W9-BYV-853

Chemistry and
Biochemistry of the
Amino Acids

Chemistry and Biochemistry of the Amino Acids

EDITED BY

G.C. Barrett
Oxford Polytechnic, UK

LONDON NEW YORK
Chapman and Hall

First published 1985 *by*
Chapman and Hall Ltd
11 New Fetter Lane, London EC4P 4EE

Published in the USA by
Chapman and Hall
733 Third Avenue, New York NY 10017

Printed in Great Britain
by J.W. Arrowsmith Ltd, Bristol

ISBN 0 412 23410 6

British Library Cataloguing in Publication Data

Barrett, G.C.
Chemistry and biochemistry of the amino acids.
1. Amino acids
I. Title
547.7′5 QD431

ISBN 0–412–23410–6

Library of Congress Cataloging in Publication Data

Main entry under title:

Chemistry and biochemistry of the amino acids.

Bibliography: p.
Includes index.
1. Amino acids. I. Barrett, G.C., 1935–
QD431.C45155 1985 574.19′245 83–15109
ISBN 0–412–23410–6 (U.S.)

Contents

Contributors

J.L. Bada
Amino Acid Dating Laboratory (A-012B)
Scripps Institution of Oceanography
University of California, San Diego
La Jolla, California 92093, USA

G.C. Barrett
Oxford Polytechnic
Headington, Oxford OX3 0BP, UK

D.A. Bender
Courtauld Institute of Biochemistry
The Middlesex Hospital Medical School
Mortimer Street
London W1P 7PN, UK

V. Cody
Medical Foundation of Buffalo Inc.
73 High Street, Buffalo
New York 14203, USA

J.S. Davies
Department of Chemistry
University College of Swansea
Singleton Park
Swansea SA2 8PP, UK

C.N.C. Drey
Robert Gordon's Institute of Technology
School of Chemistry, St Andrew Street
Aberdeen AB1 1HG, UK

M.H. Engel
School of Geology and Geophysics
830 Van Vleet Oval
The University of Oklahoma
Norman, Oklahoma 73019, USA

P.M. Hardy
Department of Chemistry
University of Exeter
Stocker Road, Exeter EX4 4QD, UK

P.E. Hare
Geophysical Laboratory
Carnegie Institution of Washington
2801 Upton Street, N.W.
Washington DC 20008, USA

S. Hunt
Department of Biological Sciences
University of Lancaster
Lancaster LA1 4YQ, UK

R.A.W. Johnstone
Department of Organic Chemistry
University of Liverpool, PO Box 147
Liverpool L69 3BX, UK

M.J. Jung
Centre de Recherche Merrell International
16 rue d'Ankara
67084 Strasbourg Cedex, France

P.J. Lea
Department of Biochemistry
Rothamsted Experimental Station
Harpenden, Herts AL5 2JQ, UK

T.H. Lilley
Department of Chemistry
The University
Sheffield S3 7HF, UK

J. Meienhofer
Chemical Research Department
Hoffmann-La Roche Inc.
Nutley, New Jersey 07110, USA

B.J. Miflin
Department of Biochemistry
Rothamsted Experimental Station
Harpenden, Herts AL5 2JQ, UK

D. Perrett
Department of Medicine
St. Bartholomew's Hospital Medical College
West Smithfield
London EC1A 7BE, UK

M.E. Rose
Department of Chemistry
Sheffield City Polytechnic
Pond Street
Sheffield S1 1WB, UK

G.A. Rosenthal
Thomas Hunt Morgan School of Biological Sciences
University of Kentucky
Lexington, Kentucky 40506, USA

P.A. St. John
Geophysical Laboratory
Carnegie Institution of Washington
2801 Upton Street, N.W.
Washington DC 20008, USA

C. Toniolo
Centro di Studi sui Biopolimeri
Istituto di Chimica Organica
Universita' di Padova
35100 Padova, Via Marzolo 1, Italy

R.M. Wallsgrove
Department of Biochemistry
Rothamsted Experimental Station
Harpenden, Herts AL5 2JQ, UK

Preface

Amino acids are featured in course syllabuses and in project and research work over a wide spectrum of subject areas in chemistry and biology. Chemists and biochemists using amino acids have many common needs when they turn to the literature for comprehensive information. Among these common interests, analytical studies, in particular, have undergone rapid development in recent years. All other chemical and biochemical aspects of amino acids – synthesis, properties and reactions, preparation of derivatives for use in peptide synthesis, racemization and other fundamental mechanistic knowledge – have been the subject of vigorous progress.

This book offers a thorough treatment of all these developing areas, and is structured in the belief that biochemists, physiologists and others will profit from access to information on topics such as the physical chemistry of amino acid solutions, as well as from thorough coverage of amino acid metabolism, biosynthesis and enzyme inhibition; and that chemists will find relevant material in biological areas as well as in the analysis, synthesis and reactions of amino acids.

It has been an important objective that comprehensive coverage of the amino acids should be contained within one volume. This has been achieved partly because much of the early 'pre-spectroscopic' chemical literature is so thoroughly covered by J.P. Greenstein and M. Winitz in their classic *Chemistry of the Amino Acids* (published by John Wiley & Sons in 1961), and partly because excellent coverage of some topic areas (such as thin layer chromatography, amino acid routine analysis) is also easily accessible. Economy has also been achieved by the conflation of some topic areas.

Grateful thanks to the publishers and their staff, and to Mary Emerson in particular for her help in the planning stages and who seemed to know just when an injection of help and encouragement was needed.

Graham Barrett
Oxford, 1984

CHAPTER ONE

Nomenclature of Amino Acids

G.C. BARRETT

1.1 INTRODUCTION

Tables 1 and 2 of the following chapter list the names, structures, and IUPAC–IUB recommendations for abbreviated names of the protein amino acids. Tables 1–21 in Chapter 4 give the same information for the non-protein amino acids, although there are no IUPAC–IUB recommendations for compounds in this category; if a non-protein amino acid cannot be named as a derivative of a protein amino acid, then it is named according to the generally accepted conventions [1].

This chapter is intended to provide a working guide to current recommendations [2] covering the nomenclature of amino acids, since it is an area in which considerable interchangeability is permitted between trivial and systematic conventions. The IUPAC recommendations [1] and the recommendations of a joint IUPAC–IUB working party [2] have been reprinted in a number of sources, including several primary journals and in the secondary literature [2].

1.2 TRIVIAL NAMES OF AMINO ACIDS

Recommendations dating from 1960 and endorsed by the working party which reported in 1975 [2] support the continuing tradition of using long-established trivial names for the protein amino acids. However, the coining of trivial names for newly-discovered α-amino acids should be avoided except for compelling reasons (e.g. if the systematic name is unreasonably cumbersome and lengthy), and these new compounds should be named as far as possible as derivatives of protein amino acids.

Some trivial names incorporate features which contradict other IUPAC recommendations. For example, norvaline $CH_3CH_2CH_2CH(\overset{+}{N}H_3)CO_2^-$ uses the prefix 'nor' to indicate 'normal' (i.e. an unbranched carbon chain). The prefix 'nor' is used in terpene nomenclature to denote replacement of all CH_3 groups attached to a ring system by H atoms.

(a) Acyl radicals

The ending 'ine' of the trivial name of an amino acid becomes 'yl' (with some exceptions) when arriving at the name for the acyl radical $RCH(NH_2)CO$–. Thus, with the exceptions aspartyl, asparaginyl, cysteinyl, glutamyl, glutaminyl, and tryptophyl, the names for the acyl radicals are in accordance with this rule: alanyl, arginyl, histidyl, cystyl, and so on. Half-cystyl is retained as the name for the acyl radical of cysteine in which the SH proton is substituted by some other grouping or the group is ionized. Di-acyl radicals of aspartic and glutamic acids are aspartoyl and glutamoyl respectively; mono-acyl radicals of these amino acid are α- or β-aspartyl or α- or γ-glutamyl, as appropriate.

(b) Esters

While 'the benzyl ester of methionine' or 'methionine benzyl ester' would be as likely to be as acceptable to most authors and editors as 'benzyl methioninate', the recommendation is for the latter style. Thus [2], 'ic acid' at the terminus of aspartic and glutamic acids, and the terminal 'e' of the other protein amino acids, is replaced by 'ate' in naming the ester of the amino acid (but 'ate' is added to tryptophan).

(c) Amides

Although recommendations corresponding to those for the naming of esters can be followed (thus, glycinamide, phenylalaninanilide, and so on), the 1974 report [2] also allows glycine amide, leucine anilide, etc. It is curious that the alternative ways of naming amino acid esters are not also recommended by the 1974 IUPAC–IUB working party.

(d) Radicals formed by links through other positions

The above recommendations refer to acylation through the α-carboxy-group. Specification of the linkages through other positions is covered in the 1974 recommendations; if the N – H proton is substituted then the letter 'o' replaces the terminal 'e' of the trivial name of the amino acid (and the terms asparto, glutamo, and tryptophano are generated for the trivial names that do not end in the letter 'e'). This would suggest 2-alanino-ethanol rather than the commonly used *N*-(2-hydroxyethyl)alanine for $HOCH_2CH_2NHCH(CH_3)CO_2H$, but the rule is useful in more complex cases. Thus, N^6-lysino- or N^ε-lysino- for the radical derived from lysine, and π-histidino- for N^π-substituted histidines, become convenient prefixes.

If positions other than the carboxy or amino groups are involved in links, i.e. C – H, S – H or S – R, or O – H, then the ending 'yl' replaces 'e' at the end of the trivial name, with the usual exceptions. Thus we have cystein-*S*-yl, threonin-O^3-yl, but aspartamid-x-yl, glutamid-x-yl, aspartic-2-yl, glutamic-x-yl, and tryptophan-1-yl.

1.3 SYSTEMATIC NOMENCLATURE FOR AMINO ACIDS

The naming of a substituted alkanoic acid recommended for a general case [1] has been used for several important amino acids. For example, γ-amino-butyric acid, GABA, is within these recommendations since the trivial name 'butyric acid' remains acceptable. The fully systematic name would be 4-aminobutanoic acid, and most higher homologous amino acids of this type have been named systematically. The best-known exception is *β*-alanine, $H_3\overset{+}{N}CH_2CH_2CO_2^-$, but this system is more widely used (e.g. leucine and *β*-leucine).

1.4 MISCELLANEOUS NOMENCLATURE RECOMMENDATIONS

(a) Numbering of atoms

Although Greek letters are still frequently used, numbering is becoming more widely adopted, but often erroneously. The carbon atom adjacent to the carboxy group is C-2 (carboxyl carbon is C-1), so 2-cyanoalanine for $NCCH_2CH(\overset{+}{N}H_3)CO_2^-$ is incorrect because it mixes the trivial nomenclature and numbering, and also because the cyano-group is on C-3 in this molecule; it is either *β*-cyano-alanine or 2-amino-3-cyano-propanoic acid.

Proline carbon atoms are numbered as in pyrrolidine, the nitrogen atom being numbered '1' and numbering then proceeds towards the carboxy group; again numbering of a trivial name crosses the trivial–systematic divide, but 3-methylproline, for example indicates widely-used nomenclature which is probably more common than Greek lettering around the ring.

The carbon atoms in the aromatic rings of phenylalanine, tyrosine, and tryptophan are numbered systematically, with 1 (or 3 for tryptophan) designating the carbon atom carrying the aliphatic chain (in which carbon atoms are labelled *α* or *β*). The histidine imidazole nitrogen atoms are designated *pros*- and *tele*- (meaning close and distant) and shown in names by using the Greek letters *π* and *τ* respectively, to indicate the nitrogen atom closest to the aliphatic chain, and that furthest from it. Alternatively, numbers may be used, with uncertainty about tautomers reflected in the numbering.

(b) The prefix 'homo'

This indicates the existence of one more CH_2 group in the side chain of the amino acid relative to the structure implied by the trivial name (e.g. homoserine, 2-amino-4-hydroxybutanoic acid, $HOCH_2CH_2CH(\overset{+}{N}H_3)CO_2^-$).

(c) Designation of stereochemistry: the *α*-carbon atom in *α*-amino acids

The absolute configuration of the *α*-carbon atom in *α*-amino acids is designated with the prefix D- or L- to indicate chemical correlation with D- or L-serine and thus with L- or D-glyceraldehyde, respectively. Using the Sequence Rule (the '*RS* System'), the common protein L-amino acids are *S*-amino-alkanoic acids (but L-cysteine is a member of the *R*-series).

An equimolar mixture of the two enantiomers of an amino acid with one chiral centre is prefixed DL- or (*RS*)-. The prefix *meso*- or its abbreviation *ms*- (using lower case italic letters) is used to denote those diastereoisomers of amino acids with more than one chiral centre which are optically-inactive because of internal compensation (e.g. *meso*-lanthionine and *ms*-cystine).

The prefixes D- or L- are placed immediately before the trivial name in derivatives of common amino acids. Examples are N^{ε}-methyl-L-lysine, 3,5-diiodo-L-tyrosine (but note: L-2-phenylglycine, L-phenylalanine, L-hydroxyproline and L-hydroxylysine).

(d) Designation of stereochemistry: configuration at chiral centres other than the α-carbon atom

The sequence rule offers the best solution to this problem, leading to names (3*S*)-L_s-threonine, (3*R*)-L_s-isoleucine, and (4*S*)-4-hydroxy-L_s-proline; the first two of these examples are allo-L-threonine and allo-L-isoleucine in more colloquial usage. The subscript 's' refers to the Fischer convention as used for amino acids; s stands for serine, the amino acid which has been configurationally correlated with D-glyceraldehyde. All other α-amino acids have been correlated with L-serine and carbohydrates have been correlated with D-glyceraldehyde. The term L_s avoids any confusion with the Fischer convention applied to carbohydrates (for carbohydrates the D- or L- refers to the highest-numbered chiral centre). The obvious improvement accompanying the full use of the *RS*-system can be seen in the unambiguous names (2*S*,3*R*)-threonine, (2*S*,3*S*)-isoleucine, and (2*S*,4*S*)-hydroxyproline for stereoisomers of these protein amino acids. Ring structures, of which the last-named amino acid is an example, have traditionally used stereochemical terms of geometrical isomerism, and this compound is then also trans-4-hydroxy-L_s-proline or erythro-4-hydroxy-L_s-proline. Allo-L_s-isoleucine could also be named threo-L_s-isoleucine.

(e) Optical rotation

Although only rarely used to specify an enantiomer in the case of common amino acids, if indication *is* given of the sign of optical rotation, it should precede all other aspects of the name of the compound (e.g. (+)-6-hydroxytryptophan or *dextro*-6-hydroxytryptophan). Depending on the conditions of the measurement, the sign of optical rotation of an L-amino acid may be + or − (and vice versa for the D-enantiomer: cf. L-serine, $[\alpha]_D^{20} = -6.83°$ ($c = 0.01\ g\,cm^{-3}$, H_2O), or $[\alpha]_D^{25} = +14.45°$ ($c = 0.009\ g\,cm^{-3}$, 1 M hydrochloric acid).

Thus, all four parameters defining the magnitude and, in some cases (such as L-serine), also the sign, of optical rotation should be specified, though the sign of rotation used should be assumed to refer to solutions of amino acids in water.*

1.5 DI(α-AMINO ACIDS)

(a) Those formed by C−N linkage

Secondary amines such as $HN\{(CH_2)_4CH(\overset{+}{N}H_3)CO_2^-\}_2$ can be considered to be derived from two amino acids and are named accordingly: this example is 6-(N^6-lysino)norleucine.

* Note that change of concentration can invert the sign of the rotation; thus, (+)-tartaric acid in water shows positive optical rotations for dilute solutions, but the rotation is negative through the wavelength range 365–589 nm for concentrated aqueous solutions [3].

(b) Those formed by C – S, S – S, and other linkages
S-(Alanin-*β*-yl) cysteine, (lanthionine), $S\{(CH_2)_2CH(\overset{+}{N}H_3)CO_2^-\}_2$ illustrates this structural type and the name to be used. Many examples of this class occur naturally, and trivial names can also be used. Djenkolic acid is $Cys_2(CH_2)$ or 3,3′-methylenethiobis(2-aminopropanoic acid), $\{^-O_2CCH(\overset{+}{N}H_3)CH_2S\}_2CH_2$.

(c) Those formed by C – C linkages
Di(allysine) is the extraordinary name in use for $^-O_2CCH(\overset{+}{N}H_3)CH_2$-$CH_2CH(CHO)CH(OH)(CH_2)_3CH(\overset{+}{N}H_3)CO_2^-$, a naturally occurring amino acid which can be considered to be the aldol formed between two molecules of the aldehyde derived from lysine. Its systematic name would be 5-(6-hydroxy-norleucin-6-yl)-6-oxonorleucine. Although a fully systematic name could be composed for this compound (norleucine ≡ 2-aminohexanoic acid), in terms of the recommended nomenclature for α-amino acids, norleucine is retained.

REFERENCES

1. *Nomenclature of Organic Chemistry* (1979) prepared for publication by J. Rigaudy and S.P. Klesney, 4th edn, Pergamon Press, Oxford, p. 193.
2. *IUPAC Information Bulletin* (1975) Appendix No. 46, September. Reproduced in *Biochemistry* (1975) **14**, 449.
3. Cf. Lowry, T.M. *Optical Rotatory Power*, Longmans, Green, and Co., London, 1935 and Dover Reprints, 1964.

CHAPTER TWO

The Protein Amino Acids

P.M. HARDY

2.1 INTRODUCTION

Amino acids are found in living organisms in both their free forms and bound by amide linkages in peptides and proteins. The diversity of structure observed in the free amino acids (several hundred are known) contrasts with the limited array of L-α-amino acids which are found in proteins and the peptides of higher organisms such as the mammals. The peptides of lower organisms, e.g. fungi and bacteria, are intermediate in the variety of their amino acids; cyclic structures, D-residues, and αβ-unsaturated residues are frequent components as well as α-hydroxy acids and non-protein amino acids.

The amino acid structures which occur in any given natural location reflect, of course, the local pathways of biosynthesis. Free amino acids arise solely through enzymic processes, but protein synthesis is a more complex procedure requiring both transfer RNAs and messenger RNA, and it takes place on a particular cellular particle, the ribosome. There is increasing evidence now that the peptides of higher organisms arise by cleavage of protein precursors, e.g. prohormones give rise to hormones, and therefore are constrained to the amino acid range found in proteins. The peptides of lower organisms are thought to originate in non-ribosomal enzymic processes, which are less discriminating: families of closely related compounds often occur.

It is possible to define protein amino acids in two ways. They may either be regarded as the components which are found in proteins as they emerge from the ribosome, i.e. the ones that are specifically coded for in the process of translation, or alternatively they may be considered as the components found in proteins which are of sufficient age for post-ribosomal modification of some residues to have taken place. For the purposes of this review, the former will be referred to as primary protein amino acids, and the latter as secondary. If modification involves the cross-linking of two amino acids, then the description tertiary protein amino acid is appropriate. In their definitive work on the amino acids in 1961, Greenstein and Winitz [1] listed twenty six protein amino acids, but six of these are now known to be secondary or tertiary amino acids. Of the twenty primary protein constituents (Table 2.1), nineteen are α-amino acids, but proline is a cyclic α-imino acid.

Although the genetic code is highly specific, non-coded amino acids can in certain circumstances be found in proteins. It has proved possible to incorporate structurally related analogues of the protein primary amino acids into bacterial

Table 2.1 The primary protein amino acids

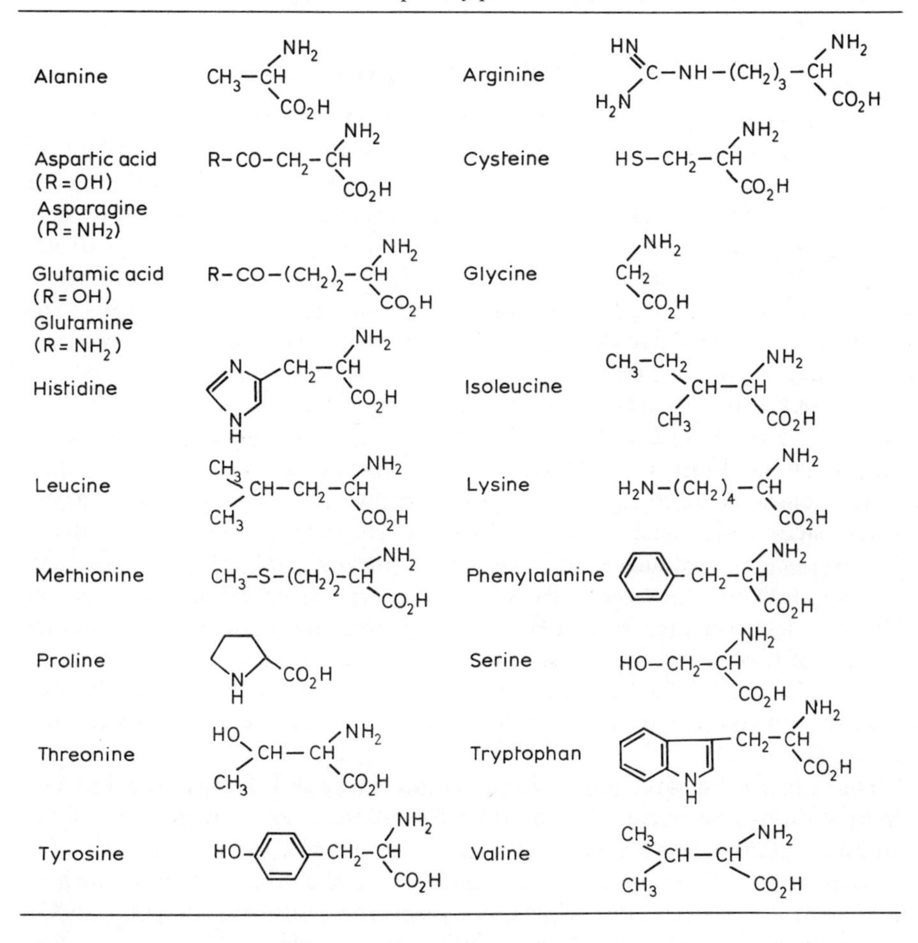

Alanine	$CH_3-CH(NH_2)-CO_2H$	Arginine	$HN{=}C(H_2N)-NH-(CH_2)_3-CH(NH_2)-CO_2H$
Aspartic acid (R = OH) Asparagine ($R = NH_2$)	$R-CO-CH_2-CH(NH_2)-CO_2H$	Cysteine	$HS-CH_2-CH(NH_2)-CO_2H$
Glutamic acid (R = OH) Glutamine ($R = NH_2$)	$R-CO-(CH_2)_2-CH(NH_2)-CO_2H$	Glycine	$CH_2(NH_2)-CO_2H$
Histidine	(imidazolyl)$-CH_2-CH(NH_2)-CO_2H$	Isoleucine	$CH_3-CH_2(CH_3)CH-CH(NH_2)-CO_2H$
Leucine	$(CH_3)_2CH-CH_2-CH(NH_2)-CO_2H$	Lysine	$H_2N-(CH_2)_4-CH(NH_2)-CO_2H$
Methionine	$CH_3-S-(CH_2)_2-CH(NH_2)-CO_2H$	Phenylalanine	(phenyl)$-CH_2-CH(NH_2)-CO_2H$
Proline	(pyrrolidine, NH)$-CO_2H$	Serine	$HO-CH_2-CH(NH_2)-CO_2H$
Threonine	$HO(CH_3)CH-CH(NH_2)-CO_2H$	Tryptophan	(indolyl, NH)$-CH_2-CH(NH_2)-CO_2H$
Tyrosine	$HO-$(phenylene)$-CH_2-CH(NH_2)-CO_2H$	Valine	$(CH_3)_2CH-CH(NH_2)-CO_2H$

protein in the absence of an adequate supply of the normal amino acid. Thus, for example, azetidine-2-carboxylic acid has been partially substituted into *E. coli* protein in place of proline, and norleucine in place of methionine [2]. However, such substitution occurs only under highly unnatural circumstances.

The structures of the primary protein amino acids were determined over the century following the characterization of glycine and leucine by Braconnet in 1820, threonine being the last of the twenty to be isolated in a pure form, in 1925. Some were prepared by chemical synthesis before being isolated from protein hydrolysates, e.g. alanine, and some initially obtained in a pure form from natural free amino acid sources, e.g. asparagine from asparagus juice. The fascinating history of the isolation and naming of the protein amino acids has been well reviewed[1,3,4], and will not be considered further here.

2.2 THE GENETICALLY CODED OR PRIMARY AMINO ACIDS

2.2.1 Physical properties

Table 2.2 outlines the more important physicochemical properties of the primary protein amino acids. Cysteine, lysine, and proline are the more water soluble of the twenty, and all except aspartic and glutamic acids and tyrosine exceed a solubility of 1 g per 100 cm^3 at 25°C. The aqueous solubility of tyrosine is too low to allow an accurate optical rotation to be measured. All the amino acids are more soluble in glacial acetic acid than water, rendering the former a useful component of solvent systems for paper and thin layer chromatography. The amino acids do not have sharp melting points but decompose over a range of several degrees above 200°C.

The p*K* of the α-carboxyl group varies from 1.71 in cysteine to 2.71 in threonine, while the p*K* of the α-amino groups extends from 8.18 in cysteine to 9.69 in alanine. In the case of fourteen of the primary amino acids, the specific optical rotations are higher in acidic solution than in simple aqueous solution. This is particularly useful in the case of alanine. The shapes of the optical rotatory dispersion curves are such as to give higher rotations at wavelengths lower than the sodium D line. In water, positive Cotton effects are observed with a peak at 216 nm or less; a change to 0.5 M HCl is accompanied by an increase in molecular rotation of 1000–1500° at the peak and a shift in the wavelength of the extremum to about 255 nm [5]. Circular dichroism curves of the primary amino acids show positive carboxyl-group Cotton effects at about 200 nm in water and 208–210 nm in acid [6].

The taste of most of the primary amino acids can be classified as either bitter or sweet, although authorities [4,7,8] vary in opinion, as assessment is essentially subjective. References to many of them can be found in which the tastes are cited as neutral or flat. Glutamic acid, methionine, and cysteine have flavours which do not fall into these categories, while lysine seems to be rather lacking in taste. On changing the configuration to the D-form, bitter amino acids become either neutral or sweet in taste, and sweet amino acids become sweeter. Protein amino acids are largely responsible for the taste of many types of foods, including soy sauce (soybean protein hydrolysed by microbial enzymes), seafood, and cheese. Many pleasant odours of cooked food are due to the interaction products of amino acids and sugars (the Maillard reaction); for example, proline and glucose give rise to the odour of newly backed bread [20].

The aliphatic members of the primary amino acids have no absorption in the ultraviolet region above 220 nm, but of course the aromatic amino acids histidine, phenylalanine, tryptophan, and tyrosine show characteristic maxima above 250 nm [9]. The zwitterionic character of the α-amino acids shows up clearly in their infrared spectra. No absorption due to the normal NH stretching frequency at 3300–3500 cm^{-1} is observed, indicating the lack of an NH_2 group. Instead, a peak near 3070 cm^{-1} due to the NH_3^+ is seen (except in proline, the NH_2^+

Table 2.2 Some physicochemical properties of the primary protein amino acids

	IUPAC–IUB abbreviations		MW	*Decomposition temperature* (°C)	*Water solubility* (g/100 g 25°C)	pK α-CO_2H	pK α-NH_3^+	pK *side chain*	*Isoelectric* pH	$[\alpha]_D^{25}$ H_2O	$[\alpha]_D^{25}$ 5 M HCl	*Taste in aqueous solution* (pH 6.0)
1. Alanine	Ala	A	89.10	297	16.5	2.34	9.69		6.01	+1.6	+13.0	sweet
2. Arginine	Arg	R	174.21	238	15.0†	2.17	9.04	12.84	10.76	+21.8	+48.1	bitter
3. Asparagine	Asn	N	132.12	236	3.1	2.02	8.60		5.41	−7.4	+37.8	bitter
4. Aspartic acid	Asp	D	133.11	270	0.5	1.88	9.60	3.65	2.77	+6.7	+33.8	bitter
5. Cysteine	Cys	C	121.16	178*	v. sol	1.71	8.18	10.28	5.02	−20.0	+7.9	sulphurous
6. Glutamic acid	Glu	E	147.14	249	0.84	2.16	9.67	4.32	3.24	+17.7	+46.8	tasty
7. Glutamine	Gln	Q	146.15	185	3.6	2.17	9.13		5.65	+9.2	+46.5	sweet
8. Glycine	Gly	G	75.07	292	25	2.34	9.60		5.97	—	—	sweet
9. Histidine	His	H	155.16	277	7.59	1.82	9.17	6.00	7.59	−59.8	+18.3	bitter
10. Isoleucine	Ile	I	131.18	284	4.12	2.36	9.68		6.02	+16.3	+51.8	bitter
11. Leucine	Leu	L	131.18	337	2.3	2.36	9.60		5.98	−14.4	+21.0	bitter
12. Lysine	Lys	K	146.19	224	v. sol	2.18	9.12	10.53	9.82	+19.7	+37.9	flat
13. Methionine	Met	M	149.22	283	3.5	2.28	9.21		5.74	−14.9	+34.6	tasty
14. Phenylalanine	Phe	F	165.20	284	2.97	1.83	9.13		5.48	−57.0	−7.4	bitter
15. Proline	Pro	P	115.14	222	162.3	1.99	10.6		6.30	−99.2	−69.5	sweet
16. Serine	Ser	S	105.10	228	5.0	2.21	9.15		5.68	−7.9	+15.9	sweet
17. Threonine	Thr	T	119.12	253	20.5	2.71	9.62		6.16	−33.9	−17.9	sweet
18. Tryptophan	Trp	W	204.23	282	1.14	2.38	9.39		5.89	−68.8	−5.7	bitter
19. Tyrosine	Tyr	Y	181.20	344	0.05	2.20	9.11	10.07	5.66	N.S.	−18.1	bitter
20. Valine	Val	V	117.15	315	8.85	2.32	9.62		5.96	+6.6	+33.1	bitter

*as HCl
†21°C

absorbing at ~2900 cm^{-1}). The carbonyl absorption of the un-ionized carboxyl group at 1700–1730 cm^{-1} is likewise replaced by a carboxylate ion absorption at 1560–1600 cm^{-1}. Two other bands associated with the NH_3^+ appear in the 1500–1600 cm^{-1} region. Most amino acids also show weak absorption at 2080–2140 cm^{-1} and 2530–2760 cm^{-1}, and a medium strength absorption at 1300 cm^{-1} [10].

2.2.2 Stereochemistry

With the exception of glycine, all the genetically coded protein amino acids are optically active, and of identical chirality at the α-carbon atom. Apart from cysteine, they are in all cases of the (*S*)-configuration. However, they are still commonly referred to as L-α-amino acids. Isoleucine and threonine have second centres of asymmetry at their β-carbon atoms; these are of the (3*S*)- and (3*R*)-configurations respectively (*2.1*) and (*2.2*).

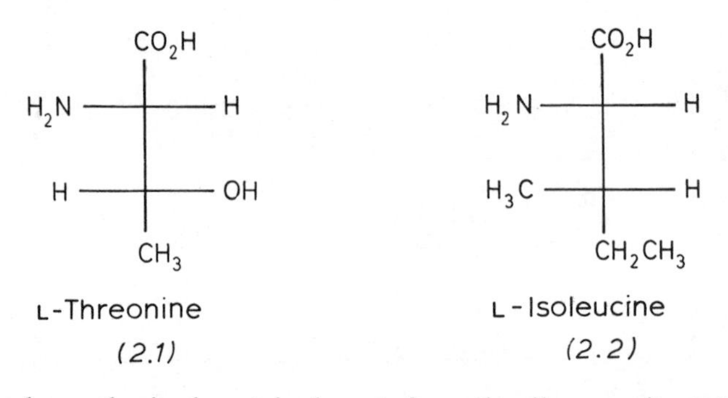

As the newly synthesized protein departs from the ribosome it contains only L-amino acids, L-proline and glycine. At present no sensitive way of assaying the configuration of the amino acids in an intact protein is known. The amide bonds must be broken either by chemical or enzymic hydrolysis before configurational studies can be made. Alkaline hydrolysis results in extensive racemization, but acidic hydrolysis is much better in this respect. However, it is still sufficient for correction to be required in order to eliminate any suspicion of occasional D-residues occurring in peptides or proteins. Carrying out the hydrolysis in tritiated hydrochloric acid and measuring the amount of tritium bound for the α-C atom for each amino acid (i.e. the tritium not readily exchangeable) allows a relatively accurate assay of the racemization occurring during hydrolysis. This can be compared to the amount of D-isomer found after coupling the hydrolysate with L-leucine or L-glutamic acid *N*-carboxyanhydride and chromatographically separating the resulting diastereoisomeric dipeptides. In general, racemization during acidic hydrolysis does not exceed 3%, but since some amino acids undergo side-chain exchange with tritiated hydrochloric acid (aspartic acid, glutamic acid, phenylalanine, tyrosine, and histidine) the method is not applicable to all residues [11]. Acid hydrolysis under these conditions also destroys tryptophan, in many cases almost completely. The prevention of this is described in the detailed discussion of protein hydrolysis in Chapter 12.

Enzymic hydrolysis may be accomplished by a combination of proteases. No single enzyme is known which is suitable for proteins containing proline, which inhibits aminopeptidases. One mixture which has advantages contains trypsin, chymotrypsin, prolidase, and aminopeptidase M bound to the polymer Sepharose [12]. Binding stabilizes the enzymes, the mixture losing no enzyme activity on storage for seven months. No hydrolysis of peptide bonds adjacent to a D-amino acid residue is observed with this 'cocktail', each such residue remaining the central one in a residual tripeptide. Quantitative comparison of the amount of each amino acid obtained with that produced by acid hydrolysis enables D-residues to be detected. Alternatively, D-amino acids can be destroyed with D-amino acid oxidase or L-amino acids with L-amino acid oxidase, and the compositional change used to monitor the configurations of the component amino acids. [13]. Use of these enzymes, however, is not free from problems. D-Lysine is not attacked by the former, or L-proline by the latter. Amino acid analysers give results at best accurate to 2–3%, which limits the accuracy of quantitative determination employing this type of instrument.

Although freshly-synthesized proteins contain only L-residues, spontaneous racemization occurs, albeit extremely slowly. The D-content of protein amino acids has been used as the basis of a method of determining their age, e.g. about 0.1% conversion of L- to D-aspartic acid occurs per year [14]. The racemization of amino acids is discussed in full in Chapter 13.

2.2.3 Distribution in proteins

The amino acid composition of proteins varies widely [15]. In some cases only a few amino acids constitute the bulk of the protein, e.g. in fish antifreeze proteins alanine and threonine may be the only amino acids present, although there is also a high carbohydrate content [16]. More typically all the amino acids are present, but there are only a few residues of some of them. Although plants and some micro-organisms are capable of the biosynthesis of all the primary amino acids, mammals are degenerate in the sense that they lack the ability to synthesize a sufficient quantity of eight of these compounds (isoleucine, leucine, lysine, methionine, phenylalanine, threonine, tryptophan and valine) from simple sources of carbon and nitrogen. Most species other than man are also unable to make histidine, and some arginine [4,8]. Metabolically related substances may be a partial replacement for the essential amino acids if present in adequate amount, e.g. tyrosine for arginine, or cystine for methionine, but in general all these amino acids have to be obtained from dietary sources. In practice, this means from plants, mainly plant protein. In ruminants micro-organisms of the digestive tract can supplement the available essential amino acids.

Plant proteins vary in composition just as do animal proteins, and their content of the essential amino acids determines their nutritional utility. The ideal amino acid composition for a protein foodstuff is attained, as might be expected, in such proteins as are found in whole eggs or human milk. The nutritionally important cereal proteins, however, are particularly deficient in lysine, and are also low in threonine and tryptophan. The limiting amino acid in soybean meal is methionine. Protein derived via micro-organisms from petroleum, on the other hand, is rich in lysine. One of the main spurs to the development of the industrial production of amino acids has been the need for the primary amino acids as dietary supplements.

Some amino acids may be supplied as their D-isomers, since natural D-amino acid oxidases can transform them into keto-acids which give the corresponding L-amino acids on stereospecific transamination. The metabolism of the amino acids is discussed in detail in Chapter 5.

2.2.4 Industrial production

The protein primary amino acids are all manufactured on an industrial scale. The production methods fall into three classes: extraction from protein hydrolysates, fermentation processes [17], and chemical synthesis [8]. Proteins were the original large scale source. Glutamic acid, the first amino acid prepared industrially because of its flavour-enhancing properties (it is normally used as monosodium glutamate or MSG), was isolated from rich sources such as wheat gliadin (43.7% L-glutamic acid) after acidic hydrolysis. However, most of the amino acids can now be more economically prepared by fermentation or synthesis, although extraction remains important for histidine, leucine, cystine, and tyrosine. Isolation of the latter two compounds from hydrolysates is, of course, facilitated by their low water solubility.

The chemical synthesis of amino acids is dealt with in Chapter 8, and only a few points pertaining to industrial practice will be mentioned here. In general, asymmetric syntheses are not at present commercially viable. In a few cases, e.g. methionine, the D-isomer is of equal nutritive value to the L-isomer, and for food supplement purposes resolution is unnecessary. However, in most cases resolution is required, the unwanted D-isomer being racemized and recycled. Preferential crystallization may be used for this in a few cases, including those of glutamic acid and threonine. Resolution by diastereoisomer formation with an optically active resolving agent usually makes use of salt formation rather than covalent bonding. Enzymic resolution using aminoacylases immobilized on carbohydrate polymers is a very advantageous method as a continuous flow system can be used. The bound enzyme is markedly more stable than in solution, allowing prolonged operation without appreciable loss of activity.

An example of the efficient combination of chemical resolution with chemical synthesis is the preparation of L-lysine from the nylon intermediate caprolactam (Fig. 2.1). In this procedure ε-DL-aminocaprolactam is resolved by the use of L-pyrrolidonecarboxylic acid, the latter of course being readily obtainable from L-glutamic acid [18]. Asparagine is prepared chemically via the β-methyl ester of

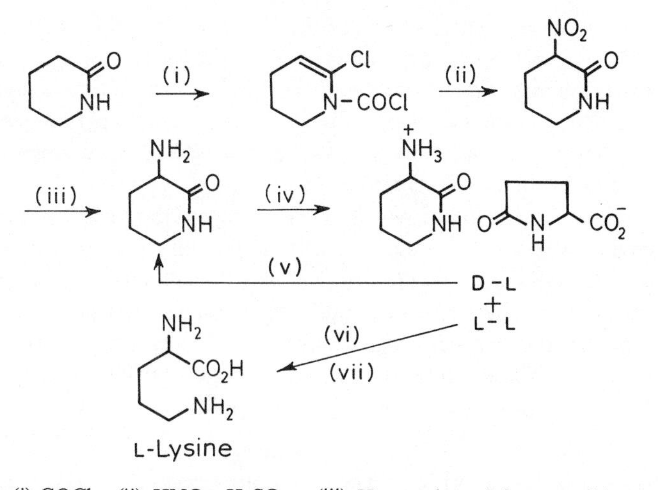

Figure 2.1 (i) $COCl_2$; (ii) HNO_3–H_2SO_4; (iii) H_2, catalyst; (iv) L-pyrrolidonecarboxylic acid; (v) regenerate D-aminocaprolactam, racemize, recycle; (vi) regenerate L-aminocaprolactam; (vii) hydrolyse.

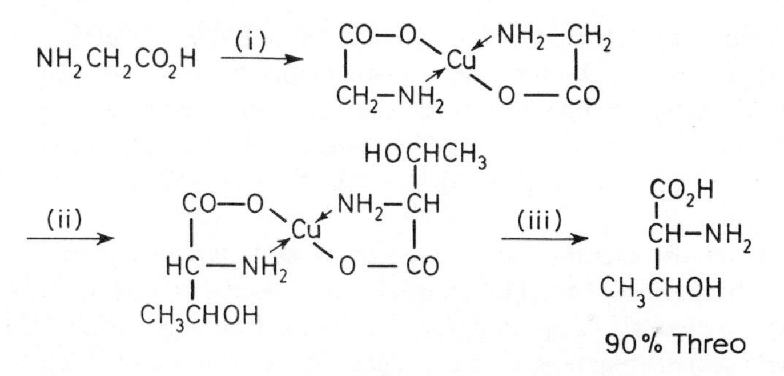

Figure 2.2 (i) Cu^{2+}; (ii) CH_3CHO, OH^-; (iii) Dowex 50, NH_4^+ form.

aspartic acid, the latter being produced at low cost by fermentation. Glycine, of course, is so readily available by chemical synthesis that other methods are unnecessary. Alkylation of glycine is a general route to other protein amino acids, and in some cases, e.g. threonine, it is used in industrial production (Fig. 2.2) [8].

The fermentation production of amino acids using micro-organisms is based on genetic and environmental modifications which overcome the normal regulatory controls, thus forcing the overproduction and excretion of primary metabolites. Consider the case of glutamic acid. All glutamate 'overproducers' are mutants having a block in the tricarboxylic acid cycle; they are deficient in α-ketoglutarate dehydrogenase, which shunts the carbon flow to glutamic acid. Even so, glutamic acid overproduction would not normally be observed due to feedback regulation. However, if the micro-organism is starved of biotin, a phospholipid-deficient cytoplasmic membrane results. This decreases the effectiveness of the membrane as a barrier to glutamic acid, allowing the amino acid to leave the cell and its biosynthesis to continue unabated [19]. The preparation of L-alanine by the fermentation method is more difficult than most amino acids, the difficulty arising from the co-existence in the system of the enzyme alanine racemase, resulting in the isolation of DL-alanine. However, L-aspartic acid can be decarboxylated to L-alanine using L-aspartate decarboxylase from bacterial sources to circumvent this problem.

Besides their use in the fortification of vegetable proteins and the massive use of glutamic acid as a flavour enhancer, synthetic primary protein amino acids find diverse uses. Alanine and glycine are used as sweeteners, for example, and synthetic mixtures of the primary protein amino acids are used in special diets, e.g. a low phenylalanine amino acid mixture is used to prevent the mental retardation effects of phenylketonuria, a metabolic inability to degrade phenylalanine in the normal way.

2.3 OTHER AMINO ACIDS FOUND IN PROTEIN HYDROLYSATES [21]

2.3.1 Secondary amino acids

No post-translational modifications of the α-amino acids without side chain functional groups, i.e. alanine, glycine, isoleucine, leucine, and valine, have been reported. Methionine and tryptophan also have not been positively identified as precursors of secondary amino acids *in vivo* (*in vitro* modifications will not be considered here). The derivatives of the other protein amino acids fall into two categories.

The first of these comprises those compounds which can be described as conjugates in the sense that a hydrogen of an OH, SH, or NH in the side chain is replaced by a glycosyl [21], phosphate [22] or sulphate grouping. All arise from enzyme mediated reactions, and in general in these cases acid hydrolysis regenerates the primary protein amino acid. A fair number of such compounds are known, glycoproteins in particular being common natural products. We can group with these other derivatives which regenerate primary amino acids on acid hydrolysis. Such compounds include side-chain amides like N^{ε}(N^{ε}-methylalanyl)- [23] and N^{ε}-(diaminopimelyl)-lysines [24], N^{5}-methylglutamine [25], and pyroglutamic acid. The latter is present at the N-terminus of a number of biologically active peptides, and is thought to arise by spontaneous cyclization of N-terminal glutamine residues. Of great interest in recent years has been the discovery of γ-carboxyglutamic acid in plasma proteins, often as a major component of some sections of the molecule. In prothrombin, for instance, it occurs at ten positions in the first 33 residues at the N-terminus [26]. Formation from glutamic acid is known to be mediated by vitamin K, and the residue is essential for calcium binding and blood coagulation functions. A number of other proteins, especially those associated with calcification, are now known also to contain this secondary amino acid [27]. Although unstable to acid, γ-carboxyglutamic acid survives alkaline hydrolysis, as does its lower homologue β-

Table 2.3 *N*-Methyl amino acids and their sources

Amino acid	*Source*
N^{ε}-Methyllysine	*Salmonella* flagellin[a,b]
N^{ε}-Dimethyllysine	Calf thymus histone[a,b]
N^{ε}-Trimethyllysine	Several histones[a,b]
N^{ε}-Trimethyl-δ-hydroxylysine	Diatom cell wall[c]
N^{ω}-Methylarginine	Calf thymus histone[a,b]
N^{ω}-Dimethylarginine N^{ω},$N^{\omega'}$-Dimethylarginine	Bovine encephalitogenic protein[a,b]
N^{τ}-Methylhistidine	Muscle proteins[d]
N-Trimethylalanine	Ribosomal protein[e]
N-Dimethylproline	Cytochrome[f]
N-Methylmethionine	Ribosomal protein[e]

[a] Paik, W.K. and Kim, S. (1975) *Adv. Enzymol.*, **42**, 227.
[b] Klagsbrun, M. and Furano, A.V. (1975) *Arch. Biochem. Biophys.*, **169**, 529.
[c] Nakajima, T. and Volcani, B.E. (1970) *Biochem. Biophys. Res. Comm.*, **39**, 28.
[d] Young, V.R. and Munro, H.N. (1978) *Fed. Proc.*, **37**, 2291.
[e] Dognin, M.J. and Wittman-Liebold, B. (1980) *Hoppe-Seyler's Z. Physiol. Chem.*, **361**, 1967.
[f] Smith, G.M. and Pettigrew, G.W. (1980) *Eur. J. Biochem.*, **110**, 123.

carboxyaspartic acid, subsequently discovered in *E. coli* ribosomal protein [28]. No function of this latter compound has yet been proposed.

The second category of secondary protein amino acids comprises those whose structures are stable to acid hydrolysis. Of this group *N*-methyl and *C*-hydroxy derivatives are the most numerous. Known *N*-methylated protein amino acids are listed in Table 2.3; histones in particular commonly contain such derivatives. All are thought to arise through the action of methyltransferases upon proteins. Hydroxylated amino acids are given in Table 2.4; 4-hydroxyproline is the most abundant, occurring in considerable amounts in hydrolysates of collagen and elastin. The two β-hydroxy aromatic amino acids listed occur in fact in proteins as their O^{β}-glycosyl derivatives; it is thought that the saccharide free amino acid is the precursor, although it has not been observed yet in this form in an intact

Table 2.4 Hydroxylated amino acids and their sources

Amino acid	*Source*
3-Hydroxyproline	Collagen[a]
4-Hydroxyproline	Gelatin[a]
3,4-Dihydroxyproline	Diatom cell wall[b]
δ-Hydroxylysine	Gelatin, wool[c]
β-Hydroxyphenylalanine	Cutinase[d]
β-Hydroxytyrosine	Cutinase[d]
β-Hydroxyglutamic acid	Calf thymus nucleoprotein[e]

[a] Bornstein, P. (1974) *Annu. Rev. Biochem.*, **43**, 567.
[b] Nordwig, A. and Pfab, P.K. (1969) *Biochim. Biophys. Acta.*, **181**, 52.
[c] Varner, J.E. (1960) *Arch. Biochem. Biophys.*, **90**, 7.
[d] Lin, T.-S. and Kolattukudy, P. (1979) *Arch. Biochem. Biophys.*, **196**, 255.
[e] Wilhelm, G. and Kunka, K.-D. (1981) *FEBS Lett.*, **123**, 141.

Table 2.5 Protein-bound derivatives of tyrosine and their sources

Amino acids	*Source*
3-Chlorotyrosine	Whelk scleroprotein[a]
3,5-Dichlorotyrosine	*Limulus* cuticle[b]
3-Bromotyrosine	Gorgonian scleroprotein[a,b]
5-Bromo-3-chlorotyrosine	Whelk scleroprotein[c]
3,5-Dibromotyrosine	Gorgonian scleroprotein[a,b]
3-Iodotyrosine	Thyroglobulin[d]
3,5-Diiodotyrosine	Thyroglobulin[d]
3,5,3′-Triiodothyronine	Thyroglobulin[e]
3,5,3′,5′-Tetraiodothyronine	Thyroglobulin[e]

[a] Hunt, S. (1972) *FEBS Letters*, **24**, 109.
[b] Welinder, S. (1972) *Biochim. Biophys. Acta.*, **279**, 491.
[c] Hunt, S. (1971) *Biochim. Biophys. Acta.*, **252**, 301.
[d] Wolff, J. and Corelli, I. (1969) *Eur. J. Biochem.*, **9**, 371.
[e] Cahnemann, H.J., Pommier, J. and Nunez, J. (1978) *Proc. Nat. Acad. Sci. USA*, **74**, 5333.

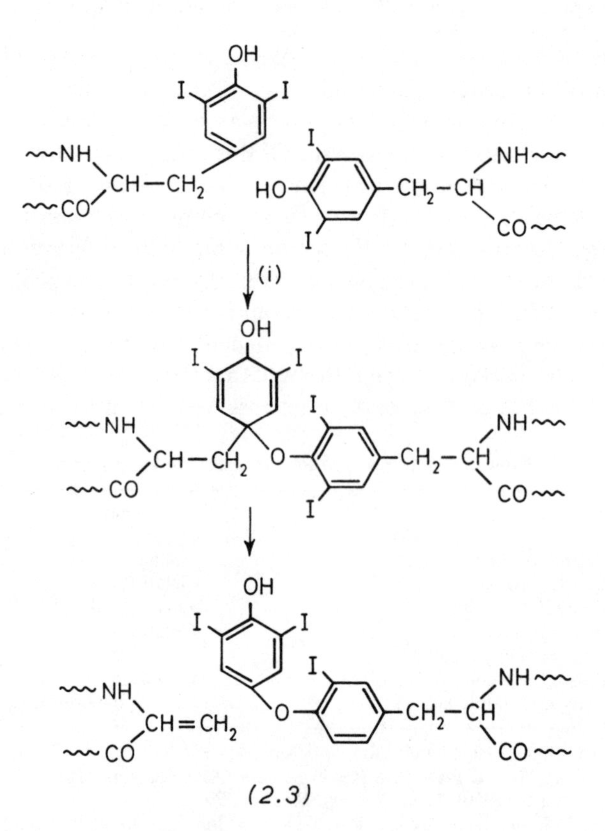

Figure 2.3 (i) Tyrosine peroxidase.

protein. Aminocitric acid, otherwise known as β-hydroxyglutamic acid, has on earlier occasions been alleged to occur in proteins [3], but now seems finally authenticated in calf thymus nucleoprotein [29].

A number of modifications of tyrosine have been identified in proteins, and are listed in Table 2.5. Although chlorotyrosines can be formed under certain conditions on hydrolysing proteins in hydrochloric acid, 3,5-dichlorotyrosine was isolated after basic hydrolysis and 3-chlorotyrosine after enzymic hydrolysis, clearly establishing that they are not artefacts. In the protein thyroglobulin iodination occurs to produce 3-iodo- and 3,5-di-iodo-tyrosines, which undergo intrachain interactions to produce tri-iodothyronine and tetraiodothyronine (*2.3*). Both iodination and the oxidative coupling occur under the influence of tyrosine peroxidase, and a β-elimination reaction leads to one of the iodotyrosine residues involved remaining as a dehydroalanine residue (Fig. 2.3).

Arginine, lysine, and cysteine are also known to undergo modification. Alkali partly degrades arginine to ornithine (*2.4*), but ornithine has been established as a genuine secondary amino acid [30]. Arginine also gives rise to citrulline (*2.5*) in

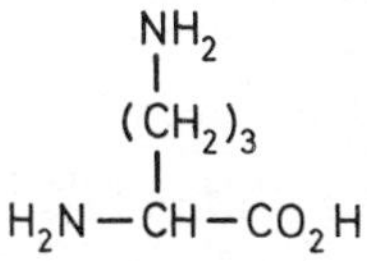

$$\begin{array}{c} NH_2 \\ | \\ (CH_2)_3 \\ | \\ H_2N-CH-CO_2H \end{array} \qquad \begin{array}{c} NHCONH_2 \\ | \\ (CH_2)_3 \\ | \\ H_2N-CH-CO_2H \end{array}$$

(*2.4*) 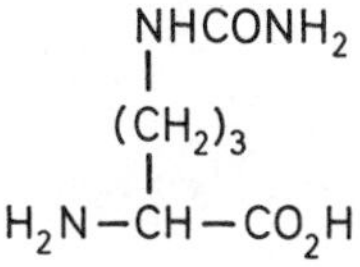 (*2.5*)

certain structural proteins [31]. Lysine is known to be involved in the cross-links important in collagen and elastin. An oxidase has been well characterized which converts lysine to allysine (α-aminoadipic acid δ-semialdehyde), and the subsequent reactions of the aldehyde grouping generate the cross-links. The reactive nature of allysine has precluded its isolation as such, but its presence in protein has been established by oxidizing to α-aminoadipic acid, or reducing to the corresponding primary alcohol. On hydrolysis in hydrochloric acid the latter forms ε-chloronorleucine [32]. Bovine teeth phosphoprotein has been shown to contain α-aminoadipic acid even though no specific oxidation step has preceded hydrolysis; it is not clear if this oxidation is enzyme-mediated or spontaneous [33]. Seleno-cysteine occurs in several enzymes, e.g. *E. coli* formate dehydrogenase, and is known to be involved in the catalytic action [34]. It is not thought that seleno-cysteine can be incorporated in the process of translation.

2.3.2 Cross-linking reactions producing tertiary amino acids

The commonest cross-link found in proteins is the disulphide bridge (cystine) formed by oxidation of the thiol groups of two residues of cysteine. This reaction is important in stabilizing the three-dimensional folding of the backbone chain in proteins, especially, as in the case of insulin, where two chains exist as a result of the post-translational excision of a central portion of the molecule. This excision, of course, is preceded by disulphide bridge formation as otherwise the correct cysteine pairing is not favoured by the natural conformation of the molecules [35]. Glutamic acid residues can also form structurally simple cross-links by forming an amide link with lysine and an ester link with serine. Several ε-(γ-glutamyl) lysine cross-links have been identified, the first being in the α-fibrous proteins of human epidermis. They arise from glutamine and lysine by the action of transglutaminase, and are important in the blood clotting process [36]. *Clostridium* proline reductase is the only protein to date in which an O^{β}-(γ-glutamyl) serine cross-link has been proposed [37].

The formation of allysine from lysine has already been mentioned, and this aldehyde is the precursor of the varied cross-links of collagen and elastin

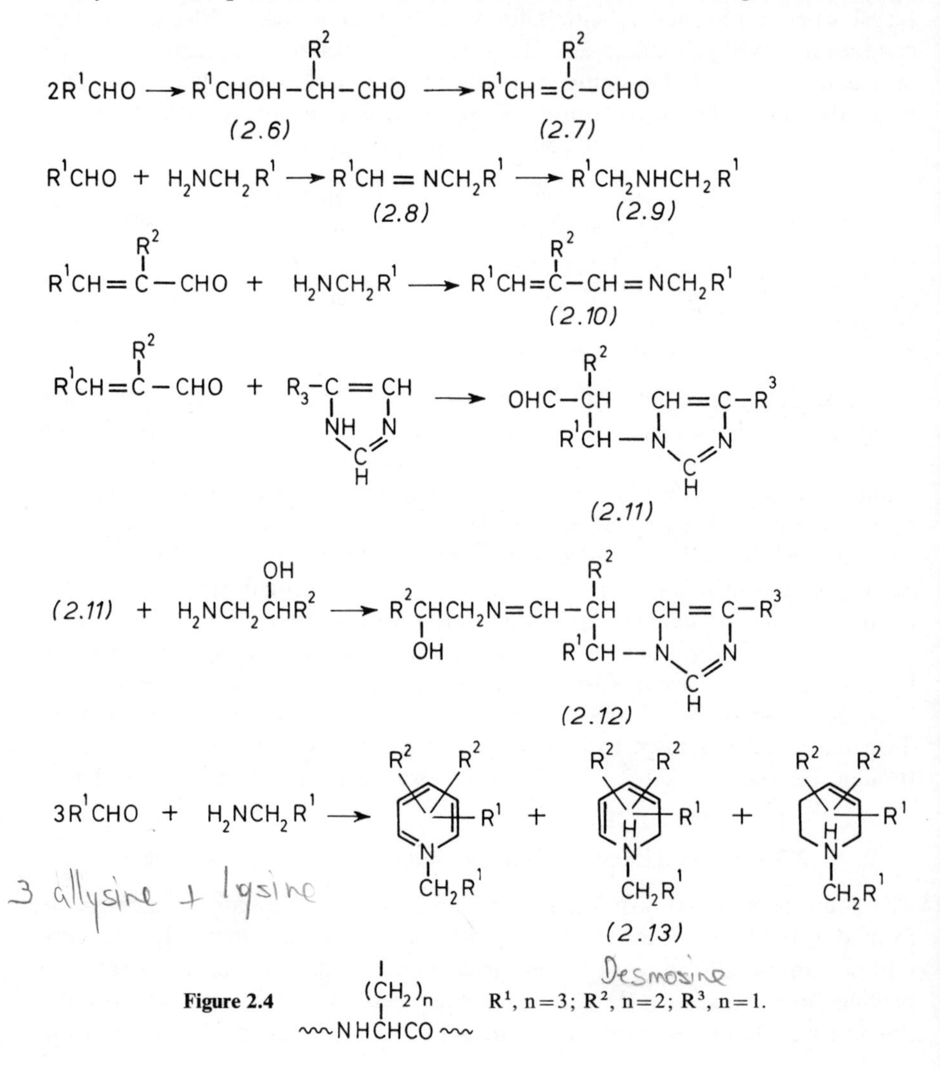

Figure 2.4 $\sim\sim NHCH(CO\sim\sim)(CH_2)_n$ R^1, n=3; R^2, n=2; R^3, n=1.

(Fig. 2.4). Two allysine residues may interact to give an aldol condensation product (*2.6*), or one may form an aldimine (*2.8*) with a lysine side-chain amino group. The reduced form of the latter, lysinonorleucine (*2.9*), has been isolated from these proteins, so natural reduction pathways do exist. The dehydrated aldol product (*2.7*) can react with lysine to give a three-residue cross-link, dehydromerodesmosine (*2.10*), at least part of which is reduced to merodesmosine. In collagen δ-hydroxylysine residues can also undergo analogous reactions, and histidine also participates, leading to aldol histidine (*2.11*) and histidine hydroxydehydromerodesmosine (*2.12*). In elastin three residues of allysine also condense with one residue of lysine in a complex reaction to produce a series of pyridine derivatives, the desmosines (*2.13*), of varying oxidation levels and substitution patterns. The identities of those cross-linking species unstable to acid hydrolysis were established after stabilization by reduction with tritiated sodium borohydride [38,39]. Pyridinoline (*2.14*), arising from two α-amino-δ-hydroxyadipic acid δ-semialdehyde residues and one hydroxylysine, has been isolated from both collagen and elastin acid hydrolysates [40]. There is some doubt as to whether or not it is an artefact of hydrolysis [41].

Tyrosine can undergo oxidative phenolic coupling in a number of proteins to give 3,3′-bityrosine and 3,3′,5′,3″-tertyrosine cross-links. Such cross-links can be produced *in vitro* by the action of peroxidases at alkaline pH, but there seems little doubt that they are not artefacts of isolation. They occur in structural proteins, e.g. elastin [42], human lens protein [43], and in adhesives produced by mussels [44]. Lanthionine (*2.15*) and lysinoalanine (*2.16*) are well known cross-links produced by the action of alkali on proteins, arising through the addition of

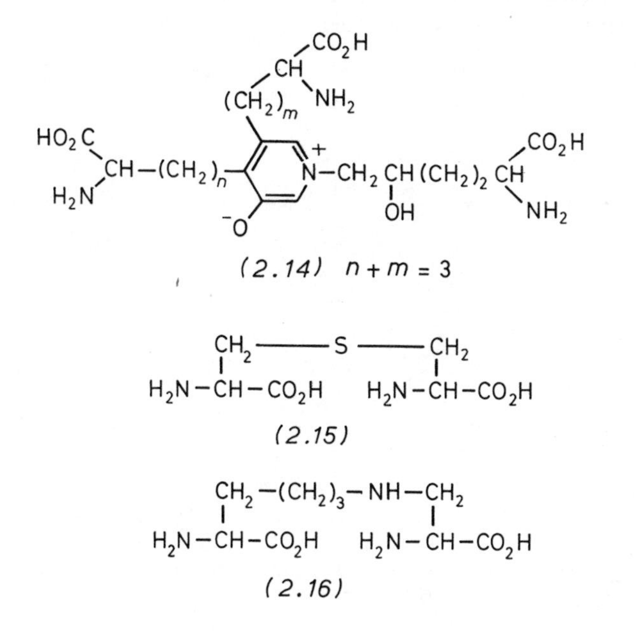

(*2.14*) $n + m = 3$

(*2.15*)

(*2.16*)

cysteine and lysine respectively to dehydroalanine residues. Recently these two cross-links have been found in the outer stratum of dermal α-keratin [45]. It is surmised that they are formed under the influence of alkaline environmental conditions, and they are therefore right on the borderline between natural products and artefacts. There is evidence that dehydroalanine residues can be formed on heating protein-bound cysteine or *O*-phosphoserine residues in the absence of alkali [46]. The existence of dehydroalanine residues at the active sites of both histidine ammonia lyase [47, 48] and phenylalanine ammonia lyase [49], enzymes which catalyse the elimination of ammonia from the free amino acids, is well established however so enzymic processes for dehydroalanine formation in proteins are clearly likely in some circumstances.

2.4 THE USES OF PRIMARY AMINO ACIDS IN ORGANIC SYNTHESIS

2.4.1 Syntheses involving temporary incorporation

Simple acidic and basic amino acids or amino acid derivatives have found some use as resolving agents. The use of L-pyrrolidonecarboxylic acid derived from L-glutamic acid to resolve a lysine precursor has already been mentioned (see Fig. 2.1), while L-tyrosine hydrazide has found more general use in resolving amino acids [50]. (*S*)-Proline has been used as the chiral inducer in the asymmetric synthesis of α-amino acids from α-keto acids and ammonia (Fig. 2.5); hydrogenation of the dioxopiperazine (*2.17*) gives the (*S*,*S*)-cyclodipeptide in essentially quantitative yield [51]. In a rather similar way proline has been used in the preparation of α,α-disubstituted α-hydroxy acids [52].

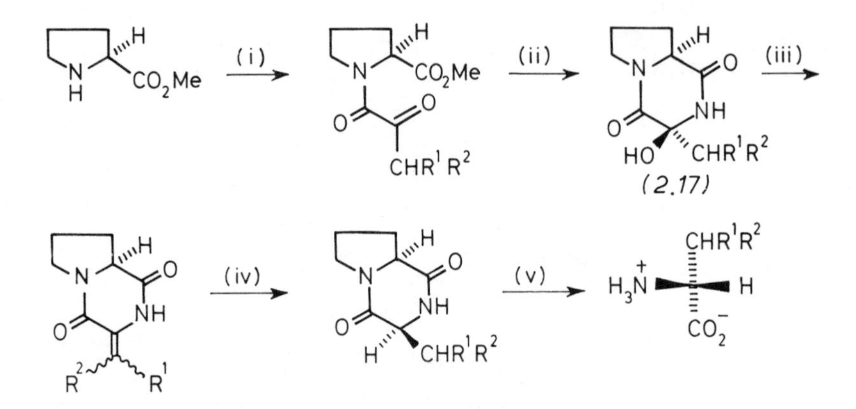

Figure 2.5 (i) $R^1R^2CHCOCO_2H$, *N,N'*-dicyclohexylcarbodiimide; (ii) NH_3, dimethoxyethane; (iii) CF_3CO_2H; (iv) Pt/H_2; (v) H^+/H_2O.

2.4.2 Amino acids as synthons

Glycine, of course, has long been used as a source of other amino acids through alkylation of either cyclic derivatives, e.g. oxazolinones [53], or protected linear compounds, e.g. $(RS)_2C{=}NCH_2CO_2Et$ [54]. Use of microbial L-serine hydroxymethyltransferase has even allowed the conversion of glycine to L-serine [55].

However, only in recent years has it been recognized that the protein amino acids are relatively low-priced compounds of high optical purity and hence economically attractive starting materials for the synthesis of a variety of natural

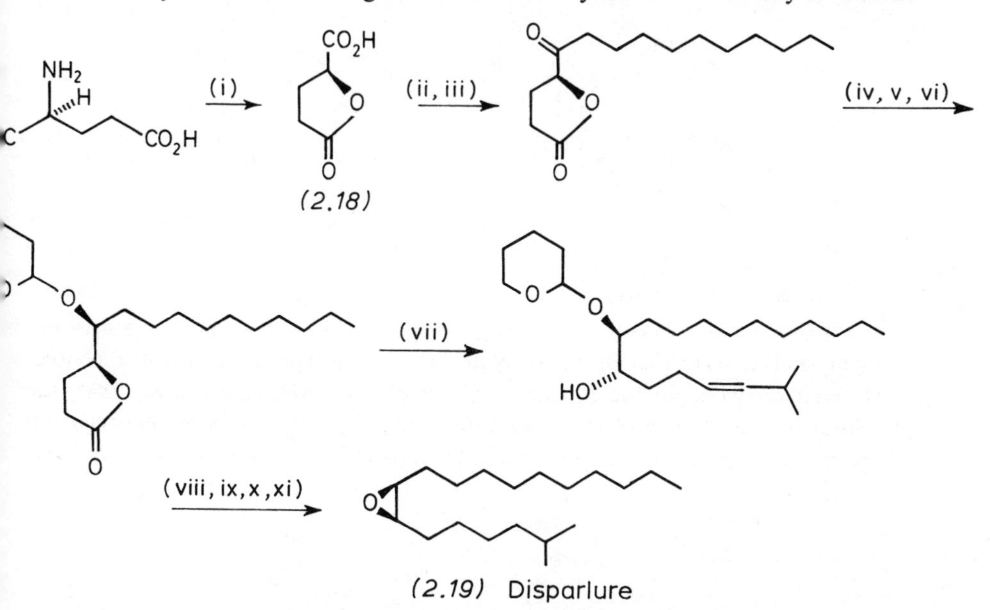

Figure 2.6 (i) HNO_2/H^+; (ii) oxalyl chloride; (iii) didecylcadmium; (iv) $NaBH_4$; (v) separate 4*S*,5*S*-isomer; (vi) dihydropyran, H^+; (vii) $Me_2CHCH\,PPh_3$; (vii) Pt/H_2; (ix) Tos-Cl/pyridine; (x) $HAc/H_2O/THF$ (2:1:1), 40°C, 5h; (xi) 0.25 M KOH/MeOH.

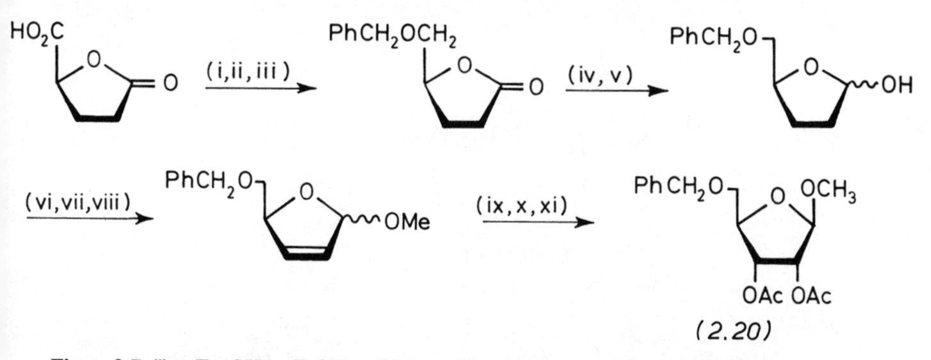

Figure 2.7 (i) TosOH, EtOH, PhH; (ii) $NaBH_4$; (iii) $PhCH_2Br/Ag_2O$; (iv) $Na/HCO_2Et/Et_2O$; (v) H^+/H_2O/dioxan; (vi) methylate; (vii) $Br_2/Et_2O/CaCO_3$; (viii) NaOMe; (ix) $KMnO_4$; (x) Ac_2O/pyridine; (xi) separate stereoisomers.

products [56]. The deamination of L-glutamic acid with nitrous acid to the lactone acid (*2.18*) has been shown to proceed with retention of configuration [57]. This product has been used as the starting point in the synthesis of a number of compounds of diverse structures. These include some insect pheromones, e.g. the sex attractant of the gypsy moth, disparlure [(*2.19*); Fig. 2.6] [58], a derivative of D-ribose [(*2.20*); Fig. 2.7] [59], some terpenes, including squalene-2,3-epoxide (*2.21*) [60], and some antileukaemic lignan lactones, e.g. (−)-isodeoxypodophyllotoxin (*2.22*) [61].

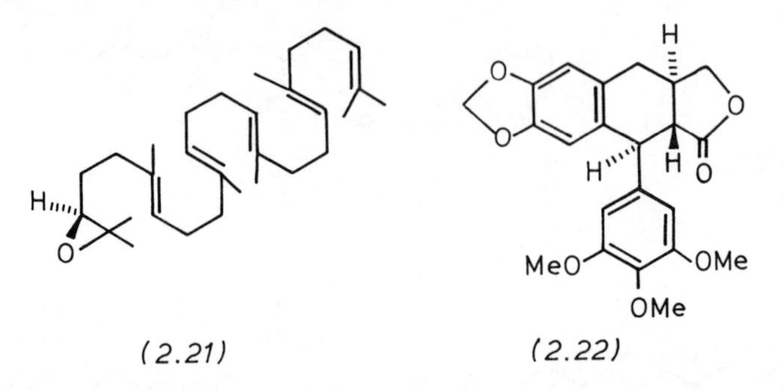

As far as the other protein primary amino acids are concerned, chiral centres have been provided in the synthesis of the nine-membered macrocycle of the streptogramin antibiotics (*2.23*) from L-cystine [62], the bark beetle aggregation pheromone (−)-ipsenol (*2.24*) from L-leucine [63], (−)-deoxoproposopinine

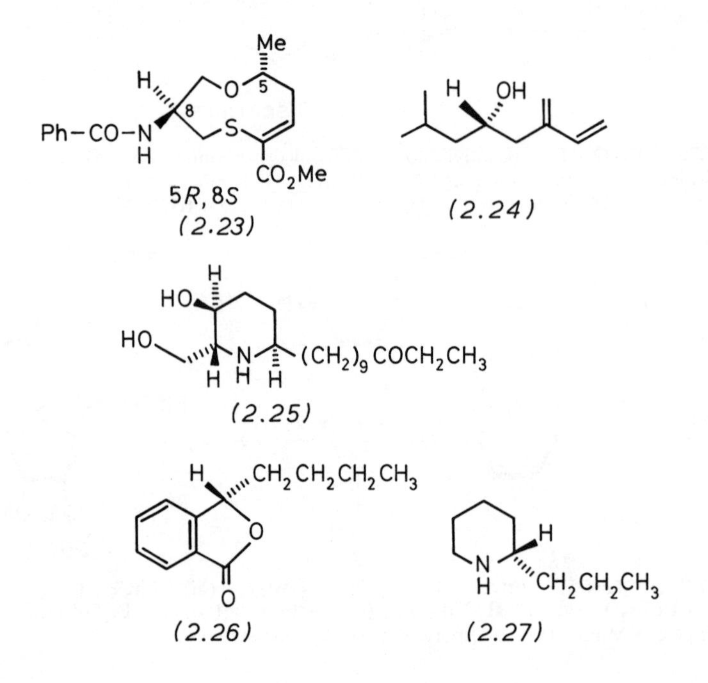

(*2.25*) from L-serine [64], a component of oil of celery (*2.26*) from L-proline [65] and (*S*)(+)-coniine (*2.27*) from L-lysine [66].

REFERENCES

1. Greenstein, J.P. and Winitz, M. (1961) *Chemistry of the Amino Acids*, (Vols 1–3), John Wiley, New York.
2. Fowden, L. and Richmond, M.H. (1963) *Biochim. Biophys. Acta.*, **71**, 459; analogue incorporation has been reviewed by Richmond, M.H. (1962) *Bacteriol. Rev.*, **26**, 398.
3. Vickery, H.B. (1972) *Adv. Protein, Chem.*, **26**, 82.
4. Meister, A. (1965) *Biochemistry of the Amino Acids*, (Vol. 1) 2nd edn, Academic Press, New York.
5. Jennings, J.P., Klyne, W. and Thomas, R.N. (1961) *J. Chem. Soc. C.*, 295.
6. Fowden, L., Scopes, P.M. and Thomas, R.N. (1971) *J. Chem. Soc. C.*, 833.
7. Solms, J., Vuataz, L. and Egli, R.H. (1965) *Experientia*, **21**, 692.
8. Kaneko, T., Izumi, Y., Chibata, I. and Itoh, T. (1974) *Synthetic Production and Utilization of Amino Acids*, John Wiley, New York.
9. Wetlaufer, D.B. (1962) *Adv. Protein. Chem.*, **17**, 304.
10. Bellamy, L.J. (1975) *The Infrared Spectra of Complex Molecules*, (Vol. I) (3rd edn) John Wiley, New York, p. 263.
11. Manning, J.M. (1970) *J. Amer. Chem. Soc.*, **92**, 7449.
12. Bennett, H.P.J., Elliott, D.F., Evans, B.E., Lowry, P.J. and McMartin, C. (1972) *Biochem. J.*, **129**, 695.
13. Riniker, B. and Schwyzer, R. (1961) *Helv. Chim. Acta.*, **44**, 658.
14. Masters, P.M., Bada, J.L. and Zigler, J.S. (1977) *Nature*, **268**, 71.
15. Tristram, G.R. and Smith, R.H. (1963) *Adv. Protein Chem.*, **18**, 227.
16. Feeney, R.E. and Yeh, Y. (1978) *Adv. Protein Chem.*, **32**, 191.
17. Yamada, K. (1972) *The Microbial Production of Amino Acids*, Kodansha and Wiley.
18. Ottenheim, H.H. and Jenneskens, P.J. (1970) *J. Agr. Food Chem.*, **18**, 1010.
19. Demain, A.L. (1980) *Naturwiss.*, **67**, 582.
20. Marimoto, T. and Johnson, J.A. (1966) *Cereal Chem.*, **43**, 627.
21. This topic has been reviewed by Wold, F. (1981) *Annu. Rev. Biochem.*, **50**, 783.
22. Phosphoproteins have been reviewed by Tabarsky, G. (1974) *Adv. Protein Chem.*, **28**, 1.
23. Chen, R. and Chen-Schmeisser, U. (1977) *Proc. Nat. Acad. Sci. USA*, **74**, 4905.
24. Braun, V., Rehn, K. and Wolff, H. (1977) *Biochemistry*, **9**, 5041.
25. Lhoest, J. and Colson, C. (1977) *Mol. Gen. Genet.*, **154**, 175.
26 Morris, H.R., Dell, A., Peterson, T.E., Sottrup-Jensen, L. and Magnussen, S. (1976) *Biochem. J.*, **153**, 663.
27. Stenflo, J. and Suttie, J.W. (1977) *Annu. Rev. Biochem.*, **46**, 157.
28. Christy, M.R., Barklay, R.M. and Koch, T.W. (1981) *J. Amer. Chem. Soc.*, **103**, 3935.
29. Wilhelm, G. and Kunka, K-D. (1981) *FEBS Lett.*, **123**, 141.
30. Allen, A.K. and Neuberger A. (1973) *Biochem. J.*, **135**, 307.
31. Steinert, P.M. and Idler, W.W. (1979) *Biochemistry*, **18**, 5664.
32. Diedrich, D.L. and Shaitman, C.A. (1978) *Proc. Nat. Acad. Sci. USA*, **75**, 3708.
33. Hiraoki, B.Y., Fukusawa, K., Fukusawa, K.M. and Harada, M. (1980) *J. Biochem. (Tokyo)*, **88**, 373.
34. Stadtmann, T.C. (1980) *Annu. Rev. Biochem.*, **49**, 93.
35. Chance, R.E., Ellis, R.M. and Bromer, W.M. (1968) *Science*, **161**, 165.
36. Falk, J.E. and Finlayson, J.S. (1977) *Adv. Protein Chem.*, **31**, 2.
37. Brain, V. and Sieglin, U. (1970) *Eur. J. Biochem.*, **13**, 336.
38. This topic has been reviewed by Gallop, P.M. and Paz, A.M. (1975) *Physiol Rev.*, **55**, 418.
39. Bernstein, P.M. and Mechanic, J.L. (1980) *J. Biol. Chem.*, **255**, 10414.
40. Deyl, Z., Macek, K., Adam, M. and Vancikova, O. (1980) *Biochim. Biophys. Acta*, **625**, 248.
41. Elsden, D.F., Light, N.D. and Bailey, A.J. (1980) *Biochem. J.*, **185**, 531.
42. Keely, F.W. and Labella, F.S. (1972) *Biochim. Biophys. Acta*, **263**, 52.
43. Garcia-Castineiras, S., Dillon, J. and Spector A. (1978) *Science*, **199**, 897.
44. DeVare, D.P. and Gruebel, R.J. (1978) *Biochem. Biophys. Res. Comm.*, **80**, 993.
45. Steinert, P.M. and Idler, N.W. (1979) *Biochemistry*, **18**, 5664.

46. Steinberg, M., Yim, C.Y. and Schwende, F.J. (1975) *Science*, **190**, 992.
47. Givot, J.L., Smith, T.A. and Abeles, R.M. (1969) *J. Biol. Chem.*, **244**, 6341.
48. Wickens, R.B. (1969) *J. Biol. Chem.*, **244**, 6550.
49. Givot, J.L. and Abeles, R.M. (1970) *J. Biol. Chem.*, **245**, 3271.
50. Vogler, K. and Lanz, P. (1966) *Helv. Chim. Acta*, **49**, 1348.
51. Bycroft, B.W. and Lel, G.R. (1975) *J. Chem. Soc. Chem. Commun.*, 988.
52. Jew, S-S., Terashima, S. and Koga, K. (1979) *Tetrahedron*, **35**, 2337.
53. Kubel, B., Gruber, P., Hurnaus, R. and Steglich, W. (1979) *Chem. Ber.*, **112**, 128.
54. Hoppe, D. and Beckmann, L. (1979) *Ann. Chem.*, 2066.
55. Ewa, M., Katimoto, T. and Chibata, I. (1979) *Appl. Environ. Microbiol.*, **37**, 1053.
56. This topic has been reviewed by Yamada, S. and Koga, K. (1975) *Yuki Gosei Kogaku Kyotai Shi*, **33**, 535.
57. Yamada, S., Taniguchi, M. and Koga, K. (1969) *Tetrahedron Lett.*, 25.
58. Cristol, S.J., Ziebarth, T.D. and Lee, G.A. (1974) *J. Amer. Chem. Soc.*, **96**, 7842.
59. Koga, K., Taniguchi, M. and Yamada, S. (1971) *Tetrahedron Lett.*, 263.
60. Yamada, S., Oh-hashi, N. and Achiura, K. (1976) *Tetrahedron Lett.*, 2561.
61. Tanioka, K. and Koga, K. (1979) *Tetrahedron Lett.*, 3315.
62. Meyers, A.I. and Amos, R.A. (1980) *J. Amer. Chem. Soc.*, **102**, 870.
63. Mori, K. (1979) *Tetrahedron*, **32**, 1101.
64. Saitoh, Y., Moriyama, Y., Takahashi, T. and Khuang-Luu, Q. (1980) *Tetrahedron Lett.*, **21**, 75.
65. Asami, M. and Mukaiyama, T. (1980) *Chem. Lett.*, 17.
66. Aketa, K-i., Terashima, S. and Yamada, S-i. (1976) *Chem. Pharm. Bull.*, **24**, 261.

CHAPTER THREE

Beta and Higher Homologous Amino Acids

C.N.C. DREY

3.1 INTRODUCTION

Prior to a review [1] on the chemistry and biochemistry of β-amino acids this area of chemistry had not been covered apart from their synthesis [2], a short review of their biochemical properties [3], and a brief account of their occurrence [4].

The burgeoning literature accumulated in the last seven years emphasizes the importance of ω-amino acids. This is particularly associated with new groups of naturally occurring peptides, antibiotics and in the synthesis of peptide analogues with increased potency and *in vivo* stability.

Reference to azetidin-2-ones, aspartic acid and glutamic acid derivatives are excluded as these have been reviewed extensively. As appropriate, mention will be made of higher azetidinones.

3.2 OCCURRENCE OF β AND ω-AMINO ACIDS

The antibiotic P168 [5] and the newly isolated intrinsic substance [6] from *Carcinus maenas* contain β-alanine. Amino acids obtained from a Mighei Type C carbonaceous chondrite include *inter alia* β-aminobutyric acids [7]. β-Alanine betaine has been isolated from southern blue whiting [8]. An unusual acid isolated from human urine has been identified as α-hydroxy-β-keto-γ-amino-butyric acid [9]. A major source of new ω-amino acids is provided by the hydrolysis of antibiotics. In recent years the very important enzyme inhibitors amastatin [10], pepstatin [11] and bestatin [12] have been found to contain (2*S*,3*R*)-3-amino-2-hydroxy-5-methylhexanoic and (AHMHA) (*3.1*), (3*S*,4*S*)-4-amino-3-hydroxy-6-methylheptanoic acid (statine) (Sta) (*3.2*) and (2*S*,3*R*)-3-amino-2-hydroxy-4-phenylbutanoic acid (AHPA) (*3.3*) respectively.

$$R-CH(NH_2)CH(OH)CO_2H$$

$$R = CH_2CHMe_2 \quad (3.1)$$

$$R = CH_2Ph \quad (3.3)$$

$$(CH_3)_2CHCH_2CH(NH_2)CH(OH)CH_2CO_2H \quad (3.2)$$

Other hydroxy ω-amino acids isolated include *threo*-γ-hydroxy-β-lysine from tuberactinomycins A and N [13], and 4-amino-3-hydroxy-2,4-dimethylbutyric acid from cleomycin [14].

Hydrolysis of iturin A provided two new β-amino acids having structures (*3.4*) and (*3.5*) [15]. Their absolute configuration has been determined. The isolation of δ-aminovaleric acid derived from rumen ciliate protozoa has been reported [16]. 2-Methylene-β-alanine previously isolated from a Red Sea sponge, has been found in an unidentified black Hawaiian Sponge [17, 18].

$$CH_3CHR[CH_2]_8CH(NH_2)CH_2CO_2H$$

R = Me (*3.4*)

R = Et (*3.5*)

3.3 SYNTHESIS OF ANALOGUES

In recent years considerable effort has been expended in research directed towards increased stability and or potentiation of activity of physiologically active peptides. Enhanced stability towards enzymic degradation may be achieved by incorporating D-residues of α-amino acids and by the inclusion of ω-amino acids which are not normally substrates for proteolytic enzymes.

(a) Bradykinin

Bradykinin has been the subject of extensive replacement studies. Recently β-alanine [19, 20] and γ-aminobutyric acid [20] analogues and fragments have been prepared by solid state and classical methods. Analogues (*3.6*) and (*3.7*) both prolonged pentobarbitol-induced sleeping time in animal experiments.

H–Gly–β-Ala–Ser–Pro–OH (*3.6*)

H–β-Ala–Pro–Pro–Gly–Phe–γ-Abu–Pro–Phe–β-Ala–OH (*3.7*)

(b) Angiotensin

In addition to previously synthesized analogues, more recently derivatives of angiotensin (*3.8*) have been prepared with enhanced resistance to enzymic degradation, but unfortunately without increased duration of action *in vivo*. Results indicate that the maximum enhancement of resistance to chymotrypsin occurred when residue four or five was increased by one carbon atom. However with or without *N*-terminal sarcosyl residues pressor activity was found to be low [21, 22].

X^1–Arg–Val–X^2–X^3–His–Pro–Phe

X^1 = Sar, X^2 = β-HTyr, X^3 = Ile

X^1 = Asp, X^2 = Tyr, X^3 = β-HIle

X^1 = Sar, X^2 = Tyr, X^3 = β-HIle (*3.8*)

(c) Corticotropin

β-Alanyl-corticotropin derivatives prepared by solid state methods have been described [23, 24].

(d) LHRH (Luliberin)

The decapeptide LHRH (*3.9*) is a key mediator in the regulation of gonadotrophin, luteinizing hormone, and follicle stimulating hormone, and considerable developmental work has been undertaken on it. The analogue L-homohis^2 LHRH prepared by solid state techniques displayed very low *in vivo* potency, a result ascribed to conformation changed [25]. A useful analogue [β-Ala^6, β-Ala-NH_2^{10}]-LHRH prepared by solid state synthesis stimulated the release of luteinizing hormone and induced ovulation. Activity at 2–5 mg gave 97% release of luteinizing hormone in ewes compared to LHRH [26].

<Glu-His-Trp-Ser-Tyr-Gly-Leu-Arg-Pro-Gly-NH_2 (*3.9*)

(e) Thyrotropin

Synthesis of an *N*-terminal tripeptide amide (*3.10*) using conventional solution methods provided a thyrotropin releasing hormone analogue with antidepressant activity [27]. A similar peptide (*3.11*) has also been reported [28].

<Glu-His-Pro-β-Ala (*3.10*)

<Glu-Gly-β-Ala (*3.11*)

(f) Enkephalins

The successful isolation and structural determination of the endogenous opiate-like peptide enkephalin from brain tissues stimulated intensive activity and to date many hundreds of analogues have been prepared. Typically work has been directed at enhancement of activity and *in vivo* stability together with elucidation of biochemical roles. A number of preparations incorporating β- and ω-amino acids have been reported. These include β-Ala^2-Met-enkephalin (*3.12*) [29], β-Ala^1-Met/Leu-enkephalin (*3.13*) [30] and tris-homo-Tyr-des-Gly-Phe-enkephalin (*3.14*) [31] prepared by solid state, or solution methods.

H-Tyr-β-Ala-Gly-Phe-Met-NH_2 (*3.12*)

H-β-Ala-Gly-Gly-Phe-D/L Met-/Leu- OH (*3.13*)

Tris-homo Tyr-Gly-Phe-Met-NH_2 (*3.14*)

(g) Bestatin

Cultures of soil streptomyces have provided broths from which potent aminopeptidase inhibitors have been isolated. These are bestatin [12] (*3.15*) (2*S*,3*R*)-3-amino-2-hydroxy-4-phenylbutanoyl-*S*-leucine and amastatin [10] (*3.16*)

(2*S*, 3*R*)-3-amino-2-hydroxy-5-methylhexanoyl-*S*-valyl-*S*-valyl-*S*-aspartic acid. Bestatin (*3.15*) inhibits aminopeptidase B and leucine aminopeptidase but not aminopeptidase A, whilst amastatin (*3.16*) inhibits the latter too but not aminopeptidase B. Both substances have become important synthetic targets for confirmation of structure and improved activities. The synthesis of the respective hydroxy-amino acids (*3.3* and *3.1*) and analogous derivatives (Section 3.4) has received great attention.

(h) Amastatin

As previously discussed in Section 3.2, amastatin (*3.16*) is a low molecular weight inhibitor of some aminopeptidases. The *N*-terminal residue of AHMHA which is essential for activity is bound to a tripeptide *C*-terminal peptide L-valyl-L-valyl-L-Asp. Workers active in the bestatin field have likewise published procedures for the synthesis and resolution of AHMHA followed by incorporation into the *C*-terminal tripeptide. The synthetic material (*3.16*) inhibited leucine aminopeptidase at levels comparable to the natural substance [32].

A mechanism for inhibition has been advanced in which the role of the β-aminohydroxyacid can be seen to have a key effect namely that it is bound irreversibly to the active site of the enzyme [33–35].

(i) Pepstatin

Pepstatin has the structure iVal-L-Val-L-Val-Sta-L-Ala-StaOH where Sta is (3*S*, 4*S*, 4*S*)-4-amino-3-hydroxy-6-methylheptanoic acid (*3.2*). First discovered by Umezawa's group [11] pepstatin has been synthesized and several analogues prepared [36–40]. It is a specific inhibitor of carboxyl proteases and binds to the active site of pepsin with an inhibitory ratio enzyme/Sta of 1:1 to 1:2 dependent on the pH. Several groups have studied binding properties with statine or derivatives. Workman and Burkitt prepared pepstatin-L[^{125}I]monoiodotyrosine methyl ester by conventional coupling techniques and found that its inhibitory properties towards pepsin were identical to pepstatin. With the aid of the radio label the workers demonstrated that the dissociation constant for the enzyme–inhibitor complex is remarkably small with a K_D of $1.31 \times 10^{-11} \pm 0.63\,\text{M}^6$ [41].

Inhibition of pepsin by pepstatin involves the rapid formation of an enzyme–inhibitor complex which undergoes transformation to a more tightly bound complex. A co-operative action of three sites on pepstatin including the 3-*S*-hydroxyl group of the third residue are required for tight-binding inhibition [42].

(j) Bleomycin

Umezawa and coworkers first reported on a structure for bleomycin in 1972 since when a number of reviews have been published [43–47]. The antibiotic, produced by a strain of *Streptomyces verticillus*, is a potent glycopeptide antitumour agent of value in the treatment of squamous cell carcinomas, lymphomas and testicular carcinomas. Clinically the antibiotic is used as a strain of some eleven active fragments of which bleomycin A is the most abundant. Its biological action is thought to involve single and double strand breaks in DNA for which a two step mechanism has been proposed involving insertion of a bithiazole moiety into the DNA helices followed by oxidation of bleomycin/Fe(II)-O_2 complex to Fe(III)-bleomycin and the formation of hydroxyl or peroxide radicals [48]. DNA 'activity' is inhibited by some first row transition metal ions including copper and zinc.

The antibiotic contains a number of unusual structural features including the hydroxyamino acids β-hydroxyhistidine and (2*S*, 3*S*, 4*R*)-4-amino-3-hydroxy-2-methyl valeric acid (AHVA). Because of the uncertainty arising from structural revisions and the importance of synthetic analogues and structural verification the group of Umezawa and Hecht have sought a total synthesis to remove the ambiguity.

(k) Miscellaneous

A β-alanine-containing analogue of oxypressin prepared (*3.17*) by solution techniques was found to have antidiuretic and uterotonic activity but no pressor effect [49].

$$\text{H-}\underline{\text{Cys-Tyr-Phe-Gln-Asn-Cys}}\text{-Pro-Leu-}\beta\text{-Ala-NH}_2$$

(3.17)

In studies with somatostatin (*3.18*) ω-amino acid derivatives have been suggested as useful analogues in the treatment of acromegaly, gastric ulcers and diabetes [50]. A ring size of between 24–33 atoms is a requirement for biological activity.

$$\text{cyclo}[\text{X}_j\text{-X}^1_k\text{-X}^2_l\text{-Phe-Phe-X}^3\text{-Lys-Thr-Phe-Thr-X}^4_m]$$

$$\text{X} = \text{HN(CH}_2)_n\text{CO},\ n = 0\text{-}4;\ \text{X}^1 = \text{Lys};\ \text{X}^2 = \text{Asn, Ala};\ \text{X}^3 = \text{Trp, D-Trp}$$

$$\text{X}^4 = \text{Ser, Gly}\quad j,k,l,m \neq 0,1$$

(3.18)

A number of publications have appeared concerned with the synthesis of edeine, an antibiotic which contains isoserine and β-phenyl-β-alanine [51]. Analogues containing β-alanine and α-phenyl-β-alanine have also been prepared [52].

3.4 SYNTHESIS OF ω-AMINO ACIDS

(a) Synthesis of ω-amino acids

In order to provide a logical framework the synthetic methods used to prepare ω-amino acids have been classified by the particular reaction employed. Syntheses published prior to 1976 have been reviewed [1].

(b) Addition to unsaturated derivatives

The versatility of nucleophilic addition of amines to α,β-unsaturated acids, esters and nitriles has led to substantial refinement of the reaction such that both regio- and stereospecific control is possible. Addition of ammonia to α-methylacrylonitrile and acrylic acid under autoclave conditions provided α-methyl-β-alanine and β-alanine respectively in high yield [53, 54]. Similarly substituted 3-aryl-3-aminopropionic acids were afforded through 1,2-addition of hydroxylamine to unsaturated acids followed by catalytic hydrogenation [55]. Russian workers prepared *N*-alkyl-β-alanine derivatives by the addition of an alkylamine (C–1–14) to appropriate unsaturated esters and reported that the corresponding quaternary hydrochloride salts make useful bacteriocides [56]. An interesting application of 1,2-aminomercuration gave access to secondary or tertiary substituted 3-aminopropanoic acid esters.

$$R^2CH{=}CHR^3CO_2Me \xrightarrow[2.\ 4\text{-}RC_6H_4NHR^1]{1.\ Hg(OAc)_2} 4\text{-}RC_6H_4NHR^1CHR^2CR^3Hg(OAc)_2CO_2R^4$$

$$\xrightarrow{NaBH_4} 4\text{-}RC_6H_4NHR^1CHR^2CHR^3CO_2R^4$$

R = H, Cl, Me R^1 = H, Me R^2 = H R^3 = H, Me R^4 = Me

Disappointingly the reductive demercuration of many derivatives gave starting materials via elimination reactions [57].

Asymmetric syntheses of β-amino acids have been achieved by addition of chiral substituted benzylamines to 1-cyanopropenes and unsaturated esters. Amino acids were isolated by hydrolysis and catalytic hydrogenolysis [58].

$$CHR^1{=}CR^2R^3 + R^4NH_2 \longrightarrow R^4NHCHR^1CHR^2R^3$$

$$\xrightarrow[2.\ Pd/C/H_2]{1.\ Hydrolysis} (R/S)\text{-}NH_2CHR^1CHR^2R^3$$

R^1 = H, Me, Ph
R^2 = H, Me, Et
R^3 = CN, CO_2Me, l-carboxymenthyl
R^4 = (*R*)-PhCHMe, (*S*)-PhCHMe

(c) Reduction of enamines

2-Alkyl and aryl-β-amino acids may be conveniently prepared by either hydride reduction or hydrogenation of 3-(*R*) or (*S*)-(α-methylbenzyl)aminoacrylates. For example starting from the enamine of (*Z*) configuration either (*R*)-3-methyl or 3-phenyl-3-aminopropionic acid may be obtained depending on the method of reduction employed [59].

Application of homogeneous chiral catalysts for selective reductions and hydrogenations has also found application in the synthesis of ω-amino acids.

Thus reduction of methyl (*Z*)-3-acetylaminoprop-2-enoic acids over a chiral rhodium biphosphine complex yielded the corresponding chiral amino esters [60].

$$(Z)\text{-AcNHCR}=\text{CHCO}_2\text{Me} \xrightarrow{H_2} \text{AcNHCHRCH}_2\text{CO}_2\text{Me}$$

R = Me, Ph

Catalysts were prepared *in situ*. The products were subsequently isolated by preparative TLC in good yield but optical yields ranged from 3–55%.

(d) Reformatski reaction

A variation of the Reformatski reaction in which a chiral Schiff base on treatment with a β-bromo ester provided 3-methyl- or 3-phenyl-3-aminopropanoic acids in 2–28% optical purity [61]. Thus:

$$R^1CH=NR^2 + BrCH_2CO_2R^3 \xrightarrow{1.\ Zn;\ 2.\ H^+;\ 3.\ H_2/Cat} NH_2CHR^1CH_2CO_2H$$

R^1 = Me, Ph
R^2 = (*R*) or (*S*)-PhCHMe
R^3 = Et, l-menthyl

(e) Rearrangements

Many applications of the Arndt–Eistert reaction have been reported for the homologation of α-amino acids. A recent example is in the preparation of benzyloxycarbonyl-L-3-aminobutyric acid and the corresponding methyl ester which were both recovered directly and in high yield after silver benzoate catalysed rearrangement of the precursor diazoketone [62]. Whilst attempting a similar conversion to *N*-tosyl-3-aminobutyric acid, the intermediate carbene was trapped and *N*-tosyl-2-methylazetidin-2-one (3.19) isolated in good yield [62].

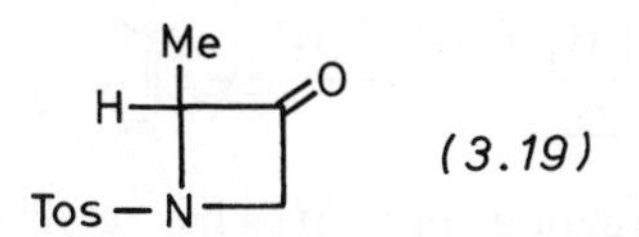

(3.19)

Other syntheses include *inter alia* β-leucine [63], and *N*-methyl-L-β-arginine [64]. The latter is a constituent of antibiotic LL-BM547B whose cyclic core structure is identical to that of tuberactinomycin A and B (cf. viomycin). The homoarginine derivative (*3.20*) was prepared as follows:

$$\text{Boc-NHCH}_2\text{CH}_2\text{CH(NMeTos)CO}_2\text{H} \xrightarrow[\text{2. CH}_2\text{N}_2;\ \text{3. PhCO}_2\text{Ag/MeOH}]{\text{1. ClCO}_2\text{Et/NMM}}$$

$$\text{Boc–NHCH}_2\text{CH}_2\text{CH(NMeTos)CH}_2\text{CO}_2\text{Me} \xrightarrow[\text{2. NH}_2\text{C(=NNO}_2\text{)OCH}_3\text{ NaOH}]{\text{1. HCL}}$$

$$\text{NH}_2\text{C(=NNO}_2\text{)NHCH}_2\text{CH}_2\text{CH(NMeTos)CH}_2\text{CO}_2\text{H} \xrightarrow[\text{2. 47\% HBr}]{\text{1. H}_2\text{/Pd}}$$

$$\text{NH}_2\text{C(=NH)NHCH}_2\text{CH}_2\text{CH(NHMe)CH}_2\text{CO}_2\text{H}$$

(3.20)

The Curtius rearrangement when applied to cyclic anhydrides gave β- and γ-amino acids in 26–50% yield [65].

$$[\text{CH}_2]_n(\text{CO})_2\text{O} + \text{Bu}_3\text{SnN}_3 \rightarrow \text{N}_3\text{CO[CH}_2]_n\text{CO}_2\text{SnBu}_3$$

$$\xrightarrow[\text{2. hydrolysis}]{\text{1. ROH}} \text{RO}_2\text{CNH[CH}_2]_n\text{CO}_2\text{H}$$

$n = 2$, R = CH_2Ph
$n = 3$, R = $CH_2C_6H_4OMe(4)$

Both *cis*- and *trans*-2-aminocyclohexanecarboxylic acids and *cis*-2-amino-4-cyclohexenecarboxylic acid and associated derivatives may be prepared from alicyclic trimethylsilyl-β-isocyanatocarboxylates. Cyclization of the deprotected β-isocyanatocarboxamides gave the corresponding 5,6-dihydrouracils [66].

(f) Miscellaneous reactions

The Knoevenagel reaction has been used to prepare 3-(cyclohex-3-enyl)-2-aminopropanoic acid [67].

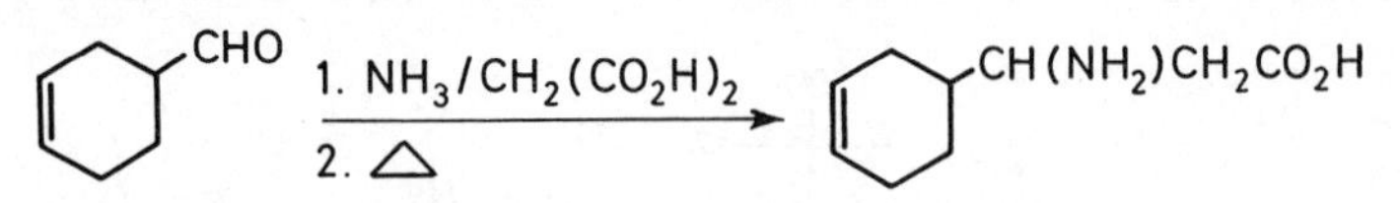

Reduction of phenylhydrazone derivatives has re-emerged as a versatile route to racemic ω-amino acids [68].

$$\text{OC(R)[CH}_2]_n\text{CO}_2\text{H} \xrightarrow{\text{PhNHNH}_2} \text{PhNHN=CR[CH}_2]_n\text{CO}_2\text{H}$$

$$\xrightarrow{\text{H}_2} (\pm)\,\text{NH}_2\text{CHR[CH}_2]_n\text{CO}_2\text{H}$$

$n = 1$, R = Ph; 30%
$n = 2$, R = naphthyl; 50%
$n = 2$, R = Ph; 95%
$n = 2$, R = 4-MeOC_6H_4-; 96%
$n = 4$, R = Ph; 100%

(g) Halogen substituted ω-amino acids

A number of fluoro-β-alanine derivatives have shown effective inhibition of γ-aminobutyric acid (GABA) transaminase in mice. The derivatives were prepared by Grignard synthesis followed by oxidation of the terminal alkene to carboxylate [69]

$$CH_2{=}CHCH_2Br \xrightarrow[\text{via Grignard reaction}]{FCH_2CN} NH_2CH(CH_2F)CH_2CH{=}CH_2$$

$$\xrightarrow[\text{2. } KMnO_4]{\text{1. t-Butoxycarbonylation}} BOC{-}NHCH(CH_2F)CH_2CO_2H$$

$$\xrightarrow[\text{2. } Et_3N]{\text{1. HCl}} NH_2CH(CH_2F)CH_2CO_2H$$

An alternative reaction provides 2-fluoro-β-alanine derivatives by an electrolytic reaction [70] using carbon electrodes. α-Chloro-β-alanine has been synthesized by Gabriel reaction with chloroacrylonitrile [71].

$$C_4H_6(CO)_2NK + CH_2{=}CClCN \xrightarrow[\text{hydroquinone}]{t\text{-BuOH}} C_4H_6(CO)_2NCH_2CHClCN$$

$$\xrightarrow{HCl} NH_2CH_2CHClCO_2H$$

(h) Phosphorus containing β-amino acids

Phosphonatoacetyl β-alanine prepared by direct coupling reactions is a potent inhibitor of DNA polymerase at extremely low concentration [72].

$$(PhO)_2P(O)CH_2CO_2H + NH_2CH_2CH_2CO_2CH_2Ph \xrightarrow[HNSu]{DCCI}$$

$$(PhO)_2P(O)CH_2CONHCH_2H_2CH_2CO_2CH_2Ph \xrightarrow{Pd/C/H_2}$$

$$(PhO)_2P(O)CH_2CONHCH_2CH_2CO_2H$$

The crystal structure and infrared spectra of β-alanylaminomethylphosphoric acid (*3.21*) have been determined. Preparation of (*3.21*) was achieved by standard methods [73].

$$ClCH_2CH_2COCl + NH_2CH_2P(O)(OH)_2 \longrightarrow NH_2CH_2CONHCH_2P(O)(OH)_2 \quad (3.21)$$

An alternative method also provided *(3.21)*:-

$$C_6H_4(CO)_2NCH_2CH_2CO_2H \xrightarrow{\text{1. } ClCO_2Et;\ \text{2. } NH_2CH_2P(O)(OH)_2;\ \text{3. } N_2H_4} (3.21)$$

Preparation of the thiophosphorus derivatives (*3.22*) was undertaken to investigate the metabolites and toxicity of thio-organophosphorus insecto-acaricides [74].

$$EtOPR(A)SCH_2COX$$
$$R = Me, OEt$$
$$A = O, S$$
$$X = \beta\text{-Ala}, \beta\text{-AlaOEt}, Val, Gly \quad (3.22)$$

Ethyl diethylenimidophosphoryl-β-aminophenyl-β-alaninate (*3.23*) has been synthesized [75].

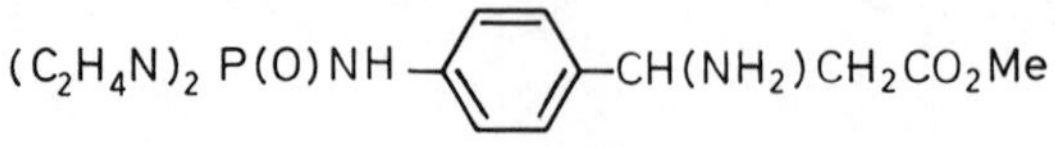

$$(C_2H_4N)_2P(O)NH-C_6H_4-CH(NH_2)CH_2CO_2Me$$
$$(3.23)$$

(i) GABA

Cyanopropionamide is readily converted to GABA amide (*3.24*) by catalytic hydrogenation [76].

$$NCCH_2CH_2CONH_2 \xrightarrow[H^+]{PtO_2/H_2/EtOH} NH_2(CH_2)_3CONH_2HCl \quad (3.24)$$

A Belgian patent discloses procedures for the preparation of α halo GABA but no data are given for its inhibiting properties [77]. A second Belgian patent describes routes to 1- and 2-substituted hydroxyl, amino and alkyl GABA derivatives. Thus (*Z*)-asparagine by dehydration to the corresponding nitrile afforded L(+)-2-amino GABA after catalytic hydrogenation [78].

(j) Methylene-β-alanine and oxo derivatives

2-Methylene-β-alanines have previously been isolated from marine sources. An oxo-analogue (*3.25*) isolated from marine sources has been synthesized by heating t-butyl 3-amino-2-methylenepropionate with 2-oxopalmitic acid in ammonia at low temperature [17].

$$NH_2CH_2C(:CH_2)CO_2{}^tBu \xrightarrow{\text{2-Oxopalmitic acid}} Me(CH_2)_{13}COCONHCH_2C(:CH_2)CO_2Me$$
$$(3.25)$$

(k) Amino acids from edeine

N-(4-Aminobutyl)-3-aminopropionic acid (*3.26*), a hydrolysis product of the antibiotic edeine (Section 3.3), has been synthesized by reaction of mono acetyl-1,4-diaminobutane with acrylonitrile followed by hydrolysis [79].

$$AcNH[CH_2]_4NH_2 + CH_2{=}CH\,CN \rightarrow AcNH[CH_2]_4NHCH_2CH_2CN$$

$$\xrightarrow[\text{2. hydrolysis}]{\text{1. }Ac_2O} AcNH[CH_2]_4NAcCH_2CH_2CO_2H \quad (3.26)$$

(l) Reduction of uracils

Development of the method of Birkofer and Storch has led to an improved procedure for the synthesis of both 2,3-diaminopropionic acid derivatives and of α-substituted-β-alanines [80].

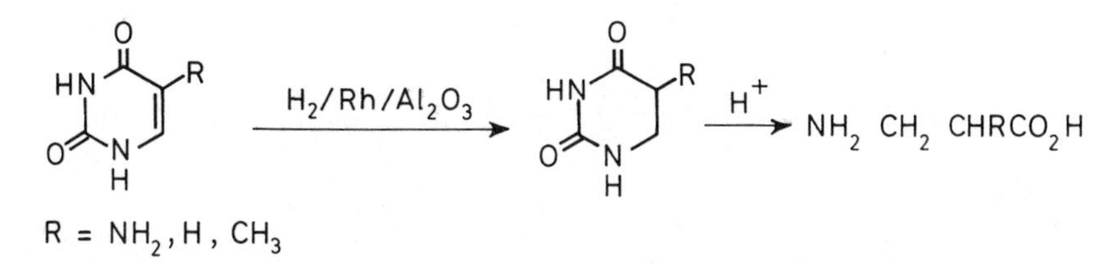

(m) Hydroxy ω-amino acids

In previous sections attention has been drawn to the occurrence of substituted 3-amino-2-hydroxyacids and to 4-amino-3-hydroxyhexanoic acid, statine (Sta), (*3.2*). All of the foregoing have been synthesized in addition to many analogues. Amastatin (*3.16*) previously isolated by Umezawa and coworkers was assigned the structure (2*S*,3*R*)-AMHA-Val-Val-Asp on the basis of chemical degradation and mass spectrometry [81]. The unique hydroxyacid AHMHA (*3.1*) has since been synthesized [82]. Principal routes to (*3.1*) included selective reduction of esters or active amides with hydride reducing agents. Rich and coworkers reduced methyl benzyloxycarbonyl-D/L-leucinate with dibal and treated the aldehyde with bisulphite and cyanide, the resulting cyanohydrin was deblocked and reprotected as the *N*-*t*-butoxycarbonyl derivative. Separation of diastereoisomeric pairs was then possible by silica gel chromatography [32].

$$\mathrm{ZNHCH}(i\text{-}\mathrm{C_4H_9})\mathrm{CO_2Me} \xrightarrow{(i)} \mathrm{Z\ NHCH}(i\text{-}\mathrm{C_4H_9})\mathrm{CHO} \xrightarrow{(ii)}$$

$$\mathrm{ZNHCH}(i\text{-}\mathrm{C_4H_9})\mathrm{CH(OH)CN} \xrightarrow{H^+} \mathrm{NH_2CH}(i\text{-}\mathrm{C_4H_9})\mathrm{CH(OH)CO_2H}$$

$$\xrightarrow{(iii)} \mathrm{Boc\text{-}NHCH}(i\text{-}\mathrm{C_4H_9})\mathrm{CH(OH)CO_2Me}$$

(*3.1*)

(i) Dibal	(a) (2*S*, 3*R*)
(ii) 1. $NaHSO_3$; 2. KCN	(b) (2*R*, 3*R*)
(iii) 1. Boc N_3; 2. CH_2N_2;	(c) (2*S*, 3*S*)
3. SiO_2	(d) (2*R*, 3*S*)

Bestatin (Section 3.2), as has already been mentioned, contains (2*S*, 3*R*)-3-amino-2-hydroxy-4-phenylbutanoic acid (AHPA) (*3.3*). Syntheses of (*3.3*) and analogues have been widely reported and are primarily based on the methodology successfully used in the preparation of AHMHA (*3.1*) [83–89].

In a recent paper Umezawa and coworkers have described a regio- and stereospecific synthesis for both (*3.1*) and (*3.3*) (±)-*threo*/*erythro* (Section 3.2) by

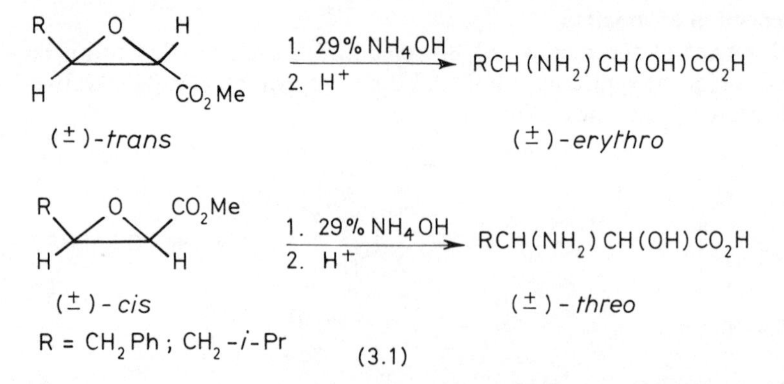

ring opening of *cis*-2,3-epoxy acids with ammonia (Reaction 3.1) [90, 91]. Quoted yields are good, however synthesis of the respective unsaturated esters required for epoxidation is not without its drawbacks.

Pepstatin (Sections 3.2 and 3.3) contains (3*S*, 4*S*)-4-amino-3-hydroxy-6-methylheptanoic acid, statine (*3.2*), which has been synthesized by a number of groups. Rich and coworkers utilized a Reformatski reaction which yielded the protected amino acid (*3.2*) directly [92].

$$\text{Boc-L-Leu-OMe} \xrightarrow[-78^\circ]{\text{Dibal}} \text{Boc-NHCH}(i\text{-}C_4H_9)\text{CHO}$$

$$\xrightarrow{\text{LiCH}_2\text{CO}_2\text{Et}} \text{Boc-NHCH}(i\text{-}C_4H_9)\text{CH(OH)CH}_2\text{CO}_2\text{Et}$$

$$\xrightarrow{\text{SiO}_2/\text{EtOAc}/C_6H_6} (3S,4S)\ \text{StaOH} + (3R,4S)\ \text{StaOH}$$

(*3.2*)

The same group also report on the synthesis of (3*S*, 4*S*)-4-amino-3-hydroxy-5-phenylpentanoic acid using similar methodology [93].

An interesting route to 4-amino-2-hydroxybutyric acid (*3.27*) is based on the dehydration of a primary amide to nitrile followed by catalytic hydrogenation [94].

$$\text{L-NH}_2\text{COCH}_2\text{CH(OH)CO}_2\text{H} \xrightarrow{\text{Ac}_2\text{O/Py}} \text{NC—CH}_2\text{CH(OAc)CO}_2\text{H}$$

$$\xrightarrow{\text{H}_2/\text{PtO}_2} \text{L-NH}_2\text{CH}_2\text{CH}_2\text{CH(OAc)CO}_2\text{H}$$

(*3.27*)

A synthesis of *threo*-γ-hydroxy-L-β-lysine a constituent of the antibiotic tuberactinomycin made use of the analogous ornithine derivative which was subject to homologation [95].

In earlier discussion reference has been made to the antibiotic bleomycin which contains (2*S*,3*S*,4*R*)-4-amino-3-hydroxy-2-methylvaleric acid. The latter's structure elegantly synthesized from L-rhamnose was thus confirmed [96], (Reaction 3.2).

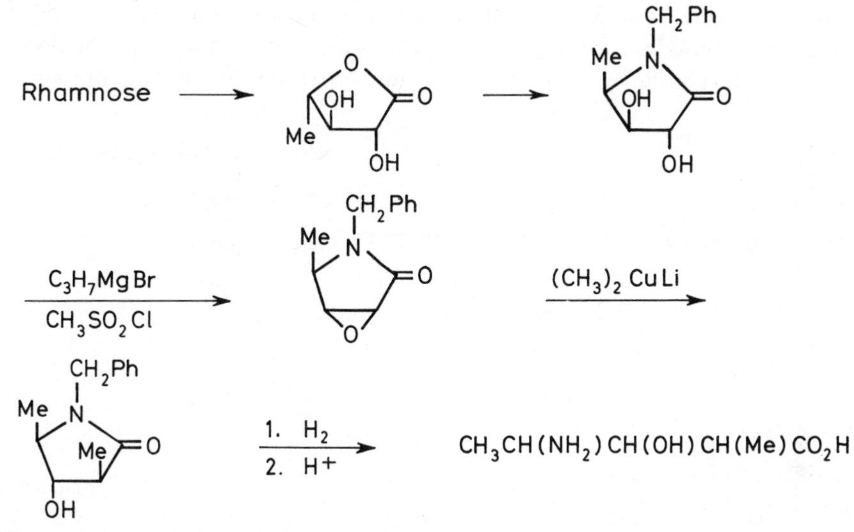

3.5 OTHER ω-AMINO ACID PEPTIDES AND DERIVATIVES

(a) Introduction

In recent years a number of trends have emerged in regard to target structures incorporating ω-amino acids. Thus the preparation of heterocyclic, secondary acylamido and phosphorus containing derivatives has featured in the search for compounds with biological action.

(b) Heterocyclic derivatives

El-Naggar *et al.* [97] recently described the synthesis of unprotected and protected amino acid and peptide derivatives of 2-aminophenazines (*3.28*) aminobenzophenazines (*3.29*) and aminophenazone derivatives (*3.30*).

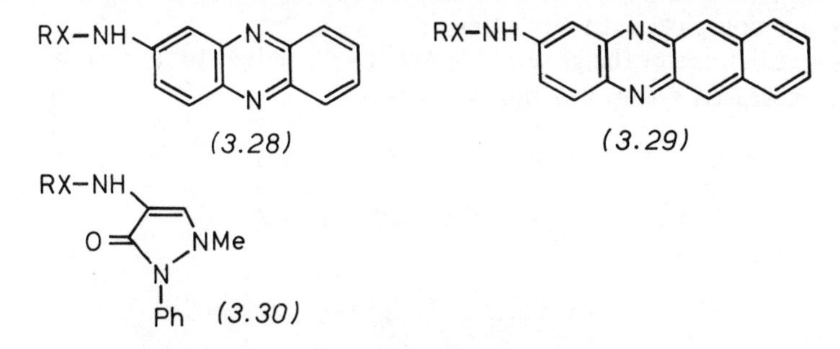

X = Gly, β-Ala, Val, Leu, (±) Phe, Ala, (±) Ser
R = H, Pth, Tos, Tos–Ala

The peptides were prepared by conventional procedures using DCCI Compound (*3.30*). X = (±) Ser, R = Tos was found to be very active against *Penicillium chrysogenum*. Similarly acridines were employed as the heterocyclic group by Wysocka-Skrzela *et al.* in a continuation of studies on the synthesis of the anticancer drug ledakrin [98]. In their compounds pyridinium salts or phenyl ethers were displaced by ω-amino esters thus providing the parent compounds (*3.31*).

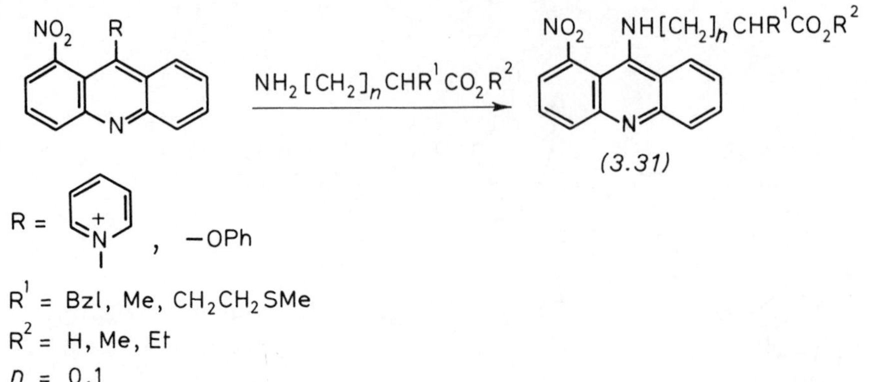

Thus the β-alanine derivative (*3.31*, $R^1 = R^2 = H$, n = 1) was recovered in 62% yield by heating *N*-(1-nitro-9-acridyl)pyridinium chloride in phenol. Reaction of DL or D and L-phenylalanine with *o*-nitrobenzoyl chloride gave the corresponding (*R*)-and (*S*)-benzodiazepines (*3.32*) [99].

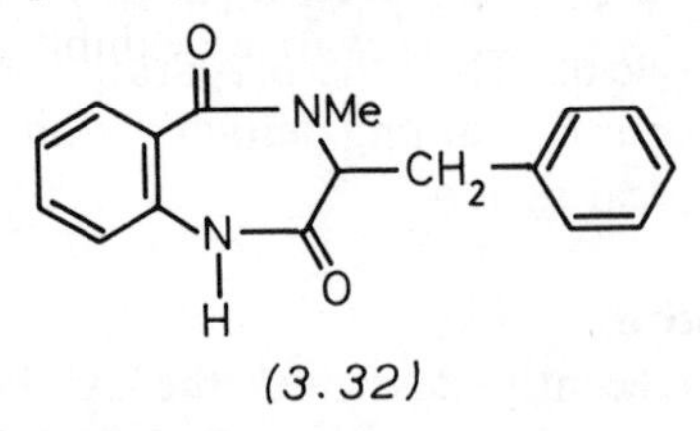

(c) ω-Amino acid amides, acids and esters

Recent patent literature reveals that many ω-amino acid amide derivatives have widespread biological activities [100].

Thus compounds of the general formula (*3.33*) reduced blood sugar levels in rats and increased serum vitamin A levels.

$$RR^1N[CH_2]_mCONH[CH_2]_nR^2 \quad (3.33)$$

R = H, Bz, $PhSO_2$, Alkoxycarbonyl, etc.

$R^1 = H$

R, R^1 = alkenedioyl, 2-COC_6H_4CO, COCH=CHCO

$R^2 = SO_3H$, OPO_3H_2

m, *n* = 2,3

Compounds of the general formulae (*3.34*) provided good protection in mice against infections of *Klebsiella pneumonia* [101].

$$R^1-(X)_m-\underset{|}{N}COR^2$$
$$[CH_2]_nCO-(X^1)_p-R^3$$

(3.34)

R^1, R^2 = lipophilic group $C \geqslant 10$
m = 0,1
n = 1,2
p = 0–3
X, X^1 = β-Ala, Gly, Ala, Arg, etc.

N-Substituted ω-aminoalkanoyl-ω-aminoalkanoic acids typified by (*3.35*) have been prepared and found to have properties promoting the secretion of the gall bladder and pancreas. ω-Amino acids used in these preparations included β-alanine, γ-aminobutyric acid and ε-aminocaproic acid [102].

$$4-MeOC_6H_4N(COC_6H_4Cl)CH_2CON(C_6H_4OMe)CH_2CH_2CH_2CO_2H$$

(3.35)

A variety of γ-aminobutyramide derivatives exhibited the ability to modify neurotropic activities [103], and lower blood sugar levels in rats [104], and NN^1-diaryl-γ-aminobutyric acid derivatives, the subject of patents, demonstrate antihepatotoxic and ulcer inhibitory properties [105].

(d) Peptides

In vitro synthesis of acetylaspartyl peptides has been demonstrated in homogenates of mouse cortex. Concerning the role of GABA it was postulated that the former may be conjugated as *N*-acetylaspartyl-GABA. Synthesis of the dipeptide by conventional methods gave a product which had no anticonvulsant or microbial activity. The corresponding succinimides formed by lactamization also had no antiepileptic properties [106]. Synthesis of β-alanyl-L-alanine and the retro-dipeptide were directed towards a study of the influence of peptides on germination of spores of *Bacillus thiaminolyticus*. The peptides were inactive [107]. In further studies on the synthesis of hindered β-amino acid peptides conventional coupling reagents were found to be ineffective. Thus with β-aminopivalic acid, carbonic mixed anhydrides provided acylated amino esters and very low yields of peptides. With DCCI, very high yields of *N*-acylurea could be recovered. Good yields of peptides were obtained with DCCI and Hbt or with the oxazinone derivative [108] (*3.36*).

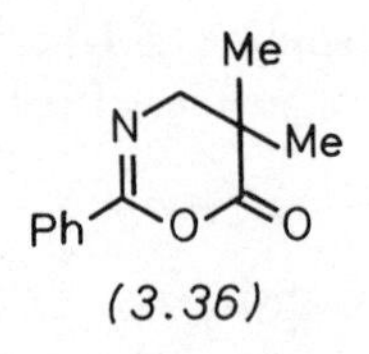

(3.36)

Miscellaneous homogeneous and mixed di- and tripeptides of β-alanine required for enzyme substrate studies have been prepared by azide and acid chloride coupling reactions [109]. Hecht and co-workers [110] in the course of very elegant synthetic studies have prepared both the *S*-tri and tetrapeptides (*3.37*) and (*3.38*), and Umezawa's group recently published syntheses for the pyrimidine moieties (*3.39*) of bleomycin and epibleomycin [111]. The synthesis of another constituent, namely L-*erythro*-β-hydroxyhistidine, has also been described [112, 113].

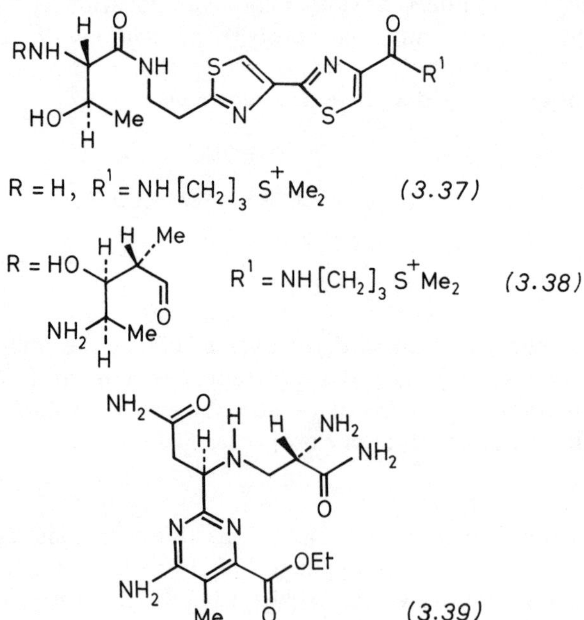

Cleomycin, a new member of the group, has been identified and its structure determined by physico-chemical methods. The major difference concerns the exchange of a threonine residue for the hydroxycyclopropyl derivative [14] (*3.40*).

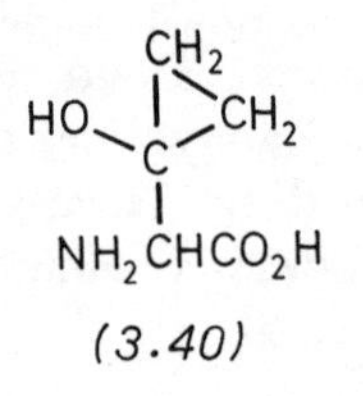

(3.40)

An associated group of antibiotics, capreomycins, viomycin (tuberactinomycin B) and tuberactinomycin S possess antitubercular activity. Structurally they are related, containing a cyclic pentapeptide and a variety of unusual amino acids [114–118].

3.6 Cyclopeptides

Growth hormone inhibitors which are 14-cyclopeptides (*3.41*) incorporating β-alanine, D-amino acids, γ-aminoisobutyric acid, cysteine and aminosulphydryl residues have been prepared [119].

$$\text{cyclo}[X^1\text{-}X^2\text{-}X^3\text{-Phe-Phe-Trp-Lys-Thr-Phe-Thr-Ser-}X^4] \quad (3.41)$$

typically $X^1 = \beta$-Ala-Gly-Cys; D-Ala-Gly-Cys; Aib-Gly-Cys
$X^2 =$ Lys, Orn
$X^3 =$ Asn, Ala
$X^4 = -NHCH_2CH_2S-$; Cys –; tBu-Cys-

The cyclodepsipeptide, protodestruxin (*3.42*), has been synthesized by conventional solution techniques. Cyclization through the ester linkage proved ineffective in combination with DCCI or the *p*-toluenesulphonate. The cyclo-derivative (*3.42*) eventually prepared by linkage at the prolyl *C*-terminus was identical in all respects to the naturally occurring material [120].

$$\text{cyclo}[\text{Pro-Ile-Val-Ala-}\beta\text{-Ala-Hmp}]$$

Hmp = D-2-hydroxy-4-methylpentanoic acid
(*3.42*)

Further to the synthesis of ($\pm$)-cyclo-tri-β-aminoisobutyryl, the (+) analogue could be isolated by cyclization of the (*S*, *S*, *S*)-tripeptide (*3.43*) using *o*-phenylenephosphochloridite in diethyl phosphate at reflux temperatures. The cyclopeptide (*3.44*) separating in a crystalline form in some 50% yield, proved to be insoluble in a variety of polar and covalent organic solvents in addition to water which is in marked contrast to the ($\pm$)-cyclotripeptide [121].

H-β-Abu-β-Abu-β-Abu-OH
(*3.43*)

$$\text{cyclo}[\beta\text{-Abu-}\beta\text{-Abu-}\beta\text{-Abu}]$$

(*3.44*)

In a continuation of investigations of high numbered cyclic amides of β-alanine, Rothe and coworkers have carried out a very careful analysis of the cyclization of the di-, tri- and tetrapeptides of β-alanine using the phosphochloridite method. Separation of cyclo-oligomers was achieved by gel filtration giving the following distribution [122].

H-[β-Ala]$_2$-OH → cyclo-[β-Ala]$_n$ n = 2,4,6,8,10,12 in 23,44,10,3,1 and 0·5% yield

H-[β-Ala]$_3$-OH $\xrightarrow{i}$ cyclo-[β-Ala]$_n$ n = 3,6,9 in 58,11 and 2·5% yield

H-[β-Ala]$_4$-OH → cyclo-[β-Ala]$_n$ n = 4,8 in 68 and 2·5% yield

i = 0·01M solution in diethyl phosphate : *o*-phenylenephosphochloridite

A study of the cyclopeptides derived from ω-lysine has been recently published. Tripeptides were prepared by conventional solution techniques using Z-*N* and Boc-*N* protection in conjunction with *p*-nitrophenyl ester couplings. Cyclization afforded the *Z*-protected derivative (*3.45*) which was deblocked to give the free ε-amino-cyclo-tripeptide (*3.46*) [123].

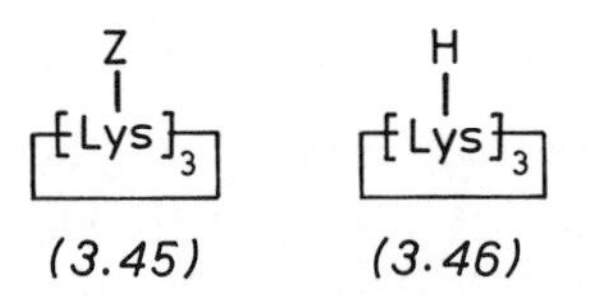

(*3.45*) (*3.46*)

A novel method of preparing cyclo-ω-amino acid lactams (*3.47*) utilized reaction of γ-butyrolactone together with the ω-amino acid in toluene. In addition to the individual lactams, depsipeptide byproducts were also isolated [124].

$NH_2[CH_2]_nCO_2H$ $\xrightarrow{\gamma\text{-butyrolactone}}$ cyclo-[$NH[CH_2]_n$–CO] n = 3,4,5

(*3.47*)

A new antiviral cytotoxic depsipeptide has been isolated from a Caribbean tunicate – didemins A, B and C (*3.48*). All three inhibit RNA and DNA viruses in addition to which they are very highly cytotoxic to L1210 cells and protect mice against P388 leukaemia and B16 melanoma. The structures were assigned by mass spectral and [^{1}H]-NMR studies. Unusual modifications of amino acids include *N*, *O*-dimethyltyrosyl, *N*-lactylleucyl residues, as well as *N*,*O*-dimethyltyrosyl, *N*-lactylprolyl and *N*,*N*-methyllactylleucyl residues [125].

MeLeu-Thr-Sta-Hip-Leu-Pro-MeTyr-O (ring closed from Thr to O)

(*3.48*)

Didemin A
Hip = Hydroxyisovalerylpropionyl
Me$_2$Tyr = *N*,*O*-diMeTyr
Sta = Statine
Didemin B
MeLeu modified by *N*-lactylPro

Didemin C
MeLeu modified by *N*-lactyl

3.7 POLYMERS

Introduction

Interest has continued in homo and heteropolymers derived from β- and higher amino acids since the area was last reviewed [1]. Polymers derived from non-α-amino acids are of interest because of their homology, properties and for comparison with other naturally occurring polymers such as collagen.

(a) β-Alanine

Hanabusa and coworkers [126, 127] have systematically examined 4-acyl-2-nitrophenolate esters as intermediates in the preparation of poly-β-alanine.

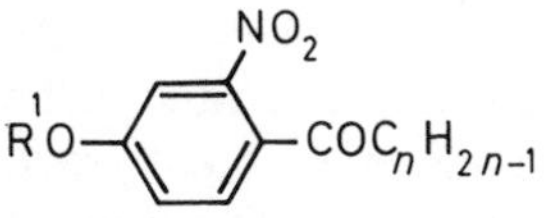

$$R^1O-C_6H_3(NO_2)-COC_nH_{2n-1} \qquad n = 1, 5, 7, 9, 11, 12, 15, 17$$

$$R^1 = H-[\beta\text{-Ala}]_m-OH \qquad m = 1, 2, 3$$

Polymerizations were found to be most efficient in solvents such as benzene or carbon tetrachloride with a 1:1 mole ratio of monomer to initiator (triethylamine) at 25–30°C. At the completion of reaction the *C*-terminal activated ester may be hydrolysed with base. Homopolymers with DPs of 96 could be recovered using the mono-β-alanyl activated ester. Addition of polar solvents such as methanol inhibited the polymerization. This was ascribed to changes in the state of aggregation and micelle formation [128].

Low molecular weight, water-soluble poly-β-alanine may be recovered from the alkoxide polymerization of acrylamide in chlorobenzene using *p*-nitrosodimethylaniline as a free-radical scavenger [129]. Previously, molecular weights of 76 000 have been found using this route [1].

(b) β-Aspartic acid

Polymerization of the β-aspartyldipeptide (*3.49*) yielded oligomers with some eight to nine residues.

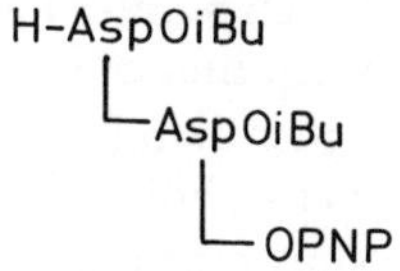

```
H-AspOiBu
  └AspOiBu
     └OPNP
```
(*3.49*)

CD and NMR spectroscopic studies suggested that ordered structures are formed when there are at least eight residues present with the most likely structure corresponding to a β-sheet form [130].

(c) ω-Amino acids

The syntheses of homopolymers derived from the amino acids $NH_2(CH_2)_nCO_2H$ where $n = 2, 3, 4$, and 5 was undertaken by the sequential solid phase approach

[131]. Polymers (DP 24, 48, 55, 48 respectively) were retrieved by treating the resin with hydrogen bromide and trifluoroacetic acid. X-ray studies show similarities to the diffraction patterns of the crystallites in silk fibroin which assume anti-parallel pleated sheet conformations. Addition of water to 2,2,2-trifluoroethanol–methanol solutions destabilized the regular conformation except for ε-aminohexanoic acid.

(d) *N*-Carboxyanhydrides

Birkofer and Modic [132] first prepared *N*-carboxyanhydrides (NCA) from *β*-amino acids which were subsequently used in polymerization studies. Further studies by Kricheldorf [133] of the copolymerization of glycine and *β*-alanine showed that both monomers had very similar reactivity ratios. In the presence of aprotic bases such as pyridine the rate of incorporation of glycine was somewhat greater than that for *β*-alanine. The polymer exhibited a random block structure. [^{15}N]-NMR allowed the unequivocal characterization of sequences as the chemical shifts, as measured upfield of nitrate ion in trifluoroacetic acid, differed by several ppm and are apparently reasonably invariant of neighbouring effects. Kricheldorf and coworkers in a further study examined the effects of amines, amine ratio and amine nucleophilicity versus basicity [134]. Their studies showed that high molecular weights could not be obtained with primary or secondary amines when the pK_a of the amine was greater than eight due to hydrogen abstraction from the NCA (*3.50*) to give a *β*-isocyanatocarboxylate anion which terminated by amine capture giving a *β*-ureido acid (*3.51*).

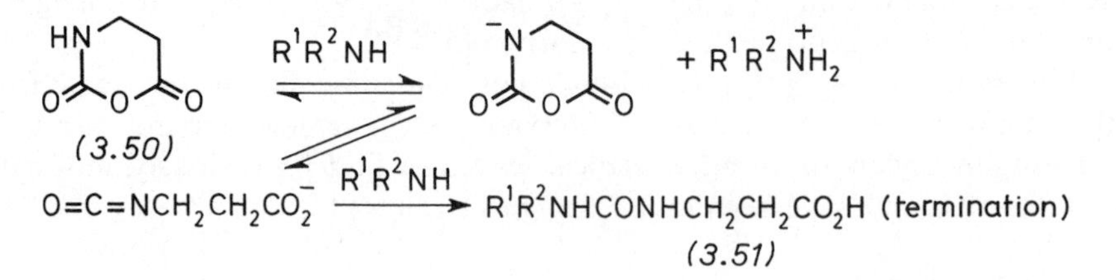

When *β*-Ala NCA was initiated with primary amines the mechanisms of reaction were similar to those observed for α-amino acid NCAs [135] with two extreme cases from a kinetic viewpoint. Thus if the initiator was more nucleophilic than the free amino terminus of the growing polymer the rate of the 'start reaction' was found to be higher than the rate of propagation. The converse with aromatic amines whose nucleophilicity is less than that of the growing amino terminus, led to a slower 'start reaction' but a higher DP. In either case the DP is a function of monomer to initiator ratio. It followed that the use of tertiary amines with a high monomer to initiator ratio should give the highest DPs.

(e) Heteropolymers

Polymerization of *N*-hydroxysuccinimido esters of alanyl-*β*-alanine and derivatives yielded block polymers [136].

H-[Ala-β-Ala]$_n$-OH

H-[Ala-Ala-β-Ala]$_n$-OH

H-[Ala-β-Ala-Ala-β-Ala]$_n$-OH

H-[Ala-Ala-DL-NHCH$_2$CHMeCO]$_n$-OH

The authors noted that *β*-alanine was not incorporated into an α-helix with alanine but into a polyproline type II helix. It was suggested that the rotational freedom around the N–C^{β} and C^{α}–C^{β} bonds could account for the findings and observations that have been made elsewhere.

Copolymerization of *β*-*p*-nitrophenyl esters of aspartic acid and *β*-alanine in dimethylformamide at room temperature for six days afforded a copolymer containing 53% mol of *β*-alanine and 43% mol of aspartic acid with a typical molecular weight of 5200 [137]. X-ray and CD studies did not provide evidence that the polymer was structured.

Further studies aimed at collagen models have led to an improved synthesis of H-(Pro-Pro-*β*-Ala)$_n$-OH via *p*-nitrophenyl active esters rather than the corresponding pentachlorophenyl esters [138, 139]. The latter are reported to be more sensitive to impurities and to the formation of diketopiperazines. With the improved route a DP over 40 could be attained in some 70% yield. [^{13}C]-NMR spectra in per-deuteriodimethyl sulphoxide gave almost identical spectra to those obtained from H-(Pro-Pro-Gly)$_n$-OH.

3.8 MODEL SYSTEMS AND REACTIONS

(a) Enzymes

It has been shown that papain-catalysed reactions with nucleophiles on an *N*-benzyloxycarbonyldipeptide derived from α-amino acids can be directed to the amide group or the terminal carboxyl. If on the other hand *N*-protected dipeptides of *β*-alanine and DL-alanine are incubated at 40°C at a pH of 4.5 with papain and aniline or phenylhydrazine, reaction is directed exclusively to the α-amino acid residue. Anilides and phenylhydrazides are subsequently recovered in very high optical purity [140].

$$\text{Z-}\beta\text{-Ala-DL-Ala} \xrightarrow[\text{papain}]{\text{RNH}_2} \text{Z-}\beta\text{-Ala-L-AlaNHR}$$

$$\text{Z-DL-Ala-}\beta\text{-Ala} \xrightarrow[\text{papain}]{\text{RNH}_2} \text{Z-L-AlaNHR}$$

R = Ph, PhNH

A group of trypsin substrates have been defined as ‘inverse-substrates’. These are categorized by the fact that it is the ester leaving group which binds to the enzyme site. Typically such ‘inverse-substrates’ are esters of *p*-amidinophenol and as such provide a method for the specific introduction of an acyl group carrying a non-specific residue into the active site of trypsin. Many *N*-acylamino acid, di- and tripeptide *p*-amidinophenolates have been prepared by the authors including both *N*-benzoyl- and *N*-benzyloxycarbonyl-protected β-alanyl-*p*-amidinophenolates [141].

Cyclo-peptides (*3.52*) containing 6-aminohexanoic acid (AHA), 11-aminoundecanoic acid (AUA), histidine and cysteine were found to accelerate the hydrolysis of *p*-nitrophenyl carboxylates [142].

[Gly-Cys-Gly-His-AHA-AUA]$_2$ (*3.52*)

In another study related to chymotrypsin a β-alanine-containing nonapeptide (*3.53*) and its cyclo-analogue (*3.54*) prepared by conventional solution techniques exhibited enhanced catalytic activity towards *p*-nitrophenyl acetate. (*3.53*) and (*3.54*) both gave sigmoid pH/k_{cat} profiles in which (*3.54*) was some 1.5 times more active than (*3.53*) [143].

Asp-β-Ala-Gly-Ser-β-Ala-Gly-His-β-Ala-Gly

(*3.53*)

cyclo (*3.53*) = (*3.54*)

(b) Nucleotide peptides

The mechanism of hydrolysis of uridyl (5′-*N*)amino acids (*3.55*) has been studied and shown to involve intramolecular catalysis with participation of the amino acid carboxyl group. This observation was confirmed by measuring the isotopic effect for the hydrolysis in alcoholic $H_2{}^{18}O$ [144].

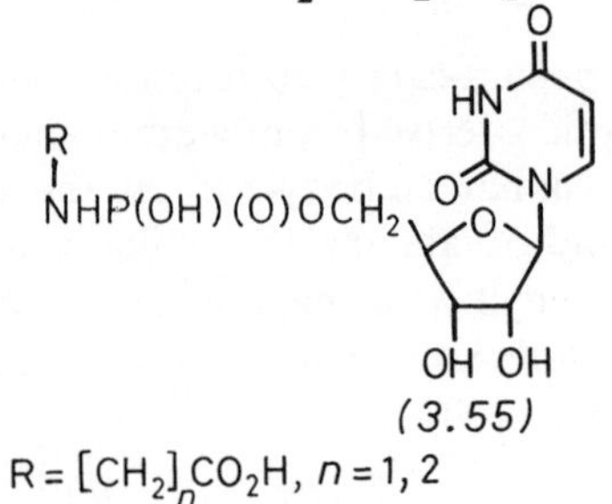

(*3.55*)

R = $[CH_2]_nCO_2H$, $n = 1, 2$

(c) Metal complexes

Metal complexes of bleomycin, epibleomycin, isobleomycin and fragments of bleomycin have been examined using ultraviolet spectroscopy and circular dichroism in attempts to throw light on their structure and mechanisms of action [145, 146].

(d) Glow discharge electrolysis

Controlled glow discharge electrolysis (CGDE) of β and ω-amino acids has been studied by means of an electrical discharge between an aqueous solution of the substrate and an electrode in contact with the solution. Reactions were carried out at 20°C with 1.0 mmol of amino acid at a current of 50 mA and 600–1500 V. Reaction pathways were thought to involve stepwise elimination of methylene groups α to the carboxyl group which were initiated by the formation of hydroxyl radicals. Analysis of intermediates formed in the reaction indicated the following reaction pathway [147].

$$\beta\text{-Ala} \xrightarrow[-H^{\bullet}]{CGDE} NH_2CH_2\dot{C}HCO_2H \xrightarrow{OH^{\bullet}} NH_2CH_2CH(OH)CO_2H$$

$$\xrightarrow{-H^{\bullet}} NH_2CH_2\dot{C}(OH)CO_2H \xrightarrow{-H^{\bullet}} NH_2CH_2COCO_2H$$

$$\xrightarrow{+OH^{\bullet}} NH_2CH_2CO_2H + CO_2$$

Similarly γ-aminobutyric acid provides *inter alia* β-alanine and glycine by an analogous pathway [148].

(e) Radical insertion

Photolysis of hydrogen azide in liquid, low molecular weight hydrocarbons provided high yields of the corresponding amines. Interpretation of the results suggested the presence of $NH(^1\Delta)$ radicals associated with an 'insertion reaction' typical of $CH_2(^1\Delta_1)$ radicals. In an extension of the reaction, photolysis of hydrogen azide was carried out in propionic acid at 25°C with a medium-pressure mercury lamp. Yield ratio of the products, alanine and β-alanine was independent of the concentration of hydrogen azide and was in the ratio 1:1.5. The following mechanism has been proposed:

$$HN_3 \xrightarrow{h\nu} NH(^1\Delta) + N_2$$

$$NH(^1\Delta) + CH_3CH_2CO_2H \rightarrow \begin{cases} CH_3CH(NH_2)CO_2H \\ NH_2CH_2CH_2CO_2H \\ NH(^3\Delta) + CH_3CH_2CO_2H \end{cases}$$

The triplet eventually decomposes to ammonia and nitrogen. In the reaction some 70% of singlet intermediate undergoes insertion. It is thought that the reaction may have prebiotic significance [149].

(f) Ortho-aminoacyl rearrangement

Brenner and coworkers demonstrated that *o*-aminoacyl rearrangement of salicylates could provide a new peptide bond [150]. Recently it has been shown that aminoacyl esters of *ortho*-acylaminophenols will also undergo a similar rearrangement leading to peptide formation. In the reaction α-amino acids react more rapidly than β and γ derivatives presumably because of a more favourable cyclic transition state [151].

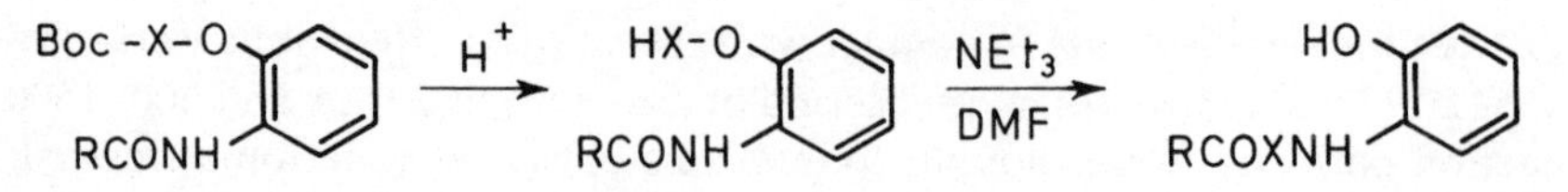

X = Ala, Gly, Leu, Phe, Pro, β-Ala
RCO = Z-Gly, Z-Cys(SBzl), Z-Gly-Ile, Z-Gly-Ala
Yields for peptides lay in the region of 50–90%.

(g) Oxazinones

It has been shown that cyclization of acyl β-amino acids and peptides leads *inter alia* to oxazin-6-ones. The latter are useful for coupling sterically hindered β-amino acids in high yield. In the course of structural investigation of the oxazinones, catalytic hydrogenation led to ring opening and the formation of β-acylaminoaldehydes in high yield.

$$\text{oxazinone}\ (R_1, R_2, R_3, R_4, R_5) \xrightarrow{H_2/Pd/C} R^1CO{-}NHCR^2R^3CR^4R^5{-}CHO$$

$R^1 = Ph$
$R^2 = R^3 = H, Me, \text{-}[CH_2]_5\text{-}$
$R^4 = R^5 = H, Me$

Reaction of the aldehydes with 2,4-dinitrophenylhydrazine gave the corresponding hydrazides which underwent elimination in acid conditions to yield the corresponding hydrazone derivatives [152].

(h) Deuteriation

δ-Aminolaevulinic acid (ALA) has been specifically labelled by pyridoxal catalysed exchange. The position of labelling was determined by NMR and lanthanide shift reagents [153].

$$\text{ALA} \ -D_2O/\text{Cat} \longrightarrow \text{ALA-5, 5-}d_2$$

$$\text{ALA} \ -15\%\ \text{DCl} \longrightarrow \text{ALA-3,3,5,5-}d_4 \xrightarrow{\text{Cat/pH 6.4}} \text{ALA-5, 5-}d_2$$

Kumarev and Almanov prepared α-deuterio derivatives of β, γ and ε-amino acids by using deuterium chloride for extended periods of time [154].

3.9 SPECTRA AND PROPERTIES

(a) Low temperature radicals

A study of the decay of X-ray generated low temperature radicals in monocrystals of β-alanine was undertaken using ESR. The rate constant for the decay of a selected signal determined at 98–188 K followed first order kinetics and was found to have an energy of activation of approximately 2.6 kcal mol^{-1} [155].

(b) Dielectric relaxation spectra

Spectra were determined for *β*-alanine at 3624, 9455, 15 350 and 35 250 Hz [156].

(c) Molal heat capacities

A range of *ω*-amino acids, $H_3\overset{+}{N}(CH_2)_nCO_2^-$ where $n = 2$–5 has been examined and molal heat capacities determined at 23–55°C using adiabatic calorimetry. The molal heat capacities were found to be associated with charge separation. The effects were quantified by comparing the difference between the amino acid and a similar uncharged molecule respectively [157].

(d) Chemical ionization

Dominant fragmentation pathways for ionizing hydrogen and methane chemical ionization mass spectroscopy of *ω*-amino acids have been compared and determined. Thus for *β*-alanine after methane chemical ionization, fragmentation of the M + 1 ion is sequential loss of water and the CH_2CO fragment; the base peak is azomethine.

$$\overset{+}{N}H_3CH_2CH_2CO_2^- \xrightarrow{CH_4/CI} NH_2CH_2CH_2CO_2\,H\overset{+}{H}$$

$$H_2NCH_2CH_2\overset{+}{C}O \xrightarrow{-CH_2CO} CH_2{=}\overset{+}{N}H_2$$

The analogous base peak $CH(CH_3){=}NH_2$ originates from 3-aminobutyric acid. Methane ionization fragmentation of GABA and 6-aminohexanoic acid proceeds by loss of water and ammonia. However in the presence of hydrogen the base peak arising from GABA is the $C_4H_5O^+$ ion of *m/e* 69, whilst that of 6-aminohexanoic acid is the $C_5H_9^+$ ion *m/e* 69 arising by further loss of carbon monoxide. Isobutane chemical ionization of the latter amino acid and 11-aminoundecanoic acid proceeds *via* the parent ion by loss of water to give the base peak consonant with lactamization [158].

(e) Configuration

X-ray analysis of *β*-alanylciliatine, a phosphonic acid dipeptide reveals that (*3.56*) exists as a monohydrated zwitterion in which conformers are stabilized by hydrogen bonding [159].

$$NH_2(CH_2)_2\,CONH\,(CH_2)_2\,PO_3H$$

(3.56)

Mugineic acid (*3.57*) copper (II) complex, a naturally occurring chelating agent in graminaceous plants has been demonstrated to exist as a distorted octahedron with one copper atom [160].

$$\text{(azetidine-2-carboxylic acid)}N{-}CH_2CH(OH)CH(CO_2H)NHCH_2CH(OH)(CO_2H)$$

(3.57)

A structural analogue of GABA (2*RS*,4*RS*)-2-hydroxy-4-aminovaleric acid (*3.58*) was shown by X-ray spectroscopy to crystallize with a triclinic space group in which the crystal structure is also stabilized by hydrogen bonding [161].

$$CH_3CH(NH_2)CH_2CH(OH)CO_2H$$

(3.58)

A combination of spectroscopic techniques utilizing X-ray, ORD, IR and far IR was brought to bear on the conformation of structures of sequential polypeptides containing L-alanine, β-alanine and DL-2-methyl-3-aminopropionic acid. The investigators concluded that β-amino acids were incorporated into a polyproline II helix type structure in which there was conformation freedom around the $N-C^{\beta}$ and $C^{\alpha}-C^{\beta}$ bonds of the β-amino acid residues [162].

Studies of the secondary structure of bradykinin despeptide analogues containing 5-aminovaleric acid in the 1–2, 2–3, and 4–5 positions in aqueous solutions were determined by CD to assess the contribution of amide bonds to biological activity [163]. It was shown that the despeptide arginylphenylalanyl analogue was devoid of activity. CD spectra demonstrated that the analogues were characterized by 3–1 and 4–1 hydrogen bonds with a salt bridge connecting the arginyl and terminal carboxyl group [163].

Further investigations relating the CD spectra of *N*-dithioethoxycarbonyl derivatives of β-amino acids have been undertaken. A negative Cotton effect when measured in chloroform or benzene at 330–350 nm is indicative of an L-residue [164, 165]. In an analogous study the ORD spectra of 3-amino-3-phenylpropionic acid has been assigned as (*S*) by comparison with (*R*) and (*S*)-3-amino-2-ethyl-2-phenylpropionic acid [166]. The absolute configuration of iturinic acid (*3.4*, *3.5*) isolated from iturin A has been determined as D by application of the dinitrophenyl chirality rule [167]. 3-Amino-3-phosphonatopropionic acid when studied using [^{1}H] and [^{31}P]-NMR spectoscopy at pH ranges between 0.5 and 13.1 gave information relating $3J_{HH}$ and $3J_{HP}$ to rotational isomerism leading to the suggestion that the most stable conformer is one in which the carboxyl group is *trans* to the phosphoric acid residue [168].

Observations of low energy conformation of a β-alanine containing cyclotetrapeptide give good correlations to the observed X-ray data [169]. The cyclic peptide cyclochlorotine (*3.59*) a potent toxin isolated from *Penicillium islandicium* containing β-phenylalanine has been determined by mass spectral sequencing [170].

In a comparative study between β-alanine, taurine and ciliating both IR and Raman spectra supported the view that the conformation around the -*N*-$C^{\alpha}H_2$-$C^{\beta}H_2$ was equivalent in each [171].

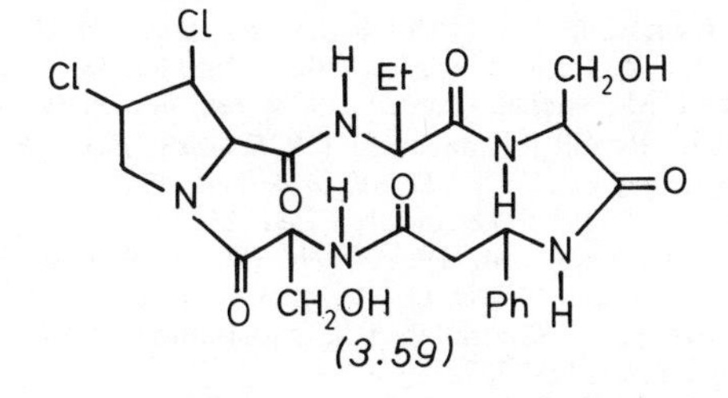

(3.59)

(f) NMR

Kricheldorf has demonstrated that in contrast to [^{13}C]-NMR, [^{15}N]-NMR provides information characterizing the sequence of the polymer prepared by copolymerization of glycine and β-alanine NCAs. Thus the chemical shifts for Gly-Gly, β-Ala-Gly, Gly-β-Ala and β-Ala-Ala differ from one another by several ppm. The copolymerization of the NCAs initiated by primary amines leads to random sequences. The following observations were made that the central nitrogen of Gly-Gly and β-Ala-Gly are insensitive to neighbours, and the latter is downfield by *ca* 5 ppm compared to the former. The nitrogen of the Gly-β-Ala bond appears *ca* 7 ppm upfield of poly-β-alanine and finally that the β-Ala-β-Ala signal is unique in that if flanked by β-alanine residues it displays a signal at *ca* 2 ppm upfield of that when β-Ala-β-Ala is flanked by glycine [133].

Gly NCA/β-Ala NCA 1:1: Initiator benzylamine

Polymer	*β-Ala-β-Ala*	*Gly-β-Ala*	*β-Ala-Gly*	*Gly-Gly (for amide group)*
$(Gly)_n$				267.3
$(\beta\text{-Ala})_n$	247.9			
$(Gly\text{-}\beta\text{-Ala})_n$		254.9	262.8	
$(Gly\text{-}Gly\text{-}\beta\text{-Ala})_n$		255.0	261.58	268.1
$(Gly\text{-}\beta\text{-Ala}\text{-}\beta\text{-Ala})_n$	245.3	255.6	262.5	
$(Gly/Ala)_n$	246.8			

REFERENCES

1. Drey, C.N.C. (1976) in *The Chemistry and Biochemistry of Amino Acids* (ed. B. Weinstein), (Vol. 4), Dekker, New York, p. 241.
2. Millar, I.T. and Springall H.D. (1966) in *Sidgwick's Organic Chemistry of Nitrogen* 3rd edn. Clarendon, Oxford, p. 207.
3. Meister, A. (1965) in *Biochemistry of the Amino Acids* (Vols. 1, 2) 2nd edn., Academic, New York.
4. Carnegie, P.R. (1963) *Biochem. J.* **89**, 459.
5. Isogai, A. and Suzukai, A. (1980) *Pept. Chem.* 125.
6. Arnould, J.M. and Frentz, R. (1975) *Comp. Biochem. Physiol. C*, **51**, 301.
7. Buhl, P.H. (1979) in *Trace Organic Analysis: New Frontiers in Analytical Chemistry* (*NBS Spec. Pub.*) Washington, D.C., **519**, 771.
8. Konsosu, M., Murakami, T., Hayashi, T. and Fuke, S. (1978) *Nippon Suisan Gakkaishi* **44**, 1165.
9. Kinuta, M. (1978) *Biochim. Biophys. Acta* **542**, 56.
10. Aoyagi, T., Tobe, H., Kojima, F. *et al.* (1978) *J. Antibiot.* **31**, 636.
11. Umezawa, H., Aoyagi, T., Morishima, H. *et al.* (1970) *J. Antibiot.* **23**, 259.
12. Umezawa, H., Aoyagi, T., Suda, H. *et al.* (1976) *J. Antibiot.* **29**, 97.

13. Teshima, T., Ando, T. and Shiba, T. (1980) *Bull. Chem. Soc. Jpn* **53**, 1191.
14. Umezawa, H., Muraoka Y., Fujii, A. *et al.* (1980) *J. Antibiot.* **33**, 1079.
15. Peypousc, F., Guinand, M., Michel, G. *et al.* (1973) *Tetrahedron* **29**, 3455.
16. Tsutsumi, W., Onodera, R. and Kandatsu, M. (1975) *Agric. Biol. Chem.* (*Japan*) **39**, 711.
17. Yunker, M.B. and Scheuer, P.J. (1978) *Tetrahedron Lett.* 4651.
18. Holm, A. and Scheuer, P.J. (1980) *Tetrahedron Lett.* 1125.
19. Okada, Y., Tsuda, Y. and Yagyu, M. (1980) *Chem. Pharm. Bull.* **28**, 310.
20. Skaric, V., Topic-Bulic, M. and Skaric, D. (1978) *Croat. Chem. Acta* **51**, 347.
21. Khosla, M.C., Stachowiak, K., Khairallah, P.A. and Bumpus, F.M. (1979) *Pept. Struct. Biol. Funct. Proc. Am. Pept. Symp. 6th.* 467.
22. Stachowiak, K., Khosla, M.C., Plucinska, K. *et al.* (1979) *J. Med. Chem.* **22**, 1128.
23. Enkoji, T. and Skibbe, M.O. (1978) *US Patent* 4 130 514, **16**, 797.
24. Amantharamaiak, G.M. and Swanandaiak, K.M. (1978) *Indian J. Chem. Sect. B.* **16**, 797.
25. Raap, A. and Kerling, K.E.T. (1981) *Recl. Trav. Chem. Pays-Bas* **100**, 62.
26. Moody, K. (1978) *Australian Patent* 497 512.
27. Gillessen, D., Stucter, R. and Trzeciak, A. (1978) *US Patent* 4 066 635.
28. Inoue, C. and Yamada, T. (1979) *Japan Kokai Tokkyo Koho Patent* 7 919 967.
29. Hudson, D., Kenner, G.W., Sharpe, R. and Szelke, M. (1979) *Int. J. Pept. Protein Res.* **14**, 177.
30. Jones, D.A., Schlatter, J.M. and Mikulec, R.A. *et al.* (1978) *Ger. Offen.* 2 735 674.
31. Hudson, D., Sharpe, R. and Szelke, M. *et al.* (1979) *Brit. U.K. Pat. Appl.* 2 00 783.
32. Rich, D.H., Moon, B.J. and Bopari, A.S. (1980) *J. Org. Chem.* **45**, 2288.
33. Subramanian, E., Swan, I.D.A. and Davies, D.R. (1976) *Biochem. Biophys. Res. Commun.* **68**, 875.
34. Workman, R.J. and Burkett, D.W. (1979) *Arch. Biochem. Biophys.* **194**, 157.
35. Bryce, G.F. and Rabin, B.R. (1964) *Biochem. J.* **20**, 509.
36. Matsushita, Y., Tone, H., Hori, S. *et al.* (1975) *J. Antibiot.* **28**, 1016.
37. Marciniszyn, J., Hartsuck, J.A. and Tang, J. (1976) *J. Biol. Chem.* **251**, 7088.
38. Liu, W.-S., Smith, S.C. and Glover, G.I. (1979) *J. Med. Chem.* **22**, 577.
39. Okada, K., Kurosawa, Y. and Nagai, S. (1979) *Chem. Pharm. Bull.* **27**, 2163.
40. Castro, B., Menard, J., Evin, G. and Coval P. (1979) *Ger. Offen.* 2 834 001.
41. Workman, R.J. and Burkett, D.W. (1980) *J. Med. Chem.* **23**, 27.
42. Rich, D.H. and Sun, E.T.O. (1980) *Biochem. Pharmacol.* **29**, 2205.
43. Takita, T., Muroaka, Y., Yoshioka, T. *et al.* (1972) *J. Antibiot.* (*Tokyo*) **25**, 755.
44. Umezawa, H. (1978) in *Bleomycin : Current Status and New Developments* (S.K. Carter, S.T. Crooke, and H. Umezawa), Academic Press, New York.
45. Takita, T. (1979) in *Bleomycin : Chemical, Biochemical and Biological Aspects*, (ed. S.M. Hecht) Springer-Verlag, New York.
46. Iitaka, Y., Nakamura, H. and Nakatani, T., *et al.* (1978) *J. Antibiot.* (*Tokyo*) **31**, 1070.
47. Oppenheimer, N.J., Rodriguez, L.O. and Hecht, S.M. (1979) *Proc. Natl. Acad. Sci. USA* **76**, 5616.
48. Umezawa, H. (1973) *Biomedicine* **18**, 459.
49. Anagnostaras, P., Cordopatis, P. and Theodoropoulous, D. (1981) *Eur. J. Med. Chem. Chim. Ther.* **16**, 171.
50. Strachan, R.G., Palevada, W.J., Veber, D.F. and Holly, F.W. (1979) *US Patent* 4 162 248.
51. Grzybowksa, J., Wojciechowska, H., Andruszkiewicz, R. and Borowski, E. (1979) *Pol. J. Sci.* **53**, 1533.
52. Grzybowska, J., Wojciechowska, H., Andruszkiewicz, R. and Borowski, E. (1979) *Pol. J. Sci.* **53**, 1267.
53. Mekhtiev, S.I., Safrov, Yu. D., Akhmedov, R.M. and Tagier, R.B. (1980) *Azerb. Khim. Zh.* 21.
54. Tokyo Fine Chemical Co. Ltd. (1980) *Japan Tokkyo Koho Patent* 31 137; 31 138.
55. Basheeruddin, K., Siddiqui, A.A., Khan, N.H. and Saleha, S. (1979) *Synth. Commun.* **9**, 705.
56. Limanov, V.E., Sobol, A.F., Schkol'nik, Ya.S. and Bobrova, G.T. (1977) *Khim.-Farm. Zh.* **11**, 41.
57. Barluenga, J., Villamana, J. and Yus, M. (1981) *Synthesis* 375.
58. Furukawa, M., Okawara, T. and Terawaki, Y. (1977) *Chem. Pharm. Bull.* **25**, 1319.
59. Furukawa, M., Okawara, T., Noguchi, Y. and Terawaki, Y. (1979) *Chem. Pharm. Bull.* **27**, 2223.
60. Achiwa, K. and Soga, T. (1978) *Tetrahedron Lett.* 1119.
61. Furukawa, M., Okawara, T., Noguchi, Y. and Terawaki, Y. (1978) *Chem. Pharm. Bull.* **26**, 260.
62. Drey, C.N.C. and Mtetwa, E. (1982) *J. Chem. Soc., Perkin Trans. 1*, 1587.
63. Sylvester, S.R. and Stevens, C.M. (1981) *Proc. West. Pharmacol Soc.* **24**, 117.
64. Nomoto, S. and Shiba, T. (1978) *Chem. Lett.* 589.

65. Kricheldorf, H.R., Schwarz, G. and Kaschig, J. (1977) *Angew. Chem.* **89**, 570.
66. Kricheldorf, H.R. (1975) *Justus Liebigs Ann. Chem.* 1387.
67. Skibarkiene, B., Kaikaris, P., Gurviciene, L. and Braziuniene, B. (1977) *Poiski Izuch Protivoopukholevykh Protivovaspalitel'Nykh Mutagennykh Veshchestv* 135 (Chem. Abs. **88**, 136 923C).
68. Khan, N.H., Siddiqui, A.A. and Kidwai, A.R. (1977) *Indian J. Chem., Sect. B* **15B**, 573.
69. Bey. P., Jung, M. and Gehart, F. (1981) *Eur. Pat. Appl.* 24 965.
70. Miyoshi, M., Matsumoto, K. and Iwasaki, T. (1978) *Japan Kokai* **78**, 90 220.
71. Mitsui Toastu Chemicals Inc. (1980) *Japan Kokai Tokkyo Koho* **80**, 154 947.
72. Von Esch, A.M., Thomas, A.M., Fairgrieve, J.S. and Seely, J.H. (1981) *US Patent* 4 272 528.
73. Cotrait, M., Dupart, E., Prigent, J. and Garrigou-Lagrange, C. (1978) *J. Mol. Struct.* **50**, 313.
74. Mastryukova, T.A., Shipov, A.E., Zhdanova, G.V. *et al.* (1980) *Izv. Akad. Nauk SSSR Ser. Khim.* 703.
75. Poskiene, R., Karparvicus, K., Puzerauskas, A. *et al.* (1976) *Izv. Akad. Nauk SSSR, Ser. Khim.* 407.
76. Kleeman, A., Martens, J. and Weigel, H. (1981) *Ger. Offen.* 2 947 825.
77. Bey, P. and Jung, M. (1979) *Belg. Patent* 870 796.
78. Yoneta, T., Shibakara, S., Seki, S. and Fukatsu, S. (1979) *Belg. Patent* 873 659.
79. Andruskiewicz, R., Grzybowska, J. and Wojciechowska, H. (1978) *Pol. J. Chem.* **52**, 2251.
80. Dietrich, R.F., Sakurai, T. and Kenyon, G.L. (1979) *J. Org. Chem.* **44**, 1894.
81. Tobe, H., Morishima, H., Naganawa, H. *et al.* (1979) *Agric. Biol. Chem.* **43**, 591.
82. Nishizawa, R., Saino T., Takita, T. *et al.* (1977) *J. Med. Chem.* **20**, 510.
83. Suda, H., Takita, T., Aoyagi, T. and Umezawa, H. (1976) *J. Antibiot.* **29**, 100.
84. Microbial Research Foundation (1978) *Neth. Appl.* 7706 503; (1979) *Fr. Demande* 2393 788.
85. Umezawa, H., Aoyagi, T., Takita, T. *et al.* (1978) *Ger. Offen.* 2 725 732.
86. Umezawa, H., Aoyagi, T. and Takita, T. *et al.* (1977) *Brit.* 1 540 019.
87. Umezawa, H., Aoyagi, T. and Takita, T. *et al.* (1979) *Japan Kokai Tokkyo Koho* 7 909 237.
88. Umezawa, H., Aoyagi, T., Takeuchi, T. *et al.* (1979), *Ger. Offen.* 2 840 636.
89. Umezawa, H., Aoyagi, T., Shirai, T. *et al.* (1980) *Ger. Offen.* 2 947 140.
90. Saino, T., Kato, K., Seya, K., *et al.* (1978) *Pept. Chem.* **16**, 5.
91. Kato, K., Saino, T., Nishizawa, R. *et al.* (1980). *J. Chem. Soc. Perkin Trans. 1*, 1618.
92. Rich, D.H., Sun, E.T. and Bopari, A.S. (1978) *J. Org. Chem.* **43**, 3624.
93. Rich, D.H., Sun, E.T. and Ulm, E. (1980) *J. Med. Chem.* **23**, 27.
94. Yoneta, T., Shibakara, S., Fukatsu, S. and Seki, S. (1978) *Bull. Chem. Soc. Japan* **51**, 3296.
95. Teshima, T., Ando, T. and Shiba, T. (1980) *Bull. Chem. Soc. Japan* **53**, 1191.
96. Ohgi, T. and Hecht, S. (1981) *J. Org. Chem.* **46**, 1232.
97. El-Naggar, A.M., Ahmed, F.S.M. and Badu, M.F. (1981) *J. Heterocycl. Chem.* **18**, 91.
98. Wysocka-Skrzela, B., Weltrowska, G. and Ledochowski, A. (1980) *Pol. J. Chem.* **54**, 619.
99. El Azzouny, A., Winter, K., Framm, J., Richter, H. and Luckner, M. (1977) *Pharmazie* **32**, 318.
100. Feuer, L., Furka, A., Sebestyen, F., Horvath, A. and Hercsel, J. *Australian Patent* 508 792.
101. Lefrancier, P., Lederer, E., Choay, J., Chedid, L. and Parant, M. (1980) *Eur. Pat. Appl.* 13 856.
102. Krastinat, W., Rapp, E. and Riedel, R. (1979) *Ger. Offen.* 2 856 753.
103. Tsybina, N.M., Ostrovskaya, R.U. and Skoldinov, A.P. (1980), *Khim-Farm. Zh.* **44**, 30.
104. Feuer, L., Furka, A., Hercsel, J., Horvath, A. and Sebestyen, F. (1980) *US Patent* 4 218 404.
105. Krastinat, W., Riedel, R. and Wolf, H. (1980) *Ger. Offen.* 2 923 698.
106. Loennechem, T., Hagan, E.A. and Aasen, A. (1979) *J. Acta Chem. Scand. Ser. B* **B33**, 387.
107. Okada, Y., Iguchi, S. Okinaka, M. *et al.* (1978) *Chem. Pharm. Bull.* **26**, 3588.
108. Drey, C.N.C. and Ridge, R.J. (1981) *J. Chem. Soc., Perkin Trans. 1*, 2468.
109. El-Naggar, A.M. Zaher, M.R. and El-Salam, A.M. (1975) *Egypt. J. Chem.* **18**, 815.
110. Levin, M.D., Subrahamanian, K., Katz, H. *et al.* (1980) *J. Am. Chem. Soc.* **102**, 1452.
111. Umezawa, Y., Morishima, H., Saito, S. *et al.* (1980) *J. Am. Chem. Soc.* **102**, 6630.
112. Takita, T., Yashioka, T., Muraoko, Y. *et al.* (1971) *J. Antibiot.* (*Tokyo*) **24**, 795.
113. Hecht, S.M., Rupprecht, K.M. and Jacobs, P.M. (1979) *J. Am. Chem. Soc.* **101**, 3982.
114. Shiba, T., Nomoto, S. and Wakamiya, T. (1976) *Experientia* **32**, 1109.
115. Nomoto, S., Teshima, T., Wakamiya, T. and Shiba, T. (1977) *J. Antibiot.* **30**, 955.
116. Nomoto, S., Teshima, T., Wakamiya, T. and Shiba, T. (1976) *Pept. Chem.* **14**, 109.
117. Nomoto, S. and Shiba, T. (1979) *Bull. Chem. Soc. Japan* **52**, 1709.
118. Teshima, T., Nomoto, S., Wakamiya, T. and Shiba, T. (1977) *J. Antibiot.* **30**, 1073.

119. Geiger, R., Obermeier, R., Wissmann, H. and Jaeger, G. (1975) *Ger. Offen.* 2416048.
120. Lee, S. and Izumiya, N. (1977) *Int. J. Pept. Protein Res.* **10**, 206.
121. Mtetwa, E. (1978) PhD Thesis (CNAA).
122. Rothe, M. and Muehlhausen, D. (1979) *Angew, Chem.* **91**, 79.
123. Iwai, M. and Okawa, K. (1980) *Kaigi Daigakko Kenhyu Hokoku* **23**, 47; (*Chem. Abs.* **94**, 66040k).
124. Blade-Font, A. (1980) *Afinidad* **37**, 445; (*Chem. Abs.* **95**, 80695e).
125. Rinehart, K.L. Jr., Gloer, J.B., Cook, J.C. Jr. *et al.* (1981) *J. Am. Chem. Soc.* **103**, 1857.
126. Hanabusa, K., Kondo, K. and Takemoto, K. (1979) *Makromol. Chem.* **180**, 307.
127. Hanabusa, K., Ohno, K., Kondo, K. and Takemoto, K. (1980) *Angew. Makromol. Chem.* **84**, 97.
128. Hanabusa, K., Kondo, K. and Takemoto, K. (1981) *Makromol. Chem. Phys.* **182**, 9.
129. Dukknenko, E.M., Komarov, V.M., Bsipenko, V.A. *et al.* (1976) *Zh. Prikl. Khim.* **49**, 181; (*Chem. Abs.*, **84**, 122385j).
130. Yuki, H., Okamoto, Y. and Doi, Y. (1979) *J. Polym. Sci. Polym. Chem. Ed.* **17**, 1911.
131. Del Pra, A., Spadon, P., Boni, R. *et al.* (1978) *Makromol. Chem.* **179**, 2707.
132. Birkofer, L. and Modic, R. (1959) *Justus Liebigs Ann. Chem.* **628**, 162.
133 Kricheldorf, H.R. (1979) *Makromol. Chem.* **180**, 147.
134. Kricheldorf, H.R. and Muelhaupt, R. (1979) *Makromol. Chem.* **180**, 1419.
135. Kricheldorf, H.R. and Schilling, G. (1979) *Makromol. Chem.* **180**, 1175.
136. Komoto, T., Minoshima, Y. and Kawai, T. (1978) *Colloid Polym. Sci.* **256**, 645.
137. Yuki, H., Okamoto, Y., Taketani, Y. *et al.* (1978) *J. Polym. Sci. Polym. Chem. Ed.* **16**, 2237.
138. Rapaka, R.S., Khaled, M.A., Urry, D.W. *et al.* (1978) *Macromolecules* **11**, 619.
139. Bhatnagar, R.S. and Rapaka, R.S. (1975) *Biopolymers* **14**, 597.
140. Abernethy, J.L., Cleary, T.S. and Kerns, B.D., Jr. (1977) *J. Org. Chem.* **42**, 3731.
141. Fujioka, T., Tanizawa, K., Nakayama, H. and Kanaoki, Y. (1980) *Chem. Pharm. Bull.* **28**, 1899.
142. Murakami, Y., Nakano, A., Matsumoto, K. *et al.* (1978) *Pept. Chem.* **16**, 157.
143. Nishi, N. (1978) *Pept. Chem.* **16**, 151.
144. Juodka, B., Sasnoukiene, S., Kazlauskaite, S. *et al.* (1981) *Bioorg. Khim.* **7**, 240.
145. Takita, T., Muroaka, Y., Nakatini, T. *et al.* (1978) *J. Antibiot.* **31**, 1073.
146. Sugiura, Y. (1979) *Biochem. Biophys. Res. Commun.* **87**, 643.
147. Harada, K. and Terasawa, J. (1980) *Chem. Lett.* 441.
148. Suzuki, S., Tamura, M., Terasawa, J. and Harada, K. (1978) *Bioorg. Chem.* **7**, 111.
149. Brenner, M., Zimmermann, G.P., Wehrmuller, J. *et al.* (1955) *Experientia* **11**, 397.
150. Sato. S., Kitamura, T. and Tsunashima, S. (1980) *Chem. Lett.* 687.
151. Mitin, Y.V. and Zapevalova, N.P. (1979) *Tetrahedron Lett.* 1081.
152. Drey, C.N.C., Ridge, R.J. and Mtetwa, E. (1980) *J. Chem. Soc. Perkin Trans. 1*, 378.
153. Lerman, C.L. and Whiteacre, E.B. (1981) *J. Org. Chem.* **46**, 468.
154. Kumarev, V.P. and Almanov, G.A. (1975) *Zh. Fy. Khim.* **49**, 1361.
155. Smith, C.J., Poole, C.P. Jr. and Farah, H.A. (1981) *J. Chem. Phys.* **74**, 993.
156. Salefran, J.L., Delbos, G., Marzat, C. and Bottreau, A.M. (1977) *Adv. Mol. Relaxation Interact. Processes* **10**, 34.
157. Cabani, S., Conti, G., Matteoli, E. and Tani, A. (1977) *J. Chem. Soc., Faraday Trans. 1*; **73**, 476.
158. Tsang, C.W. and Harrison, A.G. (1976) *J. Am. Chem. Soc.* **98**, 1301.
159. Cotrait, M., Prigent, J. and Garrigou-Lagrange, C. (1977) *J. Mol. Struct.* **39**, 175.
160. Nomato, K., Mino, Y., Ishida, T. *et al.* (1981) *J. Chem. Soc., Chem. Commun.* 338.
161. Brehm, L. and Honore, T. (1978) *Acta Crystallogr. Sect. B* 1334, 2359.
162. Komoto, T., Minoshima, Y. and Kawai, T. (1978) *Colloids Polym. Sci.* **256**, 645.
163. Hashimoto, C., Tamaki, M., Uchida, H. *et al.* (1978) *Pept. Chem.* **16**, 115.
164. Yamada, T., Kuwata, S. and Watanabe, H. (1977) *Pept. Chem.* **15**, 1.
165. Yamada, T., Kuwata, S. and Watanabe, H. (1978) *Tetrahedron Lett.* 1813.
166. Garbarino, J.A. and Nunez, O. (1981) *J. Chem. Soc. Perkin Trans. 1*, 906.
167. Nagai, U., Besson, F. and Peypoux F. (1979) *Tetrahedron Lett.* 2359.
168. Siatecki, Z. and Kozlowski, H. (1980) *Org. Magn. Reson.* **14**, 431.
169. Hall, D. and Wood, K. (1980) *J. Comput. Chem.* **1**, 368.
170. Anderegg, R.J., Biemann, K., Manmade, A. and Ghosh, A.C. (1979) *Biomed. Mass Spectrom.* **6**, 129.
171. Garrigou-Lagrange, C. (1978) *Can. J. Chem.* **56**, 663.

CHAPTER FOUR

The Non-Protein Amino Acids

S. HUNT

4.1 INTRODUCTION

In 1961 Greenstein and Winitz [1] listed some ninety amino or imino acids of non-protein origin. Scratch a reasonably well-informed biologist today and he might tell you that there are probably about two hundred natural amino acids now known. In fact this would be a wild underestimate since the tables appended to this chapter list close upon seven hundred amino and imino acids or related derivatives. The author makes no claim to have been exhaustive (though exhausted) in his literature search and the pace of discovery of new compounds makes it highly likely that many will have been overlooked. Indeed there are now so many natural amino acids known that 'rediscovery' must be becoming a common event. A comprehensive catalogue of structures, sources, chemical and biological properties is badly needed but this review does not aspire to that function. This is however perhaps the point to apologize for missed molecules, erroneous structures and spurious compounds not yet purged from the literature.

4.2 SCOPE

Organisms are wonderfully diverse in their metabolisms and the range and variety of structures listed herein reflect this diversity.

Non-protein amino acids are those amino acids which are not found in protein main chains either for lack of a specific transfer RNA and codon triplet or because they do not arise from protein amino acids by post-translational modification. In this chapter we are looking at alpha amino acids, that is to say the carboxyl group is separated from the amino or imino nitrogen by one carbon only, the α-carbon atom. Higher homologues are dealt with in another chapter. Built upon this basic structure a vast diversity of organic functional groupings can give rise to a numerous body of molecules which to a greater or lesser extent qualify as amino acids.

Many of these compounds are the end products of secondary metabolism. Their origins are diverse as are their functions although in both respects we are still more ignorant than informed. Many others arise as intermediates upon metabolic pathways or originate from the metabolism or detoxification of foreign compounds. Because of the nature of bacterial metabolism it is possible to produce many new compounds by addition of appropriate organic nutrients to culture media. Such 'artificially' generated amino acids can be found in the literature but have not been collected here. Nor has the question of whether amino acids found in meteorites are of natural origin or not been pursued. Aside from these matters I have interpreted what an amino acid is quite liberally.

4.3 SOURCES OF INFORMATION

A lot of molecules are listed in the tables which follow. Given the space available the best one can hope to do for each compound or class of compounds is to provide an entry to the relevant literature.

Many of these amino acids have been grouped before in the tables of authors' reviews. It would not seem to be a terribly effective use of space to have relisted all the primary sources that others have already cited. Accordingly I have in many cases cited secondary sources when these are likely to be readily accessible to the majority of readers. I drew heavily for information upon early reviews and several more recent ones [1–12], upon the Chemical Society Specialist Periodical Reports of Peptides and Proteins [13–25] and upon some general compilations like the CRC *Handbook of Microbiology* [26]. Not unnaturally, *Chemical Abstracts* was one rather important access route. Even so, searching for amino acids has many of the characteristics of wet fly-fishing – chuck it and chance it. Two names deserve special mention in the non-protein amino acid world; E.A. Bell and L. Fowden have probably identified more new plant amino acids than any other pair of workers and have contributed more to the review literature of this subject than anyone else [4–6, 8–10].

4.4 CLASSIFICATION AND TABULATION

Taxonomies are always to some extent artificial. Listing and classifying amino acids is no exception.

In tabulating the amino acids the broad principle has been adopted that a rough division may be made with aliphatic and other amino acids. That the benzenoid aromatics form a distinct group and that heterocyclic amino acids may be crudely separated according to their heteroatoms seems obvious while the presence of sulphur in an amino acid again would seem to distinguish it as does the presence of the analogous selenium. Probably there are sound metabolic arguments to back up these divisions in many cases. It has been convenient to segregate phosphorylated amino acids even though the parent compounds may have been drawn from several different groups and this comment applies also to a broad grouping of α-*N*-substituted amino acids containing those which are *N*-methylated, *N*-acylated or more bizarrely substituted.

4.5 NOVEL AMINO ACIDS; THEIR IDENTIFICATION, ISOLATION, CHARACTERIZATION – TWO CASE HISTORIES

Novel amino acids in bulk tissues or physiological fluids are most frequently first recognized as components of unusual behaviour upon chromatograms. Atypical

Rf values in relation to known species or an unusual colour response to ninhydrin may prompt further investigation. New compounds, as components of biologically active molecules such as depsipeptides, will emerge as a logical consequence of structural elucidation as perhaps an extra peak on an amino acid analyser trace. Encounters may thus be planned or fortuitous (hopefully serendipitously). Investigation of a metabolic pathway may have as a basis of experimental rationale the predicted isolation of a hitherto uncharacterized intermediate.

By whatever route a new amino acid is first encountered or suspected, a process of isolation and purification in sufficient quantity will be a necessary preliminary to structural characterization, proof of structure via synthesis and establishment of the conformation at chiral centres in the natural product. Modern physical techniques such as mass spectrometry and nuclear magnetic resonance spectroscopy have rendered characterization relatively simple in the majority of cases while establishment of enantiomeric type has also been made easier by the development of gas–liquid chromatographic techniques capable of separating enantiomers on a systematic basis. The following two examples of characterization of novel amino acids will give an indication of the types of technique now routinely applied.

(a) Dicysteinyldopa [27]

Behind the light-sensitive cells of the retina in the vertebrate eye is a layer called the *tapetum lucidum* whose chemistry is related to its function as a reflector of light. *Lepisosteus* the Alligator Gar (a confusingly named fish – not a reptile) has a pigmented tapetum. A desire to identify the major constituent of this yellow material led to the discovery of a novel sulphur-containing amino acid.

Molar formic acid extraction of dark-adapted tapetum gave an extract which after removal of acid *in vacuo* could be examined by two-dimensional thin layer chromatography. Under ultraviolet light the chromatograms showed at least seven major discrete components of which several were both ninhydrin and ferric chloride positive. The most intense of these spots was obtained in quantity from over a hundred grams of tissue by extraction with 1 N hydrochloric acid and absorption onto Dowex 50 ion-exchange resin. Washing the resin with a molar solution of hydrochloric acid and subsequent elution with 6 N acid gave a product which after removal of acid could be rechromatographed on Sephadex LH20 in methanol/hydrochloric acid. A major peak identified by monitoring at 305 nm was collected and further purified by preparative paper chromatography. A major ultraviolet quenching band was eluted from the paper, converted to its formate salt and rechromatographed on a column of Sephadex G25. A major peak absorbing at 305 nm was collected and freeze-dried ready for structural study.

Several factors pointed to this compound being a sulphur-containing ortho-diphenolic amino acid. Reactivity to ferric chloride and ninhydrin, an elemental analysis of $C_{15}H_{21}N_3O_8S_2$–HCOOH, infrared absorption bands at 3600–2200, 1630, 1500 and 1390 cm^{-1}, and ultraviolet absorption spectra with maxima at

303 nm in acid shifting to 316 nm at pH 7 contributed to this view. Proton nuclear magnetic resonance spectra revealed a multiplet for six methylene protons at δ 3.2–3.8, a multiplet for three methine protons at δ 4.3–4.5 and an isolated aromatic proton singlet at δ 7.20.

Reductive hydrolysis in hydriodic acid/red phosphorus yielded cysteine and dihydroxyphenylalanine (dopa) as major products leading to the formulation of Structure (*4.1*) or *S*, *S*-dicysteinyldopa composed of one molecule of dopa linked to two molecules of cysteine via thio–ether linkages.

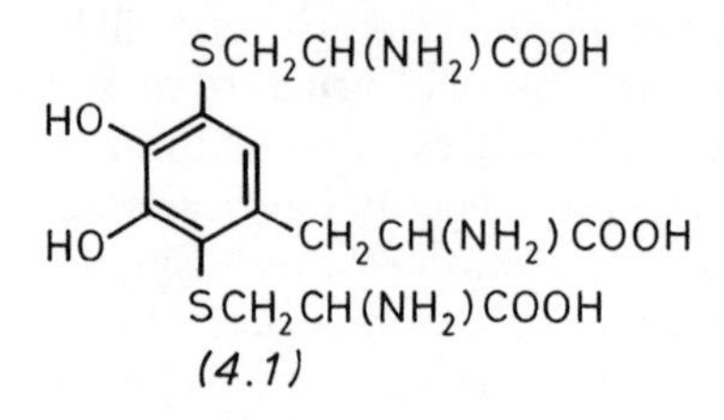

(4.1)

This structure was partly confirmed by biological synthesis. Tyrosinase oxidation of L-dopa in the presence of excess L-cysteine yielded the same amino acid together with 5- and 3-*S*-cysteinyldopa thus establishing positions 2 and 5 in the aromatic nucleus as being the most likely for substitution. Weight was added to this view by the observation that under the same conditions catechol and cysteine yield 3,5-cysteinylcatechol and 3,6-*S*,*S*-dicysteinylcatechol (*4.2*) the latter being a symmetrical structure like the putative Structure (*4.1*).

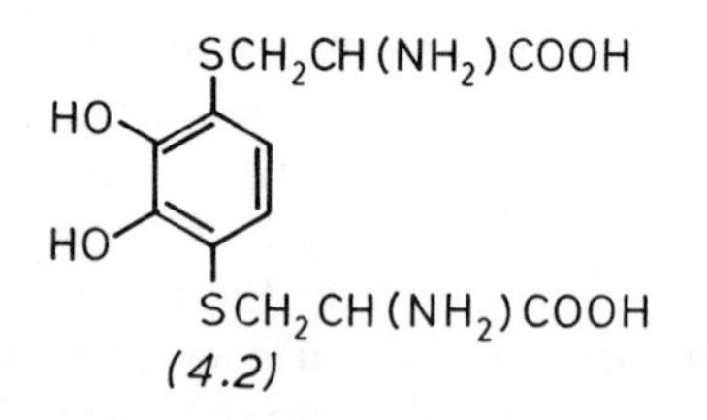

(4.2)

The absolute configuration (2*S*,2′*S*,2″*S*) was inferred from the good agreement of the specific rotation of the natural sample with that of the synthetic one prepared from L-dopa and L-cysteine.

(b) 2,4-Diamino-3-methylbutanoic acid [28]

The amino acid composition of root nodules in *Lotus* species is determined by the *Rhizobium* strain rather than the host plant. Common amino acids are present but ninhydrin-positive compounds with unusual Rf values are also evident.

Ethanol (80%) extracts of *Lotus* root nodules were hydrolysed with acid (6 N HCl), chromatographed on Amberlite IR120 ion exchange resin and the amino acid fraction converted to *N*-trifluoracetyl-n-butyl esters (TAB). Gas chromatographic/electron impact mass spectrometry analysis revealed a major component with an aspartate-related retention time of 0.86. Comparison with standard TAB amino acids suggested an unusual amino acid. High resolution mass spectrometry gave an empirical formula for the derivative of $C_{13}H_{18}O_4F_6$ isomeric with TAB-ornithine and the molecular weight of 380 was confirmed by gas chromatographic/chemical ionization mass spectrometry. Diagnostic ions in the EIMS at *m/z* 306, 279 and 166 were characteristic of TAB derivatives of basic aliphatic amino acids although there were insufficient mass spectral data to assign an unequivocal structure. The evidence however suggested an ornithine isomer and in consequence the four possible isomers were synthesized and the mass spectra compared. Such comparisons identified the unknown compound as 2,4-diamino-3-methylbutanoic acid whose configuration was established as being 2*R*,3*S* on the basis of the behaviour of the TAB amino acid, in comparison with known compounds, on the gas chromatographic stationary phase Chirasil-Val.

These two examples show slightly differing approaches to the structure determination problem. In one, nuclear magnetic resonance spectroscopy and wet chemical methods are important, in the other, the mass spectrometer provides much of the more significant information.

4.6 BIOLOGICAL SOURCES AND THE RANGE AND DIVERSITY OF NON-PROTEIN AMINO ACIDS

The compounds listed in this chapter are drawn from most groups of living organisms. What constitutes a 'non-protein' amino acid has been, as said above, interpreted quite liberally within the mentioned chemical frames of reference.

The major body of the known non-protein amino acids have as their source the plant world and micro-organisms, the latter secreting a variety of compounds into culture media [26]. Many microbiological products have antibiotic properties and many of these as well as some similar fungal products comprise unusual amino acids incorporated into more complex structures such as depsipeptides [29]. Here both D and L amino acids of common and novel character are linked by peptide bonds to each other and by other linkages to units such as hydroxy acids. Such molecules have proved a rich source of new amino acids.

In the higher plants the unusual amino acids are most frequently present in the free state or as low-molecular weight complexes in isopeptide linkage to glutamic acid. Concentrations of such compounds in plant tissues can be very high indeed but trace amounts of a great variety of compounds may also be present in a single species. A number of unusual plant and animal amino acids constitute parts of pigment structure [30–31].

On the face of things, when one looks at distributions of amino acids, micro-organisms and plants seem to be chemically rather different from animals. It is not that their basic metabolisms differ, for these are broadly similar, but we seem not to find in animals the range of secondary substances like non-protein amino acids, alkaloids, polyketides and phenolics so characteristic of plants and lower organisms. Thus the list of non-protein amino acids distributed in animal tissues and physiological fluids is more limited and less exciting. Even so this may be due to the fact that most work has concentrated upon mammals while the more diverse invertebrates have been less well investigated.

A high proportion of plant non-protein amino acids have aliphatic structures. Chains do not often exceed six carbon atoms in length although there are some quite large molecules. Diversity is attained by limited branching, hydroxyl, carboxyl and amino substitution and by the inclusion of unsaturation as allenic or allynic groupings. Other types of nitrogenous group include guanidino and cyano functions while there are numerous sulphur or selenium analogues of cysteine, cystine and methionine.

Benzenoid compounds are quite common in both plants and animals and it is therefore surprising to find that the range of aromatic amino acids based upon the benzene ring is relatively limited in comparison with the aliphatics. In view also of the diversity of halogeno aromatic compounds found in marine organisms [32] and the relative ease of substitution of the phenol ring by oxidized halide ions it is surprising that so few free halogenated phenolic amino acids are found. In fact, halogeno amino acids of any type are relatively uncommon outside the group of chlorinated bacterial products, even among the marine algae and marine invertebrates which are known to produce halogenated proteins and other secondary products [33].

In contrast to benzenoid cyclic amino acids, heterocyclics both aromatic and otherwise form a large body of non-protein amino acids. It is perhaps only to be expected that a great proportion of these will be nitrogen heterocycles although there are significant numbers which have oxygen or sulphur in the ring. A number of these nitrogen heterocycles are derived from and based upon the tryptophan indole nucleus. Other nitrogen heterocycles are closely related to proline or are homologues of this amino acid. Once again plant and bacterial sources dominate the heterocyclic scene although some interesting oxygen–nitrogen and nitrogen–sulphur heterocycles are components of animal pigments.

While there are numbers of heterocyclic imino acids, an interesting recent development has been an expansion in the limited range of aliphatic imino acids with pyruvate and amino acid-derived products such as strombine and alanopine being added to octopine as components of anoxic tissue metabolism in invertebrates [34, 35]. The related lysopine and saccharopine are plant metabolites [36–37].

4.7 BIOSYNTHETIC ORIGINS OF NON-PROTEIN AMINO ACIDS

With such diversity of structure it is not surprising that no one clear pattern of biosynthetic origin can be put forward for the non-protein amino acids. Setting aside for the moment environmental and other pressures which have selected for the maintenance of pathways leading to the accumulation of particular components it seems likely that the non-protein amino acids may have arisen in perhaps three or four general ways.

A probable route for many familiar looking products is that of modification of existing amino acids i.e. by mechanisms similar to those involved in post-translational modification of protein amino acids. That is by derivatization or simple change in structure of an existing site on a common amino acid. Dihydroxyphenylalanine [38] production from tyrosine, δ-acetylornithine [39] or *O*-acetylserine [5] synthesis in plants are well-documented examples of such modifications. The appearance of hydroxyproline [40], desmosine and isodesmosine [41] or pyridinoline [42] in mammalian urine are examples of novel amino acids which arise as post-translational modifications of amino acids actually on protein chains but which are then released by metabolic turnover to achieve free status.

Thus we can have potentially many rather simple analogues of well-known amino acids through simple post-synthetic modification of the twenty or so protein amino acids in a variety of ways. In practice certain amino acids participate in this type of modificational activity more than others and while we may encounter many products derived from lysine there may be few originating in valine. Clearly reactivities of side-chains are great determinants. The reactive thiol group of cysteine results in a relatively large range of secondary amino acids (Tables 4.13–4.15).

The formation of the α-amino–α-carboxyl grouping in the alpha amino acids takes place at initial stages in the synthesis of the amino acids by amination of keto acid. If generation of the side-chain function is one involving several enzymically catalysed steps then it is easy to see how modifications to intermediate stages can result in formation of one or even more novel amino acids. Probably many non-protein amino acids originate in this manner. At its very simplest we may see the actual intermediates in the formation of a particular amino acid as amino acids in their own right. In the absence of a metabolic role their existence may be transitory. Examples might be phosphoserine on the route to serine from 3-phosphoglycerate or glutamic-γ-semialdehyde lying between glutamate and proline [43]. Others of this type however may accumulate as does homoserine in the synthetic path to methionine or threonine from aspartate or ornithine on the route to arginine [43]. The latter synthesis has again as a precursor the amino acid glutamyl-γ-semialdehyde while the related aspartate derivative gives rise to homoserine originating in its turn from aspartylphosphate [43]. If we include all derivative intermediates of this type the list of non-protein amino acids becomes quite legally very much greater.

Failure of enzyme production for a particular step in a metabolic path may result in abnormal accumulation of amino acid intermediates. Accumulation of homocysteine in the blood and urine of homocystinurics together with other novel sulphur amino acids points to abnormalities of methionine metabolism between dietary methionine and inorganic sulphate. This can be looked upon basically as a failure to condense homocysteine with serine to form cystathione; a reaction catalysed by cystathione synthase [44, 45]. Pyroglutamic acid (5-oxoproline) may appear in abnormal urines as a result of a deficiency in 5-oxoprolinase, a component of the γ-glutamyl cycle [44].

An extension of this particular concept lies with some change in the pathway leading to a common amino acid which gives rise to a final product which is a new amino acid and which is evolutionarily selected. The best-documented example of this class of origin is that of the shikimate to chorismate route. The common amino acids phenylalanine and tyrosine are generated from shikimate (*4.3*) via chorismate (*4.4*) and prephenic acid (*4.5*). The plant amino acids β-(3-carboxyphenyl)alanine (*4.6*) and β-(3-carboxy-4-hydroxyphenyl)alanine (*4.7*)

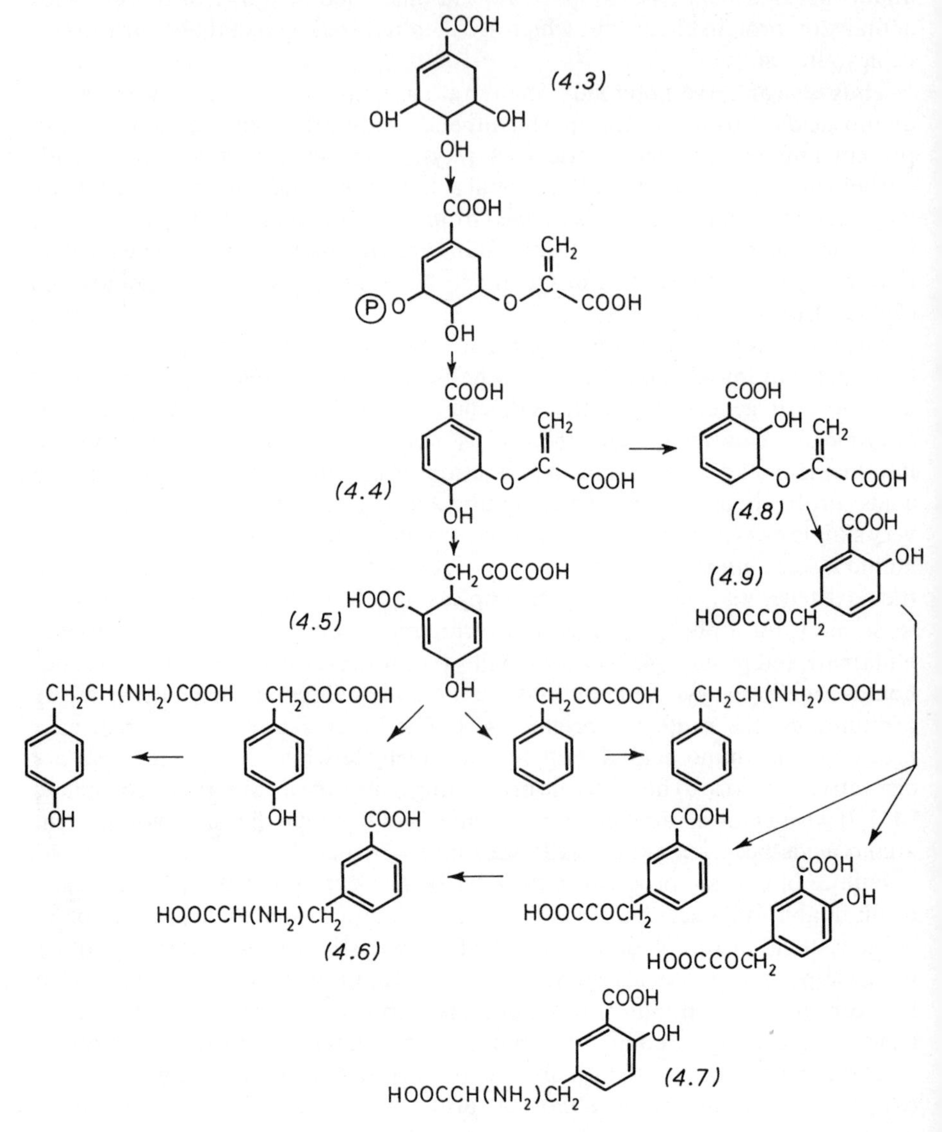

Figure 4.1

share the same route as far as chorismate but divert from it before prephenate with a conversion of chorismate to isochorismate (*4.8*) and hence to isoprephenate (*4.9*) from which the aromatic ring complete with carboxyl group is generated. The modification here may come down merely to removal of a decarboxylation stage [45, 47].

Chorismate provides a divergence point for the origin of other novel aromatic amino acids; for example *p*-aminophenylalanine in *Vigna* arises by amination of chorismate, the path thereafter returning to that for the production of phenylalanine [48]. An interesting variant on the theme is the alternative route to tyrosine from chorismate in *Pseudomonas* which takes place via the novel compound pretyrosine or arogenate (*4.10*) [49].

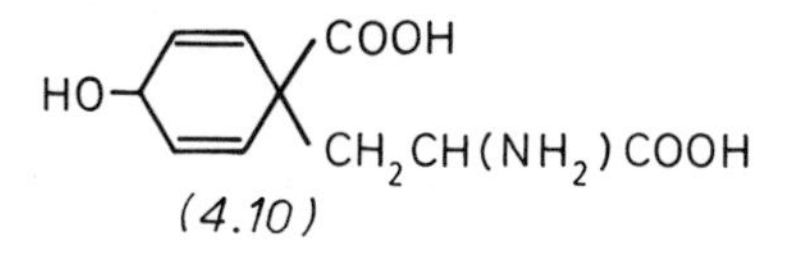

(4.10)

While many novel amino acids undoubtedly do originate from the 'tuning' of metabolic pathways and metabolites as described above it seems equally likely that such routes cannot wholly account for many of the more unusual compounds. Thus there may be a number of biosynthetic routes which like their products are themselves new and atypical. Even some non-protein amino acids whose structures seem to be close analogues of protein amino acids are products of convergence rather than parallel development. L-β-Methylaspartic acid of *Clostridium tetanomorphum* arises not from aspartate derivatization but by a novel rearrangement of L-glutamate involving 5′-deoxyadenosylcobalamin in an aerobic fermentative reaction [43]. A similar methyl derivative *erythro*-methyl-γ-glutamic acid originates in *Gleditsia triacanthus* from L-leucine by methyl group oxidation rather than from glutamic acid itself [50].

At the present time detailed accounts of biosynthesis exist for only a few groups of novel amino acids. For the major body one can only speculate informedly upon the basis of experience of routes leading to protein amino acids and related compounds. The likelihood is that even for many quite exotic structures relationships with known pathways will emerge. For example the mycosporin-like mytilins and related amino acids probably originate from condensation of known amino acids, glycine, serine and threonine with a parent diketone (*4.11*) originating from the shikimate pathway [51]. Or again strombine and alanopine derive from a NADH-coupled dehydrogenase-mediated reductive condensation

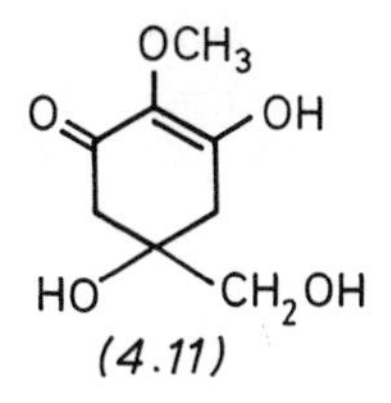

(4.11)

of pyruvate with glycine or alanine [34]; a reaction which has features in common with the formation of glutamate from α-ketoglutarate and ammonium ion by glutamate dehydrogenase and NADH.

Unusual amino acids which have been suggested to arise via quite new routes include the heterocycle azetidine-2-carboxylic acid (*4.12*) for which a putative synthetic chain from homoserine through 2,4-diaminobutanoate to 4-amino-butanoate and hence by ring closure and dehydration to the azetidine derivative has been proposed. But novel as this may seem, the suggestion has also been made that the synthesis is one made in error by the enzymes of proline synthesis (an amino acid undergoing cyclization before final reduction) [52, 53].

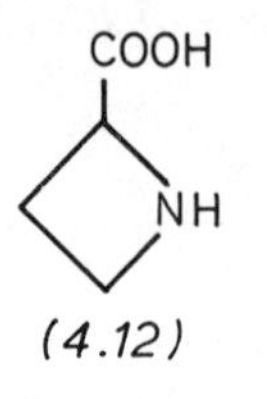

(*4.12*)

4.8 BIOLOGICAL FUNCTIONS

Novel amino acids may have many roles *in vivo* yet it is a fact that at the present time there is no function obvious for most of these compounds. An important point to bear in mind particularly in relation to plant amino acids is that while these substances often have physiological activity in animals these may be secondary actions rather than prime functions. In some cases there may be no function. As was indicated above some may originate through errors of metabolism, their presence merely illustrating and illuminating an underlying genetic abnormality; they may even be themselves the instruments of further internally directed deleterious action. Such compounds are best known for mammalian species because our knowledge of metabolic error is sparse for invertebrates or even for domestic plant species.

Function is related to the puzzling feature of non-protein amino acid distribution already mentioned in which there is an apparent disparity of occurrence and productivity in animals and plants or micro-organisms. One may speculate that this reflects a tendency in animal evolution towards increasingly efficient excretory systems while in plants the end products of metabolism are not voided but stored and exposed to a broader spectrum of enzymic secondary change. Certainly in lower, less highly evolved animals there seems to be greater versatility of low molecular weight chemical production. In this respect it is a pity that there is such a paucity of information regarding the products of nitrogen metabolism in the protozoa, the amino acid list which follows having few references relating to that group. Perhaps also the evolution of a nervous system and a physiology under delicate hormonal control has militated against the accumulation of numerous active small molecules.

An important aspect of plant metabolism is the need to conserve nitrogen and hence to limit nitrogen loss by production of secreted nitrogen compounds. Animals need not practice such economy and hence may produce and excrete nitrogenous compounds. It is therefore no accident that the excretion products of animals are one of the major sources of novel amino acids in that section of the living world.

Nitrogen required for protein synthesis in plants may be stored in large quantities as alanine, asparagine, arginine, acetylornithine, allantoin, citrulline and glutamine and in these cases it is easy to see the route by which entry to the protein synthetic path may be made. But in many cases of high levels of novel amino acid accumulation the route is less easy to see and perhaps these compounds are not intended as nitrogen stores or are unable to participate. Perhaps they may even accumulate for this very reason, having arisen through some change in a route to storage products, they have now replaced these products without a mechanism having yet evolved for their utilization.

4.9 TOXICITY

An important property of many of these products is their toxicity or ability to deleteriously affect metabolism in other organisms.

Many novel plant amino acids are structurally very similar to protein amino acids. In this light one cannot fail to be impressed by the accuracy of a translational system which has such compounds in relative abundance available and yet makes no mistakes. This is particularly so when it is remembered that the genetic code could have had the potential for incorporation of far more than twenty two amino acids into proteins. Yet there is no evidence to suggest that any of these unusual compounds find their way into the protein of their parent organism. Such incorporations are however not wholly impossible. Supply of a non-protein amino acid to a species not normally producing the compound can result in translation into a protein if it is a close enough analogue of a regular protein amino acid. A classic example of this is the incorporation of azetidine-2-carboxylic acid by *Phaseolus aureus* tissues as a replacement for proline residues in proteins [4]. *Convallaria majallis* unlike *Phaseolus* is a source of the amino acid and cannot incorporate it presumably because the enzymes of proline tRNA synthesis have extra specificity.

Incorporation of an abnormal amino acid into the proteins of a species to which the compound is foreign has obvious potential consequences for the properties of those proteins. Here there is a toxic effect which at first sight may seem to be unlikely to be exerted. Yet plants not uncommonly exude low molecular weight metabolites from roots or leaf glands or such products may slowly enter the soil from decomposing litter. Compounds from one species may then enter another via the root system. Plants may thus inhibit the development of other species in their vicinity; a well-documented phenomenon [54, 55]. Competition for nutrients and light may be weighted in favour of a species by such mechanisms operating.

Just as plants may obtain advantage over other plants by exertion of toxic effects their opportunities for survival, range expansion and success may also be affected by toxic influences upon animals. Such influences may be applied to the general metabolism or at specific physiological loci.

The leguminosae are a rich hunting ground for unusual amino acids. The seeds of this group which include peas and beans have frequent importance in the diets of a wide range of world peoples and it is therefore of some consequence that many species accumulate quite seriously toxic amino acids in their seed in appreciable quantity. This fact is made all the more important because the group is particularly useful not only as a staple food producer but also as a source of animal fodder and a soil enrichment crop via the nitrogen fixing bacteria of its root nodules.

Lathyrism is a consequence of certain types of leguminosid toxicity. Seeds of the sweet pea (*Lathyrus odoratus*) are a frequent dietary cause of connective tissue malformation in domestic animals and sometimes humans. Weakened connective tissue may even result in death from aortic rupture. β-Aminopropionitrile present in the seeds as the γ-glutamyl derivative acts as an inhibitor of lysyl oxidase on enzymes required for cross-link formation in connective tissue structural protein. The rubber-like properties of elastin and the inextensibility of collagen both depend upon such lysyl-derived cross-links. β-Aminopropionitrile originates by decarboxylation of β-cyanoalanine in the plant.

β-Cyanoalanine and its γ-glutamyl derivative are reported as neurotoxins present in another legume *Vicia sativa* and in some other *Vicia* species. Here a lesion is manifested as a neurological disorder [57] Cystathionine excretion at unusually high levels in the urine is another manifestation of this condition suggesting an interference with homocysteine production. While the mode of action of β-cyanoalanine and β-aminopropionitrile in these diseases is not wholly clear it should be noted that a link with pyridoxine and pyridoxal phosphate may be a partial basis of the toxicity since pyridoxine deficiency diminishes amine oxidase activity and produces defective elastin *in vivo* while pyridoxal phosphate is not only necessary for the conversion of cystathionine to methionine but also eliminates the neurotoxic effects of β-cyanoalanine [43–44].

Many other toxic effects in humans may be traced to plant amino acids. Hypoglycins A and B (*4.13* and *4.14*) are responsible for the hypoglycaemia which follow eating unripe Akee fruit (*Blighia sapida*) [58, 59]. Mimosine (*4.15*) causes loss of hair and wool in cattle and sheep grazing on *Leucaena leucocephala* [60] (a pity an amino acid with antidepilatory properties cannot be found) while the numerous selenium analogues of sulphur amino acids found in a variety of plants have wide ranging toxic effects upon grazing animals.

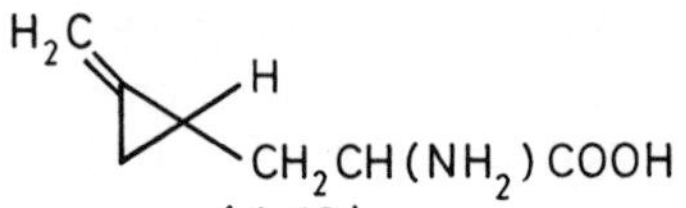

H_2C, H, $CH_2CH(NH_2)COOH$

(4.13)

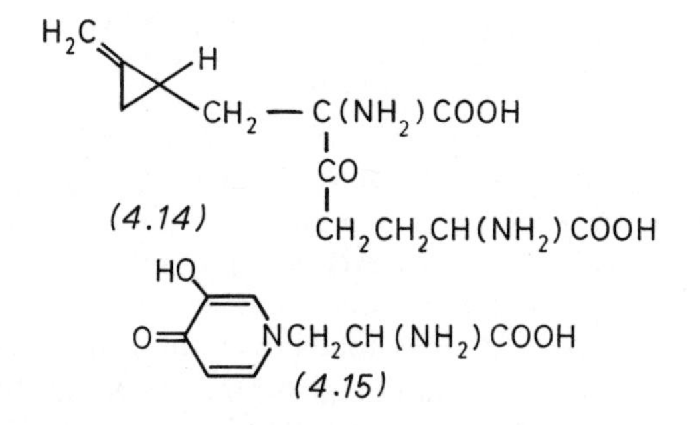

Toxic effects are also exerted in invertebrates. It is possible that insects in particular represent the original targets of some toxins since plants are prone to extensive attack by many groups and species. A static lifestyle may favour evolution of chemical defence against a highly mobile depredator. Canavanine (*4.16*) and β-hydroxy-γ-methylglutamate (*4.17*) can act as insect repellents to certain species as well as being toxic while 5-hydroxytryptophan (*4.18*) and 3,4-dihydroxyphenylalanine (*4.19*) are toxic to bruchid beetles [61, 62] and other insects.

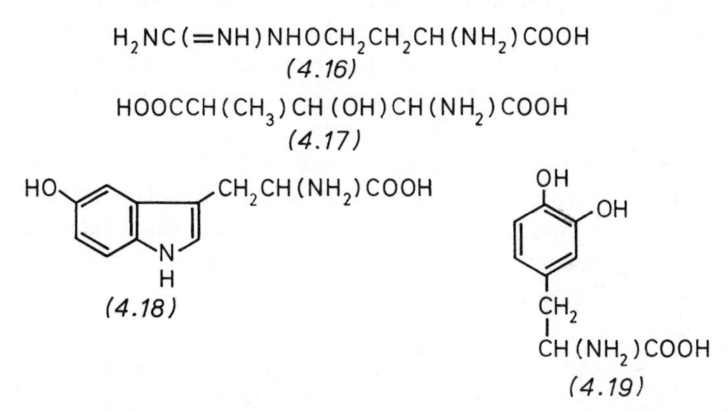

The relationship of (*4.18*) to a neurotransmitter substance may be of significance since 5-hydroxytryptamine is the major neurotransmitter of insect fore and middle gastrointestinal tract muscle [65]. Dopa is both the precursor of neurotransmitter substances and cuticle cross-linking agents in insects giving rise to dihydroxyphenylalanine and a range of quinonoid and β-substituted ketocatechols [63, 64, 66, 67].

The obvious toxicity of a such wide variety of plant products when present in high concentration leads to the speculation that some human and domestic animal non-specific diseases may originate in low concentrations of toxic amino acids more widely distributed in food plants than presently recognized. For example N^{ε}-acetyl-L-α,γ-diaminobutanoic acid (*4.20*), which can give rise to α,γ-diaminobutanoic acid, a toxin inhibiting ornithine transcarbamylase in the urea cycle, is present in small quantities in sugar beet [68].

$$CH_2(NHCOCH_3)CH_2CH(NH_2)COOH$$
(4.20)

The toxic properties of many unusual amino acid-containing products of micro-organisms are well known, some of their toxicity in mammals being acceptable in view of their even greater toxicity to other micro-organisms. Many unusual peptidic fungal toxins however are physiologically highly active in mammals, usually in a highly deleterious manner [69].

The toxic principles of poisonous toadstools, the genus *Amanita*, are cyclic peptides with unusual amino acid compositions. Although they fall into two groups the phallotoxins and amatoxins [69] they are basically similar with an indole–sulphur bridge derived from tryptophan or hydroxytryptophan and cysteine or cysteine sulphoxide bridging the cyclic structure and with unusual hydroxyamino acids such as L-dihydroxyisoleucine (*4.21*).

$$CH_2(OH)CH(OH)CH(CH_3)CH(NH_2)COOH$$
(4.21)

The toxic function in the amatoxins seems to lie in the hydroxyisoleucine component since its replacement by leucine in amanullin results in nontoxicity [69]. The toxic action here as with many of the other fungal products probably lies in the peptide's ability to fit into a symmetric site on RNA polymerase thereby blocking protein synthesis and hence explaining the slow action of the toxin.

Simpler amino acid products than the amatoxins are also responsible for the sinister reputation of fungi. Tricholomic acid (*4.22*) and muscazone (*4.23*) are the causes of a variety of lesions to vision, memory and spatial or temporal location in man, while these two from *Tricholoma muscarium* and *Amanita muscaria* together with ibotenic acid (*4.24*) from *Amanita pantherina* are potent insecticides [69]. All are based upon isoxazole.

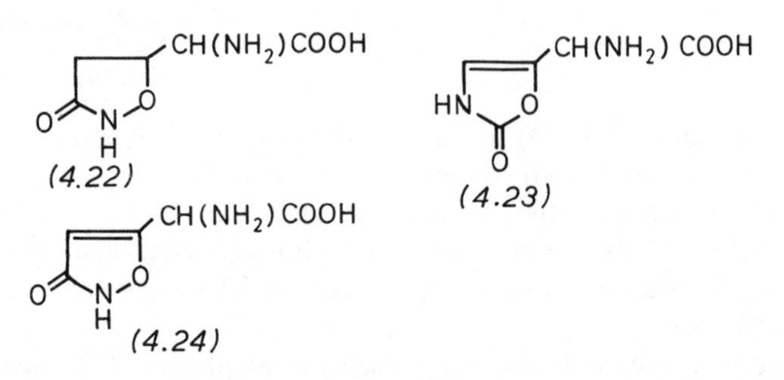

The fungi also produce non-protein amino and imino acid phytotoxins. Lycomarasmin (*4.25*) and the aspergillomarasmins (*4.26*) occur in *Fusarium* causing leaf wither in tomato by complexation of ferric ion while fusaric acid (*4.27*) from the same source causes leaf yellowing [69]. Soya leaf necrosis is caused by rhizobitoxin (*4.28*) of *Rhizobium japonicum* which blocks cystathionine conversion to homocysteine. Interference with sulphur metabolism seems to be a

$$HOOCCH_2CH(COOH)NHCH_2CH(COOH)NHCH_2CONH_2$$
(4.25)

$$HOOCCH_2CH(COOH)NHCH_2CH(COOH)NHCH_2CH(NH_2)COOH$$
(4.26)

$$HOOCCH_2CH(COOH)NHCH_2CH(COOH)NHCH_2COOH$$
(4.27)

$$HOCH_2CH(NH_2)CH_2OCH{=}CHCH(NH_2)COOH$$
(4.28)

common consequence of toxicity of the unusual amino acids. The locus of action of many antibiotics is still imperfectly understood although undoubtedly many, whose cyclic structures contain novel amino acids or protein amino acids in the D-conformation such as gramicidin or enniatin, act at the level of the bacterial cell membrane affecting ion permeability. Antibiosis is a consequence of competition and evolutionary pressure at the microbiological level [69].

The penicillins derived from penicillamine (*4.29*) act at the level of peptidoglycan biosynthesis inhibition acting as analogues of D-alanyl-D-alanine and binding to the active site of the bacterial transacylase. The cephalosporins are closely related to the penicillins [43] containing a D-α-aminoadipoyl side-chain. While the majority of antibiotics are rather elaborate molecules a few are quite simple. Azaserine (*4.30*) is one example while L-2-amino-4-(4′-amino-2′,5′-cyclohexadienyl)butanoic acid [70] is another.

$$(CH_3)_2C(SH)CH(NH_2)COOH$$
(4.29)

$$N_2CCHCOOCH_2CH(NH_2)COOH$$
(4.30)

Animals, in contrast to plants and micro-organisms, produce few toxins which contain novel amino acids. The requirements of toxins of animal origin are usually for a swift and painful response and as we have seen, biotoxic peptides of abnormal content or novel amino acids usually act upon metabolism at a slower rate. Such rapid responses are more usually achieved through the action of neurotransmitter analogues, steroid glycosides and relatively simple peptides which elicit a swift allergic response. Some molluscan venoms however are mixtures of amines, peptides and proteins and these may yet be shown to have novel constituents. Homarin (*4.31*) which is present in some cone shell venoms, has a curare-like effect [69].

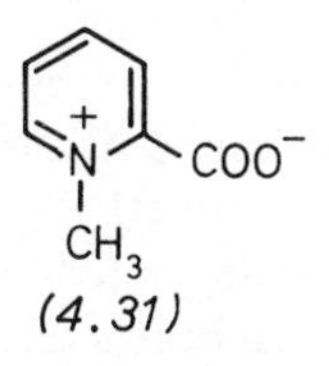

(4.31)

Some tunicates (ascidians or sea squirts) have cytotoxic, non-protein amino acid-containing depsipeptides [71] similar in many respects to depsipeptides of microbial and fungal origin but this group are atypical in many other respects of animal metabolism. For functions of non-protein amino acids in the animal kingdom we should look rather to physiological roles.

4.10 PHYSIOLOGICAL FUNCTION OF NON-PROTEIN AMINO ACIDS

As we have seen, much of the biological function of novel amino acids in plants and in micro-organisms may be associated not directly with the physiology of the organism itself but with its relationship with other organisms in its environment. We are thus dealing here with physiological (toxic), deterrent (pheromonal) or other modifying roles external to the responsible species.

The question of the physiological functions, within higher plants, of novel amino acids still remains for the majority an uncharted sea, a doubtful and disputed land. The role of such compounds in nitrogen storage has already been touched upon and clearly is of some considerable consequence to the plant; other functions are obscure.

That such compounds may be involved in physiological functions seems to be suggested by observations upon plants made at different stages of development, at different seasons and in different years or under conditions of stress. Such observations show deletions or accumulations of particular novel amino acids under different conditions although the reasons behind this are by no means clear. For example γ-hydroxy-γ-methylglutamic acid is detectable in some years in *Asplenium* but not during others [72]. Canavanine accumulations in seeds may vanish during germination [73]. There is no doubt also that osmotic control in water stress may be achieved by changes in free amino acid concentration over considerable ranges [74]. Free amino acids may also be involved in ion-binding, a factor which could be of consequence in marine and freshwater algae. Cases of the sequestration of metabolically important ions, as for example to avenic acid, have recently been established [75].

Some unusual amino acids lie upon routes to well-established plant metabolites involved in important processes. For example the plant hormone ethylene has been shown to be produced from methionine via *S*-adenosylmethionine and 1-aminocyclopropane-1-carboxylic acid [76].

In invertebrates some unusual amino acid function is associated with energy supply to tissue under anoxic conditions; strombine, alanopine and octopine [34,35] whose biosynthesis has already been referred to function in this manner to maintain a basal rate of energy production.

Certain amino acids in animals also seem to be associated with osmotic regulation. Glycine betaine for example has this function in many molluscs in addition to acting as a donor of the methyl group.

An increasing diversity of phosphorylated amino acids, less well-known than phosphocreatine, function in invertebrates as the latter compound does in the muscle of vertebrates, as phosphagens associated with the maintenance of energy transfer during contraction. Such compounds include *N*-phosphorylarginine, phosphoglycocyamine and lombricine.

Table 4.1 Some common non-protein α-amino acids and related intermediates of metabolism [43]

Amino acid	*Formula*	*Involvement*
N-Acetylglutamic acid	$HOOCCH_2\,CH_2\,\overset{NHCOCH_3}{CH}\,COOH$	Route to ornithine
N-Acetylglutamic acid semialdehyde	$OCH\,CH_2\,CH_2\,\overset{NHCOCH_3}{CH}\,COOH$	Route to ornithine
N-Acetyllysine	$CH_3\,CONH\,[CH_2]_4\,CH\,(NH_2)\,COOH$	Lysine degradation
N-Acetylornithine	$H_2N\,[CH_2]_3\overset{NHCOCH_3}{CH}\,COOH$	Route to ornithine
O-Acetylserine	$CH_3\,COO\,CH_2CH\,(NH_2)\,COOH$	Serine metabolism
S-Adenosylhomocysteine	$H\overset{Adenosyl}{\underset{+}{S}}\,CH_2CH_2\,CH\,(NH_2)COOH$	Methionine metabolism
S-Adenosylmethionine	$CH_3\,\overset{Adenosyl}{\underset{+}{S}}\,CH_2CH_2\,CH\,(NH_2)\,COOH$	Methionine metabolism
2-Aminoadipic acid	$HOOC\,[CH_2]_3\,CH\,(NH_2)\,COOH$	Lysine degradation
2-Aminoadipic acid semialdehyde	$OCH\,[CH_2]_3\,CH\,(NH_2)\,COOH$	Lysine degradation
2-Amino-3-carboxymuconic acid 6-semialdehyde	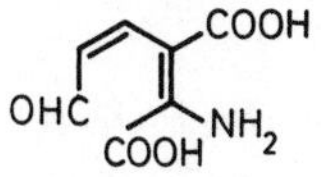	Tryptophan metabolism
2-Amino-3-ketobutanoic acid	$H_3C\,\overset{O}{\overset{\Vert}{C}}\,CH(NH_2)\,COOH$	Threonine metabolism
2-Aminomuconic acid	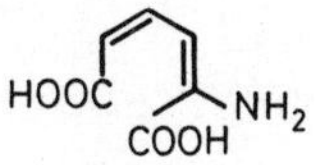	Tryptophan metabolism
2-Aminomuconic acid 6-semialdehyde	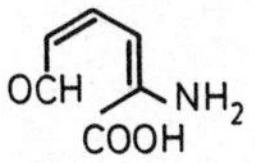	Tryptophan metabolism
Argininosuccinic acid	$H_2N\overset{HOOC\overset{}{C}H\,CH_2COOH \;|\; N \;\Vert}{C}NH\,[CH_2]_3\,CH\,(NH_2)\,COOH$	Ornithine cycle
Aspartylphosphate	Ⓟ $OOC\,CH_2\,CH\,(NH_2)\,COOH$	Aspartate to lysine route
Aspartylsemialdehyde	$OCH\,CH_2CH\,(NH_2)COOH$	Aspartate to lysine route
Betaine	$\bar{O}OC\,CH_2\,\overset{+}{N}\,(CH_3)_3$	Biosynthesis of glycine

Table 4.1 (*Contd.*)

Amino acid	*Formula*	*Involvement*
Carbamoylglutamic acid	$HOOCCHCH_2CH_2COOH$ | NH | $H_2NC=O$	Histidine degradation
Carbamoylaspartic acid (Ureidosuccinic acid)	$H_2NCONHCH(COOH)CH_2COOH$	Uridine synthesis from aspartate
Citrulline	$H_2NCONH[CH_2]_3CH(NH_2)COOH$	Ornithine cycle
Cystathionine	$CH_2CH(NH_2)COOH$ | $SCH_2CH_2CH(NH_2)COOH$	Cysteine and methionine metabolism
Cysteic acid	$HO_3SCH_2CH(NH_2)COOH$	Cysteine metabolism
Cysteinesulphinic acid	$HOOSCH_2CH(NH_2)COOH$	Cysteine metabolism
Diaminopimelic acid	NH_2 (on second CH) $HOOCCH[CH_2]_3CHCOOH$ NH_2 (on first CH)	Aspartate to lysine route
2, 3-Dihydrodipicolinic acid	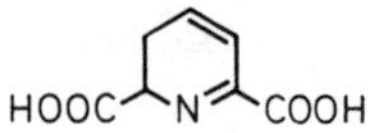 HOOC (ring N) COOH	Aspartate to lysine route
Dihydroxyphenylalanine	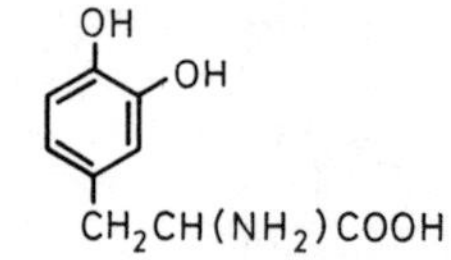 OH, OH (ring) $CH_2CH(NH_2)COOH$	Tyrosine metabolism
Dimethylglycine	$HOOCCH_2N(CH_3)_2$	Biosynthesis of glycine
Formiminoaspartic acid	$HOOCCHCH_2COOH$ | NH | $HN=CH$	Histidine degradation
Formiminoglycine	$NHCH_2COOH$ | $HN=CH$	Purine dĕgradation
Formiminoglutamic acid	$HOOCCHCH_2CH_2COOH$ | NH | $HN=CH$	Histidine degradation
Formlyaspartic acid	$HOOCCHCH_2COOH$ | NH | CHO	Histidine degradation
Formylkynurenine	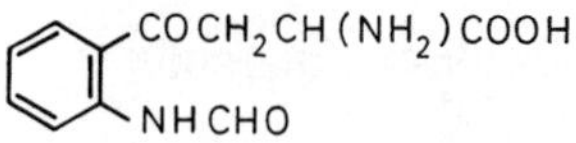 (ring) $COCH_2CH(NH_2)COOH$ $NHCHO$	Tryptophan metabolism

Table 4.1 (*Contd.*)

Amino acid	*Formula*	*Involvement*
Glutamic acid semialdehyde	$OCHCH_2CH_2CH(NH_2)COOH$	Proline metabolism
Glycocyamine	$H_2N\overset{NH}{\overset{\Vert}{C}}NH\,CH_2COOH$	Creatinine formation
Homocysteine	$HSCH_2CH_2CH(NH_2)COOH$	Methionine metabolism
Homoserine	$HOCH_2CH_2CH(NH_2)COOH$	Pathway to methionine, homocysteine or threonine
γ-Hydroxyglutamic acid	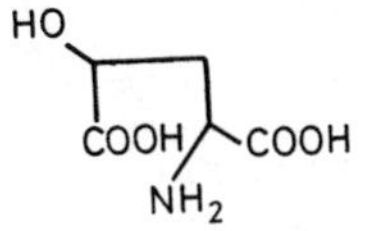 HO COOH COOH NH_2	Hydroxyproline metabolism
γ-Hydroxyglutamic acid semialdehyde	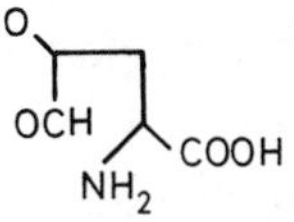 HO OCH COOH NH_2	Hydroxyproline metabolism
3-Hydroxykynurenine	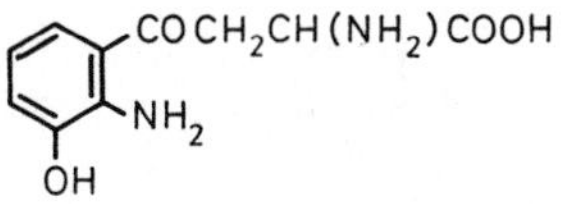 $COCH_2CH(NH_2)COOH$ NH_2 OH	Tryptophan metabolism
8-Hydroxyquinaldic acid	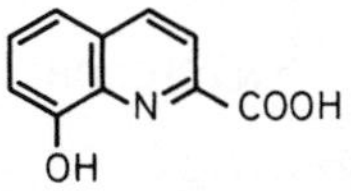 N COOH OH	Tryptophan metabolism
5-Hydroxytryptophan	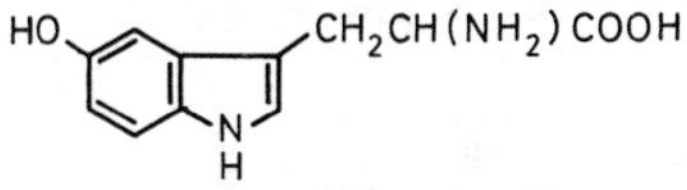 HO $CH_2CH(NH_2)COOH$ N H	Tryptophan metabolism
Kynurenic acid	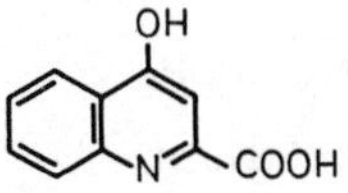 OH N COOH	Tryptophan metabolism
Kynurenine	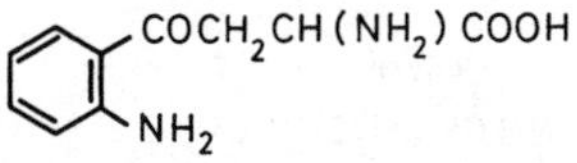 $COCH_2CH(NH_2)COOH$ NH_2	Tryptophan metabolism
Methylaspartic acid	$HOOC\overset{CH_3}{\overset{\vert}{C}H}CH(NH_2)COOH$	Glutamic acid degradation
Ornithine	$H_2N[CH_2]_3\overset{NH_2}{\overset{\vert}{C}H}COOH$	Ornithine cycle
O-Phosphohomoserine	Ⓟ$OCH_2CH_2CH(NH_2)COOH$	Pathway to threonine

Table 4.1 (*Contd.*)

Amino acid	*Formula*	*Involvement*
Phosphoserine	Ⓟ$OCH_2CH(NH_2)COOH$	Serine metabolism
Picolinic acid	N COOH	Tryptophan metabolism
Pipecolic acid	N COOH H	Lysine degradation
Δ^2-Piperideine-2-carboxylic acid	N COOH	Lysine degradation
Δ^2-Piperideine-6-carboxylic acid	N COOH	Lysine degradation
Piperideine-2, 6-dicarboxylic acid	HOOC N COOH	Aspartate to lysine route
Pyrroline-5-carboxylic acid	N COOH	Proline metabolism
Pyrroline-3-hydroxy-5-carboxylic acid	HO N COOH	Hydroxyproline metabolism
Pyrroline-4-hydroxy-2-carboxylic acid	HO N COOH	Hydroxyproline metabolism
Quinaldic acid	N COOH	Tryptophan metabolism
Quinolinic acid	COOH N COOH	Tryptophan metabolism
Quinolinic acid ribosylphosphate	COOH + N COOH RP	Tryptophan metabolism
Saccharopine	COOH $NH\,CH\,CH_2CH_2COOH$ $CH_2[CH_2]_3CH(NH_2)COOH$	Lysine degradation
Sarcosine	$HOOCCH_2NH(CH_3)$	Biosynthesis of glycine
N-Succinyl-2-amino-6-oxopimelic acid	$NHCOCH_2CH_2COOH$ $HOOCCO[CH_2]_3CH\,COOH$	Aspartate to lysine route

Table 4.1 (*Contd.*)

Amino acid	*Formula*	*Involvement*
N-Succinyl-2, 6-diaminopimelic acid	$HOOCCH(NH_2)[CH_2]_3CH(NHCOCH_2CH_2COOH)COOH$	Aspartate to lysine route
Succinylhomoserine	$HOOCCH_2CH_2COOCH_2CH_2CH(NH_2)COOH$	Pathway to methionine
Xanthurenic acid	OH, N, COOH, OH	Tryptophan metabolism

Table 4.2 Aliphatic monoaminomonocarboxylic amino acids

Amino acid	*Formula*	*Source*
O-Acetylhomoserine	$CH_3COOCH_2CH_2CH(NH_2)COOH$	Pea seedlings [6]
2-Acroleyl-3-aminofumarate	O=CH, COOH, 2, 3, NH_2, COOH	Nicotinate biosynthesis [43]
D-Alloisoleucine	$CH_3CH_2CH(CH_3)CH(NH_2)COOH$	*Bonella viridis* [104] (conjugate of a tetrapyrrole pigment bonellin)
D-Allothreonine	$CH_3CH(OH)CH(NH_2)COOH$	Peptidolipid and glycolipid in actinomycetes [3]
D-2-Aminobutanoic acid	$CH_3CH_2CH(NH_2)COOH$	Plant extracts [1]; pathological urine [1]; animal tissue [3]; bacterial extracts [3], seeds of legumes [3]; factor S in staphylomycin [79]
L-2-Aminobutanoic acid	$CH_3CH_2CH(NH_2)COOH$	
(*R*)-2-Aminobut-3-enoic acid	$CH_2{=}CHCH(NH_2)COOH$	*Rhodophyllus nidorosus* [81]
L-2-Aminobut-3-ynoic acid	$HC{\equiv}CCH(NH_2)COOH$	*Streptomyces catenulae* [82]; antibiotic FR900130
threo-L-2-Amino-3, 4-dihydroybutanoic acid	$HOCH_2CH(OH)CH(NH_2)COOH$	Micro-organisms [8]

Table 4.2 (*Contd.*)

Amino acid	*Formula*	*Source*
L-*cis*-2-Amino-3-formyl-pentenoic acid	$CH_3CH{=}C(CHO)CH(NH_2)COOH$	*Bartera fulgineoalba* [86]
2-Aminoheptanoic acid	$CH_3[CH_2]_4CH(NH_2)COOH$	*Claviceps purpurea* [26]
(2*S*)-2-Amino-4, 5-hexadienoic acid	$CH_2{=}C{=}CHCH_2CH(NH_2)COOH$	*Amanita* sp. [253, 254]
2-Aminohex-4-enoic acid	$CH_3CH{=}CHCH_2CH(NH_2)COOH$	Illamycin [26]
L-2-Aminohex-4-ynoic acid	$CH_3C{\equiv}CCH_2CH(NH_2)COOH$	New Guinea fungus [248]
2-Amino-4-hydroxyhept-6-ynoic acid	$CH{\equiv}CCH_2CH(OH)CH_2CH(NH_2)COOH$	*Euphoria longa seeds* [5]
erythro-L-2-Amino-3-hydroxyhex-4-ynoic acid	$CH_3C{\equiv}CCH(OH)CH(NH_2)COOH$	*Tricholomopsis rutilans* [85]
threo-L-2-Amino-3-hydroxyhex-4-ynoic acid	$CH_3C{\equiv}CCH(OH)CH(NH_2)COOH$	*Tricholomopsis rutilans* [85]
2-Amino-6-hydroxy-5-methyl-hex-4-enoic acid	$HOCH_2C(CH_3){=}CHCH_2CH(NH_2)COOH$	*Blighia unijugata* seeds [156, 5]; *Leucocortinarious bulbiger* (fungi) [5]
2-Amino-4-hydroxy-methylhex-5-ynoic acid	$CH{\equiv}CCH(CH_2OH)CH_2CH(NH_2)COOH$	*Euphoria longa* seeds [5]
L-*cis*-2-Amino-3-hydroxy-methylpent-3-enoic acid	$CH_3CH{=}C(CH_2OH)CH(NH_2)COOH$	*Bankera fulgineoalba* [86]
(2*S*, 3*R*)-2-Amino-3-hydroxypent-4-ynoic acid	$HC{\equiv}CCH(OH)CH(NH_2)COOH$	*Sclerotium rolfsii* [252]
2-Aminoisobutanoic acid	$CH_3C(CH_3)(NH_2)COOH$	*Paecilomyces* sp. antibiotic [2, 3] alamethicin [80]
2-Aminoisoheptanoic acid	$(CH_3)_2CHCH_2CH_2CH(NH_2)COOH$	S-520 Antibiotics [29]
2-Aminoisooctanoic acid	$(CH_3)_2CH[CH_2]_3CH(NH_2)COOH$	S-520 Antibiotics [29]
2-Amino-4-keto-3-methylpentanoic acid	$CH_3COCH(CH_3)CH(NH_2)COOH$	*Bacillus cereus* [257]
L-2-Amino-5-methyl-hex-4-enoic acid	$CH_3C(CH_3){=}CHCH_2CH(NH_2)COOH$	*Leucocortinarious* bulbiger [134]
(*S*, *S*)-2-Amino-4-methyl-hex-4-enoic acid	$CH_3CH{=}C(CH_3)CH_2CH(NH_2)COOH$	*Aesculus californica* seed [88]; *Serratia marcescens* cultures [83]

Table 4.2 (*Contd.*)

Amino acid	*Formula*	*Source*
(2*S*, 3*S*)-2-Amino-4-methyl-hex-5-enoic acid	$CH_2{=}CHCH(CH_3)CH_2CH(NH_2)COOH$	New Guinea Boletus [133] *Tricholomopsis rutilans* [249]
2-Amino-4-methyl-hex-5-ynoic acid	$CH{\equiv}CCH(CH_3)CH_2CH(NH_2)COOH$	*Euphorbia longa* seed [131]
2-Amino-4-methyl-6-hydroxyhex-4-enoic acid	$CH_2(OH)CH{=}C(CH_3)CH_2CH(NH_2)COOH$	*Aesculus californicus* seeds [5, 88]
2-Aminoneohexanoic acid	$(CH_3)_3CCH(NH_2)COOH$	Bottromycin [26]
2-Aminononanoic acid	$CH_3[CH_2]_6CH(NH_2)COOH$	S-500 Antibiotics [29]
2-Aminooctanoic acid	$CH_3[CH_2]_5CH(NH_2)COOH$	*Aspergillus* sp. [26]
2-Amino-4-oxopentanoic acid	$CH_3COCH_2CH(NH_2)COOH$	*Clostridium stricklandii* [8]
L-2-Aminopent-4-ynoic acid	$HC{\equiv}CCH_2CH(NH_2)COOH$	*Streptomyces* sp. [8]; fermentation
O-n-Butyl-L-homoserine	$CH_3[CH_2]_3OCH_2C(CH_3)(NH_2)COOH$	*Corynebacterium ethanolaminophilum* [87]
Dehydroalanine	$CH_2{=}C(NH_2)COOH$	*Alternaria mali* [77]; alternariolide – apple blotch toxin; subtilin – *Bacillus subtilis* [78]
β, *γ*-Dihydroxyisoleucine	$CH_3CH(OH)C(CH_3)(OH)CH(NH_2)COOH$	Thiostrepton [3]
(2*S*, 3*R*, 4*R*)-*γ*, *γ*-Dihydroxyisoleucine	$CH_2(OH)CH(OH)CH(CH_3)CH(NH_2)COOH$	α- and *β*-Amanatins of *Amanita phthaloides* [84]; amanin [10]
erythro-*γ*, Δ-Dihydroxy-L-leucine	$CH_3C(CH_2OH)(OH)CH_2CH(NH_2)COOH$	*Amanita phalloides*; phalloidin [3, 8]; phallacidin [69]
O-Ethyl-L-homoserine	$CH_3CH_2OCH_2C(CH_3)(NH_2)COOH$	*Corynebacterium ethanolaminopilum* [87]
Homoisoleucine	$CH_3CH_2CH(CH_3)CH_2CH(NH_2)COOH$	*Aesculus californica* [10]
α-Hydroxyalanine	$CH_3C(OH)(NH_2)COOH$	Ergotamine; ergotamimine [1]
(2*R*, 3*R*, 4*R*)-*γ*-Hydroxyisoleucine	$CH_3CH(OH)CH(CH_3)CH(NH_2)COOH$	Sterodal saponegin-yielding plants [92]
(2*R*, 3*R*, 4*R*)-*γ*-Hydroxyisoleucine	$CH_3CH(OH)CH(CH_3)CH(NH_2)COOH$	*Trigonella foenumgraecum* seeds [91]

Table 4.2 (*Contd.*)

Amino acid	*Formula*	*Source*
(2*S*, 3*R*, 4*R*)-γ-Hydroxyisoleucine	$CH_3CH(OH)CH(CH_3)CH(NH_2)COOH$	*Trigonella foenumgraecum* seeds [91]
(2*S*, 3*R*, 4*S*)-γ-Hydroxyisoleucine	$CH_3CH(OH)CH(CH_3)CH(NH_2)COOH$	γ- and ε-Amanatins of *Amantia phthalloides* [69, 84]
β-Hydroxyleucine	$(CH_3)_2CH_2CH(OH)CH(NH_2)COOH$	Antibiotics [3]
γ-Hydroxyleucine	$(CH_3)_2C(OH)CH_2CH(NH_2)COOH$	*Amanita phalloides* [3, 69]; phalloin
erythro-β-Hydroxy-L-leucine	$(CH_3)_2CH_2CH(OH)CH(NH_2)COOH$	Telomycin [3, 89]
threo-β-Hydroxy-L-leucine	$(CH_3)_2CH_2CH(OH)CH(NH_2)COOH$	*Deutzia gracilis* [3]
δ-Hydroxynorleucine	$CH_3CH(OH)CH_2CH_2CH(NH_2)COOH$	*Crotalaria juncea* seeds [255, 256]; ilamycin
β-Hydroxy-L-norvaline	$CH_3CH_2CH(OH)CH(NH_2)COOH$	Micro-organisms [8]; cycloheptamycin [145]
Δ-Hydroxynorvaline	$HOCH_2CH_2CH_2CH(NH_2)COOH$	*Canavalia ensiformis* [10]
γ-Hydroxynorvaline (2*S*, 4*R*)-(−)-	$CH_3CH(OH)CH_2CH(NH_2)COOH$	*Lathyrus odoratus* seeds [129]; **Boletus satanas** [188]
γ-Hydroxynorvaline (2*S*, 4*S*)-(+)-	$CH_3CH(OH)CH_2CH(NH_2)COOH$	
γ-Hydroxynorvaline	$CH_3CH(OH)CH_2CH(NH_2)COOH$	
α-Hydroxy-L-valine	$(CH_3)_2CHC(OH)(NH_2)COOH$	Ergotoxine; ergotinine [1, 2, 3]
β-Hydroxy-L-valine	$(CH_3)_2CH(OH)CH(NH_2)COOH$	YA56 antibiotics [250, 251]; phleomycin; bleomycin [1]; telomycin [29]
γ-Hydroxy-L-valine	$CH_2(OH)CH(CH_3)CH(NH_2)COOH$	Crown gall tumours of Kalanchoe [1, 2, 3]
γ-Keto-δ-hydroxy-L-norvaline	$HOCH_2COCH_2CH(NH_2)COOH$	**Streptomyces akiyoshiensis** *novo* [3]
β-Methyldehydroalanine **Dehydrobutyrine**	$CH_3CH{=}C(NH_2)CO_2H$	Subtilin – *B. subtilis* [78]; stendomycin [29]
β-Methyl-γ, Δ-dihydroxyisoleucine	$HOCH_2CH(OH)C(CH_3)_2CH(NH_2)COOH$	α- and β-Amanatins *of Amanita phthaloides*; amanin [10]

Table 4.2 (*Contd.*)

Amino acid	*Formula*	*Source*
β-Methylene-L-norleucine	$CH_3CH_2CH_2C(=CH_2)CH(NH_2)COOH$	*Amanita vaginata* [132]
β-Methylene-L-norvaline	$CH_3CH_2C(=CH_2)CH(NH_2)COOH$	*Lactarius helvus* [8]
α-Methylserine	$HOCH_2C(CH_3)(NH_2)COOH$	*Streptomyces* sp. [1, 3]; amicetin
L-Norvaline (2-Aminopentanoic acid)	$CH_3CH_2CH_2CH(NH_2)COOH$	*Serratia marcessens* [83]
O-Oxalylhomoserine	$HOOCCOOCH_2CH_2CH(NH_2)COOH$	*Lathyrus sativus* [9, 10]
γ-Oxonorleucine	$CH_3CH_2COCH_2CH(NH_2)COOH$	*Citrobacter freundii* [90]
Pantonine (2-Amino-3-methyl-3-hydroxymethyl-butanoic acid)	$CH_2(OH)C(CH_3)_2CH(NH_2)COOH$	*Escherichia coli cells* [1, 2]
O-n-Propyl-L-homoserine	$CH_3CH_2CH_2OCH_2C(CH_3)(NH_2)COOH$	*Corynebacterium ethanolaminophilum* [87]
O-Succinylhomoserine	$HOOCCH_2CH_2COOCH_2CH_2CH(NH_2)COOH$	Route to methionine [11]
Thermozymocidin (Myriocin) (all *cis*)	$CH_3(CH_2)_5CO(CH_2)_6CH{=}CHCH_2CH(OH)CH(OH)C(CH_2OH)(NH_2)COOH$	Thermophilic fungi [258]

Table 4.3 Aliphatic monoaminodicarboxylic amino acids

Amino acid	*Formula*	*Source*
Alanopine (2, 2′-Iminodipropanoic acid)	$HOOCCH(CH_3)NHCH(CH_3)COOH$	Fish attractant [205, 34] from *Strombus gigas*; *Crassostrea*; *Halichondria*
D-2-Aminoadipic acid	$HOOCCH_2CH_2CH_2CH(NH_2)COOH$	Cephalosporins [1]
L-2-Aminoadipic acid	$HOOCCH_2CH_2CH_2CH(NH_2)COOH$	*Vibrio cholerae* [1]; *Aspergilus oryzae* [2]; *Penicillium* [2]; *Neurospora* [2]; Higher plants [1]
2-Amino-4, 5-dihydroxy-adipic acid	$HOOCCH(OH)CH(OH)CH_2CH(NH_2)COOH$	Human urine [101]
2-Amino-4-hydroxyadipic acid	$HOOCCH_2CH(OH)CH_2CH(NH_2)COOH$	Cephalosporins [1]

Table 4.3 (*Contd.*)

Amino acid	*Formula*	*Source*
(4*R*, 2*S*)-2-Amino-4-hydroxypimelic acid	$HOOCCH_2CH_2CH(OH)CH_2CH(NH_2)COOH$	*Asplenium septentrionale* [1]; *Filicinae* [99]; *Phyllitis* [99]; *Reseda* luteola [223]
(4*S*, 2*S*)-2-Amino-4-hydroxypimelic acid	$HOOCCH_2CH_2CH(OH)CH_2CH(NH_2)COOH$	
(4*R*, 4*S*)-2-Amino-4-hydroxypimelic acid	$HOOCCH_2CH_2CH(OH)CH_2CH(NH_2)COOH$	
L-2-Amino-4-methylpimelic acid	$HOOCCH_2CH_2CH(CH_3)CH_2CH(NH_2)COOH$	*Lactarius quietus* [261]
2-Aminopimelic acid	$HOOCCH_2CH_2CH_2CH_2CH(NH_2)COOH$	*Asplenium septentrionale* [1, 2, 3]
4-Carboxyglutamic acid	$HOOCCH(COOH)CH_2CH(NH_2)COOH$	In urine [21]
4-Carboxy-4-hydroxy-2-aminoadipic acid	$HOOCCH_2C(COOH)(OH)CH_2CH(NH_2)COOH$	*Caylusea abyssinica* [100]
3, 4-Dihydroxyglutamic acid	$HOOCCH(OH)CH(OH)CH(NH_2)COOH$	Pepperwort [1] *Lepidium sativum* [130]; *Rheum rhaponticum* [2]; lettuce seed [3]; *Pandanus* sp. [94]
N^4-Ethyl-L-asparagine	$CH_3CH_2NHCOCH_2CH(NH_2)COOH$	*Ecballium elaterium* [2, 3]
4-Ethylglutamic acid	$HOOCCH(CH_2CH_3)CH_2CH(NH_2)COOH$	Seeds of *Julbernardia* [95]; *Isoberlinia*; *Brachystegia*; *Cryptosepalum*
cis-4-Ethylideneglutamic acid	CH_3 | CH ‖ $HOOCCCH_2CH(NH_2)COOH$	Distributions for 4-ethylglutamate [95, 224]; *Guilandina crista* seed [135]; *Tulipa gesneriana* [135]; *Mycena pura* [98]
3-Hydroxyasparagine	$H_2NCOCH(OH)CH(NH_2)COOH$	Human urine [3]
3-Hydroxyaspartic acid	$HOOCCH(OH)CH(NH_2)COOH$	*Azotobacter* [2]; alfalfa; clover root; urine; [3]
erythro-3-Hydroxy-L-aspartic acid	$HOOCCH(OH)CH(NH_2)COOH$	*Astralagus sinicus* seeds [8]; Phallacidine of *Amanita phalloides* [6]
threo-3-Hydroxy-L-aspartic acid	$HOOCCH(OH)CH(NH_2)COOH$	*Streptomyces* cultures [259]
N^4-Hydroxyethyl-L-asparagine	$HOCH_2CH_2NHCOCH_2CH(NH_2)COOH$	*Bryonia dioica* [2, 3]

Table 4.3 (*Contd.*)

Amino acid	*Formula*	*Source*
4-Hydroxyglutamic acid	$HOOCCH(OH)CH_2CH(NH_2)COOH$	*Phlox decussata* [6]
(2*S*, 4*R*)-4-Hydroxyglumatic acid	$HOOCCH(OH)CH_2CH(NH_2)COOH$	and other plants
(2*S*, 4*S*)-4-Hydroxyglutamic acid	$HOOCCH(OH)CH_2CH(NH_2)COOH$	
erythro-4-Hydroxyglutamic acid	$HOOCCH(OH)CH_2CH(NH_2)COOH$	Mammalian and microbial metabolism (see Table 4.1)
threo-3-Hydroxy-L-glutamic acid	$HOOCCH_2CH(OH)CH(NH_2)COOH$	Streptomycetes [8]; peptide antibiotics [93]; S-520 antibiotics [29]
4-Hydroxyglutamine	$H_2NCOCH(OH)CH_2CH(NH_2)COOH$	*Hemerocallis* [2, 3]
(2*S*, 4*R*)-4-Hydroxy-4-isobutylglutamic acid	$CH(CH_3)_2$ – CH_2 – $HOOCC(OH)CH_2CH(NH_2)COOH$ (branch on C-4)	In *Reseda odorata* flowers as the D-galactopyranose glycoside
(2*S*, 3*S*)-3-Hydroxy-4-methyleneglutamic acid	$HOOCC(=CH_2)CH(OH)CH(NH_2)COOH$	*Gleditsia caspica* [96]
(2*S*, 3*S*, 4*R*)-3-Hydroxy-4-methylglutamic acid	$HOOCCH(CH_3)CH(OH)CH(NH_2)COOH$	*Gleditsia triacanthus* and *G. caspica* seeds and seedlings [96]; Kentucky Coffee tree [260]; *Gymocladus dioicus* seeds; *Guclandina* seeds [224]
(2*S*, 4*S*)-4-Hydroxy-4-methylglutamic acid	$HOOCC(CH_3)(OH)CH_2CH(NH_2)COOH$	*Filicinae* [99]; *Caylusea abyssinica* [100]; *Phylitis scolopendrium* [3];
(2*S*, 4*R*)-4-Hydroxy-4-methylglutamic acid	$HOOCC(CH_3)(OH)CH_2CH(NH_2)COOH$	*Adiantum pedatum* [6]; *Pandanus* sp. [94]; *Reseda luteola* [223]
N^4-Methylasparagine	$CH_3NHCOCH_2CH(NH_2)COOH$	*Corallocarpus epigaeus* [10]
threo-3-Methyl-L-aspartic acid	$HOOCCH(CH_3)CH(NH_2)COOH$	*Clostridium tetanomorphum* [6]; aspartocin [6]; amphomycin [8]; glumamycin [6]
4-Methyleneglutamic acid	$HOOCC(=CH_2)CH_2CH(NH_2)COOH$	Seeds of *Julbernardia*; *Isoberlinia*; *Brachystegia*; *Cryptosepalum* [95]; *Arachis hypogaea* [2]; *Humulus lupus* [2]; *Amorpha truticosa* [2, 6]; *Mycena pura* [98]

Table 4.3 (*Contd.*)

Amino acid	Formula	Source
(2*S*)-4-Methyleneglutamic acid	$HOOCC(=CH_2)CH_2CH(NH_2)COOH$	*Phyllitis* [99]
4-Methyleneglutamine	$H_2NCOC(=CH_2)CH_2CH(NH_2)COOH$	Tulip family [1]; *Arachis hypogaea* [2]; *Humulus lupus* [2]
4-Methylglutamic acid (2*S*, 4*R*)-4-Methylglutamic acid erythro-4-Methylglutamic acid	$HOOCCH(CH_3)CH_2CH(NH_2)COOH$ $HOOCCH(CH_3)CH_2CH(NH_2)COOH$ $HOOCCH(CH_3)CH_2CH(NH_2)COOH$	*Phyllitis scolopendrium* [1, 2, 3]; *Polygala vulgaris* [6]; liliaceae [2, 3]; seeds of *Julbernardia*; *Isoberlinia*; *Brachystegia*; *Cryptosepalum* [95]; *Gleditsia triacanthus* [9]; *Lathyrus maritimus* [97] seeds [96]; *Gleditsia caspica*
4-Propylideneglutamic acid	$HOOCC(=CHCH_2CH_3)CH_2CH(NH_2)COOH$	*Mycena pura* [10, 98, 102]
Strombine (*N*-Carboxy-methylalanine)	$HOOCCH(CH_3)NHCH_2COOH$	Fish attractant [205, 34] from *Strombus gigas*; *Crassostrea*; *Halichondria*

Table 4.4 Aliphatic monocarboxylic amino acids with nitrogen in the side chain

Amino acid	Formula	Source
β-Acetamido-L-alanine (L-3-*N*-Acetyl-2, 3-diaminopropanoic acid)	$CH_2(NHCOCH_3)CH(NH_2)COOH$	*Acacia* seeds [5, 10, 106, [222]
γ-Acetamidobutyrine (L-4-*N*-Acetyl-2, 4-diaminobutanoic acid)	$CH_2(NHCOCH_3)CH_2CH(NH_2)COOH$	*Euphorbia latex* [6]; sugar beet [68]
N^6-Acetyl-allo-5-hydroxy-L-lysine	$CH_3CONHCH_2CH(OH)CH_2CH_2CH(NH_2)COOH$	*Mycobacterium* [2, 3, 6]
N^6-Acetyllysine	$CH_3CONH[CH_2]_4CH(NH_2)COOH$	*Beta vulgaris* [68]
N^5-Acetylornithine	$CH_2(NHCOCH_3)CH_2CH_2CH(NH_2)COOH$	Widely in plants [2]; *Pucinella maritima* [25]; *Phaseolus vulgaris* [114]; *Onobrychis vicifolia* [9]
Albizzine (L-2-Amino-3-ureidopropanoic acid)	$NH_2CONHCH_2CH(NH_2)COOH$	*Albizzia julibrissin* *A. lophanta* seeds and seedlings [2, 3, 9, 10, 113]; *Dialium* [107]

Table 4.4 (*Contd.*)

Amino acid	*Formula*	*Source*
β-Aminoalanine (2, 3-Diaminopropanoic acid)	$CH_2(NH_2)CH(NH_2)COOH$	Wide occurrence [5, 10] in plants and in antibiotics e.g. *Acacia*, Mimosaceae [222]; *Chainia olivacea* [26]; *Trichosanthes cucumeroides* [105]; Viomycin [1]
L-2-Amino-4-(2-aminoethoxy) butanoic acid	$NH_2CH_2CH_2OCH_2CH_2CH(NH_2)COOH$	*Streptomycetes* sp. [267]
L-2-Amino-4-(2-aminoethoxy)-*trans*-but-3-enoic acid	$NH_2CH_2CH_2OCH{=}CHCH_2CH(NH_2)COOH$	*Streptomycetes* sp. [268]
2-Amino-4-(2-amino-3-hydroxypropoxy)-*trans*-but-3-enoic acid	$HOCH_2CH(NH_2)CH_2OCH{=}CHCH(NH_2)COOH$	*Rhizobium japonicum*; nodules on *Glycine max* [265]
γ-Aminobutyrine (L-2, 4-Diaminobutanoic acid)	$CH_2(NH_2)CH_2CH(NH_2)COOH$	Antibiotics – e.g. Aerosporin, Polymixin, Circulin, Polypeptin; Comirin [1]; Coryneform bacterial mureins [110]; various higher plants e.g. *Lathyrus* sp. [2, 3, 10, 6, 9]
L-2-Amino-6-hydroxyaminocaproic acid (L-2-Amino-6-hydroxyaminohexanoic acid)	$HONH[CH_2]_4CH(NH_2)COOH$	*Mycobacterium* [2, 3, 6]
L-2-Amino-5-hydroxyaminovaleric acid	$HONH[CH_2]_3CH(NH_2)COOH$	Albomycin [3]; *Actinomyces subtropicus* [6]; Fe-deficient *Ustilago sphaerogena* [6]
O-(2-Amino-3-hydroxypropyl)homoserine	$HOCH_2CH(NH_2)CH_2OCH_2CH_2CH(NH_2)COOH$	*Rhizobium japonicum* [265]; nodules on *Glycine max*
2-Amino-3-(1-hydroxyureido) propanoic acid	$NH_2CON(OH)CH_2CH(NH_2)COOH$	*Quisqualis fructus* [109]; *Q. indica* [108]
2-Amino 3-methylaminopropanoic acid	$CH_3NHCH_2CH(NH_2)COOH$	*Cycas circinalis* seed [10]
2-Amino-4-oxalylaminobutanoic acid	$HOOCCONHCH_2CH_2CH(NH_2)COOH$	*Acacia* sp. seeds [222]; *Lathyrus* sp.

Table 4.4 (*Contd.*)

Amino acid	*Formula*	*Source*
L-Azaserine	$N_2CHCOOCH_2CH(NH_2)COOH$	*Streptomyces* sp. [2]
L-Canaline (L-4-Aminoxy-2-aminobutanoic acid)	$NH_2OCH_2CH_2CH(NH_2)COOH$	Canavanine metabolism [2]; *Astralagus suricus* [10, 112] seeds
L-Canavanine (2-Amino-4-guanidinobutanoic acid)	$NH_2C(=NH)NHOCH_2CH_2CH(NH_2)COOH$	Generally in papilionoid leguminosae [10]
5-*O*-Carbamoyl-2-amino-2-deoxy-L-xylonic acid (Polyoxamic acid)	$NH_2COOCH_2CH(OH)CH(OH)CH(NH_2)COOH$	Polyoxins [233] (N-terminus)
O-Carbamoyl-D-serine	$H_2NCOOCH_2CH(NH_2)COOH$	*Streptomyces* culture [1, 3]
N^{ε}-Carboxymethyl-L-lysine	$HOOCCH_2NH[CH_2]_4CH(NH_2)COOH$	*Sagittaria pygmacea* [281]
L-Citrulline (2-Amino-5-ureidovaleric acid)	$NH_2CONH[CH_2]_3CH(NH_2)COOH$	General distribution in plants [1, 2]; seaweeds [12]; some animal tissues [2]
β-Cyanoalanine	$NCCH_2CH(NH_2)COOH$	*Vicia* sp. [9, 6]
D-*erythro*- and L-*threo*-2, 3-Diaminobutanoic acid	$CH_3CH(NH_2)CH(NH_2)COOH$	Aspartocin [8, 111] amphomycin [8]; glumamycin [6]
2, 6-Diamino-7-hydroxyazelaic acid	$HOOCCH_2CH(OH)CH(NH_2)[CH_2]_3CH(NH_2)COOH$	Edeine A and B [8, 280] (*Bacillus brevis*)
(2*R*, 3*S*)-2, 4-Diamino-3-methylbutanoic acid	$NH_2CH_2CH(CH_3)CH(NH_2)COOH$	*Lotus tenuis* [218]
2, 4-Diaminovaleric acid	$CH_3CH(NH_2)CH_2CH(NH_2)COOH$	*Clostridium sticklandii* [8]
ε-Diazo-*δ*-oxo-L-norleucine	$N_2CHCOCH_2CH_2CH(NH_2)COOH$	*Streptomyces* sp. [2] culture media
γ, *δ*-Dihydroxyornithine	$NH_2CH(OH)CH(OH)CH_2CH(NH_2)COOH$	Echinocandin B [146]
N^G, N^G-Dimethylarginine	$(CH_3)_2NC(=NH)NH[CH_2]_3CH(NH_2)COOH$	Broad bean seeds [269] human urine [270, 272]
N^G, N'^G-Dimethylarginine	$CH_3NHC(=NCH_3)NH[CH_2]_3CH(NH_2)COOH$	Human normal and tumour urines [270, 272]; brain tissue [271]
N^6-Dimethyllysine	$(CH_3)_2N[CH_2]_4CH(NH_2)COOH$	*Reseda luteola* seeds [276, 277];

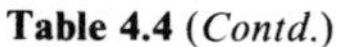

Table 4.4 (*Contd.*)

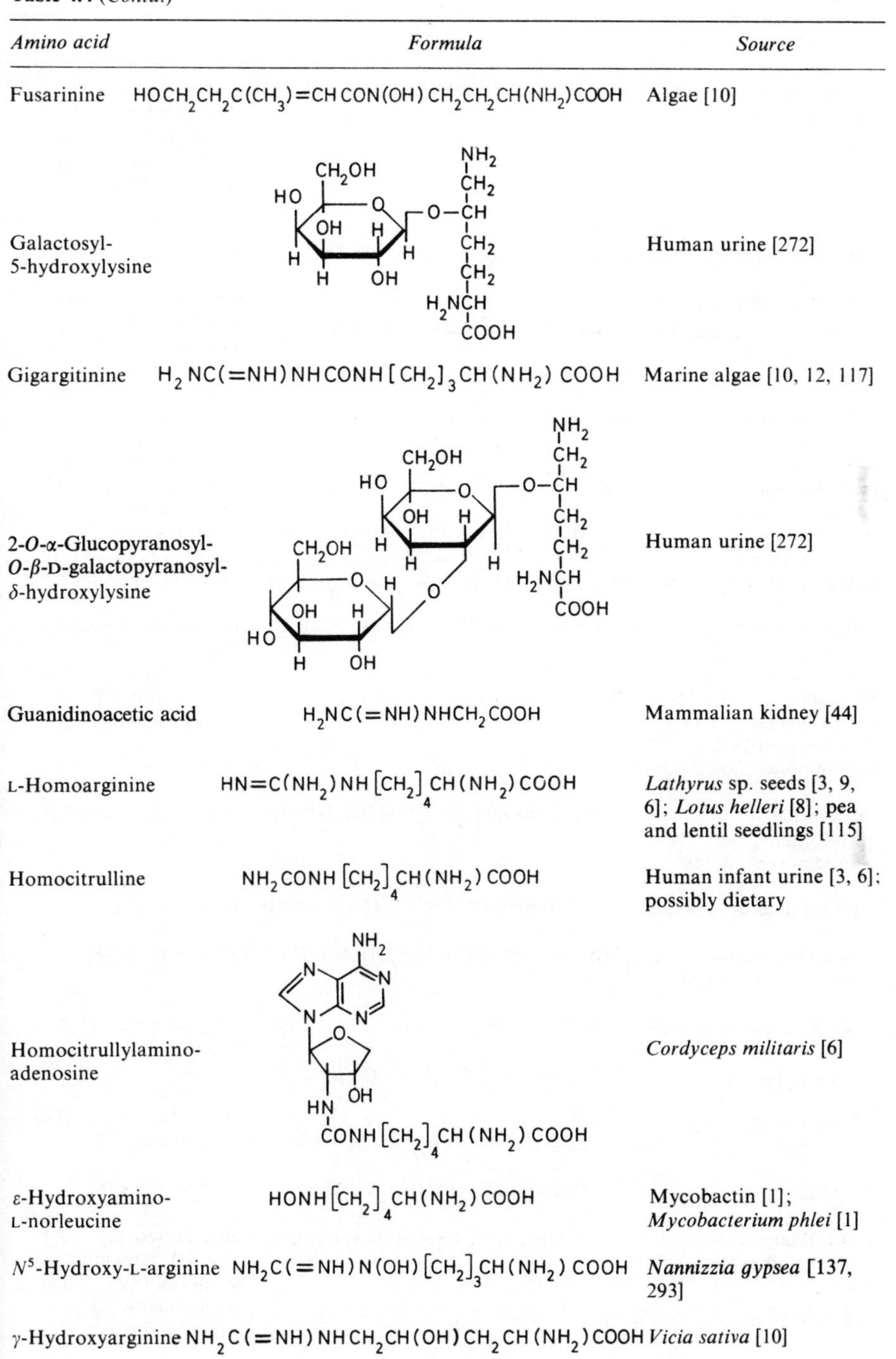

Amino acid	*Formula*	*Source*
Fusarinine	$HOCH_2CH_2C(CH_3)=CH\,CON(OH)\,CH_2CH_2CH(NH_2)COOH$	Algae [10]
Galactosyl-5-hydroxylysine		Human urine [272]
Gigargitinine	$H_2NC(=NH)NHCONH[CH_2]_3CH(NH_2)\,COOH$	Marine algae [10, 12, 117]
2-*O*-α-Glucopyranosyl-*O*-β-D-galactopyranosyl-δ-hydroxylysine		Human urine [272]
Guanidinoacetic acid	$H_2NC(=NH)NHCH_2COOH$	Mammalian kidney [44]
L-Homoarginine	$HN=C(NH_2)NH[CH_2]_4CH(NH_2)COOH$	*Lathyrus* sp. seeds [3, 9, 6]; *Lotus helleri* [8]; pea and lentil seedlings [115]
Homocitrulline	$NH_2CONH[CH_2]_4CH(NH_2)COOH$	Human infant urine [3, 6]; possibly dietary
Homocitrullylamino-adenosine		*Cordyceps militaris* [6]
ε-Hydroxyamino-L-norleucine	$HONH[CH_2]_4CH(NH_2)COOH$	Mycobactin [1]; *Mycobacterium phlei* [1]
N^5-Hydroxy-L-arginine	$NH_2C(=NH)N(OH)[CH_2]_3CH(NH_2)\,COOH$	*Nannizzia gypsea* [137, 293]
γ-Hydroxyarginine	$NH_2C(=NH)NHCH_2CH(OH)CH_2CH(NH_2)COOH$	*Vicia sativa* [10]

Table 4.4 (*Contd.*)

Amino acid	*Formula*	*Source*
erythro-γ-Hydroxy-L-arginine	$HN{=}C(NH_2)NHCH_2CH(OH)CH_2CH(NH_2)COOH$	*Polycheira rufescens* (Holothuria) [3]; *Anthopoeura japonica* (Anthozoa) [3]; Lentil and pea seedlings [115, 116]; *Vicia* sp. [6]
γ-Hydroxycitrulline	$NH_2CONHCH_2CH(OH)CH_2CH(NH_2)COOH$	*Vicia* sp. [10]
γ-Hydroxyhomoarginine	$HN{=}C(NH_2)NHCH_2CH_2CH(OH)CH_2CH(NH_2)COOH$	Distribution similar to L-homoarginine; *Tephrosieae* [274]; route to lathyrine [9, 10]
γ-Hydroxylysine	$H_2NCH_2CH_2CH(OH)CH_2CH(NH_2)COOH$	*Salvia* [10]
δ-Hydroxylysine (2, 6-Diamino-5-hydroxycaproic acid)	$NH_2CH_2CH(OH)CH_2CH_2CH(NH_2)COOH$	*Corynebacterium* [2]; *Mycobacterium phlei* [2]; *Ladino* clover seeds [118];
DL-*cis*, DL-*trans*-δ-Hydroxylysine	$NH_2CH_2CH(OH)CH_2CH_2CH(NH_2)COOH$	*Medicago sativa* [10]; Cerexins A and B [119];
L-*threo*-δ-Hydroxylysine	$NH_2CH_2CH(OH)CH_2CH_2CH(NH_2)COOH$	*Gleditsia triacanthos* [5]
γ-Hydroxyornithine	$CH_2(NH_2)CH(OH)CH_2CH(NH_2)COOH$	Pea and lentil seedlings [10]
Hypusine (N^6-(4-Amino-2-hydroxybutyl)-2, 6-diaminohexanoic acid)	$H_2NCH_2CH_2CH(OH)CH_2NH[CH_2]_4CH(NH_2)COOH$	Human urine [8]; bovine brain [8]
Indospicine (L-2-Amino-6-amidinocaproic acid)	$NH_2C(=NH)[CH_2]_4CH(NH_2)COOH$	*Indigofera spicata* [138]
δ-(*O*-Isoureido)-L-norvaline	$NH_2C(=NH)O[CH_2]_3CH(NH_2)COOH$	Bacterial culture
4-*N*-Lactyl-L-2,4-diaminobutanoic acid	$CH_2[NHCOCH(OH)CH_3]CH_2CH(NH_2)COOH$	*Beta vulgaris* [68]
α-Methylarginine	$H_2NC(=NH)NH[CH_2]_3CH(CH_3)(NH_2)COOH$	*Streptomycete* sp. [273]
N^5-Methylornithine	$CH_3NH[CH_2]_3CH(NH_2)COOH$	*Atropa belladona* [219]
N^G-Monomethylarginine	$CH_3NHC(=NH)NH[CH_2]_3CH(NH_2)COOH$	Broad bean seeds [269]; brain tissue [271]
N^6-Monomethyllysine	$CH_3NH[CH_2]_4CH(NH_2)COOH$	Human urine [8];
L-Ornithine (2, 5-Diaminovaleric acid)	$CH_2(NH_2)CH_2CH_2CH(NH_2)COOH$	Bird excreta [2]; shark liver [2]; tyrocidine [2]; gramicidin [2]; widely in higher plants [2]; red algae [12]; insects [2]

Table 4.4 (*Contd.*)

Amino acid	*Formula*	*Source*
3-*N*-Oxalyl-L-2,3-diaminopropanoic acid	$CH_2(NHCOCOOH)CH(NH_2)COOH$	*Lathyrus* sp.; *Acacia* sp. [222]; *Mimosa* sp. [262, 263, 264]; *Crotalaria incana*
L-γ-Oxalysine	$NH_2CH_2CH_2OCH_2CH(NH_2)COOH$	*Escherichia coli* [8]
Rhizobitoxin	$HOCH_2CH(NH_2)CH_2OCH{=}CHCH(NH_2)COOH$	*Rhizobium japonicum* [69, 235]
N^6-Trimethyl-L-δ-hydroxylysine	$(CH_3)_3NCH_2CH(OH)CH_2CH_2CH(NH_2)COOH$	Diatom cell wall [8]
N^6-Trimethyl-L-γ-hydroxylysine	$(CH_3)_3\overset{+}{N}[CH_2]_3CH(OH)CH(NH_2)COOH$	*Neurospora crassa* [249]
N^6-Trimethyllysine (Laminine)	$(CH_3)_3\overset{+}{N}[CH_2]_4CH(NH_2)CO\bar{O}$	*Sedum acre* [278]; *Laminaria augustata* [12]
O-Ureidohomoserine	$NH_2CONHOCH_2CH_2CH(NH_2)COOH$	Route to canavanine [5] in *Canavalia*

Table 4.5 Aliphatic diaminodicarboxylic amino acids

Amino acid	*Formula*	*Source*
2, 6-Diamino-3-hydroxypimelic acid (NH_2 *meso*, OH *threo* NH_2 L, OH *erythro*)	$HOOCCH(NH_2)[CH_2]_2CH(OH)CH(NH_2)COOH$	Bacterial cell wall [8]
2, 6-Diaminopimelic acid	$HOOCCH(NH_2)[CH_2]_3CH(NH_2)COOH$	Bacterial cell wall [1, 3] proteoglycan and free
L, L-2, 6-Diaminopimelic acid	$HOOCCH(NH_2)[CH_2]_3CH(NH_2)COOH$	*Chlorella ellipsoidea* [2]; *Pseudomonas aeruginosa* [20]; coryneform bacteria on human skin [121]; pine pollen (disputed) [122]
2, 3-Diaminosuccinic acid	$HOOCCH(NH_2)CH(NH_2)COOH$	*Streptomyces rimosus* [2, 3, 26] culture media
Dihydroxylysinonor-leucine	$HOOCCH(NH_2)[CH_2]_4NHCH_2CH(OH)[CH_2]_2CH(NH_2)COOH$	Human urine [42]; collagen degradation product

Table 4.5 (*Contd.*)

Amino acid	*Formula*	*Source*
Tabtoxinine	$HOOCCH(NH_2)CH(OH)CH_2CH_2CH(NH_2)COOH$	*Pseudomonas tabaci* [1, 2]
(Native as the lactone of the α-lactyl derivative)	$CO-CH-CH(OH)CH_2CH_2CH(NH_2)COOH$ O NH CH – CO CH_3	

Table 4.6 Miscellaneous aliphatic amino acids

Amino acid	*Formula*	*Source*
Aspergillomarasmin A	$HOOCCH_2CH(COOH)NHCH_2CH(COOH)NHCH_2CH(NH_2)COOH$	*Fusarium* sp. [11, 26]
Aspergillomarasmin B	$HOOCCH_2CH(COOH)NHCH_2CH(COOH)NHCH_2COOH$	*Fusarium* sp. [11, 26]
Avenic acid A ((2*S*, 3′*S*, 3″*S*)-*N*-[3-(3-Hydroxy-3-carboxy-propylamino)-3-carboxypropyl]homoserine)	HO, CH_2, H, COOH, H, N, H, COOH, H, N, H, COOH, OH	*Avena sativa* [75], root washings
Avenic acid B ((2*S*, 3′*S*)-*N*-(3-Hydroxy-3-carboxy-propyl)homoserine)	HO, CH_2, H, COOH, H, N, H, COOH, OH	*Avena sativa* [123] root washings
Lycomarasmin	$HOOCCH_2CH(COOH)NHCH_2CH(COOH)NHCH_2CONH_2$	*Fusarium* sp. [69, 11]
Nopaline (N^2-(1, 3-Dicarboxy-propyl)arginine)	$HOOCCH_2CH_2CHCOOH$ NH $H_2NC(=NH)NH[CH_2]_3CHCOOH$	Crown gall tumours [126], *Nicotiana tabacum*
Nopalinic acid (N^2-(1, 3-Dicarboxy-propyl)ornithine)	$H_2N[CH_2]_3CH(COOH)NHCH(COOH)CH_2CH_2COOH$	Crown gall tumours [126]; *Nicotiana tabacum*
D-Octopine (N^2-(1-Carboxyethyl)-arginine)	$H_2NCNH[CH_2]_3CHNHCHCH_3$ NH COOH COOH	Invertebrate muscle [1, 3], Sunflower crown gall tumour [124]; *Nicotiana tabacum* [126]
D-Octopinic acid (N^2-(D-1-Carboxyethyl)-L-ornithine)	$H_2N[CH_2]_3CH[NHCH(CH_3)COOH]COOH$	Crown gall tumours [124, 126]; *Nicotiana tabacum*
L-Saccharopine ((2*S*, 2′*S*)-N^6-(2′-Glutaryl)lysine)	$HOOCCH_2CH_2CH(COOH)NH[CH_2]_4CH(NH_2)COOH$	*Saccharomyces cerevisiae* [3, 127, 128]; *Reseda odorata*

Table 4.7 Phenyl, substituted phenyl, and cyclohexenyl amino acids

Amino acid	*Formula*	*Source*
(a) Benzenoid aromatic amino acids		
Actinoidic acid	(biphenyl: ring 1 OH, $CH(NH_2)COOH$; ring 2 HO, OH, $CH(NH_2)COOH$)	Ristomycin [304, 287]; actinoidin; avoparcin [408]
2-Amino-3, 4-dihydroxy-4-(4-hydroxymethyl-phenyl) butanoic acid	$HO-C_6H_4-CH(OH)CH(OH)CH(NH_2)COOH$	Echinocandine B [146]
4-*O*-[2-Amino-(4-hydroxyphenylacetic acid)]homoserine	$HOOCCH(NH_2)-C_6H_4-OCH_2CH_2CH(NH_2)COOH$	*β*-Lactam [301] antibiotics
2-Amino-4-oxo-4-(4-hydroxyphenyl)-butanoic acid	$HO-C_6H_4-COCH_2CH(NH_2)COOH$	Echinocandine [140]
2-Amino-5-(4-methoxyphenyl)-pentanoic acid	$CH_3O-C_6H_4-CH_2CH_2CH_2CH(NH_2)COOH$	Alternaridide [302]; *Alternaria mali* (apple blotch toxin)
2-Amino-3-methyl-4-hydroxy-4-(4- hydroxy-phenyl) butanoic acid	$HO-C_6H_4-CH(OH)CH(CH_3)CH(NH_2)COOH$	Nikkomycin B [141]
3-Aminomethyl-phenylalanine	$NH_2CH_2-C_6H_4-CH_2CH(NH_2)COOH$	*Combretum zeyheri* seeds [284]
4-Aminophenylalanine	$H_2N-C_6H_4-CH_2CH(NH_2)COOH$	*Vigna vexillata* [288]
2-Amino-3-phenyl-butanoic acid	$C_6H_5-CH(CH_3)CH(NH_2)COOH$	Bottromycin [1, 3]; (*Streptomyces botropensis*) [6]
*N*δ-Benzoyl-L-γ-hydroxyornithine	$C_6H_5-CONHCH_2CH(OH)CH_2CH(NH_2)COOH$	*Vicia pseudo orobus* [245]
*N*δ-Benzoyl-L-ornithine	$C_6H_5-CONH[CH_2]_3CH(NH_2)COOH$	*Vicia pseudo orobus* [245]
3-Carboxy-4-hydroxyphenylalanine	$HO(HOOC)C_6H_3-CH_2CH(NH_2)COOH$	*Reseda odorata* [142]
3-Carboxy-4-hydroxyphenylglycine	$HO(HOOC)C_6H_3-CH(NH_2)COOH$	*Neonotonia wightii* [283] seeds; *Reseda* sp. [5, 282]

Table 4.7 (*Contd.*)

Amino acid	*Formula*	*Source*
3-Carboxyphenylalanine	3-HOOC-C_6H_4-$CH_2CH(NH_2)COOH$	Iris bulbs; *Reseda*; [223] *Caesalpinia tinctoria* [284]. *Confusea abyssinca* [100]
3-Carboxyphenylglycine	3-HOOC-C_6H_4-$CH(NH_2)COOH$	Iris bulb [2, 3]; *Reseda* sp. [223, 282]; *Caylusea abyssinica* [100] leaves
2, 4-Dihydroxy-6-methylphenylalanine (Orcylalanine)	2,4-(HO)$_2$-6-CH_3-C_6H_2-$CH_2CH(NH_2)COOH$	*Agrostemma githago* seeds [234]
2, 3-Dihydroxy-phenylalanine	2,3-(HO)$_2$-C_6H_3-$CH_2CH(NH_2)COOH$	Chloridazone-degrading bacteria [294]
2, 5-Dihydroxy-phenylalanine	2,5-(HO)$_2$-C_6H_3-$CH_2CH(NH_2)COOH$	Streptomycetes [144]
3, 4-Dihydroxy-phenylalanine (dopa)	3,4-(HO)$_2$-C_6H_3-$CH_2CH(NH_2)COOH$	Widely distributed in animals; neurologically active [290] catecholamine precursor; scleroprotein cross-linking agent precursor; [63, 64, 66] melanin precursor; [30] widely distributed in plants e.g. *Vicia faba* [1, 3]; *Mucuna pruriens*, [1, 3] *Euphorbia lathyris* [2]; *Hygrocybe* sp. [289]
3, 5-Dihydroxyphenyl-glycine	3,5-(HO)$_2$-C_6H_3-$CH(NH_2)COOH$	*Euphorbia helioscopa* [143]
4-Hydroxy-3-hydroxy-methylphenylalanine	4-HO-3-$HOCH_2$-C_6H_3-$CH_2CH(NH_2)COOH$	*Caesalpinia tinctoria* seeds [150]
3-Hydroxykynurenine	3-HO-2-NH_2-C_6H_3-$COCH_2CH(NH_2)COOH$	Pathological urines [1]; as a possible cross-linking component of arthropod cuticle and other invertebrate sclerotins [63]

Table 4.7 (*Contd.*)

Amino acid	*Formula*	*Source*
3-Hydroxykynurenine-*O*-β-glycoside	(GlcO)(NH$_2$)C_6H_3–COCH$_2$CH(NH$_2$)COOH	Human eye lens [300]
4-Hydroxy-3-methoxyphenylalanine	(CH$_3$O)(HO)C_6H_3–CH$_2$CH(NH$_2$)COOH	*Cortinarius brunneus* [149]; *Pachymatisma johnstoni* [149]; Blood of Parkinsonian patients [297]
3-Hydroxymethylphenyl-alanine	(HOCH$_2$)C_6H_4–CH$_2$CH(NH$_2$)COOH	*Caesalpinia tinctoria* [150, 284]; *Iris sanguinea*
4-Hydroxymethyl-phenylalanine	HOCH$_2$–C_6H_4–CH$_2$CH(NH$_2$)COOH	*Escherichia coli* [147]
4-Hydroxy-3-nitrophenylalanine	(NO$_2$)(HO)C_6H_3–CH$_2$CH(NH$_2$)COOH	Ilamycin [26]; Rufomycin; *Streptomyces islandicus*
4-*O*-[2-Oxo-(4-hydroxyphenylacetic acid)]homoserine	HOOCCO–C_6H_4–OCH$_2$CH$_2$CH(NH$_2$)COOH	β-Lactam [301] antibiotics
2-Hydroxyphenylalanine	(OH)C_6H_4–CH$_2$CH(NH$_2$)COOH	Chloridazone-degrading bacteria [294]
3-Hydroxyphenylalanine	(HO)C_6H_4–CH$_2$CH(NH$_2$)COOH	Mammalian adrenal medulla [290, 295]; chloridazone-degrading bacteria [294]
N-Hydroxyphenylalanine	C_6H_5–CH$_2$CH(NHOH)COOH	*Harpaphe haydeniana* [299] (millipede)
3-Hydroxyphenylglycine	(HO)C_6H_4–CH(NH$_2$)COOH	*Euphorbia helioscopa* [143]
4-Hydroxyphenylglycine	HO–C_6H_4–CH(NH$_2$)COOH	Enduracidin A and B; vancomycin [285, 286]; ristocetin [287]; ristomycin [411]; actintoidin [408]; avoparcin A35512B (antibiotic); *Caesalpinia tinctoria* [284]; *Iris sanguinea* [5]

Table 4.7 (*Contd.*)

Amino acid	*Formula*	*Source*
4-Hydroxyphenylpyruvic acid oxime	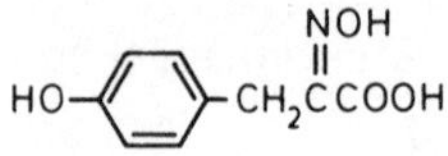 $HO-C_6H_4-CH_2C(=NOH)COOH$	*Hymeniacidon sanguinea* [220] (Porifera);
Kynurenine	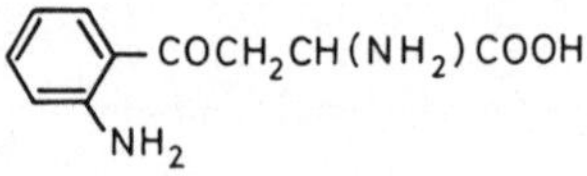 $C_6H_4(NH_2)-COCH_2CH(NH_2)COOH$	Pathological urines [1]; (tryptophan metabolism) – see Table 4.1 for this and other kynurenine derivatives
4-*O*-[2-Oximino-(4-hydroxyphenylacetic acid)]homoserine	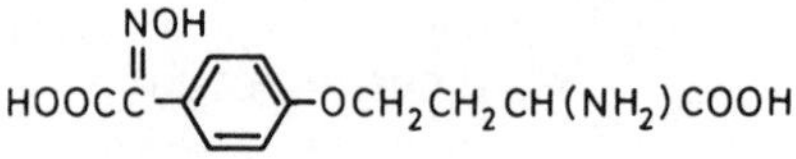 $HOOCC(=NOH)-C_6H_4-OCH_2CH_2CH(NH_2)COOH$	*β*-Lactam [301] antibiotics
L-Phenylglycine	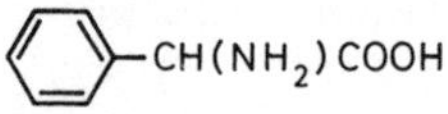 $C_6H_5-CH(NH_2)COOH$	Staphylomycins [29]; mikamycin B [16]; *Streptomyces mitakaensis* [6]; *Fagus* phloem sap [10]
L-*threo*-*β*-Phenylserine	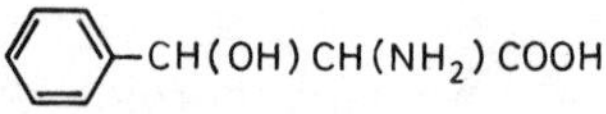 $C_6H_5-CH(OH)CH(NH_2)COOH$	*Canthium euryoides* [298]
Ristomycinic acid	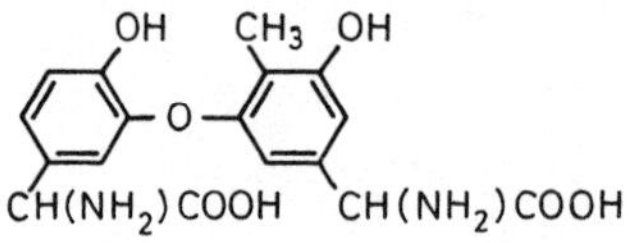 OH, CH_3, OH; two rings linked by O; $CH(NH_2)COOH$ $CH(NH_2)COOH$	Ristomycin [303, 304]
2, 4, 5-Trihydroxy-phenylalanine	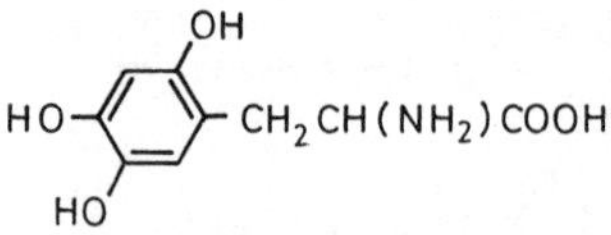 $(HO)_3C_6H_2-CH_2CH(NH_2)COOH$	*Microspira tyrosinatica* [296]
D-Tyrosine-*O*-methyl ether	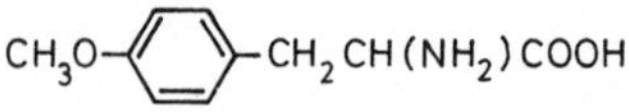 $CH_3O-C_6H_4-CH_2CH(NH_2)COOH$	Cycloheptamycin [145]
L-Tyrosine-*O*-methyl ether	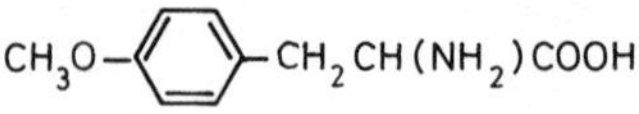 $CH_3O-C_6H_4-CH_2CH(NH_2)COOH$	Puromycin [1, 3]
Tyrosine-*O*-sulphate	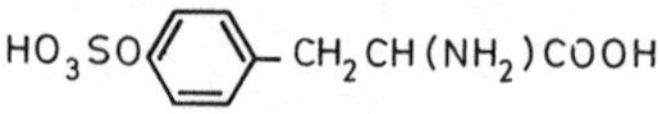 $HO_3SO-C_6H_4-CH_2CH(NH_2)COOH$	In human urine [429]

(b) Related six-membered carbon ring amino acids

Amino acid	*Formula*	*Source*
L-2-Amino-4-(4-amino- 2, 5-cyclohexadienyl)-butanoic acid	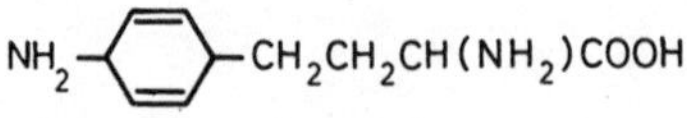 $NH_2-C_6H_6-CH_2CH_2CH(NH_2)COOH$	Stravidin [305] (*Streptomyces avidinii*)

Table 4.7 (*Contd.*)

Amino acid	*Formula*	*Source*
3-Cyclohexenylglycine	(cyclohexenyl)–CH(NH$_2$)COOH	*Streptomyces tendae* [306]
2, 5-Dihydrophenylalanine (Cyclohexa-1, 4-diene-1-alanine)	(cyclohexadienyl)–CH$_2$CH(NH$_2$)COOH	Streptomycetes [8]
Dopaquinone	(O=, O= quinone ring)–CH$_2$CH(NH$_2$)COOH	Derived from [30, 63] 3, 4-dihydroxyphenylalanine; melanin formation; sclerotization of invertebrate proteins
Mycosporine-glycine	OCH$_3$; O; NH–CH$_2$COOH; HO; CH$_2$OH	*Palythoa tuberculosa* [155] (Zoantharia)
Mytilin A	$^{-}$OOCCH$_2$–NH; +; NH–CH(COOH)CH$_2$OH; HO; CH$_2$OH	*Mytilus galloprovincialis* [51]
Mytilin B	$^{-}$OOCCH$_2$–NH; +; NH–CH(COOH)CH(OH)CH$_3$; HO; CH$_2$OH	*Mytilus galloprovincialis* [51]
Palythene	OCH$_3$; CH$_3$CH=CHN=; NH–CH$_2$COOH; HO; CH$_2$OH	*Palythoa tuberculosa* [153]
Palythine	OCH$_3$; H$_2$N; +; NH–CH$_2$COO^{-}; HO; CH$_2$OH	*Palythoa tuberculosa* [154] (Zoantharia); *Chondrus yendoi* [152] (red alga)
Palythinol	OCH$_3$; HOCH$_2$CH(CH$_3$)N=; NH–CH$_2$COOH; HO; CH$_2$OH	*Palythoa tuberculosa* [153]
Pretyrosine (Arogenate)	HO–(ring)(COOH)(CH$_2$CH(NH$_2$)COOH)	*Pseudomonas aeruginosa* [151]; *Neurospora crassa*

Table 4.8 Cycloalkane amino acids

Amino acid	*Formula*	*Source*
1-Aminocyclopropane-1-carboxylic acid	(cyclopropane) COOH, NH_2	Pears, apples and other fruit [3]; intermediate of ethylene bio-synthesis [307, 308, 309]
1-Amino-2-nitro-cyclopentanecarboxylic acid	(cyclopentane) O_2N, H_2N, COOH	*Aspergillus wenttii* [26, 204]
cis-α-(2-Carboxycyclo-propyl)glycine	HOOC CH(NH_2)COOH (cyclopropane) H H	*Aesculus parviflora* [5]
trans-α-(2-Carboxycyclo-propyl) glycine	HOOC H (cyclopropane) H CH(NH_2) COOH	*Aesculus parviflora* [5]; *Blighia sapida* [5]
trans-α-(2-Carboxy-methylcyclopropyl)-glycine	HOOCCH_2 H (cyclopropane) H CH(NH_2)COOH	*Blighia unijugata* [156]
2-Cyclopentenylglycine	(cyclopentene) CH (NH_2) COOH	*Hydrocarpus anthelminthica* seeds [301] in Flacourtiaceae [311]
α-(1-Hydroxycyclo-propyl)glycine	(cyclopropane) OH, CH (NH_2) COOH	Cleomycin [207]
Hypoglycine A (*β*-(Methylenecyclo-propyl)alanine)	H_2C= (cyclopropane) H, CH_2CH(NH_2) COOH	*Blighia sapida* seeds and unripe fruit [3, 10]
2, 4-Methanoglutamic acid	(cyclobutane) COOH, NH_2, HOOC	Seeds of *Ateleia herbert smithii* [172]
2-Methylenecyclo-hepteno-1, 3-diglycine	CH(NH_2)COOH (cycloheptene) =CH_2. CH (NH_2) COOH	*Lactarius helvus* [158]
α-(Methylenecyclo-propyl)glycine	H_2C= (cyclopropane) H, CH(NH_2) COOH	*Litchi chinensis* [157] fruit; *Billia hypoastrum* seeds; *Acer pseudo-platanus*
β-(Methylenecyclo-propyl)-*β*-methylalanine	H_2C= (cyclopropane) H, CH (CH_3) CH (NH_2) COOH	*Aesculus californicus* [131, 185]

Table 4.9 Indole amino acids

Amino acid	*Formula*	*Source*
2-Amino-3,3′-indole-prop-2-enoic acid	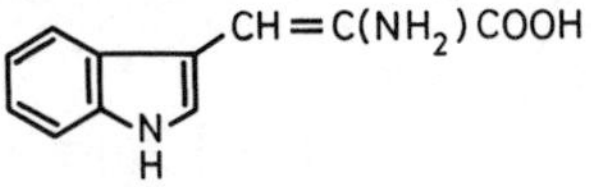 $CH{=}C(NH_2)COOH$	Telomycin [69]; Antibiotic A-128-OP
Betacyanins	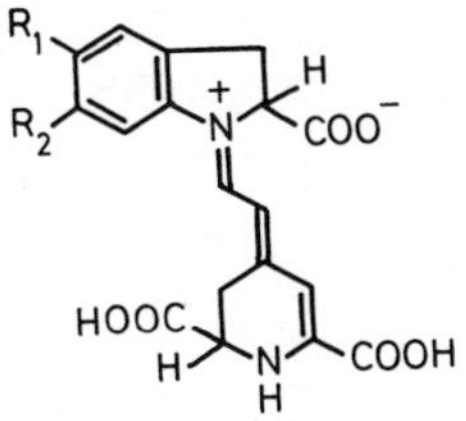 R_1, R_2, H, COO^-, HOOC, H, COOH, H	Betalains [315]; Pigments of the centrospermae
Betanin	R_2 = OH, R_1 = glucosyl	*Beta vulgaris* [315]
Amaranthin	R_2 = OH, R_1 = glucuronic acid-glycosyl	*Amaranthus tricolor* [315]
Bougainvillein-r-I	R_2 = OH, R_1 = sophorosyl	*Bougainvillea* [315]
Gomphrenin I	R_2 = OH, R_1 = glucosyl	*Gomphrena globosa* [315]

and others – also various acylated, half ester sulphated, malonylated, 3-hydroxy-3-methylglutarylated, *p*-coumaroylated, feruloylated derivatives of the sugar moiety.

Amino acid	*Formula*	*Source*
L-3-Carboxy-1, 2, 3, 4-tetrahydro-β-carboline	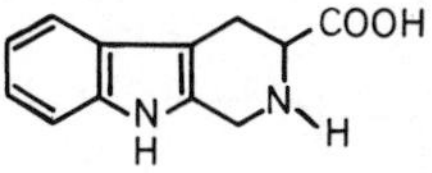 COOH, N, H	*Amanita muscaria*; *Aleurites fordii* seed [178]
Clavicipitic acid	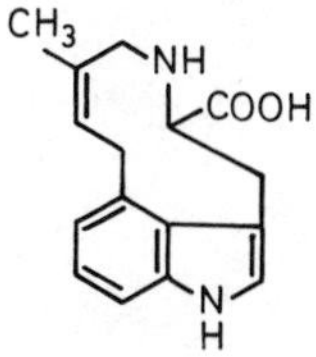 CH_3, NH, COOH, N, H	*Claviceps* sp. [164]
5, 6-Dihydroxyindole-2-carboxylic acid	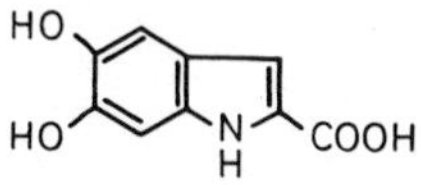 HO, HO, N, H, COOH	In sea catfish tapetum in oligomeric form; in eumelanin formation [30, 316]
1′-2′-(2, 2-Dimethyl-cyclopentano)tryptophan	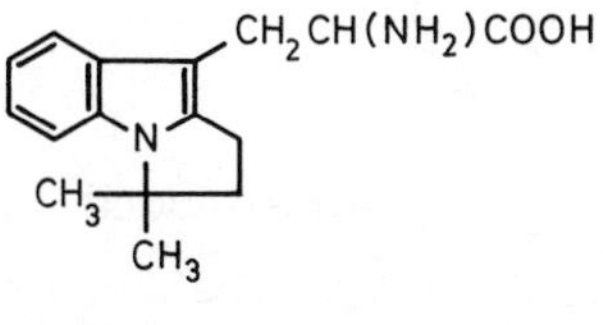 $CH_2CH(NH_2)COOH$, N, CH_3, CH_3	Ilamycin B_1 [161]
Dopachrome	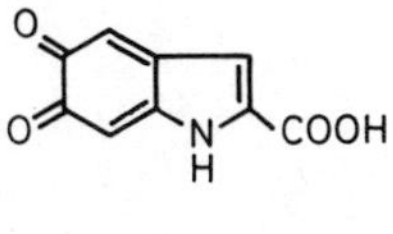 O, O, N, H, COOH	In eumelanin [30, 316] formation

Table 4.9 (*Contd.*)

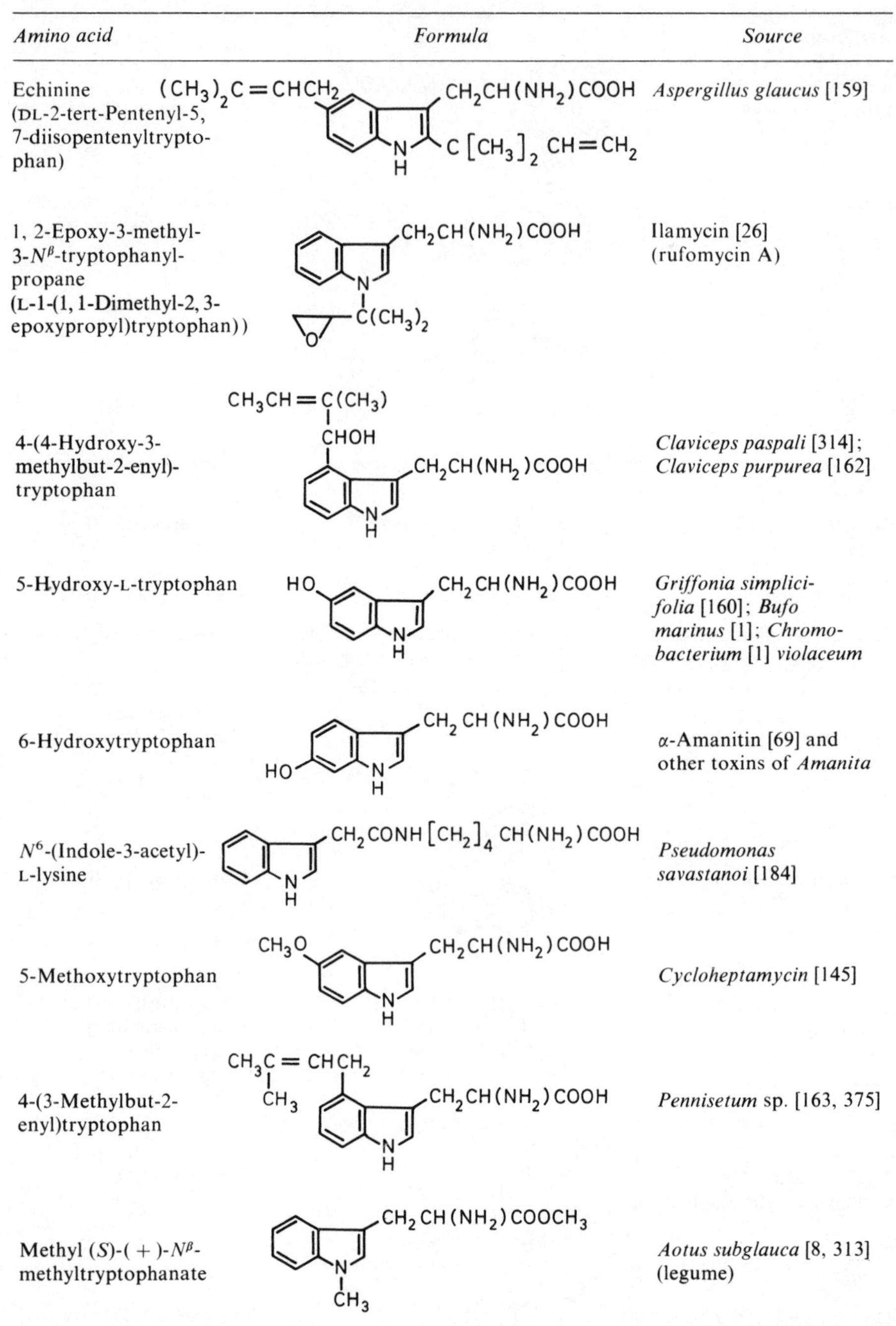

Amino acid	*Formula*	*Source*
Echinine (DL-2-tert-Pentenyl-5, 7-diisopentenyltryptophan)	$(CH_3)_2C{=}CHCH_2$; $CH_2CH(NH_2)COOH$; $C[CH_3]_2\,CH{=}CH_2$; N–H	*Aspergillus glaucus* [159]
1, 2-Epoxy-3-methyl-3-N^{β}-tryptophanyl-propane (L-1-(1, 1-Dimethyl-2, 3-epoxypropyl)tryptophan))	$CH_2CH(NH_2)COOH$; N; $C(CH_3)_2$; O	Ilamycin [26] (rufomycin A)
4-(4-Hydroxy-3-methylbut-2-enyl)-tryptophan	$CH_3CH{=}C(CH_3)$; CHOH; $CH_2CH(NH_2)COOH$; N–H	*Claviceps paspali* [314]; *Claviceps purpurea* [162]
5-Hydroxy-L-tryptophan	HO; $CH_2CH(NH_2)COOH$; N–H	*Griffonia simplicifolia* [160]; *Bufo marinus* [1]; *Chromobacterium* [1] *violaceum*
6-Hydroxytryptophan	$CH_2CH(NH_2)COOH$; HO; N–H	α-Amanitin [69] and other toxins of *Amanita*
N^6-(Indole-3-acetyl)-L-lysine	$CH_2CONH[CH_2]_4CH(NH_2)COOH$; N–H	*Pseudomonas savastanoi* [184]
5-Methoxytryptophan	CH_3O; $CH_2CH(NH_2)COOH$; N–H	*Cycloheptamycin* [145]
4-(3-Methylbut-2-enyl)tryptophan	$CH_3C(CH_3){=}CHCH_2$; $CH_2CH(NH_2)COOH$; N–H	*Pennisetum* sp. [163, 375]
Methyl (*S*)-(+)-N^{β}-methyltryptophanate	$CH_2CH(NH_2)COOCH_3$; N–CH_3	*Aotus subglauca* [8, 313] (legume)

Table 4.9 (*Contd.*)

Amino acid	*Formula*	*Source*
β-Methyltryptophan	(indole ring, N–H) $CH(CH_3)CH(NH_2)COOH$	Telomycin [3]; antibiotic A-128-OP [29]; route to Streptonigrin; in *Streptomyces floculus* [312]
3-Methyl-3-N^{β}-tryptophanylbut-1-ene (L-1-(2-Methyl-3-buten-2-yl)tryptophan)	(indole ring) $CH_2CH(NH_2)COOH$; N-substituent $CH_2{=}CHC(CH_3)CH_3$	Ilamycin [26] (rufomycin B_1)

Table 4.10 Nitrogen heterocycle amino acids

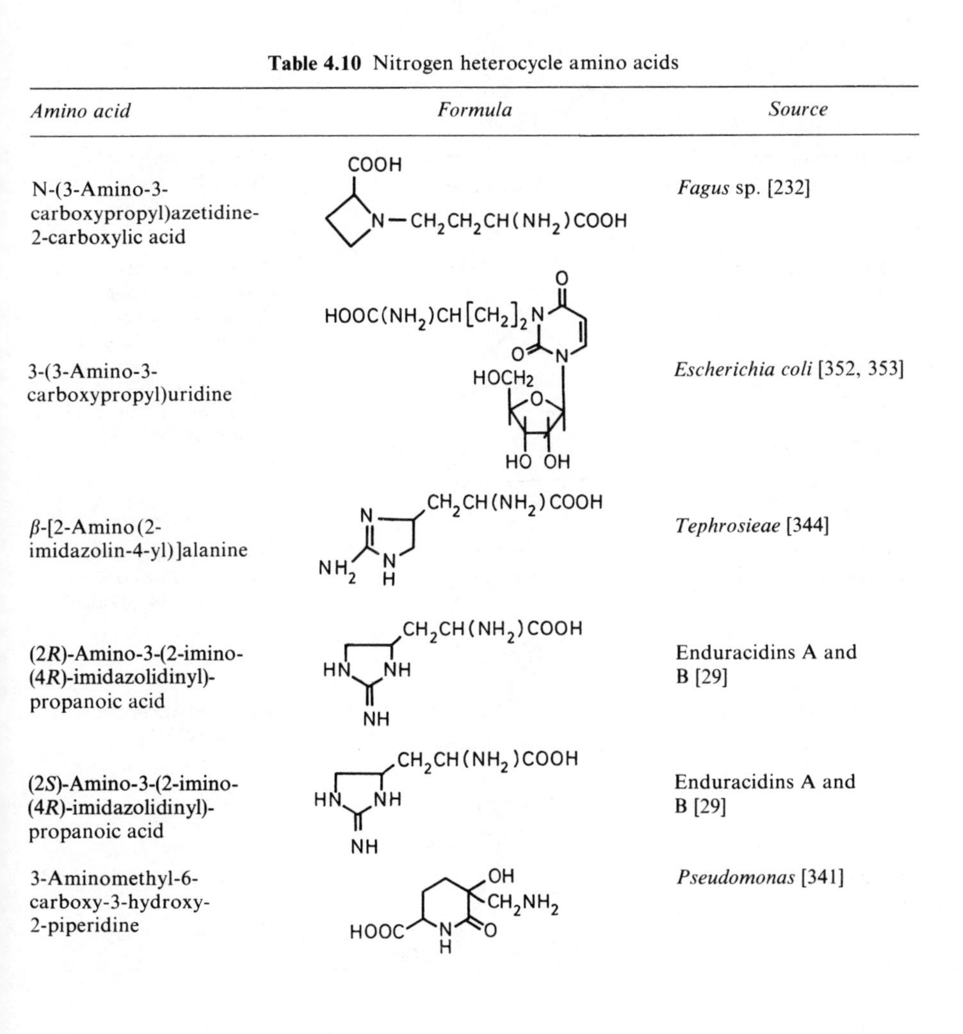

Amino acid	*Formula*	*Source*
N-(3-Amino-3-carboxypropyl)azetidine-2-carboxylic acid	COOH; N–$CH_2CH_2CH(NH_2)COOH$	*Fagus* sp. [232]
3-(3-Amino-3-carboxypropyl)uridine	$HOOC(NH_2)CH[CH_2]_2N$; O; O; N; $HOCH_2$; O; HO OH	*Escherichia coli* [352, 353]
β-[2-Amino(2-imidazolin-4-yl)]alanine	$CH_2CH(NH_2)COOH$; N; NH_2; N H	*Tephrosieae* [344]
(2*R*)-Amino-3-(2-imino-(4*R*)-imidazolidinyl)-propanoic acid	$CH_2CH(NH_2)COOH$; HN NH; NH	Enduracidins A and B [29]
(2*S*)-Amino-3-(2-imino-(4*R*)-imidazolidinyl)-propanoic acid	$CH_2CH(NH_2)COOH$; HN NH; NH	Enduracidins A and B [29]
3-Aminomethyl-6-carboxy-3-hydroxy-2-piperidine	OH; CH_2NH_2; HOOC; N H; O	*Pseudomonas* [341]

Table 4.10 (*Contd.*)

Amino acid	*Formula*	*Source*
4-Aminopipecolic acid	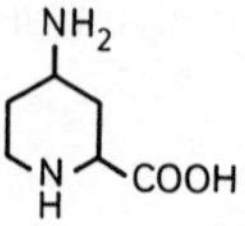 NH_2; N H; COOH	*Strophanthus scandeus* [3]
cis-3-Aminoproline	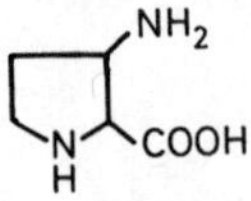 NH_2; N H; COOH	*Morchella esculenta* [323, 324]
3-Aminopyrrolidine-3-carboxylic acid (Curcubitin)	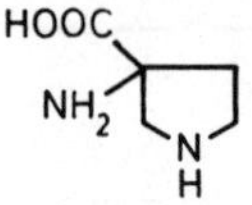 HOOC; NH_2; N H	*Curcubita moschata* [6]
Ascorbalamic acid	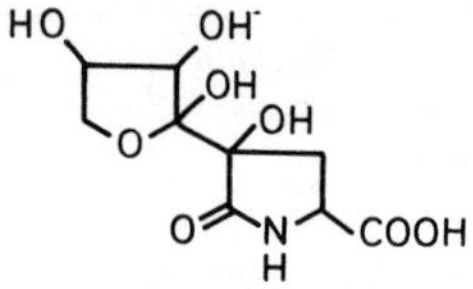 HO; OH; OH; OH; O; O; N H; COOH	Green Gram seeds [329, 330]
Azetidine-2-carboxylic acid	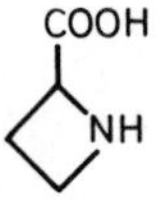 COOH; NH	Widely in plants e.g. [4] *Convallaria majalis* leaves [1], *Polygonatum officinale* [8], *Delonix majalis* [8], *Lophocladia lallemandi* [12] (alga), *Haliclona* (Porifera) [318], *Chalinopsilla* [318]
Azirinomycin	 CH_3; N; COOH	*Streptomyces aureus* [182]
L-Baikiain (4, 5-Dehydro-L-pipecolic acid)	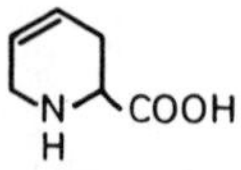 N H; COOH	*Baikiaea plurijuga* [1]; dates [1]; *Corallina officianalis* [4, 8] (alga)
Betalamic acid	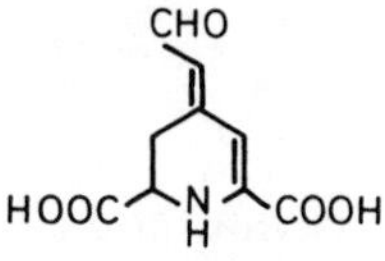 CHO; HOOC; N H; COOH	Intermediate in betalain synthesis [315]
Indicaxanthin	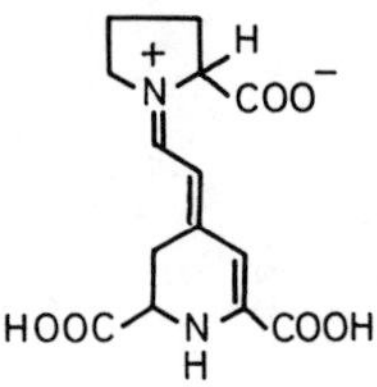 +; N; H; COO^-; HOOC; N H; COOH	*Opuntia ficas indica* [315]

Table 4.10 (*Contd.*)

Amino acid	*Formula*	*Source*
Bleomycin	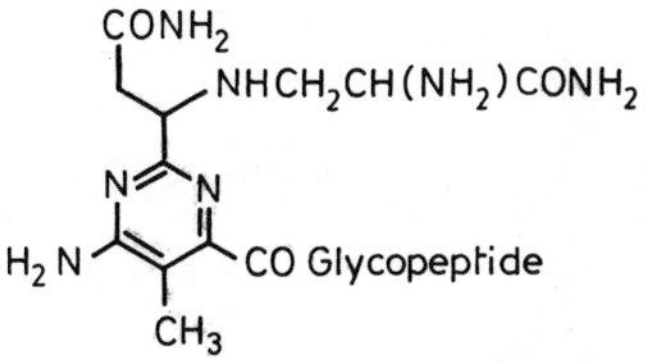 CONH$_2$; NHCH$_2$CH(NH$_2$)CONH$_2$; N; N; H$_2$N; CO Glycopeptide; CH$_3$	*Streptomyces verticillus* [217]
5-Butylpicolinic acid (Fusaric acid)	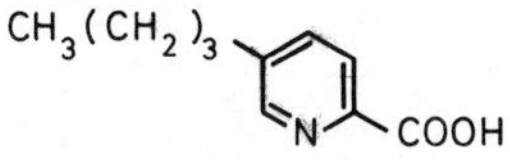 CH$_3$(CH$_2$)$_3$; N; COOH	*Fusarium lycopersii* [3]
Capreomycidine (α-(2-Iminohexahydro-pyrimid-4-yl) glycine) (2*S*, 3*S*)-(α-(2-Iminohexa-hydropyrimid-4-yl)-glycine) (2*R*S, 3*S*)-(α-(2-Iminohexa-hydropyrimid-4-yl)-glycine)	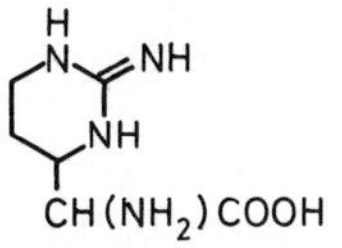 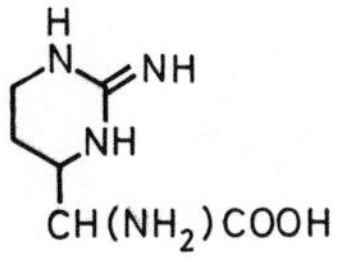 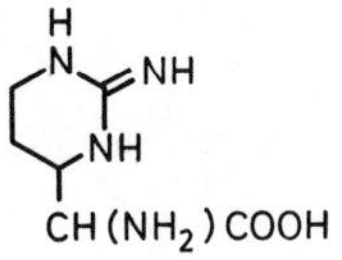 H; N; NH; NH; CH(NH$_2$)COOH — H; N; NH; NH; CH(NH$_2$)COOH — H; N; NH; NH; CH(NH$_2$)COOH	Capreomycin [8], from chymostatin [339] and from elastinal [350]
(2*S*, 4*R*)-2-Carboxy-4-acetylaminopiperidine	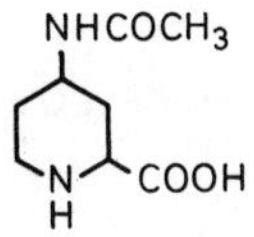 NHCOCH$_3$; N; H; COOH	*Calliandra haemato-cephala* [179]
3-Carboxymethyl-4-(2-carboxy-1-methylhexa-1, 3-dienyl) proline (Domoic acid)	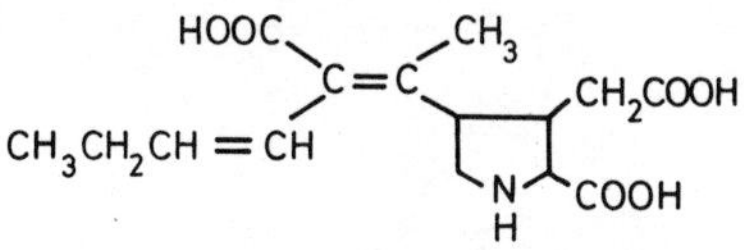 HOOC; CH$_3$; C=C; CH$_3$CH$_2$CH = CH; CH$_2$COOH; N; H; COOH	*Chondria armata* [3]; *Alsidium corallinum* [12]
3-Carboxymethyl-4-isopropenylproline (L-α-kainic acid) (L-α-allo-kainic acid)	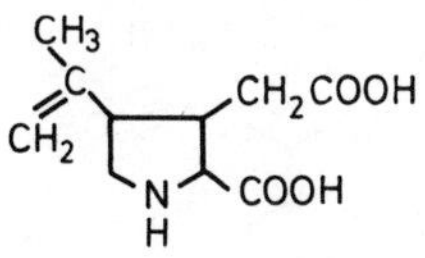 CH$_3$; C; CH$_2$; CH$_2$COOH; N; H; COOH	*Digenea simplex* [3]; *Centroceras clavulatum* [12] (marine algae)
Dehydrofusaric acid	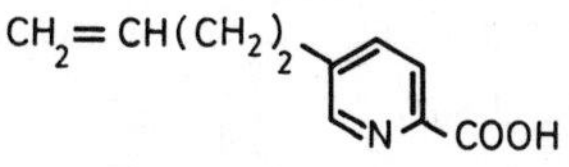 CH$_2$= CH(CH$_2$)$_2$; N; COOH	*Fusarium lycopersii* [26]

Table 4.10 (*Contd.*)

Amino acid	*Formula*	*Source*
Desmosine	HOOC(NH_2)CH[CH_2]$_2$–(pyridinium ring with (CH_2)$_3$CH(NH_2)COOH, [CH_2]$_2$CH(NH_2)COOH and N⁺–[CH_2]$_4$CH(NH_2)COOH)	Human urine [41]; product of elastin degradation
L-*trans*-2, 3-Dicarboxy-aziridine	H, COOH, HOOC, N, H, H	*Streptomyces* cultures [317]
4, 5-Dihydroxy-L-pipecolic acid	OH, HO, N, H, COOH	*Calliandra haemato-cephala* [338]
(2*S*, 4*R*, 5*R*)-4, 5-Dihydroxy-L-pipecolic acid	OH, HO, N, H, COOH	*Julbernardia paniculata* seeds [95]
(2*S*, 4*R*, 5*S*)-4, 5-Dihydroxy-L-pipecolic acid	OH, HO, N, H, COOH	*Derris eliptica* seeds [340]
(2*S*, 4*S*, 5*S*)-4, 5-Dihydroxy-L-pipecolic acid	OH, HO, N, H, COOH	*Isoberlinia* seeds [340]; *Brachystegia* seeds; *Cryptosepalum* seeds; *Derris eliptica* seeds
β-(2, 6-Dihydroxypyrimi-din-5-yl) alanine	HO, N, N, OH, $CH_2CH(NH_2)COOH$	Pea seedlings [10]
3, 4-Dihydroxyquinoline-2-carboxylic acid	OH, OH, N, COOH	*Aplysina aerophoba* [8] (Porifera)
6, 7-Dihydroxy-1, 2, 3, 4-tetrahydroisoquinoline-3-carboxylic acid	HO, HO, COOH, NH	*Mucuna mutisiana* [181, 186]
1, 3-Dimethyl-L-histidine	$CH_2CH(NH_2)CO\bar{O}$, CH_3–N⁺, N–CH_3	Red and brown algae [12] e.g. *Gracillaria secundata*

Table 4.10 (*Contd.*)

Amino acid	*Formula*	*Source*
Dipicolinic acid	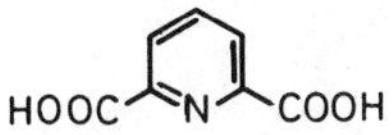 HOOC N COOH	*Bacillus megatherium* [3] spores
Discadenine	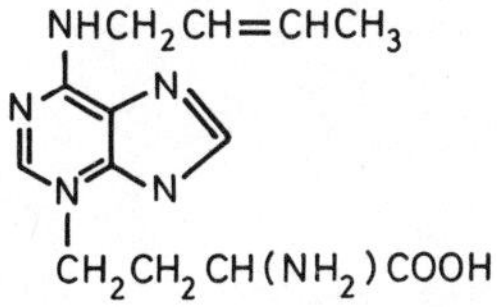 $NHCH_2CH{=}CHCH_3$ N N N N $CH_2CH_2CH(NH_2)COOH$	*Dictyostellium discoideum* [356]
Dopaxanthin	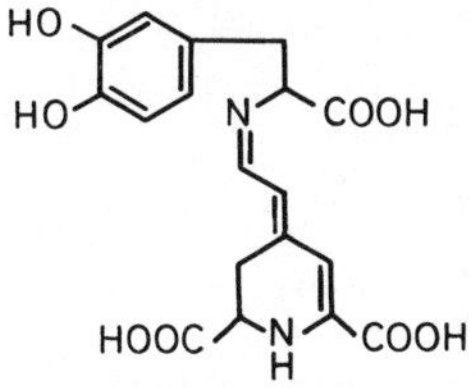 HO HO N COOH HOOC N H COOH	*Glottiphyllum longum* [315]
D-α-allo-Enduracididine	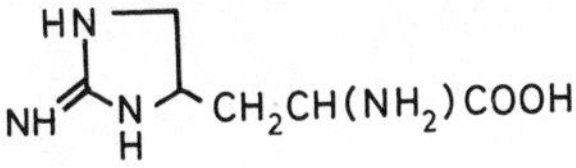 HN NH N H $CH_2CH(NH_2)COOH$	Enduracidin [345] (antibiotic)
L-α-Enduracididine	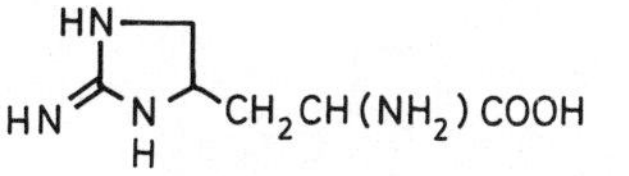 HN HN N H $CH_2CH(NH_2)COOH$	*Tephrosieae* [344]; Enduracidin [345] (antibiotic); Aspartocin; *Lonchocarpus* [5]
Ethyl hydrogen-2, 6-dipicolinate	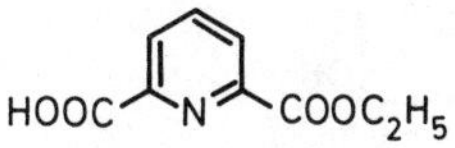 HOOC N $COOC_2H_5$	*Bacillus cerens* spores [26]
'Fluorescent Y base'	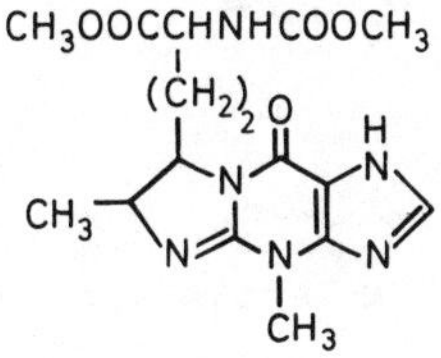 $CH_3OOCCHNHCOOCH_3$ $(CH_2)_2$ O H N CH_3 N N N N CH_3	Yeast, wheat germ and rat liver phenylalaninet RNA [357]
Histopine ((N^2-(1-Carboxyethyl)-L-histidine)	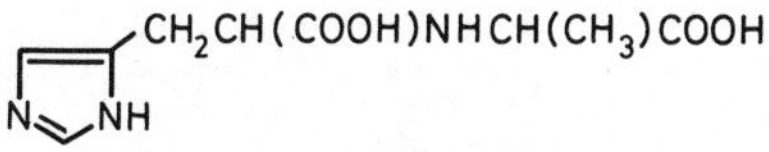 $CH_2CH(COOH)NHCH(CH_3)COOH$ N NH	Crown gall tumours [348, 349] of sunflower
Hydroxyfusaric acid	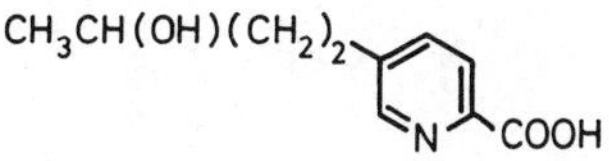 $CH_3CH(OH)(CH_2)_2$ N COOH	*Fusarium* sp. [26]
β-Hydroxy-L-histidine	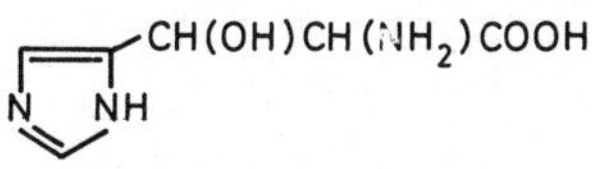 $CH(OH)CH(NH_2)COOH$ N NH	YA56 Antibiotics [346, 347], phleomycin – bleomycin group

Table 4.10 (*Contd.*)

Amino acid	*Formula*	*Source*
6-Hydroxykynurenic acid	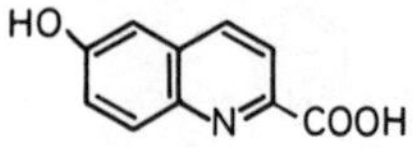	*Nicotiana* [354]; widely in plants; mammalian and avian urine
(2*S*, 5*S*, 6*S*)-5-Hydroxy-6-methylpipecolic acid	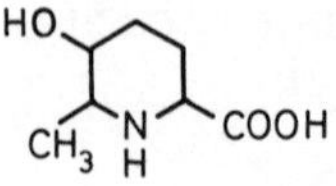	*Fagus silvatica* [167, 232, 342]
(2*S*, 5*R*, 6*S*)-5-Hydroxy-6-methylpipecolic acid	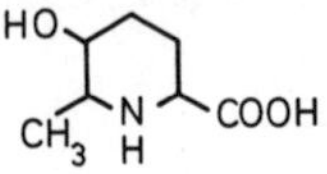	*Fagus silvatica* [167, 232, 342]
3-Hydroxy-5-methylproline	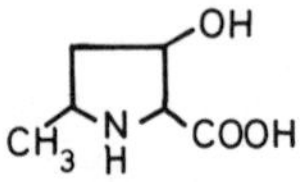	Actinomycin Z_1 [327]
cis-4-Hydroxy-methylproline	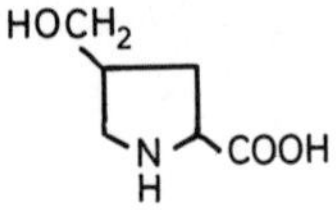	Apple wood and fruit [1, 2, 3, 4, 6]
4-Hydroxyminaline	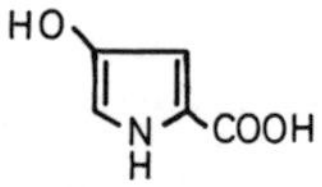	*Penicillium* and *Aspergillus* [1, 2]
3-Hydroxy-4-oxo-5-methylproline	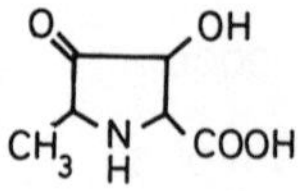	Actinomycin Z_1 [325]
3-Hydroxypicolinic acid	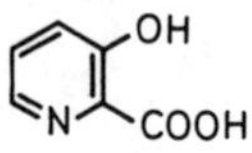	Viridogrisein; (*Streptomyces griseus*) [3]; Etamycin [214]; Staphylomycin Factor S [79]; Mikamycin B [6]
3-Hydroxypipecolic acid	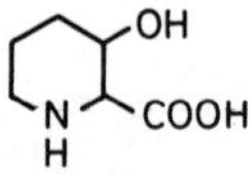	Halophyte [8] leaves, flowers and fruits
cis-5-Hydroxy-L-pipecolic acid	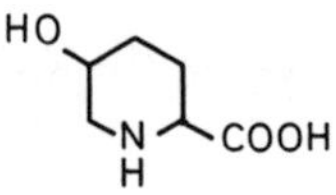	*Gymnocladus dioicus* [336]; *Morus alba* [337]; *Lathyrus japonicus* [337]; *Gleditsia triacanthes* [5]

Table 4.10 (*Contd.*)

Amino acid	*Formula*	*Source*
trans-L-4-Hydroxypipecolic acid	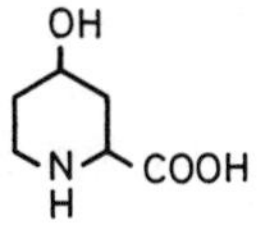 OH; N H; COOH	*Acacia pentadena* [1]; *Armeria maritima* [2]; *Peganum harmala* [8, 335]; caesalpinaceae [224, 4]; *Peltophorum* sp. [224]
trans-5-Hydroxypipecolic acid	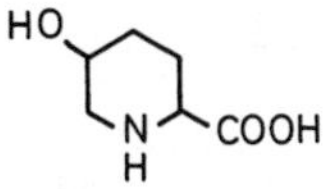 HO; N H; COOH	*Acacia* [1] sp.; *Rhapis flabelliformis* [1]; *Gleditsia triacanthos* [4, 6]; *Undaria pinnatifida* [5, 8] (alga)
(3*S*, 5*S*)-5-Hydroxypiperazic acid	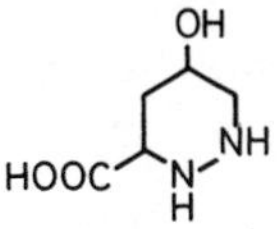 OH; NH; HOOC; N H	Monamycin [29]
(3*S*, 5*S*)-5-Hydroxy-piperidazine-3-carboxylic acid	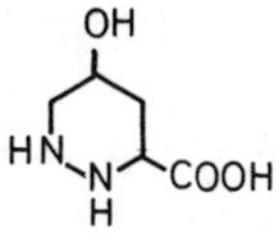 OH; HN; N H; COOH	*Streptomyces jamaicensis* [291, 407]; monamycin
D-allo-4-Hydroxyproline	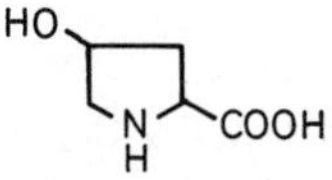 HO; N H; COOH	Etamycin [1, 4, 214]
L-allo-4-Hydroxyproline	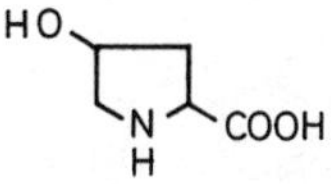 HO; N H; COOH	*Amanita phalloides* [11, 26, 69] in phalloidine; *Santalum album* leaves [1, 3]
cis-4-Hydroxy-L-proline	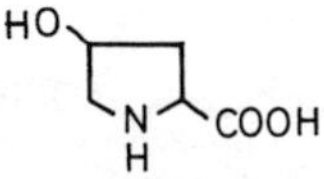 HO; N H; COOH	*Santalum album* [8] actinomycin I [6]; *Afzelia bella* seeds [221]
3-Hydroxy-L-proline (*cis* and *trans*)	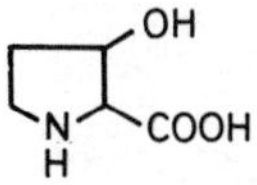 OH; N H; COOH	Telomycin [3]; *Delonix regia* seeds [320, 321] and vegetation
Isodesmosine	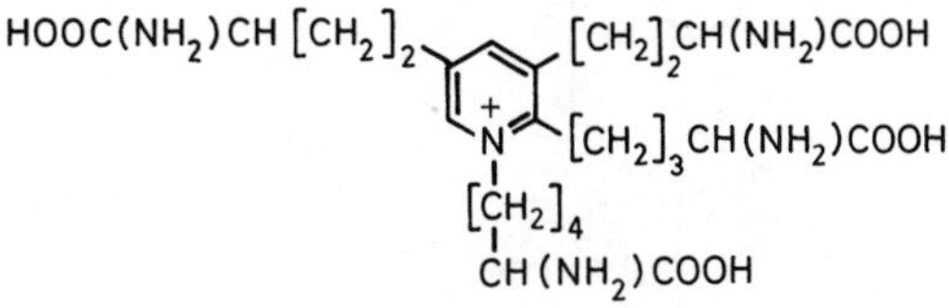 $HOOC(NH_2)CH[CH_2]_2$; $[CH_2]_2CH(NH_2)COOH$; N^+; $[CH_2]_3CH(NH_2)COOH$; $[CH_2]_4$; $CH(NH_2)COOH$	Human urine [41]; product of elastin degradation
Isowillardine (β-(2, 4-Diketo-pyrimidinyl) alanine)	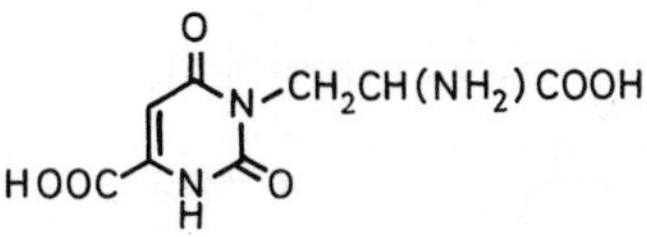 O; $N{-}CH_2CH(NH_2)COOH$; HOOC; N H; O	Pea seedlings root exudates [10]

Table 4.10 (*Contd.*)

Amino acid	*Formula*	*Source*
4-Ketopipecolic acid	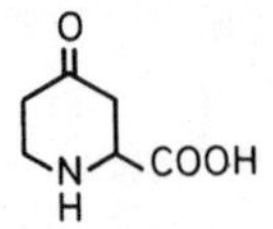 O; N H; COOH	Staphylomycin [79] (factor S) [6]; mikamycin B
4-Keto-L-proline	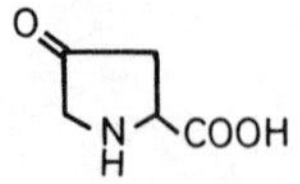 O; N H; COOH	Actinomycin V [3, 4]
Lathyrine β-(2-Aminopyrimidin-4-yl)alanine	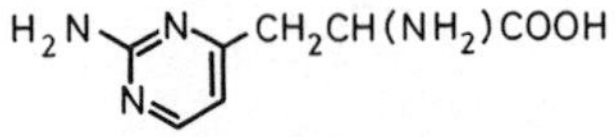 H_2N; N; N; $CH_2CH(NH_2)COOH$	*Lathyrus tingitanus* [175]; *Lathyrus japonicus* [420]; cucurbitaceae [234]
Lupinic acid (β-[6-(4-Hydroxy-3-methyl-*trans*-butenyl-amino) purin-9-yl]alanine)	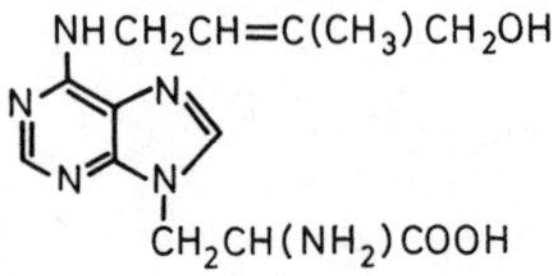 $NHCH_2CH{=}C(CH_3)CH_2OH$; N; N; N; N; $CH_2CH(NH_2)COOH$	*Lupinus augustifolius* [174] seedlings; *Lathyrus odoratus*
2, 4-Methanoproline	 N H; COOH	Seeds of *Ateleia* [172] *herbert smittii*; (leguminosae)
cis-3, 4-Methano-L-proline	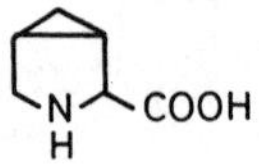 N H; COOH	*Aesculus parviflora* [8, 328]
1-Methyl-6, 7-dihydroxy-1, 2, 3, 4-tetrahydro-isoquinoline-3-carboxylic acid	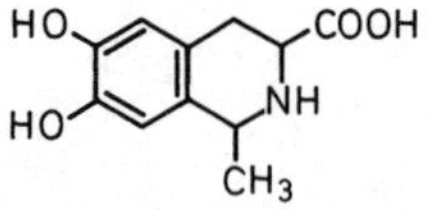 HO; HO; COOH; NH; CH_3	*Mucuna mutisiana* [183]
(1*S*, 3*S*)-1-Methyl-6, 7-dihydroxy-1, 2, 3, 4-tetrahydroisoquinoline-3-carboxylic acid	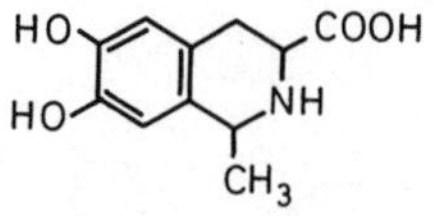 HO; HO; COOH; NH; CH_3	*Mucuna* sp. [355] (velvet bean)
1-Methyl-6-hydroxy-1, 2, 3, 4-tetrahydro-isoquinoline-3-carboxylic acid	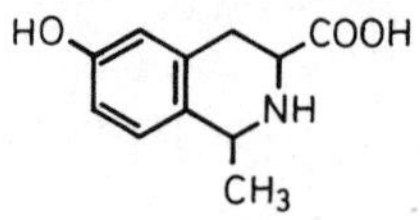 HO; COOH; NH; CH_3	*Euphorbia myrsinites* [183]
4-Methylene-DL-proline	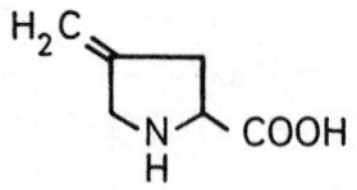 H_2C; N H; COOH	*Eribotrya japonica* seeds [3, 6, 168]; loquat seeds [170, 171]
1-Methyl-L-histidine	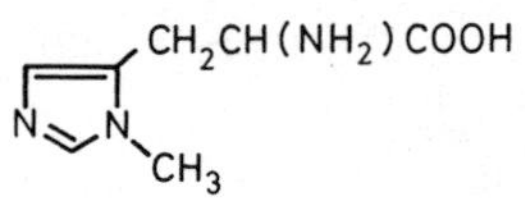 $CH_2CH(NH_2)COOH$; N; N; CH_3	Anserine [1]; human urine [1]; red and brown algae [12] e.g. *Phyllospora camosa*

Table 4.10 (*Contd.*)

Amino acid	*Formula*	*Source*
2-Methyl-L-histidine	$CH_2CH(NH_2)COOH$; N, NH; CH_3	Snake muscle [1]
3-Methyl-L-histidine	$CH_2CH(NH_2)COOH$; N, N; CH_3	Human urine [1]
(2*S*, 3*S*, 4*S*)-4-Methyl-3-hydroxyproline	CH_3, OH; N, H, COOH	Echinocandin B [146]
cis-4-Methyl-L-proline	CH_3; N, H, COOH	Antibiotic 1C1 13959 [6]
trans-4-Methyl-L-proline	CH_3; N, H, COOH	Apples [1, 3, 6]; griselimycin [319]; antibiotics from *Paecilomyces* [3]; monamycin
cis-5-Methylproline	CH_3, N, H, COOH	Actinomycin Z_5 [326]
N-Methylstreptolidine	HOOC, $CH(OH)CH_2NHCH_3$; N, NH; NH_2	Streptothricin [292]; antibiotic LL-AC541
Mimosine or Leucenol β-(1, 4-Dehydro-3-hydroxy-4-oxopyrid-1-yl)-L-alanine	HO; O=; $NCH_2CH(NH_2)COOH$	*Mimosa pudica* [1, 173]; *Leucaena glauca* [2]; *Cucurbitaceae* [234];
Minaline	N, H, COOH	Takadiastase; yeast invertase aglycone [1, 2]
Miraxanthin-1	CH_3; O=S; COOH; N; HOOC, N, H, COOH	*Mirabilis jalapa* [315]

Table 4.10 (*Contd.*)

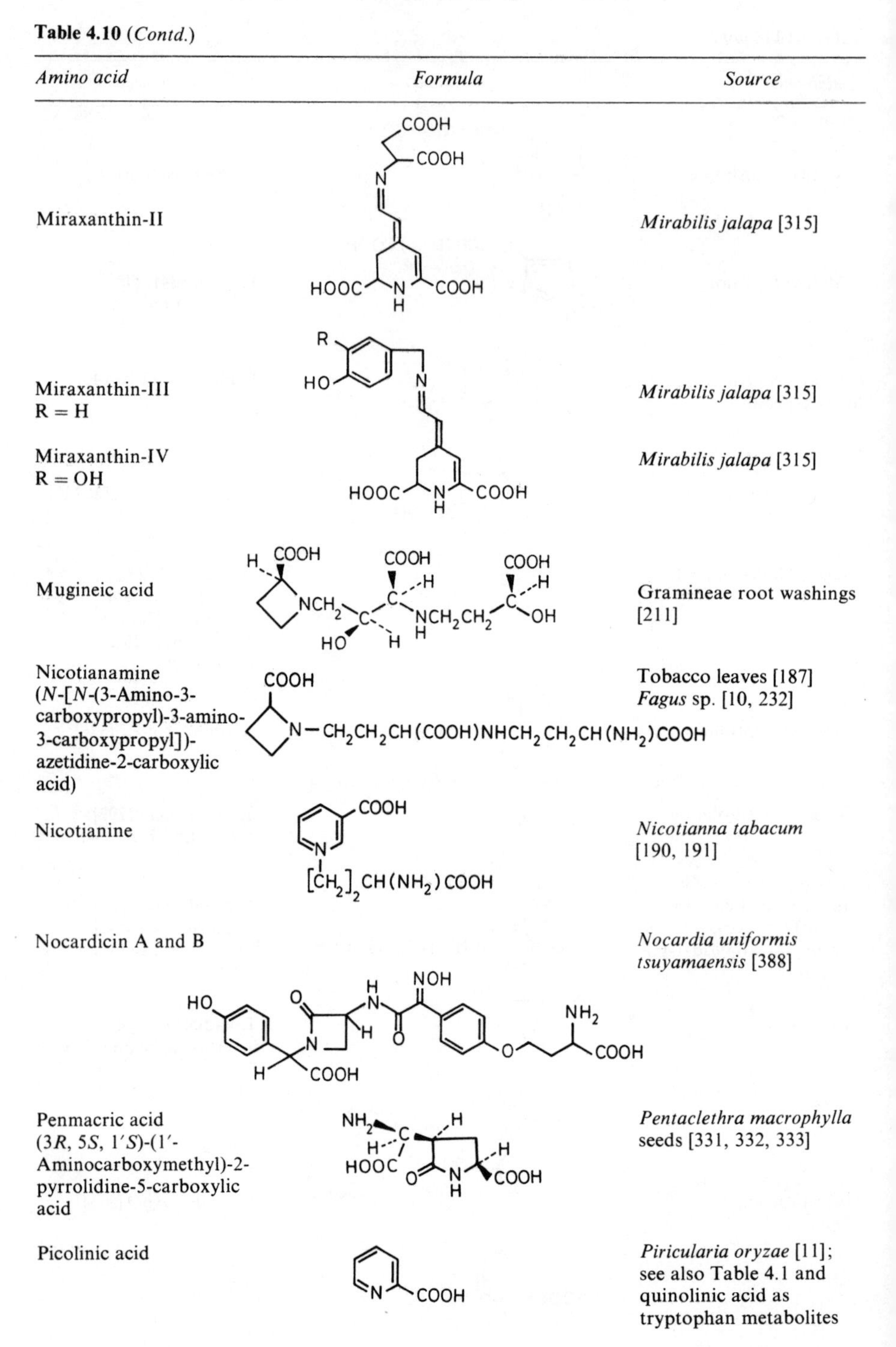

Amino acid	*Formula*	*Source*
Miraxanthin-II		*Mirabilis jalapa* [315]
Miraxanthin-III R = H		*Mirabilis jalapa* [315]
Miraxanthin-IV R = OH		*Mirabilis jalapa* [315]
Mugineic acid		Gramineae root washings [211]
Nicotianamine (*N*-[*N*-(3-Amino-3-carboxypropyl)-3-amino-3-carboxypropyl])-azetidine-2-carboxylic acid)	N−$CH_2CH_2CH(COOH)NHCH_2CH_2CH(NH_2)COOH$	Tobacco leaves [187] *Fagus* sp. [10, 232]
Nicotianine	$[CH_2]_2CH(NH_2)COOH$	*Nicotianna tabacum* [190, 191]
Nocardicin A and B		*Nocardia uniformis tsuyamaensis* [388]
Penmacric acid (3*R*, 5*S*, 1′*S*)-(1′-Aminocarboxymethyl)-2-pyrrolidine-5-carboxylic acid		*Pentaclethra macrophylla* seeds [331, 332, 333]
Picolinic acid		*Piricularia oryzae* [11]; see also Table 4.1 and quinolinic acid as tryptophan metabolites

Table 4.10 (*Contd.*)

Amino acid	*Formula*	*Source*
D-Pipecolic acid	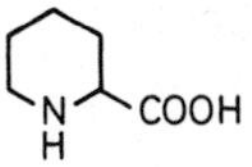 [piperidine ring, N–H, 2-COOH]	Amphomycin [6]; aspartocin [6]; glumamycin [6];
L-Pipecolic acid	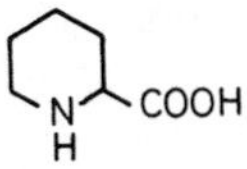 [piperidine ring, N–H, 2-COOH]	Widely in plants e.g. apples, legumes [1, 3, 12, 334]; dates; brown and red algae; neurospora; see Table 4.1
Piperazic acid (3*R*)-Piperidazine-3-carboxylic acid	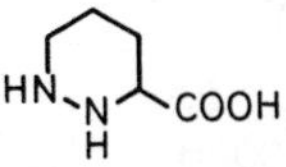 HN–NH ring, COOH	Monamycin [29] *Streptomyces jamaicensis* [291, 407]
Portulaxanthin	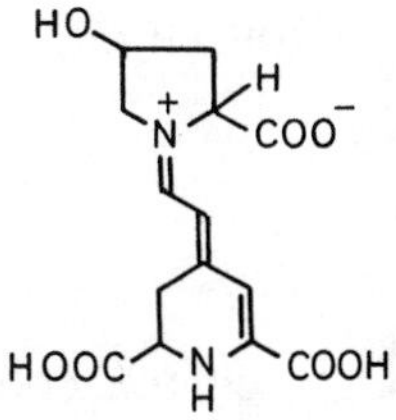 HO-substituted pyrrolidinium (N^+), H, COO^-; =CH–CH= linked to dihydropyridine with HOOC, N–H, COOH	*Portulacea grandiflora* [315]
β-(*N*-Pyrazolyl)-L-alanine (*β*-[Pyrazol-l-yl]alanine)	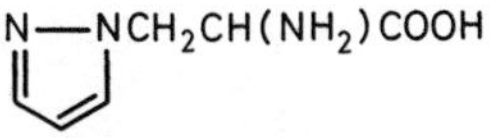 pyrazole N—$NCH_2CH(NH_2)COOH$	*Citrullus vulgaris* [3, 234]; *Cucumis sativus* seed [343]
Pyridinoline	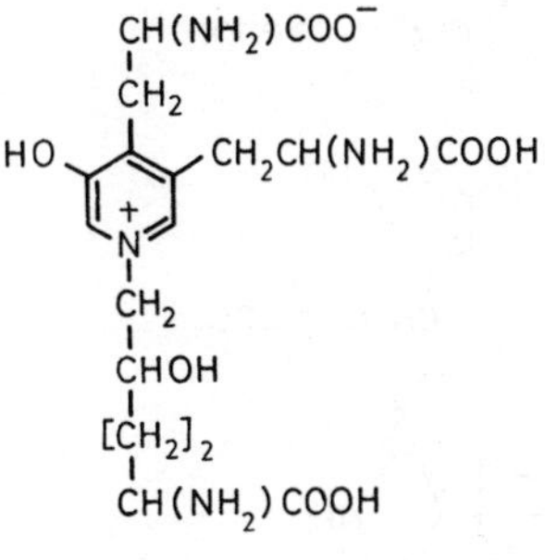 $CH(NH_2)COO^-$–CH_2– on pyridinium ring with HO and $CH_2CH(NH_2)COOH$; N^+–CH_2–CHOH–$[CH_2]_2$–$CH(NH_2)COOH$	Human urine [42]; product of collagen degradation
Pyrimine	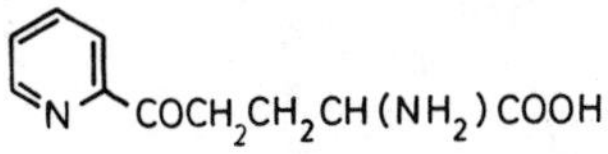 pyridin-2-yl $COCH_2CH_2CH(NH_2)COOH$	*Pseudomonas* GH [26]
Pyrrolidine-2, 4-dicarboxylic acid (*trans*-4-carboxy-L-proline)	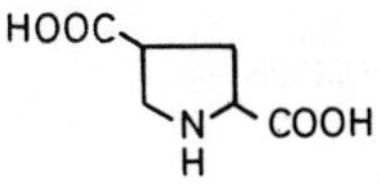 HOOC-substituted pyrrolidine, N–H, COOH	*Chondria coerulescens* [12] and other algae; *Afzelia bella* seeds [221]
Pyrrolidine-2, 5-dicarboxylic acid	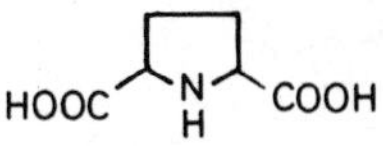 HOOC–pyrrolidine (N–H)–COOH	*Schizymeria dubyi* [12]; *Haematocelis rubens* [12, 322]

Table 4.10 (*Contd.*)

Amino acid	*Formula*	*Source*
Quinoxaline-2-carboxylic acid	N, N, COOH	Echinomycin [29]
Spinacin	H, N, CH_2, COOH, N, CH_2–NH	*Ascanthias vulgaris* liver [1]
Stendomycidine	H, N, NCH_3, N, CH_3, H_2N, COOH	Stendomycin [29, 351]
Streptolidine	HOOC, $CH(OH)CH_2NH_2$, N, NH, NH_2	Streptothricin [292]; antibiotic LL-AC541
Tetrahydrolathyrine	N, H_2N, N, H, CH_2, $CH(NH_2)COOH$	*Lonchocarpus costaricensis* [177] seeds
Trichoponauric acid	CH_3, CH_3COCH_2, N, H, COOH	In *Trichopolyn* [169]; a *Lentinus edodes* growth inhibitor from *Trichoderma polysporum*
Tuberactidine (α-(2-Imino-4-hydroxy-hexapyrimid-4-yl) glycine)	H, HO, N, NH, NH, $CH(NH_2)COOH$	Tuberactinomycin [8]
Viomycidine	HN, NH, HOOC, N, H, NH	Viomycin from *Streptomyces puniceous*; *S. floridae*; tuberactino-mycin [8, 358, 180]
Vulgaxanthin-I $R = NH_2$	R, O, COOH, N, HOOC, N, H, COOH	*Beta vulgaris* [315]
Vulgaxanthin-II $R = OH$		*Beta vulgaris* [315]

Table 4.10 (*Contd.*)

Amino acid	*Formula*	*Source*
L-Willardiine (β-Uracil-3-yl-L-alanine)	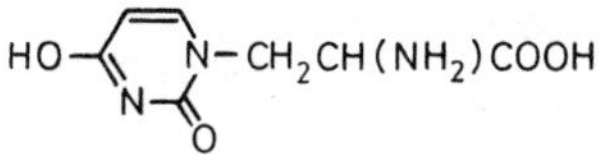 HO– (uracil ring) N–$CH_2CH(NH_2)COOH$	*Acacia* sp. seeds [176]; pea and sweet pea root exudates [206]; cucurbitaceae [234]

Table 4.11 Oxygen heterocycle amino acids

Amino acid	*Formula*	*Source*
Anticapsin (2, 3-Epoxy-4-oxo-hexahydro-L-phenylalanine)	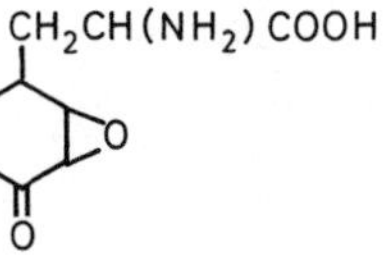 $CH_2CH(NH_2)COOH$; O	*Streptomyces griseoplanus* [192]
L-β-(3-Carboxyfuran-4-yl) alanine	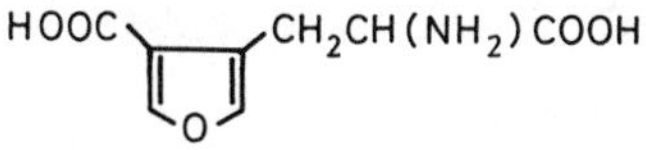 HOOC, $CH_2CH(NH_2)COOH$; O	*Phyllotopsis nidulans* [360]; *Tricholomopsis rutilans* [361]
Furanomycin (Threomycin)	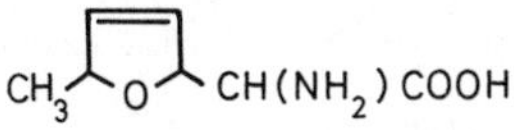 CH_3, O, $CH(NH_2)COOH$	*Streptomyces threomycetius* [26]
L-β-(2-Furoyl)alanine	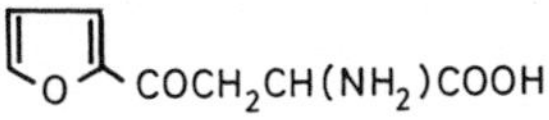 O, $COCH_2CH(NH_2)COOH$	*Fagopyrum esculentum* seeds [225, 226, 227]; probably a degradation product of ascorbalamic acid
Lycoperdic acid (β-(5(*S*)-Carboxyl-2-oxotetrahydrofuranyl)-L-alanine)	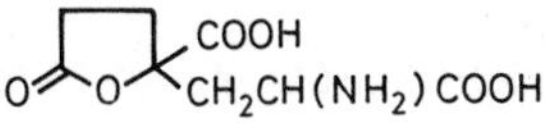 O, O, COOH, $CH_2CH(NH_2)COOH$	*Lycoperdon perlatum* [194]
Marinobufagin-3-suberoyl-L-glutamine ester	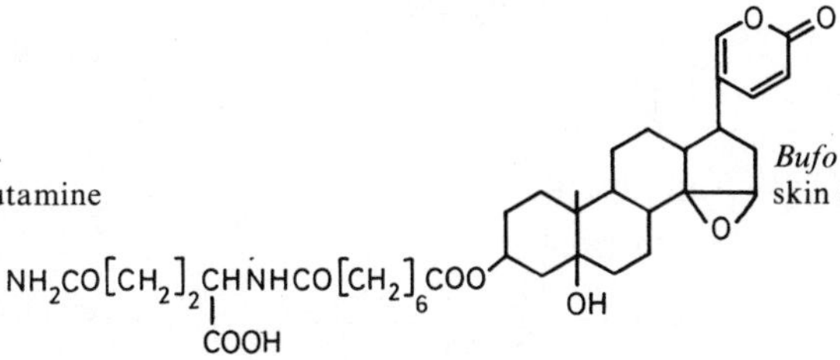 O, O; O; OH; $NH_2CO[CH_2]_2CH(COOH)NHCO[CH_2]_6COO$	*Bufo americanus* [362] skin secretion
Stizolobic acid	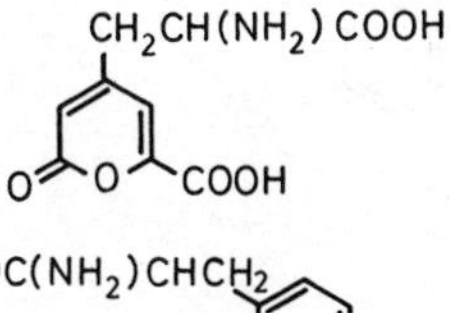 $CH_2CH(NH_2)COOH$; O, O, COOH	*Stizolobium hasjoo* seedlings [3, 10]; cucurbitaceae [234]
Stizolobinic acid	 $HOOC(NH_2)CHCH_2$; O, O, COOH	*Stizolobium hasjoo* [10]

Table 4.12 Nitrogen-Oxygen heterocycle amino acids

Amino acid	*Formula*	*Source*
2-(β-Alanyl)clavam	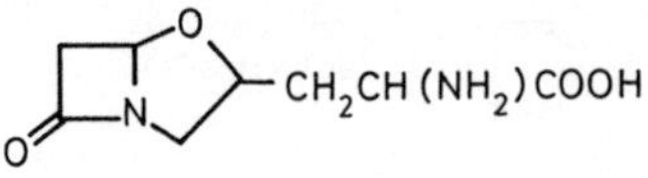 O, N, O, $CH_2CH(NH_2)COOH$	*Streptomyces clavuligerus* [363]
2-Amino-4-(isoxazolin-5-on-2-yl)butanoic acid	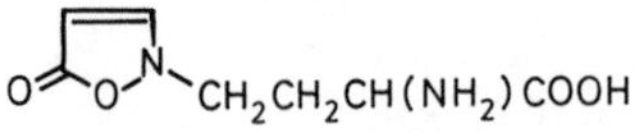 O, O, N, $CH_2CH_2CH(NH_2)COOH$	Sweet pea root exudates [206]
2-Amino-4-(isoxazolin-5′-on-3′-yl)butanoic acid	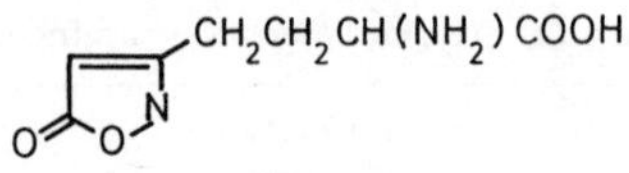 $CH_2CH_2CH(NH_2)COOH$, O, O, N	Sweet pea extracts [189]
Desaminocanavanine	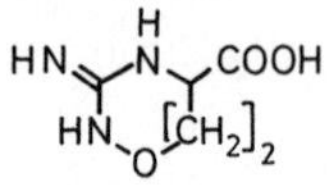 H, HN, N, COOH, HN, O, $[CH_2]_2$	Jack bean [3]
β-(3, 5-Dioxo-1, 2, 4-oxadiazolidin-2-yl)-L-alanine	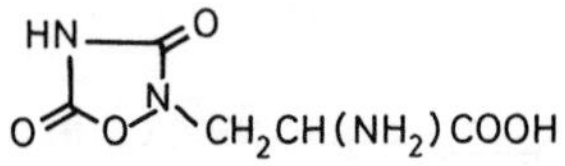 HN, O, O, O, N, $CH_2CH(NH_2)COOH$	*Quisqualis fructus* [109]; *Q. indica* [108]
β-(2-β-D-Glucopyranosyl-isoxazolin-5-on-4-yl)-alanine	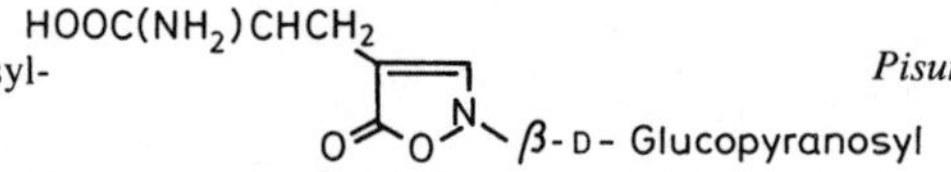 $HOOC(NH_2)CHCH_2$, O, O, N, β-D-Glucopyranosyl	*Pisum sativum* [189]
Ibotenic acid	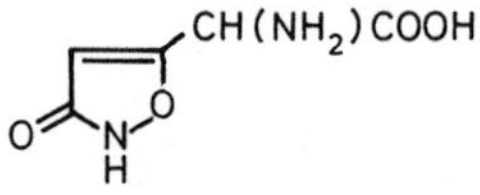 $CH(NH_2)COOH$, O, N, O, H	*Amanita* sp. [209]
β-(Isoaxazolin-5-on-2-yl)alanine	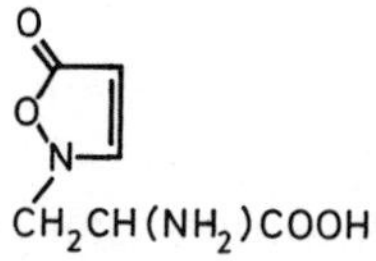 O, O, N, $CH_2CH(NH_2)COOH$	*Lathyrus odoratus* seedlings [189]; *Pisum arvense*; *Pisum sativum*; pea seedling root exudates [206]
Muscazone	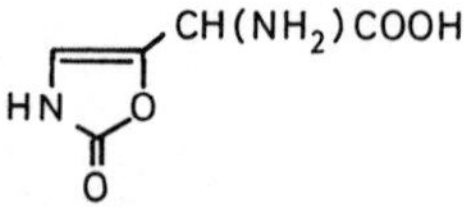 $CH(NH_2)COOH$, HN, O, O	*Amanita muscaria* [10, 11, 26]
Ommatin D	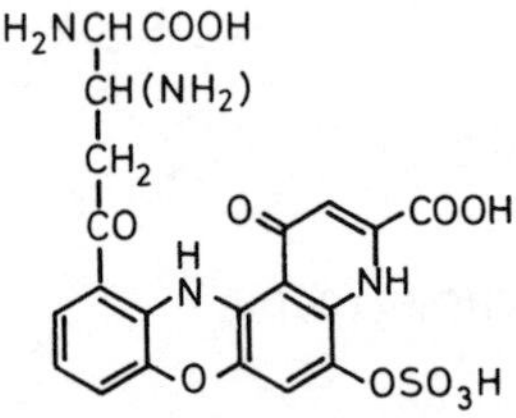 $H_2NCHCOOH$, $CH(NH_2)$, CH_2, CO, O, COOH, H, N, NH, O, OSO_3H	Insects [364]
Rhodommatin	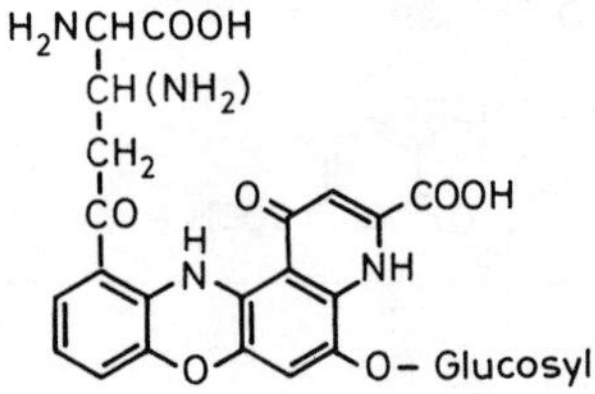 $H_2NCHCOOH$, $CH(NH_2)$, CH_2, CO, O, COOH, H, N, NH, O, O-Glucosyl	Insects [364]

Table 4.12 (*Contd.*)

Amino acid	*Formula*	*Source*
Tricholomic acid	[ring structure] $CH(NH_2)COOH$	*Tricholoma muscarium* [208]
Xanthommatin	[structure] $H_2NCHCOOH$, CH_2, CO, COOH, NH	*Calliphora erifocephala* [364]; ommochrome pigment; widely in insect and crustacea, eggs of echiurid worms, (*Urechis campo*)

Table 4.13 Aliphatic sulphur-containing amino acids

Amino acid	*Formula*	*Source*
S-(2-Acetamidoethyl)-cysteine	$CH_3CONHCH_2CH_2SCH_2CH(NH_2)COOH$	*Rozites caperata* [26]
L-Alliin (*S*-Allyl-L-cysteine sulphoxide)	$CH_2{=}CHCH_2S(=O)CH_2CH(NH_2)COOH$	*Allium sativum* [1, 6]; garlic root
S-Allyl-L-cysteine	$CH_2{=}CHCH_2SCH_2CH(NH_2)COOH$	Garlic [2, 3, 6]
S-(2-Aminoethyl)cysteine	$NH_2CH_2CH_2SCH_2CH(NH_2)COOH$	*Rozites caperata* [26]
2-Amino-3-mercaptobutanoic acid (thiothreonine)	$CH_3CH_2(SH)CH(NH_2)COOH$	Pea seedlings [366]
Arcamine	$HO_2SCH_2CH_2NHCOCH(NH_2)COOH$	*Arca zebra* (*mollusca*) [205]
S-(2-Carboxyethyl)-L-cysteine	$HOOCCH_2CH_2SCH_2CH(NH_2)COOH$	Seeds of *Albizzia julibrissin* [3]; *Acacia* sp. [2]
S-(2-Carboxylethyl)-L-cysteine sulphoxide	$HOOCCH_2CH_2S(=O)CH_2CH(NH_2)COOH$	*Acacia* sp. [370]
S-(2-Carboxyisopropyl)-L-cysteine	$HOOCCH_2CH(CH_3)SCH_2CH(NH_2)COOH$	*Acacia millefolia* seeds [3]
S-(Carboxymethyl)-cysteine	$HOOCCH_2SCH_2CH(NH_2)COOH$	Radish [10]
S-(Carboxymethyl)-homocysteine	$HOOCCH_2SCH_2CH_2CH(NH_2)COOH$	Cystathionuria [8]; homocystinuria [8]
Cystathionine	$HOOCCH(NH_2)CH_2SCH_2CH_2CH(NH_2)COOH$	Vetch [1]; *Neurospora* [1]; urine of pyridoxine-deficient rats [1]; human brain [1]; brown algae – *Heterochordaria abietina* [12]

Table 4.13 (*Contd.*)

Amino acid	*Formula*	*Source*
Cystathionine sulphoxide	$HOOCCH(NH_2)CH_2\overset{O}{\overset{\|}{S}}CH_2CH_2CH(NH_2)COOH$	Urine of cystathioninuric cases [196]
L-Cysteic acid	$HO_3SCH_2CH(NH_2)COOH$	Free in urine [2]
L-Cysteinesulphinic acid	$HO_2SCH_2CH(NH_2)COOH$	Rat brain [1, 3]; lugworm [2]; (*Arenicola cristata*); (Cysteine metabolite, Table 4.1)
Cystine disulphoxide	$HOOCCH(NH_2)CH_2\overset{O}{\overset{\|}{S}}\overset{O}{\overset{\|}{S}}CH_2CH(NH_2)COOH$	Metabolic intermediate [3]
S-(1, 2-Dicarboxyethyl)-L-cysteine	$HOOCCH_2CH(COOH)SCH_2CH(NH_2)COOH$	Bovine lens [6]
Dichrostachinic acid	$HOOCCH(OH)CH_2\overset{O}{\overset{\|}{\underset{O}{\underset{\|}{S}}}}CH_2SCH_2CH(NH_2)COOH$	*Dichrostachys glomerata* seeds [3, 6]
Dihydroalliine (*S*-Propylcysteine sulphoxide)	$CH_3CH_2CH_2\overset{O}{\overset{\|}{S}}CH_2CH(NH_2)COOH$	Onion [2]
L-Djenkolic acid	$CH_2[SCH_2CH(NH_2)COOH]_2$	Djenkol bean [1]; (*Pithecolobium lobatum*) *Albizzia lophanta* [2]; urine of djenkol eaters
Djenkolic acid suphoxide	$CH_2[\overset{O}{\overset{\|}{S}}CH_2CH(NH_2)COOH]_2$	*Acacia* sp. [370]
Felininine	$HOOCCH_2C(CH_3)_2SCH_2CH(NH_2)COOH$	Cat urine [1, 2, 3, 6]
D-Homocysteic acid	$HSO_3CH_2CH_2CH(NH_2)COOH$	*Palmaria palmata* (red alga) [367]
L-Homocysteine	$HSCH_2CH_2CH(NH_2)COOH$	*Neurospora* mutants [1]; human adrenal; abnormal urines [3, 365]
L-Homocystine	$HOOCCH(NH_2)CH_2CH_2SSCH_2CH_2CH(NH_2)COOH$	Pathological urines [3]
L-Homolanthionine	$HOOCCH(NH_2)CH_2CH_2SCH_2CH_2CH(NH_2)COOH$	Urine of homocystinurics [365]; *Escherichia coli* [3]
L-Homomethionine	$CH_3SCH_2CH_2CH_2CH(NH_2)COOH$	Cabbage [8]
S-(2-Hydroxy-2-carboxyethanethiomethyl)-cysteine	$HOOCCH(OH)CH_2SCH_2SCH_2CH(NH_2)COOH$	*Acacia georginae* seed [269]
S-(2-Hydroxy-2-carboxyethyl)cysteine	$HOOCCH(OH)CH_2SCH_2CH(NH_2)COOH$	Human urine [8]
S-(2-Hydroxy-2-carboxyethyl)-homocysteine	$HOOCCH(OH)CH_2SCH_2CH_2CH(NH_2)COOH$	Cystathionuria [8, 197]; Homocysteinuria

Table 4.13 (*Contd.*)

Amino acid	*Formula*	*Source*
S-(2-Hydroxy-2-carboxyethylthio)-homocysteine	$HOOCCH(OH)CH_2SSCH_2CH_2CH(NH_2)COOH$	Cystathionuria [8, 197]; Homocysteinuria
S-(3-Hydroxy-3-carboxy-n-propyl)cysteine	$HOOCCH(OH)CH_2CH_2SCH_2CH(NH_2)COOH$	Cystathionuria [8, 197]; Homocysteinuria
S-(3-Hydroxy-3-carboxy-n-propylthio)-homocysteine	$HOOCCH(OH)CH_2CH_2SSCH_2CH_2CH(NH_2)COOH$	Cystathionuria [8, 197]; Homocysteinuria
Isovalthine	$CH_3CH(CH_3)CH(COOH)SCH_2CH(NH_2)COOH$	Cat urine [3]; human urine in hypothyroidism, atherosclerosis, diabetes and from use of the sedative α-bromoisovalerylurea (bromural) [6]
Lanthionine (2, 2′-Thiodi(2-aminopropanoic acid))	$S[CH_2CH(NH_2)COOH]_2$	Subtilin; nisin [1, 3]; cinnamycin [2]; *Fusabacterium nucleatum* [368]; *Hypnea musciformis* [12]
S-(3-Mercaptolactic acid) cysteine	$HOOCC(CH_3)(OH)SSCH_2CH(NH_2)COOH$	Pathological urine [202]
Methionine sulphoxide	$O{=}S(CH_3)CH_2CH_2CH(NH_2)COOH$	Urine in Franconi's Syndrome [1]; red algae [8]
S-(2-Methyl-2-carboxyethyl)cysteine	$HOOCCH(CH_3)CH_2SCH_2CH(NH_2)COOH$	Urine of atherosclerotics [1] cat urine [1]; onion tripeptide [6]
S-Methyl-L-cysteine	$CH_3SCH_2CH(NH_2)COOH$	*Phaseolus vulgaris* [1]; *Neurosporra crassa* [1]; *Astragalus bisculatus* [8]; *Prodenia eridenia* [6] (haemolymph)
S-Methyl-L-cysteine sulphoxide	$O{=}S(CH_3)CH_2CH(NH_2)COOH$	Cabbage; turnip [1]; legumes; *Allium cepa* [2]
3, 3′-(2-Methylethylene-1, 2-dithiodialanine)	$HOOCCH(NH_2)CH_2SCH_2CH(CH_3)SCH_2CH(NH_2)COOH$	*Allium schoenoprasum* [201]
3-Methyllanthionine	$HOOCCH(NH_2)CH(CH_3)SCH_2CH(NH_2)COOH$	Cinnamycin [1]; nisin [1]; subtilin [1]; yeast [1]; *Phallus impudicas* [26]
S-Methylmethionine	$(CH_3)_2SCH_2CH_2CH(NH_2)COOH$	Cabbage; asparagus [1]; animals; microorganisms [2, 3]

Table 4.13 (*Contd.*)

Amino acid	*Formula*	*Source*
D-Penicillamine	$(CH_3)_2C(SH)CH(NH_2)COOH$	Penicillin [1]
S-(Prop-1-enyl)-L-cysteine	$CH_3CH{=}CHSCH_2CH(NH_2)COOH$	Chives [6]; garlic [6]
S-(Prop-1-enyl)-L-cysteine sulphoxide	$CH_3CH{=}CH\overset{O}{\overset{\Vert}{S}}CH_2CH(NH_2)COOH$	Onion [3]; garlic [6]
S-n-Propyl-L-cysteine	$CH_3CH_2CH_2SCH_2CH(NH_2)COOH$	Garlic [6]
Rhodoic acid	$HO_3SCH_2CH_2NHCH(CH_3)COOH$	Red algae e.g. [12] *Chondrus ocellatus*, *Iridaea cornucopial*, *Neodilsea yendoana*
Thiocysteine	$HSSCH_2CH(NH_2)COOH$	Minor metabolic intermediate of cysteine conversion in mammals [43]
Mixed disulphide of: *S*-(2-Carboxy-3-mercaptopropyl)-L-cysteine and 3-Mercaptoisobutanoic acid	$HOOCCH(NH_2)CH_2SCH_2CH(COOH)CH_2SSCH_2CH(COOH)CH_3$	*Asparagus officinalis* [244]
Disulphide of: *S*-(2-Carboxy-3-mercaptopropyl)-L-cysteine	$HOOCCH(NH_2)CH_2SCH_2CH(COOH)CH_2SSCH(COOH)CH_2SCH_2CH(NH_2)COOH$	*Asparagus officinalis* [244]

Table 4.14 Sulphur-containing heterocycle amino acids

Amino acid	*Formula*	*Source*
2-Amino-4-(5-carboxythiazol-2-yl)butanoic acid	$HOOCCH(NH_2)CH_2CH_2$–(thiazole ring: S, N)–COOH	*Xeronus subtomentosus* [371]
2-(1-Amino-2-methylpropyl)thiazole-4-carboxylic acid	$(CH_3)_2CHCHNH_2$–(thiazole ring: S, N)–COOH	Micrococcin P hydrolysates [6]
2-Aminomethylthiazole-4-carboxylic acid	NH_2–$CH(CH_3)$–(thiazole ring: S, N)–COOH	Thiostrepton hydrolysates [198, 199]

Table 4.14 (*Contd*)

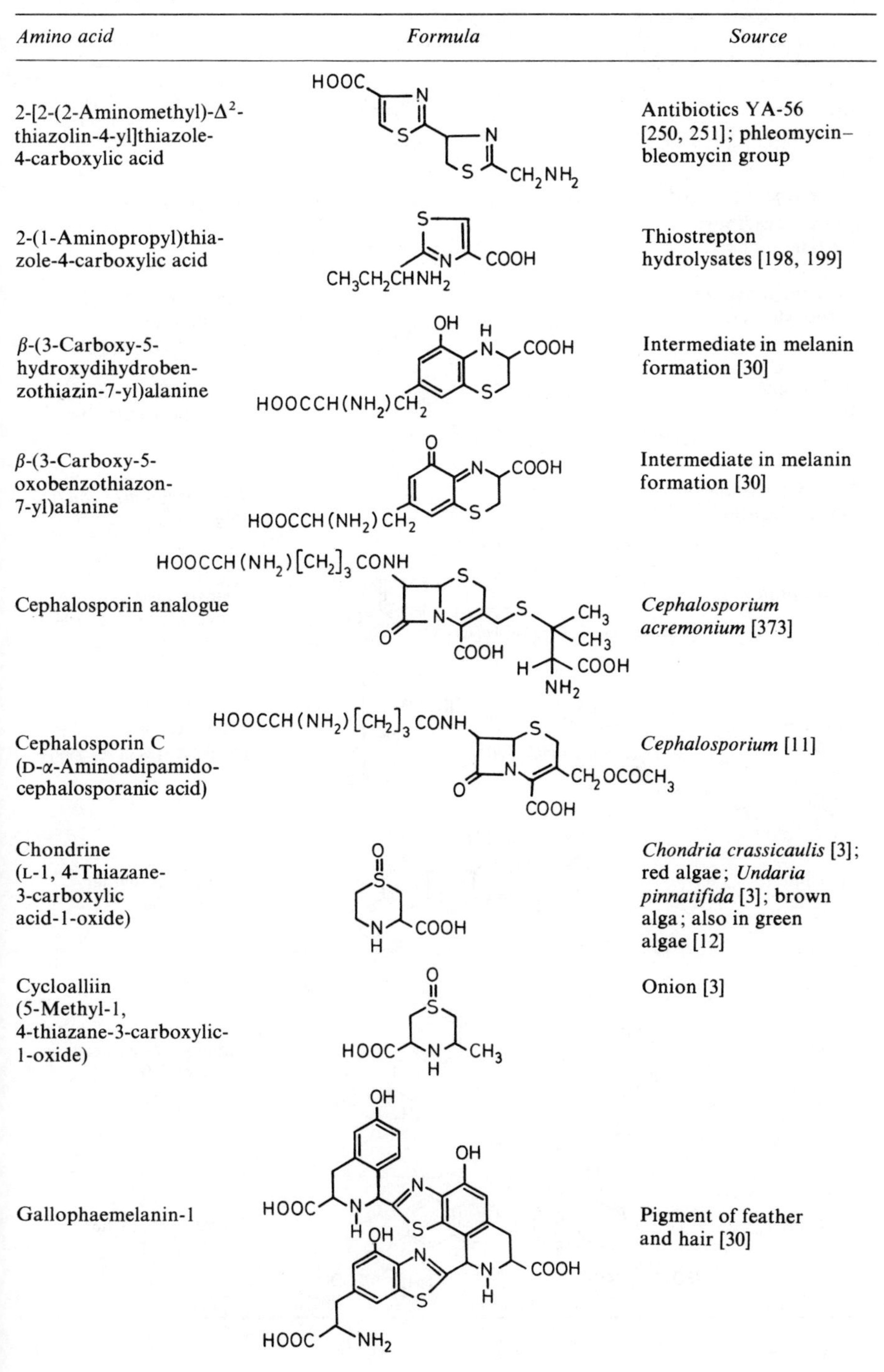

Amino acid	*Formula*	*Source*
2-[2-(2-Aminomethyl)-Δ^2-thiazolin-4-yl]thiazole-4-carboxylic acid		Antibiotics YA-56 [250, 251]; phleomycin–bleomycin group
2-(1-Aminopropyl)thiazole-4-carboxylic acid		Thiostrepton hydrolysates [198, 199]
β-(3-Carboxy-5-hydroxydihydrobenzothiazin-7-yl)alanine		Intermediate in melanin formation [30]
β-(3-Carboxy-5-oxobenzothiazon-7-yl)alanine		Intermediate in melanin formation [30]
Cephalosporin analogue		*Cephalosporium acremonium* [373]
Cephalosporin C (D-α-Aminoadipamido-cephalosporanic acid)		*Cephalosporium* [11]
Chondrine (L-1, 4-Thiazane-3-carboxylic acid-1-oxide)		*Chondria crassicaulis* [3]; red algae; *Undaria pinnatifida* [3]; brown alga; also in green algae [12]
Cycloalliin (5-Methyl-1, 4-thiazane-3-carboxylic-1-oxide)		Onion [3]
Gallophaemelanin-1		Pigment of feather and hair [30]

Table 4.14 (*Contd.*)

Amino acid	*Formula*	*Source*
Isopenicillin N (L-α-Aminoadipoyl-penicillin)	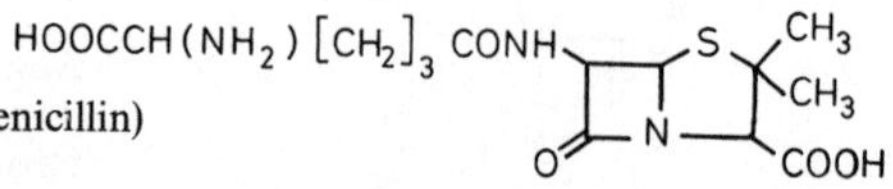 $HOOCCH(NH_2)[CH_2]_3CONH$ – S, CH_3, CH_3, N, O, COOH	*Penicillium* sp. [11]
Penicillin N (D-α-Aminoadipoyl-penicillin)	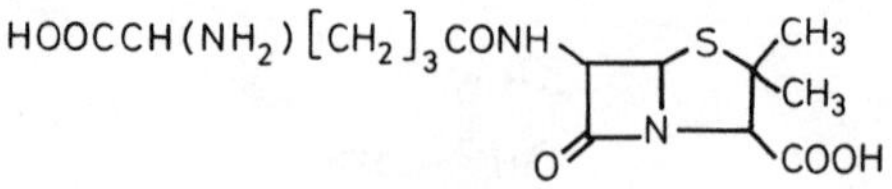 $HOOCCH(NH_2)[CH_2]_3CONH$ – S, CH_3, CH_3, N, O, COOH	*Penicillium* sp. [11]
2-Propanoylthiazole-4-carboxylic acid	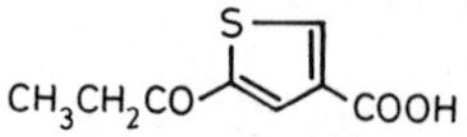 S, CH_3CH_2CO, COOH	Micrococcin P [6]; thiostreptan (hydrolysates) [198, 199]
1, 4-Thiazane-3, 5-dicarboxylic acid	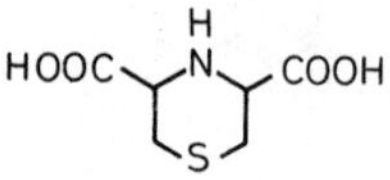 H, HOOC, N, COOH, S	Urine from cystathioninuric case [196]
(2*S*)-1, 4-Thiazane-3-carboxylic acid	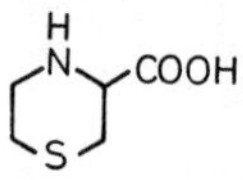 H, N, COOH, S	*Heterochordaria abietina* (alga) [372]
Thiostreptine	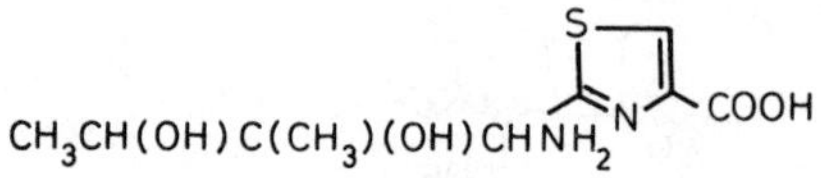 S, N, COOH, $CH_3CH(OH)C(CH_3)(OH)CHNH_2$	Thiostrepton hydrolysates [198, 199]
Thiostreptoic acid	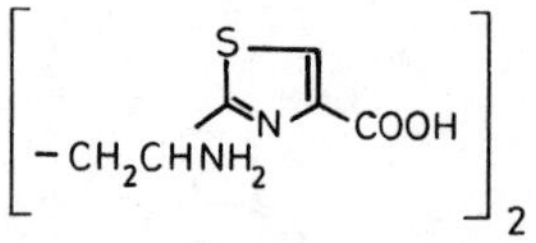 $[-CH_2CHNH_2$ – S, N, COOH$]_2$	Thiostrepton hydrolysates [198, 199]
Trichrome C	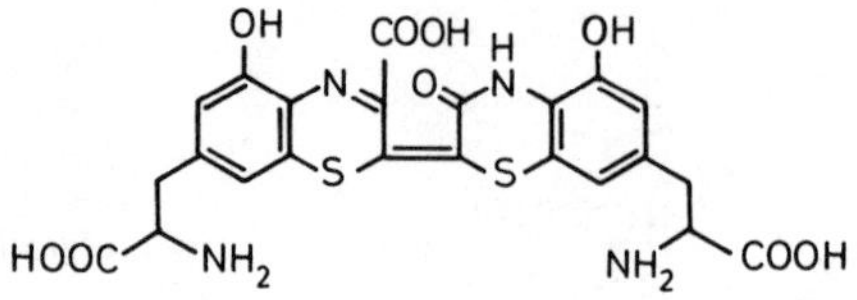 OH, COOH, H, OH, N, O, N, S, S, HOOC, NH_2, NH_2, COOH	Pigment of feather and hair [30]
Trichrome E	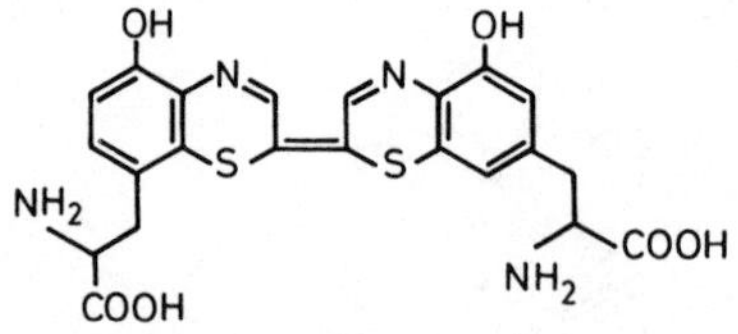 OH, OH, N, N, S, S, NH_2, COOH, COOH, NH_2	Pigment of feather and hair [30]
Trichrome F	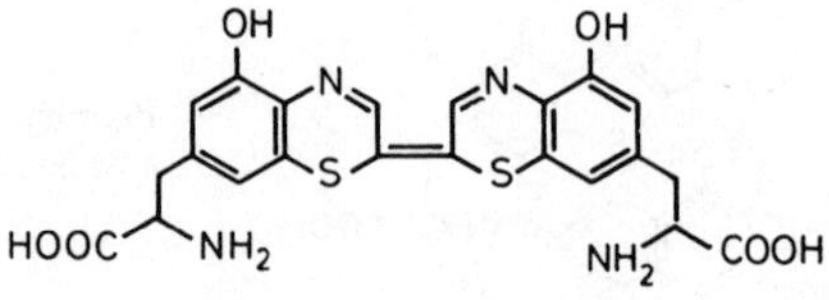 OH, OH, N, N, S, S, HOOC, NH_2, NH_2, COOH	Pigment of feather and hair [30]

Table 4.15 Cyclic amino acids with sulphur outside the ring

Amino acid	*Formula*	*Source*
Adenochromine A	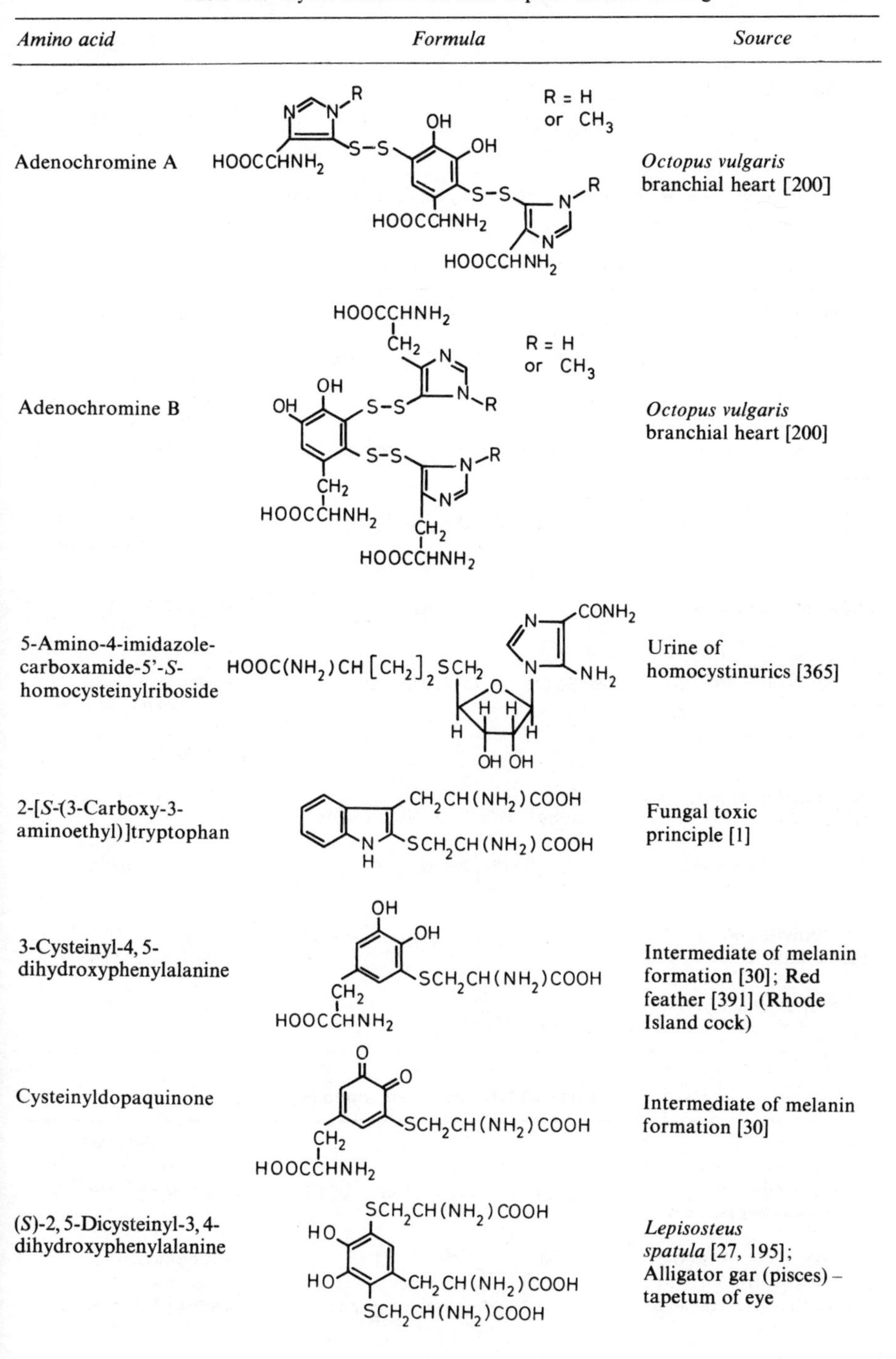	*Octopus vulgaris* branchial heart [200]
Adenochromine B		*Octopus vulgaris* branchial heart [200]
5-Amino-4-imidazole-carboxamide-5'-*S*-homocysteinylriboside		Urine of homocystinurics [365]
2-[*S*-(3-Carboxy-3-aminoethyl)]tryptophan		Fungal toxic principle [1]
3-Cysteinyl-4, 5-dihydroxyphenylalanine		Intermediate of melanin formation [30]; Red feather [391] (Rhode Island cock)
Cysteinyldopaquinone		Intermediate of melanin formation [30]
(*S*)-2, 5-Dicysteinyl-3, 4-dihydroxyphenylalanine		*Lepisosteus spatula* [27, 195]; Alligator gar (pisces) – tapetum of eye

Table 4.15 (*Contd.*)

Amino acid	*Formula*	*Source*
Thiolhistidine	$CH_2CH(NH_2)COOH$; N, NH, SH (ring structure)	Reputedly [2, 3] as the betaine ergothioneine in ergot, but of doubtful provenance

Other nucleotide derivatives of Sulphur amino acids appear in Table 4.1.

Table 4.16 Selenium-containing amino acids

Amino acid	*Formula*	*Source*
Selenocystathionine	$HOOCCH(NH_2)CH_2SeCH_2CH_2CH(NH_2)COOH$	*Astragalus* [237]; *Stanleya pumata*; *Neptunia amplexicaulis*
Selenocysteineselenic acid	$HO_2SeCH_2CH(NH_2)COOH$	*Trifolium pratense* [237]; *Lolium perenne*
Selenocystine	$[SeCH_2CH(NH_2)COOH]_2$	*Trifolium* [237]; *Lolium*; *Allium*; *Zea mays* and others
Selenohomocysteine	$HSeCH_2CH_2CH(NH_2)COOH$	*Astragalus* [237]
Selenomethionine	$CH_3SeCH_2CH_2CH(NH_2)COOH$	*Allium cepa* [237]; *Trifolium pratense*; *Lolium perense*; *Spiradela*
Selenomethionine selenoxide	$CH_3Se(=O)CH_2CH_2CH(NH_2)COOH$	*Trifolium* [237]; *Lolium*
Se-Methylselenocysteine	$CH_3SeCH_2CH(NH_2)COOH$	*Astragalus bisulcatus* [237]
Se-Methylseleno-methionine	$(CH_3)_2SeCH_2CH_2CH(NH_2)COOH$	*Astragalus* [237]; *Trifolium*; *Lolium*
Se-Propenylselenocys-teine selenoxide	$CH_3CH=CHSe(=O)CH_2CH(NH_2)COOH$	*Allium cepa* [237]

Table 4.17 Halogenated amino acids

Amino acid	*Formula*	*Source*
trans-2-Amino-5-chlorohex-4-enoic acid	$CH_3CCl=CHCH_2CH(NH_2)COOH$	*Amanita solitaria* [375]
(αS, $4S$, $5R$)-α-Amino-3-chloro-4-hydroxy-isoxazoline-5-acetic acid	Cl, OH, N, O, $CH(NH_2)COOH$ (ring structure)	Antibiotic from *Streptomyces sviceus* [383, 384, 385]

Table 4.17 (*Contd.*)

Amino acid	*Formula*	*Source*
α-Amino-3-chloro-4-hydroxyphenylacetic acid	HO–(3-Cl-C_6H_3)–CH(NH_2)COOH	Actinoidin [411]; avoparcin [408]; antibiotic LL-AV-290 [376]
(αS, 5S)-α-Amino-3-chloroisoxazoline-5-acetic acid	3-Cl-isoxazolin-5-yl–CH(NH_2)COOH	Antibiotic from *Streptomyces sviceus* [383, 384, 385]
2-Amino-4-chloro-4-pentenoic acid	CH_2=CClCH$_2$CH(NH_2)COOH	*Amanita pseudoporphyra* [374]
2-Amino-4, 4-dichlorobutanoic acid	CHCl$_2$CH$_2$CH(NH_2) COOH	*Streptomyces armentosa v. armentosa* [26]
α-Amino-3, 5-dichloro-4-hydroxyphenylacetic acid	HO–(3,5-Cl$_2$-C_6H_2)–CH(NH_2)COOH	Enduracidin [29]
L-6-Bromohypaphorine	6-Br-indol-3-yl–CH$_2$CH[N(CH$_3$)$_2$]COOH	*Pachymatisma johnstone* [380] (porifera)
N^{α}-Carboethoxyacetyl-D-4-chlorotryptophan	4-Cl-indol-3-yl–CH$_2$CH(NHCOCH$_2$COOCH$_2$CH$_3$)COOH	*Pisum sativum* seeds [8]
N^{α}-Carbomethoxyacetyl-D-4-chlorotryptophan	4-Cl-indol-3-yl–CH$_2$CH(NHCOCH$_2$COOCH$_3$)COOH	*Pisum sativum* seeds [8]
3-Chloro-β-hydroxytyrosine	HO–(3-Cl-C_6H_3)–CH(OH)CH(NH_2)COOH	Vancomycin [378, 379, 409, 410]; actinoidin [411]; ristomycin
3-Chloro-5-keto-3, 4-pyrrolidene-2-carboxylic acid	3-Cl-5-oxo-3-pyrroline-2-COOH	Antibiotic FR-900148 [386] from *Streptomyces xanthocidocus*
(3-R, 5S)-5-Chloropiperazic acid	5-Cl-hexahydropyridazine-3-COOH (H, Cl at C-5; N-H, N-H)	Monamycin [291, 407]; viomycidine; *Streptomyces jamaicensis*
5-Chloro-D-tryptophan	5-Cl-indol-3-yl–CH$_2$CH(NH_2)COOH	Longicatenamycin [377]

Table 4.17 (*Contd.*)

Amino acid	*Formula*	*Source*
cis-3, 4-Dichloroproline	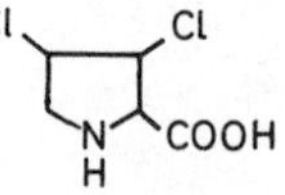 [structure: pyrrolidine ring with Cl, Cl, N–H, COOH]	Islanditoxin [382]; *Penicillium islandicum*
N-Methyl-δ, δ, δ-trichloroleucine	$CH_3CH(CCl_3)CH_2CH(NHCH_3)COOH$	*Dysidea herbacea* [231] (porifera)
Monodechlorovancomycinic acid	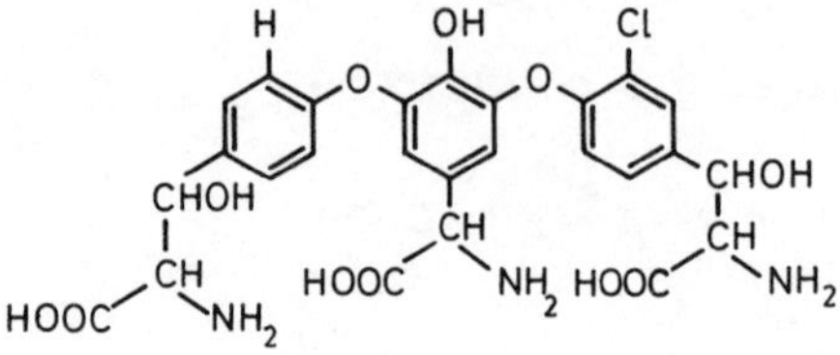 [structure: H, OH, Cl; O; CHOH; CH; HOOC; NH_2]	Actinoidin [287, 303, 304, 378]
Tetracetylclionamide	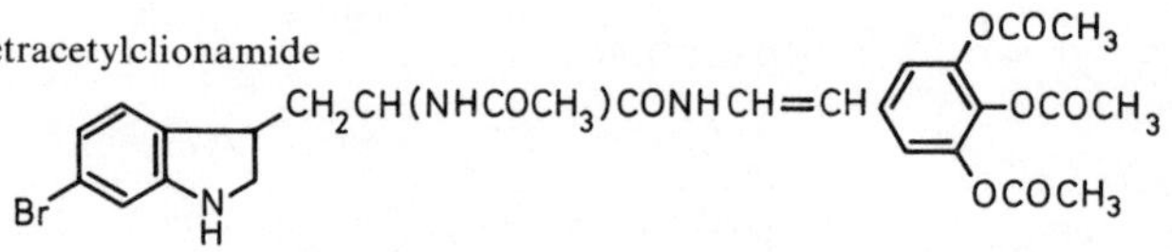 [structure: Br, N–H indoline] $CH_2CH(NHCOCH_3)CONHCH{=}CH$ [benzene ring with $OCOCH_3$, $OCOCH_3$, $OCOCH_3$]	*Cliona celata* [381] (porifera)
Thyroxine	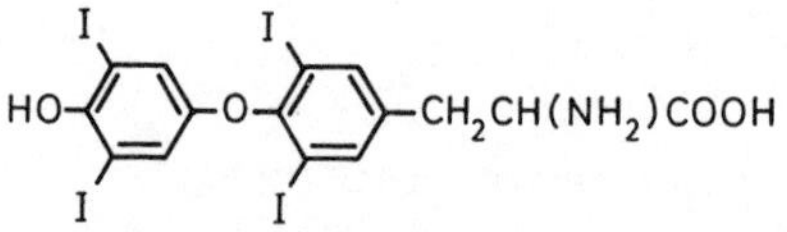 [structure: HO, I, I, O, I, I] $-CH_2CH(NH_2)COOH$	Free in serum
Triiodothyronine	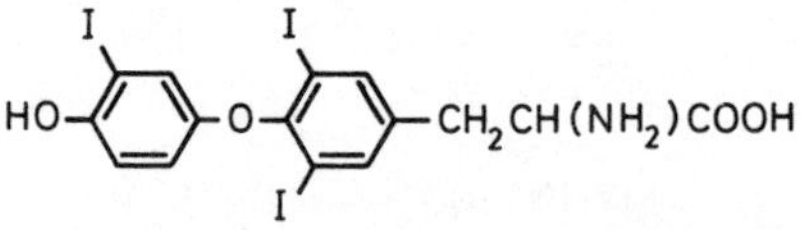 [structure: HO, I, O, I, I] $-CH_2CH(NH_2)COOH$	Free in serum
Vancomycinic acid	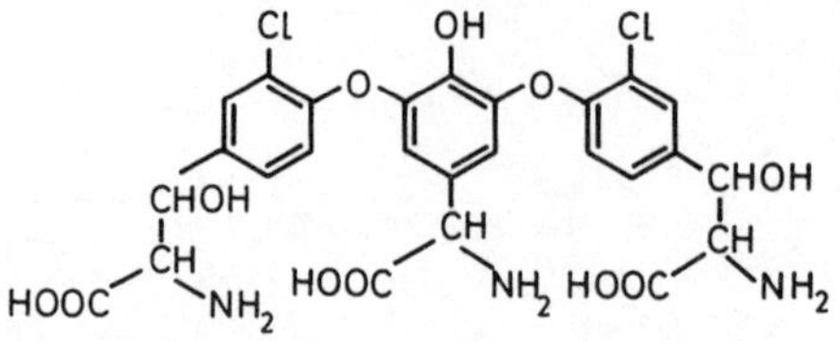 [structure: Cl, OH, Cl; O; CHOH; CH; HOOC; NH_2]	Vancomycin [287, 303, 304, 412]

Table 4.18 Alpha *N*-substituted amino acids

Amino acids	*Formula*	*Source*
(a) *N*-Alkyl		
L-Abrine	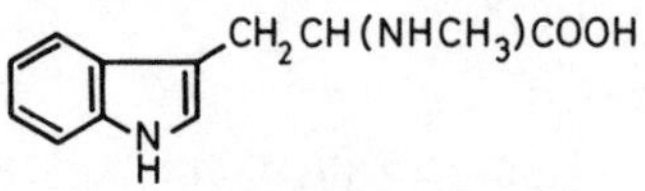 [indole, N–H] $CH_2CH(NHCH_3)COOH$	*Abrus precatorius* seeds [1]
Alanine betaine	$CH_3CH[\overset{+}{N}(CH_3)_3]COO^-$	*Dichapetalum cymosum* [392]; destruxin A and B [26]; *Limonium vulgare* wood [401]

Table 4.18 (*Contd.*)

Amino acid	*Formula*	*Source*
4-Dimethylamino-N^{α}-methyl phenylalanine	$(CH_3)_2N-C_6H_4-CH_2CH(NHCH_3)COOH$	Mikamycin B [6]
β-Dimethylamino-*N*-methyl-L-phenylalanine	$C_6H_5-CH[\overset{+}{N}(CH_3)_2]CH(NHCH_3)COOH$	Doricin [26]
N, *N'*-Dimethylcystine	$HOOCCH(NHCH_3)CH_2SSCH_2CH(NHCH_3)COOH$	Triostin C [216]
N-Dimethylglycine	$(CH_3)_2NCH_2COOH$	Metabolic intermediate [43, 44]
N-Dimethyl-L-phenylalanine	$C_6H_5-CH_2CH[N(CH_3)_2]COOH$	*Canthium euryoides* [298]
N-Dimethyltryptophan	indol-3-yl-$CH_2CH[N(CH_3)_2]COOH$	*Abrus precatorius* seeds [9]
N, O-Dimethyl-L-tyrosine	$CH_3O-C_6H_4-CH_2CH(NHCH_3)COOH$	Didemnin [71] – from *Tridideminim* sp. tunicate [393]; majusculamides
Glycine betaine	$(CH_3)_3\overset{+}{N}CH_2COO^-$	*Beta vulgaris* [1, 43, 44]; Generally in plants and animals; *Callista brevishiphonato* [389] ovary; *Atrina pedinaria* muscle [390]; *Petalonia jasmia* [8] (alga)
Histidine betaine (hercynine)	imidazol-4-yl-$CH_2CH(\overset{+}{N}(CH_3)_3)COO^-$	*Agaricus campestris* [1]; *Boletus edulis* [1]
Homoserine betaine	$HOCH_2CH_2CH[\overset{+}{N}(CH_3)_3]COO^-$	*Callista brevishiponata* [389, 390]; *Atrina pectinata* (mollusca)
Homostachydrine	piperidine ring, $\overset{+}{N}(CH_3)_2$, 2-COO^-	*Medicago sativa* [396]
N-(2-Hydroxyethyl)alanine	$CH_3CH(NHCH_2CH_2OH)COOH$	Rumen protozoan phospholipid [8]
N-(2-Hydroxyethyl)glycine	$CH_2(NHCH_2CH_2OH)COOH$	*Beta vulgaris*; [1, 43, 44]; Generally in plants and animals; *Callista brevishiphonato* [389] ovary; *Atrina pedinaria* muscle [390]; *Petalonia jasmia* [8] (alga)
4-Hydroxy-*N*-methylproline	4-HO-pyrrolidine ring, $N-CH_3$, 2-COOH	*Croton gubouga* [212] bark; *Afromosia elata* [212] wood

Table 4.18 (*Contd.*)

Amino acid	*Formula*	*Source*
4-Hydroxyphenyl-sarcosine	$HO-C_6H_4-CH(NHCH_3)COOH$	Antibiotic LL-AV-290 [376]
D-*allo*-Hydroxyproline betaine (Turicine)	HO-(pyrrolidine ring) $\overset{+}{N}(CH_3)_2$, COO^-	*Stachys sylvatica* [1]; Jack bean
L-Hydroxyproline betaine (betonicine)	HO-(pyrrolidine ring) $\overset{+}{N}(CH_3)_2$, COO^-	*Betonica officinalis* [1]
L-Isoleucine betaine	$CH_3CH_2CH(CH_3)CH[\overset{+}{N}(CH_3)_3]COO^-$	Cannabis seeds [397]
N-Methyl-L-alanine	$CH_3CH(NHCH_3)COOH$	*Dichapetalum cymosum* [392]; Destruxin A and B [26]; *Limonium vulgare* wood [401]
N-Methylaspartic acid	$HOOCCH_2CH(NHCH_3)COOH$	*Halopytis incurvus* [398]
N-Methyldehydro-phenylalanine	$C_6H_5-CH{=}CH(NHCH_3)COOH$	*Alternaria* sp. [165]; tentoxin
N^{α}-Methyl-4-dimethylallyl-L-tryptophan	$(CH_3)_2C{=}CHCH_2$ (at C-4 of indole, N-H) with $CH_2CH(NHCH_3)COOH$ at C-3	*Claviceps fusiformis* [426]
2-*N*-Methyl-4-formyl-pentanoic acid (L-γ-Formyl-*N*-methylnorvaline)	$CH_3CH(CHO)CH_2CH(NHCH_3)COOH$	Ilamycin [247]
N-Methyl-L-isoleucine	$CH_3CH_2CH(CH_3)CH(NHCH_3)COOH$	Enniatin A and B [29, 394]
N-Methyl-L-allo-isoleucine	$CH_3CH_2CH(CH_3)CH(NHCH_3)COOH$	Actinomycin $AYX_{1\alpha}$ [29]; cycloheptamycin [145]; actinomycins [29]; enniatin A [394]
N-Methyl-D-leucine	$(CH_3)_2CHCH_2CH(NHCH_3)COOH$	Griselimycin [319]
N-Methyl-L-leucine	$(CH_3)_2CHCH_2CH(NHCH_3)COOH$	Actinomycin [1, 29]; enniatin A; Majuscul-amides A and B [393]; Sporidesmolide I [29]; ilamycin B [26]; didemnin [71]; (tunicate)

Table 4.18 (*Contd.*)

Amino acid	*Formula*	*Source*
N-Methyl-3-methyl-L-leucine (*N*-Methyl-4-methyl-L-allo-isoleucine)	$(CH_3)_2CHCH(CH_3)CH(NHCH_3)COOH$	Etamycin [214]; triostin C [213, 216]
N-Methyl-*O*-methyl-L-serine	$CH_3OCH_2CH(NHCH_3)COOH$	*Mycobacterium* sp. [26]
N-Methyl-L-methionine sulphoxide	$CH_3S(=O)CH_2CH_2CH(NHCH_3)COOH$	*Centroceras clavulation* [399]; *Grateloupia turuturu* [395]; red algae
N-Methyl-5-methoxytryptophan	CH_3O-indol-3-yl-$CH_2CH(NHCH_3)COOH$	Cycloheptamycin [145]
N-Methylphenylalanine	C_6H_5-$CH_2CH(NHCH_3)COOH$	Staphylomycin [79]; factor S
N-Methyl-L-phenylglycine	C_6H_5-$CH(NHCH_3)COOH$	Etamycin [214]
N-Methylpicolinium betaine (Homarine)	1-CH_3-pyridinium-2-COO^-	Widely in invertebrates [402]; crustacea; molluscs
N-Methyl-L-serine	$HOCH_2CH(NHCH_3)COOH$	*Dichapetalum cymosum* [392]
N-Methyl-L-threonine	$CH_3CH(OH)CH(NHCH_3)COOH$	Stendomycin [29]
N-Methyl-L-valine	$(CH_3)_2CHCH(NHCH_3)COOH$	Actinomycin [1, 29]; enniatin A; majusculamides A and B [393]; sporidesmolide I. [29]; ilamycin B [26]; didemnin [71]; (Tunicate)
L-Ornithine betaine (Miokinine)	$(CH_3)_3\overset{+}{N}CH_2CH_2CH_2CH[\overset{+}{N}(CH_3)_3]COO^-$	Mammalian muscle [1]
Phenylalanine betaine	C_6H_5-$CH_2CH[\overset{+}{N}(CH_3)_3]COO^-$	*Antiaris africana* [427]
L-Proline betaine (Stachydrine)	pyrrolidinium-$\overset{+}{N}(CH_3)_2$-2-COO^-	*Stachys tubifera* [1]; Orange leaves [1] alfalfa [1]; *Arca noae* [1] (mollusca); *Medicago sativa* [396]
Sarcosine	CH_3NHCH_2COOH	Widely distributed [1, 43, 44]; animals; plants; micro-organisms; etamycin

Table 4.18 (*Contd.*)

Amino acid	*Formula*	*Source*
D-Suriname	HO-C_6H_4-$CH_2CH(NHCH_3)COOH$	*Geoffroya surinamensis* [1] (bark); *Rhatania* [1]; *Combretum zeyeri seed* [400]
Thiolhistidine betaine (Ergothioneine)	2-mercaptoimidazol-4-yl-$CH_2CH(\overset{+}{N}(CH_3)_3)COO^-$	Ergot [1]; liver [1]
2-Trimethylamino-6-oxoheptanoyl choline ester	$CH_3CO[CH_2]_3CH(\overset{+}{N}(CH_3)_3)COOCH_2CH_2\overset{+}{N}(CH_3)_3$	*Limonium vulgare* wood [215]
L-Tryptophan betaine (Hypaphorine)	indol-3-yl-$CH_2CH[\overset{+}{N}(CH_3)_3]COO^-$	*Erythrina hygaphorus* seeds [1]; *Antiaris africana* [426]
Tryptophan betaine methyl ester	indol-3-yl-$CH_2CH[\overset{+}{N}(CH_3)_3]COOCH_3$	*Abrus precatorious seeds* [8]; *Aotus subglauca* [424]
L-Valine betaine	$(CH_3)_2CHCH[\overset{+}{N}(CH_3)_3]COO^-$	*Callista brevishiponata* [389, 390]; *Atrina pectinata* (mollusca)
(b) *N*-Acyl and others		
N^α-Acetyl-L-arginine	$NH_2C(=NH_2)NH[CH_2]_3CH(NHCOCH_3)COOH$	Cattle brain [423]
O-Acetyl-*N*-(*N'*-benzoyl-L-phenylalanyl)-L-phenylalaninol	C_6H_5-$CONHCH(CH_2C_6H_5)CONHCH(CH_2C_6H_5)CH_2OCOCH_3$	*Aspergillus glaucus* [229]
N-Acetyldjenkolic acid	$HOOCCH(NH_2)CH_2SCH_2SCH_2CH(NHCOCH_3)COOH$	*Acacia farnesiana* [3, 5, 6, 370]
N-Acetylornithinine	$NH_2(CH_2)_3CH(NHCOCH_3)COOH$	Micro-organisms [3]
N-Acetyl-L-phenyl-L-phenylalaninol	$CH_3CONHCH(CH_2C_6H_5)CONHCH(CH_2C_6H_5)CH_2OH$	*Emericellopsis salmosynnemata* [230]
N-Acyl-L-glutamyl-2-phenylethylamines	$RCONHCH(CH_2CH_2CONH_2)CONHCH_2CH_2$-$C_6H_5$; R = iPr or EtCHCH$_3$	*Croton humulis* [405]
Allantoic acid	$H_2NCONHCH(COOH)NHCONH_2$	Uric acid degradation [43, 44]
N-Aminoproline	pyrrolidine-2-COOH with N-NH_2	*Vicia faba* leaves; *Linum* (flax) [10]

Table 4.18 (*Contd.*)

Amino acid	*Formula*	*Source*
Asperphenamate ((*S*)-*N*-Benzoylphenylalanine (*S*)-2-benzamido-3-phenylpropyl ester)	$C_6H_5CH_2CH(NHCOC_6H_5)COOCH_2CH(CH_2C_6H_5)NHCOC_6H_5$	*Aspergillus flavipes* [228]
N-Carbamoyl-α-(4-hydroxyphenyl) glycine	$HOC_6H_4CH(NHCONH_2)COOH$	*Vicia faba* leaves [8]
N^2, N^5-Dibenzoylornithine (Ornithuric acid)	$C_6H_5CONH[CH_2]_3CH(COOH)NHCOC_6H_5$	Avian excretion [44]
N^2, N^6-Di-(2, 3-dihydroxybenzoyl) -L-lysine	$(HO)_2C_6H_3CONH[CH_2]_4CH(COOH)NHCOC_6H_3(OH)_2$	Iron-deficient *Azobacter vinelandii* [406]
Dihydrojasmonoylisoleucine	3-oxo-2-($CH_2[CH_2]_3CH_3$)cyclopentyl-$CH_2CONHCH(COOH)CH(CH_3)CH_2CH_3$	*Gibberella fujikuroi* [8]
N^2, N^5-Diphenacetylornithine	$C_6H_5CH_2CONH[CH_2]_3CH(COOH)NHCOCH_2C_6H_5$	Avian excretion [44]
Echinulin	[structure: triprenylated indole diketopiperazine; labels HN, O, CH_3, NH, O, N, H]	*Aspergillus glaucus* [6]
N-Formyl-L-valine	$(CH_3)_2CHCH(NHCHO)COOH$	Cycloheptamycin [26]
N-Fumarylalanine	$HOOCCH=CHCONHCH(CH_3)COOH$	*Penicillium resticulosum* [26]
Hadacidin (*N*-Formyl-*N*-hydroxyglycine)	$HOOCCH_2N(OH)CHO$	*Penicillium frequentans* [3]
Hippuric acid	$C_6H_5CONHCH_2COOH$	Mammalian urinary excretion [44]
N-(Indole-3-acetyl)-aspartic acid	indol-3-yl-$CH_2CONHCH(COOH)CH_2COOH$	Etiolated pea tissue [3] and tomato treated with indole-3-acetate
Jasmonylisoleucine	3-oxo-2-($CH_2CH=CHCH_2CH_3$)cyclopentyl-$CH_2CONHCH(COOH)CH(CH_3)CH_2CH_3$	*Gibberella fujikuroi* [8, 26]
N-Malonyl-D-alanine	$CH_3CH(NHCOCH_2COOH)COOH$	*Pisum sativum* [403, 404]

Table 4.18 (*Contd.*)

Amino acid	*Formula*	*Source*
N^2-(6-Methyloctanoyl)-2, 4-diaminobutanoic acid	$H_2NCH_2CH_2CH[NHCO[CH_2]_4CH(CH_3)CH_2CH_3]COOH$	*Bacillus colistinus* [8]
N-(10-Methylundeca-2-(*cis*)4(*trans*)-dienoyl)-aspartic acid	$(CH_3)_2CH[CH_2]_4CH{=}CHCH{=}CHCONHCH(COOH)CH_2COOH$	Enduracidin A [29]
Phenaceturic acid	C_6H_5–$CH_2CONHCH_2COOH$	Mammalian urinary excretion [44]
Phenylacetylglutamine	C_6H_5–$CH_2CONHCH(COOH)CH_2CH_2CONH_2$	Mammalian urinary excretion [44]
N-Succinyl-L-diaminopimelic acid	$HOOCCH(NH_2)[CH_2]_3CH(COOH)NHCOCH_2CH_2COOH$	*Escherichia coli* [26]
N-Succinylglutamic acid	$HOOCCH_2CH_2CH(NHCOCH_2CH_2COOH)COOH$	*Bacillus megatherium* [26]
N-(Undeca-2(*cis*)-4(*trans*)-dienoyl)aspartic acid	$CH_3CH_2[CH_2]_4CH{=}CHCH{=}CHCONHCH(COOH)CH_2COOH$	Enduracidin B [29]

Table 4.19 Phosphorus-containing amino acids

Amino acid	*Formula*	*Source*
L-2-Amino-4-(methylphosphino)-butanoic acid	$CH_3P(OH)(=O)CH_2CH_2CH(NH_2)COOH$	Antibiotic SF 1293 [416. 417, 418]
D-2-Amino-5-phosphono-3-pentenoic acid	$HOP(OH)(=O)[CH_2]_3CH(NH_2)COOH$	*Streptomyces plumbeus* [7]
Creatine phosphate	$H_2O_3PNHC(=NH)N(CH_3)CH_2COOH$	Mammalian muscle [12, 43, 44]
Lombricine	$H_2NC(=NH)NHCH_2CH_2OP(=O)(OH)OCH_2CH(NH_2)COOH$	*Lumbricus terrestris* [243]
N-Phosphoglutamic acid	$HOOCCH_2CH_2CH(NHPO_3H_2)COOH$	*Pseudomonas phaseolicola* [236]
Phosphoguanidinoacetic acid (Phosphoglycocyamine)	$H_2O_3PNHC(=NH)NHCH_2COOH$	*Nereis diversicola* [44, 242]; annelid worms

Table 4.19 (*Contd.*)

Amino acid	*Formula*	*Source*
Phosphoguanidinoethyl-serylphosphate	$H_2O_3PNHC(=NH)NHCH_2CH_2OP(=O)(OH)OCH_2CH(NH_2)COOH$	Leech (annelid) muscle [44]
O-Phospho-4-hydroxypiperidine-2-carboxylic acid	4-O–$P(=O)(OH)$–OH on piperidine ring (N–H), 2-COOH	*Peltophorum* [415]
Phospholombricine	$H_2O_3PNHC(=NH)NHCH_2CH_2OP(=O)(OH)OCH_2CH(NH_2)COOH$	Annelids [243]
L-(N^s-Phosphoro) methionine-(*S*)-sulphoximine	$H_2O_3PN{=}S(=O)(CH_3){-}CH_2CH_2CH(NH_2)COOH$	*Streptomyces* sp. [414]
N-Phosphorylarginine	$H_2O_3PNHC(=NH)NH[CH_2]_3CH(NH_2)COOH$	Arthropods [241] and echinoderms
O-Phosphorylhomoserine	$H_2O_3POCH_2CH_2CH(NH_2)COOH$	*Lactobacillus casei* [3]
N-Phosphoryl-L-thalassemine	$H_2O_3PNHC(=NH)NHCH_2CH_2OP(=O)(OH)CH_2CH[N(CH_3)_2]COOH$	*Thalassema neptuni* [413] (Echiuroid worm)
L-Thalassemine	$H_2NC(=NH)NHCH_2CH_2OP(=O)(OH)OCH_2CH[N(CH_3)_2]COOH$	*Thalassema neptuni* [413] (Echiuroid worm)

Table 4.20 (a) Some gamma-glutamyl compounds

Amino acid	*Formula*	*Source*
Agaritine (*β*-*N*-(*γ*-L-Glutamyl)-4-hydroxymethyl-phenylhydrazine)	$HOCH_2$–C_6H_4–$NHNHCOCH_2CH_2CH(NH_2)COOH$	*Neurospora crassa*
N^5-(3, 4-Dihydroxy-phenyl)-L-glutamine R = OH	(HO)(HO)–C_6H_3–$NHCOCH_2CH_2CH(NH_2)COOH$	*Agaricus bisporus* [10, 26, 210]
γ-Glutamyl-2-aminoethyl-isoxazolin-5-one	isoxazolin-5-one–N–$[CH_2]_2NHCO[CH_2]_2CH(NH_2)COOH$	*Lathyrus odoratus* [206, 189]

Table 4.20 (*Contd.*)

Amino acid	*Formula*	*Source*
N-(L-γ-Glutamyl)-2-aminohexan-3-one	$H_2N(HOOC)CHCH_2CH_2CONHCH(CH_3)COCH_2CH_2CH_3$	*Russula ochroleuca* [428]
β-*N*-(γ-Glutamyl)-aminopropionitrile	$NCCH_2CH_2NHCOCH_2CH_2CH(NH_2)COOH$	*Lathyrus* sp. [2, 10]
N^5-(4-Hydroxybenzyl)-L-glutamine	$HO\text{-}C_6H_4\text{-}CH_2NHCOCH_2CH_2CH(NH_2)COOH$	Buckwheat seed [238]
N^5-(2-Hydroxybenzyl)-allo-γ-hydroxy-L-glutamine	$(2\text{-}HO)C_6H_4\text{-}CH_2NHCOCH(OH)CH_2CH(NH_2)COOH$	Buckwheat seed [238]
N^5-(1-Hydroxycyclopropyl)-L-glutamine (coprine)	cyclopropane(OH)-$NHCO(CH_2)_2CH(NH_2)COOH$	*Coprinus atramentarius* [166, 203];
N^5-Hydroxyethyl-glutamine	$HOCH_2CH_2NHCOCH_2CH_2CH(NH_2)COOH$	*Lunaria annua* [10, 240]
N^5-(4-Hydroxyphenyl)-L-glutamine R = H	$HO\text{-}C_6H_4\text{-}NHCOCH_2CH_2CH(NH_2)COOH$	*Agaricus hortensis* [148, 246]
N-[*N*-2*R*-Methylbutanoyl)-L-glutaminoyl]-2-phenylethylamine	$CH_3(CH_2)_3CONHCH(CH_2CH_2CONH_2)CONHCH_2CH_2C_6H_5$	*Croton humilis* [239]
N-[*N*-(2-Methylpropanoyl)-L-glutaminoyl]-2-phenylethylamine	$(CH_3)_2CHCONHCH(CH_2CH\ CONH_2)CONHCH_2CH_2C_6H_5$	*Croton humilis* [239]
Oxypinnatanine	dihydrofuran ring (CH_2OH)-$NHCOCH(OH)CH_2CH(NH_2)COOH$	*Staphylea pinnata* seeds; *Hemerocallis fulva* [193];
Pinnatanine (N^5-(2-Hydroxymethyl-butadienyl) allo(*threo*-γ-hydroxy-L-glutamine)	$CH_2{=}CHC(CH_2OH){=}CHNHCOCH(OH)CH_2CH(NH_2)COOH$	*Staphylea pinnata* seeds [139, 193]
Theanine (N^5-Ethyl-L-glutamine)	$CH_3CH_2NHCOCH_2CH_2CH(NH_2)COOH$	Tea leaves [6]; *Xerocomus badivus*

(**b**) Some gamma-glutamyl non-protein amino acid derivatives

Amino acid	*Source*
2-Aminohex-4-ynoic acid	*Tricholomopsis rutilans*

Table 4.20 (*Contd.*)

Amino acid	*Source*
erythro-2-Amino-3-hydroxyhex-4-ynoic acid	*Tricholomopsis rutilans*
D-Alanine	*Pisum sativum*
Hypoglycin A (as Hypoglycin B)	*Blighia sapida*
2-Methyl-cyclopropylglycine	*Litchi chinensis*; *Billia hypocastrum*; *Acer pseudoplatanus*
trans-2-(Carboxycyclo-propyl)glycine	*Blighia sapida*
Lathyrine (β-(2-Aminopyrimidin-4-yl)alanine)	*Lathyrus japonicus*
β-(Pyrazol-1-yl)-L-alanine	*Cucumis sativus* [343]
Pipecolic acid	Gleditsia sp. [10]
N-Aminoproline	*Linum* (flax) [10]
S-Methylcysteine	*Phaseolus vulgaris* [6]
S-Methylcysteine sulphoxide	*Phaseolus vulgaris* [6]
S-Allylcysteine	Garlic bulb [6]
S-Propylcysteine	Garlic bulb [6]
S-Prop-1-enylcysteine	Chive seed [6] (*Allium schoenoprasum*)
S-Prop-1-enylcysteine sulphoxide	Onion bulbs [6] (*Allium cepa*); *Santalum album* [425]
L-Djenkolic acid	*Acacia* sp. [370]
3, 3′-(2-Methylene-1, 2-thio)dialanine	Chive [10]
Lentinic acid	*Lentinus edodes* [421, 422]

Table 4.20 (*Contd.*)

Amino acid	*Source*
Se-Methylseleno-L-cysteine	In higher plants [237]
2-γ-Glutamylamino-4-(2-amino-2-carboxyethyl-selenyl)butanoic acid	*Astragalus pectinatus* [10]
2-Amino-4-(2-γ-glutamyl-amino-2-carboxyethyl-selenyl)butanoic acid	*Astragalus pectinatus* [10]
(Isomeric glutamylseleno-cystathionines) and leucine, isoleucine, valine, phenylalanine, tyrosine, alanine methionine, cysteine; from various plants. [6]	*Astragalus pectinatus* [10]

REFERENCES

1. Greenstein, J.P. and Winitz, M. (1961) *Chemistry of the Amino Acids*, (Vol. 1) John Wiley and Sons Inc. New York and London, p. 3.
2. Tschiersch, N. and Mothes, K. (1963) in *Comparative Biochemistry* (eds M. Florkin and H.S. Mason) (Vol. 5 *Constituents of Life Part C*), Academic Press, New York and London, p. 1.
3. Meister, A. (1965) *Biochemistry of the Amino Acids*, (Vol. 1) Academic Press, New York and London, p. 1; Rosenthal, G.A. (1982) *Plant Nonprotein Amino- and Imino-acids: Biological, Biochemical and Toxological Properties*, Academic Press, New York.
4. Fowden, L., Lewis, D. and Tristram, H. (1967) *Advances in Enzymology*, **29**, 89.
5. Fowden, L. and Lea, P.J. (1979) *Advances in Enzymology*, **50**, 117.
6. Fowden, L. (1964) *Annu. Rev. Biochem*, **33**, 173.
7. Park, B.K., Hirota, A., and Sakai, H. (1976) *Agric. Biol. Chem.*, **40**, 1905.
8. Bell, E.A. (1973) in *MTP International Review of Science, Organic Chemistry, Series* 1 (eds D.H. Hey and D.I. John), (Vol. 6) Butterworths, London, p. 1.
9. Bell, E.A. (1976) *FEBS Letters*, **64**, 29.
10. Bell, E.A. (1980) in *Encyclopaedia of Plant Physiology, New Series*, (eds. E.A. Bell and B.V. Charlwood) (Vol. 8 *Secondary Plant Products*), Springer-Verlag, Berlin, p. 401.
11. Turner, W.B. (1971) *Fungal Metabolites*, Academic Press, London and New York.
12. Fattorusso, E. and Piatelli, M. (1980) in *Marine Natural Products. Chemical and Biological Perspectives*, (ed. P.J. Scheur) (Vol. 3) Academic Press, New York and London, p. 95.
13. Young, G.T. (ed.) (1969) *Amino Acids, Peptides and Proteins* (*Specialist Periodical Reports*) (Vol. 1) The Chemical Society, London, p. 1.
14. Young, G.T. (ed.) (1970) *Amino Acids, Peptides and Proteins* (*Specialist Periodical Reports*) (Vol. 2) The Chemical Society, London, p. 1.
15. Young, G.T. (ed.) (1971) *Amino Acids, Peptides and Proteins* (*Specialist Periodical Reports*) (Vol. 3) The Chemical Society, London, p. 1.
16. Young, G.T. (ed.) (1972) *Amino Acids, Peptides and Proteins* (*Specialist Periodical Reports*) (Vol. 4) The Chemical Society, London, p. 1.
17. Sheppard, R.C. (ed.) (1973) *Amino Acids, Peptides and Proteins* (*Specialist Periodical Reports*) (Vol. 5) The Chemical Society, London, p. 1.
18. Sheppard, R.C. (ed.) (1975) *Amino Acids, Peptides and Proteins* (*Specialist Periodical Reports*) (Vol. 6) The Chemical Society, London, p. 1.

19. Sheppard, R.C. (ed.) (1976) *Amino Acids, Peptides and Proteins* (*Specialist Periodical Reports*) (Vol. 7) The Chemical Society, London, p. 1.
20. Sheppard, R.C. (ed.) (1977) *Amino Acids, Peptides and Proteins* (*Specialist Periodical Reports*) (Vol. 8) The Chemical Society, London, p. 1.
21. Sheppard, R.C. (ed.) (1978) *Amino Acids, Peptides and Proteins* (*Specialist Periodical Reports*) (Vol. 9) The Chemical Society, London, p. 1.
22. Sheppard, R.C. (ed.) (1979) *Amino Acids, Peptides and Proteins* (*Specialist Periodical Reports*) (Vol. 10) The Chemical Society, London, p. 1.
23. Sheppard, R.C. (ed) (1980) *Amino Acids, Peptides and Proteins* (*Specialist Periodical Reports*) (Vol. 11) The Chemical Society, London, p. 1.
24. Sheppard, R.C. (ed.) (1981) *Amino Acids, Peptides and Proteins* (*Specialist Periodical Reports*) (Vol. 12) The Chemical Society, London, p. 1.
25. Sheppard, R.C. (ed.) (1982) *Amino Acids, Peptides and Proteins* (*Specialist Periodical Reports*) (Vol. 13) The Chemical Society, London, p. 1.
26. Laskin, A.I. and Lechevalier, H.A., (eds.) (1973) *Handbook of Microbiology*, (Vol. 3, *Microbial Products*), CRC Press Inc., Cleveland, Ohio.
27. Ito, S. and Nicol, J.A.C. (1977) *Biochem. J.*, **161**, 499.
28. Shaw, G.J., Ellington, P.J. and Nixon, L.N. (1981) *Phytochemistry*, **20**, 1853.
29. James, H.A. (1973) in *MTP International Review of Science, Organic Chemistry, Series* 1, (eds. D.H. Hey and D.I. John) (Vol. 6) Butterworths, London, p. 213.
30. Prota, G. and Thomson, R.H. (1976) *Endeavour*, **35**, 32
31. Mabry, T. (1980) in *Encyclopaedia of Plant Physiology*, *New Series*, (eds. E.A. Bell and B.V. Charlwood) Vol. 8. *Secondary Plant Products*), Springer-Verlag, Berlin, p. 513.
32. Higa, T., Fujiyama, T. and Scheuer, P.J. (1980) *Comp. Biochem. Physiol.*, **65B**, 525.
33. Hunt, S., and Breuer, S.W. (1971) *Biochim. Biophys. Acta*, **252**, 401.
34. Fields, J.H.A., Eng, A.K., Ramsden, W.D., *et al.* (1980) *Arch. Biochem. Biophys.*, **201**, 110.
35. Robin, Y. and Guillou, Y. (1977) *Anal Biochem*, **83**, 45.
36. Biemann, K., Lioret, C., Asselineau, J. *et al.* (1960) *Biochim. Biophys. Acta.*, **40**, 369.
37. Darling, S. and Larsen, P.O. (1961) *Acta Chem. Scand.* **15**, 743.
38. Sekeris, C.E. and Karlson, P. (1966) *Pharmac. Rev.*, **18**, 89.
39. Brown, D.H. and Fowden, L. (1966) *Phytochemistry*, **5**, 887.
40. Smith, I. (1969) *Chromatographic and Electrophoretic Techniques.* (Vol. 1, *Chromatography*), Heinemann, London, p. 104.
41. Gunja-Smith, Z. and Boucek, R.J. (1981) *Biochem. J.*, **193**, 915.
42. Gunja-Smith, Z. and Boucek, R.J. (1981) *Biochem. J.*, **197**, 759.
43. Metzler, D.E. (1977) *Biochemistry : The Chemical Reactions of Living Cells*, Academic Press, New York.
44. White, A. Handler, P. and Smith, E.L. (1973) *Principles of Biochemistry*, McGraw Hill Kogakusha Ltd, Tokyo.
45. Perry, T.L. Hansen, S., MacDougall, L. and Warrington, P.D. (1967) *Clin. Chim. Acta*, **15**, 409.
46. Larsen, P.O., Orderka, O.F. and Floss, H.G. (1975) *Biochim. Biophys. Acta*, **381**, 397.
47. Larsen, P.O., and Wieczorkowska, E. (1975) *Biochim. Biophys. Acta*, **381**, 409.
48. Dardenne, G.A., Larsen, P.O. and Wieczorkowska, E. (1975) *Biochim. Biophys. Acta.* **381**, 416.
49. Patel, N., Pierson, D.L., and Jensen, R.A. (1977) *J. Biol. Chem.*, **252**, 5839.
50. Peterson, P. and Fowden, L. (1972) *Phytochemistry*, **11**, 663.
51. Chioccara, F., Misuraca, G., Novellino, E. and Prota, G. (1979) *Tetrahedron Lett.* 3181.
52. Sung, M.L., and Fowden, L. (1971) *Phytochemistry*, **10**, 1523.
53. Leete, E.J. (1975) *Phytochemistry*, **14**, 1983.
54. Soderquist, C.J. (1973) *J. Chem. Ed.*, **50**, 782.
55. Whittaker, R.H. and Feeney, P.P. (1971) *Science*, **171**, 757.
56. Bornstein, P. (1974) *Annu. Rev. Biochem.*, **43**, 567.
57. Ressler, C., Nelson, J. and Pfeffer, M. (1964) *Nature*, **203**, 1286.
58. Ellington, E.V., Hassall, C.H., Plimmer, J.R., and Seaforth, C.E. (1959) *J. Chem. Soc.*, 80.
59. Hassall, C.H., and Reyle, K. (1955) *Biochem. J.*, **60**, 334.
60. Hegarty, M.P., Schinkel, P.G., and Court, R.D. (1964) *Aust. J. Agric. Res.*, **15**, 153.
61. Rehr. S.S., Bell, E.A., Janzen, D.H., and Feeny, P.P. (1973) *Biochemical Systematics*, **1**, 63.
62. Rehr, S.S., Janzen, D.H., and Feeny, P.P. (1973) *Science*, **181**, 81.
63. Brunet, P.C.J. (1980) *Insect Biochem.*, **10**, 467.

64. Andersen, S.O. (1979) *Insect Biochem.*, **9**, 233.
65. Miller, T.A. (1975) in *Insect Muscle* (ed. P.N.R. Usherwood), Academic Press, London, p. 545.
66. Andersen, S.O. (1979) *Annu. Rev. Entomol.*, **24**, 29.
67. Chen, T. and Hodgetts, R.B. (1976) *Comp. Biochem. Physiol.*, **53B**, 415.
68. Fowden, L. (1972) *Phytochemistry*, **11**, 2271.
69. Barbier, M. (1976) *Introduction to Chemical Ecology*, Longmans, London.
70. Okami, Y., Kitihara, T., Hamada, H. *et al.* (1974) *J. Antibiotics*, **27**, 656.
71. Rinehart, K.L., Gloer, J.B., Cook, J. *et al.* (1981) *J. Am. Chem. Soc.* **103**, 1857.
72. Bramesfield, B. and Virtanen, A.L. (1956) *Acta. Chem. Scand.*, **10**, 688.
73. Bell, E.A. (1960) *Biochem. J.*, **75**, 618.
74. Munns, R., Brady, C.J. and Barlow, E.W.R. (1979) *Aust. J. Plant Physiol*, **6**, 379.
75. Fushiya, S., Sato, Y., Nozoe, S. *et al.* (1980) *Tetrahedron Lett.*, **21**, 3071.
76. Murr, D.P. and Yang, S.F. (1975) *Plant Physiol*, **55**, 79.
77. Okuno, T., Ishita, Y., Sawai, K. and Matsumoto, T. (1974) *Chem. Lett.*, 635.
78. Gross, E. and Kiltz, H.H. (1973) *Biochem. Biophys. Res. Commun.*, **50**, 559.
79. Vanderhaeghe, H. and Parmeutier, G. (1960) *J. Am. Chem. Soc.*, **82**, 4414.
80. Payne, J., Jakes, R. and Hartley, B.S. (1970) *Biochem. J.*, **117**, 757.
81. Dardenne, G., Casimir, J., Marlier, M. and Larsen, P.O. (1974) *Phytochemistry*, **13**, 1897.
82. Kuroda, Y., Okuhara, M., Goto, T. *et al.* (1980) *J. Antibiot.*, **33**, 125.
83. Kisumi, M., Sugiura, M., Kato, J. and Chibata, I. (1976) *J. Biochem.*, **79**, 1021.
84. Gieren, A., Narayanan, P., Hoppe, W. *et al.* (1974) *Justus Liebigs Ann. Chem.* 1561.
85. Hatanaka, S.I., Niimura, Y. and Taniguchi, K. (1973) *Z. Naturforsch.*, **28C**, 475.
86. Doyle, R.R. and Levenberg, B. (1968) *Biochemistry*, **7**, 2457.
87. Murooka, Y. and Harada, T. (1967) *Agric. and Biol. Chem.* (*Japan*)., **31**, 1035.
88. Fowden, L. and Smith, A. (1968) *Phytochemistry*, **7**, 809.
89. Sheehan, J.C., Maeda, K., Sen, A.K. and Stock, J.A. (1962) *J. Am. Chem. Soc.*, **84**, 1303.
90. Barry, G.T. and Roark, E. (1964) *J. Biol. Chem.*, **239**, 1541.
91. Fowden, L., Pratt, H.M. and Smith, A. (1973) *Phytochemistry*, **12**, 1707.
92. Hardman, R. and Abu-Al-Futuh, I.M. (1976) *Phytochemistry*, **15**, 325.
93. Shoji, J. and Sakazahi, R. (1970) *J. Antibiotics*, **23**, 418.
94. Bell, E.A., Meier, L.K. and Sorensen, H. (1981) *Phytochemistry*, **20**, 2213.
95. Shewry, P.R. and Fowden, L. (1976) *Phytochemistry*, **15**, 1981.
96. Dardenne, G.A., Casmir, J. and Sorensen, H. (1974) *Phytochemistry*, **13**, 2195.
97. Przybylska, J. and Strong, F.M. (1968) *Phytochemistry*, **7**, 471.
98. Hatanaka, S. and Katayama, H. (1975) *Phytochemistry*, **14**, 1434.
99. Meier, L. K. and Soerensen, H. (1979) *Phytochemistry*, **18**, 1173.
100. Olsen, O. and Soerensen, H. (1980) *Phytochemistry*, **19**, 1717.
101. Yuasa, S. (1978) *Biochim. Biophys. Acta*, **540**, 93.
102. Hatanaka, S., Niimura, Y., Taniguchi, K. *et al.* (1976) *Mushroom Science*, **9**, 809.
103. Larsen, P.O., Soerensen, H., Cochran, D.W. *et al.* (1973) *Phytochemistry*, **12**, 1713.
104. Pelter, A., Akela-Medici, A., Ballantine, J.A. *et al.* (1978) *Tetrahedron Lett.*, 2017.
105. Murakami, T., Mori, N. and Nagasawa, M. (1968) *Yakugaki Zasshi*, **88**, 488. (*Chem. Abs.*, **69**, 46016.)
106. Seneviratne, A.S. and Fowden, L. (1968) *Phytochemistry*, **7**, 1039.
107. Peiris, P.S. and Sirimawathie-Seneviratne, A. (1977) *Phytochemistry*, **16**, 1821.
108. Pan, P.C., Fang, S.D. and Tsai, C.C. (1976) *Sci. Sin.*, **19**, 691.
109. Takemoto, T., Koike, K., Nakajima, T. and Arihara, S. (1975) *Yakugaku Zasshi*, **95**, 448. (*Chem. Abs.* **83**, 97875.)
110. Fiedler, F. and Kandler, O. (1973) *Arch. Mikrobiol*, **89**, 51.
111. Hausmann, W.K., Borders, D.B. and Lancaster, J.E. (1969) *J. Antibiotics*, **22**, 207.
112. Inatomi, H., Inugai, F. and Murakami, T. (1968) *Chem. and Pharm. Bull.* (*Japan*), **16**, 2521.
113. Kjaer, A. and Larsen, P.O. (1959) *Acta Chem. Scand.*, **13**, 1565.
114. Zacharius, R.M. (1970) *Phytochemistry*, **9**, 2047.
115. Sulser, H. and Sager, F. (1976) *Experientia*, **32**, 422.
116. Smith, T.A. and Best, G.R. (1976) *Phytochemistry*, **15**, 1565.
117. Ito, K. and Hashimoto, Y. (1966) *Nature*, **211**, 417.
118. Kasi, T., Furukawa, K. and Sakamura, S. (1976) *Agric. Biol. Chem.*, **40**, 2489.
119. Shoji, J. and Hinoo, H. (1975) *J. Antibiotics*, **28**, 60.

120. Saleh, F. and White, P.J. (1976) *J. Gen. Microbiol.*, **96**, 253.
121. Pitcher, D.G. (1976) *J. Gen. Microbiol.*, **94**, 225.
122. Larsen, P.O. and Worris, F. (1976) *Phytochemistry*, **15**, 1761.
123. Fushiya, S., Sato, Y. and Nozoe, S. (1980) *Chem. Lett.*, 1215.
124. Kemp, J.D. (1977) *Biochim. Biophys. Res. Commun.*, **74**, 862.
125. Ahmad, I., Lahrer, F. and Stewart, G.R. (1981) *Phytochemistry*, **20**, 1501.
126. Firmin, J.L. and Fenwick, R.G. (1977) *Phytochemistry*, **16**, 761.
127. Nawaz, R. and Soerensen, H. (1977) *Phytochemistry*, **16**, 599.
128. Soerensen, H. (1976) *Phytochemistry*, **15**, 1527.
129. Fowden, L. (1966) *Nature*, **209**, 807.
130. Mueller, A L. and Uusheimo, K. (1965) *Acta Chem. Scand.*, **19**, 1987.
131. Sung, M.L. and Fowden, L., (1969) *Phytochemistry*, **8**, 1227.
132. Vervier, R. and Casimir, J. (1970) *Phytochemistry*, **9**, 2059.
133. Gellert, E., Halpern, B., and Rudzats, R. (1978) *Phytochemistry*, **17**, 802.
134. Dardenne, G., Casimir, J., and Jadot, J. (1968) *Phytochemistry*, **7**, 1407.
135. Nulu, J.R., and Bell, E.A. (1972) *Phytochemistry*, **11**, 2573.
136. Koenig, W.A., Kneifel, H., Bayer, E. *et al.* (1973) *J. Antibiotics*, **26**, 44.
137. Fischer, B., Keller-Schierlein, W., Kniefel, H. *et al.* (1973) *Arch Mikrobiol.*, **91**, 203.
138. Hegarty, M.P., and Pound, A.W. (1968) *Nature*, **217**, 354.
139. Grove, M.D., Daxenbickler, M.E., and Weisleder, D., and VanEtten, C.H. (1971) *Tetrahedron Lett.*, 4477.
140. Keller-Schierlein, W., and Joos, B. (1980) *Helv. Chim. Acta*, **63**, 250.
141. Koenig, W.A., Hass, W., Dehler, W. *et al.* (1980) *Justus Liebigs Ann. Chem.*, 622.
142. Larsen, P.O., and Kjaer, A. (1962) *Acta Chem. Scand.*, **16**, 142.
143. Müller, P., and Schütte, H.R. (1968) *Z. Naturforsch*, **23B**, 659.
144. Scannel, J.P., Pruess, D.L., Demny, T.C. *et al.* (1970) *J. Antibiotics.*, **23**, 618.
145. Godtfredsen, W.O., Vaugedal, S., and Thomas, D.W. (1970) *Tetrahedron*, **26**, 4931.
146. Keller-Juslen, C., Kuhn, M., Loosli, H.R. *et al.* (1976) *Tetrahedron Lett.*, 4147.
147. Sloane, N.H., and Smith, S.C. (1968) *Biochim. Biophys. Acta*, **158**, 394.
148. Weaver, R.F., Rajagopalan, K.V., Handler, P. *et al.* (1971) *J. Biol. Chem.*, **246**, 2010.
149. Dardenne, G., Marlier, M. and Welter, A. (1977) *Phytochemistry*, **16**, 1822.
150. Watson, R., and Fowden, L. (1973) *Phytochemistry*, **12**, 617.
151. Zamir, L.O., Jensen, R.A., Arison, B.H. *et al.* (1980) *J. Am. Chem. Soc.*, **102**, 4499.
152. Isujino, I., Yake, K., Sekikawa, I., and Hamanaka, N. (1978) *Tetrahedron Lett.*, 1401.
153. Takano, S., Uemura, D., and Hirata, Y. (1978) *Tetrahedron Lett.*, 4909.
154. Takano, S., Uemura, D., and Hirata, Y. (1978) *Tetrahedron Lett.*, 2299.
155. Ito, S. and Hirata, Y. (1977) *Tetrahedron Lett.*, 2429.
156. Fowden, L., MacGibbon, C.M., Mellon, F.A. and Sheppard, R.C. (1972) *Phytochemistry*, **11**, 1105.
157. Gray, D.O., and Fowden, L. (1962) *Biochem. J.* **82**, 385.
158. Honkanen, E., Moisio, T., Virtanen, A.L., and Melera, A. (1964) *Acta Chem. Scand.*, **18**, 1319.
159. Casnati, G. Quilico, A., and Ricca, A. (1963) *Gazz. Chim. Ital.*, **93**, 349.
160. Bell, E.A., and Fellows, L.E. (1966) *Nature*, **210**, 529.
161. Takita, T., Naganawa, H., Maeda, K., and Umezawa, H. (1964) *J. Antibiotics, (Tokyo) Ser. A.*, 90.
162. Anderson, J.A., and Saini, M.S. (1974) *Tetrahedron Lett.*. 2107.
163. Agurell, S., and Lindgren, J.A. (1968) *Tetrahedron Lett.*, 5127.
164. Robbers, J.E., and Floss, H.G. (1969) *Tetrahedron Lett.*, 1857.
165. Koncewicz, M., Mathiaparanam, P., Uchytil, T.F. *et al.* (1973) *Biochem. Biophys. Res. Comm.*, **53**, 653.
166. Lindberg, P., Bergman, R., and Wickberg, B. (1977) *J. Chem. Soc., Perkin Trans. 1*, 684.
167. Kristensen, I., Larsen, P.O., and Olsen, C.E. (1976) *Tetrahedron*, **32**, 2799.
168. Gray, D.O., and Fowden, L. (1962) *Nature*, **193**, 1285.
169. Fujita, T., Takaishi, Y., Okamura, A. *et al.* (1979) *Koen, Yoshishu-Tennen Yuki Kagobutsu Toronkai 22nd.* 424. (*Chem. Abs.*, **93**, 47 160.)
170. Burgstahler, A.W., Trollope, M.L., and Aiman, C.E. (1964) *Nature*, **202**, 388.
171. Bethell, M., Kenner, G.W., and Sheppard, R.C. (1962) *Nature*, **194**, 864.
172. Bell, E.A., Qureshi, M.Y., Pryce, R.J. *et al.* (1980). *J. Am. Chem. Soc.*, **102**, 1409.
173. Spenser, I.D., and Notation, A.D. (1962) *Can. J. Chem.*, **40**, 1374.

174. MacLeod, J.K., Summons, R.E., Parker, C.W. and Letham, D.S. (1975) *J. Chem. Soc. Chem. Commun.*, 809.
175. Lipton, S.H., and Strong, F.M. (1965) *J. Org. Chem.*, **30**, 115.
176. Kjaer, A., Knudsen, A., and Olesen, P. (1961) *Acta Chem. Scand.*, **15**, 1193.
177. Fellows, L.E., Bell, E.A., Lee, T.S., and Janzen, D.H. (1979) *Phytochemistry*, **18**, 1333.
178. Okuda, T., Yoshida, T., Shiota, N. and Nobuhara, J. (1975) *Phytochemistry*, **14**, 2304.
179. Marlier, M., Dardenne, G., and Casimir, J. (1979) *Phytochemistry*, **18**, 479.
180. Buchi, G., and Raleigh, J.A. (1971) *J. Org. Chem.*, **36**, 871.
181. Bell, E.A., and Nulu, J.R. (1971) *Phytochemistry*, **10**, 2191.
182. Stapley, E.O., Hendlin, D., Jackson, M., and Miller, A.K. (1971) *J. Antibiotics*, **24**, 42.
183. Müller, P., and Schütte, H.R. (1968) *Z. Naturforsch.*, **236**, 491.
184. Hutzinger, O., and Kosuge, T. (1968) *Biochemistry*, **7**, 601.
185. Millington, D.S. and Sheppard, R.C. (1968) *Phytochemistry*, **7**, 1027.
186. Bell, E.A., Nulu, J.R., and Cone, C. (1971) *Phytochemistry*, **10**, 2191.
187. Noria, M., Noguchi, M. and Tamaki, E. (1971) *Tetrahedron Lett.*, 2017.
188. Matzinger, P., Catalforno, P., and Engster, C.H. (1972) *Helv. Chim. Acta*, **55**, 1478.
189. Lambein, F., and Van Parijs, R. (1974), *Biochem. Biophys. Res. Commun.*, **61**, 155.
190. Noguchi, M., Sakuma, H., and Tamaki, E. (1968) *Arch. Biochem. Biophys.* **125**, 1017.
191. Noguchi, M., Sakuma, H., and Tamaki, E. (1968) *Phytochemistry*, **7**, 1861.
192. Shah, R., Neuss, N., Gorman, M., and Boeck, L.D. (1970) *J. Antibiotics.*, **23**, 613.
193. Grove, M.D., Weisleder, D., and Daxenbichler, M.E. (1973) *Tetrahedron*, **29**, 2715.
194. Rhugenda-Banga, N., Welter, A., Jadot, J., and Casimir, J. (1979) *Phytochemistry*, **18**, 482.
195. Ito, S., and Nicol, J.A.C. (1975) *Tetrahedron Lett.*, 3287.
196. Kodama, H., Ishimoto, Y., Shimoura, M. *et al.* (1975) *Physiol. Chem. Phys.*, **7**, 147.
197. Kodama, H., Ohmori, S., Suzuki, M. *et al.* (1971) *Physiol. Chem. Phys.*, **3**, 81.
198. Bodanszky, M., Fried, J., Sheehan, J.T. *et al.* (1964), *J. Am. Chem Soc.*, **86**, 2478.
199. Cross, D.F.W., Kenner, G.W., Sheppard, R.C., and Stehr, C.E. (1963) *J. Chem. Soc.*, 2143.
200. Ito, S., Nardi, G., and Prota, G. (1976) *J. Chem. Soc., Chem. Commun.*, 1042.
201. Mutikkala, E.J., and Virtaen, I. (1964) *Acta Chem. Scand.*, **18**, 2009.
202. Ampola, M., Bixby, E.M., Crawhall, J.C. *et al.* (1968) *Biochem. J.*, **107**, 16P.
203. Lindberg, P., Bergman, R., and Wickberg, B. (1975) *J. Chem. Soc., Chem. Commun.*, 946.
204. Burrow, B.F. and Turner, W.B. (1966), *J. Chem. Soc. (C)*, 255.
205. Sangster, A.W., Thomas, S.F., and Tingling, N.L. (1974) *Tetrahedron*, **31**, 1135.
206. Kuo, Y.H., Lambein, F., and Van Parijs, R. (1976) *Arch. Int. Physiol. Biochim.*, **84**, 169.
207. Kato, K., Takita, R., and Umezawa, H. (1980) *Tetrahedron Lett.*, 4925.
208. Takemoto, T. and Nakajuma, T. (1964) *Yakugaku Z.*, **84**, 1183, (*Chem. Abs.* **62**, 7859).
209. Takemoto, T., Nakajima, T. and Yokobe, T. (1964) *Yakugaku, Z.*, **84**, 1232, (*Chem. Abs.* **62**, 11 901.)
210. Kelly, R.B., Daniels, E.G., and Hinman, J.W. (1962) *J. Org. Chem.*, **27**, 3229.
211. Nomoto, K., Yoshioka, H., Takemoto, T. *et al.* (1979) *Koen, Yoshishu-Tennen Yuki Kagobutsu Toronkai 22nd.*, 619, (*Chem. Abs.*, **93**, 47 161.)
212. Morgan, J.W.W. (1964) *Chem. Ind.*, 542.
213. Shoji, K., Tori, K., and Otsuka, H. (1965) *J. Org. Chem.*, **30**, 2772.
214. Sheehan, J.C., Zachay, H.G., and Lawson, W.B. (1958) *J. Am. Chem. Soc.*, **80**, 3349.
215. Larher, F., and Hamelin, J. (1975) *Phytochemistry*, **14**, 1789.
216. Otsuka, H., and Shaji, J. (1965) *Tetrahedron*, **21**, 2931.
217. Fukuoka, T., Muraoka, Y., Fujii, A. *et al.* (1980) *J. Antibiotics*, **33**, 114.
218. Shaw, G.J., Ellingham, P.J., and Nixon, L.W. (1981) *Phytochemistry*, **20**, 1853.
219. Hedges, S.H., and Herbert, R.B. (1981) *Phytochemistry*, **20**, 2064.
220. Cimino, G., de Stefano, S., and Minale, L. (1975) *Experientia*, **31**, 756.
221. Welter, A., Marlier, M., and Dardenne, G. (1978) *Phytochemistry*, **17**, 131.
222. Evans, C.S., and Bell, E.A. (1979) *Phytochemistry*, **18**, 1807.
223. Meier, L.K., Olsen, O., and Soerensen, H., (1979) *Phytochemistry*, **18**, 1505.
224. Evans, C.S., and Bell, E.A., (1978) *Phytochemistry*, **17**, 1127.
225. Ichihara, A., Hasegawa, H., Sato, H. *et al.* (1973) *Tetrahedron Lett.*, 37.
226. Kasui, T., Kishi, Y., Sano, M., and Sakamura, S. (1973) *Agric. Biol. Chem.*, (*Japan*), **37**, 2923.
227. Couchman, R., Eagles, J., Hegarty, M.P. *et al.* (1973) *Phytochemistry*, **12**, 707.
228. Clark, A.M., and Hufford, C.D. (1978) *Phytochemistry*, **17**, 552.

229. Cox, R.E., Chexal, K.K., and Holker, J.S.E. (1976) *J. Chem. Soc., Perkin Trans. 1*, 578.
230. Argoudehis, A.D., Mizsah, S.A., and Baczynsky, L. (1976) *J. Antibiotics*, **28**, 733.
231. Zaclauskas, R., Murphy, P.T., and Wells, R.J. (1978) *Tetrahedron Lett.*, 4945.
232. Kasai, T., Larsen, P.O., and Soerensen, H. (1978) *Phytochemistry*, **17**, 1911.
233. Funayam, S., and Isono, K. (1977) *Biochemistry*, **16**, 3121.
234. Murakoshi, I., Kuramoto, H., Haginawa, J. and Fowden, L. (1972) *Phytochemistry*, **11**, 177.
235. Owens, L.D., and Thompson, J.F. (1972) *J. Chem. Soc., Chem. Commun.* 714.
236. Patil, S.S., Youngblood, P., Christiansen, P., and Moore, R.E. (1976) *Biochem. Biophys. Res. Commun.*, **69**, 1019.
237. Shrift, A. (1969) *Ann. Rev. Plant, Physiol.*, **20**, 475.
238. Koyama, M., Tsujizaki, Y., and Sakamura, S. (1973) *Agric. Biol. Chem.*, (*Japan*), **37**, 2749.
239. Stuart, K.L., McNeil, D., Kutney, J.P. *et al.* (1973) *Tetrahedron*, **29**, 4071.
240. Fosker, A.P., and Law, H.D. (1965) *J. Chem. Soc.*, 7305.
241. Durzan, D.J., and Pitel, J.A. (1977) *Insect Biochem.*, **7**, 11.
242. Florkin, M. (1969) in *Chemical Zoology*, (ed. M. Florkin and B.T. Scheer) (Vol. 4) Academic Press, New York, p. 147.
243. VanThoai, N., and Robin, Y. (1969) in *Chemical Zoology*, (ed. M. Florkin and B.T. Scheer) (Vol. 4) Academic Press, New York, p. 163.
244. Kasai, T., Hirakuri, Y., and Sakamura, S. (1981) *Phytochemistry*, **20**, 2209.
245. Hatanaka, S.I., Kaneko, S., Saito, K., and Ishida, Y. (1981) *Phytochemistry*, **20**, 2291.
246. Tsuji, H., Bando, N., Ogawa, T., and Sasaoka, K. (1981) *Biochim. Biophys. Acta.*, **677**, 326.
247. Takita, T. (1963) *Penishirin Sono Ta Koseibusshitsu. Ser. A.*, **60**, 3086, (*Chem. Abs.*, **60**, 3086.)
248. Hatanaka, S.I., Niimura, Y., and Taniguchi, K. (1972) *Phytochemistry*, **11**, 3327.
249. Rudzats, R., Gellart, E. and Halpern, B. (1972) *Biochem. Biophys. Res. Commun.* **47**, 290.
250. Ohashi, Y., Abe, H. and Ito, Y. (1973) *Agric. Biol. Chem.* (*Japan*) **37**, 2277.
251. Ohashi, Y., Abe, H. and Ito, Y. (1973) *Agric. Biol. Chem.* (*Japan*), **37**, 2283.
252. Potgieter, H.C., Vermeulen, M.M., Potgieter, D.J.J. and Strauss, H.F. (1977) *Phytochemistry*, **16**, 1757.
253. Hatanaka, S. and Kawakami, K. (1980) *Sci. Pap. Cell. Gen. Educ., Univ. Tokyo* **30**, 147, (Chem. Abs., **93**, 61 778.)
254. Chilton, W.S., and Tsou, G. (1972) *Phytochemistry*, **11**, 2853.
255. Pant, R., and Fales, H.M. (1974) *Phytochemistry*, **13**, 1626.
256. Pilbeam, D.J., and Bell, E.A. (1979) *Phytochemistry*, **18**, 320.
257. Perlman, D., Perlman, K.I., Bodansky, M. *et al.* (1977) *Bio-org. Chem.*, **6**, 263.
258. Kuo, C.H., and Wendler, N.L. (1978) *Tetrahedron Lett.*, 211.
259. Ishiyama, T., Furuta, T., Takai, M. *et al.* (1975) *J. Antibiotics*, **28**, 821.
260. Dardenne, G.A., Casimir, J., Bell, E.A., and Nulu, J.R. (1972) *Phytochemistry*, **11**, 787.
261. Hatanaka, S., Iizumi, H., Tsuji, A., and Gmelin, R. (1975) *Phytochemistry*, **14**, 1559.
262. Qureshi, M.Y., Pilbeam, D.J., Evans, C.S., and Bell, E.A. (1977) *Phytochemistry*, **16**, 477.
263. Evans, C.S., Qureshi, M.Y., and Bell, E.A., (1977) *Phytochemistry*, **16**, 565.
264. Harrison, F.L., Nunn, P.B., and Hill, R.R. (1977) *Phytochemistry*, **16**, 1821.
265. Owens, L.D., Thompson, J.F., Pitcher, R.G., and Williams, T. (1972) *J. Chem. Soc., Chem. Commun.*, 714.
266. Owens, L.D., Thompson, J.F., and Fennessey, P.V. (1972) *J. Chem. Soc., Chem. Commun.*, 715.
267. Scannell, J.P., Pruess, D.L. Ax, H.A. *et al.* (1976) *J. Antibiotics*, **29**, 38.
268. Pruess, D.L., Scannell, J.P., Kellett, M. *et al.* (1974) *J. Antibiotics*, **27**, 229.
269. Kasai, T., Sano, M., and Sakamura, S. (1976) *Agric. Biol. Chem.*, **40**, 2449.
270. Akazawa, S. (1970) *Osaka Daigaku Igaku Zasshi*, **22**, 461.
271. Nakajima, T., Matsuoka, Y., and Kakimoto, Y. (1971) *Biochim. Biophys. Acta*, **230**, 212.
272. Kakimoto, Y., and Akazawa, S. (1970) *J. Biol. Chem.*, **245**, 5751.
273. Maehr, H., Yarmchuk, L., Pruess, D.L. *et al.* (1976) *J. Antibiotics*, **29**, 213.
274. Fellows, L.E., Polhill, R.M., and Bell, E.A. (1978) *Biochem. Syst. Ecol.*, **6**, 213.
275. Larsen, P.O. (1968) *Acta Chem. Scand.*, **22**, 1369.
276. Hempel, K., Lange, H.W., and Birkhofer, L. (1968) *Z. Physiol. Chem.*, **349**, 603.
277. Hempel, K., Lange, H.W., and Birkhofer, L. (1968) *Naturwissenschaften*, **55**, 37.
278. Leistner, E., and Spenser, I.D. (1973) *J. Am. Chem. Soc.*, **95**, 4715.
279. Kaufman, R.A., and Broquist, H.P. (1977) *J. Biol. Chem.*, **252**, 7437.
280. Hettinger, T.P., and Craig, L.C. (1970) *Biochemistry*, **9**, 1224.

281. Matsutani, H., Kusumoto, S., Koizumi, R., and Shiba, T. (1979) *Phytochemistry*, **18**, 661.
282. Larsen, P.O., and Wieczorkowska, E. (1975) *Biochim. Biophys. Acta*, **381**, 409.
283. Wilson, M.F., Bholah, M.A., Morris, G.S., and Bell, E.A. (1979) *Phytochemistry*, **18**, 1391.
284. Larsen, P.O., Onderka, D.K., and Floss, H.G. (1975) *Biochim. Biophys. Acta.*, **381**, 397.
285. Rajanada, V., Norris, A.F., and Williams, D.H. (1979) *J. Chem. Soc. Perkin Trans. 1*, 29.
286. Smith, G.A., Smith, K.A., and Williams, D.H. (1975) *J. Chem. Soc. Perkin Trans. 1*, 2108.
287. Williams, D.H., Rajanada, V., and Kalman, J.R. (1979) *J. Chem. Soc., Perkin Trans. 1*, 737.
288. Dardenne, G.A., Marlier, M., and Casimir, J. (1972) *Phytochemistry*, **11**, 2567.
289. Steglich, W., and Preuss, R. (1975) *Phytochemistry*, **14**, 1119.
290. Tong, J.H., D'lorio, A., and Benoiton, N.L. (1971) *Biochem. Biophys. Res. Comm.* **43**, 819.
291. Hassal, C.H., Morton, R.B., Ogihara, Y., and Thomas, W.A. (1969) *J. Chem. Soc., Chem. Commun.*, 1079.
292. Borders, D.B., Sax, K.J., Lancaster, J.E. *et al.* (1972) *Tetrahedron*, **26**, 3123.
293. Widmer, J., and Keller-Schierlein, W. (1974) *Helv. Chim. Acta*, **57**, 657.
294. Buck, R., Eberspaecher, J., and Lingens, F. (1979) *Justus Liebigs Ann. Chen.*, 564.
295. Tong, J.H., D'lorio, A., and Benoiton, N.L. (1971) *Biochem. Biophys. Res. Comm.*, **44**, 229.
296. Lunt, D.O., and Evans, W.C. (1976) *Biochem. Soc. Trans.*, **4**, 491.
297. Fellman, J.H., Wada, G.H., and Roth, E.S. (1975) *Biochim. Biophys. Acta*, **381**, 9.
298. Boulvin, G., Ottinger, R., Pais, M., and Chiurdoglu, G. (1970) *Bull. Soc. Chim. Belg.*, **78**, 583.
299. Duffey, S.S., Underhill, E.W., and Towers, G.H.N. (1974) *Comp. Biochem. Physiol.* **47B**, 753.
300. Van-Heyningen, R. (1971) *Nature*, **230**, 393.
301. Aoki, H., and Okuhara, M. (1980) *Ann. Rev. Microbiol.*, **34**, 159.
302. Okuno, T., Ishita, Y., Sawai, K., and Matsumoto, T. (1974) *Chem. Lett.*, 635.
303. Harris, T.M., Harris, C.M., Fehlner, J.R. *et al.* (1979) *J. Org, Chem.*, **44**, 1009.
304. Harris, C.M., Kibly, J.J., Fehlner, J.R. *et al.* (1979) *J. Am. Chem. Soc.*, **101**, 437.
305. Baggaley, K.H., Blessington, B., Falshaw, C.P., and Ollis, W.D. (1969) *J. Chem. Soc., Chem. Commun.*, 101.
306. Koenig, W., Hagenmaier, H., and Dachn, U. (1975) *Z. Naturforsch.*, **30B**, 626.
307. Konze, J.R., and Kende, H. (1979) *Planta*, **146**, 293.
308. Boller, T., Herner, R.C., and Kende, H. (1979) *Planta*, **145**, 293.
309. Adams, D.O., and Yang, S.F. (1979) *Anal. Biochem.*, **100**, 140.
310. Cramer, U., and Spener, F. (1977) *Eur. J. Biochem.*, **74**, 495.
311. Cramer, U., Rehfeldt, A.G., and Spener, F. (1980) *Biochemistry.*, **19**, 3074.
312. Gould, S.J., and Darling, D.S. (1978) *Tetrahedron Lett.*, 3207.
313. Johns, S.R., Lamberton, J.A., and Sioumis, A.A. (1971) *Aust. J. Chem.*, **24**, 439.
314. Petroski, R.J., and Kelleher, W.J. (1978) *Lloydia*, **41**, 332.
315. Mabry, T.J. (1980) in *Encyclopaedia of Plant Physiology, New Series*, (eds E.A. Bell and B.V. Charlwood) (Vol. 8, *Secondary Plant Products*) Springer-Verlag, Berlin, p. 513.
316. Graham, D.G., and Jeffs, P.W. (1977) *J. Biol. Chem.*, **252**, 5729.
317. Naganawa, H., Usui, N., Takita, T. *et al.* (1975) *J. Antibiotics.*, **28**, 828.
318. Bach, B., Gregson, R.P., Holland, G.S. *et al.* (1978) *Experientia*, **34**, 608.
319. Terlain, B., and Thomas, J.P. (1971) *Bull. Soc. Chim. France.*, 2349.
320. Sung, M.L., and Fowden, L. (1968) *Phytochemistry*, **7**, 2061.
321. Ohkusu, H., and Mori, A. (1969) *J. Neurochem.*, **16**, 1485.
322. Impellizzeri, G., Mangiafico, S., Oriente *et al.* (1975) *Phytochemistry*, **14**, 1549.
323. Hatanaka, S. (1969) *Phytochemistry*, **8**, 1305.
324. Moriguchi, M., Sada, S., and Hatanaka, S. (1979) *Appl. Environ. Microbiol.*, **38**, 1018.
325. Brockmann, H., and Stahler, E.A. (1973) *Tetrahedron Lett.*, 3685.
326. Katz, E., Mason, K.T., and Mauger, A.B. (1973) *Biochem. Biophys. Res. Comm.*, **53**, 819.
327. Katz, E., Mason, K.T., and Mauger, A.B. (1975) *Biochem. Biophys. Res. Comm.*, **63**, 502.
328. Fowden, L., Smith, A., Millington, D.S., and Sheppard, R.C. (1969) *Phytochemistry*, **8**, 437.
329. Kasai, T., Kishi, Y., Sano, M., and Sakamura, S. (1973) *Agric. and Biol. Chem.* (*Japan*), **37**, 2923.
330. Couchman, R., Eagles, J., Hegart, M.P. *et al.* (1973) *Phytochemistry*, **12**, 707.
331. Welter, A., Marlier, M., and Dardenne, G. (1975) *Bull. Soc. Chim. Belg.*, **84**, 243.
332. Welter, A., Jadot, J., Dardenne, G. *et al.* (1975) *Phytochemistry*, **14**, 1347.
333. Mbadiwe, E.I. (1975) *Phytochemistry*, **14**, 1351.
334. Hatanaka, S. (1968) *Sci. Papers, Coll. Educ. Univ. Tokyo*, **17**, 219.
335. Ahmad, V.U., and Khan, M.A. (1971) *Phytochemistry*, **10**, 3339.

336. Despontin, J., Marlier, M. and Dardenne, G. (1977) *Phytochemistry*, **16**, 387.
337. Hatanaka, S., and Kaneko, S. (1977) *Phytochemistry*, **16**, 1041.
338. Marlier, M., Dardenne, G.A., and Casimir, J. (1972) *Phytochemistry*, **11**, 2597.
339. Tatsuta, K., Mikami, N., Fujimoto, K. *et al.* (1973) *J. Antibiotics*, **26**, 625.
340. Marlier, M., Dardenne, G., and Casimir, J. (1976) *Phytochemistry*, **15**, 183.
341. Taylor, P.A., Schnoes, H.K., and Durbin, R.D. (1972) *Biochim. Biophys. Acta.*, **286**, 107.
342. Kristensen, I., Larsen, P.O., and Soerensen, H. (1974) *Phytochemistry*, **13**, 2803.
343. Dunhill, P.M. and Fowden, L. (1963) *Biochem. J.*, **86**, 388.
344. Fellows, L.E., Polhill, R.M. and Bell, E.A. (1978) *Biochem. Syst. Ecol.*, **6**, 213.
345. Horii, S., and Kameda, Y. (1968) *J. Antibiotics*, **21**, 665.
346. Takita, T., Yoshioka, T., Muraoka, Y. *et al.* (1971) *J. Antibiotics*, **24**, 795.
347. Koyama, G., Nakamura, H., Muraoka, Y. *et al.* (1973) *J. Antibiotics*, **26**, 109.
348. Kemp, J.D. (1977) *Biochim. Biophys. Res. Comm.*, **74**, 862.
349. Hack, E., and Kemp, J.D. (1977) *Biochim. Biophys. Res. Comm.*, **78**, 785.
350. Okura, A., Morishima, H., Takita, T. *et al.* (1975) *J. Antibiotics*, **28**, 337.
351. Marconi, G.G. and Bodansky, M. (1970) *J. Antibiotics*, **23**, 120.
352. Ohashi, Z., Maeda, M., McClowskey, J.A., and Nishimura, S. (1974) *Biochemistry*, **13**, 2620.
353. Friedman, S., Li, H.J., Nakanishi, K., and Van Lear, G. (1974) *Biochemistry*, **13**, 2932.
354. Macnicol, P.K. (1968) *Biochem. J.*, **107**, 473.
355. Daxenbichler, M.E., Kleiman, R., Weisleder, D. *et al.* (1972) *Tetrahedron Lett.*, 1801.
356. Abe, H., Uchiyama, M., Tanaka, Y., and Saito, H. (1976) *Tetrahedron. Lett.*, 3807.
357. Nakanishi, K., Furutachi, N., Funamizu, M. *et al.* (1970) *J. Am. Chem. Soc.*, **92**, 7617.
358. Nakamiya, T., Shiba, T., Kaueko, T. *et al.* (1970) *Tetrahedron Lett.*, 3497.
359. Corbin, J.L., Karle, I.L., and Karle, J. (1970) *J. Chem. Soc., Chem. Commun.*, 186.
360. Doyle, R.R., and Levenberg, B. (1974) *Phytochemistry*. **13**, 2813.
361. Hatanaka, S., and Niimura, Y. (1975) *Phytochemistry*, **14**, 1463.
362. Shimada, K., and Nambara, T. (1979) *Tetrahedron Lett.*, 163.
363. Kellett, M., Pruess, D., and Scannell, J.P. (1980) *US Patent* 4 202 819 (*Chem. Abs.*, **93**, 130 567).
364. Prota, G. (1980) in *Marine Natural Products, Chemical and Biological Perspectives*, (ed. P.J. Scheuer) (Vol. 3) Academic Press, New York, p. 141.
365. Perry, T.L., Hansen, S., MacDougal, L., and Warrington, P.D. (1967) *Clin. Chim. Acta.*, **15**, 409.
366. Schnyder, J., and Erismann, K.H. (1973) *Experientia*, **29**, 232.
367. Laycock, M.V., McInnes, A.G., and Morgan, K.C. (1979) *Phytochemistry*, **18**, 1230.
368. Vasstrand, E., Jensen, H.B., and Miron, T. (1980) *Anal. Biochem.*, **105**, 154.
369. Ito, K. and Fowden, L. (1972) *Phytochemistry*, **11**, 2541.
370. Seneviratne, A.S., and Fowden, L. (1968) *Phytochemistry*, **7**, 1039.
371. Jadot, J., Casimir, J., and Warin, R. (1969) *Bull. Soc. Chim. Belg.*, **78**, 299.
372. Kawanchi, K., Kosura, T., Ota, S., and Susuki, T. (1977) *Nippon Suisan Gakkaishi*, **43**, 1293.
373. Kanzaki, T., Fukita, T., Kitano, K. *et al.* (1976) *Hakko Kogaku Zasshi*, **54**, 720 (*Chem. Abs.*, **36**, 28 440).
374. Hatanaka, S., Kaneko, S., Niimura, Y. *et al.* (1974) *Tetrahedron Lett.* 3921.
375. Groger, D. (1980) in *Encyclopaedia of Plant Physiology, New Series*, (eds E.A. Bell and B.V. Charlwood) (Vol. 8 *Secondary Plant Products*), Springer Verlag. Berlin, p. 128.
376. Hlavka, J.J., Bitha, P., Boothe, J.F., and Morton, G. (1974) *Tetrahedron Lett.*, 175.
377. Shiba, T., Mukunoki, Y., and Akiyama, H. (1974) *Tetrahedron Lett.*, 3085.
378. Sztariczkai, F., Harris, C.M., and Harris, T.M. (1979) *Tetrahedron Lett.*, 2861.
379. Katrukha, G.S., Diarra, B., Silav, A.B. *et al.* (1979) *Antibiotiki*, **24**, 179.
380. Raverty, W.D., Thomson, R.H., and King, T.J. (1977) *J. Chem. Soc., Perkin Trans. 1*, 1204.
381. Anderson, R.J. (1978) *Tetrahedron Lett.*, 2541 and 4651.
382. Fowden, L. (1968) *Proc. Roy. Soc. Ser. B.*, **171**, 5.
383. Martin, D.G., Chidester, C.G., Mitzak, S.A. *et al.* (1975) *J. Antibiotics*, **28**, 91.
384. Hanka, L.J., Gerpheide, S.A., Spieles, P.R. *et al.* (1975) *Antimicrobial Agents and Chemotherapy*, **7**, 807.
385. Martin, D.G., Duchamp, D.J., and Chidester, C.G. (1973) *Tetrahedron Lett.*, 2549.
386. Kuroda, Y., Okuhara, M., Goto, T. *et al.* (1980) *J. Antibiotics*, **33**, 259.
387. Kruger, G.J., Du Plessis, L.M., and Grobbelaar, N. (1976) *J. S. Afr. Chem. Inst.*, **29**, 24.
388. Kurita, M., Jomon, K., Komori, T. *et al.* (1976) *J. Antibiotics*, **29**, 1243.
389. Yasumoto, T., and Shimizu, N. (1977) *Nippon Suisan Gakkaishi*, **43**, 201, (*Chem. Abs.*, **86**, 117886.)

390. Hayashi, T., and Konosu, S. (1977) *Nippon Suisan Gakkaishi*, **43**, 343.
391. Agrup, G., Hansson, C., Rorsman, H. *et al.* (1978) *Acta Derm-Venereol.*, **58**, 269.
392. Eloff, J.N. (1980) *Z. Pflanzenphysiol.*, **98**, 403.
393. Marner, F.J., Moore, R.E., Hirotsu, K. and Clardy, J. (1977) *J. Org. Chem.*, **42**, 2815.
394. Audhya, T.K., and Russell, D.W. (1974) *J. Chem. Soc., Perkin Trans.* **1**., 743.
395. Miyazawa, K., and Ito, K. (1974) *Nippon Suisan Gakkaishi*, **40**, 655.
396. Sethi, J.K., and Carew, D.P. (1974) *Phytochemistry*, **13**, 321.
397. Bercht, C.A.L., Lonsberg, R.J.J.C., Kuppers, T.J.E.M., and Salermink, C.A. (1973) *Phytochemistry*, **12**, 2457.
398. Sciuto, S., Piatelli, M., and Chillemi, R. (1979) *Phytochemistry*, **18**, 1058.
399. Impellizzeri, G., Mangiafico, S., Oriente *et al.* (1975) *Phytochemistry*, **14**, 1549.
400. Mwanluka, K., Bell, E.A., Charlwood, B.V., and Briggs, J.M. (1975) *Phytochemistry*, **14**, 657.
401. Larker, F. and Hamelin, J. (1975) *Phytochemistry*, **14**, 205.
402. Gasteiger, E.L., Haake, P.C., and Gergen, J.A. (1960) *Ann. N.Y. Acad. Sci.*, **90**, 622.
403. Ogawa, T., Fukuda, M., and Sasaoka, K. (1973) *Biochim. Biophys, Acta*, **297**, 60.
404. Fukuda, M., Tokumura, A., Ogawa, T., and Sasaoka, K. (1973) *Phytochemistry*, **12**, 2593.
405. Stuart, K.L., McNeil, D., Kutney, J.P. *et al.* (1973) *Tetrahedron*, **29**, 4071.
406. Corbin, J.L., and Bulen, W.A. (1969) *Biochemistry*, **8**, 757.
407. Hassall, C.H., Ogihara, Y., and Thomas, W.A. (1971) *J. Chem. Soc. (C)*, 522.
408. McGahren, W.J., Martin, J.H., Morton, G.O. *et al.* (1980) *J. Am. Chem. Soc.*, **102**, 1671.
409. Williams, D.H., and Kalman, J.R. (1976) *Tetrahedron Lett.*, 4829.
410. Smith, G.A., Smith, K.A., and Williams, D.H., (1975) *J. Chem. Soc., Perkin Trans. 1*, 2108.
411. Makleit, S., Starichkai, F., and Pushkash, M. (1967) *Visn. Kiiv. Univ. Ser. Fiz. Khim.*, **7**, 155.
412. Trifonova, Z.P., Katrukha, G.S., Silav, A.B. *et al.* (1979) *Khim. Prir. Soedin.*, 875.
413. Van Thoai, N., Robin, Y., and Guillon, Y. (1972) *Biochemistry*, **21**, 3890.
414. Pruess, D.L., Scannell, J.P., Ax, H.A. *et al.* (1973) *J. Antibiotics.*, **26**, 261.
415. Evans, C.S., and Bell, E.A. (1978) *Phytochemistry*, **17**, 1127.
416. Ogawa, Y., Yoshida, H., Inoue, S., and Niida, T. (1973) *Meiji Seika Kenkyu Nempo*, **13**, 49.
417. Ogawa, Y., Yoshida, H., Inoue, S., and Niida, T. (1973) *Meiji Seika Kenkyu Nempo*, **13**, 42.
418. Ezaka, N., Amano, S., Fukushima, K. *et al.* (1973) *Meiji Seika Kenkyu Nempo*, **13**, 60.
419. Kruger, G.J., DuPleissis, L., and Grobbelaar, N. (1976) *J.S. Afr. Chem. Inst.*, **29**, 24.
420. Hatanaka, S., and Kaneko, S. (1978) *Phytochemistry*, **17**, 2027.
421. Yasumoto, K., Iwami, K., Mizusawa, H., and Mitsuda, H., *Nippon Nogei Kagaku Kaishi*, 1976, **50**, 563, (*Chem. Abs.*, **86**, 185 866.)
422. Hofle, G., Gmelin, R., Luxa, H.H. *et al.* (1976) *Tetrahedron Lett.* 3129.
423. Ohkusi, H., and Mori, A. (1969) *J. Neurochem.*, **16**, 1485.
424. Johns, S.R., Lamberton, J.A., and Sioumis, A.A. (1971) *Aust. J. Chem.*, **24**, 439.
425. Kuttan, R., Nair, N.G., Radhakrishnan, A.N. *et al.* (1974) *Biochemistry*, **13**, 4394.
426. Barrow, K.D., and Quigley, F.R. (1975) *Tetrahedron Lett.*, 4269.
427. Okogun, J.I., Spiff, A.I., and Ekong, D.E.U. (1976) *Phytochemistry*, **15**, 826.
428. Welter, A., Jadot, J., Dardenne, G. *et al.* (1976) *Phytochemistry*, **15**, 1984.
429. Tallan, H.H., Bella, S.T., Stein, W.H., and Moore, S. (1955) *J. Biol. Chem.*, **217**, 703.

CHAPTER FIVE

Metabolic and Pharmacological Studies

D.A. BENDER

5.1 NITROGEN BALANCE AND DYNAMIC EQUILIBRIUM

By the end of the last century the concept of nitrogen balance or equilibrium was well established. The measurement of nitrogenous compounds in foods, urine and faeces was technically simple, and the balance between intake and excretion had been studied by a number of workers. It was known that for a healthy adult animal or human being there was an equilibrium between the intake and excretion of nitrogenous compounds: nitrogen was ingested mainly as proteins, and excreted as a number of smaller molecules, most of which had been identified. The main excretory compound is urea; it was the synthesis of urea from ammonium carbamate by Wöhler in 1828 that led to the abandoning of the older concept of a difference between 'organic' compounds that contained some 'vital principle' and 'inorganic' compounds that did not.

It was also known that in young animals that were still growing there was a net retention of nitrogenous material in the body, in the form of new tissue proteins – the excretion was less than the intake, the condition of positive nitrogen balance. Similarly, it was known that in starvation, on feeding inadequate diets and in some pathological conditions there was a net loss of nitrogen from the body, and a loss of tissue proteins. The excretion was greater than the intake, the condition of negative nitrogen balance.

The ability of foods or proteins to support nitrogen balance is still used today as one of the methods of assessing the adequacy of protein intake and the level of protein requirements. Three major problems arise in the interpretation of results:

(i) There is a considerable degree of individual variation in protein requirements, as with all other nutrients.

(ii) There is some degree of adaptation to different levels of protein intake – one of the effects of this is that there is not a sharp change from balance to the maximum rate of nitrogen loss on a deficient diet, but a period of several days over which there is increasingly more negative balance before the maximum rate of loss is reached.

(iii) Munro and Young [1] have pointed out that determination of the precise level of protein intake that will support balance, even allowing for adaptation to the new level of intake, is difficult, because the relationship between intake and nitrogen balance is not linear. At low levels of intake, up to near that which will support balance, there is a straight line relationship, but as the intake of protein is increased, so the efficiency with which it is used for tissue protein synthesis decreases, and there is an increasing degree of curvature of the graph.

5.1.1 Essential amino acids

The name protein was coined by the Dutch chemist Mulder early in the nineteenth century, for the residual substance from casein and other sources which he regarded as the 'essential' constituent of 'organized bodies'. Later, the name was applied to a variety of nitrogenous compounds, and it was the work of Magendie, Liebig and others through the middle of the nineteenth century that showed that 'protein' was not a single compound, but a mixture of many different proteins. Liebig demonstrated that all protein was not nutritionally equivalent when he fed his experimental animals on gelatin ('the most highly nitrogenized of the elements of the diet') and showed that they were no longer able to grow. Indeed, their loss of tissue protein was so great that they died.

At the beginning of this century, the amino acids that make up the proteins were being isolated and identified. Willcock and Hopkins [3] showed that the newly discovered amino acid tryptophan was a dietary essential, and that if it was not provided in the diet animals were unable to grow or maintain nitrogen balance. This was the first of the dietary essential amino acids to be so identified, in 1906. By 1914, Osborne and Mendel [4] had studied a great many of the amino acids, and much of what we now know as the list of dietary essential amino acids was complete. Methionine was not isolated until 1922, and threonine, the last of the essential amino acids, was not identified until 1935.

The amino acids that were dietary essentials were defined initially as those that could not be synthesized in the body, and were essential for the maintenance of nitrogen balance in adults and growth in young animals or human beings. By contrast, other amino acids could be synthesized in apparently adequate amounts, provided that a source of fixed nitrogen was available – i.e., provided that the total intake of amino acid nitrogen was adequate. These were known as the non-essential or dispensable amino acids; this name was not meant to mean that they were not required in the body, but only that there was no absolute dietary requirement for them; they could be formed from a variety of common metabolic intermediates.

Later the definition of essential amino acids was modified to include those that could be synthesized in the body to some extent, but not at a rate sufficient to permit full growth, and thus arginine and histidine were added to the list.

There is also evidence that dietary histidine may be essential for the maintenance of nitrogen balance in adults. Bergstrom and coworkers [5] showed in 1970 that the nitrogen balance of uraemic patients was improved by the addition of histidine to the essential amino acid solution that was administered intravenously. Kopple and Swendseid [6] later showed that histidine was essential for both uraemic and normal subjects. When patients were fed on a semi-synthetic diet deficient in histidine, they gradually went into negative nitrogen balance. In three of the seven subjects there was also a decrease in the concentration of serum albumin. Five subjects developed a fine scaly dry skin with erythema, and there was also an increase in serum iron, with a fall in haematocrit and haemoglobin concentration. When histidine was added to the amino acid mixture, these changes were reversed; the serum iron concentration fell rapidly, and there was an increase in haematocrit, with a period of positive nitrogen balance and an increase in serum albumin concentration. The authors noted that previous studies that had apparently demonstrated the non-essential nature of histidine for the adult had all been based on relatively short studies of nitrogen balance, whereas they had kept their subjects on a low protein diet (40 g day^{-1}) for a month before the initiation of the histidine-deficient diet, which was maintained for 35 days, so that tissue reserves of the amino acid were depleted.

Two of the amino acids occupy a somewhat ambiguous position between those that are dietary essentials and those that are not. These are cysteine and tyrosine, both of which can be synthesized in the body, as far as we know in adequate amounts under normal conditions, but only from essential amino acid precursors – methionine for cysteine and phenylalanine for tyrosine. This means that although neither amino acid is itself essential, their presence in the diet will 'spare' the parent essential amino acid. This is especially important in the case of methionine and cysteine since, as discussed below, the availability of methionine is frequently a limiting factor in the utilization of dietary proteins. Because of this, it is usual to express protein composition in terms of 'sulphur amino acids' (i.e., methionine plus cysteine). Phenylalanine is rarely a limiting factor in proteins, and therefore it is not usual to consider the tyrosine content of proteins as an important factor.

5.1.2 Dynamic equilibrium

By the middle of the nineteenth century it was known that even when eating a wholly protein-free diet, an adult human being excreted nitrogenous metabolites equivalent to about 35 g of protein a day. This was termed the endogenous protein loss, and is still the basis of our calculations of minimum protein requirements. It was also known that any intake of (exogenous) proteins led to an equivalent excretion of nitrogenous end-products. Thus, there arose a belief that there were two pools of protein metabolism: separate endogenous and exogenous pools, which did not mix.

It was not until the early 1940s that this was clarified, by the work of Schoenheimer and coworkers [7]. They used the then newly discovered heavy isotopes ^{15}N and ^{13}C to follow the metabolism of administered compounds, a technique that was then revolutionary and is now a standard part of biochemical research. If the then current concept of separate endogenous and exogenous pools of amino acid metabolism was correct, it would be expected that more or less all of an administered dose of labelled amino acid would be recovered in urine or exhaled air within 24 hours. In fact, they showed that this was not so, and only about one half to one third of the administered label was recovered in this time; the remainder was sequestered in the tissue proteins, and was released only slowly, over a prolonged period. Schoenheimer's studies led him to formulate the concept of dynamic equilibrium – a state in which there is no change in the gross composition of the body or tissues, yet there is a continual change of the individual molecules that make up those organs and tissues. Such a state could not be detected without the use of isotopically labelled tracers, yet it is interesting to note that Magendie, in 1829, had stated, presumably quite intuitively and without any experimental evidence, that 'all parts of the body of man undergo an intimate movement that serves both to expel those molecules that can or ought no longer to make up the body, and to replace them with new ones' [2]. This is still one of the most succinct definitions of what we now know as dynamic equilibrium.

Our current view of dynamic equilibrium is essentially the same as that formulated by Schoenheimer. However, there was a major problem in his interpretation of the experimental results. He assumed that once a protein has been synthesized (by what was then a wholly unknown process), it remains unchanged, and what is observed as an exchange of label during tracer experiments is the result of specific excision of individual amino acids from proteins, and their replacement. We now know that there is a continual breakdown and resynthesis of proteins, and that many key regulatory enzymes have extremely short half-lives. We also know that the total catabolism and synthesis of protein is very much in excess of the 35 g day^{-1} that might be predicted from the obligatory endogenous nitrogen loss.

This means that we are now able to define nitrogen balance not just in terms of a balance between the intake and excretion of nitrogenous compounds, but also in terms of a balance between the rates of protein synthesis and catabolism. Any factor that affects one of these processes without also affecting the other will lead to a change in nitrogen balance.

We know a great deal about the mechanisms of protein synthesis, and are beginning to learn how it is regulated and controlled by hormonal and other factors – any modern text-book of biochemistry covers this subject in detail. On the other hand, we are still largely ignorant of the details of the catabolism of tissue proteins. At least three proteolytic enzymes in the lysosomes are involved. Collectively, these are known as cathepsins: cathepsin A is carboxypeptidase (EC. 3.4.17.1); cathepsin C is dipeptidyl peptidase (EC. 3.4.14.1) and cathepsin D is endopeptidase D (EC. 3.4.23.5).

The regulation of cathepsin activity, and of their release from the lysosomes is little understood. Adrenalectomy leads to an increase in the free cathepsin activity in tissues, and it has been suggested that one of the functions of corticosteroid hormones is to maintain the stability of the lysosomal membrane. Bird and coworkers [8] showed that restricting food intake to 25% of normal led to an increase in total cathepsin activity in tissues, with no change in the fraction that was free, and suggested that this was because the stress of such severe food restriction led to an increase in corticosteroid secretion, and hence to enhanced stability of the lysosomal membranes. Obled and coworkers [9] have shown that cathepsin activity is generally independent of food intake in experimental animals, although it is maximal at the beginning of the light period in the rat, which is a nocturnally feeding animal; the light period would normally be the time at which the rat might need to catabolize labile protein stores as a source of energy-yielding metabolites, since it does not feed during the day.

The concept of a dynamic equilibrium between the rates of protein synthesis and catabolism becomes important when considering the actions of the glucocorticoid hormones (cortisol in man and corticosterone in the rat). At the level of whole body physiology the response to increased secretion, or administration, of these hormones is catabolism of the so-called labile muscle protein reserves, and an increase in gluconeogenesis in the liver. However, as discussed above, cortisol seems to have no effect, or even a contrary effect, on the activity of cathepsins. The principal metabolic response to glucocorticoids at the molecular level is induction of the synthesis of two liver enzymes, tryptophan oxygenase and tyrosine aminotransferase. The increase in activity of these two enzymes leads to a depletion of the liver pools of tyrosine and tryptophan, and in turn this leads to withdrawal of these two amino acids from the circulation. This means that the availability of tyrosine and tryptophan becomes a limiting factor for protein synthesis, which is therefore reduced. However, the rate of protein catabolism continues unchecked, so that the net result is an excess of breakdown over synthesis and hence overall catabolism of tissue proteins in response to corticosteroid action. The amino acids arising from this catabolism cannot be used for new protein synthesis, because of the withdrawal of tyrosine and tryptophan, and they are therefore metabolized in the same way as other amino acids that are 'surplus' to requirements; i.e., they are deaminated and used for energy-yielding metabolism or glucose synthesis as appropriate. Thus, the induction of two enzymes of amino acid catabolism can lead to the switch from the fed to the fasting state, and an increase in the rate of gluconeogenesis, with no direct effect on the enzymes of glucose synthesis.

Nakamura and coworkers [10] have investigated the hormonal induction of tryptophan oxygenase in isolated cultured rat hepatocytes. Normally, these cells show a considerable loss of tryptophan oxygenase activity during culture, but this can be prevented by the addition of tryptophan, dexamethasone (a synthetic glucocorticoid analogue) or glucagon to the culture medium. The glucagon effect can also be produced by dibutyryl cyclic AMP, and is antagonized by insulin, indicating that it may be a normal physiological effect of the hormone. The effects of tryptophan, dexamethasone and glucagon are all additive, suggesting that they act by different mechanisms. Dexamethasone induction is inhibited by actinomycin D, and is therefore presumably an effect at the level of transcription of DNA to mRNA; the effects of both glucagon and dexamethasone are inhibited by cycloheximide, which prevents the translation of mRNA to protein, suggesting that glucagon acts at this level, while the effects of tryptophan are unaffected by any inhibitors of protein synthesis. This last is not surprising, since it is known that tryptophan acts to stabilize the oxygenase against proteolysis [11, 12].

5.1.3 Protein quality

Table 5.1 shows the amounts of the essential amino acids that are required per kg of dietary protein, for complete utilization of that protein for the synthesis of tissue proteins. This is frequently called the reference pattern of amino acids, and is a standard against which the quality of dietary proteins can be assessed. The reference pattern or ideal mixture of amino acids was determined experimentally by varying the proportions fed, and using nitrogen balance as the criterion of adequacy [13].

The essential amino acid that is present in the dietary protein in lowest amount relative to its requirement is called the limiting amino acid of that protein. It is the content of this amino acid that limits the extent to which the protein can be used for tissue protein synthesis. If a protein contains only 70% of the requirement for (methionine plus cysteine) then it will only be 70% useable, even if the other amino acids are present in excess. The remainder will be 'wasted' by being used as an energy source.

This means that knowing the essential amino acid composition of a protein it is possible to predict its nutritional value and its ability to support nitrogen balance or growth. The ratio of the proportion of the limiting amino acid that is present in a protein to the amount required is termed the Protein Score of that protein. Before the reference pattern was established, it was usual to use the amino acid composition of egg protein as a standard, since it is 100% used when fed to animals or man at low levels. The ratio of the amount of limiting amino acid present in a test protein to that in egg protein is the Chemical Score of the protein. For practical purposes, Chemical Score and Protein Score are quantitatively similar.

Oser [14] has suggested that it may be more useful to consider not just the first limiting amino acid of a protein, but the content of all the essential amino acids. He has defined an Essential Amino Acid Index, which is the geometric mean of

Table 5.1 The reference pattern of essential amino acids for man [13]

Essential amino acid	*Requirements, g amino acid per kg dietary protein*
Leucine	48
Lysine	42
Isoleucine	42
Valine	42
Phenylalanine	28
(Tyrosine	28)
Threonine	28
Methionine	22
(Cysteine	22)
Tryptophan	14

the ratio of the amount of each essential amino acid that is present to that which is required.

Neither Protein Score nor the Essential Amino Acid Index is a really satisfactory predictor of the nutritional value of dietary proteins, because both are based on chemical analysis of protein, and therefore ignore the possibility that one or more of the essential amino acids may be present in a form that is not liberated during digestion, and is therefore biologically unavailable. This is commonly a problem with lysine, which is the limiting amino acid of cereal proteins (in general meat and legume proteins are limited by methionine plus cysteine). Lysine can form links from its ε-amino group to a variety of compounds, and although many of these bonds are hydrolysed by the acid or alkali used in chemical analysis of protein, they are not hydrolysed by mammalian digestive enzymes. This means that Protein Score will overestimate the quality of the protein. Two methods are commonly used to attempt to determine chemically the extent to which ε-amino groups of lysine are free, and therefore to predict the biological availability of lysine:

(i) Fluorodinitrobenzene (FDNB) will react with free amino groups, and after hydrolysis the proportion of lysine that has been fluorodinitrophenylated can be assessed. This is a tedious procedure, and unreliable in the presence of large amounts of starch, as is the case when cereals are being analysed. Furthermore, FDNB is a very toxic reagent, and contamination of the skin can lead to the development of auto-immune disease and skin ulceration.

(ii) Acidic dyes will bind to free amino groups, and therefore the binding of dye to the protein will be a useful estimate of the extent to which lysine ε-amino groups are free. The dyes will also bind to arginine and histidine residues, and therefore this must be taken into account, for example by blocking lysine ε-amino groups with propionic anhydride and measuring the residual dye binding. The reaction is complete within a few minutes with some foods, but with others can take much longer.

Both of these methods for determining the extent to which lysine ε-amino groups are free suffer from the problem of access of the reagent to lysine residues that are buried deep within the native structure of the protein. If relatively severe methods are used to denature the protein, and so reveal all the lysine groups for reaction, it is possible that some of the bound amino groups will be liberated, thus leading to falsely high results [15].

Because chemical analysis, even when refined to include some determination of the availability of lysine and other amino acids, is not wholly satisfactory, biological assays of protein nutritional value are still essential. They are based on principles developed by the pioneering workers at the beginning of this century, and measure the ability of the test protein, fed under suitably limiting conditions, to maintain nitrogen balance or promote growth in experimental animals or man. They depend on the content of the limiting amino acid, and provide no further information.

The earliest biological measure of the nutritional value of proteins is Biological Value (BV), which is the ratio of nitrogen retained in the body to that absorbed. If digestibility is taken into account, i.e. the measure is of the proportion of dietary protein that is retained, then the value is termed Net Protein Utilization (NPU). The NPU is the BV of the protein multiplied by the digestibility. The third biological measure of protein quality is the Protein Efficiency Ratio (PER), which is the gain in weight per gram of dietary protein. This is obviously susceptible to distortion if there is any significant gain in weight that is not protein, for example adipose tissue or even oedema, and varies with the amount of food eaten.

All of these biological measures of protein quality depend on feeding the test protein at a sub-optimal level – in other words, protein must be (marginally) a limiting factor in the diet, so that the quality of the protein is important. Provision of protein in excess of requirements would mask any short-comings in the quality of that protein; this is what happens in practice in most human diets.

5.1.4 Protein–energy nutrition

In many ways, the determination of protein nutritional value, by whatever means selected, is an academic rather than a practical exercise. For most of the hungry of the world it is total food that is lacking rather than protein specifically, so that limitation of the nutritional value of that protein may be relatively unimportant. The body's first call is for energy-yielding metabolites, and therefore until energy requirements have been met, protein will be catabolized for energy-yielding metabolism, and not used for protein synthesis. Even for children in many cases, it is food energy rather than protein that is lacking.

Iyengar and coworkers [16–18] have studied the inter-relationships of protein and energy intake and requirements in both adults and children in India. They showed that for adults it was possible to achieve positive nitrogen balance with an intake of only 1.0 g protein per kg body weight day^{-1} if the daily energy intake was 232 kJ per kg body weight. However, reducing the energy intake to 186 kJ per kg body weight day^{-1} (about the degree of energy deficit that they say is common among the disadvantaged of the Indian population) led to an increase in the amount of protein required to maintain nitrogen balance to 1.2 g per kg body weight day^{-1} [16, 17]. In 1981 [18] they reported similar studies with schoolchildren all of whom were receiving the previously determined 'safe' level of protein intake (1.75 g per kg body weight day^{-1}). In order to maintain a nitrogen retention of 40 mg N per kg body weight day^{-1}, it was necessary to provide 326 kJ per kg body weight day^{-1}. Below this energy intake, the 'safe' level of protein intake was inadequate to permit even this minimal degree of nitrogen retention.

In a similar study, Garza and coworkers [19] fed six healthy medical students on diets providing the calculated safe level of protein, and an energy intake adequate for their calculated requirements. In five of their subjects this resulted in negative nitrogen balance, which was corrected by an increase in the non-protein energy intake. The mean observed change in nitrogen balance was 0.335 mg nitrogen per additional kJ consumed. At this level of energy intake, all the subjects gained weight, indicating that the energy intake was in fact in excess of their requirements for energy balance.

Reeds and coworkers [20] came to a similar conclusion from animal studies. They measured nitrogen balance in pigs maintained on minimally adequate diets, and then added either protein or non-protein energy supplements. They showed that both supplements led to an increase in nitrogen retention. Their results suggested that for each gram of additional digestible nitrogen fed to the pigs there was an increase in nitrogen retention of 0.88 g, and for each MJ of metabolizable non-protein energy provided there was an increase of 0.93 g in nitrogen retention. They also studied the metabolism of [^{14}C]-leucine during constant infusion, and showed that its oxidative metabolism, as assessed by measurement of exhaled $^{14}CO_2$, was increased by the provision of additional protein in the diet (in other words there was now an excess of protein over requirements) and was reduced by the provision of additional non-protein energy, reflecting the increased synthesis of proteins, and the reduced use of protein for energy-yielding metabolism.

Until the FAO/WHO report of 1973 [13] it was usual to consider requirements for protein and energy separately. The two can be combined by expressing the protein quality of a diet not just as the quality of that protein (generally expressed as NPU), but also as the proportion of the dietary energy that is being derived from proteins. The Net Dietary Protein : Energy Ratio (NDpE) is the NPU of the dietary protein multiplied by the energy derived from protein as a proportion of the total energy intake of the diet. Consideration of NDpE will permit a decision to be made as to whether protein (quantity and quality) is really the limiting factor, or whether the provision of additional energy sources will be beneficial.

In another sense the measurement of the nutritional value of proteins is an academic rather than a practical exercise. No significant groups of people live on a single source of protein; all eat a mixture of proteins from different sources. This means that the valuable measurements will be of the protein quality of whole diets rather than of individual foods. Such measurements as have been made show that the range between the well-fed people of the developed countries, with a considerable intake of animal proteins, and the less well-fed people of developing countries, is very small: a range of Biological Value between a low of 0.7 (very rarely as low as 0.6) and a high of 0.8.

The explanation of this has been known for many years – a judicious mixture of proteins will permit complementation between the relative surpluses of one and the deficits of another. Thus a mixture of maize protein (limited by lysine and therefore with a relative excess of methionine) and pea flour (limited by methionine and therefore with a relative excess of lysine) has a BV of 0.7, compared with values of 0.4 and 0.35 respectively for the individual proteins.

Woodham and coworkers [21, 22] have studied the practical nutritional value of a number of mixtures of different protein supplements in the diets of chickens and rats, and have compared the predicted value of the mixtures (by both

Chemical Score and Essential Amino Acid Index) with the biological usefulness determined by a chick growth assay (total protein efficiency) and NPU and PER in rats. Neither Chemical Score nor Essential Amino Acid Index could be used to predict the results in chickens, where the biological assay was a 'practical' one rather than a true determination of protein quality, since they fed the test material at above the minimal level. They showed that a number of protein mixtures with apparently unfavourable amino acid compositions scored well *in vivo*. In their studies with rats [22], they were able to predict the rank order of their protein mixtures for either PER or NPU (both of which depend on feeding the test material at a sub-optimal level) by the Chemical Score. However, they showed no agreement with the Essential Amino Acid Index. For some mixtures there was a significant discrepancy between the Chemical Score and the results obtained *in vivo*. This may have been the result of interactions between various components of the protein supplements, leading to lower availability of lysine or other amino acids than in the individual components, or it may have reflected some degree of amino acid imbalance, as discussed below.

It is obvious from the discussion above that the various measures of protein quality serve more to characterize a given protein than to define its 'quality', and moreover, each assay serves mainly to define the value of that protein for a specific purpose – the maintenance of tissues, recovery during convalescence, growth in infants, or the production of meat, milk, eggs or wool in farm animals.

As was noted above, biological assays of protein quality depend solely on the limiting amino acid, and provide no information about the other essential amino acids present. What is required is chemical analysis to provide information on the total amount of each amino acid present, with bio-assay to reveal how much of the limiting amino acid is available.

5.1.5 Amino acid imbalance and antagonism

Hier and coworkers noted in 1944 [23] that the addition of an excess of one essential amino acid would reduce the Biological Value of a protein or mixture of amino acids. There have been many reports of similar amino acid imbalance since then; Harper and coworkers [24] have reviewed the field, and have noted that in general there must be a moderately large excess of the offending amino acid, and that the total intake of protein must be only marginally adequate. Under these conditions, there is a depression of growth and a failure to maintain nitrogen balance, although chemical analysis of the protein indicates that it is adequate to support nitrogen balance. The severity of the effect varies with the nature and amount of the amino acid added in excess, and with the degree of protein deficiency. No adequate explanation has been advanced to account for this phenomenon.

In the special case of hyperphenylalaninaemia, as occurs in phenylketonuria, Hughes and Johnson [25] have suggested that the observed inhibition of protein synthesis may result from competition between phenylalanine and methionine for methionine-tRNA. The result of this would be a reduced capacity for the initiation of protein synthesis, because of lower availability of methionyl-tRNA. However, this is an extreme case of imbalance, with blood concentrations of phenylalanine some 10-fold higher than normal, so that it may represent phenylalanine intoxication rather than a simple amino acid imbalance.

Recent studies by Eggum and coworkers [26] have used a less extreme model than was used by Harper and other earlier workers to demonstrate the phenomenon of amino acid imbalance. They fed rats on egg protein at a level that was just adequate to support optimum growth, and showed that the addition of any one of the essential amino acids at a level that only doubled the amount that was initially present led to a marked reduction in the Biological Value of the protein. When they fed more than a minimally adequate amount of protein, the addition of essential amino acids had little or no effect. This suggests that for most human populations, eating more than the minimally adequate amount of protein, any imbalances of essential amino acids that may occasionally be found in dietary proteins are unlikely to be nutritionally important. The potentially important case of leucine excess is discussed below.

Harper and coworkers [24] drew a clear distinction between the apparently general phenomenon of amino acid imbalance, which can be relieved by increasing the dietary intake of the essential amino acid that was initially limiting in the test protein, and the more specific phenomenon of amino acid antagonism. Here the addition of one amino acid leads to a failure of nitrogen retention that does not respond to the addition of the limiting amino acid, but specifically to addition of the antagonistic amino acid. Two examples of such specific antagonism are known: lysine and arginine are mutually antagonistic, and the addition of an excess of one (generally arginine) will reduce the Biological Value of proteins that are not limited by lysine, yet only the addition of lysine will overcome the effect. The other antagonism occurs between the branched chain amino acids, leucine, isoleucine and valine; an excess of leucine will impair the utilization of a protein, and addition of isoleucine or valine will overcome the effect. Bender [27] showed that halving the amount of isoleucine in a complete amino acid mixture reduced the NPU from 100 to 34. When the amount of leucine was also halved, the NPU increased from 34 to 56.

No explanation can be offered for the antagonism between arginine and lysine; apart from a net basic charge at physiological pH they have little in common. For the branched chain amino acids, although again no real explanation of the phenomenon can be advanced, it is possibly important that all three share the same enzymes for the first three steps of their oxidative metabolism, so that an excess of one may disturb the catabolism of the others.

5.1.6 Measurement of the rates of protein synthesis and catabolism

Studies of nitrogen balance give an indication of the overall magnitude and direction of changes in the state of the body's protein reserves, but they only measure the results of continual synthesis and catabolism of proteins; the work of Schoenheimer and co-workers [24] cited above, showed that even in nitrogen equilibrium there is still a considerable amount of protein turnover. From the very different values of half-life of different proteins, it is obvious that the overall rate of turnover that is observed in the body is the result of a great many different rates of synthesis and catabolism in different tissues, in different cells in the same tissue, and indeed within the same cell. However, Waterlow and Stephen [28] have suggested that measurement of the overall rate of protein synthesis and catabolism in the body may give us a great deal of useful information about the mechanisms involved in disease processes and undernutrition, in the same way as measurement of basal metabolic rate, which itself is the mean of a great many different rates of respiration, has been useful in developing our understanding of physiology.

Dunlop and coworkers [29–32] have measured the rates of protein synthesis and catabolism in the brain, both using brain slices *in vitro*, and *in vivo* following the administration of radioactive amino acids as tracers. They have shown that the rates observed in tissue slices may not be a true reflection of the rates that occur *in vivo*.

Studies in experimental animals have the advantage that it is possible to administer a radioactive amino acid at relatively high specific activity, and in relatively large amounts, so that greater sensitivity is possible. It is also possible to obtain samples of tissues to follow the accumulation of labelled protein, a procedure that is not generally possible in human studies, although there have been studies of the rates of accumulation of labelled serum albumin (which is readily accessible) and even of muscle proteins, by means of serial biopsy sampling [33, 34].

In studies of protein synthesis in the brain it is possible to inject the tracer amino acid directly into the third ventricle and thus minimize the effects of extra-cerebral metabolism. Antonas and coworkers [35] have shown that under some conditions, and with some amino acids, the apparent rates of incorporation into proteins are different when the amino acid is given in this way rather than by intraperitoneal or intravenous injection. While this may reflect the metabolism of the tracer outside the brain, it is also possible that it reflects competition for uptake into the brain by other amino acids in the blood-stream. This may be an especially important factor in hyperphenylalaninaemia [25] and other conditions where there is an elevated concentration of one amino acid in the blood-stream.

Dunlop and coworkers [29, 30] showed that in the adult mouse there appear to be two major pools of protein metabolism: the larger pool has a relatively long half-life (about 9.7 days), and the smaller component (about 3.5% of the total) has a half-life of 7 h. It is likely that this rapidly changing pool of protein is concerned with metabolic regulation and the synthesis of neurotransmitters. By contrast, in young mice, where the rate of protein synthesis is very much greater than in adults, there appears to be only a single pool, with a half-life of 2 days. The fractional rate of protein synthesis in the brain of a young mouse may be as high as $2\% h^{-1}$ compared with $0.6\% h^{-1}$ in the adult. The net accumulation of protein in the brain of the developing mouse is only $0.7\% h^{-1}$ at its greatest, and it has been suggested that the rapid turnover that is occurring reflects the considerable degree of plasticity that is known to exist in the developing brain. To a very great extent the pattern of inter-neuronal connections in the central nervous system is not predetermined, but is shaped by environmental and other stimuli during development. If there is to be such plasticity, there must obviously be a considerable degree of formation, breakdown and reformation of enzymes and structural proteins.

Dunlop and coworkers [31, 32] also studied the catabolism of brain proteins, and showed that the actual proteolytic activity of the developing brain was greater than that of the adult, although there was a greater potential proteolytic capacity in the adult brain. They showed that in response to food deprivation there was a reduction in the net protein accumulation in both adult and immature mouse brains, a factor that may be important in considering the long-term *sequelae* of under-nutrition in infancy. In the adult, the reduced protein accumulation was the result of increased catabolism, while in the immature mouse there was a 33% reduction in protein synthesis and an 18% reduction in protein catabolism, i.e., a reduction in the rate of turnover of proteins, as well as a reduction in net synthesis.

(a) The choice of labelled amino acid for use as a tracer

In animal studies, it is possible to use [^{14}C]- and [^{3}H]-labelled amino acids at high specific activity, so that the administered tracer does not cause any significant alteration in the size of the tissue amino acid pools.

For human studies, it is not possible to use large amounts of radioactive material, for obvious reasons, and it is more common to use the stable isotope ^{13}C. When studies involve the amino group of the administered amino acid, there is no suitable radioactive tracer anyway, and for both human and animal studies it is necessary to use ^{15}N.

There are two different approaches to the choice of the amino acid to be used as tracer in studies of protein synthesis:

(i) An amino acid such as leucine or methionine may be used, since it is involved in relatively few metabolic pathways, and the label that is incorporated into protein will be almost entirely in the same chemical form as that in which it was administered. In such cases the measurement of the excretion of ^{15}N in urea or ammonia will indicate more or less specifically the oxidative metabolism of the amino acid that was given. The same is true when $^{13}CO_2$ or $^{14}CO_2$ in expired air is measured.

McFarlane and coworkers [34] used *guanido*-[^{14}C]-arginine as tracer for their studies of albumin synthesis; this permitted them to measure the specific activity of the tissue free arginine pool (and hence the specific activity of the precursor pool used for protein synthesis) and also that of the arginine incorporated into albumin (and hence the synthesis of new albumin), as well as the specific activity of the urea excreted. This last permitted an estimate to be made of the oxidative metabolism of arginine, since arginine is catabolized almost entirely by way of arginase action to release the guanido group as urea.

(ii) The second approach to tracer experiments is to use an amino acid that will lead to labelling of the whole amino nitrogen pool more or less uniformly, or at least that part of the pool that is represented by the non-essential amino acids. The incorporation of ^{15}N into proteins, or its excretion in urea or ammonia, can then be measured, as an index of either protein synthesis or the oxidative metabolism of amino acids in general. Here it is assumed that the fate of the label is the same as the fate of the amino nitrogen pool as a whole. For such studies it is most usual to use [^{15}N]-glycine, because of its ready availability and low cost, although other [^{15}N]-amino acids have been used.

Waterlow [36] has discussed some of the assumptions that are made in the use of tracers for the assessment of protein synthesis; they apply whether the labelling of proteins or end products of nitrogen metabolism is to be measured, and whether the tracer is given as a single dose or over a prolonged period so as to achieve a constant degree of labelling of the tissue pools (the constant infusion technique, see below).

It is assumed that the ratio tracer to (amino acid pool leading to end-products) is the same as that of tracer to (amino acid pool involved in protein synthesis). This is certainly an incorrect assumption. The tissue free amino acid pool has a considerable excess of glutamine and some of the other non-essential amino acids, as well as methionine and lysine, compared with the mixture being incorporated into proteins.

The selection of amino acids that contribute to the formation of end-products (urea and ammonia) is not the same as either the total amino acid pool or that involved in protein synthesis. More than 90% of the nitrogen excreted as ammonia comes from glutamine, and therefore ultimately from those amino acids that readily undergo de-amination – glutamate, glycine, serine and, to a lesser extent, threonine. In urea the two nitrogen atoms come from different sources. One is derived from ammonia, and therefore presumably reflects largely the de-amination pool, while the other comes from aspartate, and will therefore reflect the metabolism of those amino acids that are preferentially metabolized by transamination.

This means that the amino acids contributing to protein, ammonia and urea will be different, and possibly synthesized in different tissues of the body. It is obvious that the real situation does not fit the simple model that is used to interpret the data, which assumes that there is a single pool of amino acids in equilibrium with a single pool of protein, giving rise to a single pool of end-products. Nevertheless, Golden and Waterlow [37] obtained essentially similar results when they assessed protein synthesis by the incorporation of [^{14}C]-glycine (when the label that was incorporated into proteins was largely in glycine residues) and [^{15}N]-glycine (when the label was present in a number of amino acids, because of the ready interchangeability of the amino group of glycine through the body amino nitrogen pools). Other amino acids may give very misleading results – Taruvinga and coworkers [38] found that [^{15}N]-aspartate led to considerable labelling of urea, as might be expected, and therefore a low apparent rate of protein synthesis. By contrast, [^{15}N]-leucine contributed little label to urea and gave a high value for the apparent rate of protein synthesis.

A further problem arises with the use of leucine as tracer if the exhalation of $^{14}CO_2$ is used as an index of oxidative metabolism, because the metabolism of leucine responds very rapidly to changes in food intake [39]. Waterlow [36] cites unpublished studies in which leucine was infused intragastrically, at a constant rate, together with a constant rate of food infusion. When the infusion of food was halted, but that of leucine continued as before, there was a change in the exhalation of $^{14}CO_2$ within minutes. He suggests that leucine metabolism is only a useful measure if the intake of energy-yielding substrates is maintained at a constant level throughout the study.

It was noted above that the precursor pools from which the nitrogen atoms of urea and ammonia are derived are not the same, and therefore it is not surprising that values of the apparent rate of protein synthesis derived from measurement of the labelling of urea and ammonia differ. The mean of the apparent rates derived from measurement of these two end-products (the end-product average) gives a result that agrees well with that from studies using other techniques [40].

Fern and coworkers [40] showed that the route of administration of the labelled precursor affected the results obtained. When [^{15}N]-glycine was given orally, the rate of protein synthesis calculated on the basis of the labelling of ammonia was higher, and that derived from the labelling of urea was lower, than when the tracer was given intravenously. However, when the end-product average rate of protein synthesis was calculated, the route of administration of the tracer did not affect the results.

(b) Constant infusion techniques

Waterlow and coworkers [28, 41] have pioneered an elegant technique to permit the determination of the rate of whole body protein synthesis *in vivo*, based on a constant infusion of labelled amino acid. Theoretically, if there is a constant rate of input of labelled amino acid into a homogeneous metabolic pool that exchanges freely with a homogeneous pool of protein, the specific activity of the free amino acid will rise hyperbolically until a plateau is reached. The re-entry of labelled amino acids into the free pool from the catabolism of newly synthesized proteins is assumed to be negligible during the period of the experiment. The rate of increase of the specific activity of the free amino acid pool is determined by the rate constants of amino acid disappearance, and the height of the maximum (plateau) is determined by the degree of dilution of the infused material with unlabelled amino acid liberated from proteins by catabolism, and derived from food.

If the metabolic pools are not in fact homogeneous, the specific activity curve for any one compartment (for example, the plasma) will be complex, and it will not be possible to assign a single value for the rate constant. Nevertheless, the curve will still tend towards an identifiable plateau, and it will still be easier to fit an appropriate curve through the experimental points than is the case when only a single pulse dose of the tracer amino acid is given, when the kinetic analysis is very complex.

Knowing the rate of infusion of the labelled amino acid, and measuring the specific activity of the free amino acid pool, it is possible to calculate the rate of apparent irreversible disappearance of the amino acid from the free pool – the total amino acid flux. In the simplest model, this flux has two components, metabolism to end-products, which can be measured, and the synthesis of proteins. Hence it is possible to estimate the fraction rate of protein synthesis [42–44].

Various workers have used different labelled amino acid tracers (both [^{14}C]- and [^{15}N]-labelled) both in animals and in human beings. Different amino acids yield slightly different results, depending largely on how many true metabolic pools there are, and how many metabolic components of the flux have to be taken into account, as compared with the single pool and single metabolic flux that are assumed in the model. Furthermore, as discussed above, the pools of amino acids that are used for protein synthesis do not have the same composition as the general pool, so again discrepancies are likely to occur. In other words, although the model used to analyse the data assumes homogeneous pools of both protein and precursor amino acids, this situation is not encountered in practice.

Most workers have used essential amino acids as tracers in constant infusion studies; if a non-essential amino acid is used there is an additional complication in that synthesis of the amino acid will contribute to the dilution of the lavel, which, in the simplest model, is assumed to result only from the entry of amino acids resulting from protein catabolism and food intake [45].

Constant infusion is obviously a technique that is capable of yielding highly precise results. However, it is rather tedious to perform, and calls for considerable co-operation from the experimental subjects during the prolonged period of infusion (often in excess of 30 hours of intra-gastric or intravenous infusion of the labelled precursor). Gersowitz and coworkers [46] have described an interesting variation; instead of a truly continuous infusion, they used repeated oral administration of [^{15}N]-glycine at 3 h intervals for 60 h. They then measured the isotopic enrichment of urinary urea, as an index of the specific activity of the liver pool of arginine. From this and the degree of isotopic enrichment of arginine in albumin, they were able to calculate the rate of synthesis of albumin. They reported a rate of between 3–4% per day, which agrees well with rates reported by other workers using different methods. Their method was adequately sensitive to detect a reduction in the rate of albumin synthesis when their subjects were fed on a low protein diet for several days.

Young and coworkers [47] have measured the rates of total protein synthesis in healthy people of various ages; the rates show a progressive decline with age: premature and new-born infants synthesize 17.4 g of protein per kg body weight day^{-1}, older infants 6.9, young adults 3.0 and elderly subjects only 1.9 g per kg body weight day^{-1}.

Young and Bier [45] have reviewed a series of studies in which volunteers were fed on either the minimum amount of essential amino acids to meet their requirements (as determined by nitrogen balance studies) or a 'generous allowance'. The rate of protein synthesis was calculated by a constant infusion technique. They concluded that the rate of protein synthesis was greater when the allowance of essential amino acids was above the minimum requirement level, and suggested that the currently accepted minimum levels of intake are too low, and should be higher. This conclusion may, however, be incorrect, since it is based on an observed increase in the rate of protein synthesis, and therefore an increase in the rate of protein turnover, and we have no information on what are appropriate rates of turnover. On the other hand, we do know that the dietary requirement derived from nitrogen balance is the minimum that is required to maintain nitrogen equilibrium (provided that energy intake is adequate) and that at these levels of intake a relatively small excess of one of the essential amino acids will lead to an imbalance, and hence to an impairment of nitrogen retention [26]. To avoid the problems of imbalance, it may be desirable to increase the 'safe level' of protein intake above the minimum derived from nitrogen balance studies; this is already achieved in most communities.

(c) Urinary excretion of 3-methylhistidine

For studies in the field and in out-patients, it would be extremely useful to have an indicator of protein turnover that could readily be measured on a large number of people, preferably by a wholly non-invasive method – obviously, a urinary metabolite that was unique to skeletal muscle metabolism would be ideal for such a purpose.

In 1967, it was shown that 3-methylhistidine was a significant constituent amino acid of the two major contractile proteins of muscle, actin and myosin [48,49], and that it was formed by post-synthetic modification of the polypeptide chains. Young and coworkers [50] showed that there was no incorporation of radioactive 3-methylhistidine into proteins, and no amino acid charging of tRNA; there is no codon for 3-methylhistidine, and no specific tRNA has been identified. This means that any 3-methylhistidine that is released by the catabolism of tissue proteins will not be used for new protein synthesis, but will either be metabolized or excreted unchanged. In studies with rats, it was shown that there was quantitative recovery of labelled 3-methylhistidine in urine and faeces. In young animals this was largely as unchanged amino acid, while in older animals there was an increasing proportion of *N*-acetyl-3-methylhistidine [50].

Elia and coworkers [51] showed that the greater part of the total body content of 3-methylhistidine was in skeletal muscle – some 3.3 μmol per kg protein, compared with 2–3 $\mu mol\,kg^{-1}$ in smooth and cardiac muscle, which constitute a very much lower proportion of the total body protein than does skeletal muscle, and less than 1 $\mu mol\,kg^{-1}$ in tissues such as liver and kidney, where it is presumably in the contractile proteins of blood capillaries.

It thus appears likely that measurement of the urinary excretion of 3-methylhistidine will largely reflect the turnover of skeletal muscle protein. A number of experimental studies support this view:

(i) During a 20-day fast by obese subjects there was a progressive decline in the excretion of 3-methylhistidine, and a parallel decline in total nitrogen excretion, suggesting, in agreement with other studies, that there is progressive adaptation during prolonged starvation, to reduce the catabolism of muscle proteins as ketone concentrations increase [52].

(ii) Increasing dietary protein intake led to an increase in the excretion of 3-methylhistidine, suggesting an increase in the rate of protein turnover with increasing intake [53].

(iii) Growing rats fed on an adequate intake of protein showed a progressive increase in the excretion of 3-methylhistidine at the same time as there was an increase in body weight, and therefore muscle protein mass. Conversely, protein deprivation led to a progressive decline in 3-methylhistidine excretion together with the loss of muscle tissue [54].

These and other studies suggest that measurement of urinary 3-methylhistidine will give a useful indication of the state of muscle protein metabolism. However, at least in some species, this is not so. Harris and Milne [55–58] have shown that in cattle the excretion of radioactively labelled 3-methylhistidine is quantitative and more or less immediate, and suggest that, as in the studies cited above, the excretion of this metabolite may indeed reflect muscle protein metabolism in cattle. However, in sheep and pigs they found that the recovery of administered 3-methylhistidine was very poor, and there was a considerable accumulation of label in muscle, much of it as dipeptides, including balenine (β-alanyl-3-methylhistidine). Under these conditions, it is most unlikely that the measurement of urinary 3-methylhistidine would yield any useful information; the 'buffer' pool in muscle would be large enough to mask any changes due to changes in protein turnover.

Recent studies by Bates and coworkers [59,60] have suggested that in man also the excretion of 3-methylhistidine may not be as accurate a reflection of muscle protein metabolism as had previously been thought. Although the majority of the total body content of this amino acid is in skeletal muscle [48, 51], they have argued that the relative contributions of various tissues to the urinary excretion reflect not only the proportion of body content, but also the rate of turnover in different tissues. Using [^{14}C]-*S*-adenosylmethionine to label newly synthesized

3-methylhistidine in rats, they have shown that the rate of turnover in skeletal muscle is 1% day^{-1}, whereas in skin it is 3.2% day^{-1}, and in intestinal mucosa it is 9% day^{-1}. They have suggested that less than 40% of urinary 3-methylhistidine may originate in skeletal muscle. This casts serious doubt on the validity of 3-methylhistidine excretion as an index of muscle protein metabolism.

5.2 THE SYNTHESIS OF UREA

The pathway for the synthesis of urea, shown in Fig. 5.1, was elucidated by Krebs and Henseleit in 1932. It was the first cyclic metabolic pathway to be described, and the history of the discovery and the logical development of the concept, have been described by Krebs in an essay which forms the first chapter in a published symposium on the cycle [61].

The regulation of the urea synthesis cycle *in vivo* has been discussed by Raijman [62]; there is normally a considerable excess activity of all the enzymes of the pathway, so that there should be no limitation of the capacity to metabolize all the nitrogenous waste from protein intake and catabolism. Problems in patients with inborn errors of the enzymes of the cycle are discussed below. Although the apparent capacity of the cycle is in excess of the requirement, Schimke [63–66] has shown that the activities of all the enzymes increase markedly with increases in protein intake. The variation in arginase activity is relatively small, but the activities of the other enzymes, and also of glutamate–aspartate and glutamate–alanine aminotransferases, which are concerned with the input of nitrogen into the cycle, vary over a 10–20-fold range with variations in protein intake.

The immediate source of ammonia for the synthesis of carbamoyl phosphate remains unclear. Much certainly comes from the de-amidation of glutamine, catalysed by glutaminase, and hence ultimately from the de-amination of amino acids, as discussed above. There is also evidence that much arises from the α-amino group of glutamate, by the reaction of glutamate dehydrogenase. McGiven and Chappell [67] have shown that the rate of reaction of glutamate dehydrogenase in isolated mitochondria is not adequate to meet the observed rate of urea synthesis, partly as a result of the slow rate of entry of glutamate into the mitochondrion. However, Krebs and Lund [68] pointed out the limitations involved in studies of isolated mitochondria.

The equilibrium of glutamate dehydrogenase is such that the reaction will normally favour the formation of glutamate from 2-oxo-glutarate and ammonia rather than the de-amination of glutamate, and this will result in a low rate of ammonia formation in isolated mitochondria. *In vivo*, they say that the products of glutamate dehydrogenation will be removed more rapidly, since 2-oxo-glutarate will be a substrate for either oxidation or transamination back to glutamate; this will shift the equilibrium towards the de-amination of glutamate, so that the rate *in vivo* may be very much greater than is observed with isolated mitochondria.

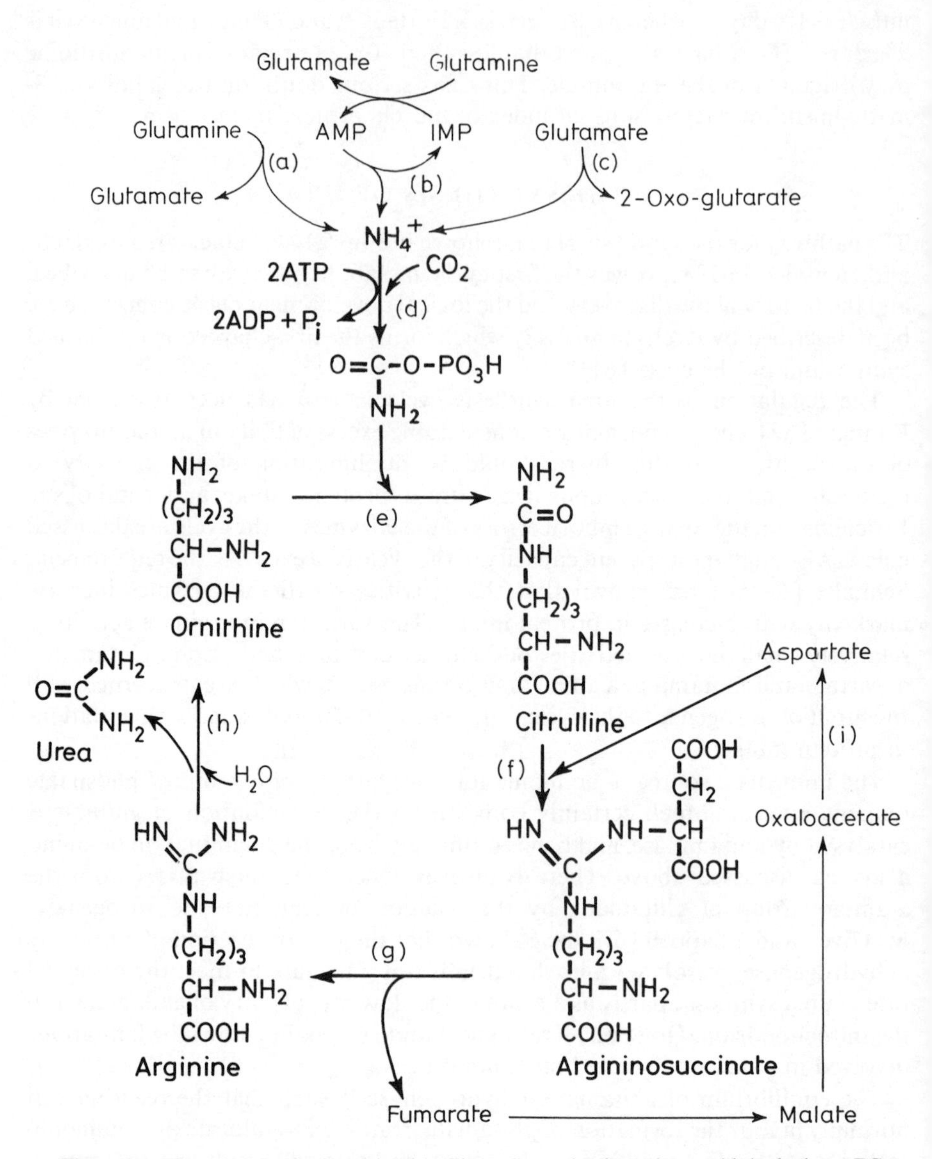

Figure 5.1 The urea synthesis cycle. (a) Glutaminase (L-glutamine amidohydrolase, EC 3.5.1.2); (b) AMP deaminase (AMP amino-hydrolase, EC 3.5.4.6); (c) glutamate dehydrogenase (L-glutamate:NAD oxidoreductase (deaminating), EC 1.4.1.2); (d) carbamoyl phosphate synthetase, EC 6.3.4.16; (e) ornithine carbamoyltransferase (carbamoyl phosphate: L-ornithine carbamoyltransferase, EC 2.1.3.3.); (f) argininosuccinate synthetase (L-citrulline:L-aspartate ligase (AMP forming), EC 6.3.4.5); (g) argininosuccinase (L-argininosuccinate arginine-lyase, EC 4.3.2.1); (h) arginase (L-arginine amidinohydrolase, EC 3.5.3.1); (i) aspartate aminotransferase (L-aspartate:2-oxoglutarate aminotransferase, EC 2.6.1.1).

McGiven and Chappell [67] proposed an alternative source of ammonia for carbamoyl phosphate synthesis – a cyclic de-amination of adenosine monophosphate to inosine monophosphate, with subsequent synthesis of AMP using glutamine as the nitrogen donor. While this may occur, Krebs and Lund [68] have suggested that it would be unable to meet the demand for ammonia for urea synthesis because of the inhibition of AMP deaminase by physiological concentrations of GTP.

The total excretion of urea is related directly to the intake of proteins; pyrimidines also give rise to urea, but the proportion of daily urea excretion from this source is normally very small.

5.2.1 The entero-hepatic circulation of urea

Urea is a small freely permeable molecule, and therefore it can diffuse from the blood-stream into the lumen of the gut. Under normal conditions there is no detectable urea in the faeces, but in germ-free animals, and in response to gut-sterilizing antibiotics such as neomycin, the faecal concentration of urea approaches that of blood. Obviously intestinal bacteria are capable of hydrolysing urea to ammonia and carbon dioxide; there is no urease activity in mammalian tissue.

Regoeczi and co-workers [69] showed that the administration of tracer amounts of [^{15}N]-urea led to a considerably longer biological half-life of the label than when [^{14}C]-urea was used. They suggested that this was because much of the ammonia released by bacterial urease activity in the gut was absorbed into the portal circulation, and was re-used in the liver, while the carbon dioxide was largely lost in expired air. This utilization of ammonia entering the liver from the gut could be by one of three routes: some may be used directly for the synthesis of urea, while the remainder may be incorporated into glutamate by the action of glutamate dehydrogenase, or into glutamine by the action of glutamine synthetase. Wrong [70] noted that these latter two reactions would permit incorporation of the nitrogen into a wide variety of amino acids and other metabolites.

This entero-hepatic circulation of urea may provide an important route of urea elimination in patients with chronic uraemia as a result of renal failure. However, despite the increase in the availability of urea in this condition, there seems to be no increase in the rate of intestinal urea hydrolysis [71, 72]. Wrong [70] suggested that this might be because the large intestine, which has the majority of the ureolytic bacteria, is poorly permeable to urea, while the small intestine, which is more freely permeable, has only a small bacterial population.

In chronic renal failure, severe restriction of dietary protein intake is necessary so as to reduce as far as possible the amount of urea that is synthesized and has to be eliminated from the body. One method of reducing the nitrogen load yet further, is to feed a diet that is virtually free from protein, together with minimally adequate amounts of the essential amino acids, and to rely on transamination of common metabolic intermediates to provide the non-essential amino acids. This will utilize much of the nitrogen arising from protein catabolism, as well as that resulting from the entero-hepatic circulation of urea [73]. Richards and coworkers [74] proposed an extension of this, to reduce the nitrogen load yet further by feeding the oxo-acids of the essential amino acids, rather than the amino acids themselves. This would permit synthesis of the essential amino acids by transamination of the oxo-acids at the expense of the nitrogen pool that would otherwise be used for urea synthesis. Only lysine, for which there is no transaminase and threonine, which is a poor substrate for transamination, need be provided as the amino acids. They demonstrated that there was a greater incorporation of label into proteins from $^{15}NH_4^+$ in subjects fed on a low protein diet than in normally fed subjects, and in uraemic patients this incorporation was greater still.

Walser and coworkers [75, 76] showed that the extent to which different essential oxo-acids were used for the synthesis of their corresponding amino acids differed. Valine, leucine, isoleucine, methionine and phenylalanine were readily synthesized by transamination, while there was less synthesis of histidine, threonine and tryptophan, and none of lysine. Since the three branched-chain amino acids together constitute some 45% of the total essential amino acid pool, this may represent a considerable potential use of ammonia. They also showed that the utilization of essential oxo-acids was greater when animals were maintained on low protein diets than when they were receiving a more or less normal intake of protein [77].

The hydroxy-acids corresponding to the oxo-acids of methionine, leucine and phenylalanine are also useable for amino acid synthesis, although that corresponding to valine is not [78]. This is of interest if it is intended to administer amino acid precursors as a part of the treatment of uraemia, since the hydroxy-acids are chemically more stable than the oxo-acids.

Although it would appear from the discussion above that the administration of essential oxo- or hydroxy-acids, together with severe restriction of protein intake, may be useful in the treatment of uraemia, the extent and usefulness of the entero-hepatic circulation of urea, both under normal and uraemic conditions, has been challenged. Varcoe and coworkers [72] have shown that little of the label from [^{15}N]-urea is incorporated into proteins, even when protein restriction is severe enough to reduce the synthesis of albumin, and they concluded that the entero-hepatic circulation of urea is probably not nutritionally significant.

5.2.2 Inborn errors of the urea synthesis cycle

As are all inborn errors of metabolism, the genetic diseases that affect the urea synthesis cycle are extremely rare, and have been discovered only recently. The first to be described (in 1958) was argininosuccinic aciduria, due to a deficiency of argininosuccinase. The first reports of ornithine carbamoyltransferase deficiency and citrullinaemia due to argininosuccinate synthetase deficiency were in 1962, carbamoyl phosphate synthetase deficiency was first reported in 1964 and arginase deficiency leading to hyperargininaemia was reported in 1969 [79]. All of these diseases are characterized by hyperammonaemia as well as accumulation in the blood, tissues and urine of the immediate substrates of the affected enzymes. In general, the accumulation of substrates is not particularly dangerous, but hyperammonaemia can prove fatal.

The mechanism of ammonia intoxication is not clear; it is not due to a change in tissue pH, since coma and even death can occur before there is any detectable change in pH. It has been suggested that the coma may be the result of synthesis of glutamate and glutamine at the expense of 2-oxo-glutarate; especially in the central nervous system this may cause so severe a depletion of the tissue pools of 2-oxo-glutarate that energy-yielding metabolism is impaired [80].

In all patients with inborn errors of the urea cycle who survive for more than a few hours after birth, there is some residual activity of the affected enzyme. Presumably a total lack of the enzyme would prove fatal within a short time after birth, although there would be no problem *in utero*, since ammonia could be passed across the placenta to the maternal circulation for metabolism. In some cases there may be some activity of the affected enzymes in tissues other than the liver, or there may be residual activity in the liver that is adequate to deal with small amounts of substrate. Many of the defects that have been investigated appear to be of the type where it is the K_m of the enzyme that is abnormal, so that there is significant activity only when a high concentration of the substrate is available. This means that there will, necessarily, be a considerable accumulation of the substrate of the affected enzyme, but the overall rate of urea synthesis will be more or less normal. It is assumed that the hyperammonaemia that accompanies nearly all inborn errors of the cycle is the result of feed-back inhibition of other enzymes because of the accumulation of one substrate. With the exception of arginase, all of the enzymes of the pathway have equilibrium constants such that it is only because of the removal of products that the cycle functions in the direction of urea synthesis [79].

The therapy of inborn errors of the urea cycle is based on as severe a restriction of protein intake as possible, with frequent small meals so as to reduce still further the amount of ammonia that must be metabolized at any time. Obviously, this will call for very delicate dietary control in young children, where there must be a balance between a protein intake low enough to prevent hyperammonaemia, and an intake adequate to permit nitrogen retention and growth.

In order to permit some degree of relaxation of strict dietary control, there have been a number of attempts to increase protein tolerance by increasing the availability of 2-oxo-glutarate for glutamine synthesis by feeding citrate [81, 82]. Aspartate has also been used [83]; while this may function as a source of 2-oxo-glutarate, it is more likely that it provides an alternative route of ammonia excretion, by the formation of asparagine, or perhaps by increasing the synthesis of pyrimidines, and so permitting some nitrogen excretion in the form of orotic acid and other intermediates of pyrimidine synthesis.

Brusilow and Batshaw [84] have reported an interesting approach to the problem of increasing protein tolerance in an infant with argininosuccinase deficiency. They fed 4–5 mmol of arginine per kg body weight day^{-1}, so as to provide a source of ornithine for the incorporation of ammonia into argininosuccinate. In this way the excretion of argininosuccinate becomes a pathway for the net excretion of nitrogen, with repletion of the pools of intermediates required for its synthesis by the administration of arginine.

The same authors [85] have also improved protein tolerance and increased the urinary excretion of nitrogen by exploiting the well-known conjugation of 'foreign substances' with amino acids. Benzoic acid is excreted largely as the glycine conjugate, hippuric acid (as well as a lesser amount of an alanine conjugate), and phenylacetic acid is excreted largely as phenylacetylglutamine. This therapy was useful in the treatment of a number of patients in hyperammonaemic hepatic coma; it led to a reduction in the circulating concentration of ammonia and a clinical improvement. It also led to increased protein tolerance in a patient with carbamoyl phosphate synthetase deficiency. Similar depletion of the body amino nitrogen pools by the administration of these and other compounds that are excreted as amino acid conjugates may also prove useful in other cases.

5.3 AMINO ACID TOXICITY

A number of essential amino acids are toxic when fed in excess. The amounts involved are very much greater than for the more general amino acid imbalance and antagonism discussed above – intakes of between 2–20% of the diet as a single amino acid have been used to demonstrate toxicity. While such amounts of amino acids are very much greater than would normally be eaten in dietary proteins, some of the effects may be magnified when the dietary protein intake is very low. Other effects may be similar to those that are seen in some of the inborn errors of amino acid metabolism, so that study of toxicity may give us useful information on the pathology of these rare genetic diseases.

5.3.1 Leucine excess and pellagra

People living on sorghum (*Sorghum vulgare*) as the dietary staple tend to develop pellagra. Gopalan and Srikantia [86] suggested that the cause might be the relative excess of leucine found in this cereal. Pellagra is due to a deficiency of niacin, which may be present in foods as nicotinic acid and nicotinamide, or may be synthesized from tryptophan. Like the protein of maize, sorghum protein is relatively deficient in tryptophan, and most varieties contain a relatively large amount of leucine, with little isoleucine or valine. Gopalan and Srikantia [87] showed that feeding experimental animals on diets that contained an excess of leucine (between 1.5–3% of the diet) led to the development of pellagra-like conditions. Initially, their findings were challenged by a number of workers who were unable to repeat their observations [88–90]. However, their principal finding has recently been confirmed; in rats that are maintained on diets that provide only minimally adequate amounts of tryptophan and niacin the addition of 1.5% leucine leads to depletion of tissue nicotinamide nucleotides [91]. This seems to be an effect on the endogenous synthesis of nicotinamide nucleotides from tryptophan rather than on the utilization of preformed niacin in the diet. It therefore seems likely that people who are living on a low level of protein nutrition, and with only a marginally adequate intake of tryptophan and niacin, may well be at risk of developing pellagra if they have an excess intake of leucine.

5.3.2 Maple syrup urine disease

Patients suffering from maple syrup urine disease, an inborn error of branched-chain amino acid metabolism, are severely mentally retarded unless the condition is treated early enough. The treatment consists of a diet that contains only minimally adequate amounts of leucine, isoleucine and valine to support maintenance and growth. As far as is known, this dietary restriction must be continued throughout life, since intermittent elevation of blood concentrations of the branched-chain amino acids, even in otherwise well-controlled patients who have developed with normal intelligence, leads to acute respiratory distress and neurological crises. The mechanism of the neurological damage in this disease is not known, but it is possible that it results, at least in part, from inhibition of uptake into the brain of tryptophan and tyrosine, as a result of competition from the branched-chain amino acids, as discussed below.

The enzyme affected in maple syrup urine disease is branched-chain oxo-acid decarboxylase (2-oxoisovalerate dehydrogenase (lipoamide), EC. 1.2.4.4). In about half of the reported cases, the disease appears to be a ‘classical’ inborn error of metabolism, with no detectable activity of the enzyme in the liver or other tissues. In the remaining cases, there is some residual activity of the enzyme, which can be stimulated by large amounts of the cofactor thiamin pyrophosphate. The mutation in these cases seems to affect the cofactor binding site of the enzyme, so that a very high concentration is required for activity, which may then be more or less normal. Such vitamin dependency diseases have been discussed by Frimpter and coworkers [92] and by Mudd [93]. These patients are fortunate, since it is not necessary to restrict their intake of branched-chain amino acids; adequate control can be achieved by a thiamin intake of the order of 10–50 mg day^{-1}, compared with a recommended daily intake for normal people of 1–2 mg day^{-1}.

It is interesting to note that in maple syrup urine disease there seems to be no evidence of pellagra [94]. This may reflect the generally good nutritional status of these patients, or it may reflect the counteracting effects of isoleucine and valine together with the elevated leucine; in these patients there is not an excess of leucine *per se*, but rather a toxic accumulation of all three branched-chain amino acids and their oxo-acids.

5.3.3 Tyrosine

Very high intakes of tyrosine, especially with a low protein diet, lead to specific signs of toxicity in rats – redness and swelling of the paws and a dark exudate around the eyes, a loss of weight and a high mortality. With an increasing intake of protein there is an increasing tolerance of tyrosine, although even in animals that are receiving high protein diets an intake of tyrosine as high as 10% of the diet will still lead to toxicity. At lower levels of tyrosine (up to about 3% of the diet) there is some development of tolerance, and the signs of toxicity disappear after a time. In rabbits and dogs a high intake of tyrosine is nephrotoxic, and in rabbits and rats there is also liver necrosis. The metabolic bases of these signs of tyrosine intoxication is not known [25].

5.3.4 Phenylalanine and phenylketonuria

Rats or other animals treated with large amounts of phenylalanine show some superficial similarities to the biochemical abnormalities that are observed in patients suffering from the inborn error of metabolism phenylketonuria, a congenital absence of phenylalanine hydroxylase – a high blood concentration of phenylalanine and a massive urinary excretion of phenylketones (phenylpyruvate, phenyllactate and phenylacetate). However, such animals also have an elevated blood concentration of tyrosine, in contrast to the abnormally low levels in phenylketonurics [95].

Fig. 5.2 shows the possible pathways of phenylalanine metabolism; normally the major pathway is by hydroxylation to tyrosine, and transamination to phenylpyruvate is a very minor route. The evidence from animals treated with high doses of phenylalanine suggests that the K_m of phenylalanine transaminase is high, so that there is little activity until there is a considerable accumulation of the substrate.

Phenylketonuria is one of the least rare of the inborn errors of amino acid metabolism, occurring in about 80 per 10^6 live births. Untreated phenylketonurics are very severely mentally retarded, but if the disease is detected soon after birth and appropriate dietary therapy (a severe restriction of phenylalanine intake) is started early enough, brain damage is minimal and the patients develop with normal intelligence. Because of this, it is now common in most developed countries to screen all infants shortly after birth to detect the disease. This is done either by detecting the presence of phenylketones in the urine or, more commonly, by measurement of the blood phenylalanine concentration, generally by means of the Guthrie bacterial inhibition test [96, 97].

The metabolic basis of the mental defect in untreated phenylketonuria is not known; however, high circulating levels of phenylalanine seem to be harmful only

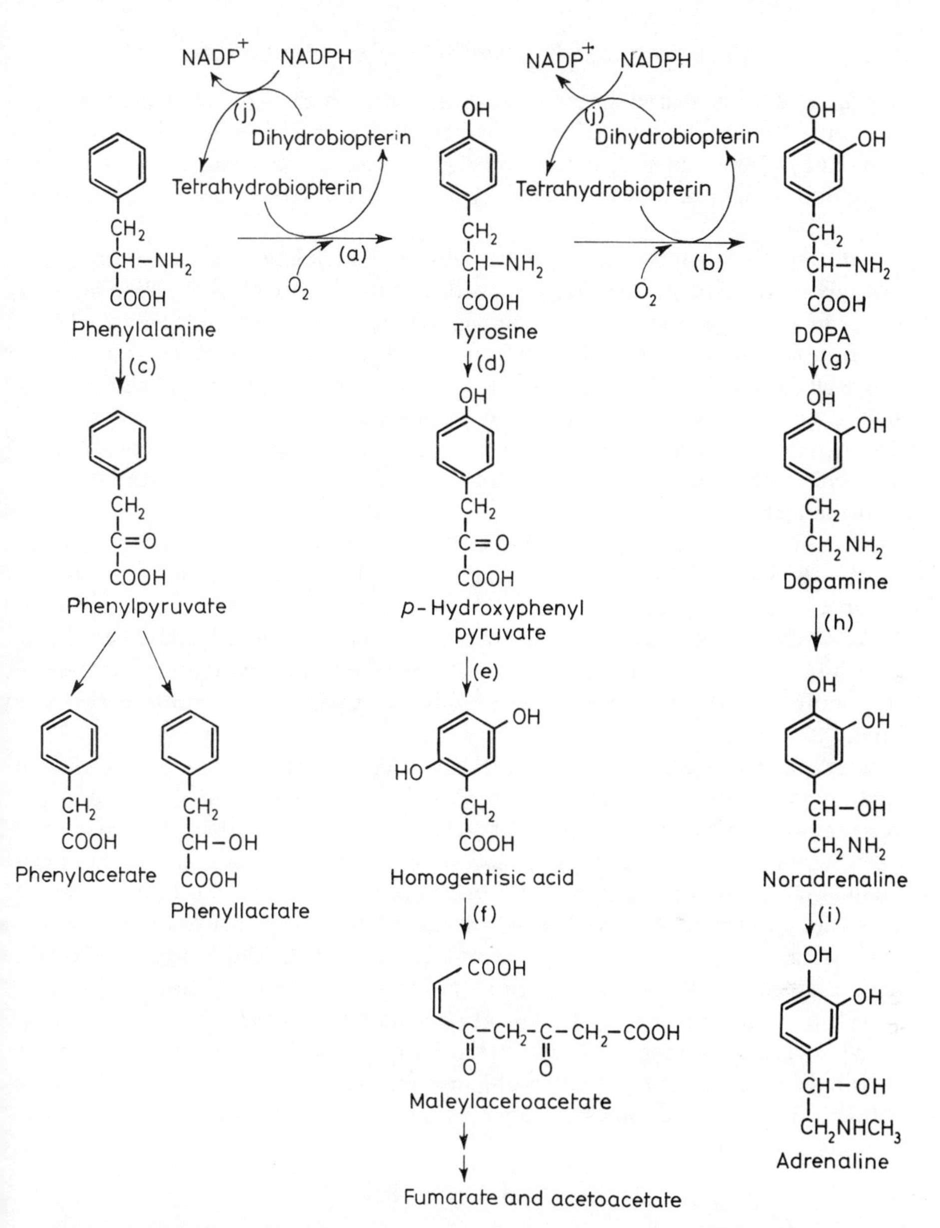

Figure 5.2 The metabolism of phenylalanine and tyrosine. (a) Phenylalanine hydroxylase (L-phenylalanine, tetrahydropteridine; oxygen oxido-reductase (4-hydroxylating), EC 1.14.16.1); (b) tyrosine hydroxylase (L-tyrosine: tetrahydropteridine: oxygen oxido-reductase (3-hydroxylating), EC 1.14.16.2); (c) phenylalanine aminotransferase (L-phenylalanine: pyruvate aminotransferase, EC 2.6.1.58); (d) tyrosine aminotransferase (L-tyrosine: 2-oxoglutarate aminotransferase, EC 2.6.1.5); (e) *p*-hydroxyphenylpyruvate oxidase (4-hydroxyphenylpyruvate: oxygen oxido-reductase (hydroxylating, decarboxylating), EC 1.13.11.27); (f) homogentisic acid oxidase (homogentisate: oxygen 1,2-oxidoreductase (decyclizing), EC 1.13.11.5); (g) aromatic amino acid decarboxylase (L-aromatic amino acid carboxy-lyase, EC 4.1.1.28); (h) dopamine-β-hydroxylase (3,4-dihydroxyphenylethylamine, ascorbate: oxygen oxidoreductase (β-hydroxylating), EC 1.14.17.1); (i) phenylethanolamine *N*-methyltransferase (*S*-adenosylmethionine: phenylethanolamine *N*-methyltransferase, EC 2.1.1.28); (j) dihydrobiopterin reductase.

during development, and the evidence available to date suggests that from the age of about 10 years the strict dietary control can be relaxed progressively. During pregnancy there must again be very strict dietary control, since maternal hyperphenylalaninaemia leads to very severe neurological defects in the fetus [98].

A number of workers have used *p*-chlorophenylalanine as an inhibitor of phenylalanine hydroxylase, together with a loading dose of phenylalanine, as a better model of phenylketonuria than simple hyperphenylalaninaemia [35, 99]. The inhibition of the hydroxylase by *p*-chlorophenylalanine is interesting; *in vitro* it is ineffective, and it must be administered for several days before there is any significant inhibition *in vivo*, suggesting that it has an effect on protein synthesis. The mechanism has not been elucidated, but does not appear to be incorporation of *p*-chlorophenylalanine into proteins in place of either phenylalanine or tyrosine [100].

p-Chlorophenylalanine also inhibits tryptophan hydroxylase *in vivo*, in the same way as it inhibits phenylalanine hydroxylase. The two enzymes have a number of similar properties, and share the same tetrahydrobiopterin cofactor and dihydrobiopterin reductase system. *p*-Chlorophenylalanine has been used clinically to reduce the synthesis of 5-hydroxytryptophan and 5-hydroxytryptamine in patients with carcinoid syndrome, so as to give symptomatic relief [101, 102].

Classical phenylketonuria involves absence of the enzyme phenylalanine hydroxylase; there have been a few reports of patients who failed to respond to restriction of phenylalanine intake, and who showed a number of neurological defects that are not seen in the classical disease. In most of these cases it has been shown that the defect is in dihydrobiopterin reductase, and it has been suggested that the additional neurological problems result from a failure of 5-hydroxytryptamine synthesis as well as the failure of phenylalanine metabolism. To date it has not proven possible to provide any suitable therapy for these patients [103]. A further two cases of atypical phenylketonuria have been described in which there was an inability to synthesize biopterin, and in these cases it has been possible to resolve the clinical problems, without restriction of phenylalanine intake, by administration of suitable precursors of biopterin [104].

5.3.5 Tryptophan

As with a number of other essential amino acids, intake of tryptophan as high as 1.5% of the diet can lead to growth depression in rats maintained on low protein diets. Higher intakes of tryptophan are toxic even when the dietary protein intake is adequate. As with tyrosine, there is some degree of adaptation to high intakes of tryptophan with time [25].

Administration of tryptophan leads to the induction of phosphoenol pyruvate carboxykinase (EC. 4.1.1.32) one of the rate-limiting enzymes of gluconeogenesis, but also leads to inhibition of the enzyme due to the formation of relatively large amounts of quinolinic acid [105]. This means that hypoglycaemia is a potential problem when relatively large doses of tryptophan are used. However, quinolinate phosphoribosyltransferase (nicotinate nucleotide: pyrophosphate phosphoribosyl transferase (carboxylating) EC. 2.4.2.19) appears to have a considerable excess capacity to metabolize quinolinate onwards to nicotinamide nucleotides (Bender, unpublished observations). There are no reports of hypoglycaemia in patients receiving up to about 12 g of tryptophan daily as antidepressant medication, and McDaniel and coworkers [106] have suggested that inhibitory concentrations of quinolinate are not achieved *in vivo*.

The administration of tryptophan also has specific behavioural effects in experimental animals, characterized by hyperactivity. It is is assumed that these effects are due to increased synthesis of 5-hydroxytryptamine, and possibly also tryptamine; they are enhanced if a mono-amine oxidase inhibitor is given before the test dose of tryptophan, so as to inhibit the catabolism of these amines [107]. The rate of synthesis of tryptophan in the central nervous system seems to depend to a great extent on the availability of tryptophan, as discussed below, and this has been exploited in psychiatry; it is believed that at least some forms of psychological depression are associated with lower than normal concentrations of 5-hydroxytryptamine in the brain, and the administration of relatively large doses of tryptophan has been shown to be beneficial in some cases [108].

5.3.6 Dihydroxyphenylalanine (dopa)

Studies of Parkinson's disease have shown that there is a loss of dopaminergic function in the central nervous system and in the relatively early stages of the disease an increase in the amount of dopamine available in the brain is beneficial. During the 1960s there was a considerable breakthrough in the treatment of this disease when therapy with dihydroxyphenylalanine (dopa), a precursor of dopamine, was introduced.

Many patients are unable to tolerate the doses of dopa that are necessary to achieve benefit. There is considerable decarboxylation of dopa to dopamine in the liver, lung, kidney and other tissues, and this systemic formation of dopamine leads to hypertensive crises, nausea, dizziness and other serious side effects [109]. A number of other side effects of dopa therapy have also been reported (many anecdotally), including such alarming effects as blackening of the urine, saliva and sweat, due to the formation of melanin from dopamine by non-enzymic oxidation and polymerization. Despite these problems, dopa therapy continues to be useful in Parkinsonism, especially when the dose can be reduced by the simultaneous administration of an inhibitor of dopa decarboxylase (aromatic L-amino acid decarboxylase, EC. 4.1.1.28) that does not cross the blood–brain barrier, and therefore reduces the extra-cerebral decarboxylation of dopa without affecting the formation of dopamine in the central nervous system.

5.3.7 Cysteine and methionine

Excessive intake of either cysteine or methionine will cause growth retardation in animals maintained on a low protein diet. Paradoxically, although a higher intake of protein reduces this effect of cysteine, it increases its toxicity.

Cysteine toxicity is manifested by cirrhosis of the liver, and with higher doses there is also necrosis of the liver and kidneys. The effect also depends to some extent on the fat content of the diet; increasing the proportion of fat increases the incidence of liver haemorrhage but reduces the severity of necrosis and cirrhosis. This has not been explained [25].

The toxicity of methionine is aggravated by vitamin B_6 deficiency, and conversely, the addition of even a relatively small excess of methionine to the diet of experimental animals leads to depletion of vitamin B_6 and hastens the development of signs of deficiency. Methionine toxicity is accompanied by hypertrophy of the kidney, as a result of dilatation of the tubules, and marked changes in the pancreatic acinar cells. Liver damage appears to be variable [25].

There is some alleviation of methionine toxicity by feeding large amounts of glycine or serine [110]. It has been suggested that this reflects a limitation of the ability to convert homocysteine to cystathionine (i.e., a partial essentiality of the non-essential amino acid serine under the abnormal conditions of methionine intoxication). Certainly either acute or chronic administration of large amounts of methionine to rats leads to a reduction by 25–50% in the circulating concentrations of glycine and serine, and even larger falls in the brain, together with a considerable increase in the tissue concentrations of cystathionine [111].

The disease of homocystinuria is an inborn error of methionine metabolism caused by a failure of cystathionine synthetase (EC. 4.2.1.22). This enzyme is pyridoxal phosphate-dependent, and about half the known cases respond to administration of very high intakes of vitamin B_6 [92, 93]. The remaining patients have a more classical type of genetic disease, and in these cases it is necessary to restrict the intake of methionine severely. At the same time, these patients must be provided with adequate amounts of cysteine, since they are now unable to form it from methionine.

Clinically, homocystinuria is characterized by abnormal fragility of the long bones, spontaneous dislocation of the lens of the eye and thrombosis, which may be fatal. In about 50% of the cases there is also mental retardation, and it has been suggested that this is the cumulative result of numerous small thromboses in the brain. All of these features, including possibly the occurrence of thrombosis, can be accounted for by abnormal synthesis of collagen, which is both more soluble than normal and less well mineralized in bone. At the same time, it is suggested that abnormal collagen provides loci in blood vessel walls at which thrombus formation is enhanced. The effect on collagen synthesis seems to be inhibition of cross-linkage as a result of inhibition of lysyl oxidase by homocysteine, homocystine and the mixed disulphide of cysteine and homocysteine [112]. Collagen synthesis is discussed below.

5.4 AMINO ACID LOADING TESTS

One of the accepted methods of testing the body's metabolic capacity is by the administration of a loading dose of the substrate of a pathway, followed by measurement of the blood and urine concentrations of metabolites. This has been an especially useful technique in studying sub-clinical vitamin undernutrition, where the excess of substrate may exceed the capacity of the pathway if the vitamin-derived cofactor is not present in adequate amounts [113].

5.4.1 The tryptophan loading test for vitamin B_6 status

Pyridoxal phosphate, which is derived from vitamin B_6, is required for the activity of kynureninase, and therefore the administration of tryptophan in cases of vitamin B_6 insufficiency will lead to an accumulation of kynurenine and hydroxykynurenine. Under these conditions there is a considerable increase in the excretion of xanthurenic and kynurenic acids, which are normally minor side-products of tryptophan metabolism. The standardization of conditions for the tryptophan load test has been discussed by Coursin [114].

While it is certainly true that the metabolism of tryptophan is disturbed in vitamin B_6 deficiency, it is not necessarily true to say that therefore any such disturbance in tryptophan metabolism is the result of vitamin B_6 deficiency. There is a considerable amount of research that apparently shows the development of vitamin B_6 deficiency in women receiving oral contraceptive steroids [115]. In general, these studies have used the tryptophan load test as an index of vitamin B_6 status, but other indices of vitamin B_6 nutrition are not affected by the steroids. Such indices include the plasma concentration of pyridoxal phosphate, the activation of aspartate aminotransferase by pyridoxal phosphate added *in vitro*, the excretion of 4-pyridoxic acid and the ability to metabolize a test dose of methionine [116]. This casts some doubt on the interpretation of an abnormal response to the tryptophan load test as due to vitamin B_6 deficiency in these cases.

Bender and Wynick [117] have shown that oestrone sulphate and glucuronide (and hence by implication a number of other oestrogen conjugates) are inhibitors of kynureninase, competing with the substrate but not with the cofactor. It was suggested that the abnormal response to the tryptophan load test in women receiving oestrogens might therefore be due to this direct inhibition of kynureninase, and not be connected with vitamin B_6 depletion. Bender and coworkers [118] have also shown that the administration of oestrogens to experimental animals leads to inhibition of kynureninase and the expected changes in tryptophan metabolism, but has no effect on blood or tissue concentrations of vitamin B_6, urinary excretion of 4-pyridoxic acid or the activation of erythrocyte aspartate aminotransferase by added pyridoxal phosphate. They suggested that abnormal responses to the tryptophan load test must be interpreted with caution, since a direct effect of a drug on kynureninase would yield results indistinguishable from drug-induced vitamin B_6 deficiency.

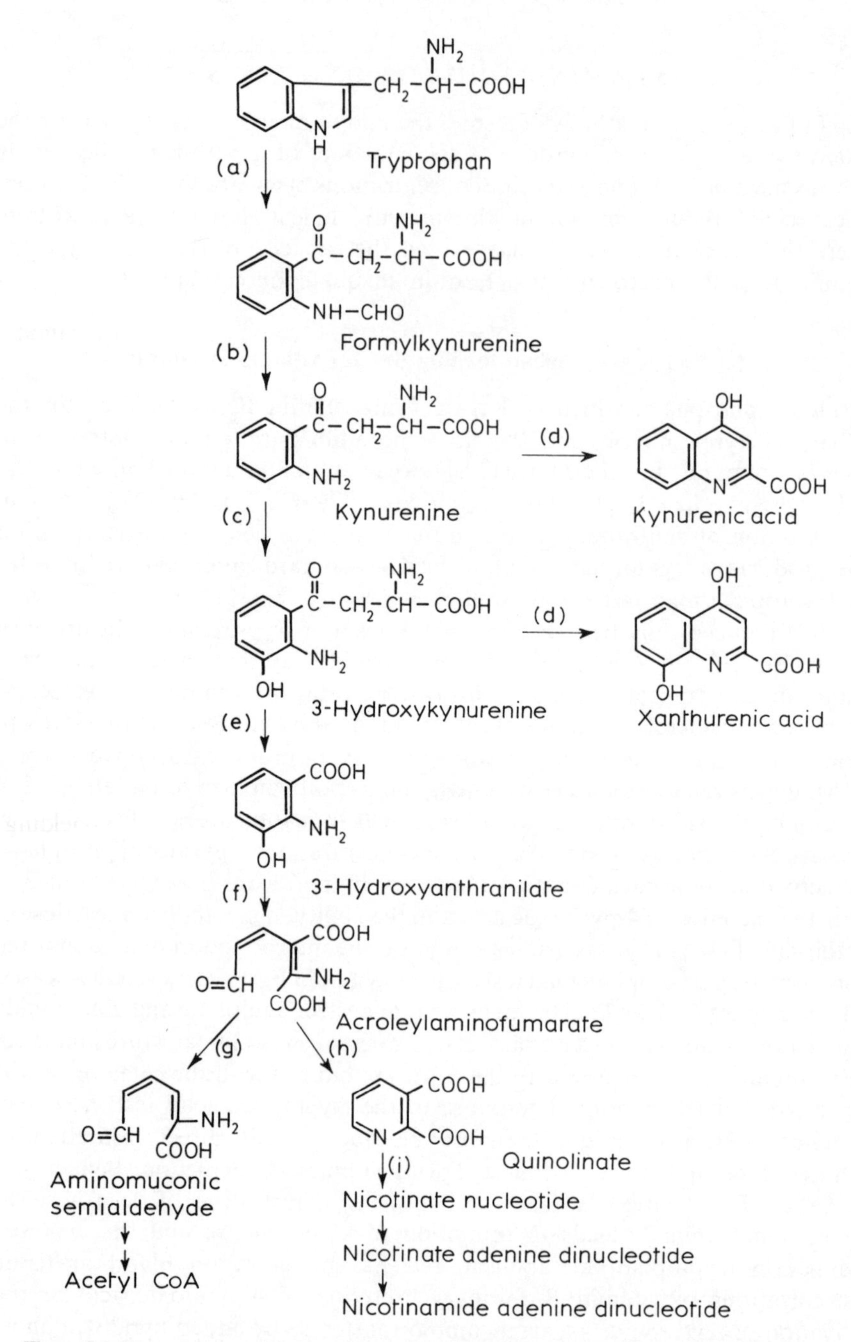

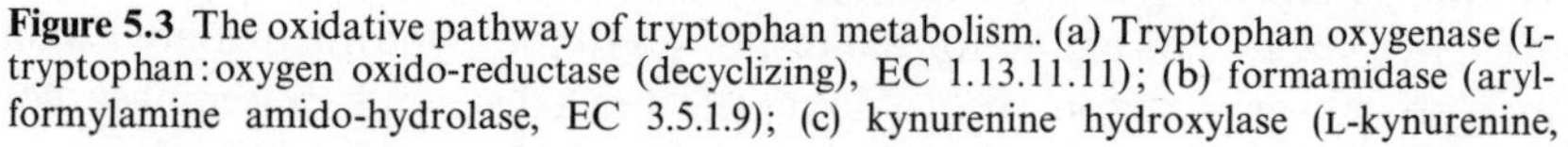

Figure 5.3 The oxidative pathway of tryptophan metabolism. (a) Tryptophan oxygenase (L-tryptophan: oxygen oxido-reductase (decyclizing), EC 1.13.11.11); (b) formamidase (aryl-formylamine amido-hydrolase, EC 3.5.1.9); (c) kynurenine hydroxylase (L-kynurenine,

Wolf and coworkers [119] have introduced a modification of the tryptophan load test, the administration of a test dose of kynurenine, to distinguish between inhibition of kynurenine metabolism (for example as a result of vitamin B_6 deficiency) and excessive formation of kynurenine. The latter would exceed the normal capacity to metabolize it by way of kynureninase, as a result of hormonal induction of tryptophan oxygenase.

5.4.2 The histidine loading test for folate status

The metabolism of histidine involves the formation of formiminoglutamate (FIGLU), which is normally metabolized to glutamate, with the formation of formimino-tetrahydrofolate. In folic acid deficiency this reaction is impaired, and after a test dose of histidine there is considerable excretion of FIGLU in the urine. This test was a standard part of the armoury of the clinical biochemist for the differential diagnosis of megaloblastic anaemia due to folate deficiency until the development of convenient and adequately sensitive methods for the determination of blood folic acid concentrations [120]. Thereafter it fell into disuse.

5.5 AMINO ACID METABOLISM IN FASTING AND STARVATION

It has long been recognized that any dietary protein in excess of the relatively small amount that is required for net new protein synthesis is catabolized as a source of metabolic energy – the average energy yield of protein is 17 MJ kg^{-1}, the same as that of carbohydrate. Furthermore, it was noted above that if the total energy intake is inadequate, protein will be used as an energy-yielding substrate as the first priority, and the body will no longer be able to maintain nitrogen balance on an otherwise adequate intake of protein.

This suggests that the body has a considerable reserve in the form of muscle proteins which might be mobilized during starvation. A number of studies of nitrogen excretion during therapeutic fasting in obese subjects show that there is considerable catabolism of muscle protein for gluconeogenesis initially, but as the circulating concentrations of ketones (acetoacetate and 3-hydroxybutyrate) increase to a level where the central nervous system can utilize them rather than glucose for its energy-yielding metabolism (after some 2–3 weeks), there is a reduction in the rate of breakdown of protein [52, 121]. Pawan and Semple [122] have attempted to minimize this loss of muscle protein during therapeutic

Figure 5.3 (*Contd.*)
NADPH: oxygen oxido-reductase (3-hydroxylating), EC 1.14.13.9); (d) kynurenine aminotransferase (L-kynurenine: 2-oxoglutarate aminotransferase (cyclizing), EC 2.6.1.7); (e) kynureninase (L-kynurenine hydrolase, EC 3.7.1.3); (f) 3-hydroxyanthranilate oxidase (3-hydroxyanthranilate: oxygen 3,4-oxido-reductase (decyclizing), EC 1.13.11.6); (g) picolinate carboxylase (aminocarboxymuconate semialdehyde carboxy-lyase, EC 4.1.1.45); (h) non-enzymic cyclization; (i) quinolinate phosphoribosyltransferase (nicotinate nucleotide: pyrophosphate phosphoribosyltransferase (carboxylating, EC 2.4.2.19).

starvation of obese patients by infusion of 3-hydroxybutyrate to achieve maximum concentrations of ketones in the blood-stream at an earlier stage than would normally occur. They showed that, although the overall loss of weight was the same when 3-hydroxybutyrate was infused as during simple starvation, there was very much less excretion of nitrogen, and therefore presumably the weight loss was almost entirely adipose tissue, with little loss of muscle protein.

Felig and Wahren [123, 124] have drawn attention to the difference between the average amino acid composition of muscle proteins and the pattern of amino acids released into the blood-stream from muscle during fasting. There is a considerable excess of alanine and glutamine relative to the proportions of these two amino acids in muscle. There is also a net uptake of amino acids across the splanchnic bed (effectively the gastro-intestinal tract and liver) during fasting, and again alanine and glutamine predominate [125]. From the few studies that have been possible in patients undergoing elective surgery, they have shown that the upper gastro-intestinal tract takes up the majority of the glutamine from the splanchnic circulation, and in fact puts out a significant amount of alanine, so that measurement of alanine removal across the splanchnic bed underestimates the hepatic uptake of this amino acid for gluconeogenesis [126].

In the rat there is also a net removal of alanine across the splanchnic bed during starvation. The jejunum takes up glutamine and ketones from the circulation as its major energy sources, and also glucose and lactate, which are released as alanine, for uptake into the liver [127].

Felig and Wahren [123, 124] have proposed a glucose–alanine cycle involving the production and release of alanine by muscle and its uptake for gluconeogenesis by the liver and kidneys. The alanine is synthesized from amino acids released by proteolysis, and also by transamination of pyruvate formed by glycolysis of muscle glycogen and of glucose taken up by the muscle from the general circulation. The cycle provides a mechanism whereby the muscle can spare glucose in starvation for use by the central nervous system, while relying on amino acid, ketone and fatty acid metabolism to meet its own energy requirements. The cycle is shown in Fig. 5.4.

Infusion of 3-hydroxybutyrate into fasting subjects leads to a considerable fall in the circulating concentration of alanine [128], suggesting that increasing ketosis during prolonged starvation may inhibit the cycle, and reduce the catabolism of proteins; the excretion of nitrogen also falls during ketone infusion, mirroring the circulating alanine concentrations.

Chang and Goldberg [129] have shown that, even during fasting, there is still a considerable uptake of glucose into skeletal muscle, and this is largely released as alanine. The source of the nitrogen for this is not only the catabolism of muscle protein, but also leucine taken up from the circulation and used as a major energy source in fasting muscle [130]. It was noted above that leucine appears to be an important energy source, and that its metabolism is very sensitive to changes in food intake [36, 39]; in atrial muscle from fed rats the addition of leucine leads to an increase in the rate of production of alanine and glutamine, presumably because of an increase in the availability of amino groups for transamination

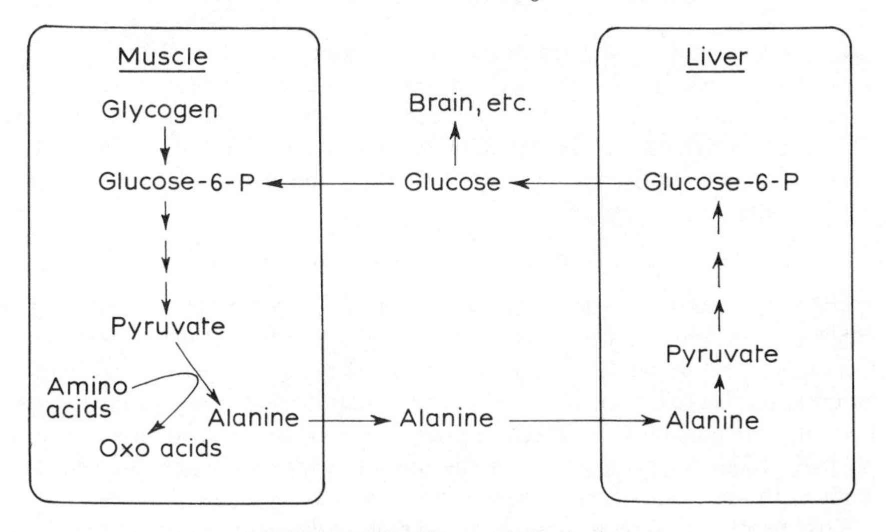

Figure 5.4 The glucose–alanine cycle.

[131]. The addition of leucine also reduces the oxidation of glucose in atrial muscle from fed or fasted rats, by inhibiting the oxidation of pyruvate, which will therefore be transaminated to alanine [132].

In man, during a four-day fast, in diabetes and immediately following surgery there is a reduction in the plasma concentration of alanine. Elia and coworkers [133] suggested that in diabetes and following surgery this is the result of increased removal by the liver for gluconeogenesis, while in fasting it is the result of reduced release from muscle with increasing ketosis.

Alone of the species that have been studied, the rabbit is incapable of significant gluconeogenesis from alanine, and it seems likely that glycine is a major means of carrying potentially gluconeogenic fragments from muscle to the liver in this species [134, 135].

5.6 TRANSPORT, EXCRETION AND UPTAKE OF AMINO ACIDS

Under normal conditions the excretion of amino acids in the urine amounts to some 3 g day^{-1}, one third each as free amino acids, small peptides and conjugates of 'foreign compounds' and some metabolic end-products. Aminoaciduria, the excretion of inappropriate amounts of amino acids, is a common finding in inborn errors of amino acid metabolism. In such cases there is generally only a single amino acid present in large amounts, that which has disturbed metabolism, so that the circulating concentration is too high for the kidney to resorb all the filtered load.

There are also a number of genetic diseases in which there is a more general aminoaciduria – rather than massive excretion of a single amino acid, there is a considerable increase in the excretion of several. Studies of two of these conditions, cystinuria and Hartnup disease (discussed below), have given us a great deal of information about the mechanisms and specificity of amino acid transport, not only in the kidney, but also in the intestinal mucosa and other tissues, including the uptake of amino acids into the brain, which may be an important factor in the regulation of brain function, as discussed below.

5.6.1 Intestinal absorption of amino acids

A major advance in the study of intestinal transport mechanisms came during the 1950s with the introduction by Wiseman [136] and others of the everted gut sac technique. The principle is that eversion of the gut will permit the sensitive mucosal surface to be bathed in an appropriate nutrient medium, and if the sac is tied off, transport into the interior of the sac seems to approximate well to the normal process of transport from the intestinal lumen (the exposed mucosal surface) to the blood-stream (the enclosed serosal surface of the sac).

Wiseman [136] showed that the uptake of amino acids across the intestinal mucosa was by a process of active transport – there was uptake against a concentration gradient and the process was dependent on a source of metabolic energy. By contrast, passive diffusion and carrier-mediated transport (facilitated diffusion) do not permit accumulation of material against a concentration gradient, and do not require a source of energy. Both facilitated diffusion and active transport display considerable specificity (including stereo-specificity), whereas passive diffusion is very much less specific, and depends largely on the lipid solubility of the permeant molecule.

The intestinal digestion of proteins results in the release of free amino acids and di- and tripeptides. The dipeptidases and tripeptidases are located inside the brush border of the mucosal epithelial cells [137], and hydrolysis of small peptides occurs intracellularly. This means that there must be distinct mechanisms for the uptake of free amino acids and small peptides into the mucosal cells. Asatoor and coworkers [138] showed that the absorption of phenylalanine from the dipeptide phenylalanyl-phenylalanine, and that of tryptophan from glycyl-tryptophan, was slower than after feeding equivalent amounts of the free amino acids. By contrast, Cook [139] showed that the uptake of glycine from glycyl-glycine was faster than that of free glycine. In both cases, the uptake from a mixture of free amino acid and dipeptide was the sum of the two separate rates of uptake, showing that they are additive, and therefore distinct, processes. There seems to be no competition between amino acid and dipeptide for uptake. In Hartnup disease, where there is a more or less total absence of the intestinal transport mechanism for tryptophan, the dipeptide transport system is unaffected, and becomes the principal mechanism of tryptophan absorption [140].

5.6.2 Renal transport of amino acids

It was noted above that much of our knowledge of the mechanisms of amino acid transport in the kidney and other tissues has come from the study of two genetic diseases which result in a relatively generalized aminoaciduria, cystinuria and Hartnup disease. There are other diseases of this kind, and collectively they are known as the aminoacidurias of renal origin. They arise as a result of a failure of the renal transport system rather than from a metabolic defect which would overload the kidney's capacity to resorb filtered metabolites.

(a) Cystinuria

The first study of cystinuria was in 1810, when Wollaston [141] isolated cystine, the first of the amino acids to be discovered, as a result of his studies of renal calculi. The amount of this amino acid can be so great that its concentration exceeds its solubility product, and crystals form in the kidney and bladder. Such cystine stones may be several centimetres long.

A number of other amino acids are also excreted in large amounts in cystinuria, notably the basic amino acids lysine, ornithine and arginine [142, 143]. Dent and Rose [144] showed that although there were very large amounts of these amino acids in the urine, the blood concentrations were normal, and suggested that the disease was due to a defect in the normal process of resorption of amino acids from the glomerular filtrate.

In cystinuria there are also abnormally high levels of diamines and polyamines, including dimethylamine, piperidine, pyrrolidine, cadaverine, putrescine and spermidine, both in blood and urine. This seems to be the result of intestinal bacterial metabolism rather than metabolism in the liver or other tissues, and it has been suggested that there might be a defect in the absorption of the dibasic amino acids from the intestine similar to that observed in the kidney. This would mean that there would be higher than normal concentrations of these amino acids in the intestinal lumen, thus providing more substrate for bacterial action. It is therefore assumed that the renal and intestinal amino acid transport systems are the same, so that a genetic mutation that affects one also affects the other [145–147].

(b) Hartnup disease

The first report of Hartnup disease was in 1956 [148], when a child suffering from pellagra was brought to the Middlesex Hospital in London. He showed a generalized aminoaciduria, with tryptophan and other neutral amino acids predominating, as well as a number of indolic metabolites of tryptophan that are not normally found in human urine. As in the case of the abnormal amines found in the urine of homocystinuric patients, these 'exotic' metabolites of tryptophan were shown to be the result of bacterial metabolism of abnormally high intestinal concentrations of tryptophan because of a failure of intestinal uptake of tryptophan. Administration of gut-sterilizing antibiotics such as neomycin led to the disappearance of these compounds from the urine, and conversely, the administration to normal subjects of tryptophan in enteric-coated capsules (so that the contents were released in the lower gut) led to the production, absorption and urinary excretion of the same abnormal metabolites as were found in Hartnup disease [149].

The defect in Hartnup disease appears to be one of the mechanisms for the transport of neutral amino acids in the kidney and intestinal mucosa. The pellagra can be accounted for by a lack of tryptophan for endogenous synthesis of nicotinamide nucleotides, and responds to the administration of either nicotinamide or tryptophan-containing dipeptides, but not, of course to the oral administration of free tryptophan.

The other feature of the first Hartnup disease patient, and of many of the others that have been reported since, was a complex of neurological signs (cerebellar ataxia resulting in an abnormal gait) which is attributed to deficient synthesis of the neurotransmitter 5-hydroxytryptamine in the central nervous system, as a result of reduced availability of tryptophan as a precursor for the synthesis.

(c) Iminoglycinuria

A number of patients have been reported with a defect of the renal transport mechanism for glycine, proline and hydroxyproline, so that these three amino acids are excreted in abnormally large amounts, although again the blood concentrations are normal. Some patients also show an impairment of the intestinal uptake of these amino acids, although in others it is normal [150].

Some degree of transient iminoglycinuria is normal in new-born infants, because the maturation of this renal transport system is a post-natal event, and development is incomplete at birth. The normal conservation of glycine, proline and hydroxyproline may not develop until about six months of age [150].

A different type of glycinuria has been described in members of one Swiss family, in which proline and hydroxyproline excretion are normal, but there is hyperglycinuria, together with glucosuria. There is no evidence of any renal tubular dysfunction, and no explanation of this condition has been advanced [151].

Yet a further defect of renal glycine transport has been reported, in which there is an elevation of the blood concentration of glycine; the administration of test doses of proline has shown that the maximum capacity for glycine and proline transport is normal in these cases, but the carrier has a reduced affinity for its substrates [152].

On the basis of studies with isolated rabbit renal tubules, Hillman and coworkers [153] proposed three separate systems for glycine transport: the glycine-imino acid system which also transports proline and hydroxyproline, a glycine and alanine system which may also transport other small neutral amino acids, and a specific glycine transport system which does not seem to carry other amino acids.

(d) Hyperdibasic aminoaciduria

A number of families have been reported in whom there is a dibasic aminoaciduria with normal or lower than normal concentrations of arginine, lysine and ornithine in the blood-stream. The excretion of cystine in these patients is normal [140].

(e) Fanconi syndrome

The renal aminoacidurias discussed above are all more or less specific, in that there is excessive excretion of chemically related amino acids; study of these conditions has permitted us to define the classes of amino acid transport systems that occur in the kidney and intestinal mucosa. By contrast, Fanconi syndrome is a generalized impairment of renal tubular transport mechanisms, with a generalized aminoaciduria. The amino acid profile in the urine resembles that in a plasma ultrafiltrate. All solutes show an abnormally high renal clearance.

The defect in Fanconi syndrome seems to be a grossly impaired ability to transport both solutes and water across the tubular membrane, and in some cases there is a similar defect in intestinal mucosal transport. The affinities of the transport systems seem to be unaffected and it has been suggested that the defect is one of the coupling between the membrane binding proteins and the metabolic energy source. In some cases there are also anatomical abnormalities of the proximal renal tubule [140].

5.6.3 Methods for studying renal transport mechanisms

The genetic diseases of renal amino acid transport have allowed us to define several different classes of amino acids that share a common transport mechanism. In cystinuria, it is the basic amino acid carrier that is defective, while in Hartnup disease the defect is in the neutral amino acid system. This latter may be divided into two systems, one which transports the large neutral amino acids (leucine, isoleucine, valine, methionine, phenylalanine, tyrosine, tryptophan and threonine), often called the L-system (for leucine), and a separate system which transports the smaller neutral amino acids (alanine, serine, cysteine, glutamine and asparagine), which is known as the A-system (for alanine). There is also a distinct acidic amino acid transport system, carrying glutamate and aspartate, as well as up to three systems that carry glycine, as discussed above.

(a) *In vivo* studies

It is possible to study the kinetics of renal amino acid transport *in vivo* by measurement of either the renal clearance or the maximum rate of transport.

The renal clearance of a substrate is inversely proportional to its rate of tubular resorption after glomerular filtration, and can be estimated by measurement of blood and urine concentrations. Two important assumptions are involved:

(i) There must be no active secretion of the amino acid into the urine. There is indeed no evidence that any of the physiologically important amino acids is actively secreted into the urine, although this does occur with a number of drugs and drug metabolites.

(ii) The amino acid must not be metabolized by the kidney itself. This is certainly not a valid assumption, and indeed glutamine is a major substrate of the kidney, both for energy-yielding metabolism and for gluconeogenesis.

Studies of renal clearance of amino acids have permitted the definition of three classes of aminoaciduria [154]:

(i) *Renal aminoaciduria* of the type discussed above, where there is a defect of a specific carrier system, so that a number of related amino acids that share the same carrier are excreted together.

(ii) *Overflow aminoaciduria* where the concentration of the amino acid in the urine is so high that the carrier mechanism is saturated, and therefore resorption from the urine is incomplete. In such cases the concentration of the affected amino acid will be abnormally high in the blood-stream as well.

(iii) *Competitive aminoaciduria* where saturation of the transport system by one amino acid will lead to an increase in the excretion of others carried by the same system, because of mutual competition for the same carrier. There is therefore some degree of combined aminoaciduria. The blood concentration of at least one of the affected amino acids will be elevated above normal.

The maximum rate at which an amino acid can be resorbed from the urine can be determined by infusion of the amino acid under study into a peripheral vein. By careful control of the rate of infusion and repeated measurement of concentrations in blood and urine, it is possible to construct a curve of excretion against blood concentration, and hence to estimate the values of the maximum rate of transport (T_m) and the affinity of the carrier system for its substrate. Two classes of inhibition of transport can be distinguished in this way:

(i) *Competitive inhibition* where the inhibition will vary with the concentration of the amino acid. There is a change in the affinity of the carrier system, but T_m is unaffected, and if the urine concentration is high enough resorption will be normal.

(ii) *Non-competitive inhibition* where the affinity of the carrier is unaffected, but the T_m is lower than normal.

The administration of a relatively small intravenous load of lysine leads to a considerable increase in the excretion of cystine, ornithine and arginine in normal subjects, and in cystinurics it also increases the already high renal clearance of these amino acids. Lester and Cusworth [155] showed that in cystinurics the renal threshold for lysine was very much lower than normal, and the T_m for lysine was less than one third of that in normal subjects or heterozygotes for the condition. This confirms that cystine and the dibasic amino acids share a common carrier system, and also suggests that the defect in this disease must be partial rather than total, since otherwise the administration of lysine would have no further effect on the excretion of cystine.

(b) *In vitro* studies

Studies of the uptake and transport of amino acids in suitable tissue preparations *in vitro* will obviously be capable of giving much useful information about the mechanisms involved and their specificity. The major problem is that of how closely the tissue preparation mirrors the normal physiological function of the kidney, which is the transport of solutes across the tubule epithelium from the urine back into the blood-stream.

Rosenberg [156] measured the uptake of amino acids into slices of kidney, together with a correction for uptake into the intercellular space using inulin, which does not enter cells. A complication that may occur is that the amino acid may be metabolized in the tissue, and therefore a number of non-metabolizable amino acid analogues (and especially α-amino-isobutyrate and cyclo-leucine (1-amino-cyclopentane carboxylate)) have been used. Here it must be assumed that the transport of non-metabolizable compounds is the same as that of physiological substrates.

The major criticism of studies with kidney slices is that they give information about the uptake of materials into the cells of the kidney as a whole; access is possible to all cell surfaces, and indeed it is likely that the most hindered access will be to the tubule epithelium, the surface across which transport is physiologically important.

Essentially the same problem arises when isolated renal tubules are used for uptake studies, although there is less complication due to the presence of a multiplicity of different cell types. However, the uptake that is measured will be largely that from the anti-luminal face of the tubule, rather than transport from the lumen to the blood-stream [157, 158].

Burg and coworkers [158] have developed an elegant technique in which renal tubule fragments are isolated and the lumen is perfused with the test solution, so that uptake from the luminal face and transport across the tubule epithelium can be studied. Obviously this is the closest approach to the physiological situation, but it is technically very demanding, and seems not to have been widely adopted.

An example of the problems involved in the interpretation of *in vitro* studies of renal transport comes from the use of phlorizin. *In vivo* in the dog, phlorizin causes lysinuria by inhibiting the net resorption of lysine, but does not affect the resorption of glycine [159]. *In vitro*, using kidney cortex slices, phlorizin has no effect on the transport of either amino acid [160]. The discrepancy seems to be due to the site of action of phlorizin; it inhibits transport across the basement membrane into the blood-stream, without affecting transport at the luminal face of the cell. This means that lysine will eventually accumulate intracellularly to such an extent that it will diffuse passively across the luminal membrane into the urine, and hence lysinuria will occur *in vivo*. By contrast, glycine, which also accumulates intracellularly, will be metabolized in the epithelial cell, so that passive diffusion will not occur to any significant extent. This means that although phlorizin affects the transport of both amino acids to the same extent, it will not appear to affect that of glycine at all [160].

5.6.4 Transport of amino acids into the brain

The uptake of amino acids into the brain may have considerable physiological importance in the control of brain function, as discussed below. Additionally, the brain provides a convenient tissue for uptake studies, since it is relatively simple to cannulate the carotid artery and jugular vein, and perform perfusion studies *in vivo*. The single pass perfusion technique, where uptake is measured after a pulse infusion of the test substrate has been widely exploited [161, 162]. In small experimental animals it is also possible to remove the brain rapidly after pulse infusion, and thus obtain a direct measurement of uptake, as opposed to extraction from the blood-stream. It is also possible to measure the efflux of substrates from the brain, by means of injection into the third ventricle and measurement of the rate of appearance in the blood-stream.

Perfusion techniques of this type measure transport across what is known as the blood–brain barrier, a useful concept which has not been localized to any specific anatomical feature. We know that there must be permeability barriers between the brain and the general circulation, on the evidence of both uptake studies and the differences in metabolic pools between brain and the general circulation. There does not seem to be any single site of this barrier, rather there are multiple membrane barriers – there is obviously some degree of control exercised by the permeability of the epithelium of the brain capillary blood vessels, and the membranes that surround the central nervous system, and finally the cell membranes of different cell types also seem to have different transport characteristics.

The composition of the free amino acid pools of the brain differs considerably from that of other tissues. Not only are there amino acids that are more or less unique to the brain, such as γ-amino-butyrate (GABA), but the brain is richer than other tissues in glutamate, glutamine, aspartate, acetylaspartate and glutathione [163].

The rate of amino acid uptake into the brain is extremely high. As much as 30–50% of the amino acid present in an infusion bolus may be removed during 15 seconds of a single pass through the brain capillary bed [161]. This occurs without any change in the net pool size in the brain, so there must be considerable counter-transport – there is a rapid exchange of amino acids across the blood–brain barrier, much of which may be carrier-mediated rather than by active transport [164].

There are also active transport systems for the efflux of amino acids from the brain; this can be distinguished from exchange transport on the basis of specificity, cross-inhibition and sensitivity to drug inhibition. It was first demonstrated by Lajtha and Toth [165, 166] who showed that the loss of label after the intra-cerebral administration of labelled amino acids occurred at different rates for different substrates. They also showed that efflux occurred against a concentration gradient, and therefore was an active process.

There have been a great many studies of amino acid uptake into slices of brain tissue. On the basis of such studies, Blasberg and Lajtha [167] defined five distinct carrier systems, for large and small neutral amino acids, β-amino acids, acidic and basic amino acids. These classes agree relatively well with those defined for renal and intestinal transport.

Studies with brain slices have defined two separate classes of carrier, those with high affinities (K_m of the order of 10 μM) and low affinity carriers with values of K_m of the order of 1 mM. In some reports such low affinity carriers are referred to as 'non-saturable' systems, since it is unlikely that they will be saturated under physiological conditions, whereas the high affinity systems will presumably be saturated, and therefore acting at their maximum rates, under normal physiological conditions [168, 169].

Single pass perfusion studies have demonstrated the presence of the same substrate-specific classes of carriers as have studies with brain slices – on the basis of self-inhibition of the uptake of labelled amino acids and cross-inhibition studies, Oldendorf and Szabo [170] proposed the existence of a single neutral amino acid carrier, with separate basic and (low capacity) acidic transport systems. The acidic amino acid carrier was not detected in early studies with single pass perfusion when the arterio-venous difference was measured as an index of brain uptake [162] because of the very low rate of extraction. Later studies [170] used the uptake of label into the brain as the index of transport, and revealed the low capacity acidic amino acid carrier.

On the basis of inhibitor sensitivity and substrate specificity, it is possible to distinguish at least three different transport systems for neutral amino acids in the brain capillary bed – the L (leucine) and A (alanine) systems described above and a separate alanine–serine–cysteine (ASC) system. The L-system has been identified on both the luminal and anti-luminal faces of the epithelial cell, while the ASC-system seems to be solely luminal and the A-system is anti-luminal, and therefore concerned with amino acid efflux rather than uptake [171, 172]. Wade and Brady [173] have suggested that the ASC-system does not transport cysteine to any significant extent in the brain capillaries, since alanine and serine did not inhibit cysteine uptake in carotid infusion studies, and inhibition of the L-system with 2-amino-bicyclo-(2, 2, 1)-heptane-2-carboxylate readily inhibited cysteine uptake, but not that of alanine or serine.

Pardridge [174] has compared the observed rates of amino acid uptake into the brain with values calculated from the affinities and maximum rates of transport determined by single pass perfusion studies. In general, there is good agreement between the observed and theoretical rates, although the uptake of phenylalanine is twice as fast as predicted, suggesting that there may be an additional carrier, as yet unidentified. The rate of tryptophan uptake is about half that predicted; this may be the result of the binding of tryptophan to serum albumin, which reduces the concentration available for transport to lower than that which was used for the calculations – the total plasma tryptophan concentration.

(a) The binding of tryptophan to serum albumin

McMenamy and Oncley [175] first demonstrated that tryptophan was bound non-covalently to serum albumin. The binding is specific, and only a limited number of analogues and metabolites will compete with tryptophan. The equilibrium is a rapid one and such that, under physiological conditions, only about 10% of plasma tryptophan is freely diffusible; the remainder is associated with albumin. Albumin binding is specific for L-tryptophan; the D-isomer does not bind to any significant extent, and indeed this has been exploited as a means of resolving racemic mixtures of tryptophan, by chromatography on columns of immobilized serum albumin [176].

Non-esterified fatty acids will displace tryptophan from albumin binding [177]. This may be a significant factor *in vivo*; the addition of palmitate over the physiological range of 0.3–1.2 mM leads to an increase in the percentage of tryptophan that is free, and changes the dissociation constant of the tryptophan–albumin complex [178]. It seems likely that fatty acids either cause a change in the conformation of albumin on binding, so altering the affinity of the tryptophan binding site, or overlap the tryptophan site to some extent.

A number of drugs, including salicylate [179], indomethacin, benzoate and clofibrate [180, 181] and several phenothiazines with anti-schizophrenia activity [182] also displace tryptophan from albumin binding. The displacement is either direct (and hence demonstrable *in vitro*) or by some means that is detectable only *in vivo*, but not when the drug is added to solutions of albumin [182]. The ability of different drugs to displace tryptophan from albumin binding is not related to the affinity of albumin for the drugs, and it therefore seems likely that, as for fatty acid displacement of tryptophan, the effect is due to either overlapping of the binding sites or a conformational change in albumin [183].

Several different methods have been used to estimate the binding of tryptophan to albumin. McMenamy and Oncley used relatively large-scale dialysis [175], while Smith and Lakatos [179] used reversed phase dialysis, with a small sac of saline suspended in a relatively large volume of serum. Bender and coworkers [178] used small scale equilibrium dialysis with a [^{14}C]-tryptophan tracer. The use of [5-^{3}H]-tryptophan leads to anomalous results, because of isotope exchange and the production of $^{3}H_2O$, although the results with tryptophan tritiated in other positions agree with those obtained with [^{14}C]-tryptophan (Bender, unpublished observations). It is known that 5-substituted tryptophan derivatives will not bind to albumin [175, 177], and the isotope exchange effect suggests that the 5-position of tryptophan is involved in the binding reaction. Other methods to assess the binding of tryptophan to albumin have depended on gel filtration, and various forms of ultra-filtration. Different workers have used temperatures ranging from 4°C, through 'room temperature' to 37°C; some have attempted to control pH and others have not. Baumann and Perry have reviewed the various methods used, and have shown that despite the wide range of conditions used, there is generally good agreement between different workers [184]. At a meeting of workers in the field held in Lausanne in July 1978, it was agreed that, as far as possible, studies should be based on the use of ultra-filtration to determine the extent of tryptophan binding to albumin, so as to permit more precise comparison between different laboratories.

The binding of tryptophan to serum albumin may be important in controlling the uptake of tryptophan into the brain. Curzon and coworkers [185–187] have shown that a number of experimental manipulations that increase plasma free fatty acids, and therefore lead to displacement of tryptophan from binding, lead to an increase in the uptake of tryptophan into the brain. Similarly, they have shown that insulin and nicotinic acid, both of which reduce free fatty acids, also increase the binding of tryptophan to albumin, and reduce its uptake into the brain. Salicylates displace tryptophan from albumin [179], and also increase brain uptake [188]. After the administration of clofibrate to rats there is a reduction in the total plasma concentration of tryptophan, but a considerable displacement from binding, and an increase in the rate of uptake into the brain [181].

Yuwiler and coworkers [189] suggested that the binding of tryptophan to albumin would not be a significant factor in the rate of uptake into the brain, since the equilibrium of binding is so rapid that tryptophan would be stripped off albumin during a single pass through the brain capillary bed. Therefore the concentration of tryptophan available for uptake would be the same as the total amount present in plasma, rather than only free tryptophan. However, this may not be so; although the albumin-bound tryptophan represents a considerable buffer pool that can replenish the free pool, and is considerably greater than the free pool, nevertheless, what is immediately available to the transport system, and is competing with other amino acids for uptake, is the 10% or so that is free. The role of albumin-bound tryptophan is to maintain this concentration at a more or less constant level despite uptake. Pardridge [174] showed that the dissociation constant of the tryptophan–albumin complex was of the same order of magnitude as the K_m of the uptake mechanism, and therefore these two would compete for the free tryptophan. In other words, the albumin-bound pool of tryptophan would be available to replenish the free pool, but would not be apparent to the uptake system.

Etienne and coworkers [190] showed that the addition of albumin to the infusion bolus in single pass perfusion studies reduced the rate of tryptophan uptake into the brain, while it had no effect on the uptake of tyrosine, which is not albumin-bound. This can be considered as further evidence of the importance of the albumin binding of tryptophan in controlling the rate of uptake into the brain, and hence of the importance of the effects of drugs that modify this binding, either directly or by modifying the plasma concentrations of free fatty acids.

(b) The physiological importance of brain amino acid uptake

For a number of amino acids that are the precursors of neurotransmitters or other important metabolic intermediates in the central nervous system, it has been suggested that uptake into the brain is a rate-limiting factor in their utilization, and that an increase in their uptake will lead to an increase in the rate of formation of the product. This is because the rate-limiting enzyme of the pathway involved is not saturated with its substrate at the normal steady state concentration in the brain, and therefore any change in the availability of substrate will lead to an increase in the rate of reaction. The physiological significance of this will depend on the extent to which the enzyme is unsaturated; if the steady state concentration of substrate is close to the K_m of the enzyme then a relatively small change in substrate availability will make little difference to the rate of reaction, while if the concentration of substrate is very much lower than the K_m of the enzyme, even a small change in substrate will result in quite a large change in the rate of reaction.

Methionine

Rubin and coworkers [191] have suggested that the availability of methionine in the brain may be a regulatory factor in the synthesis of *S*-adenosylmethionine; administration of methionine at such a level that the plasma concentration was increased, although still within the normal physiological range, led to an increase in the concentrations of methionine and *S*-adenosylmethionine in the brain. Feeding a low-protein diet, which did not affect the plasma concentration of methionine, but reduced that of tyrosine, also led to an increase in brain uptake of methionine and synthesis of *S*-adenosylmethionine. They showed that the K_m of methionine adenosyl transferase (ATP:L-methionine adenosyltransferase, EC 2.5.1.6) was 900 μM, while the concentration of methionine in the brain was in the range 30–50 μmol kg^{-1}, suggesting that the effect was the result of considerable unsaturation of the enzyme with its substrate.

Histidine

Schwartz and coworkers [192] showed that the administration of relatively large amounts of histidine to rats (500 mg per kg body weight) led to a significant increase in the brain content of histamine. At the same time there was a reduction in the activity of histidine decarboxylase (L-histidine carboxy-lyase, EC. 4.1.1.22), which they suggested was the result of feed-back repression of enzyme synthesis by its product; histamine had no effect on the activity of the preformed enzyme *in vitro*.

They reported a concentration of histidine in the brain of the order of 7 mmol kg^{-1}, and a K_m of histidine decarboxylase of 0.4 mM; under these conditions the enzyme would be either saturated or very nearly so, and it is unlikely that the availability of histidine would be a major factor in the regulation of histamine synthesis. Indeed, they cited a number of other workers who had failed to show any effect of the administration of histidine on brain histamine synthesis.

Tyrosine

Fernstrom and Wurtman [193] have suggested that the availability of tyrosine in the brain is an important factor in the regulation of the synthesis of the catecholamines, dopamine, noradrenaline and adrenaline. They reported a steady state concentration of tyrosine in the brain of the order of 83 μmol kg^{-1}, and said that this was of the same order of magnitude as the K_m of tyrosine hydroxylase (L-tyrosine, tetrahydrobiopterin: oxygen oxido-reductase (3-hydroxylating) EC 1.14.16.2). However, this was based on estimates of the K_m of the enzyme that had been determined using synthetic pterin cofactors; with the physiological cofactor, tetrahydrobiopterin, the K_m of the enzyme is nearer 7 μM, and it is therefore likely that the enzyme will be more or less fully saturated with substrate under normal conditions [194].

A further factor that may be involved is the observed inhibition of tyrosine hydroxylase by an excess of its substrate, which suggests that it is unlikely that an increase in the concentration of tyrosine in the brain will lead to an increase in the rate of catecholamine synthesis [194].

In vivo the situation may be confused yet further by the observation that tyrosine hydroxylase exists in a soluble, relatively inactive, form and a particulate, active, form. Activation of the enzyme, which is achieved by a variety of physiological inputs, not only increases the rate of reaction, but also reduces the feedback inhibition by dopamine [195].

Tryptophan

Eccleston and coworkers [196] showed that the administration of relatively large amounts of tryptophan to rats led to an increase in the synthesis of the neurotransmitter amine 5-hydroxytryptamine, as well as an increase in the production of its principal metabolite, 5-hydroxyindole-acetic acid. Fernstrom and Wurtman [197] showed that changes in plasma tryptophan within the normal physiological range also affected the synthesis of 5-hydroxy-tryptophan, and Tagliamonte and coworkers [198] showed that a number of drugs that affect the rate of 5-hydroxytryptamine synthesis do so by increasing the concentration of tryptophan in the brain.

Curzon and Green [199] showed that the fall in circulating tryptophan that follows the administration of hydrocortisone to rats (the result of induction of tryptophan oxygenase) is accompanied by a reduction in the synthesis of 5-hydroxytryptamine in the brain. Similarly, the increased protein synthesis in response to growth hormone administration leads to a reduction in the circulating concentration of tryptophan, and a reduction in the rate of synthesis of 5-hydroxytryptamine [200].

The K_m of tryptophan hydroxylase in the brain, when measured using the physiological cofactor, tetrahydrobiopterin, is of the order of 50 μM, while estimates of the concentration of tryptophan in the brain under normal conditions range between 5–25 μmol kg^{-1} [201]. The enzyme will therefore be very much unsaturated under normal conditions.

This means that the rate of synthesis of 5-hydroxytryptamine in the brain will depend to a very great extent on the availability of tryptophan; this in turn will be determined by the activity of competing pathways of tryptophan utilization and the rate of uptake into the brain. There is a great deal of evidence to suggest that factors that affect either the total concentration of tryptophan in plasma, or the extent of binding of tryptophan to serum albumin, as well as the concentrations of amino acids that compete for uptake via the L-carrier system, all have effects on the rate of 5-hydroxytryptamine synthesis [202]. However, there is also evidence that the uptake of tryptophan into individual neurons may be more closely regulated than the uptake of tryptophan across the blood–brain barrier [203].

(c) The enzymology of amino acid transport

Meister and coworkers [204–207] have proposed that amino acids may be transported across cell membranes as γ-glutamyl peptides, which are formed by transpeptidation at the outer surface, with glutathione acting as the γ-glutamyl donor, and hydrolysed at the inner surface of the membrane to release the free amino acid and 5-oxo-proline (pyroglutamic acid or 2-pyrrolidone-5-carboxylate) from cyclization of the γ-glutamyl moiety. The 5-oxo-proline is then used for the resynthesis of glutathione. They have called the pathway the 'γ-glutamyl cycle'; it is shown in Fig. 5.5.

The cycle would account for the very high rate of glutathione turnover in tissues; Tateishi and Higashi [208] have observed two separate pools in rat liver, with half-lives of the order of 1.7 and 28.5 hours. It would also explain why there is such a high concentration of glutathione in rat liver – of the order of 7–8 mmol kg^{-1}, which is far in excess of the amount required for the maintenance of reduced sulphydryl groups in proteins. The cycle also provides a role for ATP in amino acid transport, and hence explains the energy dependency of the process.

Histochemical studies have located the key enzyme of the cycle, γ-glutamyl transpeptidase, in the brush border of the proximal convoluted tubule of the kidney [209, 210], in the apical portion of the epithelial cells of the jejunal villi [211] and in the choroid plexus and brain [212] – all sites where a key enzyme of amino acid transport would be expected.

The specificity of amino acid transport seems not to be a property of γ-glutamyl transpeptidase, which will form γ-glutamyl peptides with all the protein amino acids except proline [205]. It therefore seems likely that there are specific amino acid binding sites on the outer surface of cell membranes that confer the observed specificity of transport discussed above.

The formulation of the γ-glutamyl cycle still begs the key question of amino acid transport – the process of translocation of the amino acid or the derived γ-glutamyl peptide across the cell membrane from the outer surface to the inside of the cell. It is also unable to account for the observed sodium dependence of amino acid transport, although it does account for the relatively high requirement for ATP; the action of 5-oxo-prolinase is an unusual one in that it represents effectively an ATP-dependent hydrolysis of a peptide bond [213].

There have been two reports of a patient with an apparent inborn error of the γ-glutamyl cycle, who excreted large amounts of 5-oxo-proline, and was presumably lacking 5-oxo-prolinase. The administration of amino acids at such a level as to raise the plasma concentrations 3-fold led to a doubling of the excretion of 5-oxo-proline, as well as aminoaciduria. Under basal conditions he showed no aminoaciduria, suggesting that there must be just adequate capacity

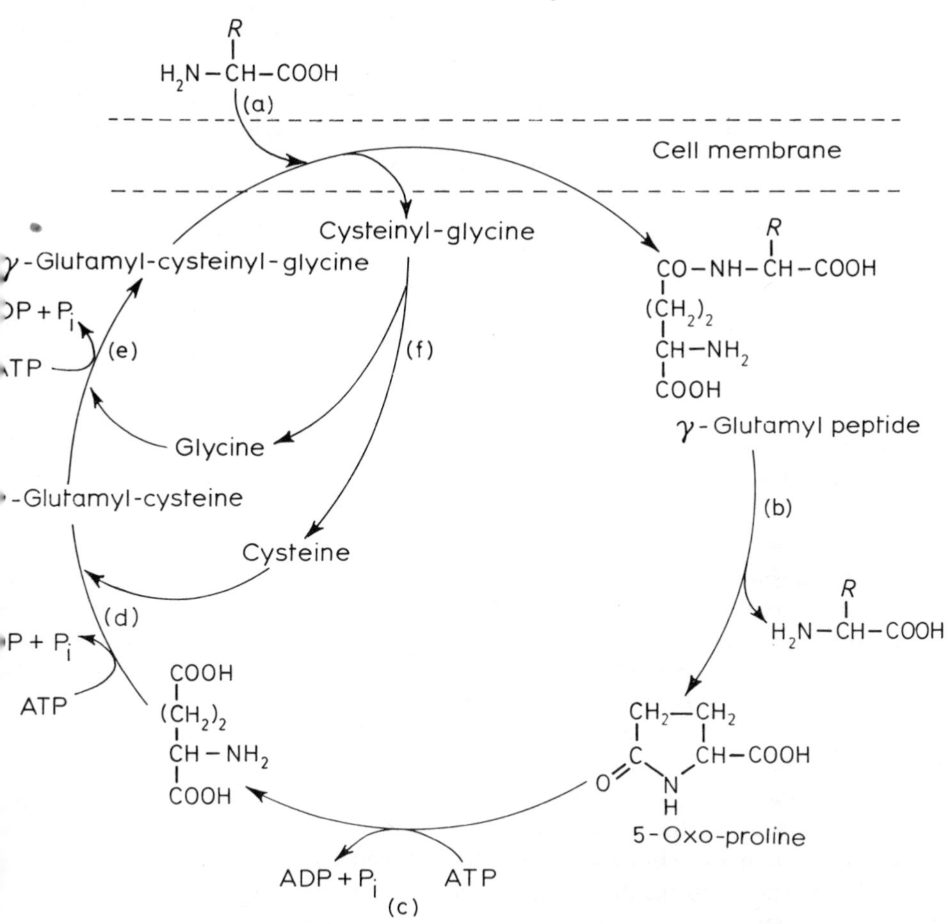

Figure 5.5 The γ-glutamyl cycle. (a) γ-Glutamyl transpeptidase ((5-glutamyl)-peptide: amino acid 5-glutamyltransferase, EC 2.3.2.2); (b) γ-glutamyl cyclotransferase ((5-glutamyl)-L-amino acid 5-glutamyltransferase (cyclizing), EC 2.3.2.4); (c)5-oxo-prolinase (5-oxo-L-proline amido-hydrolase (ATP hydrolysing), EC 3.5.2.9); (d) γ-glutamyl-cysteine synthetase (L-glutamate: L-cysteine γ-ligase (ADP)-forming, EC 6.3.2.2); (e) glutathione synthetase (γ-L-glutamate-L-cysteine: glycine ligase (ADP-forming), EC 6.3.2.3); (f) dipeptidase.

to synthesize glutathione under normal conditions despite the loss of 5-oxo-proline, but after the increasing of amino acid load this capacity was exceeded, resulting in aminoaciduria [214, 215]. Meister [204] has calculated that the amount of 5-oxo-proline excreted by this patient was approximately that which might be expected to be formed (and normally re-utilized) in the re-uptake of the normal filtered load of 0.2–0.4 mol of amino acids per day.

Two patients have been reported in whom there was a severe haemolytic anaemia associated with a deficiency of γ-glutamyl-cysteine synthetase, so that glutathione synthesis was impaired. They showed a generalized aminoaciduria, which would be expected if glutathione were important in amino acid transport [216].

One patient has been reported in whom there was an inborn error of metabolism that was not compatible with the hypothesis that the γ-glutamyl cycle is important in amino acid transport. This was a patient who had apparently no activity of γ-glutamyl transpeptidase, and showed hyperglutathionaemia and glutathionuria. However, renal resorption of amino acids appeared to be normal, and there was no aminoaciduria [217].

5.7 AMINO ACIDS FORMED BY POST-SYNTHETIC MODIFICATION OF PROTEINS

A number of amino acids are found in proteins, yet are not incorporated during protein synthesis, and have no specific tRNA. Such amino acids include phosphoserine, 3- and 4-hydroxyproline, 5-hydroxylysine, γ-carboxyglutamate and 3-methylhistidine, which was discussed above.

Phosphoserine is an important amino acid in a number of secretory proteins, including casein, and the phosphorylation occurs during the process of translocation from the cell. In a number of enzymes, serine residues are reversibly phosphorylated as a mechanism of regulation; the phosphorylation may increase or decrease the activity of the enzyme. The protein kinases and phosphoprotein phosphatases involved are generally activated and inactivated by cyclic AMP and cyclic GMP acting as second messengers to hormone stimulation. This has been reviewed by Carlson and coworkers [218].

Elucidation of the mechanisms involved in the synthesis of hydroxyproline and hydroxylysine in the maturation of collagen, and in the synthesis of γ-carboxyglutamate in prothrombin and other proteins has enabled us to define the molecular mechanisms of action of vitamins C and K respectively.

5.7.1 Hydroxyproline and hydroxylysine

Hydroxyproline and hydroxylysine are found in very few proteins: collagen and elastin (the principal proteins of connective tissue), the C_{1q} component of complement and the tail structure of acetyl cholinesterase. The most studied system has been the synthesis of collagen by isolated skin fibroblasts in tissue culture.

There are at least five different types of collagen, which differ in their hydroxyamino acid content and the degree of glycosylation of hydroxylysine residues, found in different types of connective tissue [219, 220]. Most of the hydroxyproline is hydroxylated in the 4-position, but some forms of connective tissue also contain relatively large amounts of 3-hydroxyproline.

Hydroxylation seems to be essential for the stability of the triple helical structure of collagen. Although unhydroxylated collagen molecules will form a helix *in vitro* at low temperatures, it is not stable at physiological temperatures [220].

Overall the formula of the α-chain of collagen is $(X\text{-}Y\text{-}Gly)_{333}$; about 100 of the X sites are occupied by proline, and about 100 of the Y sites by hydroxyproline, which is incorporated as proline and is hydroxylated during the post-synthetic processing of the procollagen molecule. The pro-α chains that are synthesized on the ribosome are considerably longer than the α-chains of mature collagen, and removal of the amino and carboxy terminal regions is the final stage in the maturation and export of collagen from the fibroblast. While the nascent polypeptide chain is still attached to the ribosome, appropriate proline and lysine residues are hydroxylated; three separate enzymes are involved, procollagen proline 3- and 4-hydroxylases and procollagen lysine hydroxylase [221]. This followed by glycosylation of hydroxylysine residues, formation of an initially hydrogen-bonded triple helix and proteolysis to remove the terminal peptides.

Proline and lysine hydroxylases are the two most studied of a group of hydroxylases that contain iron and are dependent on ascorbate (vitamin C) for activity. The hydroxylation of the substrate is obligatorily linked to the oxidative decarboxylation of 2-oxo-glutarate to succinate [222]. The iron must be in the Fe(II) state for activity; *in vitro* proline hydroxylase will catalyse some 15–30 cycles of enzyme activity (5–10 s) without the addition of ascorbate, but thereafter the rate of reaction falls to very low levels unless ascorbate is added. At this stage the iron has been oxidized to the Fe(III) state, and will not exchange with Fe(II) in the incubation medium, although the reduced form of the enzyme will exchange its Fe(II) with the medium. Ascorbate is specifically required for the reduction of the enzyme-bound Fe(III) and other reducing agents will not substitute. Myllylä and coworkers [223] have suggested that the oxidation of the iron of the enzyme is the result of a side reaction rather than a part of the principal action, since it is possible for a number of cycles of activity to be completed before there is any appreciable oxidation of Fe(II) and loss of activity.

The collagen molecules that are extruded from the cell spontaneously form fibrils that appear to be the same as those of mature collagen, but lack the stability and tensile strength that come with cross-linkage. Three stages are involved in the cross-linkage of collagen:

(i) There is oxidative de-amination of the ε-amino group of lysine and hydroxylysine residues to form reactive aldehydes, a reaction catalysed by lysyl oxidase. This is a copper-containing enzyme, and it is noteworthy that in copper deficiency in chickens and other animals the connective tissue is abnormally weak, and indeed a common cause of death is aortic rupture due to the low elasticity of the abnormal elastin that is formed. The enzyme is also inhibited by β-aminopropionitrile, the toxin of *Lathyrus* species peas that is responsible for the disease of lathyrism. Aortic rupture is a common occurrence in lathyrism.

Lysyl oxidase reacts poorly with isolated procollagen α-chains, and requires the triple helical structure for maximum activity [224, 225].

(ii) Intramolecular cross-links involve the aldol condensation of two of the lysine or hydroxylysine reactive aldehydes in adjacent chains of the same collagen molecule.

(iii) Inter-chain links are formed initially by Schiff base formation between hydroxylysine or lysine residues and reactive aldehydes formed from lysine, hydroxylysine or glycosylated hydroxylysine.

A number of rearrangements of these Schiff bases can then occur, to form a variety of inter-chain links, including rearrangement to form a ketone, hydration and rearrangement to form a peptide bond and a complex rearrangement that also includes the incorporation of a histidine residue in the link [220].

5.7.2 γ-Carboxyglutamate

It has long been known that in vitamin K deficiency the synthesis of prothrombin is affected, and as a result there is defective blood clotting. However, it was not until 1974 that the difference between normal and vitamin K-deficient prothrombin (now called pre-prothrombin) was elucidated. Three groups of workers [226–228] showed that a new amino acid, γ-carboxyglutamic acid, was present in prothrombin and not in pre-prothrombin. They noted that the reason that it had not been discovered before was that it is unstable to acid, and therefore would not be detectable in the acid hydrolysates that are usually used for determination of protein structure; it decarboxylates spontaneously to glutamate at low pH. All ten of the glutamate residues in the amino terminal 40-amino-acid sequence of prothrombin are normally carboxylated.

Shah and Suttie [229] showed that the carboxylation occurred after synthesis of the protein, was vitamin K-dependent, and occurred in liver microsomes. The vitamin K-dependent carboxylase was then also identified in kidney and bone, both of which contain calcium binding proteins that are rich in γ-carboxyglutamate. In bone as much as 1–2% of the total protein may be the calcium binding protein osteocalcin [230, 231].

The microsomal vitamin K-dependent carboxylase has been solubilized, and has been shown to require a peptide substrate rather than free glutamate, oxygen and either vitamin K and NADPH or the quinol of vitamin K. The immediate source of carbon dioxide is CO_2 rather than bicarbonate [232–234].

In isolated microsomes the carboxylation of glutamate residues is associated with the formation of vitamin K epoxide. Although it is not known whether this is an obligatory part of the reaction or not, since the oxidation and peroxidation of vitamin K can be uncoupled from the carboxylation of glutamate, it is known that the coumarin anti-coagulants inhibit vitamin K epoxide reductase and lead to an accumulation of the inactive epoxide. Warfarin, another widely used anti-coagulant, acts to inhibit both the reduction of vitamin K epoxide to the quinone, and also the reduction of the quinone to the quinol which is the final active form [235, 236]. The reaction sequence of glutamate carboxylation and vitamin K reduction is shown in Fig. 5.6.

The physiological function of γ-carboxyglutamate residues in proteins seems to be the chelation of calcium. Pre-prothrombin, with no γ-carboxyglutamate binds less than 1 mol of calcium per mol of protein, while prothrombin that is fully

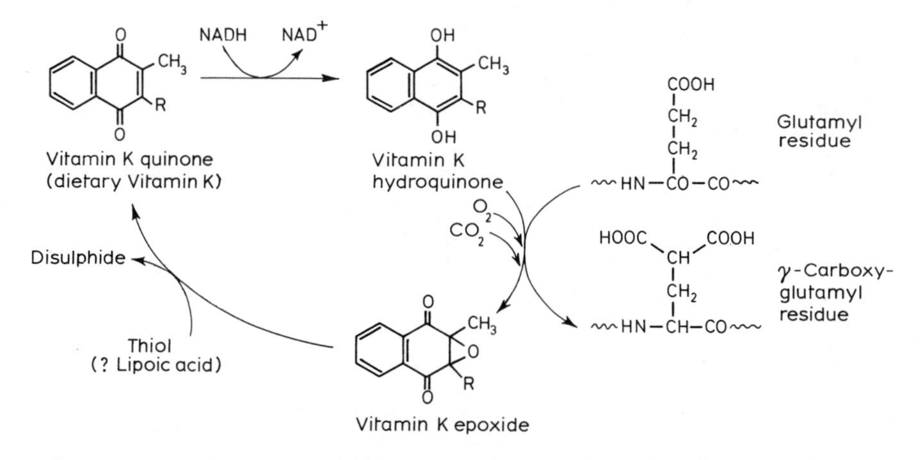

Figure 5.6 The role of vitamin K in γ-carboxyglutamate synthesis. The enzymes involved have not yet been characterized, and the thiol compound involved in the reduction of the vitamin epoxide to the quinone has not been identified, but is thought to be lipoic acid. It is this reduction which is inhibited by the vitamin K antimetabolite Warfarin.

carboxylated binds 10–12 mol of calcium per mol. This calcium appears to be essential for the binding of prothrombin onto phospholipid-rich membranes for activation to thrombin by limited proteolysis [237].

The γ-carboxyglutamate-rich protein of bone, osteocalcin, is formed in bone matrix just before mineralization, and the osteocalcin content of bone increases with increasing bone density. Hall and coworkers [238] have suggested that the fetal Warfarin syndrome of defective bone formation and mineralization in infants whose mothers took Warfarin during the first trimester of pregnancy is the result of inhibition of glutamate carboxylation, and hence defective synthesis of osteocalcin at a critical stage of bone formation.

GENERAL FURTHER READING

This review has concentrated on a limited number of areas of amino acid metabolism in which there have been considerable advances over the last two decades, and a number of well-established pathways of amino acid metabolism have been omitted or mentioned only briefly. There are two standard books that will provide much of the missing information:

Meister, A. (1965) *Biochemistry of the Amino* Acids, Academic Press, New York.
Bender, D.A. (1975) *Amino Acid Metabolism*, John Wiley & Sons Chichester.

REFERENCES

1. Munro, H.H. and Young, V.R. (1981) *Recent Advances in Clinical Nutrition*, (eds A. Howard and I. McLean-Baird) (Vol. 1) John Libby, London, p. 33.
2. Munro, H.N. (1964) in *Mammalian Protein Metabolism*, (eds H.N. Munro and J.B. Allison) (Vol. 1) Academic Press, New York, p. 1.

3. Willcock, E.G. and Hopkins, F.G. (1906) *J. Nutr.*, **35**, 88.
4. Osborne, T.B. and Mendel, L.B. (1914) *J. Biol. Chem.*, **17**, 325.
5. Bergstrom, J., Furst, P., Josephson, B. and Noree L-O. (1970) *Life Sci.*, pt 2, **9**, 787.
6. Kopple, J.D. and Swendseid, M.E. (1975) *J. Clin. Invest.*, **55**, 881.
7. Schoenheimer, R. (1946) *The Dynamic State of Body Constituents*, Harvard University Press, Cambridge, Mass.
8. Bird, J.W.C., Berg, T. and Leathem, J.H. (1968) *Proc. Soc. Exp. Biol. Med.*, **127**, 182.
9. Obled, C., Arnal, M. and Valin, C. (1980) *Brit. J. Nutr.*, **44**, 61.
10. Nakamura, T., Shinno, H. and Ichihara, A. (1980) *J. Biol. Chem.*, **255**, 7533.
11. Schimke, R.T., Sweeney, E.W. and Berlin, C.M. (1965) *J. Biol. Chem.*, **240**, 322.
12. Schimke, R.T., Sweeney, E.W. and Berlin, C.M. (1965) *J. Biol. Chem.*, **240**, 4609.
13. FAO/WHO (1973) *FAO Nutr. Rpt.* No. 52 (WHO Tech. Rpt. Ser. No. 522), FAO/WHO, Rome and Geneva.
14. Oser, B.L. (1951) *J. Amer. Diet. Ass.*, **27**, 396.
15. Hurrell, R.F. and Carpenter, K.J. (1974) *Brit. J. Nutr.*, **32**, 589.
16. Iyengar, A.K., Narasinga Rao, B.S. and Reddy, V. (1979) *Brit. J. Nutr.*, **42**, 417.
17. Iyengar, A. and Narasinga Rao, B.S. (1979) *Brit. J. Nutr.*, **41**, 19.
18. Iyengar, A.K., Narasinga Rao, B.S. and Reddy, V. (1981) *Brit. J. Nutr.*, **46**, 295.
19. Garza, C., Scrimshaw, N.S. and Young, V.R. (1977) *Brit. J. Nutr.*, **37**, 402.
20. Reeds, P.J., Fuller, M.F., Cadenhead, A. *et al.* (1981) *Brit. J. Nutr.*, **45**, 539.
21. Woodham, A.A. and Deans, P.S. (1977) *Brit. J. Nutr.*, **37**, 289.
22. Woodham, A.A. and Clarke, E.M.W. (1977) *Brit. J. Nutr.*, **37**, 309.
23. Hier, S.W., Graham, C.E. and Klein, D. (1944) *Proc. Soc. Exp. Biol. Med.*, **56**, 187.
24. Harper, A.E., Benevenga, N.J. and Wohlhuetter, R.M. (1970) *Physiol. Revs*, **50**, 428.
25. Hughes, J.V. and Johnson, T.C. (1977) *Biochem. J.*, **162**, 527.
26. Eggum, B.O., Back-Knudsen, K.E. and Jacobsen, I. (1981) *Brit. J. Nutr.*, **45**, 175.
27. Bender, A.E. (1961) *Meeting Protein Needs of Children* (Publ. no. 843), National Academy of Science, National Research Council, Washington, p. 407.
28. Waterlow, J.C. and Stephen, J.M.L. (1967) *Clin. Sci.*, **33**, 489.
29. Dunlop, D.S., van Elden, W. and Lajtha, A. (1974) *J. Neurochem.*, **22**, 821.
30. Dunlop, D.S., van Elden, W. and Lajtha, A. (1977) *J. Neurochem.*, **29**, 939.
31. Dunlop, D.S., van Elden, W. and Lajtha, A. (1978) *Biochem. J.*, **170**, 637.
32. Dunlop, D.S., van Elden, W., Plucinska, I. and Lajtha, A. (1981) *J. Neurochem.*, **36**, 258.
33. Halliday, D. and McKeran, R.O. (1975) *Clin. Sci. Molec. Med.*, **49**, 581.
34. McFarlane, A.S. (1963) *Biochem. J.*, **89**, 277.
35. Antonas, K.N., Coulson, W.F. and Jepson, J.B. (1974) *Biochem. Soc. Trans*, **2**, 105.
36. Waterlow, J.C. (1981) *Proc. Nutr. Soc.*, **40**, 317.
37. Golden, M.H.N. and Waterlow, J.C. (1977) *Clin. Sci. Molec. Med.*, **53**, 277.
38. Taruvinga, M., Jackson, A.A. and Golden, M.H.N. (1979) *Clin. Sci. Molec. Med.*, **57**, 281.
39. Garlick, P.J., Clugston, G.A. and Waterlow, J.C. (1980) *Amer. J. Physiol.*, **238**, E235.
40. Fern, E.B., Garlick, P.J., McNurlan. M.A. and Powell-Tuck, J. (1981) *Proc. Nutr. Soc.*, **40**, 88A.
41. Waterlow, J.C. (1967) *Clin. Sci. Molec. Med.*, **33**, 507.
42. Waterlow, J.C., Garlick, P.J. and Millward, D.J. (1978) *Protein Turnover in Mammalian Tissues and in the Whole Body*, Associated Scientific Publishers/North Holland, Amsterdam and New York.
43. Picou, D. and Taylor-Roberts, T. (1969) *Clin. Sci. Molec. Med.*, **36**, 283.
44. Reeds, P.J. and Lobley, G.E. (1980) *Proc. Nutr. Soc.*, **39**, 43.
45. Young, V.R. and Bier, D.M. (1981) *Proc. Nutr. Soc.*, **40**, 343.
46. Gersowitz, M., Munro, H.N., Udall, J. and Young, V.R. (1980) *Metabolism*, **29**, 1075.
47. Young, V.R., Steffee, W.P., Pencharz, P.B. *et al.* (1975) *Nature*, **253**, 192.
48. Asatoor, A.M. and Armstrong, M.D. (1967) *Biochem. Biophys. Res. Commun.*, **26**, 168.
49. Johnson, P., Harris, C.I. and Perry, S.V. (1967) *Biochem. J.*, **105**, 361.
50. Young, V.R., Alexis, S.D., Baliga, B.S. *et al.* (1972) *J. Biol. Chem.*, **247**, 3592.
51. Elia, M., Carter, A. and Smith, R. (1979) *Brit. J. Nutr.*, **42**, 567.
52. Young, V.R., Haverberg, L.N., Bilmazes, C. and Munro, H.N. (1973) *Metabolism*, **22**, 1429.
53. Nishizawa, N., Shimbo, M., Hareyama, S. and Funakiki, R. (1977) *Brit. J. Nutr.*, **37**, 345.
54. Haverberg, L.N., Deckelbaum, L., Bilmazes, C. *et al.* (1975) *Biochem. J.*, **152**, 503.
55. Milne, G. and Harris, C.I. (1978) *Proc. Nutr. Soc.*, **37**, 18A.

56. Harris, C.I. and Milne, G. (1980) *Brit. J. Nutr.*, **44**, 129.
57. Harris, C.I. and Milne, G. (1981) *Brit. J. Nutr.*, **45**, 411.
58. Harris, C.I. and Milne, G. (1981) *Brit. J. Nutr.*, **45**, 423.
59. Bates, P.C., Grimble, G.K. and Millward, D.J. (1979) *Proc. Nutr. Soc.*, **38**, 136A.
60. Bates, P.C. and Millward, D.J. (1981) *Proc. Nutr. Soc.*, **40**, 89A.
61. Krebs, H.A. (1976) in *The Urea Cycle* (eds S. Grisolia, R. Bagueña and F. Mayor), John Wiley and Sons, New York, p. 1.
62. Raijman, L. (1976) in *The Urea Cycle* (eds S. Grisolia, R. Bagueña and F. Mayor), John Wiley and Sons, New York, p. 243.
63. Schimke, R.T. (1962) *J. Biol. Chem.*, **237**, 459.
64. Schimke, R.T. (1962) *J. Biol. Chem.*, **237**, 1921.
65. Schimke, R.T., Brown, M.B. and Smallman, E.T. (1963) *Ann. N. Y. Acad. Sci.*, **102**, 587.
66. Schimke, R.T. (1964) *J. Biol. Chem.*, **237**, 3808.
67. McGiven, J.D. and Chappell, J.B. (1975) *FEBS Letters*, **52**, 1.
68. Krebs, H.A. and Lund, P. (1977) *Adv. Enz. Reg.*, **15**, 375.
69. Regoeczi, E., Erons, L., Koj, A. and McFarlane, A.S. (1965) *Biochem. J.*, **95**, 521.
70. Wrong, O.M. (1978) *Amer. J. Clin. Nutr.*, **31**, 1587.
71. Mitch, W.E. (1978) *Amer. J. Clin. Nutr.*, **31**, 1594.
72. Varcoe, A.R., Halliday, D., Carson, E.R. *et al.* (1978) *Amer. J. Clin. Nutr.*, **31**, 1601.
73. Giordiano, C. (1966) *J. Lab. Clin. Med.*, **62**, 231.
74. Richards, P., Metcalfe-Gibson, A., Ward, E.E. *et al.* (1967) *Lancet*, **ii**, 845.
75. Chow, K-W. and Walser, M. (1974) *J. Nutr.*, **104**, 1208.
76. Walser, M., Lund, P., Ruderman, N.B. and Coulter, A.W. (1973) *J. Clin. Invest.*, **52**, 2865.
77. Chow, K-W. and Walser, M. (1975) *J. Nutr.*, **105**, 119.
78. Chow, K-W. and Walser, M. (1975) *J. Nutr.*, **105**, 372.
79. Shih, V.E. (1976) in *The Urea Cycle* (eds S. Grisolia, R. Bagueña and F. Mayor), John Wiley & Sons, New York, p. 367.
80. Visek, W.J. (1972) *Fed. Proc.*, **31**, 1178.
81. Levin, B. and Russell, A. (1967) *Amer. J. Dis. Child.*, **113**, 142.
82. Sunshine, P., Lindenbaum, J.E., Levy, H.L. and Freeman, J.M. (1972) *Pediatrics*, **50**, 100.
83. Russell, A. (1969) in *Enzymopenic Anaemias, Lysosomes and Other Papers* (eds J.S. Allen, K.S. Holt, J.T. Ireland and R.S. Pollitt), (*Proc. Soc. Study Inborn Errors Metabolism*), E and S Livingstone, Edinburgh.
84. Brusilow, S.W. and Batshaw, M.L. (1979) *Lancet*, **i**, 124.
85. Brusilow, S., Tinker, J. and Batshaw, M.L. (1980) *Science*, **207**, 659.
86. Gopalan, C. and Srikantia, S.G. (1960) *Lancet*, **i**, 954.
87. Gopalan, C. and Rao, K.S.I. (1975) *Vit. and Horm.*, **33**, 505.
88. Manson, J.A. and Carpenter, K.J. (1978) *J. Nutr.*, **108**, 1883, 1889.
89. Nakagawa, I. and Sasaki, A. (1977) *J. Nutr. Sci. Vitaminol.*, **23**, 535.
90. Patterson, J.I., Brown, R.R., Linkswiler, H. and Harper, A.E. (1980) *Amer. J. Clin. Nutr.*, **33**, 2157.
91. Magboul, B.I. and Bender, D.A. (1983) *Brit. J. Nutr.*, **49**, 321.
92. Frimpter, G.W., Andelman, R.J. and George, W.F. (1969) *Amer. J. Clin. Nutr.*, **27**, 794.
93. Mudd, S.H. (1971) *Fed. Proc.*, 30, 970.
94. Dancis, J. and Levitz, M. (1978) in *The Metabolic Basis of Inherited Disease*, (eds J.B. Stanbury, J.B. Wyngaarden and D.S. Fredrickson), McGraw-Hill, New York, p. 397.
95. Agrawal, H.C., Bone, A.H. and Davison, A.N. (1970) *Biochem. J.*, **117**, 325.
96. Tourian, A.Y. and Sidbury, J.B. (1978) in *The Metabolic Basis of Inherited Disease*, (eds J.B. Stanbury, J.B. Wyngaarden and D.S. Fredrickson), McGraw-Hill, New York, p. 240.
97. Medical Research Council Steering Committee for the MRC/DHSS Phenylketonuria Register (1981) *Brit. Med. J.*, **282**, 1680.
98. Smith, I., Erdohazi, M., MacCartney, F.J. *et al.* (1979) *Lancet*, **i**, 17.
99. Lipton, M.A., Gordon, R., Guroff, G. and Udenfriend, S. (1967) *Science*, **156**, 248.
100. Kelly, C.J. and Johnson, T.C. (1978) *Biochem. J.*, **174**, 931.
101. Engelman, K., Lovenberg, W. and Sjoedsma, A. (1967) *New Engl. J. Med.*, **277**, 1103.
102. Sandler, M. (1968) *Adv. Pharmacol.*, **6B**, 127.
103. Kaufman, S., Holtzman, N.A., Milstein, S. *et al.* (1975) *New Engl. J. Med.*, **293**, 785.
104. Nixon, J.C., Lee, C.-L., Milstein, S. *et al.* (1980) *J. Neurochem.*, **35**, 898.
105. Foster, D.O., Ray, P.D. and Lardy, H.A. (1966) *Biochemistry*, **5**, 563.

106. McDaniel, H.G., Boshell, B.R. and Reddy, W.J. (1973) *Diabetes*, **22**, 713.
107. Green, A.R. and Youdim, M.B.H. (1975) *Brit. J. Pharmacol.*, **55**, 415.
108. Sourkes, T.L. (1977) *Can. Psych. Ass. J.*, **22**, 467.
109. Cotzias, G.C., Papavasiliou, P.S. and Gellene, R. (1969) *New Engl. J. Med.*, **280**, 337.
110. Sauberlich, H.E. (1961) *J. Nutr.*, **75**, 61.
111. Daniel, R.G. and Waisman, H.A. (1969) *J. Neurochem.*, **16**, 787.
112. Thier, S.O. and Segal, S. (1978) in *The Metabolic Basis of Inherited Disease* (eds J.B. Stanbury, J.B. Wyngaarden and D.S. Fredrickson), McGraw-Hill, New York, p. 1578.
113. Bender, D.A. (1980) in *Vitamins in Medicine* (eds B.M. Barker and D.A. Bender) (Vol. 1) William Heinemann Medical, London, p. 1.
114. Coursin, D.B. (1964) *Amer. J. Clin. Nutr.*, **14**, 56.
115. Rose, D.P. and Braidman, I. (1971) *Amer. J. Clin. Nutr.*, **24**, 673.
116. Leklem, J.E., Brown, R.R., Rose, D.P. and Linkswiler, H. (1975) *Amer. J. Clin. Nutr.*, **28**, 535.
117. Bender, D.A. and Wynick, D. (1981) *Brit. J. Nutr.*, **45**, 269.
118. Bender, D.A., Tagoe, C.E. and Vale, J.A. (1982) *Brit. J. Nutr.*, **47**, 609.
119. Wolf, H., Brown, R.R. and Arend, N.A. (1980) *Scand. J. Clin. Lab. Invest.*, **40**, 9.
120. Chanarin, I. (1980) in *Vitamins in Medicine*, Vol. 1. (eds B.M. Barker and D.A. Bender), William Heinemann Medical, London, p. 247.
121. Owen, O.E., Felig, P., Morgan, A.P. *et al.* (1969) *J. Clin. Invest.*, **48**, 574.
122. Pawan, G.L.S. and Semple, S.J.G. (1980) *Proc. Nutr. Soc.*, **39**, 48A.
123. Felig, P. and Wahren, J. (1974) *Fed. Proc.*, **33**, 1092–7.
124. Felig, P. (1975) *Annu. Rev. Biochem.*, **44**, 933.
125. Felig, P. and Wahren, J. (1971) *J. Clin. Invest.*, **50**, 1702.
126. Felig, P. and Wahren, J. (1971) *J. Clin. Invest.*, **50** 2703.
127. Windmueller, H.G. and Spaeth, A.E. (1978) *J. Biol. Chem.*, **253**, 69.
128. Sherwin, R.S., Hendler, R.G. and Felig, P. (1975) *J. Clin. Invest.*, **55**, 1382.
129. Chang, T.W. and Goldberg, A.L. (1978) *J. Biol. Chem.*, **253**, 3677.
130. Chang, T.W. and Goldberg, A.L. (1978) *J. Biol. Chem.*, **253**, 3685.
131. Tischler, M.E. and Goldberg, A.L. (1980) *Amer. J. Physiol.*, **238**, E487.
132. Tischler, M.E. and Goldberg, A.L. (1980) *Amer. J. Physiol.*, **238**, E480.
133. Elia, M., Ilic, V., Bacon, S. *et al.* (1980) *Clin. Sci. Molec. Med.*, **58**, 301.
134. Huigbregste, C.A., Rufa, G.A. and Ray, P.P. (1977) *Biochim. Biophys. Acta*, **499**, 99.
135. Nissim, I. and Lapidot, A. (1979) *Amer. J. Physiol.*, **237**, E418.
136. Wiseman, G. (1956) *J. Physiol.*, **133**, 626.
137. Peters, T.J. (1970) *Gut*, **11**, 720–5.
138. Asatoor, A.M., Cheng, B., Edwards, K.D.G. *et al.* (1970) *Gut*, **11**, 350.
139. Cook, G.C. (1973) *Brit. J. Nutr.*, **30**, 13.
140. Scriver, C.R. and Bergeron, C.R. (1974) in *Heritable Disorders of Amino Acid Metabolism*, (ed. W.L. Nyhan), John Wiley & Sons, New York, pp. 515–92.
141. Wollaston, W.H. (1810) *Proc. Trans. Roy. Soc. Lond.*, **100**, 223.
142. Yeh, H.L., Frankl. W., Dunn, M.S. *et al.* (1947) *Amer. J. Med. Sci.*, **214**, 507–12.
143. Stein, W.H. (1951) *Proc. Soc. Exp. Biol. Med.*, **78**, 705–8.
144. Dent, C.E. and Rose, G.A. (1951) *Quart. J. Med.*, **20**, 205–19.
145. Milne, M.D., Asatoor, A.M., Edwards, H.D.G. and Loughridge, L.W. (1961) *Gut*, **2**, 323–37.
146. Bremer, H.J., Kohne, E. and Endres, W. (1971) *Clin. Chim. Acta*, **32**, 407–18.
147. Thier, S., Segal, S., Fox, M. *et al.* (1965) *J. Clin. Invest.*, **44**, 442–8.
148. Baron, D.N., Dent, C.E., Harris, H. *et al.* (1956) *Lancet*, **i**, 421–8.
149. Jepson, J.B. (1978) in *The Metabolic Basis of Inherited Disease* (eds J.B. Stanbury, J.B. Wyngaarden and D.S. Fredrickson), McGraw-Hill, New York, pp. 1563–78.
150. Scriver, C.R. (1978) in *The Metabolic Basis of Inherited Disease* (eds J.B. Stanbury, J.B. Wyngaarden and D.S. Fredrickson), McGraw-Hill, New York, pp. 1593–1606.
151. Kaser, H., Cottier, P. and Autener, I.J. (1962) *Pediatrics*, **61**, 386–94.
152. Greene, M.L., Lietman, P.S., Rosenberg, L.E. and Seegmiller, J.E. (1973) *Amer. J. Med.*, **54**, 265–71.
153. Hillman, R.E., Albrecht, I. and Rosenberg, L.E. (1968) *J. Biol. Chem.*, **243**, 5566–71.
154. Webber, W.A., Brown, J.L. and Pitts, R.F. (1961) *Amer. J. Physiol.*, **200**, 380–6.
155. Lester, I.T. and Cusworth, D.C. (1973) *Clin. Sci.*, **44**, 99–111.
156. Rosenberg, L.E., Blair, A. and Segal, S. (1961) *Biochim. Biophys. Acta*, **54**, 479–88.

157. Burg, M. and Orloss, J. (1962) *Amer. J. Physiol.*, **203**, 327–30.
158. Burg, M.B., Grantham, J., Abramow, M. and Orloff, J. (1966) *Amer. J. Physiol.*, **210**, 1293–8.
159. Webber, W.A. (1965) *Amer. J. Physiol. Pharmacol.*, **43**, 79–87.
160. Segal, S., Blair, A. and Rosenberg, L.E. (1963) *Biochim. Biophys. Acta*, **71**, 676–87.
161. Pardridge, W.M. and Oldendorf, W.H. (1977) *J. Neurochem.*, **28**, 5–13.
162. Oldendorf, W.H. (1970) *Brain Res.*, **24**, 372–6.
163. Oldendorf, W.H. (1971) *Amer. J. Physiol.*, **221**, 1629–39.
164. Lajtha, A. (1974) in *Aromatic Amino Acids in the Brain*, CIBA Fdn Symp 22, Associated Scientific Publishers, Amsterdam, pp. 25–51.
165. Lajtha, A. and Toth, J. (1961) *J. Neurochem.*, **8**, 216–25.
166. Lajtha, A. and Toth, J. (1963) *J. Neurochem.*, **10**, 909–20.
167. Blasberg, R. and Lajtha, A. (1966) *Brain Res.*, **1**, 86–104.
168. Levi, G. and Raiteri, M. (1973) *Life Sci.*, **12**, 81–8.
169. Logan, W.J. and Snyder, S.H. (1972) *Brain Res.*, **42**, 413–31.
170. Oldendorf, W.H. and Szabo, J. (1976) *Amer. J. Physiol.*, **230**, 94–8.
171. Betz, L.A. and Goldstein, G.W. (1978) *Science*, **202**, 225–7.
172. Sershen, H. and Lajtha, A. (1979) *J. Neurochem.*, **32**, 719–26.
173. Wade, L.A. and Brady, H.M. (1981) *J. Neurochem.*, **37**, 730–4.
174. Pardridge, W.M. (1977) *J. Neurochem.*, **28**, 103–8.
175. McMenamy, R.H. and Oncley, J.L. (1958) *J. Biol. Chem.*, **233**, 1436–47.
176. Stewart, K.K. and Doherty, R.F. (1973) *Proc. Nat. Acad. Sci. U.S.A.*, **70**, 2850–2.
177. McMenamy, R.H. (1965) *J. Biol. Chem.*, **240**, 4235–43.
178. Bender, D.A., Boulton, A.P. and Coulson, W.F. (1975) *Biochem. Soc. Trans.*, **3**, 193–4.
179. Smith, H.G. and Lakatos, C. (1971) *J. Pharm. Pharmacol.*, **23**, 180–9.
180. Iwata, H., Okamoto, H. and Koh, S. (1975) *Jap. J. Pharmacol.*, **25**, 303–10.
181. Spano, P.F., Szyszka, K., Galli, C.L. and Ricci, A. (1974) *Pharmacol. Res. Commun.*, **6**, 163–73.
182. Bender, D.A. and Cockcroft, P.M. (1977) *Biochem. Soc. Trans.*, **5**, 155–7.
183. Müller, W.E. and Wollert, U. (1975) *Res. Commun. Chem. Pathol. Pharmacol.*, **10**, 565–8.
184. Baumann, P. and Perry, M. (1977) *Clin. Chim. Acta*, **76**, 223–31.
185. Knott, P.J. and Curzon, G. (1972) *Nature*, **239**, 452–3.
186. Curzon, G. and Knott, P.J. (1974) *Brit. J. Pharmacol.*, **50**, 197–204.
187. Curzon, G. (1974) *Adv. Biochem. Psychopharmacol.*, **10**, 263–71.
188. Tagliamonte, A., Biggio, G., Vargiu, L. and Gessa, G.L. (1973) *J. Neurochem.*, **20**, 909–12.
189. Yuwiler, A., Oldendorf, W.H., Geller, E. and Braun, L. (1977) *J. Neurochem.*, **28**, 1015–23.
190. Etienne, P., Young, S.N. and Sourkes, T.L. (1976) *Nature*, **262**, 144–5.
191. Rubin, R.A., Ordoñez, L.A. and Wurtman, R.J. (1974) *J. Neurochem.*, **23**, 227–31.
192. Schwartz, J.C., Lampart, C. and Rose, C. (1972) *J. Neurochem.*, **19**, 801–10.
193. Fernstrom, J.D. and Wurtman, R.J. (1975) *Amer. J. Clin. Nutr.*, **28**, 638–47.
194. Shiman, R. and Kaufman, S. (1970) *Meth. Enzymol.*, **17A**, 609–615.
195. Kuczenski, R.T. and Mandell, A.J. (1972) *J. Biol. Chem.*, **247**, 3114–22.
196. Eccleston, D., Ashcroft, G.W. and Crawford, T.B.B. (1965) *J. Neurochem.*, **12**, 493–503.
197. Fernstrom, J.D. and Wurtman, R.J. (1971) *Science*, **173**, 149–52.
198. Tagliamonte, A., Tagliamonte, P., Perez-Cruet, J. *et al.* (1971) *J. Pharm. Exp. Ther.*, **177**, 475–80.
199. Green, A.R. and Curzon, G. (1968) *Nature*, **220**, 1095–97.
200. Cocchi, D., di Giulio, A., Groppetti, A. *et al.* (1975) *Experientia*, **31**, 384–6.
201. Gal, E.M. (1974) *Adv. Biochem. Psychopharmacol.*, **11**, 1–11.
202. Bender, D.A. (1978) *Proc. Nutr. Soc.*, **37**, 159–65.
203. Parfitt, A. and Grahame-Smith, D.G. (1974) in *Aromatic Amino Acids in the Brain*, CIBA Fdn Symp 22, Associated Scientific Publishers, Amsterdam, pp. 175–96.
204. Orlowski, M. and Meister, A. (1970) *Proc. Nat. Acad. Sci. U.S.A.*, **67**, 1248–55.
205. Meister, A. (1973) *Science*, **180**, 33–9.
206. Meister, A. (1978) in *Functions of Glutathione in Liver and Kidney* (eds H. Sies and A. Wendel), Springer Verlag, Berlin, pp. 43–60.
207. Meister, A. (1979) in *Glutamic Acid : Advances in Biochemistry and Physiology* (eds L.J. Filer, S. Garattini, M.R. Kare, W.A. Reynolds and R.J. Wurtman), Raven Press, New York, pp. 69–84.
208. Tateishi, N. and Higashi, T. (1978) in *Functions of Glutathione in Liver and Kidney* (eds H. Sies and A. Wendel), Springer Verlag, Berlin, pp. 3–8.
209. Albert, Z., Orlowski, M. and Szewczuk, A. (1961) *Nature*, **191**, 767–8.

210. Glenner, G.G. and Folk, J.E. (1961) *Nature*, **192**, 338–40.
211. Albert, Z., Orlowska, J., Orlowski, M. and Szewczuk, A. (1964) *Acta Histochem.*, **18**, 78–89.
212. Okonkwo, P.O., Orlowski, M. and Green, J.P. (1974) *J. Neurochem.*, **22**, 1053–8.
213. van der Werf, P., Orlowski, M. and Meister, A. (1971) *Proc. Nat. Acad. Sci. U.S.A.*, **68**, 2982–5.
214. Jellum, E., Kluge, T., Borreson, H.C. *et al.* (1970) *Clin. Lab. Invest.*, **26**, 327–35.
215. Eldjarn, L., Jellum, E. and Stokke, O. (1972) *Clin. Chim. Acta*, **40**, 461–76.
216. Konrad, P.N., Richards, F., Valentine, W.N. and Paglia, D.E. (1972) *New Engl. J. Med.*, **286**, 557–61.
217. Schulman, J.D., Goodman, S.I., Mace, J.W. *et al.* (1975) *Biochem. Biophys. Res. Commun.*, **65**, 68–74.
218. Carlson, G.M., Bechtel, P.J. and Graves, D.J. (1979) *Adv. Enzymol.*, **50**, 41–115.
219. Grant, M.E. and Prockop, D.J. (1972) *New Engl. J. Med.*, **286**, 194–9, 242–9, 291–300.
220. Prockop, D.J., Kivirikko, K.I., Tuderman, L. and Guzman, N.A. (1979) *New Engl. J. Med.*, **301**, 13–23, 77–85.
221. Tryygvason, K., Majamaa, K. and Kivirikko, K.I. (1979) *Biochem. J.*, **178**, 127–31.
222. Cardinale, G.J. and Udenfriend, S. (1974) *Adv. Enzymol.*, **41**, 245–300.
223. Myllylä, R., Kuntki-Savolainen, E.R. and Kivirikko, K.I. (1978) *Biochem. Biophys. Res. Commun.*, **83**, 441–8.
224. Siegel, R.C. and Fu, J.C.C. (1976) *J. Biol. Chem.*, **251**, 5779–85.
225. Stassen, F.L.H. (1976) *Biochim. Biophys. Acta*, **438**, 49–60.
226. Magnusson, S., Sottrup-Jensen, L., Petersen, T.E. *et al.* (1974) *FEBS Letters*, **44**, 189–93.
227. Nelsestuen, G.L., Zytkovicz, T.H. and Howard, J.B. (1974) *J. Biol. Chem.*, **249**, 6347–50.
228. Stenflo, J., Fernlund, P., Egan, W. and Roepstorff, P.P. (1974) *Proc. Nat. Acad. Sci. U.S.A.*, **71**, 2730–3.
229. Shah, D.V. and Suttie, J.W. (1974) *Biochem. Biophys. Res. Commun.*, **60**, 1397–402.
230. Hauschka, P.V., Friedman, P.A., Traverso, H.P. and Gallop, P.M. (1976) *Biochem. Biophys. Res. Commun.*, **71**, 1207–13.
231. Lian, J.B. and Friedman, P.A. (1978) *J. Biol. Chem.*, **253**, 6623–6.
232. Esmon, C.T. and Suttie, J.W. (1976) *J. Biol. Chem.*, **251**, 6238–43.
233. Mack, D.O., Suen, E.T., Girnadot, J.M. *et al.* (1976) *J. Biol. Chem.*, **251**, 3269–76.
234. Jones, J.P., Gardner, E.J., Cooper, T.G. and Olson, R.E. (1977) *J. Biol. Chem.*, **252**, 7738–42.
235. Olson, R.E. and Suttie, J.W. (1978) *Vit Horm.*, **35**, 59–108.
236. Dam, H., Søndegaard, E. and Olson, R.E. (1982) in *Vitamins in Medicine* (eds B.M. Barker and D.A. Bender) (Vol. II) William Heinemann Medical, London, pp. 92–113.
237. Gallop, P.M., Lian, J.B. and Hauschka, P.V. (1980) *New Engl. J. Med.*, **302**, 1460–6.
238. Hall, J.G., Pauli, R. and Wilson, K. (1980) *Amer. J. Med.*, **68**, 122–40.

CHAPTER SIX

The Biosynthesis of Amino Acids in Plants

P.J. LEA, R.M. WALLSGROVE AND B.J. MIFLIN

6.1 INTRODUCTION

In the previous chapter, Bender has given a very detailed description of the metabolism of amino acids in humans. It is well known that animals are unable to synthesize the 'essential' amino acids, and most rely upon a continuous supply provided in their diet. Despite attempts by a number of industrial companies to provide a cheap supply of protein from bacterial and fungal cells, by far the most important source of essential amino acids for all animals including humans, is the plant kingdom.

In this chapter we will discuss the biosynthesis of amino acids within plants, and lay particular stress on the regulation of the synthesis of the essential amino acids. The field of amino acid biosynthesis from inorganic nitrogen has been one of the most active in plant biochemistry in the last five years, and a number of reviews [1–3] and a book [4] have been devoted to it.

6.2 AMMONIA ASSIMILATION

In contrast to animal metabolism, plants have to conserve all the available nitrogen, as this is frequently a limiting factor in growth. Ammonia is produced initially either by the reduction of nitrate [5] or the fixation of nitrogen gas [6, 7]. However, as will be seen in later sections, ammonia is frequently liberated by internal catabolic reactions and is rapidly re-assimilated to prevent both toxic action and loss to the atmosphere.

6.2.1 The glutamate synthase cycle

Ammonia is assimilated in all plant tissues via the glutamate synthase cycle as shown in Fig. 6.1.

(a) Enzymes involved

Glutamine synthetase (GS) [L-glutamate: ammonia ligase (ADP-forming) EC 6.3.1.2] catalyses Reaction 6.1:

$$\text{L-Glutamate} + \text{ATP} + NH_3 \rightarrow \text{L-glutamine} + \text{ADP} + P_i + H_2O \qquad (6.1)$$

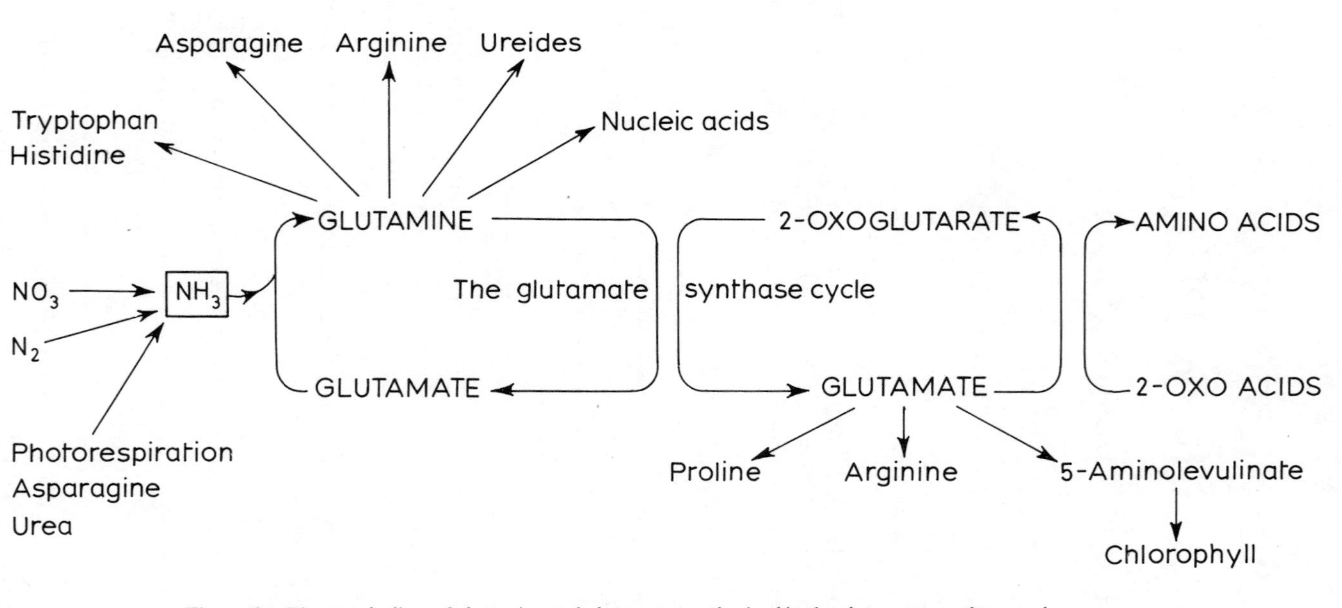

Figure 6.1 The metabolism of glutamine and glutamate synthesized in the glutamate synthase cycle.

The problems involved in assaying the enzyme have been discussed previously [1, 3]. The enzyme is universally distributed throughout the plant kingdom and is usually present in large amounts in all tissues. GS has a very high affinity for ammonia (K_m 10^{-5}–10^{-4} M), and is thus capable of preventing any build-up of toxic ammonia.

The ammonia assimilated into the amide group of glutamine is transferred to the 2-amino position of glutamate via the action of glutamate synthase, which exists as two forms in L-glutamate: NAD^+ oxidoreductase (transaminating), EC.1.4.7.14, and L-glutamate: ferredoxin oxidoreductase (transaminating), EC.1.4.7.1. These enzymes catalyse Reaction 6.2:

$$\text{2-Oxoglutarate} + \text{L-glutamine} + (\text{NADH}_2 \text{ or ferredoxin}_{\text{red}}) \rightarrow 2\,\text{L-glutamate} + (\text{NAD}^+ \text{ or ferredoxin}_{\text{ox}}) \qquad (6.2)$$

Glutamine synthetase and glutamate synthase are able to act in conjunction to form the glutamate synthase cycle (Fig. 6.1). The key point is that one molecule of glutamate continuously recycles whilst the second may be transaminated to all other amino acids or converted directly to proline and arginine. Glutamate may also be regenerated when the amide amino group of glutamine is transferred to form other compounds such as asparagine or carbamoyl phosphate.

6.2.2 Evidence for operation of the glutamate synthase cycle

Three kinds of evidence for the operation of the cycle in various plant tissues will be presented: the presence of the two requisite enzymes, labelling studies and inhibitor studies.

(a) Assimilation in photosynthetic tissue

GS has been extensively purified from leaves of *Pisum* [8,9], *Lemna* [10] and *Oryza* [11]. Two separate forms have been isolated in *Hordeum* [12], *Glycine* [13], *Oryza* [11] and *Pisum* [14] by ion exchange chromatography.

Ferredoxin-dependent glutamate synthase was originally isolated from pea leaves [15], but has since been detected in a wide range of plants [16, 17, 18]. The enzyme has been partially purified from *Vicia* [16] and to homogeneity from *Oryza* [19] and *Spinacia* [20]; it is specific for glutamine, 2-oxoglutarate and ferredoxin (although methyl viologen can also act as electron donor, as it can for the NADH-dependent enzyme [21]). The NAD-specific enzyme has recently been shown to occur in etiolated and green leaves in several species [21, 22]. The two enzymes can be separated by ion-exchange chromatography or gel filtration [21], and are immunologically distinct [23]. In etiolated leaves, the NADH enzyme comprises 10–30% of the total glutamate synthase activity, but in green leaves it is less than 3% of the total [22], and probably plays little or no role in nitrogen metabolism in such tissue.

Early [^{15}N]-labelling data obtained by Lewis and Pate [24] indicated that $^{15}NO_3^-$, [^{15}N]-glutamate and [^{15}N-amide]-glutamine were all incorporated into the amino nitrogen of amino acids suggesting the direct conversion of the amide-N of glutamine into the 2-amino position. More recently, Rhodes *et al.* [25] analysed $^{15}NH_3$ assimilation in *Lemna* under steady-state conditions and subjected the results to careful computer analysis. Two models involving two separate compartments were found to fit the data. Both had in compartment (1) the glutamate synthase cycle and in compartment (2) a second site of glutamine synthesis. The models did not exclude the possibility that 10% of the total nitrogen flux could be assimilated directly into glutamate via a second reaction. Confirmation as to the exact route of $^{15}NH_3$ assimilation was then obtained using methionine sulphoximine (MSO) and azaserine, inhibitors of GS and glutamate synthase respectively which totally inhibited incorporation. A number of other studies using MSO and azaserine have been carried out in *Lemna* [26], soybean [27] and spinach [28], all confirming the operation of the glutamate synthase cycle in leaves.

(b) Non-photosynthetic tissue

GS has been purified from rice roots [29] and isolated from the roots of a range of other plants [17, 18, 30–32]. The first report of the purified enzyme from a plant source was from dry pea seeds [33, 34]. Although GS has been demonstrated in maturing legume cotyledons [35–37] and maturing cereal grains [32, 38, 39] no further work on a purified enzyme has been published.

Glutamate synthase was first isolated from pea root tissue by Fowler *et al.* [40] and in a somewhat different form by Miflin and Lea [41]. Although the activity tends to be lower in roots than in other tissues its presence there has been confirmed [17, 18, 30–32]. There was initially some confusion over the electron donor in root systems, as extracts were able to utilize either NADH, NADPH or ferredoxin. Using antibodies raised against ferredoxin-dependent leaf glutamate synthase Susuki *et al.* [23] were able to show that although the root ferredoxin-dependent enzyme was similar to that found in the leaf it was distinct from the NADPH-dependent glutamate synthase.

Similarly, in maturing seeds the precise electron donor for glutamate synthase is unclear. A ferredoxin-dependent enzyme has been demonstrated in lupin [35] and pea [42]. An NADPH enzyme has been isolated from barley [39], maize [31, 43], pea [44] and soyabean [45]. Sodek and Da Silva [43] showed that in the developing endosperm of maize, NADPH-dependent glutamate synthase gave a well-defined peak of activity coinciding with the period of most active nitrogen accumulation.

For those non-photosynthetic tissues that have been studied in detail, NADH-dependent glutamate synthase is no more than 50% of the total activity [42,46,47]. Failure to appreciate the existence of both enzymes, and difficulties with extraction and assay, have led to reports of very low (or non-existent) glutamate synthase activity in some tissues (see Duffus and Rosie [38] for example). Many of the published levels of activity should be treated as underestimates.

Early work by Cocking and Yemm [48] and later by Yoneyama and Kumazawa [49] on feeding ^{15}N to roots is difficult to interpret, although it is possible to deduce some evidence for the GS/glutamate synthase pathway after the roots have been returned to ^{14}N [50]. Rice [51] and *Datura* [52] roots, if pretreated with MSO, are unable to assimilate either $^{15}NH_3$ or $^{15}NO_3$. A more detailed study of $^{15}NO_3$ and $^{15}NH_3$ assimilation by barley roots utilizing computer models based on Rhodes *et al.* [25] has confirmed the route of assimilation as via the glutamate synthase cycle [53]. Labelling studies with ^{15}N in developing seeds [24, 54] both suggest that nitrogen is readily transferred to protein α-amino groups via glutamine.

(c) Legume root nodules

Nitrogen-fixing nodules form on the roots of legumes as a result of infection by free-living soil micro-organisms of the genus *Rhizobium* [55]. Division of the rhizobia within the plant cytoplasm takes place at a rapid rate, until the root cells are almost filled with a modified form of the bacteria known as bacteroids, enclosed within a peribacteroid membrane. At some time during development of the nodule the bacteroids begin to reduce (fix) nitrogen using the enzyme nitrogenase and excrete ammonia. In the absence of externally supplied nitrogen, the ammonia can meet the total requirements of the plant during the complete life cycle.

Legume root nodules have been routinely separated into 'bacteroid' and 'plant cytoplasm' fractions, with little attention devoted to organelles present in the plant fraction. It is now accepted that the bacteroids do not have sufficient ammonia-assimilating enzymes to cope with the rate of nitrogen fixation [55–58]. GS is found in the cytoplasm of the plant nodule cell [59] and increases markedly during nodule development in conjunction with leghaemoglobin and nitrogenase [57]. Recent data suggest that the synthesis of at least one novel nodule-specific form of GS accounts for the rapid increase in activity but that a normal root-specific GS isoenzyme is also present in the nodules [60]. In soyabeans, the enzyme comprises 2% of the soluble protein, from where it has been purified to homogeneity and its properties studied [61].

Despite problems in assaying glutamate synthase in root nodules, large amounts of NADPH-specific enzyme have been demonstrated in lupins, *Lathyrus*, *Phaseolus* and soyabean [62]. Boland *et al.* [63] measured the activity of glutamate synthase and GS in the plant fraction of 12 different legumes; specific activities of glutamate synthase varied between 7% and 100% of GS. The enzyme has been purified extensively from lupin [64] and *Phaseolus* [65], and the molecular weights, substrate specificities and reaction mechanisms studied.

The initial ^{15}N experiments of Kennedy [66] suggested that ammonia was incorporated directly into glutamate; however, the data were later re-interpreted in terms of the GS/glutamate synthase pathway [41]. $^{13}N_2$ gas, when fed to soybean nodules, was initially rapidly incorporated into glutamine and then glutamate; the kinetics of labelling showed a typical precursor–product relationship [67]. Similar work with $^{15}N_2$ gas indicated that MSO and azaserine both blocked assimilation with accumulation of label in ammonia and glutamine respectively [68].

(d) Localization of ammonia assimilation within the plant cell

Much of the early work on the subcellular localization of enzymes of plant metabolism is of poor quality and should be treated with scepticism [69]. Studies by differential centrifugation suggest that GS [70] and glutamate synthase [15] are present in the chloroplasts. These suggestions have been confirmed by density gradient centrifugation techniques using mechanically prepared homogenates. However, the use of leaf protoplasts as starting material allows a much improved recovery of intact chloroplasts, and it is now certain that all the ferredoxin-dependent glutamate synthase is present in the chloroplast [71] as is the NADPH-dependent enzyme [47]. GS is located both in the chloroplast and cytoplasm, and the two isoenzymes have different properties [12, 72]. The ratio of cytoplasm to chloroplast GS appears to vary in a wide range of leaves examined. It has been suggested that GS is also present in the mitochondria of leaves [73], but subsequent work has been unable to substantiate this claim [74, 75].

Evidence for the localization of enzymes in non-green tissue in particular roots is still somewhat confused due to problems in obtaining good yields of intact organelles. GS was initially shown to be localized in both the plastid and cytoplasm [24, 59, 76], but a wider and more detailed examination in roots suggests that the enzyme is totally cytoplasmic [77]. However, in all studies reported, glutamate synthase appears to be located solely in the plastids [24, 59, 77].

Various techniques have been used to show that isolated intact chloroplasts assimilate ammonia into amino acids in the light via the GS/glutamate synthase pathway [78]. Furthermore, Anderson and Done [79, 80] have coupled the reactions to oxygen evolution by isolated chloroplasts, presumed to be dependent on the re-oxidation of ferredoxin by glutamate synthase.

In contrast, there is little evidence that mitochondria can assimilate ammonia, although the reverse reaction (the oxidation of glutamate) is readily catalysed. There is a small rate of assimilation under anaerobic conditions [81], but a detailed study by Wallsgrove *et al.* [74] did not suggest that mitochondria could assimilate ammonia released during photorespiration.

6.2.3 Transamination

An essential part of the scheme shown in Fig. 6.1 is that 2-oxoglutarate is regenerated from glutamate to take part in the glutamate synthase cycle. The amino group of glutamate may be transferred to other oxo acids in a series of reactions catalysed by aminotransferases. These enzymes appear to be remarkably similar in micro-organisms, animals and plants [82, 83] and do not require much further discussion in this review.

6.2.4 Summary of ammonia assimilation

Although suggestions are still being made that ammonia may be incorporated directly into the 2-position of oxo acids (in particular 2-oxoglutarate via glutamate dehydrogenase), the overwhelming evidence suggests that, in all plant tissues, the GS/glutamate synthase cycle operates (Fig. 6.1). For further references on the subject the readers should consult articles by Stewart *et al.* [84], Miflin and Lea [85] and Miflin *et al.* [86].

6.3 BIOSYNTHESIS OF AMINO ACIDS OTHER THAN GLUTAMINE AND GLUTAMATE

6.3.1 Introduction

The synthesis of the remainder of the amino acids will be dealt with in groups according to the way in which the carbon skeleton is derived following the well-established concept of amino acid families. An attempt will be made to present evidence for the suggested pathways from both *in vitro* studies and the presence of the relevant enzymes. The regulation of the synthesis of the amino acids will be discussed and some indication of the subcellular compartmentation of the reactions will be given. For further details on the subject the reader is referred to a recent book on plant amino acid metabolism [4].

6.3.2 Synthesis of amino acids derived from glutamate

Besides glutamine, ornithine, arginine and proline are the major amino acids synthesized from glutamate as shown in Fig. 6.2 [87].

(a) Enzymic evidence

Amino acid acetyltransferase, EC.2.3.1.1 (1) activity has been demonstrated in extracts of *Chlorella* [88] and *Beta* [89]. In both cases the enzyme appears to copurify with glutamate acetyltransferase EC.2.3.1.35 (2). In higher plants, enzyme (2) has a 10-fold lower K_m for glutamate than (1), but the values are very similar in *Chlorella*. The purified form of ATP: *N*-acetylglutamate kinase EC.2.7.2.8 (3) from *Chlamydomonas* [90] has been studied in detail, and the enzyme has been detected in extracts of other algae [88] and higher plants [89]. The next two enzymes in the pathway; *N*-acetyl-5-glutamyl phosphate reductase EC.1.2.1.38 (4) and *N*-acetylornithine aminotransferase EC.2.6.1.11 (5) have been little studied [87, 90]. *N*-Acetylornithine is converted to ornithine by reaction (2); it is then converted to ornithine by a series of reactions first described in the Krebs–Henseleit cycle [91]. Carbamoyl phosphate is formed from glutamine, ATP and CO_2 via the action of carbamoyl phosphate synthetase EC.6.3.5.5 (6) which has been demonstrated in several plants [92, 93, 94]. The enzyme ornithine carbamoyltransferase EC.2.1.3.3 (7) has been purified to near homogeneity (2000-fold) from peas [95], but in several plants two forms of the enzyme have been

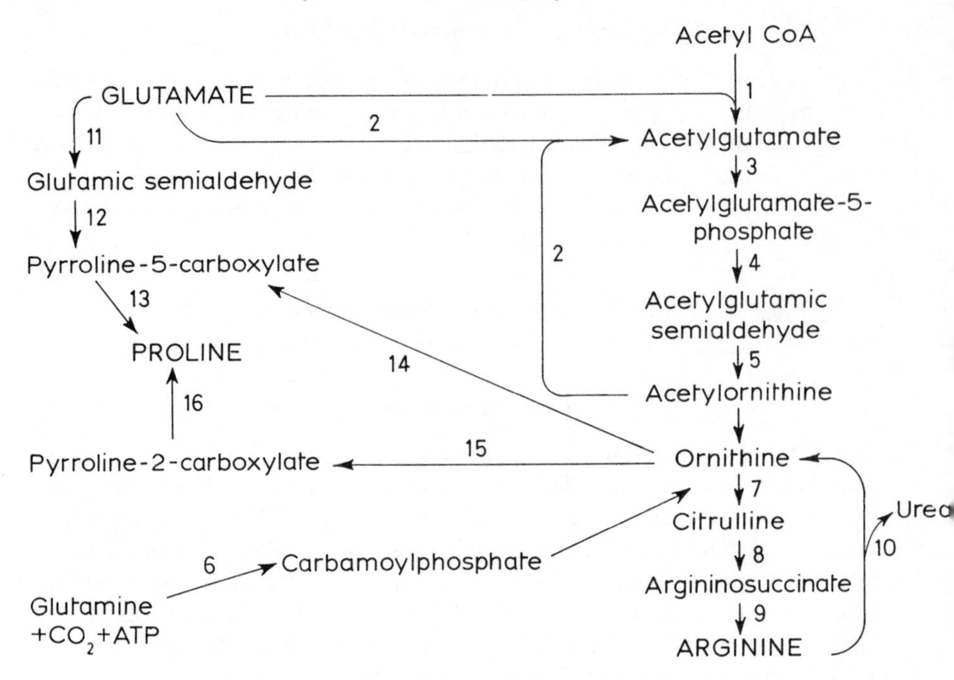

Figure 6.2 Pathway of the synthesis of proline and arginine from glutamate.

detected which differ markedly in their properties [96, 97]. Argininosuccinate synthetase EC.6.3.4.5 (8) has been demonstrated in plant extracts [98, 99] and purified 400-fold from germinating pea cotyledons [99]. The argininosuccinate so formed is subsequently cleaved via the action of argininosuccinate lyase EC.4.3.2.1 (9) to give arginine and fumarate. The enzyme has been detected in a number of cell-free extracts [98–100], but is sensitive to degradation by a protease [101]. In animals, to complete the Krebs–Henseleit cycle, arginine is broken down to yield urea via the enzyme arginase (10). Although this pathway does operate in plants, there are at least three other known mechanisms of arginine breakdown [3, 102].

In micro-organisms the biosynthesis of proline is considered to occur directly from glutamate via enzymes (11)–(13) [103], although the enzymic evidence for the first step is not convincing even in *E. coli*. The only evidence in plants that reactions 11 and 12 can take place is provided by Morris *et al.* [104]. However, extracts of a number of plants will reduce pyrroline-5-carboxylate to proline with either NADH or NADPH [105–107]. Plants will also carry out the reverse oxidation of proline with NAD [106, 107].

It can be seen from Fig. 6.2 that there are two potential methods of converting ornithine to proline depending on whether the 2-amino or 5-amino position is transaminated first. The majority of studies on ornithine transferases have not attempted to distinguish between the two possibilities [108–110], and extracts are capable of converting both pyrroline-5- and pyrroline-2-carboxylate (16) to proline [111].

(b) *In vivo* labelling studies

Many labelling studies have demonstrated that [^{14}C]-glutamate may be converted to proline and arginine [87], but these do not discriminate between the multiple pathways shown in Fig. 6.2. Specific evidence for the involvement of the acetylated pathway in arginine biosynthesis comes from the isotope competition experiments of Dougall and Fulton [112]. Using double-labelled ornithine, Mestichelli *et al.* [113] have positively shown that plants can convert ornithine to proline via pyrroline-2-carboxylate (15, 16). Although at first sight this evidence appears convincing, it has been shown that in *Neurospora* externally supplied ornithine is catabolized, rather than used as a source for arginine biosynthesis [114]. Furthermore, Mestichelli *et al.* dried their plants at 45–50°C prior to extraction, and it is known that wilting accelerates the conversion of ornithine to proline [104, 115]. This may have caused abnormal metabolism within the plant. It is clear that the statement of Mestichelli *et al.* 'that the time is ripe for a re-examination of proline biosynthesis' is a very true one.

(c) Subcellular localization

Although there have been a number of studies on the subcellular localization of arginine-metabolizing enzymes in higher plants the results have been contradictory. Enzyme (7) has been reported to be either cytoplasmic [116] or entirely chloroplastic [94], arginase (10) has been found to be mitochondrial [117] or cytoplasmic [118]. However, it is known in animals [119] and fungi [120], that there is a strict compartmentalization of arginine catabolism and anabolism. To clarify the situation Taylor and Stewart [121] re-examined the localization of arginine-metabolizing enzymes utilizing pea leaf protoplasts. Enzymes (2), (6), and (7) were localized in the chloroplast, whilst the last enzyme argininosuccinate lyase (9) was present only in the cytoplasm. Such a system would imply the transport of argininosuccinate out of the chloroplasts before conversion to arginine. In contrast, the catabolic enzymes (10) and (14) were located in the mitochondria.

(d) Regulation

Arginine is able to regulate the rate of its own synthesis, it inhibits the synthesis of acetylglutamate from acetyl CoA (1) but not from acetylornithine (2) [88, 89]. Arginine also appears to control the activity of acetylglutamate kinase (3) in a range of species studied [88–90]. The other source of carbon and nitrogen for arginine is carbamoyl phosphate, which is also utilized for pyrimidine synthesis. Carbamoyl phosphate synthetase (6) is inhibited by UMP, but this inhibition is released by ornithine, thus allowing arginine synthesis in the presence of sufficient UMP [92, 93].

Evidence for the *in vivo* regulation of proline synthesis has been provided by

Oaks and her colleagues [122]. The degree of control appears to change as a function of age in maize roots [122], and is affected by the level of water stress [115].

(e) Mutant studies

It can be seen that there are three potential routes of proline biosynthesis all initially dependent upon glutamate. It is perhaps fair to say that it has not been shown unequivocally which, if any, of these is operating at a particular point during the development of a higher plant.

Controversies regarding metabolic pathways in micro-organisms have often been resolved by the use of auxotrophic mutants. It is perhaps unfortunate that the only such well-documented plant mutant is the proline-requiring line of maize isolated by Gavazzi *et al.* [123]. Initial studies suggested that there was a block in enzyme (13) but later studies showed that pyrroline-5-carboxylate reductase was present at even higher levels in the mutant lines [124]. [^{14}C]-Labelling studies with glutamate, glutamine and ornithine indicated that they all acted as proline precursors in mutant and normal plants. The only suggestion now available to account for the proline requirement is that proline breakdown is acclerated in the mutant plant [125].

6.3.3 Synthesis of amino acids derived from aspartate

This section will be divided into asparagine taken separately and lysine, threonine, methionine and isoleucine taken as a group.

(a) Asparagine

Apart from being a constituent of proteins, asparagine is used almost universally in plants as a nitrogen storage and transport compound. The incorporation of ^{15}N into asparagine tends to be much slower than into glutamine although rapid turnover has been demonstrated [25, 126]. Most evidence suggests that ammonia is first incorporated into the amide position of glutamine and is then transferred to aspartate to yield asparagine. The synthesis of aspartate is probably via oxaloacetate from the tricarboxylic acid cycle, although in the root nodule oxaloacetate may be formed by the action of phosphoenolpyruvate carboxylase. Cyanide is rapidly converted to asparagine when fed to plants, but despite recent claims [127] this is probably a detoxification mechanism rather than a major pathway of asparagine synthesis.

Asparagine synthetase catalyses the amidation of aspartate by glutamine in an ATP-dependent reaction.

$$\text{Aspartate} + \text{glutamine} + \text{ATP} \xrightarrow{Mg^{2+}} \text{Asparagine} + \text{glutamate} + \text{AMP} + \text{PP}_i$$

The enzyme has been isolated from the cotyledons of a number of germinating seeds [35, 128–132], maize roots [133, 134] and root nodules [135]. The enzyme is notoriously unstable and measurement of activity has been prevented by the presence of inhibitors, asparaginase, and lack of chloride ions which are necessary for optimum activity [129]. Irrespective of the source, there is evidence that the enzyme is able to utilize ammonia in place of glutamine; Stulen *et al.* [134] have claimed that in maize roots the K_m for ammonia is low enough to allow direct incorporation under certain physiological conditions.

The involvement of asparagine in nitrogen transport and its subsequent metabolism have been discussed by Lea and Miflin [136]. However, a number of other compounds, in particular the ureides allantoin and allantoic acid, are also used as nitrogen transport compounds.

(b) Lysine, threonine, methionine and isoleucine

The biosynthesis of these four amino acids from aspartate is shown in Fig. 6.3. Evidence for the operation of this pathway has come from many studies discussed below and reviewed previously [1, 137, 138].

Perhaps the most interesting point in the pathway is the synthesis of lysine by higher plants from aspartate rather than via the fungal aminoadipic acid pathway [139, 140]. Isotope dilution experiments [141] have indicated that homoserine is an intermediate in threonine and methionine synthesis, that cystathionine and homocysteine (but not *O*-succinyl- or *O*-acetylhomoserine) are intermediates in methionine formation and that diaminopimelate is on the route to lysine synthesis.

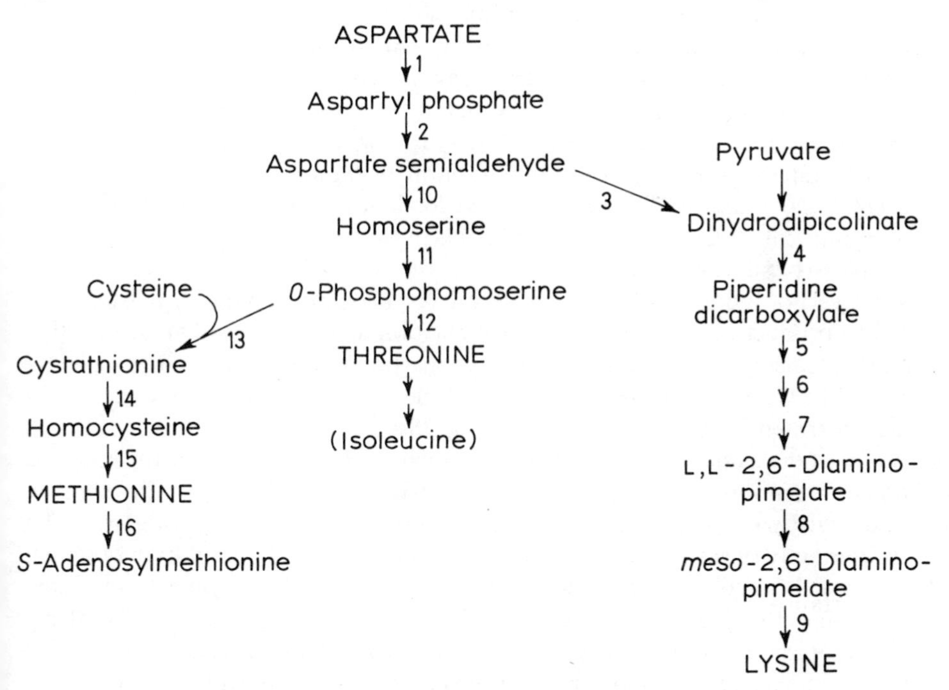

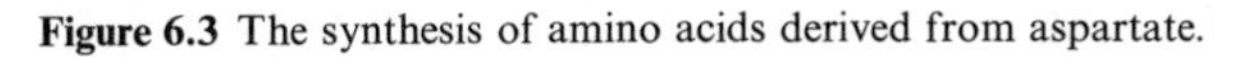
Figure 6.3 The synthesis of amino acids derived from aspartate.

Regulation of the synthesis in vitro

A number of studies have shown that lysine and threonine are both able to inhibit their own biosynthesis from [^{14}C]-labelled precursors [142–145]. Methionine synthesis from $^{35}SO_4^{2-}$ is also regulated by methionine and by the addition of lysine and threonine [146, 147]. In isolated chloroplasts lysine inhibits its own synthesis from aspartate and to a lesser extent that of threonine. Threonine, however, only inhibits its own synthesis and not that of lysine [148].

Further evidence for the regulation of methionine synthesis by lysine and threonine can be obtained by determining the effects of certain combinations of amino acids on plant growth. Lysine and threonine are not generally very inhibitory to growth on their own but together they have a marked synergistic effect. The growth inhibition can be relieved by very low levels of methionine suggesting that it is methionine synthesis that is prevented [149–152].

The enzymes involved in the pathway and their regulation in vitro

Aspartate kinase EC.2.7.2.4 (1) has been extensively studied in plants and progressively reviewed [1, 3, 137, 138]. A more coherent pattern of regulation has recently emerged and it can be now said that plants contain at least two isoenzyme forms, one regulated by threonine and one by lysine [145, 153–155]. In barley, the only plant where it has been studied in detail, there is evidence of two distinct lysine-sensitive forms [156]. The proportions of activities of the threonine- and lysine-sensitive forms appear to vary between plants and, more importantly, with developmental age. In general, the lysine-sensitive form is predominant in physiologically active, rapidly growing cells.

Although methionine has no direct effect on aspartate kinase, there is now evidence that it may regulate its own synthesis via an activated product *S*-adenosylmethionine (16). Thus the addition of 0.1 mM *S*-adenosylmethionine decreases the amount of lysine required for half-maximal inhibition of barley aspartate kinase from 340 to 48 μM [157]. In this way *S*-adenosylmethionine is unable to regulate aspartate kinase activity on its own, but in the presence of very low levels of lysine there is a very sensitive synergistic regulatory mechanism.

The product of the aspartate kinase reaction aspartyl phosphate is unstable and is probably immediately reduced by aspartate semialdehyde dehydrogenase EC.1.2.1.11 (2). The enzyme has been little studied in plants but has been demonstrated in pea, maize and barley [138].

Aspartate semialdehyde is the first branch of the pathway; on the route to lysine formation it is condensed with pyruvate to form dihydrodipicolinate, catalysed by dihydrodipicolinate synthase (3) EC.4.2.1.52 [155, 158–160]. The enzyme activity is very tightly regulated by lysine, with 50% inhibition obtained at 11 μM. Beyond this point there is no further report of enzymes from plants catalysing the postulated reactions (4–8) until the last step, catalysed by diaminopimelate decarboxylase EC.4.1.1.20 (9). The enzyme which is specific for the *meso* form of the substrate has been isolated and characterized from a number of plants [161–163].

Aspartate semialdehyde is also a precursor of threonine and methionine and is initially reduced by the action of homoserine dehydrogenase EC.1.1.1.3. (9) which, due mainly to its ease of assay in both directions, has been extensively studied and purified [164–172]. The enzyme appears to exist in at least two major forms; a high-molecular-weight form inhibited by threonine and a low-molecular-weight form insensitive to threonine but inhibited by low concentrations of cysteine [168–172]. When cereal tissues of increasing ages are extracted the degree of sensitivity to threonine appears to decrease [168–171] though this is somewhat dependent on the extraction conditions [173]. The precise physiological role of this decrease in sensitivity is not understood, but there is considerable variation between tissues studied [138]. There is no evidence that methionine has any regulatory effect on homoserine dehydrogenase.

Homoserine kinase EC.2.7.1.39 (11) catalyses the addition of a phosphate group to homoserine and has been demonstrated in barley [174] and peas [175]. Homoserine phosphate is then metabolized to threonine via the action of threonine synthase EC.4.2.99.2 (12). This enzyme is subject to an unusual control in that it is strongly activated by *S*-adenosylmethionine [174, 176, 177]. Threonine synthase from *Beta vulgaris* is stimulated 20-fold by 0.5 mM *S*-adenosylmethionine; this activation is antagonized by cysteine [176].

Detailed studies by Giovanelli, Datko and Mudd [147, 178, 179] have established that phosphohomoserine is the 2-aminobutyryl donor for cystathionine biosynthesis in plants. Cystathionine synthase (13) (EC unclassified since EC.4.2.99.9, *O*-succinylhomoserine (thiol)-lyase is not an appropriate name for the plant enzyme) has been studied in extracts of several plants [180]. The enzyme is apparently not subject to any regulatory controls. Cystathionine is cleaved by cystathionine β-lyase EC.4.4.1.8 (14) to give homocysteine, pyruvate and ammonia. The enzyme has been purified 430-fold from spinach leaves [181] and has also been shown to cleave cysteine. Conversely, a cysteine lyase purified from turnip is able to cleave cystathionine [182]; the precise physiological role of these enzymes is unclear. The final step in methionine synthesis is the methylation of homocysteine catalysed by tetrahydropteroylglutamate methyltransferase EC.2.1.1.14 (15) which has been demonstrated in a number of plant extracts [183, 184].

It has been suggested that sulphide could be incorporated directly into homocysteine without prior formation of cysteine, this would bypass reactions (13) and (14). Careful assessment of the relative contributions of the two pathways in *Chlorella* and *Lemna* have indicated that the direct sulphydration pathway is unlikely to contribute more than 10% of the homocysteine formed in plants [147, 179]. In considering methionine synthesis, it is important to note that it is not only used for protein synthesis but is a precursor of *S*-adenosylmethionine which is involved in plants in a variety of reactions (e.g. ethylene biosynthesis, methylating reactions etc. [147, 185]). The enzyme catalysing its synthesis is methionine adenosyltransferase EC.2.5.1.6 (16) which has been studied in barley [147] and pea [184, 186].

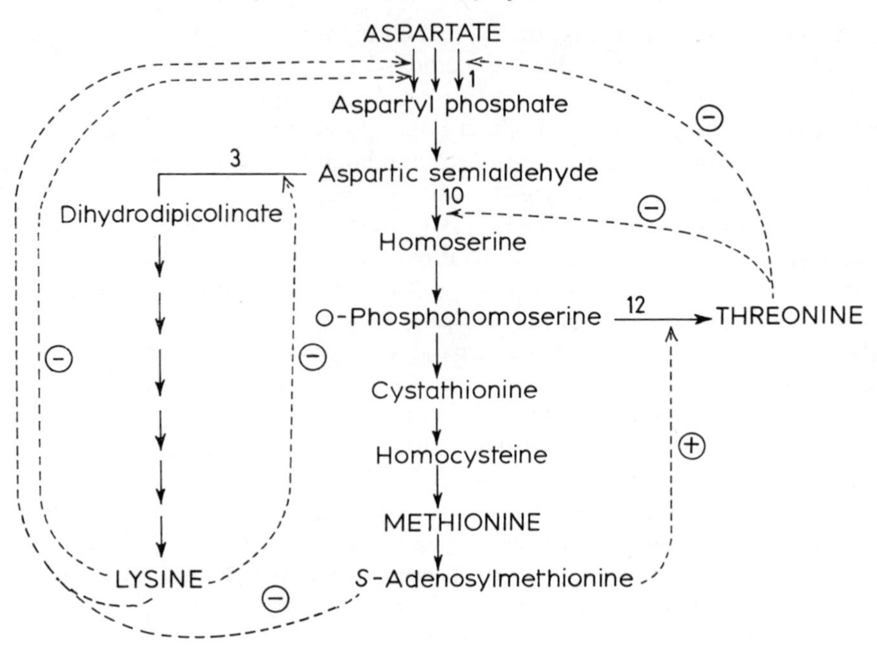

Figure 6.4 End product feedback regulation of lysine, threonine and methionine synthesis.

From the previous discussion, it is possible to build up a model of how the aspartate pathway is regulated (Fig. 6.4). Synthesis of excess lysine will shut off reaction (3) immediately but only partly reduce the synthesis of aspartyl phosphate. Carbon would continue to flow to methionine and *S*-adenosylmethionine alone until sufficient accumulated to activate threonine synthase (12) and synergistically inhibit (1). The flux to threonine in the presence of lysine and *S*-adenosylmethionine would be dependent on the threonine-sensitive aspartate kinase, once sufficient threonine accumulated it would lead to cessation of flux through (1) and (10). A subsequent fall in the lysine content could be accommodated as the inhibition of the lysine-sensitive aspartate kinase and of dihydrodipicolinate synthase would be removed, but the pathway to homoserine (10) would still be blocked by threonine. It can clearly be seen that the external addition of lysine and threonine would completely block the flow of carbon to methionine.

Subcellular localization

Isolated intact chloroplasts can carry out light-dependent synthesis of lysine, threonine, isoleucine and to a much lesser extent methionine from [^{14}C]-aspartate or [^{14}C]-malate [148], these reactions could not be demonstrated in mitochondria.

Initial studies showed that at least some of the aspartate kinase activity was present in chloroplasts [187, 188], and recent studies have shown that essentially all of the enzyme activity in green leaves is chloroplast-located [189]. Aspartate semialdehyde dehydrogenase activity has been found in chloroplasts [190], though the complete localization has not yet been shown.

Detailed studies with homoserine dehydrogenase have shown that the enzyme is located both in the chloroplast and cytoplasm [169, 191]. The chloroplastic form in peas and barley differs from the cytosol enzyme in size, affinity characteristics, regulatory properties and inhibition kinetics [169] and thus the two forms are obviously isoenzymes of entirely different nature and correspond to the two classes of enzyme established by purification studies [170–173]. Whether or not the cytosolic form plays any biosynthetic role remains to be determined.

The localization of two enzymes of the lysine branch of the pathway has been studied in leaves. Using a mechanical isolation technique Mazelis *et al.* [161] were able to show that diaminopimelate decarboxylase (9) was present in intact chloroplasts. Further studies with protoplast-derived organelles indicated that all of this enzyme (9) and all the dihydrodipicolinate synthase (3) were located in the chloroplasts [192, 193], suggesting that chloroplasts are the sole site of lysine synthesis.

Threonine synthesis would also appear to be confined to the chloroplast, as all of the leaf cells' homoserine kinase and threonine synthase are located in the chloroplast [189]. The first two enzymes of the branch leading to methionine, cystathionine synthase and cystathionine β-lyase, are also in the chloroplast (though not all of the latter enzyme), but current evidence suggests that methionine synthase and methionine adenosyltransferase are not associated with the chloroplast, or indeed with any organelle [189]. Some methionine synthase, though, may exist in the mitochondria [184]. Earlier evidence for limited methionine synthesis by isolated chloroplasts [148] should perhaps be treated with some caution.

Mutant selection studies

A number of studies have been carried out to obtain tissue culture lines or intact plants with altered feedback control in the aspartate pathway. Mutants resistant to ethionine, 5-hydroxylysine and aminoethylcysteine (AEC) have been selected which overproduce methionine [194] and lysine [195], but the enzyme mechanism has not been determined. Maize callus lines resistant to lysine and threonine have been selected [196]. One of the lines (D33) had an altered aspartate kinase with a K_i for lysine of 530 μM compared with a value of 62 μM for the parental line. Although the line had higher levels of threonine and lysine, a genetic analysis has not proved possible. More recently, a fertile maize mutant has been regenerated from callus culture, which has an elevated threonine content in the mature seed [197].

Three lines of mutant barley plants (R2501, R3004 and R3202) have been isolated at Rothamsted which are resistant to the toxic action of lysine and threonine [156, 198, 199]. Analysis of the properties of the three aspartate kinases separated by DEAE-cellulose chromatography, have shown that R2501 and R3202 both have AKII with greatly decreased sensitivity to inhibition by lysine and normal AKIII whilst R3004 has an unchanged AKII but an AKIII that is relatively unaffected by lysine. Genetic data suggested that the genes *lt1* and *lt2* are likely to be structural loci for the two aspartate kinase isoenzymes [199]. Both R2501 and R3004 showed greater than 10-fold increases in the soluble levels of threonine in the seed. None of the mutants contained higher levels of lysine, and they were not resistant to the toxic action of AEC [200]. It must be assumed that the enzyme dihydrodipicolinate synthase is still tightly regulated by lysine.

(c) Synthesis of amino acids derived from pyruvate

Alanine (by direct transamination) and the branched chain amiho acids leucine, valine, isoleucine and lysine derive some of their carbon from pyruvate; lysine, however, as it derives the majority of its carbon from aspartate has been considered in the previous section. Evidence for the pathway occurring as shown

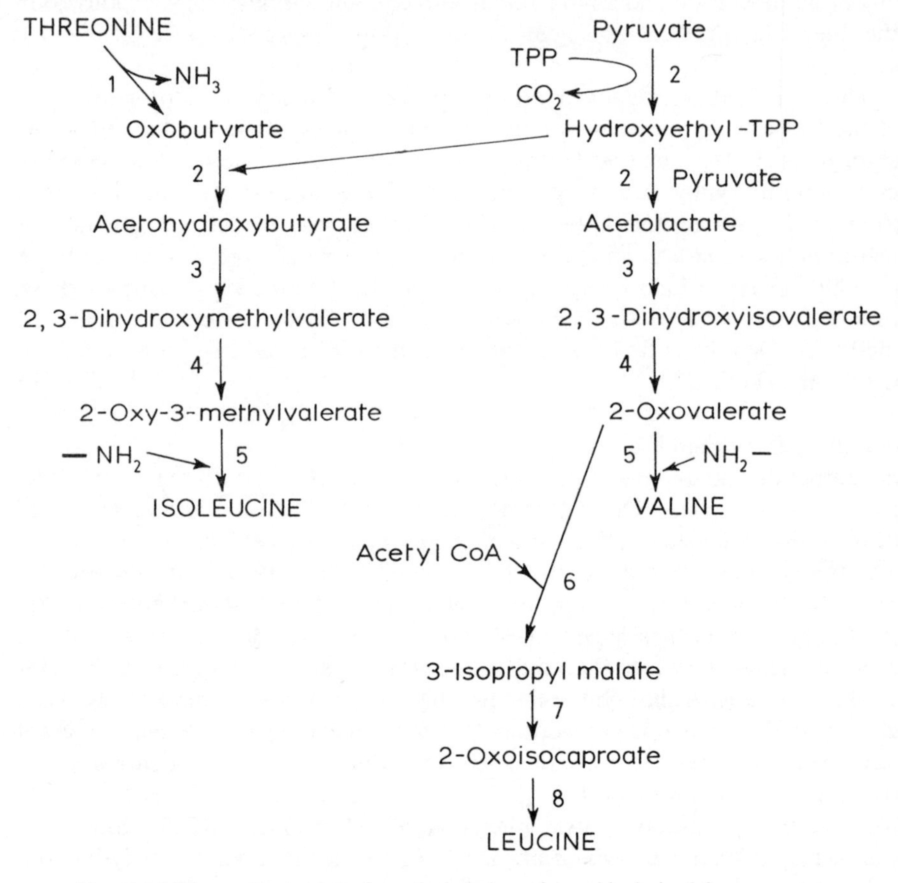

Figure 6.5 The synthesis of the branched chain amino acids derived from pyruvate.

in Fig. 6.5 has been obtained from isotope dilution experiments [141] and the feeding of labelled precursors [201].

Enzymes and their regulation

Studies with labelled precursors showed that the addition of very small amounts of exogenous leucine regulated the *in vivo* synthesis of the amino acid in maize root tips [202]; similar evidence also suggested that the synthesis of valine and isoleucine were regulated. Analysing the growth inhibitory action of the amino acids on maize seedlings, Miflin [203] showed that there was a co-operative feedback effect of leucine and valine. Similar results were also obtained with *Spirodela* [204].

Starting with threonine (whose synthesis has been described in Section 6.3.3 (b)) the first enzyme involved is threonine dehydratase EC.4.2.1.16 (1) which deaminates threonine to yield oxobutyrate. This enzyme shows complex kinetics indicating its probable allosteric nature, and is inhibited by isoleucine, this inhibition being partly reversed by valine [138, 205, 206].

Recently, an auxotrophic mutant requiring isoleucine, deficient in threonine deaminase has been isolated [207].

The first common enzyme in the pathway is acetolactate synthase EC.4.1.3.18 (2). The reaction involves the decarboxylation of pyruvate to form hydroxyethyl thiamin pyrophosphate, which is then accepted by another molecule of pyruvate or by 2-oxobutyrate. The enzyme is found in a number of plants and is co-operatively inhibited by low concentrations of leucine and valine, although both amino acids are inhibitory on their own at higher concentrations [208, 209]. The enzymes catalysing reactions (3) and (4) are acetohydroxyacid reductoisomerase EC.1.1.1.86 and dihydroxyacid dehydratase EC.4.2.1.9, the latter has been purified from spinach [210]. The final steps of valine and isoleucine synthesis are catalysed by aminotransferases (5) the specificity of which are not known [82, 83]. The synthesis of isopropylmalate (6) has been shown to be subject to inhibition by leucine in maize [211], but, in general, there is no detailed enzymology of this branch of the pathway in higher plants. However, aminotransferases active with leucine have been described [82, 83].

Subcellular localization

Evidence that at least part of the pathway occurs in chloroplasts comes from the finding that acetolactate synthase is associated with plastids in green and non-green tissue [76], and that threonine dehydratase is present in easily sedimenting particles [206]. The light-dependent synthesis of isoleucine from [^{14}C]-labelled aspartate and threonine has been demonstrated in isolated chloroplasts [148], and recently chloroplasts have been shown to synthesize valine from $^{14}CO_2$ at rates equivalent to that of whole cells [212].

(d) Synthesis of glycine, serine and cysteine

The biosynthetic routes important in the synthesis of these amino acids are summarized in Fig. 6.6. Reactions (1–5) plus (7–9) comprise the photorespiratory

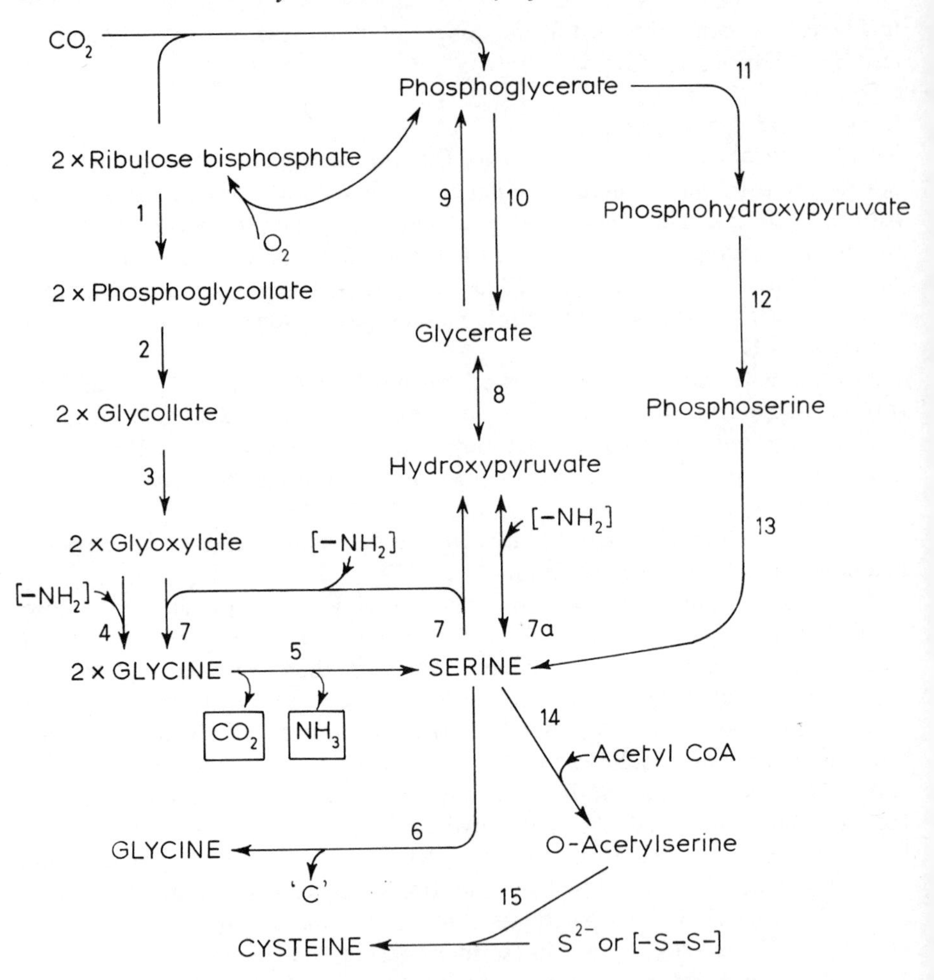

Figure 6.6 The pathway of synthesis of glycine, serine and cysteine.

pathway; these and other reactions involved in CO_2 photosynthetic reactions are outside the scope of this review [213–215].

Enzymic evidence

Glycine is formed by transamination from glyoxylate which may arise by the photorespiratory pathway or via the glyoxylate cycle. The major donors are glutamate, alanine and serine [216]. Purification of the enzymes [217] and mutant evidence [215, 218] suggest that serine:glyoxylate aminotransferase EC.2.6.1.45 (7) and glutamate:glyoxylate aminotransferase EC.2.6.1.4 (4) are separate proteins: the origin of alanine:glyoxylate aminotransferase is as yet not clear. The transamination of glycine to yield serine via reaction (7) does not take place, although the synthesis of serine from hydroxypyruvate and other amino donors (e.g. alanine using serine:pyruvate aminotransferase EC.2.6.1.57 (7a)) may occur.

The conversion of glycine to serine in the photorespiratory pathway is catalysed by system (5) [213, 214] and operates predominantly in intact green-leaf mitochondria [219]. It is similar to the reaction catalysed by rat liver mitochondria and involves the oxidative decarboxylation and de-amination of glycine to yield a C-1 fragment and equal amounts of CO_2 and NH_3. The reductant formed is either converted to ATP via the electron transport chain or is exported from the mitochondria via a malate:oxaloacetate shuttle [220]. The reaction is irreversible, has not been broken down into separate parts and is thus dependent upon intact mitochondria for assay.

Glycine may however be formed from serine by serine hydroxy-methyltransferase EC.2.1.1.1 (6). The enzyme has been detected in a number of plants [221], and purified from *Vigna* [222]. The activity is dependent on 5, 10-methylenetetrahydrofolate and pyridoxal phosphate, and subject to complex product inhibition kinetics [222]. It is not clear whether this is a separate enzyme, or a breakdown product of the complex in intact mitochondria that catalyses reaction (5). Serine may also arise from phosphoglyceric acid via phosphoglycerate phosphatase EC.3.1.1.18 (10) and glycerate dehydrogenase EC.1.1.1.29 (8). The equilibrium of the latter reaction favours glycerate formation, and is probably involved in the flow of carbon from serine during photorespiration. A third route of serine synthesis is the phosphorylated pathway starting with 3-phosphoglycerate dehydrogenase EC.1.1.9.5 (11). The enzyme has been purified 450-fold from peas [223] and is subject to inhibition by serine [224] and nucleotides [225] and activation by methionine [226]. Phosphohydroxy-pyruvate is transaminated to phosphoserine by an aminotransferase EC.2.6.1.52 (12) [223, 227], which is converted to serine by the action of phosphoserine phosphatase EC.3.1.3.3. (13) [223, 227].

Serine is acetylated to *O*-acetylserine by serine acetyltransferase EC.2.3.1.30 (14) prior to its sulphydration [228–230]. Cysteine is then formed by *O*-acetylserine (thiol)-lyase (cysteine synthase) EC.4.2.99.8 (15), which has been purified from many species [147] and two forms of the enzyme have been reported in mungbean [231] and other *Phaseolus* [232] species. There is some controversy as to whether free or bound sulphide is the natural substrate of the enzyme [233], although mutant evidence from *Chlorella* [234] suggests that it is the 'bound' pathway that is normally used.

Subcellular localization

The photorespiratory pathway is distributed between three organelles and the cytoplasm [213, 214]. Reactions (1) and (2) take place in the chloroplast, reactions (3), (4), (7) and (8) in the peroxisomes and reaction (5) is carried out only in intact mitochondria. The enzymes involved in (4), (7) and (8) have also been shown to be present in microbodies from other plant tissues [235]. Larson and Albertsson [227] have studied the localization of the phosphorylated pathway of serine synthesis and reported the presence of enzymes catalysing reactions (11), (12) and (13) in chloroplasts, although they considered that a parallel pathway could also exist outside the chloroplast.

Chloroplasts are able to catalyse the incorporation of sulphate, free sulphite and free sulphide into cysteine [233] and cysteine synthase has been shown to be localized in chloroplasts [233, 236]. However, the enzyme has also convincingly been shown to be a cytoplasm-soluble enzyme [231, 237, 238]; whether this is due to the rupture of plastids is not clear.

Regulation

The regulation of the photorespiratory pathway has been adequately discussed elsewhere [213, 214]. It is clear that the major effect on the rate of glycine and serine synthesis is the competition of O_2 and CO_2 for ribulose bisphosphate carboxylase [215, 239] (1); there is no evidence that glycine and serine are able to inhibit their own synthesis via this pathway. It is important to note that the conversion of glycine to serine produces ammonia at rates ten times faster than nitrate reduction. If this ammonia is not rapidly re-assimilated [240] the plants are unable to carry out the photorespiration process [241]. Evidence is also now available to suggest that the source of nitrogen supplied to the plant may have an influence on photorespiration rates [242].

There is evidence that cysteine can regulate its own synthesis by inhibiting the metabolism of sulphate [147]. The three main points of control appear to be: sulphate uptake [243], sulphate adenylyltransferase [244] and APS sulphotransferase [237]. In bacteria, cysteine regulates its own synthesis by feedback inhibition of (14) and repression of (15) [245]; in plants there is no evidence of such effects at physiological concentrations of cysteine [147]. The feedback regulation of sulphate reduction is however not under strict control as plants are able to emit large quantities of H_2S from the leaves by reactions that are dependent upon light and a high level of sulphate supply [246].

(e) Synthesis of aromatic amino acids

The group consisting of tryptophan, phenylalanine and tyrosine is synthesized initially by a common pathway involving the intermediate shikimic acid (Fig. 6.7). Besides leading to the formation of aromatic acids for protein synthesis, the pathway also provides precursors for a number of secondary metabolites produced in higher plants. The evidence for the existence of the proposed pathway has been reviewed by Haslam [247] and Gilchrist and Kosuge [248].

Enzymic evidence

3-Deoxy-D-arabinoheptulosonate-7-phosphate (DAHP) synthase (1) EC.4.1.2.15 has been demonstrated in a number of plants [248]. In peas, separate isoenzymes have been reported [249], but their role in the biosynthesis of the three amino acids is not clear. Reaction (2) is catalysed by 3-dehydroquinate synthase EC.4.6.1.3 [250, 251], which uses NAD as a cofactor and is stimulated

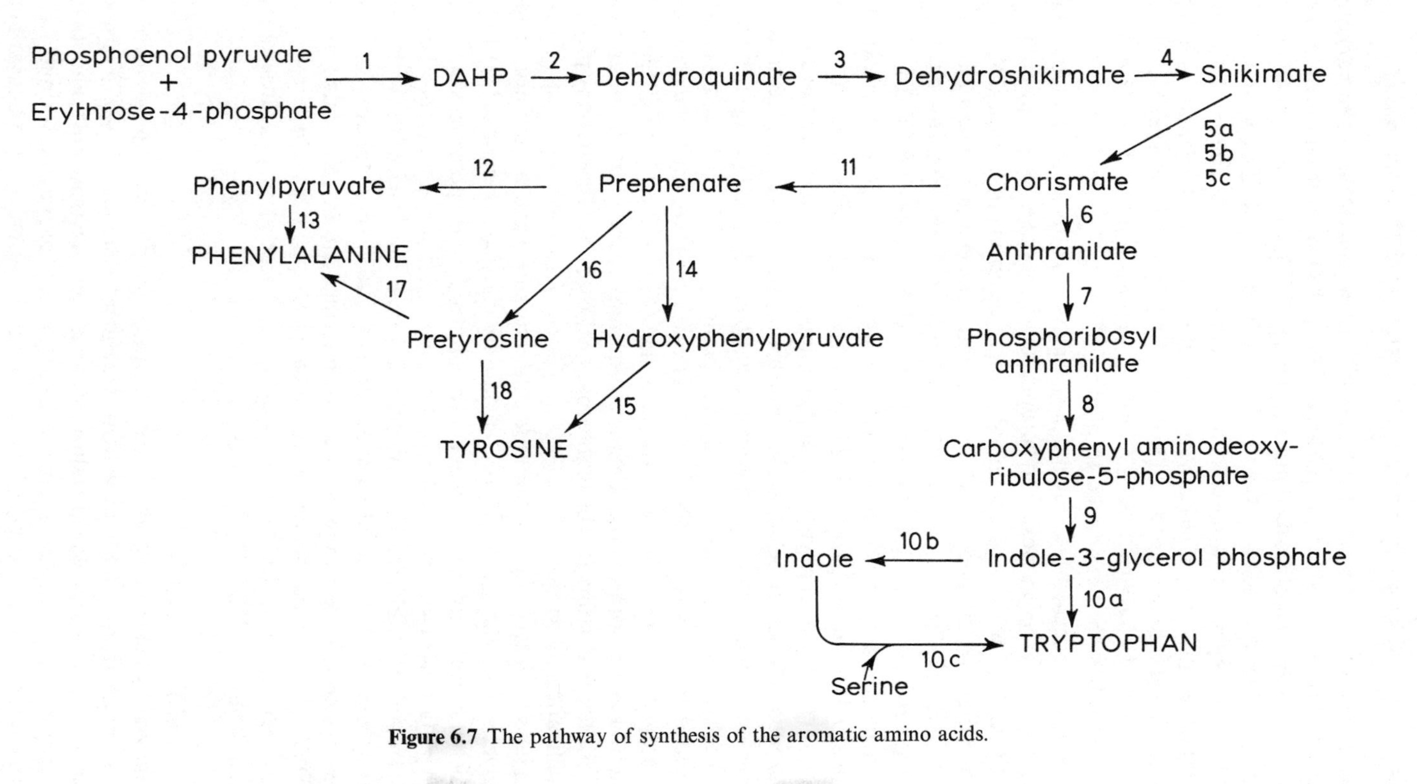

Figure 6.7 The pathway of synthesis of the aromatic amino acids.

by Cu^{2+} or Co^{2+} but not Mn^{2+} or Ca^{2+}; in fungi it occurs in an aggregate with the next four enzymes of the pathway [248]. 3-Dehydroquinate dehydratase (3) EC.4.2.1.10 has been purified independently [252], and also in an aggregate with shikimate dehydrogenase (4) EC.1.1.1.25 which is NADP-dependent [253, 254]. The latter enzyme exists in five forms in *Vicia* [255], four in barley and at least two in other plants [256]. The conversion of shikimate to chorismate is by the three enzymes shikimic kinase (5a) EC.2.7.1.71, 3-enolpyruvolylshikimate-5-phosphate synthase (5b) EC.2.5.1.19 and chorismate synthase (5c) EC.4.6.1.4. Only (5a) has been examined in plants [257].

Chorismate occupies the pivotal position in the pathway of aromatic amino acid synthesis. For the synthesis of tryptophan it is aminated by anthranilate synthase (6) EC.4.1.3.27 with a subsequent removal of a pyruvyl side chain. The enzyme is a classical glutamine-dependent reaction and has been demonstrated in a number of plants [258–260]. Anthranilate is converted to *N*-phosphoribosyl anthranilate by the action of anthranilate phosphoribosyl transferase (7) EC.2.4.2.18 which is in turn isomerized by phosphoribosyl anthranilate isomerase (8) and then converted to indole-3-glycerol phosphate (IGP) by IGP synthase (9) EC.4.1.1.48. The only detailed reports of these three enzymes are in peas and maize, where they were shown to be separable enzymes [261] and not aggregates or single multifunctional proteins as in bacteria, fungi and *Euglena* [262]. The final step in the pathway has been widely studied in plants [261, 263, 264] and is catalysed by tryptophan synthase (10) EC.4.2.1.20. The enzyme is complex and consists of two subunits A and B which show three activities (Fig. 6.7, reactions 10a, b and c). For reaction (10a) both components are required whereas the A subunit catalyses (10c) and the B, (10b). It is probable that only reaction (10a) has any physiological significance.

The syntheses of phenylalanine and tyrosine have a common step in the conversion of chorismate to prephenate by chorismate mutase (11) EC.5.4.99.5 which has been shown to exist as a number of different isoenzymes in plants [265–268]. There are two possible mechanisms by which prephenate may be converted to tyrosine. The differences are dependent upon the order of the transamination step, which may be at either reaction (15) [269] or reaction (16) [270, 271]; pretyrosine may however, also be an intermediate in phenylalanine synthesis [270, 271]. Prephenate may be connected to phenylpyruvate by prephenate dehydratase (12) EC.4.2.1.51 which is subsequently transaminated to phenylalanine [272, 273].

Subcellular localization

Isolated intact chloroplasts can synthesize phenylalanine, tyrosine and tryptophan from $^{14}CO_2$ in a light-dependent reaction [274, 275]. Studies on enzyme distribution have also shown that all enzymes from anthranilate synthase to tryptophan synthetase are present in etioplasts of pea and walnuts [276, 277]. Little work has been carried out on the localization of the phenylalanine and tyrosine enzymes, but presumably they are also present in the chloroplasts. Data on the localization of shikimate dehydrogenase are somewhat variable suggesting that the isoenzymes may exist in the chloroplast, microbodies and in the cytosol [255, 256].

Regulation

Evidence for feedback regulation occurring *in vivo* comes from studying the effects of externally applied tryptophan on the flow of carbon from shikimate [202, 258], or on the effect of tyrosine and phenylalanine on growth of barely seedlings [203]. Anthranilate synthase has been shown to be very sensitive to tryptophan inhibition. Mutant cell lines resistant to 5-methyltryptophan, with a greatly increased level of soluble tryptophan were found to have anthranilate synthase activity much less sensitive to feedback inhibition [259, 260]. Later, more detailed studies showed that, in normal plants, two isoenzymes of anthanilate synthase exist; one sensitive and one insensitive to feedback inhibition by tryptophan, with the former greatly in excess of the latter [278]. In contrast, the resistant cells had predominantly the resistant form.

In mungbeans, chorismate mutase (CM) has been separated into two forms [279, 280]. CM-1 was inhibited by phenylalanine and tyrosine, although not additively, and this inhibition was antagonized by equimolar amounts of tryptophan; CM-2 was apparently unaffected by the three amino acids or other products of the pathway. A similar control exists in oak [267], but in alfalfa there are three isoenzymes. CM-1 and CM-3 were sensitive to control by the aromatic acids, whereas CM-1 and CM-2 were inhibited by chlorogenoquinone and caffeic acid and CM-3 by ferridic acid [268].

Inhibition of the synthesis of phenylalanine, tyrosine and tryptophan from $^{14}CO_2$ by isolated spinach chloroplasts has been reported [274, 275]. However, phenylalanine and tyrosine inhibited jointly their own synthesis, and tryptophan inhibited the synthesis of all three amino acids suggesting a site of action between shikimate and chorismate.

No clear pattern emerges as to the regulation of the aromatic amino acid pathway. This may be due to the fact that the enzymes often exist in multiple forms, possibly in different locations within the cell, producing different secondary metabolites in various plants.

(f) Synthesis of histidine

Very little information regarding histidine biosynthesis in plants is available [281]. It is generally assumed that the route of synthesis is the same as in bacteria shown in Fig. 6.8; histidinol has been shown to block the incorporation of [^{14}C]-glucose into histidine [141].

The enzymes ATP-phosphoribosyl transferase (2) EC.2.4.2.17, imidazole glycerol-phosphate dehydratase (7) EC.4.2.1.19 and histidinol phosphatase (9) EC.3.1.3.15 have been demonstrated in extracts of barley, oat and pea shoots [282]. More recently, histidinol dehydrogenase (10) EC.1.1.1.23 has been partially purified from wheat germ, and shown to exist in two forms readily

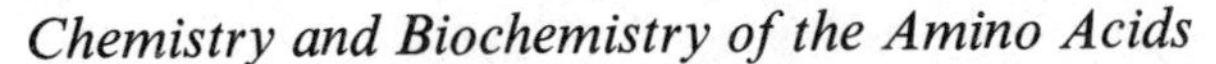

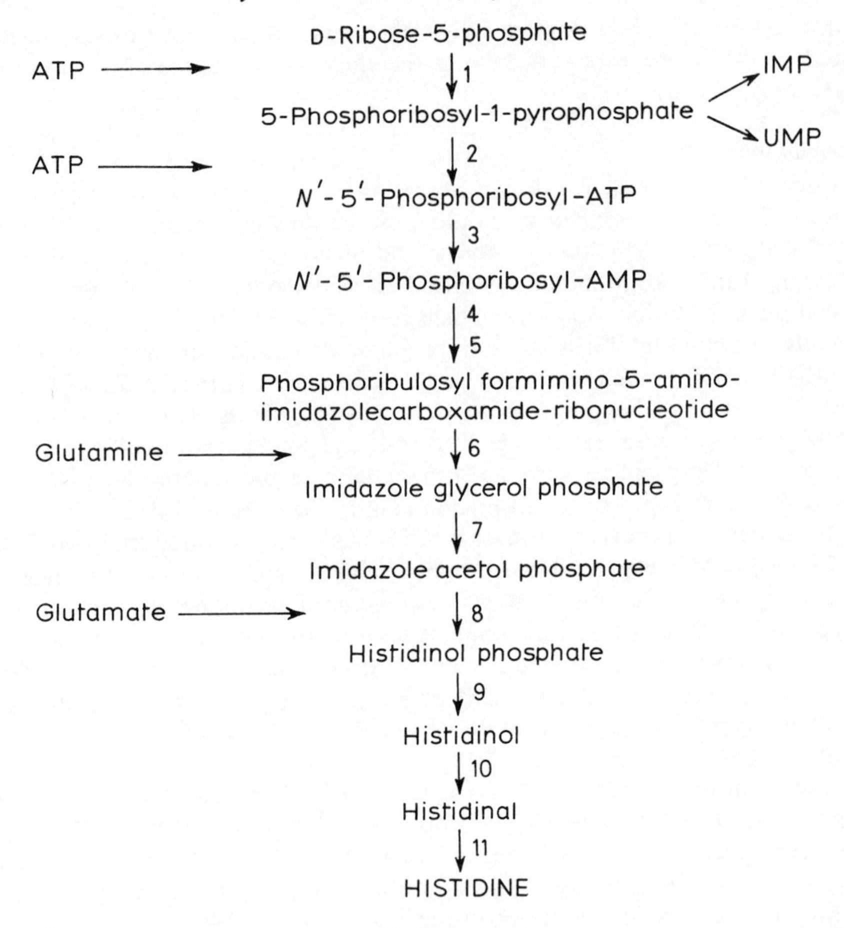

Figure 6.8 The pathway of the synthesis of histidine.

separable by DEAE-cellulose chromatography [283]. The herbicide aminotriazole (which is also a potent catalase inhibitor) inhibits reaction (7) and leads to the accumulation of imidazole glycerol (presumably formed by the action of a phosphatase); this accumulation is abolished by the addition of histidine [284, 285]. There is also some suggestive evidence for the feedback control of plant ATP-phosphoribosyl transferase by low levels of histidine [282]. A number of mutants have been isolated which require histidine for growth. The biochemistry of these mutations has not been established [286].

6.4 CONCLUSIONS

Although it is clear from the previous sections that pathways of the biosynthesis and their subsequent regulation of a number of the protein amino acids are well-documented in plants, there is still a shortage of information on the aromatic and branched-chain amino acids. Further work is required to fill the gaps in our knowledge and it is possible that the availability of mutant amino acid auxotrophs and plants with altered feedback regulation may be more fruitful lines of research than those previously followed. It is also readily apparent that the production of crop plants with elevated levels of the nutritionally important amino acids lysine, threonine and methionine, but without reduced yields would be of economic importance.

It has become clear that those amino acids essential for animal nutrition so far studied in detail are synthesized mainly, if not totally, in the chloroplasts of plants.

REFERENCES

1. Miflin, B.J. and Lea, P.J. (1977) *Annu. Rev. Plant Physiol.*, **28**, 299.
2. Lea, P.J., Wallsgrove, R.M., Mills, W.R. and Miflin, B.J. (1981) in *The Origin of Chloroplasts* (ed. J.A. Schiff), Elsevier, New York, p. 149.
3. Miflin, B.J. and Lea, P.J. (1982) in *Encyclopaedia of Plant Physiology* (ed. D. Boulter) (Vol. 14A) Springer, Berlin, p. 3.
4. Miflin, B.J., ed. (1980) *The Biochemistry of Plants* (Vol. 5. Amino Acids and Derivatives) Academic Press, New York.
5. Guerrero, M.G., Vega, J.M. and Losada, M. (1981) *Annu. Rev. Plant Physiol.*, **32**, 169.
6. Stewart, W.D.P. (1979) in *Encyclopaedia of Plant Physiology* (eds M. Gibbs and E. Latzko) (Vol. 6) Springer, Berlin, p. 457.
7. Yates, M.G. (1980) in *The Biochemistry of Plants* (ed. B.J. Miflin), Academic Press, New York, p. 1.
8. O'Neal, T.D. and Joy, K.W. (1973) *Arch. Biochem. Biophys.*, **159**, 113.
9. O'Neal, T.D. and Joy, K.W. (1975) *Plant Physiol.*, **55**, 968.
10. Stewart, G.R. and Rhodes, D. (1977) *New Phytologist*, **79**, 257.
11. Hirel, B. and Gadal, P. (1980) *Plant Physiol.*, **66**, 619.
12. Mann, A.F., Fentem, P.A. and Stewart, G.R. (1980) *FEBS Letters*, **110**, 265.
13. Stasiewicz, S. and Dunham, V.L. (1979) *Biochem. Biophys. Res. Commun.*, **87**, 627.
14. Estigneeva, Z.G., Pushkin, A.V., Radyukira, N.A. and Kretovich, V.L. (1977) *Dokl. Akad. Nauk SSR*, **237**, 962.
15. Lea, P.J. and Miflin, B.J. (1974) *Nature*, **251**, 614.
16. Wallsgrove, R.M., Harel, E., Lea, P.J. and Miflin, B.J. (1977) *J. Exp. Bot.*, **28**, 588.
17. Lee, J.A. and Stewart, G.R. (1978) *Adv. Bot. Res.*, **6**, 1.
18. Stewart, G.R. and Rhodes, D. (1978) *New Phytol.*, **80**, 307.
19. Suzuki, A. and Gadal, P. (1982) *Plant Physiol.*, **69**, 848.
20. Tamura, G., Oto, M., Hirasawa, M. and Aketagawa, J. (1980) *Plant Sci. Lett.*, **19**, 209.
21. Matoh, T., Ida, S. and Takahashi, E. (1980) *Plant cell Physiol.*, **21**, 1461.
22. Wallsgrove, R.M., Lea, P.J. and Miflin, B.J. (1982) *Planta*, **154**, 473.
23. Suzuki, A., Vidal, J. and Gadal, P. (1982) *Plant Physiol.*, **70**, 827.
24. Lewis, O.A.M. and Pate, J.S. (1973) *J. Exp. Bot.*, **24**, 596.
25. Rhodes, D., Sims, A.P. and Folks, B.F. (1980) *Phytochemistry*, **19**, 357.
26. Stewart, G.R. and Rhodes, D.A. (1976) *FEBS Letters*, **64**, 296.
27. Stewart, C.R. (1979) *Plant Sci. Lett.*, **14**, 269.
28. Woo, K.C. and Canvin, D.T. (1980) *Can. J. Bot.*, **58**, 511.
29. Kanamori, T. and Matsumoto, H. (1972) *Arch. Biochem. Biophys.*, **152**, 404.
30. Emes, M.J. and Fowler, M.W. (1978) *Planta*, **144**, 249.
31. Oaks, A., Jones, K. and Misra, S. (1979) *Plant Physiol.*, **63**, 793.
32. Oaks, A., Stulen, I., Jones, K. *et al.* (1980) *Planta*, **148**, 477.
33. Elliott, W.H. (1933) *J. Biol. Chem.*, **201**, 661.
34. Kingdon, H.S. (1974) *Arch. Biochem. Biophys.*, **163**, 429.

35. Lea, P.J. and Fowden, L. (1975) *Proc. Roy. Soc. London Ser. B*, **192**, 13.
36. Storey, R. and Beevers, L. (1978) *Plant Physiol.*, **61**, 494.
37. Sodek, L., Lea, P.J. and Miflin, B.J. (1980) *Plant Physiol.*, **65**, 22.
38. Duffus, C.M. and Rosie, R. (1978) *Plant Physiol.*, **61**, 570.
39. Miflin, B.J. and Shewry, P.R. (1979) in *Recent Advances in the Biochemistry of Cereal* (eds D. Laidman and R. Wyn Jones), Academic Press, New York, p. 239.
40. Fowler, M.W., Jessup, W. and Stephan-Saskissian, G. (1974) *FEBS Letters*, **46**, 340.
41. Miflin, B.J. and Lea, P.J. (1975) *Biochem. J.*, **149**, 403.
42. Matoh, T., Takahashi, E. and Ida, S. (1979) *Plant Cell Physiol.*, **20**, 1455.
43. Sodek, L. and Da Silva, W.J. (1977) *Plant Physiol.*, **60**, 602.
44. Beevers, L. and Storey, R. (1976) *Plant Physiol.*, **57**, 862.
45. Storey, R. and Reporter, M. (1978) *Plant Physiol.*, **61**, 494.
46. Matoh, T., Susuki, F. and Ida, S. (1979) *Plant Cell Physiol.*, **20**, 1329.
47. Matoh, T. and Takahashi, E. (1981) *Plant Cell Physiol.*, **22**, 727.
48. Cocking, E.C. and Yemm, E.W. (1961) *New Phytol.*, **60**, 103.
49. Yoneyama, T. and Kumazawa, K. (1975) *Plant Cell Physiol.*, **16**, 21.
50. Miflin, B.J. and Lea, P.J. (1976) *Phytochemistry*, **15**, 873.
51. Arima, Y. and Kumazawa, K. (1977) *Plant Cell Physiol.*, **18**, 1121.
52. Probyn, T.A. and Lewis, O.A.M. (1979) *J. Exp. Bot.*, **30**, 299.
53. Fentem, P.A., Lea, P.J. and Stewart, G.R. (1983) *Plant Physiol.*, **71**, 496.
54. Atkins, C.A., Pate, J.S. and Sharkey, P.J. (1975) *Plant Physiol.*, **56**, 807.
55. Robertson, J.G. and Farnden, K.T.F. (1980) in *The Biochemistry of Plants* (ed. B.J. Miflin), Academic Press, New York, p. 65.
56. Brown, C.M. and Dilworth, M.J. (1975) *J. Gen. Microbiol.*, **86**, 39.
57. Robertson, J.G., Farnden, K.J.F., Warburton, M.P. and Banks, J.M. (1975) *Aust. J. Plant Physiol.*, **2**, 265.
58. Stripf, R. and Werner, D. (1978) *Z. Naturforsch.*, **33C**, 373.
59. Awonaike, K.O., Lea, P.J. and Miflin, B.J. (1981) *Planta Sci. Lett.*, **23**, 189.
60. Cullimore, J.V., Lara, M., Lea, P.J. and Miflin, B.J. (1983) *Planta*, **157**, 245.
61. McParland, R.H., Guevara, J.G., Becker, R.R. and Evans, H.J. (1976) *Biochem. J.*, **153**, 597.
62. Lea, P.J., Cullimore, J.V. and Miflin, B.J. *et al.* (1982) *Israel J. Bot.*, **31**, 140.
63. Boland, M.J., Fordyce, A.M. and Greenwood, R.M. (1978) *Aust. J. Plant Physiol.*, **5**, 553.
64. Boland, M.J. (1979) *Eur. J. Biochem.*, **99**, 531.
65. Awonaike, K.O. (1980) PhD Thesis, University of London.
66. Kennedy, I.R. (1966) *Biochim. Biophys. Acta*, **130**, 295.
67. Meeks, J.C., Wolk, C.P., Schilling, N. *et al.* (1978) *Plant Physiol.*, **61**, 980.
68. Okyama, T. and Kumazawa, K. (1980) *Soil Sci. Plant Nutr.* (*Tokyo*), **26**, 109.
69. Quail, P.H. (1979) *Annu. Rev. Plant Physiol.*, **30**, 425.
70. O'Neal, T.D. and Joy, K.W. (1973) *Nature New Biol.*, **246**, 61.
71. Wallsgrove, R.M., Lea, P.J. and Miflin, B.J. (1979) *Plant Physiol.*, **63**, 232.
72. Mann, A.F., Fentem, P.A. and Stewart, G.R. (1979) *Biochem. Biophys. Res. Commun.*, **88**, 515.
73. Jackson, C., Dench, J.E., Morris, P. *et al.* (1979) *Biochem. Soc. Trans.*, **7**, 1124.
74. Wallsgrove, R.M., Keys, A.J., Bird, I.F. (1980) *J. Exp. Bot.*, **31**, 1005.
75. Nishimura, M., Douce, R. and Akazawa, T. (1980) *Plant Physiol.*, **65**, 14S.
76. Miflin, B.J. (1974) *Plant Physiol.*, **54**, 550.
77. Susuki, A., Gadal, P. and Oaks, A. (1981) *Planta*, **151**, 457.
78. Lea, P.J. and Miflin, B.J. (1979) in *Encyclopaedia of Plant Physiology* (eds M. Gibbs and E. Latzko) (Vol. 6) Springer, Berlin, p. 445.
79. Anderson, J.W. and Done, J. (1977) *Plant Physiol.*, **60**, 354.
80. Anderson, J.W. and Done, J. (1977) *Plant Physiol.*, **60**, 504.
81. Davis, D.D. and Teixeira (1975) *Phytochemistry*, **14**, 647.
82. Wightman, F. and Forest, J.C. (1978) *Phytochemistry*, **17**, 1455.
83. Givan, C. (1980) in *The Biochemistry of Plants* (ed. B.J. Miflin) (Vol. 5) Academic Press, New York, p. 329.
84. Stewart, G.R., Mann, A.F. and Fentem, P.A. (1980) in *The Biochemistry of Plants* (ed. B.J. Miflin) (Vol. 5) Academic Press, New York, p. 271.
85. Miflin, B.J. and Lea, P.J. (1980) in *The Biochemistry of Plants* (ed. B.J. Mifflin) (Vol. 5) Academic Press, New York, p. 169.

86. Miflin, B.J., Lea, P.J. and Wallsgrove, R.M. (1981) in *Current Topics in Cellular Regulation* (eds B.L. Horecker and E.R. Stadtman) (Vol. 20) Academic Press, New York, p. 1.
87. Thompson, J.F. (1980) in *The Biochemistry of Plants* (ed. B.J. Miflin) (Vol. 5) Academic Press, New York, p. 375.
88. Morris, C.J. and Thompson, J.F. (1975) *Plant Physiol.*, **55**, 960.
89. Morris, C.J. and Thompson, J.F. (1977) *Plant Physiol.*, **59**, 684.
90. Farago, A. and Denes, G. (1967) *Biochim. Biophys. Acta*, **136**, 6.
91. Cohen, P. (1981) in *Current Topics in Cellular Regulation* (eds B.L. Horecker and E.R. Stadtman) (Vol. 18) Academic Press, New York, p. 1.
92. O'Neal, T.D. and Naylor, A.W. (1976) *Plant Physiol.*, **57**, 23.
93. Ong, B.J. and Jackson, J.F. (1972) *Biochem. J.*, **129**, 583.
94. Shargool, P.D., Steeves, T., Weaver, M.G. and Russell, M. (1978) *Can. J. Biochem.*, **56**, 273.
95. Kleczkowski, K. and Cohen, P. (1964) *Arch. Biochem. Biophys.*, **107**, 271.
96. Eid, S., Waly, Y. and Abdelal, A.T. (1974) *Phytochemistry*, **13**, 99.
97. Glenn, E. and Maretzki, A. (1977) *Plant Physiol.*, **60**, 122.
98. Roubelakis, K.A. and Kliewer, W.M. (1978) *Plant Physiol.*, **62**, 340.
99. Shargool, P.D. (1971) *Phytochemistry*, **10**, 2029.
100. Walker, J.B. and Myers, J. (1953) *J. Biol. Chem.*, **203**, 143.
101. Shargool, P.D. (1975) *Plant Physiol.*, **55**, 632.
102. Bidwell, R.G.S. and Durzan, D.J. (1975) in *Historical and Current Aspects of Plant Physiology* (ed. P.J. Davies), Cornell University Press, New York, p. 152.
103. Umbarger, H.E. (1978) *Ann. Rev. Biochem.*, **47**, 533.
104. Morris, C.J., Thompson, J.F. and Johnson, C.M. (1969) *Plant Physiol.*, **44**, 1023.
105. Noguchi, M. Koiwai, A. and Tamaki, E. (1966) *Agric. Biol. Chem.*, **30**, 452.
106. Rena, A.B. and Splittstoesser, W.E. (1975) *Phytochemistry*, **14**, 657.
107. Vansuyt, G., Vallee, J.C. and Prevost, J. (1979) *Physiol. Veg.*, **17**, 95–105.
108. Seneviratne, A.S. and Fowden, L. (1968) *Phytochemistry*, **7**, 1047.
109. Mazelis, M. and Fowden, L. (1969) *Phytochemistry*, **8**, 801.
110. Lu, T.S. and Mazelis, M. (1975) *Plant Physiol.*, **55**, 502.
111. MacHolan, L., Zobac, P. and Hekelova, P. (1965) *Hoppe Seyler's Z. Physiol. Chem.*, **349**, 97.
112. Dougall, D.K. and Fulton, M.M. (1967) *Plant Physiol.*, **42**, 387.
113. Mestichelli, L.J.J., Gupta, R.N. and Spenser, I.D. (1979) *J. Biol. Chem.*, **254**, 640.
114. Davis, R.H. and Mora, J. (1968) *J. Bacteriol.*, **96**, 383.
115. Boggess, S.F., Stewart, C.R., Aspinall, D. and Paleg, L.G. (1976) *Plant Physiol.*, **58**, 398.
116. Kolloffel, C. and Stroband, H.W.J. (1973) *Phytochemistry*, **12**, 2635.
117. Kolloffel, C. and van Dyk, H.D. (1975) *Plant Physiol.*, **55**, 507.
118. Splittstoesser, W.E. (1969) *Phytochemistry*, **8**, 753.
119. Gamble, J.G. and Lehningen, A.L. (1973) *J. Biol. Chem.*, **248**, 610.
120. Weiss, R.L. and Davis, R.H. (1973) *J. Biol. Chem.*, **248**, 5403.
121. Taylor, A.A. and Stewart, G.R. (1981) *Biochem. Biophys. Res. Commun.*, **101**, 1281.
122. Oaks, A., Mitchell, I.J., Barnard, R.A. and Johnson, F.T. (1970) *Can. J. Bot.*, **48**, 2249.
123. Gavazzi, G., Racchi, M. and Torelli, C. (1975) *Theor. Appl. Genet.*, **46**, 339.
124. Bertani, A., Tonelli, C. and Gavazzi, G. (1980) *Maydica*, **25**, 17.
125. Dierks-Ventling, C. and Tonelli, C. (1982) *Plant Physiol.*, **69**, 130.
126. Bauer, A., Urquart, A.A. and Joy, K.W. (1977) *Plant Physiol.*, **59**, 915.
127. Cooney, D.A., Jayaram, H.N., Swengros, S.G. *et al.* (1980) *Int. J. Biochem.*, **11**, 69.
128. Rognes, S.E. (1970) *FEBS Letters*, **10**, 62.
129. Rognes, S.E. (1980) *Phytochemistry*, **19**, 2287.
130. Streeter, J.G. (1973) *Arch. Biochem. Biophys.*, **157**, 613.
131. Dilworth, M.F. and Dure, L. (1978) *Plant Physiol.*, **61**, 698.
132. Kern, R. and Chrispeels, M.J. (1978) *Plant Physiol.*, **62**, 815.
133. Stulen, I. and Oaks, A. (1977) *Plant Physiol.*, **60**, 680.
134. Stulen, I., Israelstam, G.F. and Oaks, A. (1979) *Planta*, **146**, 237.
135. Scott, D.B., Robertson, J. and Farnden, K.J.F. (1976) *Nature*, **262**, 703.
136. Lea, P.J. and Miflin, B.J. (1980) in *The Biochemistry of Plants* (ed. B.J. Miflin) (Vol. 5) Academic Press, New York, p. 569.
137. Miflin, B.J., Bright, S.W.J., Davies, H.M. *et al.* (1979) in *Nitrogen Assimilation in Plants* (eds E.J. Hewitt and C.V. Cutting), Academic Press, London, p. 335.

138. Bryan, J.K. (1980) in *The Biochemistry of Plants* (ed. B.J. Miflin) (Vol. 5) Academic Press, New York, p. 525.
139. Vogel, H.J. (1959) *Proc. Nat. Acad. Sci. USA*, **45**, 1717.
140. Moller, B.C. (1974) *Plant Physiol.*, **34**, 638.
141. Dougall, D.K. and Fulton, M.M. (1967) *Plant Physiol.*, **42**, 941.
142. Oaks, A. (1963) *Biochim. Biophys. Acta*, **76**, 638.
143. Dunham, V.L. and Bryan, J.K. (1971) *Plant Physiol.*, **47**, 91.
144. Henke, R.R. and Wilson, K.G. (1974) *Planta*, **121**, 155.
145. Davies, H.M. and Miflin, B.J. (1978) *Plant Physiol.*, **62**, 536.
146. Bright, S.W.J., Shewry, P.R. and Miflin, B.J. (1978) *Planta*, **139**, 119.
147. Giovanelli, J., Mudd, S.H. and Datko, A.H. (1980) in *The Biochemistry of Plants* (ed. B.J. Miflin) (Vol. 5) Academic Press, New York, p. 454.
148. Mills, W.R., Lea, P.J., and Miflin, B.J. (1980) *Plant Physiol.*, **65**, 1166.
149. Dunham, V.L. and Bryan, J.K. (1969) *Plant Physiol.*, **44**, 1601.
150. Green, C.E. and Phillip, R.L. (1974) *Crop Sci.*, **14**, 827.
151. Henke, R.R., Wilson, K.G., Mclure, J.W. and Treick, R.W. (1974) *Planta*, **116**, 333.
152. Bright, S.W.J., Wood, E.A. and Miflin, B.J. (1978) *Planta*, **139**, 113.
153. Davis, H.M. and Miflin, B.J. (1977) *Plant Sci. Letters*, **9**, 323.
154. Sakano, K. and Komamine, A. (1978) *Plant Physiol.*, **63**, 583.
155. Matthews, B.F. and Widholm, J.M. (1979) *Z. Naturforsch.*, **34C**, 1177.
156. Bright, S.W.J., Miflin, B.J. and Rognes, S.E. (1982) *Biochem. Genet.*, **20**, 229.
157. Rognes, S.E., Lea, P.J. and Miflin, B.J. (1980) *Nature*, **287**, 357.
158. Cheshire, R.M. and Miflin, B.J. (1975) *Phytochemistry*, **14**, 695.
159. Mazelis, M., Whatley, F.R. and Whatley, J. (1977) *FEBS Letters*, **64**, 197.
160. Wallsgrove, R.M. and Mazelis, M. (1981) *Phytochemistry*, **20**, 2651.
161. Mazelis, M., Miflin, B.J. and Pratt, H.M. (1976) *FEBS Letters*, **64**, 197.
162. Mazelis, M. and Crevelling, R.K. (1978) *J. Food Biochem.*, **2**, 29.
163. Sodek, L. (1978) *Rev. Bios. Bot.*, **1**, 65.
164. Bryan, J.K. (1969) *Biochim. Biophys. Acta*, **171**, 205.
165. Aarnes, H. and Rognes, S.E. (1974) *Phytochemistry*, **13**, 2717.
166. Di Marco, G. and Grego, S. (1975) *Phytochemistry*, **14**, 943.
167. Aarnes, H. (1977) *Plant Sci. Letters*, **10**, 381.
168. Matthews, B.F. and Widholm, J.M. (1979) *Phytochemistry*, **18**, 395.
169. Sainis, J., Mayne, R.G., Wallsgrove, R.M., *et al.* (1981) *Planta*, **152**, 491.
170. DiCamelli, C.A. and Bryan, J.K. (1975) *Plant Physiol.*, **55**, 999.
171. DiCamelli, C.A. and Bryan, J.K. (1980) *Plant Physiol.*, **65**, 176.
172. Walter, T.J., Connelly, J.A., Gengenboch, B.G. and Wold, F. (1979) *J. Biol. Chem.*, **254**, 1349.
173. Bryan, J.K. and Lochner, N.R. (1981) *Plant Physiol.*, **68**, 1395.
174. Aarnes, H. (1978) *Planta*, **140**, 185.
175. Thoen, A., Rognes, S.E. and Aarnes, H. (1978) *Plant Sci. Letters*, **13**, 103.
176. Madison, J.T. and Thompson, J.F. (1976) *Biochem. Biphys. Res. Commun.*, **71**, 684.
177. Thoen, A., Rognes, S.E. and Aarnes, H. (1978) *Plant Sci. Letters*, **13**, 113.
178. Giovanelli, J., Mudd, S.H. and Datko, A.H. (1974) *Plant Physiol.*, **54**, 725.
179. Giovanelli, J., Mudd, S.H. and Datko, A.H. (1978) *J. Biol. Chem.*, **253**, 5665.
180. Datko, A.H., Mudd, S.H. and Giovanelli, J. (1977) *J. Biol. Chem.*, **252**, 3436.
181. Giovanelli, J. and Mudd, S.H. (1971) *Biochim. Biophys. Acta*, **227**, 654.
182. Anderson, N.W. and Thompson, J.F. (1979) *Phytochemistry*, **18**, 1953.
183. Burton, E.G. and Sakami, W. (1969) *Biochem. Biophys. Res. Commun.*, **36**, 228.
184. Cladinin, M.T. and Cossins, E.A. (1974) *Phytochemistry*, **13**, 585.
185. Bright, S.W.J., Lea, P.J. and Miflin, B.J. (1980) in *Sulphur in Biology* (eds K. Elliott and J. Wheland) Ciba Foundation Symposium No. 72, Exerpta Medica, Amsterdam, p. 101.
186. Aarnes, H. (1977) *Plant Sci. Letters*, **10**, 381.
187. Lea, P.J., Mills, W.R. and Miflin, B.J. (1979) *FEBS Letters*, **98**, 165.
188. Wahnbaeck-Spencer, R., Henke, R.R., Mills, W.R., *et al.* (1979) *FEBS Letters*, **104**, 303.
189. Wallsgrove, R.M., Lea, P.J. and Miflin, B.J. (1983) *Plant Physiol.*, **71**, 780.
190. Wallsgrove, R.M. and Verbruggey, I. unpublished results.
191. Bryan, J.K., Lissik, E.A. and Matthews, B.F. (1977) *Plant Physiol.*, **59**, 673.
192. Wallsgrove, R.M., Lea, P.J., Mills, W.R. and Miflin, B.J. (1979) *Plant Physiol.*, **63**, 265.

193. Wallsgrove, R.M. and Mazelis, M. (1980) *FEBS Letters*, **116**, 189.
194. Sloger, M. and Owens, L.D. (1974) *Plant Physiol.*, **53**, 469.
195. Widholm, J.M. (1976) *Can. J. Bot.*, **54**, 1523.
196. Hibberd, K.A., Walter, T., Green, C.E. and Gengenbach, B.G. (1980) *Planta*, **148**, 183.
197. Hibberd, K.A. and Green, C.E. (1982) *Proc. Nat. Acad. Sci. USA*, **79**, 559.
198. Rognes, S.E., Bright, S.W.J. and Miflin, B.J. (1983) *Planta*, **157**, 32.
199. Bright, S.W.J., Kueh, J.S.H., Franklin, J. *et al.* (1982) *Nature*, **299**, 278.
200. Bright, S.W.J., Featherstone, L.C. and Miflin, B.J. (1979) *Planta*, **146**, 629.
201. Borstlap, A.C. (1975) *Acta Bot. Neerl.*, **24**, 203.
202. Oaks, A. (1965) *Plant Physiol.*, **40**, 149.
203. Miflin, B.J. (1969) *J. Exp. Bot.*, **20**, 810.
204. Bortslap, A.C. (1970) *Acta Bot. Neerl.*, **19**, 211.
205. Dougall, D.K. (1970) *Phytochemistry*, **9**, 959.
206. Bleckman, G.I., Kogan, Z.S. and Kretowich, W.L. (1971) *Biokhimiya*, **36**, 1050.
207. Sidorou, V., Menczel, L. and Maliga, P. (1981) *Nature*, **294**, 87.
208. Miflin, B.J. (1971) *Arch. Biochem. Biophys.*, **146**, 542.
209. Miflin, B.J. and Cave, P.R. (1972) *J. Exp. Bot.*, **23**, 511.
210. Kiritani, K. and Wagner, R.P. (1970) *Meth. Enzymol.*, **XVIIA**, 755.
211. Oaks, A. (1965) *Biochim. Biophys. Acta*, **111**, 79.
212. Bassham, J.A., Larsen, P.O., Lawyer, A.L. and Cornwell, K.L. (1981) in *Nitrogen and Carbon Metabolism* (ed. J.D. Bewley) Martinus Nijhoff/Dr W. Junk, The Hague, p. 135.
213. Tolbert, N.E. (1979) in *Encyclopaedia of Plant Physiology* (eds M. Gibbs and E. Latzko) (Vol. VI) Springer, Berlin, p. 338.
214. Keys, A.J. (1980) in *The Biochemistry of Plants* (ed. B.J. Miflin) (Vol. 5) Academic Press, New York, p. 359.
215. Somerville, C.R. and Ogren, W.L. (1982) *TIBS*, **7**, 171.
216. Walton, N.J. and Butt, V.S. (1981) *Planta*, **153**, 232.
217. Brock, B.L., Wilkinson, D.A. and King, J. (1970) *Can. J. Biochem.*, **48**, 486.
218. Somerville, C.R. and Ogren, W.L. (1980) *Proc. Nat. Acad. Sci. USA*, **77**, 2684.
219. Gardestrom, P., Bergman, A. and Ericson, I. (1980) *Plant Physiol.*, **65**, 389.
220. Woo, K.C. and Osmond, C.B. (1977) *Plant Cell Physiol., Special Issue.*, 315.
221. Mazelis, M. and Liu, E.S. (1967) *Plant Physiol.*, **42**, 1763.
222. Rao, D.N. and Rao, N.A. (1982) *Plant Physiol.*, **69**, 11.
223. Slaughter, J.C. and Davies, D.D. (1968) *Biochem. J.*, **109**, 743.
224. Slaughter, J.C. and Davies, D.D. (1968) *Biochem. J.*, **109**, 749.
225. Slaughter, J.C. (1970) *FEBS Letters*, **7**, 245.
226. Slaughter, J.C. (1973) *Phytochemistry*, **12**, 2627.
227. Larson, C. and Albertsson, E. (1979) *Physiol. Plant*, **45**, 7.
228. Smith, I.K. and Thompson, J.F. (1971) *Biochim. Biophys. Acta*, **227**, 288.
229. Ngo, T.T. and Shargool, P.D. (1974) *Can. J. Bot.*, **52**, 6.
230. Smith, I.K. (1977) *Phytochemistry*, 165, 1293.
231. Masada, M., Fukushima, K. and Tamura, G. (1975) *J. Biochem.*, **77**, 1107.
232. Bertagnolli, B.L. and Wedding, R.T. (1977) *Plant Physiol.*, **60**, 115.
233. Anderson, J.W. (1980) in *The Biochemistry of Plants* (ed. B.J. Miflin) (Vol. 5) Academic Press, New York, p. 203.
234. Schmidt, A., Abrams, W.R. and Schiff, J.A. (1974) *Eur. J. Biochem.*, **47**, 423.
235. Beevers, H. (1979) *Annu. Rev. Plant Physiol.*, **30**, 159.
236. Fankhauser, H., Brunold, C. and Ericsman, K.H. (1976) *Experientia*, **32**, 1494.
237. Ascano, A. and Nicholas, D.J.D. (1976) *Phytochemistry*, **16**, 889.
238. Smith, I.K. (1972) *Plant Physiol.*, **50**, 477.
239. Lorimer, G. (1981) *Annu. Rev. Plant Physiol.*, **32**, 349.
240. Keys, A.J., Bird, I.F., Cornelius, M.J., *et al.* (1978) *Nature*, **275**, 74.
241. Somerville, C.R. and Ogren, W.L. (1980) *Nature*, **286**, 257.
242. Cresswell, C.F. (1980) *S. Afr. J. Sci.*, **76**, 107.
243. Brunhold, C. and Schmidt, A. (1978) *Plant Physiol.*, **61**, 342.
244. Reuveny, Z. and Filner, P. (1977) *J. Biol. Chem.*, **252**, 1858.
245. Umbarger, H.E. (1978) *Annu. Rev. Biochem.*, **47**, 533.
246. Rennenberg, H. and Filner, P. (1982) *Plant Physiol.*, **69**, 766.

247. Haslam, E. (1974) *The Shikimate Pathway*, Halsted, New York, Toronto.
248. Gilchrist, D.G. and Kosuge, T. (1980) in *The Biochemistry of Plants* (ed. B.J. Miflin) (Vol. 5) Academic Press, New York, p. 507.
249. Rothe, G.M., Maurer, W. and Melke, C. (1976) *Ber. Dtsch. Bot. Ges.*, **89**, 163.
250. Yamamoto, E. (1977) *Plant Cell Physiol.*, **18**, 995.
251. Saijo, R. Kosuge, T. (1978) *Phytochemistry*, **17**, 223.
252. Balinsky, D. and Davies, D.D. (1961) *Biochem. J.*, **80**, 300.
253. Koshiba, T. (1978) *Biochim. Biophys. Acta*, **522**, 10.
254. Boudet, A.M., Boudet, A., and Bouyssou, H. (1977) *Phytochemistry*, **16**, 919.
255. Feierabend, J. and Brassel, D. (1977) *Z. Pflanzenphysiol.*, **82**, 334.
256. Rothe, G.M. (1974) *Z. Pflanzenphysiol.*, **74**, 152.
257. Bowen, J.R. and Kosuge, T. (1979) *Plant Physiol.*, **63**, 382.
258. Belser, W., Murphy, J.B., Delmer, D.P. and Mills, S.E. (1971) *Biochim. Biophys. Acta*, **237**, 1.
259. Widholm, J.M. (1972) *Biochim. Biophys. Acta*, **261**, 52.
260. Widholm, J.M. (1972) *Biochim. Biophys. Acta*, **279**, 48.
261. Hawkins, C.N., Largen, M.T. and Mills, S.F. (1976) *Plant Physiol.*, **57**, 101.
262. Lara, J. and Mills, S.E. (1972) *J. Bacteriol.*, **110**, 1100.
263. Delmer, D.P. and Mills, S.E. (1968) *Biochim. Biophys. Acta*, **167**, 431.
264. Nagao, R.T. and Moore, T.C. (1972) *Arch. Biochem. Biophys.*, **149**, 402.
265. Cotton, R.G.H. and Gibson, F. (1969) *Biochim. Biophys. Acta*, **156**, 187.
266. Gilchrist, D.G., Woodin, T.S., Johnson, M.S. and Kosuge, T. (1972) *Plant Physiol.*, **49**, 52.
267. Gadal, P. and Bouyssou, H. (1973) *Physiol. Plant*, **28**, 7.
268. Woodin, T.S. and Nishioka, L. (1973) *Biochim. Biophys. Acta*, **309**, 211.
269. Gamborg, O.L. and Keeley, F.W. (1966) *Biochim. Biophys. Acta*, **115**, 65.
270. Patel, N., Pierson, D.L. and Jensen, R.A. (1977) *J. Biol. Chem.*, **252**, 5839.
271. Rubin, J.L. and Jensen, R.A. (1979) *Plant Physiol.*, **64**, 727.
272. Gamborg, O.L. (1965) *Can. J. Biochem.*, **43**, 723.
273. Redkina, T.V., Uzfenskaya, Z.V. and Kretovich, W.L. (1969) *Biokhimya*, **34**, 247.
274. Bickel, H., Palme, L. and Schultz, G. (1978) *Phytochemistry*, **17**, 119.
275. Bucholz, D., Reupke, B., Bickel, H. and Schulz, G. (1979) *Phytochemistry*, **18**, 1109.
276. Grosse, W. (1976) *Z. Pflanzenphysiol.*, **80**, 463.
277. Grosse, W. (1979) *Z. Pflanzenphysiol.*, **83**, 249.
278. Carlson, J.E. and Widholm, J.M. (1978) *Physiol. Plant*, **44**, 251.
279. Gilchrist, D.G. and Kosuge, T. (1974) *Arch. Biochem. Biophys.*, **164**, 95.
280. Gilchrist, D.G. and Kosuge, T. (1975) *Arch. Biochem. Biophys.*, **171**, 36.
281. Miflin, B.J. (1980) in *The Biochemistry of Plants* (ed. B.J. Miflin) Academic Press, New York, p. 533.
282. Wiater, A., Krajewska-Grynkiewicz, K. and Kloptowski, T. (1971) *Acta Biochim. Pol.*, **18**, 299.
283. Wong, Y.-S. and Mazelis, M. (1981) *Phytochemistry*, **20**, 1831.
284. Siegel, J.N. and Gentile, A.C. (1966) *Plant Physiol.*, **41**, 670.
285. Davies, M.E. (1971) *Phytochemistry*, **10**, 783.
286. Gebhardt, C., Schnebc, V. and King, P.J. (1981) *Planta*, **153**, 81.

CHAPTER SEVEN

Enzyme Inhibition by Amino Acids and their Derivatives

M.J. JUNG

Over 240 amino acids have been isolated and characterized from bacteria, animals or plants [1]. Only twenty of them are used for protein synthesis. An additional number such as ornithine, citrulline, 5-hydroxytryptophan, dihydroxyphenylalanine and homoserine are intermediates in metabolic processes. Most of the others are much rarer and often localized in a given species where they may be secondary metabolites or, more interestingly, the result of an evolutionary defence mechanism against predators. In addition to those natural amino acids, chemists have made almost innumerable modifications of the most common amino acids. In the present review, only the amino acids either natural or man-made for which the biochemical action has been reported and could be related to the inhibition of a specific enzyme will be mentioned.

One can distinguish two broad classes of enzyme inhibitors, reversible and irreversible. Reversible inhibitors have been studied almost from the time of the 'lock and key' hypothesis of E. Fisher. They are usually substrate analogues which bind to the enzyme's active site and compete with the normal substrate [2]. They can either be refractory to the enzyme catalytic act or undergo it. The most famous example of a reversible inhibitor is probably α-methyl dopa. The compound was found to be an inhibitor of aromatic amino acid decarboxylase *in vitro* [3], later it was recognized that its antihypertensive action was due to its decarboxylation to α-methyldopamine and hydroxylation to α-methylnoradrenaline [4]. The reader is referred to the excellent review of Shive and Skinner for other examples [5]. The affinity of this type of inhibitor is usually similar to that of the substrate. As in chemical reactions, a reaction catalysed by an enzyme proceeds via a high-energy transition state. The energy is provided by minimizing the enzyme–substrate interactions, so that a substrate analogue resembling this transition state should show maximal affinity for the enzyme [6]. If the guess of what the transition state looks like has been correct and if a stable molecule approaching the geometry of this transition state can be synthesized, one can gain several orders of magnitude in binding constant. For instance, proline racemase has a 160-fold higher affinity for pyrrole-2-carboxylic acid than for proline [6]. When this binding constant reaches 10^{-9}–10^{-10} M or even lower values the notion of reversibility becomes almost irrelevant and the frontier with irreversible inhibitors very fragile. For the sake of clarity, we will define irreversible inhibitors as those forming a covalent bond with some residue of the enzyme's active site.

The chemical entity on the substrate analogue which forms the covalent bond can be built in from the beginning, for instance a halomethyl ketone. Then one speaks of affinity labelling reagents. One classical example is TPCK, the irreversible inhibitor of trypsin [7]. These compounds are usually considered as being non-selective as they may react with any nucleophile on their way to the enzyme's active site. Nevertheless, chloromethyl ketone analogues of peptides have been described showing several orders of magnitude difference in the rate of inhibition of different serine proteases [7]. The late B.R. Baker has done extensive work on this class of compounds [8]. In affinity labels the degree of selectivity resides in the affinity of the substrate analogue for the enzyme. A further degree of selectivity is achieved if the reactive chemical entity is revealed during enzyme-catalysed transformation of the substrate analogue. Compounds of this type have been called suicide-inhibitors [9], k_{cat} inhibitors [10], mechanism-based [11] or enzyme-activated inhibitors [12]. The first documented examples are probably the irreversible inhibition of aspartate aminotransferase by serine *O*-sulphate [13] and of aspartate β-decarboxylase by chloroalanine [14]. Examples of this type of inhibitor are also to be found in nature [15]. Due to their increased selectivity and duration of action, it is not surprising that a number of compounds of this class are presently being evaluated as therapeutic tools [16].

7.1 PYRIDOXAL-DEPENDENT ENZYMES

Among the enzymes having amino acids as substrates, the pyridoxal-dependent enzymes occupy a privileged position. They catalyse a great variety of reactions centred around the α-amino acid portion of the molecule: transamination to the α-keto acid, decarboxylation to the corresponding amine, in some cases oxidative decarboxylation to the corresponding aldehyde (or ketone), racemization, C_α–C_β bond breaking and β- or γ-elimination of leaving groups. The first part of these different reactions is the same for all enzymes: formation of a Schiff base between the cofactor and the amino acid. It is believed that in this Schiff base, the C–H or C–C bond which is orthogonal to the plane of the imine conjugated with the pyridoxal-ring is weakened [17]. The specificity of the enzyme resides in achieving this positioning. We will examine now case by case what happens thereafter and what amino acids inhibit each class of pyridoxal enzymes.

7.1.1 Transaminases of amino acids

Most of the common amino acids can serve as amine donors when incubated with a liver homogenate in the presence of α-ketoglutarate. The accepted mechanism is shown in Fig. 7.1(a) from the point of Schiff base formation. The α-H is abstracted by some basic residue of the enzyme active site, the α-carbanion generated is stabilized by charge delocalization over the whole pyridine nucleus. The enzyme neutralizes this charge by protonation on the methylene group of the cofactor.

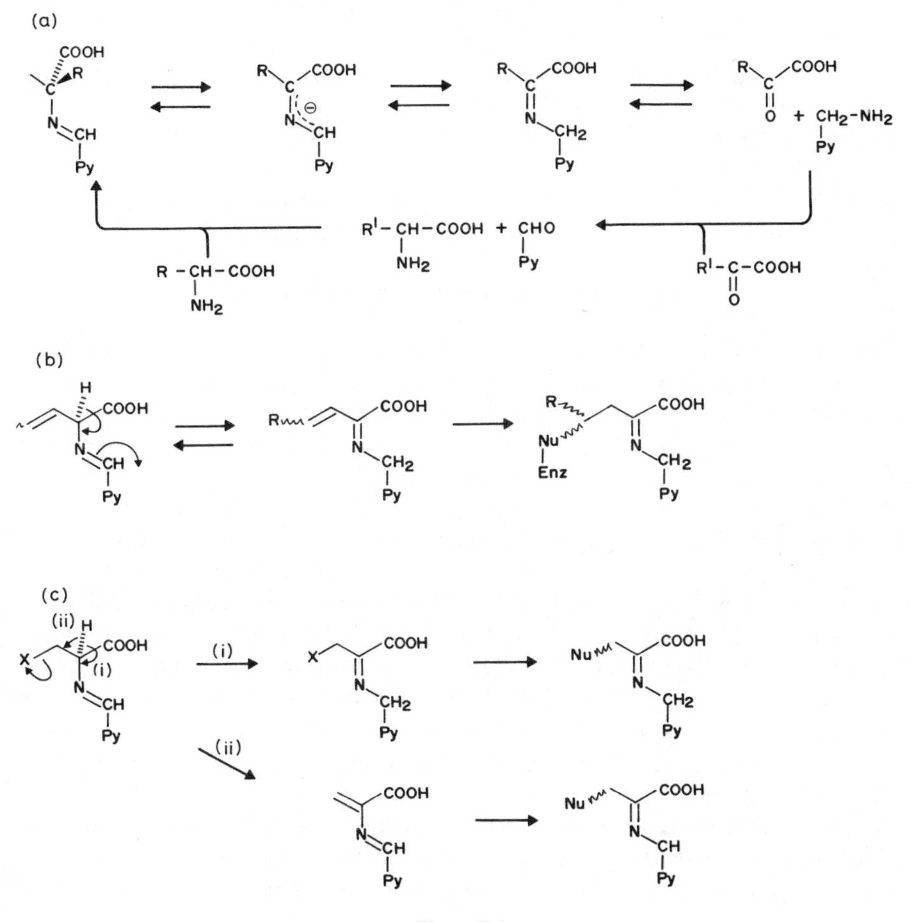

Figure 7.1

After hydrolysis of the tautomeric Schiff base, the newly formed α-keto acid leaves the active site. The enzyme is now in the pyridoxamine form and needs to be transformed back into the pyridoxal form before it can accept another molecule of amino acid. This happens by a reverse sequence of events after Schiff base formation with an α-keto acid present in excess (such as α-ketoglutarate oxaloacetate or pyruvate). Irreversible inhibitors of the suicide type of transaminases are found in nature: 2-amino-4-methoxy-*trans*-3-butenoic acid (*7.1*) [18], cycloserine (*7.2*) [15], propargylglycine (*7.3*) [19], β-cyano-L-alanine (*7.4*) [20]. Others have been synthesized: serine *O*-sulphate (*7.5*) [13], vinylglycine (*7.6*) [21], haloalanines (Fig. 7.2, 7.7) [22–24]. All these compounds generate an electrophilic species within the active site of the enzyme either through the normal prototropy as for vinylglycine or by elimination of a leaving group in β as for the haloalanines (Fig. 7.1(b) and (c)). This electrophilic species may then react

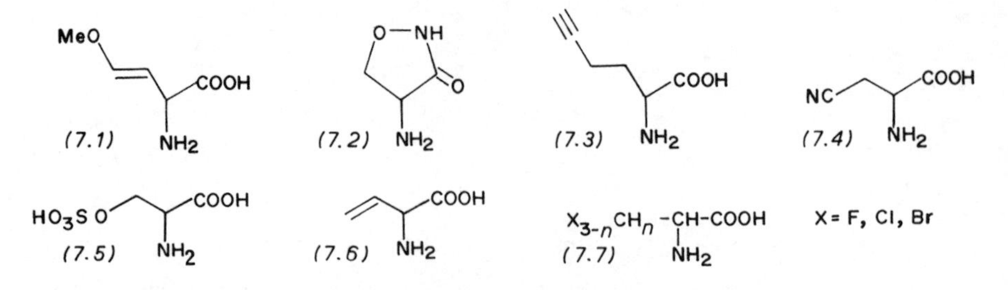

Figure 7.2 α-Amino acid transaminase inhibitors.

with a nucleophile in the active site or be hydrolysed before this reaction can occur. The efficiency of enzyme killing depends on this partition ratio which can range from 1 (one molecule of enzyme is inactivated on each turnover) to several thousands (an enzyme molecule catalyses the transamination of several thousand substrate molecules before it gets inactivated) [24].

Some enzymes of this class act on ω amino acids such as γ-amino-butyric acid (GABA), β-alanine, and the δ-amino function of ornithine. GABA aminotransferase is inhibited by 4-aminohex-5-ynoic acid (γ-acetylenic GABA) (*7.8*) and 4-amino-hex-5-enoic acid (γ-vinyl GABA) (*7.9*) by a mechanism similar to that shown in Fig. 7.1 (b) [25]. GABA transaminase (GABA-T) is also inhibited by ethanolamine *O*-sulphate (*7.10*) [26] and ω-fluoromethyl derivatives of its substrates, β-alanine and GABA (*7.11*) [27] by the mechanism shown in Fig. 7.1 (c). In addition, this enzyme is inhibited by 3-amino-4,5-cyclohexadienoic acid (gabaculine) (*7.12*), a natural compound extracted from a *Streptomyces* culture. The mechanism of inhibition does not rest on enzyme alkylation but on the generation of a transition-state analogue (Fig. 7.3). After the normal prototropy, the substrate–coenzyme adduct rearranges to the more stable anthranilic acid derivative. This compound has a high affinity for the enzyme and

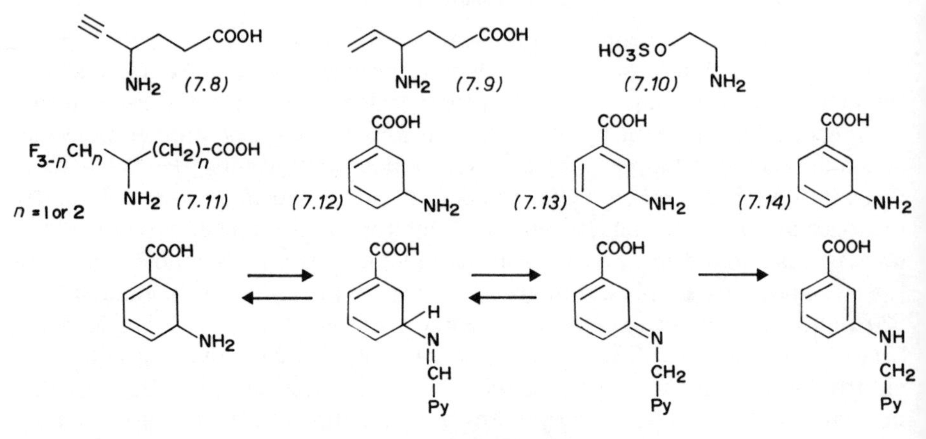

Figure 7.3 GABA-T inhibitors.

does not dissociate under non-denaturing conditions [28]. γ-Acetylenic GABA and gabaculine also inactivate orinithine δ-aminotransferase (Orn-T) by similar mechanisms. At least one of the two double bond isomers of gabaculine, 3-amino-1,4-cyclohexadienyl carboxylic acid (*7.13*) inhibits GABA-T and Orn-T by the aromatization mechanism. The third double-isomer (*7.14*) has been synthesized but no biochemical data have been reported [29].

GABA is the major inhibitory neurotransmitter in the mammalian CNS. GABA-T inhibitors when administered to animals cause an elevation of brain GABA levels resulting in anticonvulsant properties [29]. γ-Vinyl GABA is presently undergoing clinical evaluation in a number of neuropyschiatric disorders. The effect of ornithine transaminase inhibition on proline and glutamic acid biosynthesis has not yet been evaluated.

7.1.2 Amino acid racemases

These enzymes exist in mammals but they are essential in micro-organisms. D-Ala is a constituent of bacterial cell wall and is formed from L-Ala. The mechanism of action is presumably similar to that of transaminases except that reprotonation of the carbanionic intermediate occurs again on the α-carbon. Both L and D amino acids are substrates and this has been used to design inhibitors of bacterial racemases which have little effect on mammalian enzymes. D-Fluoro-alanine is an irreversible inhibitor of Ala-racemase and has been proposed as a new type of antibiotic some years ago. The compound is even more selective and less toxic if the α-hydrogen is replaced by a deuterium [30]. A number of other suicide inhibitors of this enzyme have been described: chloroalanine, *O*-carbamyl-D-serine, D-cycloserine [31].

7.1.3 Amino acid decarboxylases

This class of enzymes is ubiquitous. The amines produced play a variety of roles: chemical transmitters (biogenic amines, γ-aminobutyric acid), factors needed for cell division and differentiation (putrescine and its derivatives spermidine and spermine), autacoids in various physiological processes (histamine).

The mechanism of action is depicted in Fig. 7.4. After Schiff base formation, the carboxyl group placed trans-antiparallel with the imine-pyridoxal plane is lost, generating a carbanionic species as in α-amino acid transaminases. Normally, the enzyme catalyses protonation on the α-carbon to generate the amine after hydrolysis. Every few thousand turnovers, there is abnormal protonation on the methylene group of the cofactor leaving an aldehyde and the pyridoxamine form of the holoenzyme after hydrolysis. The incidence of this oxidative decarboxylation is higher when the enzyme is presented with an α-methyl substrate analogue (α-methyl glutamate and α-methyl dopa) [32]. In the absence of exogenous pyridoxal phosphate, the pyridoxamine form is inactive. In all cases where the stereospecificity has been documented, the protonation on the α-carbon occurs with retention of the configuration.

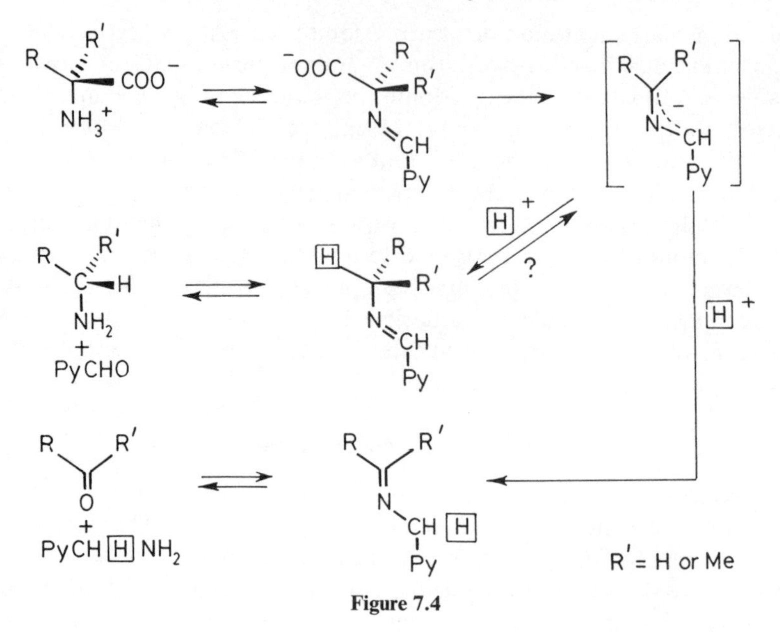

Figure 7.4

As in the case of transaminases, two types of amino acid analogue are inhibitors of decarboxylases [32]: one type has a β,γ-unsaturation which forms a Michael acceptor after oxidative decarboxylation, the other type bears a leaving group on a β-carbon. Experience has demonstrated that α-halomethyl analogues are better inhibitors than those where the halogen atom is placed on the chain carbon atom in β. The examples are numerous and are best given in the form of a table (Table 7.1).

The mechanism of action for the inhibition of aromatic amino acid decarboxylase by monofluoromethyl dopa has been documented [33]. There is stoichiometric binding of the inhibitor to the inactivated enzyme. During the process of inactivation, there is loss of CO_2 and fluorine. This compound is able to block the synthesis of catecholamines and indoleamines completely and to lower their concentrations in the peripheral sympathetic system as well as in brain. The peripheral sympathectomy results in pronounced antihypertensive effects in spontaneously hypertensive rats; the pharmacology of central amine depletion has not been reported yet. Weaker inhibitors of aromatic amino acid decarboxylase such as difluoromethyl dopa could be particularly useful to replace α-methyl hydrazinodopa (carbidopa) in the treatment of Parkinsonism. Inhibitors of ornithine decarboxylase (ODC) (in particular α-difluoromethyl ornithine) have been shown to have antitumour effects in animal models of cancer and, more recently they showed antiparasite properties [32]. The therapeutic usefulness of inhibitors of glutamate decarboxylase and histidine decarboxylase has not yet been demonstrated clinically.

So far we have only examined analogues of amino acids as inhibitors of

Table 7.1

Enzyme	*Inhibitors*	*Reference*
Ornithine decarboxylase	R, COOH, NH_2, NH_2; R: –≡, $-CHF_2$, $-CH_2F$, $-CH_2Cl$	a b c d
Glutamate decarboxylase	CH_2F, HOOC, COOH, NH_2 CH_3, HOOC, COOH, NH_2	e f
Histidine decarboxylase	CH_2F, COOH, N, NH, NH_2	g
Dopa decarboxylase	R, COOH, NH_2, HO, OH; R: –≡, $-CHF_2$, $-CH_2F$	h i j k

[a] Danzin, C. *et al.* (1980) *J. Med. Chem.*, **24**, 16–20.
[b] Metcalf, B. *et al.* (1978) *J. Amr. Chem. Soc.*, **100**, 2551–3.
[c] Kollonitsch, J. *et al.* (1978) *Nature*, **274**, 906–8.
[d] Bey, P. (1978) in *Enzyme-Activated Irreversible Inhibitors* (eds N. Seiler, M.J. Jung and J. Koch-Weser) Elsevier/North Holland, Amsterdam, 27–41.
[e] Kuo, D. and Rando, R. (1981) *Biochemistry*, **20**, 506–11.
[f] Chrystal, E. *et al.* (1979) *J. Neurochem.*, **34**, 1501–7.
[g] Garbarg, M. *et al.* (1980) *J. Neurochem.*, **35**, 1045–52.
[h] Ribereau-Gayan, G. *et al.* (1979) *Biochem. Pharmacol.*, **28**, 1331–5.
[i] Maycock, A. *et al.* (1979) in *Drug Action and Design*, (ed. Th. Kalman) Elsevier/North Holland, Amsterdam, pp. 115–29.
[j] Palfreyman, M. *et al.* (1978) *J. Neurochem.*, **31**, 927–32.
[k] Jung, M. *et al.* (1979) *Life Sci.*, **24**, 1037–42.

decarboxylases. By virtue of the reversibility of most enzymic reactions, a decarboxylase should be able to bind the produced amine, to catalyse the formation of a Schiff base with the cofactor and to abstract the pro-*R* hydrogen (the same one it would introduce in the decarboxylation process) (see Fig. 7.4). If the normal amine is replaced by an amine analogue bearing an α-substituent in the *R* configuration (vinyl, acetylene or halomethyl), one should have inhibition of the enzyme by a mechanism similar to that of amino acid transaminase inhibition. Experience again demonstrates that acetylenic amines are the best inhibitors of decarboxylases: (*R*)-4-amino-hex-5-ynoic acid inhibits bacterial glutamic acid decarboxylase, while the (*S*) enantiomer inhibits the GABA transaminase, (+)-1,4-diamino-5-hexyne inhibits mammalian ODC (stereochemistry reported, [75]), and (*R*) α-acetylenic dopamine inhibits aromatic amino acid decarboxylase [32]. The inhibition of mammalian glutamic acid decarboxylase by (*S*)-4-aminohex-5-ynoic acid cannot be explained by this mechanism. The same is true for the inhibition of histidine decarboxylase by (*R*) and (*S*) α-fluoromethyl histamine.

7.1.4 Pyridoxal-dependent enzymes catalysing β,γ-eliminations

Two very important enzymes in plant and bacterial metabolism catalyse β,γ-elimination: cystathionine γ-synthetase and methionine γ-lyase. Both enzymes are inhibited irreversibly by the natural amino acids, 2-amino-4-(2-amino-3-hydroxypropoxy)-*trans*-but-3-enoic acid (rhizobitoxine) (*7.15*) [34] and propargylglycine (*7.3*) [35]. The mechanism of action of methionine γ-lyase and the mechanism of inhibition by propargylglycine is depicted in Fig. 7.5(a) and (b). After the first α-proton abstraction and prototropic shift as in transaminases, the enzymes are able to abstract a β-proton now activated by the ketimine. In the case of cystathionine γ-synthetase, the β carbanion is stabilized by elimination of succinic acid, in the case of methionine γ-lyase by elimination of methane thiol. The intermediate would be the result of vinyl glycine transamination which, as we have seen before, inhibits Asp-transaminase. Indeed, vinylglycine (*7.6*) is a substrate of the enzyme and does not cause inactivation [36]. In these two enzymes, there is probably no nucleophile in the vicinity and the reactive Michael acceptor adds exogenous cysteine or, in the absence of cysteine, is hydrolysed to α-ketobutyrate. However, β-trifluoromethylalanine (*7.16*) has also been reported to inactivate cystathionase by elimination of a fluoride ion – generating a difluoro-substituted Michael acceptor which may be more reactive than the Michael acceptor derived from the normal substrate [37]. If the normal substrate is replaced by propargylglycine (*7.3*), the second proton abstraction leads to a conjugated allene (Fig. 7.5(b)). Allene formation has also been implicated for the inhibition of γ-cystathionase by compound (*7.17*) [37]. The mechanism of inhibition by rhizobitoxine (*7.15*) has not been studied but is presumably due to the formation of a Michael acceptor positioned differently from the one produced with the normal substrate.

A very elaborate process for the inhibition of enzymes capable of abstracting a β-proton has been described recently [38]. Compounds have been designed in which the second proton abstraction generates an allyl sulphoxide which rearranges spontaneously to form an allyl sulphenate which can alkylate a nucleophilic residue of the enzyme. Thus, 2-amino-4-chloro-5-(nitrophenylsulphinyl)pentanoic acid (*7.18*) exhibits time-dependent inhibition of both cystathionine γ-synthetase and methionine γ-lyase (Fig. 7.5(c)). The extension of this concept to inhibition of a great variety of pyridoxal enzymes has been claimed in the patent literature [39].

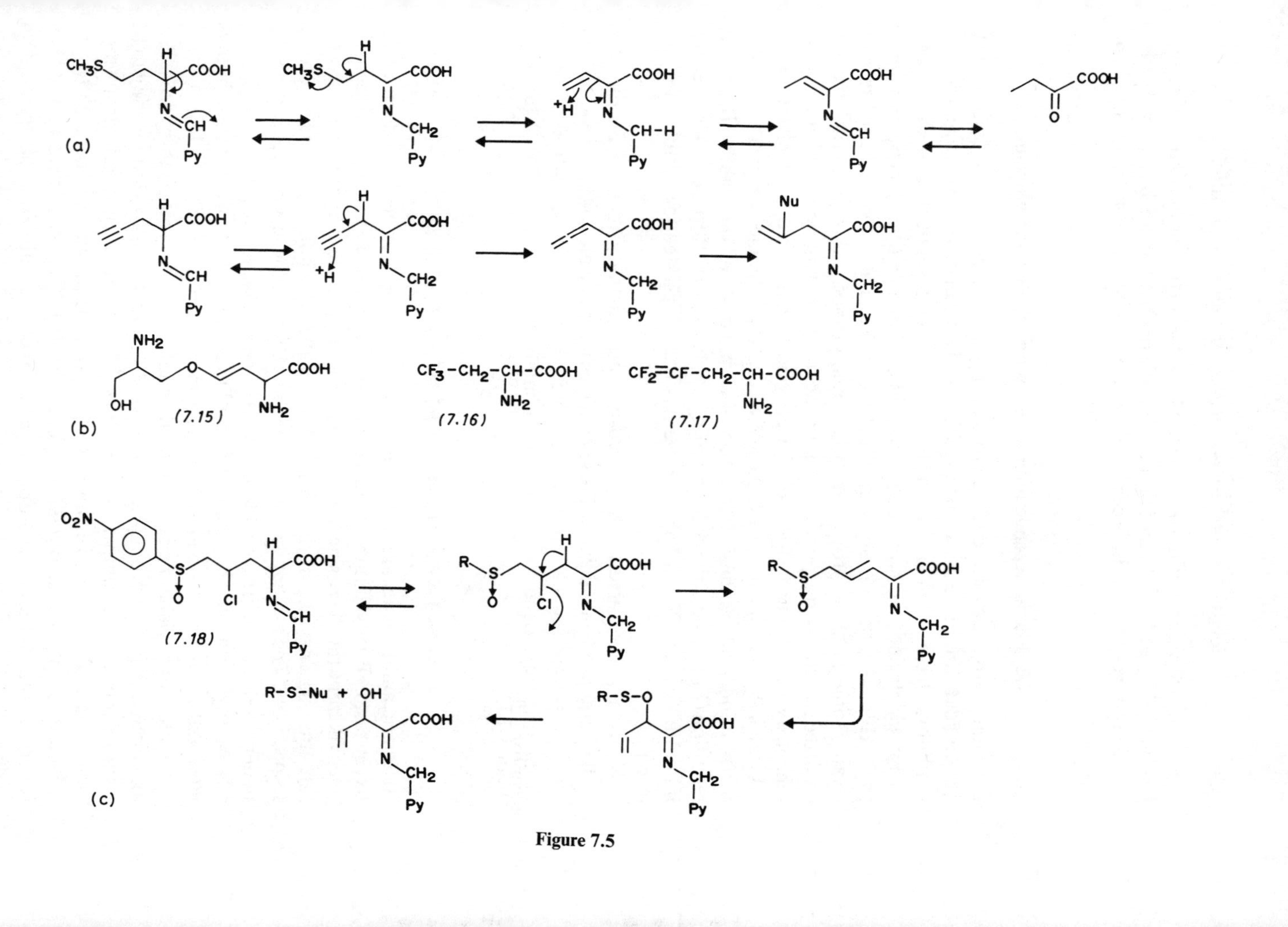

Figure 7.5

Propargylglycine inhibits bacterial and yeast growth. The antibiotic effects can be reversed by addition of methionine. Rhizobitoxin inhibits ethylene synthesis in plants and would have important applications in allowing the control of fruit ripening.

7.2 Flavin-dependent enzymes in amino acid metabolism

D-Amino acids are oxidized to α-keto acids by flavin-dependent oxidases while L-amino acids are more often oxidized by NAD-dependent dehydrogenases. No inhibitors have been reported for the NAD-dependent dehydrogenases. Propargylglycine inactivates D-amino acid oxidase in a time-dependent manner [40]. However, there is some residual enzyme activity: the alkylated enzyme retains some catalytic properties. It has been proposed that propargylglycine is oxidized to the corresponding amino acid which tautomerizes either spontaneously or by intervention of the enzyme to the conjugated allene, which alkylates the enzyme at or near the active site. Haloalanines are good substrates of D-amino acid oxidase with either β-elimination of the halide (Br, Cl) or retention (F) and yield the corresponding pyruvates [41]. Little or no irreversible inhibition occurs. No therapeutic use has been suggested for inhibition of this enzyme.

N-Methyl amino acids such as sarcosine and dimethyl glycine are de-alkylated by flavin-dependent enzymes. These enzymes resemble monoamine oxidase and, not surprisingly, are inhibited by *N*-propargyl and *N*-allyl derivatives of glycine or sarcosine just as monoamine oxidase is inhibited by propargyl- or certain allylamines [42].

7.3 PTERIDINE-DEPENDENT ENZYMES

Aromatic amino acids are hydroxylated by dihydropteridine-dependent monooxygenases. Phenylalanine hydroxylase is the enzyme which is deficient in phenylketonuria. Tyrosine hydroxylase and tryptophan hydroxylase are the rate-limiting enzymes in the biosynthesis of the catecholamines and serotonin respectively. The mechanism of oxygen insertion on the aromatic ring is poorly understood and no irreversible inhibitors are known so far [43].

α-Methyl *p*-tyrosine is a competitive inhibitor of tyrosine hydroxylase both *in vitro* and *in vivo* and is widely used as a tool in neurochemistry [44].

p-Chlorophenylalanine is a competitive inhibitor of tryptophan hydroxylase and phenylalanine hydroxylase *in vitro*. *In vivo*, its action is too long-lasting to be totally explicable by competitive inhibition [45]. More recently, dihydrophenylalanine has been isolated from a *Streptomyces* strain and has been shown to inhibit tryptophan hydroxylase competitively [46]. The *in vivo* effect has not been reported yet. 6-Fluorotryptophan, a competitive inhibitor of tryptophan hydroxylase *in vitro*, modifies sleep *in vivo* in a time course which is consistent with its serotonin-depletory effects [47].

7.4 SPECIFIC ENZYMES ACTING ON GLUTAMATE, GLUTAMINE, ASPARTATE AND ASPARAGINE

In previous sections, we have already described inhibitors of enzymes acting on the α-amino acid part of glutamate and aspartate: i.e. aminotransferases and α-amino acid decarboxylases. The ω-carboxyl functions of glutamate and aspartate are involved in a number of reactions for which some inhibitors are known.

7.4.1 Glutamate and glutamine

Glutamine synthetase and γ-glutamylcysteine synthetase (a precursor of glutathione) are inhibited by methionine sulphoximine (*7.19*), a convulsing agent found in flour bleached with nitrogen trichloride [48]. In both reactions, the γ-carboxyl function is activated toward nucleophilic attack by forming a mixed phosphoric acid anhydride (Fig. 7.6(a)). ATP, the cosubstrate and phosphate donor, is also needed for the inhibition of both enzymes by methionine sulphoximine. It could be demonstrated that methionine sulphoximine phosphate (*7.20*) is formed in the enzyme active site and does not dissociate from the enzyme (Fig. 7.6(b)). Methionine sulphoximine does not need alkylation to inhibit the enzyme and is another example of an enzyme-generated transition-state analogue (gabaculine was the first example). Only one enantiomer, L-methionine *S*-sulphoximine, is inhibitory for both enzymes. Substitution of alkyl groups on the α-carbon of methionine sulphoximine allows selection for inhibitors of glutamine synthesis. For instance, α-ethyl methionine sulphoximine (*7.21*) inhibits glutamine synthetase with about half the efficiency of methionine sulphoximine and does not inhibit γ-glutamylcysteine synthetase [49]. On the other hand, by increasing the bulkiness of the alkyl group from methyl to butyl in the sulphoximine part as in compound (*7.22*), selective inhibitors of γ-glutamylcysteine synthetase were generated [50].

Glutamine hydrolysis by glutaminase is inhibited irreversibly by azaserine (*7.23*), a bacterial metabolite with antibiotic and antitumour activity, and by 6-diazo-5-oxo norleucine (*7.24*) [51]. The mechanism of hydrolysis presumably involves protonation of the amide function before nucleophilic attack and hydrolysis. The diazo compounds upon protonation would form diazonium ions or carbenes, both extremely reactive species. The same compounds also inhibit amide transferase reactions such as asparagine synthesis catalysed by L-asparagine synthetase (L-glutamine hydrolysing) and formylglycine amidine ribotide synthesis from formylglycine amide ribotide (γ-glutamine hydrolysing) [51].

γ-Glutamyl derivatives of amino acids are hydrolysed by three types of enzyme, processes which differ in the fate of the glutamyl part which can be transferred to another amine, released as free glutamate or as oxoproline. The importance of these reactions in the transport of amino acids across membranes was revealed by the work of A. Meister [52]. L-γ-Glutamyl 2-(2-carboxyphenyl)hydrazines (anthglutin) (*7.25*) extracted from *Penicillium oxalicum* inhibits γ-glutamyl

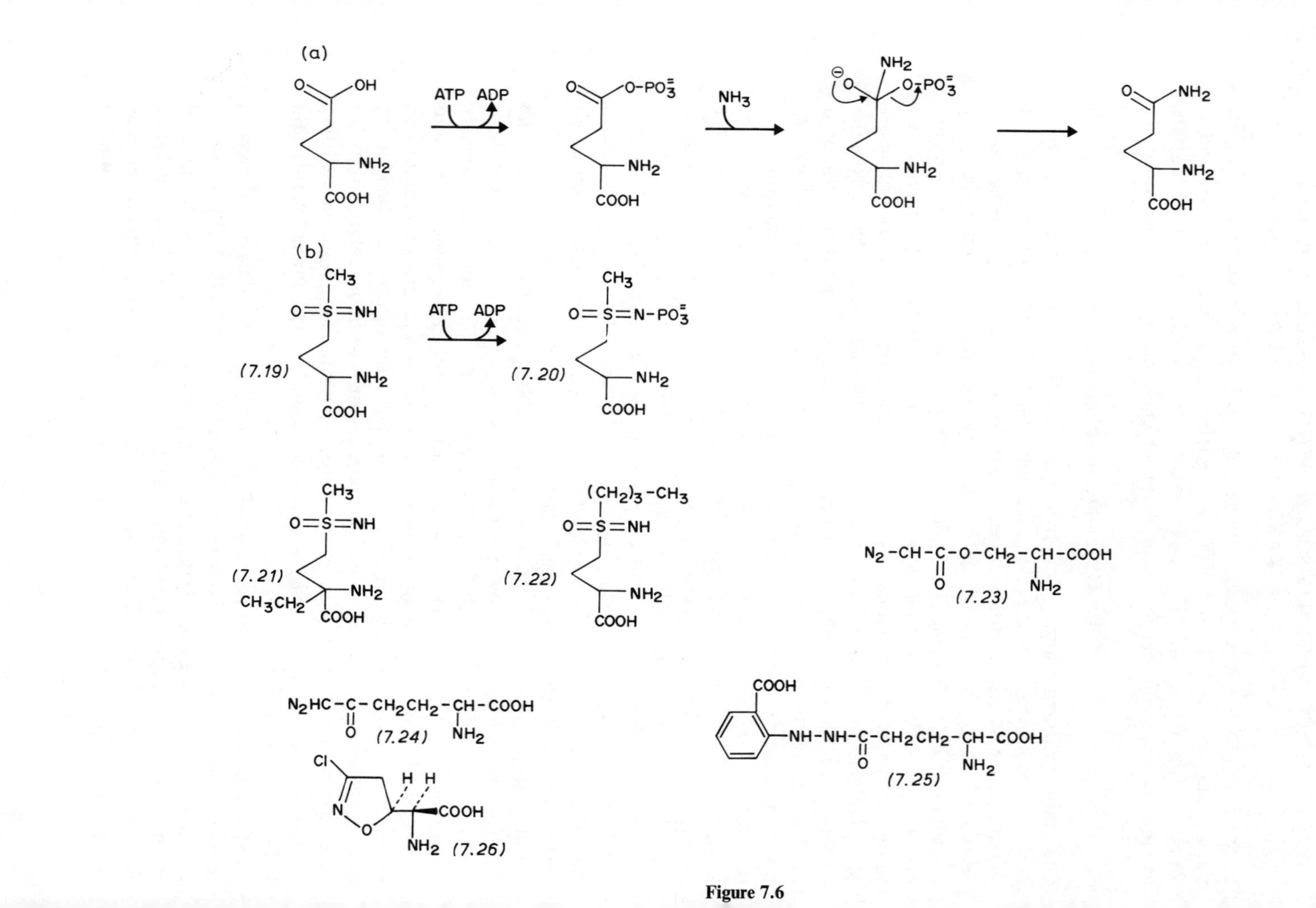

Figure 7.6

transpeptidase competitively [53]. Azaserine and 6-diazo-5-oxo-L-norleucine inhibit the enzyme irreversibly as they did for glutaminase. More recently, the fermentation product (α-S, 5-S) α-amino-3-chloro-4,5-dihydro-5-isoxazoleacetic acid (*7.26*) was found to be an irreversible inhibitor of γ-glutamyl transpeptidase, although the mechanism of action has not been elucidated [54]. The antitumour effects of this compound, of azaserin and diazo-oxonorleucine could be due at least in part to the inhibition of enzymes related to glutamate and glutamine synthesis or transport.

7.4.2 Aspartic acid and asparagine

The affinity label 2-amino 4-oxo-5-chloropentanoic acid and the suicide inhibitor 5-diazo 4-oxo-norvaline inhibit the synthesis and hydrolysis of asparagine [55]. In bacteria and plants, homoserine and aspartate are interconvertible. In *E. coli*, the transformation is catalysed by one multifunctional enzyme: aspartokinase–homoserine dehydrogenase. The dehydrogenase function is selectively inhibited by the affinity label 2-amino 4-oxo-5-chloropentanoic acid already mentioned [56].

The initial step in *de novo* synthesis of pyrimidines is the carbamoylation of aspartate by carbamoyl phosphate. This reaction is effectively blocked *in vitro* and *in vivo* by *N*-phosphonoacetylaspartate, a transition state analogue of the reaction [57]. This compound has antitumour activity in a number of experimental models and is in clinical trials for the treatment of cancer.

7.5 HYDROLYTIC ENZYMES

Nature and chemists have devised various ways of stabilizing peptide bonds against the hydrolytic action of proteinases. For instance, peptides in which one or several of the normal L-amino acids have been replaced by their D-counterparts or by an α-alkyl amino acid may retain their pharmacological activity *in vitro* and be more active than the natural peptide *in vivo* due to resistance to proteolytic breakdown [58]. In addition, there is a vast literature on inhibitors of proteolytic enzymes. In this short review, only general concepts can be given.

The cleavage of a peptide bond is believed to occur by formation of a tetrahedral transition state as depicted in Fig. 7.7. Some natural protease inhibitors (pepstatins) contain an unusual amino acid: (3*S*, 4*S*)-4-amino-3-hydroxy-6-methylheptanoic acid (statin) (*7.27*) [59]. It is believed that the extremely strong binding of pepstatin to pepsin ($K_1 = 10^{-11}$ M) is due to the fact that statin resembles the tetrahedral transition state. Cysteine-dependent proteinases are inhibited by natural peptide derivatives in which the terminal carboxyl function has been replaced by an aldehyde (leupeptin and antipain contain an arginal (*7.28*) residue as terminal amino acid) [59]. The aldehyde supposedly forms an hemithioacetal with a cysteine residue in the active site, thus binding

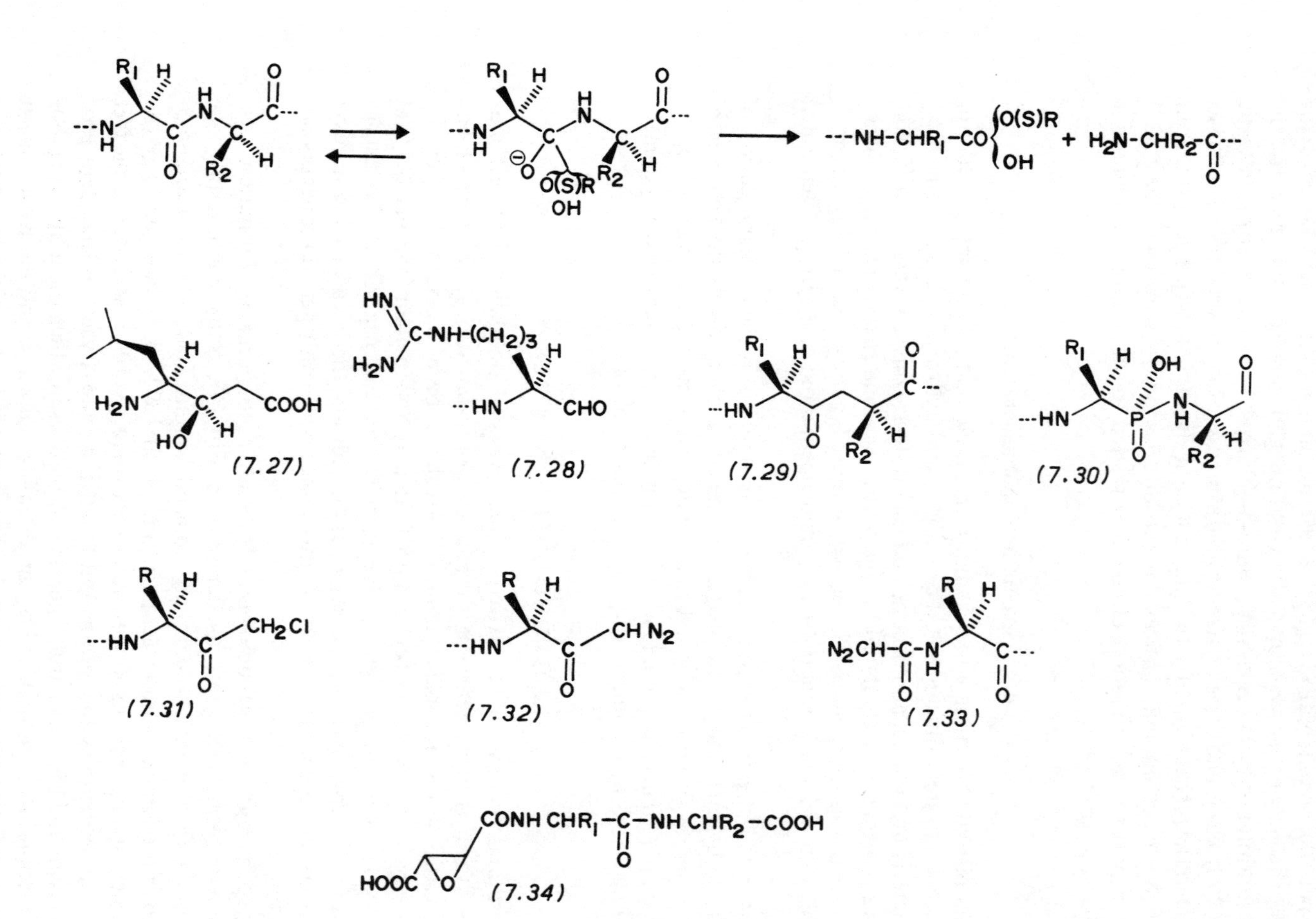

Figure 7.7

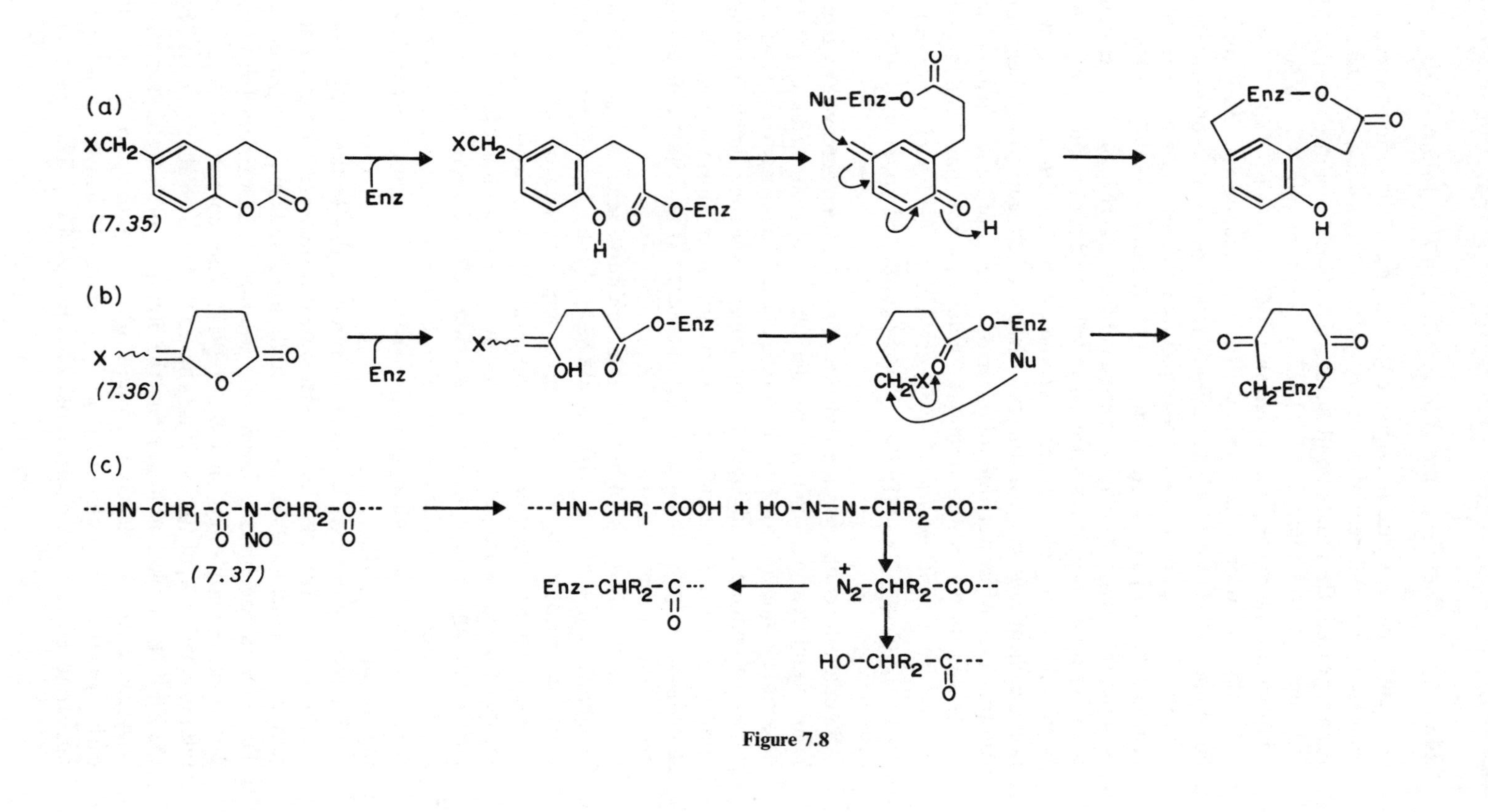

Figure 7.8

covalently the substrate analogue to the enzyme in the form of a transition-state analogue [60]. Different attempts to synthesize transition-state analogues have been reported: the amide bond has been replaced by a methylene ketone (*7.29*) [61] or a phosphonamide (*7.30*) [62]. The compounds obtained have good affinities for the corresponding enzymes ($K_I \simeq 10^{-6}$–10^{-8} M).

Peptides in which the terminal carboxyl function has been replaced by a chloromethyl ketone (*7.31*) are affinity labels of serine proteases. By properly choosing the peptide sequence, astonishing selectivity for a given serine protease can be achieved [63]. Cysteine and acid proteases are inhibited irreversibly by peptides terminated by a diazo methyl ketone (*7.32*) [64] or peptides in which the terminal amine function has been acylated with diazoacetate as in (*7.33*) [65, 66]. Both types of inhibitors need protonation or activation by a Lewis acid to become chemically reactive and can therefore be regarded as suicide inhibitors. Another type of irreversible inhibitor of cysteine proteases is found in nature: dipeptides in which the terminal amine function has been acylated with transepoxy succinic acid as in (*7.34*) are irreversible inhibitors of papain [67]. There is presumably alkylation of the protonated epoxide by the active site cysteine.

Suicide inhibition of serine proteases has also been achieved by two types of bifunctional reagent: halomethyl derivatives of coumarins (*7.35*) [68] and haloenol lactones (*7.36*) [69]. Both types have been designed in such a way that while the enzyme is acylated by opening the lactone function, an alkylating species is generated in another part of the inhibitor molecule (Fig. 7.8(a) and (b)): a methylene quinone in the case of the coumarin (Fig. 7.8(a)), a haloketone in the case of the haloenol lactone (Fig. 7.8(b)).

The different types of irreversible inhibitor seen so far are limited to exopeptidases. A concept which could be applicable to endopeptidases has been described recently [70]. By replacing the amide by a nitroso amide as in (*7.37*), a nitroso amine is generated on enzyme-catalysed hydrolysis (Fig. 7.8(c)). The nitroso amine re-arranges spontaneously to a carbonium ion, so reactive that it reacts with any nucloephile or with water before leaving the active site. Thus, chymotrypsin is inhibited irreversibly by the *N*-benzyl nitroso amide of D-phenylalanine. The L-phenylalanine derivative was hydrolysed without inhibiting the enzyme [70].

One enzyme deserves a special mention in this section: angiotensin I converting enzyme (ACE). This zinc-dependent protease has attracted much attention from medicinal chemists as its inhibition results in antihypertensive effects. The first therapeutically interesting inhibitor was captopril (D-3-mercapto-2-methylpropanoyl-L-proline) (*7.38*) [71]. This compound is a competitive inhibitor with a K_I of 10^{-8} M. It is believed that its high affinity for ACE is due to the Zn–SH interaction. Reports in the literature of its antihypertensive effects triggered a world-wide effort to find other inhibitors. So far, substituted *N*-carboxymethyl dipeptides such as *N*-1-carboxy-3-phenylpropyl-Ala-Pro (*7.39*) appear to be the best competitors of captopril [72]. The specific compound

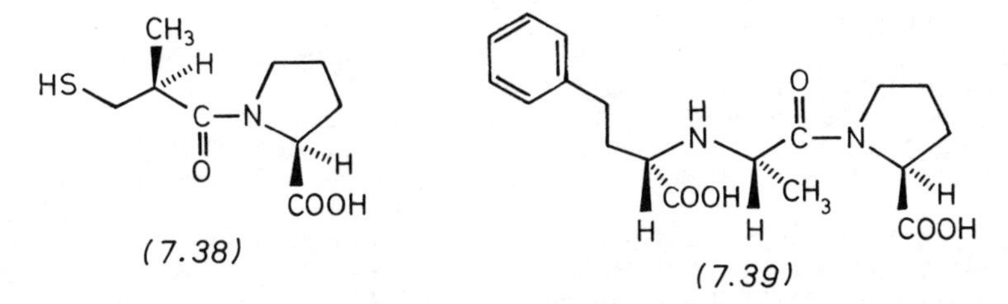

(7.38) (7.39)

mentioned has an affinity for ACE in the range of 1–2 nM (i.e. 10 times better than captopril) and its ethyl ester has a very long duration of action *in vivo* [72]. It is believed that the *N*-carboxymethyl part has the correct geometry to mimic the transition state in the hydrolysis of the corresponding tripeptide. Very few other protease inhibitors have found therapeutic applications, but it can be predicted with certainty that this will change in the near future [73, 74].

7.6 CONCLUSION

In this short and unavoidably incomplete review, I have tried to give examples of the different types of enzyme inhibitors: competitive, preformed transition state analogues, transition-state analogues generated by enzyme catalysis, irreversible inhibitors with built-in reactive groups (affinity labels) or with reactive groups revealed by enzyme activation (suicide inhibitors). Many of the compounds cited are found in nature. More are certainly present in species of animals or plants which have not yet been studied. In addition, the adaptability of bacterial metabolism and the ingenuity of chemists represent an endless source of new amino acid structures which may have interesting enzyme inhibitory properties.

REFERENCES

1. Bell, E.A. (1981) *Act. Chim.*, 22.
2. Clark, W.G. (1963) in *Metabolic Inhibitors* (eds R.M. Hochster and J.H. Quastel) (Vol. 1) Academic Press, New York, London, p. 315.
3. Sourkes, T.L. (1954) *Arch. Biochem. Biophys.*, **51**, 444.
4. Muschol, E. (1972) in *Handbuch der Expt. Pharmakologie* (eds H. Blaschko and E. Mucholl) (Vol. 33, Catecholamines) Springer-Verlag, Berlin and Heidelberg, p. 618.
5. Shive, W. and Skinner, C.G. (1963) in *Metabolic Inhibitors* (eds R.M. Hochster and J.H. Quastel), (Vol. 1) Academic Press, New York, London, p. 1.
6. Wolfenden, R. (1972) *Acc. Chem. Res.*, **5**, 10.
7. Powers, J.C. (1977) *Meth. Enzymol.*, **46**, 197.
8. Baker, B.R. (1967) *Design of Active-Site-Directed Irreversible Enzyme Inhibitors*, John Wiley and Sons, New York.
9. Abeles, R.H. and Maycock, A.L. (1976) *Acc. Chem. Res.*, **9**, 313.
10. Rando, R.R. (1974) *Science*, **185**, 320.
11. Rando, R.R. (1977) *Meth. Enzymol.*, **46**, 28.
12. Seiler, N., Jung, M.J. and Koch-Weser, J. (eds) (1978) Enzyme-Activated Irreversible Inhibitors, Elsevier/North Holland, Amsterdam.
13. John, R.A. and Fasella, P. (1969) *Biochemistry*, **8**, 4477.
14. Tate, S.S., Relyea, N.M. and Meister, A. (1969) *Biochemistry*, **8**, 5016.

15. Rando, R.R. (1975) *Acc. Chem. Res.*, **8**, 281.
16. Sandler, M. (ed.) (1980) *Enzyme Inhibitors as Drugs*, MacMillan Press, London, Basingstoke.
17. Dunathan H.C. (1971) in *Adv. Enzymol.*, **35**, 79.
18. Rando, R.R., Relyea, N. and Cheng, L. (1976) *J. Biol. Chem.* **251**, 3306.
19. Marcotte, P. and Walsh, C. (1975) *Biochem. Biophys. Res. Commun.*, **62**, 677.
20. Alston, Th. A., Porter, D.J.T., Mela, L. and Bright, H.J. (1980) *Biochem. Biophys. Res. Commun.*, **92**, 299.
21. Gehring, H., Rando, R.R. and Christen, Ph. (1977) *Biochemistry*, **16**, 4832.
22. Morino, Y. and Okamoto, M. (1973) *Biochem. Biophys. Res. Commun.*, **50**, 1061.
23. Silverman, R. and Abeles, R.H. (1976) *Biochemistry*, **15**, 4718.
24. Walsh, C., Johnston, M., Marcotte, P. and Wang, E. (1978) in *Enzyme-Activated Irreversible Inhibitors* (eds N. Seiler, M.J. Jung and J. Koch-Weser), Elsevier/North Holland, Amsterdam, p. 177.
25. Metcalf, B.W., Lippert, B. and Casara, P. (1978) in *Enzyme-Activated Irreversible Inhibitors* (eds N. Seiler, M.J. Jung and J. Koch-Weser) Elsevier/North Holland, Amsterdam, p. 123.
26. Fowler, L.J. and John, R.A. (1972) *Biochem. J.*, **130**, 569.
27. Bey, P., Jung, M.J. and Gerhart, F. *et al.* (1981) *J. Neurochem.*, **37**, 1341.
28. Rando, R.R. (1977) *Biochemistry*, **16**, 4604.
29. Jung M.J. (1980) in *Enzyme Inhibitors* (ed. U. Brodbeck) Verlag Chemie, Weinheim, p. 85.
30. Kollonitsch, J. and Barash, L. (1976) *J. Amer. Chem. Soc.*, **98**, 5591.
31. Wang, E. and Walsh, C. (1978) *Biochemistry*, **17**, 1313.
32. Jung, M.J. and Koch-Weser, J. (1981) in *Molecular Basis of Drug Action* (eds. Th.P. Singer and R. Ondaza), Elsevier/North Holland, Amsterdam, p. 135.
33. Maycock, A.L., Aster, S.D. and Patchett, A.A. (1980) *Biochemistry*, **19**, 709.
34. Owens, L.D., Thompson, J.F., Pitcher, R.G. and Williams, T. (1972) *J. Chem. Soc. Chem. Commun.*, 714.
35. Washtien, W. and Abeles, R.H. (1977) *Biochemistry*, **16**, 2485.
36. Johnston, M., Marcotte, P., Donovan, J. and Walsh, C. (1979) *Biochemistry*, **18**, 1729.
37. Alston, Th. A., Muramatsu, H., Ueda, T. and Bright, H.J. (1981) *FEBS Letters*, **128**, 293.
38. Johnston, M., Raines, R., Walsh, C. and Firestone, R.A. (1980) *J. Amer. Chem. Soc.*, **102**, 4241.
39. Merck and Co. Inc., *European Patent* 104 805, 1980.
40. Marcotte, P. and Walsh, C. (1978) *Biochemistry*, **17**, 2864.
41. Dang, T.Y., Cheung, Y.F. and Walsh, C., (1976) *Biochem. Biophys. Res. Commun.*, **72**, 960.
42. Yaouanc, J.J., Dugenet, P. and Kraus, J.L. (1979) *Pharmacol. Res. Commun.*, **11**, 115.
43. Dmitrienko, G.I., Snieckus, V. and Viswanatha, V. (1977) *Bioorganic Chem.*, **6**, 421.
44. Udenfriend, S., Zaltzmann-Nirenberg, P. and Nagatsu, T. (1965) *Biochem. Pharmacol.*, **145**, 837.
45. Chang, N., Kaufman, S. and Milstien, S. (1979) *J. Biol. Chem.*, **254**, 2665.
46. Okabayashi, K., Morishima, H., Hamada, M. *et al.* (1977) *J. Antibiot.*, **30**, 675.
47. Nicholson, A.N. and Wright, C.M. (1981) *Neuropharmacology*, **20**, 335.
48. Meister, A. (1978) in *Enzyme-Activated Irreversible Inhibitors*, (eds N. Seiler, M.J. Jung and J. Koch-Weser), Elsevier/North Holland, Amsterdam, p. 187.
49. Griffith, O.W. and Meister, A. (1978) *J. Biol. Chem.*, **253**, 2333.
50. Griffith, O.W. and Meister, A. (1979) *J. Biol. Chem.*, **254**, 7558.
51. Buchanan, J.M. (1978) in *Enzyme-Activated Irreversible Inhibitors* (eds N. Seiler, M.J. Jung and J. Koch-Weser), Elsevier/North Holland, Amsterdam, p. 277.
52. Meister, A. (1979) in *Glutamic Acid* (*Advances in Biochemistry and Biophysiology*) (eds L.J. Filer, S. Garattini and M.R. Kare), Raven Press, New York, p. 69.
53. Kinoshita, T., Watanabe, H. and Murato, S. (1981) *Bull. Chem. Soc. Jap.*, **54**, 2219.
54. Allen, L., Meck R. and Yunis, A. (1980) *Res. Commun. Chem. Pathol. Pharmacol.*, **27**, 175.
55. Jayaram, H.N., Cooney, D.A., Milman, H.A. *et al.* (1976) *Biochem. Pharmacol.*, **25**, 1571.
56. Hirth, C.G., Veron, M., Villar-Palasi, C. *et al.* (1975) *Europ. J. Biochem.*, **50**, 425.
57. Collins, K.D. and Stark, G.R. (1971) *J. Biol. Chem.*, **246**, 6599.
58. Morley, J.S. (1980) *Trends Pharmacol. Sci.*, p. 463.
59. Umezawa, H. (1972) *Enzyme Inhibitors of Bacterial Origin*, University Park Press, Baltimore, London, Tokyo, p. 14.
60. Frankfater, A. and Kuppy, T. (1981) *Biochemistry*, **20**, 5517.
61. Almquist, R.G., Chao, W.R., Ellis, M.E. and Johnson, H.L. (1980) *J. Med. Chem.*, **23**, 1392.
62. Jacobsen, N.E. and Bartlett, P.A. (1981) *J. Amer. Chem. Soc.*, **103**, 654.

63. Shaw, E. (1980) in *Enzyme Inhibitors as Drugs*, (ed. M. Sandler) MacMillan Press, London, Basingstoke, p. 25.
64. Green, G.D.J. and Shaw, E. (1981) *J. Biol. Chem.*, **256**, 1923.
65. Kanazawa, H. (1977) *J. Biochem.*, **81**, 1739.
66. Johnson, R.L. and Poisner, A.M. (1980) *Biochem. Biophys. Res. Commun.*, **95**, 1404.
67. Tamai, M., Hanada, K., Adachi, T. *et al.* (1981) *J. Biochem.*, **90**, 255.
68. Bechet, J.J., Dupaix, A. and Blagoeva, I. (1977) *Biochimie*, **59**, 231.
69. Krafft, G.A. and Katzenellenbogen, J.A. (1981) *J. Amer. Chem. Soc.*, **103**, 5459.
70. White, E.H., Jelinski, L.W., Politzer, E.R. *et al.* (1981) *J. Amer. Chem. Soc.*, **103**, 4231.
71. Ondetti, M.A., Cushman, D.W., Sabo, E.F. *et al.* (1981) in *Molecular Basis of Drug Action*, (eds. Th. P. Singer and R. Ondarza), Elsevier/North Holland, Amsterdam, p. 235.
72. Gross, D.M., Sweet, C.S., Ulm, E.H. *et al.* (1981) *J. Pharmacol. Exp. Therap.*, **216**, 552.
73. Barrett, A.J. (1980) in *Enzyme Inhibitors as Drugs*, (ed. M. Sandler) MacMillan Press, London, Basingstoke, p. 219.
74. Roques, B.P., Fournie-Zaluski, M.C., Soroca, E. *et al.* (1980) *Nature*, **288**, 286.
75. Casara, P., Danzin, C., Metcalf, B.W. and Jung, M.J. (1982) *J. Chem. Soc., Chem. Commun.*, 1190.

CHAPTER EIGHT

Synthesis of Amino Acids

G.C. BARRETT

8.1 INTRODUCTION

This chapter covers the methods available for the synthesis of amino acids, and refers the reader to other sections of this book where complementary coverage has been provided. No attempt has been made to cover large-scale fermentative production of some of the protein amino acids [1] but due emphasis is given in this chapter to the use of protein amino acids as starting materials for the synthesis of other α-amino acids.

Incidentally, Chapter 11 covers a number of reactions in which one amino acid is converted into another, but several examples are also given in this chapter (Table 8.14) as well as other uses in synthesis.

8.1.2 Use of the chapter

The recent literature on the synthesis of amino acids is a mixture of routine applications of well-established methods, useful improvements to these well-established methods, and exploration of new synthetic approaches or under-valued older methods.

Readers attracted to this chapter may therefore consult the early sections if representative applications of well-established methods are needed (Tables 8.1–8.3), to help to make a choice of route towards a specific synthetic objective, or may use the chapter as a whole to derive an overall view of current activity in this field.

Asymmetric synthesis of amino acids is undergoing considerable development and methods are collected separately (Tables 8.21–8.25) though many of the methods are variations of well-established methods and of some newly-introduced routes.

8.2 ESTABLISHED METHODS FOR THE SYNTHESIS OF α-AMINO ACIDS

The title of this section refers to standard textbook methods of general applicability for the synthesis of α-amino acids, all of which are covered briefly in Greenstein and Winitz' classic treatise on the amino acids [2–4]. The methods given here are therefore of long standing, and illustrative examples from the literature published since the coverage provided by Greenstein and

Winitz further support the continuing confidence shown in the methods, but also reveal that some useful modifications are still being discovered. The *Specialist Periodical Reports* series published by The Royal Society of Chemistry [5–18] from 1969 partly covers the gap in the coverage of the established methods of synthesis of α-amino acids, but this series tends to concentrate on the newer developments in the various subjects covered.

The methods can be grouped into three main categories:

(i) Alkylation of glycine derivatives and of compounds related to them (Reaction 8.1).

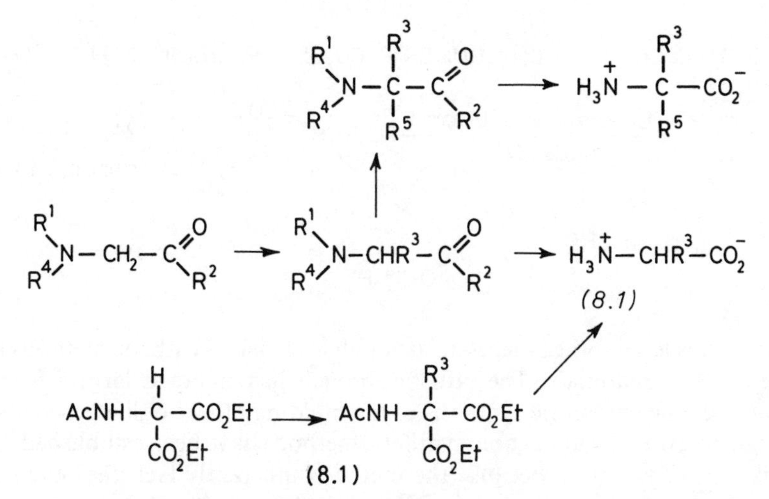

(ii) Substitution reactions in which the amino- or carboxy-function is introduced to complete the synthesis (e.g. amination of keto-, bromo-, or hydroxy-acids; carboxylation of amines; Reaction 8.2):

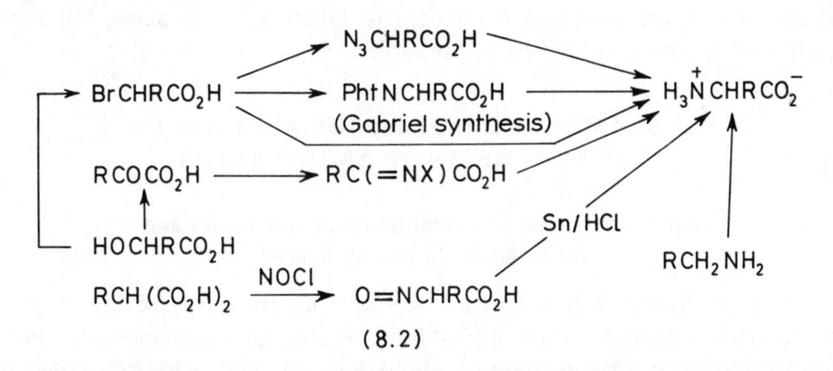

(iii) Other methods. These include long-established rearrangement reactions (Reaction 8.3) and some newer rearrangement procedures (Table 8.6). Miller's work published in 1953 [19,20] on synthesis of amino acids from methane, NH_3,

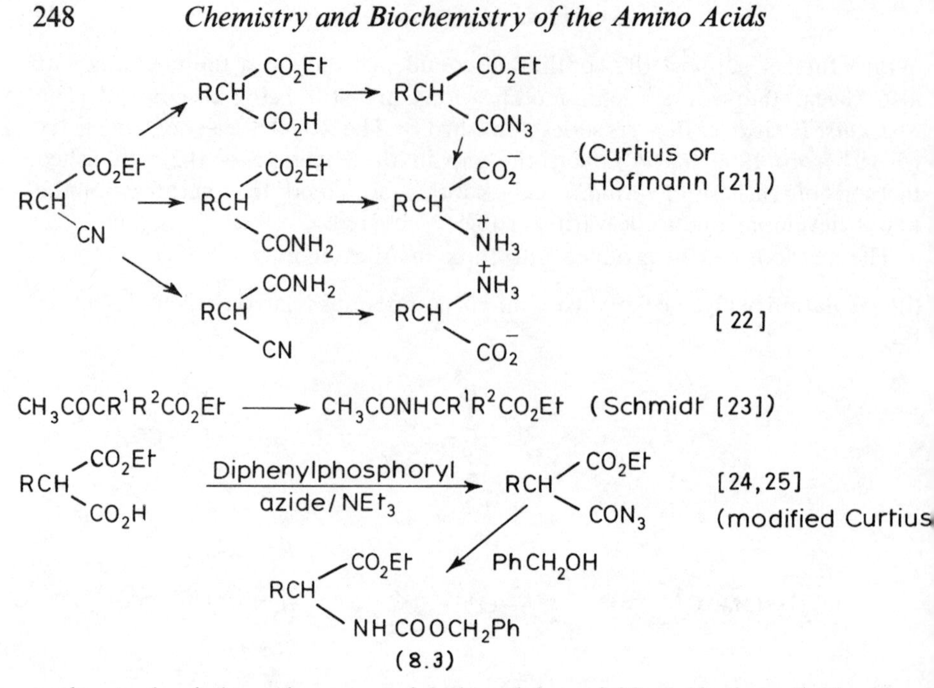

and water has led to other successful (though low-yield) synthetic methods using simple starting materials. The latter approach has emerged largely from the interest in chance prebiotic synthesis of the building blocks of life processes, but although no generally applicable synthetic methods have been established, based on this area of research (because the methods inherently lack the potential of leading to one product), a broad summary is included in this chapter of methods in this category (Table 8.11). The requirements for isotopically-labelled protein amino acids for biosynthesis studies often call for novel variations of standard synthetic methods, and coverage of this topic is also included in this chapter. However, no separate section is devoted to labelled amino acids, but representative examples are included in the tables.

8.3 EXAMPLES OF METHODS APPLIED TO THE SYNTHESIS OF AMINO ACIDS

8.3.1 Alkylation of acetamidomalonate esters and other *N*–acylaminomalonates

The route, displayed in Reaction 8.1, depends on the easy alkylation of the starting material, and no possibility of di-alkylation exists. The method remains the most frequently chosen route to amino acids carrying side-chains (*8.1*; R^3) which can withstand the hydrolysis step involved at the end of the operation (this usually involves reaction with moderately concentrated hydrochloric acid at reflux during short periods).

A typical experimental procedure based on diethyl acetamidomalonate (reference h, Table 8.1) involves condensation of the alkyl chloride with the substrate, which often gives a crystalline malonate. This is hydrolysed with refluxing dilute aqueous sodium hydroxide, and excess alkali is removed with a strong cation exchange resin (Amberlite IR-120 in H^+ form). This brings about hydrolysis to the malonic acid, which is decarboxylated by boiling in water.

Representative examples from the recent literature are given in Table 8.1. Although the use of acetamidomalonates is still most frequently involved in applications of this route, formamidomalonates, benzyloxycarbonylaminomalonates, and phthalimidomalonates have also been used and examples are included in Table 8.1. The conditions applied for the removal of the *N*-substituent, e.g. alkaline hydrolysis, hydrogenolysis or use of acetic acid saturated with HBr, and hydrazinolysis, respectively, differ from those used to cleave the acetamido group at the end of the alkylation procedure, and there may be specific reasons based on this fact, for the choice of one of these alternative starting materials.

8.3.2 Alkylation of glycine derivatives: Schiff base esters $RCH{=}NCH_2CO_2R$ and heterocyclic analogues

The di-anion formed from benzylideneglycine is a convenient starting material for this route to α-amino acids, but mixtures of mono- and di-alkylated products have been obtained:

$$PhCH{=}NCH_2CO_2H \rightarrow PhCH{=}N\bar{C}HCO_2^- \rightarrow PhCHRN{=}CRCO_2H$$
$$PhCH{=}N\bar{C}HCO_2^- \rightarrow PhCH{=}NCHRCO_2H \rightarrow PhCH{=}N\bar{C}RCO_2^- \rightarrow PhCH{=}NCR_2CO_2H$$

The problem highlighted in this example, the formation of mixtures of products, as well as the relatively low yields and initial exploration needed to establish appropriate reaction conditions in each case, indicates the limitations inherent in this method.

4-Alkylation of a 2-substituted oxazolin-5-one or the equivalent 4-alkylidenation (Ploechl–Erlenmeyer reaction) followed by catalytic hydrogenation (the 'azlactone synthesis') is the equivalent of the alkylation of a glycine Schiff base (although this and related systems have a variety of tautomeric structures available to them, in contrast to the more restricted delocalization within a Schiff base anion). Table 8.17 summarizes representative applications of these heterocyclic systems in the synthesis of α-amino acids. The oxazolinone route is probably the most fully studied of these alternatives, especially the 'azlactone synthesis', and relative merits of the use of thiazolinones and imidazolinones have not yet been established.

The initially-formed 'dehydro-amino acid' derivative can be hydrogenated to yield alanine derivatives $H_3\overset{+}{N}CH(CHR^1R^2)CO_2^-$ after alkaline hydrolysis. Use of deuterium or tritium offers a simple entry to [2,3-$^2H_2$- or -$^3H_2$]-labelled amino acids, while addition of a Grignard reagent provides aliphatic amino acids fully substituted at C-3 (e.g. t-leucine, $H_3\overset{+}{N}CH(CMe_3)CO_2^-$, reference a, Table 8.17). The scope for asymmetric hydrogenation offered by the use of chiral rhodium (I) complexes has been explored with these derivatives, and extraordinarily high enantioselectivity can be achieved (Table 8.21).

8.3.3 Alkylation of α-isocyanoacetates, α-nitroacetates, and their homologues

The use of isocyanoacetate esters, $\bar{C}\equiv\overset{+}{N}CH_2CO_2R$, in the same way as Schiff bases, has been more thoroughly explored. Useful syntheses have been reported of α-alkyl-α-amino acids and β-hydroxy-α-amino acids through alkylation of α-isocyanopropionate esters and oxazolines, respectively (Table 8.4). Other more straightforward uses of these compounds are also described in this table. Later tables describe the equivalent use of α-nitroalkanoates (Table 8.5) and azido analogues (Table 8.17).

8.3.4 Alkylation of hydantoins, and the Bücherer—Bergs synthesis

Saturated heterocyclic systems enclosing the α-amino-acyl residue −NH−CHR−CO− are formed in the Bücherer–Bergs synthesis, employing simple starting materials. The methods are particularly useful for the synthesis of α-alkyl analogues of common α-amino acids since the heterocyclic synthesis step is equally successful using ketones as starting materials. Asymmetric synthesis possibilities exist in routes to amino acids using tautomeric heterocyclic systems, where enzymic hydrolysis can be used in the final (or penultimate) ring-opening step (this topic is covered in Chapter 10).

For the hydantoin synthesis, hydantoin itself may be alkylated, which is an efficient procedure (Reaction 8.4 [26]) employing the magnesium enolate; the reagents are (i) NaOH and $PhCH_2Cl$, (ii) magnesium methyl carbonate, (iii) RHal and (iv) aq $Ba(OH)_2$. Alternatively the aldehyde corresponding to the required α-amino acid side-chain is treated with KCN and $(NH_4)_2CO_3$ and the product hydrolysed in aqueous acid.

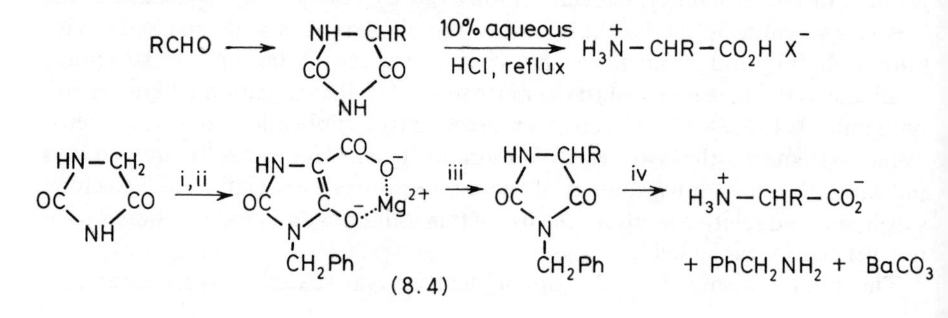

(8.4)

8.3.5 The Strecker synthesis; other condensation and substitution reactions in the synthesis of α-amino acids

The Strecker synthesis (Reaction 8.5) illustrates a well-established condensation reaction, yielding α-aminoalkanenitriles which, on hydrolysis, give the corresponding α-amino acids.

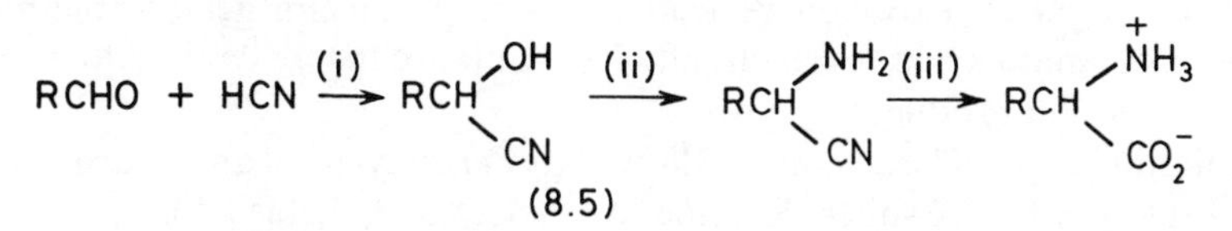

For step (i) the Zelinsky–Stadnikoff modification employs the less hazardous (since control is easier) $KCN + NH_4Cl$ as source of HCN [27]; anhydrous HCN at 0°C [28] or liquid HCN containing a trace of saturated aqueous KCN [29] leads to a pure cyanohydrin in cases where the basic conditions cause rearrangement and other side reactions of the aldehyde; NaCN/AcOH/MeOH has been used [30]. For step (ii) NH_4OH and NH_4Cl are applied to the pure cyanohydrin, or (i) and (ii) can be combined where the amino-nitrile can be formed satisfactorily in this way. For step (iii) hydrolysis with 10 M hydrochloric acid has been most commonly used (short reflux [31], or 12 h at room temperature followed by 3 h at 50°C in sensitive cases [29]).

There is a connection between this process and the Bücherer–Bergs hydantoin synthesis described in the preceding section, in that both routes use aldehydes and later stages depend on the initial formation of cyanohydrins. Recent examples of the application of the Strecker synthesis are collected in Table 8.2. Yields are variable [31] but this is still the best method for the preparation of α-amino acids from aldehydes [29]. Recently, the Ugi 'four-component condensation' has been developed into a substantial alternative method (next section).

Substitution reactions leading to amino acids which have been used over many years are based on the easy availability of halogeno-alkanoic acids and hydroxy- or epoxy- alkanoic acids. The introduction of the nitrogen function either as the amino group directly (using ammonia), or as an analogue (using an amine or hydroxylamine) leads to the corresponding amino acid. An extraordinary reaction in which $Cl_2CHCHCl_2$ reacts with aqueous NH_4OH at 160–170°C to give glycine proceeds via amino-acetonitrile (reference z, Table 8.10).

Condensation of ammonia or its derivatives with keto-acids, and other condensation methods of introducing a nitrogen function adjacent to a carboxy group (e.g. isonitrosation of 'active' methylene groups), are serviceable routes to α-amino acids. Reactions of these types are collected in Table 8.10.

Carboxylation of amines, and amination of alkanoic acids, are simple and therefore attractive synthetic routes to amino acids, but there is generally insufficient activation within the substrates to favour convenient reaction rates in these processes. Established methods (Tables 8.8 and 8.9) employ relatively drastic reaction conditions, thus leading to mixtures of products through indiscriminate substitution. The present level of exploration of these methods has been stimulated by the search for models of prebiotic synthesis of amino acids (Table 8.11), in which the objective has been to establish the various routes through which amino acids may have arisen in primeval times, from simple starting materials. Further development of the carboxylation and amination processes might be expected to lead to more promising general methods of synthesis of amino acids, in which sufficient control can be exercised to give good yields of specific products.

Lithiation of benzyl isocyanide followed by carboxylation and hydrolysis gives serviceable yields of C-phenylglycine (references a–c, Table 8.9).

$$PhCH_2\overset{+}{N}\equiv\overset{-}{C} \xrightarrow{BuLi} PhCHLi\overset{+}{N}\equiv\overset{-}{C} \xrightarrow[-80°C]{CO_2} PhCH(CO_2Li)\overset{+}{N}\equiv\overset{-}{C}$$

$$\xrightarrow{H_3O^+} H_3\overset{+}{N}-CHPh-CO_2^-$$

8.3.6 Rearrangements leading to α-amino acids (Reaction 8.3, Table 8.6)

The Curtius rearrangement was used from the time of its discovery for the synthesis of amino acids, including a notable incidental synthesis in an experiment designed to verify the Walden inversion. This, and related rearrangements (Schmidt, Hofmann, etc.) are now only rarely used for the purpose in view of the availability of efficient, more convenient, alternative general methods. Newer rearrangement reactions applied to the synthesis of amino acids include some promising routes which still await full development (Table 8.6).

8.3.7 Comparisons of standard methods of synthesis of α-amino acids

Any one of the standard methods could be adopted in a straightforward case, to give an acceptable yield of the product in a satisfactory state of purity. Comparisons of procedures under controlled conditions (i.e. the same workers working through a group of procedures in a systematic manner) are rarely found.

The standard methods using an aldehyde as starting material have been compared for the synthesis of DL-serine [32]. A variety of routes (some shown in Reaction 8.6) are available to glycollaldehyde, from which a 70% yield of DL-serine was obtained through the Zelinsky–Stadnikoff modification of the Strecker synthesis ([33]; see also Reaction 8.5 and Table 8.2). The straightforward Strecker synthesis gave only 9%, and the Bücherer–Bergs route gave a 54% yield.

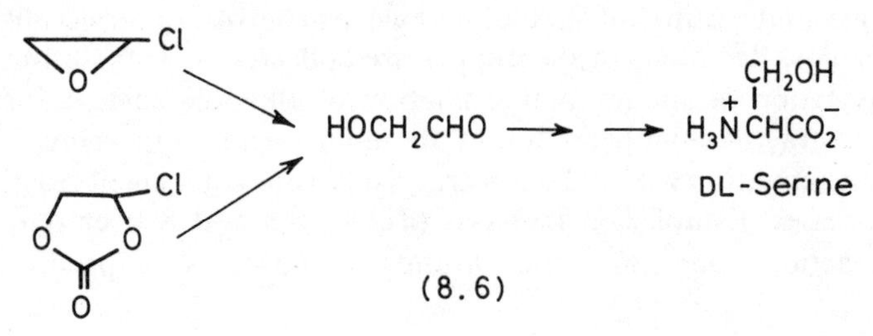

8.4 NEWER METHODS OF AMINO ACID SYNTHESIS

8.4.1 Ugi four-component condensation

This 'one-pot' approach, though a spectacular example of synthesis design (Reaction 8.7), is neither a one-stage synthesis of amino acids nor is it entirely general in character. It has therefore not replaced any of the long established methods in the field, but in certain cases there have been shown to be advantages in the Ugi condensation.

A primary amine, a ketone or aldehyde (these combine to form a Schiff base), together with a simple isocyanide, and an acid yield the corresponding N^{α}-acetyl-α-amino acid *N*-alkyl (or aryl) amide, and stripping of the different groups from the amino- and carboxy-functions leads to the required product. Where the product is an *N*-benzylamino acid anilide, successive treatment with hydrogen bromide, and hydrogen with palladized charcoal, conveniently leads to the amino acid; a good example of the superiority of this method in certain circumstances is the synthesis of α-benzylphenylalanine, a highly hindered amino acid, in 56% yield (AcOH, PhNC, and $PhCH_2N{=}C(CH_2Ph)_2$ were condensed [34]). For the synthesis of 1,4-dihydrophenylalanine, a conventional Strecker synthesis was inappropriate, while the Ugi method was quite suitable [35].

The condensation of the *cis*-furan (*8.2*), (*S*)-phenylethylamine, benzoic acid and t-butyl isocyanide gives a mixture of (α*R*,2*S*,5*S*) and (α*R*,2*R*,5*R*)-amide (*8.3*) and their enantiomers [36].

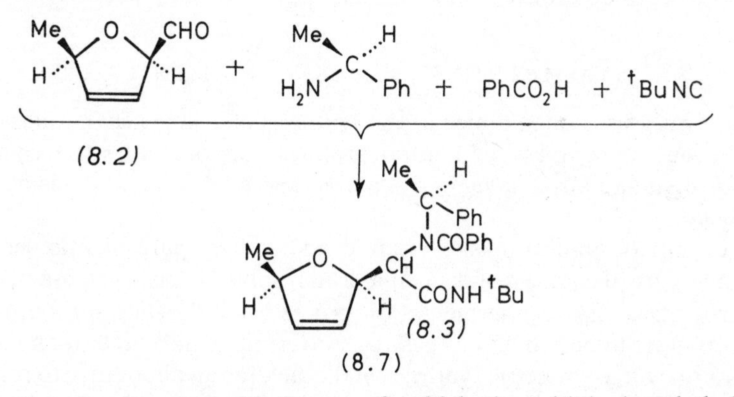

(8.7)

The reaction was repeated with the *trans*-furaldehyde and (*R*)-phenylethylamine, and stereoisomers thus obtained were de-benzylated and hydrolysed to the corresponding (2-methyl-2,5-dihydrofuran-5-yl)glycines, of which the (α*S*,2*R*,5*S*)-stereoisomer was found to be identical with (+)-furanomycin.

This recent example gives a detailed account of the use of the Ugi synthesis, also commonly known as the 4-component condensation or 4CC-synthesis. Other examples in recent literature vary the procedure only slightly, e.g. in using cyclohexyl isocyanide [37] or in using CO_2 instead of the carboxylic acid component [38]. A further example includes a device for incorporating a labile *N*-protecting group in the product [39]; the condensation conducted with *N*-

benzyloxycarbonylglycine, 9-(aminomethyl)fluorene, benzaldehyde and cyclohexyl isocyanide gave the expected product, from which the *N*-substituent was removed in high (87%) yield using 1.1 equivalents 1,8-diazabicyclo[5.4.0]undec-7-ene in pyridine at room temperature.

In addition to the expected α-amino acid derivatives, the corresponding amino-malonamide can also arise, in relative amounts depending on the proportions of reactants and solvent. The amino-malonic acid product does not incorporate the aldehyde component of the condensation mixture, but it is not formed in the absence of the aldehyde [40].

$$PhCO_2H + R^1R^2CHNH_2 + R^3CHMeCHO + {}^tBuNC \rightarrow PhCON(CHR^1R^2)CH(CONH\,{}^tBu)_2 + \text{4CC product}$$

8.4.2 Specific syntheses of amino acids not based on general methods

Where the side-chain of the required amino acid contains particular functional groups which are not easily introduced by using one of the long-established methods of amino acid synthesis, alternative strategies must be considered.

An example is the synthesis of lysine from ε-aminocaprolactam [41], since the starting material is easily prepared (and the established routes would call for side-chain protection and deprotection stages). The industrial synthesis of lysine using this approach is described in Chapter 2 (p. 12).

8.5 ASYMMETRIC SYNTHESIS OF AMINO ACIDS

Organic synthesis is often aimed at the synthesis of a single enantiomer, and as elsewhere in the synthesis of natural products, opportunities for asymmetric synthesis of amino acids have been explored and successful methods have been established.

Most synthetic applications of α-amino acids call for pure enantiomers. Some amino acids are of course readily available in resolved form (e.g. L-α-amino acids from proteins or through fermentation processes) and, while substantial ingenuity continues to be brought to bear to convert these into other L-amino acids, there is a role for asymmetric synthesis in this field, as an alternative to resolution of racemates.

A recent review [42], a potted version of the continuing coverage of the specialist periodical report *Amino acids, Peptides and Proteins* [5–18] includes 21 references to the literature before early 1978, and there is an increasing number of papers now appearing on this topic.

The earlier examples were often capable either of leading to high stereoselectivity in the synthesis of particular α-amino acids (e.g. L-aspartic acid; Reaction 8.8), or rather disappointingly low enantiomer purities associated with conformationally-mobile substrates (Reactions 8.9 and 8.10).

Table 8.21 gives brief details of other related approaches, bringing the coverage up to date (1981). Further tables give details of alternative approaches to the heterogeneous or homogeneous catalysed hydrogenation procedures to which most effort has been applied (Tables 8.22–8.24). The text which follows is devoted to specific examples deserving detailed analysis.

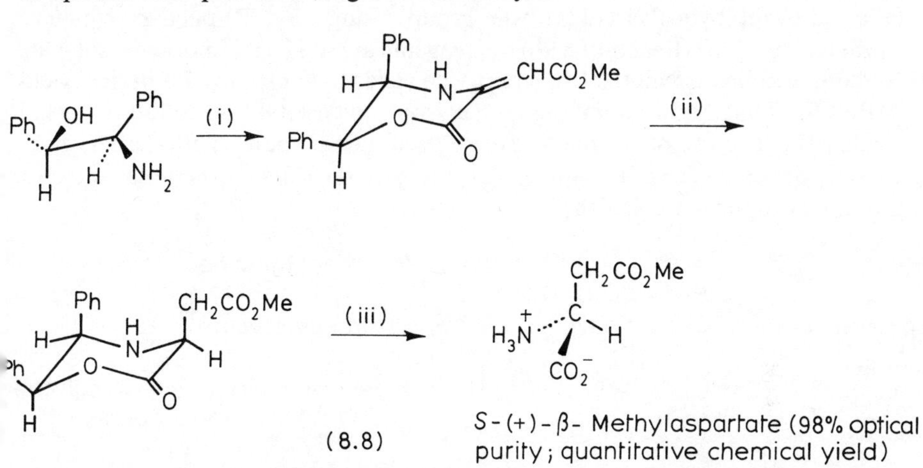

The reagents are (i) dimethylacetylenedicarboxylate; (ii) Pd/H_2 and (iii) Raney Ni/H_2 and H_3O^+ hydrolysis [43].

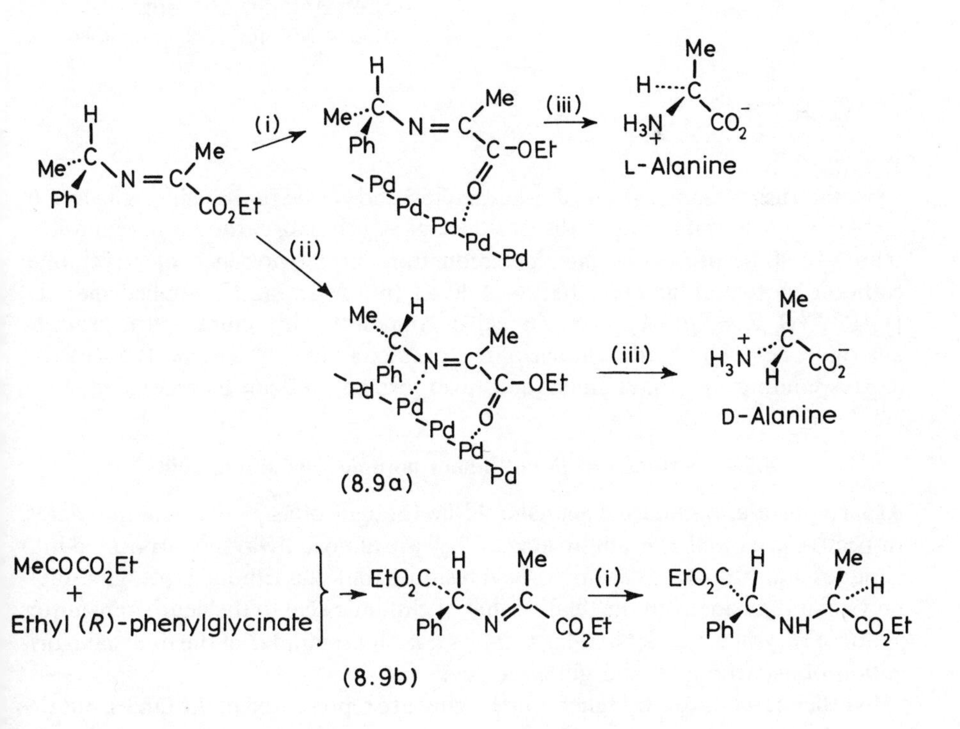

With (*R*)- or (*S*)-α-phenylethylamine as chiral reagent [44,45] (Reaction 8.9a), the reagents are (i) H_2/Pd at low temperatures, giving 60% optical purity in favour of the *S*-enantiomer from which L-alanine is obtained; but zero stereoselectivity at 17°C; (ii) H_2/Pd at temperatures above room temperature giving 43% optical purity in favour of the *R*-enantiomer at 50°C, from which D-alanine is obtained followed by (iii) hydrolysis of the ester group. Using ethyl (R)-phenyglycinate as chiral reagent [46] (Reaction 8.9b) the reagents are (i) H_2/Pd in organic solvents containing aqueous NaOH. Ethyl pyruvate yields L-alanine in 72% optical yield in Bu^{t}OH, while higher homologues (i.e. isopropyl esters) lead to lower optical yields [47]. The use of (*R*)-phenyglycine itself, rather than its ethyl ester, gives 10–62% optical yields of L-alanine, depending on the solvent (less polar solvents give higher optical yields [48]).

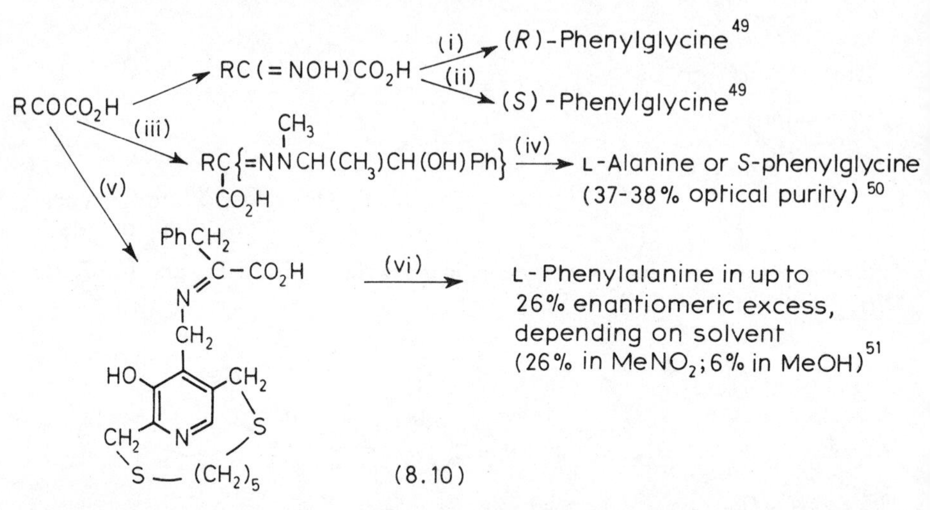

Asymmetric transamination of α-keto-acids and derivatives, using reagents (i) electrochemical reduction in the presence of strychnine, cathodic potential less than −1.40 V; (ii) electrochemical reduction in the presence of strychnine; cathodic potential greater than −1.40 V; (iii) *N*-amino-(−)-ephedrine; (iv) H_2/Pd; (v) R = $PhCH_2$ (use Zn salt); condense with chiral pyridoxamine analogue, i.e. the (−)-enantiomer of the pyridoxamine shown and (vi) $ZnClO_4$ (corresponding pyridoxal analogue is also formed and can be recovered).

8.5.1 Synthesis of β-and higher homologous amino acids

Aspartic acid and glutamic acid should be thought of as β- and γ-amino acids, respectively, as well as α-amino acids. They are almost always incorporated into peptides and proteins as α-amino-acyl residues (but glutathione is γ-L-glutamyl-L-cysteinyl-glycine), but the relationship of glutamic acid to the neurotransmitter γ-aminobutyric acid, $H_3\overset{+}{N}(CH_2)_3CO_2^-$, is a useful reminder of the dual categorization of aspartic acid and glutamic acid.

Synthetic routes to the higher homologues are represented in the tables, but the main intention in this chapter has been the coverage of the synthesis of α-amino acids in a rather thorough manner, with little attempt to cover the higher homologues except in outline. Chapter 3 deals with β-amino acids in detail.

8.6 USE OF THE TABLES

The tables are arranged to provide representative examples of the applications of closely related reactions to the synthesis of amino acids. Asymmetric synthesis methods are collected in Tables 8.21–8.25 and most of these methods are based on the reactions illustrated in the earlier tables.

The opportunity has been taken to collect together the routes to proline derivatives (Table 8.19), and some routes to β-amino acids (Table 8.20); otherwise, examples of syntheses achieved will be found either in the first five tables (established methods of synthesis), or in the following tables where newer and less frequently used methods may offer specific attraction or advantages (for example, more appropriate reaction conditions for the synthesis of amino acids with sensitive functional groups in the side-chains). The tables give leads to the recent literature, where practical details of the various routes can be found.

Table 8.1 Synthesis of α-amino acids from acylaminomalonate esters

Alkylating agent as starting material	*Synthesis achieved*	*Reference*
Diethyl acetamidomalonate		
Benzyl chloromethyl sulphide $PhCH_2SCH_2Cl$	[α-^{2}H]-*S*-Benzyl-DL-cystine (also [ββ-$^2H_2$]-and [αββ-$^2H_3$]-analogues)	a,b
(*R*)-^{2}H-Benzyl toluene-*p*-sulphonate (H, ^{2}H, C, Ph, OTs)	(2*S*,3*R*)-and (2*R*,3*R*)-[β-^{2}H]-Phenylalanine	c–e
Starting material $AcNHC(CO_2Et)_2CH_2CHO$, Fischer cyclization of appropriate phenylhydrazone	7- or 5-Chloro-DL-tryptophan	f,g
2,3-Dichloropropene	L-2-Amino-4-chloropent-4-enoic acid	h
$MeP(O)(OH)CH_2CH_2Cl$	L-2-Amino-4-(methylphosphino) butyric acid	i
N-(Toluene-*p*-sulphonyl) imidazolylmethyl esters	DL-Histidine	j
$I(CH_2)_3X$	N^5-Hydroxy-L-arginine	k
[1-^{14}C]-2-Methylthioethyl chloride	[3-^{14}C]-Methionine	l
Starting material $AcNHC(CO_2Et)_2CH_2CHO \rightarrow AcNHC(CO_2Et)_2CH_2$-$CH(SMe)S(O)Me$	(+)-(*S*)-*E*-2-Amino-4-methylthio-but-3-enoic acid ('dehydro-L-methionine')	m
Allyl chloride; then (i) add Br_2 (ii) 48% aq. HBr (iii) hydrolyse lactone	DL-γ-Hydroxyproline and allo-isomer; γ-hydroxyglutamic acid	n

Table 8.1 (*Contd.*)

Alkylating agent as starting material	*Synthesis achieved*	*Reference*
Substituted benzyl chlorides	Ring-substituted phenylalanines (3,4,5-trimethoxy-, 4-hydroxy-3,5-dimethoxy-, and 3,4-dihydroxy-5-methoxy-,[o] 4-pentafluorophenyl-[p] [4-^{18}F]-)[q]	o–q
3-Bromopropyl bromide; then substitution of side-chain bromine atom using $PhCH_2ONHAc$	Rhodotorulic acid	r
3-*N*-(Toluene-*p*-sulphonyl-*N*-benzyloxy) aminopropyl bromide	N^{δ}-Hydroxyornithine	s
1,4-Dichlorobutyne	2,6-Diaminohex-4-ynoic acid	t
$CH_3S(CH_2)_4Cl$	DL-2-Amino-6-(methylthio)heptanoic acid	u
[$^{14}C_4$]-2-Bromobutane	[$\beta,\gamma,\gamma^1,\delta$-^{14}C]-DL-Isoleucine	v
$Me_2C^2HCH_2Br$ (starting from isobutene and hydroboration with $B_2{}^2H_6$)	[γ-^{2}H]-DL-Leucine	w
	DL-α-Aminosuberic acid	x
$PhtNCH_2CH_2O^{14}CH_2CH_2Cl$	[4-^{14}C]-DL-2-Amino-4-(amino-ethoxy)butanoic acid	y
p-Methoxyphenacyl bromide	γ-Oxohomotyrosine	z
$(Bu^tO_2C)_2CHCH_2N$(piperidine ring) with diethyl *N*-benzyloxycarbonylaminomalonate	γ-Carboxyglutamic acid	aa
HO, Br, N, O, CH_2Br (isoxazole ring structure)	Analogues of ibotenic acid	bb
Alkyl halide with solid K_2CO_3/crown ether/DMF	Representative α-amino acids	cc
Butadienes with $PhO^-Na^+/PPh_3/Pd(OAc)_2$	Side-chains derived from the dimerized butadiene, e.g. $CH_2{=}CR(CH_2)_3CH{=}CHR$-$CH_2CH(\overset{+}{N}H_3)CO_2^-$	dd
$I(CH_2)_3CN$	δ-Amino-adipic acid	ee
C^2H_3COCl followed by $NaBH_4$ or NaB^2H_4 reduction, then hydrolysis	Labelled threonines	ff
$Me_2C{=}CHCH_2Cl$ then $PhCH_2SH/BF_3$	*S*-Benzyl derivative of a homologue of penicillamine	gg
Formaldehyde, diethyl oxalate	4-Methyleneglutamic acid	hh
$AcNH(CH_2)_2OCH{=}CHCl$	2-Amino-4-(2-aminoethoxy)-*trans*-but-3-enoic acid	ii
(2*S*)-Methylbutan-l-ol (via chloride)	2-Amino-4-methylhexanoic acid	jj

Table 8.1 (*Contd.*)

Alkylating agent as starting material	*Synthesis achieved*	*Reference*
Crotonaldehyde; Michael addition	→ HO–CH(–CH₂–CHMe–)C(CO₂Et)₂ (ring closed through NAc) thence to *trans*-3-methylproline,	kk
2,5,6-Trifluoro-3,4-dimethoxybenzyl bromide	DL-2,5,6-Trifluorodopa	ll
Diethyl formamidomalonate (introduced by A. Galat[mm])		
Propargyl bromide then cyclotrimerization with $^{14}C_2$-acetylene using $Ni(CO)_2(Ph_3P)_2$ as catalyst	DL-3-[2′,3′,4′,5′-$^{14}C_4$]-Phenylalanine	nn
Morpholine and formaldehyde then mono-fluoroindoles	Fluoro-substituted tryptophans	oo
Cycloheptatrienylium tetrafluoroborate	Cycloheptatrienylglycine	pp
Allenic bromides, $CH_2{=}C{=}CHCH_2Br/NaH \rightarrow$ $CH_2{=}C(-CH_2-)-CHCH_2C(CO_2Et)_2(NHCRO) \xrightarrow[-CO_2]{H_2O}$	Alka-4,5-dienoic acid and homologues	qq
	(2*S*,4*S*)/(2*R*,4*R*)-hypoglycin A	qq
5-Bromopent-3-en-1-yne	2-Aminohept-4-en-6-ynoic acid	rr
3,5-Di-*t*-butyl-4-hydroxybenzyl chloride	3,5-Di-*t*-butyltyrosine	ss
Ethyl acetamidocyanoacetate		
3-Hydroxy-4-nitrobenzyl chloride	Further elaboration leads to 5-(oxindolyl) alanine	tt
Ethyl [2-^{14}C]-*acetamidocyanoacetate*		
o-Hydroxybenzyl bromide	*o*-Hydroxy-[2-^{14}C]-DL-phenylalanine	uu
Michael addition to $ArC(=CH_2)CHO$	R–C₆H₄–substituted ring: CH–CH(OH)–N(CHO)–C(CO₂Et)₂ followed by reduction, indole ring construction en route to tryptophans	vv
2-Acetylamino-4-chloromethyl-selenazole	*β*-(2-Amino-1,3-selenazol-4-yl)-alanine	ww
Diethyl phthalimidomalonate (Sorenson method*) $PhtN(CH_2)_3\,^{13}CH_2Br$ and [^{15}N]-phthalimidomalonate	DL-[3-^{13}C, 2-^{15}N]-Lysine	xx
Diethyl benzyloxycarbonylaminomalonate		
(For a preparation of the starting material, see Cox *et al.*[yy])	*γ*-Carboxyglutamic acid	p,aa

Footnotes to Table 8.1

[a] Upson, D.A. and Hruby, V.J. (1976) *J. Org. Chem.*, **41**, 1353.
[b] Thanassi, J.W. (1971) *J. Org. Chem.*, **36**, 3019.
[c] Nagai, U. and Kobayashi, J. (1976) *Tetrahedron Lett.*, 2873.
[d] Kirby, G.W. and Michael, J. (1971) *Chem. Commun.*, 4151.
[e] Ife, R. and Haslam, E. (1971) *J. Chem. Soc. (C)*, 2818.
[f] Shiba, T., Mukunoki, Y. and Akiyama, H. (1975) *Bull. Chem. Soc. Japan*, **48**, 1902.
[g] Van Pee, K.H., Salcher, O. and Lingens, F. (1981) *Liebigs Ann. Chem.*, 233.
[h] Hatanaka, S., Kaneko, S., Niimura, Y. *et al.* (1974) *Tetrahedron Lett.*, 3931.
[i] Ogawa, Y., Tsuruoka, T., Inoue, S. and Niida, T. (1973) *Meiji Seika Kenkyu Nempo*, **13**, 42; (*Chem. Abs.*, 1974, **81**, 37806).
[j] Matsumoto, K., Miyahara, T., Suzuki, M. and Miyoshi, M. (1974) *Agric. Biol. Chem. (Japan)*, **38**, 1097.
[k] Maehr, H. and Leach, M. (1974) *J. Org. Chem.*, **39**, 1166.
[l] Pichat, L. and Beaucourt, J.P. (1974) *J. Labelled Compounds*, **10**, 103.
[m] Balenovic, K. and Delkac, A., (1973) *Rec. Trav. Chim. Pays-Bas*, **92**, 117.
[n] Lee, Y.K. and Kaneko, T. (1973) *Bull. Chem. Soc. Japan*, **46**, 2924.
[o] Sethi, M.L., Subba Rao, G. and Kapadia, G.J. (1973) *J. Pharm. Sci.*, **62**, 1802.
[p] Bosshard, H.R. and Berger, A. (1973) *Helv. Chim. Acta*, **56**, 1838.
[q] Goulding, R.W. and Palmer, A.J. (1972) *Internat. J. Appl. Radiat. Isotopes*, **23**, 133.
[r] Fuji, T. and Hatanaka, Y. (1973) *Tetrahedron*, **29**, 3825.
[s] Isowa, Y., Takashima, T., Ohmori, M., *et al.* (1972) *Bull. Chem. Soc. Japan*, **45**, 1461.
[t] Jansen, A.C.A., Kerling, K.E.T. and Havinga, E. (1970) *Rec. Trav. Chim., Pays-Bas*, **89**, 861.
[u] Lee, C. and Serif, G.S. (1970) *Biochemistry*, **9**, 2068.
[v] Pascal, G., Pichat, L. and Baret, C. (1968) *Bull. Soc. Chim. France*, 1481.
[w] Sogn, J.A., Gibbons, W.A. and Wolff, S. (1976) *Internat. J. Peptide Protein Res.*, **8**, 459.
[x] Hase, S., Kiroi, R. and Sakakibara, S. (1968) *Bull. Chem. Soc. Japan*, **41**, 1266.
[y] Liu, Y.-Y., Thom, E. and Liebman, A.A. (1978) *Canad. J. Chem.*, **56**, 2853.
[z] Keller-Schierlein, W. and Joos, B. (1980) *Helv. Chim. Acta*, **63**, 250.
[aa] Juhasz, A. and Bajusz, S. (1980) *Int. J. Peptide Protein Res.*, **15**, 154.
[bb] Hansen, J.J. and Krogsgaard-Larsen, P. (1980) *J.Chem. Soc. Perkin 1*, 1826.
[cc] Lolodziejczyk, A.M. and Arendt, A. (1980) *Pol. J. Chem.*, **54**, 1327.
[dd] Haudegoud, J.P., Chauvin, Y. and Commereuc, D. (1979) *J. Org. Chem.*, **44**, 3063.
[ee] Wolfe, S. and Jokinen, M.G. (1979) *Canad. J. Chem.*, **57**, 1388.
[ff] Fuksova, K. and Benes, J. (1977) *Radiochem. Radioanalyt. Lett.*, **30**, 187.
[gg] Dilbeck, G.A., Field, L. Gallo, A.A. and Gargiulo, R.J. (1978) *J. Org. Chem.*, **43**, 4593.
[hh] Powell, G.K. and Dekker, E.E. (1981) *Prep. Biochem.*, **11**, 339.
[ii] Keith, D.D., Yang, R., Tortora, J.A. and Weigele, M. (1978) *J. Org. Chem.*, **43**, 3713.
[jj] Gellert, E., Halpern, B. and Rudzats, R. (1978) *Phytochemistry*, **17**, 802.
[kk] Mauger, A.B., Irreverre, F. and Witkop, B. (1966) *J. Am. Chem. Soc.*, **88**, 2019.
[ll] Filler, R. and Rickert, R.C. (1981) *J. Fluorine Chem.*, **18**, 483.
[mm] Galat, A. (1947) *J. Am. Chem. Soc.*, **69**, 965.
[nn] Pichat, L., Liem, F.N. and Guermont, J.P. (1972) *Bull. Soc. Chim. France*, 4224.
[oo] Beutov, M. and Roffman, C. (1969) *Israel J. Chem.*, **7**, 835.
[pp] Hanessian, S. and Schütze, G. (1969) *J. Med. Chem.*, **12**, 347.
[qq] Black, D.K. and Landor, S.R. (1968) *J. Chem. Soc. (C)*, 283.
[rr] Black, D.K. and Landor, S.R. (1968) *J. Chem. Soc. (C)*, 288.
[ss] Teuber, H.J., Krause, H. and Berariu, V. (1978) *Annalen*, 757.
[tt] Atkinson, J.G., Wasson, B.K., Fuentes, J.J. *et al.* (1979) *Tetrahedron Lett.*, 2857.
[uu] Petitclerc, C., D'iorio, A. and Benoiton, N.L. (1969) *J. Labelled Compounds*, **5**, 265.
[vv] Hengartner, U., Batcho, A.D., Blount, J.F. *et al.* (1979) *J. Org. Chem.*, **44**, 3748.
[ww] Hanson, R.N. and Davis, M.A. (1981) *J. Heterocyclic. Chem.*, **18**, 205.
[xx] Gould, S.J. and Thiruvengadam, T.K. (1981) *J. Am. Chem. Soc.*, **103**, 6752; see also *Tetrahedron Lett.* (1982), **22**, 4255.
[yy] Cox, D.A., Johnson, A.W. and Mauger, A.B. (1964) *J. Chem. Soc.*, 5024.
* Referred to in *Tetrahedron Lett.* (1982), **22**, 4255 as the classical Sörenson method (Sörenson, S.P.L. (1903–6) *Compt. rend. trav. Carlsberg. Ser. Chem.*, **6**, 1).

Table 8.2 Strecker synthesis of α-amino acids

Aldehyde or ketone (or acetal equivalent)	*Objective of the synthesis*	*Reference*
$R_2C{=}C{=}CHCMe_2CHO$ (Unsuccessful for H in place of Me groups)	Corresponding 2-amino-alkadienoic acids	a
3,4-Dihydroxyphenylpropanone	L-α-Methyl-dopa	b
γ-Guanidinobutyraldehyde[c] 3-Cyanopropionaldehyde[d]	$[1\text{-}^{14}C]$-DL-Arginine,[c–e] also $[1\text{-}^{14}C]$-DL-glutamic acid[e] and $[1\text{-}^{14}C]$-DL-ornithine[e]	c–e
3-Carboxycyclohexanone	1-Aminocyclohexane-1,3-dicarboxylic acid (several analogues also, starting from the appropriate cyclohexanone)	f
$MeCOCH_2Ar$ + [structure: H—N, Ph, O, H, O] + NaCN 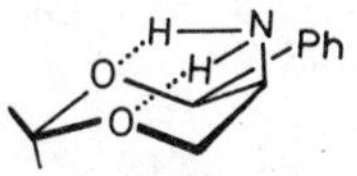 (modified Strecker synthesis, see also Table 8.22)	Asymmetric Strecker synthesis of (*S*)-α-methyl-α-amino acids (e.e. 80–90%)	g
(*S*)-$[\gamma\text{-}^2H_3]$-Isobutyraldehyde ('mild' Strecker reaction, $NH_4OH/NaCN/0°$)	(2*RS*,3*S*)-$[4,4,4\text{-}^2H_3]$-Valine	h–j
Acrylonitrile → $NCCH_2CH_2CHO$	L-Glutamic acid (industrial synthesis)	k
Acrolein, $CH_3CH{=}CHCHO$	2-Aminobut-3-enoic acid ('vinylglycine') and homologue	l
p-Hydroxybenzaldehyde	*p*-Hydroxyphenyglycine	m
Phenoxyacetaldehyde	2-Amino-3-phenyloxypropanoic acid	n
Various aldehydes	*N*-(2-Hydroxyethyl) amino acids	o
Butanone	Isovaline	p, q
Isobutyraldehyde (modified Strecker synthesis)	DL-Valine, via $Me_2CHCH(OH)CN$, which, with $SOCl_2$, gives $Me_2CHCH(OSOCl)CN$, which is reacted with NH_3	r
$MeP(O)(OH)CH_2CH_2CHO$	NH-Phosphinothricin	s
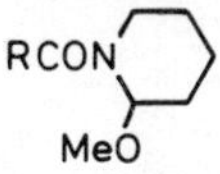 [structure: RCON, MeO]	Lysine and ornithine	t
$AcOCH_2COCH_2CO_2Et$	α-Hydroxymethyl-aspartic acid	u
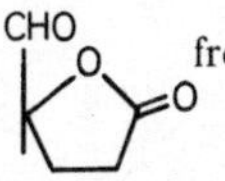 [structure: CHO, O, =O] from D-glucosamine	*erythro*-β-Hydroxy-L-histidine	v
$ClCH_2CHO$ bisulphite adduct	β-Chloroalanine	w
$ClCH_2CHO$, NH_3, Me_2CO, NaSH then HCN	DL-Cysteine	x
γ-(*p*-Methoxyphenyl)butyraldehyde	$p\text{-}MeOC_6H_4(CH_2)_3CH(\overset{+}{N}H_3)CO_2^-$ Alternamic acid	y
Glycollaldehyde, $HOCH_2CHO$	DL-Serine	z

[a] Black, D.K. and Landor, S.R. (1968) *J. Chem. Soc.* (*C*), 281.
[b] Reinhold, D.F., Firestone, R.A., Gaines, W.A. *et al.* (1968) *J. Org. Chem.*, **33**, 1209.

Footnotes to Table 8.2 (*Contd.*)

[c] Pichat, J., Guermont, J.-P. and Liem, P.N. (1968) *J. Labelled Compounds*, **4**, 251.
[d] Pichat, J., Guermont, J.-P. and Liem, P.N. (1971) *Bull. Soc. Chim. France*, 837.
[e] Mezo, I., Teplan, I. and Marton, J. (1969) *Acta Chim. Acad. Sci. Hung.*, **60**, 301.
[f] Gass, J.D. and Meister, A. (1970) *Biochemistry*, **9**, 842.
[g] Weinges, K., Graab, G., Nagel, D. and Stemmle, B. (1971) *Chem. Ber.*, **104**, 3594.
[h] Aberhardt, D.J. and Lin, L.J. (1973) *J. Am. Chem. Soc.*, **95**, 7859.
[i] Aberhardt, D.J. and Lin, L.T. (1974) *J. Chem. Soc. Perkin Trans. 1*, 2320.
[j] Hill, R.K., Yan, S. and Artin, S.M. (1973) *J. Am. Chem. Soc.*, **95**, 7857.
[k] Yoshida, T. (1970) *Chem.-lng. Tech.*, **42**, 641.
[l] Friis, P., Helboe, P. and Larsen, P.O. (1974) *Acta Chem. Scand.* (*B*), **28**, 317.
[m] Eizember, R.F. and Ammons, A.S. (1976) *Org. Prep. Proced. Internat.*, **8**, 149.
[n] Sarda, N., Grouiller, A. and Pacheco, H. (1976) *Tetrahedron Lett.*, 271.
[o] Giraud-Clenet, D. and Anatol, J. (1969) *Compt. rend.*, **268C**, 117.
[p] Baker, C.G., Shou-Cheng, J.Fu, Birnbaum, S.M. *et al.* (1952) *J. Am. Chem. Soc.*, **74**, 4701.
[q] Shou-Cheng, J.Fu and Birnbaum, S.M. (1953) *J. Am. Chem. Soc.*, **75**, 918.
[r] Davis, J.W. (1978) *J. Org. Chem.*, **43**, 3980.
[s] Gruszecka, E., Soroka, M. and Masterlerz, P. (1969) *Pol. J. Chem.*, **53**, 937.
[t] Warning, K., Mitzlaff, M. and Jensen, H. (1978) *Liebigs Annalen Chem.*, 1707.
[u] Walsh, J.J., Metzler, D.E., Powell, D. and Jacobson, R.A. (1980) *J. Am. Chem. Soc.*, **102**, 7136.
[v] Hecht, S.M., Rupprecht, K.M. and Jacobs, P.M. (1979) *J. Am. Chem. Soc.*, **101**, 3982.
[w] Nakayasu, K., Furuya, O., Inoue, C. and Moriguchi, S. (1981) *Ger. Offen.*, 3 021 566 (*Chem. Abstr.*, **95**, 25629).
[x] Martens, J., Offermanns, H. and Scherberich, P. (1981) *Angew. Chem.*, **93**, 680.
[y] Okuno, T., Ishita, Y., Sugawara, A. *et al.* (1975) *Tetrahedron Letts*, 335.
[z] Bassignani, L., Biancini, B., Bandt, A. *et al.* (1979) *Chem. Ber.*, **112**, 148.

Table 8.3 Hydantoin synthesis (Bücherer–Bergs synthesis) of α-amino acids

Aldehyde, ketone or hydantoin	*Objective of the synthesis*	*Reference*
5-Hydroxybutylhydantoin	γ-Hydroxynorleucine	a
5-Methyl-5-(3-azidopropyl) hydantoin	α-Methyl-L-arginine	b
	2-Amino-4-(3,4-dihydroxyphenyl)butyric acid ('homo-dopa') and 3,4,5-trihydroxyphenyl analogue	c
	α-Methyl-(3-allyl-4-hydroxyphenyl)-alanine and close structural analogues	d
	αα′-Di-aminopimelic acid	e
	α-Methylproline	f
	α-Methylornithine	f, g
[α-^{13}C]-Dibenzyloxybenzylmethyl ketone	[β-^{13}C]-α-Methyl-dopa	h
5-(4-Bromobutyl)hydantoin	[^{15}N]-Lysine (using $^{15}NH_3$);[i] or 'anabilysine' (by reaction with anabasine)[m]	i, m
5-(But-3-enyl)-3-phenylhydantoin	δ-Hydroxynorleucine, 5,6-dihydroxynorleucine, and hydroxylysinohydroxynorleucine	i
5,5-Dialkylhydantoins, converted into 3-(*N*-toluene-*p*-sulphonyl) derivative to facilitate hydrolysis (dilute NaOH then dilute HCl)	α-Alkyl-α-amino acids	k

Table 8.3 (*Contd.*)

Aldehyde, ketone or hydantoin	*Objective of the synthesis*	*Reference*
Aldehyde or ketone with $K^{11}CN$, NH_4Cl, $(NH_4)_2CO_3$, rapid reaction at 210°	[2-^{11}C]-Amino acids	l
	L- and D-2-(1,2:5,6-di-*O*-Isopropylidene-α-D-allofuranos-3-yl)glycine	n
Cycloalkanones	Corresponding l-amino cycloalkane-l-carboxylic acids	o
Alcohol, KCN, $(NH_4)_2CO_3$, electro-oxidation	Glycine, alanine, valine, diphenylalanine	p
4-Chlorohydantoin from [structure: HN–C(=O)–C(=O)–NH ring with C=O] (reacts with diethyl malonate, KBH_4 then $SOCl_2$)	β-Carboxy-DL-aspartic acid	q
$K^{11}CN$, standard Bücherer–Bergs procedure with indoleacetaldehyde	[2-^{11}C]-DL-Tryptophan	r

[a] Perier, C., Ronziere, M.C., Rattner, A. and Frey, J. (1976) *J. Chromatogr.*, **125**, 526.
[b] Maehr, H., Yarmchuk, L. and Leach, M. (1976) *J. Antibiotics*, **29**, 221.
[c] Winn, M., Rasmussen, R., Minard, F. *et al.* (1975) *J. Medicin, Chem.*, **18**, 434.
[d] El Masry, A.H., El Masry, S.E., Hare, L.E. and Counsell, R.E. (1975) *J. Medicin. Chem.*, **18**, 16.
[e] Arendt, A., Kolodziejczyk, A., Sokolowska, T. and Mrozowski, M. (1974) *Roczniki Chem.*, **48**, 883.
[f] Ellington, J.J. and Honigberg, I.L. (1974) *J. Org. Chem.*, **39**, 104.
[g] Abdel-Monem, M.M., Newton, N.E. and Weeks, C.E. (1974) *J. Medicin. Chem.*, **17**, 447.
[h] Ames, M.M. and Castagnoli, N. (1974) *J. Labelled Compounds*, **10**, 195.
[i] Mizon, J. and Mizon, C. (1974) *J. Labelled Compounds*, **10**, 229.
[j] Davis, N.R. and Bailey, A.J. (1972) *Biochem. J.*, **129**, 91.
[k] Hiroi, K., Achiwa, K. and Yamada, S. (1968) *Chem. Pharm. Bull.* (*Tokyo*), **16**, 444.
[l] Howes, R.L., Washburn, L.C. Wieland, B.W., *et al.* (1978) *Int. J. Appl. Radiat. Isotopes*, **29**, 186.
[m] Hardy, P.M., Hughes, G.J. and Rydon, H.N. (1977) *J.Chem. Soc. Chem. Comm.*, 759.
[n] Rosenthal, A. and Dodd, R.H., (1979) *J. Carbohydr., Nucleosides Nucleotides*, **6**, 467.
[o] Grunewald, G.L., Kuttab, S.H., Pleiss, M.A., *et al.* (1980) *J. Medicin. Chem.*, **23**, 754.
[p] Krysin, E.P., Tsodikov, V.V. and Grinberg, V.A. (1976) *Elektrokhimiya*, **12**, 1590 (*Chem. Abs.*, **80**, 162752).
[q] Henson, E.B., Gallop, P.M. and Hauschka, P.V. (1981) *Tetrahedron*, **37**, 2561.
[r] Zalutsky, M.R., Wu, J., Harper, P.V. and Wickland, T. (1981) *Int. J. Appl. Radiat. Isot.*, **32**, 182.

Table 8.4 Synthesis of α-amino acids from α-isocyano-alkanoates and α-isothiocyano-alkanotes

Isocyanoalkanoate and alkylating agent	*Objective of the synthesis*	*Reference*
Ethyl isocyanoacetate + $R^1R^2CBrCO_2R^3$	ββ-Dialkylaspartic acids	a
Ethyl isocyanopropionate + [structure: imidazole with CH_2OCOR and NTos]	α-Methylhistidine	b

Table 8.4 (*Contd.*)

Isocyanoalkanoate and alkylating agent		*Objective of the synthesis*	*Reference*
Methyl α-isocyanoalkanoates $\bar{C}\equiv\overset{+}{N}CHRCO_2Me$	ArCHO → (oxazoline: Ar, H, R, CO_2Me, O, N)	β-Hydroxy-α-alkyl-α-amino acids (predominantly *threo*) Warm aqueous ethanolic NEt_3 brings about ring opening to the *N*-formyl derivative, cleaved with 2 M HCl	c
	ArCOCl or $(RCO)_2O$ then H_2 with Pt or Pd, or $NaBH_4$[k]	β-Hydroxy-α-alkyl-α-amino acids (predominantly *erythro*)	c, d
	Aldehydes,[e,f] Ketones[g–i]	Substituted threonines or αβ-unsaturated α-amino acids	e, f
	Halides[j]	α-Alkyl and /or αα-dialkyl α-amino acids	f
Ethyl α-isocyanoacrylates $R^1R^2C{=}C(NC)CO_2Et$	Grignard reagent R^3MgX	C-tert-Alkylglycines $R^1R^2RCCH(\overset{+}{N}H_3)CO_2^-$	k
	Amines (secondary)	$R^3_2NCH{=}NC{=}CR^1R^2)CO_2Me$ through initial Michael addition then synchronous elimination and insertion into the isocyano-group	l
Methyl α-isocyanoalkanoates	Primary amine then BuLi then R^3Br then AcOH	(imidazolone: R^1, R^3, N=, NR^2, O) → $AcNHCR^1R^3CO_2H$	m
Methyl α-isocyanopropionate	3-Hydroxy-4-nitrobenzyl chloride	α-Methyl-β-(5-(oxindolyl))alanine	n
Isocyanoacetamide	Aldehydes	*threo*-β-Hydroxy-α-amino acids via *trans*-oxazolines	o
5-Alkoxyoxazoles (masked isocyanoacetates, lithio-derivatives of which form with BuLi)	Benzyl halides	α-Methyl-dimethoxydopa from the 4-methyloxazole	p
$(Me)(R)C{=}C(NC)(CO_2R^2)$	R^1X/LDA	$H_2C{=}CRCR^1(NC)CO_2R^2$ ↓ αβ-Unsaturated α-amino acids	q, r
Ethyl α-isothiocyanoacetate + aldehyde	(Ac, N, CO_2Et, R, S, O) or (N, CO_2Et, BzlS, O, R)	α-Aminoacrylates	s

[a] Bochenska, M. and Biernat, J.F. (1976) *Roczniki Chem.*, **50**, 1195.
[b] Suzuki, M., Miyahara, T., Yoshioka, R., *et al.* (1974) *Agric. and Biol. Chem., Japan*, **38**, 1709.
[c] Suzuki, M., Iwasaki, T., Matsumoto, K. and Okumura, K. (1973) *Chem. and Ind.*, 228.
[d] Suzuki, M., Iwasaki, T., Miyoshi, M., *et al.* (1973) *J. Org. Chem.*, **38**, 3571.
[e] Matsumoto, K., Urabe, Y., Ozaki, Y. *et al.* (1975) *Agric. and Biol. Chem., Japan*, **39**, 1869.
[f] Ozaki, Y., Matsumoto, K. and Miyoshi, M. (1978) *Agric. and Biol. Chem. Japan*, **42**, 1565.

Footnotes to Table 8.4 (*Contd.*)

[g] Praetorius, H.J., Flossdorf, J. and Kula, M.R. (1975) *Chem. Ber.*, **108**, 3079.
[h] Suzuki, M., Nunami, K. and Yoneda, N. (1978) *J. Chem. Soc., Chem. Commun.*, 1978, 270.
[i] Schöllkopf, U. and Meyer, R. (1981) *Liebigs Ann. Chem.*, 1469.
[j] Schöllkopf, U., Hoppe, D. and Jentsch, R. (1975) *Chem. Ber.*, **108**, 3079.
[k] Schöllkopf, U. and Meyer, R. (1975) *Angew. Chem.*, **87**, 624.
[l] Suzuki, M., Nunami, K., Moriya, T., *et al.* (1978) *J. Org. Chem.*, **43**, 4933.
[m] Schöllkopf, U., Hausberg, H.H., Hoppe, I., *et al.* (1978) *Angew. Chem.*, **90**, 136.
[n] Atkinson, J.G., Wasson, B.K., Fuentes, J.J., *et al.* (1979) *Tetrahedron Lett.*, 2857.
[o] Ozaki, Y., Maeda, S., Miyoshi, M. and Matsumoto, K. (1979) *Synthesis*, 216.
[p] Jacobi, P.A., Ueng, S.-N. and Carr, D. (1979) *J. Org. Chem.*, **44**, 2042.
[q] Kirihata, M., Tokumori, H., Ichimoto, I. and Ueda, H. (1978) *Nippon Nogei Kagaku Kaishi*, **52**, 271.
[r] Hoppe, I. and Schöllkopf, U. (1981) *Synthesis*, 646.
[s] Hoppe, D. (1973) *Angew. Chem. Internat. Edit.*, **12**, 656, 658.

Table 8.5 Synthesis of α-amino acids from α-nitroacetates and homologues

Nitroalkanoate or nitroalkenoate	*Alkylating agent*	*Objective of the synthesis*	*Reference*
Methyl nitroacetate	Aminoalkyl halides	Aminoalkyl glycines	a
Ethyl nitroacetate	3-Formylindole	3-Mercaptotryptophans and homologues	b
Ethyl nitroacetate		L- and D-2-(1,2:5,6-di-*O*-isopropylidene-α-D-allofuranos-3-yl)glycines	c
Ethyl nitroacetate or diethyl nitromalonate	1-Nitrobuta-1,3-diene	DL-Lysine (after hydrogenation and acid hydrolysis)	d
Methyl α-nitro-*ββ*-dimethylacrylates	$Me_2C{=}CHCO_2Me$ with HNO_3 then Sn-aq. HCl dil. HCl reduction	DL-Isodehydrovaline after KH treatment and dil. HCl	e
Methyl nitroacetate	Substituted 3-formylindoles, followed by $SnCl_2$ reduction ($NO_2 \rightarrow NH_2$) then H_2-Rh/phosphine	Substituted tryptophans	f

[a] Kaji, E. and Zen, S. (1973) *Bull. Chem. Soc. Japan*, **46**, 337.
[b] Vinograd, L.K., Shalygina, O.D., Kostyuchenko, N.P. and Suvorov, N.N. (1974) *Khim. geterosikl. Soedinenii*, 1236.
[c] Rosenthal, A. and Cliff, B. (1975) *J. Carbohydrates, Nucleosides, and Nucleotides*, **2**, 263.
[d] Samoilovich, T.I., Polyanskaya, A.S. and Perekalin, V.V. (1974). *Doklady Akad. Nauk. S.S.S.R.*, **217**, 1335.
[e] Baldwin, J.E., Haber, S.B., Hoskins, C. and Kruse, L.I. (1977) *J. Org. Chem.*, **42**, 1239.
[f] Hengartner, U., Batcho, A.D., Blount, J.F., *et al.* (1979) *J. Org. Chem.*, **44**, 3748.

Table 8.6 Rearrangement reactions leading to amino acids*

Name of rearrangement	*Reactions involved*	*Objective of the synthesis*	*Reference*
Neber rearrangement	Nitriles to imino-esters to aziridines to α-amino-ortho-esters	General synthesis of α-amino acids	a, b
	$PhCH_2CH_2{}^{14}CN \rightarrow Ph(CH_2)_2{}^{14}C(OMe)=NCl \rightarrow PhCH_2CH(NHZ){}^{14}C(OMe)_3$	[1-^{14}C]-Labelled phenylalanines	c
Stevens rearrangement	$R^1R^2\overset{+}{N}(CH_2CO_2Me)_2)X^-$	to *N,N*-dialkyl-aspartic acids	d
Azidoformate rearrangement	(*S*)-2-Methyl-3-phenylpropyl azidoformate	(*R*)-α-methylphenyl-alanine	e
	$Me-CH(CH_2OCON_3)-CH_2Ph \xrightarrow{h\nu}$ $Me-C(CH_2Ph)$ in ring $CH_2-O-CO-NH$		
Curtius rearrangement	Malonic acid half-esters with $Ph_2P(O)N_3/Et_3N/PhCH_2OH$	*N*-Benzyloxycarbonyl-amino acid esters	f, g
Schmidt rearrangement	HN_3/ethyl dibenzyl-acetoacetate	α-Benzylphenylalanine after hydrolysis	h
Ogura-Tsuchihashi rearrangement; α-amino acids from nitrites	β-Amino-α-methanesulphinyl vinyl sulphides to N-acetyl-α-amino acids	α-Amino acids	i, j
	$Me-S(=O)-CH_2SMe + RCN \xrightarrow{NaH} RC(NH_2)=C(SMe)(S(=O)-Me) \xrightarrow{Ac_2O} AcNH-C(R)(SMe)-COSMe \xrightarrow{(i)\ MeOH,\ (ii)\ Ni} RCH(NH_2)CO_2$		
Curtius rearrangement	$Bu^tCH(CO_2Et)CO_2H$ (from $Me_2C=C(CO_2Et)_2$ + MeMgI, etc)	*t*-Leucine	k
Hofmann rearrangement			l
	cyclopropane $(MeO_2C)_2$, H, Et $\rightarrow$ MeO_2C, H_2NCO, H, Et $\rightarrow$ MeO_2C, MeO_2C-HN, H, Et $\rightarrow$ DL-Coronamic acid		
Curtius rearrangement of $H(Ph)C=C(CONH_3)(CO_2Et)$	Addition of $PhCH_2OH$ to resulting isocyanate	*N*-Benzyloxycarbonyl-*E*-dehydrophenylalanine ethyl ester	m
Hofmann rearrangement of Boc-L-asparagine	$(CF_3CO_2IPh/DMF-H_2O)py$	L-Boc-$NHCH(CH_2NH_2)CO_2H$	n
Schmidt rearrangement	$HN_3/MeCOCHBu^tCO_2Et$	*t*-Leucine	o

Table 8.6 (*Contd.*)

Name of rearrangement	*Reactions involved*	*Objective of the synthesis*	*Reference*
Arndt-Eistert rearrangement of *N*-protected α-amino acids		β-Amino acids (see Table 8.20)	
Wolff rearrangement		β-Amino acids (see Table 8.20)	

* See also Reaction 8.3 and Table 8.22.
[a] Graham, W.H. (1969) *Tetrahedron Lett.*, 2223.
[b] Zemlicka, J. and Murata. M. (1976) *J. Org. Chem.*, **41**, 3317.
[c] Partingham, C.R. and Mertes, M.P. (1978) *J. Labelled Compounds Radiopharm.*, **14**, 223.
[d] Babayan, A.T., Kocharyan, S.T. and Ogandzhanyan, S.M. (1976) *Armyan. khim. Zhur.*, **29**, 456.
[e] Terashima, S. and Yamada, S. (1968) *Chem. and Pharm. Bull., Japan*, **16**, 2064.
[f] Yamada, S., Ninomiya, K. and Shioiri, T. (1973) *Tetrahedron Lett.*, 2343.
[g] Yamada, S., Ninomiya, K. and Shioiri, T. (1974) *Chem. and Pharm. Bull., Japan*, **22**, 1398.
[h] Barrett, G.C., Hardy, P.M., Harrow, T.A. and Rydon, H.N. (1972) *J. Chem. Soc. Perkin 1*, 2634.
[i] Ogura, K. and Tsuchihashi, G. (1974) *J. Am. Chem. Soc.*, **96**, 1960.
[j] Ogura, K., Yoshimura, I., Katoh, N. and Tsuchihashi, G. (1975) *Chem. Lett.*, 803.
[k] Miyazawa, T., Nagai, T., Yamada, T., *et al.* (1979) *Mem. Konan Univ., Sci. Ser.*, **23**, 51.
[l] Ichihara, A., Shiraishi, K. and Sakamura, S. (1977) *Tetrahedron Lett.*, 269.
[m] Nitz, T.J., Holt, E.M., Rubin, B. and Stammer, C.H. (1981) *J. Org. Chem.*, **46**, 2667.
[n] Waki, M., Kitajima, Y. and Izumiya, N. (1981) *Synthesis*, 266.
[o] Barrett, G.C. and Cousins, P.R. (1975) *J. Chem. Soc. (C)*, 2313.

Table 8.7 Synthesis of amino acids from Schiff bases and related imines*

Imine	*Method and synthetic objective*	*Reference*
Schiff bases derived from α-keto acids	Reductive alkylation at − 2.28 V at Hg cathode $PhCH_2N{=}CR^1CO_2R^2 \xrightarrow{R^3X} PhCH_2NH{-}C(R^1)(R^3){-}CO_2R^2$ H_3O^+ then H_2/Pd gives 38–86% yields	a
Oximes of α-keto acids	$RC({=}NOH)CO_2H \rightarrow RCH(\overset{+}{N}H_3)CO_2^-$ (Aluminium amalgam as reducing agent for R = alkenyl;[b] catalytic hydrogenation in a synthesis of t-leucine[c])	b c
Phenylhydrazones of α-keto acids	$RC({=}NNHPh)CO_2H \rightarrow RCH(N^+H_3)CO_2^-$ (Hydrogenation of aqueous suspensions over palladium)	d, e
Addition of Me_3SiCN to simple Schiff bases	$MeCH{=}NCHMePh \rightarrow MeCH(CN)N(SiMe_3)CHMePh \xrightarrow{H_2}$ alanine	f
Addition of Grignard reagent to $EtOCON{=}CHCCl_3$	$\rightarrow EtOCONHCHRCCl_3 \rightarrow H_3\overset{+}{N}CHRCO_2^-$	g
Addition of HCN to an aldoxime bisulphite adduct	$HON{=}CHCH_2OH \rightarrow (HONH)(NC)CH{-}CH_2OH \rightarrow$ Serine	h, i

Table 8.7 (*Contd.*)

Imine	*Method and synthetic objective*		*Reference*
Arylideneglycine derivatives	Aldehydes	β-Hydroxy-α-amino acids	j
(Salicylideneglycine)$_2$ copper (II)		Alanine, phenylalanine, in good yields	k, l
(Salicylideneglycine)$_2$Co(III)	Acrylate esters	Glutamic acid derivs	m
	Gramine methosulphate	Tryptophan	n
Schiff base from glycine t-butyl ester	LDA/benzyl or butyl halide	Phenylalanine or norleucine (some di-alkylation observed)	o
$Ph_2C=NCH_2CO_2Et$	Phase-transfer catalysed alkylation using $EtO_2CCH=CHCH_2Br$	(*R*)-5-Amino-hex-2-ene-dioic acid (analogue of D-α-aminoadipic acid)	p
$Ph_2C=NCH_2CO_2Et$	LDA/alkyl halide/THF/−78°/HMPA or phase transfer catalysis (some di-alkylation with benzylic halides)	Representative α-amino acids	q
$Ph_2C=NCH_2CN$	Phase-transfer catalysed alkylation with 'unactivated' halides	Representative α-amino acids	r
$(RS)_2C=NCH_2CO_2Et$	Bu^{t}OK and alkyl halide (some di-alkylation observed)	Representative α-amino acids	s
$PhCH=NCHRCO_2Me (R=Me)$	Alkylation with halogenomethane	α-Halogenomethyl-α-methyl-α-amino acids	t, u
$PhCH=NCH_2CO_2H$	LDA or NaH in THF (−78° or room temp respectively) then R′I	α-Alkyl-α-amino acids	v, w
$Me_2NCH=NCHRCO_2Me$ (from α-amino acids and *NN*-dimethylformamide dimethyl acetal)	Alkylation or Michael addition after carbanion formation	α-Alkyl-α-amino acids	x
$PhCH=NC(=CHMe)CO_2Me$	$LiN(SiMe_3)_2$ then RX	α-Vinyl-α-amino acids (exclusive α-alkylation)	y

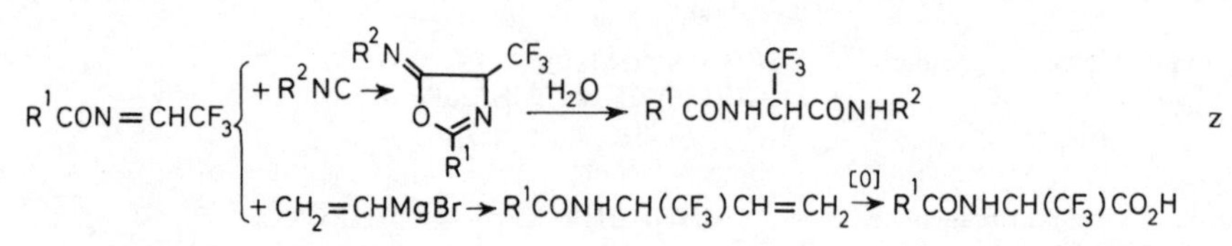

z

Imine	*Method and synthetic objective*		*Reference*
$PhCH=NCHMeCO_2Me$	*E*- or *Z*-RCH=CHBr/LDA	α-Methyl-(*E*)-3,4-dehydroglutamic acid	aa

Table 8.7 (*Contd.*)

Imine	*Method and synthetic objective*		*Reference*
$PhCH_2N{=}CR^1CO_2R^2$	Alkyl halide and cathodic reduction in DMF	α-Methyl-α-amino acids	bb
α-Keto ester, amine, dihydropyridine	Reduction by a dihydropyridine	α-Amino acid esters ($PhNHCHMeCO_2Me$ in 57% yield)	cc
$PhCH{=}NCHRCO_2Et$	Formaldehyde/LDA followed by 6M HCl	α-Hydroxymethylamino acids	dd
$BrCH_2C({=}NOH)CO_2Et$	Alkylation in the presence of Na_2CO_3, then Al amalgam	$RCH_2CH(\overset{+}{N}H_3X^-)CO_2Et$	ff

* See also Table 8.14.

[a] Iwasaki, T. and Harada, K. (1974) *J.C.S. Chem. Comm.*, 338.
[b] Drinkwater, D.J. and Smith, P.W.G. (1971) *J. Chem. Soc.* (*C*), 1305.
[c] Pospisek, J. and Blaha, K. (1977) *Coll. Czech. Chem. Commun.*, **42**, 1069.
[d] Khan, N.H. and Kidwai, A.R. (1973) *J. Org. Chem.*, **38**, 822.
[e] Tabakovic, I., Trkovnik, M. and Dzepina, M. (1977) *Croat. Chem. Acta*, **49**, 497.
[f] Nakajima, Y., Makino, T., Oda, J. and Inouye, Y. (1975) *Agric. and Biol. Chem. Japan*, **39**, 571.
[g] Kashima, C., Aoki, Y. and Omote, Y. (1975) *J. Chem. Soc. Perkin 1.*, 2511.
[h] Natta, G. and Pasquon, I. (1973) *Chimie et Industrie*, **35**, 323.
[i] Sizov, S.Y., Semenova, L.V. and Utrabin, N.P. (1978) *Prikl. Biokhim. Mikrobiol.*, **14**, 915.
[j] Belokon, Yu. N., Kuznetsova, N.I., Murtazin, R.M. and Dolgaya, M.M. (1972) *Izv. Akad. Nauk S.S.S.R., Ser. Khim.*, 2772.
[k] Nakahara, A., Nishikawa, S. and Mitani, J. (1967) *Bull. Chem. Soc., Japan.*, **40**, 2212.
[l] Wakamiya, T. Konishi, K. Chaki, H., *et al.* (1981) *Heterocycles*, **15**, 999.
[m] Belokon, Yu.N., Kuznetsova, N.I., Murtazin, R.M. and Dolgaya, M.M. (1972) *Bull. Acad. Sci. USSR*, 1288.
[n] Belokon, Yu.N., Belikov, V.M., Maksakov, V.A., *et al.* (1978) *Izvest. Akad. Nauk S.S.S.R., Ser. Khim.*, 1026.
[o] Oguri, T., Shioiri T. and Yamada, S. (1977) *Chem. Pharm. Bull.*, **25**, 2287.
[p] Allan, R.D. (1980) *J. Chem. Res.* (*S*), 392.
[q] O'Donnell, M.J., Boniece, J.M. and Earp, S.E. (1978) *Tetrahedron Lett.*, 2641.
[r] O'Donnell M.J. and Eckrich, T.M. (1978) *Tetrahedron Lett.*, 4625.
[s] Hoppe, D. and Beckmann, L. (1979) *Liebig's Ann. Chem.*, 2066.
[t] Bey, P. and Vevert, J.P. (1978) *Tetrahedron Lett.*, 1215.
[u] Bey, P., Vevert, J.P., Van Dorsselaer, V. and Kolb, M. (1979) *J. Org. Chem.*, **44**, 2732.
[v] Bey, P. and Vevert, J.P. (1977) *Tetrahedron Lett.*, 1455.
[w] Stork, G., Leong, A.Y.W. and Touzin, A. (1976) *J. Org. Chem.*, **41**, 3491.
[x] Fitt, J.J. and Gschwend, H.W. (1977) *J. Org. Chem.*, **42**, 2639.
[y] Greenlee, W.J., Taub, D. and Patchett, A.A. (1978) *Tetrahedron Lett.*, 3999.
[z] Weygand, F., Steglich, W., Ottmeier, W., *et al.* (1966) *Angew. Chem. Int. Ed.*, **5**, 600.
[aa] Bey, P. and Vevert, J.P. (1980) *J. Org. Chem.*, **45**, 3249.
[bb] Iwasaki, T. and Harada K. (1977) *J.Chem. Soc. Perkin 1*, 1730.
[cc] Nakamura, K., Ohno, A. and Oka, S. (1977) *Tetrahedron Lett.*, 4593.
[dd] Calcagni, A., Rossi, D. and Lucente, G. (1981) *Synthesis*, 445.
[ee] Gilchrist, T.L., Lingham, D.A. and Roberts, T.G. (1979) *J.C.S. Chem. Commun.*, 1089.

Table 8.8 Synthesis of amino acids by amination of carboxylic acids and related compounds

Carboxylic acid	*Amination reagent*	*Product*	*Reference*
Arylacetic acids	$MeONH_2$/LDA in THF—HMPA[a,b]; $ArSO_2ONH_2$[c]	$PhCH(NH_2)CO_2H$ (55%)	a–c
Alkanoic acids	Conc. aq. NH_4OH/contact glow discharge electrolysis	Mixtures of amino acids	d
Alkenoic acids	Aq. NH_4OH/contact glow discharge electrolysis	Acrylic acid gives 1–3% yields of alanine, serine, β-alanine, isoserine Maleic and fumaric acids give β-hydroxyaspartic acids	e
Alkanenitriles	Aq. NH_4OH/contact glow discharge electrolysis	Propionitrile gives small yields of glycine, alanine, β-alanine	f
Acetic or succinic acids	NH_3, γ-irradiation[g] or atomic nitrogen (N plasma)[h]	Amino acid mixtures	g h
Propionic acid	$NH(^1\Delta)$ insertion into CH bond (photolysis of NH_3)	Alanine and β-alanine	i
Formic acid or ammonium bicarbonate	Aq. NH_4OH, glow discharge electrolysis	Glycine; traces of five other protein amino acids	j

[a] Yamada, S., Oguri, T. and Shioiri, T. (1972) J.C.S. *Chem. Comm.*, 623.
[b] Yamada, S., Oguri, T. and Shioiri, T. (1975) *Chem. Pharm. Bull.*, **23**, 167.
[c] Hansen, J. and Krogsgaard-Larsen, P. (1980) *J. Chem. Soc., Perkin 1*, 1826.
[d] Harada, K. and Iwasaki, T. (1974) *Nature*, **250**, 426.
[e] Harada, K., Suzuki, S. and Ishida, H. (1978) *Experientia*, **34**, 17.
[f] Harada, K., Suzuki, S. and Ishida, H. (1978) *Biosystems*, **10**, 247.
[g] Ema, K. and Masuda, T. (1980) *Technol. Rep. Osaka Univ.*, **30**, 313; (*Chem. Abs.*, **93**, 168566).
[h] Margulis, M.M., Grundel, L.M. and Girina, E.L. (1980) *Dokl. Akad. Nauk. S.S.S.R., Ser. Khim.*, **251**, 639.
[i] Sato, S., Kitamura, T. and Tsunashima, S. (1980) *Chem. Lett.*, 687.
[j] Harada, K. and Suzuki, S. (1977) *Naturwiss.*, **64**, 484.

Table 8.9 Synthesis of amino acids by carboxylation of amines and related nitrogen compounds

Starting material	*Carboxylation reagent*	*Product*	*Reference*
Benzyl isocyanide	BuLi/THF then CO_2/ −78°, or $ClCO_2R$ or $(RO)_2CO$	$PhCH(\overset{+}{N}H_3)CO_2^-$	a–c
Aliphatic amines	Formic acid, contact glow discharge electrolysis	Corresponding amino acids	d
t-Butyl *N*-trimethylsilyl prop-2-ynyl carbamate	CO_2/LDA/TMEDA/ −78° then diazomethane	$BocNHCH(CO_2Me)C{\equiv}CSiMe_3$ (+allenic isomer, 20%)	e

Table 8.9 (*Contd.*)

Starting material	*Carboxylation reagent*	*Product*	*Reference*
Simple primary amines	Contact glow discharge electrolysis in aqueous solutions (oxidation by HO·)	$H_2NEt \rightarrow$ glycine, alanine, β-alanine and homologues $H_2NPr \rightarrow$ isoserine and β-alanine	f, g
Simple primary amines and diamines	γ-Radiolysis in aq. formic acid	Non-specific carboxylation gives mixtures of amino acids	h
Cyanation of simple primary amines	Contact glow discharge electrolysis in aq. NaCN	Mixtures of amino acids	i
Cyanation of α-halogeno-alkylamines	$ZNHCHClCF_3$	Trifluoroalanine	j
α-Methoxyurethanes	Phenyl isocyanide/$TiCl_4$ $MeO_2CNR^1CHR^2OMe$	$MeO_2CNR^1CHR^2CONHPh$	k

[a] Vaalburg, W., Strating, J., Waldring, M.G. and Wynberg, H. (1972) *Synthetic Commun.*, **2**, 423.
[b] Matsumoto, K., Suzuki, M. and Miyoshi, M. (1973) *J. Org. Chem.*, **38**, 2094.
[c] Oguri, T., Shioiri, T. and Yamada, S. (1975) *Chem. Pharm. Bull. Japan*, **23**, 173.
[d] Harada, K. and Iwasaki, T. (1975) *Chem. Lett.*, 185.
[e] Metcalf, B.W. and Casara, P. (1979) *J.C.S. Chem. Comm.*, 119.
[f] Terasawa, J. and Harada, K. (1980) *Chem. Lett.*, 73.
[g] Harada, K., Nomoto, M.M. and Gunji, H. (1981) *Tetrahedron Lett.*, **22**, 769.
[h] Davison, A., Barker, N.T. and Sangster, D.F. (1977) *Austral. J. Chem.*, **30**, 807.
[i] Harada, K., Suzuki, S. and Ishida, H. (1978) *Biosystems*, **10**, 247.
[j] Uskert, A., Neder, A. and Kasztreiner, E. (1973) *Magyar Kem. Folyoirat*, **79**, 333.
[k] Shono, T., Matsumura, Y. and Tsubata, K. (1981) *Tetrahedron Lett.*, **22**, 2411.

Table 8.10 Synthesis of amino acids by amination of halogeno-alkanoic acids, hydroxyalkanoic acids, αβ-epoxyalkanoic acids, keto acids, or alkenoic acids

Starting material	*Amination reagent*	*Product*	*Reference*
α-Hydroxy acids	Phthalimide/PPh_3/diethyl azodicarboxylate	α-(Phthaloylamino)acids	a
cis-Epoxysuccinic acid	NH_3	*threo*-β-Hydroxyaspartic acid	b, c
Trans-Epoxysuccinic acid	NH_3	*threo*/*erythro*-β-Hydroxyaspartic acid, formed in ratio 1:2	b, c
cis-2,3-Epoxyalkanoate esters	NH_4OH	*threo*-2-Hydroxy-3-amino-alkanoic acids	d
β-Isopropylglycidic acid	NH_3	*erythro*/*threo*-β-Hydroxyleucine	e
Pyruvic acid	$^{15}NH_3/NaBH_3CN$	[^{15}N]-Labelled α-amino acids	f
Pyruvic acid	Indoles, NH_3, tryptophanase	L-Tryptophan and analogues	g

Table 8.10 (*Contd.*)

Starting material	*Amination reagent*	*Product*	*Reference*
α-Keto acids	Electroreductive amination with NH_3/NH_4Cl at Hg electrode	Amino acids; low yields due to concurrent condensation processes	h, i
α-Hydroxy acids	γ-Irradiation in aq. NH_4OH	Corresponding amino acids	j
α-Keto acids	Benzyl carbamate/TosOH 4h in refluxing benzene	*N*-Benzyloxycarbonyl αβ-unsaturated α-amino acids ('Z-dehydro-amino acids')	k
α-Keto acids	Reductive amination with NaB^2H_4	[2-^{2}H]-3-Fluoroalanine from FCH_2COCO_2H	l
Alkenedioic acids (citraconic and itaconic acid derivatives)	Heat ammonium salts at 130–210°, hydrolyse the resulting poly(imides)	α- and β-Methylaspartic acids	m
Acrylic acid homologues	H_2NOH followed by H_2/Pd	General α-amino acid synthesis	n
α-Bromoepoxides (from ArCHO and $CHBr_3$)	$NH_3/KOH/LiNH_2$ in glyme	*C*-Aryglycines	o
α-Bromo acids	$CH_2{=}CHCHO$ → cyanohydrin → hydroxy-acid → α-bromo-acid bromide with PBr_3, then NH_3	α-Vinylglycine	p, q
α-Bromo-βγ-unsaturated alkanoates (from αβ-isomers with LDA/ − 78°/THF)	NH_3, room temperature, 90 min in DMSO	βγ-Unsaturated α-amino acids	r
2-{(*R*)-Phenylethyl} malonate	Decarboxylation and amination	2*S*-Amino-3*S*-phenyl butanoic acid	s
α-Bromoalkanoates $R^2CHBrCO_2R^1$	M^+ OCN^- then R^3OH	$R^3OCONHCHR^2CO_2R^1$ (91–99% yields of amino acids after hydrolysis in aq. HCl)	t, u
(*Z*)-$ClCH_2CX{=}CHCO_2H$ from $ClCH_2C{\equiv}CCO_2H$ and HX (X = Cl, Br, I)	NH_3	$H_3\overset{+}{N}CH_2CX{=}CHCO_2^-$ (GABA analogues)	v
Chloroacetamide	$Me_3SiCH_2NH_2$	$Me_3SiCH_2NHCH_2CONH_2$	w
Hydroxy acid 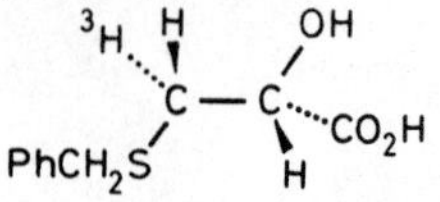 [structure: ^{3}H, H / C — C (OH, CO_2H, H) / $PhCH_2S$]	Via bromo acid + NH_3	Corresponding cysteine	x
$BrC({=}CH_2)CO_2Me$ + *trans*-2-butene $\xrightarrow{EtAlCl_2}$ $CH_2{=}CHCHMeCH_2\cdot CHBrCO_2Me$	NH_3	2-Amino-4-methyl-5-hexenoic acid	y

Table 8.10 (*Contd.*)

Starting material	*Amination reagent*	*Product*	*Reference*
$ClCH_2CCl_3$ or $Cl_2CHCHCl_2$	NH_4OH at 160–170°C	Glycine (via H_2NCH_2CN)	z
α-Chloropropionic acid	Aq. NH_4OH (10 equiv.) 70°/pressure/5 h	Alanine (76%)	aa
$MeCHFCHMeCH_2 \cdot CO_2R$ with LDA and Br_2 or $I_2 \rightarrow$ α-bromo- or -iodo acids	NaN_3 then H_2	γ-Fluoroisoleucine	bb
α-Cyano-oxiranes (R, R¹, CN epoxide) (from ketones and $ClCH_2CN$)	HF/pyridine then $NH_3 - MeOH$	β-Fluoro-α-amino-alkanoic acids via $RR^1CFCH(OH)CN$	cc
$H_2NCOCH{=}CH \cdot CO_2H$	$^{14}CH_3NH_2$	*N*-[^{14}C]-Methyl-DL-asparagine	dd

[a] Wada, M., Sano, T. and Mitsunobu, O. (1973) *Bull. Chem. Soc. Japan*, **46**, 2833.
[b] Okai, H., Imamura N. and Izumiya, N. (1967) *Bull. Chem. Soc. Japan*, **40**, 2154.
[c] Jones, C.W., Leyden, D.E. and Stammer, C.H. (1969) *Canad. J. Chem.*, **47**, 4363.
[d] Kato, K., Saino, T., Nishizawa, R. *et al.* (1980) *J. Chem. Soc. Perkin 1*, 1618.
[e] Futagawa, S., Nakahara, M., Inui, T., *et al.* (1971) *Nippon Kagaku Zasshi*, **92**, 374.
[f] Borsch, R.F., Bernstein, M.D. and Durst, H.D. (1971) *J. Am. Chem. Soc.*, **93**, 2897.
[g] Nakazawa, H., Enei, H., Okumura, S., *et al.* (1972) *FEBS Lett.*, **25**, 43.
[h] Jeffrey, E.A. and Meisters, A. (1978) *Austral. J. Chem.*, **31**, 73.
[i] Jeffrey, E.A. Johansen, C. and Meisters, A. (1978) *Austral. J. Chem.*, **31**, 79.
[j] Ema, K., Kato, T. and Shinagawa, M. (1978) *Radioisotopes*, **27**, 445.
[k] Shin, C., Yonezawa, Y., Unoki, K. and Yoshimura, J. (1979) *Tetrahedron Lett.*, 1049.
[l] Dolling, U.H., Douglas, A.W., Grabowski, E.J.J., *et al.* (1978) *J. Org. Chem.*, **43**, 1634.
[m] Harada, K. and Matsuyama, M. (1979) *Bio Systems*, **11**, 47.
[n] Basheeruddin, K., Siddiqui, A.A., Khan, N.H., and Saleha, S. (1979) *Synthetic Commun.*, **9**, 705.
[o] Compere, E.L. and Weinstein, D.A. (1977) *Synthesis*, 852.
[p] Baldwin, J.E., Haber, S.B., Hoskins, C. and Kruse, L.I. (1977) *J. Org. Chem.*, **42**, 1239.
[q] Frijis, P., Helboe, P. and Larsen, P.O. (1974) *Acta Chem. Scand.*, **28**, 317.
[r] Chari, R.V.J. and Wemple, J. (1979) *Tetrahedron Lett.*, 111.
[s] Tsuchihashi, G., Mitamura, S. and Ogura, H. (1979) *Bull. Chem. Soc. Japan*, **52**, 2167.
[t] Effenberger, F. and Drauz, K. (1979) *Angew. Chem.*, **91**, 504.
[u] Effenberger, F., Drauz, K., Forster, S. and Schöller, W. (1981) *Chem. Ber.*, **114**, 173.
[v] Allan, R.D., Johnston G.A.R. and Twitchin, B. (1980) *Austral. J. Chem.*, **33**, 1115.
[w] Fink, W. (1974) *Helv. Chim. Acta*, **57**, 1042.
[x] Aberhardt, D.J., Lin, L.J. and Chu, J.Y.-R. (1975) *J. Chem. Soc. Perkin 1*, 2517.
[y] Snider B.B. and Duncia, J.V. (1981) *J. Org. Chem.*, **46**, 3223.
[z] Inoue, M. and Enomoto, S. (1978) *Chem. Lett.*, 1231.
[aa] Ogata, Y. and Inaishi, M. (1980) *Kenkyu Hokoku—Asahi Garasu Kogyo Gijutsu Shoreikai*, **36**, 219; (*Chem. Abs.*, **95**, 43613).
[bb] Butina, D. and Hudlicky, M. (1980) *J. Fluorine Chem.*, **16**, 301.
[cc] Ayi, A.I., Remli, M. and Guedj, R. (1981) *Tetrahedron Lett.*, **22**, 1505.
[dd] Tsou, H.R. (1981) *J. Labelled Compd. Radiopharm.*, **18**, 921.

Table 8.11 Formation of amino acids from mixtures of simple reactants ('prebiotic synthesis models')

Reactants	*Energy source*	*Amino acids formed (yields are invariably low and not optimized)*	*Reference*
Methane, ammonia, H_2S, H_2O	UV irradiation	Cysteine and other amino acids	a
Methane, ammonia, water	High frequency discharge	Histidine and other amino acids	b, c
Methane, nitrogen, water, traces of ammonia	Electrical discharge	α-Hydroxy-γ-aminobutyric acid (not isoleucine as previously claimed) and other amino acids	d, e
Methane, ammonia, water, sand	930–1060°C	β-Alanine constitutes 90% of the amino acids formed at 1060° but major products at 930° are glycine (ca 96%) and alanine (ca 4%)	f
Methane or CO, ammonia, water	Volcanic lava at temperatures up to 1050°C	Amino acids and other organic compounds	g
Methane, ethane, ammonia, water	Acoustic	α-Amino-alkanenitriles	h
Carbon monoxide and ammonia, over aluminosilicates saturated with calcium, ammonium and ferrous salts	Heat	Amino acids, urea, and other organic compounds	i
Methylamine, ammonia, CO_2	n, γ-Irradiation	Glycine, alanine, lysine	j
Trimethylamine, ammonia, CO_2	n, γ-Irradiation	Also valine, γ-aminobutyric acid	j
n-Pentylamine, ammonia, CO_2	n, γ-Irradiation	Also norleucine, 6-amino-hexanoic acid	j
Formaldehyde, ammonium molybdate and phosphate, water, mineral salts	Sunlight	Mixtures of amino acids	k, s
Hydrogen cyanide, aldehydes, water	Acoustic	α-Amino acids via aldimines and α-amino-alkanenitriles	l
Formaldehyde, NH_4SCN, KH_2PO_4, $Ca(OAc)_2$, metal salts	UV irradiation	Cysteine and other amino acids	m
Formaldehyde, hydroxylamine or N_2	Ultrasound	Glycine, alanine	n, o
Formaldehyde, hydroxylamine, salt solutions	Ultrasound	Glycine, serine, aspartic acid, β-alanine	p, q

Table 8.11 (*Contd.*)

Reactants	*Energy source*	*Amino acids formed (yields are invariably low and not optimized)*	*Reference*
Hydrogen cyanide in water	[^{60}Co]-γ-Irradiation	Glycine, alanine, valine, serine, threonine, aspartic acid and glutamic acid	r
Cyanamide in water	γ-Irradiation	Arginine and other compounds	t
Hydrogen, methane, ammonia, cyanides	Electric discharge	Amino acid mixtures	u
Carbon dispersed in aqueous ammonia	Contact glow discharge electrolysis	Six amino acids, together with urea	v
Formaldehyde, hydroxylamine containing H_2O_2 and $Fe_2(SO_4)_3$	$Fe^{3+} - H_2O_2$ (chemical energy)	Glycine, serine, threonine, proline	w
Alkyl cyanides in water	[^{60}Co]-γ-Irradiation (MultiKrad doses)	Glycine from MeCN; alanine from EtCN; up to nine other amino acids	x, y
Ammonium cyanide in water	[^{60}Co]-γ-Irradiation	Condensation product releases amino acids on hydrolysis	z
Glucose, mannose or arabinose in aq. nitrate solution under O_2, N_2 or CO_2	UV irradiation	Amino acid mixtures	aa, bb, cc
Cyanamide–KNO_2 reaction products formed in aq. solutions	Aged solutions	Amino acids, nucleosides, and other organic compounds	dd
Methane, ammonia, ammonium chloride over TiO_2	Xe lamp irradiation	Glycine, alanine, serine, aspartic acid, glutamic acid	ee, ff
HCN oligomers such as di-aminomaleonitrile in water	Heat	Up to eleven amino acids formed on hydrolysis	gg, hh
Carbon vapour (carbon arc) in NH_3 at $-196°C$		Glycine, alanine, β-alanine, *N*-methylglycine, serine, and aspartic acid after hydrolysis	ii

[a] Khare, B.N. and Sagan, C. (1971) *Nature*, **232**, 577.
[b] Yuasa, S., Ishigami, M., Honda, Y. and Imahori, K. (1970) *Sci. Rep. Osaka Univ.*, **19**, 33.
[c] Yuasa, S., Yamamoto, M., Honda, Y., *et al.* (1970) *Sci. Rep. Osaka Univ.*, **19**, 7.
[d] Van-Trump, J.E. and Miller, S.L. (1972) *Science*, **178**, 859.
[e] Ferris, J.P. and Chen, C.T. (1975) *J. Am. Chem. Soc.*, **97**, 2962.
[f] Lawless, J.G. and Boynton, C.D. (1973) *Nature*, **243**, 405.
[g] Mukhin, L.M., Bondarev, V.B., Kalinichenko, V.I., *et al.* (1976) *Doklady Akad. Nauk S.S.S.R.*, **226**, 1225.
[h] Bar-Nun, A. (1975) *Origins of Life*, **6**, 109.

Footnotes to Table 8.11 (*Contd.*)

[i] Poncelet, G., van Assche, A.T. and Fripiat, J.J. (1975) *Origins of Life*, **6**, 401.
[j] Zhigunova, L.N., Manuilova, G.V. and Petryaev, E.P. (1974) *Vestsi Akad. Navuk B.S.S.R., Ser. Fiz. Energ. Navuk*, 33 (*Chem. Abstr.*, 1974, **81**, 152607).
[k] Bahadur, K., Verma, M.L. and Sing, Y.P. (1974) *Z. Allg. Mikrobiol.*, **14**, 87.
[l] Barak, I. and Bar-Nun, A. (1975) *Origins of Life*, **6**, 483.
[m] Bahadur, K. and Sen. P. (1975) *Z. Allg. Mikrobiol.*, **15**, 143.
[n] Sokolskaya, A.V. (1976) *Origins of Life*, **7**, 183.
[o] Sokolskaya, A.V. (1978) *Zhur. Obshchei Khim.*, **48**, 140.
[p] Kamaluddin, Yanagawa, H. and Egami, F. (1979) *J. Biochem., Tokyo*, **85**, 1503.
[q] Egami, F. (1981) *Origins of Life*, **11**, 197.
[r] Sweeney, M.A., Toste, A.P. and Ponnamperuma, C. (1976) *Origins of Life*, **7**, 187.
[s] Ranganayaki, S. and Srivastava, M. (1979) *Zhur. Org. Khim.*, **15**, 1124.
[t] Draganic, Z.D., Draganic, I.G. and Jovanovic, S.V. (1978) *Radiation Res.*, **75**, 508.
[u] Kalinichenko, V.I., Bondarev, V.B., Gerasimov, M.V., *et al.* (1977) *Dokl. Akad. Nauk S.S.S.R., Ser. Khim.*, **236**, 245.
[v] Harada, K. and Suzuki, S. (1977) *Nature*, **266**, 275.
[w] Matatov, Yu.I. (1980) *Zh. Evol. Biokhim. Fiziol.*, **16**, 189; (*Chem. Abs.*, **94**, 42850).
[x] Draganic, Z.D., Draganic, I.G. Shimoyama, A. and Ponnamperuma, C. (1977) *Origins of Life*, **8**, 377.
[y] Draganic, Z.D., Draganic, I.G. Shimoyama, A. and Ponnamperuma, C. (1977) *Origins of Life*, **8**, 371.
[z] Draganic, Z.D., Niketic, V., Jovanovic, S. and Draganic, I.G. (1980) *J. Mol. Evol.*, **15**, 239.
[aa] Khenokh, M.A. and Nikolaeva, M.V. (1977) *Stud. Biophys.*, **63**, 1.
[bb] Khenokh, M.A. and Nikolaeva, M.V. (1977) *Zh. Evol. Biokhim. Fiziol.*, **13**, 105; (*Chem. Abs.*, **86**, 184776).
[cc] Yanagawa, H., Kobayashi, Y. and Egami, F. (1980) *J. Biochem. Tokyo*, **87**, 359.
[dd] Wollin, G. and Ryan, W.B.F. (1979) *Biochim. Biophys. Acta*, **584**, 493.
[ee] Reiche, H. and Bard, A.J. (1979) *J. Am. Chem. Soc.*, **101**, 3127.
[ff] Dunn, W.W., Aikawa, Y. and Bard, A.J. (1981) *J. Am. Chem. Soc.*, **103**, 6893.
[gg] Moser, R.E. and Matthews, C.N. (1968) *Experientia*, **24**, 658.
[hh] Moser, R.E., Claggett, A.R. and Matthews, C.N. (1968) *Tetrahedron Lett.*, 1599.
[ii] Shevlin, P.B., McPherson, D.W. and Melius, P. (1981) *J. Am. Chem. Soc.*, **103**, 7006.

Table 8.12 Miscellaneous reactions applied to the synthesis of α-amino acids

Reaction type	*Illustrative examples*	*Product*	*Reference*
Oxo-type reaction	$R^1CH{=}CH_2 + R^2CONH_2 \xrightarrow{110\text{–}130°C} R^2CONHCH(CH_2CH_2R^1)CO_2H$ + 2CO + H_2 with $Co_2(CO)_8$ (26–80% yields)		a
Hydroformylation	$R^1CHO + CO + R^2CONH_2 \xrightarrow[EtOAc]{110\text{–}130°C} R^2CONHCHR^1CO_2H$ (21–84%; $Co_2(CO)_8$, dioxan)		a, b
Ethyl cyanoformate and active methylene compound with $ZnCl_2 - Et_3N$	$EtO_2CCN + CH_2XY \rightarrow (H_2N)(EtO_2C)C{=}CXY$	αβ-Unsaturated α-amino acids	c
Ene reaction of alkenes with *N*-arenesulphonyl-imines	$R^3C(=CHR^1)\text{–}CH(R^2)\text{–}H$ + $BuO_2C\text{–}CH{=}N\text{–}SO_2Ar$ $\xrightarrow[\text{or } SnCl_4,\ BF_3 \text{ or } AlCl_3]{120°C}$ $R^2CH{=}C(R^3)\text{–}CHR^1\text{–}CH(NHSO_2Ar)\text{–}CO_2Bu$		d

Table 8.12 (*Contd.*)

Reaction type	*Illustrative examples*	*Product*	*Reference*
Ene reaction of enols (e.g. $R^1CH=CR^2-OH$) with *N*-arenesulphonylimines	Similar to previous entry	δ-Oxo-α-amino acids	e

[a] Wakamatsu, H., Uda, J. and Yamakami, N. (1971) *Chem. Commun.*, 1540.
[b] Parnaud, J.J., Campari, G. and Pino, P. (1979) *J. Mol. Catal.*, **6**, 341.
[c] Iimori, T., Nii, Y., Izawa, T., *et al.* (1979) *Tetrahedron Lett.*, 2525.
[d] Achmatowicz, O. and Pietraszkiewicz, M. (1981) *J. Chem. Soc. Perkin 1*, 2680.
[e] Achmatowicz, O. and Pietraszkiewicz, M. (1981) *Tetrahedron Lett.*, **22**, 4323.

Table 8.13 Synthesis of α-amino acids by alkylation of derivatives of glycine and of other protein amino acids*

Glycine derivative	*Alkylation conditions*	*Synthetic objective*	*Reference*
NN-bis(Trimethysilyl)-glycine trimethysilyl ester	Base, $NaN(SiMe_3)_2$, then alkyl halide. Remove silyl groups with aq. acid	General synthesis of α-amino acids	a
	Base then aldehyde (non-enolizable)	β-Hydroxy-α-amino acids	b
	$HN(SiMe_3)_2/NaNH_2$ then $NO_2(CH_2)_4Br$ then aqueous acid	2-Amino-6-nitrocaproic acid	c
	Ketones + LDA	L-3-(3-Deoxy-1,2:5,6-di-*O*-isopropylidene-α-allofuranos-3-yl)alanine; *erythro*-β-hydroxy-α-amino acids	d
Hippuric acid (*N*-benzoylglycine)	Di/tri-anion formed using LDA/TMEDA, then add alkyl halide	Representative amino acids	e
N-Benzyloxycarbonyl-glycine ethyl ester	Ketone + LDA	*threo*-β-Hydroxy-α-amino acids	f
Copper(II) glycinate	Aldehydes	β-Hydroxy-α-amino acids	g
Formylglycine esters	Enamines, such as 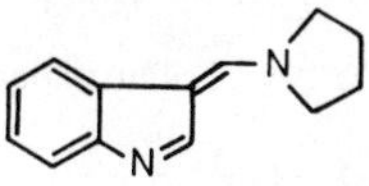	αβ-Dehydrotryptophans	h
N,N-Dibenzylglycine esters	LDA, then $MgBr_2$, Me_2CO	β-Hydroxyvaline	i
Formylaspartic acid α,β-di-*t*-butyl ester	LDA (−78°), alkyl bromide	Mixture of α- and β-alkylaspartates without racemization; (2*S*,3*R*)-β-methylaspartic acid actually obtained	j

Table 8.13 (*Contd.*)

Glycine derivative	*Alkylation conditions*	*Synthetic objective*	*Reference*
Ethoxycarbonyl (trimethylsilylethynyl)glycine methyl ester EtOCONHCH(C≡C $SiMe_3$)CO_2Me	LDA/HMPA then alkyl bromide	α-Ethynyl-α-amino acids	k,l

* See also Tables 8.7 and 8.14.

[a] Rühlmann, K. and Kuhrt, G. (1968) *Angew. Chem. Internat. Edit.*, **7**, 809.

[b] Rühlmann, K., Kaufmann, K.D. and Ickert, K. (1970) *Z. Chem.*, **10**, 393.

[c] Bayer, E. and Schmidt, K. (1973) *Tetrahedron Lett.*, 2051.

[d] Rosenthal, A. and Brink, A.J. (1976) *Carbohydrate Res.*, **46**, 289.

[e] Krapcho, A.P. and Dundulis, E.A. (1976) *Tetrahedron Lett.*, 2205.

[f] Shanzer, A., Somekh, L. and Butina, D. (1979) *J. Org. Chem.*, **44**, 3967.

[g] Otani, T.T. and Briley, M.R. (1974) *J. Pharm. Sci.*, **63**, 1253.

[h] Moriya, T., Yoneda, N., Miyoshi, M. and Matsumoto, K. (1982) *J. Org. Chem.*, **47**, 94.

[i] Scott, A.I. and Wilkinson, T.J. (1981) *J. Labelled Compd. Radiopharm.*, **18**, 347.

[j] Seebach, D. and Wasmuth, D. (1981) *Angew. Chem.*, **93**, 1007.

[k] Casara, P. and Metcalf, B.W. (1978) *Tetrahedron Lett.*, 1561.

[l] Casara, P. and Metcalf, B.W. (1979) *J. Chem. Soc., Chem. Commun.*, 119.

Table 8.14 Amino acids as starting materials for the synthesis of other amino acids*

Starting material (invariably N- and C-protected)	*Reagents and reaction conditions*	*Synthetic objective*	*Reference*
DL-Serine	Conventional substitution, cyclization	DL-Cystine	a
L-Tyrosine		L-Phenylalanine	b
L-Ornithine		L-Proline	c
L-Histidine	Bamberger cleavage of methyl ester, and catalytic hydrogenation to 2,4,5-triamino-pentanoic acid methyl ester; guanidation	(2*S*,4*R*)-Enduracidine and allo-isomer	d
L-Methionine	Chlorinolysis	2-Amino-3,4,4,4-tetrachlorobutanoic acid and 4,4,4-trichloro-analogue	e
L-Glutamic acid	via *N*-tosyl-L-glutamine ^{i}Pr ester	L-Ornithine (large-scale)	f
Phenylglycine	*N*-Trifluoroacetylation, conversion into 2-trifluoromethyloxazolin-5-one, rearrangement into 2-methoxycarbonyl homologue via 5-methoxycarbonyloxyoxazole, then to 2-phenyl-4-trifluoromethyl-5-methoxyoxazole and treatment with HBr/AcOH	Trifluoroalanine	g

Table 8.14 (*Contd.*)

Starting material (invariably N- and C-protected)	*Reagents and reaction conditions*	*Synthetic objective*	*Reference*
β-Cyano-L-alanine (from asparagine)	H_2S in pyridine	Thioasparagine	h
L-β-Chloroalanine (from L-aspartic acid via its β-hydrazide[i])	MeCH=CHSLi	*S*-(trans-Prop-l-enyl)-L-cysteine	j, k
	$NaSe(CH_2)_2NH_2$	4-Seleno-L-lysine	l
	Dibenzyl malonate	γ-Carboxy-L-glutamic acid	m
DL-β-Chloroalanine	Sodium anti-benzaldoximate	Alanosine (β-hydroxy-nitrosamino) alanine	n
O-(Toluene-*p*-sulphonyl)-L-serine	Sodium pyrimidine-2-mercaptide	*S*-(Pyrimidin-2-yl)-L-cysteine	o
	Di-*t*-butyl malonate and LDA	γ-Carboxy-L-glutamic acid	p, q
N-Boc-Amino acids	via *N*-chloro-derivatives (Bu^tOCl) thence to α-methoxy-α-amino acids and DABCO elimination	'Dehydro-amino acids' (αβ-unsaturated α-amino acids)	r
Dehydro-alanine	Michael addition with di-t-butyl malonate	γ-Carboxy-L-glutamic acid	s
4-Bromo-2-amino-butanoic acid, from L-homoserine	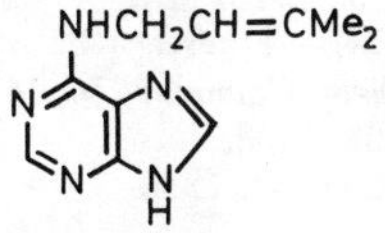 $NHCH_2CH{=}CMe_2$ (purine structure: N, N, N, N, H)	Discadenine	t, u
O-Acylserines, threonines	Ph_3P/Diethyl azo-dicarboxylate	Corresponding 'dehydro-amino acids'	v
L-Lysine	NaOCl gives the N^6,N^6-dichloro-derivative, DABCO gives the nitrile	L-α-Amino-adipic acid	x
L-Glutamic acid	$ClCO_2CH_2Ph$/LDA	γ-Carboxy-L-glutamic acid	y
L-Pyroglutamic acid	4-Bis(dimethylamino)-methylene derivative formed using $(Me_2N)_2CHOBu^t$, treated with $ClCO_2CH_2CCl_3$	γ-Carboxy-L-glutamic acid	z
L-Pyroglutamic acid	via 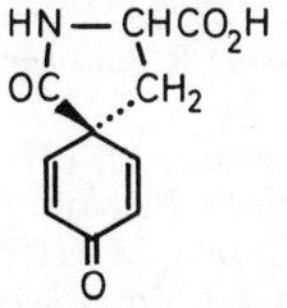 (structure: HN—$CHCO_2H$, OC, CH_2, O)	'Pretyrosine' and tyrosine	aa, bb
L-Histidine	Cleavage to δ-oxo-ornithine, then $NaBH_4$ reduction	δ-Hydroxyornithine (68% *erythro*, 32% *threo*)	cc
Pyrrolidin-2-one-5,5-bis-carboxylates	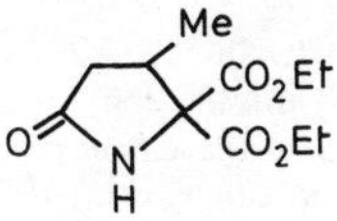 (structure: Me, CO_2Et, CO_2Et, O, N, H)	*threo*- and *erythro*-γ-Methyl-DL-glutamic acids	dd

Table 8.14 (*Contd.*)

Starting material (invariably N- and C-protected)	*Reagents and reaction conditions*	*Synthetic objective*	*Reference*
L-Asparagine	$-CONH_2 \rightarrow -COCH_2OAc \rightarrow$ [epoxide: $-C(CH_2)(CH_2OAc)-O$] $\rightarrow$ $-C(CH_2OH)(OH)(CH_2OH)$	$\gamma\delta\delta'$-Trihydroxy-L-leucine	ee
L-Asparagine	Hofmann rearrangement ($-CONH_2 \rightarrow -NH_2$)	L-2,3-Di-aminopropanoic acid	ff
L-Lysine	$h\nu/Cl_2$ then hydrolysis	*threo*-γ-Hydroxy-L-lysine	gg
L-Glutamic acid	$h\nu/Cl_2 \rightarrow \beta$-Chloro-L-glutamic acid (*erythro* : *threo* = 1:1) then heterocyclic ring construction	'AT-125' [structure: 3-chloro-4,5-dihydroisoxazol-5-yl, H, H, CO_2^-, $\overset{+}{N}H_3$]	hh

*See also Tables 8.13 and 8.19.

[a] Rambacher, P. (1968) *Chem. Ber.*, **101**, 2595.
[b] Kishi, T., Kato, Y. and Tanaka, M. (1968) *J. Agric. Chem. Soc. Japan*, **42**, 238.
[c] Ohshiro, S., Kuroda, K. and Fujita, T. (1967) *J. Pharm. Soc. Japan*, **87**, 1184.
[d] Tsuji, S., Kusumoto, S. and Shiba, I. (1975) *Chem. Lett.*, 1281.
[e] Urabe, Y., Okawara, T., Okukura, K. *et al.* (1974) *Synthesis*, 440.
[f] Gut, V. and Poduska, K. (1971) *Coll. Czech. Chem. Commun.*, **36**, 3470.
[g] Höfle, G. and Steglich, W. (1971) *Chem. Ber.*, **104**, 1408.
[h] Ressler, C. and Banerjee, S.N. (1976) *J. Org. Chem.*, **41**, 1336.
[i] Fischer, E. and Raske, K. (1907) *Ber.*, **40**, 3717.
[j] Okumura, K., Iwasaki, T., Okawara, T. and Matsumoto, K. (1972) *Bull. Inst. Chem. Res. Kyoto Univ.*, **50**, 209.
[k] Nishimura, H., Mizuguchi, A. and Mizutani, J. (1975) *Tetrahedron Lett.*, 3201.
[l] De Marco, C., Rinaldi, A., Dernini, S. and Cavallini, D. (1975) *Gazzetta*, **105**, 1113.
[m] Weinstein, B., Watrin, K.G., Loie, J.H. and Martin, J.C. (1976) *J. Org. Chem.*, **41**, 3634.
[n] Eaton, C.N., Denney, G.H., Ryder, M.A., *et al.* (1976) *J. Medicin. Chem.*, **16**, 289.
[o] Holy, A., Votruba, I. and Jost, K. (1974) *Coll. Czech. Chem. Commun.*, **39**, 634.
[p] Boggs, N.T., Gawley, R.E., Koehler, K.A. and Hiskey, R.G. (1975) *J. Org. Chem.*, **40**, 2850.
[q] Boggs, N.T., Goldsmith, B., Gawley, R.E., *et al.* (1979) *J. Org. Chem.*, **44**, 2262.
[r] Poisel, H. and Schmidt, U. (1976) *Angew. Chem. Internat. Edit.*, **15**, 294.
[s] Bajusz, S. and Juhasz, A. (1976) *Acta Chim. Acad. Sci. Hung.*, **88**, 157.
[t] Uchiyama, M. and Abe, H. (1977) *Agric. Biol. Chem.*, **41**, 1549.
[u] Seela, F. and Hasselmann, D. (1979) *Chem. Ber.*, **112**, 3072.
[v] Wojciechowska, H., Pawlowicz, R., Andruszkiewicz R. and Grzybowska, J. (1978) *Tetrahedron Lett.*, 4063.
[x] Scott, A.I. and Wilkinson, T.J. (1980) *Synth. Commun.*, **10**, 127.
[y] Zee-Cheng, R.K.-Y. and Olsen, R.E. (1980) *Biochem. Biophys. Res. Commun.*, **94**, 1128.
[z] Danishefsky, S., Berman, E., Clizbe, L.A. and Hirama, M. (1979) *J. Am. Chem. Soc.*, **101**, 4385.
[aa] Danishefsky, S., Morris, J. and Clizbe, L.A. (1981) *J. Am. Chem. Soc.*, **103**, 1602.
[bb] Danishefsky, S. Morris, J. and Clizbe, L.A. (1981) *Heterocycles*, **15**, 1205.
[cc] Mizusaki, K. and Makisumi, S. (1981) *Bull. Chem. Soc. Japan*, **54**, 470.
[dd] Mauger, A.B. (1981) *J. Org. Chem.*, **46**, 1032.
[ee] Weygand, F. and Mayer, F. (1968) *Chem. Ber.*, **101**, 2065.
[ff] Waki, M., Kitajima, Y. and Izumiya, N. (1981) *Synthesis*, 266.
[gg] Fujita, Y., Kollonitsch, K. and Witkop, B. (1965) *J. Am. Chem. Soc.*, **87**, 2030.
[hh] Silverman, R.B. and Holladay, M.W. (1981) *J. Am. Chem. Soc.*, **103**, 7357.

Table 8.15 α-Amino acids from α-hydroxyglycine and related derivatives

Starting material	*Reagent*	*Product*	*Reference*
α-Hydroxyhippuric acid (adduct of HO_2CCHO and $PhCONH_2$)	Benzene and derivatives with $MeSO_3H$	*C*-Phenylglycine and homologues (yields 20% upwards)	a–c
	Benzyl chlorides	*p*-Substituted phenylglycines	d
	Alkenes $R^1R^2C{=}CH_2$	Leucine, aspartic acid semi-aldehyde; also 2-phenyloxazines	e
	Active methylene compounds	Aspartic acid analogues	f
	RCH_2CO_2Et	*β*-Amino aspartic acids *β*-Alkyl aspartic acids	g
α-Acetoxy-*N*-acetylglycine	Phenyl ethers with $SnCl_4$	4-Methoxyphenyglycine	h
	Elimination of AcOH	'Dehydro-amino acids'	h
	RSH	α-Alkanethio glycines	h
N-Acyl-α-chloroglycine esters (from α-hydroxy analogues)	RSH, $PhNH_2$, RCO_2H, MeOH	α-Hetero-atom substituted glycines	i

[a] Ben-Ishai, D., Sataty, I. and Bernstein, Z. (1976) *Tetrahedron*, **32**, 1571.
[b] Ben-Ishai, D., Sataty, I. and Bernstein, Z. (1977) *Tetrahedron*, **33**, 3261.
[c] Schouteeten, A., Christidis, Y. and Mattioda, G. (1978) *Bull. Soc. Chim. Fr.*, 248.
[d] Ben-Ishai, D., Altman, J. and Peled, N. (1977) *Tetrahedron*, **33**, 2715.
[e] Ben-Ishai, D., Moshenberg, R. and Altman, J. (1977) *Tetrahedron*, **33**, 1533.
[f] Ben-Ishai, D., Altman, J., Bernstein, Z. and Peled, N. (1978) *Tetrahedron*, **34**, 467.
[g] Ozaki, Y., Iwasaki, T., Miyoshi, M. and Matsumoto, K. (1979) *J. Org. Chem.*, **44**, 1714.
[h] Ozaki, Y., Iwasaki, T., Horikawa, K., Miyoshi, M. and Matsumoto, K. (1979) *J. Org. Chem.*, **44**, 391.
[i] Matthies, D., Bartsch, B. and Richter, H. (1981) *Arch. Pharm.*, **314**, 209.

Table 8.16 Synthesis of α-amino acids from α-azidoalkanoates, α-diazoalkanoates, and related compounds

Reactant	*Reaction conditions*	*Product*	*Reference*
Ethyl 2-azido-2-alkenoates	Al/Hg reduction	Aliphatic α-amino acids	a
[structure: 3-azido-azetidin-2-one bearing R^1 and NR^2]	H_2/Pd	$H_2NCH(CH_2R^1)$-$CONHR^2$ (used for the synthesis of dopa and analogues)	b,c
[structure: ethyl 2,3-dioxopiperidine-4-carboxylate → (i) Br_2 (ii) NaN_3 (iii) $MeCO_3H$ → ethyl 3-azido-2-oxopyrrolidine-3-carboxylate]	Diborane (CONH → CH_2NH) then H_2/Pt followed by ester hydrolysis	(±)-Cucurbitine	d

Table 8.16 (*Contd.*)

Reactant	*Reaction conditions*	*Product*	*Reference*
Ethyl α-azido-β-hydroxypropionate	From ethyl α-bromo-β-hydroxy-propionate under phase transfer catalysis with 18-crown-6 catalyst	Serine (also some iso-serine in absence of 18-crown-6)	e
$N_3CH(CO_2Me)\bar{C}H(CO_2Me)_2$	H_2/Ni	DL-β-Carboxyaspartic acid	f
Ethyl diazoacetate N_2CHCO_2Et	MeCN, hν → oxazole (CF_3, CO_2Et, Me substituted); H_2/Pt →	$AcNHCH(CH_2CF_3)CO_2Et$	g

[a] Shin, C., Yonezawa, Y. and Yoshimura, J. (1976) *Chem. Lett.*, 1095.
[b] Ojima, I., Suga, S. and Abe, R. (1980) *Chem. Lett.*, 853.
[c] Ojima, I., Suga, S. and Abe, R. (1980) *Jpn. Kokai Tokkyo Koho*, **81**, 32443; (*Chem. Abs.*, **95**, 151158).
[d] Monteiro, H.J. (1973) *J. Chem. Soc. Chem. Commun.*, 2.
[e] Nakajima, K., Kinishi, R., Oda, J. and Inouye, Y. (1977) *Bull. Chem. Soc., Japan*, **50**, 2025.
[f] Christy, M.R., Barkley, R.M., Koch, T.H., *et al.* (1981) *J. Am. Chem. Soc.*, **103**, 3935.
[g] Steglich, W., Heninger, H.U., Dworschak, H. and Weygand, F. (1967) *Angew. Chem. Int. Ed.*, **6**, 807.

Table 8.17 Synthesis of α-amino acids via oxazolin-5-ones ('azlactones'), thiazolin-5-ones, and imidazolin-5-ones

Starting material	*Reagent*	*Product*	*Reference*
Oxazolin-5-ones (ring: R^1 at C-2, R^2, R^3 at C-4, C=O at C-5)			
4-Isopropylidene-2-phenyl-oxazolin-5-one	MeMgI/CuCl → 4-*t*-Butyl-2-phenyl-oxazolin-5-one	*t*-Leucine	a
4-Substituted oxazolinones ($R^3 = H, R^2$ = alkyl/aryl)	Base + RX then hydrolysis	αα-Disubstituted α-amino acids	b
	Michael addition to $PhSO_2C \equiv CH$ (Et_3N as base)	α-Vinyl- α-amino acids	c
	Formaldehyde	α-Hydroxymethyl-α-amino acids	d
2,4-Dimethyl-4-alkoxycarbonyl-oxazolin-5-one	ArMgX then $NaBH_4$	β-Aryl-α-methylserines (erythro-isomer predominates at low temps)	e, f
Oxazolin-5-one trimethylsilyl enol ethers (alias 5-trimethyl-silyloxyoxazoles)	Ketone + $SnCl_4$ → 2-phenyl-4-(CR^1R^2)-oxazolin-5-one →	Dehydro-amino acids $PhCONHC(=CR^1R^2)CO_2H$	g

Table 8.17 (*Contd.*)

Starting material	*Reagent*	*Product*	*Reference*
Thiazolin-5-ones			
2-Phenyl-4-substituted thiazolin-5-ones	Maleic anhydride and other electron-deficient alkenes; Michael addition then hydrolysis	αα-Dialkyl-α-amino acids	h
2-Thiothiazolidin-5-one	Ketone/then HI/P or Zn/AcOH followed by HCl/MeOH	$H_3\overset{+}{N}CH(CHR^1R^2)CO_2^-$	i
Imidazolin-5-ones			
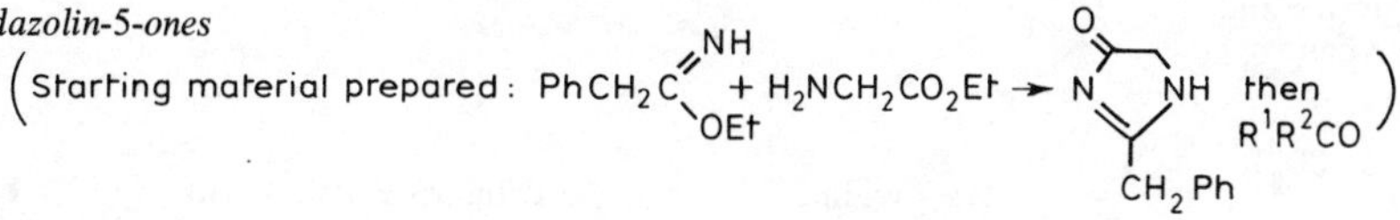 (Starting material prepared: $PhCH_2C(=NH)OEt + H_2NCH_2CO_2Et \rightarrow$ imidazolinone (4-CH_2Ph) then R^1R^2CO)			l
2-Phenyl-4-arylidene-imidazolin-5-ones	Zn/aq. KOH/reflux 1 h	*N*-Benzyl phenylalanines	j
2-Benzyl-4-alkylidene-imidazolin-5-ones	H_2/Pd	β-Methyl-leucine	k
	H_2/Pd then methylation	*threo-N*,β-Dimethyl-leucine	
Benzodiazepinones			
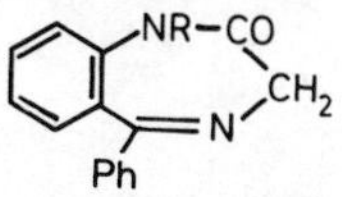 (benzodiazepinone: NR–CO, CH_2, =N, Ph)	Base, alkyl halide followed by alkaline hydrolysis	β-Fluoroalanine, α-^{2}H-α-amino acids	m

[a] Miyazawa, T., Nagai, T., Yamada, T., *et al.* (1979) *Mem. Konan Univ., Sci. Ser.*, **23**, 51; (*Chem. Abs.*, 1980, **92**, 94681).
[b] Kuebel, B., Gruber, P., Hurnaus, R. and Steglich, W., (1979) *Chem. Ber.*, **112**, 128.
[c] Steglich, W. and Wegmann, H. (1980) *Synthesis*, 481.
[d] Leplawy, M.T. and Olma, A. (1979) *Pol. J. Chem.*, **53**, 353.
[e] Pines, S.H. Karady, S. and Sletzinger, M. (1968) *J. Org. Chem.*, **33**, 1758.
[f] Pines, S.H., Karady, S., Kozlowski, M.A. and Sletzinger, M. (1968) *J. Org. Chem.*, **33**, 1762.
[g] Takagaki, H., Tanabe, S., Asaoka, M. and Takei, H. (1979) *Chem. Lett.*, 347.
[h] Barrett, G.C. and Walker, R. (1976) *Tetrahedron*, **32**, 571, 577; see also G.C. Barrett, *Tetrahedron*, 1980, **36**, 2023.
[i] Billimoria J.D. and Cook, A.H. (1949) *J. Chem. Soc.*, 2323.
[j] Devasia, G.M., and Pillai, C.R. (1975) *Tetrahedron Letters*, 4051.
[k] Kotake, H., Saito, T. and Okubo, K. (1969) *Bull. Chem. Soc. Japan*, **42**, 1367.
[l] Lehr, H., Karlan, S. and Goldberg, M.W. (1953) *J. Am. Chem. Soc.*, **75**, 3640.
[m] Decorte, E., Tosc, R., Sega, A., Sunjic, V., Ruzic-Toros, Z., Kojic-Prodic, B., Bresciani-Pahor, N., Nardin, G. and Randoccio, L. (1981) *Helv. Chim. Acta*, **64**, 1145.

Table 8.18 Aziridines and azirines in α-amino acid synthesis*†

Aziridine or azirine	*Reagents*	*Product*	*Reference*
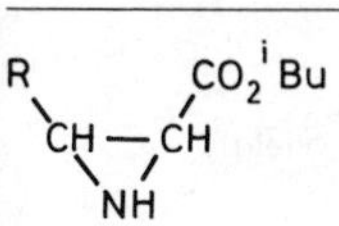 R–CH–CH–$CO_2{}^iBu$ (aziridine, NH)	70% HF/pyridine	$RCHFCH(NH_2)CO_2{}^iBu$ (Relative stereochemistry not defined but assumed to be the result of *trans*-addition)	a, b

Table 8.18 (*Contd.*)

Aziridine or azirine	*Reagents*	*Product*	*Reference*
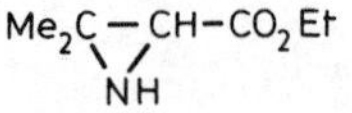 $Me_2C-CH-CO_2Et$ (aziridine, NH) (from ethoxycarbonylnitrene + ethyl *ββ*-dimethylacrylate)	AcOH then base hydrolysis	*β*-Methylthreonine	c
N-Benzyloxycarbonyl-aziridine-2-carboxylates, from *N*-benzyloxy-carbonyl-2-chloroglycine and diazomethane	$MeSH-BF_3$ $MeOH-H^+$	*S*-Methylcysteine *O*-Methylserine	d d
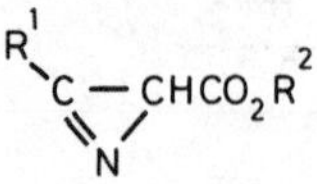 $R^1C=N-CHCO_2R^2$ (azirine) from $R^1C(N_3)=CHCO_2R^2$	HF/pyridine	*ββ*-Difluoro-*α*-amino acids	e
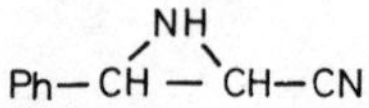 $Ph-CH-CH-CN$ (aziridine, NH)	HF/pyridine	*β*-Fluorophenylalanine	f

*See also Table 8.24.
† Azetidinones for the same purpose: see Table 8.16.
[a] Wade, T.N. and Kheribet, R. (1980) *J. Chem. Res.* (*S*), 210.
[b] Wade, T.N., Gaymard, F. and Guedj, R. (1979) *Tetrahedron Lett.*, 2681.
[c] Berse, C. and Bessette, P. (1971) *Canad. J. Chem.*, **49**, 2610.
[d] Bernstein, Z. and Ben-Ishai, D. (1977) *Tetrahedron*, **33**, 881.
[e] Wade, T.N. and Guedj, R. (1979) *Tetrahedron Lett.*, 3953.
[f] Ayi, A.I., Remli, M. and Guedj, R. (1981) *J. Fluorine Chem.*, **18**, 93.

Table 8.19 Synthesis of prolines and related imino acids*

Starting material	*Reagents and intermediates*	*Synthetic objective*	*Reference*
Pyrrole-2-carboxylic acid	$H_3PO_2 + HI$	3,4-Dehydroproline	a, b
Pyrrolidine	Via *N*-chloropyrrolidine, formed using Bu^tOCl, de-hydrochlorination (NaOMe), add HCN and hydrolyse	DL-Proline	c
L-Pyroglutamic acid (2-Oxopyrrolidine-5*S*-carboxylic acid)	Sequence $-CONH- \rightarrow -C(OEt)=N-$ with $Et_3O^+BF_4^-$ then $NaBH_4$[d]; or P_4S_{10} then Raney nickel[r]	L-Proline	d
D-Glutamic acid	via 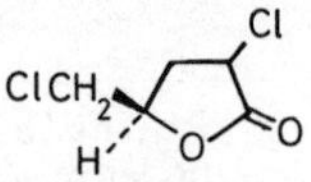 $ClCH_2$-substituted 3-chloro-γ-butyrolactone (H, Cl, O)	Hydroxy-L-proline and D-allohydroxyproline	e
Glyoxal, oxaloacetic acid, NH_3	Reaction in water at room temperature; followed by $NaBH_4$ reduction	4-Hydroxyproline (40% yield)	f

Table 8.19 (*Contd.*)

Starting material	*Reagents and intermediates*	*Synthetic objective*	*Reference*
Dieckmann cyclization Glycine methyl ester added to MeCH =$CHCO_2Me$, then $ClCO_2Me$	→ MeCH(CH_2CO_2Me)–N(CO_2Me)–CH_2CO_2Me	CH_2–CHOH, MeCH, $CHCO_2Me$, N, CO_2Me → 3-Hydroxy-5-methylproline (after hydrolysis)	g
4-Cyano-2,2-dimethyl butyraldehyde	→ (Cl, N–H, O) $\xrightarrow{OH^-}$ ($\overset{+}{N}H_2$, CO_2^-)	4,4-Dimethylproline	h
Intramolecular ene reaction	CH_2–H (2), $CHCO_2Et$, CO_2Et, CO_2Et, N, $COCF_3$ $\xrightarrow[10\ min]{170°}$ CH_2CO_2Et, CO_2Et, CO_2Et, N, $COCF_3$ → $\overset{+}{N}H_2$, CO_2^- kainic acid		i,j
4-Hydroxy-L-proline	Via dioxopiperazine; oxidation to ketone, then SF_4-HF	4,4-Difluoro-L-proline	h
trans-4-Bromoproline	Via *N*-toluene-p-sulphonyl butyl ester, reaction with purine anions	*cis*-4-(*N*-9-Adeninyl)-(*N*-,9-guanin-4-yl)-L-proline	k
cis-4-Aminoproline	Via *N*-toluene-p-sulphonyl butyl ester, construction of pyrimidine	*cis*-4-(*N*-9-Hypoxanthin-4-yl)-L-proline	l
L-Isoleucine	Via Cl–CH_2–CH_2–C(H)(CH_3)–C(H)(CO_2H)–$\overset{+}{N}H_3$ formed by photochlorination	*trans*-3-Methyl-L-proline (allo-isoleucine gives the *cis* isomer)	m
N-(Pyruvylideneglycinato)-copper(II) Michael addition to acrylonitrile, or methyl acrylate	Via RCH–CH_2, MeC, CH, OC, NH, CO, O, Cu, O, pyridine	4-Substituted prolines	n
L-Amino acid esters (norvaline, allo-isoleucine, isoleucine, and leucine)	Hofmann–Loeffler–Freytag cyclization (*N*-chloro-amino acid ester in 85% H_2SO_4; UV irradiation during 40 hours)	3- and 4-Methyl-L-prolines	o
Pyrrolidone ^{1}H, ^{2}H, C, CH_2–CH_2, CO, NH (from L-glutamic acid by enzymic decarboxylation in ^2H_2O)	$(CO_2Et)_2$ then hydrolysis and resolution	(2*S*,5*R*)-[5-^{2}H]-Proline	p

Table 8.19 (*Contd.*)

Starting material	*Reagents and intermediates*	*Synthetic objective*	*Reference*
3-Substituted pyrroline-2-carboxylic acids (ring structure with R and CO_2H)	R = Br → R = substituent; $NaBH_4$ or amine-borane	*cis*- or *trans*-3-Substituted prolines, respectively	q, r
L-3-Pyrroline-2-carboxylic acid (ring structure with CO_2H, H, NH)	F_3CCO_3H gives exo/endo-epoxide mixture; hydrolysed	Stereoisomers of 3,4-dihydroxyproline	s

* See also Tables 8.1. and 8.14.
[a] Corbella, A., Garibaldi, P., Jommi, G. and Mauri, F. (1969) *Chem. and Ind.*, 583.
[b] Scott, J.W., Focella, A., Hengartner, U.O., Parrish, D.R. and Valentine, D. (1980) *Synth. Commun.*, **10**, 529.
[c] Schmidt, U. and Poisel, H. (1977) *Angew. Chem., Internat. Edit.*, **16**, 777.
[d] Monteiro, H.J. (1974) *Synthesis*, 137.
[e] Eguchi, C. and Kakuta, A. (1974) *Bull. Chem. Soc. Japan*, **47**, 1704.
[f] Ramaswamy, S.G. and Adams, E. (1977) *J. Org. Chem.*, **42**, 3440.
[g] Mauger, A.B., Stuart, O.A., Katz, E. and Mason, K.T. (1977) *J. Org. Chem.*, **42**, 1000.
[h] Shirota, F.N., Nagasawa, H.T. and Elberling, J.A. (1977) *J. Med. Chem.*, **20**, 1176.
[i] Oppolzer, W. and Andres, H. (1978) *Tetrahedron Lett.*, 3397.
[j] Oppolzer, W. and Andres, H. (1979) *Helv. Chim. Acta*, **62**, 2282.
[k] Kaspersen, F.M. and Pandit, U.K. (1975) *J. Chem. Soc. Perkin Trans. 1*, 1617.
[l] Kaspersen, F.M. and Pandit, U.K. (1975) *J. Chem. Soc. Perkin Trans. 1*, 1798.
[m] Kollonitsch, J., Scott, A.N. and Doldouras, G.A. (1966) *J. Am. Chem. Soc.*, **88**, 3624.
[n] Casella, L., Gullotti, M., Pasini, A. and Psaro, R. (1979) *Synthesis*, 150.
[o] Titouani, S.L., Lavergne, J.-P., Viallefont, P. and Jacquier, R. (1980) *Tetrahedron*, **36**, 2961.
[p] Gramatica, P., Manitto, P., Manzocchi, A. and Santaniello, E. (1981) *J. Labelled Compd. Radiopharm.*, **18**, 955.
[q] Haeusler, J. (1981) *Liebigs Ann. Chem.*, 1073.
[r] Kleeman, A., Martens, J. and Drausz, K. (1981) *Chem.-Ztg.*, **105**, 266.
[s] Kahl, J.-U. and Wieland, T. (1981) *Liebigs Ann. Chem.*, 1445.

Table 8.20 Synthesis of β-amino acids

Reactants	*Intermediates*	*Product*	*Reference*
$EtO_2CNHCSR$ $Ph_3P\overset{+}{-}CHCO_2Et$	$EtO_2CNHCR{=}CHCO_2Et \xrightarrow[\text{(ii) HCl HOAc}]{\text{(i) } H_2/Pt}$	$H_3N^+CHRCH_2CO_2^-$	a
$PhCH_2N{=}CHPh$ $CH_2{=}\overset{+}{C}(SBu^t)OBBu_2{}^n$ (Vinyloxyboranes)	$PhCH_2NHCHPhCH_2COSBu^t \longrightarrow$	$H_3\overset{+}{N}CHPhCH_2CO_2^-$	b
Acrylonitrile + oleum		$H_3\overset{+}{N}CH_2CH(SO_3^-)CO_2H$	c
β-Alanine + oleum		$H_3\overset{+}{N}CH_2CH(SO_3^-)CO_2H$	d
$\alpha\beta$-Unsaturated esters $Hg(OAc)_2$ + secondary amine	$ArNR^1CHR^2CR^3(HgOAc)CO_2R^4 \xrightarrow{NaBH_4}$	*N*-Aryl β-amino acids	e

Table 8.20 (*Contd.*)

Reactants	*Intermediates*	*Product*	*Reference*
Wolff rearrangement	$PhtNCH^iPrCOCl + CHN_2 \rightarrow$	$PhtNCH^iPrCH_2CO_2H$ (better than 68% yield)	f
Arndt–Eistert rearrangement N^{α}-Tos-N^{δ}-Boc-ornithine	$TosNHCH(CO_2H)(CH_2)_3$-NHBoc	*β*-Lysine derivative, further elaboration giving the *β*-arginine	g
α-Cyano-alkanoic acids	Electrochemical reduction	*β*-Amino acids	h

[a] Slopianka, M. and Gossauer, A. (1981) *Synth. Commun.*, **11**, 95.
[b] Otsuka, M., Yoshida, M., Kobayashi, S., *et al.* (1981) *Tetrahedron Lett.*, **22**, 2109.
[c] Zilkha, A., Barzilay, I., Naiman, J. and Feit, B.-A. (1968) *J. Org. Chem.*, **33**, 1686.
[d] Wagner, D., Gertner, D. and Zilkha, A. (1968) *Tetrahedron Lett.*, 4479.
[e] Barluenga, J., Villamana, J. and Yus, M. (1981) *Synthesis*, 375.
[f] Sylvester, S.R. and Stevens, C.M. (1981) *Proc. West. Pharmacol. Soc.*, **24th**, 117; (*Chem. Abs.*, **95**, 133328).
[g] Nomoto, S. and Shiba, T. (1978) *Chem. Lett.*, 589.
[h] Krishnan, V., Ragapathy, K. and Udupa, H.V.K. (1978) *J. Appl. Electrochem.*, **8**, 169.

Table 8.21 Asymmetric synthesis of α-amino acids based on stereoselective hydrogenation or reduction*

Substrate	*Reagent*	*Product and stereoselectivity*	*Reference*
Di-oxopiperazines from *N*-(L-prolyl)-α-amino acid αβ-unsaturated amides	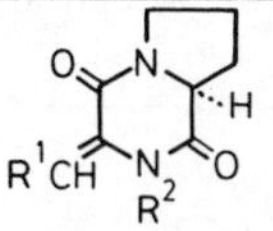 H_2/Pt	L-Phenylalanine 90% $R^1 = Ph$	a (see also ref. p of Table 8.22)
(−)-Menthyl esters of achiral imines	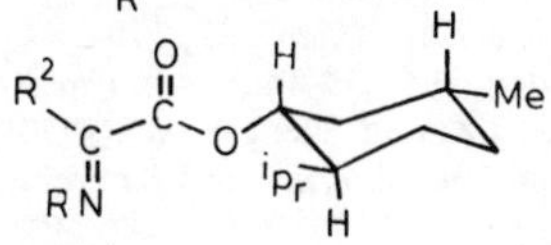	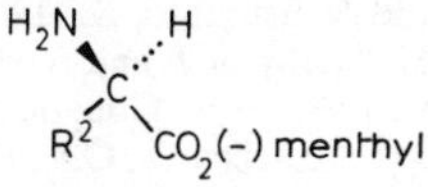	b
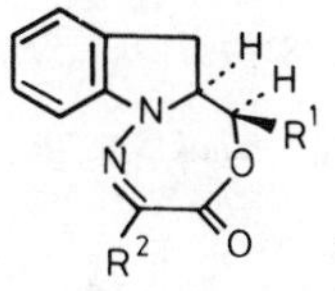	$Al/Hg/H_2O$ then Pd/H_2 then H_3O^+	96–99% $R^1 = Me$ 80–90% $R^1 = H$	c, d
4-Arylidene-oxazolin-5-ones	$H_2/Pd/(S)$-phenylethylamine	13% L-phenylalanine	e
		low from (*E*)-isomer	f
		L-dopa	g
	Reductive methanolysis in the presence of $PdCl_2$ and (*S*)-phenylethylamine; (*Z*)-4-benzylideneoxazolin-5-one gives *N*-benzoyl-L-phenylalanine methyl ester		f
Di-oxopiperazine of glycyl-L-alanine, alkylidene derivative	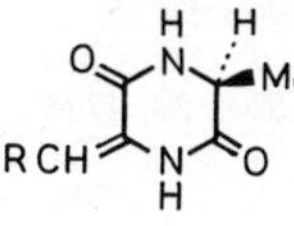 H_2/Pd	L-Alanine	h
		90–99% 85–98.4%	i, j

Table 8.21 (*Contd.*)

Substrate	*Reagent*	*Product and stereoselectivity*	*Reference*
$R^1CONHC(=CHR^2)CO_2R^3$ *N*-Acylaminoacrylic acids	Hydrogen and heterogeneous catalyst, or chiral phosphine complexed to rhodium(I)	(*R*)-$Ph_2PCHRCH_2PPh_2$ (R $= PhCH_2$ or Me_2CH) 84–99% optical purity[r]	n–r
[cyclic structure: R^1, R^2, H, N, O, O, O]	Al/Hg	D-Amino acids	k
[cyclic structure: Ph, H, H, N, $CHCO_2Me$, O, O]	Al/Hg or H_2/ Raney Ni	L-Aspartic acid 14–17%	l
Sodium bis-(*N*-salicylidene-L-glutamato) cobaltate (III) (from methyl acrylate and the glycine complex)	Electrochemical reduction	L-Glutamic acid 10–46% (100% chemical yield)	m
2-(Indolylidene)glycines	H_2/Rh(I)-chiral phosphine	(*R*)-6-Methyltryptophan in 82% e.e.	s
$AcNHC(=CHR)COOC(Ph)(H)Me$	H_2	Low optical yields	t

* See also Reactions 8.8–8.10, Table 8.22 and Fig. 8.5.

[a] Poisel, H. and Schmidt, U. (1973) *Chem. Ber.*, **106**, 3408.

[b] Harada, K. and Matsumoto, K. (1967) *J. Org. Chem.*, **32**, 1794.

[c] Corey, E.J., McCaully, R.J. and Sachdev, H.S. (1970) *J. Am. Chem. Soc.*, **92**, 2476.

[d] Corey, E.J., Sachdev, H.S., Gougoutas, J.Z. and Saenger, W. (1970) *J. Am. Chem. Soc.*, **92**, 2488.

[e] Karpeiskaya, E.I., Chel'tsova, G.S., Klabunovskii, E.I. and Kharchevnikov, A.P. (1980) *Izv. Akad. Nauk S.S.S.R., Ser. Khim.*, 1082.

[f] Godunova, L.F., Levitina, E.S., Karpeiskaya, E.I. and Klabunovskii, E.I. (1981) *Izv. Akad. Nauk S.S.S.R. Ser. Khim.*, 815.

[g] Karpeiskaya, E.I., Neupokoeva, E.S., Godunova, L.F., *et al.* (1978) *Izv. Akad. Nauk S.S.S.R., Ser Khim.*, 1368.

[h] Kanmera, T., Lee, S., Aoyagi, H. and Izumiya, N. (1979) *Tetrahedron Lett.*, 4483.

[i] Lee, S., Kanmera, T., Aoyagi, H. and Izumiya, N. (1979) *Int. J. Peptide Protein Res.*, **13**, 207.

[j] Kanmera, T., Aoyagi, H. and Izumiya, N. (1980) *Int. J. Peptide Protein Res.*, **16**, 280.

[k] Irurre Perez, J., Martin Juarez, J. and Bosch Rovita, A. (1979) *An. Quim.*, **75**, 958.

[l] Tamura, M. and Harada, K. (1980) *Bull. Chem. Soc. Japan*, **53**, 561.

[m] Belekon, Yu. N., Savel'eva, T.F. and Saporovskaya, M.B. (1977) *Izv. Akad. Nauk S.S.S.R., Ser. Khim.*, 428.

[n] Hanaki, K., Kashiwabara, K. and Fujita, J. (1978) *Chem. Lett.*, 489

[o] Takaishi, N., Ima, H., Bertelo, C.A. and Stille, J.K. (1978) *J. Am. Chem. Soc.*, **100**, 264.

[p] Masuda, T. and Stille, J.K. (1978) *J. Am. Chem. Soc.*, **100**, 268.

[q] Koettner, J. and Greber, G. (1980) *Chem. Ber.*, **113**, 2323.

[r] Bergstein, W., Kleemann, A. and Martens, J. (1981) *Synthesis*, 76.

[s] Hengartner, U., Batcho, A.D., Blount, J.F., *et al.* (1979) *J. Org. Chem.*, **44**, 3748.

[t] Sheehan, J.C. and Chandler, R.E. (1961) *J. Am. Chem. Soc.*, **83**, 4795.

Table 8.22 Alternative asymmetric syntheses of α-amino acids from Schiff bases and other imines*

Substrate		*Reagent*	*Product and stereoselectivity*	*Reference*
Glyoxylic imine (−)menthyl esters, $R^1N=CHCOO(-)$menthyl		Grignard reagent R^3MgX	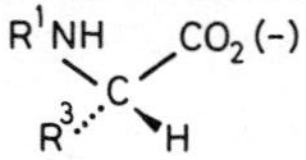 R^1NH–C(R^3)(H)–$CO_2(-)$menthyl (up to 63% e.e. for $R^3 = Me, Pr^i$)	a
Imine from glycine *t*-butyl ester and (1*S*,2*S*,5*S*)-2-hydroxy-pinan-3-one	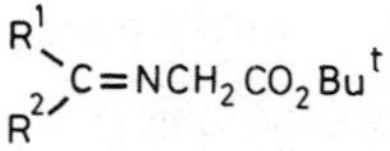 $R^1R^2C=NCH_2CO_2Bu^t$	Alkylation (LDA then alkyl halide) and acid hydrolysis	D-amino acids (50–79%)	b
Imine from DL-alanine *t*-butyl ester and (−)-menthone	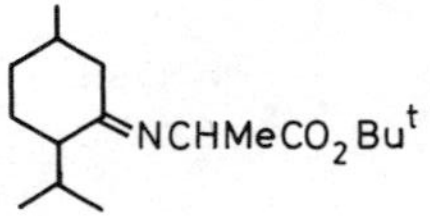 (menthone)$=NCHMeCO_2Bu^t$	Alkylation (LDA then alkyl halide) and acid hydrolysis	α-Methylphenylalanine (21% optical yield)	b
Imine from an aldehyde and a chiral benzylamine	$R^1N=CHR^2$ ($R^2 = Me$;[c–e]	HCN ('asymmetric Strecker synthesis')	Alanine; (*R*)-phenylethylamine gives D-alanine (optical yield depends on solvent)	c–e
	R^2 = various[f])	HCN, then aqueous $NaID_3$	Other amino acids (22–58% optical purity)	f
Imine from a methyl ketone and a chiral 5-amino-4-phenyl-1,3-dioxan	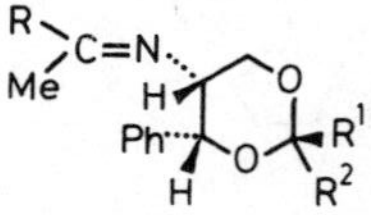 R(Me)C=N–(4-Ph-1,3-dioxan-5-yl, 2-R^1,R^2)		(*S*)-α-Methyl-α-amino acids, predominantly (80–90% optical yields)	g–j
Chiral phenylethylimine	$R^1N=CHR^2$	PhCOCN	(R)-Imine gives D-α-amino acids 15–37% optical purity	k
Chiral imine, from a benzaldehyde and an L-amino acid *t*-butyl ester	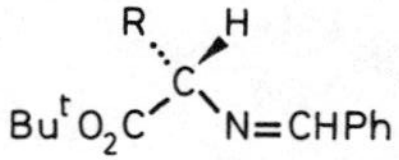 Bu^tO_2C–C(R)(H)–N=CHPh	HCN then acid hydrolysis and Bu^tOCl cleavage of the secondary amine	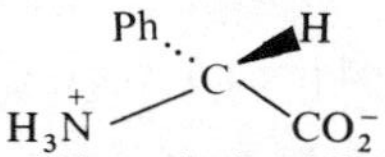 $H_3\overset{+}{N}$–C(Ph)(H)–CO_2^- D-Phenyglycine (62% optical purity using valine)	l
Chiral imine from methyl pyruvate and (*R*)-phenylethylamine		H_2/Pd	See Reaction 8.9	m

Table 8.22 (*Contd.*)

Substrate		*Reagent*	*Product and stereoselectivity*	*Reference*
Chiral imines, from a nitrile and di-isopinocamphenylborane	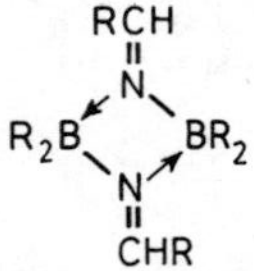	HCN (acetone cyanohydrin as a convenient source) then MeOH and hydrolysis	L-Valine (optical purity 12.4%)	n
Chiral imine from (*R*)-phenylethylamine, ethyl glyoxylate, and $Fe_2(CO)_9$	$PhCHMeN=CHCO_2Et$	Alkylation with alkyl bromides then remove phenylethyl group (H_2/Pd)	L-Phenylalanine (77% optical purity) L-Aspartic acid (78% optical purity)	o
From N^{α}-(α-keto-acyl) *N*-methyl-L-prolinamide by reaction with $NH_3/(MeOCH_2)_2$ or $MeNH_2$, then TFA ($R^1=H$) or $SOCl_2$ ($R^1=Me$)	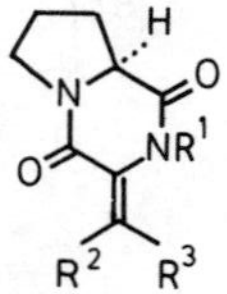	H_2/Adams' catalyst, ethanol	Corresponding L-α-amino acids ($R^1=H$) and *N*-methyl analogues ($R^1=Me$). Chiral induction >90% and ca 60% yields	p
Chiral imine formed from ethyl pyruvate and (*S*)-bornylamine	$R^1N=CMeCO_2Et$	H_2	46.5% Optical purity in the best case (use of a bulky amine)	q
Imine from dimethyl 2,3-dibromosuccinate and (*R*)-phenylethylamine	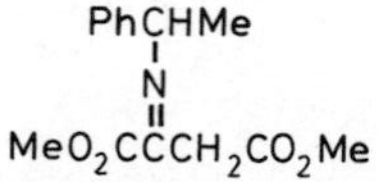 $MeO_2CC(=NCHMePh)CH_2CO_2Me$ formed together with the isomeric aziridine	Catalytic hydrogenation	D-Aspartic acid in 49% optical yield	r
Chiral imine formed from glycine *t*-butyl ester and MeCOCMePhOH	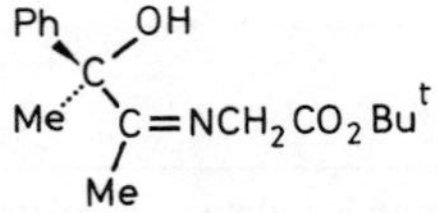 $C=NCH_2CO_2Bu^t$	Alkylation with aldehydes and potassium di-isopropylamide	*threo*-β-Hydroxy-α-amino acids of high optical purity	s

Chiral imine from (*S*)-(−)-phenylethylamine	$(Bu^tO_2C)_2CHCH_2CH{=}NCHMePh$	HCN	γ-Carboxy-L-glutamic acid (100% optically pure)	t
Chiral imine from (*S*)-phenylethylamine and adamantylacetaldehyde	PhCHMeN=CHR	HCN	(*S*)-(+)-2-Amino-1-adamantylpropionic acid	u
Chiral imine from (*S*)-phenylethylamine and pivalaldehyde	$PhCHMeN{=}CHBu^t$	HCN, H_2/Pd, then hydrolysis and hydrogenolysis	L-2-*t*-Butylglycine ('L-*t*-leucine')	v
Chiral imine from (*S*)-phenylethylamine and Me_3CCH_2CHO	$PhCHMeN{=}CHCH_2CMe_3$	HCN then hydrolysis to amide and hydrogenolysis	γ-Methyl-D-leucinamide, enantioselectively de-amidated with leucine aminopeptidase	w
α-Keto-acid		H_2/Pd in the presence of (*S*)-phenylethylamine	L-Alanine in 91% e.e. from pyruvic acid	x
Imines derived from *N*-amino-anabasine	H, N, N, $N{=}CMeCO_2Et$	Zn/HCl	L-Alanine (ca 40% e.e.)	y

* See also Reactions 8.9 and 8.10 and Table 8.2.

[a] Fiaud, J.C. and Kagan, H.B. (1971) *Tetrahedron Lett.*, 1019.
[b] Yamada, S., Oguri, T. and Shioiri, T. (1976) *J.C.S. Chem. Commun.*, 136.
[c] Oguri, T., Shioiri, T. and Yamada, S. (1977) *Chem. Pharm. Bull.*, **25**, 2287.
[d] Harada, K. and Fox, S.W. (1964) *Naturwiss.*, **51**, 106.
[e] Patel, M.S. and Worsley, M. (1970) *Canad. J. Chem.*, **48**, 1881.
[f] Harada, K. and Okawara, T. (1973) *J. Org. Chem.*, **38**, 707.
[g] Weinges, K. and Stemmle, B. (1973) *Chem. Ber.*, **106**, 2291.
[h] Weinges, K., Graab, G., Nagel, D. and Stemmle, B. (1971) *Chem. Ber.*, **104**, 3594.
[i] Weinges, K., Brune, G. and Droste, H. (1980) *Liebigs Ann. Chem.*, 212.
[j] Weinges, K., Klotz, K.P. and Droste, H. (1980) *Chem. Ber.*, **113**, 710.
[k] Harada, K. and Okawara, T. (1973) *Bull. Chem. Soc. Japan*, **46**, 191.
[l] Yamada, S. and Hashimoto, S. (1976) *Chem. Letters*, 921.
[m] Yamada, S. and Hashimoto, S. (1976) *Tetrahedron Lett.*, 997.
[n] Diner, U.E., Worsley, M., Lown, J.R. and Forsythe, J. (1972) *Tetrahedron Lett.*, 3145.
[o] Chenard, J.Y., Commereuc, D. and Chauvin, Y. (1972) *J.C.S. Chem. Commun.*, 750.
[p] Bycroft, B.W. and Lee, G.R. (1973) *J.C.S. Chem. Commun.*, 988.
[q] Kiyooka, S., Takeshima, K., Yamamoto, H. and Suzuki, K. (1976) *Bull. Chem. Soc. Japan*, **49**, 1897.
[r] Harada, K. and Nakamura, I. (1978) *Chem. Lett.*, 9.
[s] Nakatsuka, T., Miwa, T. and Mukaiyama, T. (1981) *Chem. Lett.*, 279.
[t] Oppliger, M. and Schwyzer, R. (1977) *Helv. Chim. Acta*, **60**, 43.
[u] Do, K.Q., Thanei, P., Caviezel, M. and Schwyzer, R. (1979) *Helv. Chim. Acta*, **62**, 956.
[v] Fauchere, J.L. and Petermann, C. (1980) *Helv. Chim. Acta*, **63**, 824.
[w] Fauchere, J.L. and Petermann, C. (1981) *Int. J. Pept. Protein Res.*, **18**, 249.
[x] Hiskey, R.G. and Northrop, R.C. (1961) *J. Am. Chem. Soc.*, **83**, 4798.
[y] Kost, A.N., Sagitullin, R.S. and Yurovskaya, M.A. (1966) *Chem. and Ind.*, 1496.

Table 8.23 Asymmetric alkylation and asymmetric addition reactions

Starting material	*Reagents*	*Product and stereoselectivity*	*Reference*
Chiral Schiff base from glycine *t*-butyl ester and (+)-pinanone	LDA/alkyl halide	D-Amino acid t-butyl esters in low optical yields (see also Table 8.22)	a
N-Formyl-*N*-(*S*)-(phenylethyl) amino-acetonitrile $HCONRCH_2CN$	MeI/NaH in DMF	Hydrogenation (H_2/Pd) and hydrolysis yields D-alanine (2–18% optical yields)	b
Menthyl isocyano-propionate NC–CHMe–CO_2(-) menthyl	NaH and acrylonitrile	$NCCH_2CH_2C(NC)(Me)CO_2$ (−) menthyl- → ornithine	c
(+)-Bornyl, (−)-menthyl etc esters of isocyano-acetic acid $NCCH_2CO_2R$	MeI/OH^-/phase transfer conditions	L-Alanine in 48% optical purity	d
Amidines of DL-amino acid methyl esters formed using *O*-methyl-L-prolinol	LDA/alkyl halide	α-Alkyl-α-amino acids (highest enantiomer excess 50%)	e
α-Phthalimidoacrylate $PhtNC(=CH_2)CO_2H$	$PhCH_2SH$ with acrylonitrile—quinine copolymer	*N*-Phthaloyl-*S*-benzyl-L-cysteine in low optical yield	f
L-Alanine dioxopiperazine di-*O*-methyl ether (93-95% optically pure)	BuLi and an alkyl bromide	D-α-Methyl-α-amino acids (>90% diastereoselectivity)	g–j
	BuLi and a ketone	D-α-Methylserines (41–74% enantiomer excesses)	k
Dehydroalanyl-L-prolinamide derivative	MeSH addition (basic conditions)	Predominance of D-(*S*-methyl)cysteine moiety in the adduct	l
(*S*)-Phenylethylamine and dimethyl fumarate or dimethyl maleate		R^1NH–C(H)(CO_2R^2)–$CH_2CO_2R^2$ L-Aspartic acid favoured	m
(+)-Enantiomer of K bis (*N*-salicylidene glycinato) cobaltate(III)	Addition of acetaldehyde	D-Threonine and D-allothreonine (19–46 and 55–64% optical yields, respectively)	n
N-Chloroacetyl-*N*-cyanomethyl-(*R*)-benzylamines $RN(CH_2CN)COCH_2Cl$	β-Lactam formation through NaH, then hydrolysis and H_2/Pd	L-Aspartic acid (up to 67% optical yield)	o

Table 8.23 (*Contd.*)

Starting material	*Reagents*	*Product and stereoselectivity*	*Reference*
C-Carboxylation of isocyanides via lithium aldimines	$R^1N=CR^2Li$ $R^1=(R)$-phenylethyl CO_2 or $ClCO_2Et$	Representative amino acids; low	p
Chiral 1-substituted 4-methyl-3-imidazolin-5-ones (from chiral isocyanopropionamides)	H, N, N—C, Ph, Me, Me, OH	L-α-Methylphenylalanines (>95%) H, N, N—C, Ph, Me, Me, O, CH_2, Ph	q

[a] Oguri, T., Kawai, N., Shioiri, T. and Yamada, S. (1978) *Chem. Pharm. Bull.*, **26**, 803.
[b] Harada, K., Tamura, M. and Suzuki, S. (1978) *Bull. Chem. Soc. Japan*, **51**, 2171.
[c] Kirihata, M., Mihara, S., Ichimoto, I., and Ueda, H. (1978) *Agric. Biol. Chem.*, **42**, 185.
[d] Laangstroem, B., Stridsberg, B. and Bergson, G., (1979) *Chem. Scripta*, **13**, 49.
[e] Kolb, M. and Barth, J. (1979) *Tetrahedron Lett.*, 2999.
[f] Kobayashi, N. and Iwai, K. (1980) *J. Polymer Sci., Polymer Lett. Ed.*, **18**, 417.
[g] Schöllkopf, U., Hartwig, W. and Groth, U. (1979) *Angew. Chem.*, **91**, 922.
[h] Schöllkopf, U., Hartwig, W., Groth, U., and Westphalen, K.O. (1981) *Liebigs Ann. Chem.*, 698.
[i] Schöllkopf, U., Groth, U., and Deng, C. (1981) *Angew. Chem.*, **93**, 793.
[j] Schöllkopf, U. and Groth, U. (1981) *Angew. Chem.*, **93**, 1022.
[k] Schöllkopf, U., Hartwig, W. and Groth, U. (1980) *Angew. Chem.*, **92**, 205.
[l] Schmidt, U. and Öhler, E. (1976) *Angew. Chem. Internat. Edit.*, **15**, 42.
[m] Nakajima, Y., Oda, J. and Inoue, Y. (1975) *Agric. and Biol. Chem. Japan*, **39**, 2065.
[n] Belokon, Yu.N., Belikov, V.M., Vitt, S.V., *et al.* (1975) *J.C.S. Chem. Commun.*, 86.
[o] Okawara, T. and Harada, K. (1972) *J. Org. Chem.*, **37**, 286.
[p] Hirowatari, N. and Walborsky, M.H. (1974) *J. Org. Chem.*, **39**, 604.
[q] Schöllkopf, U., Hausberg, H.H., Segal, M., *et al.* (1981) *Liebigs Ann. Chem.*, 439.

Table 8.24 Miscellaneous asymmetric synthesis approaches to α-amino acids

Reaction type		*Products and stereoselectivity*	*Reference*
Carbene insertion into chiral amines	$R^2NH_2 + :CR^1CO_2R^3$	$R^2NHCHR^1CO_2R^3$; low optical yields	a
Asymmetric hydroformylation and hydrocarbonylation of enamides	H_2/CO using HRh·(CO)(PPh_3) with a chiral phosphine as catalyst, *N*-vinylsuccinimide as representative enamide	HCOCHMeN (CO, CH_2, CH_2, CO) L-Alanine esters in low optical yield	b
(*R*)-Phenylethylamine and alkyl 2,3-dibromopropionates	Aziridines CH_2—$CHCO_2R$, N, CHMePh	Hydrolysis and hydrogenation yields D-serine, 28–39% optical purity	c
meso-$(MeO_2CCHBr)_2$ *erythro*-$MeCHBrCHBrCO_2Me$		L-aspartic acid; *S*(−) phenylethylaziridines give L-threonine containing 15% *erythro*, isomer	d,e

Table 8.24 (*Contd.*)

Reaction type			*Products and stereoselectivity*	*Reference*
Ugi synthesis (chiral amine, an aldehyde, an isocyanide, and a carboxylic acid)	$R^1NH_2 + R^2CHO + R^3NC + R^4CO_2H$		$R^4CONR^1CHR^2CONHR^3$ L-Valine (75%) using (*R*)-α-ferrocenylamine D-Amino acids using *N*-amino-L-proline	f, g h
Enantioselective decarboxylation of α-alkyl-α-aminomalonic acids within a chiral cobalt complex	H_2N, R, C, ^-O_2C, CO_2^-	Chiral cobalt complex anion	L-Alanine. Optical purity 30%; better results can be achieved using Yoshikawa's method for releasing the amino acids from the complex	i, j
Neber rearrangement of (−)-menthyl *N*-chloroimidates	See Table 8.6 for details		L-Phenylglycine (27%), L-alanine (33%), L-phenylalanine (62%), L-leucine (75% optical yields, respectively)	k
D-Mannose oxime O, O, H, H, HO, H, CH=NOH, H, O, O	$Bu^tOCOCHO + H_2C=CH_2$ 75°/65 bar/17 h/$CHCl_3$ then acid hydrolysis		L-5-Oxoproline in 54% e.e. O, +, N, H_2, H, CO_2^-	l

[a] Nicoud, J.F. and Kagan, H.B. (1971) *Tetrahedron Lett.*, 1540.
[b] Becker, Y., Eisenstadt, A. and Stille, J.K. (1980) *J. Org. Chem.*, **45**, 2145.
[c] Harada, K. and Nakamura, I. (1978) *J. Chem. Soc. Chem. Commun.*, 522.
[d] Harada, K. and Nakamura, I. (1978) *Chem. Lett.*, 1171.
[e] Harada, K. and Nakamura, I. (1979) *Chem. Lett.*, 313.
[f] Urban, R. and Ugi, I. (1975) *Angew. Chem.*, **87**, 67.
[g] Urban, R. and Ugi, I. (1975) *Angew. Chem. Int. Ed.*, **14**, 61.
[h] Achiwa, K. and Yamada, S. (1974) *Tetrahedron Lett.*, 1799.
[i] Job, R.C. and Bruice, T.C. (1974) *J. Am. Chem. Soc.*, **96**, 809.
[j] Ajioka, M., Yano, S., Matsuda, K. and Yoshikawa, S. (1981) *J. Am. Chem. Soc.*, **103**, 2459.
[k] Nogami, Y., Kawazoe, Y. and Taguchi, T. (1973) *Yokugaku Zasshi*, **93**, 1058.
[l] Vasella, A. and Voeffray, R. (1981) *J. Chem. Soc., Chem. Commun.*, 97.

Table 8.25 Asymmetric synthesis of β-amino acids by the hydrogenation of a derivative of β-aminoacrylic acid and by the alkylation of imines

Starting material	*Reaction system*	*Product*	*Reference*
(*Z*)−$AcNHCR=CHCO_2Me$	H_2/Chiral biphosphine-Rh(I)Cl		a
(*Z*)−$R^1NHCR^2=CHCO_2R^3$ (R^1 = (*R*)-2-phenylethyl; R^3 = Et)		(*S*)-β-Amino acids; 2–28% optical purity	b, c

Table 8.25 (*Contd.*)

Starting material	*Reaction system*	*Product*	*Reference*
(R^1 = alkyl or benzyl; R^3 = (−)-menthyl)		Low optical purity	b, c
Addition of a chiral benzylamine to MeCH=CHCN	Hydrolysis and H_2/Pd treatment to cleave the benzylamine moiety	$H_3\overset{+}{N}CHMeCH_2CO_2$ in moderate optical yields	d
Imines $R^1CH=NR^2$ in Reformatzki reaction	$BrCH_2CO_2$-(−)menthyl/Zn	$R^1CH(NHR^2)CH_2CO_2$-menthyl; (*S*)-β-amino acids in 2–28% optical yield when R^2 = (*R*)-PhCHMe	e

[a] Achiwa, K. and Soga, T. (1978) *Tetrahedron Lett.*, 1119.
[b] Furukawa, M., Okawara, T., Noguchi, Y., and Terawaki, Y. (1979) *Chem. Pharm. Bull.*, **27**, 2223.
[c] Furukawa, M., Okawara, T., Noguchi, Y. and Terawaki, Y. (1978) *Chem. Pharm. Bull.*, **26**, 260.
[d] Furukawa, M., Okawara, T. and Terawaki, Y. (1977) *Chem. Pharm. Bull.*, **25**, 1319.

REFERENCES

1. Kaneko, T., Izumi, Y., Chibata, I. and Itoh T. (1974) *Synthetic Production and Utilization of Amino Acids*, Wiley, New York.
2. Greenstein, J.P. and Winitz, M. (1961) *Chemistry of the Amino Acids*, (Vol. 1–3) Wiley, New York.
3. Cook, A.H. (1944) *Ann. Repts. Chem. Soc.*, **41**, 120.
4. Rydon, H.N. (1949) *Ann. Repts. Chem. Soc.*, **46**, 184.
5. Jones, J.H. (1969) *Amino Acids, Peptides and Proteins*, (Vol. 1) The Royal Society of Chemistry, London, p. 1.
6. Bycroft, B.W. (1970) *Amino Acids, Peptides and Proteins*, (Vol. 2) The Royal Society of Chemistry, London, p. 1.
7. Bycroft, B.W. (1971) *Amino Acids, Peptides and Proteins*, (Vol. 3) The Royal Society of Chemistry, London, p. 1.
8. Bycroft, B.W. (1972) *Amino Acids, Peptides and Proteins*, (Vol. 4) The Royal Society of Chemistry, London, p. 1.
9. Hardy, P.M. (1974) *Amino Acids, Peptides and Proteins*, (Vol. 5) The Royal Society of Chemistry, London, p. 1.
10. Barrett, G.C. (1975) *Amino Acids, Peptides and Proteins*, (Vol. 6) The Royal Society of Chemistry, London, p. 1.
11. Barrett, G.C. (1976) *Amino Acids, Peptides and Proteins*, (Vol. 7) The Royal Society of Chemistry, London, p. 1.
12. Barrett, G.C. (1976) *Amino Acids, Peptides and Proteins*, (Vol. 8) The Royal Society of Chemistry, London, p. 1.
13. Barrett, G.C. (1978) *Amino Acids, Peptides and Proteins*, (Vol. 9) The Royal Society of Chemistry, London, p. 1.
14. Barrett, G.C. (1979) *Amino Acids, Peptides and Proteins*, (Vol. 10) The Royal Society of Chemistry, London, p. 1.
15. Barrett, G.C. (1981) *Amino Acids, Peptides and Proteins*, (Vol. 11) The Royal Society of Chemistry, London, p. 1.
16. Barrett, G.C. (1981) *Amino Acids, Peptides and Proteins*, (Vol. 12) The Royal Society of Chemistry, London, p. 1.
17. Barrett, G.C. (1982) *Amino Acids, Peptides and Proteins*, (Vol. 13) The Royal Society of Chemistry, London, p. 1.
18. Barrett, G.C. (1983) *Amino Acids, Peptides and Proteins*, (Vol. 14) The Royal Society of Chemistry, London, p. 1.

19. Miller, S.L. (1955) *J. Am. Chem. Soc.*, **77**, 2351.
20. Miller, S.L. and Urey, H.C. (1953) *Science*, **117**, 528.
21. Cook, A.H. (1944) *Ann. Repts. Chem. Soc.*, **46**, 123.
22. Gagnon, P.E., Gaudry, R. and King, F.E. (1944) *J. Chem. Soc.*, **13**, 1677.
23. Barrett, G.C., Hardy, P.M., Harrow, T.A. and Rydon, H.N. (1972) *J. Chem. Soc. Perkin Trans. 1*, 2634.
24. Ninomiya, K., Shioiri, T. and Yamada, S. (1974) *Chem. and Pharm. Bull., Japan*, **22**, 1398.
25. Ninomiya, K., Shioiri, T. and Yamada, S. (1973) *Tetrahedron Lett.*, 2343.
26. Finkbeiner, H. (1965) *J. Org. Chem.*, **30**, 3414.
27. Zelinsky, N. and Stadnikoff, C. (1906) *Ber.*, **39**, 1722.
28. Ultee, M.A.J. (1909) *Rec. Trav. Chim.*, **28**, 248.
29. Black, D.K. and Landor, S.R. (1968) *J. Chem. Soc.* (*C*), 281.
30. Weinges, K., Graab, G., Nagel, D. and Stemmle, B. (1971) *Chem. Ber.*, **104**, 3594.
31. Kendall, E.C. and McKenroe, B.F. (1941) *Org. Synth., Coll.* (Vol. 1) 2nd edn, Wiley, New York, p. 21.
32. Bassignani, L., Biancini, B., Brandt, A., *et al.* (1979) *Chem. Ber.*, **112**, 148.
33. Geipel, H., Gladede, J. Hilgetag, K.P. and Gross, H. (1965) *Chem. Ber.*, **98**, 16.
34. Maia, H.L., Ridge, B. and Rydon, H.N. (1973) *J. Chem. Soc. Perkin Trans. 1*, 98.
35. Scholz, D. and Schmidt, U. (1974) *Chem. Ber.*, **107**, 2295.
36. Semple, J.E., Wang, P.-C., Lysenko, Z. and Joullie, M.M. (1980), *J. Am. Chem. Soc.*, **102**, 7505.
37. Divanford, H.R., Lysenko, Z., Wang, P.C. and Joullie, M.M. (1978) *Synth. Commun.*, **8**, 269.
38. Haslinger, E. (1978) *Monatsh. Chem.*, **109**, 749.
39. Hoyng, C.F. and Patel, A.D. (1981) *J. Chem. Soc., Chem. Commun.*, 491.
40. Gieven, A., Dederer, B., George, G., Marquarding, D. and Ugi, I. (1977) *Tetrahedron Lett.*, 1503.
41. Chiha, M., Stefanikova, E. and Zvakova, A. (1978) *Chem. Listy*, **72**, 1066.
42. Asimon, J.W.P. and Seguin, R.P. (1979) *Tetrahedron*, **35**, 2797.
43. Vigneron, J.P., Kagan, H. and Horeau, A. (1968) *Tetrahedron Lett.*, 5681.
44. Harada, K. and Yoshida, T. (1970) *Chem. Commun.*, 1071.
45. Harada, K. and Yoshida, T. (1973) *J. Org. Chem.*, **38**, 4366.
46. Harada, K. and Kataoka, Y. (1978) *Chem. Lett.*, **8**, 791.
47. Harada, K. and Kataoka, Y. (1978) *Tetrahedron Lett.*, 2103.
48. Harada, K. and Tamura, M. (1979) *Bull. Chem. Soc., Japan*, **52**, 1227.
49. Jubault, M., Raoult, E., Armand, J. and Boulares, L., (1977) *J.C.S. Chem. Commun.*, 250.
50. Takahashi, H., Neguchi, H., Tomita, K. and Otomasu, H. (1978) *Yakugaku Zasshi*, **98**, 618.
51. Kuzuhara, H., Komatsu, T. and Emoto, S. (1978) *Tetrahedron Lett.*, 3563.

CHAPTER NINE

Protected Amino Acids in Peptide Synthesis

J. MEIENHOFER

9.1 INTRODUCTION

Modern peptide synthesis is predominantly concerned with synthesis of biologically active peptides and studies on peptide conformation. For these purposes suitably protected amino acids are needed. Temporary protection of the α-amino group of one component and of the α-carboxyl group of the other component permits directed coupling to produce the desired sequence and eliminates the dipolar character of the starting zwitterionic amino acids. Side-chain functional groups also frequently need protection. Although a large number of novel protective groups have been developed during the past 20 years, the main methods of peptide bond formation for use in routine synthesis have largely remained unchanged.

The major part of this chapter, therefore, deals with protective groups and a relatively shorter part with special amino acid derivatives needed for coupling. It is mainly a reference work quoting the recent literature which describes the preparation of a multitude of protected amino acids available for synthesis. Some recent trends and a few leads on potential future developments have also been included. An apology is due to the reader for the frequent use of abbreviations and crowded tables.

9.2 PROTECTIVE GROUP TACTICS

Temporary protection of α-amino and α-carboxyl groups of a growing peptide chain is a typical feature of peptide synthesis. The side-chains of trifunctional amino acids e.g. ω-amino, ω-carboxy, thiol, hydroxy, indolyl, guanidino, and carboxamido groups also frequently need to be protected. Protection is obligatory for amino and thiol groups ('minimal protection' [1, 2]). Blocking of other functionalities is often advantageous or necessary. 'Maximal protection' [3] may provide optimal safeguard from undesired side reactions; it is required in solid phase synthesis [4, 5].

A large number of amino protective groups (> 200), suitable for peptide synthesis, have been developed, whereas considerably fewer (*ca* 70) are available for thiol protection, and still smaller numbers for carboxyl, hydroxyl, and other groups [2, 6–17]. Basic requirements for blocking groups to be suitable are

(i) complete stability during synthesis, and (ii) quantitative cleavage by readily available reagents without detriment to peptide bonds, chiral centres or regenerated functional groups. Essential side-chain protective groups (e.g. for $-NH_2$ or -SH) must be completely stable throughout repetitive cleavage of α-protective groups for subsequent coupling. The required high cleavage selectivities during synthesis, coupled with the need for quantitative deblocking of the final product, render the design of effective sets of complementary protecting groups a formidable challenge. The majority of peptide syntheses have been carried out, with the use of combinations of benzyl- and *t*-butyl-derived blocking groups, once the required protected amino acids became commercially available (see Reactions 9.1 and 9.2, Table 9.1; see also ref. 16, pp. 229–45). System 1, (Table 9.1) developed by Schwyzer [18] provides perfect selectivity of cleavage of N^{α}-benzyloxycarbonyl groups (N^{α}–Z) by catalytic hydrogenolysis, to which the *t*-butyl-type side-chain protection (Boc, O^tBu, ^tBu) is completely stable. With Cys- or Met-containing peptides catalytic hydrogenolysis of N^{α}–Z may be carried out in liquid ammonia as a solvent [19, 20]. Completed peptides are deprotected by mild acidolysis, e.g. by trifluoroacetic acid (TFA). This system is close to ideal except that N^{α}–Z hydrogenolysis may become sluggish with large peptides, especially if these are poorly soluble.

With the reverse system (N^{α}-*t*-butyloxycarbonyl and benzyl-type side-chain protection) (system 2, Table 9.1) [5, 21, 22] the cleavage selectivity is not absolute but a matter of relative stability toward acidolysis, which has created considerable problems in solid phase synthesis [5, 23–26]. However, in general N^{α}-Boc acidolysis proceeds well, even with large and sparingly soluble intermediates. A disadvantage, created by such a scheme, is the use of strong acid treatment for the deprotection of the completed peptide, e.g. by liquid HF [27], which may give rise to a variety of side reactions [28]. Nevertheless, this combination of α-*t*-butyl- and benzyl-type side-chain protection is presently the most frequently used tactic in both solid phase and solution synthesis.

System 3 (Table 9.1) [29, 30] provides excellent cleavage selectivity of N^{α}-9-fluorenylmethyloxycarbonyl (Fmoc) groups by base (e.g. piperidine), to which the

Table 9.1 Protective group combinations used in peptide synthesis[a]

Main-chain protection	*Repetitive cleavage*	*Side-chain protection*	*Product deprotection*
(1) Benzyloxycarbonyl[b]	H_2/Pd	*t*-Butyl type	TFA (Mild acidolysis)
(2) *t*-Butyloxycarbonyl[c]	TFA	Benzyl type	HF (Strong acidolysis)
(3) 9-Fluorenylmethyl-oxycarbonyl[d]	Base (Piperidine)	*t*-Butyl type	TFA (Mild acidolysis)

[a] For a detailed discussion of these and other protective group combinations, see Fauchere and Schwyzer in ref, 16, p. 203.
[b] Refs 18–20
[c] Refs 21–28; 36; 37.
[d] Refs 29–33; 101–115.

t-butyl-type side-chain protection is fully stable, and works efficiently in solid phase synthesis. This system is compatible with the 4-alkyloxybenzyl ester [31] and related peptide to polymer anchorage, cleavable by mild acid (TFA), and obviates the use of liquid HF. In solution phase synthesis, side reactions have been observed, e.g. a slow removal of the Fmoc protective group by the primary or secondary amines which are the nucleophilic components of coupling reactions [32, 33].

In selecting one or a set of protective groups for peptide synthesis, the most important consideration should be their facilities for complete cleavage from the target peptide, especially from large peptides possessing multiple sets of each protective group. Several reported failures [34–36] to obtain biologically active material from large synthetic peptides have been attributed to incomplete deprotection of the final product. For example, deprotection of the eight acetamidomethyl (Acm) thiol protecting groups from a synthetic 129-peptide analogous to lysozyme provided an overall cleavage of 75% [34]. Assuming an approximately equal average cleavage yield for all of the eight Acm groups, no more than 10% (0.75^8) recovery of fully SH-deprotected peptide could be theoretically expected.

In the recent synthesis of ribonuclease A [21, 37] a yield of 6.6% of fully active enzyme was obtained from the protected peptide indicating an approximate 93% average cleavage of each of the 33 protective groups. It is clear from these examples that substantially improved cleavage, i.e. quantitative cleavage, will have to be obtained if syntheses of yet larger proteins should be attempted by conventional strategies.

Sections 9.3–9.6 on different types of protective groups are therefore categorized according to the conditions needed for their removal, rather than by structure. Literature for their preparation is given in the footnotes to Tables 9.2–9.6.

9.3 AMINE PROTECTIVE GROUPS

Notwithstanding an abundance of amine protective groups suitable for peptide synthesis [13, 15–17], new groups as well as novel reagents for their preparation continue to be developed apace. However, in the vast majority of peptide syntheses the benzyloxycarbonyl group (**1**) [38] (Table 9.2A), and the *t*-butyloxycarbonyl group (**9**) [39, 40] (Table 9.3A), are used in the vast majority of applications.

(a) Groups cleaved by strong acid or catalytic reduction

The benzyloxycarbonyl and related groups (Table 9.2A) are preferentially cleaved by strong acid treatment or reduction. Complete cleavage under acidic conditions may be obtained by liquid HF [27], 2 N HBr–AcOH [41–43], methanesulphonic acid (MSA) [44], trifluoromethanesulphonic acid (TFMSA) [45], and related procedures (refs 5, p. 70; 13, pp. 51–64; 16, pp. 211–23). Carbonium ion

Table 9.2 Protective groups cleaved by strong acid treatment [liq. HF (anisole)[a], HBr-AcOH[b], MSA[c], TFMSA[d]] and by reduction (H_2/Pd[e], in liq. NH_3[f], transfer-H_2[g], Na/liq. NH_3[h], electrochem.[i])

No.	Name	Abbreviation	Structure	Stability[j]	Other cleavage agents[j,k]
	A Amino group protection				
(1)	Benzyloxycarbonyl[l,m]	Z	C_6H_5–CH_2OCO–	WA, B, hν	BBr_3, BTFA, H_2–$K_3[Co(CN)_5]$ Me_3SiI
(2)	2-Chlorobenzyloxycarbonyl[n]	Z(2Cl)	2-Cl-C_6H_4–CH_2OCO–	WA	BBr_3
(3)	2-Bromobenzyloxycarbonyl[n]	Z(2Br)	2-Br-C_6H_4–CH_2OCO–	WA, B	BBr_3
(4)	2-Nitrobenzyloxycarbonyl[n]	Z($2NO_2$)	2-NO_2-C_6H_4–CH_2OCO–	WA	hν, 320 nm
	B Carboxyl group protection				
(5)	Benzyl ester[o,p]	OBzl	–OCH_2–C_6H_5	WA	Me_3SiI, HO^-, BBr_3, BTFA, H_2–$K_3[Co(CN)_5]$
	C Hydroxyl group protection (serine[q], threonine[q], tyrosine[r])				
(6)	Benzyl (ether)[p,q,r]	Bzl	H_2C(–)–C_6H_5	WA, B H_2–$K_3[Co(CN)_5]$	Me_3SiI, BTFA, BBr_3
	D Thiol group protection				
(7)	Benzyl (thioether)[s]	Bzl	H_2C(–)–C_6H_5	WA, Iodine-ox. $(SCN)_2$	Electrolyt.–liq. NH_3
(8)	3,4-Dimethylbenzyl[t]	Bzl(3,4-Me_2)	H_2C(–)–C_6H_3-3,4-$(CH_3)_2$	WA	

[a] Ref. 27.
[b] Refs 41; 42. For procedural variants, see ref. 43.
[c] Ref. 44.
[d] Refs 45; 46.
[e] Ref. 38.
[f] Ref. 19.
[g] Refs 47; 48.
[h] Refs 49; 50; 51.
[i] Ref. 52.
[j] B = Base; BBr_3 = boron tribromide, ref. 163; BTFA, Pless, J., Baner, W (1973) *Angew. Chem. Int. Ed.*, **12**, p. 147; H_2–$K_3[Co(CN)_5]$, Losse, G., Stiehl, H.U. (1981) *Z. Chem.* 188; $h\nu$ = photolysis; Me_3Sil = trimethylsilyl iodide ref. 73, WA = weak acid treatment.
[k] For other cleavage procedures, see ref. 15, index.
[l] Rev.: refs 11, pp. 56–61; 13, pp. 47–91; 15, pp. 239–45; 16, pp. 15–9; 17, Vol. 9, pp. 317–20.
[m] Synth.: benzyl chlorocarbonate ref. 13, pp. 47–50, pp. 65–9; other reagents refs 12, p. 192; 13, p. 50; 15, p. 224. Cleav.: ref. 13: pp. 51–64.
[n] Synth., Cleav.: refs 13; pp. 78–91; 59, p. 326.
[o] Rev.: refs 11, pp. 196–8; 13, pp. 348–71; 15, p. 172; 16, pp. 116–7. Synth.: $RCOO^-Cs^+$ + Bzl–Br, refs 159; 160; Bzl – OH in $SOCl_2$, ref. 167; dibenzyl sulphite, refs 168; ref 13, pp. 348–53. Cleav.: refs 13, 253–4; 47; 73; 163; 164; 169; 170.
[p] For solid phase synthesis several halobenzyl protective groups have been recommended, see refs 5, pp. 190–1, pp. 208–9; 23; 24–26.
[q] Rev.: refs 13, pp. 576–9, pp. 591–8; 15, pp. 29–31; 16, pp. 180–1, p. 185. Synth.: Boc – Ser – OH (or Boc – Thr – OH) + Bzl–Br in DMF/NaH, refs 193; 194. Cleav.: ref. 15, pp. 29–31.
[r] Rev.: refs 13, pp. 613–7, pp. 628–33; 15, pp. 97–8; 16, pp. 193–4. Synth.: Tyr–copper complex + Bzl–Br refs 195; 196. Cleav.: ref. 15, pp. 97–8. For acyl type protective groups, see ref. 16, pp. 191–2.
[s] Rev.: refs 11, pp. 241–71; 13, pp. 736–44; 15, pp. 195–7; 16, pp. 138–49. Synth.: Cy_2 + Na/liq. NH_3 + Bzl–Cl, refs 219; 220; CySH in liq. NH_3 + Bzl–Cl, refs 221; 13, pp. 736–44, 836–7. Cleav.: ref. 15, pp. 195–7.
[t] Refs 222–226, ref. 15, p. 196.

scavengers added to minimize alkylation of side-chain functional groups, e.g. of Lys, Arg, Tyr, Met, His and Trp, include anisole, thioanisole, dithioethane, indoles and others, see ref. 5, pp. 72–5. *m*-Cresol has been recommended recently [46]. Reductive cleavage can be carried out by catalytic hydrogenolysis (H_2/Pd) [19, 38], transfer hydrogenation [47, 48], sodium in liquid ammonia [49–51], and electrochemical procedures [52]. Introduction of the benzyloxycarbonyl group into amino acids is generally carried out with benzyl chlorocarbonate under Schotten–Baumann conditions [38], although more than a dozen other reagents have since been recommended (footnote m, Table 9.2). Few procedures in the literature describe larger scale (> 0.5 mol) preparations of protected amino acids. A recent need for kilogram amounts of highly pure and chirally homogeneous Z-Lys(Boc)-OH required modification of literature procedures [53–55].

$$\text{HCl H-Lys-OH} \xrightarrow{(i)} \text{C}_6\text{H}_5\text{-CH=N}[\text{CH}_2]_4\text{C(H)(NH}_2\text{)COOH} \xrightarrow{(ii)} \text{Z-Lys-OH} \xrightarrow{(iii)} \text{Z-Lys(Boc)-OH DCHA} \quad (9.1)$$

The conditions are (i) PhCHO, 2 N LiOH, 2–4°, 17h; (ii) (a) $PhCH_2OCOCl$, aq. NaOH, EtOH, – 20 to 5°, 1h; (b) concentrated HCl, 50°, then 3h, *ca* 20°; (iii) $PhOCOOC(CH_3)_3$, DMSO, $(CH_3)_2NC(NH)N(CH_3)_2$, 50°, 5h, DCHA. Overall yield 61%. The Schotten-Baumann N^{α}-benzyloxycarbonylation of the azomethine (ii, Reaction 9.1) [56] had to be carried out at – 20° to obtain a high yield (82%) of pure material, and in order to prevent freezing, ethanol was used as an internal 'antifreeze'. Best yields (92%) of N^{ε}-*t*-butylcarbonylation (iii) were produced with the reagent *t*-butylphenyl phenylcarbonate [57, 58]. A number of other literature procedures may possibly need modification to be suitable for scaled-up preparation of homogeneous protected peptides.

The 2-chloro- (**2**, Table 9.2), 2-bromo- (**3**, Table 9.2), and 2-nitro- (**4**, Table 9.2) benzyloxycarbonyl groups are representative of a large number of derivatives of **1** (Table 9.2) (ref. 13, pp. 78–91) with a range of modified cleavage conditions. The greater acid stability of **2** and **3** (ref. 59, p. 326) made these and related groups suitable for N^{ε}-amino protection of lysine in solid phase synthesis [23–26]. Group **4** may also be cleaved by photolysis [60].

(b) Groups cleaved by mild acid treatment

A range of conditions have been described for the mild acid cleavage of the N^{α}-*t*-butyloxycarbonyl (Boc) group (**9**, Table 9.3) and groups with similar or higher acid lability, **10–16** (Table 9.3A). Preferred reagents for Boc-group cleavage include: (i) trifluoroacetic acid (20°, 1 h) [61, 62] where thiophenol may act as a scavenger of *t*-butylcarbonium ions to prevent Met or Trp alkylation [63], (ii) formic acid [64], (iii) HCl-solvent (e.g. EtOAc) [39, 40], (iv) toluenesulphonic acid [65], and (v) 0.4 M BF_3–AcOH [66, 67]. Recommended reagents for the preparation of Boc-amino acids are 'Boc-ON', 2-(*t*-butyloxycarbonyloximino)-2-phenylacetonitrile (Structure *9.1*) [68–70], and di-*t*-butyl dicarbonate (Structure

Table 9.3 Protective groups cleaved by mild acid treatment [TFA (cold)[a], formic Acid[b], HCl/solvent[c], TosOH/solvent[d], BF_3 AcOH[e]]

No.	*Name*	*Abbreviation*	*Structure*	*Stability*[f]	*Other cleavage agents*[f]
A	*Amino group protection*				
(9)	*t*-Butyloxycarbonyl[g,h]	Boc	$(CH_3)_3COCO-$	Red., B	Me_3SiI[i] $AlCl_3/PhOCH_3$ in CH_2Cl_2/CH_3NO_2[j]
(10)	Adamantyloxycarbonyl[k]	Adoc	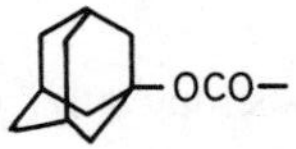 $-OCO-$	Red., B	
(11)	4-Methyloxybenzyloxycarbonyl[l]	Z(OMe)	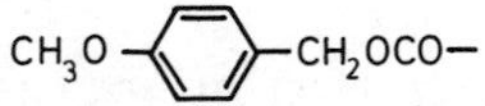 CH_3O $-CH_2OCO-$		H_2/Pd
(12)	2-(3,5-Dimethyloxyphenyl)propyl(2)oxycarbonyl[m]	Ddz	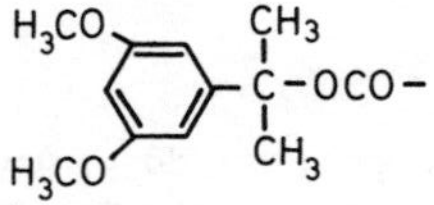 H_3CO, H_3CO; CH_3, $-C-OCO-$, CH_3	Red.	hν, 5% TFA/CH_2Cl_2
(13)	2-(4-Biphenylyl)propyl(2)oxycarbonyl[n]	Bpoc	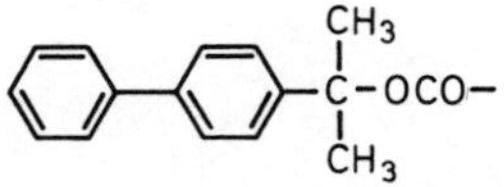 CH_3, $-C-OCO-$, CH_3		80% AcOH, H_2/Pd 75% $ClCH_2COOH/CH_2Cl$
(14)	Triphenylmethyl (trityl)[o]	Trt	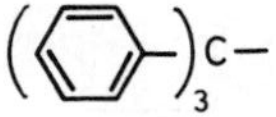 $(C_6H_5)_3C-$		AcOH, Red.
(15)	2-Nitrophenylsulphenyl[p]	Nps	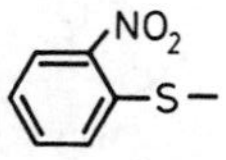 NO_2, $-S-$		Nucleophiles, thiols, desulphuration Ph_3P_4ROH[q]
(16)	3-Nitro-2-pyridinesulphenyl[r]	Npys	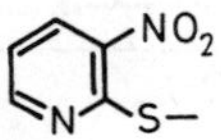 NO_2, N, $-S-$	TFA Formic acid	0.2 N HCl/dioxane $Ph_3P + ROH$, Py(O)SH[q]
B	*Carboxyl group protection*				
(17)	*t*-Butyl ester[s]	O[t]Bu	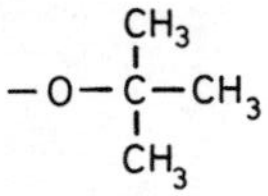 $-O-C(CH_3)_2-CH_3$	Red., B, N_2H_4	Me_3SiI[i]

Table 9.3 (*Contd.*)

No.	*Name*	*Abbreviation*	*Structure*	*Stability*[f]	*Other cleavage agents*[f]
C	*Hydroxyl group protection* (serine[t], threonine[t], tyrosine[u])				
(18)	*t*-Butyl (ether)[t,u]	^{t}Bu	$-C(CH_3)_3$	Red., B	Me_3SiI[i] HBr/AcOH

[a] Refs 61; 62.
[b] Ref. 64.
[c] Refs 38; 39.
[d] Ref. 65.
[e] Refs 66; 67.
[f] B = Base; hν = photolysis; Red = reduction; TFA = trifluoroacetic acid.
[g] Rev.: refs 11, pp. 58–9; 13, pp. 117–54; 15, pp. 232–4; 16, pp. 31–6.
[h] Synth.: Boc – ON = C(CN)Ph [Boc – ON], refs 68–70; $(Boc)_2O$, di-*t*-butyl dicarbonate, refs 71, 72. Exp.: refs 13, pp. 123–6, pp. 130–2; 15, pp. 232–3. Other reagents: refs 12, pp. 192–3; 13, pp. 117–23; 14, pp. 30–4; 16, p. 32. Cleav.: refs 13, pp. 126–9; 15, pp. 232–3.
[i] Refs 73; 198–200.
[j] Ref. 74.
[k] Synth.: Adoc–chloride. Exp.: ref. 13, pp. 139–41.
[l] Synth.: Z(OMe)–azide, refs 13, pp. 71–2; 15, p. 241. Exp.: ref. 13, pp. 87–91. Cleav.: ref. 13: pp. 72–3.
[m] Synth.: Ddz–azide refs 79; 80; 81. Exp.: ref. 13, pp. 150–3.
[n] Synth.: Bpoc–azide (DCHA salts). Exp.: refs 13, pp. 141–9; 17, Vol. 9, p. 321.
[o] Synth.: Trt–chloride, refs 13, pp. 266–76; 15, pp. 273–4; 16, p. 44.
[p] Rev.: refs 11, p. 72; 13, pp. 203–22; 15, pp. 283–4; 16, pp. 40–1; 17, Vol. 11, p. 324. Synth.: Nps–chloride, ref. 13, pp. 203–6, pp. 218–9. Cleav.: ref. 13, pp. 207–22.
[q] Ref. 97.
[r] Synth.: Npys–chloride, refs 98; 99; 100.
[s] Rev.: refs 11, p. 196; 13, pp. 390–8; 15, pp. 168–9; 16, pp. 106–10. Synth.: RCOOH + isobutene – H_2SO_4 – Et_2O, refs 171; 172; 173; transesterification: AcO^{t}Bu + AA in 60% aq. $HClO_4$ ref. 174. Exp.: ref. 13, pp. 390–8. Cleav.: ref. 13, pp. 395–6.
[t] Rev.: refs 11, p. 97; 13, pp. 579–84; 15, pp. 26–7; 16, pp. 181–2, p. 185. Synth.: Z – AA – OBzl ($4NO_2$) + isobutene – H_2SO_4 in CH_2Cl_2 → Z – AA(^{t}Bu) – OBzl($4NO_2$), H_2 – Pd, ref 201. Exp.: refs 13, pp. 579–84, pp. 591–8; 15, pp. 26–7. Cleav.: ref 13, p. 584; 15, p. 27.
[u] Rev.: refs 13, pp. 619–25; 15, pp. 96–7. Exp.: ref. 13, pp. 628–33. Cleav.: ref. 13, pp. 624–5. For other ether and acyl protective groups, see ref. 16, pp. 191–6.

9.2) [71, 72]. (For compilations of other reagents see ref. 14, pp. 30–4; footnote h, Table 9.3.)

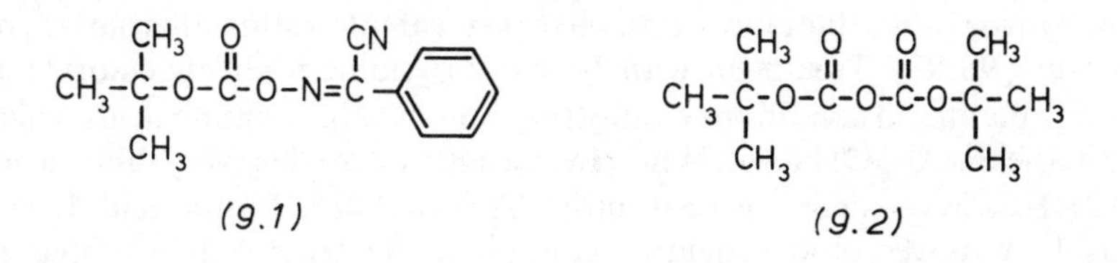

The somewhat more acid stable adamantyloxycarbonyl group (**10**, Table 9.3A) is introduced into amino acids using adamantyl chlorocarbonate [75]. It is suitable for N^G-arginine protection [76], and its lipophilicity improves the solubility of the protected amino acids in organic solvents.

Preparation of N^α-4-methyloxybenzyloxycarbonyl [Z(OMe)] amino acids is carried out with Z(OMe)-azide [77] or -piperidyl(1)carbonate [78]. The Z(OMe)-group (**11**, Table 9.3A) is more acid sensitive than Boc and is cleaved by trifluoroacetic acid (TFA) at 0° or by reduction.

The reagent 2-(3,5-dimethyloxyphenyl)propyl(2)oxycarbonyl (Ddz) azide is used to prepare N^α-Ddz-amino acids [79] which are considerably more acid sensitive than Boc-amino acids and may be cleaved by 5% TFA in CH_2Cl_2, or by photolysis using a mercury lamp. Several larger peptides have been synthesized using N^α-Ddz-amino acids [80–83].

The N^α-2-(4-biphenylyl)propyl(2)oxycarbonyl (Bpoc) group (**13**, Table 9.3A) is introduced into amino acids with the use of Bpoc-azide or 2-(4-biphenylyl)propyl(2)carbonate [84–86]. Cleavage (ref. 16, pp. 35–36, pp. 212–215, pp. 230–233) is carried out in 80% AcOH, 75% chloroacetic acid in CH_2Cl_2, or catalytic hydrogenolysis. TFA may give rise to Trp or Tyr alkylation. N^α-Bpoc-amino acids have been used in many laboratories both in solid phase and synthesis in solution [87, 88]. Relative rates of Ddz and Bpoc cleavage in 80% acetic acid are 1400 and 3000 compared to Boc = 1.

The N^α-trityl (Trt) group (**14**, Table 9.3A) may be introduced into free amino acids using trityl chloride [89] in the presence of diethylamine [90] or into amino acid esters, followed by saponification with strong alkali. Mild acidolysis (e.g. by aqueous AcOH) of the Trt group is 21 000 times faster than Boc, thus providing excellent selectivity towards groups **9** to **13**. This lability, together with solvent dependent selectivity was utilized in the synthesis of insulin by the Ciba-Geigy group [91, 92].

Similar cleavage selectivity is provided by the N^α-2-nitrophenylsulphenyl (Nps) group (**15**, Table 9.3A) which is readily introduced into amino acids by 2-nitrophenylsulphenyl chloride at pH 7–8 [93]. Since Nps-amino acids are not very stable they are stored as their DCHA salts. Very mild acidolysis (e.g. by 2 equiv. HCl in ether) is obtained within a few minutes at 20°. Numerous other reagents for cleavage include nucleophiles [94], thiols [95, 96], and triphenylphosphine under neutral conditions [97]. Nps-Amino acids have been widely used in peptide synthesis in solution. The novel, related, N^{α}-3-nitro-2-pyridinesulphenyl (Npys) group (**16**, Table 9.3A), suitable for the protection of amino, hydroxy and thiol functions, offers several interesting alternatives to the Nps group [98, 99]. Treatment with 3-nitro-2-pyridinesulphenyl chloride under Schotten–Baumann conditions readily provides N^{α}-Npys-amino acids which are stable in the free COOH form. While **16** is readily cleaved by very dilute acid, e.g. 0.1–0.2 N HCl in dioxane it is resistant to TFA and 88% formic acid. It may be selectively removed under neutral conditions by triphenylphosphine or 2-pyridinethiol-1-oxide [97] without affecting groups such as **1, 6, 9, 13** or **17** (Tables 9.2 and 9.3). When activated by tertiary phosphines in the presence of RCOOH (carboxyl component), peptide bonds are formed via oxidation–reduction condensation [100]. Examples of solid phase and synthesis in solution have been described [98]. The Npys group appears promising for further use.

(c) Groups cleaved by base treatment

Protective groups cleaved by base treatment have become of interest in recent years. Preferred cleavage of the N^{α}-9-fluorenylmethyloxycarbonyl (Fmoc) group (**19**, [12, 29, 30, 101–111], Table 9.4), is obtained via a β-elimination process using piperidine or morpholine in suitable solvents (e.g. DMF; 5 min, 20°) or by liquid ammonia. The strong UV absorption of **19** permits monitoring of both cleavage and peptide bond formation. Cleavage selectivity is excellent with *t*-butyl type and related base stable side-chain protective groups. This factor presents an advantage in solid phase synthesis by eliminating the need for repetitive acidolysis in each cycle. The published procedures for the preparation of Fmoc-amino acids [101–106] using Fmoc-chloride in a Schotten–Baumann reaction generate products with sizeable (up to 10%) contamination of di- and tripeptide. Recently, it was found that the use of $LiOH/Li_2CO_3$ or $LiOH/Na_2CO_3$ almost completely suppresses the side product formation [112]. The outstanding acid stability of **19** made it possible to prepare stable crystalline symmetrical anhydrides of Fmoc-amino acids which can be stored at room temperature [107]. They react with amino acids and peptides within 2–3 minutes. This permits the use of defined quantities of homogeneous coupling agents in solid phase synthesis (rather than the complex systems in carbodiimide couplings, ref. 5, pp. 122–39). Several larger peptides have been synthesized by Fmoc-solid phase synthesis [29, 30, 108, 113–115].

The N^{α}-methylsulphonylethyloxycarbonyl (Msc) group (**20**, Table 9.4) may be cleaved as **19** by piperidine and morpholine [116], but complete scission requires several hours. The preferred reagent is 1 N NaOH in methanol which cleaves the group in less than 5 seconds [116–122]. The preparation of Msc-amino acids is carried out with the use of Msc-4-nitrophenylcarbonate or Msc-chlorocarbonate [116, 117]. The Msc group proved to be useful in chemical modifications of insulin [118–121], and cytochrome C [122].

Table 9.4 Protective groups cleaved by base (β-elimination)[a] (piperidine or morpholine[b], liq. NH_3[c], 1 N NaOH/MeOH)

No.	Name	Abbreviation	Structure	Stability[d]	Other cleavage agents[d]
	A Amino group protection				
(**19**)	9-Fluorenylmethyloxycarbonyl[e]	Fmoc	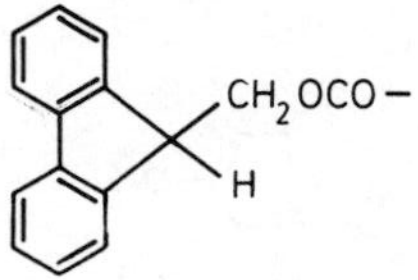 9-fluorenyl(H)–CH_2OCO-	SA (liq. HF)	Piperazine, H_2/Pd
(**20**)	Methylsulphonylethyloxycarbonyl[f]	Msc	$CH_3SO_2CH_2CH_2OCO-$	SA, H_2/Pd	$NaOCH_3/CH_3OH$
	B Carboxyl group protection				
(**21**)	2-(4-Toluenesulphonyl)ethyl ester[g]	OTse	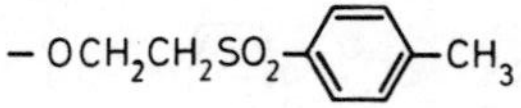 $-OCH_2CH_2SO_2-C_6H_4-CH_3$	SA (liq. HF)	
(**22**)	9-Fluorenylmethyl ester[h]	OFm	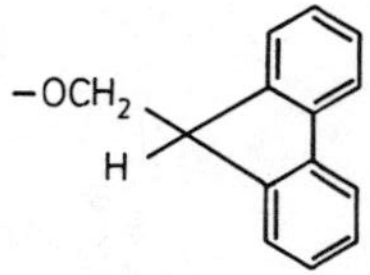 $-OCH_2$–9-fluorenyl(H)	SA	

[a] For a compilation, see ref. 16, pp. 223–9.
[b] Refs 12, 101–103.
[c] Ref. 49. It is important to work at the boiling point ($-33°$) of liq. NH_3.
[d] SA = Strong acid.
[e] Rev.: refs 13, pp. 94–6, 15, pp. 225–6, 16, pp. 21–4. Synth.: 9-Fluorenylmethyl chlorocarbonate: refs 101–106, 112. Exp.: refs 29, 30, 108–111, 113–115. Cleav.: refs 29, 30, 101–103, 106, 111.
[f] Rev.: refs 15, p. 228, 16, pp. 21–4. Synth.: Methylsulphonylethyl chlorocarbonate or Msc/NOSu: refs 117, 118. Cleav.: refs 117–123.
[g] Rev.: refs 13, pp. 343–4, 15, p. 167, 16, pp. 122–3. Synth.: $RCOOH + HOCH_2CH_2SO_2C_6H_4\text{-}4\text{-}CH_3$ + DCC in Py: ref. 15, p. 167. Cleav.: ref. 16, p. 225.
[h] Ref. 177.

Table 9.5 Protective groups cleavable by chemical reduction ($Zn-AcOH$ or $MeOH$[a], electrochemical red.[b], thiols[c])

No.	Name	Abbreviation	Structure	Stability[d]	Other cleavage agents[d]
	A Amino group protection				
(**23**)	2,2,2-Trichloroethyloxycarbonyl[e]	Troc	Cl_3CCH_2OCO-	SA (liq. HF)	H_2/Pd, $NaBH_4/Co^{II}Pc$
(**24**)	Piperidyl(l)oxycarbonyl[f]	Pipoc	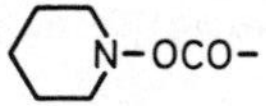 N–OCO–	A (liq. HF)	$H_2/Pd-AcOH$ Na – dithionite
(**25**)	4-Picolyloxycarbonyl[g]	Picoc	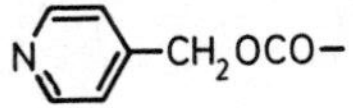 N ... $-CH_2OCO-$	SA (liq. HF)	H_2/Pd Tributylphosphine–pH 10
(**26**)	Dithiasuccinoyl[h]	Dts	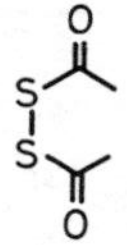 O, S, S, O	SA (6 N HCl, 110°), B, hν	$NaBH_4$, 1 N NaOH
	B Carboxyl group protection				
(**27**)	2,2,2-Trichloroethyl ester[i]	OTre	$-OCH_2CCl_3$	SA	HO^-
(**28**)	4-Picolyl ester[j]	OPic	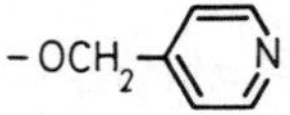 $-OCH_2-$ ... N	SA	H_2/Pd, Na/liq. NH_3, OH^-
	C Hydroxyl group protection (serine[k], threonine[k], tyrosine[l])				
(**29**)	2,2,2-Trichloroethyl (ether)[k]	Tre	CH_2CCl_3	A	
(**30**)	4-Picolyl (ether)[l]	Pic	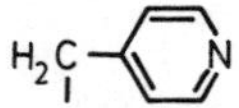 H_2C- ... N	A	H_2/Pd
	D Thiol group protection				
(**31**)	4-Picolyl (thioether)[m]	Pic	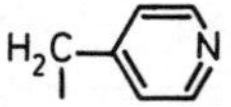 H_2C- ... N	A, 2 N HBr – AcOH, OH^-	
(**32**)	Diphenyl-4-pyridylmethyl[n]	Dppm	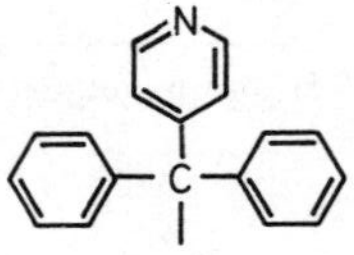	A	$Hg(OAc)_2/AcOH$ pH 4, $I_2-80\%$ AcOH

(33) *t*-Butylsulphenyl (disulphide)[o]	S^{t}Bu	$S-C(CH_3)_3$	A (but not liq. HF), B	$NaBH_4$, Na_2SO_3, Na – liq. NH_3 Tributylphosphine/TFE

[a] Refs 123, 124.
[b] Ref. 52.
[c] β = Mercaptoethanol, dithioethane, dithiothreitol, mercaptoacetic acid, *N*-methylmercaptoacetamide, 2-mercaptopyridine, etc.
[d] A = acid; B = base; hν = photolysis; $NaBH_4/Co^{II}Pc$ = cobalt-phthazocyanin (Eckert, H., Kiesel, Y. (1980) *Synthesis*, 947); SA = strong acid.
[e] Rev.: refs 13, pp. 103–5, 15, pp. 226–7, 16, pp. 26–7. Synth.: 2,2,2-trichloroethyl chlorocarbonate, refs 129, 130. Cleav.: refs 123–128.
[f] Rev. refs 13, pp. 155–9; 15, pp. 237–8; 16, p. 39. Synth.: 2,4,5-trichlorophenyl piperidyl (1) carbonate, ref. 131. Cleav.: refs 132; 13, pp. 157–9.
[g] Rev.: refs 15, p. 247; 16, p. 26; 17, vol. 10, pp. 313–14. Synth.: picolyl-4-nitrophenylcarbonate, ref. 133.
[h] Rev.: refs 5, p. 114–6; 15, p. 266; 16, p. 14; 17, vol. 10, pp. 312–13. Synth.: $RNH_2 \rightarrow EtOC = SNHR \rightarrow + ClSCCl \rightarrow DTS - NR$ = 1,2,4-dithiazolidine-3, 5-dione, refs 134, 135. Cleav.: refs 136, 137.
[i] Synth.: RCOOH + $HOCH_2CCl_3$ + DDC in Py or TosOH – $C_6H_5CH_3$, reflux, refs 123; 125; 178; 13, pp. 340–1; 15, p. 165.
[j] Rev.: refs 13, pp. 327–33; 15, pp. 117–8, 16, pp. 118–9, p. 318. Synth.: RCOOH + picolyl alcohol/DCC or Pic–Cl, ref. 179. 'Handle Method', see ref. 180. Cleav.: refs 13, p. 330; 15, pp. 177–8.
[k] Refs 15, p. 26; 202.
[l] Synth.: Tyr – Ni complex + Pic/Cl + NaOH, refs 203–205. Exp.: refs 15, p. 99; 16, pp. 172–3, pp. 239–40.
[m] Synth.: CySH + Picolyl chloride in liq/NH_3, ref. 205. Cleav.: refs 15, p. 198; 16, p. 140.
[n] Synth.: CySH + $HOCPh_2Py/BF_3 - Et_2O$ – AcOH, refs 226; 15, p. 202; 16, pp. 138–9.
[o] Synth.: Cy_2 + tBuSH in NaOH/MeOH, refs 227, 228, Rev.: refs 13, pp. 789–96; 15, p. 214; 16, pp. 141–2, pp. 161–3; 229–231.

(d) Groups cleaved by chemical reduction

Several interesting groups, cleavable by chemical reduction (Table 9.5), have been developed, although these have not yet found the attention they appear to merit. Some of these are suitable for the protection of amino, carboxy, hydroxy, and thiol functions. Cleavage procedures include treatment with zinc powder in acetic acid, methanol or tetrahydrofuran [123], electrochemical reduction [52], and reduction by thiols, e.g. β-mercaptoethanol, dithiothreitol, etc. Most of these groups are resistant to cleavage by strong acid treatment.

The N^α-2,2,2-trichloroethyloxycarbonyl (Troc) group (**23**, Table 9.5) was developed by Woodward *et al.* [123] for use in a cephalosporin synthesis and has been useful recently as a hydrazide N'-protective group in the synthesis of ribonuclease A [21, 37]. Cleavage conditions are: Zn (12 mol equiv) in AcOH (48 h, followed by EDTA in 5% Na_2CO_3, 68%) [124]; Zn in aq. THF (pH 4.2, 30 min, 86%, or pH 5.5–7.2, 18 h, 96%) [125]; – 1.70 V, 0.1 M $LiClO_4$, (85%) [126, 127]. The Troc group [e.g. Lys(Troc)] might also be cleaved by catalytic hydrogenolysis [128]. Introduction of Troc into amino acids is carried out with 2,2,2-trichloroethyl chlorocarbonate under Schotten-Baumann conditions [129, 130].

The N^α-piperidyl(1)oxycarbonyl (Pipoc) group (**24**, Table 9.5) is introduced similarly using the corresponding Pipoc chloride [131]. Reported cleavage conditions are: catalytic hydrogenolysis in acetic acid; electrolysis (200 mA, 1 N H_2SO_4, 20°, 90 min); $Na_2S_2O_4$ in AcOH (20°, 5 min, 93%); Zn–AcOH (20°, 10 min, 94%) [131, 132]. The Pipoc group is completely stable to strong acid treatment including liquid HF.

The N^α-4-picolyloxycarbonyl (Picoc) group (**25**, Table 9.5) [133] is similarly stable towards acid, and these groups are therefore highly recommended for N^ε-lysine protection. An additional advantage is improved solubility of Picoc-lysine peptides in organic solvents. Cleavage is obtained by treatment with Zn in AcOH (25°, 1.5 h, 100%), or by catalytic hydrogenolysis. Introduction of **25** into amino acids is carried out with picolyl-4-nitrophenyl carbonate [133].

The N^α-dithiasuccinoyl (Dts) group (**26**) [134, 135] was advocated for application to orthogonal protection schemes where each class of protective group can be removed in any order and in the presence of all other classes, an almost unrealistic goal in peptide synthesis. The Dts-amine protection may be cleaved by thiolytic reduction [136, 137] (see ref. 5, pp. 114–6). It is stable to strong acid and mild base treatment. Unfortunately, Dts-amino acids cannot be prepared directly from free amino acids; a two-step procedure via an intermediate ethoxythiocarbonyl derivative and reaction with chlorocarbonylsulphenyl chloride is required [134, 135]. In spite of this handicap the Dts group might prove to be useful.

(e) New amine protective groups

A selection of the considerable number of novel amine protective groups proposed for peptide synthesis in recent years is presented below (see *Specialist Periodical Reports*, ref. 17, for more information).

The 6-methylene-2-trifluoromethylchromone (Tcroc) group (Structure *9.3*) has been proposed recently [138, 139] and used for a synthesis of somatostatin in combination with the 2-methyleneanthraquinone (Maq) ester group [191]. Introduction into lysine via its copper complex was achieved by treatment with Tcroc-Cl. Group (*9.3*) is stable to TFA and cleaved by neat *n*-propylamine (25°, 2–40 min). Hydrazine may also be used but not branched-chain primary amines.

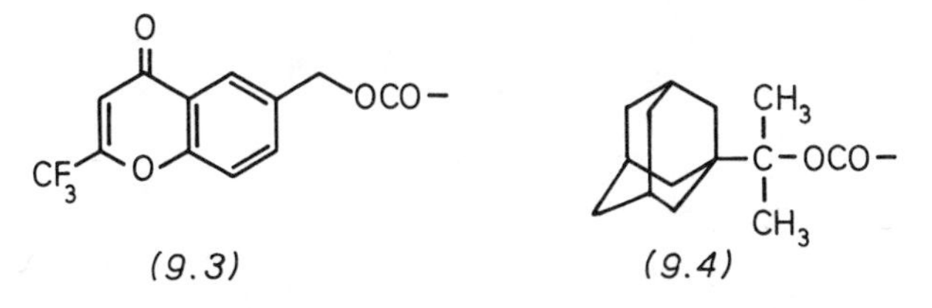

(9.3) (9.4)

The 2-(adamantyl)propyl(2)oxycarbonyl (Adpoc) group [140] (Structure *9.4*) is readily introduced into amino acids using the stable crystalline reagents Adpoc-phenylcarbonate or Adpoc-oximino-2-phenylacetonitrile to form well crystallized derivatives. Cleavage by 5% TFA/CH_2Cl_2(25°, 5 min) proceeds *ca* 10^3 times faster than Boc. The group is suitable for solid phase synthesis and synthesis in solution.

The 2-(trimethylsilyl)ethyloxycarbonyl (Teoc) group (Structure *9.5*) for amine protection [141] is introduced by reaction of amino acids with Teoc chloride or azide. Cleavage is carried out with tetraethylammonium fluoride in polar organic solvents, e.g. DMF, CH_3CN, or by acidolysis (TFA). The group is stable to H_2/Pd and base treatment (e.g. piperidine, NH_3). Recently, 2-(trimethylsilyl)ethyl-4-nitrophenyl carbonate has been recommended for introduction of the Teoc group as well as an alternative cleavage using a suspension of $ZnCl_2$ (15 equiv.) in CH_3NO_2 or TFA (25°, 2 h) [142]. A side reaction during both deblocking procedures was observed, i.e. formation of up to 80% of *β*-aspartyl derivatives from -Asp(O^{t}Bu)-peptides [142].

(9.5) (9.6)

The trichloro-*t*-butyloxycarbonyl (Tcboc) group (Structure *9.6*) can be cleaved by the supernucleophilic Co^{I}phthalocyanine anion [143]. Introduction is carried out by reaction of amino acids with the stable Tcboc-chlorocarbonate under Schotten-Baumann conditions. The group is stable to alkali and conditions of *t*-butyl group acidolysis.

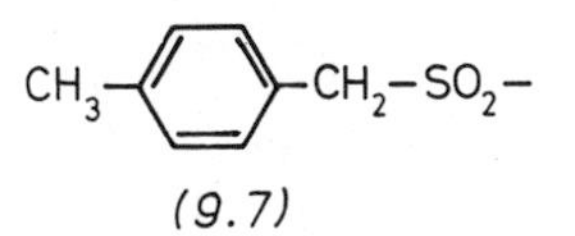

(9.7)

The 4-toluenemethylsulphonyl (Pms) group (Structure *9.7*) has been used for the protection of the ε-amino group of lysine [144]. Lys(Pms) is prepared via the copper complex by reaction with Pms-Cl. The group is stable to TFA, dilute HCl and hydrogenolysis and is deblocked by HF/anisole (20°, 1 h). The Pms group has been useful in a synthesis of neurotensin in solution and a solid phase synthesis of serum thymic factor.

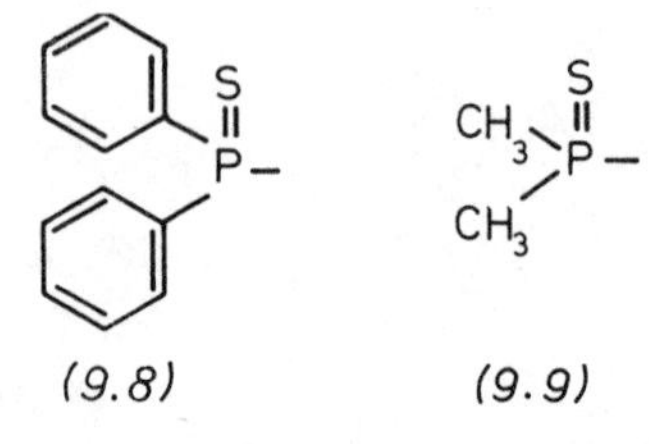

(9.8) (9.9)

Diphenyl- and dimethylthiophosphinyl groups [145–147] (Structures *9.8*, *9.9*) are prepared by reaction of amino acids with the respective thiophosphinyl chloride, i.e. $Ph_2\overset{\overset{O}{\|}}{P}S\text{–}Cl$ or $Me_2\overset{\overset{O}{\|}}{P}S\text{–}Cl$. They are cleaved by 1 M triphenylphosphine/2 N HCl (20°, 5 min). Since carbonium ions are not formed these groups are suitable for use in solid phase synthesis of Trp-containing peptides.

The Azo-Tac group [148] (Structure *9.10*) is cleaved under very mild conditions, i.e. warming with 95% EtOH or EtOAc/1% AcOH. Introduction into amino acids is carried out using 4-phenylazophenylsulphonyl isocyanate. Its utility in peptide synthesis remains to be studied.

$C_6H_5\text{–}N{=}N\text{–}C_6H_4\text{–}SO_2NHCO\text{–}$ $\qquad$ $CH_2{=}CH\text{–}OCO\text{–}$

(9.10) (9.11)

The vinyloxycarbonyl (Voc) group [149] (Structure *9.11*) can be specifically cleaved using one equivalent of bromine while leaving Boc groups intact. The group may also be cleaved by a variety of acidic treatments (HBr/AcOH, HCl/dioxane followed by warm EtOH, etc.). Vinyl chlorocarbonate is used for the preparation of Voc-amino acids. Their usefulness in synthesis needs to be established.

The anthrylmethyloxycarbonyl group [150] (Structure *9.12*) is distinguished by specific nucleophilic displacement reactions, i.e. rapid cleavage by mercaptide ion in DMF (25°, 1–2 min or −20°, 2 h). Acidolysis (TFA) is also rapid. Introduction into amino acids is carried out with the 4-nitrophenyl carbonate. The utility of this group in peptide synthesis still remains to be studied.

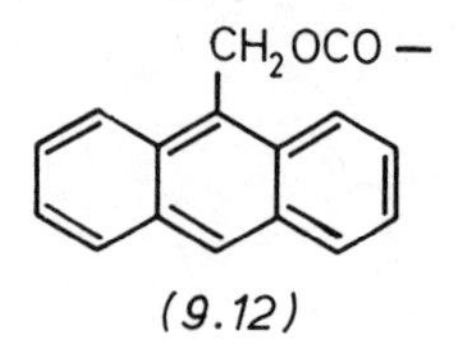

(9.12)

9.4 CARBOXYL GROUP PROTECTION

In contrast to the large selection of amine protective groups, the number of carboxyl blockers suitable for peptide synthesis is relatively small. One reason is that the alkaline saponification of simple alkyl esters (OMe, OEt, OBzl) is limited to very small peptides. However, several useful novel groups have been developed recently.

(a) Methods for ester formation

For classical methods of ester introduction into amino acids, see refs 11, pp. 183–91; 16, pp. 101–36; 151 and 152. Newer methods of ester formation suitable for peptide synthesis include Reactions 9.2 to 9.5, see also ref. 15, pp. 154–6 for a compilation.

$$\text{RCOOH} + \text{R'OH} \xrightarrow{\text{(i)}} \text{RCOOR'} \quad (9.2)$$

The conditions [153–156] are (i) DCC/4-dimethylaminopyridine, Et_2O, 25°, 1–24 h, 70–95% (racemization may occur [157, 158]).

$$\underset{\text{(or Peptide)}}{\text{PG-HNCHRCOOH}} \xrightarrow{\text{(i)}} \text{PG-HNCHRCOO}^-\,\text{Cs}^+ \xrightarrow{\text{(ii)}} \text{PG-HNCHRCOOR'} \quad (9.3)$$

The conditions [160] are (i) Cs_2CO_3, pH 7; (ii) R′X, DMF (HMPA preferred [159]), 6 h. R′ = Me, 80%; Bzl, 70–90%; Bzl($2NO_2$), 90%; Bzl(OMe), 70%; Trt, 40–60%, ᵗBu, 14%; MePac, 80%. X = I, Br, Cl. PG = Protective group.

$$\text{PG-HNCHRCOOH} \xrightarrow{\text{(i)}} \text{PG-HNCHRCOOR'} \quad (9.4)$$

The conditions [161] are (i) $Me_2NCH(OR')_2$ (DMF-alkyl acetal) 25–80°, 1–36 h, 80–95%; R′ = Me, Et, Bzl, SBu.

$$\text{RCOOH} + \text{R'OH} \xrightarrow{\text{(i)}} \text{RCOOR'} \quad (9.5)$$

The conditions [162] are (i) ᵗBuNC, 0–20°, 24 h, 36–98%; R′ = Me, Et, ᵗBu.

Examples of newer methods of cleavage are shown in Reactions 9.6 to 9.8, see also ref. 15, pp. 156–7.

$$\text{PG-HNCHRCOOR'} \xrightarrow{\text{(i,ii)}} \text{H}_2\text{NCHRCOOH} \quad (9.6)$$

The conditions are (i) Me_3SiI, 1–4 equiv., $CDCl_3$ (NMR monitoring), 25–50°, 5–10 min; (ii) MeOH, 3–16 equiv.; R′ = Bzl, Et, Me; PG = Boc, Z (also cleaves benzyl ether of Tyr, 50°, 1 h). Neutral conditions; no carbonium ion formation.

$$\text{PG-HNCHRCOOR}' \xrightarrow{(i)} \text{H}_2\text{NCHRCOOH} \quad (9.7)$$

The conditions [163] are (i) BBr_3 in CH_2Cl_2, $-10°$, 1 h→25°, 2 h, 60–85%; R′ = Bzl, Et, Me, ^{t}Bu; PG = Boc, Z; OEt, OMe, O^{t}Bu, Bzl (side chain ethers).

$$\text{PG-HNCHRCOOCH}_2\text{CH}_2\text{SiMe}_3 \xrightarrow{(i)} \text{PG-HNCHRCOOH} + \text{CH}_2{=}\text{CH}_2 + \text{F-SiMe}_3 \quad (9.8)$$

The conditions [164] (i) 0.15 M $[Et_4N]^+F^-$ 5–10 equiv. in anhydrous DMF or DMSO, 20–30°, 10–60 min; PG = Boc, Bpoc, side-chain *S*-Trt, *S*-Acm, ^{t}Bu.

The preparation of amino acid methyl and ethyl esters and their cleavage by aqueous base has been described in detail (ref. 6, pp. 925–32; ref. 9, derivatives). The decline of their use in peptide synthesis may continue in view of the recent development of direct procedures for hydrazide [165] and azide formation [166], see Section 9.8a.

(b) Esters cleaved by strong acid or catalytic reduction

Cleavage by reduction or strong acid treatment has made the benzyl ester (OBzl) group (**5**, Table 9.2B) very useful in peptide synthesis, see Table 9.1. A large variety of procedures for the preparation of amino acid benzyl esters have been developed [153–162, 167, 168] (see Table 9.2B; Reactions 9.2–9.5; ref. 15, pp. 154–8, 171–2). Furthermore, novel methods of cleavage have recently been explored including rapid catalytic transfer hydrogenolysis [47, 48, 169, 170] and cleavage under neutral conditions using trimethylsilyl iodide, Reaction 9.6. Although the OBzl group is frequently used for the side-chain protection of Asp and Glu in solid phase synthesis, derivatives such as the 4-bromo, and 4-chlorobenzyl esters are preferred for their increased acid stability [23–26].

(c) Esters cleaved by mild acid treatment

The *t*-butyl ester (O^{t}Bu) group [171, 172] (**17**, Table 9.3B) is generally stable to nucleophilic attack i.e. by weak base or reduction. It is cleaved by mild acid treatment under a variety of experimental conditions (e.g. HCl in organic solvents, trifluoroacetic acid with or without diluents, ref. 16, pp. 219–21), or by treatment with stronger acids (e.g. HBr/AcOH [41–43], liquid HF [27]) for final deprotection (see ref. 13, pp. 395–6). Addition of carbonium ion scavengers is often necessary [63]. Amino acid *t*-butyl esters are prepared from free [171, 173] or N^α-protected [172] amino acids by reaction with isobutene in organic solvents (e.g. Et_2O) in the presence of H_2SO_4 or TosOH. Acid catalysed transesterification using *t*-butyl acetate has also been used [174].

(d) Esters cleaved by β-elimination

The 2-(4-toluenesulphonyl)ethyl ester (OTse) group [175] (**21**, Table 9.4B) can be cleaved at pH 11.5 in aqueous–organic medium [176], or by potassium cyanide in dioxane/H_2O [175]. Similar reactivity, more closely resembling the rapid cleavage under mildly basic conditions (e.g. treatment with piperidine in organic solvent), might be expected of amino acid 9-fluorenylmethyl esters (OFm, **22**, Table 9.4B). This potentially promising class of compounds [177] has not yet been studied in sufficient detail.

(e) Esters cleaved by chemical reduction

Cleavage of the 2,2,2-trichloroethyl ester (OTre) group [123, 178] (**27**, Table 9.5B) has been carried out by Zn in AcOH [124] or THF buffer at pH 4–7 [125], or alternatively by electrochemical reduction at − 1.65V [126, 127]. For the preparation of these compounds N^{α}-protected amino acids are condensed with 2,2,2-trichloroethanol using DCC in Py [123] or TosOH in Bzl-OH [178]. Pyridyl-4-methyl (4-picolyl, OPic) esters (**28**, Table 9.5B) are obtained by reaction of N^{α}-Z-amino acids with picolyl alcohol and DCC or picolyl chloride [179]. These acid stable compounds may be cleaved by catalytic hydrogenolysis [19, 38], sodium in liquid ammonia [49–51], electrochemical reduction [52] and by basic hydrolysis. The basic site in amino acid picolyl esters permits their extraction into an acidic medium, a procedure that has become known as the 'handle method of peptide synthesis' [180].

(f) Esters cleaved by other methods

Among several esters cleaved differently, the phenyl (OPh) and phenacyl (OPac,–OCH_2COPh) esters have been used in peptide synthesis. Rapid cleavage of phenyl esters [181] is obtained by reaction with peroxide anion at pH 10.5 [182, 183]. Alternatively, conversion to hydrazides proceeds swiftly [184]. Phenyl esters are prepared by reaction of N^{α}-protected amino acids with phenol in the presence of DCC [185] or with the reagent benzotriazolyloxytris(dimethyl-amino)phosphonium hexafluorophosphate [186]. Amino acid phenacyl esters (refs 13, pp. 344–7; 15, p. 162; 16, pp. 124–5) are readily prepared using phenacyl bromide [187]. They are cleaved by treatment with thiophenoxide in DMF [187], zinc in AcOH [188], or catalytic hydrogenolysis [189], and are stable to strong acid hydrolysis, including liquid HF.

Potentially promising ester groups that would seem to merit more frequent use include, among others, the diphenylmethyl (benzhydryl) ester (ref. 13, pp. 385–90), the trimethylsilyl ester (ref. 13, pp. 599–691; ref. 16, pp. 104–5), the trimethylsilylethyl ester cleaved by fluoride ion [164, 190], the 2-methyleneanthraquinone ester, cleavable by reduction or photolysis [191], and the recently explored 2,4-dimethyloxybenzyl ester cleavable by mild acid treatment or catalytic hydrogenolysis [192].

9.5 HYDROXYL GROUP PROTECTION

The protection of hydroxyl groups is not mandatory during peptide synthesis in solution, see for example the synthesis of ribonuclease A [21, 37]. However, a considerable number of undesired side reactions due to free hydroxyl groups have been observed (see ref. 16, pp. 169–201), especially with tyrosine. Therefore, hydroxyl side-chain protection is required in solid phase synthesis and frequently desirable for synthesis in solution.

(a) Established groups

The benzyl ether (Bzl) group (**6**, Table 9.2C) has been used for permanent side-chain protection of Ser, Thr, and Tyr. Cleavage from final protected peptides is most often carried out by strong acid treatment, see Table 9.2. For smaller peptides reductive cleavage is an efficient alternative. In a synthesis of β-endorphin, Na in liquid NH_3 cleavage was superior to liquid HF acidolysis [28]. The benzyl ether protection is introduced into Boc-serine or -threonine by reaction with benzyl bromide in DMF-NaH [193, 194]. For the synthesis of Tyr(Bzl) see refs [195, 196].

The *t*-butyl ether (^{t}Bu) group [197] (**18**, Table 9.3C) for side-chain protection of Ser, Thr, and Tyr is cleaved by mild acid treatment, i.e. 50% trifluoroacetic acid in CH_2Cl_2 at 20°. Trimethylsilyl iodide can also be used for its quantitative cleavage (in $CHCl_3$, 25°, < 0.1 h) [73, 198–200]. For the introduction of the ^{t}Bu group, N^α-Z-AA-OBzl(4NO_2) is treated with isobutene in CH_2Cl_2 in the presence of H_2SO_4, followed by catalytic hydrogenolysis [197, 201].

The 2,2,2-trichloroethyl ether (Tre) group [202] (**29**, Table 9.5C) and the 4-picolyl ether (Pic) group [203] (**30**) can be cleaved by chemical reduction. They are resistant to strong acids. While potentially interesting, the *O*-Tre ether protection has not yet found frequent use. Introduction of the Tre ether group into amino acids should be obtainable with the use of Tre-bromide [202]. Tyr(Pic) has proven to be of value in syntheses by the 'handle method' [180]. For its preparation, the Ni-complex of tyrosine is treated with picolyl chloride in the presence of EtOH and NaOH, followed by EDTA to remove Ni [204, 205].

(b) Other hydroxyl protective groups

O-Acyl protective groups for serine and threonine, e.g. *O*-acetyl, *O*-benzyloxycarbonyl, or *O*-4-toluenesulphonyl, as well as the tetrahydropyranyl [206] and the 1-benzyloxycarbonylamino-2,2,2-trifluorethyl (Ztf) groups [207] have all been replaced by the more suitable groups **6**, **18**, and **30**, described above (see ref. 16, pp. 169–200). However, the trimethylsilyl ether (Tms) group [208] may merit more frequent application as a temporary *O*-protective group for serine and threonine (see ref. 15, pp. 39–43 for introduction and cleavage conditions). Recently developed *O*-protective groups of considerable promise include the dimethylcarbamoylbenzyl ether [209], cleavable by hydrogenolysis, and 3-nitro-2-pyridinesulphenyl (Npys) group [98, 99], cleaved by 0.5 N HCl/dioxane (30 min) or triphenylphosphine (1 equiv. in CH_2Cl_2/MeOH, 1:1, 25°, 30 min) (not recommended for Tyr protection).

Preparation of Boc-Ser(Me)-OH, and Boc-Thr(Me)-OH by direct methylation of the N^α-*t*-butyloxycarbonylamino acid [210] and convenient preparations of a variety of *O*-alkyl ethers of tyrosine have been described [211]. For the protection of the phenolic functionality of tyrosine in solid phase synthesis the *O*-2-bromobenzyloxycarbonyl group [26] has been found most useful [212, 213]. The 2,4-dinitrophenyl (Dnp) ether protective group for tyrosine [214] is stable to liquid HF, TFA, and Et_3N and can be cleaved quantitatively by 2-mercapto-ethanol. Introduction into tyrosine is readily carried out with 2,4-dinitrofluorobenzene.

9.6 PROTECTION OF SULPHYDRYL GROUPS

Many sulphydryl protective groups (*ca* 70) have been described [215–218] (see also ref. 5, pp. 233–47), however, the search for more efficient protection of cysteine continues. Deprotection to form free thiols or disulphide bonds requires specific reagents, such that only a few sulphydryl protective groups fit (partially) into the deprotection schemes of Tables 9.2–9.5.

(a) Established groups

The benzyl thioether group (7, Tables 9.2, 9.6) [49, 219] has been the workhorse of thiol protection in syntheses of neurohypophyseal hormones and hundreds of analogues. Since it is quite resistant even to rather strong acid treatment (though partially cleaved by prolonged treatment with liquid HF), the preferred cleavage has been by reduction with Na/liq. NH_3. However, side reactions with larger peptides during that procedure led to a decreased use of **7** and its eventual replacement by the *S*-acetamidomethyl and *S*-*t*-butyl groups. *S*-Benzylcysteine is prepared by treatment of CySH in liquid NH_3 with Bzl-Cl [219–221]. For solid phase synthesis, the ring-substituted 3,4-dimethylbenzyl (Dmb) thioether group (**8**, Tables 9.2, 9.6) [222–224] (or 4-methylbenzyl [225, 226]) have been very useful. Group **8** is stable to TFA/CH_2Cl_2. Complete cleavage is obtained by HF/anisole treatment (0°, 10 min.) Introduction into cysteine is carried out using Dmb-Br/aq. EtOH/Et_3N.

Three sulphydryl protective groups listed in Table 9.6, i.e. the 4-picolyl (Pic) thioether (**31**) [205], the diphenyl-4-pyridylmethyl (Dppm) thioether (**32**) [227] and the *t*-butylsulphenyl (S^tBu) (**33**) [228] groups, are quite resistant to acid and base treatment and can be cleaved by chemical reduction under conditions very similar to those for the corresponding amino, carboxy, and hydroxy protective groups. Preparation of **31** is accomplished by reaction of cysteine with picolyl chloride in liq. NH_3 [205]. Group **32** is introduced into cysteine with the use of diphenyl-4-pyridylmethanol/BF_3-Et_2O/AcOH [227], and group **33** is prepared by treatment of cysteine with *t*-butylmercaptan/2 N NaOH/MeOH [229]. Group **32** is also cleaved by $Hg(OAc)_2$/AcOH at pH 4 [227] or by treatment with iodine in 80% AcOH. The *t*-butylsulphenyl group (**33**) can also be cleaved by $NaBH_4$ [229], Na/liq. NH_3 [237] or Bu_3P/TFE (trifluoroethanol) and has recently been used in syntheses of several larger peptides [230–232]. It would seem to merit more frequent use.

Table 9.6. Sulphydryl protective groups

No.	Name	Abbreviation	Structure	Preparation[a]	Cleavage[b]	Stability[c]
(7)	Benzyl thioether	Bzl	H_2C– (phenyl)	Cy_2 + Bzl-Cl/Na/liq. NH_3[d]	Na/liq. NH_3[e]; HF/anisole, 25°, 1 h[f]; electrolysis/liq. NH_3[g]; electrochem redn: – 2.8 V; DMF/$R_4N^+X^-$[h]	Iodine, WA, $(SCN)_2$
(8)	3,4-Dimethylbenzyl thioether	Bzl (3,4-Me_2)	H_2C– (phenyl) –CH_3, CH_3	CySH + 3,4-dimethyl-benzyl-Br, Et_3N[i]	HF/anisole[i]	WA
(31)	4-Picolyl thioether	Pic	H_2C– (pyridyl, N)	CySH + Pic-Cl/liq. NH_3[j]	Zn/AcOH[j]; electrochem. redn.; 0.25 M H_2SO_4[j]	A, 2 N HBr/AcOH, OH^-
(32)	Diphenyl-4-pyridyl-methyl thioether	Dppm	(pyridyl, N)(phenyl)$_2$C–	CySH + $HOCPh_2Py/BF_3$ – Et_2O/AcOH[k]	Zn/AcOH[k]; $Hg(OAc)_2$/AcOH, pH 4 + H_2S[k]; electrochem. redn[k]; iodine/AcOH → SS	A, 2 N HBr/AcOH
(33)	*t*-Butylsulphenyl	S[t]Bu	S–C(CH_3)$_3$	Cy_2 + [t]BuSH/NaOH/ MeOH[l]	$NaBH_4$[l]; thiols[m,n]; Na_2SO_3[n]; tributylphosphine/TFE[o]; Na/liq. NH_3[p]	A (but not to HF), B
(34)	Benzhydryl thioether[q] (Diphenylmethyl)	Bzh	HC(–(phenyl))$_2$	CySH + $HOCHPh_2$/TFA or HBr/AcOH[r] CySH + $HOCHPh_2$/ BF_3 – Et_2O/AcOH[s]	HF/anisole, 0°, 45 min; TFA + 2.5% phenol (70°) or anisole scavengers[r]; Na/liq. NH_3[t]; Nps-Cl + DTT or $NaBH_4$[u]; $(SCN)_2$ in TFA/AcOH → SS[v]; $ClSCO_2CH_3$ → Scm[w]	$Hg(OAc)_2$, Iodine, OH^-, H_2NNH_2, WA

(35)	Triphenylmethyl thioether[x]	Trt	$-C(C_6H_5)_3$	CySH + $HOCPh_3$/TFA[r] CySH + $HOCPh_3$/ $BF_3 - Et_2O$/AcOH[s]	Hcl/AcOH or HBr/TFA[y,z]; $Hg(OAc)_2$ or $AgNO_3$/EtOH + H_2S[aa]; Na/liq. NH_3[t]; electrochem. − 2.6 V[h]; Nps-Cl + DTT or $NaBH_4$[u]; $ClSCO_2CH_3$ → Scm[w]; $(SCN)_2$/TFA → SS[v,bb]; iodine/MeOH → SS[cc,dd]	OH^-
(36)	t-Butyl thioether[ee]	ᵗBu	$-C(CH_3)_3$	CySH + ᵗBuOH reflux[ff]	$Hg(OAc)_2/H_2O$ pH 4 + H_2S[gg]; $Hg(OCOCF_3)_2$/anisole[hh]; Nps − Cl + $NaBH_4$ or thiols[ii]; HF/anisole, 20°, 30 min[jj]	A, OH^-, Na/liq. NH_3, Iodine
(37)	Acetamidomethyl aminothio-acetal[kk]	Acm	$-H_2CNHCOCH_3$	CySH + $HOCH_2NHCOCH_3$/ HCl, pH 0.5, 25°[ll]	$Hg(OAc)_2/H_2O$ pH 4 + H_2S[ll]; Nps-Cl + $NaBH_4$ or thiols[mm]; $ClSCO_2CH_3$ → Scm[w]; iodine/80% aq. AcOH[nn]	SA (liq. HF), Zn/AcOH, Electrochem. redn, H_2NNH_2, Na/liq. NH_3
(38)	Ethylcarbamoyl thioester[oo]	Ec	$-CONHCH_2CH_3$	CySH + C_2H_5NCO, 20°, 70 h[pp]	1 N NaOH, 20°, 20 min[pp]; NH_3 or H_2NNH_2/MeOH[pp]; Na/liq. NH_3[pp]; $Hg(OAc)_2/H_2O$/MeOH + H_2S[qq]; $AgNO_3/H_2O$/MeOH[gg]	SA
(39)	Carboxymethyl-sulphenylthio-carbonate[rr]	Scm	$-SCOOCH_3$	Z − Cys − OH + $ClSCOOCH_3$/ MeOH (X = Bzh, Trt, Bzl, Acm, ᵗBu)[ss]	See text, Reaction 9.9	A (TFA) B (Et_3N)

[a] TFA = trifluoroacetic acid.
[b] DTT = dithiothreitol; Nps = 2-nitrophenylsulphenyl; SS = disulphide bond; TFE = trifluoroethanol.
[c] A = acid; B = base; Et_3N = triethylamine; SA = strong acid; WA = weak acid.
[d] Refs 219–221.

Footnotes to Table 9.6 (*Contd.*)

[e] Refs 49–51.
[f] Ref. 233.
[g] Ref. 234.
[h] Ref. 52.
[i] Refs 222, 223.
[j] Ref. 205.
[k] Ref. 227.
[l] Ref. 229.
[m] Ref. 228.
[n] Ref. 235.
[o] Ref. 236.
[p] Ref. 237.
[q] Refs 11, pp. 251–71; 13, pp. 744–58; 15, pp. 199–202, 16, pp. 138–49.
[r] Ref. 238.
[s] Ref. 239.
[t] Ref. 240.
[u] Ref. 241.
[v] Ref. 242.
[w] Ref. 243.
[x] Refs 11, pp. 258–71; 13, pp. 749–58; 15, pp. 201–2; 16, pp. 138–49.
[y] Ref. 244.
[z] Refs 238, 246.
[aa] Ref. 66.
[bb] Refs 247, 248.
[cc] Ref. 249.
[dd] Refs 250–253.
[ee] Refs 11, pp. 241–71; 13, p. 760; 15, p. 203; 16, p. 143; 17, vol. 10, pp. 316–7, vol. 11, pp. 329–30; 254–260.
[ff] Ref. 260.
[gg] Refs 255, 256
[hh] Refs 258, 259.
[ii] Ref. 260.
[jj] Ref. 233.
[kk] Refs 11, pp. 282–6; 13, pp. 770–3; 15, pp. 206–7; 16, pp. 151–9; 261–266.
[ll] Refs 261, 262.
[mm] Refs 241, 265.
[nn] Refs 249–251.
[oo] Refs 11, pp. 288–95; 13, pp. 785–88; 15, pp. 212–13; 16, pp. 15–61; 268–272.
[pp] Ref. 268.
[qq] Ref. 269.
[rr] Refs 5, pp. 239–43; 11, pp. 299–300; 13, pp. 827–34; 15, pp. 215–6; 16, pp. 147–9, p. 164.
[ss] Refs 243, 273, 274.

The benzhydryl (Bzh) or diphenylmethyl thioether group (**34**, Table 9.6) and the closely related triphenylmethyl (trityl, Trt) group (**35**) are of interest because selective cleavage of the latter can be obtained. For the introduction of the Bzh group, cysteine is treated with diphenylmethanol in TFA, in HBr/AcOH [238], or in BF_3-Et_2O/AcOH [239]. The group is stable to acid, alkali, iodine, mercuric acetate and hydrazine. Acid-catalysed cleavage can be obtained only at elevated temperatures in the presence of scavengers such as phenol or anisol [238]. Na/liq. NH_3 [240], or treatment with 2-nitrophenylsulphenyl chloride (Nps-Cl) followed by reduction with dithiothreitol (DTT) or sodium borohydride [241], afford efficient cleavage of **34**. Treatment with thiocyanogen ($(SCN)_2$) in TFA/AcOH leads to direct disulphide bond formation [242]. Carboxymethylsulphenyl chloride converts **34** into the stable intermediate carboxymethylsulphenyl (Scm) group [243].

The trityl thioether group, introduced similarly [238, 239], is cleaved by acidolysis (HCl/aq. AcOH) [244] or by heavy-metal ions [$Hg(OAc)_2$/EtOH or $AgNO_3$/EtOH/Py, followed by H_2S] [66, 244]. These cleavage procedures are selective for an *S*-trityl thioether in the presence of Bzh, which is important for the stepwise formation of disulphide bonds, e.g. in insulin [245]. However, *S*-trityl group acidolysis is not quantitative [238]. Best yields are obtained using repeated treatment with HBr/TFA in the presence of 2-ethylphenol [246]. Cleavage of trityl thioethers may also be carried out by Na/liq. NH_3 [240], electrochemical reduction [52], reaction with Nps-Cl followed by DTT or $NaBH_4$ [241], conversion into the carboxymethylsulphenyl group [243], and S–S-bond formation by treatment with $(SCN)_2$ [242, 247] or iodine [248, 249]. The latter procedure has been employed in the total synthesis of human insulin [250, 251] and other peptides [252, 253].

The *t*-butyl (^{t}Bu) thioether group (**36**, Table 9.6) [254] has been used increasingly since its quantitative cleavage under mild conditions has been developed, using mercuric acetate [$Hg(OAc)_2$/pH 4, 25°] [255–257] or mercuric trifluoroacetate [$Hg(OCOCF_3)_2$] [258, 259], both followed by reduction (H_2S, mercaptoethanol, etc.) of the intermediate mercaptide. Quantitative cleavage can also be achieved by treatment with 2-nitrophenylsulphenyl chloride [241, 260]. The intermediate *S*-Nps derivatives are readily converted into thiols by dithiothreitol, mercaptoethanol, sodium borohydride, etc., see ref. 260 for a recent detailed report. Cleavage may also be obtained by HF/anisole, albeit under forcing conditions [233]. For the preparation of *S*-*t*-butylcysteine in high yield, the amino acid is refluxed in a mixture of 2 N HCl and *t*-butyl alcohol [260]. The *t*-butyl thioether group merits increased use and provides an advantageous alternative to the acetamidomethyl (Acm) group by its simplicity of preparation, greater resistance to racemization, and freedom from thiazolidine contamination [260].

The acetamidomethyl (Acm) group (**37**, Table 9.6) has been widely used both in solid phase synthesis and in solution since its development in 1969 by Hirschmann, Veber *et al.* [261–263]. Acetamidomethylcysteine is prepared in moderate yield by reaction by CySH with acetamidomethanol at pH 0.5. Among several related groups the benzamidomethyl (Bam) group may be more conveniently prepared [264]. The Acm group is cleaved by aqueous mercuric acetate at pH 4, and subsequent hydrogen sulphide treatment; for water-insoluble peptides $Hg(OAc)_2$/50% aqueous AcOH may be used. Heavy metal cleavage may not always be complete in larger peptides (see refs 16, pp. 152–9 and 265). Cleavage is also achieved by reaction with 2-nitrophenylsulphenyl chloride and subsequent reduction (DTT, $NaBH_4$) [241, 266], by conversion to the carboxymethylsulphenyl intermediate using $ClSCO_2CH_3$ [243], and by iodine oxidation for direct generation of disulphide bonds [249–251]. Disulphides may also be generated by the reaction of thiocyanogen, sulphenyl halides, sulphenyl thiocyanates, etc. with the Acm group. Iodolysis may not always effect complete cleavage in large molecules that contain a number of Acm groups [34–36, 267] (see also discussion on p. 299 of this chapter).

The ethylcarbamoyl (Ec) group (**38**, Table 9.6) [268, 269] is readily introduced by treatment of cysteine with ethyl isocyanate. The *S–N* acyl shift typical for most thioester protective groups does not occur with **38**. The *S*-Ec group is cleaved by a variety of mildly basic conditions [268] (see Table 9.6), as well as by silver nitrate or mercuric acetate treatment [269]. Hofmann and collaborators used the Ec sulphydryl protection in their approaches to ribonuclease T_1 synthesis in solution [270]. It would seem to merit more frequent use in synthesis of medium-size peptides, and seems to be suitable for solid phase synthesis [271, 272].

The carboxymethylsulphenyl (Scm) group (**39**, Table 9.6) [273, 274] represents a stable disulphide intermediate that can be utilized to generate unsymmetrical or symmetrical disulphide bonds, Reaction 9.9.

$$R^1{-}S{-}H + R^2{-}S{-}S{-}\overset{O}{\overset{\|}{C}}{-}OCH_3 \xrightarrow{(i)} R^1{-}SS{-}R^2 + COS + CH_3OH \qquad (9.9)$$

The conditions [273, 274] are (i) CH_2Cl_2 or CH_3OH, catalysed by trace Et_3N. The Scm group is formed from cysteine thioethers and acetamidomethyl residues by oxidation with carboxymethylsulphenyl chloride [243, 273, 274]. The Scm group has been used in the directed formation of an unsymmetrical disulphide bond in the total synthesis of human insulin by the Ciba–Geigy group [250, 251]. The recently proposed acid stable benzyloxycarbonylsulphenyl (SZ) group [275] might prove to be similarly useful.

(b) Other sulphydryl protective groups

A long known group, i.e. *S*-4-methyloxybenzyl [Bzl(OMe)] [276], has recently been used in several synthetic projects with variable success [36, 226, 277–279]. This group is readily introduced into cysteine by treatment of the amino acid with 4-methyloxybenzyl bromide (ref. 13, pp. 743–4). Yajima and Fujii [21, 37] have used this group for the protection of all six cysteine residues in repeated syntheses of ribonuclease A. Cleavage conditions were carefully optimized [21] and involved the use of 1 M trifluoromethanesulphonic acid/thioanisole in TFA plus *m*-cresol as a cation scavenger. Cleavage can also be achieved by $Hg(OCOCF_3)_2$ followed by H_2S [258, 259]. The Bzl(OMe) group may merit increased use.

The 3-nitro-2-pyridinesulphenyl (Npys) group [98, 99] (see **16**, Table 9.3) is also suitable for sulphydryl protection [280, 281]. Boc-Cys (Npys)-OH was prepared by treatment with Npys-Cl/CH_2Cl_2 in the presence of Et_3N. Cleavage was carried out with 10% β-mercaptoethanol/DMF/0.01% Et_3N, with aqueous dithiothreitol (pH 8.5), or with TFA/CH_2Cl_2/0.5% ethanedithiol. It was used in a synthesis of lysine-vasopressin [280].

The *S*-acetyl and *S*-benzoyl groups can be cleaved by heavy metal salts (e.g. $Hg(OAc)_2$, $Hg(OCOCF_3)_2$) [258, 259], as described by Photaki and collaborators [192] who also used improved procedures for the preparation of X-Cys(Bz)-OH (X = Nps, Trt, Boc) under non-aqueous conditions. This may encourage increased use of these long known protective groups, although the danger of *S–N* acyl shifts under basic conditions still remains.

The *S*-2-nitrobenzyl group has recently been proposed as a photo-cleavable thiol protective group [282]. It is introduced into cysteine by reaction with 2-nitrobenzyl chloride/H_2O/MeOH plus Et_3N. Deblocking at 350 nm in dimethylsulphide/semicarbazide HCl produced mixtures of CysH/Cy_2-containing peptides requiring subsequent reduction. The group was used to synthesize a tuftsin analogue.

For lists of protected cysteine derivatives, see refs. 9; 13, pp. 836–45; 14, pp. 60–5.

9.6 PROTECTIVE GROUPS FOR ARGININE, HISTIDINE, METHIONINE AND AMIDE SIDE-CHAIN FUNCTIONALITIES

The protection of these side-chain functionalities is not obligatory for peptide synthesis in solution.

Guanido function of arginine

The guanido function of arginine has been protected by: (i) the nitro group, (ii) arylsulphonyl group, (iii) acyl groups, (iv) groups formed by bifunctional aldehydes and ketones, and (v) simple protonation. Each may give rise to side reactions during deblocking even with ω-deprotected derivatives and during activation (δ-lactam formation). The choice of protection depends in each case on the synthetic strategy. The search for a universally applicable guanidine protective group continues. For reviews see refs 5, pp. 169–75; 13, pp. 506–537; 14, pp. 67–8; 16, pp. 60–70; for a list of compounds see ref. 9.

The N^G-nitro group [283] is cleaved by reductive procedures, i.e. H_2/Pd [283, 284], transfer hydrogenation [47, 48], electrochemical reduction [285, 286], Zn [287], $SnCl_2$ [288], and, recently reported, $TiCl_3$ [289]. Improved procedures for the preparation of nitroarginine have been developed [288, 290].

N^G-4-Toluenesulphonylarginine [291], prepared by reaction with tosyl chloride [292], is cleaved by Na/liq. NH_3 [291] or by HF/anisole [293], and has been widely used in synthesis in solution and in solid phase procedures. Recently, the N^G-4-methyloxybenzenesulphonyl (Mbs) [294] and mesitylene-2-sulphonyl [295] groups, cleavable by methanesulphonic acid [44], trifluoromethanesulphonic acid [45] and anhydrous HF have been utilized. Mbs was successfully used by Yajima and Fujii in syntheses of ribonuclease A [21, 37].

While monoacylation of the guanido function did not provide significant advantages over the nitro or tosyl groups, diacylation of the ω and δ positions has provided protective groups useful for peptide synthesis, such as Boc-Arg(Z_2)-OH [296, 297], Z(OMe)-Arg(Z_2)-OH [298], Z-Arg(Adoc$_2$)-OH [299], Z(OMe)-Arg (Adoc$_2$)-OH [299], Z-Arg(Boc$_2$)-OH [300]. For protein modification, bifunctional aldehydes and ketones, e.g. butanedione [301] or 1,2-cyclohexanedione [302, 303] have been used, the latter in a semisynthesis of human insulin [304].

To circumvent the protection of the guanidine moiety, N^δ-protected ornithine may be used instead. After deprotection at the end of synthesis, the ornithine residue is converted to arginine by treatment with 1-guanyl-3,5-dimethylpyrazole [305]. This strategy has been used successfully in several syntheses of complex peptides [306–308].

Histidine

Histidine has often been used in peptides without protection of the imidazol (N^{im}) function. Among a number of N^{im} protective groups the most frequently used are the benzyl [51, 309–311], 2,4-dinitrophenyl [312, 313], 4-toluenesulphonyl [44, 45, 314, 315], and *t*-butyloxycarbonyl [316–318], see also refs 5, pp. 179–89; 13, 537–64; 14, pp. 61, 66–67; 16, pp. 70–80; and 9 (derivatives). None of these protective groups is satisfactory and considerable degrees of racemization, to which histidine is particularly prone, have often been observed [319, 320]. Very recently, it was shown that racemization during carbodiimide activation is caused by internal base catalysis of N^α,$N^{im}(\tau)$-protected histidine derivatives [320] and that it can be completely eliminated by blocking the N^{im}-πnitrogen [321–323]. The N^{im}-π-benzyloxymethyl group has been recommended. It is readily prepared from Boc-His(τBoc)-OMe by reaction with benzyloxymethyl chloride as shown in Reaction 9.10.

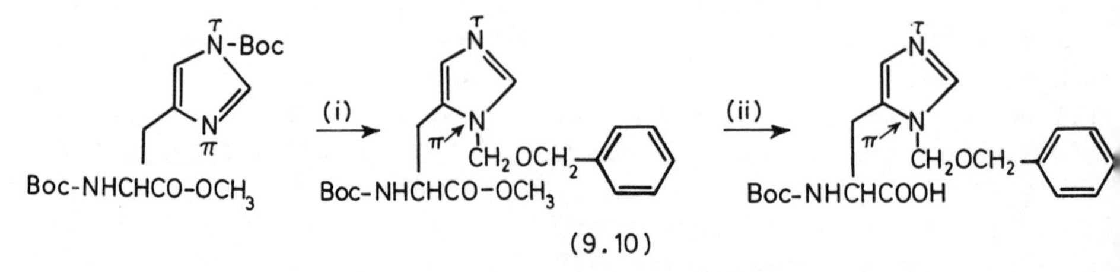

(9.10)

The conditions [221, 322] are (i) $PhCH_2OCH_2$ Cl/Et_2O, 20° then Et_3N/MeOH (62%); (ii) aqueous NaOH (78%). τ = tele (distant), π = proc (close). The group is stable to nucleophilic and basic reagents and to TFA. Rapid cleavage occurs with saturated HBr/TFA (25°, 1.5 h). The group is also cleaved by H_2/Pd, and its efficacy has been tested with syntheses of two tripeptides [321].

Thioether group of methionine

The thioether group of methionine is prone to two side reactions in peptide synthesis, i.e. formation of diastereoisomeric sulphoxides and *S*-alkylation to form sulphonium salts, see refs 5, pp. 223–33; 13, pp. 728–34; 14, p. 69; and 9 (derivatives). Nevertheless, methionine has frequently been left unprotected. Protection as its sulphoxides by treatment of methionine with H_2O_2 [324] and final reduction with thioglycollic acid or dithiothreitol have been explored long ago. Recently, more efficient procedures for the reduction have been developed. In the solid phase synthesis of a heptapeptide the methionine d-sulphoxide residue was cleanly reduced by treatment with liquid HF/anisole in the presence of 2-mercaptopyridine (10 equiv., 0°, 10 min, 63% yield) during cleavage from the resin and deprotection [325], thus superseding the previously recommended aqueous *N*-methylmercaptoacetamide [326, 327]. Complete reduction of Met(O) to Met was also obtained by treatment with HF/dimethylsulphide-4-cresol (25:65:10;0°, 1 h) followed by deprotection and cleavage from resin with HF/4-cresol (9:1;0°,1 h) [318]. Iodide in TFA, Reaction 9.11 [329], was used to reduce the methionine sulphoxide residues in substance P, motilin, and human calcitonin synthesized by conventional methods in solution [329].

$$R{-}S(O)CH_3 + 2I^- + 2H_3O^+ \xrightarrow{TFA} R{-}SCH_3 + 3H_2O + I_2\,[+HSCH_2COOH \text{ as } I_2 \text{ scavenger}] \qquad (9.11)$$

Where $R - S(O)CH_3$ represents Met(O) peptide.

Methionine sulphonium salts, arising e.g. from carbonium ions during deprotection, can be reconverted into methionine by thiolysis [330, 331] or by moderate heat [332].

Indole functionality of tryptophan

The indole functionality of tryptophan is often left unprotected in peptide synthesis. However, because of its sensitivity to oxidation under acidic conditions, the addition of scavengers and/or reducing agents is essential especially during deprotection, see refs 5, pp. 217–23; 13, pp. 565–8; 14, pp. 69; 16, pp. 82–4; and 9 (derivatives). The N^i-formyl group is the only protective group of tryptophan that has been used in synthesis [333–335]. It is stable to acid and base conditions during solid phase synthesis. Cleavage has been carried out using (i) 1 M NH_4HCO_3 (pH 9, 25°, 24 h), (ii) liq. NH_3/hydroxylamine (100 equiv., 2 h) [334], or (iii) H_2NNH_2-H_2O/DMF (1:10, 25°, 48 h) [335], (iv) HF/1,2-ethanedithiol (95:5; 0°, 10 min) [336]. N^α-Formylation (5–8%) was observed during cleavage procedures (i) and (ii).

Amide protective groups

Amide protective groups have been used infrequently in peptide synthesis. The advantages of protecting amide groups are confined to achieving (a) improved solubility of hydrophilic peptides in organic solvents [337, 338] or (b) improved crystallization which would facilitate purification [339], see refs 5, pp. 199–208; 13, pp. 711–27; 16, pp. 49–58; and 9 (derivatives). Two N^{ω}-amide protective groups are recommended (ref. 16, pp. 53, 56). The 4,4′-dimethyloxybenzhydryl [Bzh(OMe)$_2$] [340] group is introduced by reaction of Z-Gln-OH or Z-Asn-OH with 4,4′-dimethyloxybenzhydryl group/AcOH/H_2SO_4 (trace). It is stable toward H_2/Pd, and cleaved by TFA/anisole (25°, 2–3 h or 72°, 5 min) or HBr/AcOH/indole (25°, 5–10 min). The 2,4-dimethyloxybenzyl[(Bzl(OMe)$_2$] group [341, 342] is introduced by reaction of Z-Glu-OBzl or Z-Asp-OBzl with 2,4-dimethyloxybenzylamine/DCC-HOBt followed by OH^-. It is stable toward H_2/Pd, 1 N MeOH/HCl, and cleaved by TFA (72°, 1 h or 28°, 30–72 h) or HF (3 h).

9.8 PROTECTED AMINO ACIDS FOR PEPTIDE BOND FORMATION

(a) Methods of peptide synthesis

The major methods of peptide bond formation continue [343] to be the active ester [344], azide [345, 346], carbodiimide [347–349] and mixed anhydride [350–353] procedures. The most recent reviews of these and other [354] methods of peptide synthesis in solution were published in 1979 [343]. In 1980 an exhaustive up-to-date account on solid phase peptide synthesis [5] was published. Other methods reviewed in the same volume include liquid phase synthesis [355], polymeric reagents [356], four component synthesis [357], oxidation–reduction condensation [358], repetitive methods [359], and partial synthesis [360]. In volume 5 [361] recombinant DNA synthesis [362], acidolytic deprotection [363], side reactions [364, 365], sequence dependence of racemization [366], α,β-dehydropeptides [367], and unusual amino acids in peptide synthesis [368] have been reviewed.

The most complete description of synthetic methods has been presented in the encyclopaedic Houben–Weyl (Part II) [369] which covers the literature up to early 1974; see active esters, pp. 1–102; carbodiimide method, pp. 103–17; mixed and symmetrical anhydrides, pp. 169–270; and azide method, pp. 296–322. Many experimental procedures for the preparation of protected amino acids and peptides are described in full detail.

According to a 1975 survey (ref. 354, pp. 97–9) dicyclohexylcarbodiimide was by far the most frequently used reagent in solid phase synthesis, followed by active esters (10% of DCC). In solution synthesis, active esters were most often used, followed by DCC alone and DCC/HOBt or HOSu (~50% of active ester), then mixed anhydrides (~25%) and azides (~10%). This pattern was very similar to those of two earlier surveys compiled in 1968 [370] and in 1960–62 [371]. Although the use of established methods of peptide synthesis is clearly preferred, the need for new and improved methodology is widely recognized.

The azide method is the least prone to racemization [372] and for this reason it was used by Yajima and Fujii for the condensation of 33 fragments in their successful synthesis of ribonuclease [21, 37]. Yet the method has not been used frequently, perhaps because peptide hydrazide preparation by ester hydrazinolysis tended to be slow and inconvenient. Recently, several of these intermediates have become more easily accessible. Peptide acids can be directly converted into hydrazides by DCC/HOBt condensation with hydrazine (1.0 to 1.2 equivalents) [165]. This procedure is much milder than hydrazinolysis and avoids side reactions. Even more conveniently, reaction of diphenylphosphoryl azide with N^{α}-protected amino acid or peptide carboxylates in DMF in the presence of base produces the respective azide [1], Reaction 9.12 [166].

$$\text{R-COO}^- + \text{N}_3\overset{\text{O}}{\overset{\|}{\text{P}}}(\text{OPh}_2) \rightarrow \left[\begin{array}{c}\text{R-CO-O} \\ \text{N}_3\text{-}\text{P}(\text{OPh})_2 \\ | \\ \text{O}^-\end{array}\right] \rightarrow \text{R-CON}_3 + \text{HO}\overset{\text{O}}{\overset{\|}{\text{P}}}(\text{OPh})_2 \quad (9.12)$$

Moreover, $N_3PO(OPh)_2$ may be added to a mixture of amine and carboxyl components [373] as a coupling agent to form the acyl azide *in situ*. Increased use of azide segment condensation may thus be encouraged.

The mixed anhydride method would also seem to deserve more use [353]. In a recent study [374], the heat stability of mixed anhydrides of Boc-amino acids with isobutyl chlorocarbonate was found to be much higher than indicated in the literature and their propensity for wrong opening (formation of isobutyloxycarbonyl derivatives) appeared to be lower. The method was therefore adapted for use in solid phase synthesis.

Symmetrical anhydrides of Boc-amino acids, generated *in situ*, have been used with advantage in solid phase synthesis for over ten years [375, 376]. If they are prepared with the use of water-soluble carbodiimide [e.g. *N*-(3-dimethylaminopropyl)-*N'*-ethylcarbodiimide HCl], many symmetrical anhydrides of Z- and Boc-amino acids are sufficiently stable for isolation and storage [377]. Symmetrical anhydrides of N^{α}-Fmoc-amino acids are also very stable. They have been used successfully in solid phase synthesis [107].

Efforts to keep racemization as low as possible during segment condensation in synthesis of large peptides continue to be of great importance. Both the azide method [378] and the use of additives (e.g. HOBt) in carbodiimide coupling, which generally suppress racemization, occasionally fail to do so [372, 379]. Utilizing the photochemical properties of the 5-bromo-7-nitroindolinyl (Bni) group, a procedure for segment coupling with minimal racemization was recently developed [380]. Irradiation at 360 nm of a mixture of N^{α}-protected peptide-Bni-1-amide and amine component in tetramethylurea/toluene under nitrogen resulted in 70–95% coupling, with liberation of 5-bromo-7-nitroindoline, Reaction 9.13.

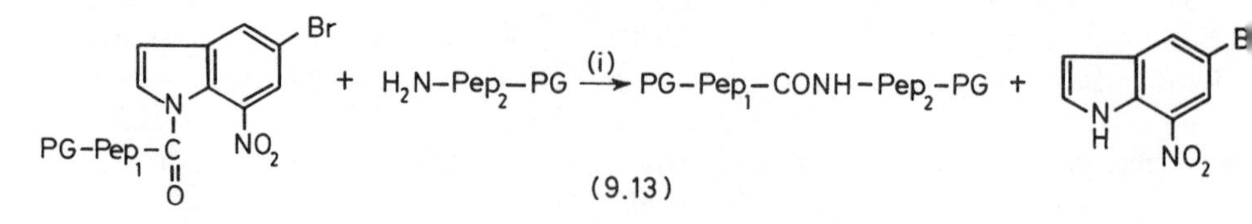

The conditions [380] are (i) $Me_2NCONMe_2$/toluene, *ca* 360 nm, 25°, 1–3 h, 70–95%, PG = protective group. Introduction of the Bni group is carried out by treatment of N^{α}-Fmoc-amino acids with 5-bromo-7-nitroindoline in $SOCl_2$/toluene (40–70°, 2–3 h, 99.5%) followed by removal of the Fmoc group by brief exposure to piperidine.

Until recently, determination of racemization during coupling required time consuming enzyme digestion studies [381, 382]. Their accuracy was not always satisfactory. Several years ago, a gas chromatographic (GC) amino acid enantiomer separation using a chiral stationary phase was developed, which allows rapid determination of racemization during peptide synthesis [383, 384]. The accuracy of this method has been greatly improved by hydrolysis of the peptide in deuterium chloride followed by GC separation and determination of the D/L ratio by mass spectrometry [385].

(b) Sources of references of amino acid derivatives

Amino acid derivatives required for use in carbodiimide or mixed anhydride coupling are simply N^{α}- and side chain-protected free acids. For azide coupling, hydrazide precursors may readily be obtained from free acids by treatment with DCC/HOBt/N_2H_4 [165] or azides can be prepared directly using diphenylphosphoryl azide [166] (see Section 9.8a). However, preparation of active esters requires additional effort, i.e. carboxyl preactivation, typically with the use of DCC. In all instances, the extent of side-chain protection for each synthesis will necessarily depend on the strategy (maximal or minimal protection, see p. 1) to be used.

The recent review of the major methods [343] (Section 9.8a) includes 13 tables listing properties (mp., $[\alpha]_D$) and references (up to 1977) of active esters [344] of N^{α}-Z- and Boc-amino acids [ONp, ONp(o), OTcp(2,4,5), OPcp, OPfp, ONSu, OPip, and OQc]. A selected list of amino acids which are useful in peptide synthesis with literature coverage up to early 1974 is compiled in ref. 9. The Houben–Weyl Part II [369] covers the literature up to the same time. Experimental procedures for the preparation of many protected amino acids are described in full detail. Lists of active esters of protected amino acids are compiled on pp. 27–35 (ONp); pp. 37–8 [ONp(o)]; pp. 45–8 [OTcp(2,4,5)]; pp. 53–5 (OPcp); pp. 57–8 (OPfp); p. 61 [Ph(SO_2MeO]; p. 67 (OQ, OQc); pp. 135–6 (OPip); pp. 161–6 (ONSu). For experimental procedures in Houben–Weyl concerning preparations useful for carbodiimide coupling, see pp. 103–17, for mixed anhydride coupling, pp. 169–270, and for azide coupling, pp. 296–323.

Any comprehensive search from 1974 to date should consult the *Specialist Periodical Reports* [17] starting with Vol. 7, Chapter 13, Appendix II. Also very useful are the *Chemical Abstracts* [386] and the *Japanese Peptide Information* [387].

With the commercial availability of a wide selection of N^{α}- and N^{ω}-protected benzyloxycarbonyl- and *t*-butyloxycarbonylamino acid derivatives, the majority of compounds used in routine peptide synthesis are now produced industrially rather than in research laboratories. However, reliable grade specifications are not generally established for these product lines. It is, therefore, important to carefully analyse commercially produced, protected amino acids prior to use for their chemical and optical purity. Analytical HPLC provides a powerful tool for rapid assessment of chemical homogeneity [384, 388, 389]. Gas chromatography on Chirasil-Val (a valine-containing polysiloxane) columns permits rapid and accurate determination of the chiral purity of protected amino acid derivatives [383–385] (Section 9.8a). Nevertheless, many novel amino acid derivatives continue to be developed in basic research laboratories.

9.9 CONCLUSION

The variety of protected amino acids of proven or potential value in peptide synthesis has dramatically increased since Greenstein and Winitz published their authoritative work in 1961 [6]. During this period striking advances have been made in the synthesis of biologically active peptides. Notwithstanding, formidable challenges still remain. Concerning protective groups, the achievement of quantitative final deprotection continues to be a major concern. Coupling of large peptide segments in good yields and with minimal racemization still presents problems. New methodology such as enzymic synthesis of peptides [390] will have different protective group requirements. To meet these challenges, the preparation of novel protected amino acids for use in peptide synthesis will continue to be of great importance.

REFERENCES

1. Hirschmann, R. (1971) in *Peptides 1969* (ed. E. Scoffone), North-Holland Publ., Amsterdam, p. 138.
2. Finn, F.M., Hofmann, K. (1976) in *The Proteins* (eds H. Neurath, R.L. Hill) (Vol. 2) 3rd edn., Academic Press, New York, pp. 105–253; see also Inman, J.K. in ref. 15, p. 253.
3. Wünsch, E. (1967) *Z. Naturforsch. Teil B*, **22**, 1269.
4. Merrifield, R.B. (1963) *J. Am. Chem. Soc.*, **85**, 2149.
5. Barany, G., Merrifield, R.B. (1980) in *The Peptides, Analysis, Synthesis, Biology* (eds E. Gross, J. Meienhofer) (Vol. 2) Academic Press, New York, p. 1.
6. Greenstein, J.P., Winitz, M. (1961) *Chemistry of the Amino Acids* (Vol. 3), Wiley, New York.
7. Schröder, E., Lübke, K. (1965) *The Peptides*, (Vol. 1) Academic Press, New York, pp. 1–270.
8. Bodanszky, M., Ondetti, M.A. (1966) *Peptide Synthesis*, Wiley, New York.
9. Fletcher, G.A., Jones, J.H. (1972) *Int. J. Peptide Protein Res.*, **4**, 347.

10. Fletcher, G.A., Jones, J.H. (1975) *Int. J. Peptide Protein Res.*, **7**, 91.
11. McOmie, J.F.W., ed. (1973) *Protective Groups in Organic Chemistry*, Plenum Press, New York.
12. Carpino, L.A. (1973) *Accts. Chem. Res.*, **6**, 191.
13. Wünsch, E., ed. (1974) *Synthese von Peptide* in *Houben-Weyl Methoden der Organischen Chemie*, (ed. E. Müller,) (Vol. 15, Part I) 4th edn., Thieme, Stuttgart.
14. Bodanszky, M., Klausner, S., Ondetti, M.A. (1976) *Peptide Synthesis*, Wiley, New York.
15. Greene, T.W. (1981) *Protective Groups in Organic Synthesis*, Wiley, New York.
16. Gross, E., Meienhofer, J. (eds) (1981) *The Peptides, Analysis, Synthesis, Biology* (Vol. 3), Academic Press, New York.
17. *Amino-acids, Peptides and Proteins, Specialist Periodical Reports*, The Chemical Society (Vol. 1) 1969 (Lit. of 1968), to (Vol. 11), 1981 (Lit. of 1978).
18. Schwyzer, R. (1961) in *Protides of the Biological Fluids*, Elsevier, Amsterdam, p. 27.
19. Felix, A.M., Jimenez, M.H., Meienhofer, J. (1980) *Org. Synth.*, **59**, 159.
20. Felix, A.M., Jimenez, M.H., Mowles, T., Meienhofer J. (1978) *Int. J. Peptide Protein Res.*, **11**, 329 (erratum, 12, 184).
21. Yajima, H., Fujii, N. (1981) *J. Am. Chem. Soc.*, **103**, 5867.
22. Sakakibara, S. (1977) in *Peptides* (eds M. Goodman and J. Meienhofer, J.), Wiley, New York, p. 436.
23. Erickson, B.W., Merrifield, R.B. (1973) *J. Am. Chem. Soc.*, **95**, 3750.
24. Yamashiro, D., Li, C.H. (1972) *Int. J. Peptide Protein Res.*, **4**, 181.
25. Yamashiro, D., Li, C.H. (1973) *J. Am. Chem. Soc.*, **95**, 1310.
26. Yamashiro, D., Li, C.H. (1973) *J. Org. Chem.*, **38**, 591.
27. Sakakibara, S. (1971) in *Chemistry and Biochemistry of Amino Acids, Peptides and Proteins* (ed. B. Weinstein) (Vol. 1) Dekker, New York, p. 51.
28. Tzougraki, C., Makofske, R.C., Gabriel, T.F. *et al.* (1980) *Int. J. Peptide Protein Res.*, **15**, 377.
29. Chang, C.D., Felix, A.M. Jimenez, M.H., Meienhofer, J. (1980) *Int. J. Peptide Protein Res.*, **15**, 485.
30. Atherton, E., Gait, M.J., Sheppard, R.C., Williams, B.J. (1979) *Bioorg, Chem.*, **8**, 351.
31. Wang, S.S. (1973) *J. Am. Chem. Soc.*, **95**, 1328.
32. Bodanszky, M., Deshmane, S.S., Martinez, J. (1979) *J. Org. Chem.*, **44**, 1622.
33. Bodanszky, A., Bodanszky, M., Chandramouli, B. *et al.* (1980) *J. Org. Chem.*, **45**, 72.
34. Kenner, G.W., Ramage, R., Galpin, I.J. (1979) in *Peptides, Structure and Biological Function* (eds E. Gross, and J. Meienhofer), Pierce Chemical Co., Rockford, Illinois, p. 431.
35. Rocchi, R., Guggi, A., Salvadori, S. *et al.* (1978) in *Peptides 1978* (eds I.Z. Siemon, and G. Kupryszewski), Wroclaw University Press, Poland, p. 649.
36. Ivanov, V.T., Tsetlin, V.I., Mikhaleva, I.I. *et al.* (1978) in *Peptides 1978* (eds I.Z. Siemon, and G. Kupryszewski Wroclaw University Press, Poland, p. 41.
37. Yajima, H., Fujii, N. (1983) in *The Peptides : Analysis, Synthesis, Biology* (eds E. Gross, and J. Meienhofer) (Vol. 5) Academic Press, New York, p. 65.
38. Bergmann, M., Zervas, L. (1932) *Ber. Dtsch. Chem. Ges.*, **65**, 1192.
39. Anderson, G.W., McGregor, A.C. (1957) *J. Am. Chem. Soc.*, **79**, 6180.
40. McKay, F.C., Albertson, N.F. (1957) *J. Am. Chem. Soc.*, **79**, 4686.
41. Ben-Ishai, D., Berger, A. (1952) *J. Org. Chem.*, **17**, 1564.
42. Ben-Ishai, D., Berger, A. (1954) *J. Org. Chem.*, **19**, 62.
43. Meienhofer, J. (1962) *Chimia*, **16**, 396.
44. Yajima, H., Kiso, Y., Ogawa, H. *et al.* (1975) *Chem. Pharm. Bull.* (*Tokyo*) **23**, 1164.
45. Yajima, H., Fujii, N., Ogawa, H., Kawatani, H. (1974) *J. Chem. Soc., Chem. Commun.*, 107.
46. Fujii, N., Yajima, H. (1981) *J. Chem. Soc. Perkin Trans. 1*, 831.
47. Felix, A.M., Heimer, E.P., Lambros, T.J. *et al.* (1978) *J. Org. Chem.*, **43**, 1494.
48. El Amin, B., Anantharamaiah, G.M., Royer, G.P., Means, G.E. (1979) *J. Org. Chem.*, **44**, 3442.
49. Sifferd, R.H., du Vigneaud, V. (1935) *J. Biol. Chem.*, **108**, 153.
50. du Vigneaud, V., Patterson, W.I. (1935) *J. Biol. Chem.*, **109**, 97.
51. du Vigneaud, V., Behrens, O.K. (1937) *J. Biol. Chem.*, **117**, 27.
52. Mairanovsky, V.G. (1976) *Angew. Chem. Int. Ed. Eng.*, **15**, 281.
53. Schwyzer, R., Rittel, W. (1961) *Helv. Chim. Acta.*, **44**, 161.
54. Wünsch, E., Zwick, A. (1964) *Chem. Ber.*, **97**, 3305.
55. Zervas, L., Hamalidis, C. (1965) *J. Am. Chem. Soc.*, **87**, 99.
56. Scott, J.W., Parker, D., Parrish, D.R. (1981) *Synth. Commun.*, **11**, 303.

57. Carpino, L.A., Carpino, B.A., Giza, C.A. *et al.* (1964) *Org. Synth.*, **44**, 20.
58. Scott, J.W., Parker, D. (1980) *Org. Prep. Proc. Int.*, **12**, 242.
59. Erickson, B.W., Merrifield, R.B. (1976) in *The Proteins* (eds H. Neurath, and R.L. Hill) (Vol. 2) 3rd edn, Academic Press, New York, p. 255.
60. Amit, B., Zehavi, U., Patchornik, A. (1974) *J. Org. Chem.*, **39**, 192.
61. Kappeler, H., Schwyzer, R. (1960) *Helv. Chim. Acta.*, **43**, 1453.
62. Schwyzer, R., Rittel, W. (1961) *Helv. Chim. Acta.*, **44**, 159.
63. Lundt, B.F., Johansen, N.L., Vølund, A., Markussen, J. (1978) *Int. J. Peptide Protein Res.*, **12**, 258.
64. Nitecki, D.E., Halpern, B. (1969) *Austral. J. Chem.*, **22**, 871.
65. Goodacre, J., Ponsford, R.J., Stirling, I. (1975) *Tetrahedron Lett.*, 3609.
66. Hiskey, R.G., Adams, J.B. (1966) *J. Org. Chem.*, **31**, 2178.
67. Schnabel, E., Klostermeyer, H., Berndt, H. (1971) *Justus Liebigs Ann. Chem.*, **749**, 90.
68. Itoh, M., Hagiwara, D., Kamiya, T. (1974) *Tetrahedron Lett.*, 4393.
69. Itoh, M., Hagiwara, D., Kamiya, T. (1977) *Bull. Chem. Soc. Jap.*, **50**, 718.
70. Itoh, M., Hagiwara, D., Kamiya, T. (1980) *Org. Synthesis*, **59**, 95.
71. Tarbell, D.S. Yamamoto, Y., Pope, B.M. (1972) *Proc. Nat. Acad. Sci. USA*, **69**, 730.
72. Moroder, L., Hallett, A., Wünsch, E. *et al.* (1976) *Hoppe-Seyler's Z. Physiol. Chem.*, **357**, 1651.
73. Lott, R.S., Chauhan, V.S., Stammer, C.H. (1979) *J. Chem. Soc., Chem. Commun.*, 495.
74. Tsuji, T., Kataoka, T., Yoshioka, M. *et al.* (1979) *Tetrahedron Lett.*, 2793.
75. Haas, W.L., Krumkalns, E.V., Gerzon, K. (1966) *J. Am. Chem. Soc.*, **88**, 1988.
76. Jager, G., Geiger, R. (1970) *Chem. Ber.*, **103**, 206.
77. Weygand, F., Hunger, K. (1962) *Chem. Ber.*, **95**, 1.
78. Jones, J.H., Young, G.T. (1966) *Chem. Ind.*, 1722.
79. Birr, C., Lochinger, W., Stahnke, G., Lang, P. (1972) *Justus Liebigs Ann. Chem.*, **763**, 162.
80. Birr, C. (1973) *Justus Liebigs Ann. Chem.*, 1652.
81. Birr, C., Nassal, M., Pipkorn, R. (1979) *Int. J. Peptide Protein Res.*, **13**, 287.
82. Birr, C., Weigand, K., Turan, A. (1981) *Biochim. Biophys. Acta*, **670**, 421.
83. Voss, C., Birr, C. (1981) *Hoppe-Seyler's Z. Physiol. Chem.*, **362**, 717.
84. Sieber, P., Iselin, B. (1968) *Helv. Chim. Acta*, **51**, 614.
85. Sieber, P., Iselin, B. (1969) *Helv. Chim. Acta*, **52**, 1525.
86. Riniker, B., Kamber, B., Sieber, P. (1975) *Helv. Chim. Acta*, **58**, 1086.
87. Mojsov, S., Merrifield, R.B. (1981) *Biochemistry*, **20**, 2950.
88. Danho, W., Sasaki, A., Bullesbach, E. *et al.* (1980) *Hoppe-Seyler's Z. Physiol. Chem.*, **361**, 747.
89. Helferich, B., Moog, L., Junger, A. (1925) *Ber. Dtsch. Chem. Ges.*, **58**, 872.
90. Zervas, L., Theodoropoulos, D.M. (1956) *J. Am. Chem. Soc.*, **78**, 1359.
91. Sieber, P., Kamber, B., Hartmann, A. *et al.* (1974) *Helv. Chim. Acta*, **57**, 2617.
92. Sieber, P., Kamber, B., Hartmann, A. *et al.* (1977) *Helv. Chim. Acta*, **60**, 27.
93. Zervas, L., Borovas, D., Gazis, E. (1963) *J. Am. Chem. Soc.*, **85**, 3660.
94. Kessler, W., Iselin, B. (1966) *Helv. Chim. Acta*, **49**, 1330.
95. Fontana, A., Marchiory, F., Moroder, L., Scoffone, E. (1966) *Tetrahedron Lett.*, 2985.
96. Stern, M., Warshawsky, A., Fridkin, M. (1979) *Int. J. Peptide Protein Res.*, **13**, 315.
97. Matsueda, R., Maruyama, H., Kitazawa, E. *et al.* (1975) *J. Am. Chem. Soc.*, **97**, 2573.
98. Matsueda, R., Walter, R. (1980) *Int. J. Peptide Protein Res.*, **16**, 392.
99. Matsueda, R., Theodoropoulos, D., Walter, R. (1979) in *Peptides, Structure and Biological Function* (eds E. Gross and J. Meienhofer), Pierce Chemical Co., Rockford, Illinois, p. 305
100. Mukaiyama, T., Matsueda, R., Ueki, M. (1980) in *The Peptides, Analysis, Synthesis, Biology* (eds E. Gross, E. and J. Meienhofer) (Vol. 2) Academic Press, New York, p. 383.
101 Carpino, L.A., Han, G.Y. (1970) *J. Am. Chem. Soc.*, **92**, 5748.
102. Carpino, L.A., Han, G.Y. (1972) *J. Org. Chem.*, **37**, 3404.
103. Carpino, L.A., Han, G.Y. (1973) *J. Org. Chem.*, **38**, 4218.
104. Carpino, L.A., Williams, J.R., Lopusinski, A. (1978) *J. Chem. Soc., Chem. Commun.*, 450.
105. Meienhofer, J., Waki, M., Heimer, E.P. *et al.* (1979) *Int. J. Peptide Protein Res.*, **13**, 35.
106. Chang, C.-D., Waki, M., Ahmad, M. *et al.* (1980) *Int. J. Peptide Protein Res.*, **15**, 59.
107. Heimer, E.P., Chang, C.-D., Lambros, T., Meienhofer, J. (1981) *Int. J. Peptide Protein Res.*, **18**, 237.
108. Chang, C.-D., Meienhofer, J. (1978) *Int. J. Peptide Protein Res.*, **11**, 246.
109. Atherton, E., Fox, H., Harkiss, D. *et al.* (1978) *J. Chem. Soc., Chem. Commun.*, 537.

110. Atherton, E., Logan, C.J., Sheppard, R.C. (1981), *J. Chem. Soc., Perkin Trans. 1*, 538.
111. Atherton, E., Bury, C., Sheppard, R.C., William, B.J. (1979) *Tetrahedron Lett.*, 3041.
112. Bolin, D., Meienhofer, J., unpublished.
113. Colombo, R. (1981) *Experientia*, **37**, 798.
114. Colombo, R. (1981) *J. Chem. Soc., Commun.*, 1012.
115. Colombo, R. (1982) *Int. J. Peptide Protein Res.*, **19**, 71.
116. Tesser, G.I. (1974) in *Peptides* 1974 (ed Y. Wolman) Wiley, New York, p. 54.
117. Tesser, G.I., Balvert-Geers, I.C. (1975) *Int. J. Peptide Protein Res.*, **7**, 295.
118. Geiger, R., Geisen, K., Summ, H.D., Langer D. (1975) *Hoppe-Seyler's Z. Physiol. Chem.*, **356**, 1635.
119. Geiger, R. (1976) *Chem. Ztg.*, **100**, 111.
120. Geiger, R., Obermeyer, R., Tesser, G.I. (1975) *Chem. Ber.*, **108**, 2758.
121. Friesen, H.J. (1980) in *Insulin, Chemistry, Structure and Function of Insulin and Related Hormones*, (eds D. Brandenburg and A. Wollmer), Walter de Gruyter & Co., p. 125.
122. Boon, P.J., Tesser, G.I., Nivard, R.J.F. (1979) *Proc. Nat. Acad. Sci. USA*, **76**, 61.
123. Woodward, R.B., Heusler, K., Gosteli, J. *et al.* (1966) *J. Am. Chem. Soc.*, **88**, 852.
124. Fujii, N., Yajima, H. (1981) *J. Chem. Soc., Perkin Trans. 1*, 804.
125. Just, G., Grozinger, K. (1976) *Synthesis*, 457.
126. Semmelhack, M.F., Heinson, G.E. (1972) *J. Am. Chem. Soc.*, **94**, 5139.
127. Kasafirek, E. (1972) *Tetrahedron Lett.*, 2021.
128. Yajima, H., Watanabe, H., Okamoto, M. (1971) *Chem. Pharm. Bull.*, **19**, 2185.
129. Carson, J.F. (1981) *Synthesis*, 268.
130. Windholz, T.B., Johnston, D.B.R. (1967) *Tetrahedron Lett.*, 2555.
131. Stevenson, D., Young, G.T. (1969) *J. Chem. Soc. C*, 2389.
132. Coyle, S., Keller, O., Young, G.T. (1979) *J. Chem. Soc., Perkin Trans. 1*, 1459.
133. Veber, D.F., Palaveda, W.J., Lee, Y.C., Hirschmann, R. (1977) *J. Org. Chem.*, **42**, 3286.
134. Barany, G., Merrifield, R.B. (1977) *J. Am. Chem. Soc.*, **99**, 7363.
135. Barany, G., Fulpius, B.W., King, T.P. (1978) *J. Org. Chem.* **43**, 2930.
136. Barany, G., Merrifield, R.B. (1979) *Anal. Biochem.*, **95**, 160.
137. Barany, G. (1979) in *Peptides, Structure and Biological Functiin* (eds E. Gross and J. Meienhofer), Pierce Chemical Co., Rockford, Illinois, p. 313.
138. Kemp, D.S., Bolin, D.R., Parham, M.E. (1981) *Tetrahedron Lett.*, 4575.
139. Kemp, D.S., Hanson, G. (1981) *J. Org. Chem.*, **46**, 4971.
140. Voelter, W., Kalbacher, H. (1981) in *Peptides 1980* (ed K. Brunfeldt), Scriptor, Copenhagen, p. 144.
141. Carpino, L.A., Tsao, J.H., Ringsdorf, H. *et al.* (1978) *J. Chem. Soc., Chem. Commun.*, 358.
142. Wünsch, E., Moroder, L., Keller, O. (1981) *Hoppe-Seyler's Z. Physiol. Chem.*, **362**, 1289.
143. Eckert, H., Listl, M., Ugi, I. (1978) *Angew. Chem. Int. Ed. Eng.*, **17**, 361.
144. Fukuda, T., Kitada, C., Fujino, M. (1978) *J. Chem. Soc., Chem. Commun.*, 220.
145. Ueki, M., Ikeda, S., Tonegawa, F. (1977) in *Peptides* (eds M. Goodman and J. Meienhofer), Wiley, New York, p. 546.
146. Ueki, M., Inazu, T., Ikeda, S. (1979) *Bull. Chem. Soc. Jpn*, **52**, 2424.
147. Ikeda, S., Tonegawa, F., Shikano, E. *et al.* (1979) *Bull. Chem. Soc. Jpn*, **52**, 143.
148. Weinstein, B., Steiner, P.A. (1979) in *Peptides, Structure and Biological Function* (eds E. Gross and J. Meienhofer), Pierce Chemical Co., Rockford, Illinois, p. 329.
149. Olafson, R.A., Yamamoto, Y.S., Mancowicz, D.J. (1977) *Tetrahedron Lett.*, 1563.
150. Kornblum, N., Scott, A. (1977) *J. Org. Chem.*, **42**, 399.
151. Haslam, E. (1979) *Chem. Ind. (London)*, 610.
152. Haslam, E. (1980) *Tetrahedron*, **36**, 2409.
153. Steglich, W., Hofle, G. (1969) *Angew. Chem. Int. Ed. Eng.*, **8**, 981.
154. Höfle, G., Steglich, W., Vorbrüggen, H. (1978) *Angew. Chem. Int. Ed. Eng.*, **17**, 569.
155. Wang, S.S. (1975) *J. Org. Chem.*, **40**, 1235.
156. Hassner, E., Alexanian, V. (1978) *Tetrahedron Lett.*, 4475.
157. Atherton, E., Benoiton, N.L., Brown, E. *et al.* (1981) *J. Chem. Soc. Chem. Commun.*, 336.
158. Wang, S.S., Wang, B.S.H., Tam, J.P., Merrifield, R.B. (1981) in *Peptides, Synthesis, Structure, Function* (eds D.H. Rich and E. Gross), Pierce Chemical Co., Rockford, Illinois, 197.
159. Pfeffer, P.E., Silbert, L.S. (1976) *J. Org. Chem.*, **41**, 1373.

160. Wang, S.S., Gisin, B.F., Winter, D.P. *et al.* (1977) *J. Org. Chem.*, **42**, 1286.
161. Brechbühler, H., Büchi, H., Hatz, E. *et al.* (1965) *Helv. Chim. Acta*, **48**, 1746.
162. Rehn, D., Ugi, I. (1977) *J. Chem. Res. Synop.*, 119.
163. Felix, A.M. (1974) *J. Org. Chem.*, **39**, 1427.
164. Sieber, P. (1977) *Helv. Chim. Acta*, **60**, 2711.
165. Wang, S.S., Kulesha, I.D., Winter, D.P. *et al.* (1978) *Int. J. Peptide Protein Res.*, **11**, 297.
166. Shioiri, T., Yamada, S. (1974) *Chem. Pharm. Bull.*, **22**, 849.
167. Brenner, M., Huber, W. (1953) *Helv. Chim. Acta*, **36**, 1109.
168. Ramachandran, J., Li, C.H. (1963) *J. Org. Chem.*, **28**, 173.
169. Anwar, K., Spatola, A.F. (1980) *Synthesis*. 929.
170. Anwar, K., Spatola, A.F. (1981) *Tetrahedron Lett.*, 4369.
171. Roeske, R. (1959) *Chem. Ind. London*, 1121.
172. Anderson, G.W., Callahan, F.M. (1960) *J. Am. Chem. Soc.*, **82**, 3359.
173. Roeske, R. (1963) *J. Org. Chem.*, **28**, 1251.
174. Taschner, E., Chimiak, A., Bator, B., Sokolowska, T. (1961) *Justus Liebigs Ann. Chem.*, **646**, 134.
175. Miller, A.W., Stirling, C.J.M. (1968) *J. Chem. Soc. C*, 2612.
176. Ludescher, U., Schwyzer, R. (1972) *Helv. Chim. Acta*, **55**, 2052.
177. Bodanszky, M., Bednarek, M., Bodanszky, A., Tolle, J.C. (1981) in *Peptides* 1980 (ed K. Brunfeldt), Scriptor, Copenhagen, p. 93.
178. Carson, J.F. (1979) *Synthesis*, 24.
179. Camble, R., Garner, R., Young, G.T. (1969) *J. Chem. Soc. C*, 1911.
180. Fletcher, G.A., Gosden, A., Nash, P.P., Young, G.T. (1973) in *Peptides 1972* (eds H. Hanson and H.D. Jakubke), Elsevier, New York, p. 65.
181. Kenner, G.W., Seely, J.H. (1972) *J. Am. Chem. Soc.*, **94**, 3259.
182. Galpin. I.J., Handa, B.K., Hudson, D. *et al.* (1976) in *Peptides 1976* (ed A. Loffet), editions de l'Universite de Bruxelles, Belgium, p. 247.
183. Choudhury, A.M., Kenner, G.W., Moore, S. *et al.* (1976) in *Peptides 1976* (ed A. Loffet), Editions de l'Universite de Bruxelles, Belgium, p. 257.
184. Meienhofer, J., Studer, R.O., Felix, A.M., Gillessen, D., unpublished.
185. Galpin, I.J., Hardy, P.M., Kenner, G.W. *et al.* (1979) *Tetrahedron*, **35**, 2577.
186. Castro, B., Evin, G., Selve, C., Seyer, R. (1977) *Synthesis*, **6**, 413.
187. Stelakatos, G.C., Paganous, A., Zervas, L. (1966) *J. Chem. Soc. C*, 1191.
188. Hendrickson, J.B., Kendall, C. (1970) *Tetrahedron Lett.*, 343.
189. Tailor-Papadimitrou, J., Yovanidis, C., Paganous, A., Zervas, L. (1967) *J. Chem. Soc. C*, 1830.
190. Sieber, P., Andreatta, R.H., Eisler, K. *et al.* (1977) in *Peptides* (eds M. Goodman, M. and J. Meienhofer), Wiley, New York, p. 543.
191. Kemp, D.S., Reczek, J. (1977) *Tetrahedron Lett.*, 1031.
192. Stelakatos, G.C., Solomos-Avaridis, C., Karayannakis, P. *et al.* (1981) in *Peptides 1980* (ed K. Brunfeldt), Scriptor, Copenhagen, p. 133.
193. Mizoguchi, T., Lewin, G., Woolley, D.W., Stewart, J.M. (1968) *J. Org. Chem.*, **33**, 903.
194. Sugano, H., Miyoshi, M. (1976) *J. Org. Chem.*, **41**, 2352.
195. Chillemi, D., Sareo, K., Scoffone, E. (1957) *Gazetta Chem. Ital.*, **87**, 1356.
196. Morley, J.S. (1967) *J. Chem. Soc. C*, 2410.
197. Beyerman, H.C., Bontekoe, J.S. (1962) *Recl. Trav. Chim Pays-Bas*, **81**, 691.
198. Jung, M.E., Lyster, M.A. (1977) *J. Org. Chem.*, **42**, 3761.
199. Jung, M.E., Lyster, M.A. (1977) *J. Am. Chem. Soc.*, **99**, 968.
200. Jung, M.E., Lyster, M.A. (1978) *J. Chem. Soc., Chem. Commun.*, 315.
201. Wünsch, E., Jentsch, J. (1964) *Chem. Ber.*, **97**, 2490.
202. Lemieux, R.U., Drieguez, H. (1975) *J. Am. Chem. Soc.*, **97**, 4069.
203. Macrae, R., Young, G.T. (1975) *J. Chem. Soc., Perkin Trans. 1*, 1185.
204. Gosden, A., Stevenson, D., Young, G.T. (1972) *J. Chem. Soc., Chem. Commun.*, 1123.
205. Gosden, A., Macrae, R., Young, G.T. (1977) *J. Chem. Res., Synop.* 22, Microfiche 0317.
206. Iselin, B., Schwyzer, R. (1956) *Helv. Chim. Acta*, **39**, 57.
207. Weygand, F., Steglich, W., Fraunburger, F. *et al.* (1968) *Chem. Ber.* **101**, 923.
208. Hirschmann, R., Schwam, H. Strachan, R.G. *et al.* (1971) *J. Am. Chem. Soc.*, **93**, 2746.
209. Chauhan, V.S., Ratcliffe, S.J., Young, G.T. (1980) *Int. J. Peptide Protein Res.*, **15**, 96.
210. Chen, F.M.F., Benoiton, N.L., (1979) *J. Org. Chem.*, **44**, 2299.

211. Kolodziejczyk, A.M., Manning, M. (1981) *J. Org. Chem.*, **46**, 1944.
212. Peña, C., Stewart, J.M., Paladini, A.C. *et al.* (1975) in *Peptides : Chemistry, Structure and Biology* (eds R. Walter and J. Meienhofer), Ann Arbor Sci. Publ., Ann Arbor, Michigan, p. 523.
213. Stewart, J.M., Pena, C., Matsueda, G.R., Harris, K. (1976) in *Peptides* 1976 (ed A. Loffet), Editions de l'Universite de Bruxelles, Belgium, p. 285.
214. Fridkin, M., Hazum, E., Tauber-Finkelstein, M., Shaltiel, S. (1977) *Arch. Biochem. Biophys.*, **178**, 517.
215. Hiskey, R.G., Rao, V.R., Rhodes, W.G. (1973) in *Protective Groups in Organic Chemistry* (ed. J.F.W. McOmie), Plenum Press, New York, p. 235.
216. Hiskey, R.G. (1981) in *The Peptides, Analysis, Synthesis, Biology* (eds E. Gross and J. Meienhofer) (Vol. 3), Academic Press, New York, p. 137.
217. Wolman, Y. (1974) in *The Chemistry of the Thiol Group* (ed. S. Patai) (Vol. 15/2), Wiley, New York, p. 669.
218. Photaki, I. (1973) in *The Chemistry of Polypeptides* (ed P.G. Katsoyannis), Plenum Press, New York, p. 59.
219. Du Vigneaud, V., Audrieth, L.F., Loring, H.S. (1930) *J. Am. Chem. Soc.*, **52**, 4500.
220. Wood, J.L., Du Vigneaud, V. (1939) *J. Biol. Chem.*, **130**, 109.
221. Hope, D.B., Morgan, C.D., Wälti, M. (1970) *J. Chem. Soc. C*, 270.
222. Yamashiro, D., Noble, R., Li, C.H. (1972) in *Chemistry and Biology of Peptides* (ed J. Meienhofer), Ann Arbor Sci. Publ., Ann Arbor, Michigan, p. 197.
223. Yamashiro, D., Noble, R. Li, C.H. (1973) *J. Org. Chem.*, **38**, 3561.
224. Hruby, V.J., Upson, D.A., Agarval, N.S. (1977) *J. Org. Chem.*, **42**, 3552.
225. Erickson, B.W., Merrifield, R.B. (1973) *J. Am. Chem. Soc.*, **95**, 3750.
226. Live, D.H., Agosta, W.C., Cowburn, D. (1977) *J. Org. Chem.*, **42**, 3556.
227. Coyle, S., Hallett, A., Munns, M.S., Young, G.T. (1981) *J. Chem. Soc. Perkin Trans. 1*, 522.
228. Weber, U., Hartter, P., Flohe, L. (1970) *Hoppe-Seyler's Z. Physiol. chem.*, **351**, 1384.
229. Wünsch, E., Spangenberg R. (1971) in *Peptides* 1969 (ed. E. Scoffone), North-Holand Publ. Co., Amsterdam, p. 30.
230. Moroder, L., Gemeiner, M., Goehring, W. *et al.* (1981) *Biopolymers*, **20**, 17.
231. Moroder, L., Gemeiner, M., Goehring, W. *et al.* (1981) in *Peptides 1980* (ed. K. Brunfeldt), Scriptor, Copenhagen, p. 121.
232. Van Rietschoten, J., Granier, C., Rochat, H. *et al.* (1975) *Eur. J. Biochem.*, **56**, 36.
233. Sakakibara, S., Shimonishi, Y., Kishida, Y. *et al.* (1967) *Bull. Chem. Soc. Jpn*, **40**, 2164.
234. Ives, D.A. (1969) *Can. J. Chem.*, **47**, 3697.
235. Inukai, N., Nakano, K., Murakami, M. (1968) *Bull. Chem. Soc. Jpn.*, **41**, 182.
236. Seely, J.H., Ruegg, U., Rudinger, J. (1973) in *Peptides 1972* (eds H. Hanson and H.D. Jakubke), North-Holland, Amsterdam, p. 86.
237. Wieland, T., Abe, K.J., Birr, C. (1977) *Justus Liebigs Ann. Chem.*, 371.
238. Photaki, I., Papadimitriou, J.T., Sakarellos, C. *et al.* (1970) *J. Chem. Soc. C*, 2683.
239. Hiskey, R.G., Adams, Jr., J.B. (1965) *J. Org. Chem.*, **30**, 1340.
240. Zervas, L., Photaki, I. (1962) *J. Am. Chem. Soc.*, **84**, 3887.
241. Fontana, A. (1975) *J. Chem. Soc., Chem. Commun.*, 976.
242. Hiskey, R.G., Tucker, W.P. (1962) *J. Am. Chem. Soc.*, **84**, 4794.
243. Hiskey, R.G., Muthukumaraswamy, N., Vunnam, R.R. (1975) *J. Org. Chem.*, **40**, 950.
244. Hiskey, R.G., Mizoguchi, T., Igeta, H. (1966) *J. Org. Chem.*, **31**, 1188.
245. Hiskey, R.G., Li, C.D., Vunnam, R.R. (1975) *J. Org. Chem.*, **40**, 950.
246. Zahn, H., Danho, W., Klostermeyer, H. *et al.* (1969) *Z. Naturforsch.*, **24B**, 1127.
247. Hiskey, R.G., Mizoguchi, T., Smithwick, E.L. (1967) *J. Org. Chem.*, **32**, 97.
248. Kamber, B. (1971) *Helv. Chim. Acta*, **54**, 398.
249. Kamber, B. (1971) *Helv. Chim. Acta*, **54**, 927.
250. Sieber, P., Kamber, B., Hartmann, A. *et al.* (1974) *Helv. Chim. Acta*, **57**, 2617.
251. Sieber, P., Kamber, B., Hartmann, A. *et al.* (1977) *Helv. Chim. Acta*, **60**, 27.
252. Kamber, B., Rittel, W. (1968) *Helv. Chim. Acta*, **51**, 2061.
253. Jones, Jr., D.A., Mikulec, R.A., Mazur, R.H. (1973) *J. Org. Chem.*, **38**, 2865.
254. Callahan, F.M., Anderson, G.W., Paul, R., Zimmerman, J.E. (1963) *J. Am. Chem. Soc.*, **85**, 201.
255. Felix, A.M., Jimenez, M.H., Meienhofer, J. (1977) in *Peptides* (eds M. Goodman and J. Meienhofer), Wiley, New York, p. 532.
256. Felix, A.M., Jimenez, M.H., Mowles, T., Meienhofer, J. (1978) *Int. J. Peptide Protein Res.*, **11**, 329.

257. Felix, A.M., Jimenez, M.H., Mowles, T., Meienhofer, J. (1978) *Int. J. Peptide Protein Res.*, **12**, 184.
258. Fujino, M., Nishimura, O. (1976) *J. Chem. Soc., Chem. Commun.*, 998.
259. Nishimura, O., Kitada, C., Fujino, M. (1978) *Chem. Pharm. Bull.*, **26**, 1576.
260. Pastuszak, J.J., Chimiak, A. (1981) *J. Org. Chem.*, **46**, 1868.
261. Hirschmann, R., Nutt, R.F., Veber, D.F. *et al.* (1969) *J. Am. Chem. Soc.*, **91**, 507.
262. Veber, D.F., Milkowski, J.D., Varga, S.L. *et al.* (1972) *J. Am. Chem. Soc.*, **94**, 5456.
263. Milkowski, J.D., Veber, D.F., Hirschmann, R. (1980) *Org. Synth.*, **59**, 190.
264. Chakravarty, P.K., Olson, R.K. (1978) *J. Org. Chem.*, **43**, 1270.
265. Williams, B.J., Young, G.T. (1979) in *Peptides, Structure and Biological Function* (eds E. Gross and J. Meienhofer), Pierce Chemical Co., Rockford, Illinois, p. 321.
266. Moroder, L., Marchiori, F., Borin, G., Scoffone, E. (1973) *Biopolymers*, **12**, 493.
267. Trudelle, Y., Caille, A. (1977) *Int. J. Pept. Protein Res.*, **10**, 291.
268. Guttman, S. (1966) *Helv. Chim. Acta*, **49**, 83.
269. Storey, H.T., Beacham, J. Cernosek, S.F. *et al.* (1972) *J. Am. Chem. Soc.*, **94**, 6170.
270. Romovacek, H., Dowd, S.R., Kawasaki, K. *et al.* (1979) *J. Am. Chem. Soc.*, **101**, 6081.
271. Hammerström, K., Lunkenheimer, W., Zahn, H. (1970) *Makromol. Chem.*, **133**, 41.
272. Rosamond, J.D., Ferger, M.F. (1976) *J. Med. Chem.* **19**, 873.
273. Brois, S.J., Pilot, J.F., Barnum, H.W. (1970) *J. Am. Chem. Soc.*, **92**, 7629.
274. Kamber, B. (1973) *Helv. Chim. Acta*, **56**, 1370.
275. Nokihara, K., Berndt, H. (1978) *J. Org. Chem.*, **43**, 4893.
276. Akabori, S., Sakakibara, S., Shimonishi, Y., Nobuhara, Y. (1964) *Bull. Chem. Soc. Jpn*, **37**, 433.
277. Fujii, H., Yajima, H. (1975) *Chem. Pharm. Bull.*, **23**, 1596.
278. Yajima, H., Ogawa, H., Fujii, N., Funakoshi, S. (1977) *Chem. Pharm. Bull.* **25**, 740.
279. Van Rietschoten, J., Muller, E.P., Granier, C. (1977) in *Peptides* (eds M. Goodman and J. Meienhofer), Wiley, New York, p. 552.
280. Ridge, R.M., Matsueda, G.R., Haber, E., Matsueda, R. (1981) in *Peptides, Synthesis, Structure, Function* (eds D.H. Rich and E. Gross), Pierce Chemical Co., Rockford, Illinois, p. 213.
281. Matsueda, R., Kamura, T., Kaiser, E.T., Matsueda, T.R. (1981) *Chem. Lett.*, 737.
282. Hazum, E., Gottlieb, P., Amit, B., Patchornik, A., Fridkin, M. (1981) in *Peptides* 1980 (ed. K. Brunfeldt), Scriptor, Copenhagen, p. 105.
283. Bergmann, M., Zervas, L., Rinke, H. (1934) *Hoppe-Seyler's Z. Physiol. Chem.*, **224**, 40.
284. Hofmann, K., Peckham, W.D., Rheiner, A. (1956) *J. Am. Chem. Soc.*, **78**, 238.
285. Clubb, M.E., Scopes, P.M., Young, G.T. (1960) *Chimia*, **14**, 373.
286. Scopeσ, P.M., Walshaw, K.B., Welford, M., Young, G.T. (1965) *J. Chem. Soc.*, 782.
287. Pless, J., Guttmann, S. (1978) in *Peptides* 1966 (eds H.C. Beyerman, A. van de Linde and W. Massen van den Brink), North-Holland Publ., Amsterdam, p. 50.
288. Hayakawa, T., Fujiwara, Y., Noguchi, J. (1967) *Bull. Chem. Soc. Jpn*, **40**, 1205.
289. Friedinger, R.M., Hirschmann, R., Veber, D.F. (1978) *J. Org. Chem.*, **43**, 4800.
290. Lenard, J. (1967) *J. Org. Chem.*, **32**, 250.
291. Schwyzer, R., Li, C.H. (1958) *Nature*, **182**, 1669.
292. Ramachandran, J., Li, C.H. (1962) *J. Org. Chem.*, **27**, 4006.
293. Mazur, R.H., Plume, G. (1968) *Experientia*, **24**, 661.
294. Nishimura, O., Fujino, M. (1976) *Chem. Pharm. Bull.*, **24**, 1568.
295. Yajima, H., Takeyama, M., Kanaki, J. *et al.* (1978) *Chem. Pharm. Bull.*, **26**, 3752.
296. Gros, C., de Garilhe, M.P., Costopanagiotis, A., Schwyzer, R. (1961) *Helv. Chim. Acta*, **44**, 2042.
297. Losse, G., Rüger, C. (1973) *Z. Chem.*, **13**, 344.
298. Weygand, F., Nintz, E. (1965) *Z. Naturforsch.*, **20B**, 429.
299. Jäqer, G., Geiger, R. (1970) *Chem. Ber.*, **103**, 1727.
300. Grønvald, F.C., Johansen, N.L., Lundt, B.F. (1981) in *Peptides 1980* (ed. K. Brunfeldt), Scriptor, Copenhaqen, p. 111.
301. Yankeelov, J.A., Jr. (1970) *Biochemistry*, **9**, 2433.
302. Toi, K., Bynum, E., Norris, E., Itano, H.A. (1967) *J. Biol. Chem.*, **242**, 1036.
303. Patthy, L., Smith, E.L. (1975) *J. Biol. Chem.*, **250**, 557.
304. Morihara, K., Oka, T., Tsuzuki, H. *et al.* (1979) in *Peptides, Structure and Biological Function* (eds E. Gross and J. Meienhofer), Pierce Chemical Co., Rockford, Illinois, p. 617.
305. Bannard, R.A., Casselman, A.A., Cockborn, W.F., Brown, G.M. (1958) *Can. J. Chem.*, **36**, 1541.
306. Bodanszky, M., Ondetti, M.A., Birkhimer, C., Thomas, P.L. (1964) *J. Am. Chem. Soc.*, **86**, 4452.
307. Borin, G., Toniolo, C., Moroder, L. *et al.* (1972) *Int. J. Peptide Protein Res.*, **4**, 37.

308. Cosand, W.L., Merrifield, R.B. (1977) *Proc. Nat. Acad. Sci. USA*, **74**, 2771.
309. Theodoropoulos, D. (1956) *J. Org. Chem.*, **21**, 1550.
310. Windridge, G.C. Jorgenson, E.C. (1971) *J. Am. Chem. Soc.*, **93**, 6318.
311. Khosla, M.C., Smeby, R.R., Bumpus, F.M. (1972) in *Chemistry and Biology of Peptides* (ed. J. Meienhofer), Ann Arbor Sci. Publ., Ann Arbor, Michigan, p. 227.
312. Shaltiel, S., (1967) *Biochem. Biophys. Res. Commun.*, **29**, 178.
313. Shaltiel, S., Fridkin, M. (1970) *Biochemistry*, **9**, 5122.
314. Sakakibara, S., Fujii, T. (1969) *Bull. Chem. Soc. Jpn*, **42**, 1466.
315. Fujii, T., Kimura, T., Sakakibara, S. (1976) *Bull. Chem. Soc. Jpn*, **49**, 1595.
316. Schnabel, E., Stoltefuss, J., Offe, H.A., Klanke, E. (1971) *Justus Liebigs Ann. Chem.*, **743**, 57.
317. Yamashira, D., Blake, J., Li, C.H. (1972) *J. Am. Chem. Soc.*, **94**, 2855.
318. Pozmev, V.F. (1978) *Zh. Obshch. Khim.*, **48**, 476.
319. Veber, D.F. (1975) in *Peptides, Chemistry, Structure, Biology* (ed. R. Walter and J. Meienhofer), Ann Arbor Sci. Publ., Ann Arbor, Michigan, p. 307.
320. Jones, J. H., Ramage, W.I., Witty, M.H. (1980) *Int. J. Peptide Protein Res.*, **15**, 301.
321. Brown, T., Jones, J.H. (1981) *J. Chem. Soc., Chem. Commun.*, 648.
322. Jones, J.H., Ramage, W.I. (1978) *J. Chem. Soc., Chem. Commun.*, 472.
323. Fletcher, A.R., Jones, J.H., Ramage, W.I., Stachulski, A.V. (1979) *J. Chem. Soc., Perkin Trans. 1*, 2261.
324. Iselin, B. (1961) *Helv. Chim. Acta*, **44**, 61.
325. Yamashiro, D. (1982) *Int. J. Peptide Protein Res.*, **20**, 63.
326. Houghten, R.A., Li, C.H. (1978) *Int. J. Peptide Protein Res.*, **11**, 325.
327. Houghten, R.A., Li, C.H. (1979) *Anal. Biochem.*, **98**, 36.
328. Tam, J.P., Heath, W.F., Merrifield, R.B. (1982) *Tetrahedron Lett.*, **23**, 2939.
329. Beyerman, H.C. Izeboud, E., Kranenburg, P., Voskamp, D. (1979) in *Peptides, Structure and Biological Functiin* (eds E. Gross and J. Meienhofer), Pierce Chemical Co., Rockford, Illinois, p. 333.
330. Naider, F., Bohak, Z. (1972) *Biochemistry*, **11**, 3208.
331. Jones, W.C., Rothgeb, T.M., Gurd, F.R. (1976) *J. Biol. Chem.*, **251**, 7452.
332. Noble, R.L., Yamashiro, D., Li, C.H. (1976) *J. Am. Chem. Soc.*, **98**, 2324.
333. Previero, A., Coletti-Previero, M.A., Cavadore, J.C. (1967) *Biochim. Biophys. Acta*, **147**, 453.
334. Yamashiro, D., Li, C.H. (1973) *J. Org. Chem.*, **38**, 2594.
335. Ohno, M., Tsukamoto, S., Sato, S., Izumiya, N. (1973) *Bull. Chem. Soc. Jpn*, 3280.
336. Matsueta, G.R. (1982) *Int. J. Peptide Protein Res.*, **20**, 26.
337. König, W., Geiger, R. (1972) *Chem. Ber.*, **105**, 2872.
338. Bajusz, S., Fauszt, I. (1972) *Acta Chim. Acad. Sci. Hung.*, **75**, 419.
339. Bajusz, S., Turan, A., Fauszt, I. (1972) in *Chemistry and Biology of Peptides* (ed. J. Meienhofer), Ann Arbor Sci. Publ., Ann Arbor, Michigan, p. 325.
340. König, W., Geiger, R. (1970) *Chem. Ber.*, **103**, 2041.
341. Weygand, F., Steglich, W., Bjarnason, J. *et al.* (1968) *Chem. Ber.*, **101**, 3623.
342. Pieta, P.G., Cavallo, P.F., Marshall, G.R. (1971) *J. Org. Chem.*, **36**, 3966.
343. Gross, E. and Meienhofer, J. (eds) (1979) *The Peptides, Analysis, Synthesis, Biology* (Vol. 1) Academic Press, New York.
344. Bodanszky, M. (1979) in *The Peptides, Analysis, Synthesis, Biology* (eds E. Gross and J. Meienhofer) (Vol. 1), Academic Press, New York, p. 106.
345. Curtius, T. (1890) *Ber. Dtsch. Chem. Ges.*, **23**, 3023.
346. Meienhofer, J. (1979) in *The Peptides, Analysis, Synthesis, Biology* (eds E. Gross and J. Meienhofer) (Vol. 1), Academic Press, New York, p. 197.
347. Khorana, H.G. (1953) *Chem. Rev.*, **53**, 145.
348. Sheehan, J.C., Hess, G.P. (1955) *J. Am. Chem. Soc.*, **77**, 1067.
349. Rich, D.H., Singh, J. (1979) in *The Peptides, Analysis, Synthesis, Biology* (eds E. Gross and J. Meienhofer) (Vol. 1), Academic Press, New York, p. 241.
350. Wieland, T., Berhard, H. (1951) *Justus Liebigs Ann. Chem.*, **572**, 190.
351. Boissonnas, R.A. (1951) *Helv. Chim. Acta*, **34**, 874.
352. Vaughan, J.R., Jr. (1951) *J. Am. Chem. Soc.*, **73**, 3547.
353. Meienhofer, J. (1979) in *The Peptides, Analysis, Synthesis, Biology* (eds E. Gross and M. Meienhofer) (Vol. 1), Academic Press, New York, p. 263.
354. Jones, J.H. (1979) in *The Peptides, Analysis, Synthesis, Biology* (eds E. Gross and J. Meienhofer) (Vol. 1), Academic Press, New York, p. 65.

355. Mutter, M., Bayer, E. (1980) in *The Peptides, Analysis, Synthesis, Biology* (eds E. Gross, J. Meienhofer) (Vol. 2), Academic Press, New York, p. 285.
356. Fridkin, M. (1980) in *The Peptides, Analysis, Synthesis, Biology* (eds E. Gross, J. Meienhofer) (Vol. 2) Academic Press, New York, p. 333.
357. Ugi, I. (1980) in *The Peptides, Analysis, Synthesis, Biology* (eds E. Gross, J. Meienhofer) (Vol. 2) Academic Press, New York, p. 365.
358. Mukaiyama, T., Matsueda, R., Ueki, M. (1980) in *The Peptides, Analysis, Synthesis, Biology* (eds E. Gross, J. Meienhofer) (Vol. 2), Academic Press, New York, p. 383.
359. Kisfaludy, L. (1980) in *The Peptides, Analysis, Synthesis, Biology* (eds E. Gross and J. Meienhofer) (Vol. 2), Academic Press, New York, p. 417.
360. Sheppard, R.C. (1980) in *The Peptides, Analysis, Synthesis, Biology* (eds E. Gross and J. Meienhofer) (Vol. 2), Academic Press, New York, p. 441.
361. Gross, E., Meienhofer, J. (eds) (1983) *The Peptides, Analysis, Synthesis, Biology* (Vol. 5), Academic Press, New York,
362. Goeddel, D., Wetzel, R. (1983) in *The Peptides, Analysis, Synthesis, Biology* (Vol. 5), Academic Press, New York, p. 1.
363. Yajima, H., Fujii, N. (1983) in *The Peptides, Analysis, Synthesis, Biology* (Vol. 5), Academic Press, New York, p. 65.
364. Bodanszky, M., Martinez, J. (1983) in *The Peptides, Analysis, Synthesis, Biology* (Vol. 5), Academic Press, New York, p. 111.
365. Bodanszky, M., Martinez, J. (1981) *Synthesis*, 333.
366. Benoiton, N.L. (1983) in *The Peptides, Analysis, Synthesis, Biology* (Vol. 5), Academic Press, New York, p. 217.
367. Izumiya, N., Noda, K., Shimohigashi, Y. (1983) in *The Peptides, Analysis, Synthesis, Biology* (Vol. 5), Academic Press, New York, p. 285.
368. Roberts, D.C., Vellaccio, F. (1983) in *The Peptides, Analysis, Synthesis, Biology* (Vol. 5), Academic Press, New York, p. 341
369. Wünsch, E. (ed) (1974) *Synthese von Peptiden* in *Houben-Weyl, Methoden der Organischen Chemie* (ed E. Muller) 4th edn, (Vol. 15, Part II), Thieme, Stuttgart.
370. Jones, J.H. (1970) in *Amino acids, Peptides and Proteins* (*Specialist Periodical Report* 5) (Vol. 3), The Chemical Soceity. London, p. 143.
371. Rydon, H.N. (1962) *R. Inst. Chem.*, Lect. Ser. No. 5.
372. Kemp, D.S. (1979) in *The Peptides, Analysis, Synthesis, Biology* (eds E. Gross and J. Meienhofer) (Vol. 1), Academic Press, New York, pp. 360–2.
373. Klausner, Y.S., Bodanszky, M. (1972) *Synthesis*, 453.
374. Fuller, W.D., Marr-Leisy, D., Chaturvedi, N.C. *et al.* (1981) in *Peptides, Synthesis, Structure, Function* (eds D.S. Rich and E. Gross) Pierce Chemical Co., Rockford, Illinois, p. 201.
375. Hagenmeier, H., Frank, H. (1972) *Hoppe-Seyler's Z. Physiol. Chem.*, **353**, 1973.
376. Lemaire, S., Yamashiro, D., Behrens, C., Li, C.H. (1977) *J. Am. Chem. Soc.*, **99**, 1577.
377. Chen, F.M., Kuroda, K., Benoiton, N.L. (1978) *Synthesis*, 928.
378. Sieber, P., Riniker, B., Brugger, M., Kamber, B., Rittel, W. (1970) *Helv. Chim. Acta*, **53**, 2135.
379. Felix, A.M., Meienhofer, J. *et al.* (1982) unpublished.
380. Pass, S., Amit, B., Patchornik, A. (1981) *J. Am. Chem. Soc.*, **103**, 7674.
381. Hofmann, K., Woolner, M.E., Spühler, G., Schwartz, E.T. (1958) *J. Am. Chem. Soc.*, **80**, 1486.
382. Finn, F.M. and Hofmann, K. (1976) in *The Proteins* (eds H. Neurath and R.L. Hill) (Vol. 2), 3rd edn, p. 179.
383. Frank. H., Nicholson, G.J., Bayer, E. (1978) *J. Chromatogr.* **167**, 187.
384. Meienhofer, J. (1981) *Biopolymers*, **20**, 1761.
385. Kusumoto, S., Matsukura, M., Shiba, T. (1981) *Biopolymers*, **20**, 1869.
386. *Chemical Abstracts*, Amino Acids, Peptides and Proteins, Organic Chemistry Section, No. 34.
387. Sakakibara, S., Seto, Y. (eds) (1975 to 1982) *Peptide Information*, (Vol. 1 to Vol. 8), Peptide Institute, Protein Research Foundation, 476 Ina, Minoh, Osaka 562, Japan.
388. Gabriel, T.F., Michalewsky, J., Meienhofer, J. (1981) in *Perspectives in Peptide Chemistry* (eds A. Eberle, R. Gieger and T. Wieland), Karger, Basel, p. 195.
389. Stein, S. (1981) in *The Peptides, Analysis, Synthesis, Biology* (eds E. Gross and J. Meienhofer), (Vol. 4), Academic Press, New York, p. 185.
390. Widmer, F., Breddam, K., Johansen, J.T. (1981) in *Peptides 1980* (ed K. Brunfeldt), Scriptor, Copenhagen, p. 46.

CHAPTER TEN

Resolution of Amino Acids

G.C. BARRETT

10.1 INTRODUCTION

Most of the purposes for which amino acids are needed call for resolved samples of the chiral amino acids. These are available either (i) from a natural source (ii) through asymmetric synthesis, or (iii) through resolution of a DL-amino acid obtained by synthesis. There are major limitations of supply involved in the dependence on a natural source for uncommon amino acids, and asymmetric synthesis has yet to be developed into a reliable general process. In any case, synthesis followed by resolution is usually a more attractive proposition for the supply of novel or uncommon amino acids in required amounts since, even if the other methods are at first sight more convenient, these also require separation stages and checks of optical purity. Assignment of absolute configuration is also a necessary step in the resolution of novel DL-amino acids; spectroscopic methods used for configurational assignments are discussed in Chapter 19.

10.2 GENERAL APPROACHES TO THE RESOLUTION OF DL-AMINO ACIDS

The major alternative approaches to the resolution of DL-amino acids which have been developed over many years are as follows.

(a) Conversion of DL-amino acids into diastereoisomeric salts or derivatives, separation, and liberation of the separated enantiomers

This, the method based on the classical Pasteur principle, continues to provide successful procedures in the amino acid field, even though it involves a 'trial and error' element in the selection of suitable reagents and solvents which may be tedious and time consuming.

(b) Chromatographic separation using a chiral stationary phase

Some aspects of this technique have been in use for many years (e.g. partition chromatography on cellulose), while other aspects are currently undergoing intensive investigation but have not yet been handed over to the preparative organic chemist for routine use.

(c) Use of enzymes

Where an amino acid oxidase is used, only one enantiomer of the DL-amino acid can be recovered (and the procedure is not a resolution, rather an enantioselective reaction), but where, for example, a protease is used as the catalyst for an amide-forming reaction, both enantiomers can be recovered separately. An advantage of enzyme-catalysed reactions whose specificity can be relied on, is the fact that the absolute configuration of the recovered amino acid enantiomer is known, although implicitly only L-enantiomers can be obtained by the use of a D-amino acid oxidase. Assumptions cannot be made, however, about the specificity of an enzyme towards analogues of common α-amino acids (e.g. α-substituted α-amino acids).

(d) Asymmetric transformations

Several examples have been established which allow for the racemization of the 'unwanted' enantiomer during, or after, the collection of the required enantiomer. This approach offers an economical resolution since, in principle, it allows a DL-amino acid to be converted entirely into one enantiomer.

(e) Preferential crystallization of one enantiomer from a racemate

Many examples of this procedure are now established, but there is, as yet, no firm understanding of the link between molecular structure and propensity towards crystallization in a form suitable for resolution by this technique. Although this is an important technique, it also involves a 'trial and error' element for any new resolution.

(f) Enantioselective processes other than enzyme-catalysed reactions: preferential destruction of one enantiomer of a DL-amino acid

Although there are some promising developments under these categories, based on enantiomeric discrimination by chiral catalysts mimicking enzymes, there is no suitable alternative to enzyme-catalysed reactions for preferential destruction of one enantiomer. Photodegradation of one enantiomer of a DL-amino acid and corresponding radiolytic methods, are being explored.

10.3 METHODS FOR THE RESOLUTION OF DL-AMINO ACIDS

10.3.1 Resolution of DL-amino acids through diastereoisomeric salt formation

The fractional crystallization of the pair of diastereoisomeric salts, formed between (i) a DL-amino acid and a chiral acid or a chiral metal complex (Table 10.1); (ii) an *N*-acyl-DL-amino acid and a chiral amine (Table 10.2); (iii) a DL-amino acid ester and a chiral acid (Table 10.2) constitutes the commonly used resolution procedure.

Relatively few examples of the separation of the diastereoisomeric salts formed between the DL-amino acid itself and a resolving agent have been described (Table 10.1). This straightforward two-stage procedure (fractional crystallization followed by recovery of the separate enantiomers and resolving agent) is restricted mainly to amino dicarboxylic acids or diaminocarboxylic acids.

Although *N*-acylation of the DL-amino acid and de-acylation of the separated enantiomers adds stages to the overall process, these are easily carried out and there is the advantage that the acyl derivatives are usually easily crystallized, thus assisting the isolation of a pure product. The alternative method in which an amino acid ester and an optically-active acid are combined to form a pair of diastereoisomeric salts may be generally less suitable since amino acid esters are capable of self-condensation and less easily purified.

Examples of procedures described in the recent literature are listed in Table 10.1, and Wilen's reviews can be consulted for detailed descriptions of theory and practice of resolution procedures [1,2]. Tables 10.1 and 10.2 are used to collect a considerable amount of material into a convenient form for readers seeking an overall view. However, this conceals a number of points of interest.

Table 10.1 Resolution of DL-amino acids through diastereoisomeric salt formation

DL-Amino acid	*Resolving agent*	*Reference*
Lysine	Optically active acid	a, b
Histidine	Optically active acid	c
Glutamic acid	Quinine (half salt)	d
Aspartic acid	Optically active base	e
Amino acids with neutral side-chains	Camphor-10-sulphonic acid	f–h
	Δ^4-Cholesten-3-one-6-sulphonic acid	i
	(+)-2-Hydroxylimonene-1-sulphonic acid	j
o-Tyrosine	1,1′-Binaphthyl-2,2′-di-*O*-phosphoric acid	k
	Tartaric acid	l–n
Lysine	Dodecatungstophosphoric acid and potassium $(-)_{546}$-[Co(EDTA)]	o
β-(2-Pyridyl)alanine	Tartaric acid	p
Histidine	The L-histidine complex precipitates first from a solution of DL-histidine and $(-)$-Co(EDTA)$^-$ in acidified aqueous ethanol	q
Aspartic and glutamic acid	(L-Argininato)Cu$(ClO_4)_2$	r
α-Methyl-*α*-amino acids	Alkaloid	s, t

[a] Berg, C.P. (1936) *J. Biol. Chem.*, **115**, 9; Boyle, P.H. (1971) *Quart. Rev.*, **25**, 323.
[b] Kearley, F.J. and Ingersoll, A.W. (1951) *J. Am. Chem. Soc.*, **73**, 4604.
[c] Pyman, F.L. (1911) *J. Chem. Soc.*, **99**, 1386.
[d] Radke, F.H., Fearing, R.B. and Fox, S.W. (1954) *J. Am. Chem. Soc.*, **76**, 2801.
[e] Harada, K. (1964) *Bull. Chem. Soc. Japan*, **37**, 1383.
[f] Wheeler, G.P. and Ingersoll, A.W. (1951) *J. Am. Chem. Soc.*, **73**, 4604.
[g] Betti, M. and Mayer, M. (1908) *Ber.*, **41**, 2071.
[h] Gronowitz, S., Sjogren, I., Wernstedt, L. and Sjoberg, B. (1965) *Arkiv Kemi*, **23**, 129.
[i] Triem, G. (1938) *Ber.*, **71**, 1522.
[j] Traynor, S.G., Kane, B.J., Betkonski, M.F. and Hirschy, L.M. (1979) *J. Org. Chem.*, **44**, 1557.
[k] Garnier, A., Collet, A., Faury, L., *et al.* (1981) in *Enantiomers, Racemates, and Resolutions*, (eds J. Jacques, A. Collet and S.H. Wilen), Wiley, New York, p. 262 and p. 333.
[l] Beyerman, H.C. (1959) *Rec. Trav. Chim.*, **78**, 134.
[m] Shafi'ee, A. and Hite, G. (1969) *J. Med. Chem.*, **12**, 266.
[n] Peck, R.L. and Day, A.R. (1969) *J. Heterocyclic Chem.*, **6**, 181.
[o] Gillard, R.D., Mitchell, P.R. and Roberts, H.E. (1968) *Nature*, **217**, 949.
[p] Veselova, L.N. and Chaman, E.S. (1973) *Zhur. obshchei Khim.*, **43**, 1637.
[q] Gillard, R.D., Mitchell, P.R. and Wieck, C.F. (1974) *J. Chem. Soc., Dalton*, 1035.
[r] Sakurai, T., Yamauchi, O. and Nakahara, A. (1976) *J. Chem. Soc., Chem. Commun.*, 553.
[s] Almond, H.R., Manning, D.T. and Niemann, C. (1962) *Biochemistry*, **1**, 243.
[t] Bollinger, F.W. (1971) *J. Med. Chem.*, **14**, 373.

Table 10.2 Resolution of DL-amino acids as *N*-acyl derivatives, or as esters, through diastereoisomeric salt formation

DL-Amino acid derivative	*Resolving agent*	*Reference*
N-Benzoyl	Brucine, strychnine	a
N-Thiobenzoyl	Brucine or morphine	b
N-Formyl	Brucine[c], quinine[j]	c,j
N-Acetyl	Dehydroabietylamine (half-quantities)	d
N-Benzyloxycarbonyl	(−)Ephedrine, phenylethylamine	e
	L-Tyrosine hydrazide	k
N-Benzyl	(−)Ephedrine	f
N-Toluene-*p*-sulphonyl	Brucine	g
N-{2-(Prop-2-en-4-onyl)}	Quinine	h
N-(*o*-Nitrophenyl)sulphenyl	Optically active base	l
N-*t*-Butyloxycarbonyl	(*R*)-α-Methyl-(*p*-nitrophenyl)ethylamine[m]	m, n
	(*R*)- or (*S*)-α-Phenylethylamine[n]	
Ethyl ester	Dibenzoyltartaric acid[g,i]	g, i, o
	Tartaric acid[o]	

[a] Fischer, E. (1899) *Ber.*, **32**, 2451.
Benzoyl-DL-phenylalanine with one half-equivalent of strychnine: Pope, W.J. and Peachey, S.J. (1899) *J. Chem. Soc.*, **75**, 1066.
Pope, W.J. and Gibson, C.S. (1912) *J. Chem. Soc.*, **101**, 939.

[b] Barrett, G.C. and Cousins, P.R. (1975) *J. Chem. Soc., Perkin Trans I*, 2313.

[c] *m*-Tyrosine: Sealock, R.R., Speeter, M.E. and Schwert, R. (1951) *J. Am. Chem. Soc.*, **73**, 5386.

[d] 2-(4-Hydroxyphenyl)glycine: Palmer, D.R. (1972) *Ger. Offen.* 2147620 (*Chem. Abs.*, 1972, **77**, 34938). Holdrege, C.T. (1974) U.S. Pat. 3796748 (*Chem. Abs.*, 1974, **80**, 121 327).

[e] 3,4,5-Trimethoxyphenylglycine: Schmidt, G. and Rosenkranz, H. (1976) *Annalen*, 124. Various amino acids using (+)-ephedrine: Wong, C.H. and Wang, K.T. (1978) *Tetrahedron Lett.*, 3813.
γγ-Di-*t*-butyl-γ-carboxyglutamic acid: Maerkl, W., Oppliger, M. and Thanei, P. and Schwyzer, R. (1977) *Helv. Chim. Acta*, **60**, 798.
Various amino acids using (−)-ephedrine: Oki, K., Suzuki, K., Tuchida, S., *et al.* (1970) *Bull. Chem. Soc. Japan*, **43**, 2554.
The salt of the *N*-benzyloxycarbonyl-D-amino acid separates first if there is a β-methyl group (also in the case of phenylalanine), otherwise the L-enantiomer separates first (except tyrosine, tryptophan and glutamic acid): Kinoshita, H., Shintani, M., Saito, T. and Kotake, H. (1971) *Bull. Chem. Soc. Japan*, **44**, 286.
Various amino acids (phenylethylamine): Felder, E., Pitre, D. and Boveri, S. (1970) *Z. Physiol. Chem.*, **351**, 943.
ββ-Dimethylalloisoleucine (ephedrine): Lehr, H., Karlan, S. and Goldberg, M.W. (1953) *J. Am. Chem. Soc.*, **75**, 3640.

[f] *threo*-β-Hydroxyaspartic acid: Liwschitz, Y., Edlitz-Pieffermann, Y. and Haber, A. (1967) *J. Chem. Soc.* (*C*), 2104.

[g] t-Leucine: Jaeger, D.A., Broadhurst, M.D. and Cram, D.J. (1979) *J. Am. Chem. Soc.*, **101**, 717.

[h] β-Fluoroalanine: Gal, G., Chemerda, J.M., Reinhold, D.F. and Purick, R.M. (1977) *J. Org. Chem.*, **42**, 142.

[i] 2-(4-Hydroxyphenyl)glycine: Lorenz, R.R. (1974) *Ger. Offen.* 2345302 (*Chem. Abs.* 1974, **80**, 133823).
Leigh, T. (1977) *Chem. and Ind.*, 36.

[j] Coronamic acid: Ichihara, A., Shitaishi, K. and Sakamura, S. (1977) *Tetrahedron Lett.*, 269.

[k] Homoserine: Curran, W.V. (1981) *Prep. Biochem.*, **11**, 269 (the salt of the D-enantiomer crystallizes first from hot ethanol).
α-Aminosuberic acid: Hase, S., Kitoi, R. and Sakakibara, S. (1968) *Bull. Chem. Soc. Japan*, **41**, 1266.
Original method applied to representative amino acid derivatives: Vogler, K. and Lanz, P. (1966) *Helv. Chim. Acta*, **49**, 1348.

[l] Konig, von J., Novak, L. and Rudinger, J. (1965) *Naturwiss.* **52**, 453.

[m] Pyrrol-3-ine-5-carboxylic acid: Kahl, J-U. and Wieland, T. (1981) *Liebigs Ann. Chem.*, 1445.

[n] β-(2-Thienyl)alanine: Lipkowski, A.W. and Flouret, G. (1980) *Pol. J. Chem.*, **54**, 2225.

[o] *C*-Arylglycines: Clark, J.C., Phillips, G.H. and Steer, M.R. (1976) *J. Chem. Soc., Perkin Trans I*, 475.

Fischer's original papers (Table 10.2, reference a) include an observation that the resolution of DL-aspartic acid as its *N*-benzoyl derivative can be arranged to allow the (−)-acid salt to separate out when using half the amount of brucine needed to neutralize both carboxy-groups. With double this amount of brucine, the (+)-acid salt separates first from the saturated solutions. Pope and Peachey (Table 10.2, reference a) found that one-half the equivalent amount of strychnine gave a satisfactory resolution in the case of *N*-benzoyl-DL-alanine.

Leigh (Table 10.2, reference i) used the principle of half-neutralization, applied by Fischer, in the reverse sense when he showed that DL-amino acid esters could be resolved with di-benzoyl tartaric acid either as half-acid salts (1:1 ratio of amine to di-acid) or as neutral salts (2:1 ratio) to bring out the chosen enantiomer of the amino acid ester.

Wong and Wang (Table 10.2, reference e) have described a variation of the standard procedure for the resolution of *N*-benzyloxycarbonyl-DL-amino acids in which (±)-ephedrine is used for the formation of mixed diastereoisomeric salts which, when seeded with the particular chiral salt, provides the required enantiomer. With further separations the procedure amounts to the simultaneous resolution of two pairs of racemates.

The recrystallization of diastereoisomeric salts until constant specific rotation is attained is the most commonly used method for monitoring the progress of a resolution, and provided that solid solution formation does not complicate the course of the crystallization process, this is usually a reliable guide. A fully detailed account of the resolution of *N*-acetyl-DL-β-(1-naphthyl)alanine with (−)-phenylethylamine is given in reference 1, p. 408; specific rotations are small ($\leq 6°$) and the sign gives no sure indication which enantiomer is being handled in the form of its salt. Barrett and Cousins (Table 10.2 reference b) describe the use of circular dichroism for monitoring the resolution of *N*-thiobenzoylamino acids, a technique which, in contrast with optical rotation data, *does* reveal directly the absolute configuration of the amino acid component in each separate fraction.

The resolution need not normally involve repeated recrystallization of the diastereoisomeric salt or derivative. It is unusual for a largely resolved amino acid derivative to show eutectic behaviour in which the residual racemate will crystallize first from supersaturated solutions. For example, *N*-acetyl-8-methyl-L-tryptophan of 82% enantiomeric excess gives fully resolved (99.8% e.e.) material through a recrystallization from aqueous ammonia (Hengartner *et al.* p. 434 of reference 1), and it is clearly better to split the partly-resolved *N*-acetylamino acid salt crops and to recrystallize the amino acid derivative to full configurational purity in this case.

10.3.2 Resolution of DL-amino acids through diastereoisomeric derivative formation

A simple extension of the principle of diastereoisomeric salt formation described in the previous section is the reaction of an optically-active acylating agent or alcohol with a DL-amino acid, yielding diastereoisomeric pairs of *N*-acyl or ester derivatives, respectively, which may be separated into their components by fractional crystallization or other physical methods (e.g. chromatography) and the D- and L-amino acids released by hydrolysis. Again, the method has been thoroughly investigated and depends upon the discovery of a suitable separation procedure in each case, as the hydrolysis stage either does not lead to full recovery of the 'resolving agent', or may return it in a form which is unsuitable for re-use.

Examples of recently described resolutions are summarized in Table 10.3.

Table 10.3 Resolution of DL-amino acids through diastereoisomeric derivative formation

Resolving agent and derivative formed	*Separation technique*	*Reference*
Use of reactions of the amino-group		
Carbomenthyloxyacetyl chloride	Fractional crystallization	a
Use of reactions of the carboxy-group		
Menthol		b
(+)*trans*-1-Cyclohexanedicarboxylic anhydride	Fractional crystallization	c
(*R*)-AcNHCHPhCH$_2$OH	Chromatography on silica gel	d
Use of reactions at both amino- and carboxy-groups		
D-HOCH(CMe$_2$CH$_2$OH)CO$_2$H → 2,5-dioxomorpholines	Fractional crystallization	e
(+)-Camphor-10-sulphonyl chloride after esterification with *p*-nitrobenzyl chloride	Partition chromatography	f
(−)-α-Methoxy-α-methyl-1-naphthaleneacetyl chloride after methyl ester formation	HPLC	g
Derived 2-trifluoromethyloxazolin-5-one with dimethyl-L-glutamate followed by hydrolysis of the resulting dipeptide (80% chemical and optical yields before recrystallization to 100% optical purity in the case of 'L-t-leucine', $H_3\overset{+}{N}CHBu^tCO_2^-$)	Recrystallization	h

[a] Witkiewicz, K., Rulko, F. and Chabudzinski, Z. (1974) *Rocz. Chem.*, **48**, 651.
[b] Belikon, V.M., Saveleva, T.F. and Safonova, E.F. (1971) *Izv. Akad. Nauk SSSR, Ser. Khim.*, 1461.
[c] Murakami, K., Katsuka, N., Takano, K., *et al.* (1979) *Nippon Kagaku Kaishi*, 765.
[d] Santoso, S., Kemmer, T. and Trowitzch, W. (1981) *Liebigs Ann. Chem.*, 658.
[e] Marieva, T.D., Kopelovich, V.M. and Gunar, V.I. (1979) *Khim. Prir. Soedin.*, 106.
[f] Furukawa, H., Mori, Y., Takeuchi, Y. and Ito, K. (1977) *J. Chromatogr.*, **136**, 428.
[g] Goto, J., Hasegawa, M., Nakamura, S., *et al.* (1977) *Chem. Pharm. Bull.*, **25**, 847.
[h] Steglich, W., Frauendorfer, E. and Weygand, F. (1971) *Chem. Ber.*, **104**, 687.

The number of stages through which the DL-amino acid must pass in the diastereoisomeric derivative procedure is smaller than in the procedure based on diastereoisomeric salt mixtures, and a wider range of physical methods is available for separating the diastereoisomers. However, the diastereoisomeric derivative procedure may be at least as labour-intensive as the other methods if the recovery of the chiral reagent in a re-usable form is also undertaken.

10.3.3 Chromatographic resolution of DL-amino acids and their derivatives

Two spots noticed on paper chromatograms [3] of DL-amino acid derivatives were recognized to be the two enantiomers, towards which the chiral stationary phase offered slightly different partition coefficients. This property of cellulose can be exploited on a preparative scale; a recent example [4] describes the resolution of DL-tryptophan on a column of powdered cellulose. The same resolution has been achieved by affinity chromatography using the chiral stationary phase formed through reacting bovine serum albumin with succinoylaminoethyl-agarose (separation is complete, D-tryptophan emerging first from the column) [5]. Paper impregnated with alginic acid and carrying precipitated silica gel provides a chiral ion-exchange medium for the resolution of DL-amino acids [6].

A chiral stationary phase synthesized by treating chloromethylated poly(styrene) or poly(acrylamide) with an L- or D-amino acid ester, then hydrolysing and co-ordinating to copper(II) ions, is suitable for ligand-exchange chromatographic resolution of DL-amino acids. The method has been studied in considerable detail [7–10], and resolution efficiency seems rather low in most cases due to slow ligand exchange, but even so, complete resolution can be achieved. A promising variation introduced by Kurganov and Davankov reverses the procedure, using *N N N′N′*-tetramethyl-(*R*)-propane-1,2-diamine-copper(II) as the solute for the resolution of DL-amino acids in water–acetonitrile (5:1) over reversed phase silica gel [9].

Resolution of an α-amino acid ester as its hexafluorophosphate salt by liquid–liquid chromatography based on the selective complexation principle has been developed to the point where clean, quantitative separation of the enantiomers can be achieved [11–14]. Aqueous $NaPF_6$ or $LiPF_6$ on Celite constitutes the stationary phase, and a chloroform solution of a chiral macrocyclic 22-crown-6-ether containing the DL-amino acid ester is the mobile phase (*10.1*). Complex formation involves the protonated amino-group of the substrate, and analogous enantiomer discrimination shown by cyclo-(L-Pro-Gly)$_n$ ($n = 3$ or 4) involving the carbonyl groups of the cyclic hexa- or octa-peptide [15] might also be exploited in the resolution of amino acid esters.

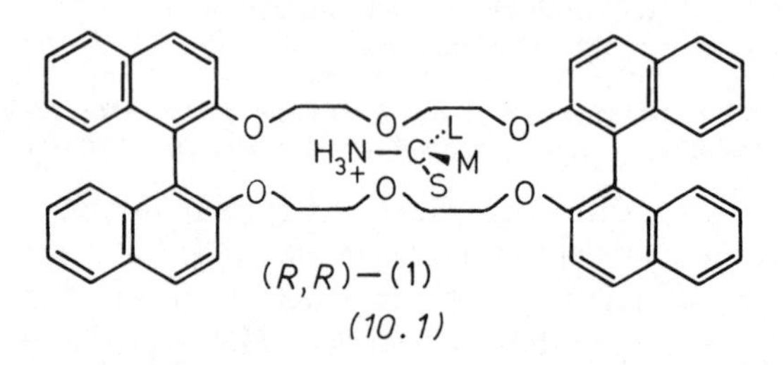

(*10.1*)

Gas–liquid chromatographic resolution has been practised mostly on the analytical scale (see Chapter 16, p. 469) but supports coated with relatively involatile trifluoroacetyl-L,L-dipeptide cyclohexyl esters [16, 17] can, in principle, be used to resolve an *N*-perfluoroalkanoyl-DL-amino acid ester on a preparative scale. The alternative approach, in which the amino acid is converted into a pair of volatile diastereoisomers through reaction with an optically-active acylating agent [18, 19] after esterification, or conversely through esterification with an optically-active alkanol followed by *N*-acylation, can also be the basis of resolution by preparative GLC.

10.3.4 Use of enzymes for the resolution of DL-amino acids

(a) Acylases

Hog renal acylase immobilized on DEAE–cellulose support has been proposed [20] as an alternative to the well known use of this enzyme for the catalysed hydrolysis of an *N*-acyl-L-amino acid in the presence of its D-enantiomer. A full

$$(DL)\text{-}RCONHCHRCO_2H \rightarrow (D)\text{-}RCONHCHRCO_2H + (L)\text{-}H_3\overset{+}{N}CHRCO_2^-$$

account of the classical procedure using hog renal acylase in solution has been given by Greenstein [21], and other workers [22–26]. The advantage of using the immobilized enzyme is the enhanced efficiency associated with the removal of reaction products as they form, thus avoiding competitive inhibition of the enzyme by the L-amino acid which is produced. Practical advantages are also associated with the use of the immobilized enzyme, not only the ease of separation of the enzyme at the end of a resolution, using the batchwise approach, but also because a continuous flow process can be used. A recent account of this variation [27] for the enantioselective hydrolysis of *N*-acetyl-DL-methionine in a column procedure advocates a flow rate of $3\,ml\,h^{-1}$ through a $1 \times 4\,cm$ column of the immobilized enzyme at 37°C for a 0.02 M solution of the acetamido-acid in a pH 7 buffer.

A D-amino acylase from *Streptomyces olivacens* has been described for the resolution of *N*-acetyl-DL-amino acids using the same principle [28].

(b) Proteases

An alternative approach employing enzymes exploits the catalysis by proteases of the hydrolysis of amides and esters of L-amino acids, leaving their D-enantiomers unchanged:

$$(DL)\text{-}R^2NHCHRCOR^3 \rightarrow (D)\text{-}R^2NHCHRCOR^3 + (L)\text{-}R^2NHCHRCO_2H$$

As well as the high level of enantiomeric discrimination involved with enzymic processes, the work-up procedure after the use of a protease is usually particularly straightforward since the unchanged D-amino acid derivative can be extracted from the reaction mixture using a water-immiscible solvent, while the *N*-acyl-L-amino acid can then be extracted in the same way after acidifying the reaction mixture.

Examples of the use of proteases include subtilisin Carlsberg (hydrolysis of *N*-benzyloxycarbonyl-L-amino acid methyl esters [29, 30]), leucine amino-peptidase (γ-methyl-leucinamide $H_2NCH(CH_2CMe_3)CONH_2$; L-isomer hydrolysed to the corresponding amino acid [31]), chymotrypsin (hydrolysis of esters of *N*-acylphenylalanines and ring-substituted derivatives [32, 33], methyl esters of α-methylphenylalanine and α-methyltryptophan [34], only the L($\equiv$S) isomer being hydrolysed, methyl (2*RS*,4*S*)-2-acetylamino-4-methylhexanoate; the ester group of the 2*S*,4*S*-diastereoisomer is hydrolysed preferentially [35], and 5-fluoro-DL-tryptophan methyl ester [36]).

The use of papain in several resolution procedures is based on the specificity of this protease for the catalysed hydrolysis of amide linkages involving an L-amino acid moiety. In this reversal of the normal function of this enzyme, it is used to catalyse the formation of amides from achiral amines and an *N*-acyl-L-amino acid, to provide a reaction mixture containing the *N*-acyl-L-amino acid amide together with the unchanged *N*-acyl-D-amino acid. Some reaction with the D-isomer does occur, however, but this has been minimized by studies in which a substituted phenylhydrazine as the achiral amine has been identified as capable of almost 100% selectivity:

$$(DL)-R^2NHCHRCO_2H + H_2NNH-C_6H_4-F \rightarrow (L)-R^2NHCHRCONHNH-C_6H_4-F + (D)-R^2NHCHRCO_2H$$

While phenylhydrazine gives only 88.2% *N*-acyl-L-amino acid phenyl-drazides in this way, *p*-fluorophenylhydrazine gives 99.9% selectivity [37]. A similar exploration of the selectivity involved when arylamines are used in this process has been described [38, 39]. Aniline gives a crystalline L,L-mono-anilide through reaction with bis(*N*-benzyloxycarbonyl)-DL-di-aminopimelic acid in the presence of papain, emphasizing the fact that maximum selectivity is not of overriding importance if the required enantiomer can be easily separated from small amounts of racemate through crystallization [40].

Carboxypeptidase A has been applied to the resolution of α-methylvaline and α-methylalanine, exploiting the enantioselectively-catalysed hydrolysis of their *N*-trifluoroacetyl derivatives [41, 42]. These reactions, though much slower than those of the corresponding derivatives of the natural amino acids, are convenient laboratory procedures for the resolution of important α-alkyl and other α-substituted α-amino acids. A further example is the digestion of *N*-chloroacetyl-3-phenyloxy-2-aminopropanoic acid with carboxypeptidase A, to give the free (*S*)-amino acid and unchanged (*R*)-derivative [43,44].

Trypsin releases D-lysine from DL-lysine ethyl ester [43, 44], and converts 2-phenylthiazolin-5-ones (*10.2*) derived from DL-arginine or lysine into the corresponding *N*-thiobenzoyl-L-amino acids [45]:

$$(DL)-C_6H_5CSNHCHRCO_2H \rightarrow (10.2) \rightleftharpoons C_6H_5\text{-thiazole(R)(OH)}; \quad (10.2) \rightarrow (L)-C_6H_5CSNHCHRCO_2H$$

(10.2)

This is an efficient process in the sense that in principle, *all* the DL-amino acid used as starting material can be converted into the L-enantiomer. Other examples of asymmetric transformations are described elsewhere in this chapter, based as in the case of the thiazolinone on easy tautomerization [46, 47], through which the 'D'-isomer (on which the enzyme cannot operate) is equilibrated with its enantiomer, replenishing that which is hydrolysed to the product.

(c) Amino acid oxidases, some from snake venom

The use of D-amino acid oxidases for the 'resolution' of DL-α-amino acids is described in many standard textbooks (references 21 and 48, for example) and is the most commonly used enzymic procedure for the resolution of α-amino acids in which an enantiomer is 'destroyed' (and therefore cannot be said to truly resolve a DL-amino acid, i.e. provide the two separate enantiomers).

$$(\mathrm{DL})\text{-}H_3\overset{+}{N}CHRCO_2^- \xrightarrow{\text{D-amino acid oxidase}} (\mathrm{L})\text{-}H_3\overset{+}{N}CHRCO_2^- + O{=}\overset{R}{\overset{|}{C}}CO_2^-$$

Generally, the reaction is not 100% selective but the rate difference is very large, the rate being substantially greater for the D-isomer. There are relative differences in rates through a series of protein D-α-amino acids, with catalysed oxidation being relatively slow for D-glutamic acid, lysine, and aspartic acid, but fast for D-proline, alanine, methionine, and tyrosine [48].

The method is best suited for the resolution of natural amino acids and near relatives, and cannot be applied to α-substituted analogues. Its rapid nature has been exploited for the resolution of [^{11}C]-labelled phenylalanine, the short half-life of the [^{11}C]-isotope calling for rapid methods of synthesis (Strecker synthesis from $PhCH_2CHO$ and $K^{11}CN$ in 40 min, including purification) and for resolution (by either L- or D- amino acid oxidase within 35 min, the immobilization and column technique discussed above for resolution by immobilized acylases facilitating the procedure [49]).

(d) Use of intact organisms or whole cells; other enzymes

There are benefits arising from the use of intact organisms (bacteria) or whole cells maintained in nutrient media, for the production of particular enantiomers of naturally occurring amino acids. These benefits are associated with the opportunities for large-scale working and the ability to introduce a resolution stage early in a chemical synthesis.

Immobilized bacteria (*Pseudomonas striata*) act as the source of dihydropyrimidinase, an enzyme capable of the enantioselective hydrolysis of DL-hydantoins [50]:

$$(\mathrm{DL})\text{-}\overset{\displaystyle HN{-}CHR}{\underset{\displaystyle NH}{OC\quad\;\, CO}} \longrightarrow (\mathrm{L})\text{-}\overset{\displaystyle HN{-}CHR}{\underset{\displaystyle NH}{OC\quad\;\, CO}} + (\mathrm{D})\text{-}HO_2CNHCHRCO_2H$$

DL-Hydantoins also yield *N*-carbamyl-D-amino acids through hydantoinase-catalysed hydrolysis [51], while an unclassified strain of bacterium has been used for the conversion of DL-5-(indol-3-ylmethyl)hydantoin into L-tryptophan [52]. Formation of L-lysine from DL-α-amino-ε-caprolactam has been studied using slurries of *Achromobacter obae* [53]. α-Amino-nitriles, like hydantoins, are intermediates in classical methods of chemical synthesis of DL-amino acids, and mutant strains of *Brevibacterium* catalyse their hydrolysis into L-α-amino acids [54].

An example of the use of an intact organism for the production of D-α-amino-adipic acid, through 'destruction' of the L-enantiomer when cultured on a medium containing the DL-amino acid, employs *Pseudomonas putida* [55].

Other enzymes which have been used include benzylpenicillinase, catalysing the enantioselective hydrolysis of *N*-phenylacetylamino acids, a substrate also for *E. coli*.

10.3.5 Asymmetric transformations

Tautomerizable racemates can be encouraged to yield more than their 50% content of the desired enantiomer or diastereoisomer, if the tautomerism involves a chiral centre. α-Amino acids (but not α-di-alkyl-α-amino acids) can be resolved through procedures in which the unwanted enantiomer is continuously racemized while the desired enantiomer is removed [21]. In enzyme-catalysed procedures applied to tautomerizable substrates (e.g. 2-phenylthiazolin-5-ones [45]), the product need not be continuously removed, but in resolutions of substrates employing physical methods such as crystallization of one enantiomer, the removal of the desired enantiomer is essential to displace an equilibrium which the unwanted enantiomer restores through tautomerization.

Examples of both the enzyme-mediated asymmetric transformation and the preferential crystallization procedure have been given in preceding sections, but these are limited to those amino acids dictated by either the enzyme specificity or by a tendency to crystallize preferentially as pure enantiomers. An example [56] of this process employing the classical Pasteur principle is described in detail here since it illustrates the general possibilities in this area. Reaction 10.1 outlines the basis of a method for the conversion of a DL- or L-phenylglycine into its D-enantiomer through transient Schiff base formation in the presence of (+)-tartaric acid:

$$\left\{\begin{matrix} \text{(DL- or L-)}\ H_3\overset{+}{N}CHArCO_2R \\ HO_2CCH(OH)CH(OH)CO_2^- \end{matrix}\right\} \underset{\text{ethanol/70°C}}{\xrightarrow{\text{PhCHO (1 equiv.)}}} \left\{\begin{matrix} \text{(D)-PhCH=NCHArCO}_2\text{R} \\ -H^+ \downarrow\uparrow +H^+ \\ PhCH\text{=}\bar{N}\text{=}CArCO_2R \\ -H^+ \uparrow\downarrow +H^+ \\ \text{(L)-PhCH=NCHArCO}_2\text{R} \end{matrix}\right.$$

$\downarrow$

(D)-Amino-ester hydrogen (+)-tartrate crystallizes out at 60°C

(10.1)

In an optimum case, the conversion of DL-*p*-hydroxyphenylglycine into its D-enantiomer, better than 90% yields of almost 100% optically-pure material were obtained by permitting crystallization at 21°C (after the first crystals had appeared at 60°C) over a period of 1–7 days. The general applicability of the method is suggested by its use for the resolution of DL-methionine, as the methyl ester, using anisaldehyde and seeding with the hydrogen (+)-tartrate salt of L-methionine methyl ester. However, the lower yield in this case (58% of the L-enantiomer was obtained after 44 h) indicates that a considerable amount of effort may be needed to optimize the procedure for a particular amino acid [56].

The same principle underlies the conversion of a racemic Schiff base into an equilibrium mixture of enantiomers through treatment of the lithium enolate with a chiral acid (Reaction 10.2) [54]:

$$\underset{[R]=[S]}{PhCH{=}NCHPhCO_2Me} \rightarrow PhCH{=}NCPh{=}C\begin{matrix} OLi \\ OMe \end{matrix} \rightarrow \underset{[R]\neq[S]}{PhCH{=}NCHPhCO_2Me} \quad (10.2)$$

An example of the preferential crystallization of one enantiomer, applied to the synthesis of L-lysine from DL-α-amino-ε-caprolactam [57,58], exploits the interconvertibility of the nickel(II) chloride complexes (L-aminocaprolactam)$_3$NiCl$_2$ $\rightleftharpoons$ (D-aminocaprolactam)$_3$NiCl$_2$ in ethanol solution containing sodium ethoxide. Seeding of a supersaturated solution with the L-enantiomer brings about separation of the L-α-amino-ε-caprolactam–nickel(II) chloride complex in 97% yield during 5 h. L-Lysine is obtained from the resolved complex through treatment with hydrochloric acid in methanol [57,58].

N-Acyl-DL-amino acids yield oxazolinones with acetic anhydride, in an equilibrium process in which tautomerization of the heterocycle involves release and return of the α-proton. A particularly favourable example, *N*-acetyl-DL-leucine dissolved in acetic acid containing 10% acetic anhydride at 110°C, deposited approximately 70% of the L-enantiomer of optical purity 92.6% during slow cooling to 40°C over 6.5 h [59], after inoculation at 100°C with *N*-acetyl-L-leucine. *N*-Butyroyl-DL-proline, however, gave crystals of optical purity only 61.4%, when melted with about 5% of its weight of acetic anhydride at 100 °C, cooled to 70 °C, and seeded with *N*-*n*-butyroyl-L-proline [59].

10.3.6 Preferential crystallization of one enantiomer from a racemic mixture

A number of examples have been established, for which induction of the crystallization of one enantiomer from a solution of a DL-amino acid or one of its derivatives has been achieved. *N*-Acetyl derivatives of DL-dopa or α-methyldopa can be resolved through seeding saturated solutions of their di-*n*-butylamine and hydrazine salts, respectively, with the L-enantiomers [60]. *N*-Acyl-DL-prolines

[61,62] and β-(methylene-3,4-dioxyphenyl)-α-methylalanine hydrochloride [63] have also been resolved through methods based on the same principle.

Early work [64] involving *N*-acyl derivatives of α-amino acids demonstrated that some of these compounds crystallize as a true racemic mixture (i.e. individual crystals of the racemate consist entirely of one enantiomer or the other) and can be resolved by seeding a supersaturated solution with either enantiomer. This causes the crystallization of the enantiomer chosen as the seed, and can be very effective in suitable cases (the method is used in the production, now discontinued on a large scale, of L-glutamic acid from acrylonitrile, conversion into 3-cyanopropanal and Strecker synthesis, and resolution by seeding the amino acid hydrochloride [65–67]). The recent demonstration [37] that the preferential crystallization procedure with *N*-acylated α-amino acids can be combined with asymmetric transformation (the 'unwanted' enantiomer is continuously racemized in solution as the required enantiomer crystallizes out) illustrates another promising aspect of the method; this has been discussed in the preceding section.

The method depends on the availability of the required enantiomer to provide seed crystals, and is only feasible with true racemic mixtures. Therefore, the large-scale synthesis of amino acid enantiomers whose resolution has already been achieved, and which can be converted into a suitable derivative, is the main area in which this method shows clear advantages. The practical details of the procedure are very straightforward. The rate of deposition of the crystals from a supersaturated solution or melt depends on a number of factors, but the supply of nuclei for crystal growth is important. Around 5% by weight of the desired enantiomer is added and the mixture cooled slowly over six or more hours, or maintained at constant temperature over this period in typical procedures.

DL-Amino acid arenesulphonate salts $H_3N^+CHRCO_2H\ ArSO_3^-$ have received particularly detailed attention since there appear to be many common amino acids whose arenesulphonate salts tend to crystallize in the true racemate form. The earliest report in this area [68] described the resolution of DL-alanine benzenesulphonate from 97% aqueous acetone, after seeding with one enantiomer. Later papers from Yamada's group [69, 70] include reports of preferential crystallization of a wide range of examples, including variations in the structure of the arenesulphonate moiety. Current work [71] is aimed at probing the physical aspects to discover optimum conditions, in parameters such as temperature, rate of cooling, concentration, solvent, and degree of stirring. It may be an objective to secure the maximum recovery of both enantiomers, especially since the 'unwanted' enantiomer can be racemized (a time-honoured method [21] being heating with acetic anhydride in acetic acid, followed by hydrolysis), and the racemate may then be resolved once again. The preferential crystallization procedure with DL-serine *m*-xylenesulphonate dihydrate seeded with the L-enantiomer deposits crystals of this enantiomer until the composition of the solution is favourable for the spontaneous crystallization of the D-enantiomer [71].

Purvis reported the preferential crystallization of D-glutamic acid from a saturated solution of DL-glutamic acid seeded with L-aspartic acid [72]. A detailed study of the phenomenon has been carried out [73], and it appears that, as well as crystallizing at slower rates than when seeded with D-glutamic acid, the product is contaminated with the aspartic acid.

Racemic threonine to which 10% L-glutamic acid was added, gave elongated prisms of D-threonine from saturated solutions, with an enantiomeric excess approaching 94% under conditions favouring rapid crystallization [74, 75]. The L-enantiomer separates next, as a fine powder [76]. Stereoselective adsorption of L-glutamic acid onto crystals of L-threonine slows their growth, and allows crystals of the D-enantiomer to predominate. As this would imply, it was found [74, 75] that dissolved L-glutamic acid inhibits the dissolution of L-threonine from crystalline DL-threonine, the remaining crystals being enriched in L-threonine. More recent studies [76] show that crystallization of DL-asparagine from solutions containing L-glutamic acid, L-aspartic acid, or L-glutamine leads to different crystal habits for the L- and D-enantiomers.

10.3.7 Enantioselective processes other than enzyme-catalysed reactions: preferential destruction of one enantiomer of a DL-amino acid

Preferential hydrolysis of the D-enantiomer of DL-leucinamides and DL-phenylalaninamides has been reported for aqueous solutions passing through a column of polystyrene carrying L-hydroxyprolyl substituents complexed with copper (II) ions [77]. This type of enantioselective reaction, coupled with a means of separating the reaction products, could offer a valuable continuous flow procedure, though clearly, further development is needed and the method must compete favourably with viable immobilized enzyme systems already well established.

There have been many investigations over the past twenty years of the effects of chiral radiation which might degrade one enantiomer of a DL-amino acid at a faster rate than the other. This interest is not merely an extension of a logical study of the photochemistry of these compounds, but was stimulated by the need to test various theories intended to account for the predominance of L-amino acids in terrestrial biology.

One of the first reports that there could be some basis to this theory was the greater degree of degradation of the D-enantiomer in aqueous solutions of DL-tyrosine exposed to [^{90}Sr]-β-radiation [78, 79], and (among other similar studies) partial resolution of DL-amino acids through irradiation with right or left circularly-polarized light [80, 81]. DL-Leucine, photolysed with 212.8 nm right circularly-polarized light was degraded to an extent of 59%, and found to contain a small (1.98%) enantiomeric excess of the L-enantiomer [81].

However, some claims have been disputed; [^{32}P]-β-radiolysis of DL-tryptophan showed asymmetric bias [82, 83] which could not be repeated [84]. DL-Leucine treated in this way over a period of time which led to the complete destruction of DL-tryptophan, was largely unchanged and was still an equimolar mixture of the enantiomers, within experimental error [85]. There are nevertheless indications from careful spectroscopic studies, that the radiation of this type does interact more effectively with one of the two enantiomers. Left circularly polarized light (284 nm) is absorbed and re-emitted by D-tryptophan in methanol with appreciably higher fluorescence efficiency than is shown by its enantiomer [86], and a chiral discrimination energy of 0.6 kcal mol^{-1} for the equilibrium constants of excimer formation under irradiation by non-polarized isotropic ultraviolet light has been established for this amino acid [87].

None of the studies described here indicates that there might be sufficient enantiomeric discrimination to expect that a practical 'resolution' procedure might be developed, based on the photochemical destruction of one enantiomer. The step to be taken in exploiting the differences in interactions of enantiomers with radiation for this purpose is considerable, and still a long way into the future.

REFERENCES

1. Jacques, J., Collet, A. and Wilen, S.H. (1981) *Enantiomers, Racemates, and Resolutions*, Wiley, New York.
2. Jacques, J., Collet, A. and Wilen, S.H. (1977) *Tetrahedron*, **33**, 2725.
3. Dalgleish, C.E. (1952) *J. Chem. Soc.*, **137**, 3940.
4. Handes, L.V., Kido, R. and Schmaeler, M. (1974) *Prep. Biochem.*, **4**, 47.
5. Stewart, K.K. and Doherty, R.F. (1973) *Proc. Nat. Acad. Sci. USA*, **70**, 2850.
6. El Din Awad, A.M. and El Din Awad, O.M. (1974) *J. Chromatogr.*, **93**, 393.
7. Davankov, V.A., Rogozhin, S.V. and Semechkin, A.V. (1974) *J. Chromatogr.*, **91**, 493.
8. Zolotarev, Yu.A., Myasoedov, N.N., Penkina, V.I., *et al.* (1981) *J. Chromatogr.*, **207**, 231.
9. Kurganov, A.A. and Davankov, V.A. (1981) *J. Chromatogr.*, **218**, 559.
10. Tapuhi, Y., Miller, N. and Karger, B.L. (1981) *J. Chromatogr.*, **205**, 325.
11. Sousa, L.R., Hoffman, D.H., Kaplan, L. and Cram, D.J. (1974) *J. Am. Chem. Soc.*, **96**, 7100.
12. Helgeson, R.C., Timko, J.M., Moreau, P., *et al.* (1974) *J. Am. Chem. Soc.*, **96**, 6762.
13. Cram, D.J. and Cram, J.M. (1974) *Science*, **183**, 803.
14. Helgeson, R.C., Koga, K., Timko, J.M. and Cram, D.J. (1973) *J. Am. Chem. Soc.*, **95**, 3021.
15. Deber, C.M. and Blout, E.R. (1974) *J. Am. Chem. Soc.*, **96**, 7566.
16. Howard, P.Y. and Parr, W. (1974) *Chromatographia*, **7**, 283.
17. Gil-Av, E. and Feibush, B. (1967) *Tetrahedron Lett.*, 3345.
18. Iwase, H. and Murai, A. (1974) *Chem. Pharm. Bull., Japan*, **22**, 1455.
19. Nambara, T., Goto, J., Toguchi, K. and Iwata, T. (1974) *J. Chromatogr.*, **100**, 180.
20. Barth, T. and Maskova, H. (1971) *Coll. Czech. Chem. Commun.*, **36**, 2398.
21. Greenstein, J.P. and Winitz, M. (1961) *Chemistry of the Amino Acids*, Wiley, New York.
22. Baker, C.G., Fu, Shou-Cheng J., Birnbaum, S.M., *et al.* (1952) *J. Am. Chem. Soc.*, **74**, 4701.
23. Fu, Shou-Cheng J. and Birnbaum, S.M. (1953) *J. Am. Chem. Soc.*, **75**, 918.
24. Baldwin, J.E., Kruse, L.I. and Cha, J.-K. (1981) *J. Am. Chem. Soc.*, **103**, 942.
25. Keller-Schierlein, W. and Joos, B. (1980) *Helv. Chim. Acta*, **63**, 250.
26. Keith, D.D., Tortora, J.A. and Yang, R. (1978) *J. Org. Chem.*, **43**, 3711.
27. Kuhlmann, W., Halwachs, W. and Schnegerl, K. (1980) *Chem.-ing. Tech.*, **52**, 607.
28. Sugie, M. and Suzuki, H. (1980) *Agric. Biol. Chem.*, **44**, 1089.
29. Berger, A., Smolarsky, M., Kurn, N. and Bosshard, H.R. (1973) *J. Org. Chem.*, **38**, 457.
30. Bosshard, H.R. and Berger, A. (1973) *Helv. Chim. Acta*, **56**, 1838.
31. Fauchere, J.L. and Petermann, C. (1981) *Int. J. Pept. Protein Res.*, **18**, 249.
32. Matta, M.S., Kelley, J.A., Tietz, A.J. and Rohde, M.F. (1974) *J. Org. Chem.*, **39**, 2291.
33. Tong, J.H., Petitclerc, C., D'Iorio, A. and Benoiton, N.L. (1971) *Canad. J. Biochem.*, **49**, 877.
34. Anantharamaiah, G.M. and Roeske, R.W. (1982) *Tetrahedron Lett.*, **23**, 3335.
35. Bernasconi, S., Corbella, A., Garibaldi, P. and Jommi, G. (1977) *Gazzetta*, **107**, 95.
36. Gerig, J.T. and Klinkenborg, J.C. (1980) *J. Am. Chem. Soc.*, **102**, 4267.

37. Abernethy, J.L., Albano, E. and Comyns, J. (1971) *J. Org. Chem.*, **36**, 1580.
38. Abernethy, J.L., Howell, F.G., Ledesma, A., *et al.* (1975) *Tetrahedron*, **31**, 2659.
39. Mohrig, J.R. and Shapiro, S.M. (1976) *J. Chem. Educ.*, **53**, 586.
40. Arendt, A., Kolodziejczyk, A., Sokolowska, T. and Szufler, E. (1974) *Rocz. Chem.*, **48**, 635.
41. Turk, J., Panse, G.T. and Marshall, G.T. (1975) *J. Org. Chem.*, **40**, 953.
42. Sarda, N., Grouiller, A. and Pacheco, H. (1976) *Tetrahedron Lett.*, 271.
43. Purdie, J.E., Demayo, R.E., Seely, J.H. and Benoiton, N.L. (1972) *Biochim. Biophys. Acta*, **268**, 523.
44. Monsan, P. and Duran, G. (1978) *Biochim. Biophys. Acta*, **523**, 477.
45. Coletti-Previero, M.A. and Axelrud-Cavadore, C. (1975) *Biochem. Biophys. Res. Commun.*, **62**, 844.
46. Wilschowitz, L., Höfle, G., Steglich, W. and Barrett, G.C. (1970) *Tetrahedron Lett.*, 169.
47. Barrett, G.C. (1980) *Tetrahedron*, **36**, 2083.
48. Fessenden, R.A. and Fessenden, J.S. (1979) *Organic Chemistry*, Wadsworth, Boston, Mass, p. 867.
49. Casey, D.L., Digenis, G.A., Wesner, D.A., *et al.* (1981) *Int. J. Appl. Radiat. Isot.*, **32**, 325.
50. Yamada, H., Shimizu, S., Shimiada, H., *et al.* (1980) *Biochimie*, **62**, 395.
51. Shimizu, S., Shimada, H., Takahashi, S., *et al.* (1980) *Agric. Biol. Chem.*, **44**, 2233.
52. Sano, K., Yokozeki, K., Eguchi, C., *et al.* (1977) *Agric. Biol. Chem.*, **41**, 819.
53. Fukumura, T. (1977) *Agric. Biol. Chem.*, **41**, 1321, 1327.
54. Arnaud, A., Gaizy, P. and Janageas, J.-C. (1980) *Bull. Soc. Chim. France, Part 2*, 87.
55. Chang, Y.-F. and Massey, S.C. (1980) *Prep. Biochem.*, **10**, 215.
56. Clark, J.C., Phillipps, G.H. and Steer, M.R. (1976) *J. Chem. Soc., Perkin 1*, 475.
57. Sifniades, S., Boyle, W.J. and Van Peppen, J.F. (1976) *J. Am. Chem. Soc.*, **98**, 3738.
58. Boyle, W.J., Sifniades, S. and Van Peppen, J.F. (1979) *J. Org. Chem.*, **44**, 4841.
59. Yamada, S., Hongo, C. and Chibata, O. (1980) *Chem. and Ind.*, 539.
60. Yamada, S., Yamamoto, M. and Chibata, I. (1975) *J. Org. Chem.*, **40**, 3360.
61. Hongo, C., Shibazaki, M., Yamada, S. and Chibata, I. (1976) *J. Agric. Food Chem.*, **24**, 903.
62. Felder, E. and Pitre, D. (1977) *Farmaco Ed. Sci.*, **32**, 123.
63. Yamada, S., Hongo, C., Yamamoto, M. and Chibata, I. (1976) *Agric. Biol. Chem.*, **40**, 1425; (*Chem. Abs.*, 1965, **62**, 13233).
64. Secor, R.M. (1963) *Chem. Rev.*, **63**, 297.
65. Kaneko, T., Izumi, Y., Chibata, I. and Itoh, T. (eds) (1974) *Synthetic Production and Utilization of Amino Acids*, Wiley, New York.
66. Watanabe, T. and Noyori, G. (1967) *J. Chem. Soc. Japan, Ind. Chem. Sect.*, **70**, 2164, 2167, 2170, 2174.
67. Watanabe, T. and Noyori, G. (1969) *J. Chem. Soc. Japan, Ind. Chem. Sect.*, **72**, 1080, 1083.
68. Chibata, I., Yamada, S., Yamamoto, M. and Wada, M. (1968) *Experientia*, **24**, 638.
69. Yamada, S., Yamamoto, M. and Chibata, I. (1973) *J. Org. Chem.*, **38**, 4408.
70. Yamada, S., Hongo, C. and Chibata, I. (1978) *Agric. Biol. Chem.*, **42**, 1521.
71. Hongo, C., Yamada, S. and Chibata, I. (1981) *Bull. Chem. Soc. Japan*, **54**, 1905, 1911.
72. Purvis, J.L. (1957) *U.S. Pat.* 2,790,001 (*Chem. Abs.*, (1957) **51**, 13911).
73. Yamamoto, H., Hasegawa, H. and Harano, Y. (1981) *J. Chem. Eng. Japan.*, **14**, 59.
74. Addadi, L. and Lahav, M. (1978) *J. Am. Chem. Soc.*, **100**, 2831.
75. Addadi, L. and Lahav, M. (1979) *Pure Appl. Chem.*, **51**, 1269.
76. Addadi, L., Gati, E. and Lahav, M. (1981) *J. Am. Chem. Soc.*, **103**, 1251.
77. Yamskov, I.A., Berezin, B.B. and Davankov, V.A. (1980) *Makromol. Chem., Rapid Commun.*, **1**, 125.
78. Garay, A.S. (1968) *Nature*, **219**, 338.
79. Garay, A.S. (1978) *Nature*, **271**, 186.
80. Norden, B. (1977) *Nature*, **266**, 567.
81. Flores, J.J., Bonner, W.A. and Massey, G.A. (1977) *J. Am. Chem. Soc.*, **99**, 3622.
82. Darge, W., Laczo, I. and Thiemann, W. (1976) *Nature*, **261**, 522.
83. Darge, W., Laczo, I. and Thiemann, W. (1979) *Nature*, **281**, 151.
84. Bonner, W.A., Blair, N.E. and Flores, J.J. (1979) *Nature*, **281**, 150.
85. Blair, N.E. and Bonner, W.A. (1980) *J. Mol. Evol.*, **15**, 21.
86. Tran, C.D. and Fendler, J.H. (1979) *J. Am. Chem. Soc.*, **101**, 1285.
87. Tran, C.D. and Fendler, J.H. (1980) *J. Am. Chem. Soc.*, **102**, 2923.

CHAPTER ELEVEN

Reactions of Amino Acids

G.C. BARRETT

11.1 INTRODUCTION

Reactions of amino acids are covered in this chapter in a systematic manner; for recent general coverage see references 1 and 2. Sections in other parts of this book also deal with certain reactions of amino acids mainly concerned with synthesis, biosynthesis, racemization, colorimetry, and other analytical operations. Chapter 13 deals with reactions leading to derivatives employed in peptide synthesis.

In principle, the side-chain of an amino acid can incorporate any one (or several) of the various organic functional groups, each with its characteristic reaction profile. To keep the discussion of side-chain reactions to a reasonable level while fulfilling the aims of this book, the coverage is mainly concerned with the protein amino acids and other biologically important examples. However, a thorough coverage of the recent literature is given for reactions of the amino- and carboxy-groups within the amino acid context.

11.2 REACTIONS UNDERGONE BY α-AMINO ACIDS DURING SAMPLE PREPARATION

This section is intended to list some of the chemical changes which may be undergone by α-amino acids under conditions used for the preparation of samples for analysis, for physico-chemical studies, or for a variety of studies in food science and biochemistry. An accompanying chapter in this book describes one aspect of this topic (Chapter 12: Degradation of amino acids accompanying *in vitro* protein hydrolysis), but other aspects (pyrolysis during sample preparation, reactions occurring during preparation and storage of solutions, reactions occurring between constituents of nutrient solutions prepared for biochemical and related studies) are briefly listed here. In many cases, further discussion can be found in later sections of this chapter.

11.2.1 Pyrolysis

Decomposition of α-amino acids at 850°C gives the hydrocarbon derived from the side-chain, together with CO, CO_2, NH_3 and HCN (the latter in relatively large amounts, from proline or glutamic acid) [3]. Ammonia and the C-1–C-4 primary amines $C_nH_{2n+1}NH_2$, and dimethylamine and diethylamine, are formed by heating glycine with alumina at 240°C [4], but with some basic manganous carbonate present, other amino acids are also formed from glycine in this way [5]. Imidazoles are released from histidine and 3-methylhistidine at 770°C [6], while the potent mutagen (*11.1*) is one of the products of the dry distillation of lysine [7]. Pyrolysis of phenylalanine gives 3-phenylpyridine and 2-amino-5-phenylpyridine [8]. Earlier amino acid pyrolysis studies [9–11] suggest that phenylalanine gives toluene, styrene, benzene, benzonitrile, 2-phenylethylamine, water and CO_2 as volatile pyrolysis products; clearly, pyrolysis–GLC analysis of amino acids is unlikely to give unambiguous results [12]. Higher temperatures (*ca* 1000°C) provide HCN as the major pyrolysis product [13].

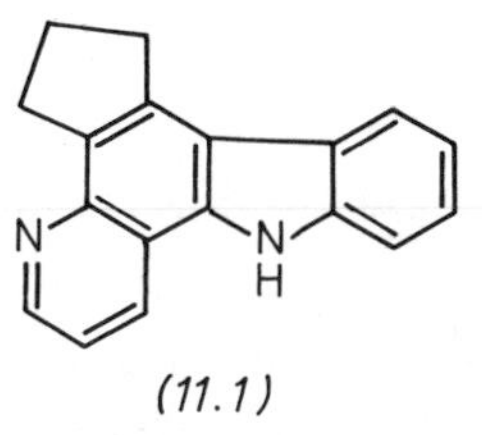

(11.1)

The interest in this area of chemistry stems not only from its relevance to food and pharmaceutical science and the generation of polycyclic hydrocarbons during combustion (e.g. tobacco [14]), but also to provide a basis for accounting for the proportions of amino acids and other organic compounds in geological samples and in meteorites [15].

Aspartic acid gives succinic acid as the major (12%) volatile product, together with dimethylmaleic anhydride, during pyrolysis through the temperature range 350–650°C [16]. Cysteine and methionine give the same thermolysis products obtained from other simple aliphatic amino acids at 850°C under N_2, as noted above, with the sulphur content appearing as CS_2 and COS [17].

Reactions of amino acids which occur at lower temperatures are discussed later in this chapter.

11.2.2 Solution reactions

Hydroxyalkyl amino acids (serine and threonine, for example) are converted into the alkyl analogues in aqueous formic acid at 100°C for 16h [18]. With benzaldehyde or a reducing sugar, amino acids generate HCN at 40–80°C in a pH 5.4 buffer, if the solution is aerated [19]. Formation of pyrroles and furans occurs at higher temperatures (105–140°C) between simple aldopentoses and aldohexoses over long reaction periods (the Maillard reaction) [20]. For example, phenylalanine and glucose yield 3-phenylfuran [21], while lysine gives ε-(2-formyl-5-hydroxymethylpyrrol-1-yl)-L-norleucine (*11.2*) [20]; ketoses also give pyrroles and furans (e.g. rhamnose or fructose give 5-methylfurfural and 2,5-dimethyl-4-hydroxyfuran-3(2*H*)-one with alanine at pH 3.5 [22]).

N-Nitroso-L-proline forms in solutions containing sodium nitrite and either L-citrulline or L-arginine under simulated human stomach conditions; yields are

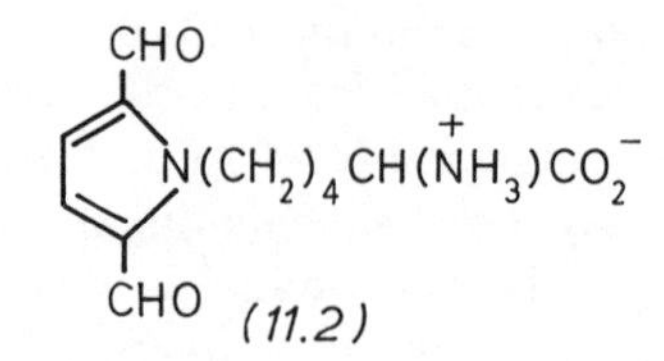

(11.2)

27.1% and 0.1% respectively from the two amino acids [23]. These observations, with others mentioned later, are relevant to current concern over the adventitious formation of potent carcinogens in the presence of nitrites. Proline, pipecolic acid, and 5-hydroxypipecolic acid are formed from the reaction of $Na_2Fe(CN)_5NO$ with ornithine, lysine, and 5-hydroxylysine, respectively [24].

11.2.3 Photolysis and radiolysis

This extensive topic has been established relatively recently, its importance arising both for routine reasons (stability of analytical standards, e.g. tryptophan solutions [25]) and because the amino acid content of geological fossils, or even extra-terrestrial samples (e.g. meteorites) can be interpreted in various contexts. Selective photodestruction of amino acids has been studied, the results include the observation that when all the α-amino acid content of a mixture has been photodegraded, *ca* 80% of β- and γ-amino acids remain unaffected [26]. The greater part of the literature on the photochemistry of the amino acids concerns the aromatic and heteroaromatic protein amino acids, particularly tryptophan and tyrosine. While much of the work seeks the finer details of aspects such as fluorescence and phosphorescence (an interesting observation is that excimer formation equilibrium constants are different for the enantiomers of tryptophan and its derivatives [27]), chemical changes continue to be reported. Cyclization of tryptophan to the hydroperoxide through photolysis in aqueous solution is an early stage in the formation of formylkynurenine and derived pigments [28–30]. Endoperoxide formation through dye-sensitized photo-oxidation of histidine derivatives [31] illustrates further the major emphasis on side-chain reactivity in these studies.

Radiolysis studies have been extensive, again because of the possible biological relevance of the process. It has been suggested (the Vester–Ulbricht theory) that the predominance of the L-amino acids is a consequence of enantiomeric discrimination in prebiotic radiolysis. Although this has been considered to have been proved experimentally in recent years, for instance, by the 19% enrichment of the L-enantiomer through [^{32}P]-β-irradiation of DL-tryptophan, the result could not be repeated [32, 33]. Similarly, the 20–30% destruction of DL-leucine by irradiation over long periods has no asymmetric bias [34]. The opposite approach, in which L-amino acids have been shown to undergo racemization as they are degraded by [^{60}Co]-irradiation in neutral (but not in acid) aqueous solutions keeps open the possibility that the amino acids found in meteorites, although all racemic, might once have been either D or L [35, 36]. The other major area of study under this heading covers radiation-induced reactions of amino acid side-chain functional groups; most of the results of these studies are more broadly significant, rather than holding special interest in amino acid chemistry, and are not given comprehensive treatment here. Tyrosine in aqueous solutions may be converted into dopa (hydroxylation of the phenolic grouping) or into dityrosine (oxidative coupling of the phenolic grouping) through γ- or X-irradiation [37–39].

11.2.4 Ultrasonic treatment

The relative proportions of amino acids in a mixture are altered by ultrasonic breakdown of aspartic acid, alanine and alloisoleucine in particular; the fact that there are increases in the proportions of glycine, glutamic acid, leucine and isoleucine indicates some of the products of this treatment. Serine, threonine, proline and valine appeared to be unaffected by this treatment, but clearly its use in preparing samples for amino acid analysis may introduce errors [40].

11.3 REACTIONS OF THE AMINO- AND CARBOXY-GROUPS

These reactions may be grouped together as: (i) reactions of the amino-group, both primary (amino acids) and secondary (imino acids); (ii) reactions of the carboxy-group; (iii) reactions depending on both amino- (or imino-) and carboxy-groups; (iv) reactions at the α-carbon atom in α-amino acids; (v) reactions of side-chains.

While the coverage is intended to be thorough, particularly as far as the recent literature is concerned, some emphasis is given to reactions which might explain some problematical observations arising from routine analytical and preparative work employing amino acids.

11.3.1 The amino group

N-Acylation and related reactions are brought about in straightforward ways, although the proximity of the carboxy-group accounts for the reactions which ensue (e.g. oxazolinone formation, Fig. 11.1) under some conditions. Optimized conditions for *N*-acetylation have been studied [41]. Thioacetic acid gives *N*-acetylamino-acids but then reacts with these (Fig. 11.1) to give 2-methyl-5-acetylthio-thiazoles. *N*-Arenesulphonylation reactions (i.e. reactions leading to toluene-*p*-sulphonyl derivatives and their analogues) are also easily brought about using the arenesulphonyl chloride and aqueous alkali. Dansyl derivatives (5-dimethylaminonaphthalene-1-sulphonyl) are particular examples of derivatives of this class which are important in amino acid analysis because of their fluorescence.

N-Thioacylation reactions are usually accomplished by aminolysis of thionesters of various types (Reaction 11.1) in alkaline media or in pyridine, providing derivatives for use both in heterocyclic synthesis and in configurational assignments based on circular dichroism measurements (Chapter 19):

$$H_3\overset{+}{N}CHR^1CO_2^- \xrightarrow{(i)} R^2OCSNHCHR^1CO_2H$$
$$\xrightarrow{(ii)} R^3SCSNHCHR^1CO_2H$$
$$\xrightarrow{(iii)} R^3CSNHCHR^1CO_2H$$
$$\xrightarrow{(iv)} [R^3NHCSNHCHR^1CO_2H] \rightarrow \text{(HN—CHR}^1\text{, C=S, C=O, N–R}^3\text{ ring)}$$

(11.1)

The reagents are (i) xanthate esters R^2OCSSR^3 [42]; (ii) CS_2 then alkyl halide [43], or esters, $R^2OCSSBu^t$ [42]; (iii) carboxymethyl dithio-esters, $R^3CSSCH_2CO_2H$ ($R^3 = Ph$ [44], $R^3 = Me$ [45]) or related compounds R^3CSSR^2 ($R^2 = CH_2CN$ [46], $CH_2CH_2N^+Me_3$ [47], Me [46]; (iv) R^3 NCS [48], or R^3 NHCSCl [49].

N-Acylamino acids have been mentioned elsewhere (in Chapter 8, Synthesis of amino acids) as starting materials for the synthesis of oxazolin-5-ones, which not only have uses in the synthesis of other amino acids but also in other aspects of synthesis (Fig. 11.1). *N*-Thiobenzoylamino acids are sources of corresponding thiazolin-5-ones whose reactivity profile differs in many ways from that of the oxazolinones [50], but they too can be used for the synthesis of amino acids since they are readily alkylated at the 4-position (Fig 11.1).

Treatment of *N*-acyl or thioacylamino acids with thioacetic acid provides thiazole derivatives [51], and corresponding amides yield 5-(*N*-trifluoroacetylamino)thiazoles with trifluoroacetic anhydride [52]; overall, amino acids act as a source for thiazoles substituted at the 5-position by *O*, *S*, or *N*-functional groups.

Reactions with nitrous acid in dilute aqueous solutions (to yield the corresponding *hydroxy acid* and nitrogen; formerly used as an analytical procedure for amino acids) or in solutions containing a hydrohalogen acid (to yield the corresponding halogeno acid) are well understood. *De-aminative bromination* of amino acids in this way usually involves inversion of configuration at the carbon atom carrying the amino-group (D-leucine gives (*R*)-Me_2-$CHCH_2CHBrCO_2H$ [53]; L-aspartic acid and its *β*-methyl ester using $NaNO_2/NaBr$ give (*S*)-bromosuccinic acid derivatives [54]). However, where retention of configuration has been observed (*β*-methyl and *β*-dimethyl aspartates [54]; 3,5-dichloro-L-tyrosine [55], conversion of L-histidine methyl ester

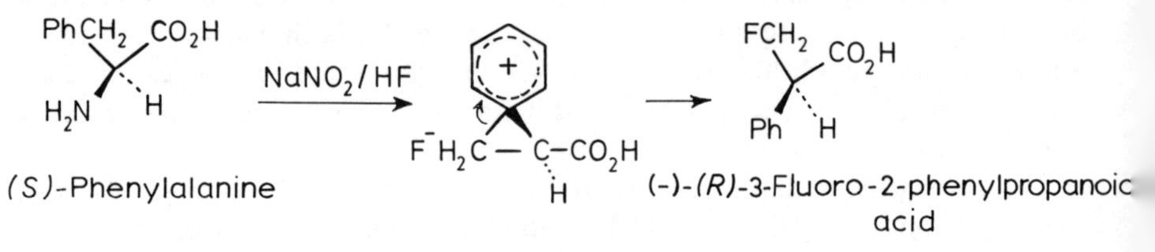

into the 2-chloro acid with $NaNO_2/HCl$ [56]) it is considered to be due to neighbouring group participation by the carboxy group [56–58]. Alanine and α-amino-butyric acid give corresponding *α-fluoro acids* with $NaNO_2/HF/py$, with retention of configuration [57, 58], while *β-fluoroalkanoic acids* are formed from phenylalanine, tyrosine, and threonine (mixtures of α- and β-fluoroalkanoic acids are formed from valine and isoleucine) as a result of an accompanying stereospecific 1,2-shift of the β-functional group [57, 58].

Imino acids yield *N-nitroso-compounds* by reaction with nitrous acid, a fact which has been known for many years but which has recently become of considerable interest in view of the mutagenic properties ascribed to these compounds, and the use of sodium nitrite as a preservative for meat. Recent chemistry of the nitrosation reaction has been reviewed [59, 60], and mention has been made earlier in this chapter of the conversion of citrulline and arginine into *N*-nitrosoproline by nitrous acid under simulated human stomach conditions [23]. However, there is a good deal of controversy in this area; one point of view suggests that there is only a small likelihood that proline can be converted *in vivo* into the nitroso-compound (this derivative occurs in several familiar sources, e.g. in malt, accompanied in some malts by *N*-nitrososarcosine [61]).

Nitrosyl tetrafluoroborate has been advocated [62] for the preparation of *N-nitroso-imino acids* from imino acids (proline and related compounds), and also for the conversion of acetylamino acids into *N-nitroso-N-acetylamino acids* [63].

$$CH_3CONHCHRCO_2H \longrightarrow (CH_3CO)(O{=}N)NCHRCO_2H$$

These have also been proved to be mutagenic, but rearrange into α-diazoalkanoic acids and give 2-methoxy- and 2-hydroxyalkanoic acids in methanol and water, respectively, this presumably reducing their potential hazard to life processes.

Conversion of the amino-group into the diazo group ($H_2NCHR\text{-} \rightarrow N_2CR\text{-}$) [64] has been accomplished for phenylalanine by reaction with isopentyl nitrite This study was directed at the establishment of stereospecific routes to *cis*-and *trans*-cinnamic acids, formed from the diazoalkanoic ester by reaction with boron trifluoride diethyletherate or sodium ethoxide respectively [64].

Schiff bases, formed in neutral or alkaline aqueous solution by reaction of an amino acid with an aldehyde or ketone, may be isolated as cyclohexylammonium salts and are moderately stable as such. In the presence of a reducing agent (sodium borohydride) the Schiff base gives the corresponding *N*-alkylamino acid [65]. Formaldehyde and formic acid gives the *N*-methyl derivative of an imino acid, while formaldehyde alone under basic conditions converts an amino acid into the *N-hydroxymethyl* derivative. This last-mentioned reaction is the basis of the 'formol' titration analysis of proteins, which depends on the reduced basicity of the *N*-hydroxymethyl derivative. The addition and condensation reactions of polyfunctional amino acids and *N*-acetylamino acids with formaldehyde have been studied by [^{13}C]-NMR spectroscopy [66]. Side-chain functional groups which react rapidly (but incompletely) with formaldehyde are NH_2, guanidinyl, OH, indolyl, and imidazolyl; the SH-group reacts rapidly and completely; the α-amino group reacts slowly but completely [66]. Condensation reactions of asparagine, threonine, histidine, and tryptophan with formaldehyde lead to cyclic derivatives (some examples are given in Table 11.1 at the end of this chapter).

Schiff bases prepared from amino acids by condensation with aldehydes and ketones under mild conditions, undergo regioselective and stereoselective cycloaddition reactions with alkenes, as well as Michael addition reactions (Fig. 11.1) [67–69].

Mono-N-alkylation of an amino acid, as mentioned above (ref. 65) is best brought about with the Schiff base as intermediate, since straightforward methods (an alkyl halide with base) lead to *betaines*, $R_3\overset{+}{N}CHR'CO_2^-$. An alternative method avoiding the possibility of racemization involves an *N*-benzyloxycarbonylamino acid, an alkyl iodide, and sodium hydride [70, 71] with hydrogenolysis as the final step to remove the benzyloxycarbonyl group. *N-Oxides of di-alkylamino acids* are formed by reacting the tertiary amine with H_2O_2 in AcOH, and are cleaved into the secondary amine when heated with toluene-*p*-sulphonyl chloride in pyridine [72]:

$$R_2\underset{\displaystyle O^-}{\overset{+}{N}}CHR'CO_2^- \longrightarrow R_2\underset{\displaystyle OTs}{\overset{+}{N}}CHR'CO_2^- \longrightarrow R_2NH + R'CHO + CO_2$$

The di-alkylamino acids themselves are formed in good yield from amino acids and formaldehyde with concurrent catalytic hydrogenation [73].

'*Dinitrophenylation*' of amino acids using 2,4-dinitrochlorobenzene ('DNP chloride') is a well-known textbook example of nucleophilic substitution of a de-activated benzene ring, and the same principle is employed in the synthesis of many analogous derivatives (e.g. *N*-(5-nitro-6-methylpyrid-2-yl)amino acids from the fluoronitropyridine [74, 75]). Other arylation reactions of conventional types can be applied successfully; one particularly effective method uses 2,4,6-trinitrobenzenesulphonic acid as the reagent for preparing *N*-(2,4,6-trinitrophenyl)amino acids [76]. *o*-Quinones react with the amino-group of amino acids [77]:

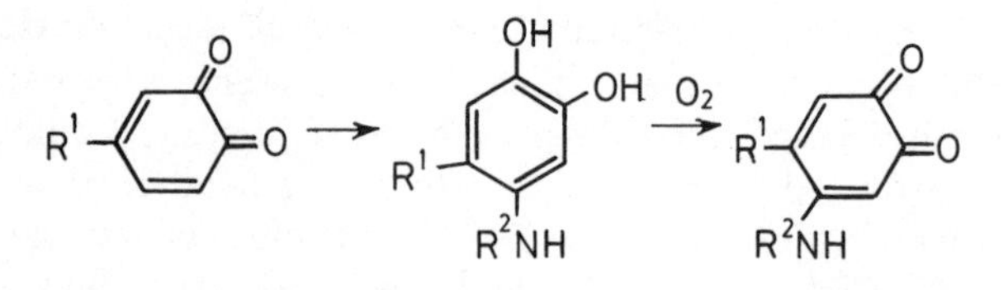

This is relevant to some biosynthetic routes, since *o*-quinones are feasibly available *in vivo*. In the case of cysteine, the SH-group is the exclusive site of addition with an *o*-quinone. Cysteine and *N*-acetyldopaquinone ethyl ester give the protected 5-(*S*-cysteinyl)dopa, which is the biosynthetic precursor of the pigments of the phaeomelanin family, present in chicken feathers [78, 79].

7-Chloro-4-nitrobenzo-2-oxa-1,3-diazole [80] reacts with the amino-group of amino acids to give fluorescent derivatives. Many other standard reactions of amino acids involving the amino-group as nucleophile can be located in the earlier literature, examples being the formation of *N-pyrrolylalkanoic acids* from condensation with 2,5-diethoxytetrahydrofuran [81], and formation of guanidines with *S*-methylisothiosemicarbazide hydriodide [82, 83].

The amino-group of an amino acid is converted into the *azido-group* by treatment with a sulphonyl azide under mild conditions:

$$H_2NCH(CH_2CHMe_2)CO_2^- + CF_3SO_2N_3 \xrightarrow[\text{pH 9, 12 h}]{H_2O-CH_2Cl_2} N_3CH(CH_2CHMe_2)CO_2^-$$

There is no loss of optical activity in this diazo-transfer reaction [84].

Conversion of the amino-group into *isocyanide* ($NH_2\text{-} \rightarrow \bar{C} \equiv \overset{+}{N}\text{-}$) can be accomplished via the *N*-formyl derivative through dehydration with $POCl_3$ [85]. Reduction of the isocyanide with $Bu^n{}_3$ SnH and α-azoisobutyronitrile gives the alkanoate corresponding to the original amino acid [85]:

$$HCONHCHRCO_2Bzl \longrightarrow \bar{C}{\equiv}\overset{+}{N}CHRCO_2Bzl \longrightarrow RCH_2CO_2Bzl$$

Preparation of *hydrazino acids* from α-amino acids can be accomplished in a multi-stage procedure [86]:

$$PhCH_2NHCHR^1CO_2R^2 \xrightarrow{HNO_2} PhCH_2N(NO)CHR^1CO_2R^2 \xrightarrow{Zn\text{-}AcOH\text{-}Ac_2O}$$

$$PhCH_2N(NHAc)CHR^1CO_2R^2 \xrightarrow{H_2\text{-}Pd/C} AcNHNHCHR^1CO_2R^2 \xrightarrow{6\,M\,HCl} H_3\overset{+}{N}NHCHR^1CO_2^-$$

N,N-Dichloroamino-acids have been prepared [87] from the amino acid using Bu^tOCl in MeOH at 0°C (these derivatives are liable to explode).

Although the nitrogen atom of a peptide bond is chlorinated by *t*-butyl hypochlorite, the amino-group of an amino acid is little affected by the reagent [88].

Isothiocyanates are formed from amino acids by reaction with CS_2 and ethyl chloroformate [89].

11.3.2 The carboxy group

Mild oxidizing agents, such as aqueous sodium hypochlorite or aqueous *N*-bromosuccinimide, cause decarboxylation of amino acids with concurrent oxidation to give aldehydes (the Strecker degradation) [90].

$$H_3\overset{+}{N}CHRCO_2^- \xrightarrow{[O]} H_2\overset{+}{N}{=}CR + CO_2 + H_2O \xrightarrow{H_2O} NH_3 + RCHO$$

A kinetic study [91] of the sodium hypochlorite oxidation of tryptophan under pseudo-first-order conditions leading to 3-indoleacetaldehyde shows that only the un-protonated tryptophan is converted into the aldehyde. Several other reagents are effective in this reaction, silver(II) picolinate being particularly valuable since it appears to give quantitative yields of the aldehyde from the amino acid [92]. The reaction has been used [93] to convert (2*S*,4*S*)-[5-^{13}C]-leucine into (2*RS*)-[4-^{13}C]-valine via [4-^{13}C]-isovaleric acid through α-halogenation and amination (see Table 5.14, Chapter 5). Lead tetra-acetate can be used to bring about the same conversion of derivatives of amino acids, soluble in organic solvents, into the carbonyl compounds (an example is given in Fig. 11.1). Since the sequence *N*-acyl-amino acid → oxazolinone → 4,4-disubstituted oxazolinone $\xrightarrow{Pb(OAc)_4}$ ketone is a practical procedure [94], an amino acid can be regarded as an acyl synthon.

α-Amino acids are converted into aldehydes by treatment with *N*-sulphinylaniline O=S=NPh in aqueous media (effectively a variant of the Strecker degradation) [95]; glycine is exceptional in this reaction in that it is oxidized to formic acid. α-Keto-acids in the absence of a catalyst bring about the same oxidative decarboxylation process, being themselves converted into the corresponding amino acids. Thus, the odour of benzaldehyde is detected above a boiling solution of phenylglycine ($H_3NCHPhCO_2^-$) with pyruvic acid, and alanine is formed [96]. Decarboxylation is avoided under the milder transamination conditions associated with catalysis by pyridoxal in the presence of metal ions, or by enzymes, mimicking the metabolic biosynthesis–degradation system involving α-amino acids and α-keto-acids. Chloramine-T in aqueous solution brings about the oxidation of amino acids to nitriles and CO_2 in quantitative yields in many cases [97], but the reaction is not useful in the analysis of amino acids since no better than ± 5% reproducibility is found with threonine, histidine, and arginine. Potassium permanganate gives aldehydes and ammonia (*N*-acyl derivatives of proline and N^{δ}-benzyloxycarbonyl-ornithine are converted into amides by this reagent [98]). Decarboxylation of tryptophan can be brought about by heating its copper(I) or zinc chelate [99].

The familiar ninhydrin reaction [90], which can also be brought about with other 1,2-dicarbonyl compounds (e.g. alloxan; used originally in the Strecker degradation of α-amino acids) brings about the same general reaction, though in the former case the formation of the nitrogen-containing condensation product (*11.3*) rather than free ammonia accounts for the exploitation of this reaction in colorimetric analysis of amino acids. With phenylalanine, a fluorescent side-product (*11.4*) formed in the ninhydrin reaction has been shown to be the result of condensation of phenylacetaldehyde formed in the reaction with further amino acid and ninhydrin [100, 101]:

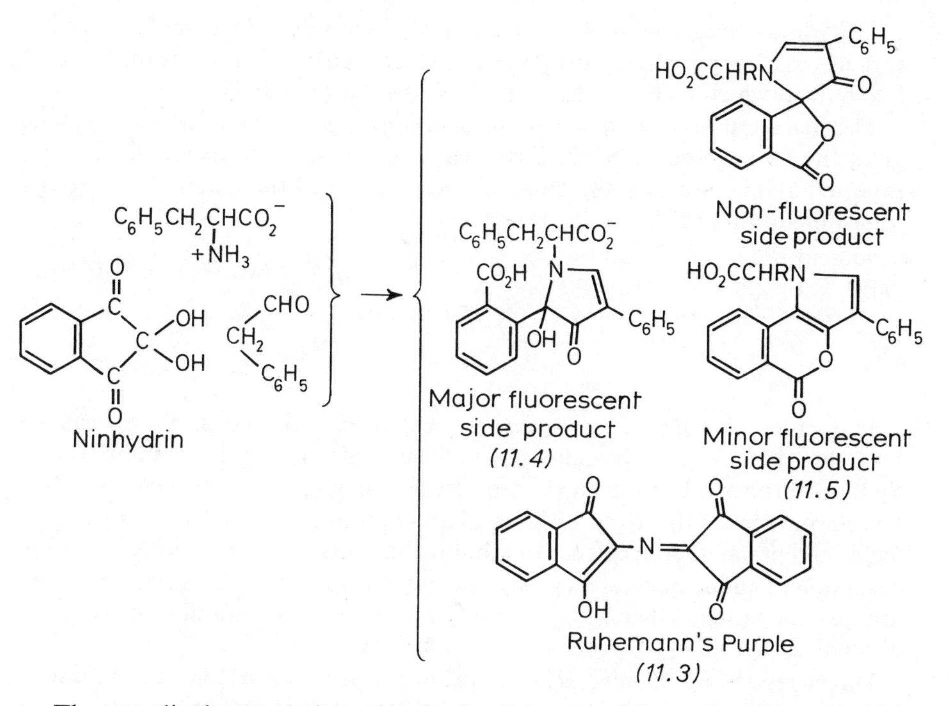

These studies have culminated in the development of the reagent fluorescamine (Chapter 20, Colorimetry and fluorimetry of amino acids), which is capable of condensing directly with amino acids (and other primary amines) to give this family of fluorescent compounds (*11.4*). Imino acids react with fluorescamine to give amino-enones [102]:

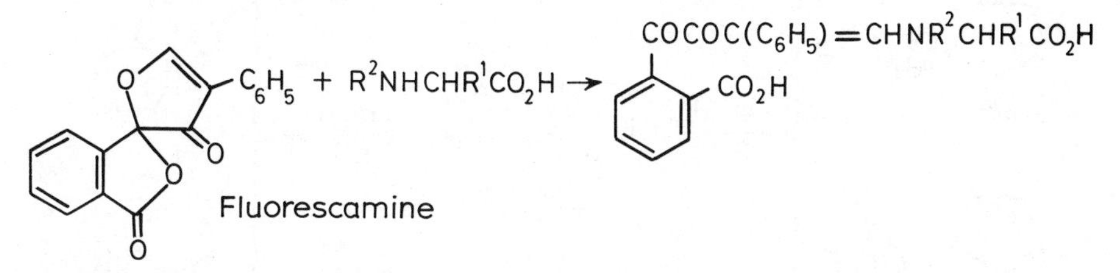

o-Phthalaldehyde gives the iso-indole (*11.6*) by condensation with a thiol and an amino acid [103]; thus, the basis of all the current colorimetric procedures for amino acids are summarized under this section, involving either amino- or carboxy-group reactions.

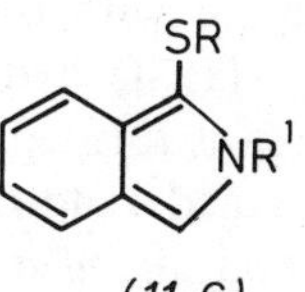

(11.6)

Dehydroascorbic acid reacts with amino acids to give several products, one of which is tris(2-deoxy-2-L-ascorbyl)amine, which on air oxidation produces a blue free radical which is fairly stable in aqueous solutions [104].

Thermal decarboxylation of α-amino acids through heating with an aryl ketone gives the corresponding Schiff base, which on hydrolysis gives not only the expected amine, but also the amine corresponding to the ketone (i.e. through transamination) [105].

$$\mathrm{Ar(R^1)C{=}N{-}CR^2R^3{-}CO_2H \xrightarrow{-CO_2} Ar(R^1)C{\cdots}\bar{N}{\cdots}CR^2R^3 \xrightarrow{H_2O} ArCOR^1 + R^3CHR^2NH_2 \;\; \text{or} \;\; R^2COR^3 + ArCHR^1NH_2}$$

Reduction of *amino acids to amino alcohols* without racemization can be brought about using diborane-dimethyl sulphide in tetrahydrofuran [106]. Sodium borohydride in ethanol is also effective but requires the carboxy-group to be esterified first [106, 107]. Lithium aluminium hydride is inferior in terms of yields and practical convenience to lithium dimethoxyaluminium hydride for the reduction of amino acid esters [108]. Curious solvent effects have been reported for the reduction of *N*-benzyloxycarbonyl-L-proline isopropylamide with lithium aluminium hydride [109] (results shown in Reaction 11.2).

Direct reduction of *amino acids* to *amino aldehydes* cannot be accomplished, but the indirect route via the oxidation of *N*-protected amino alcohols, using pyridinium dichromate, has been established [145]. The *N*-Boc-amino alcohols used in this study were prepared by reduction of corresponding *N*-Boc-amino acids with BH_3 in THF [145].

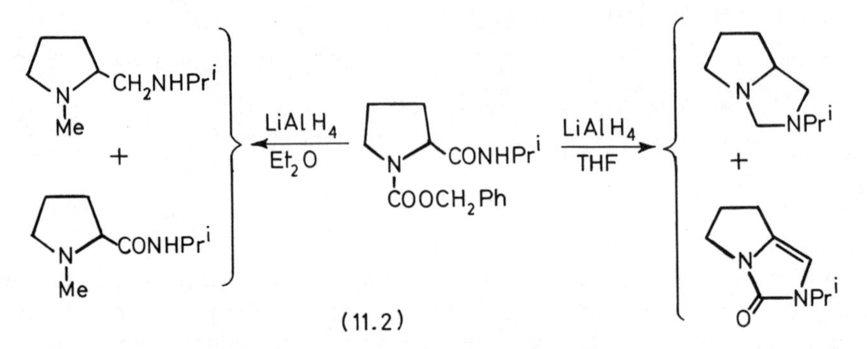

(11.2)

The carboxy-group in *N*-protected amino acids is converted into the *symmetrical anhydride* on treatment with one-half equivalent of the carbodiimide $EtN = C = N(CH_2)_3NMe_2$ (as its hydrochloride [110]), and the report [111] that a stable *O*-acyl-isourea is formed in this way seems to be incorrect [112]. *Cyclic anhydrides* are formed readily from *N*-protected aspartic and glutamic acids (Reaction 11.3) and *mixed anhydrides* are formed between the other common *N*-protected amino acids and simple anhydrides.

By treatment of *N*-benzyloxycarbonyl-L-glutamic anhydride with diazo methane, the diazoketone $ZNHCH(CH_2CH_2CO_2Me)COCHN_2$ is formed [113], not the isomer as previously claimed (this may be prepared from the α-methyl ester and diazomethane). Diazoketones of *N*-benzyloxycarbonyl- or *N*-butoxycarbonyl-amino acids are available through the reaction of the mixed anhydride or carbodi-imide adduct with diazomethane [114].

All the other standard reactions of the carboxy-group (*esterification, amidation, acid chloride* formation) are undergone by amino acids, with due regard to the need for *N*-protection. The more esoteric functional groups (e.g. hydroxamic acids) are formed from the carboxy-group or its esters by standard procedures. Friedel–Crafts acylation of arenes is a convenient route to chiral aryl α-amino-alkyl ketones (for recent references see refs 115, 116). The reaction of *N*-phthaloylaspartic anhydride with benzene in the presence of $AlCl_3$ leads to the expected product (opposition by the phthalimido-group of the build-up of charge at the neighbouring carbon atom) [117]:

PhtN–CH–CO–O–CO–CH_2 (cyclic anhydride) ⟶ PhtNCH(CH_2–COPh)CO_2H

(11.3)

11.3.3 Reactions depending on both amino- and carboxy-groups

This section discusses reactions which are undergone by α-amino acids but not by analogous amines or carboxylic acids. An example (Reaction 11.4) is the alkylation of copper or cobalt complexes of glycine by aldehydes, an aldol-type reaction yielding threonine with the correct relative stereochemistry and applied to large-scale production of this amino acid [118]. The reagents are (i) CH_3CHO, OH^- and (ii) ion-exchange resin. The chelation of both amino- and carboxy-groups of an α-amino acid leads to a stable arrangement in which the α-carbon atom is activated sufficiently towards electrophilic attack to promote these reactions. Under strongly alkaline conditions, glycine itself reacts with benzaldehyde to give β-phenylserine [119], thus demonstrating that metal chelation is not alone responsible for the activation of the α-carbon.

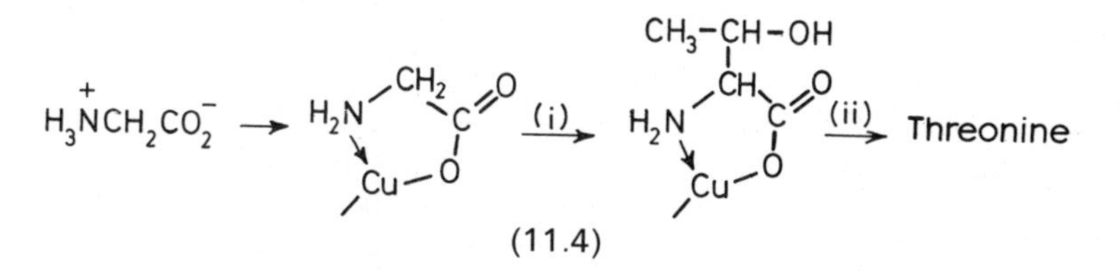

(11.4)

α,ω-Di-amino acids may be specifically ω-acylated when their α-amino group is chelated to copper(II) ions in this way; this device is used for the preparation of

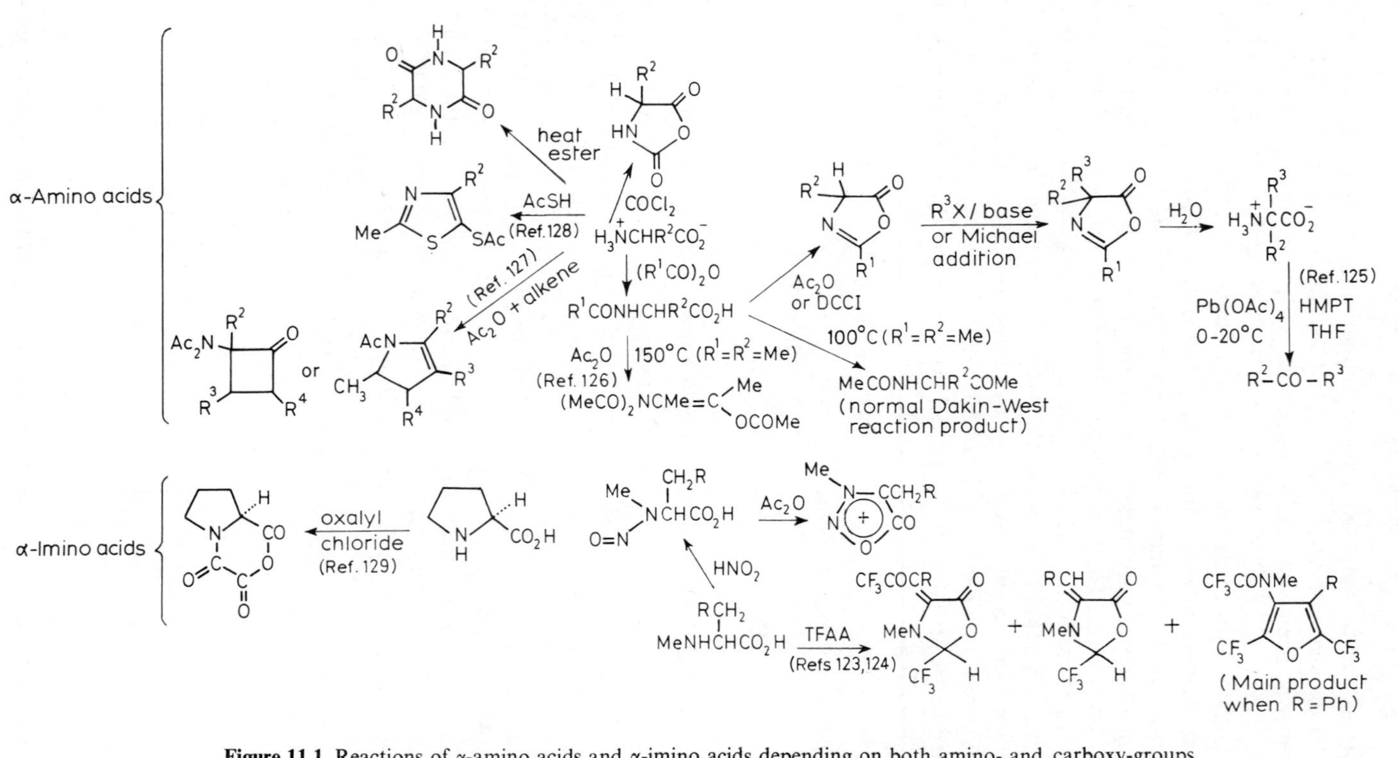

Figure 11.1 Reactions of α-amino acids and α-imino acids depending on both amino- and carboxy-groups.

N^{ε}-acyl-lysine for use in peptide synthesis. In any case, the α-amino-group is less nucleophilic in an α-amino acid than in a simple primary alkylamine due to the effect of the adjacent carboxy-group, and in certain α,ω-di-amino acids there is sufficient reactivity difference to permit selective ω-acylation without metal chelation (an example is given in Table 11.1).

Many of the reactions of amino acids have important implications in food science, and an early observation, that amino acids and simple sugars react to give heterocyclic compounds (Maillard reaction) has been developed to cover all the ways in which the protein α-amino acids or their degradation products might react with other food constituents under simple conditions. An earlier section describes the pyrolysis reactions undergone by α-amino acids, and some reactions described under the heading *Solution reactions* are also relevant to the possible chemical changes accompanying cooking processes. Gentle heating of amino acids to about 200°C causes polymerization to give oligopeptides, as should be expected for a bifunctional compound *X*---*Y*, in which groups *X* and *Y* are mutually reactive. Thermal polymerization of mixtures of α-amino acids (e.g. glycine, glutamic acid, and tyrosine [120]) does not lead to totally randomly sequenced oligopeptides when conducted at 180°C, reflecting different reactivities of the amino- and carboxy-groups in the various α-amino acids. Formation of di-oxopiperazines (cyclo-dimerization) and cyclo-oligomerization can also take place via the heating of amino acids (the melting-points of α-amino acids are frequently unreliable indices for characterization purposes, and depend on rate of heating, to some extent). In fact, di-oxopiperazine formation is usually the predominant result of the gentle heating of α-amino acids, and is invariably the major pathway open to α-amino acid esters after they have been released from their salts or stored in neutral solutions.

Most of the other reactions of amino acids involving both amino and carboxy groups are concerned with conversions into heterocyclic systems enclosing the =NCHRCO- grouping, and examples are collected together in Fig 11.1.

Some of the reactions collected in Fig. 11.1 are best performed with the *N*-acyl- or thioacyl amino acid as starting material. In the case of carbamates derived from α-amino acids, including the widely used *N*-benzyloxycarbonyl derivatives $PhCH_2OCONHCHRCO_2H$, it was assumed until recently that cyclization to the oxazolin-5-one was not possible. However, under appropriate conditions (formerly thought to cause cyclization to aziridinones [121]), 2-benzyloxy-oxazolin-5-ones can be obtained [122]. Reagents for this purpose include phosgene, thionyl chloride, or phosphorus oxychloride in THF at −20 to −30°C, triethylamine being added to maintain neutral conditions.

11.3.4 Substitution α to the carboxy-group in amino acids

There are few reactions in which the amino acids themselves undergo substitution α to the carboxy-group, though *N*-protected amino acids and other compounds carrying the =NCHRCO– grouping (particularly cyclic compounds, Fig. 11.1) can show such reactions.

The condensation of glycine with benzaldehyde in strongly alkaline media to give β-phenylserine has been mentioned earlier (p. 365). In the β-amino acid series, with β-alanine as a representative example, the introduction of the sulphonic acid function has been demonstrated [130] using oleum at room temperature; a high yield of 'α-sulpho-β-alanine' is obtained in this way.

α-Substitution of *N*-substituted glycine and *N*-substituted glycine esters now constitutes an important method for the synthesis of α-amino acids, in which Schiff bases of glycine esters, or *N*-trimethylsilylglycine derivatives, undergo alkylation after anion formation involving a powerful base. These methods are described in detail in Chapter 8 (Tables 8.7,8.13 and 8.14). An alternative approach employs *N*-benzoylhydroxyglycine ($PhCONHCH(OH)CO_2H$), in which the hydroxy-group is substituted by alkyl or aryl groups after α-carbonium ion formation with $MeSO_3H$, or first substituted by Cl using $SOCl_2$ [131]; details are given in Table 8.15 of Chapter 8.

α-Substitution of higher homologues of glycine (=NCHRCO– → =NCRR′CO–) should also be feasible in the same way, although the principle has only recently been demonstrated for the αβ-di-*t*-butyl ester of *N*-formyl-L-aspartic acid (Chapter 8, Table 8.13) [132] using LDA as base, and alkylation with an alkyl bromide. In this example, a mixture of α-and β-alkylation products was obtained.

α-Chlorination of 2-phenyloxazolin-5-one and substitution of the halogen by the acetylthio-grouping ($-Cl \rightarrow -SCOCH_3$) was accomplished some time ago [133], though in poor yield, and a similar reaction with the corresponding thiazolinone seemed to be successful [133]. The recent results reported for the preparation of α-chloroglycine derivative $R^1CONHCHClCO_2R^2$ and the corresponding nucleophilic substitution stages, show that this is a more efficient route to α-heteroatom-substituted glycines (Chapter 8, Table 8.15) [131].

Indirect routes are also represented in alkylation of oxazolinones and thiazolinones (Fig. 11.1) followed by hydrolysis to provide αα-di-alkyl analogues of amino acids. Standard methods of amino acid synthesis (Chapter 8, Tables 8.1–8.3) can be used for the preparation of αα-di-alkyl-α-amino acids.

11.3.5 Reactions of side-chains of protein amino acids and other naturally occurring amino acids

The protein amino acids include side-chain functional groups of simple types (hydroxy; carboxy; amino; thiol/disulphide; methylthio; phenyl and 4-hydroxyphenyl) and more complex types (guanidino; indolyl; imidazolyl). While all these groups show their normal reactivity profile when in the presence of the amino and carboxy groups of the amino acids, there are several further distinctive reactions which are shown by these functional groups in the structural context of an α-amino acid which are given more prominence in this section. Aspects of the side-chain chemistry of these amino acids are collected in Table 11.1.

Under conditions commonly used to hydrolyse proteins (6 M hydrochloric acid at 120°C) tryptophan reacts with cystine to give 2-(2-amino-2-carboxyethylthio)tryptophan (*alias* tryptathionine) as a transient intermediate en route to β-3-oxoindolylalanine and cysteine [134, 135]. This can account for the inaccurate quantitative amino acid analysis of proteins containing these residues, see also Chapter 12, p. 376.

Oxidation of amino acids with hydrogen peroxide or H_2O_2-$CuSO_4$ affects side-chains of the common amino acids differently; ornithine gives β-alanine via 4-amino-butyric acid [136] and proline gives 3-hydroxyproline [136], while, more predictably, methionine gives a mixture of sulphoxide and sulphone [137]. Alternative reagents for the oxidation of the methionine sulphide grouping include $NaIO_4$, *N*-chlorosuccinimide in aqueous media, $NaBrO_3$, tribromocresol, and chloramine-T [137]. Trichloroisocyanuric acid oxidizes methionine to the sulphone rapidly, but all the other common amino acids are degraded in various ways with this reagent [138]. Tetrachloro-auric acid oxidizes methionine to the sulphoxide (Au(III) $\rightarrow$ Au(I); stereospecific formation of the (*S*)-sulphoxide) by way of a methionine–$AuCl_3$ complex [139]. Aqueous chlorine or bromine reacts with cysteine, cystine, alanine-3-sulphinic acid, and cystine *SS*-dioxide to give cysteic acid and cysteinylcysteic acid [140]. Aerial oxidation of methionine to its sulphoxide is catalysed by bisulphite [141]. Further examples of oxidation reactions affecting amino acid side-chains are given in Table 11.1.

Two diastereoisomers of the spirolactone (*11.7*) are obtained through treatment of tryptophan with t-butyl hydroperoxide and $FeSO_4$ [142]

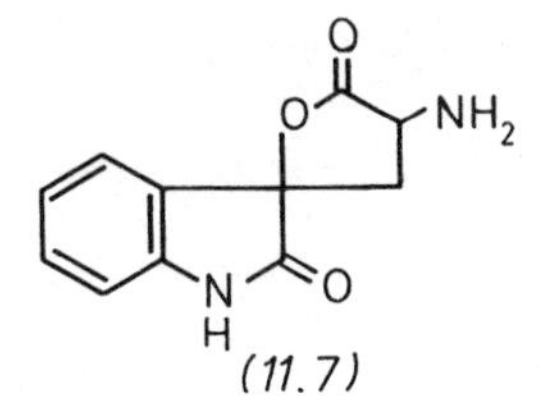

(11.7)

Nitrosation of the indole nitrogen atom of tryptophan, rather than the α-nitrogen atom, has been established using [^{15}N]-NMR spectroscopy [143].

Anchimeric assistance by nearby functional groups of the hydrolysis of the thiolester grouping in *NS*-diacetylcysteinamide and *N*-acetyl-*S*-benzoylcysteinamide is revealed by rates twenty times greater than expected on the basis of p*K* values [144]. The rate for *N*-benzyloxycarbonyl-*S*-acetyl-L-cysteinyl-L-threonine ethyl ester is enhanced by *ca* 100-fold for the same reason [144].

Table 11.1 Some reactions involving α-amino acid side-chains*

Amino acid	*Reagents and products*	*Reference*
Methionine and cysteine	Usual nucleophilic reactions of the SMe or SH groups (the methionine sulphur atom is second only to the thiolate anion among the common nucleophilic functional groups, in its nucleophilicity)	a–c
Cysteine, serine, threonine with formaldehyde	Thiazolidine- or oxazolidine-carboxylic acids via Schiff bases 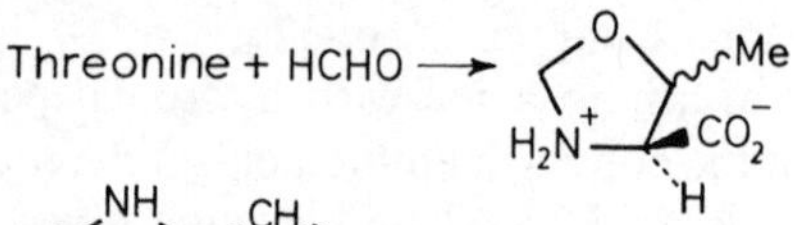 Threonine + HCHO →	d, e
Histidine with formaldehyde	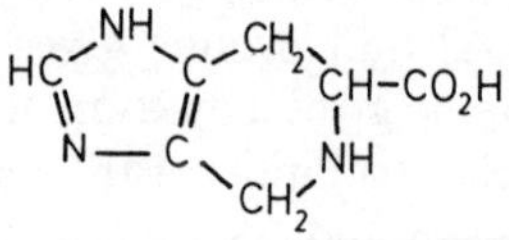	f
Tryptophan with 4-dimethyl-amino-benzaldehyde in MeOH containing HCl	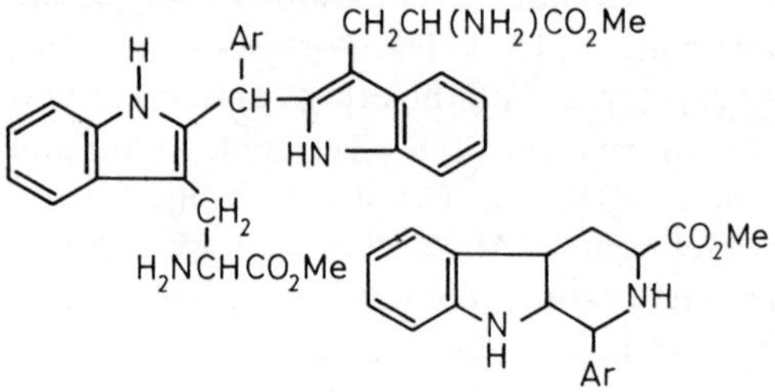	g
L-Tryptophan with 3-acetoxy-xanthine	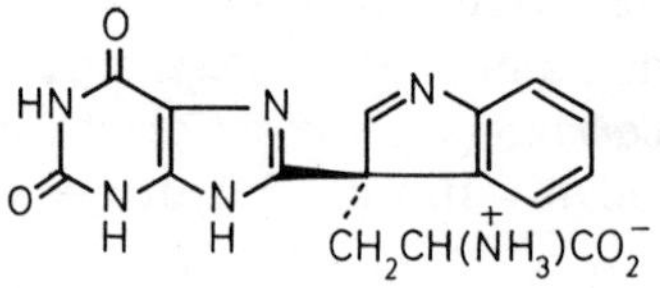	h
Tryptophan with nucleic acids (photo-initiated reactions)	Similar products	i, j
Dopa undergoes oxidative cyclization to dopachrome	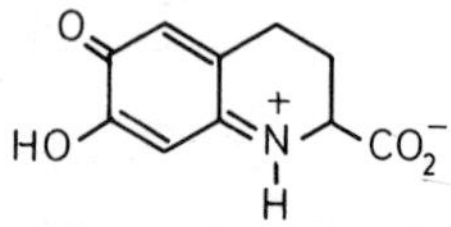	k
Arginine; conversion into ornithine (Na/NH_3 with 10 equiv. AcOH)	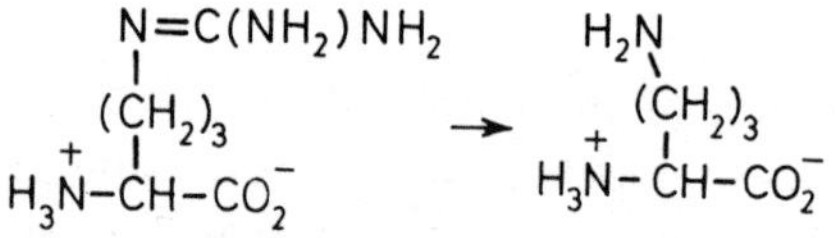	l
Arginine: quantitative modification of guanidine moiety into 5-nitropyrimidin-2-yl with nitromalonaldehyde	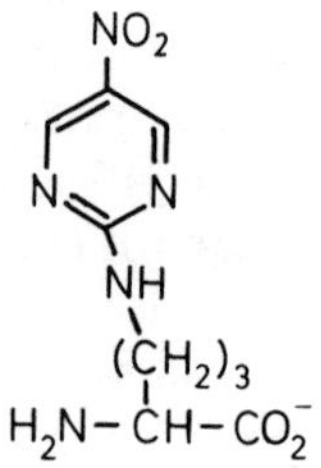	m

Table 11.1 (*Contd.*)

Amino acid	*Reagents and products*	*Reference*
N-Acetyl-L-tryptophan with ᵗBuOCl	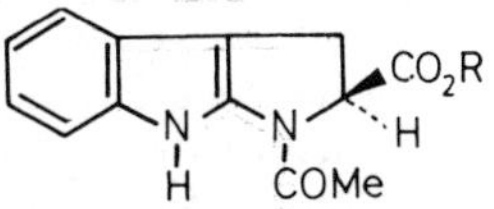 [structure: CO_2R, H, N–H, COMe]	n
L-Tryptophan with Koshland's reagent (2-hydroxy-5-nitrobenzyl bromide)	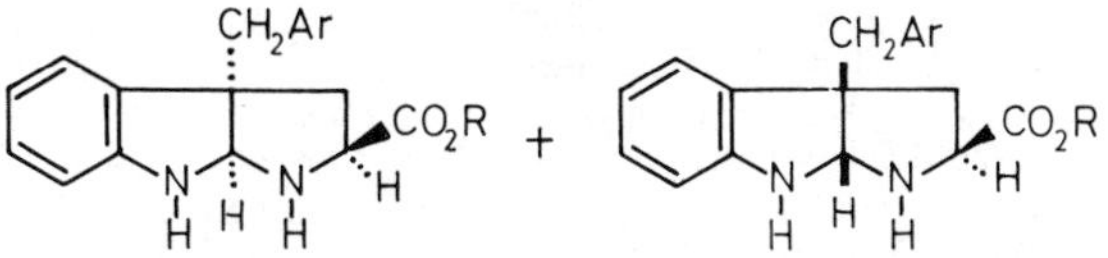 [structures: CH_2Ar, CO_2R, H, H H H + CH_2Ar, CO_2R, H, H H H]	o, p
Serine derivatives; substitution of – OH by – Cl (*erythro* via the *trans*-oxazoline, *threo* through direct S_N1 reaction)	*β*-Chloro-*α*-amino acids with retention of configuration (e.g. with *erythro-/threo-β*-arylserines)	q
Dehydration of L-asparagine and L-glutamine to *β*-cyano-L-alanine and *γ*-cyano-L-butyric acid, respectively	Phosgene in dioxan acting on the *N*-benzyloxycarbonyl derivative[r]; DCCI acting on the NPS-amino acid[s]	r, s
Hydroxylation of proline	Fe^{2+}/O_2/ascorbic acid/EDTA	t
Hydroxylation of proline	H_2O_2, O_2^-, or HO· (leading to *cis/trans* mixtures of 3- and 4-hydroxyprolines)	u
Hydroxylation of lysine	H_2O_2, O_2^-, or HO·(leading to 5-hydroxylysine)	
α,ω-Di-amino acids	Selective *ω*-acetylation of the *ω*-amino-group using *p*-nitrophenyl esters at pH 11 (but no selectivity with *α,β*- or *α,γ*-di-amino acids)	v
Dehydration of serine, threonine	Treatment of *N*-Boc or Pht amino acid esters with Ph_3P and diethyl azodicarboxylate	w
Dehydration of serine, threonine	Reaction of the amino acid with disuccinimido carbonate (leads to the succinimido esters of dehydroalanine and Z-dehydrobutyrine, respectively)	x
3-Amino-azetidinones and azirine-carboxylates from serine derivatives (L-serine *O*-benzylhydroxamate[y])	$Ph_3P/Et_3N/CCl_4$[y] or Ph_3P and diethyl azodicarboxylate[z,aa] (tendency towards aziridine formation with certain *N*-protecting groups[aa])	y, aa
O-Alkylation of tyrosine	Selective *O*-alkylation without *N*-alkylation (alkyl halide, 2 equiv. NaOH in DMSO)	bb
N,N'-Diacetylcystine + acetaldehyde under irradiation (UV < 320 nm)	*N,S*-Diacetylcysteine formed	cc
Smiles rearrangement of *S*-Dnp-cysteine (treatment with a base)	Equilibrium favouring *N*-Dnp-cysteine set up in methanol	dd
Protected tryptophan, oxidation with Fremy's salt	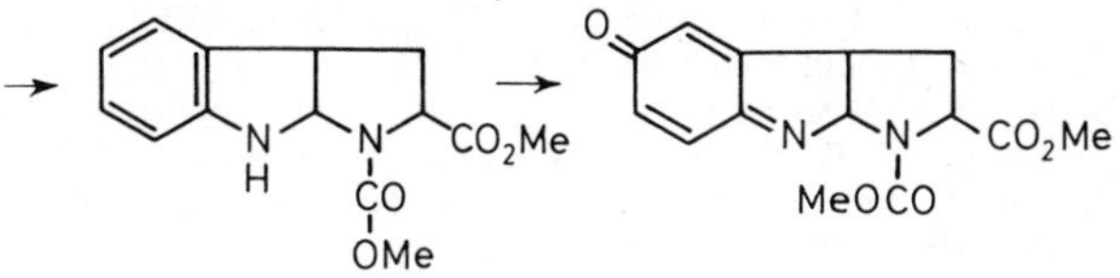 → [structure: N–H, N–CO–OMe, CO_2Me] → [structure: O, N, N–MeOCO, CO_2Me]	ee

Table 11.1 (*Contd.*)

Amino acid	*Reagents and products*	*Reference*
Cysteine with dopa or with histidine in boiling constant-boiling HBr	Substitution through *S* at phenolic moiety (position 6) and at the imidazole moiety (position 2) respectively	ff

* Table 8.14, Chapter 8 contains further examples of the coversion of one amino acid into another.
[a] Rogers, G.A., Shaltiel, N. and Boyer, P.D. (1976) *J. Biol. Chem.*, **251**, 5711.
[b] Barrett, G.C. (1979) in *Comprehensive Organic Chemistry* (eds D.H.R. Barton and W.D. Ollis) (Vol. 3) (ed. D.N. Jones) (Pergamon Press, Oxford, p. 3.
[c] Barrett, G.C. (1979) in *Comprehensive Organic Chemistry* (eds D.H.R. Barton and W.D. Ollis) (Vol. 3) (ed. D.N. Jones) Pergamon Press, Oxford, p. 33.
[d] Wolfe, S., Militello, G., Ferrari, C., *et al.* (1979) *Tetrahedron Lett.*, 3913.
[e] Szilagyi, L. and Gyorgydeak, Z. (1979) *J. Am. Chem. Soc.*, **101**, 427.
[f] Hrncir, S., Kopoldova, J., Veres, K. *et al.* (1978) *J. Labelled Comp. Radiopharm.*, **15**, 47.
[g] Pindur, U. (1978) *Arch. Pharm.*, **311**, 615.
[h] Stohrer, G., Salemnick, G. and Brown, G.B. (1973) *Biochemistry*, **12**, 5084.
[i] Reeve, A.E. and Hopkins, T.R. (1980) *Photochem. Photobiol.*, **31**, 223.
[j] Reeve, A.E. and Hopkins, T.R. (1980) *Photochem. Photobiol.*, **31**, 413.
[k] Young, T.E., Griswold, J.R. and Hulbert, M.H. (1974) *J. Org. Chem.*, **39**, 1980.
[l] Bland, J.S. and Keana, J.F.W. (1971) *Chem. Commun.*, 1024.
[m] Signor, A., Bonora, G.M., Biondi, L. *et al.* (1971) *Biochemistry*, **10**, 2748.
[n] Ohno, M., Spande, T.F. and Witkop, B. (1970) *J. Am. Chem. Soc.*, **92**, 343.
[o] Loudon, G.M., Portsmouth, D., Lukton, A. and Koshland, D.E. (1969) *J. Am. Chem. Soc.*, **91**, 2792.
[p] McFarland, B.G., Inoue, Y. and Nakanishi, K. (1969) *Tetrahedron Lett.*, 857.
[q] Pines, S.H., Kozlowski, M.A. and Karady, S. (1969) *J. Org. Chem.*, **34**, 1621.
[r] Wilchek, M., Ariely, S. and Patchornik, A. (1968) *J. Org. Chem.*, **33**, 1258, 4072.
[s] Chimiak, A. and Pastuszak, J.J. (1971) *Chem. Ind.*, 427.
[t] Bade, M. and Gould, B.S. (1968) *Biochim. Biophys. Acta*, **156**, 425.
[u] Trelstad, R.L., Lawley, K.R. and Holmes, L.B. (1981) *Nature*, **289**, 310.
[v] Leclerq, J. and Benoiton, N.L. (1968) *Canad. J. Chem.*, **46**, 1046.
[w] Andruszkiewicz, R., Grzybowska, J. and Wajciechowska, H. (1981) *Pol. J. Chem.*, **55**, 67.
[x] Ogura, H., Sato, O. and Takeda, K. (1981) *Tetrahedron Lett.*, **22**, 4817.
[y] Mattingly, P.G. and Miller, M.J. (1981) *J. Org. Chem.*, **46**, 1557.
[z] Townsend, C.A. and Nguyen, L.J. (1981) *J. Am. Chem. Soc.*, **103**, 4582.
[aa] Bose, A.K., Sahu, D.P. and Manhas, M.S. (1981) *J. Org. Chem.*, **46**, 1229.
[bb] Solar, S.L. and Schumaker, R. (1966) *J. Org. Chem.*, **31**, 1996.
[cc] Weber, A.L. (1981) *J. Mol. Evol.*, **17**, 103.
[dd] Kondo, H., Moriuchi, F. and Sunamoto, J. (1981) *J. Org. Chem.*, **46**, 1333.
[ee] Hino, T., Taniguchi, M. and Nakagawa, M. (1981) *Heterocycles*, **15**, 187.
[ff] Ito, S., Inoue, S., Yamamoto, Y. and Fujita, K. (1981) *J. Med. Chem.*, **24**, 673.

REFERENCES

1. Jones, J.H. (1979) in *Comprehensive Organic Chemistry* (eds D.H.R. Barton and W.D. Ollis) (Vol. 2) (ed. I.O. Sutherland) Pergamon Press, Oxford, p. 815.
2. Law, H.D. (1974) in *Rodd's Chemistry of Carbon Compounds*, Supplements to Vols IC, ID 2nd edn, (ed. M.F. Ansell) Elsevier, Amsterdam, p. 314.
3. Haider, N.F., Patterson, J.M., Moors, M. and Smith, W.T. (1981) *J. Agric. Food Chem.*, **29**, 163.
4. Ivanov, C. and Slavcheva, N. (1977) *Dokl. Bolg. Akad. Nauk*, **30**, 727.
5. Ivanov, C. and Slavcheva, N. (1977) *Origins of Life*, **8**, 13.
6. Smith, R.M., Solabi, G.A., Hayes, W.P. and Stretton, R.J. (1980) *J. Anal. Appl. Pyrolysis*, **1**, 197.
7. Yamaguchi, K., Iitaka, Y., Shido, K. and Okamoto, T. (1980) *Acta Crystallogr., Sect. B*, **36**, 176.
8. Tsuji, K., Yamamoto, T., Zenda, H. and Kosuge, T. (1978) *Yakugaku Zasshi*, **98**, 910 (*Chem. Abs.*, **89**, 163925).
9. Ratcliff, M.A., Medley, E.E. and Simmonds, P.G. (1974) *J. Org. Chem.*, **39**, 1481.

10. Lien, Y.C. and Nawar, W.W. (1974) *J. Food Sci.*, **39**, 914.
11. Shulman, G.P. and Simmonds, P.G. (1968) *Chem. Commun.*, 1040.
12. Merritt, C. and Robertson, D.H. (1967) *J. Gas Chromatogr.*, **5**, 96.
13. Johnson, W.R. and Kang, J.C. (1971) *J. Org. Chem.*, **36**, 189.
14. Patterson, J.M., Chen, W.Y. and Smith, W.T. (1971) *Tobacco Sci.*, **15**, 41.
15. Olafsson, P.G. and Bryan, A.M. (1971) *Geochim. Cosmochim. Acta*, **35**, 327.
16. Fort, A.W., Patterson, J.M., Small, R., *et al.* (1971) *J. Org. Chem.*, **41**, 3697.
17. Patterson, J.M., Shille, C.-Y. and Smith, W.T. (1976) *J. Agric. Food Chem.*, **24**, 988.
18. Subbaraman, A.S., Kazi, Z.A. and Choughuley, A.S.U. (1979) *Indian J. Biochem. Biophys.*, **16**, 253.
19. Lehmann, G. and Zinsmeister, H.D. (1979) *Z. Lebensm.-Untersuch. Forsch.*, **169**, 357.
20. Nakayama, T., Hayase, F. and Kato, H. (1980) *Agric. Biol. Chem.*, **44**, 1201.
21. Misselhorn, K., and Bruckner, H. (1974) *Chem. Mikrobiol., Technol. Lebensmitteln*, **3**, 25.
22. Shaw, P.E. and Berry, R.E. (1977) *J. Agric. Food Chem.*, **25**, 641.
23. Ishibashi, T. and Kawabata, T. (1981) *J. Agric. Food Chem.*, **29**, 1098.
24. Beck, M.T., Katho, A. and Drozsa, L. (1981) *Inorg. Chim. Acta*, **55**, L55.
25. Kenny, M., Lambe, R.F., O'Kelly, D.A. and Darragh, A. (1980) *Clin. Chem.* (*Winston-Salem, NC*), **26**, 1511.
26. Levi, N. and Lawless, J.G. (1978) *Anal. Biochem.*, **90**, 796.
27. Tran, C.D. and Fendler, J.H. (1980) *J. Am. Chem. Soc.*, **102**, 2923.
28. Sun, M. and Zigman, S. (1979) *Photochem. Photobiol.*, **29**, 893.
29. Nakagawa, M., Kato, S., Kataoka, S. and Hino, T. (1979) *J. Am. Chem. Soc.*, **101**, 3136.
30. Nakagawa, M., Kato, S., Kodata, S., *et al.* (1981) *Chem. Pharm. Bull.*, **29**, 1013.
31. Ryang, H.S. and Foote, C.S. (1979) *J. Am. Chem. Soc.*, **101**, 6683.
32. Bonner, W.A., Blair, N.E. and Flores, J.J. (1979) *Nature*, **281**, 150.
33. Darge, W., Laszlo, I. and Thiemann, W. (1979) *Nature*, **281**, 151.
34. Blair, N.E. and Bonner, W.A. (1980) *J. Mol. Evol.*, **15**, 21.
35. Bonner, W.A., Blair, N.E. and Lemmon, R.M. (1979) *Origins of Life*, **9**, 279.
36. Bonner, W.A., Blair, N.E. and Lemmon, R.M. (1979) *J. Am. Chem. Soc.*, **101**, 1049.
37. Lynn, K.R. and Purdie, J.W. (1976) *Int. J. Radiat. Phys. Chem.*, **8**, 685.
38. Boguta, G. and Dancewicz, A.M. (1978) *Stud. Biophys.*, **73**, 11.
39. Boguta, G. and Dancewicz, A.M. (1981) *Nukleonika*, **26**, 11.
40. Katz, B.J. and Man, E.H. (1979) *Geochim. Cosmochim. Acta*, **43**, 1567.
41. Dymicky, M. (1980) *Org. Prep. Proced. Int.*, **12**, 207.
42. Barrett, G.C. and Martins, C.M.O.A. (1972) *J. Chem. Soc., Chem. Commun.*, 638.
43. Sjoberg, B., Fredga, A. and Djerassi, C. (1959) *J. Am. Chem. Soc.*, **81**, 5002.
44. Kurzer, F. and Lawson, A. (1962) *Org. Synth.* **42**, 100.
45. Mross, G. and Doolittle, R.F. (1977) in *Advanced Methods in Protein Sequence Determination* (ed. S.B. Needleman), Springer Verlag, Mannheim.
46. Previero, A. and Pechere, J.F. (1970) *Biochem. Biophys. Res. Commun.*, **40**, 549.
47. Barrett, G.C. and Leigh, P. (1975) *FEBS Lett.*, **57**, 19.
48. Edman, P. (1950) *Acta Chem. Scand.*, **4**, 283.
49. Walter, W. and Becker, R.F. (1972) *Liebigs Ann. Chem.*, **755**, 145.
50. Barrett, G.C. (1980) *Tetrahedron*, **36**, 2023.
51. Barrett, G.C. and Khokhar, A.R. and Chapman, J.R. (1969) *Chem. Commun.*, 818.
52. Barrett, G.C. (1978) *Tetrahedron*, **34**, 611.
53. Yankeelov, J.A., Fok, K.F. and Carothers, D.J. (1978) *J. Org. Chem.*, **43**, 1623.
54. Koga, K., Tzuoh, M.J. and Yamada, S. (1978) *Chem. Pharm. Bull., Tokyo*, **26**, 278.
55. Murakami, Y., Koga, K. and Yamada, S. (1978) *Chem. Pharm. Bull., Tokyo*, **26**, 307.
56. Beyerman, H.C., Maat, L., Noordam, A. and van Zon, A. (1977) *Rec. Trav. Chim. Pays-Bas*, **96**, 222.
57. Faustini, F., De Munari, S., Panzeri, A. *et al.* (1981) *Tetrahedron Lett.*, **22**, 4533.
58. Keck, R. and Retey, J. (1980) *Helv. Chim. Acta*, **63**, 769.
59. Lijinsky, W., Keefer, L. and Loo, J. (1970) *Tetrahedron*, **26**, 5137.
60. Bonnett, R. and Nicolaidou, P. (1977) *Heterocycles*, **7**, 637.
61. Pollock, J.R.A. (1981) *J. Inst. Brew.*, **87**, 356.
62. Nagasawa, H.T., Fraser, P.S. and Yuzon, D.L. (1973) *J. Med. Chem.*, **16**, 583.
63. Chow, Y.L. and Polo, J. (1981) *J. Chem. Soc., Chem. Commun.*, 297.
64. Takamura, N., Mizoguchi, T. and Yamada, S. (1973) *Tetrahedron Lett.*, 4267.
65. Quitt, P., Hellerbach, J. and Vogler, K. (1963) *Helv. Chim. Acta*, **46**, 327.

66. Tome, D. and Naulet, N. (1981) *Int. J. Peptide Protein Res.*, **17**, 501.
67. Grigg, R., Jordan, M. and Malone, J.F. (1979) *Tetrahedron Lett.*, 3877.
68. Grigg, R. and Kemp, J. (1980) *Tetrahedron Lett.* **21**, 2461.
69. Grigg, R., Kemp, J., Malone, J. and Tangthougkam, A. (1980) *J. Chem. Soc. Chem. Commun.*, 648.
70. McDermott, J.R. and Benoiton, N.L. (1973) *Canad. J. Chem.*, **51**, 1915.
71. Okamoto, K., Abe, H., Kuromizu, K. and Izumiya, N. (1974) *Mem. Foc. Sci., Kyushu Univ., Sect. C*, **9**, 131.
72. Ikutani, Y. (1971) *Bull. Chem. Soc. Japan*, **44**, 271.
73. Ikutani, Y. (1968) *Bull. Chem. Soc. Japan*, **41**, 1679.
74. Talik, Z. and Brekiesz-Lewandowska, B. (1967) *Rocz. Chem.*, **41**, 2095.
75. Talik, T. and Talik, Z. (1968) *Bull. Acad. Polon. Sci., Ser. Sci. Chim.*, **16**, 13.
76. Harmeyer, J., Sallman, H.-P. and Ayoub, L. (1968) *J. Chromatogr.*, **32**, 258.
77. Pierpoint, W.S. (1969) *Biochem. J.*, **112**, 609.
78. Prota, G., Scherillo, G., Napolano, F. and Nicolous, R.A. (1967) *Gazzetta*, **97**, 1451.
79. Prota, G., Scherillo, G. and Nicolous, R.A. (1968) *Gazzetta*, **98**, 495.
80. Ghosh, P.B. and Whitehouse, M.W. (1968) *Biochem. J.*, **108**, 155; see also p. 588 of this book.
81. Gloede, J., Poduska, K., Gross H. and Rudinger, J. (1968) *Coll. Czech. Chem. Commun.*, **33**, 1307.
82. Gante, J. (1968) *Chem. Ber.*, **101**, 1195.
83. Nowak, K. (1969) *Rocz. Chem.*, **43**, 231.
84. Zaloom, J. and Roberts, D.C. (1981) *J. Org. Chem.*, **46**, 5173.
85. Barton, D.H.R., Bringmann, G. and Motherwell, W.B. (1980) *Synthesis*, 68.
86. Achiwa, K. and Yamada, S. (1975) *Tetrahedron Lett.*, 2701.
87. Vit, J. and Barer, S.J. (1976) *Synth. Commun.*, **6**, 1.
88. Matsushima, A., Yamazaki, S., Shibata, K. and Imada, Y. (1972) *Biochim. Biophys. Acta*, **271**, 243.
89. Terent'ev, A.P., Rukhadze, E.G. and Dunina, V.V. (1969) *Dokl. Akad. Nauk SSSR, Ser. Khim.*, **187**, 138.
90. McCaldin, D.J. (1960) *Chem. Rev.*, **60**, 39.
91. Rausch, T., Hofmann, B. and Hilgenberg, W. (1981) *Z. Naturforsch. B, Anorg. Chem., Org. Chem.*, **36B**, 359.
92. Clarke, T.G., Hampson, N.A., Lee, J.B. *et al.* (1970) *J. Chem. Soc.* (*C*), 815.
93. Sylvester, S.R., Lan, S.Y. and Stevens, C.M. (1981) *Biochemistry*, **20**, 5609.
94. Lohmar, R. and Steglich, W. (1978) *Angew. Chem. Int. Ed.*, **17**, 450.
95. Taguchi, T., Morita, S. and Kawazoe, Y. (1975) *Chem. Pharm. Bull.*, **23**, 2654.
96. Jones, J.H., in *Comprehensive Organic Chemistry* (eds D.H.R. Barton and W.D. Ollis) (Vol. 5) (ed. P.G. Sammes), Pergamon Press, Oxford, p. 825.
97. Mahadeveppa, D.S. and Gade, N.M.M. (1977) *J. Indian Chem. Soc.*, **54**, 534.
98. Miramatsu, I., Motoki, Y., Yabuchi, K. and Komachi, H. (1977) *Chem. Lett.*, 1253.
99. Kometani, T., Suzuki, T., Takahashi, K. and Fukumoto, K. (1974) *Synthesis*, 131.
100. Weigele, M., Blount, J.F., Tengi, T.P., *et al.* (1972) *J. Am. Chem. Soc.*, **94**, 4052.
101. Weigele, M., De Bernardo, S.L., Tengi, J.P., *et al.* (1972) *J. Am. Chem. Soc.*, **94**, 5927.
102. Toome, V., Wegrzynski, B. and Dell, J. (1976) *Biochem. Biophys. Res. Commun.*, **71**, 598.
103. Simons, S. and Johnson, D.F. (1978) *J. Org. Chem.*, **43**, 2886.
104. Hayashi, T. and Namiki, M. (1979) *Tetrahedron Lett.*, 4467.
105. Al-Sayyab, A.F. and Lawson, A. (1968) *J. Chem. Soc.* (*C*), 406.
106. Poindexter, G.S. and Meyers, A.I. (1977) *Tetrahedron Lett.*, 3527.
107. Kubota, M., Nagase, O. and Yajima, H. (1981) *Chem. Pharm. Bull.*, **29**, 1169.
108. Rothgery, E.F. and Hohnstedt, L.F. (1971) *Inorg. Chem.*, **10**, 181.
109. Kiyooka, S., Goto, F. and Suzuki, K. (1981) *Chem. Lett.*, 1429.
110. Chen, F.M.F. and Benoiton, N.L. (1978) *Synthesis*, 928.
111. Bates, H.S., Jones, J.H. and Witty, M.J. (1980) *J. Chem. Soc. Chem. Commun.*, 773.
112. Benoiton, N.L. and Chen, F.M.F. (1981) *J. Chem. Soc., Chem. Commun.*, 543.
113. Clarke, C.T. and Jones, J.H. (1977) *Tetrahedron Lett.*, 2367.
114. Penke, B., Czombos, J., Balaspiri, L., *et al* (1970) *Helv. Chim. Acta*, **53**, 1057.
115. McClure, D.E., Arison, B.H., Jones, J.H. and Baldwin, J.J. (1981) *J. Org. Chem.*, **46**, 2431.
116. Buckley, T.F. and Rapoport, H. (1981) *J. Am. Chem. Soc.*, **103**, 6157.

117. Reifenrath, W.G., Bertelli, D.J., Micklus, M.J. and Fries, D.S. (1976) *Tetrahedron Lett.*, 1959.
118. Kaneko, T., Izumi, Y., Chibata, I. and Itoh, T. (eds) (1974) *Synthetic Production and Utilization of Amino Acids* Wiley, New York, p. 197.
119. Reference 1, p. 825.
120. Hartman, J., Brand, M.C. and Dose, K. (1981) *Biosystems*, **13**, 141.
121. Miyoshi, M. (1973) *Bull. Chem. Soc. Japan*, **46**, 212, 1489.
122. Jones, J.H. and Witty, M.J. (1979) *J. Chem. Soc. Perkin 1*, 3203.
123. Hess, U. and Koenig, W.A. (1981) *Liebigs Ann. Chem.* 1606.
124. Hess, U. and Koenig, W.A. (1980) *Liebigs Ann. Chem.* 611.
125. Lohmar, R. and Steglich, W. (1978) *Angew. Chem. Int. Ed.*, **17**, 450.
126. Zav'yalov, S.I. and Ezhova, G.I. (1977) *Izv. Akad. Nauk S.S.S.R., Ser. Khim.*, 219.
127. Texier, F. and Yebdri, O. (1975) *Tetrahedron Lett.*, 855.
128. Barrett, G.C., Khokhar, A.R. and Chapman, J.R. (1969) *Chem. Commun.*, 818.
129. Hearn, W.R. and Worthington, R.E. (1967) *J. Org. Chem. Chem.*, **32**, 4072.
130. Wagner, D., Gertner, D. and Zilkha, A. (1968) *Tetrahedron Lett.*, 4479.
131. Matthies, D., Bartsch, B. and Richter, H. (1981) *Arch. Pharm.*, **314**, 209.
132. Seebach, D. and Wasmuth, D. (1981) *Angew. Chem.*, **93**, 1007.
133. Rae, I.D. and Umbrasas, B.N. (1971) *Austral. J. Chem.*, **24**, 2729.
134. Ohta, T. and Nakai, T. (1978) *Biochim. Biophys. Acta*, **533**, 440.
135. Ohta, T. and Nakai, T. (1976) *Biochim. Biophys. Acta*, **420**, 258.
136. Gruber, H.A. and Mellon, E.F. (1975) *Analyt. Biochem.*, **66**, 78.
137. Fujii, N., Sasaki, T., Funakoshi, S., *et al.* (1978) *Chem. Pharm. Bull.*, **26**, 650.
138. Atassi, M.Z. (1973) *Tetrahedron Lett.*, 4893.
139. Bordignon, E., Cattalini, L., Natile, G. and Scatturin, A. (1973) *J. Chem. Soc., Chem. Commun.*, 878.
140. Gordon, P.G. (1973) *Austral. J. Chem.*, **26**, 1771.
141. Inoue, M. and Hikoya, H. (1971) *Chem. Pharm. Bull. Japan*, **19**, 1286.
142. Stoehrer, G. (1976) *J. Heterocyclic Chem.*, **13**, 157.
143. Bonnett, R., Holleyhead, R., Johnson, B.L. and Randall, E.W. (1975) *J. Chem. Soc., Perkin I*, 2261.
144. Clark, D.G. and Cordes, E.H. (1973) *J. Org. Chem.*, **38**, 270.
145. Stanfield, C.F., Parker, J.E. and Kanellis, P. (1981) *J. Org. Chem.*, **46**, 4799.

CHAPTER TWELVE

Degradation of Amino Acids Accompanying *in vitro* Protein Hydrolysis

S. HUNT

12.1 INTRODUCTION

It often happens that a technique undergoes a rapid development based upon a few straightforward principles and relatively easily applied methods. Later developments most frequently are concerned with increasing sophistication of instrumentation and may fail to recognize fully inadequacies of protocol or insufficiencies in knowledge of basic principles involved in method.

Amino acid analysis is an example of this principle. The instrumentation and methods of separation have been well-developed and explored, yet the chemical principles underlying the hydrolysis of proteins to free amino acids have been examined surprisingly little, when one considers that perhaps millions of protein hydrolyses and amino acid analyses must be performed, world-wide, annually. A central problem remains, however, in providing a rational explanation for the losses of amino acids which occur during hydrolysis.

That certain amino acids undergo irreversible change during protein hydrolysis has been common knowledge for many years [1]. Thus most workers know that tryptophan is largely destroyed during acid hydrolysis and that cystine is converted in part to cysteic acid via cysteine and the sulphinic acid and sulphone. Serine, threonine and tyrosine are also well known to be partially destroyed and corrections are often (but not always) made by careful workers for these losses. What fewer recognize is that other amino acids besides those mentioned can suffer degradation to a greater or lesser degree while unidentified peaks appearing on analyser chromatograms frequently arise from the fragments of amino acid degradation and may also interfere with accurate quantitation of known components. The majority of workers have little knowledge of the nature of these by-products and indeed such knowledge is sparse. Nor are the consequences of protein hydrolysis taking place in the presence of other non-amino acid organic species widely appreciated. The type and purity of protein being hydrolysed is therefore of some consequence.

12.2 PROTEIN PURITY

Obviously for the simplest of situations in which a pure protein species, salt-free and without prosthetic or other conjugated groups is subjected to hydrolysis, only the interaction between each amino acid species and the hydrolysis reagent is of concern in the initial stages. The presence of salts however and particularly anions such as phosphate or sulphate or appreciable amounts of cations such as sodium, potassium or calcium may influence rates of hydrolysis and the formation of secondary products. The presence of iron or copper in protein combined with other circumstances may cause serious difficulties in the production of unusual artefacts and in the loss of certain amino acids. If organic groups are present, of types quite commonly accompanying protein such as carbohydrate composed perhaps of neutral sugars, hexosamines, uronic acids and neuraminic acids, then complex reactions in the hydrolysis medium will almost certainly take place. These will involve reactions between amino acids and carbohydrate degradation products. Groups such as porphyrin and other heterocyclics or lipids will themselves be released and broken down in unpredictable or little-understood ways, and these too will interact with free amino acids.

12.3 METHODS OF HYDROLYSIS

The majority of protein amino acid analyses are performed upon hydrolysates prepared in hot mineral acid media. For some purposes organic acids may be used, as for example in hydrolyses designed to recover tryptophan as an analysable component or to produce selective release of certain amino acids in a partial hydrolysis. Acetic acid may be used for the latter purpose to selectively release aspartic acid [2]. A few analyses are performed in alkaline solution usually with the same objective of releasing tryptophan intact.

12.3.1 Acid hydrolysis

While sulphuric acid has been used as a hydrolysis medium, the acid of almost universal choice is hydrochloric acid at a concentration of 6 M (constant boiling to be accurate is 5.8 M). Lower grade commercial hydrochloric acid preparations may be appreciably contaminated with amino acids, ammonia and inorganic ions of iron, copper, ammonia and bromine as well as numerous other trace materials. Amino acids and ammonia will certainly lead to erroneous quantitation in a modern highly sensitive analytical system while bromine and metal ions will enhance the decomposition or derivatization of certain amino acids. It is of paramount importance therefore that acid for hydrolysis should be of the very highest quality and should be kept well-sealed for this use only. Similarly water used for dilution of the acid should be at least distilled and deionized. Double distillation is desirable.

Hydrolysis vessels should be scrupulously cleaned with acid and distilled water and the quality of the glass is as we shall see of some importance. Old scratched glass surfaces should not be used *ad infinitum.* In general the most suitable containers for hydrolysis are either pyrex test-tubes which have been necked so that they can be flame-sealed, or glass stoppered test-tubes of the type supplied by Quickfit. The stoppers of these can usually be wired to provide a perfect seal during hydrolysis. Protein weights commonly hydrolysed these days are of the order of one milligram or less and this makes the addition of an excess of 6 M acid easier. Five hundred volumes of acid in relation to protein, on a weight–volume basis is probably ideal. Excess protein in relation to acid leads to poorer recovery of amino acids partly because of incomplete hydrolysis and partly because of enhanced decomposition.

Acid and protein are well-mixed to ensure even dispersion and any additives used to improve recovery of certain amino acids are added at this stage if they have not been incorporated already into the acid. The matter of additives is discussed below but, for example, methionine recovery can be improved by addition of mercaptoethanol. Obviously any extra reagents have to be of the same high quality as the acid itself.

Since oxidation probably plays an important part in amino acid decomposition it is usual either to remove air from the hydrolysis tube before sealing it, either by use of vacuum pump or by passing oxygen-free nitrogen through the mixture immediately prior to sealing. If the tube is to be evacuated then bumping can be avoided by freezing the sample in liquid nitrogen or solid carbon dioxide/acetone and allowing it to thaw while the vacuum is applied. Ideally nitrogen may then be admitted to fill the vacuum and the tube sealed.

Hydrolyses are usually performed at temperatures between 100°C and 110°C, the favourite being 105°C. Times of hydrolysis vary and do depend very much upon the protein. Twenty hours is average. Some proteins are very difficult to bring to complete hydrolysis and combinations in peptide linkage such as Leu–Leu, Ile–Leu, Ile–Val and Val–Gly may be very tardy in severance. In contrast aspartic acid is released very rapidly. Thus some amino acids may spend far longer in free solution than others and hence their opportunities for secondary degradation will be much enhanced.

Following hydrolysis acids must be removed. It is to be recommended that this be done as quickly as possible preferably on a rotary evaporator. While many operators evaporate *in vacuo* over potassium hydroxide pellets in a dessicator this slower procedure once again results in losses of certain amino acids as a more and more concentrated solution forms.

12.3.2 Other acid reagents leading to complete hydrolysis

Three organic acids have been used to hydrolyse protein while retaining tryptophan undergraded. Mercaptoethane-sulphonic acid [3] and *p*-toluene-sulphonic acid [4] are used at 3 N concentration while 4 N-methane-sulphonic acid has also been applied with success [5]. The disadvantage of toluene-sulphonic acid lies in the fact that it cannot be used if carbohydrate is present at levels above 50%.

Using these acids, evaporation is neither feasible nor necessary and hydrolysates may be simply neutralized with sodium hydroxide and applied to the column either as an aliquot directly or in diluted form. Once again hydrolyses are performed in sealed tubes at between 100 and 120°C for periods of usually up to 24 hours. Recoveries of amino acids are good.

12.3.3 Partial acid hydrolyses

Sequencing studies require the production of peptide fragments, and while enzymic and special reagent amino acid specific protocols are the most usual routes, partial acid hydrolysis may prove a suitable method from time to time. Molar to tenth molar hydrochloric acid or 0.3 M acetic acid [6] at 100–110°C under reflux, *in vacuo* or under nitrogen are the sort of conditions which might be applied perhaps for extended periods and it might be thought that these would be mild enough to cause few problems of degradation and artefact production. This is however not necessarily so as we shall see.

12.3.4 Alkaline hydrolysis

Alkaline hydrolysis provides a route to the recovery of tryptophan on a quantitative basis though other amino acids suffer partial or complete destruction. The methodology does however present technical and manipulative difficulties in so much that special reaction vessels must be employed and the reagent itself must be removed before analysis.

The reagent usually employed is barium hydroxide at 4 N concentration although 4.2 N sodium hydroxide has also been employed with rather more reproducible results [7, 8]. Hydrolysis is carried out at 110°C for between 16 and 70 hours depending upon the protein and the tryptophan content. It is even more important, if anything, to remove oxygen here than it is for acid hydrolyses. The barium hydroxide is added in solid form and only dissolves completely at 110°C. Ordinary glass reaction tubes are unsuitable as the glass etches badly, releasing silicates and promoting side reactions. One solution is to use (expensive) tubes of Vycor glass which have a high silica content; another, applied in my own laboratory, is to use plastic vials (scintillation vial inserts serve nicely and are cheap and expendable) held inside thick walled glass bottles with metal screw tops (McCartney bottles), having neoprene rubber seals. There have been no accidents so far.

Following hydrolysis the alkali must be neutralized and if barium hydroxide has been used the barium ions removed. Sodium hydroxide hydrolyses may be simply neutralized with hydrochloric acid although dilution will be necessary before application to the analyser column.

Barium ions may be removed either by precipitation as the carbonate using a stream of carbon dioxide or as the sulphate. It is better to add sodium sulphate than sulphuric acid as traces of sulphuric acid may prove embarrassing should the solution require concentration. The major practical disadvantages of barium hydroxide hydrolysis are the difficulty of removing all the barium, the losses incurred in adsorption of amino acids to the precipitated barium salts and the bulking of liquid samples which are a consequence of washing precipitates.

Most amino acids are severely affected by alkaline hydrolysis, although in the right hands tryptophan is reasonably stable, as is leucine with which tryptophan can be compared for correlation with the results of an acid hydrolysis.

12.4 CHANGES IN AMINO ACIDS AS A CONSEQUENCE OF HYDROLYSIS

12.4.1 Racemization

Amino acids of proteins probably undergo at least three types of change, which take place at several different stages of the hydrolysis protocol. Modification at its simplest is represented by racemization [9, 10, 11]. While this presents no serious problem under most circumstances it is important to realize that a prolonged period in hot acid may result in production of diastereoisomers from amino acids with two or more chiral centres where physical properties are so sufficiently different as to result in their separation by an ion-exchange resin column. Unidentified peaks may then arise from such products as alloisoleucine originating by racemization of leucine. Such effects may lead to erroneous quantitation.

The majority of changes experienced by amino acids during hydrolysis are however of a more radical chemical nature. These may arise at the time of hydrolysis with cleavage of labile bonds other than the peptide bonds; for example disulphide bond cleavage or loss of phosphate from phosphorylated amino acids. Progressive reactions such as the oxidative degradation of tryptophan may occur throughout hydrolysis while some amino acids may be altered in the post-hydrolysis period of sample concentration. An important class of artefact production and reduction in apparent yield lies in the ability of certain amino acids to form esters during and after hydrolysis.

12.4.2 Ester formation

The analysis of proteins bound to connective tissue proteoglycans, which are highly sulphated heteropolysaccharides, may be complicated by the inorganic sulphate which is released during acid hydrolysis [12]. If post-hydrolytic evaporation is not carried out rapidly by rotary evaporation then the sulphate esters of serine, threonine and tyrosine may form quite readily and if collagen is also a component of the hydrolysed protein then the sulphate esters of hydroxyproline and possibly hydroxylysine could conceivably be added to this list.

Amino acid sulphate esters will appear on most analyser chromatograms in the vicinity of cysteic acid. Similar effects may arise if proteins are hydrolysed in the presence of inorganic sulphates used in protein purification and incompletely removed by dialysis, or if proteins such as protamine or histone, commonly prepared as their sulphate salts, are hydrolysed [13, 14]. It is likely also that phosphate, whether it be derived from phosphate buffers, nucleic acid or phosphorylated polysaccharides, will present similar problems although information is lacking in this area.

Organic esters may also form at the drying stage of hydrolysate production. Thus glutamic acid and serine and/or threonine react readily to form *O*-(γ-L-glutamyl)-L-serine (*12.1*) and *O*-(γ-L-glutamyl)-L-threonine [15]. While concrete evidence is lacking, one may surmise that glutamic acid might react in a similar manner with tyrosine and with other hydroxy compounds commonly present in hydrolysates such as, for example, sugars.

$$HOOC(NH_2)CHCH_2OCOCH_2CH(NH_2)COOH$$

(12.1)

12.4.3 Elimination of simple substituents

Asparagine and glutamine are lost completely in the first hour or so of acid hydrolysis with the release of ammonia and the formation of aspartic and glutamic acids [16]. While this particular degradation is well-recognized and can be coped with by estimating the ammonia released, too few workers are prepared to attempt to assay the proportion of aspartic acid and glutamic acid residues present as their amides. Yet asparagine and glutamine are very different to β-aspartyl and γ-glutamyl for the purposes of understanding the role of side chains in chain conformation or biological activity.

A number of proteins, particularly those concerned with calcium-binding activity, contain phosphate groups. These too are sensitive to the acid conditions normally applied to full hydrolysis of proteins.

O-Phosphoserine and *O*-phosphothreonine are present in such materials as the non-collagenous protein of bone [17, 18]. These may be released and estimated by hydrolysis in 4 M hydrochloric acid at 105°C for 6 hours. Thus while these phosphorylated amino acids are labile during the full period of 6N acid hydrolysis they are in relative terms stable and moderately easily isolated.

γ-Glutamyl phosphate occurs in proteins such as collagen but the greater lability of this amino acid to hydrolysis led to a greater delay in its identification [19]. Reductive cleavage prior to hydrolysis yields α-amino-δ-hydroxyvaleric acid (*12.2*).

$$HOCH_2CH_2CH_2CH(NH_2)COOH$$

(12.2)

It is possible that phosphorylated amino acids, as well as those mentioned, exist undetected perhaps including the esters of hydroxyproline, hydroxylysine and tyrosine. The latter has in fact been recently found in proteins [72].

Tyrosine-*O*-sulphate is another amino acid derivative of limited distribution in proteins [20, 21], which escaped detection for many years because of its acid lability. The release of sulphate from tyrosine in fibrinogen is complete after four minutes at 93°C in 1 M hydrochloric acid. The ester is stable however to alkaline hydrolysis and tyrosine-*O*-sulphate may be obtained from fibrinogen by heating in 0.2 M barium hydroxide at 125°C for 24 hours.

If a protein amino group, either in the *N*-terminal alpha position or at the ε position of lysine, is *N*-acylated then this substituent will have a similar lability to the peptide bond itself. Thus *N*-acetylated terminal amino acids [22] or the *N*-terminal *N*-formylmethionine of bacterial protein [23] will be lost during acid hydrolysis. In a similar manner the cyclic internal amide pyroglutamic acid (*12.3*) which forms a terminal residue in certain proteins [24], will ring open in the hot acid to give free glutamic acid.

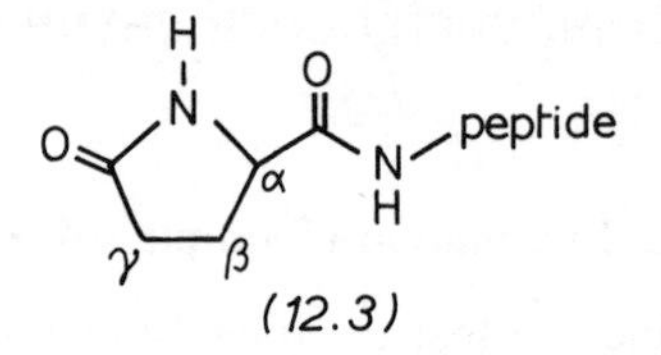

(12.3)

The quantitation of glutamic acid is attended by a number of difficulties as we have seen because of its ability to form derivatives modifying its behaviour in the native protein. Gamma-carboxyglutamate (*12.4*) is a protein amino acid of late and recent detection [25] occurring in calcium-binding proteins of mineralizing tissues and systems under calcium modulation. The second carboxyl group is acid sensitive but survives hydrolysis in 2 M potassium hydroxide at 106°C for 24 hours under nitrogen [25].

$$(HOOC)_2CHCH_2CH(NH_2)COOH$$

(12.4)

12.4.4 Degradation leading to loss of amino acid structure

The types of hydrolytic change described above involve conversion of amino acids to other chiral forms, formation of simple derivatives or conversely loss of substituents leading to formation of common protein amino acids. However, in very many cases amino acids may decompose further leading to products not immediately recognizable as originating from the parent amino acid.

Of the twenty or so common protein amino acids those most universally recognized as being seriously degraded in this way are serine, threonine, tyrosine and tryptophan [1, 26]. Of these, tryptophan is the most seriously affected by acid hydrolysis, although it is by no means always completely destroyed during hydrochloric acid hydrolysis as is so often stated. Even so, losses under normal conditions are usually so severe as to render quantitation impracticable.

(a) Tryptophan

A number of techniques of hydrolysis are available to overcome the problem of tryptophan loss and some of these have been alluded to above.

Pure protein preparations lacking sulphur-containing amino acids, carbohydrate and other non-protein constituents, hydrolysed in the complete absence of oxygen with highly purified hydrochloric acid, can give good yields of tryptophan. Addition of thioglycollic acid [27] to the 6 N hydrochloric acid also improves tryptophan recovery. But substantial improvements result from use of the strong organic acids described above, while a further improvement can be achieved by addition of 3-(2-aminoethyl)indole [5] as a 'catalyst' to these acids; for example at 0.2% concentration in 4 N methanesulfonic acid. Most of these techniques are however still sensitive in some degree to the presence of carbohydrate.

Although the hydrolysis of proteins to yield tryptophan intact has been quite intensively investigated from the aspect of methodology the exact nature of the decomposition process, its mechanisms and products, has been surprisingly little studied particularly in the latter respect.

Tryptophan degradation products are usually observable in the basic region of the amino acid analyser chromatogram and may interfere with the integration of other amino acids, particularly lysine and histidine. The nature of these products is, however, not altogether clear. There are two views of tryptophan degradation, one of these being that the amino acid is attacked by the acid, the other that there is reaction with decomposition products of other amino acids and sugars. The latter view is partially supported by studies which show that for conditions normally used for protein acid hydrolysis, even in the presence of oxygen, tryptophan alone (2.0×10^{-4} M) only follows decomposition kinetics with rate constants of the order of $10^{-5}\ s^{-1}$: too slow to decompose more than one third of the amino acid in 24 hours [28].

The decomposition of tryptophan in hydrochloric acid at 100°C has been shown to proceed by a free-radical autoxidation mechanism with the amino acid being protonated at either the 1 or the 3 position before autoxidation [28]. The reaction is catalysed by impurities in the soda glass of the tubes commonly used for hydrolysis reactions. The reaction is probably a four stage process of protonation, initiation, propagation and termination (12.1).

Protonation	$Trp + H^+ \rightleftharpoons TrpH^+$
Initiation	$TrpH^+ \longrightarrow Trp^{+\bullet}$
Propagation	$Trp^{+\bullet} + O_2 \longrightarrow TrpO_2^{+\bullet}$
Termination	$2Trp^{+\bullet} \longrightarrow$ Molecular products
	$Trp^{+\bullet} + TrpO_2^{+\bullet} \longrightarrow$ Molecular products
	$2TrpO_2^{+\bullet} \longrightarrow$ Molecular products

(12.1)

The relation between decomposition rate and pH for tryptophan at 100°C has, however, some unexpected features. In the range pH 2–7 maximum decomposition occurs at pH 5.3 the reaction following the same rate order as that noted for strongly acid solutions [29]. It would seem that protonation at two points is a factor, in this free-radical autoxidation process, taking place in the indole and the carboxyl segments. The reaction initiator, derived from the soda glass of the reaction vessel, is proposed to interact with the carboxylate ions. This process further actuates the protonation of the indole segment, producing a more reactive species and closer contact with the initiator.

While alkali is quite frequently used to release tryptophan from protein, on a quantitative basis, decomposition occurs here also. Tryptophan degradation follows similar rate order kinetics in 1 M sodium hydroxide at 100°C to those for acid decomposition [30]. Here also decomposition involves a free-radical autoxidation, but it is impurity in the alkali and not solution pH which influences stability of the amino acid. Sodium hydroxide commonly contains traces of transition elements while barium hydroxide is more easily purified. This may explain the greater stability of tryptophan in solutions of the latter base. An intermediate in alkali degradation may be the hydroperoxide (*12.5*).

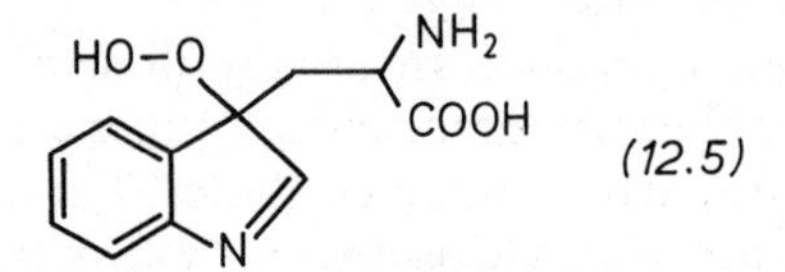

It is still not clear, however, what further products may be formed from oxidative degradation of tryptophan under hydrolytic conditions, although the production of such molecules as kynurenine [31] and anthranilic acid seem feasible in view of what is assumed to be the consequence of autoxidation namely homolysis of the indolyl N–H bond [29].

A rather surprising and hitherto unobserved effect is that of tryptophan concentration upon its degradation in isolation. At concentrations above 0.025 M tryptophan decomposition in pure 6 M hydrochloric acid at 105°C is low, but below this concentration increases rapidly (Fig. 12.1) to a maximum at 5×10^{-5} M with the formation of a complex mixture of heterocyclic compounds (Fig. 12.2) which include several quinoline carboxylic acids [32] (Figs 12.3, 12.4 and 12.5). Ring expansion is hence apparently possible and might lend support to the notion of ring opening to kynurenine and subsequent re-closure.

A clue to the process of oxidative degradation may come from the effect of *tert*-butyl hydroperoxide upon tryptophan in the presence of ferrous ions and 1 N sulphuric acid. Here the major reaction product is the spirolactone (*12.6*) [33].

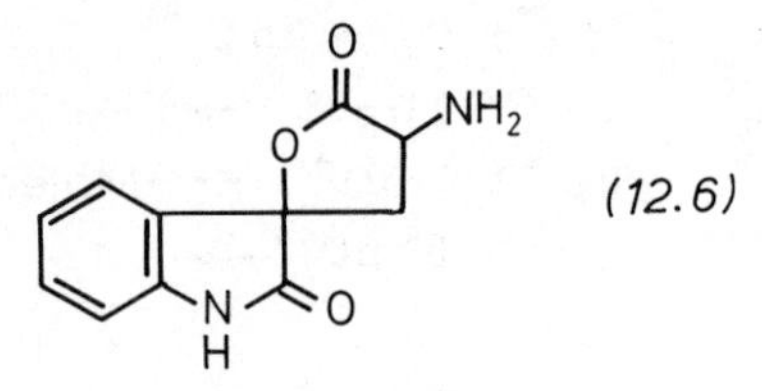

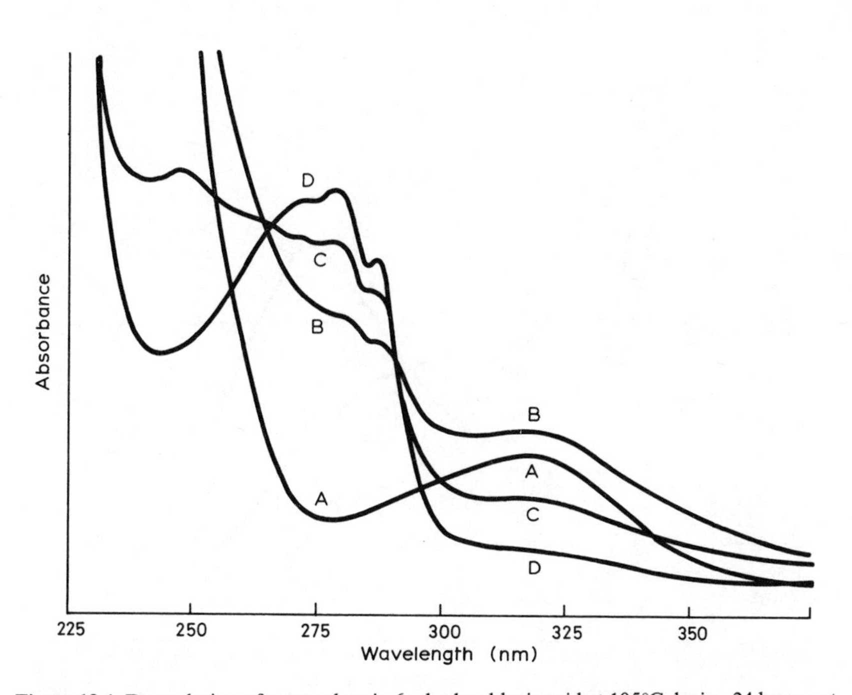

Figure 12.1 Degradation of tryptophan in 6 N hydrochloric acid at 105°C during 24 hours, at different tryptophan concentrations, shown as changes in the ultraviolet absorption spectra of the solutions. Spectra were normalized to equivalent concentrations to bring them all into the same absorbancy range. Hydrolysed tryptophan concentrations: A, 5×10^{-6} M; B, 2.5×10^{-4} M; C, 5×10^{-4} M; D, 2.5×10^{-3} M.

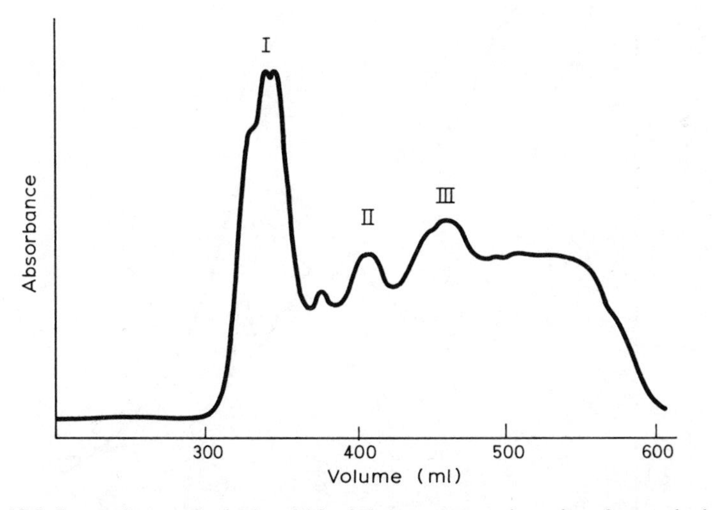

Figure 12.2 Separation on Sephadex G15 of the concentrated product from a hydrolysis solution containing 4.0 mg tryptophan in 50 ml 6 N hydrochloric acid. Hydrolytic conditions were 110°C for 24 hours, without attempt to exclude air, in a sealed vessel having approximately 200 ml dead space. The column was 2.2 × 150 cm and was eluted with 0.02 M acetic acid. Unchanged tryptophan elutes as a broad peak centred at 800 ml elution volume. Roman numerals refer to peaks whose absorption spectra follow in Figs 12.3–12.5.

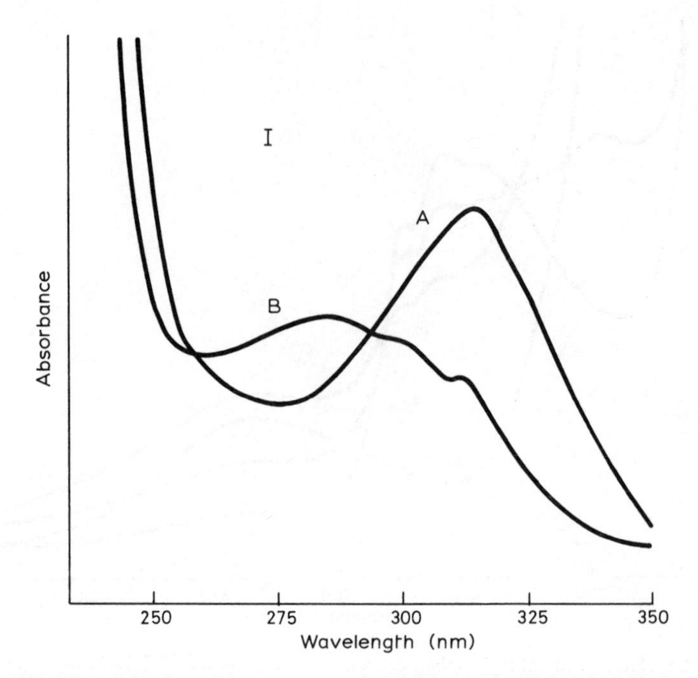

Figure 12.3 Ultraviolet absorption spectrum of peak I from Fig. 12.2. A in 0.02 M acetic acid. B at pH 12.0 in sodium hydroxide. The spectrum resembles that of quinoline-4-carboxylic acid.

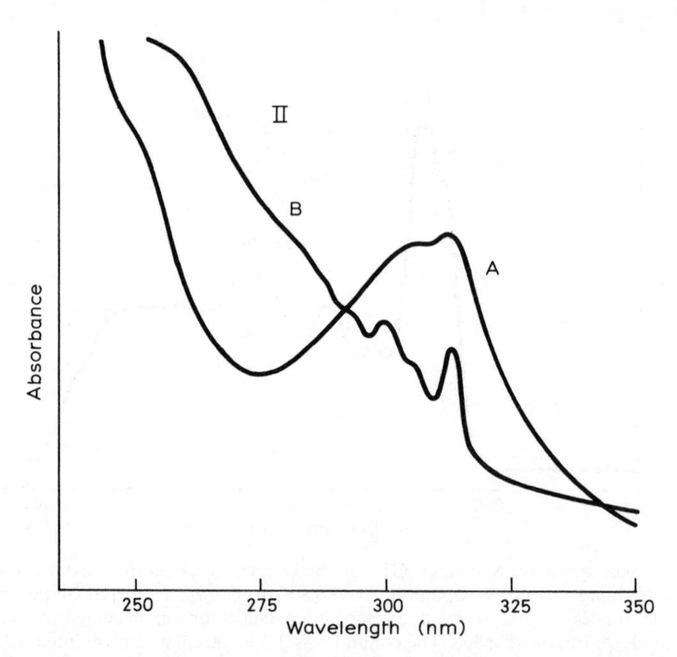

Figure 12.4 Ultraviolet absorption spectrum of peak II from Fig. 12.2; A in 0.02 M acetic acid. B at pH 12.0 in sodium hydroxide. The spectrum resembles that of unsubstituted quinoline.

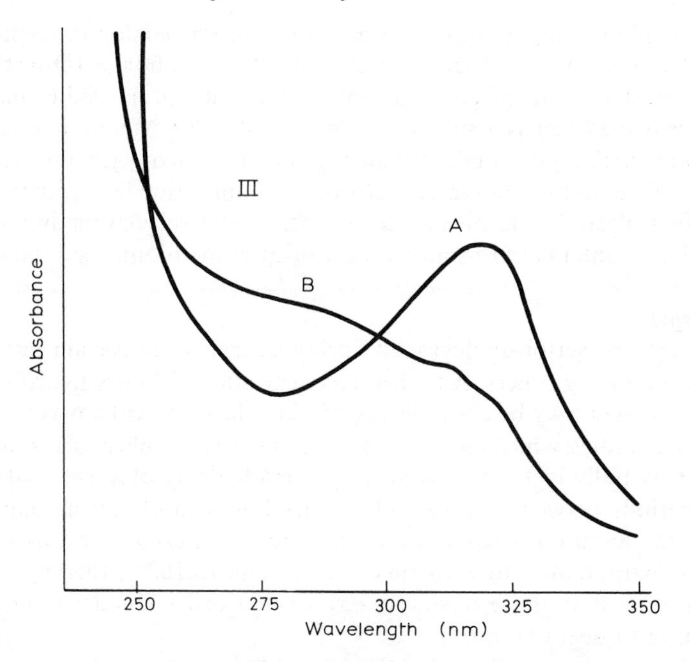

Figure 12.5 Ultraviolet absorption spectrum of peak III from Fig. 12.2. A in 0.02 M acetic acid. B at pH 12.0 in sodium hydroxide. The spectrum resembles that of quinoline-2-carboxylic acid (quinaldinic acid).

One major product of tryptophan degradation in protein hydrolysis has been firmly identified and the mechanism involved at least partly elucidated.

Beta-3-oxindolylalanine (*12.7*) was first noted in hydrolysates of the toxic cyclic peptide of *Amanita phalloides* [34]. In this peptide tryptophan and cysteine form a bridging group [35] (*12.8*). In other proteins which contain tryptophan and cystine 6 N hydrochloric acid hydrolysis yields oxindolylalanine and cysteine via the formation of 2-(2-amino-2-carboxyethylthio)tryptophan (tryptathionine) (*12.9*) as transient intermediate [36, 37].

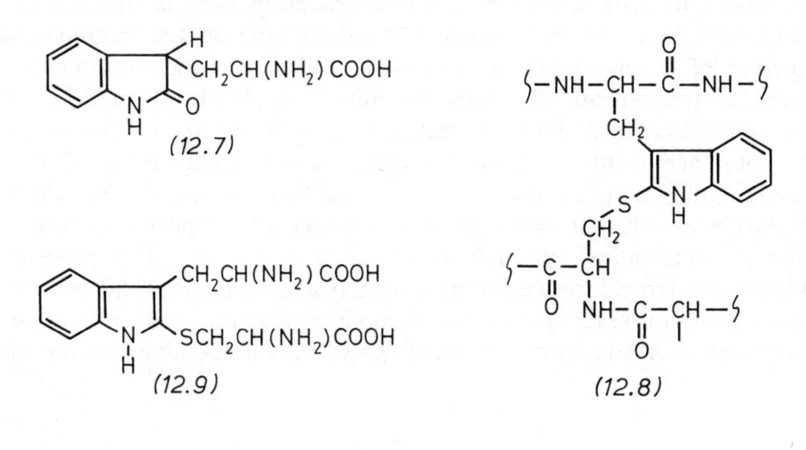

Other sulphur compounds can bring about the conversion of tryptophan to oxindolylalanine and the observation that dimethyl sulphoxide (DMSO) [38] in hydrochloric acid is an effective reagent for the reaction has led to methionine sulphoxide being effectively substituted for DMSO [39]. Methionine sulphoxide forms under hydrolysis conditions in the presence of oxygen through partial oxidation of methionine and there is an obvious connection here with tryptophan losses in hydrolysis. While most proteins contain relatively few methionines some do have high contents of this amino acid or even methionine sulphoxide itself.

(b) Tyrosine

Tyrosine may be seriously degraded during hydrolysis of certain proteins and hardly at all during others. A typical recovery after 22 hours hydrolysis in 6 N hydrochloric acid may be about 90% [40], but the only real answer for serious quantitative studies, which require knowledge of exact numbers of residues of this and other partially lost amino acids in protein hydrolysates, is to sample after various periods between twelve and seventy hours to obtain an extrapolated value for the amino acid content. Usually however, even with this technique, recoveries extrapolated to zero time rarely achieve 100%, thereby indicating reactions other than those of simple zero or first order kinetics reported to be contributing to degradation.

Loss of tyrosine is usually laid at the door of halogenation [41], however in all probability the situation is more complicated than this. It is certainly true however that traces of chlorine and bromine in commercial batches of even high quality hydrochloric acid, coupled with oxygen in the hydrolysis medium, will lead readily to halogenation of a phenolic ring. In the case of tyrosine we can then expect to obtain mono and dichlorotyrosines presumably substituted at 3 and 5 with the corresponding bromotyrosines and monochloromonobromotyrosine [41]. These compounds can, however, occur naturally in proteins and obviously care must be taken not to assume an artefactual origin in all circumstances.

Tyrosine may be degraded by other routes in aqueous media at elevated temperatures. In alkali, not unexpectedly, there is severe degradation with a major product being *p*-hydroxybenzaldehyde [42]. Natural halogenated amino acid residues in proteins suffer similar degradation [43] in alkali with the formation of halogenophenolic aldehyde. Such degradations are accompanied by production of stoichiometric quantities of oxalic acid and presumably this suggests a mechanism of oxidation involving firstly formation of 4-hydroxyphenylserine (*12.10*) (or its halogeno derivatives) and further oxidation on the α or β carbon atom with subsequent fission to the aldehyde, oxalic acid and ammonia. Other complex reactions of tyrosine and derived phenolic aldehyde may take place in alkali in view of the well-known ability of phenols to form non-specific polymerization products under such conditions. The presence of carbohydrates derived from glycoproteins may also complicate this picture, the aldehydic function reacting with the phenolic ring again producing polymeric materials. An interesting consequence of applying alkaline conditions to proteins

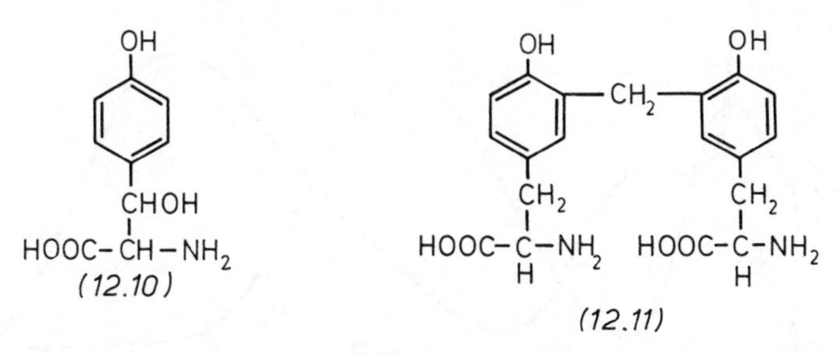

containing diiodotyrosine is the formation from the freed amino acid of small amounts of thyroxine [43]. It is not clear whether other halogeno derivatives or tyrosine itself participate in the formation of aryl ethers.

Under acid conditions, apart from the formation of halogenotyrosines, other remarkable artefacts may result. The report [44] of the unusual amino acid bis-methylene dityrosine (*12.11*) in the rubber-like protein abductin owed its origin to the consequences of heating tyrosine with methionine in the presence of ferric ions in 6 M hydrochloric acid [45]. This molluscan protein contains a large number of tyrosine and methionine (or methionine sulphoxide) residues [46, 47] and binds iron rather tenaciously. The bridging methylene group derives from methionine, although the exact mechanism of formation is not absolutely clear.

In weak acid media such as 0.3 M acetic acid, or even in water at pH 5–6 at 105°C, remarkable changes to the tyrosine molecule occur over long periods. Surprisingly again a major product seems to be *p*-hydroxybenzaldehyde but now a diverse array of other components are formed and can be separated (Figs 12.6 to 12.11) [48]. A number of these would seem from their spectra to be phenolic aldehydes or ketones but these have not yet been identified. The significance of

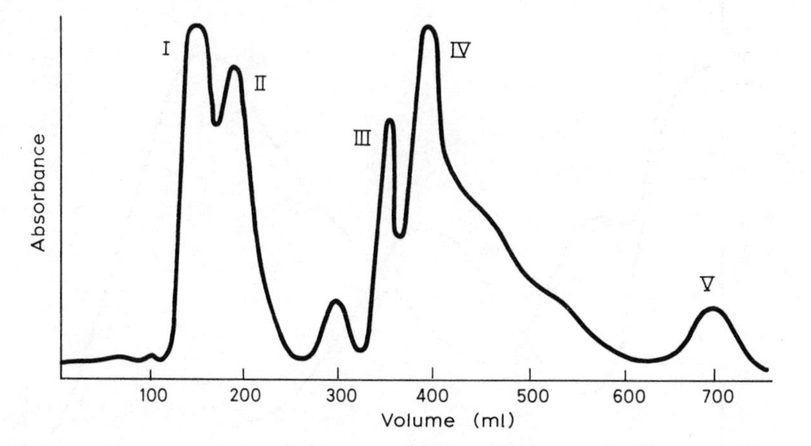

Figure 12.6 Separation on Sephadex G15 of the concentrated product from a solution of 0.03 g tyrosine in 200 ml 0.3 M acetic acid, heated at 105°C for 22 days under nitrogen in a sealed glass vessel. Column 1.6 × 86 cm, and eluted with 0.02 M acetic acid. Roman numerals refer to peaks whose absorption spectra follow in Figs 12.7–12.11.

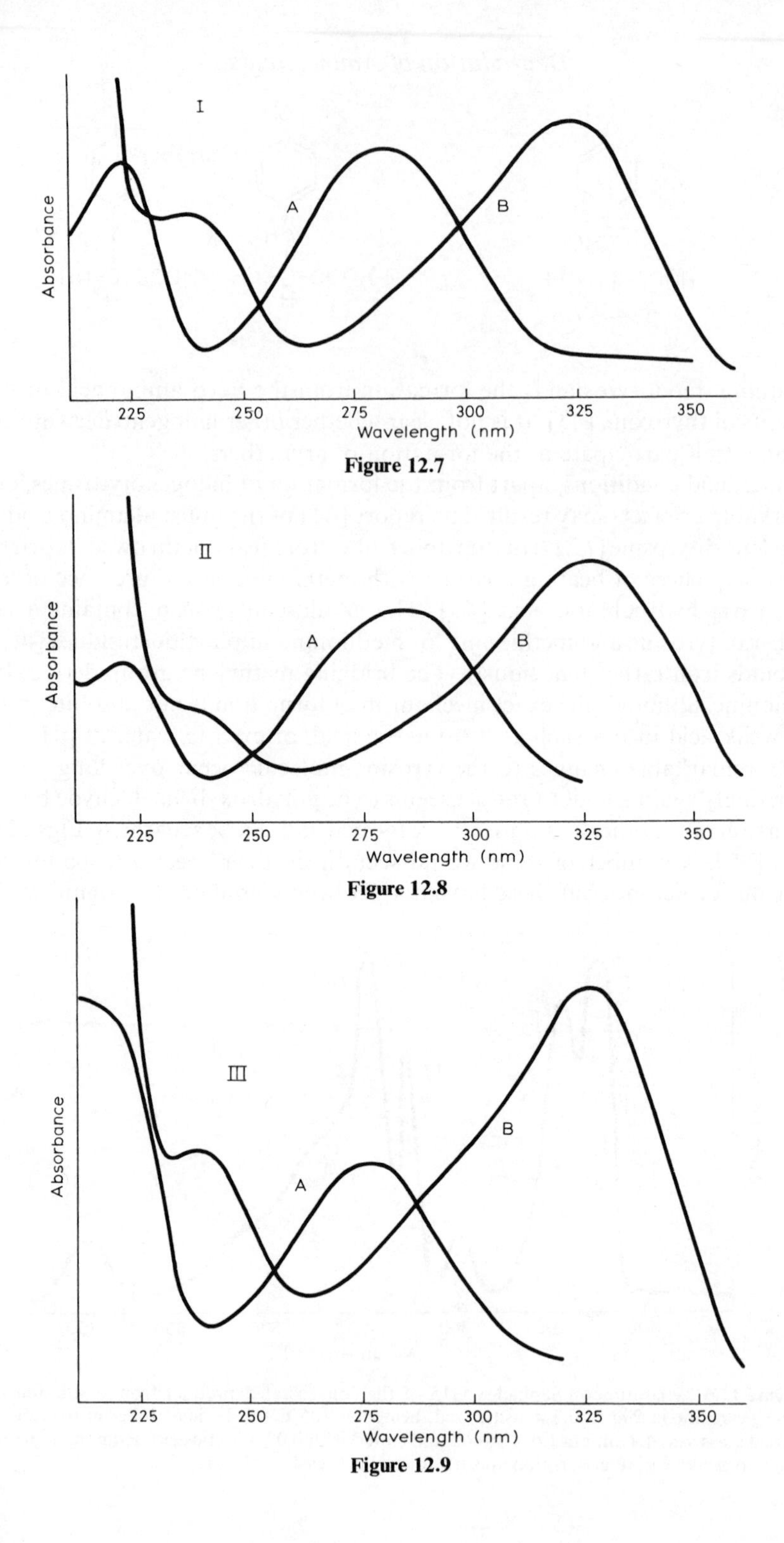

Figure 12.7

Figure 12.8

Figure 12.9

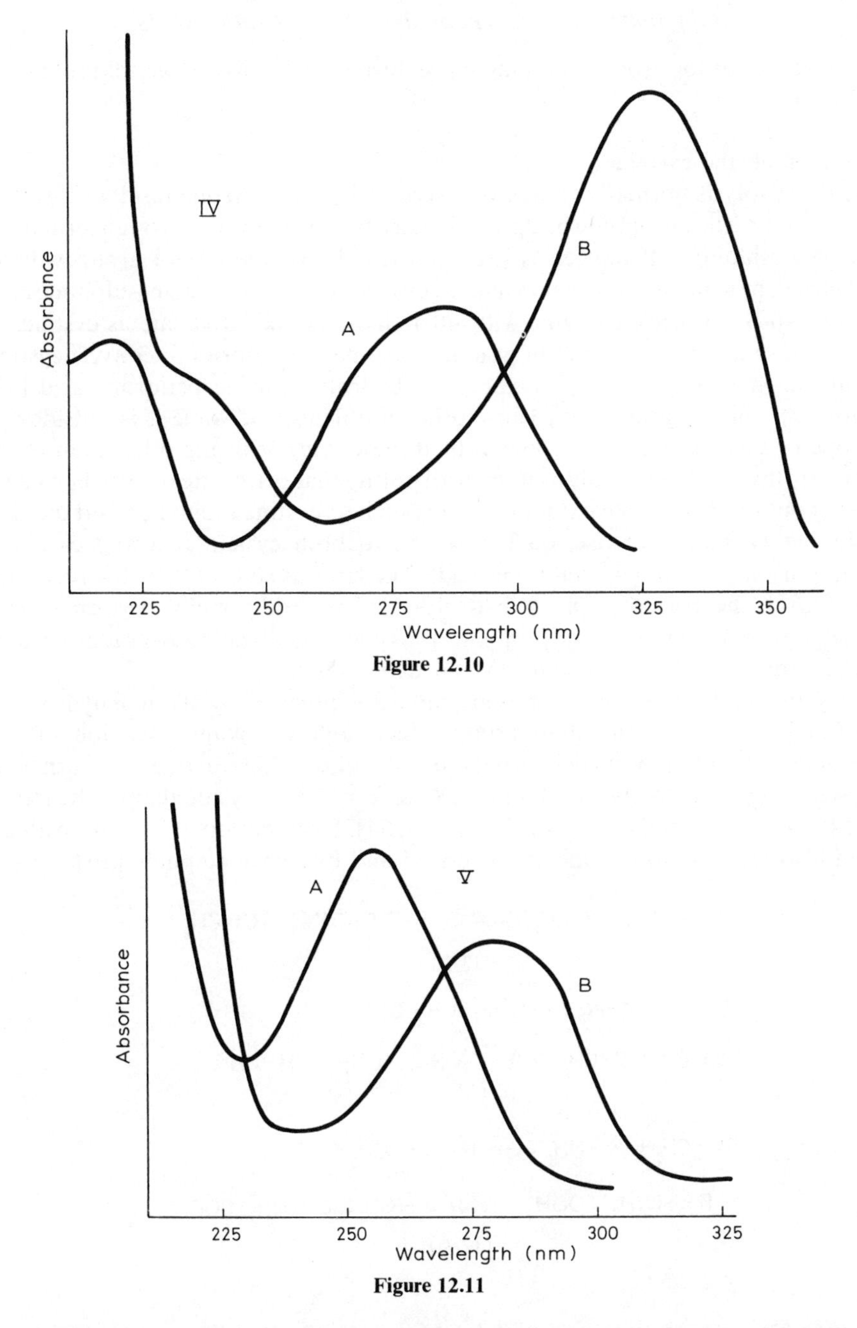

Figure 12.10

Figure 12.11

Figures 12.7–12.11 Ultraviolet absorption spectra of peaks I, II, III, IV and V from Fig. 12.6. A in 0.02 M acetic acid and B at pH 12.0 adjusted with sodium hydroxide. Peak IV is tentatively identified as 4-hydroxybenzaldehyde and peak V as 4-hydroxybenzoic acid. Peak III has a spectrum with features of 4-hydroxyacetophenone. Preliminary studies of peaks I and II using NMR suggest a dimer structure based upon phenolic aldehydes or ketones.

such reactions for processes taking place during partial hydrolysis of proteins is obvious.

(c) Cystine and cysteine

Acid hydrolysis normally causes very serious losses of cystine residues through severance of the disulphide bridges [1] or, as discussed above, through formation of tryptathionine. If oxygen is present then the released thiol group will be oxidized, to a greater or lesser extent, to cysteic acid (some cysteinesulphinic acid may also be formed). Cysteine will suffer the same oxidative fate as cystine.

Because of this it is common practice in some laboratories to convert cystine and cysteine to cysteic acid, before acid hydrolysis, using performic acid [1]. However this technique may cause other problems and while it is outside the scope of this chapter to consider protein modifying reagents other than those directly involved in hydrolysis it is worth noting that performic oxidized proteins may contain incomplete oxidation states such as cystine monoxide and dioxide [49] which may give rise, on hydrolysis, to both cysteic acid and cysteine-sulphinic acid (alanine-3-sulphinic acid). The latter is stable to 6 M hydrochloric acid over the period of most hydrolyses [50]. Additionally performic acid oxidation contributes to tryptophan, tyrosine and histidine degradation and could, for example, give rise to di- and trityrosines.

Cystine and cysteine recoveries are radically impaired by alkaline hydrolysis. Beta-elimination of the thiol group which follows alkaline scission of the disulphide bond can yield dehydroalanine (Reaction 12.2) reacting in its turn with cysteine or lysine to give lanthionine (Reaction 12.3) or lysinoalanine (Reaction 12.4) via a nucleophilic addition [51, 52, 53, 54]. Degradation of cysteine in alkali can also give rise to alanine, pyruvic acid and hydrogen sulphide [16].

$$RCH-CH_2S-S-CH_2R \rightleftharpoons {}^{-}S-S-CH_2-CHR + RC{=}CH_2 + H^+ \qquad (12.2)$$

$${}^{-}S-S-CH_2-CHR \longrightarrow S^{\circ} + {}^{-}SCH_2-CHR$$

$$RC{=}CH_2 + {}^{-}S-CH_2-CHR \xrightarrow{(+H^+)} RCH-CH_2-S-CH_2-CHR \qquad (12.3)$$

$$RC{=}CH_2 + R'-NH_2 \longrightarrow R'-NH-CH_2-CHR$$

$$R = -(NH_2)COOH, \qquad R' = -(CH_2)_4CH(NH_2)COOH \qquad (12.4)$$

(d) Methionine and methionine sulphoxide

A number of proteins apparently contain methionine sulphoxide (*12.12*) in the native molecule as well as methionine itself [47, 55, 56, 57]. Methionine sulphoxide however is reported to be reduced to methionine during anaerobic acid hydrolysis [58], although under conditions where oxygen has not been carefully excluded methionine may be oxidized both to sulphoxide and then the sulphone (*12.13*). Methionine sulphoxide is however stable to alkaline hydrolysis and hence may be estimated in this manner [47].

$$CH_3\overset{O}{\overset{\|}{S}}CH_2CH_2CH(NH_2)COOH$$

(12.12)

$$CH_3-\underset{O}{\underset{\|}{\overset{O}{\overset{\|}{S}}}}-CH_2CH_2CH(NH_2)COOH$$

(12.13)

The danger of losing methionine or the sulphoxide through reaction with tryptophan or tyrosine has been discussed above.

(e) Serine and threonine

We have already considered apparent losses of serine and threonine through ester formation. Both of these amino acids may be lost through other causes. There seems to be considerable argument as to the kinetics of degradation of serine or threonine during hydrolysis. Zero, first and second order kinetics have all been applied to their degradation in different proteins [1,40]. For many cases a linear decrease of the amino acids with time seems to apply and recoveries of about 90 and 95% after 24 hours in 6 N acid seem common but cannot be relied upon. For example, extensive decomposition of both amino acids arises during hydrolysis of *Bacillus subtilis* α-amylase, perhaps due to the presence of calcium [59].

Although serine and threonine losses are well accepted consequences of acid hydrolysis there seems not to be a clear identification of degradation products other than a stoichiometric release of ammonia. One may surmise the formation of carboxylic, and perhaps keto, acids.

Under alkaline conditions (but see below) serine can form dehydroalanine (Reaction 12.2) and in a similar manner threonine gives 3-methyldehydroalanine (*12.14*) which either do, or have the potential to, react as described above for dehydroalanine originating from cystine, with cysteine, serine, lysine, threonine, tyrosine, histidine, ornithine and arginine at thiol, hydroxyl amino or imino groups [51, 52, 53, 54]. Other reported products of alkaline hydrolysis arising from serine and threonine degradation are glycine, alanine (from both) and α-aminobutyric acid (from threonine).

$$CH_3CH{=}C(NH_2)COOH$$

(12.14)

(f) Other amino acids

Aspartic acid, glutamic acid, lysine, arginine and proline have all been variously reported to suffer loss during acid hydrolysis [16]. Involvement of glutamic acid in ester formation has already been mentioned and while similar reactivity of aspartic acid has not been recorded there seems to be no real reason why this should not occur.

The ε-amino group of lysine reacts rather readily with a number of functional groupings; reaction with dehydroalanine has already been mentioned. Formation of Schiff bases with aldehydic functions, such as might arise from other amino acid degradations or sugar decomposition products is possible and is discussed below. The deaminative oxidation of the lysyl side chain to give a lysine aldehyde may be possible under certain conditions and this in its turn could form Schiff bases with other amino groups. These suggestions are largely speculative in relation to acid hydrolysis although established *in vivo* for such proteins as collagen [60].

Bradykinin has a rather simple nonapeptide structure involving three proline residues, only two of which were detected in initial hydrolyses. One proline seems to be destroyed although the reason for this is not clear [1]. Alkaline hydrolysis in addition to its effects discussed above removes the ureido grouping from arginine to form ornithine as well as giving rise to citruline [16] while phenylalanine is extensively degraded to benzaldehyde [61].

12.5 DEGRADATIONS AS A CONSEQUENCE OF THE PRESENCE OF ORGANIC RESIDUES IN ADDITION TO AMINO ACIDS

The most serious source of amino acid degradation during hydrolysis, other than the reactions described above, lies in the presence in many proteins of conjugated oligo or polysaccharides.

To discuss the details of amino acid–sugar interaction is beyond the scope of this chapter, but the formation of secondary reaction products in acid or alkaline media from amino acids and sugars is a matter of some consequence.

Humin is a convenient collective noun used to describe the soluble and insoluble coloured condensation products which commonly form during the acid hydrolysis of glycoproteins. A major contributory factor in the formation of these products is the presence of tryptophan and amino sugars or sialic acid, but other amino acids also react with neutral as well as amino sugars to give yellow or brown products [62].

Under the conditions of strong acid hydrolysis neutral aldohexoses released from linkage to one another and to protein in glycoprotein will degrade progressively, to hydroxymethylfurfural and thence to laevulic acid and other products. Pentose sugars will give furfural itself. Since amino groups react with aldehydo groupings it is likely that formation of complexes between sugars or sugar decomposition products will take place, although the exact nature of reactions taking place in strong acid is probably complex and by no means clear. Reactions may take place during the removal of acid or afterwards. Reactions between amino acid amino groups and aldehydes take place most easily in neutral or weakly alkaline solution [63] (Reaction 12.5) forming Schiff bases (*12.15*) which are themselves unstable to acid but which may in the case of the sugar derivatives then undergo Amadori rearrangement to a more stable product (*12.16*) [64]. An alternative view may be that a glycosylamine (*12.17*) forms as an

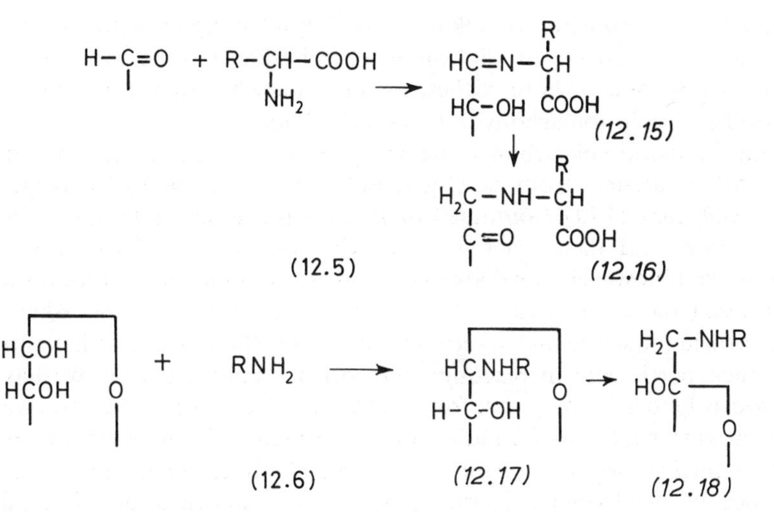

intermediate again undergoing Amadori rearrangement to a stable derivative (*12.18*) [65] (Reaction 12.6).

While the browning or Maillard type of reactions described above seem to be base-catalysed, e.g. reaction of methionine or tryptophan with glucose has maximal rate at pH 11 [66], reactions may also take place at lower pH. Reactions between rhamnose or fructose and alanine or γ-amino butyric acid at pH 3.5 give rise to pyrroles and furans [67]. The Maillard reaction between glucose and lysine in water at 105° yields ε-(2-formyl-5-hydroxymethylpyrrol-1-yl)-L-nor-leucine (*12.19*) [68].

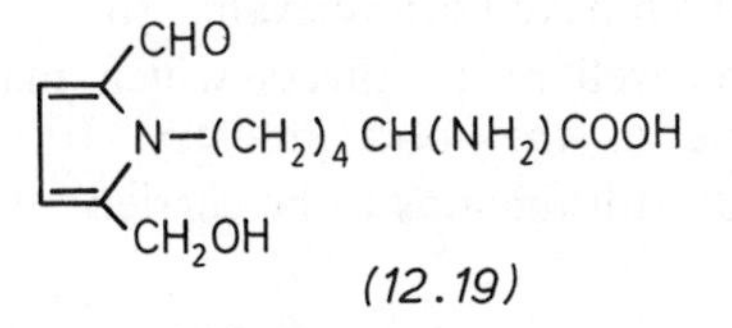

The production of brown products in protein– and amino acid–sugar mixtures, such as arise from hydrolysis, has been attributed to reactions taking place between lysine residues and sugars [69]. Initial amine–aldehyde (Maillard) products Amadori rearranged to *N*-substituted 1-amino-1-deoxy-2-ketoses could further degrade in the 1:2 enol form by 2:3 enolization to give materials absorbing in the 325 to 350 nm range.

Cysteine has also been suggested to react with sugars forming a secondary thiazoline ring although this is partially speculative [64]. Formaldehyde and pyruvic acid which may arise from either carbohydrate or amino acid degradation during hydrolysis can react with cysteine giving rise to thiazolidine-4-carboxylic acid and 2-methylthiazolidine-2,4-dicarboxylic acid [49]. Such compounds however undergo ring opening quite readily at pH 7.0 to release cysteine.

Another consequence of alkaline hydrolysis if applied to glycoproteins containing *O*-glycosidically linked oligosaccharide may be the loss of either serine or threonine or both of these amino acids when they are involved in the glycosidic protein to carbohydrate linkage region.

Beta elimination of carbohydrate will give rise to a dehydroalanine residue in the protein arising from seryl-glycoside or 3-methyldehydroalanine from threonylglycoside [70]. Continued alkaline hydrolysis may release these residues in this form and their reactivities to other amino acids have already been mentioned. If however a mild alkali-treated glycoprotein of this type is then acid hydrolysed the α,β-unsaturated acids will be converted to the α-keto acids, pyruvic acid for serine and α-ketobutyric acid for threonine. The keto grouping may then participate in reactions with other amino acids. As we have seen previously keto acids may arise from amino acid degradation in several ways and may provide routes to formation of new products from other amino acids. Thus a further degradation of tryptophan identified as taking place in the presence of α-keto acids is the formation by condensation of tetrahydroharman-1,3-dicarboxylic acid and its homologues (*12.20*) [71].

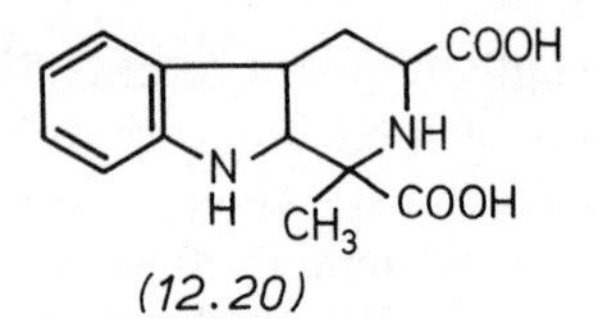

(12.20)

This brief survey emphasizes the complexity of secondary reactions which may take place in acid or alkaline media during protein hydrolysis. Clearly there are many problems here which have both relevance to the question of quantitative amino acid analysis as well as to effects which may arise during protein processing for example in the food industry. It is apparent that much fundamental basic study still remains to be carried out.

REFERENCES

1. Blackburn, S. (1968) *Amino Acid Determination, Methods and Techniques*, Marcel Dekker Inc., New York.
2. Ingram, V.M. (1963) *Methods in Enzymology*, **6**, 225 and 831.
3. Penke, B., Ferenczi, R. and Kovacs, K. (1974) *Anal. Biochem.*, 60, 45.
4. Liu, T.Y. and Chang, Y.H. (1971) *J. Biol. Chem.*, **246**, 2842.
5. Simpson, R.J., Neuberger, M.R. and Liu, T.Y. (1976) *J. Biol. Chem.*, **251**, 1936.
6. Partridge, S.M. and Davis, H.F. (1950) *Nature*, **165**, 62.
7. Hugli, T.E. and Moore, S.J. (1972) *J. Biol. Chem.*, **247**, 2828.
8. Pon, N.G., Schnackerz, K.D., Blackburn, M.W., Chatterjee, G.C. and Noltmann, E.A. (1970) *Biochem.*, **9**, 1506.
9. Piez, K.A. (1954) *J. Biol. Chem.*, **207**, 77.
10. Lockhart, I.M., Abraham, E.P. and Newton, G.G.F. (1955) *Biochem. J.*, **61**, 534.
11. Hamilton, P.B. and Anderson, R.A. (1955) *J. Biol. Chem.*, **213**, 249.
12. Dziewiatkowski, D.D., Riolo, R.L. and Hascall, V. (1972) *Anal. Biochem.*, **50**, 442.
13. Murray, K. and Milstein, C. (1967) *Biochem. J.*, **105**, 491.

14. Murray, K. (1968) *Biochem. J.*, **110**, 155.
15. Ikawa, M. and Snell, E.E. (1961) *J. Biol. Chem.*, **236**, 1955.
16. Hill, R.L. (1965) *Adv. Prot. Chem.*, **20**, 37.
17. Spector, A.R. and Glimcher, M.J. (1973) *Biochim. Biophys. Acta*, **303**, 360.
18. Cohen-Solal, L., Lian, J.B., Kossiva, D. and Glimcher, M.J. (1979) *Biochem. J.*, **177**, 81.
19. Cohen-Solal, L., Cohen-Solal, M. and Glimcher, M.J. (1979) *Proc. Natl. Acad. Sci. USA*, **76**, 4327.
20. Bettelheim, F.R. (1954) *J. Amer. Chem. Soc.*, **76**, 2838.
21. Jevons, F.R. (1963) *Biochem. J.*, **89**, 621.
22. Narita, K. (1958) *Biochim. Biophys. Acta*, **28**, 184.
23. Ochoa, S., and Mazumder, R. (1974) in *The Enzymes*, (ed. P.D. Boyer) (Vol. 10), 3rd edn., Academic Press, New York, p. 1.
24. Blomback, B. (1967) *Methods in Enzymology*, **11**, 398.
25. Hauschka, P.V., Liau, J.B. and Gallop, P.M. (1975) *Proc. Natl. Acad. Sci. USA*, **72**, 3925.
26. Noltmann, E.A., Mahowald, T.A. and Kuby, S.A. (1962) *J. Biol. Chem.*, **237**, 1146.
27. Matsabura, H. and Sasaki, R. (1969) *Biochem. Biophys. Res. Commun.*, **35**, 175.
28. Stewart, M. and Nicholls, C.H. (1972) *Aust. J. Chem.*, **25**, 39.
29. Stewart, M. and Nicholls, C.H. (1973) *Aust. J. Chem.*, **26**, 1205.
30. Stewart, M. and Nicholls, C.H. (1972) *Aust. J. Chem.*, **25**, 1595.
31. Finley, J.W. and Friedman, M. (1973) *J. Agric. Food. Chem.*, **21**, 33.
32. Hunt, S., unpublished.
33. Stohrer, G. (1976) *J. Heterocyclic Chem.*, 157.
34. Cornforth, J.W., Dalgliesh, C.E. and Neuberger, A. (1951) *Biochem. J.*, **48**, 598.
35. Ciegler, A. (1973) in *Handbook of Microbiology*, (eds A.I. Laskin and H.A. Lechvalier) (Vol. 3), CRC Press, Cleveland, p. 540.
36. Nakai, J. and Ohta, J. (1976) *Biochim. Biophys. Acta*, **420**, 258.
37. Ohta, T. and Nakai, T. (1978) *Biochim. Biophys. Act*, **533**, 440.
38. Savige, W.E. and Fontana, A. (1977) *Methods in Enzymol.*, **47E**, 442.
39. Savige, W.E. and Fontana, A. (1980) *Int. J. Peptide Protein Res.*, **15**, 285.
40. Tristram, G.E. (1966) *Techniques in Amino Acid Analysis*, Technicon Instruments Co. Ltd, Chertsey, UK, p. 61.
41. Sanger, F. and Thompson, E.O.P. (1963) *Biochim. Biophys. Acta*, **71**, 468.
42. Hunt, S., unpublished.
43. Pitt-Rivers, R. (1948) *Biochem. J.*, **43**, 223.
44. Andersen, S.O. (1967) *Nature*, **216**, 1029.
45. Andersen, S.O., personal communication.
46. Kelly, R.E. and Rice, R.B. (1967) *Science*, **155**, 208.
47. Kikuchi, Y. and Tamiya, N. (1981) *J. Biochem.*, **89**, 1975.
48. Hunt, S., unpublished.
49. MaClaren, J.A., Savige, W.E. and Sweetman, B.J. (1965) *Aust. J. Chem.*, **18**, 1655.
50. Eager, J.E. (1966) *Biochem. J.*, **100**, 37C.
51. Walsh, R.G., Nashef, A.S. and Feeney, R.E. (1979) *Int. J. Pept. Prot. Res.*, **14**, 290.
52. Friedman, M. (1977) in *Protein Crosslinking, Nutritional and Medical Consequences*, (ed. M. Friedman) Plenum Press, New York, p. 1.
53. Finley, J.W. and Friedman, M. (1977) in *Protein Crosslinking, Nutritional and Medical Consequences*, (ed. M. Friedman), Plenum Press, New York, p. 123.
54. Gorss, E. (1977) in *Protein Crosslinking, Nutritional and Medical Consequences*, (ed. M. Friedman), Plenum Press, New York, p. 131.
55. Hudson, B.G. and Spiro, R.G. (1972) *J. Biol. Chem.*, **247**, 4299.
56. Brewer, B.H., Keutmann, H.T., Potts, J.T., Reisfeld, R.A., Schlenter, R. and Munson, P.L. (1968) *J. Biol. Chem.*, **243**, 5739.
57. Sdelstein, R.S. and Kuehl, W.M. (1970) *Biochemistry*, **9**, 1355.
58. Morihara, K. (1964) *Bull. Chem. Soc. Jpn.*, **37**, 1781.
59. Junge, J.M., Stein, E.A., Neurath, H. and Fischer, E.H. (1959) *J. Biol. Chem.*, **234**, 556.
60. Bailey, A.J. (1978) *J. Clin. Path., Suppl.*, **31, 12**, 49.
61. Hunt, S., unpublished.
62. James, L.B. (1972) *J. Chromatogr.*, **68**, 123.
63. Takahashi, K. (1977) *J. Biochem.*, **81**, 395.
64. Pigman, W. (1957) *The Carbohydrates*, Academic Press, New York.

65. Jevons, F.R. (1964) in *Symposium on Foods* (eds H.W. Schultz and A.F. Anglemier) Avi Publ. Co., Westpoint, Connecticut, p. 153.
66. Dworschak, E. and Orsi, F. (1977) *Acta Aliment Acad. Sci. Hung.*, **25**, 641.
67. Shaw, P.E. and Berry, R.E. (1977) *J. Agric. Food Chem.*, **25**, 641.
68. Nakayama, T., Hayase, F. and Kato, H. (1980) *Agric. Biol. Chem.*, **44**, 1201, (*Chem. Abs.* (1981) **93**, 95 648).
69. Berrens, L. and Bleumink, E. (1966) *Biochim. Biophys. Acta*, **115**, 507.
70. Zimm, A.B., Plantner, J.J. and Carlson, D.M. (1977) in *The Glycoconjugates*, (eds M.I. and W. Pigman) (Vol. 1), Academic Press, New York, p. 69.
71. Chu, N.T. and Clydesdale, F.M. (1976) *J. Food Sci.*, **41**, 891.
72. Hunter, T. (1982) *Trends in Biochem. Sci.*, **7**, 246.

CHAPTER THIRTEEN

Racemization of Amino Acids

J.L. BADA

13.1 INTRODUCTION

A century ago it was first observed that amino acids underwent racemization when heated in strongly acidic and basic solutions. It soon became well established that the optically active amino acids isolated from biological materials could be converted into a racemic mixture by a variety of rather vigorous treatments. In the early part of the 20th century the first observations of amino acid racemization in peptides and proteins in alkaline solutions at elevated temperatures were reported. The interpretation of these measurements was complicated by peptide-bond hydrolysis and the realization that racemization rates probably depended on whether amino acids were at terminal or internal positions in the peptide. The earlier work on amino acid racemization in various systems was extensively reviewed in 1948 by Neuberger [1].

While racemization of amino acids at extreme pH values and elevated temperatures has been known for 100 years it has only been within the last 10–12 years that it was shown that racemization also takes place at neutral pH at rates which are comparable with those in dilute acid and base [2, 3]. Also racemization was detected in fossils and the metabolically stable proteins in living mammals, and it was suggested that racemization might form the basis of a dating method which could be used to determine the age of amino acids in these systems [4–8].

Racemization is a reversible first order reaction which can be written as

$$\text{L-Amino acid} \underset{k'_i}{\overset{k_i}{\rightleftharpoons}} \text{D-Amino acid} \qquad (13.1)$$

where k_i is the first order rate constant for the interconversion of amino acid enantiomers. The kinetic equation [4, 5] for this reaction is

$$\ln\left\{\frac{1+(\mathrm{D/L})}{1-K'(\mathrm{D/L})}\right\} - \ln\left\{\frac{1+(\mathrm{D/L})}{1-K'(\mathrm{D/L})}\right\}_{t=0} = (1+K')k_i t \qquad (13.2)$$

where (D/L) is the amino acid enantiomeric ratio at a particular time t and $K' = k_i/k'_i = 1/K_{eq}$. For amino acids with one centre of asymmetry, $K_{eq} = 1.0$ since $k_i = k'_i$, while for diastereoisomeric amino acids (i.e. isoleucine, threonine and 4-hydroxyproline) K_{eq} is different from unity. The K_{eq} value at pH 7.6 for 4-hydroxyproline [9] is 0.8, while that for isoleucine [10] is 1.3. The $t = 0$ term in

Equation (13.2) is needed to account for the fact that the initial D/L ratio in the system under investigation may not be exactly 0 and because some slight racemization may take place during sample processing, for example during protein hydrolysis [4, 5]. When the extent of racemization is small, i.e. D/L < ~0.15, racemization can be considered an irreversible first order reaction and Equation (13.2) can thus be simplified to

$$\ln(1 + \mathrm{D/L}) - \ln(1 + \mathrm{D/L})_{t=0} = k_i t \qquad (13.3)$$

Investigations of the racemization of free amino acids in buffered aqueous solutions indicate that the reaction obeys the expected reversible first-order kinetics [4, 11]. However, for amino acids in some peptides and proteins and in certain types of fossils, notably carbonates, the racemization kinetics are more complex [4, 5, 12].

The rate of racemization depends upon the particular amino acid (the racemization rate equals $2k_i$). Rates are also greatly influenced [4, 8] by temperature, pH, ionic strength, metal ion chelation, etc. In Table 13.1 are listed the racemization half-lives (i.e. the time required to reach a D/L ratio of 0.33) for several amino acids at pH 7.6 and 25° and 100°C. As can be seen the half-lives range from a few days for serine at 100°C to ~40 000 years for isoleucine at 25°C. The racemization rates at neutral pH are about 2–3 times and ~10 times slower [14] than in 6 M HCl and 2 M NaOH, respectively.

In this chapter the racemization of amino acids in various systems is discussed. The present survey is not meant to be a comprehensive review of the published literature since the 1948 survey by Neuberger. Rather, the significant results in certain areas are summarized with particular emphasis being placed on aqueous solution studies.

Table 13.1 Racemization half-lives[a] of several amino acids at pH 7.6 and 25° and 100°C

Amino acid	*25°C (yrs)*	*100°C (days)*
Serine[b]	~400 (?)	~4 (?)
Aspartic acid	3 500	35
Alanine	12 000	120
Isoleucine	~40 000	~300

[a] Taken from Bada [2].
[b] Estimated from measurements [13] at 122°C and assuming that the ratio of the rates of racemization of serine and aspartic acid is the same at all temperatures.

13.2 AMINO ACIDS IN AQUEOUS SOLUTION

The kinetics of racemization of amino acids in aqueous solution are the simplest to investigate. Conditions such as temperature, pH, ionic strength and buffer

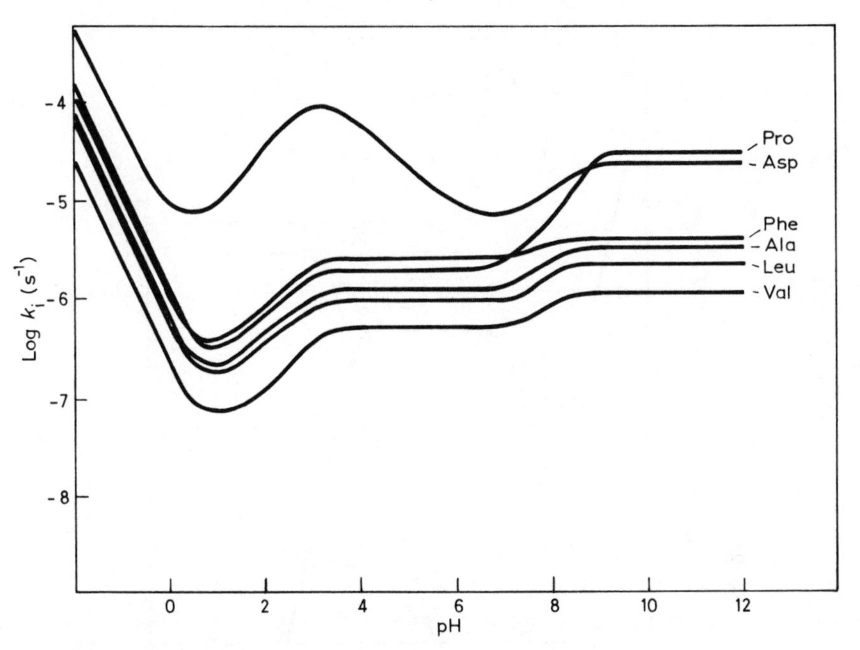

Figure 13.1 Rate of interconversion (plotted as log k_i) against pH for several amino acids at 142°C. Taken from Bada and Shou [15].

concentrations can be easily controlled and from the amino acid pK_as the distribution of the ionic species as a function of pH can be ascertained.

The kinetics of racemization of several amino acids at 142°C as a function of pH are shown in Fig. 13.1. As can be seen log k_i from monocarboxylic amino acids shows three distinct regions: an acid catalysed region for pH < 1, a region called Plateau I between pH ~ 3 and pH ~ 6.5 where k_i is independent of pH and a region between pH 9 and 12 called Plateau II where k_i is again independent of pH. In addition to the pH regions described above for monocarboxylic amino acids, the kinetic curve for aspartic acid has a pH rate maximum at about pH 3. The pH versus rate curves for the racemization of the basic and β-substituted amino acids have not been established, but are probably similar to those shown in Fig. 13.1.

The following equation [3, 15] has been found to describe the rate of interconversion of the enantiomers of monocarboxylic amino acids (A) between pH 0 and 12:

$$k_{obs}[A] = k_i^{H^+}[A^{+0}][H^+] + k_i^{+0}[A^{+0}][OH^-] + k_i^{+-}[A^{+-}][OH^-] \quad (13.4)$$

$$\text{where } A^{+0} = R\text{-}\underset{\substack{| \\ ^{+}NH_3}}{CH}\text{-}COOH \quad \text{and } A^{+-} = R\text{-}\underset{\substack{| \\ ^{+}NH_3}}{CH}\text{-}COO^-$$

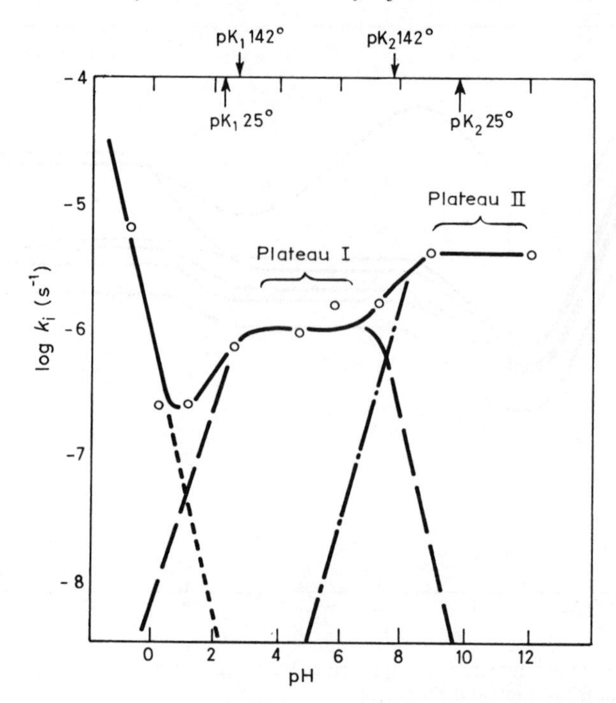

Figure 13.2 A plot of $\log k_i$ against pH for alanine at 142°C showing the various reactions given in Equation 13.4. Short dashed lines indicate k_i for acid catalysed reaction; long dashed lines, k_i of A^{+0}; and short and long dashed line, k_i of A^{+-}. Arrows indicate the values of the p*K*s at 25°C and 142°C. Taken from Bada and Shou [15].

(For dicarboxylic acids the corresponding equation is slightly more complex [3].) The various k_i values are the rate constants for the interconversion of the D and L enantiomers of the indicated ionic species of amino acids. Using the above equation and the pK_a values extrapolated to the appropriate temperature, the k_i values of the various ionic species as a function of pH can be calculated [8, 3, 15]. In Figs 13.2 and 13.3, the racemization kinetics of alanine and aspartic acid, respectively, are broken down into their component ionic reactions. For monocarboxylic amino acids at neutral pH the principal ionic species undergoing racemization is the one in which both the amino acid and carboxyl groups are protonated (i.e. A^{+0}). For aspartic acid (and probably glutamic acid as well) the reaction is more complex because this amino acid has two carboxyl groups; at neutral pH the principal species undergoing racemization is the one in which the β-carboxyl group and the amino group are protonated but the α-carboxyl group is unprotonated (asp^{0+-}).

The racemization results at elevated temperatures have important implications concerning the pH dependence of amino acid racemization at lower temperatures [15]. As can be seen from Fig. 13.2, the racemization reaction is effectively independent of pH in the Plateau I region (i.e. between pH 3 and pH 6)

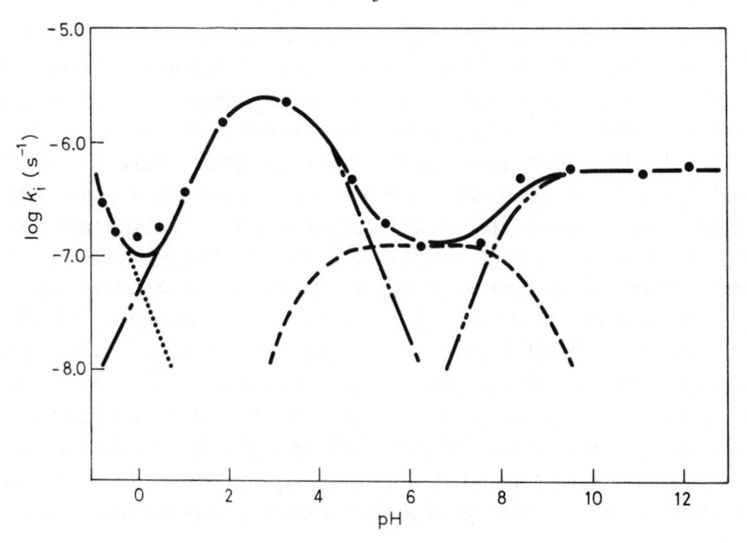

Figure 13.3 A plot of log k_i versus pH for aspartic acid at 100°C showing the reactions of the various ionic species. Dots indicate k_i of acid catalysed reaction; short and long dashes, k_i of asp^{0+0}; short dashes, k_i of asp^{0+-}; one long and two short dashes indicate k_i of asp^{-+-}. Based on data given by Bada [3].

at elevated temperatures. The width of Plateau I for each amino acid depends on the values of pK_1 and pK_2. The width of the Plateau I region will increase towards basic pH with decreasing temperatures. This is because pK_1 decreases only slightly with decreasing temperature whereas pK_2 increases substantially with decreasing temperatures. For example, at 142°C, alanine has a pK_1 of 2.6 and $pK_2 = 7.77$ while at 25°C $pK_1 = 2.35$ and $pK_2 = 9.87$. Thus, although the width of Plateau I for alanine only extends from pH 3 to ~6.5 at 142°C, this will increase from pH 3 to ~8.5–9 at 25°C. Natural systems such as lakes, rivers, the oceans and sedimentary environments generally have temperatures of the order of 0–25°C and pH values in the range of 6–9. In this temperature and pH region, it is apparent that the racemization rate of free amino acids would be effectively independent of pH.

The mechanism of amino acid racemization, as first proposed by Neuberger [1], involves the removal of the α-hydrogen of an amino acid resulting in the formation of a planar carbanion. The carbanion mechanism is supported by deuterium exchange studies [4]. At 135°C and pD ≅ 7.5, the rate of exchange (k_{exch}) of the α-hydrogen of L-isoleucine with deuterium was found to be equal to the rate of epimerization of L-isoleucine, which is the expected result if a carbanion intermediate is involved in the reaction [16].

The stability of the carbanion intermediate of a particular amino acid ionic species depends upon the electron withdrawing and resonance stabilizing capacities of the amino and carboxyl groups [1,3,15]. The more effective a substituent is in stabilizing the intermediate carbanion, the more rapid the rate of racemization. The calculated k_i for the various ionic forms of the different amino acids [3, 15] indicate that $k^{+0}_i > k^{+-}_i$. This result is consistent with the carbanion mechanism since the protonated carboxyl group greatly enhances carbanion intermediate stability by both electron withdrawal and resonance effects.

These results have demonstrated that the ionic species plays a critical role in determining the rate of racemization. Even though a particular ionic species may be present in only trace quantities at a certain pH, the racemization rate of that particular ionic species may be much greater than that of other species. Thus the observed racemization occurs primarily in the more reactive species. For example, racemization of the ionic form A^{0-} (i.e. $R-CH(NH_2)-COO^-$) is not observed until 1–2M NaOH solutions are used [14]. Since the $-NH_3^+$ group has such a large electron withdrawing capacity relative to $-NH_2$, the racemization rate of the protonated amino group species is several orders of magnitude greater than that of the unprotonated species. Only in highly basic solutions where the concentration of the protonated amino group species is very low does the racemization of the unprotonated amino group species become the dominant reaction.

Besides the effect which ionic species has on the stability of the carbanion, the R substituents of the various amino acids are important in determining the relative order of racemization rates of the amino acids themselves [3, 14, 15]. The more strongly electron withdrawing a particular R group the more stable the carbanion and thus the faster the rate of racemization. The electron withdrawing ability of a particular substituent is measured by the quantity σ^*, the polar substituent or Taft constant which is an indicator of how well a substituent can stabilize a negatively charged carbon atom. Plots of log k_i against σ^* for the ionic species A^{+0} and A^{+-} are given in Fig. 13.4. These results demonstrate that the electron withdrawing capacity of the R substituents of the various amino acids is

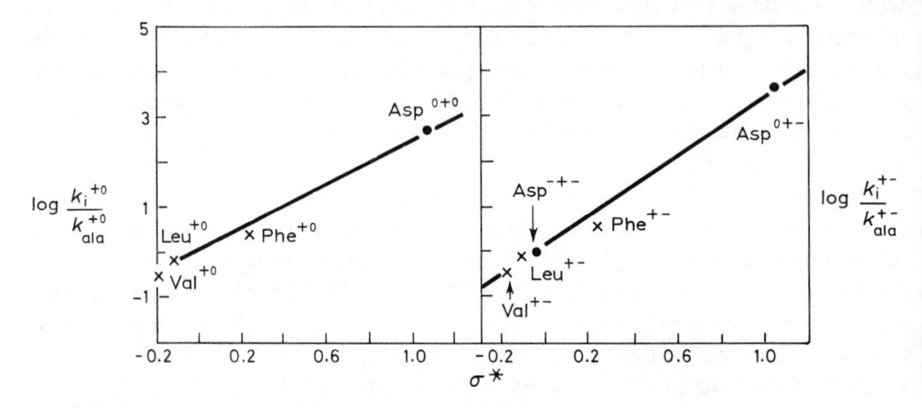

Figure 13.4 A plot of $\log k_i/k_{ala}$ versus σ^* for the ionic forms A^{+0} and A^{+-} at 142°C. The k values relative to alanine are used, since the σ^* for this amino acid is 0.0. The lines shown were obtained by a least-squares fit of the data. The correlation coefficient (R) is 0.996 for the A^{+0} results and 0.995 for A^{+-}. Based on data given in Shou [14]. The σ^* values were taken from Barlin and Perrin [17] and Taft [18].

the principal factor in determining the relative racemization rates, provided the comparison is made for the same ionic species. Amino acids that have R groups which are highly electron withdrawing (e.g. serine) have the fastest racemization rates at neutral pH while those which have substituents that are electron donating (e.g. valine) have the slowest rates.

Based on the strong correlation between the k_i values for the ionic species and the σ^* values of the amino acid R-groups, it seems reasonable to assume that σ^* values can be used as a basis for predicting how changes in the ionic speciation, or substituents, will affect relative racemization rates. The predicted relative racemization rates are given in Table 13.2. These relative racemization rates will be used in interpreting the results obtained in studies of amino acid racemization in various systems.

In aqueous solution amino acid racemization exhibits [3, 10] weak general acid–general base catalysis at both pH 2 and 7.8. Increasing the buffer concentrations by more than a factor of 10 results in only a 20% increase in the rate of racemization. Also increasing the ionic strength of the solution causes a slight increase in the racemization rate [10].

By far the most dramatic effect on racemization rates results from the presence of metal ions in dilute alkaline solutions [20, 21]. Chelation of amino acids by positively charged metal ions greatly stabilizes the carbanion intermediate resulting in an enhanced rate of racemization. In 0.1 M NaOH, amino acids in some metal ion complexes undergo detectable racemization [20] in a few hours at 34°C. Most of the studies on amino acid racemization in metal ion complexes

Table 13.2 Relative racemization rates for various amino acid substituents predicted from σ^* values [17, 18] (given in parentheses). Adapted in part from Kemp [19]. The symbol ~ is used to denote a peptide bond.

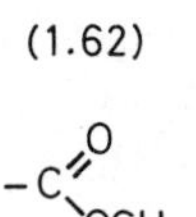

X–C(R)(H)–Y Generalized amino acid structure

X–: $-\overset{+}{N}H_3$ > $-NH-C(=O)H$ ≅ ~NH–C(=O)~ > $-NH_2$

(3.76) (1.62) (?) (0.62)

Y–: $-COOH$ ≅ $-C(=O)OCH_3$ ≅ $-C(=O)NH_2$ ≅ ~C(=O)NH~ >> COO^-

(2.08) (2.00) (1.08) (?) (-1.06)

R–: $-CH_2-COOH$ > $-CH_2-OH$ ≅ H > $-CH_2-C_6H_5$ > CH_2-COO^- ≅ CH_3 > $-CH(CH_3)_2$

(1.05) (0.56) (0.49) (0.22) (-0.06) (0.0) (-0.19)

have been carried out at pH > 9; thus it is not known whether metal ion catalysis is significant at neutral pH where the formation of metal ion–amino acid complexes would be less favourable than in dilute alkaline solutions.

In dilute alkaline solution containing metal ions, the formation of Schiff bases between the amino group of the amino acid and an active carbonyl group in, for example, pyridoxal or pyruvate, results in an enhanced rate of amino acid racemization [21]. When dilute alkaline solutions containing Cu^{2+} were heated for a few hours at 50–60°C no racemization of alanine was observed [22]. With the addition of pyruvate to the system, however, racemization was measurable under the same conditions [23]. It is interesting to note that pyridoxal dependent enzymes operate by Schiff base formation, which greatly activates the α-hydrogen of the active site amino acids [24]. The catalytic effect resulting from Schiff base formation has been exploited by a variety of enzymic systems in living organisms.

13.3 AMINO ACID DERIVATIVES

In the field of peptide synthesis the racemization of various amino acid derivatives can place severe limitations on the practicality of any particular synthetic pathway. Obviously racemization is to be avoided during peptide synthesis since the resultant peptides would have reduced chiral integrity. It is thus important to have an understanding of the factors that might facilitate or inhibit amino acid racemization. This subject has been extensively discussed by Kemp [19].

As can be seen from Table 13.2 the formation of various amino acid derivatives can have both an accelerating and inhibitory effect on racemization. The formation of certain amino acid derivatives effectively locks the amino acid into a form equivalent to that of an ionic species which normally would be present only in minute quantities at a certain pH. For example, amino acid esters can be considered as roughly equivalent to the conjugate acid form (A^{+0}), a species which has a very fast racemization rate but only a minor abundance at neutral pH. Likewise, *N*-substituted amino acids should have reactivities generally similar to A^{0-} at pH 7.

Studies of the racemization of amino acid derivatives have been mainly confined to non-aqueous solutions. There have been only a limited number of studies in aqueous solutions, mainly because hydrolysis and decomposition reactions of the amino acid derivatives complicate the system. However, studies of the racemization of various amino acid derivatives in aqueous solution could be an important area for future research. Certain amino acid derivatives such as *N*-acylamino acid amides may be good model compounds for studying the variables which affect the racemization of peptide bound amino acids [25]. That this is feasible was demonstrated [26] by the observation that the rate of racemization of *N*-acyl-L-serine amide in 0.2 M NaOH at 24°C was found to be comparable to the racemization rate of seryl residues of polyserine. There have been no investigations of the racemization rates of *N*-acylamino acid amides in the neutral pH region, however.

13.4 PEPTIDES, PROTEINS AND DIKETOPIPERAZINES

The formation of the peptide bond can have a variety of effects on the rate of amino acid racemization. In peptides, the carboxyl group has a form roughly analogous to $-$COOH (see Table 13.2). For free amino acid [3, 15] racemization $k_i^{COOH}/k_i^{COO^-} \cong 10^4$, so peptide-bound amino acids might be expected to racemize much faster than the corresponding free amino acids at neutral pH. On the other hand, the amino group of peptide bound amino acids can no longer be protonated and this should greatly retard racemization. It is difficult, however, to predict the relative magnitude of these two effects.

Amino acids at terminal positions in peptides would be expected to have racemization rates which differ from those of internally bound amino acids. For *N*-terminal amino acids the amino group can be protonated while the carboxyl group is equivalent to that of the interior peptide-bound residues. Since the NH_3^+ group greatly enhances racemization one would predict that *N*-terminal racemization rates should be greater than internal racemization rates at neutral pH. In contrast, amino acids at *C*-terminal positions might be expected to racemize more slowly than internal amino acids at neutral pH since negatively charged carboxyl groups should greatly destabilize the carbanion intermediate.

Studies of amino acid racemization in di-and tripeptides provide a basis for determining the relative rate differences between *N*-terminal, *C*-terminal and internally bound amino acids. Investigations by Kriausakul and Mitterer [27] reported that at pH $\sim$ 6 and 152°C *N*-terminal isoleucine epimerizes faster than *C*-terminal isoleucine. Smith and de Sol [13], on the other hand, reported that in studies of racemization of 37 different dipeptides at pH 7.6 and 122°C, in general *C*-terminal amino acids racemized faster than *N*-terminal amino acids. The interpretation of these results is complicated by the fact that rate constants were not determined for the various dipeptide ionic species. In addition subsequent investigations [28] demonstrated that in the previous elevated temperature dipeptide experiments, there was extensive conversion of the original dipeptide into diketopiperazines. Hydrolysis of the diketopiperazines during the course of the experiment yielded both the original dipeptide and an inverted dipeptide product. Since the formation of diketopiperazines and inverted dipeptides was not considered in earlier dipeptide studies, the interpretation of the results is difficult.

In elevated temperature studies of dipeptides in which the extent of diketopiperazine formation and thus dipeptide inversion is minimal [28] it is possible to make estimates of the relative racemization rates at the *C*- and *N*-terminal positions. In investigations of the epimerization of glycylisoleucine and isoleucylglycine it was found [28] that at 131°C and pH = 7.6–9.5 the ratio $k_i^{N-\text{term}}/k_i^{C-\text{term}}$ had a value of 15–30; the ratio showed a decreasing trend with increasing pH. It should be noted, however, that at this temperature and pH range only a small fraction of the dipeptide exists as the species in which the amino group is protonated. At lower temperatures, a substantially higher fraction of the dipeptide would exist as the protonated amino group species at neutral pH. Thus at lower temperatures the ratio of $k_i^{N-\text{term}}/k_i^{C-\text{term}}$ could be substantially greater than those measured in the neutral pH region at 131°C. The principal conclusion that can be drawn from these studies is that at pH ~ 7.5–9.5, *N*-terminal amino acids racemize faster than those at the *C*-terminal position; the rate enhancement at the *N*-terminal position is likely in the range ~ 15–100.

Investigations of amino acid racemization of tripeptides provide information on the relative rates of racemization of amino acids at terminal and internal positions in the peptide [29]. As in the case of the dipeptide studies, these investigations are complicated by the formation of cyclic intermediates and hydrolysis of the tripeptide into a variety of fragments. In the studies of tripeptides containing one leucyl and two glycyl residues, it was found that at 131°C the ratio $k_i^{N-\text{term}}/k_i^{\text{int}}$ was of the order of 5–8 at neutral pH. No detectable racemization was observed in the tripeptide where leucine was at the *C*-terminal position which provides further evidence that *N*-terminal and internal amino acid residues are racemized more rapidly than *C*-terminal residues in the neutral pH region.

With these considerations in mind it is not hard to rationalize why amino acids at either the *N*-terminal or internal positions of peptides and proteins racemize faster than free amino acids at neutral pH. Investigations [28] of the isoleucylglycine at 131°C and pH 7.6 indicate that in the dipeptide isoleucine epimerizes ~ 20 times faster than free isoleucine. In the tripeptide glycylleucylglycine at 131°C and pH 7.6 the internally bound leucine racemizes at a rate which is 2–4 times faster than the rate of free leucine racemization [29]. In studies with a variety of peptides and proteins it has been found that at neutral pH, the racemization rates of a number of amino acids are ~ 2–4 times those of the corresponding free amino acids [5, 27, 30]. Since the activation energies for racemization of protein bound and free amino acids are not well established it is difficult to extrapolate these elevated temperature results to low temperatures. The detection of aspartic acid racemization in a number of metabolically stable proteins in living mammals [31–34] suggests that at least for this amino acid, the rate of racemization in proteins is faster than that estimated [2] for free aspartic acid at 37°C and pH 7.6.

The effect which pH has on the racemization rates of peptide and protein bound amino acids is much more poorly understood than for free amino acids. The rates of racemization in proteins are apparently more or less independent of pH in the range 6–9, but for pH > 9–10 racemization rates increases with increasing pH and appear to be first order in hydroxide ion concentration [12]. In 0.1 M NaOH the rate of racemization of protein bound amino acids is about 10 times that measured at neutral pH [12]. The first order $[OH^-]$ dependence of racemization rates in alkaline solutions is consistent with a mechanism wherein hydroxide ion abstracts the α-hydrogen from the amino acid residues. The observation that the racemization of protein-bound amino acids is independent of pH in the range pH 6–9 is more difficult to explain. One possibility is that for pH values less than about 9 the abstraction of the α-hydrogen by water is more significant than that by hydroxide ion.

As was the case for free amino acids, the electron withdrawing capacities of the various R-substituents appear to be the principal factor which determines the relative racemization rates for amino acids bound in the interior positions of peptides and proteins [4, 5, 12, 15]. In a number of proteins heated at elevated temperature at both neutral pH and in dilute basic solutions the relative order of racemization rates are Asp $>$ Ala $\simeq$ Glu $>$ Leu which is consistent with the σ^* values of the various substituents [12, 15, 35]. There are some exceptions to this pattern which have been attributed to intramolecular interactions of neighbouring amino acid residues. For example, Engel *et al.* [36] showed that in the peptide hormone β_p-MSH aspartic acid was only slowly racemized in 0.1 M NaOH at 97°C; the aspartic acid racemization rate was comparable to that of hydrophobic amino acids. Engel *et al.* suggested that this might be due to neighbouring group effects which somehow retarded the racemization of aspartic acid. However, in the peptide hormone that was investigated by Engel and coworkers aspartic acid was at the *N*-and *C*-terminal positions; the peptide did not contain any internal aspartic acid residues [36]. In 0.1 M NaOH, the amino and carboxyl groups for terminal amino acid residues would both be unprotonated which would have a great inhibitory effect on racemization rates in comparison to internally bound residues (see Table 13.2). Thus the apparent slow racemization of aspartic acid in β_p-MSH in 0.1 M NaOH is simply due to the retarding effects of the $-NH_2$ and $-COO^-$ substituents and is not the result of intramolecular neighbouring group interactions.

The effects which a neighbouring group may have on racemization rates of peptide bound amino acids are not established at the present time. The similarity of the relative rates of racemization in several different proteins suggest that neighbouring group effects are not large [12]. Certain residues in peptides and proteins may be very prone to racemization, however [37]. For example, one factor which could enhance racemization rates of glutamyl or aspartyl residues, is the change in the pK_a of the β-carboxyl group which is associated with some enzymic reactions. At physiological pH the unprotonated β-carboxyl group is normally the prevalent ionic species, but in bovine trypsin the pK_a of aspartyl-102 is altered such that at neutral pH the protonated carboxyl group species is the predominant form [38]. This change in ionic character should have an enhancing effect (see Table 13.2) on the racemization of that particular residue [8, 37]. Unfortunately the detection of racemization of a single amino acid residue in a protein is difficult since in the analysis of a total protein hydrolysate, enhanced racemization of a single amino acid residue would be obscured.

Although the racemization of many amino acids has been investigated in various proteins and peptides, others, such as the β-substituted amino acid, have not been extensively studied [36, 39]. This has been partially due to the lack of routine analytical methods which can be used to separate the enantiomers of these amino acids. Moreover, the carbanions of β-substituted amino acids have a tendency to undergo β-elimination, which complicates the racemization kinetics of these amino acids [40].

Amino acids in cyclopeptides (i.e. diketopiperazines) might be expected to have racemization rates similar to amino acids bound in the interior positions of peptides. However, there appear to be some rate enhancing effects which arise from conformational properties of diketopiperazines. Amino acids in diketopiperazines have long been known to undergo rapid racemization in dilute alkaline solutions at room temperature [1]. Recent studies [28, 29] have demonstrated that even in neutral pH range the rate of amino acid racemization in diketopiperazines is more rapid than *N*-terminal amino acid residues in various di- and tripeptides. Only at pH values less than about 5 is the racemization of *N*-terminal amino acid residues in dipeptides greater than in the diketopiperazines. An explanation for the enhanced rate of racemization in diketopiperazines has been advanced by Gund and Veber [25] who used molecular orbital calculations to show that the reactivity of the amino acid α-hydrogen was affected by the *cis-trans* configuration around the peptide bond in both diketopiperazines and peptides.

It is interesting to note that in cyclo-(L-prolyl-L-phenylalanyl), the proline residue is racemized in a matter of minutes when the diketopiperazine is treated with 0.5 N NaOH at room temperature; the phenylalanyl residue retains its chirality, however [41, 42]. In studies of racemization in various proteins in dilute alkaline solutions phenylalanine was always found to be more rapidly racemized than proline [12]. The enhanced racemization of proline in the cyclopeptide is apparently attributable to conformational factors [42].

13.5 RACEMIZATION IN VARIOUS SYSTEMS ON THE EARTH

Over the last several years investigation of amino acid racemization in various geochemical and biochemical systems has been an area of active research. There has been an increasing use of amino acid racemization in dating fossil materials as well as for determining the ages of living mammals. There have also been studies on the use of racemization for estimating the temperature histories of various environments. The possible role played by racemization in the aging process in living mammals has also been discussed as has been the possible nutritional consequences of amino acid racemization in various food proteins. The results obtained in these various investigations have been extensively discussed in several reviews [4–8, 37, 43, 44]. Only a brief summary will be given here.

The determination of a fossil's age using amino acid racemization is complicated by the fact that the extent of racemization in a fossil is dependent on both time, t, and the value of k_i, the rate of interconversion of the D and L enantiomers. The value of k_i has been found to be primarily a function of the temperature of the environment where the fossil was found [15,45]. In order to determine the age of the fossil using racemization the value of k_i for the fossil locality must be determined. A calibration procedure has been developed to evaluate k_i at any particular locality [4,5,8]. In this procedure a sample of known age from the area of interest is analysed for its extent of amino acid racemization. The age and measured enantiomeric ratios of this calibration sample are substituted into Equation 13.2 and an *in situ* value of k_i is thus determined. This k_i value represents the average integrated value over the age of the calibration sample. Any variations in the temperature, or other environmental parameters, of the locality throughout the past are evaluated when the calibration constant, k_i, is determined. Once k_i has been established for a site and for a certain type of fossil, this value can then be used to date other fossils of that type in the same general area [8,45,46]. The only limitation in using this calibration procedure is that the average temperature to which the calibration sample has been exposed must be similar to that of the sample which is being dated. One additional complication is that a k_i value must be determined for each type of fossil at a particular locality since it has been found [8] that the rate of racemization in bone, shell, and wood is different.

The calibration procedure has been used to determined the racemization ages of a variety of fossil materials [8]. This application has been found to be particularly important in dating fossils which are too old to be dated by the conventional ^{14}C method, that is, ones with ages greater than $\sim$40 000 years. Since radiometric methods such as uranium series and potassium argon dating do not yield reliable ages for fossils such as bones and shells, the racemization technique has been found to be useful in dating an interval of geological time which was previously difficult to date accurately.

Initially the investigations of amino acid racemization concentrated on fossil materials but recently racemization has also been detected in the metabolically stable proteins of long-lived mammals [31–34]. Enamel, dentine and the proteins present in the nucleus of the ocular lens are metabolically inert and are thus incubated at $\sim$ 37°C throughout a mammal's lifespan. Under these conditions detectable racemization of aspartic acid occurs. The results obtained for racemization in human dentine are shown in Fig. 13.5. No racemization of other amino acids such as phenylalanine, alanine and leucine was detected which is the expected result since aspartic acid has the fastest racemization rate of these amino acids.

The racemization of amino acids thus not only provides the basis for a geochronological method but it can also be used as a biochronological tool. This new aging method may be particularly useful for determining the ages of humans whose ages are not well documented [31] or for aging certain mammals which are difficult or impossible to age by other methods [47].

Besides the chronological applications, amino acid racemization in various systems on Earth may have important implications. For example, once amino acids are isolated from the enzymic reactions which maintain their optical

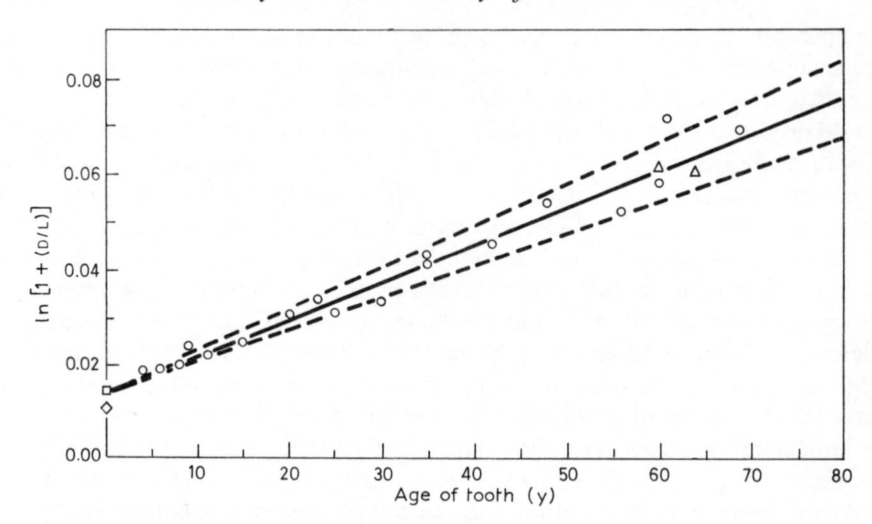

Figure 13.5 A plot, in the form of Equation 13.3, of aspartic acid racemization in dentine isolated from human teeth of various ages. The slope (i.e., $7.87 \times 10^{-4}\,y^{-1}$) is the k_{asp} value in human dentine. The dashed lines are the k_{asp} value calculated for a ± 0.7°C difference in body temperature (the temperature range for humans). The symbols are defined as: O, individuals of known age; Δ, exact ages of individuals are unknown, but were estimated by dentists; □ and ◇, albumin and bovine tendon collagen, respectively, determined for controls. Taken from Helfman and Bada [31].

activity, racemization would rapidly (on the geological time scale) take place and convert the amino acids into a racemic mixture. Since the bulk of the amino acids on the Earth are contained in reservoirs outside the biosphere it would appear that although the amino acids in living organisms are characterized by the presence of only L-amino acids all the amino acids on the Earth are racemized to some degree [8]. In fact, averaged over the entire Earth, the amino acid enantiomeric ratio may be closer to being racemic.

The biochemical implications of racemization in living organisms have only begun to be investigated. Racemization in proteins may induce structural changes which in turn affect the functionality of the protein [33, 34]. Thus, racemization may play some role in the aging process in certain organisms such as long-lived mammals [8, 37]. The recent discovery [48] of an enzyme which methylates and apparently inverts D-aspartyl residues in various mammalian tissues indicates that organisms have developed a mechanism for eliminating or repairing racemized residues. That such a 'repair' system should exist, strongly suggests that racemization in peptides and proteins has detrimental biological consequences.

Racemization may also have important nutritional implications for mammals. Alkaline treated proteins such as casein, textured soy protein and corn flour are becoming increasingly used in commercial foods. The alkali treatment (usually exposure to 0.1–0.5 N NaOH for several hours at temperatures of 40–80°C) results in significant racemization of amino acids such as aspartic and glutamic acids, phenylalanine and alanine [12, 49]. Racemization in these treated proteins may reduce their nutritional value [12]. For example, alkaline treated proteins have been found to have reduced digestibility by proteolytic enzymes such as pepsin, trypsin, and chymotrypsin [50–52]. This is due to the fact that these enzymes are highly stereoselective and cannot hydrolyse peptide bonds which contain D-amino acid residues. Even if racemized amino acid residues can be enzymically hydrolysed there is the possibility that D-amino acids are unusable or at least have a reduced usability [12]. In order for D-amino acids to satisfy the nutritional requirements in mammals they must first be oxidatively deaminated to α-keto acids which are then reaminated stereoselectively to L-amino acids. Because of the necessity to first convert D-enantiomers into L-enantiomers prior to metabolism D-amino acids are apparently more slowly utilized than L-amino acids. It seems likely that racemization can have deleterious nutritional consequences especially if the racemized amino acid residues are some of the essential amino acids. With alkali treated proteins becoming an increasingly larger component of the human diet it seems highly desirable to evaluate what effects racemized amino acids have on the nutritional value of foods.

13.6 CONCLUSIONS

During the last decade numerous investigations have helped establish the kinetics and mechanism of racemization of both free and peptide-bound amino acids. These investigations have provided the basis for the development of a chronological tool based on racemization. Although these chronological applications have been fairly extensively investigated, the geochemical and biochemical implications of racemization are only now beginning to be investigated.

Also, while the racemization reaction of many amino acids has been fairly thoroughly studied, others such as the basic and β-substituted amino acids have received only limited attention. With the development of routine analytical techniques for determining the enantiomeric composition of each of the various amino acids [53], it should become possible to extend our knowledge of amino acid racemization to various systems on the Earth.

REFERENCES

1. Neuberger, A. (1948) *Adv. Protein Chem.*, **4**, 298.
2. Bada, J.L. (1971) *ACS Adv. Chem. Ser. No.* **106**, 309.
3. Bada, J.L. (1972) *J. Am. Chem. Soc.*, **94**, 1371.
4. Bada, J.L. and Schroeder, R.A. (1975) *Naturwissenschaften*, **62**, 71.
5. Schroeder, R.A. and Bada, J.L. (1976) *Earth Sci. Rev.*, **12**, 347.
6. Williams, H.M. and Smith, G.G. (1977) *Origins of Life*, **8**, 91.
7. Vicar, J. (1977) *Chemicke Listy*, **71**, 160.
8. Bada, J.L. (1982) *Interdispl. Sci. Revs*, **7**, 30.
9. Finley, T.H. and Adams, E. (1970) *J. Biol. Chem.*, **245**, 5248.

10. Smith, G.G., Williams, K.M. and Wonnacott D.M. (1978) *J. Org. Chem.*, **43**, 1.
11. Dungworth, G., Vincken, N.J. and Schwarz, A.W. (1973) in *Advances in Organic Geochemistry* (eds B. Tissat and F. Bienner), Technip, Paris, p. 689.
12. Masters, P.M. and Friedman, M. (1980) *ACS Symposium Series No.* 123, 165.
13. Smith, G.G. and de Sol, B.S. (1980) *Science*, **207**, 765.
14. Shou, M.-Y. (1979) Ph.D. Thesis, Scripps Institution of Oceanography, University of California, San Diego.
15. Bada, J.L. and Shou, M.-Y. (1980) in *Biogeochemistry of Amino Acids* (eds P.E. Hare, T.C. Hoering and K. King), Wiley, New York, p. 235.
16. Cram, D.J. (1964) in *Fundamentals of Carbanion Chemistry*, Academic Press, New York.
17. Barlin, C.B. and Perrin, D.D. (1966) *Quart. Rev. Chem. Soc., Lond.*, **20**, 75.
18. Taft, R.W. (1956) in *Steric Effects in Organic Chemistry* (ed. M.S. Neuman), Wiley, New York, p. 556.
19. Kemp, D.S. (1979) in *The Peptides; Analysis, Synthesis and Biology* (eds E. Gross and J. Meienhofer) (Vol. 1) Academic Press, New York, p. 315.
20. Buckingham, D.A., Marzilli, L.G. and Sargeson, A.M. (1967) *J. Am. Chem. Soc.*, **89**, 5133.
21. Pasini, A. and Cassella, L. (1974) *J. Inorg. Nucl. Chem.*, **36**, 2133.
22. Gillard, R.D., O'Brien, P., Norman, P.R. and Phipps, D.A. (1977) *J. Chem. Soc. Dalton*, **1977**, 1988.
23. Gillard, R.D. and O'Brien, P. (1978) *J. Chem. Soc. Dalton*, **1978**, 1444.
24. Snell, E.E., Braunstein, A.E., Severin, E.S. and Torchinsky, Y.M. (eds) *Pyridoxal Catalysis: Enzymes and Model Systems*, Interscience, New York.
25. Gund, P. and Veber, D.F. (1979) *J. Am. Chem. Soc.*, **101**, 1885.
26. Bohak, Z. and Katchalski, E. (1963) *Biochemistry*, **2**, 228.
27. Kriausakul, N. and Mitterer, R. (1978) *Science*, **201**, 1011.
28. Steinberg, S. and Bada, J.L. (1981) *Science*, **213**, 544.
29. Steinberg, S. (1982) Ph.D. thesis, Scripps Institution of Oceanography, University of California, San Diego.
30. Smith, G.G. and Evans, R.C. (1980) in *Biogeochemistry of Amino Acids* (eds P.E. Hare, T.C. Hoering and K. King), Wiley, New York, p. 257.
31. Helfman, P.M. and Bada, J.L. (1976) *Nature*, **262**, 279.
32. Masters, P.M., Bada, J.L. and Zigler, J.S. (1977) *Nature*, **268**, 71.
33. Masters, P.M., Bada, J.L. and Zigler, J.S. (1978) *Proc. Natl. Acad. Sci. USA*, **75**, 1204.
34. Garner, W.H. and Spector, A. (1978) *Proc. Natl. Acad. Sci. USA*, **75**, 3618.
35. Friedman, M. and Masters, P.M. (1982) *J. Food Sci.*, **47**, 760.
36. Engel, M.H., Sawyer, T.K., Hadley, M.E. and Hruby, V.J. (1981) *Anal. Biochem.*, **116**, 303.
37. Helfman, P.M., Bada, J.L. and Shou, M.-Y. (1977) *Gerontology*, **23**, 419.
38. Koeppe, R.H. and Stroud, R.M. (1976) *Biochemistry*, **15**, 3450.
39. Engel, M.H., and Hare, P.E. (1981) *Carnegie Inst. Wash. Yearbook*, **80**, 400.
40. Whitaker, J.R. (1980) *ACS Symposium Series No. 123*, p. 145.
41. Ott, H., Frey, A.J. and Hofmann, A. (1963) *Tetrahedron*, **19**, 1675.
42. Vicar, J. and Blaha, K. (1973) *Collection Czechoslov. Chem. Commun.*, **38**, 3307.
43. Dungworth, G. (1976) *Chem. Geol.*, **17**, 135.
44. Pautet, F. (1980) *Pathol. Biol.*, **25**, 325.
45. Belluomini, G. (1981) *Archaeometry*, **23**, 125.
46. Bada, J.L. (1981) *Earth Planet. Sci. Lett.*, **55**, 292.
47. Bada, J.L., Brown, S. and Masters, P.M. (1980) in *Age Determinations of Toothed Whales and Sirenians* (eds W.F. Perrin and A.C. Myrick), International Whaling Commission, Special Issue 3, Cambridge, England, p. 113.
48. McFadden, P.N. and Clarke, S. (1982) *Proc. Natl. Acad. Sci. USA*, **79**, 2460.
49. Masters, P.M. and Friedman, M. (1979) *J. Agr. Food Chem.*, **27**, 507.
50. Hayashi, R. and Kameda, I. (1980) *J. Food Sci.*, **45**, 1430.
51. Hayashi, R. and Kameda, I. (1980) *Agr. Biol. Chem.*, **44**, 891.
52. Friedman, M., Zahnley, J.C. and Masters, P.M. (1981) *J. Food Sci.*, **46**, 127.
53. Engel, M.H. and Hare, P.E. (1981) *Carnegie Inst. Wash. Yearbook.*, **80**, 394.

CHAPTER FOURTEEN

Ion-Exchange Separation of Amino Acids

P.E. HARE, P.A. St. JOHN AND M.H. ENGEL

14.1 INTRODUCTION

Whenever new methods are developed for the quantitative analysis of amino acids the standard for comparison is nearly always the classic ion-exchange chromatography method developed in the 1950s by such workers as Stanford Moore, William H. Stein, D. Spackman, and Paul Hamilton [1,2,3]. When new analytical methods are compared to the classical ion-exchange method, too often claims of higher sensitivity and/or faster analysis times are made by comparison with ion-exchange systems 15–20 years old. Improvements in ion-exchange resins, use of small bore columns and the development of high-pressure injection valves and pumps have enabled researchers using the classic ion-exchange method of amino acid analysis to decrease the analysis time from 24 hours in 1958 to less than 30 minutes, and to improve sensitivity by more than six orders of magnitude. Currently, ion-exchange chromatography of amino acids is just one of many techniques available in the rapidly expanding field of high performance liquid chromatography (HPLC).

Many of the current innovations in HPLC were first developed and used in the ion-exchange chromatography of amino acids and later adapted and often improved upon for other HPLC applications. These innovations include spherical narrow-size-range resin beads, gradient elution, high-pressure pumps, flow-through photometers, and post-column reaction systems.

For detection, amino acids are generally derivatized to form a reaction product that will absorb light at some particular wavelength. Fluorescent derivatives absorb at one wavelength and re-emit light at a longer wavelength. Fluorescence is generally more sensitive than absorption, and several excellent fluorescent derivatives of amino acids can be made for analytical purposes. In preparative chromatography for micromole and larger levels of amino acids a refractive index detector may be useful. Absorption of UV light at 190 nm has also been reported but is difficult to use and not very sensitive.

14.2 ION-EXCHANGE RESINS

Amino acid ion-exchange chromatography is almost exclusively performed on strongly-acidic sulphonated polystyrene/divinylbenzene copolymers. These chromatographic resins are manufactured by the process of suspension polymerization. This process results in solid spherical particles of resin gel. The manufacturing process requires three steps. First, the neutral polystyrene/divinylbenzene beads are synthesized. Second, the beads are sulphonated with sulphuric acid, chlorosulphonic acid, or similar reagents. Third, the sulphonated beads are separated into size fractions by the process of water elutriation.

Resin particles have no true pores in the usual mechanical sense of the word. Resin porosity is 'apparent porosity' which refers to the size of the ion that can diffuse into the polymer matrix. Apparent porosity is a function of the extent of polymer chain crosslinking and seldom exceeds about 40 Å even in weakly crosslinked polymers. Diffusion into the bead matrix is a rate-limiting step in the ion exchange process; therefore, small beads with large surface areas and short diffusion paths are favoured. With the availability of high-pressure metering pumps and stainless steel columns, the trend in chromatographic media has been toward smaller and smaller bead diameters. Eight or nine micron diameter beads are now commonly in use. Smaller diameter resins have found limited use. Difficulties in manufacturing make such very small bead resins scarce and expensive, and equipment to take advantage of their capabilities has been lacking.

The collective works of Moore and Stein [1,2] and of Hamilton [3] set standards against which subsequent efforts have been compared. Over the past three decades, resin production has been tailored to match existing elution methodologies. Only in recent years has there been an increase in the level of experimentation with chromatographic materials themselves.

Performance of ion-exchange resin is influenced by resin matrix characteristics and hence by monomer composition and purity. The extent of sulphonation and the chemical approach to sulphonation influences the relative performance of the product. Efficiency is dependent on particle size and particle size distribution. Resin characteristics can be optimized for each area of the amino acid chromatogram. Resin production is thus a series of trade-offs calculated to yield a material with acceptable performance over the entire range of analyte characteristics. Low process yields, lengthy sizing procedures, and batch-to-batch variability account for the expense of modern resins. Polystyrene/divinylbenzene cation exchangers are extremely durable. They will withstand concentrated acids, bases, organic solvents, and strong ionic strength environments. For example 6 M hydrochloric acid is used to remove heavy metal ions from the resin matrix, hot concentrated sulphuric acid is used to strip away tenacious protein contamination (by brief exposure to the acid), and 1 M HCl and 0.6 M NaOH are used for routine cleaning of the resins. Chromatographic columns can frequently be cleaned *in situ* by pumping alternately 0.2 M NaOH containing 1 g dm^{-3} EDTA and pH 3.25 citrate elution buffer and elevating the column temperature to 80–90°C. Oxidants, such as peroxides and hypohalites, oxidizing acids, and nitro-compounds, such as picric acid, should be avoided. Such resin durability should be contrasted with that of silica-based supports that are limited in usable pH range and offer limited reclamation potential if contaminated. For these reasons the cost effectivity of modern cation exchangers is usually much greater than that of silica-based supports.

Polystyrene-based cation exchangers will shrink or swell depending on the environment to which they are exposed. The degree of polymer crosslinking determines the rigidity of the polymer lattice and consequently the relative degree of swelling. For example, a 1% crosslinked resin has a wet/dry diameter ratio of 2.5, while a 10% crosslinked resin swells only by a factor of 1.4. Elution methodologies that utilize ionic strength gradients will necessarily cause the column resin bed to shrink as the ionic strength of the solution increases. Such shrinkage is frequently erroneously assigned to resin settling. Shrinking and swelling will eventually cause some degree of ordering in a randomly packed resin bed. Extended re-equilibration is an unfortunate side effect of ionic strength gradient elution.

14.3 POST-COLUMN AND PRE-COLUMN DERIVATIZATION

Because amino acids at analytical levels must be derivatized for detection it becomes a matter of choice whether the derivatization process precedes or follows the column separation. The trend in recent years has been towards pre-column derivatization in which the sample mixture of free amino acids is reacted to form suitable derivatives. The amino acid derivatives are then injected onto the column and eluted directly into the detector.

The post-column reaction system requires an additional reagent(s) to be continuously metered into the effluent stream from the analytical column. This is usually done with a metering pump but can also be accomplished effectively with a pressure system consisting of a reagent reservoir pressurized with 10–20 psi of nitrogen or helium. The reagent flow rate is controlled by the applied gas pressure. The principal advantage of the post-column reaction system is that an underivatized mixture of free amino acids is injected onto the column for separation. In pre-column derivatization there is always the possibility of incomplete reaction in forming the individual derivatives as well as the question of stability of the derivatives during chromatography. Derivative stability and completeness of reaction is not as important in post-column systems. Reproducibility is achieved by maintaining uniform column temperatures and flow rates and standardizing on known concentrations of amino acids.

Although this chapter emphasizes the use of post-column reaction systems with ion-exchange columns, this is not to minimize the usefulness of pre-column derivative systems for amino acid analysis. For analysis at less than picomole (10^{-12} moles) levels of amino acids certain fluorescent derivatives such as dansyl [4, 5] and *o*-phthaldialdehyde/2-mercaptoethanol (OPA) [6, 7, 8] appear to offer significant advantages. For example, ubiquitous trace amounts of amino acids and ammonia in the mobile phases do not generally interfere with the chromatograms of the fluorescent derivatives and are not detected in the flow fluorimeter. In contrast, in post-column systems, trace amounts of amino acids in the mobile phase react with the post-column reagent and interfere with the baseline stability during the chromatographic run. Amine impurities are revealed by rising baselines and buffer change peaks and interfere particularly in the region where the basic amino acids elute. The use of high purity reagents in making up the buffers reduces, but does not eliminate, the problem. This problem of amino acid impurities in the eluting buffers is the limiting factor in the sensitivity of the analysis for amino acids. At analysis levels above 10 picomoles of individual amino acids it is possible to obtain good results for even single column analysis. Below 10 picomoles the elution of basic amino acids and ammonia impurities (which collect on the column during the elution of the low pH buffers) interferes with the analysis of the basic amino acids in the sample to such an extent that it is usually necessary to go to a two-column system that analyses the neutral and acidic amino acids on one column and the basics on another [9].

14.4 COMMERCIAL EQUIPMENT FOR AMINO ACID ANALYSIS

Commercial amino acid analysers have evolved relatively little over the last 25 years. Automatic sample injection, data handling, and improved column

Figure 14.1 Combination High Performance Liquid Chromatograph amino acid analyser with fluorescence detector.

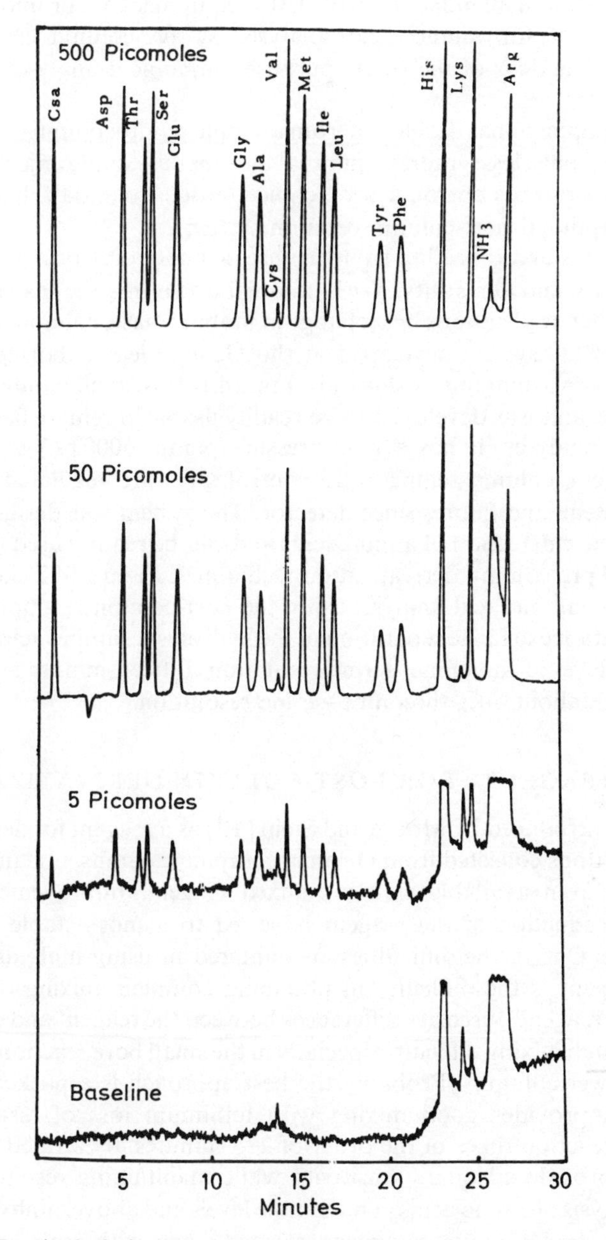

Figure 14.2 Routine hydrolysate amino acid separations utilizing *o*-phthaldialdehyde fluorimetric post-column detection.

configurations and resins have been added but the resulting instrument is still difficult to use for applications other than amino acids.

On the other hand, modern liquid chromatography has evolved rapidly over the last 10 years and manufacturers of HPLC equipment are promoting the use of their equipment for amino acid analysis by pre-column *or* post-column techniques! The days of the single purpose amino acid analyser appear to be numbered.

HPLC equipment has developed from a simple pump-column–UV detector to sophisticated pulseless gradient pumps, automatic sample injectors, optional column temperature control, a wide choice of detectors, data handling equipment, and optional post-column reaction systems.

Such systems are appealing from the user's standpoint primarily because of their flexibility and versatility. With minimal downtime the instrument can be shifted to other applications by a change in mobile phase, column, or detector. A compact, HPLC system developed at the Geophysical Laboratory, Carnegie Institution of Washington, is shown in Fig. 14.1. It is small enough to fit into a carrying case and was developed to be readily usable in remote field areas as an amino acid analyser. It has a high-pressure pump (6000 psi), a step-gradient solvent system, column temperature control, pressure regulated post-column reaction system, and fluorescence detector. The system was designed primarily for fluorescence detection of amino acids and can be readily used for both post-column and pre-column derivatization techniques. Figure 14.2 shows a routine ion-exchange amino acid analysis using the post-column reaction system. The speed of a cation exchange separation of the hydrolysate amino acids is limited by the desired level of threonine–serine resolution. Fifteen-minute separations are possible with about 50% threonine–serine resolution.

14.5 REAGENTS FOR POST-COLUMN DERIVATIZATION

Ninhydrin, introduced by Moore and Stein [10] as a reagent for detecting amino acids in fractions collected from chromatographic columns, remains in wide use today. It is even available as a pre-mixed reagent. Improvements since the original introduction of the reagent have led to a more stable and sensitive formulation. One of the difficulties encountered in using ninhydrin as a post-column reagent is the difficulty in obtaining complete mixing of the column effluent and reagent. Viscosity differences between the reagent and effluent buffer make thorough mixing difficult, especially in the small bore reaction systems used with the newer columns. Probably the best approach is a packed-bed reactor [11], which provides good mixing with minimum loss of resolution. With ninhydrin, reaction times of the order of 2–3 minutes at elevated temperatures (*ca* 120°C) provide adequate sensitivity while maintaining resolution. For the routine analysis of amino acids at nanomole levels and above, ninhydrin provides a convenient post-column reagent system that reacts with proline and hydroxyproline as well as with the α-amino acids.

Over the years other reagents have been proposed for post-column detection of amino acids. The sensitivity and convenience of fluorimetric methods have made such methods popular for high sensitivity amino acid analysis. Commercial systems are available for adding one or more post-column reaction systems to virtually any existing HPLC system.

One of the most popular post-column reagents for amino acids is *o*-phthaldialdehyde (OPA) which also can be used for pre-column derivatives. Unlike ninhydrin, the OPA reagent is a dilute aqueous solution and mixes readily with the column effluent. The signal-to-noise ratio of this system appears to be superior to other post-column systems resulting in enhanced sensitivity.

14.6 IMINO ACIDS AND SECONDARY AMINES

Ortho-phthaldialdehyde-2-mercaptoethanol reagent (OPA) reacts only with primary amines. Additional chemical manipulation must be performed if the imino acids and other amines are to be determined with OPA. In his first description of the OPA reagent Roth [12] noted that proline and hydroxyproline could be converted to reactive fragments by treatment with sodium hypochlorite or chloramine-T. St. John [13, 14] investigated this process and developed an automated post-column reaction system which was used in the AMINCO Aminalyser, a commercial amino acid analyser, which featured fluorescence detection. Drescher and Lee [15] utilized chloramine-T in this system to determine proline in hydrolysates of fixed stained protein bands from polyacrylamide electrophoresis gels, and in other protein hydrolysates. The review of fluorimetric amino acid analysis with OPA by Lee and Drescher [16] is recommended reading, covering the time period up to 1978. Böhlen and Mellet [17] studied the use of NaOCl for the determination of proline and hydroxyproline. They modified an existing commercial analyser in such a way as to add the oxidant to the post-column effluent during the proline elution interval only. Thomas [18] and Ishida *et al.* [19] have reported procedures whereby the oxidant is added to the post-column continuously during the entire separation.

Figure 14.3 shows a schematic diagram of a proline detection system. The procedure involves the addition of the oxidant post-column (either during the elution interval only or during the entire separation) followed by a delay coil where the conversion reaction occurs. The reaction effluent is then mixed with OPA as in conventional fluorescence systems, passes through a second delay coil, and is measured in the flow fluorimeter.

A typical system would be constructed of microbore mixing tees and 0.3 mm bore Teflon tubing. The oxidant (for example, chloramine-T, 5 mM in 0.2 M K borate, pH 10.5) is added with a metering pump. The mixture passes through coil A (1 m, 55°C) and is then mixed with OPA in the second tee (1 g dm^{-3} OPA, 1 ml dm^{-3} mercaptoethanol, 0.2 M K borate, pH 10.5). The effluent then passes through coil B (2 m, ambient temperature) and into the flow cell. Solution temperature at the flow cell is typically 35°C. Flow rate ratios for

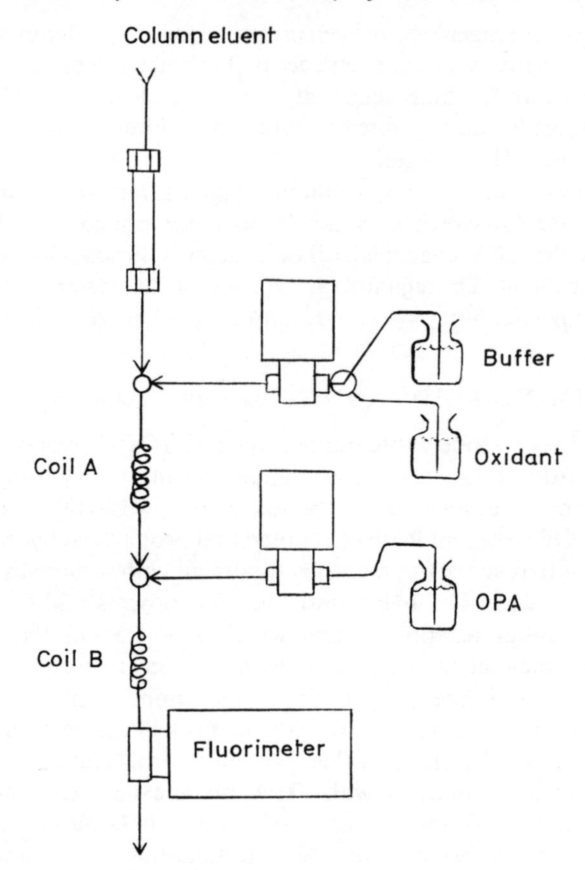

Figure 14.3 Schematic flow diagram of a post-column reaction system for imino acid determination.

column:oxidant:OPA are about 2:3:3. Procedures for the constant addition of oxidant as described by Thomas [18] and Ishida *et al.* [19] require that a balance be maintained between imino acid response and amino acid response. Considerable care is required in optimizing oxidant concentration, reaction pH, and temperature. When optimized the procedure provides proline response, and very much improved cystine response, presumably via oxidation to cysteic acid.

Use of valving methods whereby the oxidant is added to the effluent only during the elution interval of the selected imino acids permits the optimization of reaction conditions without concern for the destruction of amino acids. A microcomputer controlled diverter valve on the inlet side of the oxidant feed pump can be used to inject oxidant at will without disturbing system flow. Alternatively the oxidant can be added during the first elution buffer, for instance aspartic through cystine, and a non-oxidizing buffer added during the remainder of the separation. This yields proline response, and improved cystine response.

The post-column reaction systems outlined can be easily adapted to a variety of analytical problems. Moye and St. John [20] have shown the herbicide Glyphosate, *N*-(phosphonomethyl)glycine, can be easily split into a reactive fragment, post column. Similarly, Moye *et al.* [21] have utilized such procedures for the post-column conversion of *N*-methylcarbamates and related compounds to methylamine for OPA fluorimetric detection. Epinephrine likewise can be efficiently converted to a reactive fragment for fluorescence detection [22]. Such procedures frequently have the added benefit of background reduction because of the more complex chemical requirements for a positive fluorescence response.

14.7 CHIRAL SEPARATIONS

Interest in the resolution of amino acids into their respective D- and L-enantiomers has grown in recent years, resulting in a number of new methods using both gas and liquid chromatographic techniques. Since amino acid enantiomers cannot be resolved in achiral systems it is necessary to add a second chiral centre either as an integral part of the derivative to form a diastereoisomer or in the chromatographic system itself. Both approaches have proved successful, with the development of a number of novel separations. Chromatographic separation and resolution of underivatized amino enantiomers on both ion-exchange and reversed-phase columns using chiral mobile phases have been described [23, 24]. Post-column reaction with OPA requires chiral mobile phases with free amino groups. Dilute aqueous solutions of L-proline and cupric acetate in sodium acetate buffers can be used as the mobile phase on ion-exchange columns as well as on C-18 reversed-phase columns. On reversed-phase columns substituted L-amino acids such as *N*,*N*-dipropyl-L-alanine have been found to give excellent separation of all of the common amino acid enantiomers [25].

Efficient enantiomer separations have also been described for dansyl derivatives of amino acids including the dual column separation of the amino acid derivatives followed by the separation of the enantiomers on a separate column with a chiral mobile phase [4].

With the present rate of progress in this area it seems not too far into the future that amino acid analyses will routinely include quantitative D- and L-concentrations of each amino acid. Such information is already of practical interest in geochemical age-dating methods using amino acid racemization reactions. Other areas of medicine, biochemistry, and nutrition are showing an increased interest in the analytical determination of D- and L-amino acids.

14.8 PREPARATIVE CHROMATOGRAPHIC SEPARATIONS

The high exchange capacity of ion exchange columns makes them ideal for scaling up from analytical techniques to preparative scale chromatography. Hirs *et al.* [26] showed the use of anion and cation exchange columns and volatile acids to prepare milligram amounts of pure amino acids. Aliquots from the fractions collected can be analysed for identification of eluted amino acids as well as their purity. Preliminary data from our laboratory show promising results using OPA fluorescence detection with a stream splitter for directing the bulk of the column effluent to a fraction collector.

14.9 CONCLUSION

Ion exchange chromatography with post-column reaction continues to be a popular method for amino acid analyses. Improvement in ion-exchange resins, column packing techniques and detection instrumentation have all contributed to make ion-exchange chromatography an analytical method for amino acids that is as rapid and as sensitive as any other chromatographic method available. Post-column reaction systems for amino acid analysis require that the mobile phase buffers be as free as possible from amine and ammonia contaminants. Eluting contaminants react and interfere with the baseline and sample components. Pre-column derivatization generally entails fewer problems with contaminants in the mobile phase.

Future analytical approaches will continue to evolve along with both the pre-column and post-column concepts. New developments with packing materials, solvent and reagent purities, new or improved derivatives, and detection instrumentation will ensure that the routine analysis of the amino acids as well as their D- and L-enantiomer concentrations will be possible on sub-picomole levels of samples. Liquid chromatography promises to continue to be an exciting and useful analytical tool, not only for amino acids and their related products, but for a wide range of organic as well as inorganic products.

REFERENCES

1. Moore, S. and Stein, W.H. (1954) *J. Biol. Chem.*, **211**, 893.
2. Spackman, D., Stein, W.H. and Moore, S. (1958) *Anal. Chem.*, **30**, 1190.
3. Hamilton, P.B. (1963) *Anal. Chem.*, **35**, 2055.
4. Tapuhi, Y., Miller, N. and Karger, B.L. (1981) *J. Chromatogr.*, **204**, 325.
5. Tapuhi, Y., Schmidt, D.E., Lindner, M. and Karger, B.L. (1981) *Anal. Biochem.*, **115**, 123.
6. Hill, D.M., Walters, F.H., Wilson, T.D. and Stuart, J.D.S. (1979) *Anal. Chem.*, **51**, 1338.
7. Liebezeit, G. and Dawson, R. (1981) *J. High Resolut. Chromatogr., Chromatogr. Commun.*, **4**, 354.
8. Larson, B.R. and West, F.G. (1981) *J. Chromatogr. Sci.*, **19**, 259.
9. Hare, P.E. (1977) in *Methods in Enzymology* (eds. C.H.M. Hirs and S.N. Timasheff) (Vol. XLVII Part E), Academic Press, New York.
10. *Moore, S. and Stein, W.H.* (1951) *J. Biol. Chem.*, **192**, 663.
11. Huber, J.F.K., Jonker, K.M. and Poppe, H. (1980) *Anal. Chem.*, **52**, 2.
12. Roth, M. (1971) *Anal. Chem.*, **43**, 880.
13. St. John, P.A. (1975) *Aminco Lab. News*, **31**, Spring.
14. St. John, P.A. (1976) *Aminco Reprint 541*, available from American Research Products Corp., Beltsville, Md.
15. Drescher, D.G. and Lee, K.S. (1978) *Anal. Biochem.*, **84**, 559.
16. Lee, K.S. and Drescher, D.G. (1978) *Int. J. Biochem.*, **9**, 457.
17. Böhlen, P. and Mellet, M. (1979) *Anal. Biochem.*, **94**, 313.
18. Thomas, A.J. (1981) Chapter 2 in *Amino Acid Analysis* (ed. J.M. Rattenbury), J. Wiley, New York.
19. Ishida, Y., Fujita, T. and Asai, K. (1981) *J. Chromatogr.*, **204**, 143.

20. Moye, H.A. and St. John, P.A. (1980) *ACS Symp. Series No.* 136, *Pesticide Analytical Methodology* (eds J. Harvey Jr and G. Zweig).
21. Moye, H.A., Scherer, S.J. and St. John, P.A. (1977) *Analytical Letters*, **10**, 1049.
22. St. John, P.A. (1976) Unpublished data.
23. Hare, P.E. and Gil-Av, E. (1979) *Science*, **204**, 1226.
24. Gil-Av, E., Tishbee, A. and Hare, P.E. (1980) *J. Am. Chem. Soc.*, **102**, 5115.
25. Weinstein, S., Engel, M.H. and Hare, P.E. (1982) *Anal. Biochem.*, **121**, 370.
26. Hirs, C.H.W., Moore, S. and Stein, W.H. (1954) *J. Am. Chem. Soc.*, **76**, 6063.

CHAPTER FIFTEEN

Liquid Chromatography of Amino Acids and their Derivatives

D. PERRETT

15.1 INTRODUCTION

Since the classical studies of Moore and Stein [1,2] the separation of amino acids on cation exchange resins has been the method for their quantitation in both hydrolysates and physiological fluids. Subsequently the ease with which this could be accomplished was greatly improved by the development of an automatic analyser [3]. However over the last few years the technologies associated with high performance liquid chromatography (HPLC) have been applied with increasing success to the determination of amino acids and their derivatives. Much early work in HPLC involved investigating and 're-discovering' much that was already well known to users and manufacturers of amino acid analysers. For example the comprehensive studies on the chromatography of amino acids performed by Hamilton [4–6] in the late 1950s are reflected a decade later in the HPLC literature. Similarly it is only in the 1980s that HPLC manufacturers and users are advocating post-column derivatization to enhance both selectivity and sensitivity and developing a theoretical understanding of the mechanisms involved. Modern HPLC equipment has only recently reached the level of sophistication common in amino acid analysers ten years ago. It should be stressed that cation-exchange chromatography of amino acids represents one specific application of the general analytical technique of liquid column chromatography and that high performance (or high pressure) liquid chromatography is simply the terminology applied to the technique as currently optimized.

At present, cation-exchange chromatography with ninhydrin detection still offers a good allround compromise to those wishing to quantitate all amino acids at nanomolar levels routinely and reliably. This technique is discussed in detail in Chapter 14 of this book and elsewhere [7]. The overwhelming success of the original system meant that there was little impetus to develop other methods but with the growing availability of HPLC equipment increasing numbers of both general and specific amino acid separations are being published which utilize the equipment's capabilities.

15.2 GENERAL ASPECTS OF LIQUID CHROMATOGRAPHY

Since HPLC is currently the fastest growing analytical method and is now available in many laboratories, it is probably necessary only to give a general outline of the equipment and methodologies involved in the context of this chapter. Those requiring further information should consult either introductory texts [8, 9] or for more detail and specific reference there are an increasing number of standard texts [10, 11, 12]. The current status of our theoretical understanding of the chromatographic processes has been summarized by Knox [13] and Said [14].

Although the heart of a HPLC system is the column or more correctly the column packing, the modern system is a complex of advanced mechanical and electronic components. Figure 15.1 shows a schematic representation of the components usually found in a modern system.

The ideal chromatographic pump should give an accurate pulse-free flow and possess both minimal dead volume and rapid solvent changeover characteristics. Usually pumps are a compromise of these requirements being either constant pressure gas-driven pumps or constant volume electro-mechanical pumps. The latter pumps comprise either syringe-driven units or most commonly piston pumps coupled to one or more pulse suppressing mechanisms. Most modern HPLC pumps will operate satisfactorily at pressures in the range 100–5000 psi. The pumps either operate isocratically or pump solvent of varying composition formed in a low-pressure gradient former. High-pressure gradient formation requires at least two pumping units and to generate a gradient their individual outputs are varied electronically and the two flow streams are mixed immediately before the column.

The column usually consists of a stainless steel tube containing the packing material, which is retained by fine metal frits or sintered discs at either end of the column. Analytical columns are usually 4–5 mm i.d. and 10–30 cm in length. Such a column is capable of separating up to 1 mg of applied material but more often nanomolar amounts are applied. Higher sensitivity and lower solvent consumption is achieved by using micro-bore columns. These columns with i.d. < 2 mm are becoming increasingly available but require careful redesign of most other parts of the chromatographic system. Much larger diameter columns (2–5 cm) are used for preparative scale chromatography but require suitably scaled pumps and other equipment. Samples are introduced onto the column via either a valve (loop) injector or for maximum efficiency a syringe injector; both methods allow sample volumes from 1 μl to over 1 ml to be applied to the column. Increasingly, users with many repetitive analyses to perform are using automatic injection systems.

The eluate from the column passes into the detector(s) via a minimum length of micro-bore PTFE tubing. A large number of detectors are available giving varying degrees of sensitivity and selectivity. The most frequently used detectors are based on the absorption of UV light and employ either the lines of the

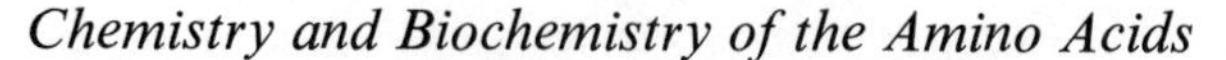

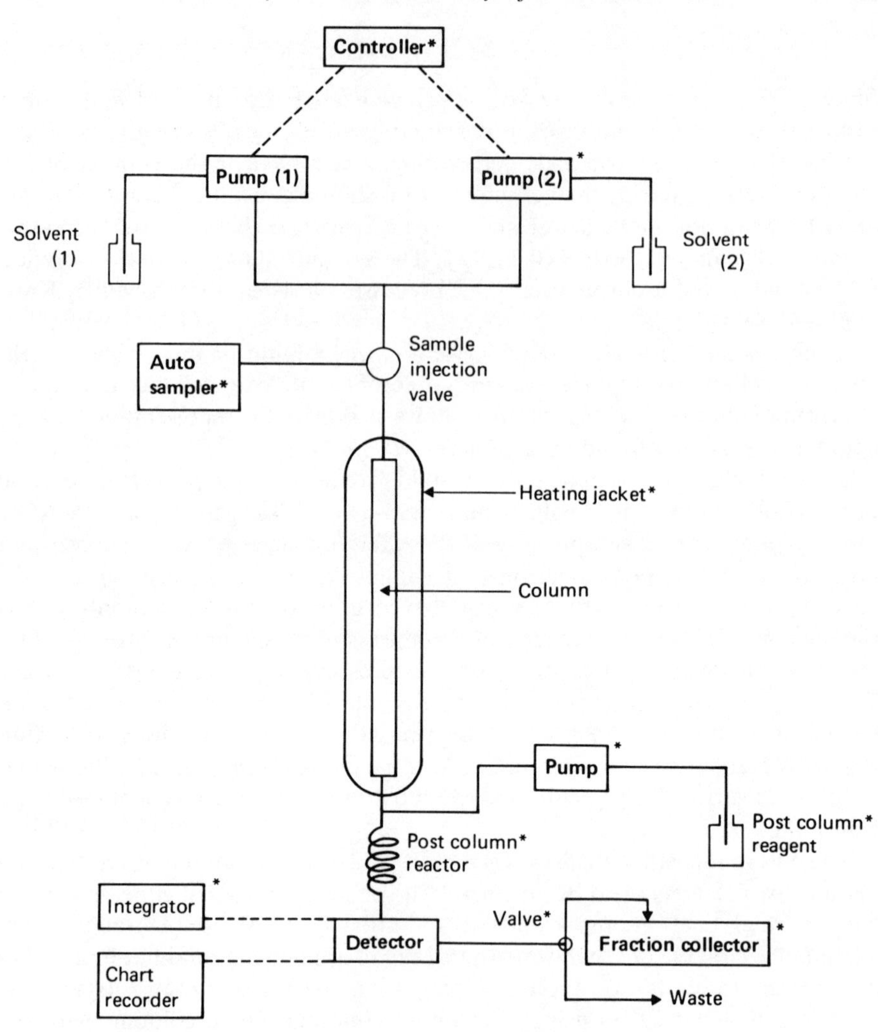

Figure 15.1 Schematic diagram of the components of a modern HPLC system. The items marked* are not found in all systems but are necessary for many of the applications described here.

mercury spectrum, particularly 254 nm, or variable wavelengths obtained from a deuterium source. Other detectors operating in the visible spectrum are available and are particularly useful when utilized with post-column reactions. Improved sensitivity, often of the order of 100–1000 fold, can be obtained by employing fluorescence detectors to measure either natural or chemically-induced fluorescence. Similar increases in sensitivity can be obtained by using an electrochemical detector to determine suitable electroactive compounds. These three detectors are the ones most commonly employed in work relating to amino acids. Although other detectors are available their application in amino acid work has been limited. The levels of sensitivity are generally better than 1 nmol injected onto the column for the UV detector and both the electrochemical and fluorescence detector can often detect less than 1 pmol of a suitable compound. All detectors are limited by the overall noise of the system and quoted S/N (signal to noise ratio) are often unobtainable in real work. To maintain chromatographic resolution cell volume should be small ($< 10\,\mu l$).

No high sensitivity universal detectors have yet been developed although some of low to moderate sensitivity can be purchased. Much work is currently being devoted to the design of linkages for the direct introduction of HPLC eluates into mass spectrometers to provide high sensitivity detection combined with structural information. The first commercially made linkages are now available. Since most HPLC detectors are non-destructive it is, of course, easy to isolate chromatographic peaks for mass spectral analysis. Similarly radioactivity incorporated into samples can be determined either by collecting fractions for off-line counting or by counting on-line but this method often gives poor counting efficiency. A more detailed account of the types of detectors available for HPLC and their mode of operation has been given by Scott [15].

Detection systems can be coupled in series. A non-destructive detector such as a UV detector could be followed by a post-column reaction with fluorescent detection giving additional sensitivity and specificity. The output from the detector is usually plotted on a chart-recorder. Peaks are usually identified on the basis of their elution times compared to known standards. In cases of difficulty additional information is obtained by co-chromatography with authentic material under varying elution or detection conditions. Where quantitation is required peak heights or areas can be determined manually from the chart, or the detector could also be coupled to an electronic integrator which would supply a digital output. In either case accurate quantitation can only be achieved by relating the unknown peak to standards of the same compound.

It is the nature of the column packing material that most affects the chromatography. The early amino acid analysers employed totally porous sulphonated ion-exchange resins (diameter $> 30\,\mu m$) and therefore required long columns (1 m), excessive analysis times (20 h) and gradient elution to separate the protein amino acids. Hamilton, Bogue and Anderson [6] demonstrated that considerable improvements could be achieved using smaller particles. Resins of smaller diameter allow quicker diffusion into and out of the pores within the body of the resin. Particle sizes employed in commercial systems gradually decreased to $17\,\mu m$, then $13\,\mu m$ and now $7\,\mu m$ is common. Shorter columns, often only 15 cm long still give complete resolution of the common amino acids. The other approach, adopted in HPLC to reduce band broadening, is to coat layers of modifier onto inert cores. This approach was pioneered by Horvath, Preiss and Lipsky [16] with their pellicular packing consisting of physically covered glass beads of $50\,\mu m$ diameter. Later chemically bonded phases, which are more stable to both physical and chemical degradation, were introduced and are now the most popular form of HPLC packing. For example a mono-molecular coating of ion-exchange material can be bonded via Si–C or Si–O–C linkages onto small diameter silica particles. The available packings, which may be either spherical or rough-shaped silica particles, have decreased in size from 10 μm to 5 μm and 3 μm packings are now available. The smaller packings give higher efficiencies and increased plate numbers for the same size column but at the expense of increased column back-pressure. Typically a 100×5 mm column containing a 5 μm microparticulate silica gel and eluted at an optimum flow rate of 1 ml min^{-1} should give a plate count (N) of 4000–6000 and a back-pressure of approximately 1000 psi depending on the type of packing, the solvent system and the equipment used.

Chromatographic mechanisms may be classified as adsorption (liquid–solid), partition (liquid–liquid or bonded phase), ion-exchange or steric exclusion. To effect these mechanisms column packings with a variety of surface coatings are available and some are listed in Table 15.1. Packings for adsorption chromatography may be either weakly acidic, e.g. silica, or weakly basic, e.g. alumina and they give excellent separation of un-ionized compounds when eluted with relatively non-polar solvents such as hexane modified with a few percent ethanol. The compounds are eluted in the order of increasing polarity. Packings for partition work now usually consist of an organic stationary phase bonded onto silica and are described as normal phase (i.e. polar stationary phase eluted with a non-polar solvent) or reversed phase (i.e. non-polar stationary phase eluted with a polar solvent). Reversed phase HPLC using octadecylsilane (C_{18}, ODS) coatings is probably the most popular HPLC mode at present.

Ion-exchange separations can be performed on suitably modified silica and both strong and weak varieties of cation and anion exchangers are available. Resolution of fully ionized compounds is accomplished by changing the pH and/or the molarity of the buffer. The popularity of ion-pair or soap chromatography, which allows the separation of ionized compounds on normal or reverse phase systems under conditions when they would frontally elute, is increasing. In reversed phase ion-pair chromatography, the mobile phase usually contains a

Table 15.1 Commonly available HPLC packings

Type	*Other names*	*Bonded group*
Unmodified	Silica	None
	Alumina	None
Reversed phase	C_{18}, ODS or octadecyl	$Si\text{–}(C_{17}H_{34})CH_3$
	C_8 or octyl	$Si\text{–}(C_7H_{14})CH_3$
	C_5	$Si\text{–}(C_4H_8)CH_3$
	TMS	$Si\text{–}(CH_3)_3$
Anionic	Amino	$Si\text{–}C_3H_6-NH_2$
Cationic	Sulphonated	$Si\text{–}C_3H_6-SO_3H$
Others	Cyano	$Si\text{–}C_3H_6-CN$
	Phenyl	$Si\text{–}C_6H_5$

low concentration (1–5 mM) of a suitable counter-ion such as octylsulphonate or tetrabutylammonimum ion depending upon whether cationic or anionic species are to be resolved. The retention of the solutes is a function of the counter-ion concentration, pH, the type of stationary phase and the nature of the counter-ion. The exact mechanisms involved vary and are, at present, the subject of some discussion but the subject has been reviewed recently by Hearn [17].

Steric exclusion chromatography on HPLC materials is of little importance with regard to low molecular weight molecules such as amino acids but is of growing importance in the related areas of protein and peptide separations.

Although a single mechanism may appear to explain a chromatographic separation it is more often the case that a number of different interactions occur between the solute, the solvent and the stationary phase during chromatography. Although modern coated silica packings have a high percentage covering it is not always possible to react all the silanol sites. These unreacted sites can play a major part in the chromatographic interactions. Attempts to eliminate such sites, particularly on reversed phase materials involve the 'capping' of the un-reacted silanol groups with trimethylchlorosilane, the smallest monochlorosilane available; such action may produce useful chromatographic changes.

Although all the packing types are available in differing particle sizes from a growing number of manufacturers, a number of authors have published methods for the coating of silicas to give either the common surfaces or customized surfaces [18–20]. The care and maintenance of HPLC columns has been thoroughly covered by Rabel [21].

15.3 DETECTION OF AMINO ACIDS

Few amino acids possess intrinsic physical properties that allow sensitive and/or specific detection with current HPLC detectors. Only phenylalanine, tyrosine and tryptophan possess strong UV chromophores with absorption maxima at 257 nm, 277 nm and 280 nm respectively. At lower wavelengths (< 210 nm) most amino acids absorb strongly and some modern UV detectors are capable of stable operation at these wavelengths. Schuster [22] has reported a procedure for the separation of 20 amino acids plus several vitamins on a $-NH_2$ column eluted with a stepwise water/acetonitrile gradient. Careful control of column temperature was essential to maintain resolution but the analysis time of 27 min for 28 compounds was good. The system could not separate asparagine, glutamine and glycine. Operation at these low wavelengths is difficult since many other compounds, including solvents and their impurities, can also absorb at these wavelengths and cause baseline problems. Because of possible interference the method can only be applied to relatively pure samples and its application to intravenous solutions and beverages, where the sample matrix is relatively simple and the concentration of amino acids is high, was demonstrated.

More specificity and more sensitivity can be achieved by the use of fluorescence spectra to determine compounds separated by HPLC. Only tyrosine and tryptophan exhibit significant emission when excited at 254 nm and the specific determination of these amino acids using such means is discussed below.

15.4 DERIVATIZATION REAGENTS

Fortunately, since the inherent physical properties of amino acids are poor with regard to HPLC detection, the amino group is able to easily undergo derivatization to readily detectable compounds. This derivatization can be performed either pre- or post-column.

The ideal reagent for LC derivatization reactions should fulfil the following criteria:

(i) It should be specific for the group of interest, in this case $-NH_2$.
(ii) It should react to the same extent with all compounds of interest and without artefact formation.
(iii) The products should be capable of detection with improved preferably high sensitivity by standard LC detectors.
(iv) The reaction should be rapid, preferably taking place at room temperature, and in aqueous solution.

To date no reagent for the determination of amino acids satisfies all these conditions entirely.

15.4.1 Ninhydrin

Ninhydrin, the classic reagent for detecting amino acids, has been used for 25 years in automated amino acid analysers. Ruhemann's purple has the well known E_{max} at 570 nm characteristic of primary amines and the yellow adduct formed with secondary amines notably proline and hydroxyproline has E_{max} at 440 nm. Sensitivity, when applied to monitoring small-bore cation exchange columns, can be as low as 100 pmol [23]. Although it is applicable to all amino acids the arduous reaction conditions, i.e. 100°C at 15 min, when generated in post-column flow systems cause significant loss of resolution and sensitivity. Attempts to overcome these disadvantages have included shorter reaction times giving non-steady state conditions or generating higher temperatures for the reaction by pressurizing the reaction coil. Although systems are usually monitored at both 570 nm and 440 nm with modern high stability instruments, detection at 400 nm alone gives similar sensitivity. Such an approach would reduce equipment costs particularly by allowing single channel integration. Unfortunately ninhydrin also reacts readily with ammonia giving high reagent blanks, a rising baseline and other baseline artefacts. Neither pre-column derivatization nor the use of ninhydrin with particulate HPLC packings has been reported.

15.4.2 Dansyl chloride

Dansyl chloride (Dns-Cl, 1-dimethylaminonaphthalene-5-sulphonyl chloride) was introduced as a derivatizing agent for amino groups [24] in 1951. The

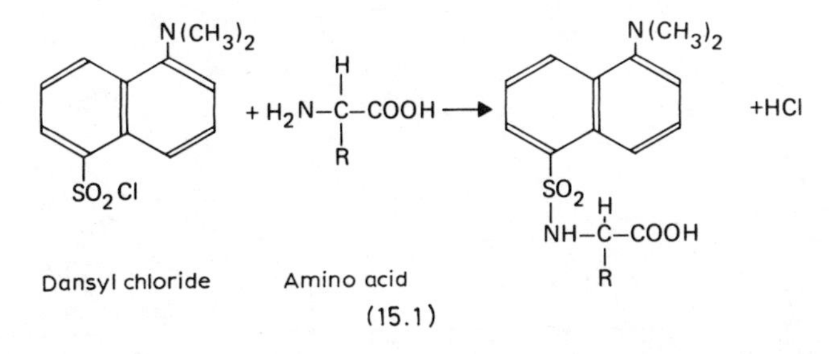

(15.1)

primary reaction (15.1) is shown above. Early studies demonstrated that it can be far from ideal as a LC reagent. Using the classical condition of derivatization (acetone/H_2O 2:1, pH 9 bicarbonate buffer), the percentage conversion of some amino acids, e.g. alanine, was reported to be dependent on the molar ratio of Dns-Cl to amino acid [25] due to the possibility of two competing reactions forming dansylsulphonic acid and dansyl amide. Recently Tapuhi *et al.* [26] have thoroughly investigated the conditions of this reaction and described conditions that are independent of the DnsCl:amino acid ratio over a 1000-fold range coupled with maximum yields. Optimum conditions were acetonitrile:water (1:2), lithium carbonate buffer (40 mmol dm^{-3}, pH 9.5) and a DnsCl:amino acid molar ratio of 1:10, the reaction at room temperature requires 35 min. Most amino acids yield monodansyl derivatives in yields of greater than 90%. Only asparagine for unknown reasons gives a poor yield (78%) unless long reactions times are employed. Lysine, histidine, tyrosine and cystine form didansyl derivatives. Care should be taken with the final products since the derivatives are photosensitive [27]. Dansyl derivatives of amino acids can be measured either by their absorption at 254 nm or by fluorescence (excitation max. = 385 nm, emission max. = 460–495 nm). The related compound dabsyl chloride has also been employed to derivatize amino acids to give reaction products that can be separated on an ODS column and quantitated by monitoring at 425 nm [28].

15.4.3 Fluorescamine

The fluorimetric determination of phenylalanine with ninhydrin in the presence of a dipeptide is a well-known reaction, the chemistry of which was investigated in detail by Samejima, Dairman and Udenfriend [29]. From their understanding a method involving the condensation of ninhydrin and phenylacetaldehyde with primary amines to give highly fluorescent products was evolved [30]. The formation and structure of the product were investigated and defined by Weigele *et al.* [31] who then designed a new highly sensitive reagent, fluorescamine, which would react not only with amino acids (Reaction 15.2) but also peptides [32,33]. This reagent is manufactured and marketed under the name of Fluram by Hoffmann–La Roche. Fluorescamine is hydrolysed by aqueous solvents and solutions are usually prepared in acetone. The optimum pH for the reaction with the primary amine group of amino acids is *ca* 9 whilst for peptides it is *ca* 7.5, and

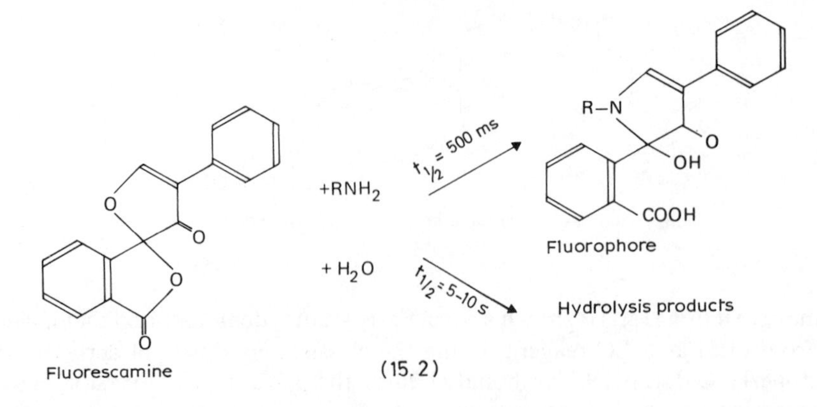

(15.2)

the excitation and emission wavelengths are 390 nm and 475 nm respectively. The reaction is rapid ($t_{1/2} = 0.1$–0.5 s) and unreacted fluorescamine is also hydrolysed ($t_{1/2} = 5$–10 s). Secondary amino acids such as proline unfortunately do not react with fluorescamine but oxidation with *N*-chlorosuccinimide yields reactive amines [34].

15.4.4 *Ortho*-phthalaldehyde (OPA)*

This reagent was first applied to the quantitation of amino acids by Roth [35] although it had previously been used for the analysis of glutathione. OPA reacts with amino acids in the presence of 2-mercaptoethanol (2-ME) in alkaline solution to form highly fluorescent adducts (excitation max. = 340 nm, emission max. = 455 nm). As with fluorescamine the imino acids do not react to give fluorescent products under the standard conditions and cystine and lysine have low fluorescent yields. Imino acids can be oxidized, to compounds that will react with OPA, by sodium hypochlorite or chloramine-T (sodium *N*-chloro-*p*-toluenesulphonamide) [35,36]. *N*-Chlorosuccinimide is not employed since in alkaline solution it gives a high background fluorescence. Lysine response was found to be markedly improved when BRIJ35, a common additive to amino acid analyser buffers, was present [37]. The sensitivity of the reagent towards cystine and cysteine can be improved by performic acid oxidation to cysteic acid [37, 38]. The chemistry of the reaction between primary amines, ME and OPA, has been studied by Simons and Johnson [39]. Using OPA, *t*-butyl mercaptan and *n*-propylamine these workers isolated a crystalline adduct 1-(2-methyl-2-propylthio)-2-propylisoindole. On the basis of this information they postulated the following scheme for the reaction (15.3). Compared to fluorescamine the reaction is relatively slow and dependent on the amino acid of interest requiring up to 5 min at room temperature to procede to completion [40].

The same authors [41] also showed that replacement of 2-ME with ethanethiol gave a fluorophore of improved sensitivity and stability. A finding confirmed by Cronin, Pizzarello and Gandy [42], who showed that both methanethiol and

*o-phthalaldehyde is also spelt o-phthaldialdehyde in many texts.

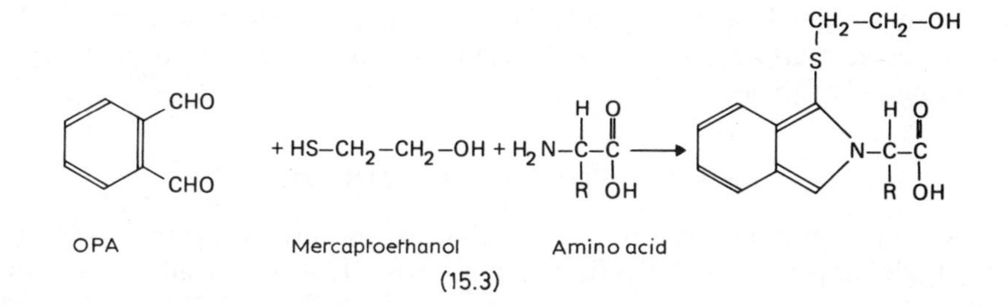

(15.3)

ethanethiol gave an approximately 10% increase in sensitivity and a further increase could be obtained by reaction at 70°C but no change to the fluorescent characteristics was found. This increase in sensitivity was not found by Lee and Drescher [38], who also commented on the stronger odour of ethanethiol. Both of the research groups Cronin, Pizzarello and Gandy [42] and Simons and Johnson [41] found that the stability of the adducts was improved by the use of thiols other than 2-ME. 2-ME adducts had been shown to be stable at room temperature in ethanol or pH 8.5 borate buffer for up to 2.5 h but decomposed 30–50% after 24 h. The use of OPA has been comprehensively reviewed by Lee and Drescher [38]. Since OPA requires no organic solvents and is also substantially cheaper than ninhydrin an amino acid analyser requires only 25–40% (£500) of the reagent costs per annum of a conventional system.

15.5 PRE-COLUMN DERIVATIZATION

The simplest approach to derivatization is to form the derivative prior to injection onto the column. It is economical on reagents and produces little, if any, dilution of the sample since the reaction mixture can be concentrated by evaporation or selective extraction. A wide range of reaction conditions are allowable since the time course of the reaction is not restricted by the instrumentation. All the requirements for successful derivatization, which were listed earlier, still apply with the following additional considerations.

(i) If multiple products are expected they should be capable of satisfactory chromatographic resolution, i.e. the reaction product for each amino acid should still be resolvable.
(ii) The reaction products should be stable.
(iii) The derivatization reagent should either not itself be detectable after the reaction or be easily resolved from peaks of interest.

It may be desirable to include an internal standard in the sample prior to derivatization but such a compound must accurately reflect all changes that occur to the amino acids and be readily resolved.

Pre-column reactions usually cause extensive changes to the chromatographic properties of the sample compounds. The loss of a free $-NH_2$ moiety in the amino acid molecule following derivatization means that traditional cation-exchange separations are no longer applicable. The resultant change in polarity means the carboxylic acid group becomes more significant and different modes of separation must be applied.

15.6 POST-COLUMN TECHNIQUES

Post-column reactions do not require changes to chromatographic systems already developed for underivatized compounds. Other major advantages are that:

(i) Artefact formation is unimportant provided it does not cause unacceptable increases in background or competing non-detectable products.
(ii) The reaction does not have to go to completion.
(iii) The reaction products do not need to be well characterized or capable of chromatographic separation.
(iv) The derivatives do not require to be stable for long periods of time.

Disadvantages can be:

(i) The need to adjust the conditions of the column effluent to give the required reaction conditions can cause a reduction in sensitivity by dilution.
(ii) The loss of chromatographic resolution in the flow reactor.
(iii) Increases in baseline noise due to reagent mixing.
(iv) Increased equipment and reagent costs.

Item (ii) is potentially the most critical since the peak broadening effects of all operations during chromatography are additive in the following manner; where σ can be assessed by calculation or from the observed peak width (4σ).

$$\sigma^2_{\text{total}} = \sigma^2_{\text{injector}} + \sigma^2_{\text{column}} + \sigma^2_{\text{detector}} + \sigma^2_{\text{tubing}} \ldots$$

Excessive mixing in the reactor can spoil a critical separation and it is necessary to optimize all these parameters to minimize re-mixing. The kinetics of the derivatization reaction are important in determining whether a reaction detector can be employed and what type of reactor can be used. Three different types of reactor have been employed.

(i) Tubular or capillary reactors consist of a micro-T-piece at the column exit, into one arm of which is pumped the reagent and the other arm is connected to the detector by a length of micro-bore tubing usually in the form of a spiral of PTFE tubing. The theoretical aspects of tubular reactors following classical flow dynamics have been discussed by Deelder *et al.* [43]. With rapid reaction kinetics such as occur with fluorescamine tubular reactors are simple and relatively efficient. Frei [44] found that for the reaction of fluorescamine with the peptide oxytocin in a 500 μl volume reaction coil and a reaction time of 20s the peak broadening produced in the reactor was an acceptable 7%. Increasing the reaction time to 1 min increased this to approximately 28%, a figure which could be unacceptable for poorly resolved compounds and for such reactions one of the other two reactor designs would be preferable.

(ii) Bed reactors consist of a column packed with inert glass beads and band broadening can be predicted from theory equivalent to that of chromatography columns [43]. For reaction times of 1 min or more such reactors can reduce band broadening by over 500% compared to tubular reactors.

(iii) For even longer reaction times (i.e. in excess of 5 min, as with ninhydrin), it is necessary to use flow-segmented systems. The column eluate is mixed with a segmented stream of reagent prior to colour development. Segmentation may be accomplished by any substance immiscible with the eluant/reagent mixture. Although oils and organic solvents can be employed the most common substances are air or other gases, e.g. N_2 as in the well-known Technicon AutoAnalyzer system. Because of the air-segmentation and the usual need to de-bubble before detection, the dynamics of such detectors are complex. Theoretical treatments have been provided by Snyder [45]. Techniques which do not de-bubble the stream before passage through the flow cell and then electronically correct the disturbance to the signal (i.e. bubble-gating) have been developed and are employed in the Chromspek amino acid analyser.

15.7 GENERAL SEPARATION OF AMINO ACIDS

Both pre- and post-column derivatization techniques have been successfully applied to amino acid analysis with OPA as the most favoured reagent. Attempts at pre-column derivatization with fluorescamine have proved disappointing since two fluorescent products are separated by reversed phase/ion pair chromatography. McHugh *et al.* [46] suggested that this was due to the formation of 5- and 6-membered lactones from the corresponding acid alcohol. The ratio of acid alcohol to lactone was dependent on steric effects and lactone ring size and in the extreme case with 4-aminobutyric acid and peptides, when the carboxylic acid is some distance from the primary amine no lactone can be formed. Derivatization with dansyl chloride has generally proved unsatisfactory for the quantitation of amino acids in complex mixtures and the majority of applications have been concerned with structural studies on proteins and these are discussed in a later section.

Pre-column techniques using OPA were introduced by Hill *et al.* [47] and Lindroth and Mopper [48]. Both groups separated the OPA-amino acids on ODS-columns with fluorescence detection. Lindroth and Mopper [48] found that using an isocratic system separation of 23 amino acid/OPA derivatives in 50 min was possible but they were unable to achieve adequate baseline separation and concentrated on gradient elution systems. Using a convex gradient going from citrate/phosphate pH 7.7 with 20% methanol to 70% methanol they separated 25 amino acid derivatives in 22 min, on a 200 × 4.6 mm C_{18} column (RP-18, 5 μm). Similar conditions were used by Hill *et al.* [47] who produced a separation of 20 amino acid derivatives in 40 min with a nearly linear phosphate/acetonitrile gradient. The order of elution resembled that found with cation-exchangers, but there are many fine differences and Hill *et al.* [47] commented that the order can be varied by changing the modifier used. Neither procedure totally separated threonine and glycine. Lindroth and Mopper demonstrated an absolute sensitivity (excitation = 330 nm, emmision = 418 nm) of 50 fmol but baseline noise and drift were excessive at such sensitivity. More realistic routine analyses were possible in the range 10–100 pmol, with a precision of $\pm 5\%$ provided the reaction time (2 min) before injection was carefully controlled. Hill *et al.* [47] showed that the use of a lower excitation wavelength 229 nm with a 470 nm cut-off emission filter increased sensitivity 6-fold. Both groups showed satisfactory applications of their method to trace amino acid determination in physiological fluids. Gardner and Miller [49] produced essentially similar separations with a single pump two-step solvent system. Recently Jones, Paabo and Stein [50] with a view to using the method for picomole quantitation of peptides following exopeptidase hydrolysis, improved the methodologies significantly. They evaluated three different HPLC column types (ODS, ODS with ion-pairing and CN). An Ultrasphere ODS (250 × 4.6 mm) column and a complex tetrahydrofuran: methanol: aqueous sodium acetate pH 5.9 gradient gave the best separation of 26 amino acid derivatives including threonine and glycine, in 33 min. The response of lysine and hydroxylysine was found to be improved by the addition of sodium dodecyl sulphate to the derivatizing reagent. Maximum sensitivity (50 fmol) was only obtainable with the more advanced of the Schoeffel fluorimeters tested; the FS970 was 10-fold more sensitive than the FS950 (Fig. 15.2).

Recently Joseph and Davies [163] have shown that between 0.4 and 1.0 V OPA/amino acid derivatives are electroactive. The above chromatographic systems can therefore be monitored using an electrochemical detector at 0.4–0.6 V with pmol sensitivity.

The major disadvantages of what is otherwise a simple and sensitive reagent are (i) the poor response of cyst(e)ine, lysine and hydroxylysine, (ii) the total lack of reaction with the imino acids proline and hydroxyproline and (iii) the decay in fluorescence with time of the OPA/amino acid derivatives.

With the development of fluorescent reagents, researchers were quick to use them as replacements for ninhydrin in the post-column reactors of standard amino acid analysers. Samejima, Dairman and Udenfriend [30] used the phenylacetaldehyde/ninhydrin system prior to the introduction of fluorescamine [32] for post-column monitoring and significant improvements in sensitivity were claimed. Georgiadis and Coffey [51] developed a single column system to separate the protein amino acids. Amino acids were resolved on a 500 × 3 mm column of traditional cation-exchange resin using a step-wise citrate gradient and detected with fluorescamine. A very similar system was also used by Voelter and Zech [52], who claimed a 100- to 1000-fold increase in sensitivity compared to using a conventional amino acid analyser. Such claims were tested by Benson and Hare [23] who showed that given the same column dimensions there was little

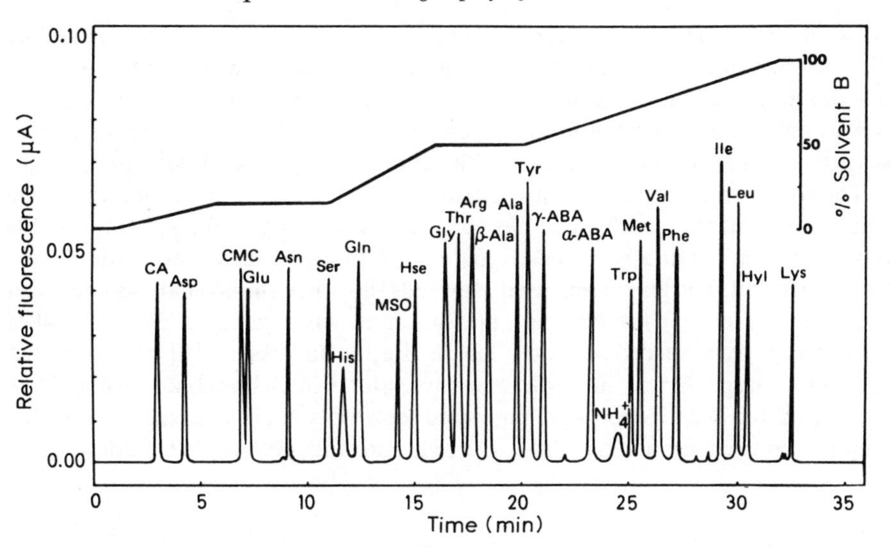

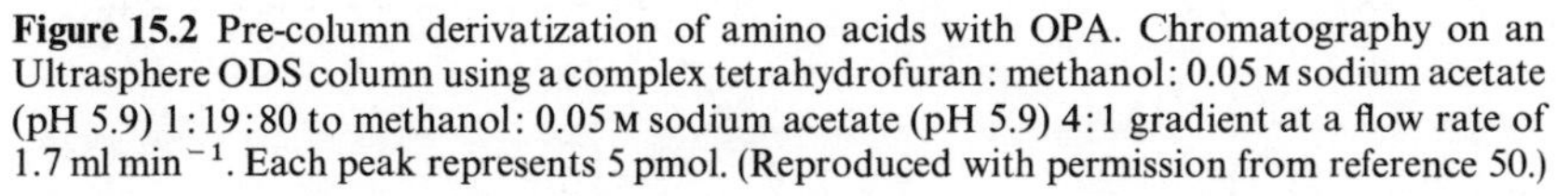
Figure 15.2 Pre-column derivatization of amino acids with OPA. Chromatography on an Ultrasphere ODS column using a complex tetrahydrofuran : methanol : 0.05 M sodium acetate (pH 5.9) 1 : 19 : 80 to methanol : 0.05 M sodium acetate (pH 5.9) 4 : 1 gradient at a flow rate of 1.7 ml min^{-1}. Each peak represents 5 pmol. (Reproduced with permission from reference 50.)

difference in sensitivity between the two detection systems, both ninhydrin and fluorescamine could be employed at pmol levels. Since fluorescamine is almost instantaneously hydrolysed by water, solutions must be prepared in acetone. Therefore when it is mixed with the hot eluate from an amino acid analyser the efficiency of the fluorescamine may be reduced by hydrolysis whilst the boiling of the acetone causes additional baseline noise. The use of acetone means that a solvent-resistant reagent pump is also necessary; to overcome this, isopropanol was introduced as the solvent because it can be pumped using a standard peristaltic pump [52]. Because of these difficulties and the fact that fluorescamine is extremely expensive particularly when used as a post-column reagent, *o*-phthalaldehyde has become the preferred reagent in this detection mode also.

OPA was first applied to monitoring chromatographic systems by Roth and Hampai [36] in 1973 and has been used successfully by Benson and Hare [23] and Cronin, Pizzarello, and Gandy [42] among others. The non-reactivity of the imino acids was overcome by the introduction of chloramine-T (*N*-chloro-*p*-toluenesulphonamide) or sodium hypochlorite to the column eluate prior to addition of the OPA reagent during the time periods when the imino acids elute. Such arrangements are fraught with technical difficulties since they require precise electronic control of solenoid valves. The switching in of another reagent causes baseline disturbances and reduction in operational sensitivity. Even so, amino acid analyser manufacturers were quick to offer a fluorescence option to their systems and at least one dedicated fluorescent system was introduced. A major advantage of OPA is its poor response to ammonia and these systems do not suffer from the problem often found in ninhydrin systems of serious baseline disturbances in the region of basic amino acids; the so-called ammonia plateau due to ammonia in the column buffers. One attraction of the reagent was that as well as detecting amino acids at high sensitivity it could also measure with equal sensitivity other primary amines, e.g. catecholamines [54] and polyamines [55].

The continuous mixing of hypochlorite with the eluate from a cation exchange column was introduced by Thomas [56]. The sensitivity of the system towards proline was substantially increased and an 8-fold improvement towards cystine was also found. The loss in sensitivity towards the other amino acids was between 5% and 25% except for histidine and arginine when gains of 36% and 40% respectively were recorded. Later Ishida, Fujita and Asai [57] also reported continuous hypochlorite infusion to be successful. Both Thomas and Ishida *et al.* confirmed that the reaction gave improved sensitivity at elevated temperatures, a finding reported earlier by Cronin, Pizzarello and Gandy [42]. These three

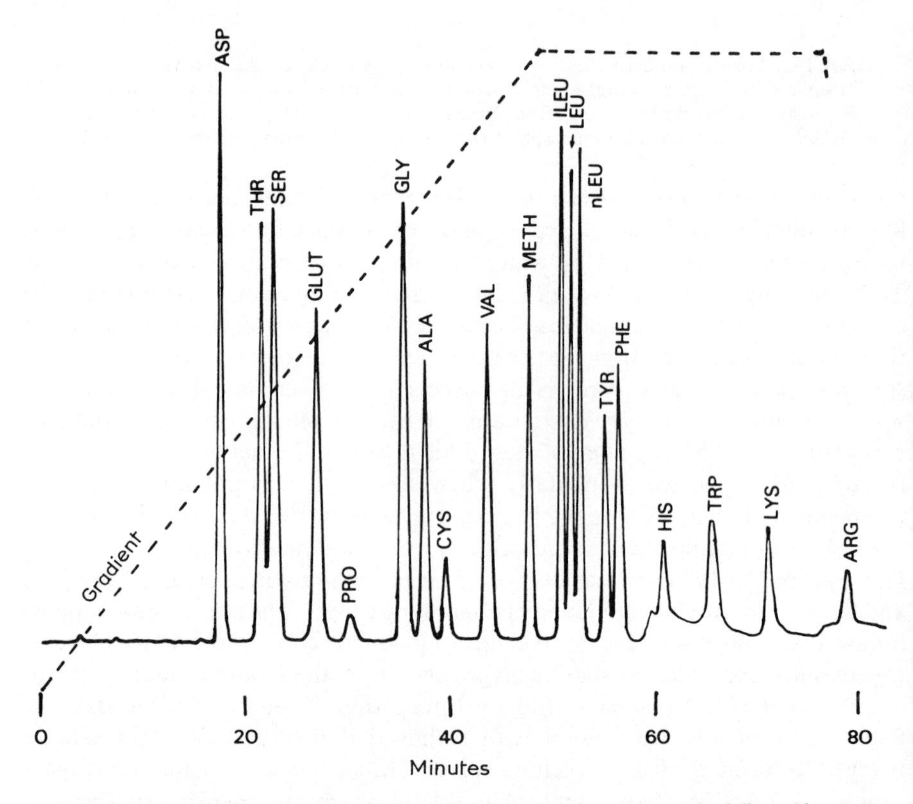

Figure 15.3 Separation of amino acids by cation exchange chromatography followed by post-column derivatization with *o*-phthalaldehyde with continuous reaction with hypochlorite. Conditions: Waters amino acid system 9 μm cation exchange column, 0.4 × 25 cm, gradient as indicated is constant ionic strength (0.2 M Na) and pH gradient from 2.9 to 9.4. All amino acids 1 nmol except proline which is 5 nmol. Fluorimeter set at × 4 minimum; ex: 338 nm; em: 425 nm.

groups unfortunately found differing reaction temperature (35, 55 and 100°C) to be optimal and different reaction times were also employed. A full investigation of these parameters and optimization of the reaction parameters would seem important. Even so sensitivities of 10–50 pmol were reported for these new OPA systems. A chromatogram obtained using such a system in the author's laboratory is shown in Fig. 15.3.

This continuous post-column reaction with hypochlorite suggests that the pre-column system could be improved by including hypochlorite in the derivatization mixture prior to addition of OPA/ME. Whether the oxidation products of proline and hydroxyproline could be satisfactorily chromatographed and resolved from the other amino acid OPA derivatives awaits investigation. If not it may prove possible to improve the chromatographic separation of amino acids by using a micro-particular cation exchanger rather than the 'tried and tested' porous cation exchange resins; either way amino acid analysis can only gain from a meeting of the older techniques with the newer approaches of HPLC.

The picomole levels of sensitivity of the newer techniques place critical demands on both the analyst, his equipment and their reagents if satisfactory trace analysis is to be accomplished. Factors of importance to the chromatographic system in trace analysis have been reviewed [58]. For example good detector design and a high signal to noise ratio for both the detector and flow system are necessary. If trace compounds are to be quantitated in the presence of large amounts of others it is better, if possible, to arrange that a minor peak of interest runs before any major components rather than sits on a tail. Both Hamilton and Nagy [59] and Thomas [56] gave instructions with regard to buffer preparation for trace amino acid analysis and with the reduction of extraneous contaminants particular care is necessary to avoid contamination from fingers. (A single finger print can deposit up to 10 pmol of some amino acids into a sample!)

15.8 CHROMATOGRAPHY OF INDIVIDUAL AMINO ACIDS

In theory it is possible for any amino acid to be measured following resolution from all other amino acids by HPLC but unfortunately many groups of amino acids are difficult to resolve. In particular, methods are required for a number of amino acids of biological importance which occur in physiological fluids in picomole amounts. Therefore in practice the analysis of a single amino acid within a complex matrix usually takes either of two approaches:

(i) A non-specific detection system is coupled with high resolution (specific) chromatography; or

(ii) A less critical chromatography step is used with a detection principle or reagent specific to a characteristic group of the amino acid.

In both systems the chromatographic separation is usually isocratic which gives both simpler equipment requirements and increased sensitivity since the avoidance of buffer changes results in improved baseline noise levels.

15.8.1 Gamma-amino-butyric acid

One of the earliest such applications was the measurement of gamma-amino-butyric acid (GABA) in human cerebrospinal fluid. GABA has been implicated as a possible inhibiting neurotransmitter and it may play a role in some neurological disorders but its concentration (240 pmol ml^{-1}) is below that determinable by an amino acid analyser using ninhydrin. Glaeser and Hare [60] introduced a system which involved separation on a 660 × 9 mm column of cation-exchanger followed by post-column reaction with OPA and fluorescence detection. The

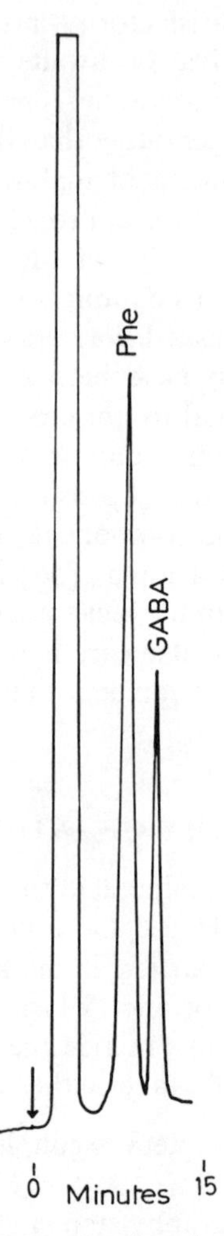

Figure 15.4 Rapid separation of gamma-amino-butyric acid (GABA). Column 100 × 5 mm, 7 μm sulphonated polystyrene resin eluted with 0.1 M sodium citrate buffer pH 5.0 at 0.6 ml min^{-1}. Ambient temperature. Post-column detection with *o*-phthalaldehyde and Aminco Fluoro-colorimeter (Excitation Wratten 7–60, Emission Wratten 2A, Range × 10). Sample 250 pmol each GABA, Phe, Asp*, Gly*, Ala*, Val*, Leu*; *indicates frontal elution.

eluting lithium citrate buffers were adjusted so that no interfering amino acids eluted at the same time (90 min) as GABA. Sensitivity was of the order of 50 pmol. Subsequently the same group modified their system to use a 500 × 2 mm and this increased sensitivity to 1 pmol whilst reducing analysis time to 25 min [61]. By coupling three columns into the system by using switching values the productivity of the system was improved. The new system was successfully applied to human whole blood and amniotic fluid [62]. A similar method for GABA developed in the author's laboratory is shown in Fig. 15.4.

Using the same approach Meek [63] developed assays for taurine, GABA and 5-hydroxytrytophan and used the GABA assay to measure the activity of glutamic acid decarboxylase. For each amino acid, the concentration of aqueous $NaClO_4$ mobile phase was found that specifically eluted the amino acid from a 250 × 2.1 mm column of Aminex A-5 cation exchanger. For example taurine was separated from cysteic acid, phosphoethanolamine and hypotaurine in a chromatographic run of 9 min using 0.1 M $NaClO_4$ pH 2.2 buffer at a flow rate of 0.4 ml min^{-1}.

There is increasing evidence that taurine plays an important part in human metabolism as a possible neurotransmitter or as a membrane stabilizer. This increasing interest has led to a number of HPLC assays. Like Meek [63], Lee and Drescher [38] used a similar post-column technique to quantitate taurine. Wheler and Russell [64], modifying the method of Stuart *et al.* [65], combined pre-column derivatization of taurine with OPA followed by separation on an ODS column. Using an acetonitrile gradient they were able to separate seven acidic amino acids from taurine and claimed a maximum sensitivity of 1 pmol.

15.8.2 Imino acids

4-Chloro-7-nitrobenzofurazan (NBD-Cl) can react with both primary and secondary amines but the reaction with secondary amines such as hydroxyproline and proline occurs one order of magnitude faster than for primary amines (Reaction 15.4). The reaction product exhibits strong light absorption at *ca* 400 nm, the exact extinction coefficient and wavelength being solvent dependent. Fluorescent detection is also possible with emission maxima at *ca* 520 nm. However with increasing solvent polarity the fluorescent quantum efficiency decreases 40–80-fold and for HPLC with aqueous solvents fluorescence detection may not be more sensitive than UV detection.

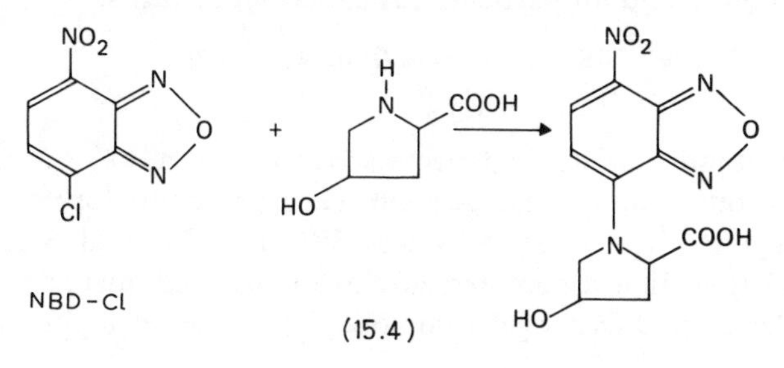

Ahnoff *et al.* [66] used these selective conditions to quantitate hydroxyproline in collagen hydrolysates. Derivatization was performed by incubation with NBD-Cl in methanol solution at 60°C for exactly 3 min. The NBD-hydroxyproline derivative was separated from other amino acid derivatives and the hydroxyl and methoxy side-products by chromatography on a C-18 column (150 × 4.5 mm) using KH_2PO_4 pH 1.9/acetonitrile buffer (80:20) with heptane sulphonate as the ion-pair agent. A similar approach for the analysis of proline should also be possible.

15.8.3 The sulphur amino acids

Although it is possible to determine these amino acids by the usual pre- and post-column techniques involving detection of the amino group, the presence of sulphur allows specific detection of either the –SH group in the case of cysteine or homocysteine or the oxidized sulphur in their disulphides and methionine. Determination of disulphides usually involves reduction to the corresponding sulphydryl compounds followed by determination of the –SH group for which a number of colorimetric, titrimetric and polarographic methods have been developed. Where the sample matrix may contain other thiols it is necessary to employ chromatography to improve specificity.

Thiols are electroactive compounds that can be quantitated at high sensitivity when the eluate from the HPLC column is passed through an electrochemical detector (ECD). Additional selectivity can be obtained by the choice of suitable electrode material and working potential. A number of different ECD designs [67, 68], employing various electrode configurations and electrode materials have been published and marketed. Electrode materials such as glassy carbon, carbon paste and noble metals have all been employed. Although it is possible to determine thiols directly at a high potential (< 1.0 V) on a glassy carbon electrode [69], Rabenstein and Saetre [70] introduced an ECD based on the characteristic reaction (15.5) which in their detector occurred on the surface of a mercury-pool. These authors successfully applied their detector to the measurement of the naturally occurring sulphur amino acids, cysteine and homocysteine [71, 72]. They also applied the detector to the determination of the unnatural amino acid D-penicillamine (beta, beta-dimethylcysteine, 3-mercaptovaline) in physiological fluids [73]. D-Penicillamine is an important drug in the treatment of Wilson's disease, cystinuria, and, in particular, rheumatoid arthritis.

$$Hg + 2RSH \longrightarrow Hg(RS)_2 + 2H^+ + 2e \tag{15.5}$$

In these methods, amino acids were separated on a 300 × 2 mm column of Zipax SCX strong cation-exchange resin. Their procedure for penicillamine has been modified by Bergstrom, Kay and Wagner [74] and Kucharczyk and Shahinian [75] so that a commercially available gold–mercury cell could be employed. Perrett and Drury [69] employed this same cell to quantitate another thiol-amino acid drug captopril, following reversed phase chromatography. In all these systems the high specificity of the detector allows the use of simple and rapid chromatographic separations yet produces high (1 pmol) sensitivity. The use of a gold surface alone to detect penicillamine has been reported [76] and a more complex separation is shown in Fig. 15.5.

Recently two pre-column derivatization procedures have been published. Reed *et al.* [77] overcame the problem of oxidation and/or disulphide exchange by initially forming the *S*-carboxymethyl derivative of the free-thiol by reaction with iodoacetic acid. The free amino group was then coupled with Sanger's reagent (1-fluoro-2,4-dinitrobenzene) which absorbs strongly at 365 nm. The resulting derivatives were separated on a 3-aminopropylsilane column. Difficulties were encountered in finding a suitable column for the separation of the N-DNP compounds which since the procedure is not specific, could include every amino acid. Reeve, Kuhlenkamp, and Kaplowitz [78] used the classic Ellman's reagent

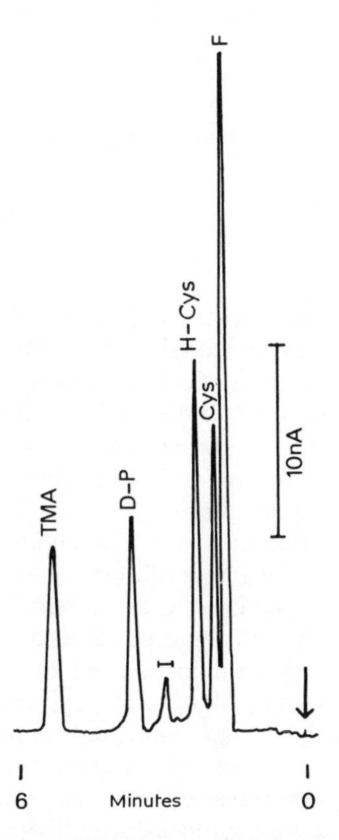

Figure 15.5 Separation of thiol-amino acids. Column 150 × 5 mm, 3 μm ODS–Hypersil eluted with 0.1 M KH_2PO_4 pH 2.0/methanol (94:6) at 1.0 ml min^{-1}. Electrochemical detector with gold electrode + 0.80 V *v* Ag/AgCl. Sample: cysteine, homocysteine, D-penicillamine (D–P) and thiomalic acid (TMA), 30 pmol each; I = impurity F = frontal peak.

to form a mixed disulphide with cysteine and then separated the 5-thio-2-nitrobenzoic acid–cysteine mixed disulphide by reverse-phase HPLC. Nanomole sensitivity was achieved by monitoring the derivatives at 280 nm. Ellman's reagent has also been employed as a post-column reagent to quantitate thiols [79].

A more specific and sensitive reaction (15.6) involves the derivatization of the –SH group with bimanes [80]. This new class of reagents reacts selectively with thiols producing highly fluorescent compounds (excitation max. = 359 nm, emission max. = 388–520 nm). Newton, Dorian and Fahey [81] employed monobromobimane (mBBr) to derivatize free thiols not only in standard solution but also in crude blood cell extracts. Using a 5 μm C-18 column (150 × 4.6 mm) they separated 19 thiols including cysteine, homocysteine, glutathione and cysteamine in less than 30 min, using a combination of isocratic and gradient elution. Using a filter fluorimeter it was possible to quantify as little as 2 pmol of cysteine.

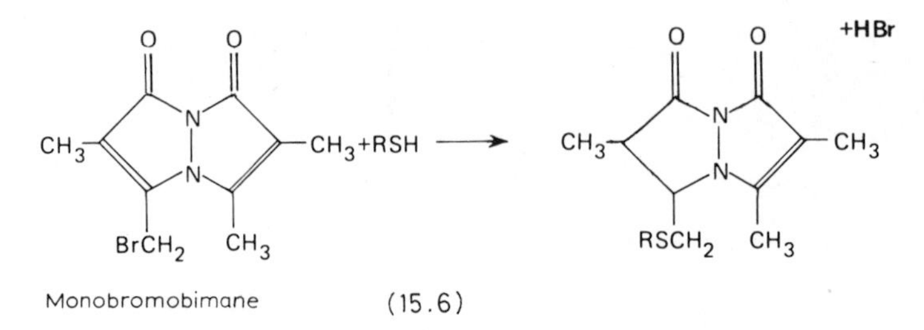

(15.6)

Another pre-column fluorescent reagent for thiols, dansylaziridine (5-dimethylaminonaphthalene-l-sulphonylaziridine) was introduced by Lankmayr *et al.* [82, 83] for the quantitation of cysteine and D-penicillamine. The reaction yield is maximal at pH 8.2 with a 3-fold molar reagent excess and reaction time of 1 h at 60°C. The derivatives can be separated by reversed phase HPLC and detected either by UV absorption at 254 or 345 nm or with much increased sensitivity by fluorescence monitoring (excitation max. = 345 nm, emission max. = 540 nm). The detection limit was 12 pmol for D-penicillamine.

The methods discussed so far are specific for free thiols, fewer methods capable of detecting both oxidized and reduced sulphur amino acids are available. Disulphides can be electrolytically reduced before analysis of the free thiol [73]. Eggli and Asper [84] included a reduction cell into the flow stream prior to a gold/mercury cell. Fowler and Robins [85] employed the bleaching of a black solution of H_2PtCl_2–KI by sulphur compounds to monitor the eluate from an amino acid analyser. Recently Werkhoven-Goewie *et al.* [86] modified this approach to produce a positive fluorescence yield rather than a negative colour response. Divalent palladium can quench the fluorescence of suitable ligands such as calcein upon complex formation. If a reactive sulphur group is mixed with this complex the palladium is displaced and the calcein fluoresces. Using this chemistry in a post-column detector it was possible to quantitate sulphur amino acids following chromatography on a cation-exchange column. A significant increase (40%) in sensitivity followed the incorporation of Zn(II) at a final concentration of 10 μmol dm^{-3} into either the LC eluant or the reagent stream. Absolute sensitivity towards thiols varied with the structure of thiol, typical sensitivities found were cysteine 3 pmol and methionine 20 pmol injected.

15.8.4 Phenylalanine, tyrosine and tryptophan

These amino acids are linked together in terms of their chromatographic characteristics and specific methodologies have been developed since the plasma levels of the aromatic amino acids are abnormal in various metabolic disorders whilst tryptophan levels can be changed in a variety of neurological diseases. These three amino acids are readily separated, at acid pH, on either cation exchange columns or reversed phase columns. Although it is possible to detect them by UV absorption following chromatography and simple systems have been developed, in general, levels of sensitivity are too low for most applications. Specificity is also limited since samples of biological origin contain many other compounds with similar spectral characteristics. In order to resolve tryptophan from a large number of other compounds in plasma, which absorb at 254 nm Krstulovic *et al.* [88] had to employ a methanol gradient to achieve satisfactory resolution whereas with a fluoresence monitor only a single peak corresponding to tryptophan was observed. Because of the limitations of UV detection, Necker, Delisi, and Wyatt [87] resorted to a post-column reaction to gain the necessary sensitivity and selectivity for phenylalanine but fluorescence monitoring at 280 nm (excitation) and 330 nm (emission) gave sufficient sensitivity and selectivity to measure tyrosine and tryptophan in acid extracts of plasma. With isocratic elution the limits of detection were 5 and 1 pmol for tyrosine and tryptophan respectively.

Analytical methods for tryptophan have usually been designed to quantitate not only the parent amino acid but also its important metabolites such as serotonin and kynurenine in plasma, urine and tissue extracts. Satisfactory resolution is easily obtained by either cation-exchange or reversed phase HPLC, followed by either UV [89], native fluorescence [90] or EC detection [91]. Both Morita *et al.* [90] and Koch and Kissinger [91] found it necessary to employ enrichment techniques to concentrate the tryptophan metabolites prior to injection into the HPLC. Selectivity with the ECD is gained by the choice of a suitable working potential for the glassy carbon electrode, e.g. for hydroxyindoles the potential should be + 0.6 V and for indoles it should be + 1.0 V.

15.8.5 Histidine and the methylhistidines

Interest in the analysis of this group of amino acids is particularly concerned with the rapid estimation of 3-methylhistidine, which is considered to be an index of the rate of myofibrillar protein catabolism. The chromatographic properties of histidine, 1-methylhistidine and 3-methylhistidine are similar and resolution has often proved difficult on conventional amino acid analysers. Their separation on cation exchange columns with *o*-phthalaldehyde as a post-column reagent has been employed by a number of workers [92, 93]. Friedman, Smith and Hancock [94] modified this approach by resolving the three amino acids on a C_{18} column using an aqueous solution of the ion-pairing agent sodium hexanesulphonate (5 mmol dm^{-3}) prior to fluorescent detection.

Nakamura and Pisano [95] had described conditions under which fluorescamine reacts specifically with some imidazoles including histidine and 3-methylhistidine but not 1-methylhistidine. Wassner, Schlitzer and Li [96] reacted biological extracts with fluorescamine at 80°C in 0.67 M HCl to give strongly fluorescent products, which are probably structurally different from the characteristic derivatives formed at pH 6–9 (see above). Histidinol was included as an internal standard. The fluorescent products were separated by reversed phase chromatography using an acetonitrile gradient. Urine was shown to contain a number of other fluorescent species which frontally eluted while histidine and 3-methylhistidine eluted at 10 and 14 min respectively. Sensitivity exceeded 1 pmol injected onto the column which is some 100-fold better than ninhydrin. The specificity of the reaction has recently been improved by the incorporation of formaldehyde into the reaction to suppress the fluorescence derived from histidine and histamine and an accurate, non-chromatographic, assay for 3-methylhistidine developed [97].

15.9 RESOLUTION OF DL-AMINO ACIDS

The resolution of amino acid enantiomers has both theoretical interest and commercial importance. In principle amino acid enantiomers can be separated by various chromatographic methods including HPLC either by (i) direct resolution using chiral metal chelates, which can be loaded onto an ion-exchange resin or added to the mobile phase or (ii) after conversion to diastereoisomeric derivatives.

During the last decade much progress has been made in the direct resolution (i.e. without derivatization) of enantiomers by LC and the subject now commands a vast and complex literature that has recently been the subject of three excellent reviews [98–100]. In outline, for stereoselective resolution it is necessary for an optically active molecule to interact with three fixed points on a stationary phase simultaneously. In Fig. 15.6 this is represented diagrammatically and shows only the D-form being retained whereas the spatial orientation of the L-form prevents retention. All forms of molecular interaction, in particular hydrogen bonding, contribute to the three points of interaction. Various stationary phases have been designed which will allow such stereo-specific interactions to occur.

Crown ethers can interact strongly and specifically with amines and amino acids [101]. These ethers are generally bi-naphthyl derivatives which can be grafted onto silica or styrene polymers and the efficiency of the stereo-resolution

(a)

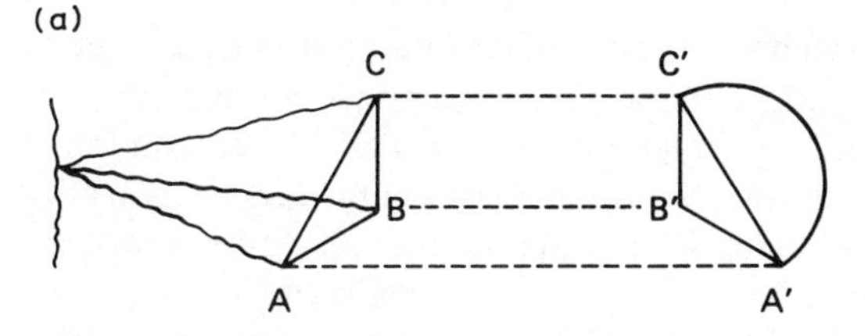

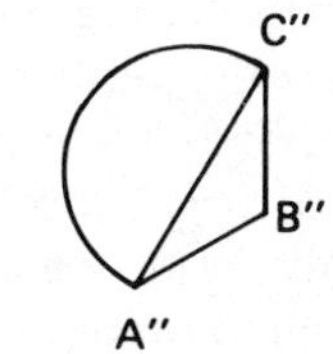

Bonded chiral support

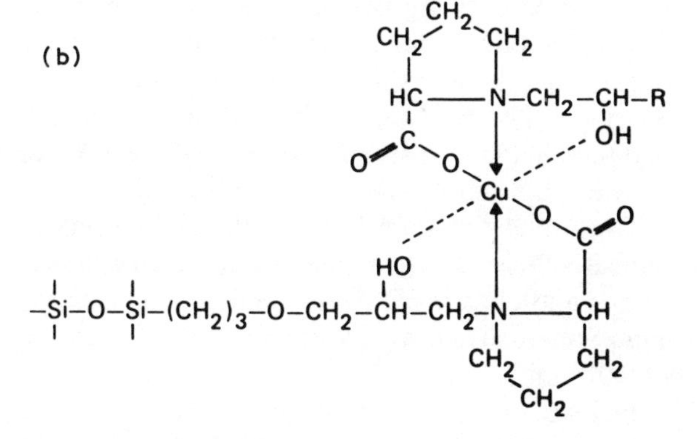

Figure 15.6 (a) Diagrammatic illustration of the 3-point rule showing only the binding of the D-chiral substrate is allowable; (b) proposed structure of the bidentate complex chemically bonded to the silica gel.

is governed by the orientation of the dinaphthyl units since the amino acids fit into the crown ether ring.

If an asymmetric sorbent forms a multidentate complex with a transition metal, which can also complex the solute, a process of ligand exchange may be able to occur. This technique was introduced simultaneously by Humbel, Vonderschmitt and Bernauer [102], Snyder, Angelici and Meck [103] and Rogozhin and Davankov [104]. Although many amino acids have been tested as the chiral graft when bonded onto acrylamide resins with a variety of metal ions, e.g. Co(II), Ni(II), Davankov *et al.*'s formulation [105] of an optically-pure ring-structured amino acid (proline or hydroxyproline) and Cu(II) is still the most efficient.

The early separations [105] required some 9 h to achieve the resolution of D-proline on an asymmetric resin containing L-hydroxyproline–Cu(II) fixed complexes. In recent studies the chromatographic times have been substantially reduced and the resolution of most amino acid pairs can be accomplished in less than one hour [106]. Ammonia is the most widely used eluant and the chromatographic efficiency is a function of the degree of copper saturation and the temperature of the column; lower temperatures increase the efficiency of the resolution. The quality of the final separation is directly dependent on the optical purity of the graft amino acid. Even with optimized chromatographic parameters the runs were still relatively slow and other groups have prepared more efficient packings based on HPLC technology. Lefebvre, Audebert and Quivoron [107] used very hydrophilic packings composed of a porous gel formed by pearl co-polymerization of acrylamide and methylene bisacrylamide. The diameter of the resin so produced was 10–20 μm and the L-amino acid (usually proline) was then grafted through reaction with formaldehyde. Using a 30 × 0.5 cm column of this material optimum flow rates were high (4–5 ml min^{-1}) and plate counts up to $N = 3500$ were obtained. With water as the eluant excellent resolution of test mixtures comprising DL-phenylalanine and DL-tryptophan were obtained.

Further improvements in efficiency have been obtained by bonding chiral phases directly onto silica gels as with other HPLC packings. Gubitz, Jellenz and Santi [108] bonded L-proline to silica via 3-glycidoxypropyl-trimethyloxysilane and then loaded the support with Cu(II) ions. In another publication [109] these workers optimized the factors affecting selectivity and the efficiency of the separation, e.g. pH, molarity of the mobile phase. Preliminary chromatograms indicated that excellent and rapid separations could be achieved. Foucault, Caude and Oliveros [110] bonded directly a chiral phase via C–Si linkages onto 7 μm Spherosil and were able to completely resolve DL-tryptophan with a HETP of 0.07 mm.

Instead of using chiral species bonded to the packing it is possible to incorporate the chiral species into the mobile phase. Lindner *et al.* [111] resolved

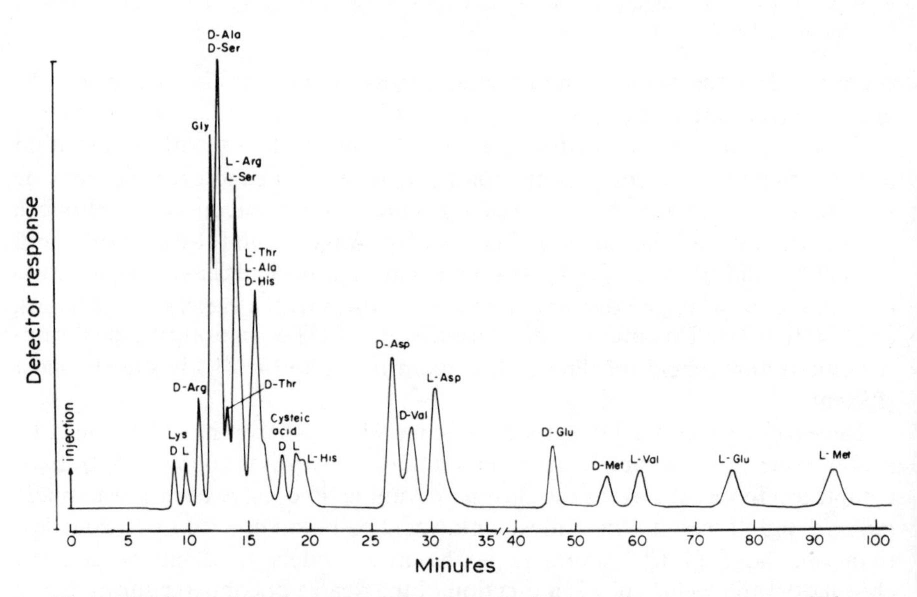

Figure 15.7 Separation of a mixture of amino acids into enantiomers on a reversed phase column using the chiral additive *N,N*-di-*n*-propyl-L-alanine and cupric acetate in the eluant. (Reproduced with permission from reference 112.)

with high efficiency the D- and L-isomers of dansyl amino acids. This separation was achieved by the addition of the counterion L-2 alkyl-octylethylenetriamine plus Zn(II) or Cd(III) to the mobile phase used to elute a C_8 column. Recently Weinstein, Engel and Hare [112] used a new chiral mobile phase consisting of *N*,*N*-di-*n*-propyl-L-alanine and cupric acetate and were able to resolve the D- and L-enantiomers of all the common amino acids. Amino acid mixtures were first fractionated into three groups on a cation exchange column and each group was then resolved on a reversed phase C_{18} column using an acetonitrile/water eluate containing the chiral compounds. High sensitivity was achieved by the use of post-column derivatization with *o*-phthalaldehyde and fluorimetric detection.

The formation and chromatographic separation of diastereoisomeric derivatives can also be used to resolve DL-amino acids but unless mild derivatization conditions are employed racemization can occur. Non-polar amino acid esters have been resolved on silica columns [113, 114]. The formation of diastereoisomeric dipeptides which can be readily resolved by reversed phase HPLC has also been successfully employed. The procedure of Manning and Moore [115] by which a neutral or acidic amino acid was converted into a dipeptide by coupling with *N*-carboxy-L-leucine anhydride (L-Leu-NCA) and an aromatic or basic amino acid with *N*-carboxy-L-glutamic acid anhydride (L-Glu-NCA), was originally intended for use with amino acid analysers but although a simple procedure its use was restricted by difficulties with the synthesis and storage of the reagents. Mitchell *et al.* [116] replaced these unstable reagents with *tert*-butoxycarbonyl-L-leucine *N*-hydroxysuccinimide ester (Boc-Leu-OSu) for forming the leucyl-dipeptides. Takaya, Kishida and Sakakibara [117] used this modification to resolve amino acids by HPLC and showed that savings in analytical time could be achieved. Additionally increased sensitivity was obtained by monitoring at 210 nm rather than using ninhydrin. Applying their procedure to commercial amino acid samples they showed that the optical purity could be between 98.92 and 99.99%. Nachtmann [118] used a similar approach to quantify the levels of the toxic isomer L-penicillamine in commercial preparations of the drug D-penicillamine. Penicillamine like cysteine and other thiols would complex the Cu(II) ions used in the ligand exchange systems.

Another novel reagent, 2,3,4,6-tetra-*O*-acetyl-β-D-glucopyranosyl isothiocyanate (GITC) was described by Nimura, Ogura and Kinoshita [119]. GITC reacts readily with enantiomeric amino acids at room temperature to form derivatives which are readily resolved on ODS columns with aqueous methanol mobile phases; DL-mixtures of at least five amino acids could be separated simultaneously in 40 min.

15.9.1 Dansyl amino acids

Because of the formation of side products methods for the separation of dansyl amino acids have been most successful where the derivatives are in a simple matrix such as occur in the *N*-terminal analysis of proteins. Numerous systems for the identification of dansyl amino acids by TLC have been published but the need for quantitative analysis is best met by LC. Bayer *et al.* [120] were among the first to attempt the transfer of TLC methods to HPLC. They employed normal phase separation on silica gel columns and achieved the separation of the 20 common protein amino acids by the use of two columns in 30 min. The comparative efficiencies of three different types of ODS packings were investigated by Hsu and Currie [121] and isocratic separation of most but not all the derivatives was achieved on a Partisil PAC column. With fluorescent detection a sensitivity of 100 fmol was attained. Wilkinson [122] reported a single column separation on an ODS phase using an acetonitrile/aqueous phosphate buffer gradient. Maximum retention for the dansyl amino acids occurred in the pH range 3–4 and decreased above the pK of the dimethyl amino group (pK = 4.07) but retention times were not entirely predictable from polarity considerations. Although an excellent separation was achieved the use of aqueous buffers severely quenched the fluorescent response and the quantum yield was only one tenth that obtained in organic solvents [123]. Wilkinson also used UV absorbance at 250 nm and found a detection limit of 500 pmol on a 5 μm ODS column.

Recently Kobayashi and Imai [124] have demonstrated an exciting new approach to the detection of fluorescent species separated by HPLC and illustrated this method by the detection of dansyl amino acids at femtomole levels. Instead of the normal excitation of the fluorophores by an external light source they employed a chemiluminescence reaction to generate the necessary photons within a post-column flow reactor. The chemiluminescence was generated by the reaction of bis (2,4,6-trichlorophenyl)oxalate with hydrogen peroxide (Reaction 15.7). Such a detection system reduces problems associated with light source instability and background stray light at the photomultiplier. The intensity and half-life of the chemiluminescence reaction was found to be solvent-sensitive with non-polar solvents being significantly better and ethyl acetate was the chosen solvent. Monitoring the reaction system at 510 nm (a

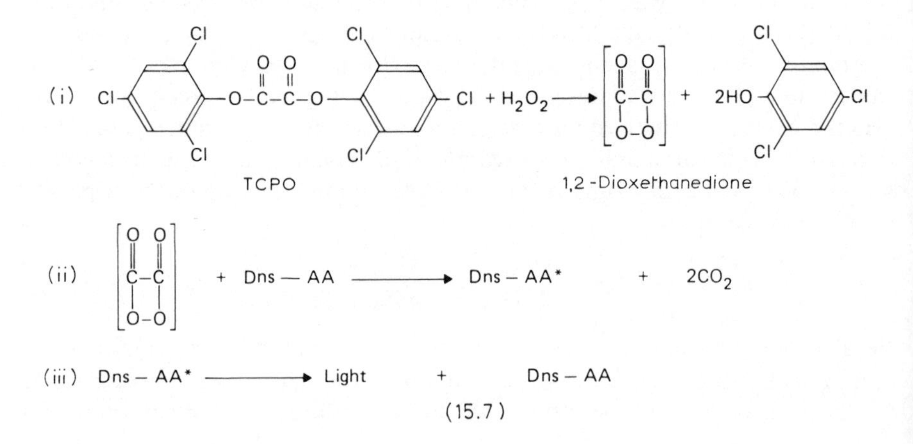

(15.7)

Schoeffel FS970 without a lamp was used) as little as 50 fmol of representative dansyl amino acids were detected following separation on a Bondapak C_{18} column. Calculation suggests that 10 ng of protein could now be sufficient for N-terminal analysis. A later paper by the same authors [125] increased the sensitivity to 25 fmol when chemiluminescence detection was applied to the determination of catecholamines.

15.9.2 Phenylthiohydantoin-amino acids

Most workers involved in the determination of the amino acid sequence of peptides and proteins now employ the classical Edman degradation procedure to sequentially cleave the N-terminal amino acids. This involves the reaction of the N-terminal amino acid with phenylisothiocyanate under alkaline conditions (pH 8–9) to form a phenylthiocarbamyl derivative (Reaction 15.8). This PTC-peptide is then treated with anhydrous trifluoroacetic acid which causes cyclization and the release of the N-terminal amino acid as an anilinothiazolinone derivative. The

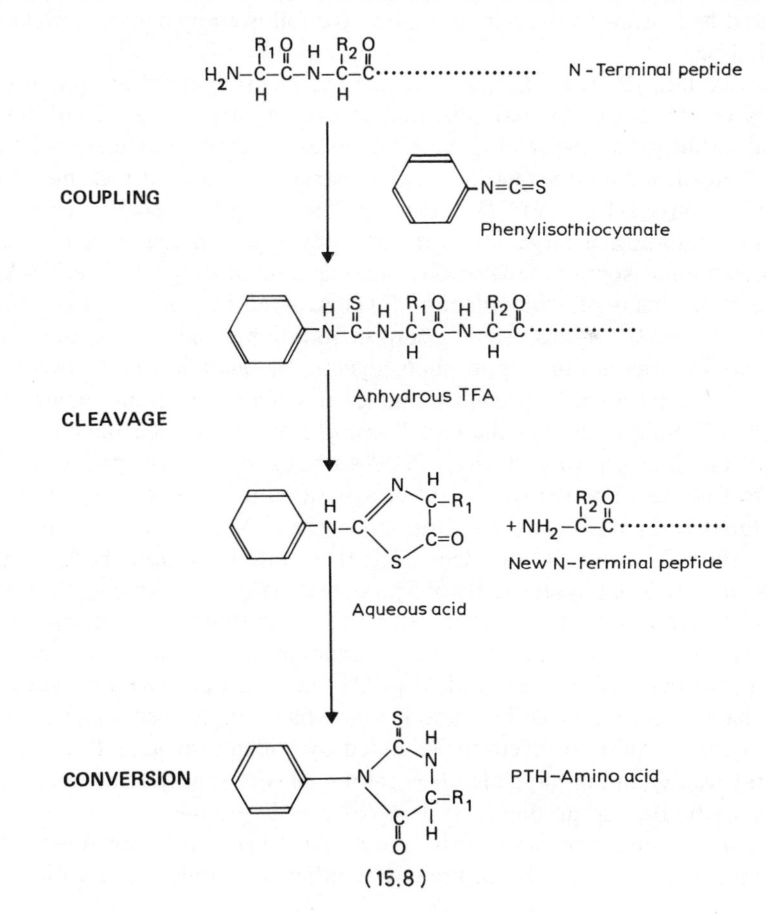

(15.8)

next N-terminal amino acid of the protein is now available for the next coupling cycle. Reaction of the anilinothiazolinone with aqueous acid produces a phenylthiohydantoin-amino acid (PTH-AA).

This cycle can now be performed automatically using commercial equipment. These automatic sequencers have enhanced the speed and predictability of accomplishing the Edman degradation whilst lowering the amounts of protein initially required to less than 100 pmol. The analytical requirement is therefore to identify each N-terminal amino acid with high sensitivity (< = 1 nmol) with good and reliable resolution and accurate quantitation. The PTH-AA separation should be at least as fast as the sequencer operates. Because of partial and side reactions as the number of cycles increases it becomes increasingly difficult to uniquely identify the PTH-AA from the N-terminal of interest among the large number of PTH-AAs found in each sample. This situation generally starts to develop after about 15 cycles and so causes the need for excellent chromatographic and reproducible resolution. The 20 PTH-AAs have been commonly identified using gas–liquid chromatography, thin-layer chromatography and back-acid hydrolysis to the parent amino acid followed by conventional amino acid analysis.

Since the late 1970s HPLC has become the most accepted approach to the analysis of PTH-AAs as derivatization is not required, lack of volatility or thermal stability is unimportant, sensitivity is satisfactory, total analysis time can be less then 20 min and the method is non-destructive. Following the initial work [126–128] on the HPLC of PTH-AAs in 1973, which utilized both reversed phase and silica columns, a large literature has developed, much of it devoted to perfecting simple isocratic separations capable of separating all the PTH-AAs in a single run. This is difficult since the PTH-AAs fall into at least two different polarity groups, the hydrophobic group including proline, the leucines, valine, phenylalanine, methionine, tryptophan, glycine, and alanine and the hydrophilic group including tyrosine, lysine, threonine, and the acidic amino acids.

In 1976 Zimmerman, Appella and Pisano [129] introduced the use of ODS columns for the separation of the PTH-AAs but soon improved this approach [130] so that all 20 derivatives could be separated in a single 20 min run. They were eluted from a 25 × 0.46 cm Dupont Zorbax ODS column maintained at 62°C with an acetonitrile gradient in 0.01 M sodium acetate buffer pH 4.5. Monitoring at 254 nm a sensitivity of 5 pmol was claimed. Following these initial successes a large number of other workers have published modifications and improvements to the system. The majority have employed octadecylsilane (ODS) columns but octyl [131], phenylalkyl [132] and cyanopropyl [133] stationary phases have also been used. The mobile phase has usually been aqueous acetate buffers with methanol or acetonitrile added by phosphoric acid, trifluoroacetic acid and triethylamine have also bee employed. Elution systems have ranged from isocratic or step gradients to complex profile gradients.

The elution behaviour of the basic and acidic PTH-AAs is dependent on pH, buffer concentration, and the nature of the buffering anion. The retention of the

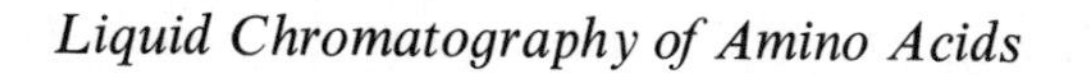

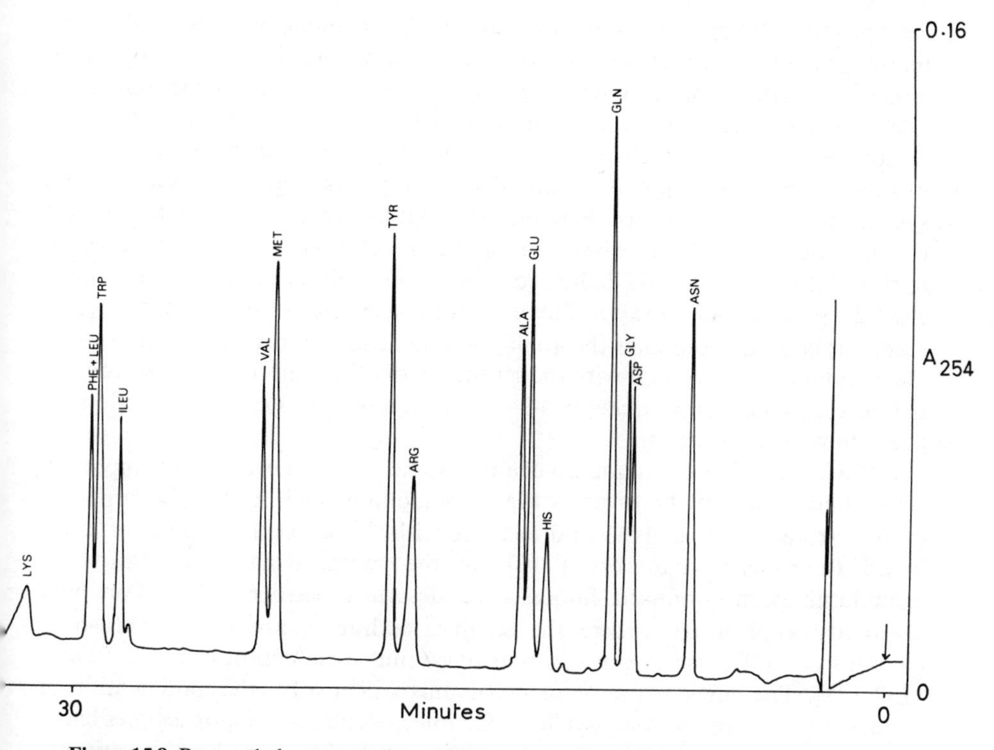

Figure 15.8 Reversed phase separation of PTH-amino acids. Column 5 μm Ultrasphere ODS eluted with 10–40% acetonitrile/tetrahydrofuran in 10 mM triethylamine phosphate pH 3.0. Flow rate 1.5 ml min^{-1}. Sample approx. 1 nmol each PTH-AA. (Courtesy Dr. P.J. Lowry, Pituitary Hormone Lab., St. Bartholomew's Hospital, London.)

acidic compounds depends on their degree of ionization; for example PTH-Glu and PTH-Asp co-elute at pH 4.65 but are well separated at pH 4.50. Buffer concentration varies the elution order of the basic derivatives, in particular changing from 7.8 mM to 9.6 mM acetate can cause PTH-Arg and PTH-His to elute 10% faster. Annan [134] has used diethylenetriamine to control ionization and dichloroacetic acid to ion-pair the basic derivatives during chromatography on Bondapak C_{18} columns. Methylthiohydantoin-amino acids can be generated in some sequencing procedures and their analysis by HPLC has been reported [135].

The progress in the field of separating phenylthiohydantoin-amino acids by HPLC means that the chromatography with its immediate quantitation, reproducibility and good resolution is now faster than the sequencer.

15.10 CONCLUSIONS

The automatic amino acid analyser with its cation-exchange columns and ninhydrin detection has for twenty five years been a valuable workhorse in many laboratories dealing with protein chemistry and clinical analysis. Technically many commercial amino acid analysers now lag behind the sophistication and level of automation found in the equipment evolved for HPLC because this larger market allows higher spending on development. Because of higher sales HPLC hardware is also relatively cheaper and with the current awareness by the equipment manufacturers of the possibilities of pre- and post-column reaction systems we can expect to see them introduce amino acid analysis systems based on their equipment. The flexibility and sensitivity of this approach should appeal to those who do not want a dedicated amino acid analyser and the cheaper price should appeal to all workers. There is still a need for a better derivatization reaction; one that combines the ability of ninhydrin to react with both primary and secondary amino acids with the sensitivity of OPA and the speed of reaction of fluorescamine and produces products that can be readily separated following pre-column derivatization.

HPLC, itself, will also undergo changes some of which are already apparent. In particular sensitivity towards ideal compounds such as the fluorophores derived from amino acids should increase with the wider availability of laser-based fluorimetric monitors [136] or the more general application of chemiluminescence-induced fluorescence described earlier [124]. For high sensitivity applications where the sample loading is not great micro-bore columns can offer useful gains in sensitivity plus a reduction in reagent costs [137, 138]. The efficient use of such columns will involve the modification of much existing equipment so that flowcells, injectors and the response times of the electronics, etc., do not lower the resolution attainable. Very high sensitivities have been obtained by the coupling of micro-bore columns with laser-induced fluorescence detection [139]. The coupling of micro-bore columns with their low solvent flow rates should allow the efficient coupling to other forms of detector such as the universal flame ionization detector and the direct linkage with a mass spectrometer [139]. The LC–MS may then rival the GC–MS in analytical power. Over the next few years it should be possible for amino acid analysis using HPLC technologies to operate routinely at picomole or even sub-picomole levels but instrument constraints will be replaced by fresh constraints, in particular those resulting from chemical and biological contamination, and this could prove limiting to picomolar amino acid analysis.

15.11 ADDENDUM

Since the body of this chapter was completed, the growth of HPLC has continued unabated and many publications on the analysis of amino acids by the technique have appeared. This note outlines some of the more significant publications and developments.

Recent advances in HPLC have been summarized in a major review in *Analytical Chemistry* [141] and a comprehensive treatise on current techniques edited by Simpson [142] appeared at the end of 1982. The continuing growth of interest means that a rapidly increasing number of commercial instruments and, in particular, column packing materials are now available. Many published methods do not readily transfer from one packing to another of nominally the same material; a number of authors have therefore attempted to rationalize this and publish comparative data [143, 144] to aid the analyst. The performance of column types, such as microbore, wide-bore and standard, differing with regard to efficiency and speed has been critically evaluated by McCoy and Pauls [145]. The related topic of the advantages of 3 μm over 5 μm packings and the role of flow rate in fast analyses has also occupied some workers [146, 147]. All these studies have concentrated on reversed phase packings and columns indicating the continuing dominance of this separation mode in LC.

There has been increased interest in the liquid chromatography of amino acids in applied situations; this has been spurred on by protein chemists wishing for amino acid analysis on nanograms of isolated material. For such analyses *o*-phthalaldehyde plus 2-mercaptoethanol (OPA/ME) with its high sensitivity towards amino acids but poor reactivity with ammonia, is increasingly used. The problems of contamination when working at picomolar concentrations have ready been referred to but now Bohlen and Schroeder [148] have reviewed the problems and presented guide-lines for the successful routine determination of amino acid compositions on 100 pmol of peptide. In order to increase the sensitivity of analysis, a micro-bore column has been coupled to an optimized post-column reactor by Kucera and Umagat [149]. In order to develop their system they also investigated the kinetics of the reaction of the OPA/ME reagent with amino acids using their micro-post-column reactor as a flow injection system. Although the reaction was complete within 9 s, like earlier authors they found that the fluorescence then decayed whereas with ethanethiol the fluorescence continued to develop for a number of minutes. Investigation of other thiols showed that 3-mercaptopropionic acid gave a stable fluorophore after a 9 s reaction, this finding would appear to be of greater importance to those using pre-column derivatization since, to avoid the errors caused by fluorescence decay, very accurate timing has proved essential.

In order to overcome the problem of fluorescence decay when using pre-column derivatization with OPA/ME, an automatic derivatization device has been described [150] and a major manufacturer (Waters) has found it possible to use its micro-processor controlled sample injector to simultaneously inject sample and reagent onto the column via a pre-column reaction device. Using automatic control of reagent volumes and reaction time it is possible to reduce the analytical variation for glycine, the amino acid with the most labile reaction product, from approximately 10% to less than 2%. Pre-column derivatization now has the facility to remove some of the problems of contamination and allows the attainment of the highest possible sensitivities with acceptable reproducibility.

The use of OPA/ME for amino acid analysis in clinical chemistry continues to increase, with most HPLC manufacturers now offering post-column detection facilities. A number of recent publications [151–154] have used pre-column derivatization with reversed phase separation to quantitate the 30 + amino acids found in plasma. All found the systems reproducible, accurate (given precise timings) and in particular economical, but generally values obtained were somewhat lower than those from traditional amino acid analysers.

In the search for even higher sensitivity the potential of chemi-luminescence detection [124] has not received further attention but the original authors have suggested an improved reagent [155]. They have shown that the reaction of bis(2,4-dinitrophenyl)oxalate with hydrogen peroxide was 10-fold more sensitive towards dansyl amino acids than their original reagent (2,4,4-tri-chlorophenyl) oxalate. Another detection system with the potential for high sensitivity is electrochemical detection. The electrochemical detection of OPA/2ME derivatives has already been referred to but more recently underivatized amino acids have been measured at a copper electrode in the 10–100 pmol range [156] and at a platinum electrode [164].

The major disadvantage of OPA/ME continues to be its inability to react with the imino acids unless they are oxidized, a condition incompatible with pre-column derivatization. A reagent which partially overcomes this problem has now been described. The use of 4-chloro-7-nitrobenzofurazan (NBD-Cl) for the detection of imino acids has been described [66, 157] but this reagent can also be used to detect other amino acids [158]. In general the reaction conditions were excessive for post-column detection, Imai and Watanabe [159] found that the fluoro-derivative (NBD-F) reacts one order of magnitude faster with both primary and secondary amines. In another publication [160] they demonstrated the separation of derivatized NBD-amino acids in 70 min on a reversed phase gradient system with a detection limit of about 10 fmol. Unfortunately the use of this promising reagent may be restricted since NBD-Cl has been shown to be both cytotoxic and mutagenic [161] and the same may apply to NBD-F. The synthesis of ammonium 4-chloro-7-sulphobenzofurazan [161] and later ammonium 4-fluoro-7-sulphobenzofurazan [162] for the specific detection of proline and hydroxyproline may point the way to new safer reagents for the detection of all amino acids. Pfeifer and Hill have recently reviewed some aspects of gradient elution of amino acids from both ion exchange and reversed columns [165].

REFERENCES

1. Moore, S. and Stein, W.H. (1951) *J. Biol. Chem.*, **192**, 663.
2. Moore, S. and Stein, W.H. (1954) *J. Biol. Chem.*, **211**, 893.
3. Spackman, D.H., Stein, W.H. and Moore, S. (1958) *Anal. Chem.*, **30**, 1190.
4. Hamilton, P.B. (1958) *Anal. Chem.*, **30**, 914.
5. Hamilton, P.B. (1960) *Anal. Chem.*, **32**, 1779.
6. Hamilton, P.B., Bogue, D.C. and Anderson, R.A. (1960) *Anal. chem.*, **32**, 1782.
7. Rattenbury, J.M. (ed.) (1981) *Amino Acid Analysis*, Ellis Horwood, Chichester, England.
8. Perrett, D. (1979) in *Techniques in the Life Sciences B215* (eds H.L. Kornberg, J.C. Metcalfe, D.H. Northcote, C.I. Pogson and K.F. Tipton), Elsevier/North Holland, Amsterdam.
9. Knox, J. (ed.) (1978) *High Performance Liquid Chromatography*, Edinburgh University Press.

10. Kirkland, J.J. (1971) *Modern Practice of Liquid Chromotography*, Wiley, New York.
11. Snyder, L.R. and Kirkland, J.J. (1979) *Introduction to Modern Liquid Chromatography*, 2nd edn, Wiley, New York.
12. Bristow, P.A. (1976) *Liquid Chromatography in Practice*, HETP, Wilmslow, England.
13. Knox, J. (1977) *J. Chromatogr. Sci.*, **15**, 352.
14. Said, A.S. (1981) *Theory and Mathematics of Chromatography*, Huthig, Germany.
15. Scott, R.P.W. (1976) *J. Chromatogr.* (Library Vol. 11) Elsevier, Amsterdam.
16. Horvath, C.G., Preiss, B.A. and Lipsky, S.R. (1967) *Anal. Chem.*, **39**, 1422.
17. Hearn, M.T.W. (1980) *Adv. Chromatogr.*, **18**, 59.
18. Cox, G.B., Loscombe, C.R., Slucutt, M.J. *et al.* (1976) *J. Chromatogr.*, **117**, 269.
19. Little, C.J., Dale, A.D., Whatley, J.A. and Evans, M.B. (1979) *J. Chromatogr.*, **171**, 431.
20. Little, C.J., Whatley, J.A., Dale, A.D. and Evans, M.B. (1979) *J. Chromatogr.*, **171**, 435.
21. Rabel, F.M. (1980) *J. Chromatogr. Sci.*, **18**, 394.
22. Schuster, R. (1980) *Anal. Chem.*, **52**, 617.
23. Benson, J.R. and Hare, P.E. (1975) *Proc. Nat. Acad. Sci. USA*, **72**, 619.
24. Weber, G. (1951) *Biochem. J.*, **51**, 155.
25. Neadle, D.J. and Pollitt, R.J. (1965) *Biochem. J.*, **97**, 607.
26. Tapuhi, Y., Schmidt, D.E., Lindner, W. and Karger, B.L. (1981) *Anal. Biochem.*, **115**, 123.
27. Seiler, N. (1970) *Methods of Biochemical Analysis* (ed. D. Glick), **18**, Interscience, New York, p. 259.
28. Lin, J.K. and Wang, C.H. (1980) *Clin. Chem.*, **26**, 579.
29. Samejima, K., Dairman, W. and Udenfriend, S. (1971) *Anal. Biochem.*, **42**, 222.
30. Samejima, K., Dairman, W. and Udenfriend, S. (1971) *Anal. Biochem.*, **42**, 237.
31. Weigele, M., Blount, J.F., Tengi, J.P. *et al.* (1972) *J. Am. Chem. Soc.*, **94**, 4052.
32. Weigele, M., DeBarnardo, S.L., Tengi, J.P. and Leimgruber, W. (1972) *J. Am. Chem. Soc.*, **94**, 5927.
33. Udenfriend, S., Stein, S., Bohlen, P. *et al.* (1972) *Science*, **178**, 1667.
34. Felix, A.M. and Terkelson, G. (1973) *Anal. Biochem.*, **56**, 610.
35. Roth, M. (1971) *Anal. Chem.*, **43**, 880.
36. Roth, M. and Hampai, A. (1973) *J. Chromatogr.*, **83**, 353.
37. Schwabe, C. and Catlin, J.C. (1974) *Anal. Biochem.*, **61**, 302.
38. Lee, K.S. and Drescher, D.G. (1978) *Int. J. Biochem.*, **9**, 457.
39. Simons, S.S. and Johnson, D.F. (1976) *J. Am. Chem. Soc.*, **98**, 7098.
40. Svedas, V-J.K., Galaev, I.J., Borisov, I.L. and Berezin, I.V. (1980) *Anal. Biochem.*, **101**, 188.
41. Simons, S.S. and Johnson, D.F. (1977) *Anal. Biochem.*, **82**, 250.
42. Cronin, J.R., Pizzarello, S. and Gandy, W.E. (1979) *Anal. Biochem.*, **93**, 174.
43. Deelder, R.S., Kroll, M.G.F., Beeren, A.J.B. and van den Berg, J.H.M. (1978) *J. Chromatogr.*, **149**, 669.
44. Frei, R.W. (1979) *J. Chromatogr.*, **165**, 75.
45. Synder, L.R. (1976) *J. Chromatogr.*, **125**, 287.
46. McHugh, W., Sandmann, R.A., Haney, W.G. *et al.* (1976) *J. Chromatogr.*, **124**, 376.
47. Hill, D.W., Walters, F.H., Wilson, T.D. and Stuart, J.D. (1979) *Anal. Chem.*, **51**, 1338.
48. Lindroth, P. and Mopper, K. (1979) *Anal. Chem.*, **51**, 1667.
49. Gardner, W.S. and Miller, W.H. (1980) Anal Biochem., **101**, 61.
50. Jones, B.N., Paabo, S. and Stein, S. (1981) *J. Liquid Chromatogr.*, **4**, 565.
51. Georgiadis, A.G. and Coffey, J.W. (1973) *Anal. Biochem.*, **56**, 121.
52. Voelter, W. and Zech, K. (1975) *J. Chromatogr.*, **112**, 643.
53. Davidson, E., Fowler, S. and Poole, B. (1975) *Anal. Biochem.*, **69**, 49.
54. Davis, T.P., Gehrke, C.W., Gehrke, C.W. Jnr. *et al.* (1978) *Clin. Chem.*, **24**, 1317.
55. Marton, L.J. and Lee, P.L.Y. (1975) *Clin. Chem.*, **21**, 1721.
56. Thomas, A.J. (1981) in *Amino Acid Analysis* (ed. J.M. Rattenbury) Ellis Horwood, Chichester, England, p. 36.
57. Ishida, Y., Fujita, T. and Asai, K. (1981) *J. Chromatogr.*, **204**, 143.
58. Kirkland, J.J. (1974) *Analyst*, **99**, 859.
59. Hamilton, P.B. and Nagy, B. (1972) *Space Life Sciences*, **3**, 432.
60. Glaeser, B.S. and Hare, T.A. (1975) *Biochem. Med.*, **12**, 274.
61. Hare, T.A. and Manyam, N.V.B. (1980) *Anal. Biochem.*, **101**, 349.
62. Grossman, M.H., Hare, T.A. and Manyam, N.V.B. (1979) *Fed. Proc.*, **38**, 375.

63. Meek, J.L. (1976) *Anal. Chem.*, **48**, 375.
64. Wheler, G.H.T. and Russell, J.T. (1981) *J. Liquid Chromatogr.*, **4**, 1281.
65. Stuart, J.D., Wilson, T.D., Hill, D.W. *et al.* (1979) *J. Liquid Chromatogr.*, **2**, 809.
66. Ahnoff, M., Grundevik, I., Arfwidsson, A. *et al.* (1981) *Anal. Chem.*, **53**, 485.
67. Rucki, R.J. (1980) *Talanta*, **27**, 147.
68. Hanekamp, H.B. and Nieuwkerk, H.J. van (1980) *Anal. Chim. Acta*, **121**, 13.
69. Perrett, D. and Drury, P.L. (1982) *J. Liquid Chromatogr.*, **5**, 97.
70. Rabenstein, D.L. and Saetre, R. (1977) *Anal. Chem.*, **49**, 1036.
71. Saetre, R. and Rabenstein, D.L. (1978) *Anal. Biochem.*, **90**, 684.
72. Saetre, R. and Rabenstein, D.L. (1978) *J. Agric. Food Chem.*, **26**, 982.
73. Saetre, R. and Rabenstein, D.L. (1978) *Anal. Chem.*, **50**, 276.
74. Bergstrom, R.F., Kay, D.R. and Wagner, J.G. (1981) *J. Chromatogr.*, **222**, 445.
75. Kucharczyk, N. and Shahinian, S. (1981) *J. Rheumatol.*, **8**, Suppl. 7, 28.
76. Kreuzig, F. and Frank, J. (1981) *J. Chromatogr.*, **218**, 615.
77. Reed, D.J., Babson, D.J., Beatty, P.W. *et al.* (1980) *Anal. Biochem.*, **106**, 55.
78. Reeve, J., Kuhlenkamp, J. and Kaplowitz, N. (1980) *J. Chromatogr.*, **194**, 424.
79. Beales, D., Finch, R., McLean, A.E.M. *et al.* (1981) *J. Chromatogr.*, **226**, 498.
80. Kosower, E.M., Pazhenchevsky, B. and Hershkowitz, E. (1978) *J. Am. Chem. Soc.*, **100**, 6516.
81. Newton, G.L., Dorian, R. and Fahey, R.C. (1981) *Anal. Biochem.*, **114**, 383.
82. Lankmayr, E.P., Budna, K.W., Muller, K. and Nachtmann, F. (1979) *Z. Anal. Chem.*, **295**, 371.
83. Lankmayr, E.P., Budna, K.W., Muller, K. *et al.* (1981) *J. Chromatogr.*, **222**, 249.
84. Eggli, R. and Asper, R. (1978) *Anal. Chim. Acta*, **101**, 253.
85. Fowler, B. and Robins, A.J. (1972) *J. Chromatogr.*, **72**, 105.
86. Werkhoven-Goewie, C.E., Niessen, W.M.A., Brinkman, U.A.Th. and Frei, R.W. (1981) *J. Chromatogr.*, **203**, 165.
87. Necker, L.M., Delisi, L.E. and Wyatt, R.J. (1981) *Clin. Chem.*, **27**, 146.
88. Krstulovic, A.M., Brown, P.R., Rosie, D.M. and Champlin, P.B. (1977) *Clin. Chem.*, **23**, 1984.
89. Tarr, J.B. (1981) *Biochem. Med.*, **26**, 330.
90. Morita, I., Masujima, T., Yoshida, H. and Imai, H. (1981) *Anal. Biochem.*, **118**, 142.
91. Koch, D.D. and Kissinger, P.T. (1977) *J. Chromatogr.*, **164**, 441.
92. Ward, L.C. (1978) *Anal. Biochem.*, **88**, 598.
93. Nakamura, H., Zimmerman, C.L. and Pisano, J.J. (1979) *Anal. Biochem.*, **93**, 423.
94. Friedman, Z., Smith, H.W. and Hancock, W.S. (1980) *J. Chromatogr.*, **182**, 414.
95. Nakamura, H. and Pisano, J.J. (1976) *Arch. Biochem. Biophys.*, **177**, 334.
96. Wassner, S.J., Schlitzer, J.L. and Li, J.B. (1980) *Anal. Biochem.*, **104**, 284.
97. Murray, A.J., Ballard, F.J. and Tomas, F.M. (1981) *Anal. Biochem.*, **116**, 537.
98. Krull, I.S. (1978) *Adv. Chromatogr.*, **16**, 175.
99. Davankov, V.A. (1980) *Adv. Chromatogr.*, **18**, 139.
100. Audebert, P. (1979) *J. Liquid Chromatogr.*, **2**, 1063.
101. Cram, D.J., Helgeson, R.C., Sousa, L.R. and Timko, J.M. (1975) *Pure Appl. Chem.*, **43**, 327.
102. Humbel, F., Vonderschmitt, D. and Bernauer, K. (1970) *Helv. Chem. Acta*, **53**, 1983.
103. Snyder, R.V., Angelici, R.J. and Meck, R.B. (1972) *J. Am. Chem. Soc.*, **94**, 2660.
104. Rogozhin, S. and Davankov, V. (1971) *J. Chromatogr.*, **60**, 280.
105. Davankov, V.A., Rogozhin, S.V., Semechkin, A.V. and Sachkova, T.P. (1973) *J. Chromatogr.*, **82**, 359.
106. Davankov, V.A., Zolotarev, Y.A. and Kurganov, A.A. (1979) *J. Liquid Chromatogr.*, **2**, 1191.
107. Lefebvre, B., Audebert, R. and Quivoron, C. (1978) *J. Liquid Chromatogr.*, **1**, 761.
108. Gubitz, G., Jellenz, W. and Santi, W. (1981) *J. Liquid Chromatogr.*, **4**, 701.
109. Gubitz, G., Jellenz, W. and Santi, W. (1981) *J. Chromatogr.*, **203**, 377.
110. Foucalt, A., Caude, M. and Oliveros, L. (1979) *J. Chromatogr.*, **185**, 345.
111. Lindner, W., Lepage, J.N., Davies, G. *et al.* (1979) *J. Chromatogr.*, **185**, 323.
112. Weinstein, S., Engel, M.H. and Hare, P.E. (1982) *Anal. Biochem.*, **121**, 370.
113. Furukawa, H., Sakakibara, E., Kamei, A. and Ito, K. (1975) *Chem. Pharm. Bull.*, **23**, 1625.
114. Furukawa, H., Mori, Y., Takeuchi, Y. and Ito, K. (1977) *J. Chromatogr.*, **136**, 428.
115. Manning, J.M. and Moore, S. (1968) *J. Biol. Chem.*, **243**, 5591.
116. Mitchell, A.R., Kent, S.B.H., Chu, I.C. and Merrifield, R.B. (1978) *Anal. Chem.*, **50**, 637.
117. Takaya, T., Kishida, Y. and Sakakibara, S. (1981) *J. Chromatogr.*, **215**, 279.
118. Nachtmann, F. (1980) *Int. J. Pharmaceutics*, **4**, 337.

119. Nimura, N., Ogura, H. and Kinoshita, T. (1980) *J. Chromatogr.*, **202**, 375.
120. Bayer, E., Grom, E., Kaltenbegger, B. and Uhmann, R. (1976) *Anal. Chem.*, **48**, 1106.
121. Hsu, K.-T. and Currie, B.L. (1978) *J. Chromatogr.*, **166**, 555.
122. Wilkinson, J.M. (1978) *J. Chromatogr. Sci.*, **16**, 547.
123. Chen, R.F. (1967) *Arch. Biochem. Biophys.*, **120**, 609.
124. Kobayashi, S. and Imai, K. (1980) *Anal. Chem.*, **52**, 424.
125. Kobayashi, S., Sekino, J., Honda, K. and Imai, K. (1981) *Anal. Biochem.*, **112**, 99.
126. Graffeo, A.P., Haag, A. and Karger, B.L. (1973) *Analyt. Lett.*, **6**, 505.
127. Frank, G. and Stubert, W. (1973) *Chromatographia*, **6**, 522.
128. Zimmerman, C.L., Pisano, J.J. and Appella, E. (1973) *Biochem. Biophys. Res. Comm.*, **55**, 1220.
129. Zimmerman, C.L., Appella, E. and Pisano, J.J. (1976) *Anal. Biochem.*, **75**, 77.
130. Zimmerman, C.L., Appella, E. and Pisano, J.J. (1977) *Anal. Biochem.*, **77**, 569.
131. Abrahamsson, M., Groningsson, K. and Castensson, S. (1978) *J. Chromatogr.*, **154**, 313.
132. Henderson, L.E., Copeland, T.D. and Oroszlan, S. (1980) *Anal. Biochem.*, **102**, 1.
133. Johnson, N.D., Hunkapiller, M.W. and Hood, L.E. (1979) *Anal. Biochem.*, **100**, 335.
134. Annan, W.D. (1981) in *Amino Acid Analysis* (ed. R.M. Rattenbury), Ellis-Horwood, Chichester, p. 66.
135. Horn, M.J., Hargrave, P.A. and Wang, J.K. (1979) *J. Chromatogr.*, **180**, 111.
136. Yeung, E.S. and Sepaniak, M.J. (1980) *Anal. Chem.*, **52**, 1465A.
137. Scott, R.P.W. and Kucera, P. (1979) *J. Chromatogr.*, **185**, 27.
138. Novotny, M. (1980) *Clin. Chem.*, **26**, 1474.
139. Folestad, S., Johnson, L., Josefsson, B. and Galle, B. (1982) *Anal. Chem.*, **54**, 925.
140. Krien, P., Devant, G. and Hardy, M. (1982) *J. Chromatogr.*, **251**, 129.
141. Majors, R.E., Barth, H.G. and Lochmuller, C.H. (1982) *Anal. Chem.*, **54**, 323R.
142. Simpson, C.F. (ed.) (1982) *Techniques in Liquid Chromatography*, Wiley–Heyden, London.
143. Goldberg, A.P. (1982) *Anal. Chem.*, **54**, 342.
144. Landy, J.S., Ward, J.L. and Dorsey, J.G. (1983) *J. Chromatogr. Sci.*, **21**, 49.
145. McCoy, R.W. and Pauls, R.E. (1982) *J. Liquid Chromatogr.*, **5**, 1869.
146. Cooke, N.H.C., Archer, B.G., Olsen, K. and Berick, A. (1982) *Anal. Chem.*, **54**, 2277.
147. Mellor, N. (1982) *Chromatographia*, **16**, 359.
148. Bohlen, P. and Schroeder, R. (1982) *Anal. Biochem.*, **126**, 144.
149. Kucera, P. and Umagat, H. (1983) *J. Chromatogr.*, **255**, 563.
150. Venema, K., Leever, W., Bakker, J.O. *et al.* (1983) *J. Chromatogr.*, **260**, 371.
151. Hill, D., Burnworth, L., Skea, W. and Pfeifer, R. (1982) *J. Liquid Chromatogr.*, **5**, 2369.
152. Turnell, D.C. and Cooper, J.D.H. (1982) *Clin. Chem.*, **28**, 527.
153. Griffin, M., Price, S.J. and Palmer, T. (1982) *Clin. Chim. Acta*, **125**, 89.
154. Hogan, D.L., Kraemer, K.L. and Isenberg, J.I. (1982) *Anal. Biochem.*, **127**, 17.
155. Honda, K., Sekino, J. and Imai, K. (1983) *Anal. Chem.*, **55**, 940.
156. Kok, W.Th., Brinkman, U.A.Th. and Frei, R.W. (1983) *J. Chromatogr.*, **256**, 17.
157. Roth, M. (1978) *Clin. Chim. Acta*, **83**, 273.
158. Yoshida, H., Sumida, T., Masujima, T. and Imai, H. (1982) *J. High Res. Chromatogr. and Chromatogr. Comm.*, **5**, 509.
159. Imai, K. and Watanabe, Y. (1981) *Anal. Chim. Acta*, **130**, 377.
160. Imai, K., Watanabe, Y., and Toyo'oka, T. (1982) *Chromatographia*, **16**, 214.
161. Andrew, J.L., Ghosh, P., Ternai, B., and Whitehouse, M.W. (1982) *Arch. Biochem. Biophys.*, **214**, 386.
162. Imai, K., Toyo'oka, T. and Watanabe, Y. (1983) *Anal. Biochem.*, **128**, 471.
163. Joseph, M.H. and Davies, P. (1983) *J. Chromatogr.*, **277**, 125.
164. Polta, J.A. and Johnson, D.C. (1983) *J. Liq. Chromatogr.*, **6**, 1727.
165. Pfeifer, R.F. and Hill, D.W. (1983) *Adv. Chromatogr.*, **22**, 37.

CHAPTER SIXTEEN

Gas–Liquid Chromatographic Separation of Amino Acids and their Derivatives

M.H. ENGEL AND P.E. HARE

16.1 INTRODUCTION

The quantitative determination of amino acid composition and, in many instances, optical purity is critical for a number of disciplines, including investigations of naturally occurring free amino acids, peptides and proteins, as well as the design and synthesis of synthetic analogues [1]. Over the past twenty years, advances in the development of gas chromatographic procedures for amino acid analyses have provided investigators with sensitive, rapid and, to some extent, less costly alternatives to the conventional methods of ion-exchange chromatography developed by Moore, Stein and Hamilton [2, 3, 4]. While marked improvements in ion-exchange chromatography for amino acid analyses have resulted in reductions in analysis times as well as in improved resolution and sensitivity (for examples see references 5 and 6), a principal advantage of gas chromatography has been the facility with which it can be combined with mass spectrometry for amino acid identification and, in the event of possible co-elutions, confirmation of purity (for examples see references 7 and 8).

A summary of recent advances in the analysis of amino acids with gas chromatography is presented below. The refinement of procedures for the separation of the respective D- and L-enantiomers of complex mixtures of protein amino acids will be stressed, as these techniques provide information with respect to the optical purity of amino acids, in addition to overall abundances. The quantitative determination of amino acid abundances and stereochemistry with gas chromatography is dependent on the initial steps of sample preparation and derivatization. A discussion of the procedures for sample preparation (acid hydrolysis, desalting) and derivatization (esterification, acylation) precedes a review of the gas chromatographic techniques presently available for amino acid analyses.

16.2 SAMPLE PREPARATION AND DERIVATIZATION

16.2.1 Hydrolysis

Improvements in sensitivity for the analysis of amino acids with gas chromatography (< picomole levels) necessitate the continuous monitoring of water, acids, solvents and reagents that are used in sample preparation and derivatization. Water, solvents and, when practical, reagents should be distilled in glass. Hydrochloric acid is often double or triple distilled in glass [9] to remove trace amino acid contaminants.

The preparation of peptide and protein samples for amino acid analysis with gas chromatography begins with the complete hydrolysis of the amide bonds. Samples are placed in an excess of 6 N HCl in Pyrex tubes, sealed under an inert atmosphere e.g. N_2 or vacuum and hydrolysed for 22 h at 110°C. Hydrolysis with 6 N HCl, however, destroys tryptophan. While substitution of 3 N mercaptoethanesulphonic acid (MESA) for 6 N HCl reduces tryptophan decomposition during hydrolysis [10], MESA is more appropriate for the hydrolysis of samples to be analysed with ion-exchange chromatography. Unlike ion-exchange chromatography, where the pH of MESA hydrolysates are adjusted with NaOH and then injected directly onto the ion-exchange column, the preparation of hydrolysates for gas chromatography requires that the hydrolysates be evaporated to dryness prior to derivatization. Mercaptoethanesulphonic acid, however, is not amenable to evaporation, as side reactions may occur [11]. While preliminary experiments have resulted in the isolation of amino acids from MESA with ion-exchange chromatography prior to evaporation and derivatization for gas chromatographic analyses (Engel, unpublished results), hydrolysis with 6 N HCl remains the method of choice for the preparation of peptides and proteins for gas chromatographic analyses of their amino acid constituents. In addition to MESA, alternative procedures (e.g. the addition of 1,2-ethanedithiol to 6 N HCl to protect tryptophan) to minimize hydrolytic losses of cysteine, histidine, tryptophan and methionine have been reported [12, 13, 14].

In addition to the selection of a suitable acid for hydrolysis, the heating time and temperature selected for hydrolysis should be carefully evaluated. The complete hydrolysis of peptides and proteins is achieved by heating samples in 6 N HCl for 22 h at 110°C. Hydrolysis for 22 h at 110°C may, however, partially racemize amino acids [15, 16]. Acid hydrolysis for 24 h at 100°C or 15 min at 150°C has, in some instances, been reported to minimize partial racemization during hydrolysis [16]. Regardless of the conditions selected, it is usually necessary to correct for a small amount of interconversion (racemization) that occurs during hydrolysis to ensure the quantitative determination of the actual extent of racemization (i.e. D/L values) of amino acid constituents of peptides and proteins with gas chromatographic techniques [17]. Kusumoto *et al.* [18] have recently reported a novel approach for the elimination of a correction factor for partial racemization that is attributed to hydrolysis. Substitution of 6 N DCl for

6 N HCl for hydrolysis resulted in the exchange of D for H at amino acid chiral centres. With combined gas chromatography–mass spectrometry (GC–MS), Kusumoto *et al.* avoided the minor contributions of partial racemization during hydrolysis by monitoring selected ions of the amino acid derivatives one mass unit lower than the deuterium substituted ions.

16.2.2 Desalinization

Many samples that have been hydrolysed, or are being analysed for free amino acids, often contain organic and inorganic impurities e.g. salts, carbohydrates, lipids and, in the case of free amino acid analyses, peptides and proteins. Column chromatography with a cation-exchange resin e.g. Dowex 50-X8 (50 to 100 mesh) is routinely performed to remove impurities from amino acid extracts prior to derivatization. For the hydrolysates of, for example, soils and sediments, the removal of Si^{4+}, Al^{3+}, Fe^{3+} and Ca^{2+} is often accomplished with an anion-exchange column prior to the removal of organic impurities by cation-exchange [19, 20, 21]. For the quantitative determination of amino acid abundances, it is recommended that the desalting procedure that is selected be tested for the recovery of standard mixtures of amino acids. There have been numerous reports of amino acid losses during cation-exchange (for examples see references 22, 23 and 24), although most losses experienced with Dowex 50-X8 resin can often be minimized by acidifying samples (pH 2.5–3.0) prior to cation-exchange [7]. When a detector that is selective for nitrogen [25, 26] is substituted for the less specific flame ionization detector, purification of hydrolysates with cation-exchange columns can, in some instances, be avoided. Non-nitrogen containing organic compounds will be detected at reduced sensitivity; inorganic salts that precipitate during esterification can be removed by filtration or centrifugation. Although this procedure is adequate for the preparation of samples for qualitative analyses (one of the authors, P.E.H., routinely uses this procedure for the gas chromatographic determination of D/L values of amino acid constituents of protein and peptide hydrolysates), desalting with cation-exchange resins is preferred for the quantitative determination of amino acid abundances with gas chromatography [27].

16.2.3 Derivatization

Successful analysis of amino acids with gas chromatography is dependent on the synthesis of derivatives that are stable, yet volatile. The initial step in the preparation of amino acid derivatives for analysis with gas chromatography is esterification. To ensure quantitative esterification, amino acid extracts must be free from residual water. A final drying step prior to derivatization is accomplished by the addition of a small quantity of CH_2Cl_2 to the samples, followed by evaporation at 40° to 60°C under a stream of N_2. A variety of alcohols have been used for esterification, including methanol, propan-1-ol, isopropyl alcohol, butan-1-ol and isobutyl alcohol, as well as some optically pure alcohols e.g. (+)-butan-2-ol, (+)-octan-2-ol (99% pure, Norse Laboratories, Santa Barbara, CA). To drive the esterification reaction to completion, the alcohols are acidified (2–4 N HCl) prior to esterification. Alcohol acidification is performed by one of three methods: (i) Direct bubbling of dry HCl gas into the alcohol, with normality determined by weight (for example see reference 28). (ii) The addition of concentrated HCl to concentrated H_2SO_4, with the resultant HCl gas passed through concentrated H_2SO_4 prior to bubbling into the alcohol [29]. (iii) The addition of an appropriate amount of acetyl chloride to the alcohol (at 0°C) to yield 3.0 N HCl (for examples see references 7 and 30). In some instances, ethanethiol has been added to the acidified alcohol as an antioxidant [31].

An excess of the acidified alcohol is added to the dry amino acid residues. The vial is sealed with a Teflon-lined cap and is placed in an oil bath or heating block at the level of the miniscus, to promote refluxing. A variety of heating conditions, ranging from 3 h at 100°C [17] to 1 h at 110°C [32] to 3.5 min at 150°C [29] have been used for esterification. Heating times and temperatures may vary with respect to the alcohol selected for esterification, the acidity of the alcohol, the determination of reaction rates for amino acid standards and whether enantiomeric analyses are planned. Subsequent to heating, the samples are evaporated under N_2. Evaporation may be done at 0°C to avoid losses of volatile amino acid esters.

The amino acid esters are next acylated by the addition of acetic anhydride, trifluoroacetic anhydride (TFAA), pentafluoropropionic anhydride (PFPA) or heptafluorobutyric anhydride (HFBA) along with an appropriate solvent e.g. CH_2Cl_2, ethyl acetate. In some cases, 2,5-di-*tert*-butylhydroxytoluene (BHT) has been added as an antioxidant [29]. Larsen and Thornton [32] have reported, however, that the addition of BHT during acylation with HFBA did not increase recoveries of histidine or methionine. As with esterification, time and temperature parameters for acylation vary from study to study, ranging from 2 h at room temperature [33] to 10 min at 150°C [34]. After acylation, the samples are evaporated to dryness and the resultant *N(O,S)*-acylamino acid esters are redissolved in a suitable solvent for injection onto the gas chromatographic column. At the time of sample injection, it is common to co-inject an aliquot of the acylating reagent to ensure the complete conversion of histidine to the di-acyl derivative [35].

In earlier reports, Gehrke and co-workers [36, 37] reported a one step synthesis of the trimethylsilyl derivatives of the protein amino acids by reaction with bis-(trimethylsilyl)trifluoroacetamide (BSTFA) for 2.5 h at 150°C. The limited stability of the trimethylsilyl derivatives, and the occurrence of multiple peaks [29] however, curtailed further investigation of these derivatives for general amino acid analyses. The *N(O,S)*-acylamino acid esters have emerged as the derivatives of choice for the analysis of amino acids with gas chromatography.

16.3 GAS–LIQUID CHROMATOGRAPHY OF AMINO ACIDS

16.3.1 Non-chiral separations

In the past twenty years, marked improvements in the quantitative analysis of amino acids with gas chromatography have occurred. Many of these advances can be attributed to the pioneering experiments of Charles W. Gehrke and his associates at the University of Missouri. The synthesis of stable, yet volatile, *N*(*O*,*S*)-trifluoroacetyl-*n*-butyl esters [38,39], the selection and development of appropriate phases and column parameters for packed and capillary columns, the development of more sensitive and selective detectors, have all contributed to improvements in analysis time, resolution, sensitivity and quantitation of complex mixtures of protein amino acids. Cram *et al.* [40] provide an extensive review of recent advances in gas chromatographic instrumentation. Summaries of earlier research on the non-chiral separation of protein amino acids with gas chromatography are provided by Husek and Macek [41] and Gehrke *et al* [42]. Some examples of non-chiral separations of amino acid derivatives with packed and capillary gas chromatographic columns are briefly described below.

While research continues on the dual column [38] and single column [39] separation of the *N*-TFA-*n*-butyl esters of the protein amino acids (for example see reference 42) a number of alternative procedures for non-chiral separations have emerged. The synthesis and subsequent separation of *N*(*O*, *S*)-heptafluorobutyryl isobutyl esters of protein amino acids on packed and capillary columns have received widespread attention (for examples see references 7, 32, 34, 43, 44 and 45). With the exception of tryptophan, MacKenzie and Tenaschuk [34] have resolved all of the protein amino acids to baseline in ~40 min, using a 3.1 m × 2.0 mm i.d. glass column packed with SE-30 on 100–120 mesh chromosorb W.H.P. (carrier gas = He; temperature programme = 2 min isothermal at 100°C, then 100° to 250°C at 4°C min^{-1}). A chromatogram [34] is shown in Fig. 16.1. Following the procedure of Roach *et al.* [35], reproducible molar response values (RMR) were determined for histidine by simultaneous injection of an aliquot of acylating reagent with the sample, to convert histidine to the di-acyl derivative [34]. Larsen and Thornton [32] have reported a slightly modified version of the procedure of MacKenzie and Tenaschuk [34] that has resulted in the separation of 3-methylhistidine from the other protein amino acids.

The synthesis of *N*-HFB-isobutyltryptophan (di-acyl) has been reported to result in the occurrence of side products, possibly affecting tryptophan RMR values [46]. The resolution of *N*-HFB-isobutyl ester derivatives of tryptophan from the other *N*-HFB-isobutyl esters of the protein amino acids with packed columns has, however, been reported (for example see reference 7). Additional separations of protein amino acids with SE-30 packed columns have been reported. Good separations of the protein amino acid *N*(*O*,*S*)-HFB-*n*-propyl esters [29,47] as well as of the *N*(*O*,*S*)-HFB-isopropyl esters [48] have been achieved with SE-30 packed columns.

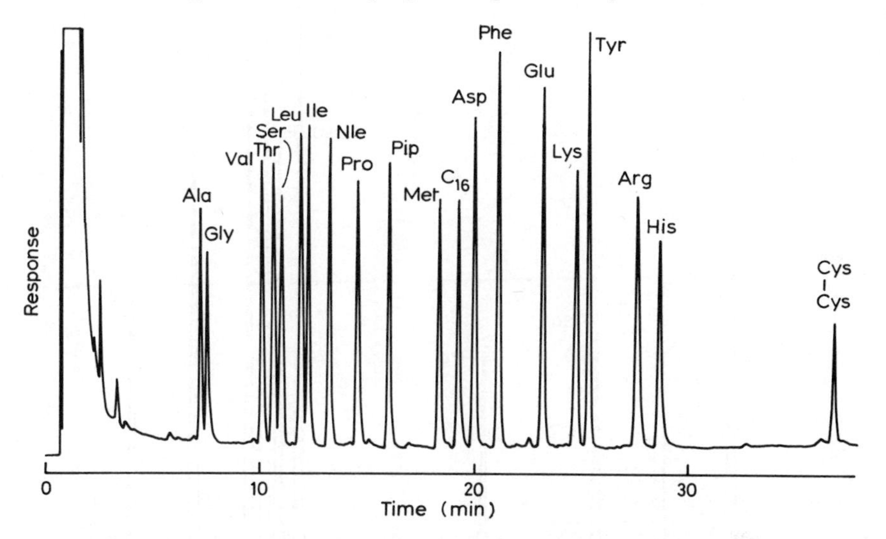

Figure 16.1 Separation of the $N(O,S)$-heptafluorobutyrylisobutyl esters of amino acid standards. The column employed was a Pyrex glass column (3.1 m × 2.0 mm i.d.) that was packed with 3% SE-30 on 100–120 mesh Chromosorb W HP; the oven temperature programme was 2 min isothermal at 100°C, then 100° to 250°C at 4°C min^{-1}; inlet temperature, 250°C; detector (FID) temperature, 300°C, the carrier gas flow rate was 25 ml min^{-1}. (Courtesy of S.L. MacKenzie and D. Tenaschuk, Prairie Regional Laboratory, National Research Council of Canada, Saskatoon, Saskatchewan, Canada and Elsevier Science Publishers, B.V., Amsterdam.)

Advances in the sensitivity and selectivity of detectors have resulted in the increased usage of capillary columns for non-chiral amino acid separations. While the amount of sample (load limit) that can be injected onto a capillary column is less than a packed column, the efficiency, resolution and, in many cases, shortened analysis times achieved with capillary columns have proved advantageous for many investigators (for examples see references 28, 45, 49 and 50). The load limitations of capillary columns can, when necessary, be compensated for by the substitution of more sensitive detectors e.g. electron capture detectors (ECD), nitrogen–phosphorus detectors (NPD), mass spectrometers (MS) for the traditional flame ionization detector (FID). Cram *et al.* [40] and references therein provide a review of the detectors currently available for gas chromatographic analysis.

Stainless steel, glass and, more recently, fused silica WCOT (wall coated open tubular) and SCOT(support coated open tubular) capillary columns coated with SE-30, OV-101, for example, have been used to resolve $N(O,S)$-HFB-n-butyl esters, $N(O,S)$-HFB-isobutyl esters and N-acetyl-n-propyl esters of the protein amino acids [7, 26, 28, 45]. A chromatogram of N-HFB-isobutyl esters of amino acids separated with a stainless steel capillary column coated with OS-138 is shown in Fig. 16.2. Bengtsson *et al.* [45] have reported the detection of $N(O,S)$-HFB-isobutyl ester derivatives of amino acids at femtomole levels using a

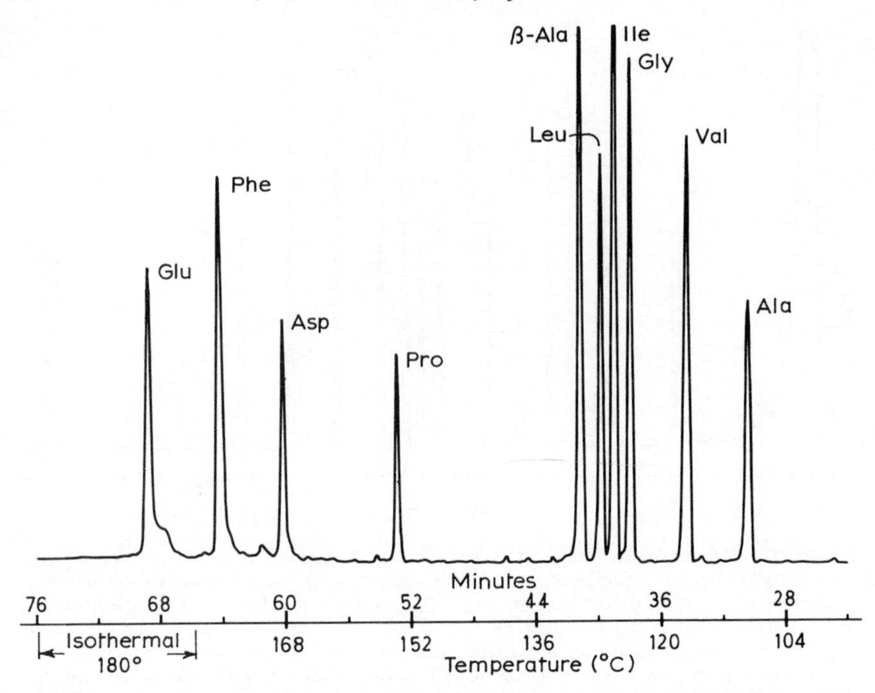

Figure 16.2 Gas chromatogram of *N*-heptafluorobutyryl isobutyl esters of amino acid standards. An OS-138 50 ft × 0.01 in i.d. SCOT stainless steel capillary column was used; the carrier gas (He) flow rate was 3.5 ml min^{-1}; the oven temperature programme was 80°C isothermal for 15 min, then 80° to 180°C at 2°C min^{-1}; The programme was finished isothermally at 180°C; the inlet temperature was 200°C.

combination of capillary columns and ECD and MS detectors. Recent developments in combined gas chromatography–mass spectrometry i.e. selected ion monitoring (SIM) and chemical ionization (CI) have not only enhanced sensitivity but have also provided improved methods for determining the identity as well as the purity of trace amounts of amino acids in biological and geological samples [7, 45, 51, 52, 53, 54]. A combination of gas chromatographic and mass spectrometric techniques has also been used to separate and identify phenylthiohydantoin (PTH) amino acids obtained from the Edman degradation of peptides and proteins [55, 56, 57, 58].

While routine procedures are now available for the separation and identification of the protein amino acids with gas chromatography, quantitative determinations for some of the protein amino acids are still problematical. Research continues on the modification of sample derivatization and column selection for the quantitative gas chromatographic analyses of cysteine, cystine, methionine, histidine, arginine and tryptophan individually, as well as in the presence of the other protein amino acids [13, 14, 42, 44]. As discussed in Chapter 14, new developments in ion-exchange chromatography afford alternatives for the rapid, single column, quantitative analyses of picomole and sub-picomole quantities of the protein amino acids. The development of HPLC–MS procedures will provide the additional methods of sample identification and determination of purity that are now routine for GC–MS [59].

16.3.2 Chiral separations

Previous examples of the separation of protein amino acids with gas chromatography have been limited to non-chiral separations. A number of techniques, however, have been developed for the separation of the respective D- and L-enantiomers of complex mixtures of protein amino acids. These procedures afford a method for the simultaneous determination of optical purity, in addition to relative and absolute amino acid abundances. The original procedures developed for the resolution of complex mixtures of amino acids into their respective D- and L-enantiomers involved the introduction of a second, optically pure asymmetric centre to the carboxyl or amine functional groups of the amino acids, resulting in diastereoisomers that, owing to different physical properties, e.g. melting point, boiling point, could be resolved on optically inactive stationary phases [60, 61, 62, 63, 64, 65, 66, 67].

The synthesis of amino acid diastereoisomers with optically pure alcohols e.g. (+)-butan-2-ol, (+)-octan-2-ol, menth-1-ol have received much attention [68, 69]. While menth-1-ol [70] and (+)-octan-2-ol [71] have been used for the esterification and subsequent separation of the D- and L-enantiomers of individual amino acids, (+)-2-butyl esters have been the derivatives of choice for the separation of the complex mixtures of amino acid enantiomers.

The procedures that have been described for the derivatization of amino acids for non-chiral separations are also used for the preparation of the amino acid diastereoisomers. The resultant *N(O,S)*-TFA-(+)-2-butyl or *N(O,S)*-PFP-(+)-2-

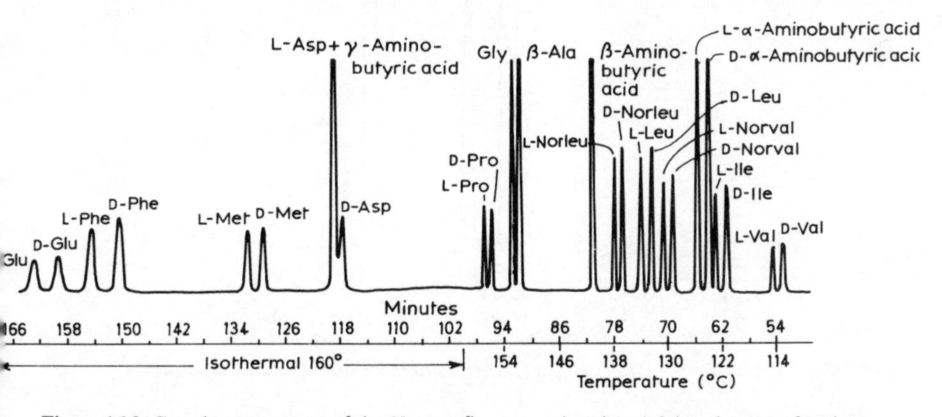

Figure 16.3 Gas chromatogram of the *N*-pentafluoropropionyl-(+)-2-butyl esters of amino acid D,L-standards. A Carbowax-20 M 61 m × 0.5 mm i.d. WCOT nickel capillary column was used; the carrier gas (He) flow rate was 11.5 ml min^{-1}; the oven temperature programme was 80°C isothermal for 20 min, then 80° to 160°C at 1°C min^{-1}; the programme was finished isothermally at 160°C; the inlet temperature was 150°C.

butyl esters are relatively stable, yet are of suitable volatility for gas chromatographic analyses. Stainless steel, nickel, glass and fused silica capillary columns coated with polar phases e.g. Carbowax 20 M, Ucon 75-H-90,000 have been used to separate complex mixtures of the D- and L-enantiomers of acidic and neutral amino acids. A chromatogram of the *N*-PFP-(+)-2-butyl ester derivatives of the acidic and neutral amino acids is shown in Fig. 16.3. A chromatogram showing the separation of the *N*-TFA-(+)-2-octyl esters of the D- and L-enantiomers of valine and alanine is shown in Fig. 16.4. While (+)-octan-2-ol is excellent for the derivatization and separation of individual or simple mixtures of amino acids, a substantial number of peak overlaps and variance in elution order have occurred

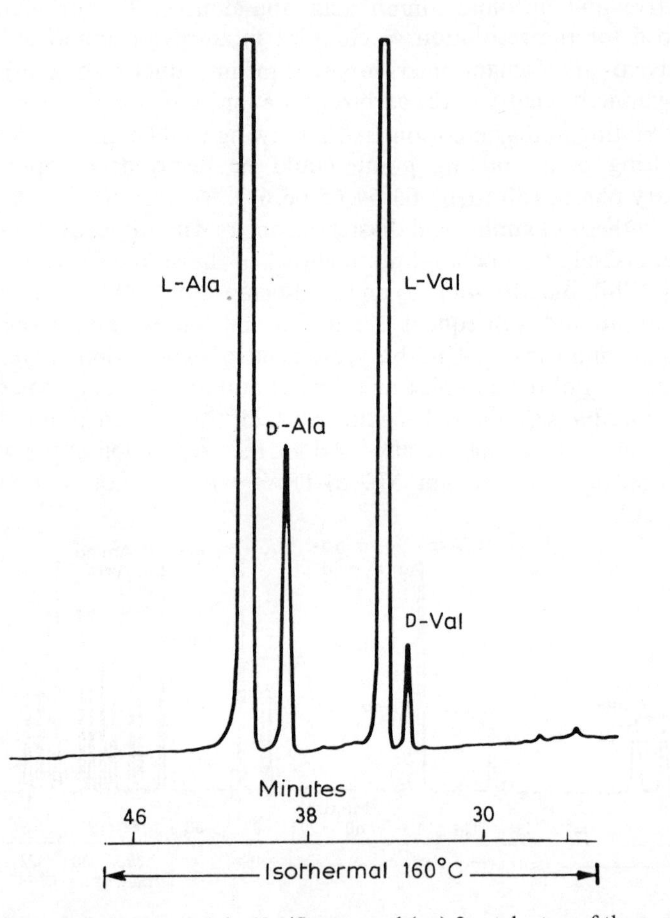

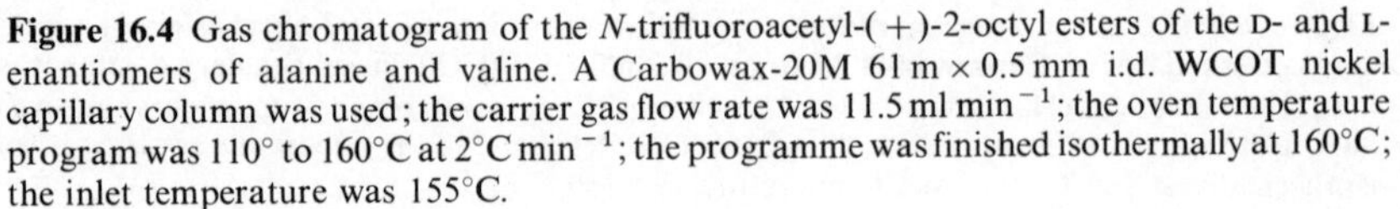
Figure 16.4 Gas chromatogram of the *N*-trifluoroacetyl-(+)-2-octyl esters of the D- and L-enantiomers of alanine and valine. A Carbowax-20M 61 m × 0.5 mm i.d. WCOT nickel capillary column was used; the carrier gas flow rate was 11.5 ml min^{-1}; the oven temperature program was 110° to 160°C at 2°C min^{-1}; the programme was finished isothermally at 160°C; the inlet temperature was 155°C.

when complex amino acid mixtures of the (+)-2-octyl ester derivatives were analysed with gas chromatography (G.E. Pollock, personal communication).

The hydroxy amino acid diastereoisomeric derivatives, such as serine, undergo on-column decomposition on Carbowax 20 M columns. To make these derivatives stable for analysis with Carbowax 20 M, a second acylation step is performed in which the PFP or TFA groups on the side chain hydroxyl moieties are replaced with an acetyl group [20,64,72].

In addition to determinations of the occurrence and abundance of D-amino acids in recent and ancient samples of biological and geological interest, the separation of diastereoisomeric derivatives of amino acids with gas chromatography provides a method for the determination of the optical purity of synthetic peptides. Figure 16.5 shows chromatograms of the *N*-PFP-(+)-2-butyl esters of acidic and neutral amino acid constituents of the hydrolysates of synthetic {Nle[4]}-α-melanocyte stimulating hormone (α-MSH) prior and subsequent to heat–alkali treatment [72].

While the synthesis and subsequent gas chromatographic separation of diastereoisomeric derivatives of the respective enantiomers of complex amino acid mixtures have received much attention, the technique has a number of disadvantages. (i) The optically pure alcohols e.g. (+)-butan-2-ol necessary for esterification are very time-consuming to prepare and, when purchased from commercial suppliers, very expensive (~ $40–50 per g). (ii) A single chromatographic separation of a complex mixture of *N*-PFP-(+)-2-butyl amino acid esters takes more than 2 h. (iii) The basic amino acids (lysine, histidine, arginine, ornithine) and tryptophan are not amenable to routine separation with this procedure. (iv) In addition to correcting for racemization during hydrolysis, a correction for the ~1.0% impurity of the alcohols must be made. (v) Diastereoisomeric fractionation has been reported to occur during esterification [73, 74]. Although an alternative, less expensive approach in which the second

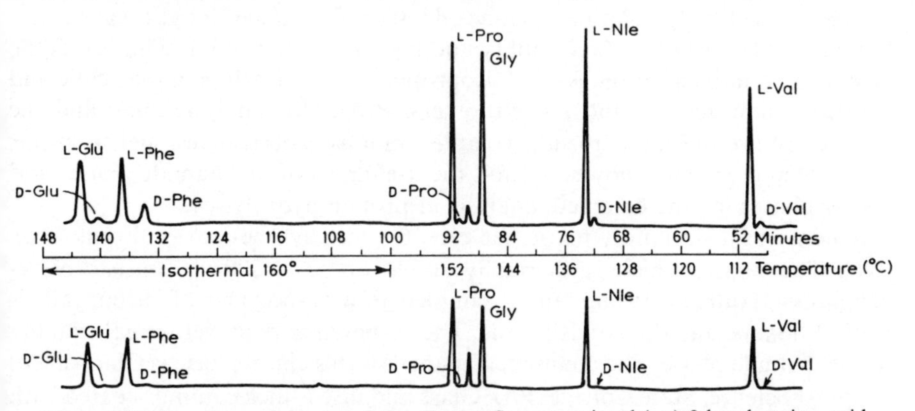

Figure 16.5 Gas chromatograms of the *N*-pentafluoropropionyl-(+)-2-butyl amino acid esters of the acid hydrolysates of [Nle[4]]-α-MSH prior (bottom) and subsequent (top) to heat–alkali treatment (0.1 N NaOH at 100°C for 10 min); the column conditions are the same as in Fig. 16.3.

asymmetric centre is introduced to the amine functional groups of the amino acid enantiomers has been pursued (i.e. *N*-TFA-L-prolyl amino acid methyl esters, see reference 75), the major breakthroughs in the gas chromatographic separation of the D- and L-enantiomers of the protein amino acids have come with the development of optically active stationary phases for capillary gas chromatographic columns.

The development of optically active (chiral) stationary phases for capillary column gas chromatography is largely the result of the pioneering experiments of E. Gil-Av and his colleagues at the Weizmann Institute of Science, Rehovot, Israel. An elegant review of the early experiments that led to the development of chiral stationary phases has been written by Gil-Av [73]. The first separations of *N*-TFA esters of amino acid enantiomers were accomplished with a chiral stationary phase that consisted of *N*-trifluoroacetyl-L-isoleucine lauryl ester [76]. The asymmetric centre was provided by L-isoleucine, while esterification with lauryl alcohol (dodecan-1-ol) rendered the phase non-volatile.

Improvements in the resolution of the *N*-acyl ester derivatives of amino acid D- and L-enantiomers were achieved by the substitution of dipeptides such as *N*-trifluoroacetyl-L-valyl-L-valine isopropyl ester in the chiral stationary phases [77]. The synthesis of chiral stationary phases consisting of *N*-TFA-dipeptide cyclohexyl esters [78, 79, 80] led to the development of routine procedures for the separation of *N*-acyl ester derivatives of complex mixtures of acidic and neutral amino acid enantiomers. Lengthy analysis times (>2 h) and the unsuccessful elution of the less volatile amino acid derivatives (such as lysine, histidine, ornithine, arginine, tryptophan), however, were a consequence of the limited thermal stabilities ($\sim 100°–110°$C) of many of the initial chiral phases, e.g. *N*-TFA-L-valyl-L-valine cyclohexyl ester [78].

The synthesis of diamide chiral stationary phases e.g. *N*-docosanoyl-L-valine *tert*-butylamide [81, 82, 83] and of mixed chiral stationary phases e.g. *N*-octadecanoyl-L-valyl-L-valine cyclohexyl ester and *N*-docosanoyl-L-valine *tert*-butylamide [8, 74] of lower volatility and higher thermal stability improves the resolution obtained for the *N*-TFA-isopropyl ester derivatives of the acidic and neutral amino acid D- and L-enantiomers. Prolonged analysis times and the unsuccessful resolution of the less volatile amino acid derivatives, such as lysine, histidine and arginine, however, limit the usefulness of the diamide phases and mixed phases for the complete analysis of protein hydrolysates.

A new chiral stationary phase has recently been synthesized by Frank *et al.* [84]. The phase, which is thermally stable up to $\sim 200°$C, consists of *N*-propionyl-L-valine *tert*-butylamide coupled to a co-polymer of carboxyalkyl-methylsiloxane and dimethylsiloxane. The co-polymer renders thermal stability to the diamide phase. The commercial name for this chiral phase is Chirasil-Val (Applied Science, State College, PA). Glass and fused silica columns coated with Chirasil-Val resolve, to baseline, the *N*(*O*, *S*)-PFP-isopropyl esters of the D- and L-enantiomers of the twenty common protein amino acids in ~ 35 min [84, 85]. Excellent resolution has also been obtained for the *N*(*O*,*S*)-PFP-*n*-propyl [31] and the *N*(*O*,*S*)-TFA-isopropyl esters of the common protein amino acids. A chromatogram of the *N*(*O*,*S*)-PFP-isopropyl esters of the D- and L-enantiomers of a complex mixture of amino acids that were resolved with a fused silica column coated with Chirasil-Val is shown in Fig. 16.6. The separation characteristics of the Chirasil-Val phase have recently been improved with the addition of 15% phenyl groups to the polysiloxane co-polymer [86].

Frank *et al.* [31, 87] have modified the general procedures for esterification and acylation that were described above, to improve the recoveries of histidine, arginine, tryptophan and cysteine. Ethanethiol was added to isopropanol/2N HCl to prevent oxidation of tryptophan and to convert cystine to cysteine. The heating conditions for esterification were 1.0 h at 110°C. The temperature for acylation (PFPA/ethyl acetate) was raised from 110° to 150°C (10 min) to ensure the quantitative acylation of arginine. Subsequent to acylation

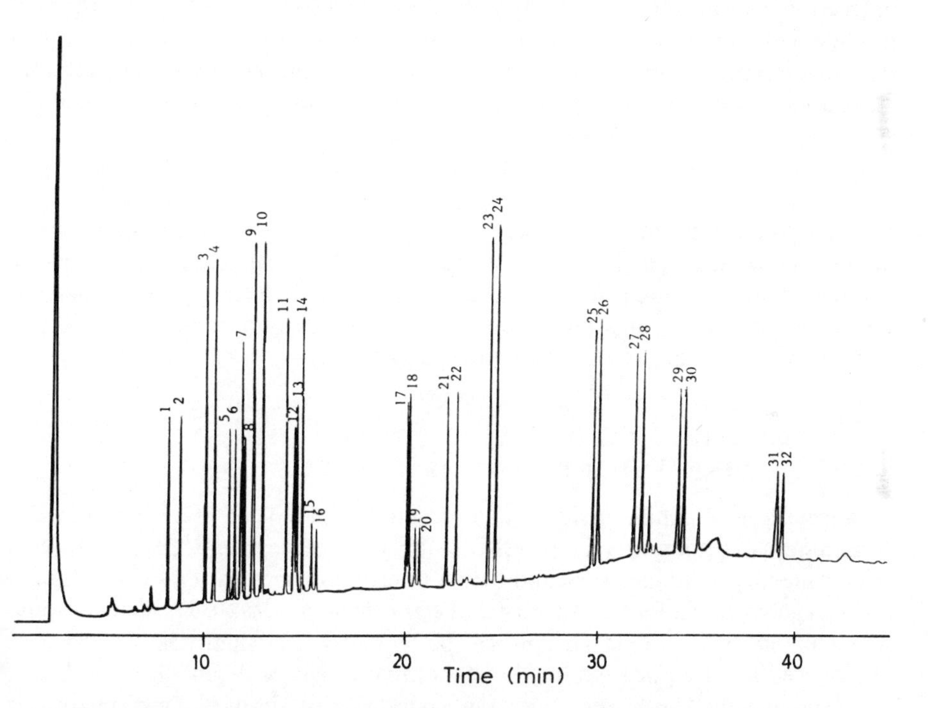

Figure 16.6 Gas chromatogram of the *N*(*O*,*S*)-pentafluoropropionyl-isopropyl esters of D- and L-amino acid standards. A Chirasil-Val 30 m × 0.25 mm i.d. WCOT fused silica column was used; the oven temperature programme was 90°C isothermal for 4 min, then 90° to 190°C at 4°C min^{-1}; the numbers on the chromatogram correspond to the following amino acids: (1) D-alanine; (2) L-alanine; (3) D-valine; (4) L-valine; (5) D-threonine; (6) L-threonine; (7) glycine; (8) D-allo-isoleucine; (9) D-isoleucine; (10) L-isoleucine; (11) D-leucine; (12) D-proline; (13) L-proline; (14) L-leucine; (15) D-serine; (16) L-serine; (17) D-aspartic acid; (18) L-aspartic acid; (19) D-cysteine; (20) L-cysteine; (21) D-methionine; (22) L-methionine; (23) D-phenylalanine; (24) L-phenylalanine; (25) D-tyrosine; (26) L-tyrosine; (27) D-ornithine; (28) L-ornithine; (29) D-lysine; (30) L-lysine; (31) D-tryptophan; (32) L-tryptophan. (Courtesy of D. Wulff, Applied Science Division, State College, PA.)

with PFPA, the sample is taken to dryness, and benzene and diethyl pyrocarbonate (10:1, v:v) are added to the sample, which is then heated for 10 min at 150°C. This procedure converts the D- and L-enantiomers of histidine to the N^{im}-ethyloxycarbonyl derivatives [88], which can be resolved with capillary columns coated with Chirasil-Val.

Using D-amino acids as internal standards, Frank *et al.* [31,85,87] have developed a procedure for the quantitative analysis of the 20 protein amino acids. Prior to sample clean-up and derivatization, a protein hydrolysate is taken to dryness and redissolved in water. Two identical aliquots are sampled from the solution. One of the aliquots is desalted, derivatized and analysed for its enantiomeric composition, while the other aliquot is spiked with a standard solution of D-amino acids prior to work up. D-Alanine is used as the internal standard for glycine. Frank *et al.* [87] have derived a series of equations and correction factors for the quantitative determination of protein amino acid abundances based on a comparison of the peak areas of the amino acid derivatives in the sample with the peak areas of the D-amino acid internal standards. In the simple case, where the sample consists entirely of L-amino acids,

$$X_a = M_D \frac{A_s}{A_T}$$

where X_a is the amount of amino acid, A_s is the area of the L-enantiomer, A_T is the area of the D-amino acid internal standard and M_D is the actual amount of the standard that was added. The only assumption is that of linearity versus detector response for the D- and L-enantiomers. For a detailed explanation of this procedure, the reader is referred to Frank *et al.* [87].

16.4 AMINO ACID ANALYSIS WITH COMBINED GAS CHROMATOGRAPHY–MASS SPECTROMETRY

Confirmation of amino acid gas chromatographic peak identities and determination of peak purity, i.e. whether a single chromatographic peak represents two or more co-eluting compounds, is routinely performed by the substitution of a mass spectrometer for the detector of the gas chromatograph. The conventional mode of electron impact (EI) mass spectrometry, in which the amino acid derivatives are fragmented by direct bombardment with an electron beam, however, usually results in the very low abundance or absence of molecular ions [89]. Because of the high molecular weights of the *N(O, S)*-acyl alkyl ester derivatives prepared for gas chromatography, it is preferable, when possible, to substitute chemical ionization (CI) for EI for combined gas chromatography–mass spectrometry analyses (for examples see references 53, 71).

Munson and Field [90] introduced CI–MS as a useful alternative to EI–MS for the structural determination and subsequent identification of organic compounds. CI–MS involves the introduction of a reaction gas, usually methane, into the ionization chamber (source) of a mass spectrometer at a pressure of 1 Torr. The primary ions of methane, the major fragments being CH_4^+ and CH_3^+, react rapidly to produce CH_5^+ and $C_2H_5^+$, which comprise ~90% of the secondary ions generated [90]. The ion $C_3H_5^+$ is produced in smaller quantities [90, 91]. The stable ions (CH_5^+ and $C_2H_5^+$) generated from methane react with the small quantities of organic compounds introduced into the ionization chamber to produce a spectrum of characteristic ions for each compound.

Unlike EI–MS, the principal ions generated by CI–MS are of higher molecular weight. The CH_5^+ and $C_2H_5^+$ ions add a proton to the *N(O, S)*-acyl alkyl esters of the amino acids. The resulting $m + 1$, i.e. the characteristic quasiparent ion [90] is often detected for amino acid *N(O, S)*-acyl alkyl ester derivatives [71]. Conversely, the parent is small, if present at all, in the EI–MS spectra of the amino acid derivatives. The amino acid derivatives react with the $C_2H_5^+$ ion to form characteristic $m + 29$ adducts. In some instances, $m + 41$ ($C_3H_5^+$) adducts are detected. The $m + 29$ and $m + 41$ adducts are commonly reported for oxygen and nitrogen containing organic compounds [90]. The reaction occurs at the nitrogen heteroatom and to a lesser extent, at the oxygen atoms present in the *N(O, S)*-acyl alkyl ester amino acid derivatives.

In GC–CIMS, the use of methane, for example, as the carrier gas for the gas chromatograph, as well as for the reactant gas for CI, eliminates the need for a splitter and a molecular separator [92, 93]. Mass spectra of glutamic acid obtained with GC–CIMS, in which methane was the carrier and the reactant gas, are shown in Fig. 16.7. Note the presence of the $m + 1$ quasiparent ion (m/z 406) as well as the $m + 29$ adduct (m/z 434). The instrument employed for this GC–CIMS analysis was a Finnigan 9500 gas chromatograph coupled to a Finnigan 3300 mass spectrometer (quadrupole) equipped with a 6110 series data system [94]. A Carbowax 20 M column, 150 ft × 0.02 in i.d. SCOT made of stainless steel was used for this analysis. The column programme consisted of an initial isothermal period of 20 min at 90°C; then 90° to 150°C at 2°C min^{-1}. The run was finished isothermally at 150°C. The GC inlet temperature was 150°C. The carrier gas (CH_4) flow rate was 8.5 ml min^{-1}. The MS ionization chamber was 150°C. The emission current was 1 milliamp. The methane ions were generated by 150 eV.

16.5 CONCLUSIONS

In the past few years there have been hundreds of publications describing gas chromatographic procedures for the study of amino acids, a tribute to the popularity of the technique. In the present review we have chosen to highlight some of the past as well as recent advances in gas chromatographic analyses of amino acids. While many procedures have been developed for the non-chiral gas chromatographic separation of protein amino acids, the development [76] and refinement [84] of chiral stationary phases for the gas chromatographic analyses of amino acids provide the investigator with a method for the simultaneous determination of absolute abundances and enantiomeric composition. It is

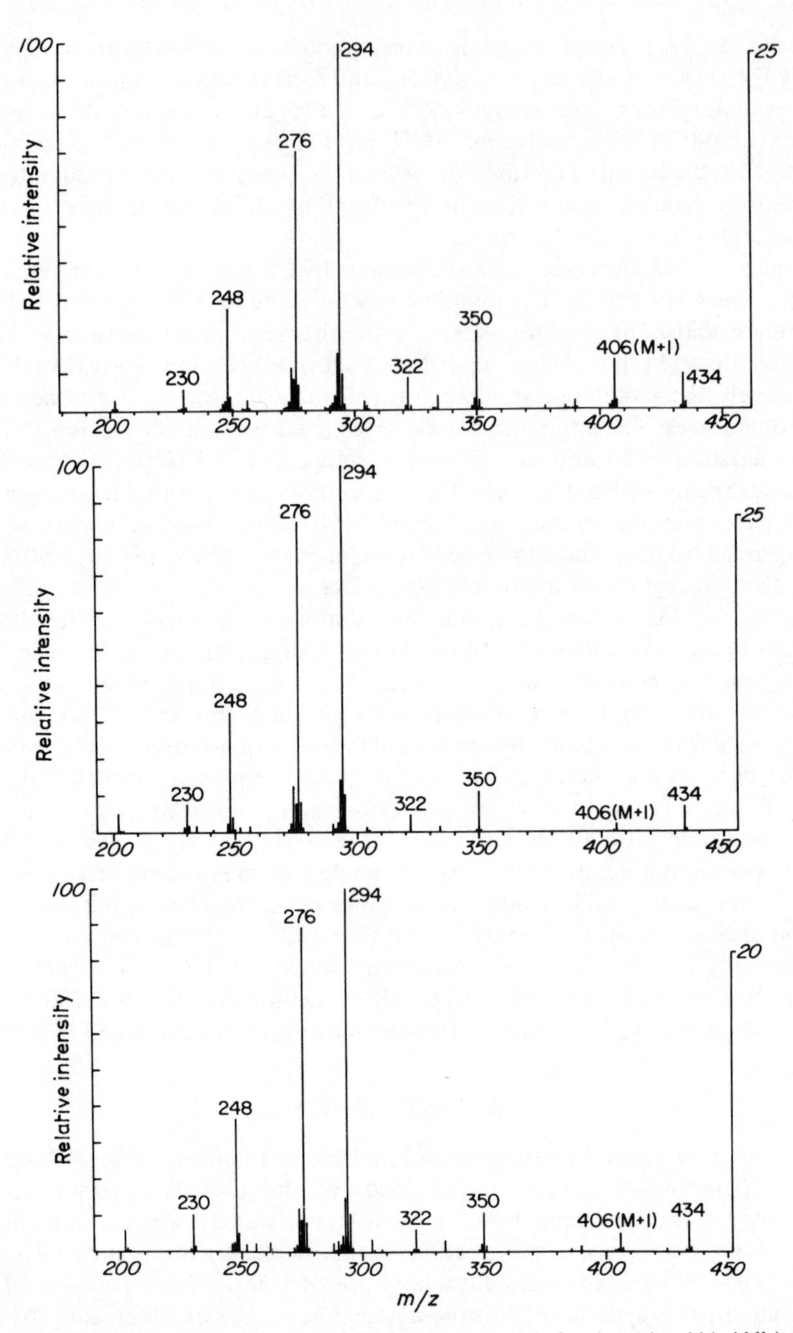

Figure 16.7 Mass spectra of glutamic acid standard (top) and of D-glutamic acid (middle) and L-glutamic acid (bottom) in the water extract of the Murchison meteorite; operating conditions for the gas chromatograph and mass spectrometer are given in the text. (Reproduced by permission from reference 94.)

possible that future investigations of the occurrence and abundance of amino acids in protein and peptide hydrolysates with chiral stationary phases will result not only in the determination of absolute abundances, but in the discovery of quantities of 'D' amino acids in systems previously thought or assumed to be comprised entirely of the 'L' enantiomers. The combination of gas chromatography with mass spectrometry provides a powerful method for confirmation of the identity and purity of common amino acids, in addition to providing information with respect to the elucidation of structure of uncommon non-protein amino acids.

Recent advances in ion-exchange chromatography have been reviewed in Chapter 14. Our experience to date has been that the techniques of ion-exchange chromatography and gas chromatography complement each other with respect to amino acid analyses. Ion-exchange provides a rapid method for the determination of amino acid abundances in a sample, with minimal preparation time e.g. desalting, derivatization. Gas chromatography affords a method for the complete enantiomeric analysis of complex mixtures of amino acids, in addition to the confirmation of identity with GC–MS. It is anticipated that future advances in ion-exchange chromatography and gas chromatography will continue to provide investigators with a selection of analytical techniques for the analysis of amino acids.

REFERENCES

1. Sawyer, T.K., Sanfilippo, P.J., Hruby, V.J., Engel, M.H., Heward, C.B., Burnett, J.B. and Hadley, M.E. (1980) *Proc. Nat. Acad. Sci. USA*, **77**, 5754.
2. Moore, S. and Stein, W.H. (1951) *J. Biol. Chem.*, **192**, 663.
3. Moore, S. and Stein, W.H. (1954) *J. Biol. Chem.*, **211**, 893.
4. Hamilton, P.B. (1963) *Anal. Chem.*, **35**, 2055.
5. Benson, J.R. and Hare, P.E. (1975) *Proc. Nat. Acad. Sci. USA*, **72**, 619.
6. LePage, J.N., Lindner, W., Davies, G., Seitz, D.E. and Karger, B.L. (1979) *Anal. Chem.*, **51**, 433.
7. Bengtsson, G. and Odham, G. (1979) *Anal. Biochem.*, **92**, 426.
8. Smith, G.G. and Wonnacott, D.M. (1980) in *The Biogeochemistry of Amino Acids* (eds P.E. Hare, T.C. Hoering and K. King, Jr.) Wiley, New York, p. 203.
9. Wolman, Y. and Miller, S.L. (1971) *Nature*, **234**, 548.
10. Penke, B., Ferenczi, R. and Kovacs, K. (1974) *Anal. Biochem.*, **60**, 45.
11. Pierce Chemical Company (1982) *General Catalogue*, p. 116.
12. Gehrke, C.W. and Takeda, H. (1973) *J. Chromatogr.*, **76**, 77.
13. Felker, P. (1976) *Anal. Biochem.*, **76**, 192.
14. Finlayson, A.J. and MacKenzie, S.L. (1976) *Anal. Biochem.*, **70**, 397.
15. Hare, P.E. and Hoering, T.C. (1973) *Carnegie Inst. Wash. Year Book 72*, p. 690.
16. Engel, M.H. and Hare, P.E. (1982) *Carnegie Inst. Wash. Year Book 81*, p. 422.
17. Engel, M.H., Zumberge, J.E. and Nagy, B. (1977) *Anal. Biochem.*, **82**, 415.
18. Kusumoto, S., Matsukura, M. and Shiba, T. (1981) *Biopolymers*, **20**, 1869.
19. Cheng, C.-N., Shufeldt, R.C. and Stevenson, F.J. (1975) *Soil Biol. Biochem.*, **7**, 143.
20. Pollock, G.E., Cheng, C.-N. and Cronin, S.E. (1977) *Anal. Chem.*, **49**, 2.
21. Pollock, G.E. and Kvenvolden, K.A. (1978) *Geochim. et Cosmochim. Acta*, **42**, 1903.
22. Cancalon, P. and Klingman, J.D. (1974) *J. Chromatogr. Sci.*, **12**, 349.
23. Krutz, M. (1975) *Z. Anal. Chem.*, **273**, 123.
24. Ogino, H. and Nagy, B. (1982) *G.S.A. Abstracts with Programs*, **14**, *addenda.*
25. Kolb, B. and Bischoff, J. (1974) *J. Chromatogr. Sci.*, **12**, 625.

26. Adams, R.F., Vandemark, F.L. and Schmidt, G.J. (1977) *J. Chromatogr. Sci.*, **15**, 63.
27. Gehrke, C.W. and Leimer, K. (1970) *J. Chromatogr.*, **53**, 195.
28. Poole, C.F. and Verzele, M. (1978) *J. Chromatogr.*, **150**, 439.
29. March, J.F. (1975) *Anal. Biochem.*, **69**, 420.
30. Frank, H., Bimboes, D. and Nicholson, G.J. (1979) *J. Chromatographia*, **12**, 168.
31. Frank, H., Rettenmeir, A., Weicker, H., Nicholson, G.J. and Bayer, E. (1982) *Anal. Chem.*, **54**, 715.
32. Larsen, T.W. and Thornton, R.F. (1980) *Anal. Biochem.*, **109**, 137.
33. Zumberge, J.E., Engel, M.H. and Nagy, B. (1980) in *The Biogeochemistry of Amino Acids* (eds P.E. Hare, T.C. Hoering and K. King, Jr.) Wiley, New York, p. 503.
34. MacKenzie, S.L. and Tenaschuk, D. (1979) *J. Chromatogr.*, **171**, 195.
35. Roach, D., Gehrke, C.W. and Zumwalt, R.W. (1969) *J. Chromatogr.*, **43**, 311.
36. Gehrke, C.W., Nakamoto, H. and Zumwalt, R.W. (1969) *J. Chromatogr.*, **45**, 24.
37. Gehrke, C.W. and Leimer, K. (1971) *J. Chromatogr.*, **57**, 219.
38. Gehrke, C.W., Kuo, K.C. and Zumwalt, R.W. (1971) *J. Chromatogr.*, **57**, 209.
39. Gehrke, C.W. and Takeda, H. (1973) *J. Chromatogr.*, **76**, 63.
40. Cram, S.P., Risby, T.H., Field, L.R. and Yu, W.-L. (1980) *Anal. Chem.*, **52**, 324R.
41. Husek, P. and Macek, K. (1975) *J. Chromatogr.*, **113**, 139.
42. Gehrke, C.W., Younker, D.R., Gerhardt, K.O. and Kuo, K.C. (1979) *J. Chromatogr. Sci.*, **17**, 301.
43. MacKenzie, S.L. and Tenaschuk, D. (1974) *J. Chromatogr.*, **97**, 19.
44. Bonvell, S.I. and Monheimer, R.H. (1980) *J. Chromatogr. Sci.*, **18**, 18.
45. Bengtsson, G., Odham, G. and Westerdahl, G. (1981) *Anal. Biochem.*, **111**, 163.
46. MacKenzie, S.L. and Hogge, L.R. (1977) *J. Chromatogr.*, **132**, 485.
47. Yoneda, T. (1980) *Anal. Biochem.*, **104**, 247.
48. Kirkman, M.A., Burrell, M.M., Lea, P.J. and Mills, W.R. (1980) *Anal. Biochem.*, **101**, 364.
49. Ettre, L.S. and March, E.W. (1974) *J. Chromatogr.*, **91**, 5.
50. Jonsson, J., Eyem, J. and Sjoquist, J. (1973) *Anal. Biochem.*, **51**, 204.
51. Coutts, R.T., Jones, G.R. and Liu, S.-F. (1979) *J. Chromatogr. Sci.*, **17**, 551.
52. Nagy, B., Engel, M.H., Zumberge, J.E., Ogino, H. and Chang, S.Y. (1981) *Nature*, **289**, 53.
53. Matthews, D.E., Starren, J.B., Drexler, A.J., Kipnis, D.M. and Bier, D.M. (1981) *Anal. Biochem.*, **110**, 308.
54. Okano, Y., Kataoka, M., Miyata, T., Morimoto, H., Takahama, K., Hitoshi, T., Kase, Y., Matsumoto, I. and Shinka, T. (1981) *Anal. Biochem.*, **117**, 196.
55. Pisano, J.J. (1975) *Mol. Biol. Biochem. Biophys.*, 8, *Protein Sequence Determination*, 2nd edn, p. 280.
56. Van Eerd, J.-P. (1976) *Anal. Biochem.*, **71**, 612.
57. Dwulet, F.E. and Gurd, F.R.N. (1977) *Anal. Biochem.*, **82**, 385.
58. Fairwell, T. and Brewer, H.B. (1980) *Anal. Biochem.*, **107**, 140.
59. Arpino, P.J. and Guiochon, G. (1979) *Anal. Chem.*, **51**, 682A.
60. Charles, R., Fischer, G. and Gil-Av, E. (1963) *Israel J. Chem.*, **1**, 234.
61. Pollock, G.E., Oyama, V.I. and Johnson, R.D. (1965) *J. Gas Chromatogr.*, **3**, **174.**
62. Pollock, G.E. and Oyama, V.I. (1966) *J. Gas Chromatogr.*, **4**, 126.
63. Pollock, G.E. (1967) *Anal. Chem.*, **39**, 1194.
64. Pollock, G.E. and Kawauchi, A.H. (1968) *Anal. Chem.*, **40**, 1356.
65. Bonner, W.A., Van Dort, M.A. and Flores, J.J. (1974) *Anal. Chem.*, **46**, 2104.
66. Halpern, B. and Westley, J.W. (1965) *Biochem. Biophys. Res. Comm.*, **19**, 361.
67. Manning, J.M. and Moore, S. (1968) *J. Biol. Chem.*, **243**, 5591.
68. Gil-Av, E. and Nurok, D. (1974) *Advances in Chromatography*, (Vol. 10) (eds. J.C. Giddings and R.A. Keller), Marcel Dekker, New York, p. 99.
69. Williams, K.M. and Smith, G.G. (1978) *Origins of Life*, **8**, 91.
70. Hasegawa, M. and Matsubara, I. (1975) *Anal. Biochem.*, **63**, 308.
71. Engel, M.H. (1980) Ph.D thesis, The University of Arizona, Tucson, AZ.
72. Engel, M.H., Sawyer, T.K., Hadley, M.E. and Hruby, V.J. (1981) *Anal. Biochem.*, **116**, 303.
73. Gil-Av, E. (1975) *J. Mol. Evol.*, **6**, 131.
74. Smith, G.G. and Wonnacott, D.M. (1980) *Anal. Biochem.*, **109**, 414.
75. Hoopes, E.A., Peltzer, E.T. and Bada, J.L. (1978) *J. Chromatogr. Sci.*, **16**, 556.
76. Gil-Av, E., Feibush, B. and Charles-Sigler, R. (1966) *Tetrahedron Lett.*, 1009.
77. Gil-Av, E., Feibush, B. and Charles-Sigler, R. (1967) in *Gas Chromatography* (ed. A.B. Littlewood), Inst. Petroleum, London, p. 227.

78. Nakaparksin, S., Birrell, P., Gil-Av, E. and Orò, J. (1970) *J. Chromatogr. Sci.*, **8**, 177.
79. Parr, W., Yang, C., Bayer, E. and Gil-Av, E. (1970) *J. Chromatogr. Sci.*, **8**, 591.
80. Koenig, W.A. and Nicholson, G.J. (1975) *Anal. Chem.*, **47**, 951.
81. Charles, R., Beitler, U., Feibush, B. and Gil-Av, E. (1975) *J. Chromatogr.*, **112**, 121.
82. Bonner, W.A. and Blair, N.E. (1979) *J. Chromatogr.*, **169**, 153.
83. Blair, N.E. and Bonner, W.A. (1980) *Origins of Life*, **10**, 255.
84. Frank, H., Nicholson, G.J. and Bayer, E. (1977) *J. Chromatogr. Sci.*, **15**, 174.
85. Frank, H., Rettenmeir, A., Weicker, H. Nicholson, G.J. and Bayer, E. (1980) *Clinica Chimica Acta*, **105**, 201.
86. Wulff, D. (1982) *Pittsburgh Conference, Abstracts*, Atlantic City, New Jersey, p. 791.
87. Frank, H., Nicholson, G.J. and Bayer, E. (1978) *J. Chromatogr.*, **167**, 187.
88. Moodie, I.M. (1974) *J. Chromatogr.*, **99**, 495.
89. Lawless, J.G. and Romiez, M.P. (1974) *Advances in Mass Spectrometry*, **6**, 143.
90. Munson, M.S.B. and Field, F.H. (1966) *J. Am. Chem. Soc.*, **88**, 2621.
91. Arsenault, G.P. (1972) in *Biochemical Applications of Mass Spectrometry* (ed. G.R. Waller), Wiley-Interscience, New York, p. 817.
92. Arsenault, G.P., Dolhun, J.J. and Biemann, K. (1970) *Chem. Comm.*, 1542.
93. Wagaman, K.L. and Smith, T.G. (1971) *J. Chromatogr. Sci.*, **9**, 241.
94. Engel, M.H. and Nagy, B. (1982) *Nature*, **296**, 837.

CHAPTER SEVENTEEN

Mass Spectrometry of Amino Acids and their Derivatives

R.A.W. JOHNSTONE AND M.E. ROSE

17.1 INTRODUCTION

The term 'mass spectrometry' encompasses a number of analytical techniques, the common denominator of which is the ionization of molecules and separation and detection of the ions according to mass-to-charge ratio (m/z). Mass spectrometry has been described in considerable detail elsewhere [1]. Applications to amino acids and their derivatives have been reviewed previously [2–7]. In particular, consecutive reviews by Vetter [6, 7] have covered thoroughly work carried out in the twenty years up to 1978. Two series of biennial reports on mass spectrometry [8,9] and one series of annual reviews on amino acids, peptides and proteins [10] continue to provide coverage of mass spectrometry of amino acids.

Previous reviews of the mass spectrometry of amino acids [2–7] have been classified largely according to the derivative employed for analysis. This article breaks with tradition to give the reader a different perspective. After a discussion of fragmentation pathways, the chapter is divided into sections on the different uses to which mass spectrometry can be put. To some extent the categories are artificial, but generally the subject matter can be grouped clearly into structure elucidation of isolated amino acids, qualitative and quantitative analyses of mixtures, differentiation of isomeric amino acids, identification of amino acids cleaved from proteins as an aid to peptide sequencing, and determination of thermochemical quantities. This organization allows ready assessment of the relative utilities of mass spectrometry and other analytical techniques for a given application. Furthermore, it is the more logical approach for those readers having a particular problem and wishing to learn how it might be solved.

A review of this size cannot be comprehensive in its coverage. Emphasis is placed on fundamental aspects and examples taken from the recent literature. For the same reason, this article will concentrate on protein amino acids with some reference to other, representative amino acids of biological importance.

17.2 FRAGMENTATION PATHWAYS

Unlike other spectroscopic methods for structure determination, mass spectrometry does not measure a well-defined property of a molecule [1]. The appearance of a mass spectrum depends not only on the structure of the sample, but also on the method of sample introduction, the initial energy distribution in the molecule, the method of ionization, and the design and condition of the mass spectrometer (particularly the pressure and temperature of the ion source and the time between ionization and detection). The uncertainty in the measurement hinders quantitative prediction of the mass spectra of novel compounds. Also, the behaviour of compounds subjected to mass spectrometry is not thoroughly understood, the structures and electronic states of ions being largely unknown. Thus, structural analysis of unidentified compounds by mass spectrometry relies on data accumulated, collated and classified from observations on compounds of known structure. Such a store of knowledge is used empirically to rationalize mass spectra in terms of proposed ion structures and the fragmentations they are observed to undergo. Accordingly, the first section of this review outlines fragmentation pathways proposed for amino acids and their derivatives.

The speculative nature of such discussions must be borne in mind – the ion structures, assumed to be formed with minimal change from ground-state neutral molecules, may not in fact be correct but they are useful concepts nonetheless.

17.2.1 Amino acid esters

The mass spectral behaviour of these derivatives is quite simple, allowing their fragmentation to be rationalized [11–14] when few previous data were available. The experimentally observed mass spectral behaviour of amino acid ethyl esters has been shown to be in reasonable agreement with theoretically computed probabilities of bond scissions [15]. The two main fragmentation pathways occur through rupture of the carbon–carbon bonds α to the amine group ('α-cleavage' as shown in Reaction 17.1). The principal driving force of the two reactions is thought to be the stability of the resulting immonium ions. Generally, the peak due to ion *a* is very prominent in mass spectra of ester derivatives, especially for amino acids lacking additional functional groups in the side-chain, R, (see Fig. 17.1(a)). When the side-chain consists of two or more aliphatic carbon atoms, ion *a* ejects an alkene by hydrogen migration from the γ or δ position and affords an ion at m/z 30 (Reaction 17.2) [6, 13]. For a case such as leucine ethyl ester, in which there is a δ-hydrogen atom, a six-centre (McLafferty) rearrangement [16] may be postulated to explain the presence of a peak at m/z 44 (Reaction 17.3 and Fig. 17.1 (a)). This type of rearrangement also explains the further decomposition

$$\left[\mathrm{R} \wr \underset{\substack{|\\ \mathrm{NH_2}}}{\mathrm{CH}} \wr \mathrm{COOR'}\right]^{+\bullet} \longrightarrow \underset{(a)}{\mathrm{R{-}CH{=}\overset{+}{N}H_2}} \quad\text{or}\quad \underset{(b)}{\mathrm{H_2\overset{+}{N}{=}CH{-}COOR'}} \;\text{or}\; \underset{(c)}{\mathrm{R^+}}$$

$M^{+\bullet}$

(R' = H, alkyl)

(17.1)

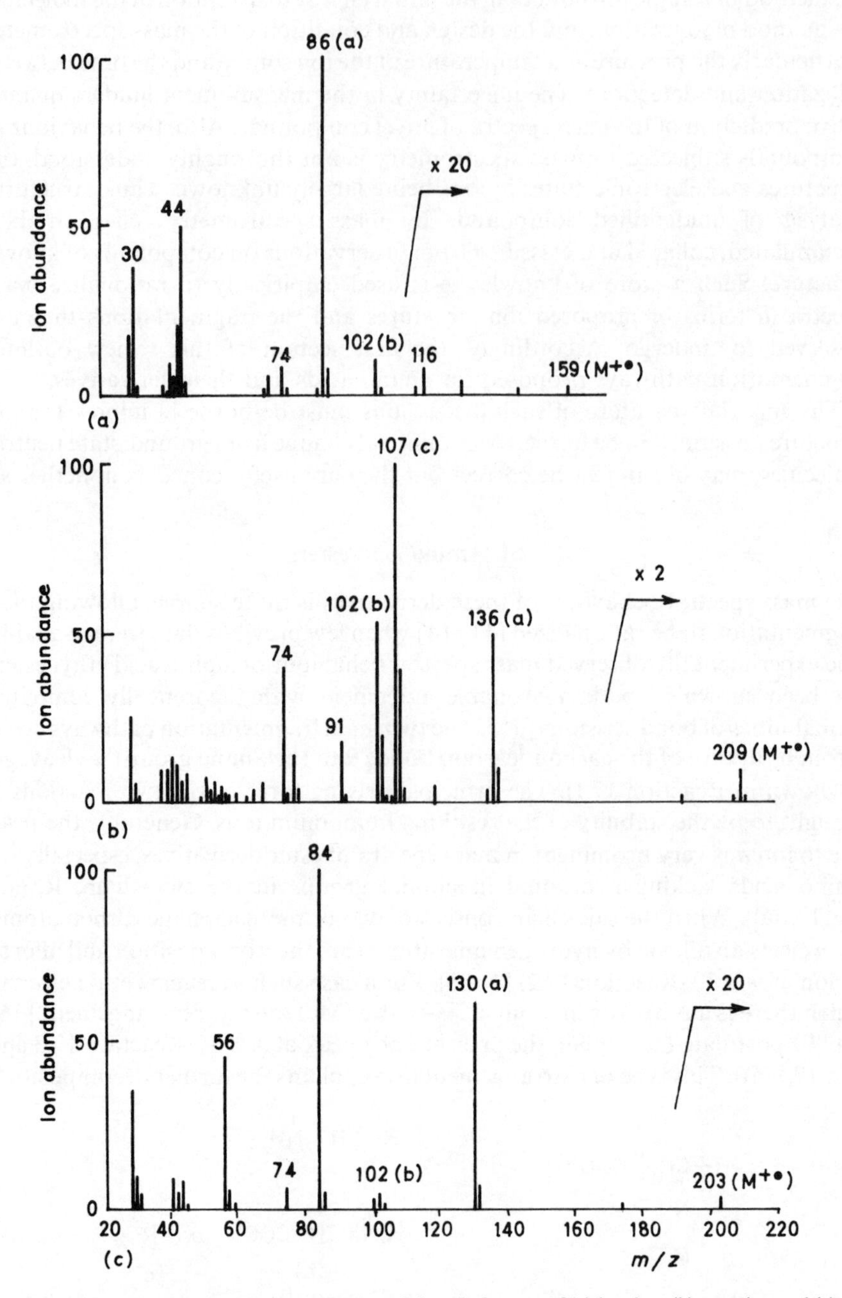

Figure 17.1 Electron impact mass spectra of the ethyl esters of (a) leucine, (b) tyrosine and (c) glutamic acid.

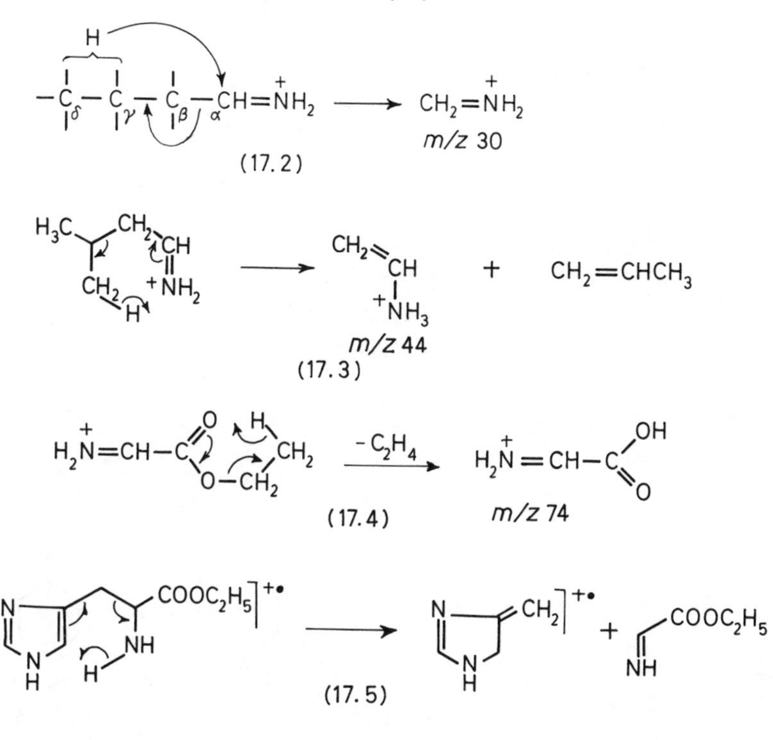

of ion *b* when the ester group, R′, is ethyl or larger (Reaction 17.4). Amino esters with aromatic groups in the side-chain, $H_2NCH(CH_2Ar)COOR'$, afford much more abundant ions *b* and *c* than do the aliphatic esters. This results from the stability of the neutral species, $ArCH_2^{\cdot}$, ejected in forming ion *b* and of the ion, $ArCH_2^+$ (*c*), in which the charge is presumably delocalized into the aromatic ring. The mass spectrum of tyrosine ethyl ester exemplifies these features. (Fig. 17.(b)). Histidine affords not $ArCH_2^+$ but $ArCH_3^{+\cdot}$ and that by a six-centre hydrogen transfer (Reaction 17.5).

The presence of functional groups in the side-chain gives rise to fragmentation pathways in addition to those already described or to some processes that compete with those shown in Reaction 17.1. Examples of the former situation are found in aspartic and glutamic acid diesters. Compare the electron impact mass spectra of the ethyl esters of leucine and glutamic acid (Fig. 17.1 (a), (c)). Whereas the leucine derivative affords very few ions apart from those shown in Reactions 17.1–17.4 the diester of glutamic acid shows additional and abundant ions at *m/z* 84 and 56 due to loss of CH_3CH_2OH and then CO from the side-chain of ion *a*. Additional fragment ions are caused by loss, principally from ions *a* but also from molecular ions, of water from hydroxyl substituents, hydrogen sulphide from thiols, methanethiol from thiomethyl groups, and $R'O^{\cdot}$, R′OH, $^{\cdot}COOR'$ and (R′OH + CO) from side-chains with COOR′ ester groups.

In the case of threonine ethyl ester, the ions *b* at *m/z* 102 are absent. A

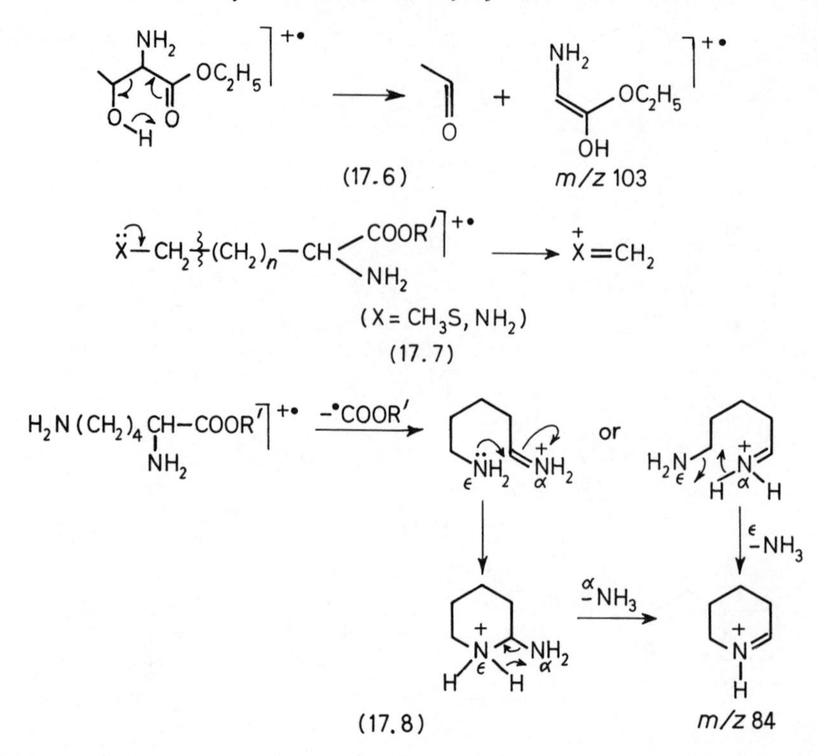

competing six-centre rearrangement of the molecular ion takes precedence, affording an ion at m/z 103 as shown in Reaction 17.6. The presence of sulphur or amino groups in side-chains tends to enhance decompositions which compete with the primary fragmentations shown in Reaction 17.1. A facile cleavage for methionine esters affords a prominent peak at m/z 61 due to the relatively stable $CH_3\overset{+}{S}=CH_2$ ion. Lysine ethyl ester undergoes a similar cleavage to yield $CH_2=\overset{+}{N}H_2$ from the ε-amino group (Reaction 17.7). Another noteworthy feature of the spectrum of lysine ethyl ester is that ejection of ammonia from ion *a* affords the base peak of its electron impact spectrum. Usually, amines do not readily eliminate NH_3, preferring fragmentation routes such as α-cleavage which give stable immonium ions. The elimination of either NH_2 group as ammonia from lysine derivatives is rationalized by formation of a cyclic immonium ion (Reaction 17.8). The volatile arginine derivatives (*17.1*), formed by condensation

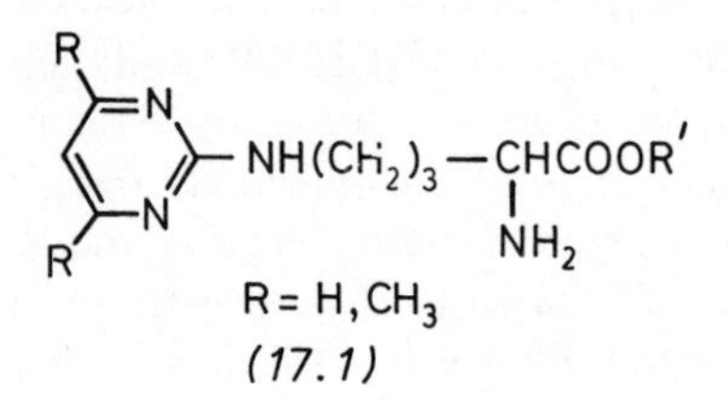

(17.1)

with β-dicarbonyl compounds, also readily eliminate NH_3 from their $(M - {}^{\cdot}COOR')^+$ ions (*a*) [17, 18]. To a lesser extent, ammonia is ejected from the molecular ion of lysine esters and from ion *a* of cysteine esters.

When a heteroatom (N,O,P,S) occurs α to the amine group, the normal α-cleavage reactions are again suppressed because of a competing reaction [19]. The carbon–heteroatom bond is preferentially cleaved with concomitant hydrogen migration and charge retention on either fragment. This is illustrated for a derivative of proline in Reaction 17.9, showing relative abundances in the 12 eV electron impact mass spectrum [19].

Of all the α-amino acid esters, that of proline exhibits the simplest fragmentation. The mass spectrum is comprised almost totally of $(M - {}^{\cdot}COOR')^+$ ions. Because of its cyclic structure, other α-cleavages do not lead to fragmentation. Esters of β- and γ-amino acids have also been investigated [13]. Generally, the loss of the ester function is not a significant process for these compounds, since it does not lead to immonium ion stabilization (Reaction 17.10). The fragmentation is dominated by α-cleavage of the amino group so that, for instance, ethyl γ-aminobutyrate affords an electron impact mass spectrum with base peak at *m/z* 30 (Reaction 17.10). Another biologically important γ-amino acid is the plant hormone, indole-3-acetic acid. Structures such as this, being vinylogous to α-amino acids, behave like their α-counterparts in that loss of the alkoxycarbonyl

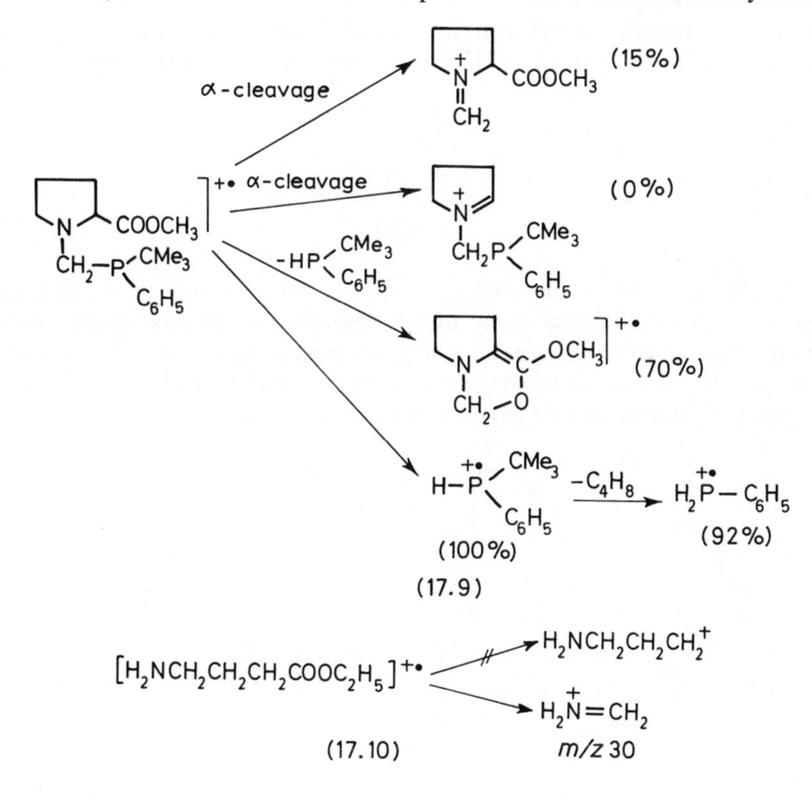

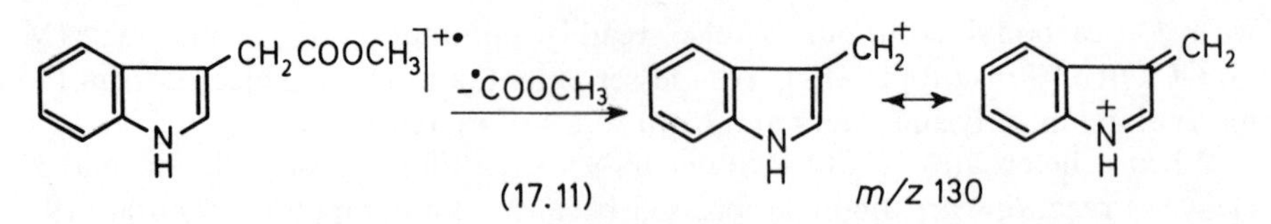

(17.11)

group is very facile. In other words, the double bond allows the charge to be delocalized, at least in part, onto the nitrogen atom (Reaction 17.11).

In summary, the behaviour of amino esters is determined largely by the amino group (α-cleavage reactions). For structures containing additional functionalities of low ionization energy such as aromatic, ester, hydroxyl or amino groups, sulphur or phosphorus, fragmentation strongly reflects these parts of the molecule. Mass spectrometry of amino acid esters has been thoroughly reviewed [2, 4, 6, 13, 14].

17.2.2 Free amino acids

The amino group is again the strongest influence on the mass spectrometric behaviour of this class of compounds (α-cleavage). Thus, it is not surprising that free amino acids follow much the same fragmentation pathways as the esters (Fig. 17.1, R′ = H) [20–22].

Variation in fragmentation with ring size has been investigated [23] for cyclic α-amino acids (*17.2*). Ejection of HOOC˙ is observed only for the larger rings. In

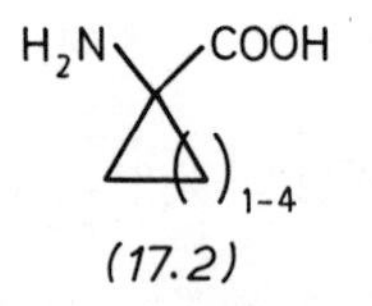

(*17.2*)

the cyclopropane and cyclobutane derivatives, fragmentation of the ring itself becomes dominant. Interaction of amino and acid groups sometimes leads to elimination of water from molecular ions of amino acids. The process is most pronounced in 1,2-disubstituted aromatic compounds in which case it is termed an *ortho*-effect. An example [24] is shown in Reaction 17.12.

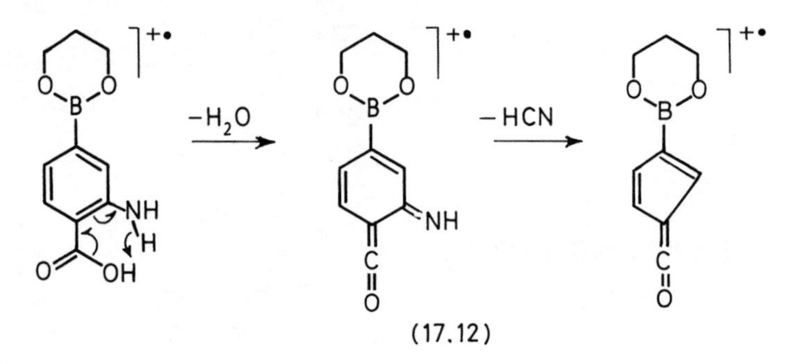

(17.12)

17.2.3 *N*-Acetylamino esters and acids

Mostly the same cleavages occur for the amides as for the amines but additional fragmentations are observed as a result of the presence of the acetyl group (Reaction 17.13) [22, 25–27]. Products of primary α-cleavages, ions *d* and *e*, tend to eject ketene and afford the familiar ions *a* and *b*. With aromatic *N*-acetyl derivatives, formation of ion *c* ($ArCH_2^+$, or $ArCH_3^{+\bullet}$ for histidine) is favoured just as it is for the amino acid esters. Fragmentation to $CH_3CO^{\bullet +}$ (*m*/*z* 43) is a facile process for all *N*-acetylated derivatives. Whilst most studies have employed methyl or ethyl esters, the 4-nitrobenzyl [28, 29], *n*-propyl [30] and amyl [27] esters have been described.

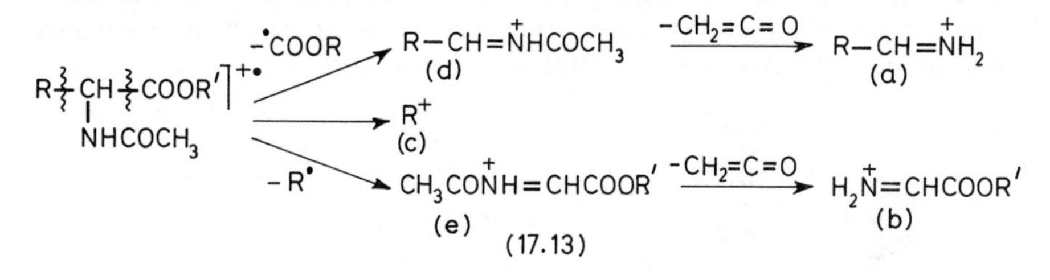

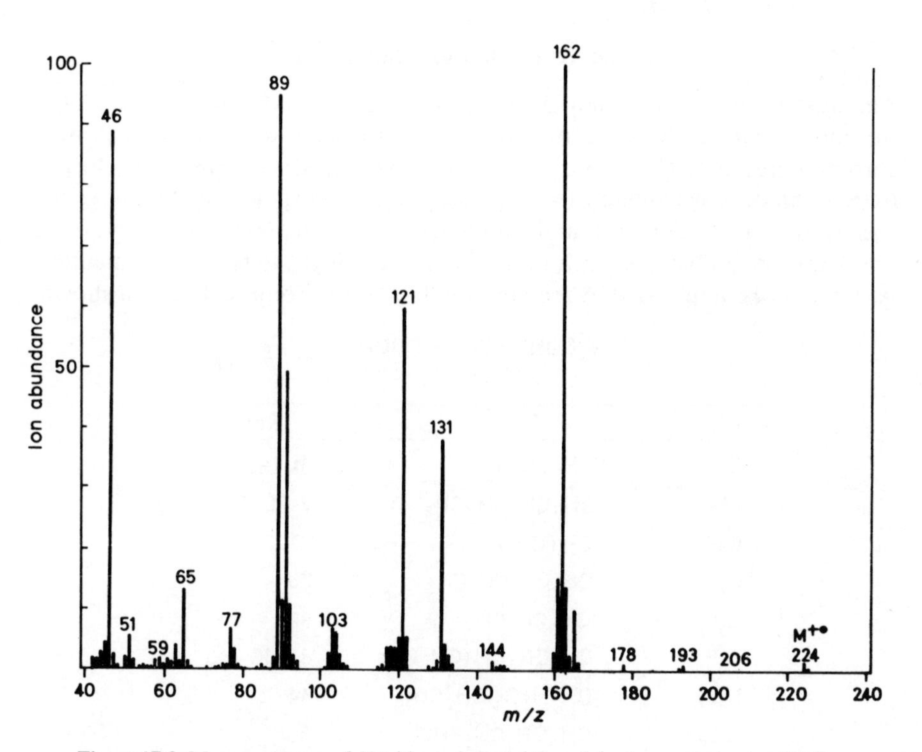

Figure 17.2 Mass spectrum of *N*-trideuterioacetylphenylalanine methyl ester (*17.3*).

As an illustration of the care required in assigning fragmentation pathways, consider the case of the electron impact mass spectrum of *N*-acetylphenylalanine methyl ester. The base peak of the spectrum occurs at *m/z* 162 corresponding to $(M-59)^+$. Incorrectly, this peak might be taken to be ion *d* (loss of the alkoxycarbonyl group from $M^{+\cdot}$). In fact, elimination of the elements of acetamide is responsible for the peak as shown by the mass spectrum of the trideuterioacetylated analogue (*17.3*), illustrated in Fig. 17.2. Since the base peak does not shift from *m/z* 162 after isotope labelling, the deuterium atoms must be ejected with the neutral fragment (Reaction 17.14). Determination of the elemental composition of the ion at *m/z* 162 by accurate mass measurement confirms the result. The process is common to the analogous derivatives of tyrosine, tryptophan and histidine, presumably as a result of resonance stabilization in the $(ArCH{=}CHCOOR')^{+\cdot}$ product ion.

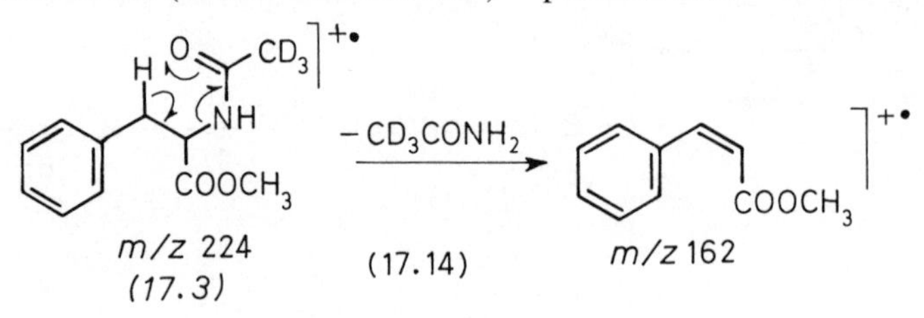

17.2.4 *N*-Perfluoroacylamino esters

Amongst derivatives of amino acids, the *N*-perfluoroacylamino esters (*17.4*) are the most volatile, allowing all protein amino acids to be analysed by gas chromatography (GC; see previous chapter) and hence combined gas chromatography/mass spectrometry (see below). For the purposes of obtaining mass spectra which correlate well and simply with structure, *N*-perfluoroacyl methyl esters are preferable because, unlike the larger alkyl esters, the fragmentation pattern is not complicated by reactions of the alkoxy group such as that shown in

$$R''CONH-CHR-COOR'$$
(*17.4*)

R″	R′	Ref.
CF_3	CH_3, C_2H_5	31–33
CF_3	$CH_3CH_2CH_2CH_2$	7, 30, 32, 34–36
C_2F_5	$CH(CF_3)_2$	37, 38
C_2F_5	$CH_3CH_2CH_2CH_2$	35
C_3F_7	$CH_3CH_2CH_2$	39
C_3F_7	$CH_3CH_2CH(CH_3)$	30, 40–43
C_3F_7	$(CH_3)_2CHCH_2CH_2$	44
C_6F_5	$CH_3CH_2CH_2CH_2$	35

Reaction 17.4. However, having been developed largely for gas chromatographic analyses, the butyl esters have taken precedence owing to their superior GC properties. Therefore, the largest store of knowledge has been built up for *N*-trifluoroacetyl *n*-butyl esters [7, 34]. Since the various alkyl ester groups have little effect on general mass spectrometric behaviour [44, 45] and the mass spectra of *N*-perfluoroacyl esters have been discussed in great detail elsewhere [7, 33, 34], a brief treatment with reference to *N*-trifluoroacetylamino *n*-butyl (TAB) esters suffices here.

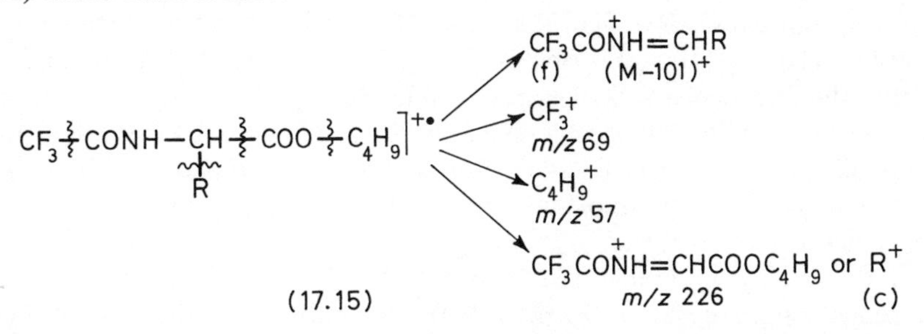

(17.15)

As shown in Reaction 17.15, the TAB derivatives of amino acids undergo α-cleavage yielding $(M-101)^+$ ions, which frequently give rise to the base peaks of electron impact spectra, and R^+ ions in the case of derivatives with aromatic side-chains ($R = ArCH_2$). Other direct cleavages result in ions at m/z 69 (CF_3^+) and 57

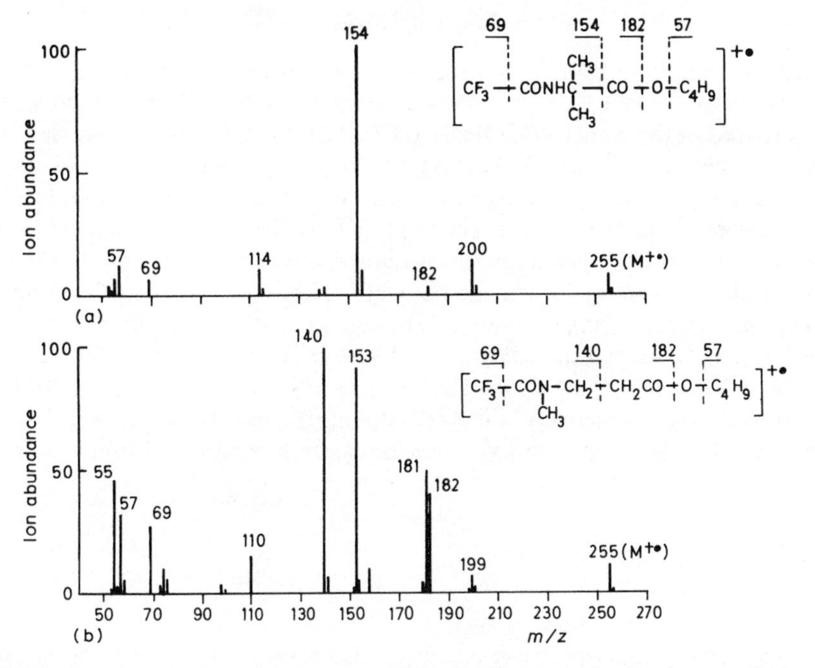

Figure 17.3 Electron impact mass spectra of the *N*-trifluoroacetyl *n*-butyl esters of (a) α-aminobutyric acid and (b) *N*-methyl-*β*-alanine.

($C_4H_9^+$). A double hydrogen rearrangement leads to expulsion of C_4H_7 from the butyloxy group of the molecular ion, affording $(M-55)^+$ ions. Functional groups in the side-chain increase the number of available fragmentation routes and most of the extra processes are analogous to those observed for amino esters as discussed above. For example, the ion *f* of the diester of glutamic acid ejects C_4H_8, C_4H_9OH and $(C_4H_9OH + CO)$ from the side-chain $COOC_4H_9$ group. Also, the perfluoroacyl esters of methionine readily afford the $CH_2=\overset{+}{S}CH_3$ ion.

The presence of an *N*-methyl group further facilitates formation of the $(M-101)^+$ ion, but causes few significant changes in fragmentation behaviour. An ion at *m/z* 110 ($C_3H_3NF_3^+$) is observed only for *N*-methylated compounds (see Fig. 17.3(b)) and, hence, is of diagnostic value [34,45]. The dominating influence of α-cleavage of amines and amides is again clear when the derivatives of β- and γ-amino acids are examined [34,45]. For such compounds, loss of the COOR′ group is a relatively unimportant process. Figure 17.3 compares the electron impact mass spectra of the *N*-trifluoroacetyl *n*-butyl esters of α-aminobutyric acid and *N*-methyl-β-alanine. The former affords $(M-55)^+$ and $(M-101)^+$ ions as expected, but in the latter, the $(M-101)^+$ ion is absent. As shown in the figure, the major peak at *m/z* 140 for the *N*-methyl-β-alanine derivative corresponds to an α-cleavage of the amide. The peaks at *m/z* 181 and 153 are due to ejection from the $M^{+\bullet}$ ion of C_4H_9OH and $(C_4H_9OH + CO)$ respectively [34,45].

17.2.5 Trimethylsilyl derivatives

Trimethylsilylation also is best suited to chromatographic studies and has the advantage of being a fast, one-step derivatization for all groups commonly encountered in this area (NH, OH, SH, COOH). Many workers have described the mass spectrometry of TMS derivatives of amino acids [35,46–59], and *N*-acylglycines [60]. Of these publications, two extensive ones provide much information on fragmentation pathways [55,59]. The TMS derivatives of amino acid alkyl esters [35,59] have also been studied, but this type of derivative has lost the one clear advantage of trimethylsilylation: that of a one-step reaction.

Typical of many TMS derivatives, the amino acids readily afford $(M-15)^+$ ions by loss of a methyl radical from a TMS group, and ions at *m/z* 73, 75, 88, 90 and 147[$(CH_3)_3Si^+$, $(CH_2)_2Si=\overset{+}{O}H$, $(CH_3)_3\overset{+}{Si}=NH$, $(CH_3)_3Si\overset{+\bullet}{O}H$ and $(CH_3)_3Si-\overset{+}{O}=Si(CH_3)_2$, respectively]. The remaining fragmentation tends to be analagous to that of the amino alkyl esters (compare Reaction 17.1 and 17.16). The

$$\left[(CH_3)_3Si-NH-\underset{R}{CH}\text{-}COOSi(CH_3)_3\right]^{+\bullet} \longrightarrow (CH_3)_3Si-\overset{+}{N}H=CHR \quad \text{or} \quad (CH_3)_3Si-\overset{+}{N}H=CH-COOSi(CH_3)_3 \ \text{or}\ R^+ \qquad (17.16)$$

electron impact mass spectrum of the TMS derivative of serine is typical of α-amino acid derivatives and is shown in Fig. 17.4.

Under the silylation conditions, arginine is degraded to ornithine so that the

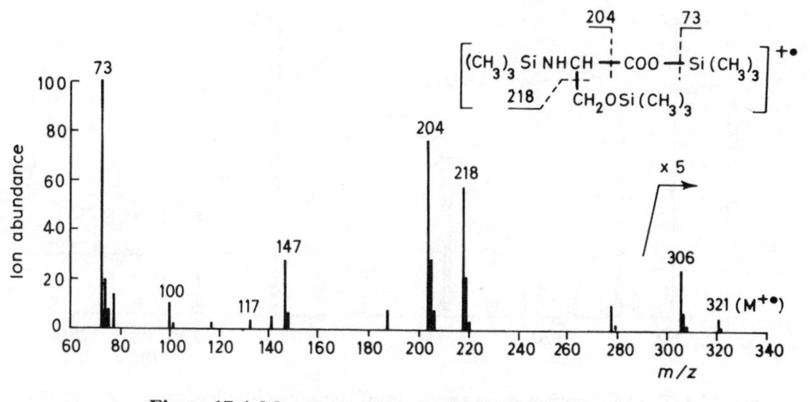

Figure 17.4 Mass spectrum of trimethylsilylated serine.

end product is N^2,N^5,N^5,O-tetra-trimethylsilylornithine [55]. Trimethylsilylated N-methyl amino acids [55], β-amino acids [55, 59] and ω-amino acids [51, 54] have been reported. Their major fragmentations (α-cleavage at the amine group) lead to ions, $(CH_3)_3Si\overset{+}{N}(CH_3)=CHR$ and $(CH_3)_3Si\overset{+}{N}H=CH_2$, respectively.

17.2.6 Thiohydantoins

The fragmentation pathways of phenylthiohydantoin (PTH) derivatives have been established [61–69]. Because the amino group is incorporated into a diazolidine ring (Reaction 17.17, $R' = C_6H_5$), fragmentation is entirely different from that already discussed for acyclic compounds. It is unfortunate that many fragmentation pathways do not reflect the character of the side-chain, R, as

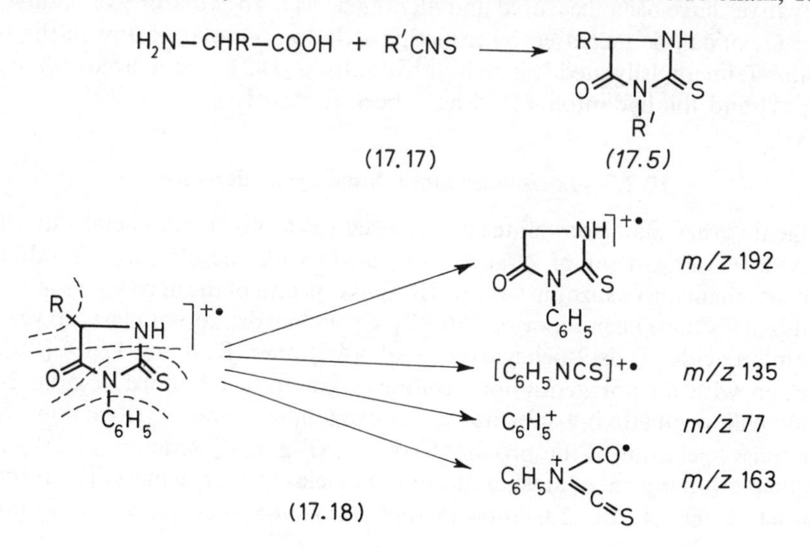

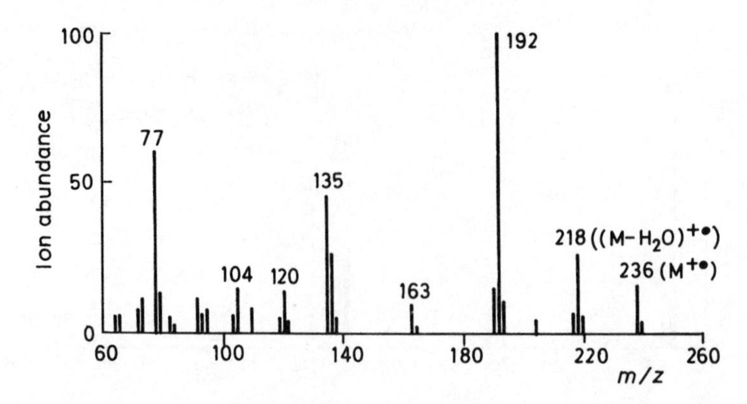

Figure 17.5 Electron impact mass spectrum of the phenylthiohydantoin of threonine.

shown in Reaction 17.18 and illustrated in Fig. 17.5 for the PTH derivative of threonine. Only an $(M - H_2O)^{+\bullet}$ ion indicates the nature of the side-chain. As the ionization energy of functionalities (where these may be considered essentially independent of the total structure) in the side-chain decreases, the spectra become more diagnostic so that sulphur-containing and aromatic amino acids, for instance, show considerable fragmentation triggered from the side-chain rather than from the thiohydantoin nucleus. The mass spectral behaviour of PTH-amino acids is described more thoroughly elsewhere [63].

Other thiohydantoins have been investigated, and their fragmentation resembles that of their PTH analogues. Major variants are 4-bromophenylthiohydantoins $\{(17.5), R' = BrC_6H_4\}$ [70–72], methylthiohydantoins $\{(17.5), R' = CH_3\}$ [65, 66, 73–76] and thiohydantoins $\{(17.5), R' = H\}$ [77–80]. The major metastable ions in spectra of PTH and methylthiohydantoin derivatives have been measured and identified [81]. The volatility of thiohydantoins (*17.5*) can be increased by trimethylsilylation of polar groups in the side-chain. Trimethylsilylated phenylthiohydantoins [82], methylthiohydantoins [83, 84] and thiohydantoins [85] have been reported.

17.2.7 Oxazolones and related cyclic derivatives

Reagents other than isocyanates can be used to convert amino acids into cyclic derivatives. Treatment of *N*-acylamino acids with dicyclohexylcarbodiimide converts them into oxazolin-5-ones. The mass spectra of the derivatives of simple amino acids have been reported [86, 87]. Cyclodehydration of benzyloxycarbonylamino acids yields 2-benzyloxyoxazol-5(4H)-ones (Reaction 17.19) [88]. By reaction with dichlorotetrafluoroacetone, α-amino acids afford volatile 2-bis-(chlorodifluoromethyl)-4-substituted-1,3-oxazolidin-5-ones (Reaction 17.20). The mass spectrum of the proline derivative (Fig. 17.6, Reaction 17.21) shows fragmentation typical of most of the α-amino acids [89]. The mass spectrometric characteristics of the 2-trifluoromethyl-3-oxazol-5-ones (*17.6*) of a limited

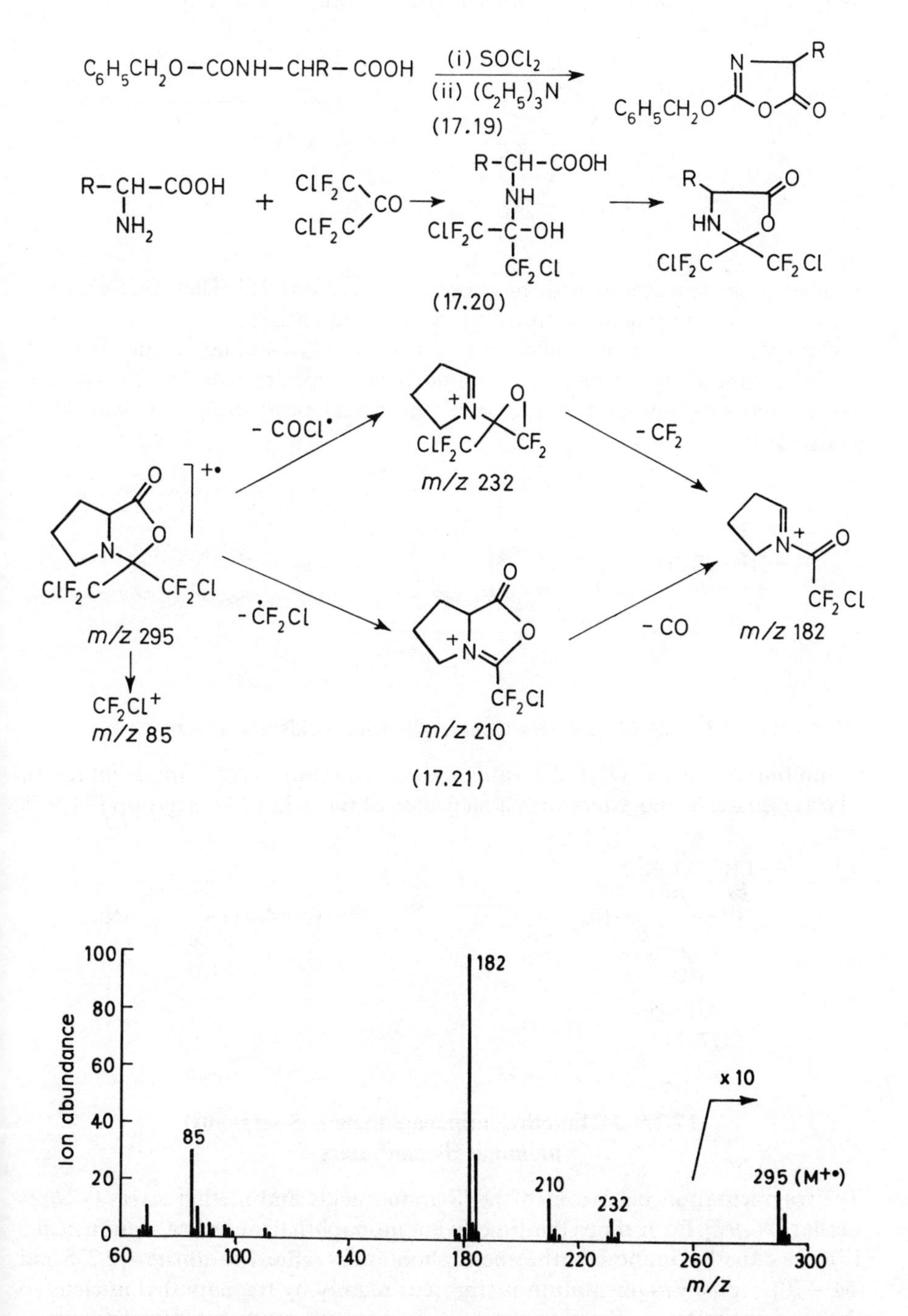

Figure 17.6 Mass spectrum of the 2-bis-(chlorodifluoromethyl)-1,3-oxazolidin-5-one of proline.

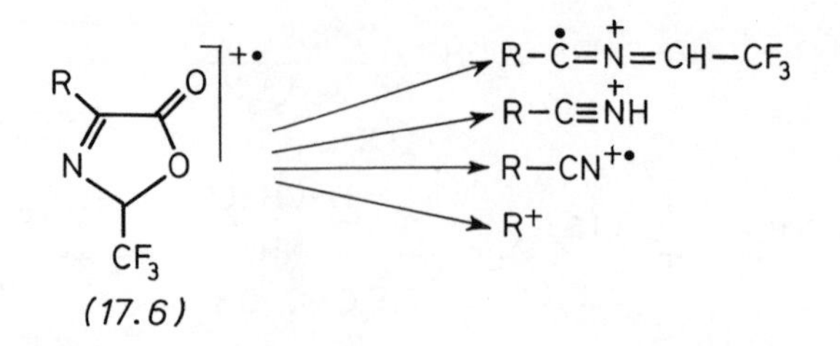

(17.6)

number of protein amino acids have been reported [90,91]. They are formed by reaction of α-amino acids with trifluoroacetic anhydride.

Oxazoles can be produced by reaction of α-amino acids with *o*-naphthoquinone (Reaction 17.22); a limited mass spectrometric study has been carried out [92]. A general review of the mass spectrometry of oxazoles is available [93].

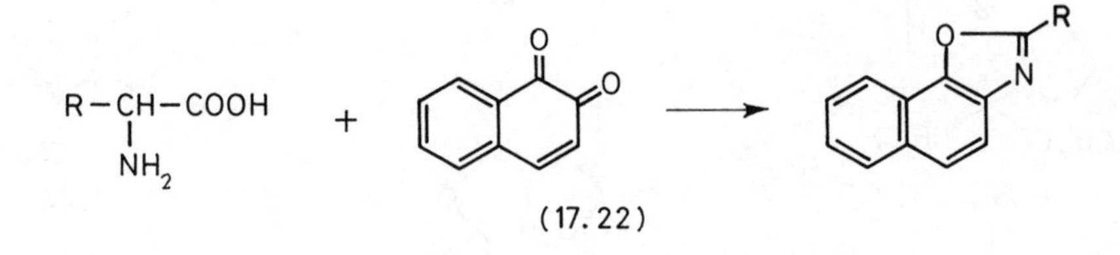

(17.22)

17.2.8 *N*-(2,4-Dinitrophenyl)amino acids and esters

Commonly known as DNP derivatives, these compounds (*17.7*) fragment rather like amino acids and esters with a facile loss of the acid or ester group [94,95].

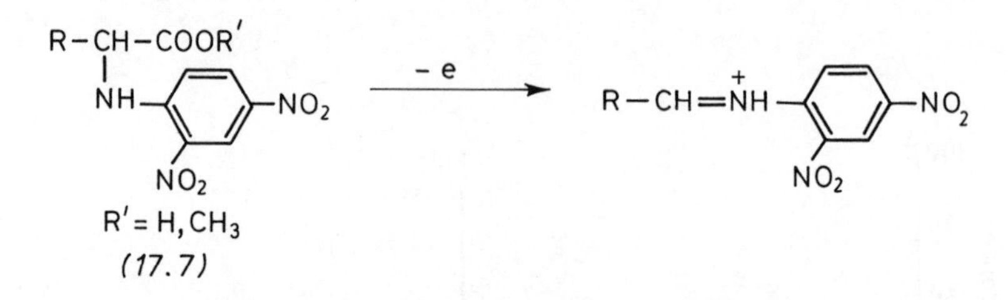

(17.7)

17.2.9 1-Dimethylaminonaphthalene-5-sulphonyl amino acids and esters

The fragmentation behaviour of DNS-amino acids and methyl esters (*17.8*) is similar [96,97]. Both afford the dimethylaminonaphthalene radical-cation at m/z 171, the dimethylaminonaphthalenesulphonic acid radical-cation at m/z 235 and $(M - R^{\cdot})^+$ ions. Fragmentation is triggered mainly by the naphthyl nucleus so that the character of the side-chain, R, is not well represented in their mass spectra.

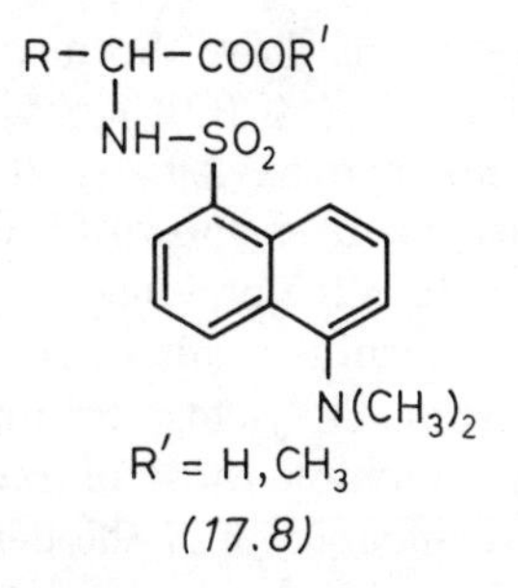

(17.8)

17.2.10 Fluorescamine derivatives

Reaction of α-amino acids with fluorescamine leads, after heat treatment, to spirolactones (*17.9*), the mass spectra of which have been reported [98–101]. Fragmentation of these derivatives is complex and governed largely by that portion of the molecule derived from fluorescamine.

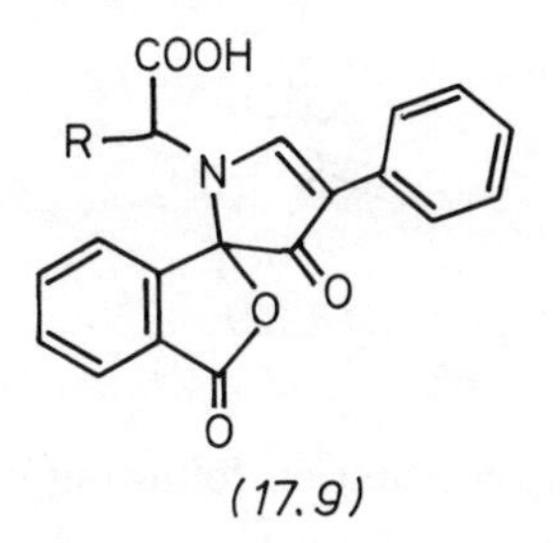

(17.9)

17.2.11 Miscellaneous derivatives

Mass spectra have been reported for derivatives with *N*-acyl groups other than acetyl: *N*-formyl [102], *N*-succinoyl [103], *N*-benzoyl [35, 104] and *N*-nitrobenzoyl [105]. Phthaloylamino acids [106, 107] show facile α-cleavage to afford relatively stable immonium ions, $(M - {}^{\bullet}COOH)^+$, and much fragmentation analogous to that of *N*-acetylamino acids. Reaction of amino acids or peptides with 1-fluoro-2,6-dinitrobenzene or 2-fluoro-3-nitropyridine, followed by reduction and cyclization affords pyrazine derivatives well suited to mass spectrometric analysis [108] (see Reaction 17.26 later). *N*-alkyl- and *N*-aralkyloxycarbonyl derivatives of amino acids, such as (*17.10*), tend to decompose thermally. Derivatives (*17.10*) fragment mass spectrometrically as shown, with facile formation also of $C_6H_5CH_2^+$ and $(M - C_6H_5CH_2O^{\bullet})^+$ ions [109].

$$C_6H_5CH_2OCONHCHRCOOH \xrightarrow[-{}^{\bullet}COOH]{-e} C_6H_5CH_2\,O{-}C{=}O\ \ {}^{+}NH{=}CHR \xrightarrow{-CO_2} C_6H_5CH_2{-}\overset{+}{N}H{=}CHR$$

(17.10)

Two single-step derivatizations have been proposed [110, 111]. Firstly, dimethylformamide dialkylacetals, $\{HC(OR')_2N(CH_3)_2\}$, react with both amino and carboxyl groups (but not hydroxyl) to yield *N*-dimethylaminomethylene alkyl esters (*17.11*) which fragment as shown [110, 112]. Secondly, reaction of amino acids with pivalaldehyde in the presence of trimethylanilinium hydroxide followed by pyrolysis of the resulting salt in the injector of a combined gas chromatograph/mass spectrometer, yields the imine esters (*17.12*) [111]; as expected, their mass spectra resemble those of free amino alkyl esters.

The mass spectrometric behaviour of *N*-thiocarbonyl esters of amino acids (*17.13*) is very similar to that of the free amino esters [113]. Finally, *N*-dithiocarbamates [114], triazines [115] and amides $(H_2NCHRCONH_2)$ [116] derived from amino acids have been investigated by mass spectrometry.

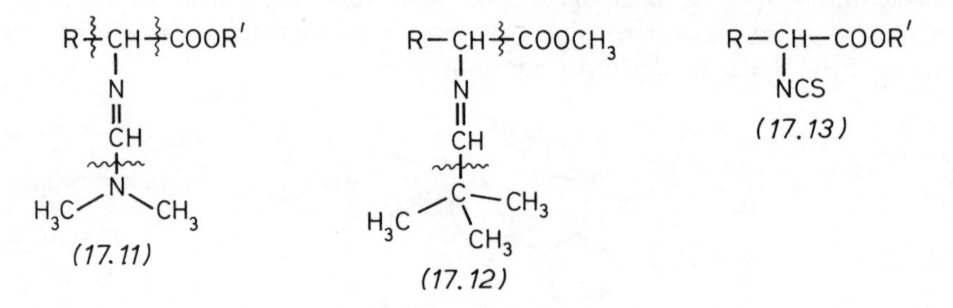

17.2.12 Fragmentation following cationization

Several methods of ionization give rise not to molecular ions, $M^{+\cdot}$, but to protonated molecules, $(M + H)^+$ (see below). The latter ions are more stable or are formed with less ro-vibronic energy than their radical-cation, $M^{+\cdot}$, counterparts so that, generally, they undergo less fragmentation. Also, fewer reaction channels are open to $(M + H)^+$ ions. Whilst radical-cations eliminate radicals or neutral molecules when they fragment, the even-electron $(M + H)^+$ ions decompose almost exclusively by ejection of neutral molecules to give other even-electron ions.

Protonation of amino acids and esters is assumed to occur at their basic sites and the small excess of internal energy of the resulting species allows fragmentation only if a particularly stable neutral molecule is ejected and/or a stable ion is formed [117–119]. Free amino acids eliminate water and carbon monoxide consecutively or (apparently) simultaneously [118, 120, 121]. The ions formed, *a* (Reaction 17.23), can undergo further fragmentation (some ions are formed by ejection of H_2O from any carboxyl or hydroxyl groups, CH_3SH from the

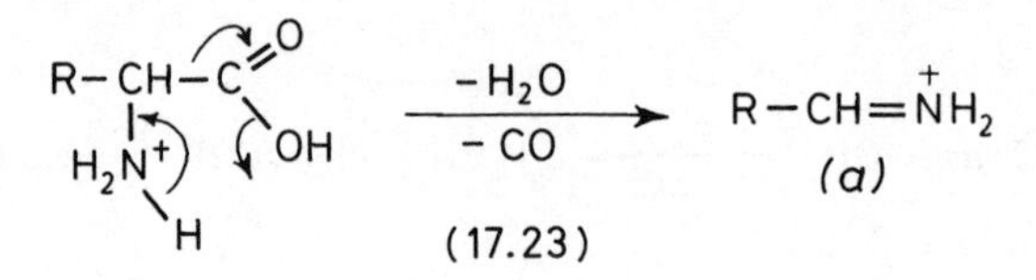

methionine side-chain, and so on). Ammonia is ejected from $(M + H)^+$ ions of amino acids with heteroatoms in the side-chain [121]. This is rationalized by postulation of anchimeric assistance and formation of cyclic ions [117, 121]. Fragmentations of protonated sulphur-containing amino acids [122], cyclic and acyclic α-amino acids, α,ω-diamino acids, ω-amino acids and their esters [121] have been discussed.

In some situations, other cationized species, $(M + X)^+$, may be formed, where $X = NH_4, C_2H_5, C_3H_5$, Na, K, etc. Such adduct ions fragment in much the same way as protonated molecules [7, 121, 123]. The same general principles can be applied to protonated derivatives of amino acids, e.g. *N*-perfluoroacyl esters [7, 30, 36], TMS derivatives [56], phenyl- [66, 68, 69], diphenylindonyl- [124], and methyl-thiohydantoins [66], thiohydantoins [78–80], fluorescamine derivatives [100] and *N*-benzoyl esters [104].

17.3 STRUCTURE ELUCIDATION OF ISOLATED AMINO ACIDS

In this section, purified samples are considered. Mixtures of amino acids, and amino acids in a chemical matrix, are discussed in the following section. Having achieved a pure sample of an unknown amino acid, the method of analysis depends on several factors, as for example, the nature of the sample (thermal stability and volatility), the information required, the amount of sample and, in particular, the instrumentation available to the investigator. It may be unfortunate, but in practice this last consideration may be the overriding one, especially with expensive instruments like mass spectrometers. It is rare that mass spectrometry alone is able to provide an unambiguous structure for an unknown sample. After mass spectrometry, a structure (or a number of structures) which is consistent with the given spectrum may be proposed. Therefore, other analytical techniques should be employed to gain supplementary evidence of structure.

Most mass spectrometers use electron impact (EI) ionization. However, there are obstacles to obtaining useful EI mass spectra from free amino acids, even from the simplest amino acid, glycine [125]. These problems stem mainly from the zwitterionic character of free amino acids which imparts low volatility to these compounds. Also, polar groups enhance intermolecular and intramolecular reactions prior to ionization. For instance, glutamic acid lactamizes, and cysteine and arginine are pyrolysed. The yellow pigment (*17.14*) from South American butterflies was erroneously assigned Structure (*17.15*) because of thermal cyclization in the heated inlet system [126, 127].

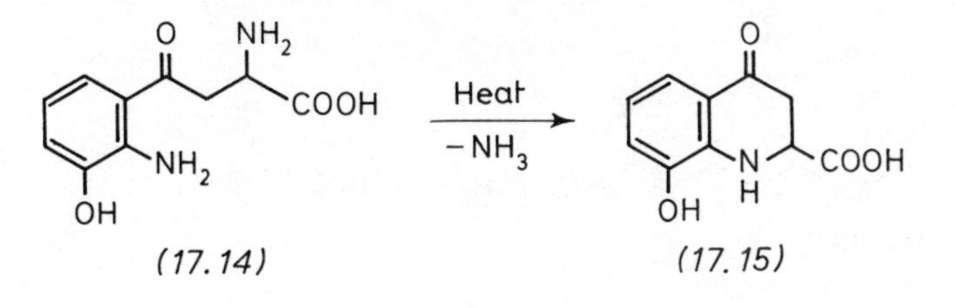

(17.14) (17.15)

The utility of mass spectrometry for amino acids is improved either by derivatization of the sample [1, 128, 129] or by application of so-called 'mild' ionization methods [1, 130–132]. Whilst it seems likely that any isolated, underivatized amino acid may be analysed by the newer methods of ionization, these are not always accessible. Hence, the uses of derivatives and EI mass spectrometry are discussed first. The wisest policy for analysis is to apply minimal derivatization commensurate with the available method(s) of ionization. Extensive derivatization can lead to manipulative losses, unnecessarily complicated mass spectra, long analysis times and the possibility of unknowingly changing the original structure.

17.3.1 Electron impact mass spectrometry

Because of the widely different chemical properties of different amino acids, no single derivative is optimal for all samples. Amino acids with apolar side-chains can be analysed directly or as simple esters by EI mass spectrometry whereas polar side-chains can impart thermal instability, involatility and cause excessive, non-diagnostic fragmentation.

By using the principles of fragmentation established above, mass spectrometry has aided the identification of very many novel amino acids and a complete list here is not possible. One study of note [133] was the elucidation by mass spectrometry of the structure of lysopine (*17.16*) from crown gall tissue. The compound was examined as its ethyl ester. Mass spectrometry complemented UV and IR spectroscopy for the identification of 5-methoxyindole-3-acetic acid (*17.17*) [134]. The identification of cross-linking amino acids in elastin and collagen has been successful with the help of mass spectrometry [135–140]. The structure of lysinonorleucine from elastin was confirmed [135] by the mass spectrum of its ethyl ester (*17.18*). For this compound an unusually abundant doubly-charged ion was observed, corresponding to loss of both ethoxycarbonyl groups, $H_2\overset{+}{N}=CH(CH_2)_4NH(CH_2)_4CH=\overset{+}{N}H_2$. More complex cross-linking species require more extensive derivatization: the reduced ($NaBD_4$) product (*17.19*) from collagen was examined as the *N,O*-trifluoroacetyl methyl ester [136].

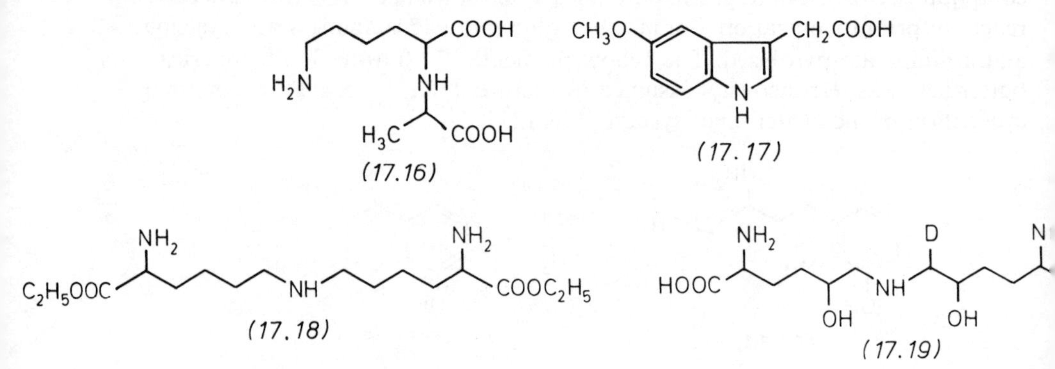

The unusual alicyclic amino acid (*17.20*) from seeds of *Aesculus californica* was identified by spectroscopy including mass spectrometry of the underivatized substance [141]. *N*-(2-Furoylmethyl)glycine is a natural product arising as a result of reaction between glycine and glucose [142]. The mass spectrum of the TMS derivative of the human metabolite, hawkinsin (*7.21*), aided its structure determination [143]. Two other sulphur-containing amino acids (*17.22*) have been isolated from the urine of a baby with probable inherited error in valine metabolism and characterized by mass spectral analysis and synthesis [144]. Column chromatography of the hydrolysate of the antibiotic, gardimycin, was used to isolate amino acids later identified as lanthionine and *β*-methyllanthionine (*17.23*) following esterification and mass spectrometry [145]. Amino acid analysis, TLC, electrophoresis and mass spectrometry have confirmed the presence of aminocitric acid in calf thymus ribonucleoprotein [146]. It has been noted [147] that amino acids separated by polyamide thin-layer chromatography can be subjected to mass spectrometry without elution from the plate. The 'spot' is scraped off and the absorbent is placed on a direct insertion probe for mass spectrometry whereupon the sample is desorbed and evaporated.

Comprehensive lists of novel amino acids, identified at least in part by mass spectrometry, continue to appear [7, 8, 10].

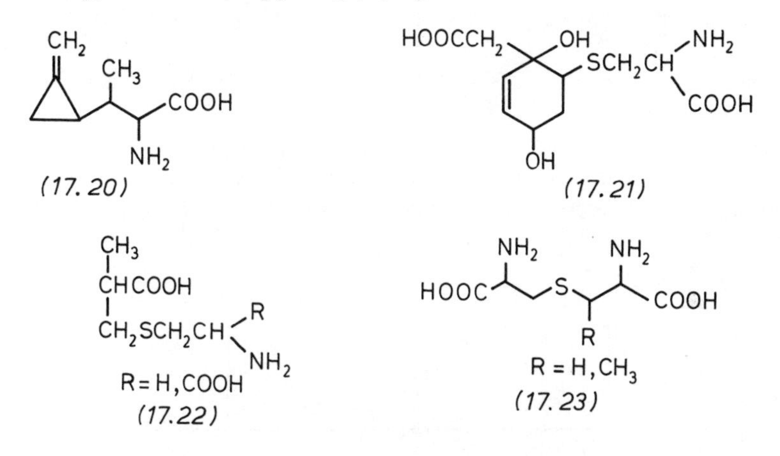

17.3.2 Mild methods of ionization

The EI mass spectra of amino acids and many derivatives often afford a very low or negligible abundance of the important molecular ion as seen in Figs 17.1, 17.2, 17.4, 17.6 and 17.7(a) (although there are exceptions such as PTH derivatives, Fig. 17.5). The use of mild methods of ionization allows underivatized amino acids to be analysed and provides direct information on molecular weight. The most useful mild methods of ionization for amino acids are chemical ionization [1, 130, 148], fast atom bombardment [149–151], field desorption [152–154], and 'in-beam' methods [132, 155–157]. Of these, chemical ionization (CI) is currently the most widespread.

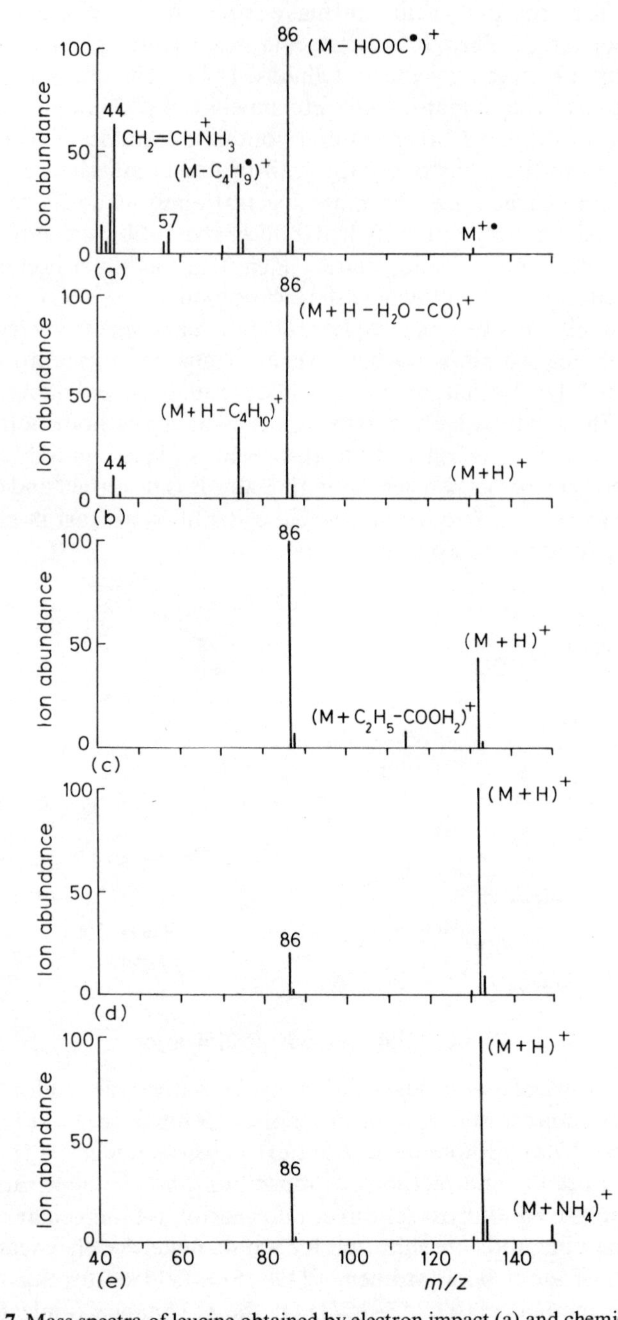

Figure 17.7 Mass spectra of leucine obtained by electron impact (a) and chemical ionization by hydrogen (b) methane (c) isobutane (d) and ammonia (e).

For *chemical ionization*, a reactant gas (G), frequently hydrogen, methane, isobutane or ammonia, at relatively high pressures is ionized by electron impact to primary ions, $G^{+\cdot}$. Through collisions with neutral molecules, G, thermally equilibrated secondary ions, $(G+H)^+$ or $(G-H)^+$, are formed. These reactant gas ions collide with sample molecules, M, and a number of reactions can occur, the principal product in the case of amino acids being the protonated molecule, $(M+H)^+$ (Reaction 17.24). Ions from the sample one mass unit greater than the

$$M + (G+H)^+ \longrightarrow (M+H)^+ + G \quad (17.24)$$

molecular weight are formed. These 'quasi-molecular' $(M+H)^+$ ions are produced with little excess of internal energy and have larger relative abundances than the molecular ions, $M^{+\cdot}$, formed by EI. This difference is illustrated in Fig. 17.7 for leucine. By changing the reactant gas, the degree of fragmentation can be controlled to some extent. Generally, hydrogen [118] and helium [30] cause more fragmentation than methane [30, 117, 118] whilst isobutane, in some cases, affords $(M+H)^+$ ions and virtually no fragment ions [120, 131]. The use of ammonia also leads to little fragmentation of amino acids [119]. Thus, the CI reactant gas can be chosen to suit the information required from the mass spectrum [1]: viz. molecular weight determination, structure-determining fragmentation or both. When interpreting CI mass spectra of unknown amino acids, the possible occurrence of cluster ions such as $(4M+H)^+$ should be borne in mind [158].

Whilst many underivatized amino acids are amenable to CI mass spectrometry, the use of derivatives is mandatory or advantageous in some situations. (i) If the sample is to be presented to the mass spectrometer via a gas chromatograph, then the latter instrument determines the volatility required (see later). (ii) Mass spectrometry may not be the primary method of analysis and hence its requirements take on a subsidiary role. This situation is met with digests from proteins, so that CI mass spectrometry has been applied to fluorescamine [100], and various thiohydantoin and thiazolinone [64, 66, 69, 78–80, 124] derivatives. A related method, Penning ionization, has also been applied to PTH derivatives [68]. (iii) If negative ion, rather than positive ion, CI is required, then derivatization with an electronegative group (*N*-pentafluorobenzoyl, *N*-nitrobenzyl and so on) increases sensitivity and enhances the abundance of negative quasi-molecular ions [1, 105, 159]. (As yet, the application of negative ion mass spectrometry to amino acids has been limited [28, 29, 105, 160–163] but this situation is likely to change.) (iv) Most importantly, some amino acids decompose during heating to effect vaporization. No matter how mild the method of ionization, if thermal decomposition occurs, no quasi-molecular ions will be formed. Examples of this last situation are provided by citrulline and arginine.

Another mild method of ionization is *field ionization* (FI) [1, 130, 131, 152, 154].

Molecules approaching a surface of high curvature maintained at a very high positive potential experience perturbations of their electric fields (molecular orbitals) and quantum tunnelling of a valence electron from the molecule to the anode occurs to give $M^{+\cdot}$ ions. The molecular ions are not formed in highly excited ro-vibronic states and do not fragment excessively. Sometimes $(M + H)^+$ ions are formed because ionization occurs near the anode in regions relatively dense with sample molecules and ion/molecule reactions can occur leading to proton transfer. (Molecular and quasi-molecular ions are distinguishable by changing the experimental conditions.) The general popularity of FI [152, 154] has not been mirrored in the amino acid field because it suffers from the same disadvantage as chemical ionization, namely, evaporation of the amino acid is necessary for ionization. Thermally unstable compounds tend to degrade just as they do in EI and CI mass spectrometry. For example, the EI, FI and CI mass spectra of creatine (*17.24*) reflect only the lactam (*17.25*) because of facile thermal cyclization [131]. Few reports of FI mass spectrometry of amino acids and their derivatives have appeared [114, 131, 164, 165].

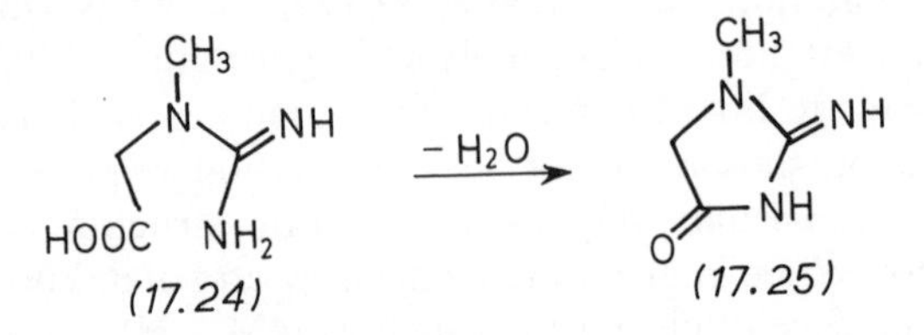

Involatile and heat-sensitive amino acids may be analysed by several newer methods of ionization by using either of two principles: (i) special vaporization conditions to prevent thermal degradation or (ii) ionization in the solid phase. For thermally labile compounds the rate of heating rather than the absolute temperature appears to control the competition between evaporation and decomposition. Rapid rates of heating (10–5000°C s^{-1}) favour evaporation mainly because there is less time for the molecules to transfer energy into unstable vibrational states and yet the molecules can gain sufficient kinetic energy to evaporate [132]. When a sample is heated very rapidly and evaporated directly into the electron beam by use of an elongated direct insertion probe, the method is termed *'in-beam' electron impact ionization.* Analogously, deposition of the vaporized sample directly into the plasma of reactant gas ions for CI is termed *'in-beam'* or *desorption chemical ionization.* Free amino acids, including arginine, afford useful 'in-beam' EI spectra with molecular weight information [166]. The more useful in-beam CI method produces very good spectra of arginine [166–172] and creatine [167, 168, 172] (see Fig. 17.8(c)). Despite the frequently transitory nature of in-beam mass spectra, several studies have proved its utility [119, 155–157, 168–172], especially when using an inert probe tip, of Teflon for example, to reduce irreversible adsorption.

A simple variant of the rapid heating method which produces very impressive positive and negative ion mass spectra of many biological compounds, including

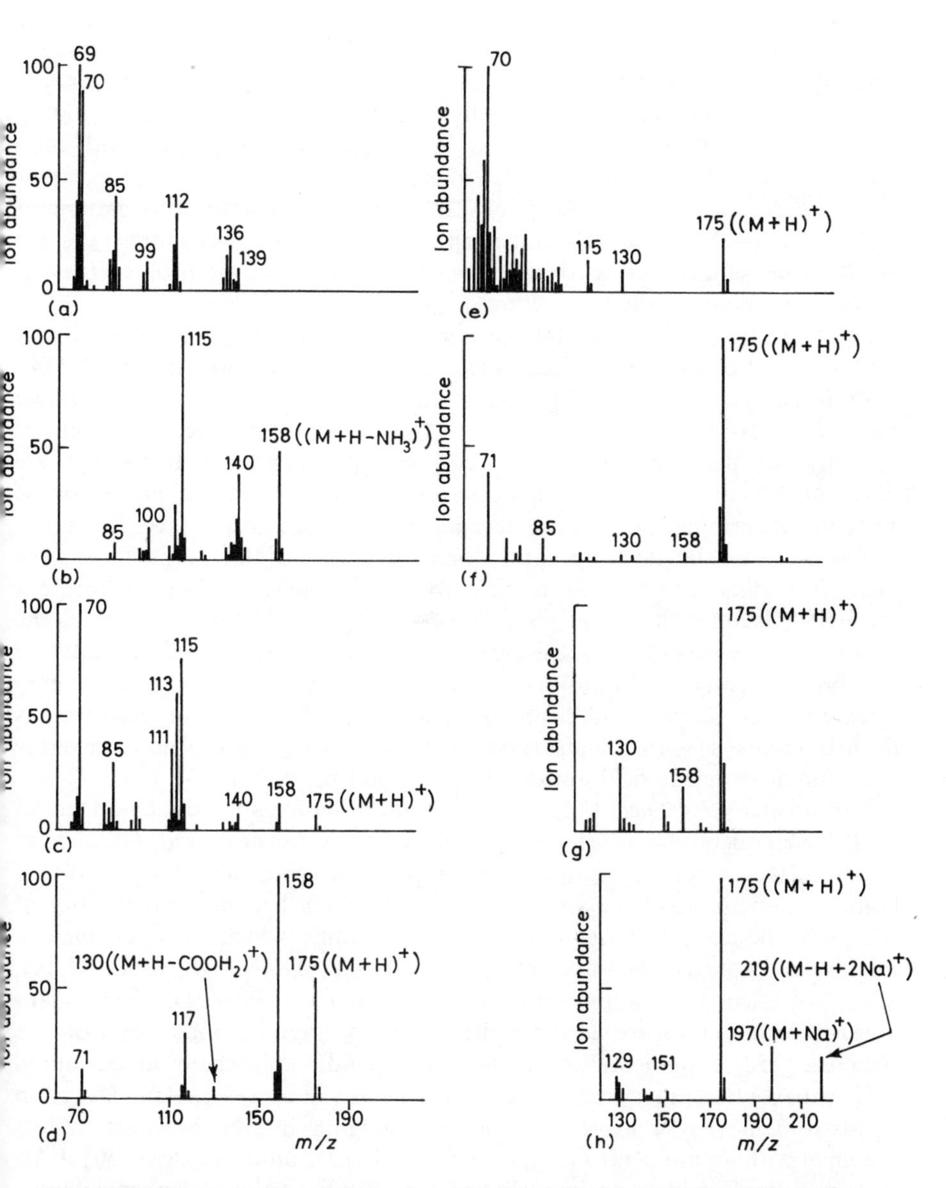

Figure 17.8 Mass spectra of arginine (mol. wt 174) obtained by (a) electron impact ionization, (b) chemical ionization by isobutane, (c) in-beam chemical ionization by isobutane (d) field desorption, (e) fast atom bombardment, (f) rapid heating for liquid chromatography/mass spectrometry, (g) secondary ion mass spectrometry, and (h) [^{252}Cf] radionuclide ionization. Figures 17.8 (c), (e), (f) and (h) have been taken with kind permission of the publishers of references 171, 149, 236 and 188 respectively.

amino acids [173] (Fig. 17.8(f)) is best described as a method of analysing mixtures by combined liquid chromatography/mass spectrometry and is discussed later.

Field desorption (FD) [152–154] is a simple modification of field ionization in which the sample is coated onto the anode of a field ionization mass spectrometer, later to be desorbed during ionization by the same principle as for FI (see above). Ionization and desorption can occur at ambient temperature or with mild heating. As with FI, $M^{+\cdot}$ or $(M + H)^+$ ions are formed, the latter being usual for amino acids, but field desorption sometimes gives rise also to other cationized molecules $(M + X)^+$ (X = Na, K, Li). Such ions can be induced by adding salts, e.g. NaCl, to the sample, just as the addition of an acid to a zwitterion such as an amino acid enhances the production of $(M + H)^+$ ions [174].

Field desorption is a well-established and sensitive technique for analysis and determination of molecular weights of free amino acids and their salts [131, 174–181]. It has also been applied to phenylthiohydantoins [67] and fluorescamine derivatives [101], but its strength lies in the fact that the molecular weight of virtually any underivatized amino acid may be ascertained. Its failings are practical difficulties resulting in poor reproducibility and the preclusion of structure determination because there are so few fragment ions. The latter problem can be alleviated by raising the temperature of the sample to observe thermal fragments or by inducing desorbed quasi-molecular ions to fragment through collision with neutral gas molecules, i.e. by 'collisional activation'. Some translational energy of the quasi-molecular ions can be converted into internal ro-vibronic energy by ion/molecule collisions. The increased vibrational energy often initiates fragmentation, the product ions of which are observed as metastable peaks [1, 182]. This powerful technique has been applied successfully to amino acids (using field ionization) [164] and peptides [183].

Fast atom bombardment [149–151] does not suffer from the disadvantages of the FD method, but because few reports of its use have yet appeared, it is less easy to assess. It is practically simple. The sample, as a solid, colloid or solution is bombarded with rapidly moving atoms (Ar, He or Xe). The sudden deposition of energy in the sample causes ionization in a manner which is as yet unclear. Positive and negative ions, normally $(M + X)^+$ and $(M - H)^-$ where X = H, Na, K, etc., are formed with approximately equal efficiency. Unlike FD, the abundant quasi-molecular ions are accompanied by many fragment ions diagnostic of structure [150]. Fragmentation is similar to that induced by chemical ionization. It is probable that any underivatized amino acids will be amenable to fast atom bombardment (FAB) mass spectrometry, judging by the excellent results obtained with amino acids [151], peptides and other biomolecules [150]. FAB mass spectra of histidine, arginine (Fig. 17.8(e)), glycine, valine and phenylalanine have been published [149, 151].

The rapid, localized heating caused within a solid sample by bombardment with laser beams, radionuclide fission products or ions can also cause ionization [1]. *Laser desorption* requires rather sophisticated instrumentation, but spectra of free amino acids have been reported [184–187]. Generally, abundant cationized molecules are obtained in the positive ion mode and $(M - H)^-$ ions in the negative ion mode. When a sample is deposited onto a thin nickel foil, the back of which is coated with a radionuclide [^{252}Cf], fission products penetrate the nickel foil and lead to localized 'hot spots' sufficient to cause rapid vaporization and ionization, but not decomposition of the sample [188, 189]. Both positive and negative molecular or quasi-molecular ions are generated, generally accompanied by fragment ions [190]. Mass spectra with large peaks for $(M + H)^+$ ions have been recorded for arginine (Fig. 17.8(h)) and cystine with this method [188, 189]. Unfortunately, its radiochemical hazards limit its use. The mass spectra it affords have been compared to those obtained by laser desorption [185].

A substance to be analysed by *secondary ion mass spectrometry* (ion bombardment) is coated onto a surface, frequently metallic, and bombarded by ions having large kinetic energies, such as Ar^+. Charge exchange reactions lead to sample ions diagnostic of structure [1, 191, 192]. Again, abundant positive and negative molecular or quasi-molecular ions of free amino acids are produced [193–198]. Secondary ion mass spectrometry also can be coupled with liquid chromatography for analysis of mixtures of underivatized amino acids [197, 199] (see below).

17.3.3 Summary

No one ionization technique is always superior for identification of isolated amino acids. In most cases, judicious manipulation of a sample through derivatization will make it suitable for the technique(s) available to the worker. The capabilities of many methods of ionization for intractible compounds can be assessed to some extent by reference to Fig. 17.8 showing mass spectra of arginine obtained in different ways. Recourse to several methods of ionization, if available, is advisable, since many ionization techniques yield complementary information. A good example of the use of several spectroscopic techniques is provided by the structural elucidation of alamethicin I and II [200].

17.4 QUALITATIVE ANALYSIS OF MIXTURES

Identification of amino acids in matrices by mass spectrometry can be carried out in two distinct ways:

(i) Using a direct insertion probe (a non-chromatographic inlet), and
(ii) with combined gas- or liquid-chromatography/mass spectrometry to separate components prior to mass spectrometry.

17.4.1 Non-chromatographic inlets

Direct insertion probes and hot and cold inlets have little potential for fractionating samples unless the components differ considerably in volatility. If this is the case, temperature programming of the inlet will effect at least partial separation. In general, components of mixtures remain largely unseparated from each other and from impurities. Since mass spectra of mixtures are usually too complex to be interpreted unambiguously, it is reasonable to ask if there is any justification for use of direct analysis of mixtures except in specialized cases. Its major advantages over methods based on chromatographic inlets are speed of analysis, avoidance of losses on passage through chromatographic columns, and free choice of ionization method, leading to less emphasis on derivatization. A gas chromatographic analysis can take an hour or more and requires volatile derivatives. Liquid chromatography/mass spectrometry may not require volatile derivatives, but is at a relatively early stage of development (see below).

As an example of a special case for clinical diagnosis, underivatized amino acids can be determined in human biological fluids by chemical ionization [201–203]. On the other hand, it has been shown that, under CI conditions, free amino acids can associate [204]. Such association reactions could lead to erroneous conclusions about the composition of mixtures. A further potential pitfall is suppression, when the abundances of ions created from a mixture are not a true reflection of the composition of the unionized mixture because one component suppresses the ionization of others. Characterization of a mixture of 'difficult' amino acids (arginine, citrulline, cysteine, histidine and tryptophan) as their dimethylaminomethylene alkyl esters has been effected with a direct insertion probe and temperature-programming of the ion source [112]. Secondary ion emissions from mixtures of free amino acids have been investigated [205, 206].

Underivatized amino acid mixtures can be examined also by FD mass spectrometry [177, 200, 207]. Because of the complexity of mass spectra of mixtures, the use of mild methods of ionization like FD, which yield only a small number of mass peaks per component, is advantageous. Paradoxically, identification of a novel compound becomes more difficult as its number of peaks decreases. This problem can be solved by using metastable ion analysis [1, 208–210]. In this method, one component is separated from all others by setting the first analyser to pass only ions of a single mass unique to that component. All others are 'defocussed' and hence, do not contribute further. Collisional activation [182] of the selected ions causes fragmentation and a diagnostic mass spectrum is recorded by scanning the second analyser. The technique is best used in conjunction with a mild method of ionization like FI [164] or FD. Ideally, if only $M^{+\cdot}$ ions are produced, each one will be unique (except for leucine and isoleucine) at least for the common protein amino acids. Mixtures of dansyl amino acids have been examined using a similar principle and electron impact with low electron beam energy [211].

The direct probe method is fast and works well with simple mixtures, but there is a number of disadvantages: mixtures of isomeric compounds cannot be separated by mass selection, artifact peaks, suppression effects, intermolecular associations can occur, and fragment ions of one component and molecular ions of another may be coincident in mass. In fact, it is not advisable to use this method to identify low molecular weight components in the presence of several higher molecular weight ones. Despite its sophistication, the method is not likely to become a routine method in the amino acid field because of well-established methods like ion-exchange chromatography for which sample involatility is no barrier.

17.4.2 Combined gas chromatography/mass spectrometry (GC/MS)

Both being gas phase methods, gas chromatography and mass spectrometry are readily made compatible [1,212,213]. Modern, computerized GC/MS and its capabilities have been explained in depth with reference to a mixture of derivatized amino acids and dipeptides in urine [1]. With computer control, mass spectra can be recorded repetitively every few seconds (or milliseconds if necessary) whilst compounds elute from the GC column. In this way, each component is analysed in turn. Computerized systems are mandatory for thorough qualitative analysis by GC/MS. Its use for identification of amino acids is so widespread that comprehensive coverage is not feasible here, but is given elsewhere [8]. GC/MS lends itself so readily to quantitative studies that the majority of publications belong more properly in the section following this one.

Whilst the mass spectra of alkyl esters of amino acids are more readily interpreted in terms of structural features than those of more complex derivatives, the free amine group imparts poor gas chromatographic properties. For GC/MS, more extensive derivatization is applied because the requirements for GC take precedence. The most popular derivatives are the highly volatile *N*-perfluoroacylamino *n*- or *i*-butyl esters [7, 34], which render even arginine amenable to GC/MS. Trimethylsilylation is less favoured for several reasons. It is difficult to attain conditions which convert all amino acids to a single product because of the widely differing reactivities of the groups derivatized. Also, once formed, TMS derivatives are prone to hydrolysis and complex rearrangements during mass spectrometry; like most other derivatizations for GC/MS, trimethylsilylation does not succeed with arginine. While some of the newer derivatives show promise (e.g. oxazolidinones [89]), many procedures are suitable only for the simpler amino acids [91,110,111]. Therefore, no particular advantages accrue from the use of derivatives other than *N*-perfluoroacylamino esters unless they are intended for specialized procedures like negative ion chemical ionization.

The plant hormone, indole-3-acetic acid [214], and its 4-chloro analogue [215,216] have been identified in Scot's Pine and peas respectively. Amino acids are formed in simulated prebiotic syntheses [87,217] and extraterrestrially [217]. In many situations, the resolution provided by packed GC columns is not sufficient to separate all amino acid derivatives and capillary columns are used to increase resolution. Separation of *N*,*O*-heptafluorobutyroylamino i-butyl esters in a clinical study [41] demonstrates the use of high efficiency GC columns. The GC/MS combination has the significant advantage over simple chromatographic methods that it provides information on the structure of uncommon or unexpected amino acids. Some representative analyses are of γ-carboxyglutamic [218] and β-carboxyaspartic acid [219], *N*-methylleucine, *N,O*-dimethyltyrosine and statine (4-amino-3-hydroxy-6-methylheptanoic acid) [207], *N*-acylglycines [220], *N*-nitrosoamino acids [221, 222], aminophosphonic acids [223], and ω-amino acids [54].

Characterization of natural metabolites [42, 224, 225] in human urine by GC/MS is playing an important role in diagnosis and characterization of diseases, especially inherited metabolic disorders [226]. The diagnosis of some mentally retarded patients as having tyrosinosis was disproved by analysis of their urine for amino acids and carboxylic acids as TMS derivatives [227]. An inborn error of leucine metabolism gives rise to *N*-isovalerylglutamic acid as a minor but diagnostic metabolite [228]. Following administration of [^{15}N]- or [^{2}H]-labelled substances (ammonium chloride, amino acids, etc.), their metabolic fates can be traced by GC/MS because mass spectrometry has the ability to distinguish by their characteristic masses compounds labelled with stable isotopes [229–232].

Finally, as a warning, it should be noted that, during derivatization to *N*-trifluoroacetyl *n*-butyl esters, a mixed disulphide was formed from cystine and homocystine in mixtures of amino acids [233].

17.4.3 Combined liquid chromatography/mass spectrometry (LC MS)

This method combines the advantages of the direct insertion probe method and GC/MS inasmuch as volatile derivatives are not necessary and components are separated prior to mass spectrometry. Unfortunately, coupling (interfacing) of the two techniques is not simple [213, 234]. Three approaches to the combination show particular promise and these have been discussed in detail [1].

The effluent from the LC column can be heated extremely rapidly to vaporization and the majority of the solvent removed by molecular diffusion through a jet separator [173, 235, 236]. Remaining droplets, which have been charged by thermal and frictional ionization during passage through the interface, carry the eluates and pass into the mass spectrometer. The droplets impinge onto a heated metallic probe whereupon they vaporize and some of the resulting molecules appear as ionized species. Ion/molecule reactions occur to produce $(M + H)^+$ and $(M - H)^-$ ions. This technically simple instrument provides mass spectra as good as, if not better than, field desorption for free amino acids as shown for arginine, the base peak of which is due to $(M + H)^+$ ions (Fig. 17.8(f)) [236]. The sensitivity for full mass spectra is 1–10 ng, a value comparable to that for GC/MS. Since the cost of this instrument is no more than that of a GC/MS system, it has considerable potential.

The LC eluate can be applied continuously to a moving belt. The belt is passed into a vacuum where solvent is removed by an infra-red heater, after which the solute remaining on the belt is transported mechanically into the ion source of a mass spectrometer and is evaporated by a powerful heater. When the belt protrudes right into the ion source, the solute evaporates near the electron beam, affording 'in-beam' EI or CI mass spectra. This system has been developed into a useful (and commercially available) instrument [237] but has been applied little to amino acids. When the solute on the belt is bombarded with argon ions, secondary ion mass spectra of amino acids are produced [197, 199]. A similar system with laser bombardment of the substances on the belt is feasible [187].

In another commercial LC/MS system [213, 237], a splitter directs just a small proportion of the LC effluent into a chemical ionization mass spectrometer. The solvent, or another injected gas, is used as CI reactant gas. Alternatively, this 'direct liquid introduction' method is compatible with the smaller flow rates of capillary liquid chromatographic columns without splitting. Using the latter technique, it was possible to analyse glutamic acid [238].

In the amino acid field, no real problems have yet been solved by LC/MS, but the future looks promising. Gradient elution and buffers, at least volatile ones, are acceptable for present instruments. Improvements in micro-LC methods will aid considerably the LC/MS combination.

17.5 QUANTITATIVE ANALYSIS

Quantification by mass spectrometry is not straightforward, basically because mass spectrometric analyses are not entirely reproducible. To estimate the amount of a substance, an internal standard must be added to the sample to be measured, so that ratios and not absolute values of ion abundance may be determined. The response of the mass spectrometer can be calibrated with different, known amounts of the substance under investigation against a fixed amount of internal standard, before the true sample and a known amount of internal standard is analysed. A calibration graph of sample and internal standard ratios provides a quantitative measure.

The measurement of sample and standard is usually performed by 'selected ion monitoring'. Rather than recording a whole mass spectrum, the mass spectrometer is tuned to monitor just one mass peak, or a small number of peaks in turn. In this way, an increase in sensitivity of 2–3 orders of magnitude over full mass spectral scanning is gained because, for the entire time that the compound(s) of interest resides in the source, the mass spectrometer is recording only selected ions for that compound and not the gaps between peaks or regions of the mass spectrum which contain few, if any, peaks. Measurement of a full spectrum can be attained with about 1 ng of sample but, in the selected ion monitoring mode, detection limits are of the order 0.1–100 pg. The fewer peaks (compounds) that are monitored, and the greater the percentage of ion current carried by each peak, the better the sensitivity.

The method is not only sensitive, but also highly specific. Consider an impurity which resides in the ion source at the same time as the compound to be measured (e.g. a co-eluting impurity in GC/MS). As long as the compound of interest has at least one unique mass peak, that *m/z* value can be monitored and the impurity will not be detected at all. Co-eluting compounds are, in effect, separated by mass. The mass spectrometric conditions can be chosen to minimize interference. For example, ammonia as a chemical ionization reactant gas does not ionize hydrocarbons, so much lipid 'background' can be ignored by its use. Mild methods of ionization have a special role in quantitative studies. Such methods (positive ion and negative ion CI especially) afford only a small number of peaks per sample so that there is less chance of interfering ions from the chemical matrix. Also, because one ion, frequently the quasi-molecular ion, carries a large proportion of the total ion current, sensitivity is very high when that ion is monitored. It should be borne in mind that, if the efficiency of ionization is lower for the mild method of ionization, then overall sensitivity may be better with EI ionization, even by measuring a peak with less percentage of total ion current. Since the number and abundance of background ions increase with decreasing mass, selected ions should be of high mass (preferably over 250), again favouring mild methods of ionization because they concentrate the ion current in the higher mass regions.

Careful derivatization can enhance quantitative mass spectrometry considerably [35]. The type of fragmentation undergone is of no real concern, but a derivative which affords at high mass one large peak is favoured. Perfluoroacyl groups do add considerable mass but, on EI, these derivatives afford few large peaks at high mass suitable for monitoring. Thus, the use of *N*-perfluoroacylamino esters with CI and not EI is recommended. Negative ion chemical ionization, with proper electronegative derivatizing reagents [1, 105, 159, 239], can achieve extremely good sensitivity and specificity. For example, 5 pg of indole-3-acetic acid is detectable as its pentafluorobenzyl ester by GC/MS and negative ion CI with ammonia reactant gas [239].

If the internal standard is added to the crude sample (e.g. biological tissue) and behaves identically to the compound of interest during work-up and derivatization, it compensates for accidental losses, and incomplete extractions and derivatizations. The best internal standards are homologues, enantiomers or stable isotope-labelled analogues. If the compound of interest co-elutes with its isotopically labelled internal standard during GC/MS their different masses allow both to be detected by mass spectrometry without mutual interference. By choosing a homologue affording ions common to the compound of interest, or an enantiomer, just one mass peak need be monitored to measure both compounds. Of course, the internal standard and compound of interest must not reside in the ion source at the same time for they will not be distinguished. This 'single ion monitoring' mode is advantageous because it is the most sensitive technique available and one which any mass spectrometer can perform without recourse to sophisticated, expensive computerization.

Quantitative mass spectrometry has been described in detail in two books [1, 240], and a review [241].

17.5.1 Non-chromatographic inlets

This method suffers from all the disadvantages mentioned in the previous section and, because of the compounding difficulties of quantitative measurement, it cannot be recommended except for pure compounds and simple mixtures or when speed of analysis of routine (repetitive) samples is of the essence. As examples, blood specimens can be screened for amino acids using [^{2}H]- or [^{15}N]-internal standards and CI mass spectrometry [201, 242]. Rapid screening of drugs [243] and metabolites [202] is possible and the [^{14}N]/[^{15}N] isotope abundance ratio in simple amino acids can be measured by FI and collisional activation mass spectrometry [164].

17.5.2 Gas chromatography/mass spectrometry

Along with the application of new ionization techniques to amino acids, quantification by GC/MS constitutes the main area of expansion for application of mass spectrometry to amino acids. Selected ion monitoring during GC/MS allows detection and quantification at the fg to ng level. A few recent, specimen applications will illustrate the utility of the method.

Employing chemical ionization with methane, 3-methylhistidine can be measured rapidly in biological samples [244]. Precise assays in the pmol and nmol range were realized with selected ion monitoring of $(M + H)^+$ ions of derivatives of 3-methylhistidine and a dideuteriated internal standard. Also GC/MS with a capillary GC column and *N*-trifluoroacetyl propyl ester derivatives can be used for determination of 3-methylhistidine isolated from urine by ion-exchange or charcoal column chromatography [245]. A third method for assay of the same substance has been reported (*N*-heptafluorobutyroyl derivative) [246]. Capillary GC/MS was again used for determination in urine and muscles of l-methylhistidine as its *N*-trifluoroacetyl propyl ester [247].

The quantitative GC/MS method readily handles the less usual amino acids. Taurine (*17.26*) in plasma has been determined by stable isotope analysis [248], and assays of N^4-(2-acetamido-2-deoxy-β-D-glucopyranosyl)-L-asparagine [249] and *N*-(phosphonoacetyl)-L-aspartic acid [250] have been described. Nicotinic acid (*17.27*) has been determined by selected ion monitoring against the 2-methyl analogue using TMS derivatives [251]. Plasma levels of the drug, phenylalanine mustard, have been measured in patients by use of CI and the derivative shown (*17.28*) [252]. The plasma level peaked between 0.7 and 2.3 h then dropped to very low levels over 24 h. *N*-Acylglycines in urine have been assayed [253].

$HO_3SCH_2CH_2NH_2$

(17.26)

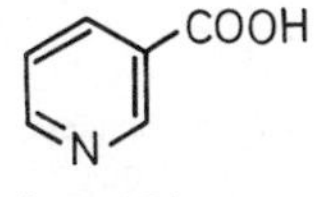

COOH, N

(17.27)

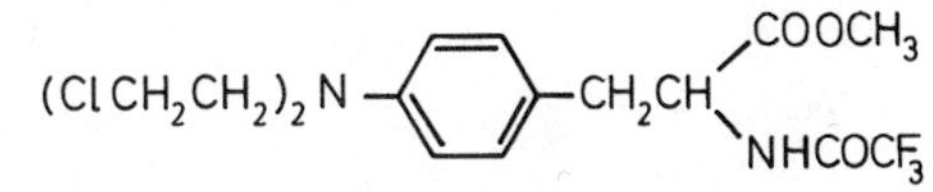

$(ClCH_2CH_2)_2N$–C$_6$H$_4$–$CH_2CH(COOCH_3)(NHCOCF_3)$

(17.28)

Some interesting and useful internal standards have been used. To take advantage of the most sensitive and simplest quantitative procedure of single ion monitoring, D-amino acids can be used as internal standards for natural L-amino acids. Derivatization with a chiral reagent yields disastereoisomers which can be separated on a normal (achiral) GC column. Alternatively, a chiral stationary phase can be used to separate conventional derivatives of the enantiomers. As *N*-perfluoroacyl *n*-butyl esters, *S*-methyl-L-cysteine from haemoglobin and the added D-enantiomer (internal standard) can be resolved completely on a capillary column of Chirasil-Val [254, 255]. The separation is illustrated in Fig. 17.9 for single ion monitoring at *m/z* 288 for the *N*-trifluoroacetyl *n*-butyl esters of both *S*-methyl-L- and D-cysteine (200 pg of each) [254]. The ions monitored are $(M + H)^+$ ions, corresponding to the base peaks in the isobutane CI spectra. The method is three times more sensitive than when monitoring $(M + H)^+$ ions at *m/z* 288 and 291 for the *N*-trifluoroacetyl *n*-butyl esters of *S*-methylcysteine and d_3-*S*-methylcysteine (internal standard) [254]. Using these methods [254–256], it was possible to estimate *S*-methyl-L-cysteine down to 0.1 nmol g^{-1} of haemoglobin in rats and humans exposed to methylating agents. The detection limit was in the low pg range with good signal to noise ratio [255]. The only disadvantage of the use of enantiomeric internal standards is that a correction factor has to be applied to the results if any racemization occurs during work-up [254]. One further report [257] describes the use of capillary GC/MS for determining the degree of amino acid alkylation of blood proteins. Analysis of tyrosine employing 3-hydroxyphenylalanine as internal standard achieved $\pm 5\%$ precision at a level of 80 nmol ml^{-1} in human plasma [258]. Synthesis of amino acids with two ^{18}O atoms in the carboxyl group has been reported [259]. The labelled compounds are suitable as internal standards because loss of

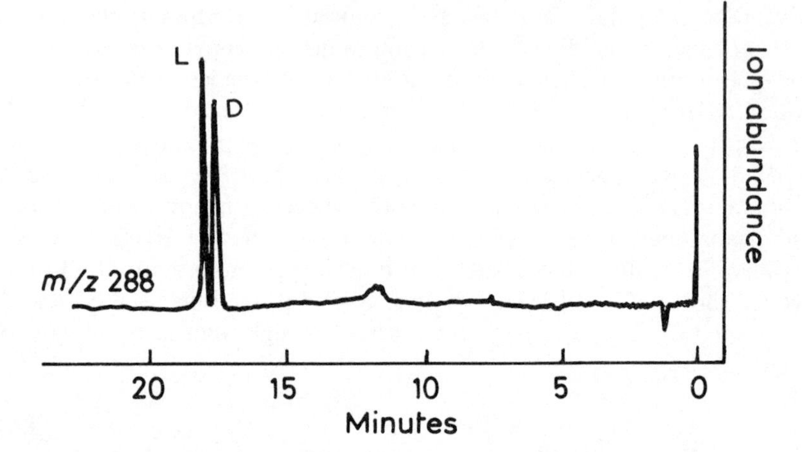

Figure 17.9 Single ion monitoring trace (*m/z* 288) of *N*-trifluoroacetyl *n*-butyl esters of L- and D-*S*-methylcysteine (200 pg of each). Reproduced from reference 254 by kind permission of the publishers.

^{18}O by exchange was slow (but pH dependent). For assay of indole-3-acetic acid, tetra- and penta-deuteriated analogues are useful [260, 261]. Radioimmunoassay is a cheaper, simpler alternative for assay of indole-3-acetic acid, but unlike GC/MS, in some cases it is only reliable after chromatographic removal of interfering substances [261].

The isotope ^{15}N has widespread use in the amino acid field for tracer studies and determining turn-over rates [262–265]. The pool size and turn-over rate of glycine in neonates has been determined by administration of [^{15}N]-glycine [262, 266]. Also, [^{13}C, ^{15}N]-glycine has been used as an internal standard for the determination of glycine in brain tissue [267]. Another stable isotope, ^{34}S, has been employed to measure bound cysteine and methionine as *N*-trifluoroacetyl *n*-butyl esters [268]. Stable isotopic enrichment of individual amino acids in plasma can be measured very accurately by GC/MS and CI, irrespective of the choice of isotope: ^{15}N, ^{2}H, ^{13}C, ^{18}O [231].

Three recent publications have described assays of 4-aminobutyric acid in the brain [269–271]. The *in vivo* flux of leucine has been determined by administration to dogs of a trideuteriated leucine and use of heptadeuteriated leucine as internal standard. The deuteriated and unlabelled leucines were assayed by selected ion monitoring at three m/z values [272]. Dosage with heptadeuterio-L-phenylalanine [273] and pentadeuterio-L-tryptophan [274] provides an *in vivo* measurement of the phenylalanine and tryptophan hydroxylating systems respectively when coupled with selected ion monitoring of the phenolic products.

Fourteen urinary α-amino acids have been determined quantitatively in one GC/MS analysis with isobutane CI [275]. The $(M+H)^+$ ion of each *N*-trifluoroacetylamino *n*-butyl ester and its [^{13}C]-analogue (internal standard) was monitored by appropriate switching under computer control. The total analysis time was 35 min and detection limits were near 1 ng. Finally, glass capillary GC/MS has been employed for the detection of amino acids in biological microenvironments. For most of the *N*-heptafluorobutyroyl isobutyl esters, the detection limit was under 1 pg using selected ion monitoring [276].

17.5.3 Liquid chromatography/mass spectrometry

This newer technique has not yet been applied to true quantitative studies of amino acids. The above selected ion monitoring methods are applicable and preliminary studies with one of the simpler LC/MS systems indicates that, when rapidly switching the analyser to measure the $(M + H)^+$ ions for a set of twenty free amino acids, the detection limit would be in the pg range [236].

17.6 DIFFERENTIATION OF ISOMERIC COMPOUNDS

Mass spectra of isomers frequently show fewer differences than the day-to-day variation in recording of the spectra. Leucine and isoleucine methyl esters can be distinguished only by differences in some peak ratios, but other derivatives

[34, 55, 92] or examination of metastable ions [81] provide a sure basis for their distinction. Of more interest is the differentiation between enantiomers and for this GC/MS is an excellent tool. Since GC is the essential feature of the resolution, it is reported in more detail in the previous chapter. Basically, there are two approaches, (i) separation on a chiral stationary phase and (ii) derivatization with a chiral reagent and separation of the resulting diastereoisomers on a conventional GC column. The mass spectra of the isomers may still be identical, or nearly so, but they are recorded independently because they are separated chromatographically.

Enantiomeric *N*-perfluoroacylamino acid isopropyl esters can be separated on *N*-trifluoroacetyl-*S*-phenylalanyl-*S*-leucine cyclohexyl esters [277] or *N*-propionyl-L-valine-*t*-butylamide polysiloxane [278]. A very useful application of chiral phases for determination of *S*-methyl-L-cysteine has already been discussed [254, 255]. Complementing GC/MS studies, the constituent amino acids of the peptide antibiotics, suzukacillin [279] and alamethicins [200], were shown to be L-isomers by GC on chiral phases. The optical purity of amino acid residues in proteins can be assessed by hydrolysis with $^2HCl/^2H_2O$. Any molecule undergoing inversion is then labelled with ^2H. Subsequent analysis of the mixture of amino acids as *N*-perfluoroacyl esters is performed by GC/MS with a chiral column and selected ion monitoring of proteo- and deuterio-forms of both (separated) enantiomers. From relative ion abundances, the degree of inversion and thus the true D/L ratio can be calculated very accurately, forming a useful method of differentiating D-amino acids initially present in a protein from those formed during hydrolysis [280–282]. Twelve amino acids can be measured using two separate GC/MS analyses and racemization levels up to 10% were measured under common protein hydrolysis conditions [282].

The advent of chiral stationary phases for LC [283] should be noted because of the potential for separation and mass spectral analysis of underivatized enantiomeric amino acids by LC/MS.

Derivatization of the amino or carboxyl groups can be used to convert enantiomeric pairs into mixtures of diastereoisomers. Reaction of amino acids with *N*-trifluoroacetyl-L-prolyl chloride [35, 284, 285], or synthesis of (−)-menthyl esters [35, 39], (+)-3-methyl-2-butyl esters [286] and *N*-(-*S*-α-methoxy-α-trifluoromethylphenylacetyl) derivatives [287] serve to produce diastereoisomers suitable for analysis by GC/MS. The possibility that derivative formation may not go to completion can be accounted for by a quantitative GC/MS measurement using CI and an additional and isotopically-labelled chiral reagent [285].

17.7 AMINO ACIDS RELEASED DURING PROTEIN SEQUENCE DETERMINATION

As this area has been well reviewed [7, 73, 288, 289], only a brief discussion is given here. A recent and sensible treatment [289] of this topic is recommended reading. The most familiar form of sequence determination is the (automated) Edman method [290] which releases anilinothiazolinones and these are re-arranged by acid to PTH derivatives (*17.5*, $R' = C_6H_5$). As the sequence study progresses the product PTH derivatives are formed in an increasingly complex matrix due to incomplete reactions and by-products. Established methods of identifying the PTH-derivatives are by paper, thin-layer, gas and high pressure liquid chromatography (HPLC), back hydrolysis to the amino acid and conventional analysis, and mass spectrometry. Of these, the most expensive and complex is mass spectrometry so, to compete with the other methods, it must offer significant advantages. Mass spectrometry does provide a more specific structural identification and the ability to analyse thiazolinones directly [69, 291], but not a simple quantitative assay of the released amino acid derivatives [64, 65, 72, 74, 75]. The nearest approach to a routine mass spectral analysis is one using chemical ionization [69]. The fragmentations of PTH derivatives were outlined above [63].

Since the primary Edman products, the anilinothiazolinones, thermally rearrange to PTH derivatives in the ion source of a mass spectrometer, the acid catalysed rearrangement step may be omitted [291]. This is advantageous for those derivatives (tryptophan, serine, threonine, aspargine, glutamine) undergoing decomposition or side-reactions under the acidic conditions required for formation of PTH derivatives. Whilst the discussion below concentrates on thiohydantoins it should be remembered that it applies equally to the thiazolinones.

With their different volatilities, PTH derivatives of the common amino acids evaporate from a direct insertion probe at different temperatures. Therefore, it is necessary to take many mass spectra whilst the sample is heated to be sure of a correct identification.

With the exception of leucine and isoleucine, all common PTH derivatives have a different molecular weight. Therefore, mild methods of ionization simplify identification because they afford mainly $(M + H)^+$ ions and few fragment ions. Chemical [64, 69], Penning [68], field desorption [67, 153] and low energy electron impact [65] ionization have been applied. There seems to be little reason for favouring FD or Penning ionization over CI, and the low energy electron impact method suffers from poor sensitivity.

The use of 4-bromo- and 4-chlorophenylisothiocyanate in alternation as peptide coupling reagents has been proposed [292]. The characteristic isotope ratios for the halogens reduce ambiguity in the identification of the resulting PTH derivatives. Use of 4-bromophenylisothiocyanate on its own has also been suggested for the same reason [70–72].

Following successful application of HPLC methods for separating and quantifying PTH derivatives, the appropriate eluate fraction can be collected and subjected to mass spectrometry for unambiguous identification [293]. This approach should be used with caution because the choosing of the relevant fraction could pre-empt the result. To prevent this, every collected fraction would have to be laboriously analysed, or an on-line LC/MS system used, taking mass spectra every few seconds. However, the off-line method is useful for distinguishing between closely or co-eluting PTH derivatives [293]. By the same token, the TLC method could be coupled with mass spectrometry without the need for eluting the sample from the plate [147]. A developed TLC plate can be 'scanned' by a beam of argon ions to produce secondary ion mass spectra wherever a 'spot' occurs. This too could be of help for complex chromatograms. Similarly, it was to be expected that there should be an extension from GC to GC/MS for analysis of PTH derivatives. For this, the PTH amino acids must be trimethylsilylated to make all but arginine of sufficient volatility for the analysis [82]. Alternatively, the PTH derivatives can be characterized by their thermal decomposition products using a pyrolysis/GC/MS system [294]. Replacement of phenyl by methyl isothiocyanate affords the more volatile methylthiohydantoin (MTH) derivatives. These can be handled by direct insertion probe [65, 73, 75] or by GC/MS after trimethylsilylation [83, 84]. Likewise, the thiohydantoins released from the *C*-terminus of a peptide after reaction with ammonium thiocyanate can be analysed by direct insertion probe [77–80] or GC/MS [85].

The use of 2-*p*-isothiocyanatophenyl-3-phenylindone as coupling reagent affords ITH amino acids [124]. Given that these compounds have excellent properties for HPLC and UV detection (sensitivity 5 pmol), application of CI mass spectrometry was probably an unnecessary complication. Both methods of detection were applied to sequence determination of phospholipase in cobra venom [124].

Not only are thiazolinones converted to PTH derivatives in the mass spectrometer, but so too are the initial coupling products of Edman degradations, *N*-methyl- or phenylthiourea peptides (*17.29*; Reaction 17.25). The pyrolysis products afford mass spectra identical to those of the corresponding PTH derivatives [295, 296].

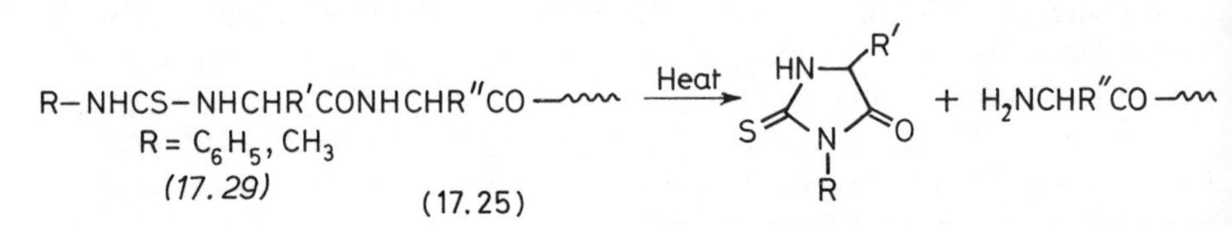

Field desorption is capable of producing molecular weight information on underivatized peptides. In a variant of the usual procedure for combining Edman sequencing with mass spectrometry, small portions of the peptide sample are subjected to field desorption before and after each Edman cycle. The difference in molecular weight characterizes the residue lost [297–299]. The PTH derivatives are analysed with the same instrument as a check. The method will be limited to small peptides with present restrictions on ionization and detection of high mass compounds, but it is reported to be suitable for simultaneous sequencing of peptides in mixtures [297–299].

As a routine means of identifying and quantifying PTH derivatives, mass spectrometry does not compete well with HPLC. It is best used to provide complementary evidence when simpler methods have failed to provide an unequivocal identification. It is, of course, the best method to use when uncommon amino acid residues are found because it provides direct evidence of structure. An example of this is provided by the identification of γ-carboxyglutamic acid residues in prothrombin as the methyl esterified PTH derivative [300]. Recently [301], a computer program has been introduced to aid sequence determination by Edman degradation in conjunction with mass spectrometry.

Following reaction of a peptide with l-fluoro-2,6-dinitrobenzene or 2-fluoro-3-nitropyridine, reduction and acid-catalysed cleavage released the *N*-terminal residue as a piperazine derivative (Reaction 17.26) which is readily characterized by EI mass spectrometry. The reaction cycle can be repeated on the separated peptide residue for sequence analysis [108].

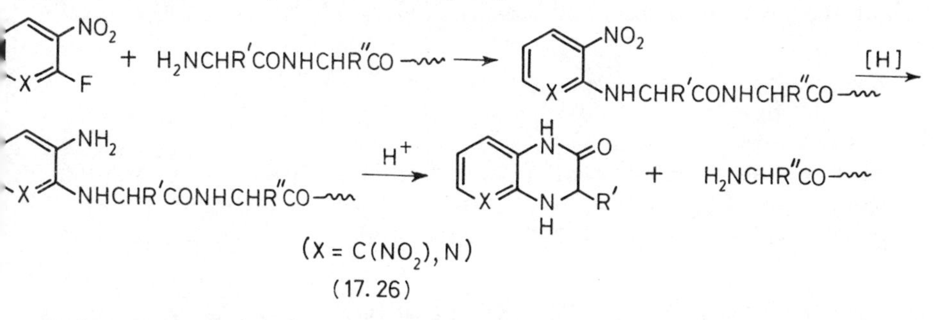

End group analysis is frequently performed by dansylation or by treatment of peptides with 1-fluoro-2,4-dinitrobenzene. The dinitrophenylamino acids are normally analysed by chromatography, but a rapid and sensitive alternative involves mass spectrometry of the acids [95] or methyl esters [94]. Dansyl derivatives are normally identified by chromatographic techniques as well. Mass spectrometry is not well suited to these compounds, but it has been applied nonetheless [96, 97]. Analysis of mixtures of dansyl amino acids was performed using low energy electron impact ionization [302], and also a metastable ion technique [211]. Mass spectrometry as a tool for end group analysis of proteins has been reviewed [288].

17.8 DETERMINATION OF THERMOCHEMICAL QUANTITIES

Ionization energies of free amino acids have been determined, by extrapolation of ionization efficiency curves, and found to be very similar to those of the corresponding amines [21]. This finding is consistent with the fact that fragmentation of amino acids is directed largely by the amino group (see section on fragmentation pathways). Distinction between $M^{+\cdot}$ and $(M+H)^+$ ions generated by field ionization of the amino acid (*17.30*) can be made on the basis of the appearance energy of each [164], the quasi-molecular ion having the lower

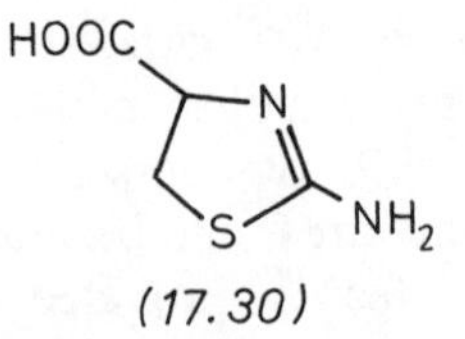

(17.30)

appearance energy. The proton affinities of amino acids can be measured by chemical ionization. At the high pressures used for CI, proton transfer equilibria are established through many ion/molecule reactions [1, 303]. By measuring the thermodynamic quantities of equilibria between an amino acid, M, and a reference base, B, of known proton affinity (Reaction 17.27), values for the simple amino acids were determined [303]. The thermodynamics of solvation and association reactions of protonated amino acids at high pressure have also been studied [204]. By using the method of ion cyclotron resonance spectroscopy [1], both the gas-phase acidity and basicity of glycine have been measured [304].

$$(M + H)^+ + B \rightleftharpoons M + (B + H)^+$$

(17.27)

For an interesting discussion of fragmentation mechanisms, centred on observations of the mass spectrometric behaviour of methionine and seleno-methionine, the reader is referred elsewhere [305].

17.9 CONCLUSION

Qualitative examination of isolated amino acids by conventional electron impact mass spectrometry has become routine. Recent emphasis in publications follows two major strands.

(i) Development and application of mild methods of ionization with a view to making any free amino acid, its conjugates and derivatives amenable to mass spectrometry.
(ii) Precise quantitative detection of known amino acids, mainly by GC/MS, in the clinical, biochemical and geochemical fields.

The objective of the first strand has largely been met. Fast atom bombardment mass spectrometry, combining the benefits of field desorption and electron impact in a relatively simple instrument, should become a particularly important tool for identifying novel amino acids. The almost inevitable marriage of FAB with a moving belt LC/MS system will be a very significant development for analysis of mixtures of underivatized amino acids. One instrument combining LC/MS with a mild method of ionization already shows much promise in this field [236].

Detection and quantification of derivatized amino acids by selected ion monitoring and GC/MS reaches impressive degrees of sensitivity and specificity.

By using capillary columns and choosing carefully the derivative and method of ionization, detection limits under 10^{-11} g are attained.

REFERENCES

1. Rose, M.E. and Johnstone, R.A.W. (1982) *Mass Spectrometry for Chemists and Biochemists*, Cambridge University Press, Cambridge.
2. Biemann, K. (1962) in *Mass Spectrometry*, McGraw-Hill, New York, Ch. 7.
3. Biemann, K. (1963) in *Mass Spectrometry of Organic Ions* (ed. F.W. McLafferty), Academic Press, New York, p. 544.
4. Budzikiewicz, H., Djerassi, C. and Williams, D.H. (1964) *Structure Elucidation of Natural Products by Mass Spectrometry*, (Vol. 2) Holden-Day, San Francisco, Ch. 26.
5. Jones, J.H. (1968) *Quart. Rev.*, **22**, 302.
6. Vetter, W. (1972) in *Biochemical Applications of Mass Spectrometry* (ed. G.R. Waller), Wiley-Interscience, New York, p. 387.
7. Vetter, W. (1980) in *Biochemical Applications of Mass Spectrometry* (eds G.R. Waller and O.C. Dermer) (First Supplementary Volume) Wiley-Interscience, New York, p. 439.
8. *Mass Spectrometry*, The Chemical Society, London, published every odd year. See chapters on natural products and chromatography/mass spectrometry.
9. Burlingame, A.L. *et al.*, *Anal. Chem.* (Mass Spectrometry), published every even year.
10. *Amino-acids, Peptides and Proteins*, The Chemical Society, London, published annually, see Chapter 1.
11. Andersson, C.-O. (1958) *Acta Chem. Scand.*, **12**, 1353.
12. Biemann, K., Seibl, J. and Gapp, F. (1959) *Biochem. Biophys. Res. Commun.*, **1**, 307.
13. Biemann, K., Seibl, J. and Gapp, F. (1961) *J. Am. Chem. Soc.*, **83**, 3795.
14. Andersson, C.-O., Ryhage, R., Ställberg-Stenhagen, S. and Stenhagen, E. (1962) *Arkiv Kemi*, **19**, 405.
15. Saito, M., Yamamoto, M. and Hirota, K. (1969) *Nippon Kagaku Zasshi*, **90**, 663.
16. McLafferty, F.W. (1963) *Mass Spectrometry of Organic Ions*, Academic Press, New York, p. 336.
17. King, T.P. (1966) *Biochemistry*, **5**, 3454.
18. Vetter-Diechtt, H., Vetter, W., Richter, W.J. and Biemann, K. (1968) *Experientia*, **24**, 340.
19. Kostyanovsky, R.G., Voznesensky, V.N., Kadorkina, G.K. and El'natanov, Yu.I. (1980) *Org. Mass Spectrom.*, **15**, 412.
20. Biemann, K. and McCloskey, J.A. (1962) *J. Am. Chem. Soc.*, **84**, 3192.
21. Junk, G. and Svec, H.J. (1963) *J. Am. Chem. Soc.*, **85**, 839.
22. Heyns, K. and Grützmacher, H.F. (1963) *Justus Liebigs Ann. Chem.*, **667**, 194.
23. Coulter, A.W. and Fenselau, C.C. (1972) *Org. Mass Spectrom.*, **6**, 105.
24. Longstaff, C. and Rose, M.E. (1982) *Org. Mass Spectrom.*, **17**, 508.
25. Andersson, C.-O., Ryhage, R. and Stenhagen, E. (1962) *Arkiv Kemi*, **19**, 417.
26. Fölsch, G., Ryhage, R. and Stenhagen, E. (1962) *Arkiv Kemi*, **20**, 55.
27. Manusadzhyan, V.G. and Varshavskii, Ya.M. (1964) *Izv. Akad. Nauk Arm. SSR Khim.*, **17**, 137.
28. Brown, C.L. and Chan, C.L. (1976) *J. Am. Chem. Soc.*, **98**, 2682.
29. Brown, C.L. and Chan, C.L. (1978) *Adv. Mass Spectrom.*, **7A**, 371.
30. Padieu, P., Desgres, J., Maume, B.F. *et al.* (1978) *Adv. Mass Spectrom.*, **7B**, 1604.
31. Manhas, M.S., Hsieh, R.S. and Bose, A.K. (1970) *J. Chem. Soc.* (*C*), 116.
32. Gelpi, E., Koenig, W.A., Gibert, J. and Oro, J. (1969) *J. Chromatogr. Sci.*, **7**, 604.
33. Prox, A. and Schmid, J. (1969) *Org. Mass Spectrom.*, **2**, 105.
34. Leimer, K.R., Rice, R.H. and Gehrke, C.W. (1977) *J. Chromatogr.*, **141**, 121.
35. Iwase, H. and Murai, A. (1977) *Anal. Biochem.*, **78**, 340.
36. Kingston, E.E. and Duffield, A.M. (1978) *Biomed. Mass Spectrom.*, **5**, 621.
37. Bertilsson, L. and Costa, E. (1976) *J. Chromatogr.*, **118**, 395.
38. Wolfensberger, M., Redweik, U. and Curtius, H.-Ch. (1979) *J. Chromatogr.*, **172**, 471.
39. March, J.F. (1975) *Anal. Biochem.*, **69**, 420.
40. Bengtsson, G. and Odham, G. (1979) *Anal. Biochem.*, **92**, 426.
41. Desgres, J., Boisson, D. and Padieu, P. (1979) *J. Chromatogr.*, **162**, 133.
42. Desgres, J., Boisson, D., Veyrac, F. *et al.* (1979) *Rec. Dev. Mass Spectrom. Biochem. Med.*, **2**, 377.

43. MacKenzie, S.L. and Hogge, L.R. (1977) *J. Chromatogr.*, **132**, 485.
44. Felker, P. and Bandurski, R.S. (1975) *Anal. Biochem.*, **67**, 245.
45. Lawless, J.G. and Chadha, M.S. (1971) *Anal. Biochem.*, **44**, 473.
46. Baker, K.M., Shaw, M.A. and Williams, D.H. (1969) *Chem. Commun.*, 1108.
47. VandenHeuvel, W.J.A., Smith, J.L., Putter, I. and Cohen, J.S. (1970) *J. Chromatogr.*, **50**, 405.
48. VandenHeuvel, W.J.A., Smith, J.L. and Cohen, J.S. (1970) *J. Chromatogr. Sci.*, **8**, 567.
49. VandenHeuvel, W.J.A. and Cohen, J.S. (1970) *Biochim. Biophys. Acta*, **208**, 251.
50. Bergström, K., Gürtler, J. and Blomstrand, R. (1970) *Anal. Biochem.*, **34**, 74.
51. Bergström, K. and Gürtler, J. (1971) *Acta Chem. Scand.*,**25**, 175.
52. Laseter, J.L., Weete, J.D., Albert, A. and Walkinshaw, C.H. (1971) *Anal. Letters*, **4**, 671.
53. Mischer, G. (1972) *Z. Anal. Chem.*, **262**, 81.
54. Marık, J., Capek, A. and Králíčk, J. (1976) *J. Chromatogr.*, **128**, 1.
55. Leimer, K.R., Rice, H.R. and Gehrke, C.W. (1977) *J. Chromatogr.*, **141**, 355.
56. Budzikiewicz, H. and Meissner, G. (1978) *Org. Mass Spectrom.*, **13**, 608.
57. Yamada, S. (1978) *GC/MS News*, **6**, 76.
58. Clay, K.L. and Murphy, R.C. (1979) *J. Chromatogr.*, **164**, 417.
59. Iwase, H., Takeuchi, Y. and Murai, A. (1979) *Chem. Pharm. Bull.*, **27**, 1307.
60. Fennessey, P.V. and Tjoa, S.S. (1980) *Org. Mass Spectrom.*, **15**, 202.
61. Wulfson, N.S., Stepanov, V.M., Puchkov, V.A. and Zyakun, A.M. (1963) *Izv. Akad. Nauk SSSR, Ser. Khim.*, 1524.
62. Weygand, F. (1968) *Z. Anal. Chem.*, **243**, 2.
63. Hagenmaier, H., Ebbighausen, W., Nicholson, G. and Vötsch, W. (1970) *Z. Naturforsch.*, **25b**, 681.
64. Fales, H.M., Nagai, Y., Milne, G.W.A. *et al.* (1971) *Anal. Biochem.*, **43**, 288.
65. Sun, T. and Lovins, R.E. (1972) *Anal. Biochem.*, **45**, 176.
66. Fairwell, T. and Brewer, H.B. Jr. (1973) *Fed. Proc.*, **32**, 648.
67. Schulten, H.-R. and Wittmann-Liebold, B. (1976) *Anal. Biochem.*, **76**, 300.
68. Laudenslager, J.B. and Theard, L.P. (1978) *Adv. Mass Spectrom.*, **7B**, 1388.
69. Fairwell, T. and Brewer, H.B. Jr. (1980) *Anal. Biochem.*, **107**, 140.
70. Tschesche, H. and Wachter, E. (1970) *Eur. J. Biochem.*, **16**, 187.
71. Weygand, F. and Obermeier, R. (1971) *Eur. J. Biochem.*, **20**, 72.
72. Tschesche, H., Schneider, M. and Wachter, E. (1972) *FEBS Letters*, **23**, 367.
73. Richards, F.F. and Lovins, R.E. (1972) *Meth. Enzymol.*, **25**, 314.
74. Richards, F.F., Barnes, W.T., Lovins, R.E. *et al.* (1969) *Nature*, **221**, 1241.
75. Fairwell, T., Barnes, W.T., Richards, F.F. and Lovins, R.E. (1970) *Biochemistry*, **9**, 2260.
76. Lindeman, J. and Lovins, R.E. (1976) *Anal. Biochem.*, **75**, 682.
77. Suzuki, T., Matsui, S. and Tuzimura, K. (1972) *Agric. Biol. Chem. (Tokyo)*, **36**, 1061.
78. Suzuki, T., Song, K-D., Itagaki, Y. and Tuzimura, K. (1976) *Org. Mass Spectrom.*, **11**, 1061.
79. Okada, K. and Sakuno, A. (1978) *Org. Mass Spectrom.*, **13**, 535.
80. Okada, K. and Itagaki, Y. (1978) *Koenshu—Iyo Masu Kenkyukai*, **3**, 249.
81. Sun, T. and Lovins, R.E. (1972) *Org. Mass Spectrom.*, **6**, 39.
82. Burgus, R., Butcher, M., Amoss, M. *et al.* (1972) *Proc. Natl. Acad. Sci. USA*, **69**, 278.
83. Vance, D.E. and Feingold, D.S. (1970) *Anal. Biochem.*, **36**, 30.
84. Lamkin, W.M., Weatherford, J.W., Jones, N.S. *et al.* (1974) *Anal. Biochem.*, **58**, 422.
85. Rangarajan, M., Ardrey, R.E. and Darbre, A. (1973) *J. Chromatogr.*, **87**, 499.
86. Grahl-Nielsen, O. and Solheim, E. (1975) *J. Chromatogr.*, **105**, 89.
87. Grahl-Nielsen, O. and Solheim, E. (1975) *Anal. Chem.*, **47**, 333.
88. Jones, J.H. and Witty, M.J. (1979) *J.C.S., Perkin 1*, 3203.
89. Liardon, R., Utt-kuhn, U. and Husek, P. (1979) *Biomed. Mass Spectrom.*, **6**, 381.
90. Weygand, F., Steglich, W. and Tanner, H. (1962) *Annalen*, **658**, 128.
91. Ferrito, V., Borg, R., Eagles, J. and Fenwick, G.R. (1979) *Biomed. Mass Spectrom.*, **6**, 499.
92. Jayasimhulu, K. and Day, R.A. (1980) *Biomed. Mass Spectrom.*, **7**, 7.
93. Traldi, P., Vettori, U. and Clerici, A. (1980) *Heterocycles*, **14**, 847.
94. Penders, Th. J., Copier, H., Heerma, W. *et al.* (1966) *Rec. Trav. Chim. Pays-Bas*, **85**, 216.
95. Studier, M.H., Moore, L.P., Hayatsu, R. and Matsuoka, S. (1970) *Biochem. Biophys. Res. Commun.*, **40**, 894.
96. Marino, G. and Buoncore, V. (1968) *Biochem. J.*, **110**, 603.
97. Seiler, N., Schneider, H.H. and Sonnenberg, K.-D. (1971) *Anal. Biochem.*, **44**, 451.

98. Pritchard, D.G., Schnute, W.C. Jr. and Todd, C.W. (1975) *Biochem. Biophys. Res. Commun.*, **65**, 312.
99. Shieh, J.-J., Leung, K. and Desiderio, D.M. (1976) *Org. Mass Spectrom.*, **11**, 479.
100. Shieh, J.-J., Leung, K. and Desiderio, D.M. (1977) *Anal. Letters*, **10**, 575.
101. Murray, K.E. and Ingles, D.I. (1979) *Chem. Ind., Lond.*, 476.
102. Heyns, K. and Grützmacher, H.F. (1961) *Z. Naturforsch.*, **16b**, 293.
103. DeJongh, D.C., Faus, G., Nayar, M.S.B. *et al.* (1976) *Biomed. Mass Spectrom.*, **3**, 191.
104. Höfle, G., Höhne, G., Respondek, J. and Schwarz, H. (1977) *Org. Mass Spectrom.*, **12**, 477.
105. Stapleton, B.J. and Bowie, J.H. (1976) *Org. Mass Spectrom.*, **11**, 429.
106. Aplin, R.T. and Jones, J.H. (1967) *Chem. Commun.*, 261.
107. Aplin, R.T. and Jones, J.H. (1968) *J. Chem. Soc. (C)*, 1770.
108. Johnstone, R.A.W., Povall, T.J. and Entwistle, I.D. (1975) *J.C.S., Perkin 1*, 1424.
109. Aplin, R.T., Jones, J.H. and Liberek, B. (1968) *J. Chem. Soc. (C)*, 1011.
110. Thenot, J.-P. and Horning, E.C. (1972) *Anal. Letters*, **5**, 519.
111. Williams, K.M. and Halpern, B. (1973) *Anal. Letters*, **6**, 839.
112. Harman, I. and Hesford, F.J. (1974) *Biomed. Mass Spectrom.*, **1**, 115.
113. Halpern, B., Close, V.A., Wegmann, A. and Westley, J.W. (1968) *Tetrahedron Letters*, 3119.
114. Szafranek, J., Blotny, G. and Vouros, P. (1978) *Tetrahedron*, **34**, 2763.
115. Maekawa, K., Taniguchi, E. and Kuwano, E. (1978) *Org. Mass Spectrom.*, **13**, 4.
116. Kasai, T., Furukawa, K. and Sakamura, S. (1979) *J. Fac. Agric., Hokkaido Univ.*, **59**, 279.
117. Milne, G.W.A., Axenrod, T. and Fales, H.M. (1970) *J. Am. Chem. Soc.*, **92**, 5170.
118. Tsang, C.W. and Harrison, A.G. (1976) *J. Am. Chem. Soc.*, **98**, 1301.
119. Gaffney, J.S., Pierce, R.C. and Friedman, L. (1977) *J. Am. Chem. Soc.*, **99**, 4293.
120. Meot-Ner, M. and Field, F.H. (1973) *J. Am. Chem. Soc.*, **95**, 7207.
121. Weinkam, R.J. (1978) *J. Org. Chem.*, **43**, 2581.
122. Cooper, A.J.L., Griffith, O.W., Meister, A. and Field, F.H. (1981) *Biomed. Mass Spectrom.*, **8**, 95.
123. Leclerque, P.A. and Desiderio, D.M. (1973) *Org. Mass Spectrom.*, **7**, 515.
124. Nasimov, I.V., Levina, N.B., Shemyakin, V.V. *et al.* (1979) *Methods Pept. Protein Sequence Anal., Proc. Int. Conf., 3rd*, 475.
125. Majer, J.R., Al-Ali, B.I. and Azzouz, A.S.P. (1981) *Org. Mass Spectrom.*, **16**, 147.
126. Brown, K.S. Jr. (1965) *J. Am. Chem. Soc.*, **87**, 4202.
127. Brown, K.S. and Becker, D. (1967) *Tetrahedron Letters*, 1721.
128. Knapp, D.R. (1979) *Handbook of Analytical Derivatization Reactions*, Wiley, New York.
129. Vouros, P. (1980) *Pract. Spectrosc.*, **3** (Mass Spectrom., Part B), 129.
130. Milne, G.W.A. and Lacey, M.J. (1974) *CRC Crit. Rev. Anal. Chem.*, **4**, 45.
131. Fales, H.M., Milne, G.W.A., Winkler, H.U. *et al.* (1975) *Anal. Chem.*, **47**, 207.
132. Davies, G.D. Jr. (1979) *Accounts Chem. Res.*, **12**, 359.
133. Biemann, K., Lioret, C., Asselineau, J. *et al.* (1968) *Biochim. Biophys. Acta*, **40**, 369.
134. Lerner, A.B., Case, J.D., Biemann, K. *et al.* (1959) *J. Am. Chem. Soc.*, **81**, 5264.
135. Franzblau, C., Faris, B. and Papaioannou, R. (1969) *Biochemistry*, **8**, 2833.
136. Mechanic, G., Gallop, P.M. and Tanzer, M.L. (1971) *Biochem. Biophys. Res. Commun.*, **45**, 644.
137. Tanzer, M.L. and Mechanic, G. (1970) *Biochem. Biophys. Res. Commun.*, **39**, 183.
138. Tanzer, M.L., Housley, T., Berube, L. *et al.* (1973) *J. Biol. Chem.*, **248**, 393.
139. Robins, S.P., Shimokomaki, M. and Bailey, A.J. (1973) *Biochem. J.*, **131**, 771.
140. Robins, S.P. and Bailey, A.J. (1977) *Biochem. J.*, **163**, 339.
141. Millington, D.S. and Sheppard, R.C. (1968) *Phytochemistry*, **7**, 1027.
142. Lipton, S.H. and Dutky, R.C. (1972) *J. Agric. Food Chem.*, **20**, 235.
143. Niederwieser, A., Matasovic, A., Tippett, P. and Danks, D.M. (1977) *Clin. Chim. Acta*, **76**, 345.
144. Truscott, R.J.W., Malegan, D., McCairns, E. *et al.* (1981) *Biomed. Mass Spectrom.*, **8**, 99.
145. Zerilli, L.F., Tuan, G., Turconi, M. and Coronelli, C. (1977) *Ann. Chim. (Rome)*, **67**, 691.
146. Wilhelm, G. and Kupka, K.D. (1981) *FEBS Letters*, **123**, 141.
147. Kraft, R., Otto, A., Makower, A. and Etzold, G. (1981) *Anal. Biochem.*, **113**, 193.
148. Richter, W.J. and Schwarz, H. (1978) *Angew. Chem.*, **17**, 424.
149. Surman, D.J. and Vickerman, J.C. (1981) *J.C.S. Chem. Commun.*, 324.
150. Barber, M., Bordoli, R.S., Sedgwick, R.D. and Tyler, A.N. (1981) *Nature*, **293**, 270.
151. Surman, D.J. and Vickerman, J.C. (1981) *J. Chem. Res. (S)*, 170.
152. Beckey, H.D. (1977) in *Principles of Field Ionisation and Field Desorption Mass Spectrometry*, Pergamon Press, Oxford.
153. Schulten, H.-R. (1977) *Meth. Biochem. Analysis*, **24**, 313.

154. Schulten, H.-R. (1979) *Int. J. Mass Spectrom. Ion Phys.*, **32**, 97.
155. Horning, E.C., Mitchell, J.R., Horning, M.G. *et al.* (1979) *Trends Pharm. Sci.*, **1**, 76.
156. Ohashi, M. (1979) *Shitsuryo Bunseki*, **27**, 1.
157. Cotter, R.J. (1980) *Anal. Chem.*, **52**, 1589A.
158. Gaffney, J.S., Pierce, R.C. and Friedman, L. (1977) *Int. J. Mass Spectrom. Ion Phys.*, **25**, 439.
159. Budzikiewicz, H. (1981) *Angew. Chem. Int. Ed.*, **20**, 624.
160. Tannenbaum, H.P., Roberts, J.D. and Dougherty, R.C. (1975) *Anal. Chem.*, **47**, 49.
161. Hunt, D.F., Stafford, G.C. and Devine, C. (1975) *23rd Annual Conference on Mass Spectrometry and Allied Topics*, Houston, Texas, p. 600.
162. Voigt, D. and Schmidt, J. (1978) *Biomed. Mass Spectrom.*, **5**, 44.
163. Smit, A.L.C. and Field, F.H. (1977) *J. Am. Chem. Soc.*, **99**, 6471.
164. McReynolds, J.H. and Anbar, M. (1977) *Anal. Chem.*, **49**, 1832.
165. Heinen, H.J., Gerlich, H.H. and Beckey, H.D. (1975) *J. Phys. (E)*, **8**, 877.
166. Ohashi, M., Nakayama, N., Kudo, H. and Yamada, S. (1976) *Shitsuryo Bunseki*, **24**, 265.
167. Hunt, D.F., Shabanowitz, J., Botz, F.K. and Brent, D.A. (1977) *Anal. Chem.*, **49**, 1160.
168. Hansen, G. and Munson, B. (1980) *Anal. Chem.*, **52**, 245.
169. Beuhler, R.J., Flanigan, E., Greene, L.J. and Friedman, L. (1974) *J. Am. Chem. Soc.*, **96**, 3990.
170. Beaugrand, C. and Devant, G. (1980) *Adv. Mass Spectrom.*, **8B**, 1806.
171. Cotter, R.J. (1979) *Anal. Chem.*, **51**, 317.
172. Hansen, G. and Munson, B. (1978) *Anal. Chem.*, **50**, 1130.
173. Blakley, C.R., Carmody, J.J. and Vestal, M.L. (1980) *J. Am. Chem. Soc.*, **102**, 5931.
174. Keough, T. and DeStefano, A.J. (1981) *Anal. Chem.*, **53**, 25.
175. Winkler, H.U. and Beckey, H.D. (1972) *Org. Mass Spectrom.*, **6**, 655.
176. van der Greef, J., Nibbering, N.M.M., Schulten, H.-R. and Lehmann, W.D. (1978) *Z. Naturforsch. Anong. Chem. Org. Chem.*, **33b**, 770.
177. Winkler, H.U., Linden, H.B. and Beckley, H.D. (1978) *Adv. Mass Spectrom.*, **7A**, 104.
178. Keough, T., DeStefano, A.J. and Sanders, R.A. (1980) *Org. Mass Spectrom.*, **15**, 351.
179. Gol'denfel'd, V., Bondarenko, R.N. and Golovatyi, V.G. (1973) *Inst. Exp. Technol.*, **16**, 852.
180. Yatsimirskii, K.B., Golovatyi, V.G., Grigor'eva, A.C. *et al.* (1976) *Proc. Acad. Sci. USSR (Phys. Chem. Sect.)*, **226**, 35.
181. Winkler, H.U. and Beckey, H.D. (1973) *Org. Mass Spectrom.*, **7**, 1007.
182. McLafferty, F.W. (1979) *Phil. Trans. R. Soc. London, Ser. A*, **293**, 93.
183. Matsuo, T., Matsuda, H., Katakuse, I. *et al.* (1981) *Anal. Chem.*, **53**, 416.
184. Kistemaker, P.G., Lens, M.M.J., Van der Peyl, G.J.Q. and Boerboom, A.J.H. (1980) *Adv. Mass Spectrom.*, **8A**, 928.
185. Krueger, F.R. and Schueler, B. (1980) *Adv. Mass Spectrom.*, **8A**, 918.
186. Kupka, K.-D., Hillenkamp, F. and Schiller, Ch. (1980) *Adv. Mass Spectrom.*, **8A**, 935.
187. Hardin, E.D. and Vestal, M.L. (1981) *Anal. Chem.*, **53**, 1492.
188. Torgerson, D.F., Skowronski, R.P. and Macfarlane, R.D. (1974) *Biochem. Biophys. Res. Commun.*, **60**, 616.
189. Macfarlane, R.D. (1980) in *Biochemical Applications of Mass Spectrometry* (eds G.R. Weller and O.C. Dermer) (First Supplementary Volume) Wiley-Interscience, New York, p. 1209.
190. Chait, B.T., Agosta, W.C. and Field, F.H. (1981) *Int. J. Mass Spectrom. Ion Phys.*, **39**, 339.
191. Benninghoven, A., Evans, C.A., Jr., Powell, R.A. *et al.* (eds) (1979) *Secondary Ion Mass Spectrometry, SIMS-II*, Springer Series in Chemical Physics (Vol. 9) Springer-Verlag. New York.
192. Winograd, N. and Garrison, B.J. (1980) *Accounts Chem. Res.*, **13**, 406.
193. Benninghoven, A., Jaspers, D. and Sichtermann, W. (1976) *Appl. Phys.*, **11**, 35.
194. Benninghoven, A., Jaspers, D. and Sichtermann, W. (1978) *Adv. Mass Spectrom.*, **7B**, 1433.
195. Grade, H., Winograd, N. and Cooks, R.G. (1977) *J. Am. Chem. Soc.*, **99**, 7725.
196. Liu, L.K., Busch, K.L. and Cooks, R.G. (1981) *Anal. Chem.*, **53**, 109.
197. Benninghoven, A., Eicke, A., Junack, M. *et al.* (1980) *Org. Mass Spectrom.*, **15**, 459.
198. Kloeppel, K.D. and Von Buenau, G. (1981) *Int. J. Mass Spectrom. Ion Phys.*, **39**, 85.
199. Smith, R.D., Burger, J.E. and Johnson, A.L. (1981) *Anal. Chem.*, **53**, 1603.
200. Pandey, R.C., Cook, J.C., Jr., and Rinehart, K.L., Jr. (1977) *J. Am. Chem. Soc.*, **99**, 8469.
201. Mee, J.M.L. (1980) *Am. Lab. (Fairfield, Conn.)*, **12**, 55.
202. Issachar, D. and Yinon, J. (1980) *Adv. Mass Spectrom.*, **8B**, 1321.
203. Issachar, D. and Yinon, J. (1976) *Clin. Chim. Acta*, **73**, 307.

204. Neot-Mer, M. and Field, F.H. (1974) *J. Am. Chem. Soc.*, **96**, 3168.
205. Tamaki, S., Benninghoven, A. and Sichtermann, W. (1979) *Springer Ser. Chem. Phys.*, **9**, 127.
206. Benninghoven, A. and Sichtermann, W. (1981) *Int. J. Mass Spectrom. Ion Phys.*, **38**, 351.
207. Rinehart, K.L. Jr., *et al.* (1981) *Pure Appl. Chem.*, **53**, 795.
208. Bente, P.F. III, and McLafferty F.W. (1980) *Pract. Spectrosc.*, **3**, (Mass Spectrom., Part B), 253.
209. McLafferty, F.W. (1980) *Accounts Chem. Res.*, **13**, 33.
210. Russell, D.H., McBay, E.H. and Mueller, T.R. (1980) *Int. Lab.*, 49.
211. Addeo, F., Malorni, A. and Marino, G. (1975) *Anal. Biochem.*, **64**, 98.
212. McFadden, W.H. (1973) *Techniques of Combined Gas Chromatography/Mass Spectrometry*, Wiley-Interscience, New York.
213. McFadden, W.H. (1979) *J. Chromatogr. Sci.*, **17**, 2.
214. Sandberg, G., Andersson, B. and Dunberg, A. (1981) *J. Chromatogr.*, **205**, 125.
215. Engvild, K.C., Egsgaard, H. and Larsen, E. (1980) *Physiol. Plant.*, **48**, 499.
216. Heikes, D.L. (1980) *J. Assn. Offic. Anal. Chem.*, **63**, 1224.
217. Oró, J. and Nooner, D. (1979) in *Practical Mass Spectrometry. A Contemporary Introduction* (ed. B.S. Middleditch), Plenum Press, New York, p. 327.
218. Matsu-Ura, S., Yamamoto, S. and Makita, M. (1981) *Anal. Biochem.*, **114**, 371.
219. Christy, M.R., Barkley, R.M., Koch, T.H. *et al.* (1981) *J. Am. Chem. Soc.*, **103**, 3935.
220. Ramsdell, H.S. and Tanaka, K. (1980) *J. Chromatogr.*, **181**, 90.
221. Roeper, H. and Heyns, K. (1980) *J. Chromatogr.*, **193**, 381.
222. Ishibashi, T., Kawabata, T. and Tanabe, H. (1980) *J. Chromatogr.*, **195**, 416.
223. Huber, J.W. III (1978) *J. Chromatogr.*, **152**, 220.
224. Lawson, A.M. (1975) *Clin. Chem.*, **21**, 803.
225. Jellum, E. (1977) *J. Chromatogr.*, **143**, 427.
226. Williams, K.M. and Halpern, B. (1973) *Austral. J. Biol. Sci.*, **26**, 831.
227. Haraguchi, S., Shinka, T., Hisanaga, N. *et al.* (1977) *Koenshu—Iyo Masu Kenkyukai*, **2**, 107.
228. Lehnert, W. (1981) *Clin. Chim. Acta*, **116**, 249.
229. Kodama, H., Samukawa, K., Maki, I. *et al.* (1980) *Koenshu—Iyo Masu Kenkyukai*, **5**, 233.
230. Curtius, H.-C., Völlmin, J.A. and Baerlocher, K. (1973) *Anal. Chem.*, **45**, 1107.
231. Matthews, D.E., Ben-Galim, E. and Bier, D.M. (1979) *Anal. Chem.*, **51**, 80.
232. Summons, R.E. and Osmond, C.B. (1981) *Phytochemistry*, **20**, 575.
233. Coffin, R.D. and Thompson, R.M. (1977) *J. Chromatogr.*, **138**, 223.
234. McFadden, W.H. (1980) *J. Chromatogr. Sci.*, **18**, 97.
235. Blakley, C.R., Carmody, J.J. and Vestal, M.L. (1980) *Anal. Chem.*, **52**, 1636.
236. Blakley, C.R., Carmody, J.J. and Vestal, M.L. (1980) *Clin. Chem.*, **26**, 1467.
237. Games, D.E. (1980) *Anal. Proc. (London)*, **17**, 110.
238. Yoshida, Y., Yoshida, H., Tsuge, S. *et al.* (1980) *J. High Resolut. Chromatogr. Chromatogr. Commun.*, **3**, 16.
239. Epstein, E. and Cohen, J.D. (1981) *J. Chromatogr.*, **209**, 413.
240. Millard, B.J. (1978) *Quantitative Mass Spectrometry*, Heyden, London.
241. Lehmann, W.D. and Schulten, H.-R. (1978) *Angew. Chem. Int. Edn.*, **17**, 221.
242. Mee, J.M.L., Korth, J., Halpern, B. and James, L.B. (1977) *Biomed. Mass Spectrom.*, **4**, 178.
243. Yoshizumi, H. Tatematsu, M., Tatematsu, A. *et al.* (1981) *Shitsuryo Bunseki*, **29**, 89.
244. Matthews, D.E., Starren, J.B., Drexler, A.J. *et al.* (1981) *Anal. Biochem.*, **110**, 308.
245. Cotellessa, L., Marcucci, F., Cani, D. *et al.* (1980) *J. Chromatogr.*, **221**, 149.
246. Hough, L.B., Khandelwal, J.K., Morrishow, A.M. and Green, J.P. (1981) *J. Pharmacol. Methods*, **5**, 143.
247. Mussini, E., Cotellessa, L., Colombo, L. *et al.* (1981) *J. Chromatogr.*, **224**, 94.
248. Irving, C.S. and Klein, P.D. (1980) *Anal. Biochem.*, **107**, 251.
249. Maury, P. (1979) *J. Lab. Clin. Med.*, **93**, 718.
250. Strong, J.M., Kinney, Y.E., Branfman, A.R. and Cysyk, R.L. (1979) *Cancer Treat. Rep.*, **63**, 775.
251. Sautebin, L., Galli, G., Puglisi, L. *et al.* (1981) *Pharmacol. Res. Commun.*, **13**, 141.
252. Pallante, S.L., Fenselau, C., Mennel, R.G. *et al.* (1980) *Cancer Res.*, **40**, 2268.
253. Gregersen, N., Keiding, K. and Koeluraa, S. (1979) *Biomed. Mass Spectrom.*, **6**, 439.
254. Bailey, E., Farmer, P.B. and Lamb, J.H. (1980) *J. Chromatogr.*, **200**, 145.
255. Farmer, P.B., Bailey, E., Lamb, J.H. and Connors, T.A. (1980) *Biomed. Mass Spectrom.*, **7**, 41.
256. Farmer, P.B., Bailey, E. and Connors, T.A. (1980) *Adv. Mass Spectrom.*, **8B**, 1227.
257. Vollner, L. (1980) *GSF-Ber. O*, **599**, *Inst. Oekol. Chem., Kolloq.*, **44**.

258. Sjoequist, B. (1979) *Biomed. Mass Spectrom.*, **6**, 392.
259. Murphy, R.C. and Clay, K.L. (1979) *Biomed. Mass Spectrom.*, **6**, 309.
260. Magnus, V., Bandurski, R.S. and Schulze, A. (1980) *Plant Physiol.*, **66**, 775.
261. Pengelly, W.L., Bandurski, R.S. and Schulze, A. (1981) *Plant Physiol.*, **68**, 96.
262. Lapidot, A. and Nissim, I. (1980) *Adv. Mass Spectrom.*, **8B**, 1142.
263. Nissim, I., Yudkoff, M., Yang, W. *et al.* (1981) *Anal. Biochem.*, **114**, 125.
264. Lapidot, A. and Nissim, I. (1980) *Metabolism*, **29**, 230.
265. Robinson, J.R., Starratt, A.N. and Schlahetka, E.E. (1978) *Biomed. Mass Spectrom.*, **5**, 648.
266. Amir, J., Reisner, S.H. and Lapidot, A. (1980) *Pediatr. Res.*, **14**, 1238.
267. Lapin, A. and Karobath, K. (1980) *J. Chromatogr.*, **193**, 95.
268. White, R.H. (1981) *Anal. Biochem.*, **114**, 349.
269. Holdiness, M.R., Justice, J.B., Salamone, J.D. and Neill, D.B. (1981) *J. Chromatogr.*, **225**, 283.
270. Maroni, F., Bianchi, C., Tanganelli, S. *et al.* (1981) *J. Neurochem.*, **36**, 1691.
271. Hasegawa, Y., Ono, T. and Maruyama, Y. (1981) *Jap. J. Pharmacol.*, **31**, 165.
272. Haymond, M.W., Howard, C.P., Miles, J.M. and Gerich, J.E. (1980) *J. Chromatogr.*, **183**, 403.
273. Trefz, F.K., Erlenmaier, T., Hunneman, D.H. *et al.* (1979) *Clin. Chim. Acta*, **99**, 211.
274. Curtius, H.C., Farner-Wegmann, H., Niederwieser, A. and Rey, F. (1980) *Dev. Biochem.*, **16**, 281.
275. Finlayson, P.J., Christopher, R.K. and Duffield, A.M. (1980) *Biomed. Mass Spectrom.*, **7**, 450.
276. Bengtsson, G. Odham, G. and Westerdahl, G. (1981) *Anal. Biochem.*, **111**, 163.
277. Koenig, W.A., Parr, W., Lichtenstein, H.A. *et al.* (1970) *J. Chromatogr. Sci.*, **8**, 183.
278. Frank, H., Nicholson, G.J. and Bayer, E. (1977) *J. Chromatogr. Sci.*, **15**, 174.
279. Jung, G., König, W.A., Leibfritz, D. *et al.* (1976) *Biochim. Biophys. Acta*, **433**, 164.
280. Kuzumoto, S., Matsukura, M. and Shiba, T. (1980) *Pept. Chem.*, **18th**, 59.
281. Frank, H., Woiwode, W., Nicholson, G.J. and Bayer, E. (1979) in *Stable Isotopes—Proceedings of the Third International Conference* (eds E.R. Klein and P.D. Klein), Academic Press, New York, p. 165.
282. Liardon, R., Ledermann, S. and Ott, U. (1981) *J. Chromatogr.*, **203**, 385.
283. Wanatabe, N., Ohzeki, H. and Niki, E. (1981) *J. Chromatogr.*, **216**, 406.
284. Dabrowiak, J.C. and Cooke, D.W. (1971) *Anal. Chem.*, **43**, 791.
285. Wiecek, C., Halpern, B., Sargeson, A.M. and Duffield, A.M. (1979) *Org. Mass Spectrom.*, **14**, 281.
286. König, W.A., Rahn, W. and Eyem, J. (1977) *J. Chromatogr.*, **133**, 141.
287. Gal, J. and Ames, M.M. (1977) *Anal. Biochem.*, **83**, 266.
288. Lovins, R.E. (1979) *Pract. Spectrosc.*, **3** (Mass Spectrom., Part A), 19.
289. Biemann, K. (1980) in *Biochemical Applications of Mass Spectrometry* (eds. G.R. Waller and O.C. Dermer) (First Supplementary Volume) Wiley-Interscience, New York, p. 469.
290. Niall, H.D. (1973) *Methods Enzymol.*, **27**, 942.
291. Fairwell, T. and Lovins, R.E. (1971) *Biochem. Biophys. Res. Commun.*, **43**, 1280.
292. Murai, A. and Takeuchi, Y. (1975) *Bull. Chem. Soc. Jpn*, **48**, 2911.
293. Hagins, D.M., Henke, J. and Lovins, R.E. (1977) *Immunochemistry*, **14**, 697.
294. Merritt, C., Jr, DiPietro, C., Robertson, D.H. and Levy, E.J. (1974) *J. Chromatogr. Sci.*, **12**, 668.
295. Ellis, S., Fairwell, T. and Lovins, R.E. (1972) *Biochem. Biophys. Res. Commun.*, **49**, 1407.
296. Fairwell, T., Ellis, S. and Lovins, R.E. (1973) *Anal. Biochem.*, **53**, 115.
297. Hong, Y.-M., Shimonishi, Y., Matsuo, T. *et al.* (1979) *Pept. Chem.*, **17th**, 47.
298. Matsuo, T., Katakuse, I., Matsuda, H. *et al.* (1980) *Shitsuryo Bunseki*, **28**, 169.
299. Shimonishi, Y., Hong, Y.-M., Kitagishi, T. *et al.* (1980) *Eur. J. Biochem.*, **112**, 251.
300. Fernlund, P., Stenflo, J., Roepstorff, P. and Thomsen, J. (1975) *J. Biol. Chem.*, **250**, 6125.
301. Matsuo, T., Matsuda, H. and Katakuse, I. (1981) *Biomed. Mass Spectrom.*, **8**, 137.
302. Addeo, F., Malorni, A., Marino, G. and Randazzo, G. (1974) *Biomed. Mass Spectrom.*, **1**, 363.
303. Meot-Ner, M., Hunter, E.P. and Field, F.H. (1979) *J. Am. Chem. Soc.*, **101**, 686.
304. Locke, M.J., Hunter, R.L. and McIver, R.T. Jr (1979) *J. Am. Chem. Soc.*, **101**, 272.
305. Bentley, T.W. (1977) in *Mass Spectrometry* (ed. R.A.W. Johnstone) (Vol. 4) The Chemical Society, London, p. 50.

CHAPTER EIGHTEEN

Nuclear Magnetic Resonance Spectra of Amino Acids and their Derivatives

G.C. BARRETT and J.S. DAVIES

This chapter covers the influence of structure, solvent, and other parameters on the nuclear magnetic resonance (NMR) spectral characteristics of amino acids (*18.1*;$R^1 = H$, $R^3 = OH$) and their derivatives.

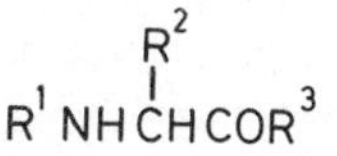

$$R^1NH\overset{R^2}{\overset{|}{C}}HCOR^3$$

(*18.1*)

The term *influence of structure* is taken to include the influence of conformation as well as steric and electronic factors of the groups R^1, R^2, and R^3, on the chemical shifts of atoms of the –NH–CH–CO– grouping.

The term *influence of solvent and other parameters* is taken to include the variations in chemical shift values which accompany changes in pH or in solvent polarity, as well as the results of shift reagent studies and other binding processes involving amino acids and their derivatives.

Less space has been allocated in this chapter to compilations of chemical shift values for amino acid side-chains (*18.1*;R^2) and other routine information, but standard chemical shift data have been compiled in the Appendix section.

18.1 AQUEOUS SOLUTIONS OF AMINO ACIDS

In aqueous solutions, the state of ionization of the amino- and carboxy-groups of an amino acid (*18.1*;$R^1 = H_2^+$ or H, $R^3 = O^-$ or OH, respectively) has a substantial effect on the resonances of nearby atoms in both [^{1}H]- and [^{13}C]-NMR spectra. These effects are partly the consequence of electrostatic effects and partly reflect changes in conformational equilibria which may accompany changes in pH. For ^{1}H spectra the chemical shift changes are greatest for the α-protons with the effects on the side-chain diminishing fairly rapidly with the number of intervening bonds. Exceptions to this behaviour may be expected for side-chains bearing functional groups. As the pH is raised, dissociation of the carboxyl group produces a negative charge density which increases the shielding of all protons [1]; α-protons shift upfield by about 0.4 ppm. A somewhat larger shielding increase occurs upon removal of protons from amino groups, since $-NH_3^+$ is quite potent in reducing local electron density.

Taking dopa (L-3,4-dihydroxyphenylalanine, *18.2*) as an example [2], two separate ABC multiplets are seen in the [^{1}H]-NMR spectrum (for 0.02 mol dm^{-3} solutions in ^2H_2O at 25°C, between p^{2}H 2.5–12.5 in the presence of 0.1 mol dm^{-3} KNO_3). One of the multiplets arises from the two closely coupled alkyl protons of the side-chain, and the other multiplet arises from the closely coupled aryl protons.

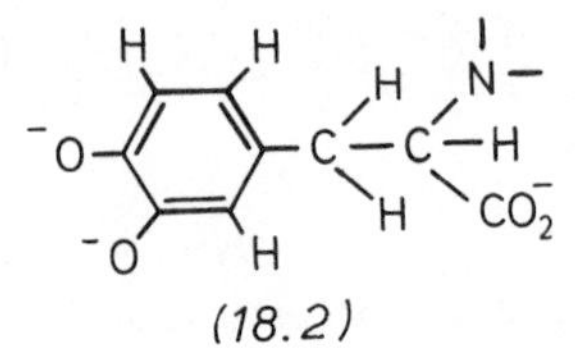

(18.2)

The chemical shift of each proton varies considerably with the dissociation of the phenolic and ammonium groupings as p^{2}H is varied. Since the protons of the aliphatic part of the molecule are less affected at high p^{2}H than the aryl protons, it is clear that the ammonium grouping, $-NH_3^+$, is less acidic than the phenolic hydroxy groups (Fig. 18.1).

Deprotonation of primary aliphatic amines causes a shift in the resonance positions of the γ-protons [3], and some of the observed shift of the aryl proton resonances for dopa (Fig. 18.1) is caused by deprotonation of the $-NH_3^+$ group; and similarly, dissociation of OH groups is likely to account for part of the shift seen for the alkyl protons as p^{2}H is raised [2].

The variation in the spin-spin coupling constants for the dopa multiplets, as p^{2}H is varied, is only slight. This indicates that the populations of different rotamers remain almost constant over a wide range of p^{2}H, for aqueous solutions of dopa [2].

p*K* Values 8.77 and 9.81 determined for dopa from these [^{1}H]-NMR data are likely to be more reliable than those determined by other techniques (most commonly UV spectra, or EMF-titrations) since it is the only procedure among those available for the purpose, where unavoidable oxidation of the amino acid introduces no error into the results (chemical shift values are independent of concentration).

A similar conclusion, that aqueous solutions of amino acids adopt conformer populations whose proportions differ only slightly over wide ranges of pH, has been drawn from [^{1}H]-NMR spin-spin coupling constant data for γ-aminobutyric acid [4], $H_3\overset{+}{N}CH_2CH_2CH_2CO_2^-$. γ-Amino-β-hydroxybutyric acid, however, adopts predominantly *trans-trans* and *gauche-trans* conformations in aqueous solutions at neutral pH, while the *trans-trans* form predominates at other pH values [4].

270 MHz [^{1}H]-NMR spectra of appropriately deuteriated DL-histidine and L-histidine at p^{2}H 8.2 in ^2H_2O show that the lower field resonance for the β-

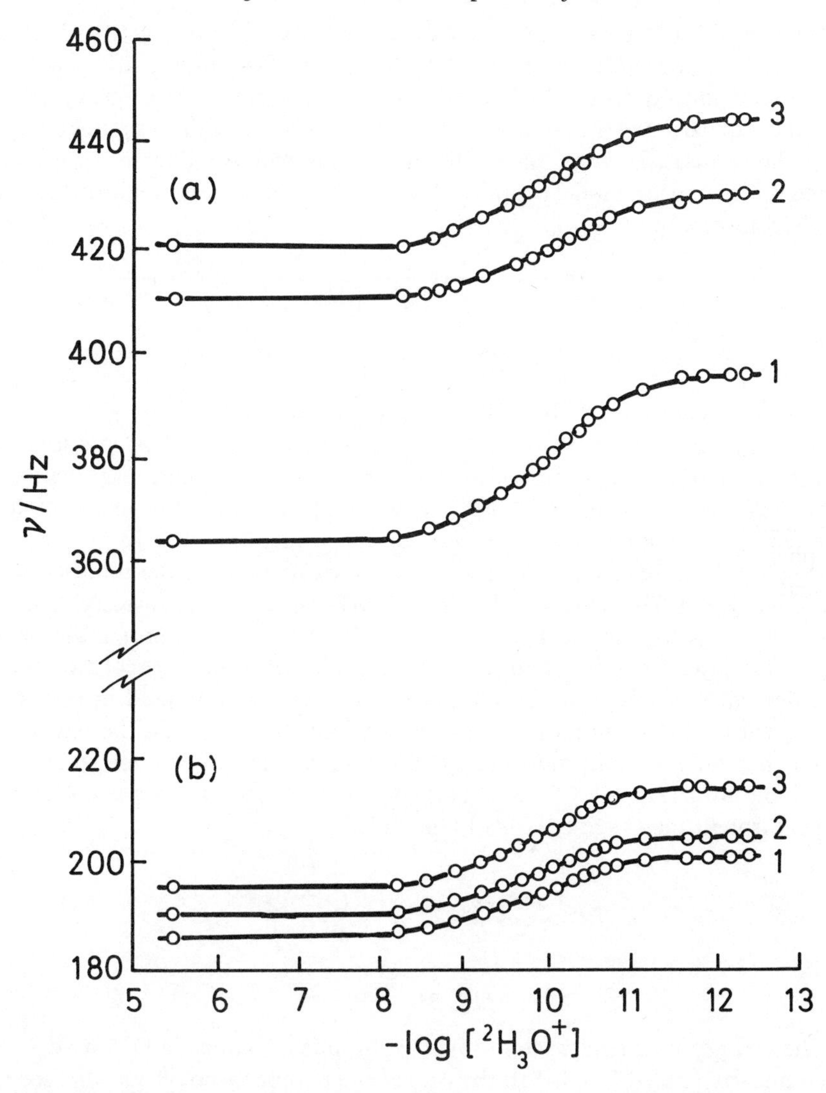

Figure 18.1 p^{2}H Dependence of the chemical shifts of (a) the alkyl and (b) the aryl protons of dopa. (Reproduced from reference 2 with the permission of authors and publishers.)

methylene protons originates in the *pro-R* proton, while the *pro-S* proton is responsible for the higher-field signal [5]. [^{1}H]-NMR and CD studies have been used with L-histidine and its derivatives [6] to reveal the considerable dependence of the side-chain conformation of this amino acid on the state of protonation of amino- and carboxy-groups.

The simpler amino acids leucine, isoleucine, and valine have been subjected to variable pH NMR studies [7, 8] and variable temperature studies [7, 8]. In leucine, the C^{α}–C^{β} bond rotamer populations at 45°C differ by less than 5% from those at 25°C [8], with the preferred conformation carrying the side-chain in a *gauche* relationship to the NH_2 group, and *trans* to the carboxy-group [9]. A 400 MHz spectrum of the β and γ protons in conjunction with computer analysis [10] shows that the couplings in the free amino acid are different from those found for leucine in a model peptide, where the side-chain is essentially locked in a single conformation:—

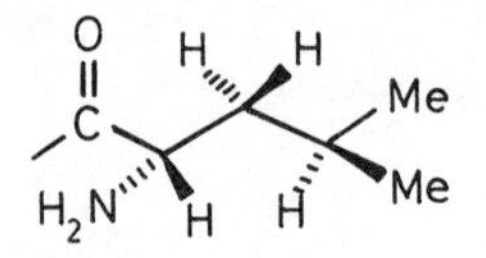

Variable temperature [^{1}H]-NMR studies of arginine and lysine have provided corresponding information on protonation behaviour [11]. Lysine adopts an intramolecularly hydrogen-bonded structure in aqueous solutions, judging by these variable temperature NMR studies [11], and arginine exists in four different zwitterionic forms at pH values ≤ 1, 8, 11–12, and ≥ 13 [11].

Pogliani *et al.* [12] have presented a detailed account of the interpretation of 270 MHz [^{1}H]-NMR and [^{13}C]-NMR spectra of *N*-acetyl-L-serine and *N*-acetyl-*O*-phospho-L-serine and *N*-acetyl-*O*-phospho-DL-threonine. Phosphorylation of serine residues in polypeptides yields mixtures of mono- and di-anions at physiological pH values (p^{2}H values 4–8 were used in this work [12]). These NMR studies have established that the charge on the phosphate group has only a small influence of NMR parameters. The conformation in which the atoms H^{α}–C^{α}–C^{β}–O–P adopt a planar W-type arrangement predominates through the range p^{2}H 4–14 (*18.3*) [12].

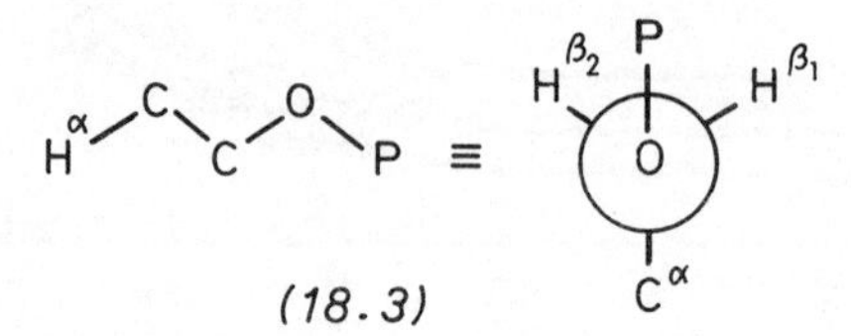

(*18.3*)

The NH proton chemical shifts (δ) of acyl-L-phenylalanine in DMSO-d_6 have been linearly related [13, 14] to the pK_a of the parent carboxylic acids used for acylation through the formula $\delta = -0.23\,pK_a + 9.19$. The α-proton chemical shifts were independent of the acyl group but resonated from highfield to lowfield in the order DMSO-d_6, methanol, trifluoroethanol, dichloroethanoic acid, trifluoroethanoic acid.

α-Aminoisobutyric acid has been subjected to detailed [^{13}C]-NMR study, and pH variation of resonances of carbonyl carbon, C^{α}, and C^{β} has been described [15]. Figure 18.2 and Table 18.1 reveal the relatively larger influence of pH on the carbonyl carbon resonance.

These are representative examples of thorough exploration of structural information available from [^{13}C]-NMR spectra for amino acids in aqueous

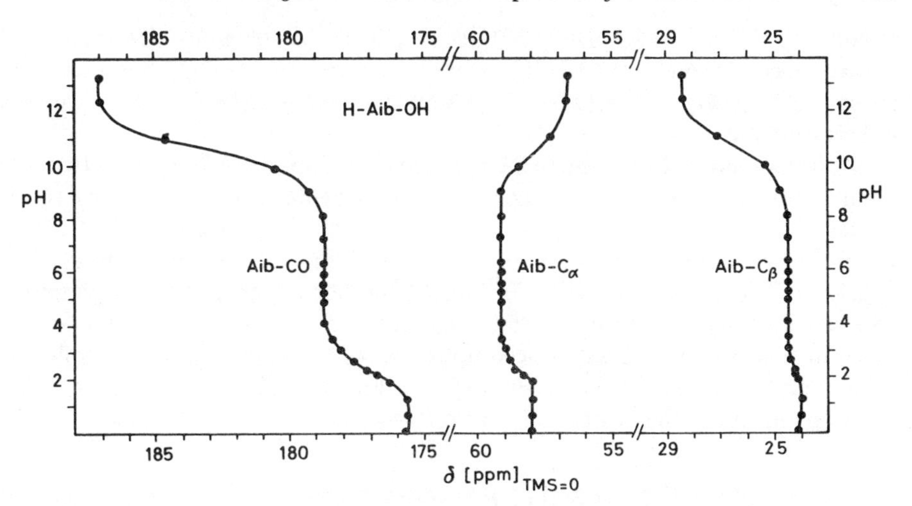

Figure 18.2 [^{13}C]-NMR of 2-methylalanine as a function of pH (100 mg ml^{-1} in water, 303 K, δ (TMS)$_{ext}$ = 0 ppm). Reproduced with permission of authors and publishers from reference 15.

Table 18.1 [^{13}C]-NMR Chemical shift variation (Δ in ppm) as a result of variation in p^{2}H for representative amino acids in which the α-carbon atom is secondary (glycine), tertiary (alanine) or quaternary (α-aminoisobutyric acid). Reproduced with permission of authors and publishers from ref. 15.

	$\Delta(\delta_{anion} - \delta_{zwitterion})$	$\Delta(\delta_{zwitterion} - \delta_{cation})$
Carbonyl carbon		
Glycine	−6.0	−3.5
Alanine	−9.0	−5.0
α-Amino-isobutyric acid	−7.0	−3.5
α-Carbon		
Glycine	−3.5	−1.5
Alanine	−1.5	−2.0
α-Amino-isobutyric acid	+2.0	−1.0
β-Carbon		
Glycine	–	–
Alanine	−4.0	−1.0
α-Amino-isobutyric acid	−4.0	−1.0

Negative sign indicates shift to higher field; positive sign indicates shift to lower field.

solutions. While there are several further studies of these types, showing a similar depth of exploration, most of the accumulated results can be summarized within a few generalizations:

(i) C^α and C^β resonances of amino acids are shifted upfield by protonation of either the $-NH_2$ or the $-CO_2^-$ groups [16]. For C^γ–C^ω carbon atoms, protonation causes very small upfield (≤ -1 ppm) shifts. For C^α the shift is -3.1 ± 0.1 (secondary C^α) or -1.4 ± 0.4 (tertiary C^α); for C^β the shift is -4.4 ± 0.7

(secondary C^{β}) or -3.2 ± 0.4 ppm (tertiary C^{β}). These values are observed as a consequence of protonation of NH_2, and the protonation of $-CO_2^-$ has a smaller effect, -2.6 ± 0.4 ppm for CO, -1.9 ± 0.6 for C^{α}, -0.7 ± 0.4 for C^{β}, and very small upfield shifts for C^{γ}–C^{ω}.

(ii) [^{2}H]-isotope shifts of up to 0.9 ppm have been observed for [^{13}C]-NMR of amino acids in ^2H_2O solutions, compared with chemical shift values in ^1H_2O [17].

(iii) Protonation of the α-amino group shifts the CO resonance by -8.3 ± 0.7 ppm. Characteristic signals for various carbon atoms in twenty common amino acids [18] have been reported, together with the suggestion that analysis of amino acid mixtures might be accomplished by [^{13}C]-NMR. Signals unique to particular protein amino acids form the basis for their identification in aqueous solutions in which they form $< 1\%$ of the solution.

18.2 OTHER SOLVENTS

NMR Studies of amino acid derivatives (*18.1*) which are soluble in organic solvents, have provided useful information on hydrogen bonding behaviour of amide carbonyl and amide NH moieties. Such results are typical of a much broader range of studies of amino acid derivatives (*18.1*) using other spectroscopic techniques, with the general objective of establishing the behaviour of particular amino acids as residues in peptides and proteins.

Dimerization of *N*-acetyl-L-alanine methyl ester (*18.1*:$R^1 = CH_3CO$, $R^2 = CH_3$, $R^3 = OCH_3$) in non-polar solvents is different from the hydrogen bonding pattern established for the corresponding *N*-methylamide [19,20].

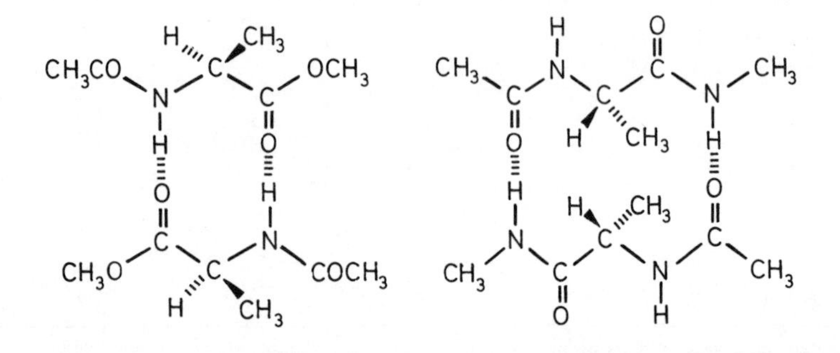

This result illustrates the different behaviour of a specific pair of compounds, one of which (the *N*-methylamide) is a model for the behaviour of the alanine residue within a peptide. Mizumo and coworkers have reported the analogous self-association of a series of *N*-acetyl amino acid *N*,*N*-dimethylamides in CCl_4 by [^{1}H]-NMR [21], and have also employed infrared spectroscopy in the same work. In fact the IR technique is particularly suitable for obtaining this type of information, and a good deal of work has been published from the early 1960s in the IR area [22]. [^{13}C]-NMR Studies [23] on a series of *N*-benzoyl amino acids in deuteriochloroform show that self-association occurs largely through two COOH – O=C hydrogen bonds and does not involve intermolecular hydrogen bonding to the NH proton.

Protonation behaviour of the imidazole group of a histidine residue has been studied in a range of H_2O or ^2H_2O-containing solvents, based on [^{1}H]-NMR spectroscopy of *N*-acetyl-L-histidine-*N*-methylamide [24], complementing the information obtained for the free amino acid in aqueous solutions [25]. The relevance of the use of amino acid derivatives in less polar media, for this sort of objective, lies in the fact that amino acid residues in peptides and proteins can be located within either hydrophobic or exposed environments, depending on the tertiary structures available overall to the molecules. The roles of functional groups in amino acid side-chains, in the physiological function of proteins, includes their participation in hydrogen bond formation and in other structure-conserving functions.

A compilation of characteristic [^{13}C]-NMR chemical shifts for a series of amino acid derivatives has been published [26] and Climie and Evans [27] give [^{13}C]-NMR data for several derivatives of cysteine, histidine, lysine, methionine and serine. [^{13}C]-NMR data for N^{α}-Boc-derivatives of methyl esters of these amino acids are given in Appendix 18.2, as well as side-chain derivatives [(*S*-, *N*- or *O*-(2-cyanoethyl), -(2-aminocarbonylethyl), -(2-methoxycarbonylethyl), -methyl)] with or without the N^{α}-Boc and methyl ester groups. This study reveals the relatively small sensitivity to pH changes, and solvent variation, of the [^{13}C]-chemical shift values of the constituent carbon atoms of these protein amino acids. The side-chain substituent causes less than 1 ppm shift in the chemical shifts of the carbon atoms of the parent amino acid, the greatest effect being felt by the β-carbon atom.

This study incidentally illustrates the use of partially decoupled spectra (i.e. full noise decoupling and off resonance proton decoupling) in unambiguous assignment of resonances to particular atoms, based on the creation of singlet or partially coupled multiplets.

The *cis*-orientation of the amide bond contributes significantly to the conformer population for *N*-acylated imino acids. The protein amino acids, proline and 4-hydroxyproline, are more correctly classified as imino acids (*18.1*; R^1 = alkyl), and provide sites in polypeptides at which the otherwise invariable progression of *trans*-amide linkages through the three-dimensional structure can be interrupted by a *cis*-amide bond. [^{1}H]-NMR studies of acyclic *N*-benzoyl-*N*-alkylamino acids [28] have been reported and some [^{13}C]-NMR data for these compounds were included in the same paper. Separate signals for the *cis*- and *trans*- forms

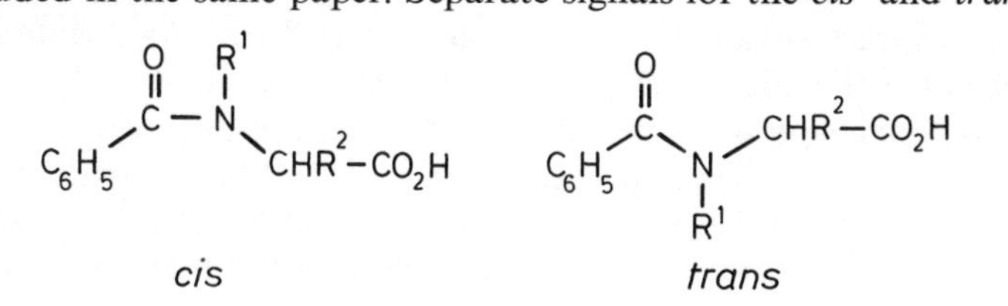

were seen in the [^{1}H]-NMR spectra of all members of a range of these compounds (except *N*-benzoylazetidine-2-carboxylic acid). In many cases, about equal proportions of *cis*- and *trans*-isomers existed in solutions of these derivatives in C^2HCl_3; benzoyl derivatives of *N*-methylamino acids with a branch at the β-carbon atom (*N*-methylisoleucine and *N*-methylvaline) actually existed preferentially in the *cis*-conformation, probably as a result of the steric interaction between the *N*-methyl group and the amino acid side-chain (the phenyl ring is rotated out of the plane of the amide group by the *N*-methyl substituent). Benzoylproline adopted predominantly the *trans* form, as did benzoylpipecolic acid, as judged by [^{13}C]-NMR data from this study [28]. This conclusion is based on the C^β and C^γ resonances adopting a characteristic *cis-trans* pattern.

Since absolute intensities of ^{13}C resonances will depend on nuclear Overhauser enhancement and spin lattice relaxation effects, quantitative analysis of *cis-trans* ratios of this type by [^{13}C]-NMR spectroscopy will involve fairly wide error limits. However, taking into account the confirmatory evidence provided by [^{1}H]-NMR spectra, the precision of the NMR method for configurational analysis of *N*-acyl amino acids is sufficiently high to provide figures assignable to two significant figures (Table 18.2):

Table 18.2 *cis* : *trans* Ratios of benzoylated amino acids in C^2HCl_3 at ambient temperature

	MeGly	*MeAla*	*MeVal*	*MeNor Val*	*MeLeu*	*MeIle*	*Pro*	*Pipe*	*Az*
cis : *trans* ([^{1}H]-NMR)	42:58	47:53	57:43	51:49	46:54	54:46	(25:75)*	(35:65)*	0:100
cis : *trans* ([^{13}C]-NMR)	38:62	43:57	63:37	48:52	49:51	55:45	(12:88)	(22:78)	0:100
Average (approx.)	40:60	45:55	60:40	50:50	48:52	55:45	20:80	25:75	0:100

*Peak overlap and separation of conformers into axial and equatorial forms prevents precise measurement.

Earlier and concurrent studies of *N*-acylimino acids include detailed [^{1}H]- and [^{13}C]-NMR study of *N*-acylsarcosine esters [29] and proline analogues [30].

Unexpected conclusions can be arrived at as a result of NMR studies of amino acid solutions, and solutions of amino acid derivatives. Strong hydrogen bonding forces identified by [^{1}H]-NMR in concentrated aqueous solutions of L-proline indicate some ordering of the solute, such as a ladder-like stacking arrangement [31]. A consequence of this result is that concentrated aqueous solutions of certain amino acids and imino acids might be expected to show some physical behaviour characteristic of a hydrophilic colloid. Aggregation of *O*-alkyltyrosines in alkaline ^2H_2O (p^{2}H 13.1) has also been identified [32] by [^{1}H]-NMR. This conclusion is arrived at indirectly, through the unexpected finding that the second most abundant conformer is actually the most crowded, and therefore least favoured, form. A similar conclusion for *N*-acetyl-L-phenylalanine ethyl ester and the corresponding *N*-methylamide [33], that the most crowded conformer (*18.4*) is actually the predominant conformation for the ethyl ester in non-polar solvents, has been drawn by use of the Pachler equations for analysing NMR data in terms of conformations.

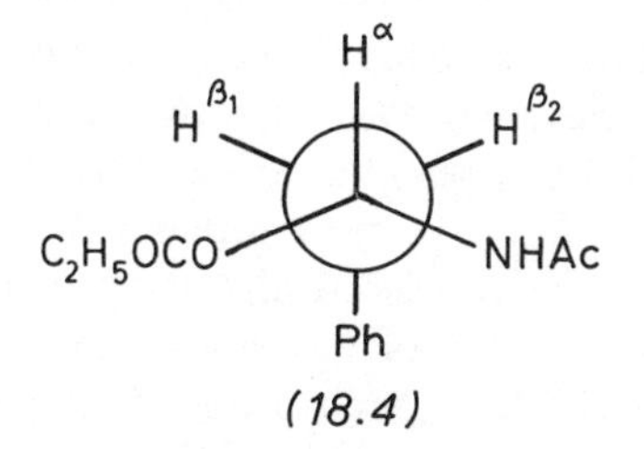

(*18.4*)

The sensitivity to solvent polarity of positions of conformational equilibria of these phenylalanine derivatives is also made clear from [^{1}H]-NMR studies and the interpretation of variations in vicinal coupling constants $^3J_{\alpha\beta_R}$ and $^3J_{\alpha\beta_S}$ for specifically deuteriated derivatives (2*S*,3*R*)- and (2*R*,3*R*)-*N*-acetyl-L-phenylalanine esters and amides [33, 34]. The relative amount of the conformer (*18.5*) increases as solvent polarity increases:

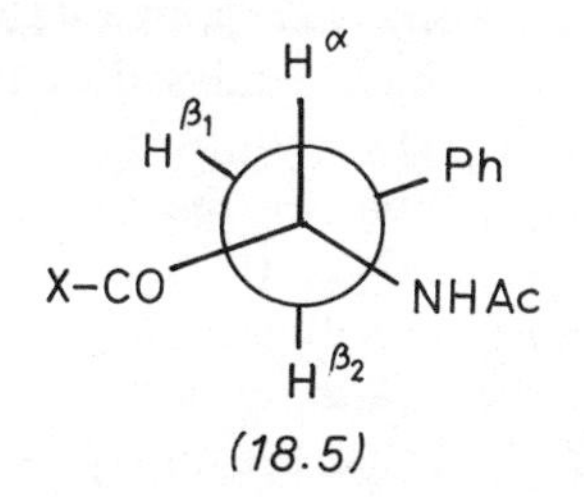

(*18.5*)

[^{1}H]-, [^{13}C]- and [^{15}N]-NMR spectra [35] of *N*-nitroso-*N*-alkylamino acids have provided useful information on the configurations of these derivatives in solution. The conformational change between *Z*- and *E*-forms as in (*18.6*)

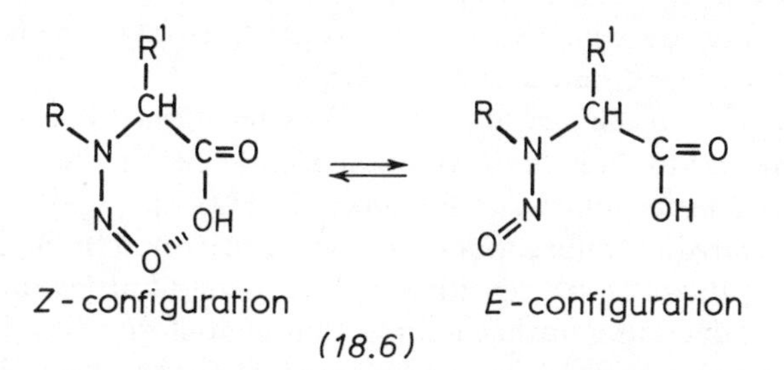

(*18.6*)

is dependent on the side-chain alkyl groups. Freshly dissolved *N*-nitroso-*N*-alkylamino acids seem to exist mainly in the *Z*-configuration; NMR signals for the *E*-configuration slowly appear, but the degree of isomerization depends upon the nature of R and R^1. At equilibrium many derivatives show a 1:1 isomeric ratio of *E*- and *Z*-forms.

18.3 NMR STUDIES BASED ON ^{15}N, ^{17}O AND OTHER NUCLEI

Natural abundance [^{15}N]-NMR studies are still in their pioneering phase, but many of the studies so far reported use amino acids as representative examples.

In water and in other protic solvents α- and ω-amino acids show an upfield shift with increase of pH in the resonance for the nitrogen atom when in the α-position [36]. There seems to be little dependence of the ^{15}N chemical shifts on structure which might be of value in analysis, but another study [37] has claimed that the ^{15}N resonance is sensitive to electronic and conformational changes. In particular cases, such as a study of the behaviour of the arginine side-chain, information not available by the more routine [^{1}H]- and [^{13}C]-NMR methods can be obtained. Thus, pH variation of the ^{15}N resonances was first established for arginine using the labelled compound $H_2{}^{15}N{-}^{13}C(={}^{15}NH)NH{-}(CH_2)_3CH(NH_2)CO_2H$ in H_2O and ^2H_2O [38]. Later studies using ^{15}N at natural abundance [39] contributed further information to the understanding of the pH dependence of the ^{15}N resonances in this amino acid and also revealed the unique contribution of the [^{15}N]-NMR method in a study of the complex formation parameters for L-arginine with Cl^-, $PO_4{}^{3-}$ and ATP [39]. The differential rates of N–H exchange were also established by Roberts and co-workers [40], who showed that base-catalysed exchange at the side-chain NH group is twice as fast as at the NH_2 groups of the guanidino group. A further specialized application [41] establishes the site of the nitro-group in nitro-arginine methyl ester hydrochloride as the side-chain imine nitrogen atom.

Only amine nitrogen signals were observable [35] at natural abundance in the [^{15}N]-NMR of *N*-nitroso-*N*-alkylamino acids. With a sample of *N*-nitroso-sarcosine enriched with 50% [^{15}N] at the nitroso nitrogen, the spectra afforded ^{15}N chemical shifts for the *Z*-isomer (in *18.6*) at +136 and −158 ppm respectively, and for the *E*-isomer at +138 and −165 ppm, respectively.

[^{17}O]-NMR studies of representative amino acids, both at natural abundance [42] and for samples enriched by up to 17% with ^{17}O, have been described [43]. Few useful features on which structural assignments might be based could be discerned beneath the broad absorption peaks.

In an alternative to the use of [^{15}N]-NMR mentioned above [39], [^{35}Cl]-NMR has been used to study the interaction of chloride ions with the amino acids arginine, lysine and histidine, as a function of pH [44].

While chemical shift values alone may yield information, in those cases of nuclei whose resonances are sensitive to local structural features and whose effects can be understood, further information in conformational terms can be derived from coupling, e.g. between ^{15}N and ^{13}C or ^{15}N and ^{1}H [37].

18.4 SHIFT REAGENT STUDIES AND RELATED BINDING PHENOMENA

Only a minority of the studies reported under this heading fall within the conventional uses of lanthanide shift reagents. *N*-Acylproline esters in C^2HCl_3 are concluded to exist mainly (*ca* 60%) in the half-chair conformation (C^α up, C^β down, alkoxycarbonyl group up) and *ca* 40% in the envelope conformation with C^γ down [45]. Similar studies have been performed with L-proline, with L-valine [46], and with L-azetidine-2-carboxylic acid [47], which exists in the puckered conformation with a dihedral angle of 25° (*18.8*).

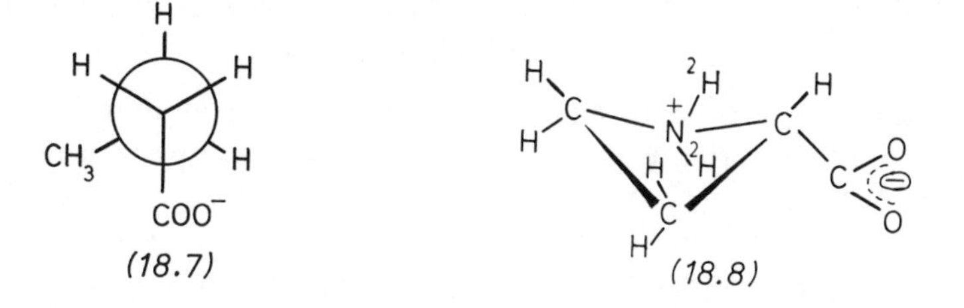

(18.7)
(18.8)

Lanthanide-induced shifts in the proton and ^{13}C resonances of sarcosine (*N*-methylglycine), demonstrate that along the lanthanide series the complexes are isostructural with respect to the amino acid ligands and the coupling constants are invariant [48]. The sarcosine ligand is in the folded *cis* conformation (*18.7*).

Binding properties of metal ions to amino acids is one of the topics within the major group of NMR studies of binding phenomena associated with amino acids. The pH dependence of the ^{13}C chemical shift values for binding by γ-carboxyglutamic acid of Ca^{2+} in competition with lanthanide cations has been interpreted to show that both γ-carboxy groups of this amino acid are involved in the binding process [49]. A related study of Mn^{2+} complexes [50], in which selective broadening of [^1H]-NMR peaks in ^2H_2O in the order $H^\alpha > H^\beta > H^\gamma$ reveals the preferred binding sites of the aliphatic amino acids, and an application of Pd(II) complexes studying rotamer populations about the C^α–C^β bond in these amino acids [51], also fall within this topic area. Binding of methylmercury (II) cations to tryptophan [52] has been studied by NMR, and represents a straightforward use of the technique for establishing the attractive interactions which may occur between biologically important species.

Some of the investigations in the preceding paragraph represent the gathering of information on binding phenomena which have an obvious relevance to possible physiological roles for amino acids. The remaining topic deals with interactions between amino acids and other organic compounds, equally encouraged by the likely importance of the results in biological contexts. One example, the binding of either D- or L-tryptophan to human serum albumin [53], is in itself not a novel system for study by spectroscopic methods, since other routine physical methods (e.g. UV spectroscopy) are well established in such studies but the use of [^1H]-NMR [53] has suggested that the amino group and the benzo moiety of this amino acid are the sites of the interaction. L-Histidine seems to undergo a stacking-type interaction with the base moiety of adenosine-5′-monophosphate, since [^1H]-NMR spectra indicate an increasing upfield shift of the resonances of the imidazole protons at positions 2 and 4, in the presence of increasing amounts of the nucleoside [54]. [^{1}H]-NMR data support the interpretation that a small red shift in the visible absorption spectra of Rose Bengal in the presence of histidine is due to an interaction involving the imidazole ring of the amino acid [55].

18.5 NMR STUDIES OF AMINO ACIDS, USING NON-STANDARD TECHNIQUES

Amino acids have proved to be unusually attractive compounds for use by research groups exploring extensions of NMR spectroscopy into non-routine techniques. A substantial amount of work in this category uses solid samples [56] giving results which, in some cases, partly complement the information available from X-ray crystal analysis. For example [57, 58], proton magnetic relaxation experiments with common aliphatic amino acids have been interpreted in terms of conformational reorientation of $-CH_3$ and $-NH_3^+$ groups, and similar [^{1}H]-NMR identification of NH_2 rotation in solid L-glutamic acid has been described [59]. Dimethyl sulphoxide solutions of *N*-acetyl-D-alloisoleucine have been studied to obtain the full panoply of scalar coupling constants and NT_1 values (relaxation rates) for [^{13}C]-atoms, indicating the adoption of a mixture of conformations. The crystal conformation of this derivative is the same as that which is predominant in the solutions [60, 61]. There is clearly considerable potential for proton relaxation spectroscopy for studying the dynamic behaviour of certain amino acid residues within polypeptides.

High resolution [^{13}C]-NMR of powdered solid samples (about 100 mg are needed) of glycine, alanine, and valine suggest that single sharp resonances are not shown for carbon atoms bonded to nitrogen [62]. Doublets are seen for the carbon atom in the grouping $C\text{–}\overset{+}{N}H_3$, with a separation of *ca* 100 Hz, arising from the failure of 'magic angle rotation' to average all solid state interactions. Whatever the physical basis of the results, it is possible that empirical structural assignments might be obtained by this NMR technique [62, 63]. While the isopropyl methyl groups of valine give ^{13}C resonances separated by 1.2 ppm for aqueous solutions [64], the separation is larger in the solid state (2.6 ppm) due to restrictions on internal rotation of the molecule [63]. Further examples of this unexpected peak splitting phenomenon for solid samples of amino acids, which is not seen in the corresponding solution spectra, have been reported by Frey and Opella for tyrosine and leucine [63].

Appendix 18.1 NMR Data for protein amino acids

Amino acid	[^{13}C]-NMR data (25°C) (p^{2}H 6.5–7.0)* (δ) (relative to TMS)		[^{1}H]-NMR data† (^2H_2O) (for zwitterionic forms) No. of hydrogens	Chemical shift (δ) (relative to DSS)	J (Hz)
Alanine	C_0	176.2	—		
C_β	C_α	51.5	1	3.78	$J_{\alpha\beta}$ 7.25
$N-C_\alpha-C_0$	C_β	17.2	3	1.49	
Arginine	C_0	175.0	—		
$C_\delta N-C_\epsilon-N$	C_α	55.3	1	(3.74) 3.18	
C_γ	C_β	28.4	2	1.87	Not available
C_β	C_γ	24.8	2	1.6	
$N-C_\alpha-C_0$	C_δ	41.5	2	3.20	
	C_ε	157.5	—		
Asparagine	C_0	174.1	—		
$C_\beta-C_\gamma ONH_2$	C_α	52.4	1	4.00	$J_{\alpha\beta}$ 5.0, 7.1
$N-C_\alpha-C_0$	C_β	35.7	2	{2.92, 2.87}	$J_{\beta\beta'}$ 16.95
	C_γ	175.3	—		
Aspartic acid	C_0	173.95	—		
$C_\beta-C_\gamma O_2H$	C_α	51.7	1	4.08	$J_{\alpha\beta}$ 5.2, 6.2
$N-C_\alpha-C_0$	C_β	35.7	2	(2.70) 3.01	
	$C_{\gamma 0}$	175.1	—		
Cysteine	C_0	173.9	—		
C_β	C_α	57.0	1	3.82	$J_{\alpha\beta}$ 4.7, 5.2
$N-C_\alpha-C_0$	C_β	25.8	2	(3.08) 2.93 (3.05)	$J_{\beta\beta'}$ 14.9
Cystine $(Cys)_2$	C_0	181.2	—	—	
	C_α	55.8	2	3.56	$J_{\alpha\beta}$ 4.0, 8.0
	C_β	44.6	4	{2.98, 2.80}	$J_{\alpha\beta}$ 15.5
Glutamine	C_0	164.75	—	—	
$C_\gamma-C_\delta O_2H$	C_α	55.35	1	3.77	$J_{\alpha\beta}$ 5.95
C_β	C_β	27.40	2	2.16	$J_{\beta\gamma}$ 7.50
$N-C_\alpha-C_0$	C_γ	32.05	2	2.43	
	$C_{\delta 0}$	178.55	—	—	
Glutamic acid	C_0	175.6	—		
$C_\gamma C_\delta O_2H$	C_α	55.7	1	3.75	$J_{\alpha\beta}$ 6.25
C_β	C_β	28.1	2	2.12	$J_{\beta\gamma}$ 7.50
$N-C_\alpha-C_0$	C_γ	34.5	2	2.49	
	$C_{\delta 0}$	182.3	—	—	
Glycine	C_0	173.5	—		
$N-C_\alpha-C_0$	C_α	42.5	2	3.56	

Appendix 18.1 (*Contd.*)

Amino acid	[^{13}C]-NMR *data* (25°C) (p^{2}H 6.5–7.0)* (δ) (*relative to TMS*)		[^{1}H]-NMR *data*† (^2H_2O) (*for zwitterionic forms*) *No. of hydrogens*	*Chemical shift* (δ) (*relative to DSS*)	*J* (Hz)
Histidine 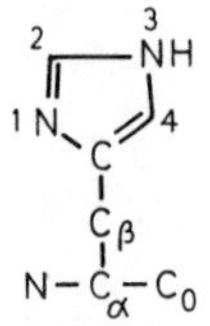	C_0	173.7	—		
	C_α	54.7	1	3.97	$J_{\alpha\beta}$ 5.1, 7.5
	C_β	27.85	2	(3.19) (3.14) 3.16	$J_{\beta\beta'}$ 15.65
	C_2	136.05	1	7.72	$J_{2,4}$ 0.9
	C_4	131.5	1	7.03	$J_{4\beta}$ 0.3–0.6
	C_5	116.95	—		
Isoleucine 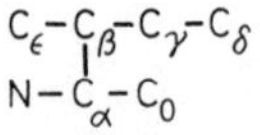	C_0	174.4	—		
	C_α	59.7	1	3.66	$J_{\alpha\beta}$ 3.9
	C_β	36.0	1	(1.5) 1.97	$J_{\beta\varepsilon}$ 7.0
	C_γ	24.6	2	1.35	$J_{\gamma\delta}$ 7.3
	C_δ	11.2	3	} 1.00	
	C_ε	14.85	3		
Leucine 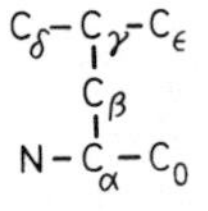	C_0	175.8	—		
	C_α	53.6	1	3.70	$J_{\alpha\beta}$ 5.0, 9.9
	C_β	39.9	2	(1.69) (1.73)	$J_{\beta\beta'}$ 14.5
	C	24.3	1	1.71	
	C_δ }	21.1	3	} 0.96	
	C_ε }	22.2	3		
Lysine 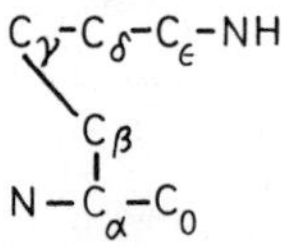	C_0	175.5	—		
	C_α	55.7	1	(3.73) 3.37	$J_{\alpha\beta}$ 6.2
	C_β	31.1	2	(1.87) 1.69	$J_{\beta\gamma}$ 7.3
	C_γ	22.7	2	1.43	$J_{\gamma\delta}$ 7.3
	C_δ	27.6	2	1.69	$J_{\delta\varepsilon}$ 7.2
	C_ε	40.3	2	2.95	
Methionine 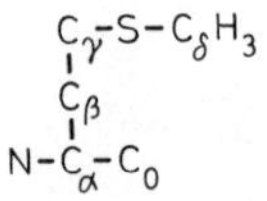	(Alkaline solution)				
	C_0	182.7	—		
	C_α	56.1	1	3.80	$J_{\alpha\beta}$ 5.75, 6.85
	C_β	30.6	2	(2.12) 2.60	$J_{\beta\gamma}$ 7.3
	C_γ	34.6	2	(2.56) 2.12	
	C_δ	15.0	3	2.12	
Phenylalanine 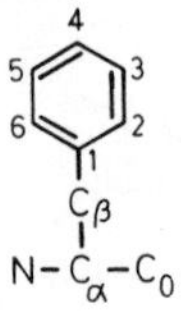	C_0	174.3	—		
	C_α	56.5	1	3.97	$J_{\alpha\beta}$ 5.4, 7.9
	C_β	36.9	2	3.20	$J_{\beta\beta'}$ 14.7
	C_1	135.9	—		
	$C_{2,6}$	130.0	2	} 7.36	
	$C_{3,5}$	130.0	2		
	C_4	128.3	1		
Proline 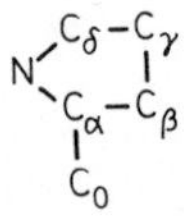	C_0	175.2	—		
	C_α	62.0	1	4.08	$J_{\alpha\beta}$ 7.5
	C_β	29.9	2	(2.11) (2.31) 2.06	
	C_γ	24.6	2	2.06	
	C_δ	46.9	2	(3.34) (3.40)	$J_{\delta\delta'}$ 11.5

Appendix 18.1 (*Contd.*)

Amino acid	[^{13}C]-NMR *data* (25°C) (p^{2}H 6.5–7.0)* (δ) (*relative to TMS*)		[^{1}H]-NMR *data*† (^2H_2O) (*for zwitterionic forms*) *No. of hydrogens*	*Chemical shift* (δ) (*relative to DSS*)	*J* (Hz)
Serine	C_0	175.1	—		
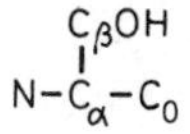 $C_\beta OH$	C_α	59.1	1	3.94	$J_{\alpha\beta}$ 3.7, 9.2
N–C_α–C_0	C_β	63.0	2	3.94	$J_{\beta\beta'}$ 11.8
Threonine	C_0	173.9	—		
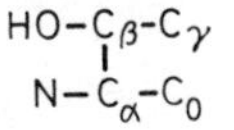 HO–C_β–C_γ	C_α	61.5	1	3.58	$J_{\alpha\beta}$ 4.9
N–C_α–C_0	C_β	67.1	1	4.23	$J_{\beta\gamma}$ 6.6
	C_γ	20.6	3	1.32	
Tryptophan	C_0	174.7	—		
	C_α	56.0	1	4.04	$J_{\alpha\beta}$ 5.0, 7.9
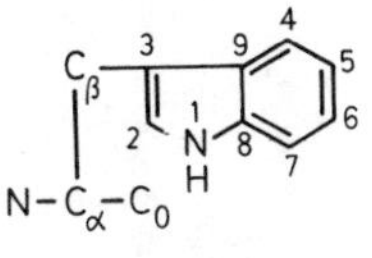	C_β	28.2	2	(3.45) (3.31) 3.38	$J_{\beta\beta'}$ 15.5
	C_2 127.6 C_3 108.8		1—	7.31	$J_{2\beta}$ 0.7, 0.3–0.4
	C_4 124.5 C_5 120.5		1,1	7.72; 7.28	
	C_6 114.25 C_7 128.9		1,1	7.55	
N–C_α–C_0	C_8 138.6 C_9 121.8		—		
Tyrosine	C_0	175.0	—		
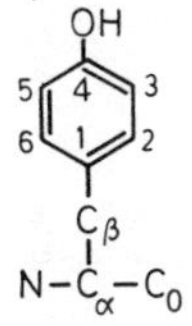	C_α	57.3	1	3.93	$J_{\alpha\beta}$ 5.7, 8.2
	C_β	37.5	2	(3.05) (3.17)	$J_{\beta\beta}$ 15.0
	C_1	N.R.	—		
	$C_{2,6}$	130.5	2	7.19 (80°)	J_{Ar} 8.9
N–C_α–C_0	$C_{3,5}$	117.5	2	6.88 (80°)	
	C_4	156.3	—		
Valine	C_0	174.6	—		
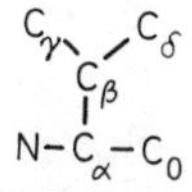 C_γ, C_δ	C_α	60.7	1	3.62	$J_{\alpha\beta}$ 4.4
C_β	C_β	29.35	1	2.29	$J_{\beta\gamma}$ 6.9, 7.0
N–C_α–C_0	C_γ	18.25	3	1.06	
	C_δ	17.0	3		

* [^{13}C]-NMR Data from ^{13}*C-N.M.R. Spectral Data* by Bremser, W., Ernst, L. and Frank, B. (1978) Verlag Chemie; Surprenant, H.L., Sarneski, J.E., Key, R.R., Byrd, J.T. and Reilley, C.N. (1980) *J. Magn. Res.*, **40**, 231. (Chemical shifts have all been converted to TMS scale.)

† Data recorded using sodium 2,2-dimethyl-2-silapentane-5-sulphonate (DSS) as internal standard. (For TMS add 0.47 ppm). Data based on results of Mandel, M. (1965) *J. Biol. Chem.*, **240.** (100 MHz) but supplemented with information and data from Roberts, G.C.K. and Jardetsky, O. (1970) *Adv. in Protein Chemistry*, **24**, 447; Bovey, F.A. (1972) in *High Resolution N.M.R. of Macromolecules*, Academic Press, p. 247, (220 MHz); see also Bak, B. *et al.* (1968) *J. Mol. Spectrosc.*, **26**, 78. Data in brackets indicate major deviations from results reported in Mandel's paper.

Appendix 18.2 [13C]-NMR chemical shifts for $BOC{-}NH{-}{}^{2}CH({}^{3}CH_2X){-}{}^{1}CO_2Me$ in C^2HCl_3 (TMS as internal reference)

Compound	Chemical shift δ[a]													
	Resonances common to all amino acid derivatives							*Side-chain resonances 'X'*						
L-Cysteine	C-1	C-2	C-3	$(\underline{CH_3})_3C$	$(CH_3)_3C$	NHC=O	OCH_3	SCH_2	$SCH_2\underline{CH_2}$	CN	C=O	OCH_3	SCH_3	CH
Boc-Cys-OMe	170.79[b] s	54.98 d	27.30 t	28.27 q	80.33 s	155.07 s	52.64 q							
CH_2CH_2CN Boc-Cys-OMe	171.05 s	53.49 d	34.57 t	28.27 q	80.45 s	155.00 s	52.77 q	28.77 t	18.27 t	117.89 s				
$CH_2CH_2CONH_2$ Boc-Cys-OMe	171.63 s	53.68 d	34.83 t	28.34 q	80.40 s	155.26 s	52.64 q	28.34 t	36.00 t		173.72 s			
$CH_2CH_2CO_2Me$ Boc-Cys-OMe	171.38 s	53.49 d	34.70† t	28.34 q	80.26 s	155.07 s	52.51 q	27.67 t	34.57† t		172.03 s	51.80 q		
CH_2CONH_2 Boc-Cys-OMe	171.45 s	53.44 d	35.54 t	28.32 q	80.58 s	155.42 s	52.80 q	36.04 t			171.45 s			
CH_2CO_2Me Boc-Cys-OMe	171.38 s	53.29 d	35.03 t	28.34 q	80.35 s	155.06 s	52.28† q	33.86 t			170.47 s	52.45† q		
CH_3 Boc-Cys-OMe	171.70 s	53.30 d	36.90 t	28.34 q	80.20 s	155.13 s	52.51 q						16.25 q	
$CH_2CH(OH)CO_2Me$ Boc-Cys-OMe	171.35 s	52.76 d	35.32 t	23.03[c] q		173.33 s	52.35 q	36.90 t			170.42 s	52.35 q		70.35 71.05
L-Histidine								C′-2	C′-4	C′-5	NCH_2	$\underline{CH_2}CN$	CN	
Boc-His-OMe	172.74 s	53.88 d	29.83 t	28.34 q	80.07 s	155.65 s	52.32 q	135.37 d	134.07 s	116.14 d				
CH_2CH_2CN Boc-His-OMe	172.55 s	56.68 d	30.35 d	28.40 q	79.74 s	155.59 s	52.19 q	136.87 d	138.75 s	116.27 d	42.57 t	20.54 t	116.72 s	
L-Lysine								C-4	C-5	C-6	NCH_2	$\underline{CH_2}CN$	CN	
Boc-Lys-OMe	173.39 s	53.29 d	28.92 m	28.40 q	79.87 s	155.91 s	52.91 q	22.68 t	32.30 t	42.11 t				
CH_2CH_2CN Boc-Lys-OMe	173.26 s	53.49 d	29.44 t	28.34 q	79.94 s	155.39 s	52.19 q	22.94 t	32.63 t	48.81 t	45.10 t	18.72 t	118.61 s	
L-Serine								OCH_2	CH_2CN	CN				
Boc-Ser-OMe	171.50 s	56.02 d	63.30 t	28.34 q	80.39 s	155.85 s	52.58 q							
CH_2CH_2CN Boc-Ser-OMe	171.10. *	53.70 *	71.36 t	28,33 q	80.26 s	155.37 *	52.40 *	65.96 t	18.85 t	119.00 s				

[a] Chemical shift downfield from TMS in C^2HCl_3 solution.
[b] Multiplicity in off-resonance proton decoupled spectra; s singlet d doublet m multiplet, etc.

Appendix 18.3 [^{13}C]-NMR Chemical shifts for $R{-}NH{-}{}^{2}CH(-{}^{3}CH_2{-}X){-}{}^{1}CO{-}Y$ (R = H, or H_2 with $\overset{+}{N}$) (Y = OH, O^-, or OMe)

Compound	*pH*	*Chemical shift δ[a]*											
		Common resonances			*Side-chain resonances*								
L-Cysteine		C-1	C-2	C-3	SCH_2	CH_2CN	CN	C=O					
CH_2CH_2CN Cys	7.0	174.24[b] s	55.24 d	33.67 t	28.54 t	19.83 t	122.06 s						
CH_2CH_2CN Cys	14.0	182.37	56.68	38.47	28.41	19.80	122.20						
CH_2CH_2CN Cys	1.0	171.90	54.01	32.83	28.86	19.96	121.99						
CH_2CONH_2 Cys	1.0	170.86 s	52.97 d	32.92 t	35.74 t								
CH_2CO_2H Cys	14.0	*	55.83 d	38.64 t	38.02 t								
CH_2CO_2H Cys	1.0	171.03†	52.92	32.80	34.73			174.95†					
L-Histidine					C′-2	C′-4	C′-5	N_τ-CH_2	N_π-CH_2	CH_2CN	CN	C=O	OCH_3
CH_2CH_2CN His-OMe	7.0	174.52 s	55.67 d	29.47 t	*	*	119.77 d	43.14 t		20.49 t	119.77 s		54.12 q
CH_2CH_2CN His	1.0	170.33 s	52.72 d	25.94 t	136.50 *	128.57 *	122.03 *	45.77 t		20.02 t	118.74 *		
CH_2CO_2H His	4.5	173.40† s	54.35 d	26.72 t	136.91 d	128.81 s	122.54 d	52.66 t	50.52 *			172.09† s	
CH_2CO_2H His	14.0	183.41 s	57.03 d	34.00 *	139.14 d	138.34 s	199.48 d	50.92 t	—			176.52 s	

Appendix 18.3 (*Contd.*)

Compound	*pH*	*Chemical shift δ*[a]								
		Common resonances			*Side-chain resonances*					
CH_2CO_2H His	1.0	170.82† s	52.53 d	25.85 t	137.57 d	127.92 s	123.12 d	50.69 t		170.66† s
$(CH_2CO_2H)_2$ His	4.0	172.47† s	53.13 d	25.13 t	139.27 d	130.41 s	123.26 d	52.18 t	50.10 t	171.98 s
$(CH_2CO_2H)_2$ His	1.0	*	51.83 d	24.53 t	140.10 *	130.16 *	123.73 *	51.02 t	48.75 t	*
L-Lysine					C-4	C-5	C-6	NCH_2	C=O	
CH_2CO_2H Lys	7.2	175.50†	55.34	25.98	22.36	30.71	47.86	50.03	172.24†	
CH_2CO_2H Lys	1.0	172.57†	53.33	25.74	22.26	29.98	47.97	47.00	172.57†	
L-Methionine					C-4	SCH_2	SCH_3	C=O		
CH_2CO_2H Met	7.0	173.34 s	53.70 d	26.14	38.38 t	48.34 t	24.57 q	169.03 s		
CH_2CO_2H Met	1.0	172.30 s	52.91 d	25.83 t	38.28 t	47.78 t	24.70 q	168.80 s		

[a] Chemical shift referenced to dioxan = 67.46.
[b] Multiplicity in off-resonance proton decoupled spectra; s singlet, d doublet, etc.
*Not observed. †Assignment could be reversed. From Climie, I.J.G. and Evans, D.A. (1982) *Tetrahedron*, **38**, 697, reproduced by permission of the authors and publishers.

REFERENCES

1. Bovey, F.A. (1972) in *High Resolution NMR of Macromolecules*, Academic Press, p. 247.
2. Jameson, R.F., Hunter, G. and Kiss, T. (1980) *J. Chem. Soc., Perkin Trans. 2*, 1105.
3. Sudmeier, F.L. and Reilley, C.N. (1964) *Anal. Chem.*, **36**, 1698.
4. Tanaka, K., Akatsu, H., Ozaki, Y. *et al.* (1978) *Bull. Chem. Soc. Japan*, **51**, 2654.
5. Kainosho, M., Ajisaka, K., Sawasa, S. *et al.* (1979) *Chem. Lett.*, 395.
6. Tran, T., Lintner, K., Toma, F. and Fermandjian, S. (1977) *Biochim. Biophys. Acta*, **492**, 245.
7. Tikhonov, V.P. and Kostromina, N.A. (1977) *Teor. Eksp. Khim.*, **13**, 496.
8. Fischman, A.J., Wyssbrod, H.R., Agosta, W.C. *et al.* (1977) *J. Am. Chem. Soc.*, **99**, 2953.
9. Fischman, A.J., Wyssbrod, H.R., Agosta, W.C. and Cowburn, D. (1978) *J. Am. Chem. Soc.*, **100**, 54.
10. Abraham, R.J., Jackson, J.T. and Thomas, W.A. (1980) *Org. Magn. Res.*, **14**, 543.
11. Tikhonov, V.P. and Kostromina, N.A. (1978) *Ukr. Khim. Zhur.*, **44**, 451.
12. Pogliani, L., Zeissow, D. and Krüger, C. (1979) *Tetrahedron*, **35**, 2867.
13. Shimohigashi, Y., Inoue, M., Kato, T. *et al.* (1981) *Memoirs of the Faculty of Sci. Kyushu Univ.*, **13**, 135.
14. Shimohigashi, Y., Inoue, M., Kato, T. *et al.* (1979) *Tetrahedron Lett.*, 1327.
15. Leibfritz, D., Haupt, E., Dubischar, N. *et al.* (1982) *Tetrahedron*, **38**, 2165.
16. Batchelor, J.G., Feeney, J. and Roberts, G.C.K. (1975) *J. Magn. Reson.*, **20**, 19.
17. Led, J.J. and Petersen, S.B. (1979) *J. Magn. Reson.*, **33**, 603.
18. Svergun, V.I., Tarabakin, S.V. and Panov, V.P. (1980). *Khim.-Farm. Zh.*, **14**, 104; (*Chem. Abs.*, **92**, 193790).
19. Asukura, T., Kanio, M. and Nishioka, A. (1979) *Biopolymers*, **18**, 467.
20. Asukura, T. and Nishioka, A. (1980) *Bull. Chem. Soc. Japan*, **53**, 490.
21. Mizumo, K., Nishio, S. and Shindo, Y. (1979) *Biopolymers*, **18**, 693.
22. Surveyed (1965–1982) in *Amino-acids, Peptides and Proteins*, (*Specialist Periodical Reports*) (Vols 1–13), The Chemical Society.
23. Fong, C.W. and Grant, H.G. (1981) *Australian J. Chem.*, **34**, 1869.
24. Tanokura, M., Tasumi, M. and Miyazawa, T. (1978) *Chem. Lett.*, 739.
25. Weinkam, R.J. and Jorgensen, E.C. (1973) *J. Am. Chem. Soc.*, **95**, 6084.
26. Schweizer, B., Scheller, D. and Losse, G. (1979) *J. Prakt. Chem.*, **321**, 1007.
27. Climie, I.J.G. and Evans, D.A. (1982) *Tetrahedron*, **38**, 697.
28. Davies, J.S. and Thomas, W.A. (1978) *J. Chem. Soc., Perkin Trans. 2*, 1157.
29. Kricheldorf, H.R. and Schilling, G. (1977) *Makromol. Chem.*, **178**, 3115.
30. Higashijima, T., Tasumi, M. and Miyazawa, T. (1977) *Biopolymers*, **16**, 1259.
31. Schobert, B. and Tschesche, H. (1978) *Biochim. Biophys. Acta*, **541**, 270.
32. Menger, F.M. and Jerkunica, J.M. (1977) *Tetrahedron Lett.*, 4569.
33. Kobayashi, J. and Nagai, U. (1977) *Tetrahedron Lett.*, 1803.
34. Kobayashi, J. and Nagai, U. (1978) *Biopolymers*, **17**, 2265.
35. Chow, Y.L. and Polo, J. (1981) *Org. Magn. Reson.*, **15**, 200.
36. Kricheldorf, H.R. (1979) *Org. Magn. Reson.*, **12**, 414.
37. Blomberg, F. and Rueterjans, H. (1978) *Jerusalem Symp. Quantum Chem. Biochem.*, **11**, 231.
38. London, R.E., Walker, T.E., Whaley, T.W. and Matwiyoff, N.A. (1977) *Org. Magn. Reson.*, **9**, 598.
39. Kanamori, K., Cain, A.H. and Roberts, J.D. (1978) *J. Am. Chem. Soc.*, **100**, 4979.
40. Yavari, I. and Roberts, J.D. (1978) *Biochem. Biophys. Res. Commun.*, **83**, 635.
41. Gust, D., Dirks, G. and Pettit, G.R. (1979) *J. Org. Chem.*, **44**, 314.
42. Valentine, B., St. Amour, T., Walter, R. and Fiat, D. (1980) *Org. Magn. Reson.*, **13**, 232.
43. Valentine, B., St. Amour, T., Walter, R. and Fiat, D. (1980) *Org. Magn. Reson.*, **13**, 413.
44. Jonsson, B. and Lindman, B. (1977) *FEBS Letters*, **78**, 67.
45. DeTar, D.F. and Luthra, N.P. (1979) *J. Org. Chem.*, **44**, 3299.
46. Mossoyan, J., Asso, M. and Beuliau, D. (1980) *Org. Magn. Reson.*, **13**, 287.
47. Inagaki, F., Takahashi, S., Tasumi, M. and Miyazawa, T. (1975) *Bull. Chem. Soc. Japan*, **48**, 1590.
48. Elgavish, G.A. and Reuben, J. (1981) *J. Magn. Reson.*, **42**, 242.
49. Sperling, R., Furie, B.C., Blumenstein, M. *et al.* (1978) *J. Biol. Chem.*, **253**, 3898.
50. Tiezzi, E. and Valensin, G. (1976) *Biofizika*, **21**, 401.
51. Vestnes, P.I. and Martin, R.B. (1980) *J. Am. Chem. Soc.*, **102**, 2906.
52. Svejda, P., Maki, A.H. and Anderson, R.R. (1978) *J. Am. Chem. Soc.*, **100**, 7138.
53. Monti, J.P., Sarrazin, M., Briand, C. and Crevat, A. (1977) *J. Chim. Phys. Phys. Chim. Biol.*, **74**, 942.

54. Maotsch, H.H. and Neurohr, K. (1978) *FEBS Letters*, **86**, 57.
55. Sidorowicz, A. (1978) *Stud. Biophys.*, **73**, 185.
56. Pande, U. and Guptia, R.C. (1977) *Indian J. Pure and Appl. Phys.*, **15**, 777.
57. Andrew, E.R., Hinshaw, W.S., Hutchins, M.G. *et al.* (1976) *Mol. Phys.*, **32**, 795.
58. Andrew, E.R., Hinshaw, W.S., Hutchins, M.G. and Sjoeblom, R.O.I. (1977) *Mol. Phys.*, **34**, 1695.
59. Ganapathy, S., McDowell, C.A. and Raghunathan, P. (1980) *J. Magn. Reson.*, **40**, 1.
60. Niccolai, N., Miles, M.P., Hehir, S.P. and Gibbons, W.A. (1978) *J. Am. Chem. Soc.*, **100**, 6528.
61. Niccolai, N., Miles, M.P., Hehir, S.P. and Gibbons, W.A. (1980) *J. Am. Chem. Soc.*, **102**, 1412.
62. Groombridge, C.J., Harris, R.K., Packer, K.J. *et al.* (1980) *J. Chem. Soc., Chem. Commun.*, 174.
63. Frey, M.H. and Opella, S.J. (1980) *J. Chem. Soc., Chem. Commun.*, 474.
64. Horsley, W., Sternlicht, H. and Cohen, J.S. (1970) *J. Am. Chem. Soc.*, **92**, 680.

CHAPTER NINETEEN

The Optical Rotatory Dispersion and Circular Dichroism of Amino Acids and their Derivatives

C. TONIOLO

19.1 INTRODUCTION

In the last 25 years optical rotatory dispersion (ORD) and circular dichroism (CD) have found a secure place among the physical techniques used by organic and biological chemists in solving their research problems. For an introduction to ORD–CD theory, nomenclature, instrumentation, and experimental application involving a wide variety of chromophores, the reader is referred to the vast body of published review articles, monographs and chapters of specialized books (see references 1–37).

However, no prior review devoted exclusively to the chirospectroscopic [38] properties of amino acids has yet appeared. This chapter surveys the literature of the subject to October 1981. Chromophoric derivatives of α- (and β-)amino, α-carboxyl, and reactive side-chain groups of amino acids are also discussed. Finally, recent developments (CD in the vacuum-ultraviolet (VUV) and in the infrared (IR) spectral regions, magnetic CD, and induced CD) are reported.

Peptides and metal complexes have not been considered. The superiority of CD over ORD, both in sensitivity and in ease of interpretation, is well established, so that emphasis has been directed towards CD studies.

19.2 CHIROSPECTROSCOPIC PROPERTIES OF AMINO ACIDS

19.2.1 Amino acids and their derivatives with CD maxima only in the far-ultraviolet region

In 1965 Legrand and Viennet [39] were the first to report the far-UV (250–185 nm) CD properties of a number of free, C_α-alkylated L_s-α-amino acids (the L_s configuration at C-2 in α-amino acids is usually but not always (*S*) in the Cahn–Ingold–Prelog convention [40]) as a function of pH. A strong, *positive* Cotton effect (CE) (band 1), which shifts to higher wavelengths either at pH 1 or pH 13, is observed in the range 200–204 nm at neutrality. The intensity of this band is enhanced by increasing side-chain length and decreasing pH; conversely, side-chain branching reduces the magnitude of the band.

Since 1968 many other experimental and theoretical papers have appeared dealing with the CD properties of amino acids, either underivatized or as their ester derivatives, with CD maxima only in the far-UV region [41–87]. The most remarkable finding for the C_α-alkylated α-amino acids was the observation of an additional weak, *negative* CE (band 2), centred at 250–245 nm at acidic pH (Fig. 19.1). This band, with the single exception of L-Ala, disappears at neutral and alkaline pH. The intensity of band 2 for L-Ala in 1 M NaOH is substantially greater than that in 1 M HCl. The effect produced by lengthening the side-chain is similar to that observed with band 1, although not as pronounced. Bands 1 and 2 occur also in α-amino acid alkyl esters; in these compounds band 2 usually increases in intensity at alkaline pH and with branching of the alkyl group of the ester moiety.

In contrast, L-Pro, a N_α-monoalkylated cyclic α-amino acid exhibits a sigmoidal-type CD curve (a medium-intensity, positive CE near 210 nm, followed by a strong, negative CE near 190 nm) *in the* 220–185 *nm region* at pH 7 and in 1,1,1,3,3,3-hexafluoropropan-2-ol solution. Both bands decrease in intensity as

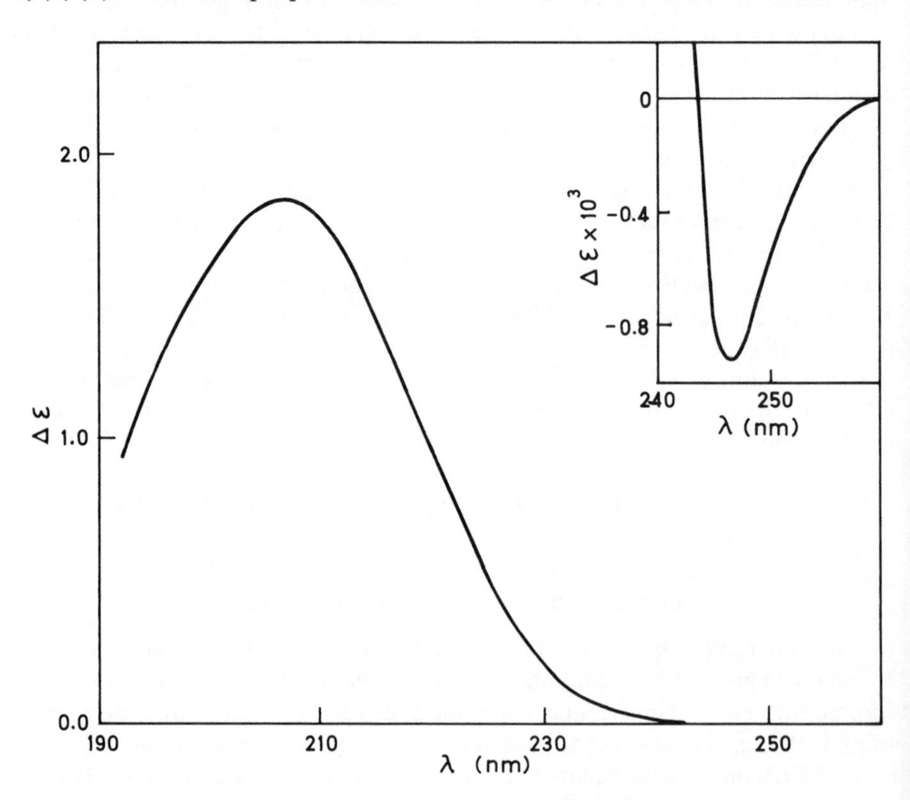

Figure 19.1 The CD spectrum of L-Val in 0.1 N HCl.

the dielectric constant and protonating power of the solvent both decrease. At 99% acetonitrile and 1% pH 7 buffered water, the positive band disappears and the shorter-wavelength band red-shifts and decreases considerably in intensity. However, neither all ring-substituted prolines (e.g. *trans*-3-methyl-L-proline), nor all N_α-monoalkylated cyclic α-amino acids (e.g. L-pipecolic acid) show the double-humped CD curve below 220 nm at neutrality characteristic of L-Pro [48].

In acidic and alkaline solutions the CD spectrum of L-Pro resembles those of N_α-unalkylated α-amino acids (a positive CE at about 215 nm and a weaker, negative CE at 250 nm). These features are also shown by the CD curves of L-Pro alkyl esters.

For N_α-monoalkylated acyclic L-α-amino acids a single positive CD maximum has been reported, at about 205 nm in water and about 210 nm in acidic solution [68, 69].

In acidic solution the intense, positive CD bands of the N_α,N_α-dimethylated derivatives of L-Ala, L-Val, and L-Leu occur in approximately the same positions as the parent amino acids, but the ellipticities are reduced somewhat [53]. In alkaline solution the bands are at about 8 nm longer wavelength than the parent compounds.

At pH 1 the CD curves for the α-trimethylammonium acids derived from L-Ala and L-Val present a single band at 213 nm with the same sign as the configurationally related amino acids [70]. No CD absorption was observed corresponding to the CD band in amino acids in the 235–250 nm region. In the zwitterionic form a blue shift to 205 nm of the positive CE occurs. The two methyl ester derivatives have a dichroic band at 213 nm irrespective of the pH.

The development of a successful theoretical treatment of the electronic transitions of carboxylic acids or alkyl esters is far behind that of the carbonyl (ketone) transitions [88]. The carbonyl group exhibits a low-intensity, low-energy transition near 290 nm, due to an $n \to \pi^*$ transition. When $-OR$ ($-OH$) is attached to the carbonyl group, interaction of the non-bonding electron on the ether oxygen with the π-orbital of the carbonyl group raises the energy of the anti-bonding π-orbital and splits the bonding π-orbital into two new orbitals of increasing energy, π_1 and π_2, the latter being essentially a non-bonding orbital located mainly on the ether oxygen (Fig. 19.2). The two transitions of lowest energy of the $-COOR$ ($-COOH$) chromophore then belong to $n \to \pi_3^*$ and $\pi_2 \to \pi_3^*$. The $n \to \pi_3^*$ transition is predicted to appear at shorter wavelength than the corresponding $n \to \pi^*$ transition of the carbonyl chromophore. In fact, a weak absorption is observed to occur near 210 nm in the UV spectra of carboxylic acids and alkyl esters, which fulfils all the criteria used in the identification of an $n \to \pi^*$ transition [88, 89]. This absorption band is superimposed upon the end absorption of a much stronger band with its maximum in the VUV. The general features of the UV absorption spectra of C_α-alkylated α-amino acids resemble those of carboxylic acids [55, 90].

Problems in assigning bands 1 and 2 in the CD curves of C_α-alkylated α-amino

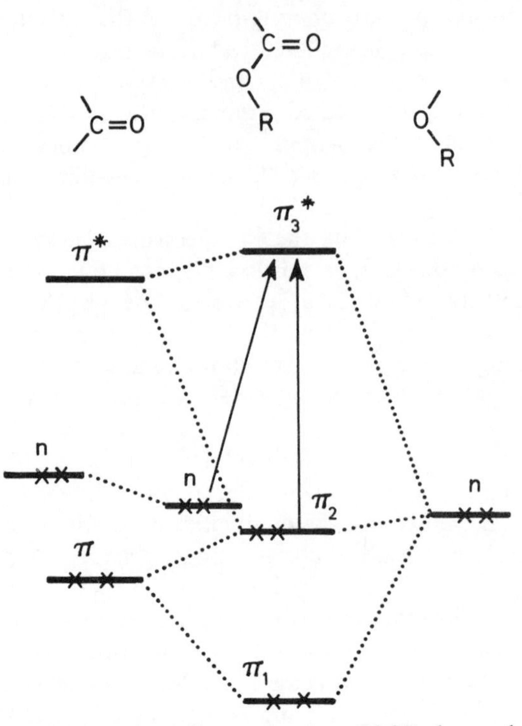

Figure 19.2 Energy level diagram for the $-COOR$ chromophore.

acids and esters in the 250–200 nm region received much attention during the last 13 years. The presence of two bands could be indicative of two different electronic transitions or of a single transition, representing either two conformational isomers or two different states of solvation. Compared with carboxylic acids and esters, the interpretation of the CD spectra of their α-amino derivatives at alkaline pH is complicated by the contribution of optically active transitions involving the non-bonding electrons of the unprotonated amino group [44–46, 66].

Anand and Hargreaves [41] assigned band 2 of α-amino acids at acidic pH to the $n \rightarrow \pi^*$ transition of the $-COOH$ group and the more intense band 1 to the $\pi \rightarrow \pi^*$ transition of the same chromophore. On the other hand, Polonsky [57] ascribed band 2 of unprotonated π-amino acid esters to a charge-transfer transition of an electron from a non-bonding orbital of the nitrogen atom to the π^* anti-bonding orbital of the $-COOR$ group, and band 1 to the $n \rightarrow \pi^*$ $-COOR$ transition.

However, at present, many authors [44–46, 49, 66, 67, 70] are attributing bands 1 and 2 of α-amino acids (esters) both to the same ($n \rightarrow \pi^*$) transition of the $-COOH$ ($-COOR$) chromophore, but arising from different isomers present in the conformational equilibrium mixtures. The fact that α-amino acid esters containing an N_α,N_α-dialkylated amino group still show band 2, argues

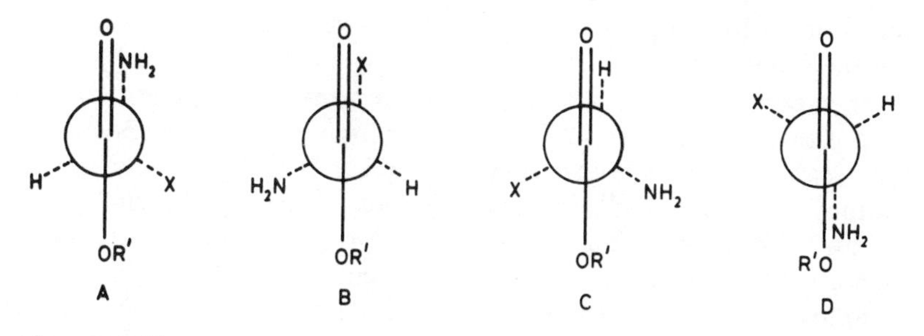

Figure 19.3 Four conformers of L-α-amino acids (X = H) or esters (X = alkyl group).

against intra- or intermolecular hydrogen-bonded conformers being the only factors responsible for this band [45, 46, 66]. The α-amino substituted acyclic acids (esters) should rotate freely about the C – C bond which connects the asymmetric α-carbon atom to the – COOH (– COOR) chromophore (Fig. 19.3). In three low-energy conformers (A, B and C) one of the bonds at C_α is *syn*-periplanar to the carbonyl group of the – COOH (– COOR) moiety. Conformer D represents an example of the *anti*-periplanar position. For a detailed discussion of the conflicting assignments of bands 1 and 2 to specific conformers the reader should consult the most recent references [57, 66, 67].

Jorgensen [47] was the first to propose a sector rule for α-amino acids which is based on the octant rule for ketones [91] (Fig. 19.4a). This rule is directly applicable to structures with the symmetrical – COO^- chromophore and relates the sign and amplitude of the 210 nm n→π* CE to the conformation and absolute configuration of C_α-alkylated acyclic and monocyclic α-amino acids. Subsequently, other groups [49, 51, 57] have discussed revised sector rules for

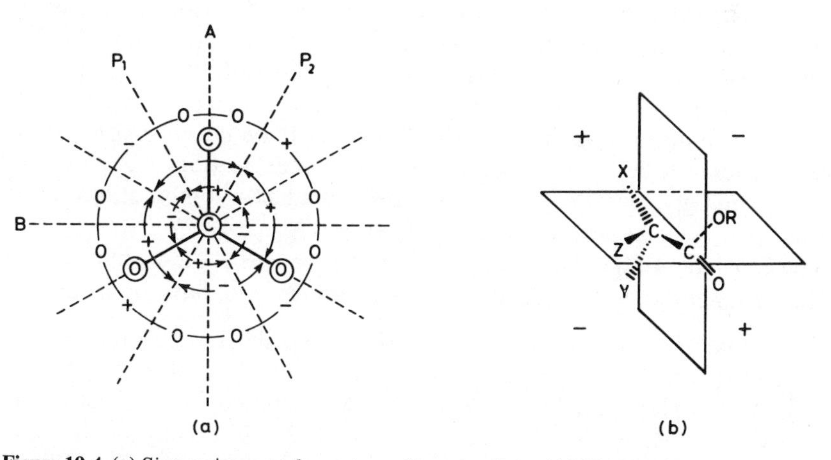

Figure 19.4 (a) Sign assignment for sectors of an α-amino acid [47]. (b) Sign assignment for back octants of an α-substituted acid (R = H) or ester (R = alkyl group) [66].

α-amino acids. More recently, Craig and coworkers [66] in particular on the basis of the observation that the ground state symmetry of the carboxyl $n \rightarrow \pi^*$ transition should be the same as that of the carbonyl chromophore of ketones, applied the ketone octant rule [91] as such to α-amino acids or esters (Fig. 19.4b).

Snyder *et al.* [55] and Richardson and Ferber [63] carried out theoretical calculations on zwitterionic L-Pro to explain *inter alia* the sigmoidal-type curve in the 220–185 nm region. The two CEs both originate with $n \rightarrow \pi^*$ transitions localized on the $-COO^-$ moiety of the molecule.

The following two conclusions also emerged from this survey. First, the observed CD bands in the far-UV of C_α-alkylated α-amino acids and their alkyl esters appear in the UV absorption spectra as weak absorptions, part of the 'tail' on the long-wavelength side of the band which has its maximum in the VUV. In this case, therefore, CD proved to be of considerable help in the elucidation of electronic absorption spectra. Secondly, many configurational assignments have been based on the positive shift observed in the optical rotation of an L_s-α-amino acid on acidification (for details of the Lutz–Jirgenson Rule see reference 15).

In the past, such correlations have been made from measurements at a single wavelength (the sodium D-line at 589 nm), but the related techniques ORD and (particularly) CD have now made possible more sound correlations of configurations based on the CE of the $-COOH$ chromophore near 210 nm (band 1). In particular, *C_α-alkylated L_s-α-amino acids give a positive CE*. However, if unusual conformational constraints and/or interfering chromophores are present, safe configurational assignments should be supported by the examination by CD of at least one additional chromophoric derivative (see Section 19.3).

The CD properties of many other types of α-amino acids containing additional chromophores, but absorbing only below 250 nm have been described. They have ethylenic [48, 81, 82, 92, 93], acetylenic [48], hydroxyl [39, 42, 48, 60, 69, 71, 72, 81], ether [71, 72], thiol [39, 48, 61, 83], thioether [39, 48, 49, 53, 56, 61, 67, 82–84], seleno-ether [61], sulphoxide [48, 85], carboxyl [39, 42, 48, 49, 54, 57, 62, 66, 67, 69, 73, 94, 95], carboxamido [39, 42, 48, 54, 71, 72, 94, 95], glucopyranosylamido [71, 72], benzamido [96], tetrazolyl [73–75], imidazolyl [39, 76–78, 94], amino [39, 42, 48, 59, 97], guanido [39, 42, 48], ureido [48, 98] moieties linked to C_β, C_γ, or C_δ atoms (however, α-amino acids containing the thiol or the thioether chromophore linked to the C_β atom will be discussed in some detail in the next section). C_α-deuterio- [87] and C_α-vinyl-glycines [80] have also been studied. Combinations of these chromophores have been examined [48, 81, 82, 95].

The CD data of a number of α-amino acid derivatives at the α-amino and/or α-carboxyl functions, with absorption only below 250 nm, have been reported:

(i) Formyl [94], acetyl [57, 58, 85, 92–94, 97, 99–116], isobutyroyl [58], isovaleroyl [58], 3,3-dimethylbutyroyl [108], pivaloyl [58, 108], long-chain acyl [117], methacryloyl [52, 58], benzoyl [58, 96], carbamoyl [94, 98], benzyloxy-carbonyl, [74, 75], *tert*-butyloxycarbonyl [118, 119], chloroalkylsulphonyl [49], methanesulphonyl [57], sultam [49], hydroxy [57], alkylidene [120, 121] derivatives at the α-amino function.

(ii) Amido and alkylamido derivatives at the α-carboxyl function [54, 99, 100, 102, 104–114, 116, 122].

(iii) Intramolecularly-linked α-amino and α-carboxyl derivatives, such as hydantoins [98, 123, 124], oxazolin-5-ones (azlactones) [115], and oxazolidin-5-ones [57].

Finally, the CD maxima of a β-amino acid, 2-pyrrolidine-acetic acid, have been listed [86].

19.2.2 Amino acids and their derivatives with circular dichroism maxima in the near-ultraviolet region

Aromatic (with the exception of His) and sulphur-containing α-amino acids are the only ones which exhibit CD maxima in the near-UV region (i.e. in the UV region above 250 nm).

Monoalkylated benzene rings have three major transitions above 185 nm: a weak band, termed ^1L_b, in the 270–245 nm region, which shows fine structure, corresponding to a symmetry-forbidden $\pi \rightarrow \pi^*$ transition, and two stronger bands near 210 nm (^1L_a) and 190 nm (^1B). The 210 nm band is also forbidden but probably involves a contribution from the first allowed $\pi \rightarrow \pi^*$ transition which overlaps it and is centred at a shorter wavelength. The CD spectra of (*S*)-Phe and its derivatives (N_α-acetyl and N_α-*tert*-butyloxycarbonyl, and $-COOMe$, $-COOEt$, and $-CONH_2$) in the near-UV region were recorded by many groups [11, 20, 39, 67, 77, 104, 109, 125–143]. The signs and intensities of the vibronic bands vary, sometimes greatly, among these compounds and on changing solvent polarity, concentration (aggregation), temperature, and pH. However, all CD spectra revealed similar vibrational perturbations arising from the benzyl moiety ($0-0$, $0+180\ cm^{-1}$, and $0+520\ cm^{-1}$ transitions, all starting progressions with $930\ cm^{-1}$ spacing to shorter wavelengths) [140].

Interestingly, if compared to the ^1L_b CE of (*S*)-Phe: (i) the sign for (*S*)-Phg, the lower homologue of Phe, is inverted, and the intensity is greatly enhanced [125, 129, 133, 141], (Fig. 19.5), and (ii) in (*S*)-homo Phe, the higher homologue of Phe, it is absent [130]. All the above results were analysed in terms of conformational equilibria. In fact, it is well-established that in conformationally-mobile systems, the observed CD curve is composed of the population-weighted contributions of the CEs of all rotameric species present [144]. In some (*S*)-Phe derivatives an additional, negative CD band, the intensity of which is greatly enhanced in solvents of low polarity, is visible at 235–240 nm [104, 132, 138]. Its origin is still unclear. In the far-UV region (*S*)-Phe and its lower and higher homologues exhibit a positive CE at 205–220 nm [7, 11, 20, 39, 57, 77, 104, 109, 125, 129–134, 137, 139, 143, 145]. In addition to the ^1L_a benzyl-ring transition these molecules possess an optically active carboxyl chromophore which absorbs in a spectral region close to the ^1L_a transition. For Phe and Phg it is now generally agreed that this CE cannot merely be ascribed to a summation of the contribution of both chromophores, but rather, to the result of their interaction. For the higher

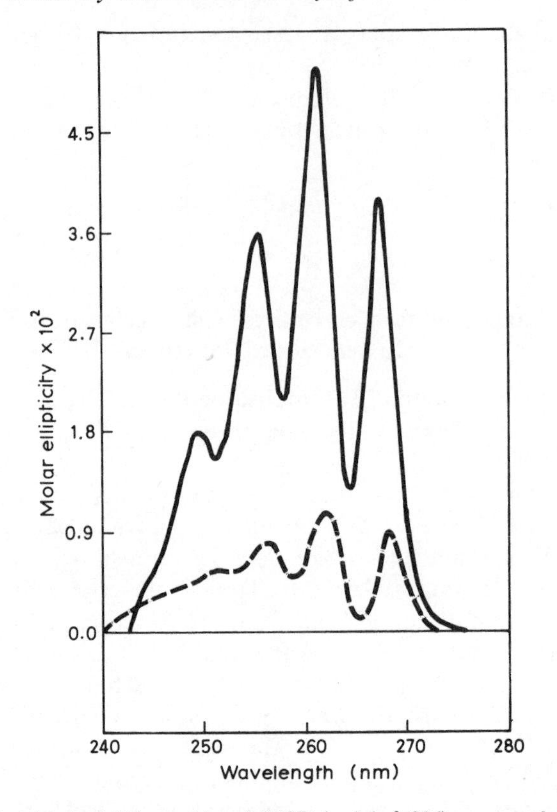

Figure 19.5 The CD (—) (this work) and MCD (----) (ref. 286) spectra of D-Phg in water above 240 nm. The MCD spectrum was measured in the laboratory of Professor C. Djerassi at the Department of Chemistry, Stanford University.

homologue of Phe, conversely, Sakota *et al.* [130] attributed its CD maximum at 207 nm only to the n→π* carboxyl transition. It was also reported that the CD of N_α-acetyl (*S*)-Phe amide below 250 nm changes dramatically from water to dioxane [109, 134]. Finally, a third (positive) CD band, centred at about 200 nm, was also reported for (*S*)-Phe, (*S*)-Phg and their derivatives [39, 104, 133, 145]. Other analogues and derivatives of Phe (C_α-methylated Phg [129], N_α, N_α-dimethylated Phg and their esters and amides [53, 129], Phe and Phg hydantoins [123] and N_α-carboxymethyl derivatives [69], *p*-chloro Phe [143], β,γ-dehydro Phe [130], *threo*-3-phenyl Ser [143], N_α-trimethylammonium Phe [70], and N_α-hydroxy Phe methyl ester [57]) were also investigated by CD.

In the UV spectra of un-ionized *p*-alkylated phenols a transition near 275 nm (^1L_b) is seen, accompanied by two other transitions at lower wavelengths, at about 225 nm (^1L_a) and 190 nm (^1B). The fine structure of the 275 nm band is evident at low temperatures. Upon ionization of the phenolic group the 275 nm band shifts to about 295 nm and the 225 nm band to about 240 nm. In the CD spectra of (*S*)-Tyr and its derivatives (N_α-acetyl, N^α-carboxymethyl, N_α-stearyl,

N_α-(ω-chlorobutylsulphonyl), butanesultam, and $-COOEt$, $-COOnHex$, and $-CONH_2$) three CEs, centred near 275, 225, and 200 nm, are visible [6, 7, 11, 20, 33, 36, 37, 39, 49, 67, 69, 77, 127, 133, 134, 139, 146–152]. The intensity of the ^1L_b CE of Tyr, higher than the corresponding one of Phe, was explained in terms of the availability of the non-bonding orbitals of the phenolic $-OH$ for mixing with the electric-dipole transition moment of the aromatic ring. As previously discussed for Phe, the signs and intensities of the vibronic bands in the near-UV region vary significantly among these compounds and on changing solvent polarity, temperature, and concentration (aggregation), and in the presence of hydrogen-bonding agents. When Tyr derivatives are cooled at very low temperatures, a number of vibronic transitions become well resolved [151, 152]. The 0–0 transition (its position changes from 282 nm to 289 nm according to the solvent) starts an intense progression with 800 cm^{-1} spacing to shorter wavelengths. A much weaker progression involves a 1250 cm^{-1} spacing. Ionization of the phenolic $-OH$ shifts the 275 nm CD band to a longer wavelength (to about 290 nm) [6, 7, 30, 36, 37, 39, 127, 133, 147, 148].

The 225 nm CE is substantially more intense than the 275 nm CE. At alkaline pH it red-shifts to about 240 nm without a significant change in ellipticity. A drastic spectral change in the 250–210 nm region is caused in N_α-acetyl (*S*)-Tyr amide by a variation in the solvent polarity [109]. This phenomenon, similar to that above described for the Phe analogue, was explained by Cann [109] as arising from the onset of an intramolecularly hydrogen-bonded folded form (called the C-7 form [153]) in the solvents of low polarity. Acetylation of the phenolic oxygen of Tyr derivatives does not induce any major change in the CD spectra [33, 149]. The main effect is to blue-shift the ^1L_a band by 5 nm, but the rotational strengths are comparable. The substitution of a *p*-methoxy group for the *p*-hydroxy group of Tyr does not alter the vibronic bands in the near-UV region; instead, the wavelength position of the 0–0 band of *O*-methyl Tyr is much less shifted by hydrogen-bonding acceptor solvents than is the case for Tyr [150]. Also, in *O*-methyl Tyr derivatives aggregation, which is responsible for a marked change in the CD properties, is minimized. The only difference in the CD spectra of (*S*)-Tyr and its C_α-methylated derivative is the sign of the ^1L_a, which is inverted [154]. The effect of pH is essentially the same in both molecules. The *ortho* and *meta* isomers of (*S*)-Tyr have CD patterns above 210 nm which are comparable to those of the *para* isomer (Tyr) [155, 156]. However, the sign of the 275 nm CE in the *ortho* [144, 155, 156] and *meta* [144, 156] compounds are opposite to that of Tyr, whereas the sign of the 225 nm CE of the *meta* isomer, which is seen only as a shoulder on a large band that reaches its maximum somewhere below 203 nm, is the same as that of Tyr [156]. The CE of the ^1L_a band of the *ortho* isomer is much stronger than that of the corresponding band in Tyr, and is again of opposite sign [155, 156].

The CD properties of many analogues of Tyr, (β-(2,3-dihydroxyphenyl) alanine [155], β-(3,4-dihydroxyphenyl)alanine, also called 3-hydroxy Tyr or dopa, and its two *O*-monomethylated derivatives [77, 143, 157], *threo*- and *erythro*-3-(3,4-

dihydrophenyl) Ser and their *O*-mono- and dialkylated derivatives [143], *p*-hydroxy Phg [158], the lower homologue of Tyr, and its amide [158] and hydantoins [123], and 3,4,5-trimethoxy Phg and its N_α-carbobenzoxy derivative [136, 137]) were also examined. The optical rotatory powers of (*S*)-Tyr [156] its *ortho* [156] and *meta* [156] isomers and its C_α-methylated analogue [154] and *p*-hydroxy Phg [158] and its amide [158] were calculated as a function of conformation.

In contrast to Tyr and Phe, the ^1L_a transition of Trp does not occur in the far-UV. This complicates the interpretation of Trp spectra, for the ^1L_a transition is broad, intense, lacking in clear vibrational structure, and sensitive in its wavelength position to environmental conditions. As a result, it often overlaps the ^1L_b transition, its long-wavelength tail sometimes extending beyond the long-wavelength end of the ^1L_b. This latter transition is, instead, similar to the Tyr ^1L_b, having comparable dipole strength and vibronic structure. Below 235 nm two absorptions (^1B) are evident in the UV spectrum of Trp, which are centred at about 220 nm and 195 nm, respectively. The former has somewhat greater intensity and more discernible fine structure than the latter. The effect of changing the state of charge of the compound and the refractive index of the solvent on the ^1B bands of Trp derivatives was also assessed. The CD properties of (*S*)-Trp and some of its derivatives (N_α-acetyl, $-COOMe$, $-COOEt$, $-CONH_2$) were investigated in detail in the 300–185 nm region [23, 28, 39, 77, 133, 134, 146–152, 159–162].

CD curves were recorded also at very low temperatures and in different solvents [77, 134, 152, 162]. The spectra were grouped into four classes: (i) ^1L_b bands intense; (ii) ^1L_a bands intense; (iii) both ^1L_b and ^1L_a bands intense, and (iv) fine structure whose origin was not identified [162]. Both the 0–0 and $0+850\,cm^{-1}$ ^1L_b transitions occur together (near 290 and 283 nm, respectively) and have the same CD sign. A number of ^1L_a transitions were identified, but their relative intensities varied greatly. The rotatory strengths at 77 K in Trp derivatives, as for the Phe and Tyr derivatives discussed above, are substantially greater than at room temperature, which appears to indicate a high degree of conformational mobility at room temperature [152, 162]. The far-UV CD bands of (*S*)-Trp and its derivatives were investigated as a function of nature of solvent and pH [39, 133, 159, 160]. Four types of spectra were found [160]. All have intense, positive ellipticity near 223 nm, but have differing patterns of zero or negative ellipticity, in one or more bands, below about 215 nm. Four bands in the region of the longer wavelength ^1B band are sufficient to describe almost every spectrum, while only one broad band is needed in the region of the shorter wavelength ^1B band. The possibility that parts of the experimental CD spectra arise from coupling with the $n\rightarrow\pi^*$ and $\pi\rightarrow\pi^*$ transitions of carbonyl, amide, or ester chromophores was discounted. To explain the observed CD spectrum below 250 nm in dioxane, the possible occurrence of the C-7 intramolecularly hydrogen-bonded folded form [153] was suggested for N_α-acetyl (*S*)-Trp amide [159]. Relatively few Trp derivatives modified at the indole moiety have been investigated by CD [77, 163, 164]. Only minor changes in the spectrum accompany the introduction of an $-OH$ group into the 5-position of Trp [77].

The interpretation of the CD spectra of β-(2-oxindolyl)-(*S*)-alanine [163], an oxidation product of (*S*)-Trp, and the tricyclic isomers of (*S*)-Trp [164] are complicated by tautomeric effects, the onset of additional chiral carbon atoms (diastereoisomerism), and disappearance of the indole aromaticity.

A major point that must be addressed in every CD study of cystine concerns which stereochemical and electronic features of the molecule play the dominant role in determining the signs and intensities of the CEs. In general terms, this role can be assigned to: (i) the inherent chirality within the CSS′C′ moiety (M, left-handed, and P, right-handed forms), (ii) the occurrence of preferred, staggered rotamers around the $C_\alpha - C_\beta$ and $C_\beta - S$ bonds, and (iii) the existence of at least two closely spaced transitions in the near-UV region. The different solid state CD spectra of L-cystine (positive CE at 287 nm, negative CE at 243 nm) and its dihydrochloride (negative CE at 272 nm) in the near-UV region were first interpreted [165, 166] as arising from the opposite chirality of the disulphide moiety in the two cases (right screw for the former, left screw for the latter, as determined by X-ray diffraction). Subsequently, it was suggested that the 'abnormal' CD curve of L-cystine is the result of an exciton splitting due to the characteristic array of the disulphides in the crystal [167]. The interpretation of the single, positive CD band between 240 and 210 nm (at 212 nm for L-cystine, while at 225 nm for its dihydrochloride) is even more difficult because of the contribution of the $-COOH$ ($-COO^-$) chromophore in that region. L-Cystine in water at neutral pH has a negative band near 225 nm (much weaker than those in the solid state), which is scarcely sensitive to acidification [4, 6, 7, 33, 34, 39, 48, 146, 165, 168–170].

These solutions are known to contain a mixture of P and M conformers. Temperature-dependence and time-dependence studies supported the view that partial cancellation of the net CD intensity by the oppositely signed P and M spectra had occurred. Although it appears certain that solutions of L-cystine contain a mobile equilibrium between conformers, the principal source of its optical activity is still in dispute. Some authors have concluded that vicinal or peripheral effects can account entirely for the observed near-UV CD properties and that these properties are diagnostic of conformational features external to the disulphide moiety [169, 171]; others, conversely, have attributed variations in CD to a displacement of the P$\rightleftarrows$M disulphide equilibrium [34, 146]. The CD properties of N_α,N'_α-diformyl-L-cystine [169], N_α,N'_α-diacetyl-L-cystine and its dimethylamide [105, 168, 172], N_α,N'_α-di(tetra, hexa)methyl-L-cystines [68, 169], L-cystine dialkyl esters [169], L-homocystine [7], D-penicillamine disulphide [169], and mixed disulphides between L-Cys (or D-penicillamine) and various alkyl and arylalkyl thiols [169, 173] were also examined in detail, as a function of pH and temperature.

The CD properties of D- and L-selenocystine were recently investigated at different pHs [174]. All bands above 230 nm were assigned to transitions within the Se$-$Se chromophore. Predominant contributions to CD from the chiral centres outside the diselenide group or from the diselenide chirality itself seem to depend on the state of ionization of the molecule.

Among the number of papers which have dealt with the chirospectroscopic properties of α-amino acids containing the thiol and thioether chromophores linked to the C_β atom, such as Cys, penicillamine, and their *S*-alkylated derivatives [26, 39, 48, 56, 61, 83, 128, 168, 175] only one [26] has described in detail the occurrence of a CE above 250 nm. In fact, a weak, negative CE is seen at about 260 nm at acidic and neutral pHs for (*S*)-Cys and (*S*)-Cys(Me), which shifts to longer wavelength at alkaline pH. Interestingly, in the N_α-acetyl Cys derivative at pH 1 the sign of this CE is reversed. In Met the sulphur atom is separated from the asymmetric C_α atom by one more methylene group than in Cys(Me), and it is significant that the CE at 260 nm is not observed.

Finally, the CD data for a few types of α-amino acid derivatives with synthetically introduced chromophores absorbing in the 300–250 nm region have been reported. In the N_α-nucleotidyl amino acids the sign of the CE at 280–250 nm is reversed on going from aliphatic to aromatic residues [176–178]. A similar effect is operative in the N_α-furoyl derivatives for the 255 nm CD band [179]. The CD spectra of several N_α-acetoacetyl amino acids have been examined as a function of solvent polarity, extent of enolization, nature of side chain, and N_α-alkylation [180, 181]. The CD curve of N_α-dimedonyl L-Val methyl ester in dioxane shows an intense, negative CE at 273 nm [115].

Among the derivatives at the α – COOH group the selenophenyl ester of L-Phe has a positive CD band at 283 nm, whereas the corresponding thiophenyl ester has a band at 275 nm [182]. The three structural elements of the – CO – X – Ar (X = Se, S) grouping, and possible interactions among them, may participate in the optically active chromophore. α-Amino acid benzylamides and phenylethyl esters with a chiral carbon atom next to the aromatic ring exhibit a CD band at about 260 nm with the vibronic fine structure characteristic of benzyl derivatives [92, 93].

The CD spectra of the nucleotidyl derivatives of Cys and its higher homologue homocysteine (at the – SH function) have been described [183, 184]. Using CD, an S→N shift of the blocking group in the Cys derivatives has been detected [184].

19.3 CHIROSPECTROSCOPIC PROPERTIES OF CHROMOPHORIC DERIVATIVES OF AMINO ACIDS

Historically, the concept of a chromophoric derivative was introduced in the electronic spectroscopies (e.g. ORD and CD) to define the case of a 'transparent' functional group which is transformed into a derivative absorbing in a more accessible spectral range [2]. Clearly, 'transparent' is a relative term and depends largely on the status of the instrumental art (i.e. penetration in the UV region). However, with the rapidly increasing availability of CD instruments capable of extending the investigation to the VUV region (Section 19.4), this strict definition (but not the utility of the approach) of a chromophoric derivative now appears to be obsolete. In this article I consider as a chromophoric derivative of an amino acid a group, synthetically introduced in the α- (or β-) NH_2, α-COOH, or side-chain functionalities, which exhibits at least one CD maximum above 300 nm, i.e. in the spectral region where no undesirable complications, arising from superposition with the absorptions of the parent compound, are found.

In addition, the new chromophore should be intrinsically optically inactive and, hopefully, have a relatively small coefficient of absorption (ε) so that its $\Delta\varepsilon/\varepsilon$ factor (corresponding to the signal-to-noise ratio) would be high. Such an approach is of interest since:

(i) It offers information of the spectral behaviour of new chromophores. In fact, in a number of instances CD measurements have uncovered hidden transitions, which are not detectable by ordinary electronic absorption spectroscopy.

(ii) It is particularly attractive in quantitative analysis.

(iii) It is quite suitable for determination of relative configurations. However, the determination of absolute configuration is a more difficult matter, since free rotation is possible around the bond(s) connecting the newly introduced chromophore to the chiral carbon atom. In practice, general predictions of configuration can often be made. This means that either the proportion of rotamers is fairly constant or that one of them is always predominant in the conformational equilibrium mixture, whatever the structure of the rest of the molecule may be [2, 21, 23, 25].

19.3.1 Chromophoric derivatives of the α- (or β-) NH_2 group

The two most extensively investigated families of chromophoric derivatives at the α-amino group of α-amino acids contain either the thiocarbonyl or the nitroaromatic moiety: among the former the alkylthio-thiocarbamoyl (also called dithiourethane, dithiocarbamate, dithiocarbalkoxy, dithioalkoxycarbonyl or alkylthiocarbonothioyl) [69, 185–193], alkoxythiocarbamoyl (or thionocarbalkoxy) [186], thiobenzoyl [194–200], and phenylthioacetyl [200] derivatives should be mentioned, while among the latter the 2,4-dinitrophenyl [201–210], 3,5-dinitro-2-pyridyl [201], 3-nitro-2-pyridyl [201], 5-nitro-2-pyridyl [201], *o*-nitrophenylthio [211, 212], and 4-nitrophthaloyl [213] derivatives.

The position and signal-to-noise ratio of the longer-wavelength CE associated with the $n \rightarrow \pi^*$ transition of the thioamide chromophore are suitable for a CD study; however, its sign largely depends upon the nature of solvent, alkylation of the amino group and state of the α-carboxy group, pointing to the need for caution in basing configurational assignments only on this family of derivatives, where different rotational isomers and differently solvated species coexist in the equilibrium mixtures. In addition, Barrett, in his extensive study of the CD properties of the *N*-thiobenzoyl derivatives [194–199] was able to show that the penultimate α-amino acid residue in a peptide chain also contributes to the CD, directly by perturbation of the electronic transition and indirectly by influencing conformational equilibria involving the substituted *N*-terminal residue [198].

A systematic study of the CD spectra of *N*-(2,4-dinitrophenyl) (DNP) derivatives of α-amino acids was carried out by Kawai, Nagai and coworkers [202–207]. According to their spectral patterns these derivatives were classified into four groups: (i) di-derivatives from α,ω-diamino acids, (ii) derivatives of aromatic, aliphatic unsaturated and sulphur-containing amino acids; (iii) derivatives of cyclic amino acids; (iv) derivatives of other amino acids. While CD spectra of group (iv) compounds have weak bands above 300 nm, each of the other groups exhibits the characteristic pattern of CD spectra of much higher intensity in this region. Several rules (e.g. di-DNP chirality rule, DNP-aromatic rule) relating the sign of the CE near 400 nm to the absolute configuration of the parent α-amino acid were found. The CD properties of *N*-(3-nitro-2-pyridyl) derivatives of α-amino acids were also reported, in comparison with those of the 5-nitro-2-pyridyl, 3,5-dinitro-2-pyridyl and 2,4-dinitrophenyl analogues [201]. It was concluded that the most useful among this group of chromophoric derivatives is the 3-nitro-2-pyridyl owing to the fact that the CD bands above 300 nm have much more favourable signal-to-noise ratios, and, in general, the spectra do not present complications arising from the presence of multisignate curves. However, also in this case the presence of α,ω di-derivatives, aromatic or cyclic amino acid side chains have a marked effect on the CD curves. These general observations, coupled with the chemical properties of the 3-nitro-2-pyridyl amino moiety, make this derivative the most suitable thus far proposed for assigning the absolute configuration of the *N-terminal* amino acid in peptides. Preliminary investigations of the solvent-dependent CEs of the *N*-(*o*-

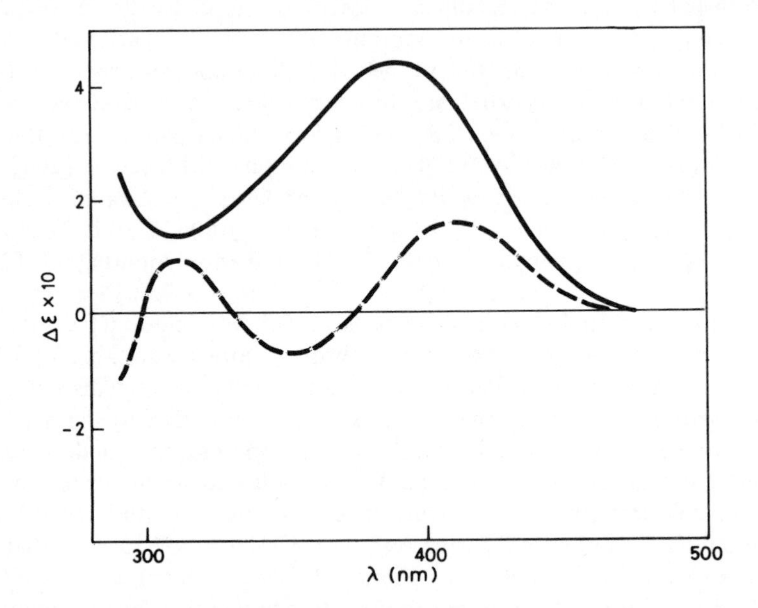

Figure 19.6 The CD spectra of N_α-NPS-L-Ala-O$^-$ dicyclohexylammonium salt in 2,2,2-trifluoroethanol (——) and dioxane (------) above 290 nm.

nitrophenylthio) (or *o*-nitrophenylsulphenyl, NPS) derivatives of α-amino acids (Fig. 19.6) were performed [211, 212]. The CD spectrum of N_α-(4-nitrophthaloyl)-*p*-dimethylamino-L-phenylalanine methyl ester shows two negative bands near 400 nm and 350 nm, respectively, which were attributed to the onset of an intramolecular charge-transfer complex [213].

The CD curves of *N*-phthaloyl-α-amino acids suffer temperature-induced modifications, suggesting various degrees of rotational mobility as the temperature is changed; in addition, concomitant blue-shift and fine structural changes, as well as the nature of the bisignate curves themselves, greatly complicate the interpretation of the results [65, 187, 188, 214, 215].

The sign of the CD near 320 nm ($\pi \rightarrow \pi^*$ transition of the enolimine tautomer) of the *N*-salicylidene derivatives of α-amino acids and esters (formed *in situ*) correlates with their absolute configurations (salicylidenimino chirality rule) [216]. The nature of solvent strongly influences the position of the enolimine ⇆ ketoamine tautomeric equilibrium, substantially perturbing the CD patterns [115, 217]. The presence of an aromatic side chain in the α-amino acid residue induces sign reversal of the 320 nm CE. The CD properties of α-amino acid derivatives of 5-chloro-2-hydroxybenzophenone were also examined and found similar to those of the salicylaldehyde derivatives [218].

The effects of >N−N=O group conformation, ring geometry, perturbing substituents, and intramolecularly hydrogen-bonding upon the $n \rightarrow \pi^*$ CD band of *N*-alkylated *N*-nitrosoamino acids were discussed [219–223]. Stereochemical conclusions can be made with confidence in many cases, although no sector rule, as yet published, successfully correlates all the available chirospectroscopic data in this series of compounds.

A series of papers dealing with the CD properties of *N*-(pyrrolin-4-one) derivatives of α-amino acids and esters were recently published [224–232]. These compounds were prepared *in situ* using furan-3-one derivatives as reactants. Two CEs are seen above 300 nm. It was suggested that the sign of the second (from the visible) CE can be safely used for configurational assignments (pyrrolinone chirality rule) [228].

Other chromophoric derivatives at the α-NH_2 group of α-amino acids and esters, the chirospectroscopic properties of which were described, include: (i) phenylhydrazone derivatives of *N*-acetoacetyl [233], (ii) *N*-acetoacetylidene [115], (iii) *N*-(2-pyrazinoyl) [234], (iv) *N*-(3-methyl-2-quinoxaloyl) [235], (v) *N*-[5(6)-benzofuroxanoyl] [236], (vi) *N*-(1-dimethylaminonaphthalene-5-sulphonyl) (dansyl) [21], (vii) 3-hydroxypyridinium [237], (viii) isothiocyanate [238], and (ix) reaction products with *o*-phthaldialdehyde in the presence of 2-mercaptoethanol [239, 240].

Finally, a few chromophoric derivatives at the β-NH_2 group of β-amino acids, basic components of a number of antibiotics, were also examined by CD: (i) *N*-salicylidene [217], (ii) *N*-(2,4-dinitrophenyl) [241, 242], and (iii) ethylthio-thiocarbamoyl [243, 244]. In case (ii) an extension of the DNP-aromatic rule was suggested for configurational assignments, while in case (iii) the sign of the CE near 340 nm in chloroform or benzene of the *N*-substituted β-amino acid dicyclohexylammonium salts was considered.

19.3.2 Chromophoric derivatives of the α-COOH group

In contrast to the α-NH_2 group, only one type of chromophoric derivative of the α-COOH group of α-amino acids was examined by CD. A negative maximum at about 300 nm is shown in 95% ethanol by the *o*-nitrophenyl esters of *N*-protected L-Val and L-Ala residues [245]. A more detailed study of the CD properties of a larger number of chromophoric derivatives of the α-COOH function of α-amino acids should be rewarding.

19.3.3 Intramolecularly-linked α-NH_2 and α-COOH chromophoric derivatives

The 3-methyl-2-thiohydantoins derived from α-amino acids show a CE at 300–315 nm which is positive for the L-series and negative for the D-series [246, 247]. The location of the CD band in correspondence to a low-intensity UV absorption and its red-shift on going from more polar to less polar solvents suggests that it is associated with an $n \rightarrow \pi^*$ transition within the chromophore. This CE does not exhibit inversion of sign on going from aqueous to organic solutions, as has been reported for other $-NH-C(=S)-$ containing compounds (see above). Hence, the use of the sign of this CE was proposed as a simple and rapid method for determining the absolute configurations of *sequences* of α-amino acids in naturally occurring peptide compounds [246].

A large number of 3-phenyl-2-thiohydantoins was also examined by CD [123, 186, 215, 248]. These derivatives suffer the disadvantage of racemizing more easily than their 3-methyl analogues. The sign of the 315–330 nm CD band of all 3-phenyl-2-thiohydantoins of identical configuration is the same in all solvents tested; moreover, as in the case of their 3-methyl analogues, positive and negative CEs correspond to L- and D-configurations, respectively. The anomalous behaviour of the 3-phenyl-2-thiohydantoins from α,α-dialkyl α-amino acids was also investigated.

19.3.4 Chromophoric derivatives of side-chain functional groups

Scattered reports of the CD properties of α-amino acids containing a single chromophoric derivative, located in their side-chain, have appeared in the literature.

Only Cys derivatives (at the $-SH$ functional group) were investigated rather extensively. The CD bands near 320 nm of the 3-hydroxy-6-methyl-2-pyridyl derivatives of Cys and its higher homologue (homocysteine), due to the pyridyl chromophore, are both negative despite the two α-carbons having opposite configuration (L for Cys and D for homocysteine) [249]. The CD spectrum of the adduct of L-Cys to 3,4-dimethoxy-β-nitrostyrene exhibits a rather intense positive CE near 300 nm, thus demonstrating a surprisingly high degree of stereoselectivity of addition (a new centre of asymmetry is generated during the reaction) [250]. The 3-nitro-2-pyridyl derivative of L-Cys shows a positive CD band at about the same wavelength of the UV absorption band (365 nm) [251]; since the CE of the 3-nitro-2-pyridyl derivative shows a much more favourable $\Delta\varepsilon/\varepsilon$ ratio when compared with the 5-nitro- [251], 3,5-dinitro- [252], and 2,4-dinitrophenyl- [251] analogues, it was concluded that the 3-nitro-2-pyridyl is the most useful for CD studies among the nitroaryl derivatives. Using this chromophoric derivative, configurational assignments and quantitative estimation of the $-SH$ function were performed. The sign of the 320 nm CD band of the methylthiocarbamoyl derivative of Cys is solvent-dependent [246, 253], as shown in other thiocarbonyl compounds (see above); therefore, it is not automatically transferable from water to organic or aqueous–organic solutions. Using this CD band, which presents a high $\Delta\varepsilon/\varepsilon$ ratio ($n \rightarrow \pi^*$ transition of the dithiocarbamate chromophore), various analytical applications in water were described. No definite advantages of other alkylthiocarbamoyl derivatives over the methylthiocarbamoyl were observed. For a more detailed discussion of the chirospectroscopic properties of the nitroaryl and thiocarbamoyl derivatives of Cys the reader is referred to reference 25.

Among the aromatic α-amino acids, only the CD curve of a chromophoric derivative of L-Tyr was published: the *p*-azobenzenearsonate derivative has a small negative ellipticity near 415 nm at pH 7.4 [254]. An extensive investigation of the CD properties of chromophoric derivatives of aromatic α-amino acid residues will prove to be of interest.

Finally, *N*-formyl-D-ferrocenylalanine has a positive CD band at 448 nm in ethanol [255].

19.4 RECENTLY DEVELOPED TECHNIQUES

19.4.1 Circular dichroism in the vacuum ultraviolet and in the infrared regions

Until recently, all CD experiments on amino acids and their derivatives were restricted to the visible and UV regions of the spectrum. As such, CD is limited with respect to the number and type of transitions that can be studied. However, in the last ten years a few laboratories have extended the CD technique to the VUV (below 185 nm) (VUCD) and to the IR (IRCD) regions, thereby providing access to new electronic and vibrational transitions.

Snyder *et al.* [55] measured the VUCD to 160 nm of 1,1,1,3,3,3-hexafluoropropan-2-ol solutions of five alkyl α-amino acids as zwitterions. In the VUV the spectra of L-Ala, L-Val, L-Leu, L-Ile, and L-Pro have a negative maximum which increases in intensity with the size of the side chain. This band falls at 168 nm for Ala, Leu, and Pro, but it is red-shifted to 172 nm in the β-branched Val and Ile. It has been associated with the first $\pi \rightarrow \pi^*$ transition of the carboxylate anion in analogy with the isoelectronic amide group. This assignment is confirmed by the VUV absorption spectra of these amino acids which show a strong band at 166 nm.

The VUCD to 167 nm of Ac-L-Ala-NHMe, measured in aqueous solution, did not reveal any new maxima [113]. The $\pi \rightarrow \sigma^*$ transition centred on the carbonyl oxygen, expected at 175 nm on the basis of the spectra of unordered polypeptides, is not seen in the VUCD of the Ala model compound. The solvent can bind more strongly to the non-bonding electrons of this diamide so that the $n \rightarrow \sigma^*$ transition should be shifted to higher energies.

Nafie and coworkers investigated the IRCD of a series of α-amino acids in both D_2O and mulls in halocarbons [256–258]. In D_2O they identified two bands of opposite sign (a positive signal at 2949 cm^{-1} and a negative signal at 2987 cm^{-1}) in the IRCD of L-Ala that was attributed to chirally perturbed $-CH_3$ stretching modes. The significantly smaller optical activity of L-Ser was related partially to cancellation of IRCD intensities of different conformers of the side-chain $-CH_2OH$ group. In mulls a larger number of vibrations, including the NH and OH stretching modes, could be studied. However, correlation of liquid- and solid-phase spectra is difficult; in view of the sensitivity of IRCD to the stereochemistry and molecular environment, it is not surprising that solution and mull spectra are rather different.

It is clear that more VUCD and IRCD investigations should be performed to broaden the scope and to increase the efficiency of measurements, and to compare theoretical and experimental results.

19.4.2 Magnetic circular dichroism

CD induced in an electronic transition by an external magnetic field (MCD) has recently been employed to investigate α-amino acids and their derivatives. As a result of the separate physical principles of the two phenomena, CD is more sensitive to subtle aspects of molecular stereochemistry while MCD essentially provides information about the electronic nature of the component chromophores.

Neither Pro nor Cys (SH) exhibit a significant MCD above 200 nm [127,259]. In Ala only a band at 220 nm is present [259]. The side-chain chromophore of His shows a band at 210–230 nm [259], but gives no signal above 240 nm [127, 260, 261].

The only protein α-amino acids which have MCD bands above 250 nm are cystine, Phe, Tyr, and Trp. In cystine the disulfide chromophore generates a weak negative band at 260–270 nm in the entire pH range [127, 170, 260–263]. Conversely, in its N_α, N'_α-diacetyl, di-methylamide derivative a positive signal is seen, but only at pH near 7 [172]. This could result from the existence of a non-zero magnetic moment, and from the sign of the **B** term which is present regardless of the degeneracy of the ground and the excited states.

Below 250 nm the MCD spectrum of Phe exhibits a very intense band at about 210 nm which changes little with pH [262, 264]. The MCD properties of L-, D-, and DL-Phe and a number of their derivatives above 250 nm were observed in aqueous and organic solutions and in films at room temperature and at 77 K [127, 135, 260–262, 264–267]. The weak, negative and multiple bands occurring in the 250–270 nm region show a remarkable pH dependence [135, 264, 267]. In this region at the longer wavelengths it is possible to find a parallelism between the CD and 'true' MCD spectra (obtained by correcting the observed MCD for intrinsic CD), while it becomes difficult at shorter wavelengths because of cancellation from CD bands of opposite sign [135]. The expected inversion symmetry upon reversal of the magnetic field was not found [135]. Freezing the internal rotation about the $C_\alpha - C_\beta$ bond by decreasing the temperature makes the population of the most stable rotamer prominent and restores the inversion symmetry. The D-enantiomer and the racemate display spectral characteristics closely similar to those of L-Phe itself [127, 135, 266]. The MCD spectrum of Phg, the lower homologue of Phe, shows sign inversion of the bands at 250–270 nm compared to Phe (Fig. 19.5) [266]. Also, Phg, which lacks the methylene group between the aromatic ring and the amino acid moiety as in Phe, exhibits inversion symmetry upon reversal of the magnetic field.

At neutral pH the MCD curves of Tyr and Ac-Tyr-OMe consist of an intense, negative band at 275 nm accompanied by a shoulder at 278–280 nm, which correspond to partially resolved vibrational components of the ^1L_b absorption band [127, 260, 262, 265, 268]. The UV band near 223 nm (^1L_a transition) gives rise to the intense, positive MCD band at about 225 nm [262]. The MCD spectrum of Tyr changes markedly on ionization of the phenolic $-OH$ group, generating a negative band at 293–295 nm with twice the magnitude of that of the 275 nm band of the protonated form [127, 268, 269]. The isosbestic point is seen at 279 nm ($pK_a = 9.9$). At pH 12 an additional, strong and positive band is visible at 243 nm [269]. The MCD curve of the model derivative arsanilazotized Ac-Tyr-NH_2 was compared to that of carboxypeptidase A arsanilazotized at Tyr in position 248 [270].

The MCD properties of Trp and its derivatives Ac-Trp-X (X = OH, OMe, OEt, NH_2) have been extensively investigated [127, 260–263, 265, 271–278]. The spectra show two oppositely signed bands which are the result of the mixing of both ^1L_a and ^1L_b electronic transitions by the magnetic field [261, 273]. The longer wavelength ^1L_b transition exhibits two partially resolved positive MCD vibronic bands at 292 nm and 285 nm, whereas the ^1L_a band gives a broad negative band near 267 nm with some vibrational fine structure. The intensity of the 292 nm band is a linear function both of concentration and magnetic field strength. The D-isomer and the amino and carboxyl substituted derivatives of Trp display MCD characteristics closely similar to those of L-Trp itself, both in regard to sign, wavelength maxima and band intensities. The 292 nm band intensity is unaffected by pH and the conformation of the molecule. Solvent perturbation studies indicate that the environment of the indole chromophore minimally affects the intensity of the 292 nm MCD band, but significantly shifts the wavelength maximum.

Trp is the only naturally occurring α-amino acid that gives positive MCD bands in the near-UV region. Furthermore, the 292 nm band is intense and almost completely free of overlapping contributions by bands associated with side chains of other α-amino acid residues. This fortuitous situation suggested the possibility of using the 292 nm MCD band as a rapid, direct, accurate, sensitive and non-destructive analytical technique to determine the Trp content in proteins [127, 260, 261, 263, 275, 278]. In order to keep the potentially interfering contribution of Tyr to the 292 nm Trp band at a minimum, it is obviously necessary to carry out the measurements at a pH below 8, where the phenolic chromophore is un-ionized (see above). The error due to overlapping contributions from the negative Tyr MCD at neutral pH values is comparatively small, but it cannot be neglected in those cases for which the Tyr/Trp ratio is high. In addition, the change in the observed intensity at 292 nm between pH 7 and pH 13 can be used to calculate the Tyr/Trp ratio. In order to test the applicability of the technique Trp determinations were carried out on a number of proteins whose molecular weight and amino acid composition were known. The resulting values were in close agreement with those given in the literature. The molar intensity per Trp residue of these proteins and model compounds has served to quantitate the Trp content of freshly purified proteins [262] and synthetic copolypeptides [272].

MCD has also provided an effective means for following the course of the reaction of free Trp and its derivatives with various oxidizing reagents [127, 274, 276–278] and 2-nitrophenylsulphenyl chloride [127, 278]. Sequential addition of oxidizing reagents rapidly abolishes the 292 nm band; consequently, this approach was also used to determine the content of Trp residues in proteins. The complexity of the oxidative processes precluded a detailed characterization of all reaction products by MCD. In contrast, reaction of 2-nitrophenylsulphenyl chloride with free Trp, which results in the formation of the 2-substituted derivative, results in the elimination of the 292 nm band but generates a new positive band at 303 nm. In this instance, both the disappearance of the substrate and the appearance of the product can be quantitated by MCD.

The potential for MCD in elucidating excited states of the amino acid chromophores has certainly not been exhausted. Significant advances await only improved instrumentation for the far-UV and VUV regions.

19.4.3 Induced circular dichroism

Induced CD arises when an electronic transition, which is itself optically inactive in a given molecule, is rendered optically active by interaction with a dissymmetric environment. This phenomenon is dependent on the distance and relative geometry of the interacting groups.

The most significant case of induced CD is met when an α-amino acid residue carrying a chromophoric group (and showing no CD in the wavelength region of absorption of that chromophore when free in solution) becomes non-covalently immobilized on a protein molecule. The specific binding of haptenic (DNP, TNP, and DNS) N_ε-substituted Lys residues and N_α-substituted Gly residues to anti-hapten antibodies (immunoglobulins) was found to prevent the hapten from tumbling freely in the site and, consequently, to generate characteristic induced Cotton effects [279–285]. In particular, the results showed that the binding of the hapten to either homologous antibodies, cross-reacting heterologous antibodies or myeloma proteins gives rise to readily distinguishable CD spectra. Induced CD, therefore, appears to provide a highly discriminating technique for probing the fine structure of antibody combining sites. Using this technique it was also shown that the conformation of the sites in various fragments derived from myeloma proteins is identical with that of the intact protein.

The interaction between bovine and human serum albumins and thyroid hormones (thyroxine and triiodothyronine) and diiodo-Tyr was investigated by CD [286–288]. An induced Cotton effect was observed at around 320 nm, the absorption wavelength region of the iodophenolic amino acids. In fact, even in the enantiomeric form, these amino acids gave no detectable CD above 240 nm when free in solution. It was concluded that the induced CD was due to the binding of these ligands to the proteins. The suggested involvement of a Lys residue in the binding sites of human serum albumin was confirmed by the CD generated at 320 nm in thyroxine upon addition of L-Lys [288].

With the increased sensitivity of modern instrumentation a considerable extension of induced CD can be expected.

REFERENCES

1. Djerassi, C. (1960) *ORD Applications to Organic Chemistry*, McGraw-Hill, New York.
2. Djerassi, C. (1964) *Proc. Chem. Soc.*, 314.
3. Velluz, L., Legrand, M. and Grosjean, M. (1965) *Optical CD : Principles, Measurements, and Application*, Academic Press, New York.
4. Velluz, L. and Legrand, M. (1965) *Angew. Chem. Internat. Edit.*, **4**, 838.
5. Crabbé, P. (1965) *ORD and CD in Organic Chemistry*, Holden Day, San Francisco, California.
6. Beychok, S. (1966) *Science*, **154**, 1288.
7. Beychok, S. (1967) in *Poly-α-Amino Acids* (ed. G.D. Fasman), Dekker, New York, p. 293.
8. Snatzke, G., (ed.) (1967) *ORD and CD in Organic Chemistry*, Heyden, London, UK.
9. Crabbé, P. and Klyne, W. (1967) *Tetrahedron*, **23**, 3449.
10. Schellman, J.A. (1968) *Accts. Chem. Res.*, **1**, 144.
11. Goodman, M. and Toniolo, C. (1968) *Biopolymers*, **6**, 1673.
12. Snatzke, G. (1968) *Angew. Chem. Internat. Edit.*, **7**, 14.
13. Urry, D.W. (1968) *Ann. Rev. Phys. Chem.*, **19**, 477.
14. Velluz, L. and Legrand, M. (1969) *Bull. Soc. Chim. Fr.*, 1785.
15. Jirgenson, B. (1969) *ORD of Proteins and Other Macromolecules*, Molecular Biology, Biochemistry and Biophysics Series, 5, (eds A. Kleinzeller, G.F. Springer and H.G. Wittmann), Springer-Verlag, Berlin, W. Germany.
16. Deutsche, C.W., Lightner, D.A., Woody, R.W. and Moscowitz, A. (1969) *Ann. Rev. Phys. Chem.*, **20**, 407.
17. Crabbé, P. (1971) in *Determination of Organic Structures by Physical Methods*, (eds F.C. Nachod and J.J. Zuckerman) (Vol. 3) Academic Press, New York, Chapter 3.

18. Crabbé, P. (1971) *An Introduction to the Chiroptical Methods in Chemistry*, Impresas Offsali, Mexico City, Mexico.
19. Bush, C.A. (1971) in *Physical Techniques in Biological Research* (ed. G. Oster) (Vol. 1A) 2nd edn, Academic Press, New York, p. 347.
20. Toniolo, C. (1971) *Farmaco, Ed. Sci.*, **26**, 741.
21. Toniolo, C. and Signor, A. (1972) *Experientia*, **28**, 753.
22. Snatzke, G. and Eckhardt, G. (1972) in *Experimental Methods in Biophysical Chemistry*, (ed. C. Nicolau) Wiley, New York, p. 67.
23. Barrett, G.C. (1972) in *Elucidation of Organic Structures by Physical and Chemical Methods* (eds K.W. Bentley and G.W. Kirby) (Vol. IV, part I) 2nd edn, Wiley, New York, p. 515.
24. Ciardelli, F. and Salvadori, P. (eds) (1973) *Fundamental Aspects and Recent Developments in ORD and CD*, Heyden, London, UK.
25. Toniolo, C. (1973) *Int. J. Sulfur Chem.*, **8**, 89.
26. Toniolo, C. and Fontana, A. (1974) in *The Chemistry of Functional Groups : The Chemistry of the Thiol Group* (ed. S. Patai), Wiley Interscience, New York, p. 355.
27. Schellman, J.A. (1975) *Chem. Rev.*, **75**, 323.
28. Fontana, A. and Toniolo, C. (1976) in *Progress in the Chemistry of Organic Natural Compounds* (eds W. Herz, H. Grisebach, and G.W. Kirby) (Vol. 33) Springer-Verlag, Berlin, W. Germany, p. 309.
29. Toniolo, C. (1977) in *The Chemistry of Functional Groups : The Chemistry of Cyanates and Their Thio-Derivatives* (ed. S. Patai), Interscience, New York, p. 153.
30. Peggion, E., Cosani, A., Terbojevich, M. and Palumbo, M. (1979) in *Optically Active Polymers* (ed. E. Sélégny) Reidel, Dordrecht, Holland, p. 231.
31. Bayley, P. (1980) in *An Introduction to Spectroscopy for Biochemists*, (ed. S.B. Brown) Academic Press, New York, p. 148.
32. Bewley, Th.A. and Yang, J.T. (1980) in *Hormonal Proteins and Peptides*, (ed. C.H. Li) (Vol. IX) Academic Press, New York, p. 175.
33. Sears, D.W. and Beychok, S. (1973) in *Physical Principles and Techniques of Protein Chemistry*, Part C (ed. S.J. Leach) Academic Press, New York, p. 445.
34. Kahn, P.C. (1979) in *Methods in Enzymology*, (eds C.H.W. Hirs and S.N. Timasheff) (Vol. 61, part H) Academic Press, New York, p. 339.
35. Blaha, K. (1981) in *Perspectives in Peptide Chemistry*, (eds A. Eberle, R. Geiger and Th. Wieland), Karger, Basel, Switzerland, p. 272.
36. Goodman, M., Davis, G.W. and Benedetti, E. (1968) *Accts. Chem. Res.*, **1**, 275.
37. Beychok, S. (1968) *Ann. Rev. Biochem.*, **37**, 437.
38. Barrett, G.C. (1970) *Chem. Br.*, **6**, 230.
39. Legrand, M. and Viennet, R. (1965) *Bull. Soc. Chim. France*, 679.
40. Cahn, R.S., Ingold, C.K. and Prelog, V. (1966) *Angew. Chem. Internat. Edit.*, **5**, 385.
41. Anand, R.D. and Hargreaves, M.K. (1968) *Chem. Ind. (London)*, 880.
42. Katzin, L.I. and Gulyas, E. (1968) *J. Am. Chem. Soc.*, **90**, 247.
43. Yamada, S., Achiwa, K., Terashima, S. *et al.* (1969) *Chem. Pharm. Bull.*, **17**, 2608.
44. Toniolo, C. (1970) *J. Phys. Chem.*, **74**, 1390.
45. Craig, J.C. and Pereira, W.E., Jr. (1970) *Tetrahedron Letters*, 1563.
46. Craig, J.C. and Pereira, W.E., Jr. (1970) *Tetrahedron*, **26**, 3457.
47. Jorgensen, E.C. (1971) *Tetrahedron Letters*, 863.
48. Fowden, L., Scopes, P.M. and Thomas, R.N. (1971) *J. Chem. Soc. C*, 833.
49. Snatzke, G. and Doss, S.H. (1972) *Tetrahedron*, **28**, 2539.
50. Rinaudo, M. and Domard, A. (1979) in *Optically Active Polymers* (ed. E. Sélegny) Reidel, Dordrecht, Holland, p. 253.
51. Webb, J., Strickland, R.W. and Richardson, F.S. (1973) *Tetrahedron*, **29**, 2499.
52. Nishihara, K., Nishihara, H. and Sakota, N. (1973) *Bull. Chem. Soc. Jpn*, **46**, 3894.
53. Hawkins, C.J. and Lawrance, G.A. (1973) *Aust. J. Chem.*, **26**, 1801.
54. Coleman, J.E. and Handschumacher, R.E. (1973) *J. Biol. Chem.*, **248**, 1741.
55. Snyder, P.A., Vipond, P.M. and Johnson, W.C., Jr. (1973) *Biopolymers*, **12**, 975.
56. Toniolo, C., Bonora, G.M. and Scatturin, A. (1975) *Gazz. Chim. Ital.*, **105**, 1063.
57. Polonsky, T. (1975) *Tetrahedron*, **31**, 347.
58. Nishihara, H., Nishihara, K., Uefuji, T. and Sakota, N. (1975) *Bull. Chem. Soc. Jpn.*, **48**, 553.
59. Silaev, A.B., Maevskaya, S.N., Trifonova, Zh.P. *et al.* (1976) *J. Gen. Chem. USSR*, **45**, 2287.

60. Kristensen, I., Larsen, P.O. and Olsen, C.E. (1976) *Tetrahedron*, **32**, 2799.
61. Craig, J.C., Lee, S.Y.C., Zdansky, G. and Fredga, A. (1976) *J. Am. Chem. Soc.*, **98**, 6456.
62. Rinaudo, M. and Domard, A. (1976) *J. Am. Chem. Soc.*, **98**, 6360.
63. Richardson, F.S. and Fe..er, S. (1977) *Biopolymers*, **16**, 387.
64. Flores, J.J., Bonner, W.A. and Massey, G.A. (1977) *J. Am. Chem. Soc.*, **99**, 3622.
65. Bernasconi, S., Corbella, A., Gariboldi, P. and Jommi, G. (1977) *Gazz. Chim. Ital.*, **107**, 95.
66. Craig, J.C., Lee, S.Y.C. and Fredga, A. (1977) *Tetrahedron*, **33**, 183.
67. Korver, O. and Liefkens, Th.J. (1980) *Tetrahedron*, **36**, 2019.
68. Shoji, J. (1973) *J. Antibiot.*, **26**, 302.
69. Miyazawa, T. (1980) *Bull. Chem. Soc. Jpn*, **53**, 2555.
70. Gacek, M. and Undheim, K. (1973) *Tetrahedron*, **29**, 863.
71. Coduti, Ph.L., Gordon, E.C. and Bush, C.A. (1977) *Anal. Biochem.*, **78**, 9.
72. Shen, T.Y., Li, J.P., Dorn, C.P. *et al.* (1972) *Carbohydr. Res.*, **23**, 87.
73. Van, T.T., Kojro, E. and Grzonka, Z. (1977) *Tetrahedron*, **33**, 2299.
74. Grzonka, Z. Gwizdala, E. and Koflek, T. (1978) *Pol. J. Chem.*, **52**, 1411.
75. Grzonka, Z., Kojro, E., Palacz, Z. *et al.* (1977) in *Peptides* (eds M. Goodman and J. Meienhofer) Wiley, New York, N.Y., p. 153.
76. Tran, T., Lintner, K., Toma, F. and Fermandjian, S. (1977) *Biochim. Biophys. Acta*, **492**, 245.
77. Verbit, L. and Heffron, P.J. (1967) *Tetrahedron*, **23**, 3865.
78. Jäckle, H. and Luisi, P.L. (1981) *Biopolymers*, **20**, 65.
79. Myer, Y.P. and Barnard, E.A. (1971) *Arch. Biochem. Biophys.*, **143**, 116.
80. Friis, P., Helboe, P. and Larsen, P.O. (1974) *Acta Chem. Scand.*, **B28**, 317.
81. Zamir, L.O., Jensen, R.A., Arison, B.H. *et al.* (1980) *J. Am. Chem. Soc.*, **102**, 4499.
82. Balenovic, K. and Deljac, A. (1973) *Rec. Trav. Chim. Pays-Bas*, **92**, 117.
83. Jung, G., Ottnad, M. and Rimpler, M. (1973) *Eur. J. Biochem.*, **35**, 436.
84. Jung, G., Ottnad, M., Voelter, W. and Breitmaier, E. (1972) *Z. Anal. Chem.*, **261**, 328.
85. Hermann, P., Willhardt, I., Blaha, K. and Fric, I. (1971) *J. Prakt. Chem.*, **313**, 1092.
86. Cassal, J.M., Fürst, A. and Meier, W. (1976) *Helv. Chim. Acta*, **59**, 1917.
87. Ringdahl, B., Craig, J.C., Keck, R. and Retey, J. (1980) *Tetrahedron Letters*, 3965.
88. Closson, W.D. and Haug, P. (1964) *J. Am. Chem. Soc.*, **86**, 2384.
89. Caswell, L.R., Howard, M.F. and Onisto, T.M. (1976) *J. Org. Chem.*, **41**, 3312.
90. Inagaki, T. (1973) *Biopolymers*, **12**, 1353.
91. Moffitt, W., Woodward, R.B., Moscowitz, A. *et al.* (1961) *J. Am. Chem. Soc.*, **83**, 4013.
92. Santoso, S., Kemmer, T. and Trowitzsch, W. (1981) *Liebigs Ann. Chem.*, 642.
93. Santoso, S., Kemmer, T. and Trowitzsch, W. (1981) *Liebigs Ann. Chem.*, 658.
94. Gravenmade, E.J. and Vogels, G.D. (1970) *Anal. Biochem.*, **32**, 286.
95. Welter, A., Marlier, M. and Dardenne, G. (1975) *Bull. Soc. Chim. Belg.*, **84**, 243.
96. Kawai, M., Nagai, U. and Katsumi, M. (1975) *Tetrahedron Letters*, 3165.
97. Kang, S., Minematsu, Y., Waki, M. *et al.* (980) *Mem. Fac. Sci., Kyushu Univ., Ser. C*, **12**, 179.
98. Gravenmade, E.J., Vogels, G.D. and Van Pelt, C. (1969) *Rec. Trav. Chim. Pays-Bas*, **88**, 929.
99. Ivanov, V.T. (1977) in *Peptides* (eds M. Goodman and J. Meienhofer) Wiley, New York, N.Y., p. 307.
100. Crippen, G.M. and Yang, J.T. (1974) *J. Phys. Chem.*, **78**, 1127.
101. Mattice, W.L. (1973) *J. Am. Chem. Soc.*, **95**, 5800.
102. Mattice, W.L. (1974) *Biopolymers*, **13**, 169.
103. Mattice, W.L. and Harrison, W.H., III (1975) *Biopolymers*, **14**, 2025.
104. Simmons, N.S., Barel, A.O. and Glazer, A.N. (1969) *Biopolymers*, **7**, 275.
105. Takagi, T., Okano, R. and Miyazawa, T. (1973) *Biochim. Biophys. Acta*, **310**, 11.
106. Madison, V. and Schellman, J. (1970) *Biopolymers*, **9**, 511.
107. Madison, V. and Schellman, J. (1970) *Biopolymers*, **9**, 569.
108. Faulstich, H. and Wieland, Th. (1973) in *Peptides* 1972 (eds H. Hanson and H.D. Jakubke) North-Holland, Amsterdam, p. 312.
109. Cann, J.R. (1972) *Biochemistry*, **11**, 2654.
110. Madison, V. and Kopple, K.D. (1980) *J. Am. Chem. Soc.*, **102**, 4855.
111. Radding, W., Ueyama, N., Gilon, C. and Goodman, M. (1976) *Biopolymers*, **15**, 591.
112. Ivanov, V.T., Kostetskii, P.V., Meschcheryakova, E.A. *et al.* (1973) *Khim. Prirod. Soedin.*, **9**, 363.
113. Johnson, W.C., Jr. and Tinoco, I., Jr. (1972) *J. Am. Chem. Soc.*, **94**, 4389.
114. Madison, V. and Schellman, J. (1970) *Biopolymers*, **9**, 65.

115. Di Bello, C., Filira, F. and Toniolo, C. (1972) *Biopolymers*, **10**, 2283.
116. Pradelles, Ph., Morgat, J.L., Fermandjan, S. *et al.* (1977) *Coll. Czech. Chem. Commun.*, **42**, 79.
117. Sakamoto, K. and Hatano, M. (1980) *Bull. Chem. Soc. Jpn*, **53**, 339.
118. Toniolo, C., Palumbo, M. and Benedetti, E. (1976) *Macromolecules*, **9**, 420.
119. Benedetti, E., Ciajolo, A., Di Blasio, B. *et al.* (1979) *Int. J. Peptide Protein Res.*, **14**, 130.
120. Okita, K., Matsui, Y. and Sakota, N. (1970) *Nippon Kagaku Zasshi*, **91**, 1179.
121. Takahashi, H. and Otomasu, H. (1978) *Chem. Pharm. Bull.*, **26**, 466.
122. Gacek, M., Undheim, K. and Håkansson, R. (1977) *Tetrahedron*, **33**, 589.
123. Poupaert, J., Claesen, M., Degelaen, J. *et al.* (1977) *Bull. Soc. Chim. Belg.*, **86**, 465.
124. Suzuki, T., Igarashi, K., Hase, K. and Tuzimura, K. (1973) *Agr. Biol. Chem.*, **37**, 411.
125. Verbit, L. and Heffron, P.J. (1968) *Tetrahedron*, **24**, 1231.
126. Parello, J. and Péchère, J.F. (1971) *Biochimie*, **53**, 1079.
127. Holmquist, B. and Vallee, B.L. (1973) *Biochemistry*, **12**, 4409.
128. Toniolo, C. and Bonora, G.M. (1976) *Can. J. Chem.*, **54**, 70.
129. Klyne, W., Scopes, P.M., Thomas, R.N. and Dahn, H. (1971) *Helv. Chim. Acta*, **54**, 2420.
130. Sakota, N., Okita, K. and Matsui, Y. (1970) *Bull. Chem. Soc. Jpn*, **43**, 1138.
131. Toniolo, C. and Bonora, G.M. (1974) *Gazz. Chim. Ital.*, **104**, 843.
132. Peggion, E., Palumbo, M., Bonora, G.M. and Toniolo, C. (1974) *Bioorg. Chem.*, **3**, 125.
133. Legrand, M. and Viennet, R. (1966) *Bull. Soc. Chim. France*, 2798.
134. Shiraki, M. (1969) *Sci. Pap. Coll. Gen. Educ., Univ. Tokyo*, **19**, 151.
135. Komiyama, T. and Miwa, M. (1979) *Chem. Phys. Letters*, **65**, 136.
136. Schmidt, G. and Rosenkranz, H. (1976) *Liebigs Ann. Chem.*, 129.
137. Schmidt, G. and Rosenkranz, H. (1976) *Liebigs Ann. Chem.*, 124.
138. Weinryb, I. and Steiner, R.F. (1969) *Arch. Biochem. Biophys.*, **131**, 263.
139. Menendez, C.J. and Herskovits, T.T. (1970) *Arch. Biochem. Biophys.*, **140**, 286.
140. Horwitz, J., Strickland, E.H. and Billups, C. (1969) *J. Am. Chem. Soc.*, **91**, 184.
141. Takagi, S., Nomori, H. and Hatano, M. (1974) *Chem. Letters*, 611.
142. Brady, H. Ryan, J.W. and Stewart J.M. (1971) *Biochem. J.*, **121**, 179.
143. Hegedüs, B., Krassó, A.F., Noack, K. and Zeller, P. (1975) *Helv. Chim. Acta*, **58**, 147.
144. Snatzke, G., Kajtar, M. and Werner-Zamojska, F. (1972) *Tetrahedron*, **28**, 281.
145. Cann, J.R., Stewart, J.M., London, R.E. and Matwiyoff, N. (1976) *Biochemistry*, **15**, 498.
146. Takagi, T. and Ito, N. (1972) *Biochim. Biophys. Acta*, **257**, 1.
147. Beychok, S. and Fasman, G.D. (1964) *Biochemistry*, **3**, 1675.
148. Simmons, N.S. and Glazer, A.N. (1967) *J. Am. Chem. Soc.*, **89**, 5040.
149. Pflumm, M.N. and Beychok, S. (1969) *J. Biol. Chem.*, **244**, 3973.
150. Strickland, E.H., Wilchek, M., Horwitz, J. and Billups, C. (1972) *J. Biol. Chem.*, **247**, 572.
151. Horwitz, J., Strickland, E.H. and Billups, C. (1970) *J. Am. Chem. Soc.*, **92**, 2119.
152. Strickland, E.H., Wilchek, M., Horwitz, J. and Billups C. (1970) *J. Biol. Chem.*, **245**, 4168.
153. Toniolo, C. (1980) *CRC Crit. Rev. Biochem.*, **9**, 1.
154. Goux, W.J., Cooke, D.B., Rodriguez, R.E. and Hooker, T.M., Jr., (1974) *Biopolymers*, **13**, 2315.
155. Buck, R., Eberspächer, J. and Lingens, F. (1979) *Liebigs Ann. Chem.*, 564.
156. Hooker, T.M., Jr. and Schellman, J.A. (1970) *Biopolymers*, **9**, 1319.
157. Van Heerden, F.R., Brandt, E.V. and Roux, D.G. (1980) *Phytochemistry*, **19**, 2125.
158. Snow, J.W. and Hooker, T.M., Jr. (1974) *J. Am. Chem. Soc.*, **96**, 7800.
159. Marche, P.,Montenay-Garestier, T., Hélenè, C. and Fromageot, P. (1976) *Biochemistry*, **15**, 5730.
160. Auer, H.E. (1973) *J. Am. Chem. Soc.*, **95**, 3003.
161. Myer, Y.P. and Mac Donald, L.H. (1967) *J. Am. Chem. Soc.*, **89**, 7142.
162. Strickland, E.H., Horwitz, J. and Billups, C. (1969) *Biochemistry*, **8**, 3205.
163. Strickland, E.H., Wilchek, M. and Billups, C. (1973) *Biochim. Biophys. Acta*, **303**, 28.
164. Taniguchi, M. and Hino, T. (1981) *Tetrahedron*, **37**, 1487.
165. Kahn, P.C. and Beychok, S. (1968) *J. Am. Chem. Soc.*, **90**, 4168.
166. Imanishi, A. and Isemura, T. (1969) *J. Biochem.* (*Tokyo*), **65**, 309.
167. Ito, N. and Takagi, T. (1970) *Biochim. Biophys. Acta*, **221**, 430.
168. Coleman, D.L. and Blout, E.R. (1968) *J. Am. Chem. Soc.*, **90**, 2405.
169. Casey, J.P. and Martin, R.B. (1972) *J. Am. Chem. Soc.*, **94**, 6141.
170. Lam-Thanh, H. and Fermandjan, S. (1977) *J. Chim. Phys. Physicochim. Biol.*, **74**, 361.
171. Strickland, R.W., Webb, J. and Richardson, F.S. (1974) *Biopolymers*, **13**, 1269.
172. Lam-Thanh, H., Lintner, K., Monnot, M. *et al.* (1978) *J. Chim. Phys. Physicochim. Biol.*, **75**, 755.

173. Ottnad, M., Ottnad, C., Hartter, P. and Jung, G. (1975) *Tetrahedron*, **31**, 1155.
174. Ringdahl, B., Craig, J.C., Zdansky, G. and Fredga, A. (1980) *Acta Chem. Scand.*, **B 34**, 735.
175. Adriaens, P., Meesschaert, B., Frère, J.M. *et al.* (1978) *J. Biol. Chem.*, **253**, 3660.
176. Gromova, E.S., Tyaglov, B.V. and Shabarova, Z.A. (1971) *Biochim. Biophys. Acta*, **240**, 1.
177. Tyaglov, B.V., Gromova, E.S., Zenin, S.V. *et al.* (1975) *Mol. Biol.* (*Moscow*), **9**, 524.
178. Tyaglov, B.V., Zenin, S.V., Gromova, E.S. *et al.* (1976) *Mol. Biol.* (*Moscow*), **10**, 279.
179. Gacek, M. and Undheim, K. (1972) *Acta Chem. Scand.*, **26**, 2655.
180. Toniolo, C., Filira, F. and Di Bello, C. (1971) *Biopolymers*, **10**, 2275.
181. Sabri, S.S., El-Abadelah, M.M. and Zaater, M.F. (1977) *J. Chem. Soc., Perkin Trans. 1*, 1356.
182. Blaha, K., Fric, I. and Jakubke, H.D. (1967) *Collect. Czech. Chem. Commun.*, **32**, 558.
183. Rekunova, V.N., Rudakova, I.P. and Yurkevich, A.M. (1975) *J. Gen. Chem. URSS*, **45**, 2047.
184. Kroeger, M. and Cramer, F. (1977) *Bioorg. Chem.*, **6**, 431.
185. Ripperger, H. (1969) *Tetrahedron*, **25**, 725.
186. Djerassi, C., Wolf, H. and Bunnenberg, E. (1962) *J. Am. Chem. Soc.*, **84**, 4552.
187. Pracejus, H. and Winter, S. (1964) *Chem. Ber.*, **97**, 3173.
188. Briggs, W.S. and Djerassi, C. (1965) *Tetrahedron*, **21**, 3455.
189. Ripperger, H. and Schreiber, K. (1965) *Tetrahedron*, **21**, 407.
190. Yamada, S., Ishikawa, K. and Achiwa, K. (1965) *Chem. Pharm. Bull.*, **13**, 892.
191. Isono, K., Asahi, K. and Suzuki, S. (1969) *J. Am. Chem. Soc.*, **91**, 7490.
192. Ishikawa, K., Achiwa, K. and Yamada, S. (1971) *Chem. Pharm. Bull.*, **19**, 912.
193. König, W.A., Pfaff, K.P., Bartsch, H.H. *et al.* (1980) *Liebigs Ann. Chem.*, 1728.
194. Barrett, G.C. (1965) *J. Chem. Soc.*, 2825.
195. Barrett, G.C. (1966) *J. Chem. Soc.*, 1771.
196. Barrett, G.C. (1967) *J. Chem. Soc. C.*, 1.
197. Barrett, G.C. and Khokhar, A.R. (1969) *J. Chem. Soc. C*, 1120.
198. Barrett, G.C. (1969) *J. Chem. Soc. C*, 1123.
199. Barrett, G.C. (1975) *J. Chem. Soc. Perkin Trans. 1*, 2313.
200. Bach, E., Kjaer, A., Dahlbom, R. *et al.* (1966) *Acta Chem. Scand.*, **20**, 2781.
201. Toniolo, C., Nisato, D., Biondi, L. and Signor, A. (1972) *J. Chem. Soc. Perkin Trans. 2*, 1179.
202. Kawai, M., Nagai, U. and Kobayashi T. (1974) *Tetrahedron Letters*, 1881.
203. Kawai, M., Nagai, U. and Katsumi, M. (1975) *Tetrahedron Letters*, 2845.
204. Nagai, U. and Kani, Y. (1977) *Tetrahedron Letters*, 2333.
205. Kawai, M. and Nagai, U. (1977) *Tetrahedron Letters*, 3889.
206. Kawai, M., Nagai, U., Katsumi, M. and Tanaka, A. (1978) *Tetrahedron*, **34**, 3435.
207. Nagai, U., Taki, N. and Kawai, M. (1981) *Chem. Pharm. Bull.*, **29**, 1750.
208. Bettoni, G., Catsiotis, S. and Franchini, C. (1977) *Farmaco, Ed. Sci.*, **32**, 367.
209. Gabbay, E.J., Sanford, K. and Baxter, S. (1972) *J. Am. Chem. Soc.*, **94**, 2876.
210. Nakayama, H., Tanizawa, K., Kanaoka, Y. and Witkop, B. (1980) *Eur. J. Biochem.*, **112**, 403.
211. Okahashi, K. and Ikeda, S. (1979) *Int. J. Peptide Protein Res.*, **13**, 462.
212. Toniolo, C., to be published.
213. Moser, P. (1968) *Helv. Chim. Acta*, **51**, 1831.
214. Wolf, H., Bunnenberg, E. and Djerassi, C. (1964) *Chem. Ber.*, **97**, 533.
215. Auterhoff, H. and Hansen, J.G. (1970) *Pharmazie*, **25**, 336.
216. Smith, H.E., Burrows, E.P., Marks, M.J. *et al.* (1977) *J. Am. Chem. Soc.*, **99**, 707.
217. Smith, H.E. and Records, R. (1966) *Tetrahedron*, **22**, 813.
218. Bettoni, G., Tortorella, V., Hope, A. and Halpern, B. (1975) *Tetrahedron*, **31**, 2383.
219. Snatzke, G., Ripperger, H., Horstmann, Ch. and Schreiber, K. (1966) *Tetrahedron*, **22**, 3103.
220. Gaffield, W., Keefer, L. and Lijinsky, W. (1972) *Tetrahedron Letters*, 779.
221. Liberek, B., Ciarkowski, J., Plucinska, K. and Stachowiak, K. (1976) *Tetrahedron Letters*, 1407.
222. Polonsky, T. and Prajer, K. (1976) *Tetrahedron*, **32**, 847.
223. Gaffield, W., Lundin, R.E. and Keefer, L.K. (1981) *Tetrahedron*, **37**, 1861.
224. Toome, V., Wegrzynski, B. and Reymond, G. (1976) *Biochem. Biophys. Res. Commun.*, **69**, 206.
225. Toome, V., Wegrzynski, B. and Dell, J. (1976) *Biochem. Biophys. Res. Commun.*, **71**, 598.
226. Toome, V. and Wegrzynski, B. (1978) *Biochem. Biophys. Res. Commun.*, **85**, 1496.
227. Kovacs, K.L. (1979) *Biochem. Biophys. Res. Commun.*, **86**, 995.
228. Toome, V. and Wegrzynski B. (1980) *Biochem. Biophys. Res. Commun.*, **92**, 447.
229. Toome, V. and Reymond, G. (1975) *Biochem. Biophys. Res. Commun.*, **66**, 75.
230. Toome, V., De Bernardo, S. and Weigele M. (1975) *Tetrahedron*, **31**, 2625.

231. Keith, D.D., De Bernardo, S. and Weigele M. (1975) *Tetrahedron*, **31**, 2629.
232. Keith, D.D., Tortora, J.A., Ineichen, K. and Leimgruber, W. (1975) *Tetrahedron*, **31**, 2633.
233. El-Abadelah, M.M., Sabri, S.S., Owais, W.M. and Tabba, H.D. (1977) *Chem. Ind.* (*London*), 200.
234. El-Abadelah, M.M., Sabri, S.S., Jarrar, A.A. and Abu Zarga, M.H. (1979) *J. Chem. Soc. Perkin Trans. 1*, 2881.
235. El-Abadelah, M.M., Sabri S.S., Nazer, M.Z. and Zaater, M.F. (1976) *Tetrahedron*, **32**, 2931.
236. El-Abadelah, Anani, A.A., Khan, Z.H. and Hassan, A.M. (1980) *J. Heterocyclic Chem.*, **17**, 213.
237. Gronnenberg, T. and Undheim, K. (1972) *Acta Chem. Scand.*, **26**, 2267.
238. Halpern, B., Patton, W. and Crabbé, P. (1969) *J. Chem. Soc. B*, 1143.
239. Romanov, V.V., Voskova, N.A. and Shvachkin Yu.P. (1980) *Khim. Prirod. Soedin.*, 132.
240. Voskova, N.A., Romanov, V.V., Sumbatyan, N.V. *et al.* (1980) *Bioorg. Khim.*, **6**, 731.
241. Nagai, U., Besson, F. and Peypoux, F. (1979) *Tetrahedron Letters*, 2359.
242. Nagai, U., Kawai, M., Yamada, T. *et al.* (1981) *Tetrahedron Letters*, 653.
243. Yamada, T., Kuwata, S. and Watanabe, H. (1978) *Tetrahedron Letters*, 1813.
244. Kuwata, S., Yamada, T., Shinogi, T. *et al.* (1979) *Bull. Chem. Soc. Jpn*, **52**, 3326.
245. Bodansky, M., Fink, M.L., Funk, K.W. *et al.* (1974) *J. Am. Chem. Soc.*, **96**, 2234.
246. Toniolo, C. (1970) *Tetrahedron*, **26**, 5479.
247. Suzuki, T. and Tuzimura, K. (1973) *Agr. Biol. Chem.*, **37**, 689.
248. Knabe, J. and Urbahn, C. (1971) *Liebigs Ann. Chem.*, **750**, 21.
249. Undheim, K. and Ulsaker, G.A. (1973) *Acta Chem. Scand.*, **27**, 1059.
250. Jung, G., Fouad, H. and Heusel, G. (1975) *Angew. Chem. Internat. Edit.*, **14**, 817.
251. Toniolo, C., Nisato, D., Biondi, L. and Signor, A. (1972) *J. Chem. Soc. Perkin Trans. 2*, 1182.
252. Nisato, D., Marzotto, A., De Pieri, G. and Signor, A. (1971) *Gazz. Chim. Ital.*, **101**, 805.
253. Toniolo, C. and Jori, G. (1970) *Biochim. Biophys. Acta*, **214**, 368.
254. Conway-Jacobs, A., Schechter, B. and Sela, M. (1970) *Biochemistry*, **9**, 4870.
255. Pospisek, J., Toma, S., Fric, I. and Blaha, K. (1980) *Collect. Czech. Chem. Commun.*, **45**, 435.
256. Diem, M., Gotkin, P.J., Kupfer, J.M. *et al.* (1977) *J. Am. Chem. Soc.*, **99**, 8103.
257. Diem, M., Gotkin, P.J., Kupfer, J.M. and Nafie, L.A. (1978) *J. Am. Chem. Soc.*, **100**, 5644.
258. Diem, M., Photos, E., Khouri, H. and Nafie, L.A. (1979) *J. Am. Chem. Soc.*, **101**, 6829.
259. Gabriel, M., Larcher, D., Rinnert, H. *et al.* (1972) *C.R. Acad. Sci. Ser. B*, **275**, 861.
260. Barth, G., Records, R., Bunnenberg, E. *et al.* (1971) *J. Am. Chem. Soc.*, **93**, 2545.
261. Barth, G., Voelter, W., Bunnenberg, E. and Djerassi, C. (1972) *J. Am. Chem. Soc.*, **94**, 1293.
262. Bayer, E., Bacher, A., Krauss, P. *et al.* (1971) *Eur. J. Biochem.*, **22**, 580.
263. Barth, G., Bunnenberg, E. and Djerassi, C. (1972) *Anal. Biochem.*, **48**, 471.
264. Gabriel, M., Larcher, D., Rinnert, H. *et al.* (1972) *C.R. Acad. Sci. Ser. B*, **275**, 935.
265. Holmquist, B. (1975) in *Protein Nutritional Quality of Foods and Feeds* Part 1 (ed. Friedman, M.), Dekker, New York, N.Y. p. 463.
266. Toniolo, C., unpublished results.
267. Komiyama, T. and Miwa, M. (1980) *Int. J. Quantum Chem.*, **18**, 527.
268. Gabriel, M., Larcher, D., Rinnert, H. and Thirion, C. (1973) *C.R. Acad. Sci. Ser. B*, **276**, 39.
269. Gabriel, M., Larcher, D., Rinnert, H. and Thirion, C. (1973) *FEBS Letters*, **35**, 148.
270. Johansen, J.T., Livingston, D.M. and Vallee, B.L. (1972) *Biochemistry*, **11**, 2584.
271. Sutherland, J.C. and Low, H. (1976) *Proc. Natl. Acad. Sci. USA*, **73**, 276.
272. Nagy, J.A., Powers, S.P., Zweifel, B.O. and Scheraga, H.A. (1980) *Macromolecules*, **13**, 1428.
273. Sprinkel, F.M., Shillady, D.D. and Strickland, R.W. (1975) *J. Am. Chem. Soc.*, **97**, 6653.
274. Heinrich, C.P., Noack, K. and Wiss, O. (1972) *Biochem. Biophys. Res. Commun.*, **49**, 1427.
275. McFarland, T.M. and Coleman, J.E. (1972) *Eur. J. Biochem.*, **29**, 521.
276. Eckstein, H., Barth, G., Linden, R.E. *et al.* (1974) *Liebigs Ann. Chem.*, 990.
277. Witzemann, V., Koberstein, R., Sund, H. *et al.* (1974) *Eur. J. Biochem.*, **43**, 319.
278. Holmquist, B. (1971) *Fed. Proc., Fed. Amer. Soc. Exp. Biol.*, **30**, 1179, Abs. No. 737.
279. Reid, V., Glaser, M., Kennett, R. and Singer, S.J. (1971) *Proc. Natl. Acad. Sci. USA*, **68**, 1184.
280. Glaser, M. and Singer, S.J. (1971) *Proc. Natl. Acad. Sci. USA*, **68**, 2477.
281. Rockey, J.H., Dorrington, K.J. and Montgomery, P.C. (1971) *Nature* (*London*), **232**, 192.
282. Rockey, J.H., Montgomery, P.C., Underdown, B.J. and Dorrington, K.J. (1972) *Biochemistry*, **11**, 3172.
283. Inbar, D., Givol, D. and Hochman, J. (1973) *Biochemistry*, **12**, 4541.

284. Merz, D.C., Litman, G.W. and Good, R.A. (1974) *Proc. Natl. Acad. Sci. USA*, **71**, 1940.
285. Holowka, D.A. and Cathou, R.E. (1976) *Biochemistry*, **15**, 3379.
286. Okabe, N., Manabe, N., Tokuoka, R. and Tomita, K. (1975) *J. Biochem. (Tokyo)*, **77**, 181.
287. Okabe, N., Manabe, N., Tokuoka, R. and Tomita, K. (1976) *J. Biochem. (Tokyo)*, **80**, 455.
288. Tokuoka, R., Okabe, N. and Tomita, K. (1980) *J. Biochem. (Tokyo)*, **87**, 1729.

FURTHER READING

A list of recent references (1981–1983) on the CD of amino acids and their derivatives is given below for the relevant sections.

Section 19.2.1

1. Benedetti, E., Di Blasio, B., Pavone, V. *et al.* (1981) *J. Biol. Chem.*, **256**, 9229.
2. Lam-Thanh, H., Juy, M., Schneider, C. *et al.* (1981) *J. Chim. Phys. Physicochim. Biol.*, **78**, 695.
3. Ando, S., Itaya, T., Nishikawa, H. and Takiguchi, H. (1981) *Fukuoka Daigaku Rigaku Shuho*, **11**, 21.
4. Dungan, J.M. III and Hooker, T.M. Jr. (1981) *Macromolecules*, **14**, 1812.
5. Delaney, N.G., Lasky, J. and Madison, V. (1981) in *Peptides : Synthesis, Structure, Function*, (eds D.H. Rich and E. Gross) Pierce Chem. Co., Rockford, Illinois, p. 303.
6. Delaney, N.G. and Madison, V. (1982) *J. Am. Chem. Soc.*, **104**, 6635.
7. Bush, C.A., Dua, V.K., Ralapati, S. *et al.* (1982) *J. Biol. Chem.*, **257**, 8199.
8. Rolli, H., Lüscher, I.F., Schneider, C.H. *et al.* (1982) *Helv. Chim. Acta*, **65**, 1965.
9. Prasad, K.U., Trapane, T.L., Busath, D. *et al.* (1982) *Int. J. Peptide Protein Res.*, **19**, 162.
10. Matsuura, H., Hasegawa, K., and Miyazawa, T. (1982) *Bull. Chem. Soc. Jpn*, **55**, 1999.
11. Madison, V. and Delaney, N.G. (1983) *Biopolymers*, **22**, 869.
12. Kozlowski, H., Decock-le-Reverend, B., Delaruelle, J.L. *et al.* (1983) *Inorg. Chim. Acta*, **78**, 31.
13. Overberger, C.G. and Shalati, M.D. (1983) *Eur. Polym. J.*, **19**, 1055.
14. Prosyanik, A.V., Fedoseenko, D.V., Chervin, I.I., Nasibov, Sh. S. and Kostyanovskii, R.G. (1983) *Proc. Acad. Sci. USSR, Chem. Sect.*, **270**, 164.
15. Kimura, H. and Stammer, C.H. (1983) *J. Org. Chem.*, **48**, 2440.
16. Yang, W.W.Y., Overberger, C.G. and Venkatachalam, C.M. (1983) *J. Polymer Sci., Polymer Chem. Edit.*, **21**, 1643.
17. Müller, J.C., Toome, V., Pruess, D.L., Blount, J.F. and Weigele, M. (1983) *J. Antibiot.*, **36**, 217.
18. Marcellet-Sauvage, J., Marcellet, M. and Loucheux, C. (1983) *Macromolecules*, **16**, 1564.
19. Malon, P., Pancoska, P., Budesinsky, M., Hlavacek, J., Pospisek, J. and Blaha, K. (1983) *Coll. Czech. Chem. Commun.*, **48**, 2844.

Section 19.2.2

1. Isogai, A., Suzuki, A., Hagashikawa, S. *et al.* (1981) in *Peptide Chemistry 1980* (ed. Okawa K.) Protein Res. Foundation, Osaka, p. 125.
2. Matsuura, H., Hasegawa, K. and Miyazawa, T. (1982) *Bull. Chem. Soc. Jpn*, **55**, 1999.
3. Rekunova, V.N., Gurevich, V.M. and Yurkevich, A.M. (1982) *J. Gen. Chem. URSS*, **52**, 1475.
4. Welsh, E.J., Frangou, S.A., Morris, E.R. *et al.* (1983) *Biopolymers*, **22**, 821.
5. Overberger, C. G. and Lu, C.X. (1983) *J. Polymer Sci., Polymer Lett. Edit.*, **22**, 193.
6. Nettsold, Kh., Smirnov, V.V., Razzhivin, V.P., Gromova, E.S., and Shabarova, Z.A. (1983) *J. Gen. Chem. USSR*, **53**, 562.
7. Sisido, M., Egusa, S. and Imanishi, Y. (1983) *J. Am. Chem. Soc.*, **105**, 4077.
8. Almog, R. (1983) *Biophys. Chem.*, **17**, 111.
9. Penke, B., Zarandi, M., Toth, G.K., Kovacs, K., Fekete, M., Telegdy, G. and Pham. P. (1983) in *Peptides 1982* (eds K. Blaha and P. Malon) de Gruyter, Berlin, p. 569.
10. Zamir, L.O., Tiberio, R., Jung, E. and Jensen, R.A. (1983) *J. Biol. Chem.*, **258**, 6486.
11. Welsh, E.J., Frangou, S.A., Morris, E.R., Rees, D.A. and Chavin, S.I. (1983) *Biopolymers*, **22**, 821.

12. Jung, G., Carrera, C., Brückner, H. and Bessler, W.G. (1983) *Liebigs Ann. Chem.*, 1608.

Section 19.3.1

1. Casella, L. and Gullotti, M. (1981) *J. Am. Chem. Soc.*, **103**, 6338.
2. Harris, C.H. and Harris, T.M. (1982) *J. Am. Chem. Soc.*, **104**, 363.
3. Kawai, M., Nagai, U. and Tanaka, A. (1982) *Bull. Chem. Soc. Jpn*, **55**, 1213.
4. Kawai, M. and Nagai, U. (1982) *Bull. Chem. Soc. Jpn*, **55**, 1327.
5. Hagenmaier, H. and König, W.A. (1982) in *Chemistry of Peptides and Proteins* (eds W. Voelter, E. Wünsch, Yu. Ovchinnikov and V. Ivanov) de Gruyter, Berlin, p. 187.
6. Voskova, N.A., Romanov, V.V., Korshunova, G.A. and Shvachkin, Yu.P. (1982) in *Chemistry of Peptides and Proteins* (eds W. Voelter, E. Wünsch, Yu. Ovchinnikov, and V. Ivanov) de Gruyter, Berlin, p. 373.
7. Casella, L. and Gullotti, M. (1983) *J. Am. Chem. Soc.*, **105**, 803.
8. Casella, L. and Gullotti, M. (1983) *Inorg. Chem.*, **22**, 2259.
9. Müller, J.C., Toome, V., Pruess, D.L., Blount, J.F. and Weigele, M. (1983) *J. Antibiot.*, **36**, 217.
10. Smith, H.E. (1983) *Chem. Rev.*, **83**, 359.

Section 19.3.4

1. Shimojima, Y., Hayashi, H., Ooka, T. *et al.* (1982) *Tetrahedron Letters*, 5435.

Section 19.4.1

1. Lal, B.B., Diem, M., Polavarapu, P.L. *et al.* (1982) *J. Am. Chem. Soc.*, **104**, 3336.
2. Freedman, T.B., Diem M., Polavarapu, P.L. and Nafie, L.A. (1982) *J. Am. Chem. Soc.*, **104**, 3343.
3. Bowman, R.L., Kellerman, M. and Johnson, W.C. Jr. (1983) *Biopolymers*, **22**, 1045.

Section 19.4.2

1. Lam-Thanh, H., Juy, M., Schneider, C. *et al.* (1981) *J. Chim. Phys. Physicochim. Biol.*, **78**, 695.
2. Rinnert, H., Maigret, B. and Thirion, C. (1981) *J. Phys. Chem.*, **85**, 3420.

Section 19.4.3

1. Okabe, N., Okamura, M., Tokuoka, R. and Tomita, K. (1981) *J. Chem. Soc. Jpn.*, **54**, 3790.

General

1. Toome, V. and Weigele, M. (1981) in *Peptides: Analysis, Synthesis, Biology* (eds E. Gross and J. Meienhofer) (Vol. 4) Academic Press, New York, p. 85.
2. Barth, G. and Djerassi, C. (1981) *Tetrahedron*, **37**, report No. 118, 4123.

CHAPTER TWENTY

Colorimetric and Fluorimetric Detection of Amino Acids

G.A. ROSENTHAL

The functional groups common to all amino acids and most substituent groups for amino acids exhibit little absorption of ultraviolet (UV) light from 210 to 300 nm where UV absorption spectrometry is most conveniently conducted. Of course, protein amino acids as well as the many aromatic and heterocyclic non-protein amino acids that are not affected by this limitation can be analysed by their UV absorption spectra. Given this limitation, amino acid detection is dependent primarily upon colorimetry and more recently fluorimetry.

20.1 COLORIMETRY

The most widely applied colorimetric assay for amino acids relies upon ninhydrin (triketohydrindene hydrate)-mediated colour formation. This reagent reacts with an amino acid to produce CO_2, NH_3, and the lower aldehyde of the original amino acid and ultimately yields a chromophore known as Ruhemann's purple that actually varies in colour from blue to purple. The exact chemical nature of the ninhydrin reaction has not been unequivocally established but it has been the subject of considerable study. A most detailed review of the ninhydrin reaction is that of McCaldin [1] but the work of Lamothe and McCormic [2] is also noteworthy.

Ninhydrin forms a coloured complex with compounds other than amino acids; e.g. primary and secondary amines, amino alcohols, ammonia, and all ammonium salts but not with tertiary or aromatic amines. The resulting Ruhemann's purple has an absorbance maximum (λ_{max}) that centres around 570 nm, and this wavelength has become the accepted standard for ninhydrin-based analysis of amino acids. Ruhemann's purple forms in substoichiometric yields that vary somewhat for the different amino acids but reproducible colour formation is obtained if a suitable reducing agent is present to block oxidative side reactions that otherwise deplete the colour. Reduction of the ninhydrin solution is usually achieved with stannous or titanous chloride but cyanide or hydrazine are also effective. The intrinsic colour formed by reaction of ninhydrin with an α-amino acid is not constant; tryptophan and cystine giving low colour yields while lysine and ornithine (possibly owing to the reactivity of their ω-NH_2 group) produce a more intense coloration (Table 20.1).

Ninhydrin also reacts with imino acids but a yellow chromogen results that possesses a broad absorption centred around 440 nm. Cysteine also forms a yellow chromogen with ninhydrin; this protein amino acid is usually pre-oxidized to cysteic acid which then produces Ruhemann's purple with ninhydrin. The presence of both amino and imino acids in many biological samples analysed by automated amino acid analysers accounts for the use of dual wavelength colorimetric detection at 440 and 570 nm. Single wavelength detection for amino acids is possible at 420 nm [4]. There is no reason to limit the analytical wavelength in amino acid detection to 440 and 570 nm because many amino acids have other maximum absorption wavelengths for their ninhydrin-produced chromophore. Charlwood and Bell [5] have extended this concept by employing multiple spectrophotometers as a means of comparing the ratio of the absorbance of the Ruhemann's purple formed after 3 min at 570 nm to the absorbances at 405, 416, 434, and 475 nm, as well as to the absorbance at 570 nm after 15 min of reaction time. The output from these multiple colorimeters is then computerized to create an information bank in which not only the column retention time but

Table 20.1 Colour yield with ninhydrin on molar basis relative to leucine[a]

Compound	*Colour Yield*	*Compound*	*Colour Yield*
Aspartic acid	0.94	Carnosine	0.93
Threonine	0.94	Citrulline	1.04
Serine	0.95	Creatinine	0.027
Proline (440 mμ)	0.225	Cysteic acid	0.99
Glutamic acid	0.99	α,ε-Diaminopimelic acid	1.24
Glycine	0.95	(per $2NH_2$)	
Alanine	0.97	Ethanolamine	0.91
Valine	0.97	Felinine	0.95
Half-cystine	0.55	Glutamine	0.99
Methionine	1.02	Glucosamine	1.03
Isoleucine	1.00	Glutathionine (oxidized, half)	0.93
Leucine	1.00	Glycerophosphoethanolamine	0.50
Tyrosine	1.00	Hydroxylysine	1.12
Phenylalanine	1.00	Hydroxyproline (400 mμ)	0.077
Ammonia	0.97	Methionine sulphone	1.02
Lysine	1.10	Methionine sulphoxide	0.98
Tryptophan	0.94	1-Methylhistidine	0.88
Arginine	1.01	3-Methylhistidine	0.86
α-Aminoadipic acid	0.96	Ornithine	1.12
β-Alanine	0.50	Phosphoethanolamine	0.43
Anserine	0.78	Sarcosine	0.28
Asparagine	0.95	Taurine	0.88
β-Aminoisobutyric acid	0.44	Urea	0.314
γ-Aminobutyric acid	1.01	Butyrine	1.02

[a] Heating time 15 min at 100°, read at 570 nm. (Reproduced from Greenstein and Winitz [3] with permission of John Wiley & Sons. Inc.)

also the absorbance ratios for each eluted compound are recorded. The data can then be retrieved for comparison with newly analysed samples and are an invaluable aid in the identification of the hundreds of non-protein amino acids present in biological materials.

Several modifications of the basic ninhydrin reagent have been described. Particularly useful ones involve the addition of metallic ions such as cadmium or copper or organic solvents such as ethanolamine that impart distinctive coloration to the treated amino acid. The copper–ninhydrin reagent developed by Moffat and Lytle [6] is probably the best known polychromatic reagent for amino acid detection. A solution containing 0.2% (w/v) ninhydrin in absolute ethanol: glacial acetic acid: 2,4,6-collidine (50:10:2, v/v) is mixed with 1% (w/v) ethanolic $Cu(NO_3)_2 \cdot 3H_2O$ such that 25 parts of the first solution are present per 1.5 parts of the ethanolic cupric nitrate. The resulting reagent is well suited for visualization of amino acid spots resulting from two-dimensional paper partition chromatography. Certain reagents other than ninhydrin can react with particular amino acids or certain R groups to generate chromophores. For example, the Sakaguchi reaction for the guanidino group of arginine yields an orange-red chromophore. The details of and reagents for these reactions have been compiled in *Data for Biochemical Research* by Dawson *et al.* [7], (see also Brenner *et al.* [8]).

20.2 FLUORIMETRY

Amino acid detection by ninhydrin has been most effectively employed in conjunction with thin-layer, ion-exchange, and especially paper partition chromatography as well as in direct amino acid colorimetric assay. One of the few deficiencies of the colorimetric method is its limited sensitivity; generally, 5 to 10 nanomoles of amino acid are required for detection. Except for certain non-specific analytical methods such as gas chromatography with flame ionization detection, amino acid determination relies upon reaction of the α-NH_2 group with a given reagent to form a desired derivative. The property of these derivatives of paramount importance is the sensitivity with which they can be detected. Recent interest in and the increased application of fluorescence detection primarily reflects the enhanced sensitivity possible in the detection of fluorescent derivatives formed by reaction with the α-NH_2 group.

Fluorescence refers to the secondary emission of light, generally for a period of 10^{-9} s but it can involve 10^{-4} s, by a compound after it has been 'excited' by the absorption of light of an appropriate wavelength. The absorption of photons of visible or ultraviolet radiation can raise the electrons of certain compounds from their stable, ground state to higher, excited states. The return of these excited electrons to their original ground state is accompanied by dissipation of some of this excess energy via the emission of radiant energy (usually of a longer wavelength than that of the excitation radiation). Thus, in fluorescence detection a given substance is usually excited by a particular monochromatic light (λ_{ex}) and the emitted light of longer wavelength (λ_{em}) is then analysed by a fluorimeter.

(a) Fluorescamine

The application of fluorimetry to amino acid analysis can be properly viewed by a consideration of fluorescamine (fluram™ *20.1*) or 4-phenylspiro[furan-2(3*H*)yl 1′-phthalan]-3,3′-dione.

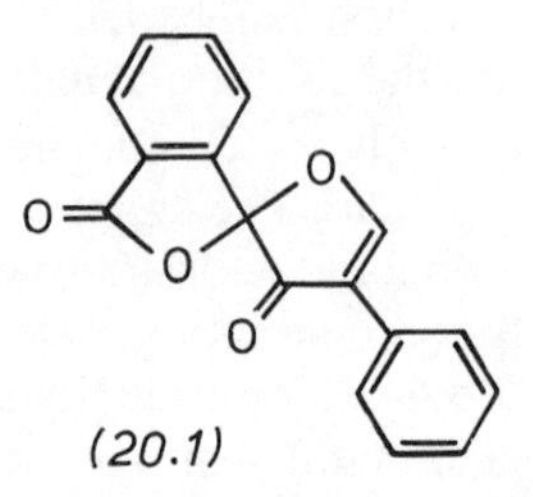

(20.1)

The discovery of this fluorescence reagent evolved from the finding of McCaman and Robins [9] that ninhydrin reacted with phenylalanine to yield a fluorescent compound. Additional details of the reaction were elucidated by Samejima *et al.* [10] who discovered that it was phenylacetaldehyde and not the amino acid *per se* that formed a fluorescent compound with ninhydrin. Samejima *et al.* [11] then developed their initial finding into a ninhydrin-based fluorimetric amino acid assay that has practical application both in automated amino acid analysis and in thin-layer chromatography. The critical elucidation of the structure of these fluorophores was achieved by Weigele *et al.* [12]; these efforts ultimately culminated in the first preparation of fluorescamine [13].

Fluorescamine lacks intrinsic fluorescence, but it can react with amino acids very rapidly ($t_{1/2} = 100$–500 ms) at room temperature and alkaline pH ($>$ pH 9) to yield highly fluorescent pyrrolinone derivatives* (*20.2*).

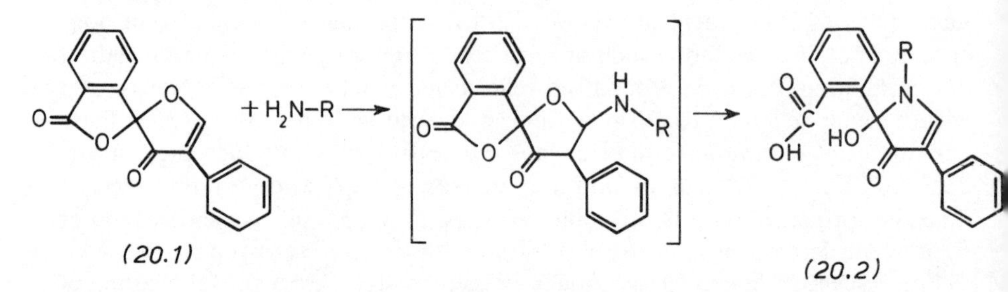

(20.1) (20.2)

The unreacted reagent is hydrolysed in seconds ($t_{1/2} = 5$ to 10 s) to yield water soluble and non-fluorescent compounds [14]. The kinetics of the reaction of fluorescamine with primary amines and the breakdown of unreacted reagent have been elucidated. There is a correlation between the reaction rate, which varies with the amine tested, and the formation of fluorescent compound [15].

Fluorescamine routinely detects as little as 50 to 100 picomoles of amino acid.

*Peptidyl materials have maximal fluorescence at a pH near 7 where amino acids exhibit minimal fluorescence. This factor provides a means for differential determination of these compounds [16].

Table 20.2 Peak areas and constants of a standard mixture of amino acids[a]

Amino acid	*Fluorescence (peak area)*	*Constant*[b]
Aspartic acid	169	0.60
Threonine	203	0.72
Serine	220	0.78
Glutamic acid	186	0.66
Glycine	271	0.96
Alanine	203	0.72
Half cystine	234	0.83
Valine	265	0.94
Methionine	259	0.92
Isoleucine	321	1.14
Leucine	338	1.20
Norleucine	282	1.00[b]
Tyrosine	296	1.05
Phenylalanine	302	1.07
Lysine	357	1.01
Histidine	191	0.54
Glycine amide	353	1.00[b]
Arginine	212	0.60

[a] See original for experimental details.
[b] The constant value expresses the fluorescence intensity of the fluorescamine derivative of the acidic and neutral amino acids relative to norleucine and the basic amino acids relative to glycine amide. (Reproduced with permission from the research of Stein *et al.* [14]).

It has been applied to amino acid detection in thin-layer chromatography [17, 18, 19], in high performance liquid chromatography [14], in ion-exchange chromatography [16, 20] and in direct fluorimetric analysis (Table 20.2). In general, fluorescamine is part of a post-column detection system but its functionality can be extended by forming the fluorescent derivative prior to compound separation. However, high performance liquid chromatography (HPLC) separation of the preformed fluorescamine–amino acid complex results in a dual peak for many of the compounds tested [21]. These two complexes, termed the acid alcohol (*20.3*) and the lactone (*20.4*) derivative, exhibit identical fluorescence characteristics.

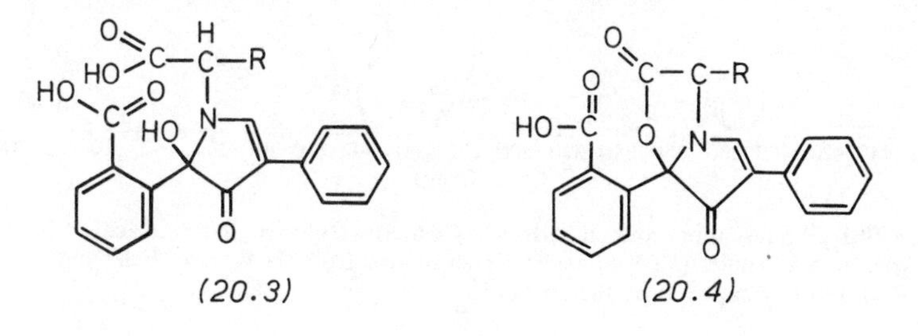

With certain amines, steric and other factors retard or prevent formation of the lactone derivative. Undesirable lactonization of the fluorescamine–amino acid complex can be suppressed by chemical modification of the amino acid prior to reaction with fluorescamine but this results in a two-step treatment procedure.

Amino acids separated by thin-layer chromatographic procedures can be detected by spraying with a solution of fluorescamine in acetone [22]. Triethylamine is effective in stabilizing the fluorescent compound; if either dimethylformamide or dimethylsulphoxide are used as the solvent, triethylamine is not required [23]. Sprayed fluorescamine detects 1 μg of various acids that have been separated by thin-layer chromatography.

Imai *et al.* [24] reacted the tested compounds with fluorescamine prior to thin-layer chromatographic separation and then sprayed twice with 10% triethanolamine in chloroform. Detection of 100 picomoles appeared routine and as little as 10 picomoles was sufficient for certain tested compounds. Fluorescamine was found superior to dansyl chloride in that each amino acid yielded a single

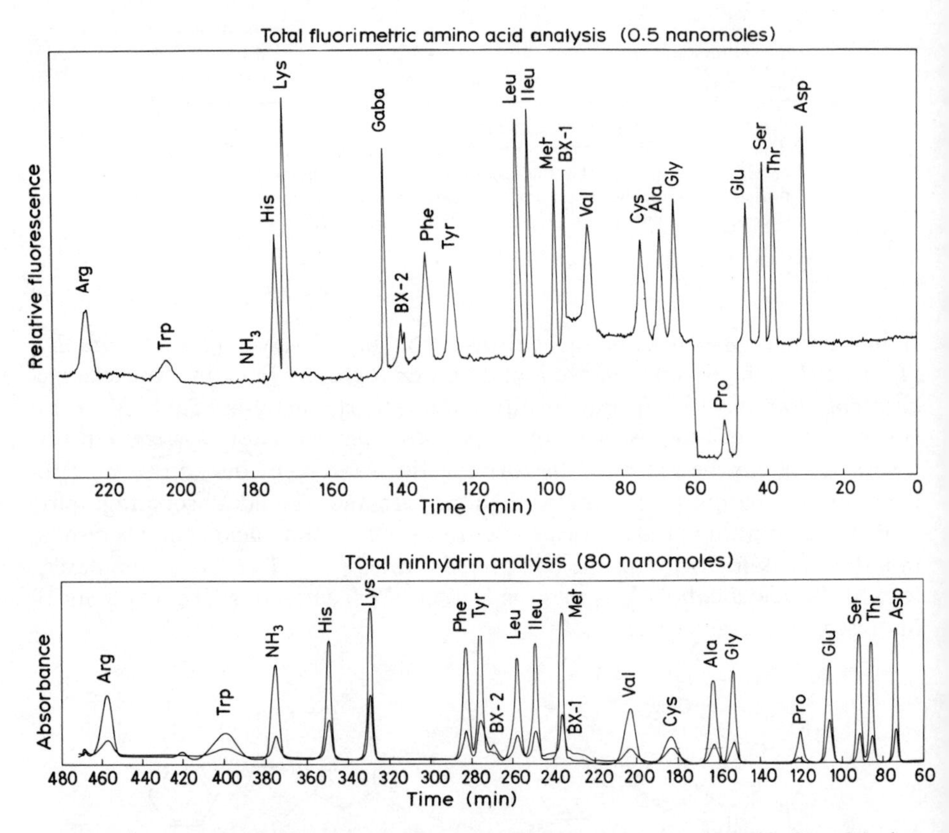

Figure 20.1 Fluorescamine and ninhydrin-based analysis of various amino acids. See original for experimental details. (Reproduced with permission from the work of Felix and Terkelsen [37]. Photo kindly supplied by the authors.)

spot with fluorescamine and the reaction was considerably faster. As one might expect, variations on the basic theme have proliferated. For example, Nakamura and Pisano [25] espoused a system in which the compounds are first applied to the thin-layer plate and then derivatized at the origin with fluorescamine prior to separation. A detection limit of 10 picomoles was indicated for this procedure.

An interesting application of fluorescamine was developed by Stein *et al.* [26] who used this reagent to detect amino acids obtained from acid hydrolysis of microgram quantities of proteins isolated by polyacrylamide gel electrophoresis. In a somewhat similar application, fluorescamine-reacted proteins subjected to electrophoresis were easily visualized *in situ* by being viewed under UV light [27]. Direct extraction of the protein from the gel matrix is unnecessary. Ninhydrin-based detection runs into difficulty with such an approach because of the ammonia generated from the gel during hydrolysis. The lack of appreciable fluorescence when fluorescamine reacts with ammonia circumvents this difficulty (see Fig. 20.1).

(b) *o*-Phthalaldehyde

Another reagent that produces fluorogenic amino acid derivatives that has gained widespread interest is *o*-phthalaldehyde. This reagent, in conjunction with 2-mercaptoethanol, is most often used in post-column detection of amino acids separated by conventional automated amino acid analysis. The potential value of *o*-phthalaldehyde came to be appreciated as a result of the effort of Roth [28] who studied the reaction of *o*-diacetylbenzene with alanine which produced a bright blue fluorogenic compound. *o*-Phthalaldehyde proved to be a superior reagent for amino acid detection. The impetus for testing the latter reagent developed from its successful use in a fluorimetric assay of glutathione [29].

The overall reaction is a ternary process that can be viewed as:

$$C_6H_4(CHO)_2 + R{-}NH_2 + R'{-}SH \longrightarrow \text{1-(S-R')-isoindole-N-R} + 2H_2O$$

where R′SH is an added thiol, usually 2-mercaptoethanol. The reaction produced for all amino acids studied has an λ_{ex} of 340 nm and an λ_{em} of 445 nm. Decomposition of the *o*-phthalaldehyde–amino acid complex as well as the reagent itself is significant after 24 h. Direct application of *o*-phthalaldehyde in the fluorimetric detection of amino acids and amines of a natural material was reported by Cronin and Hare [30], who sought to analyse certain constituents of the Murchison meteorite, a type II carbonaceous chondrite (Fig. 20.2 (a), (b)). These workers categorized the reaction of the compounds studied into three groups based on their fluorescence relative to glycine. The first contained substituents on the α-amino group, e.g. sarcosine or proline; these do not fluoresce and thus go undetected. The second, exemplified by cystine and 2-aminoiso-

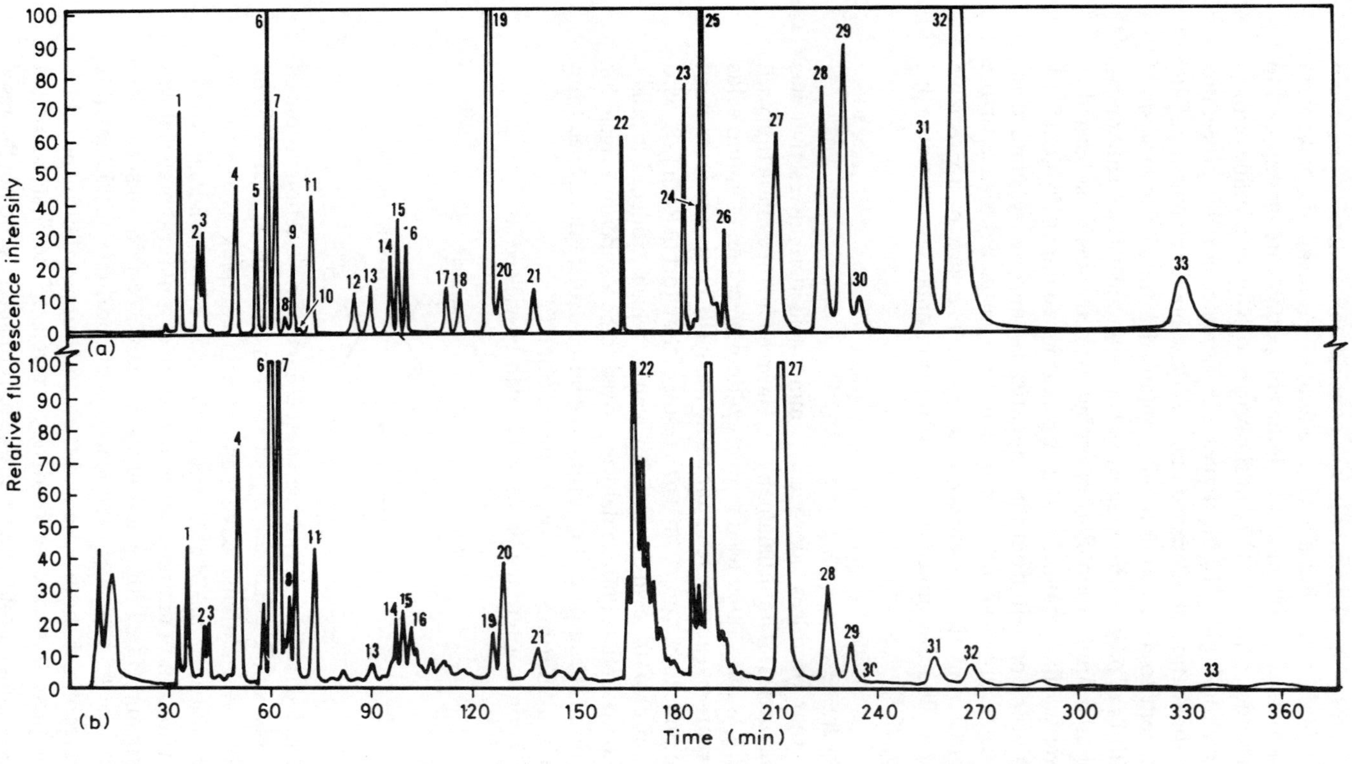

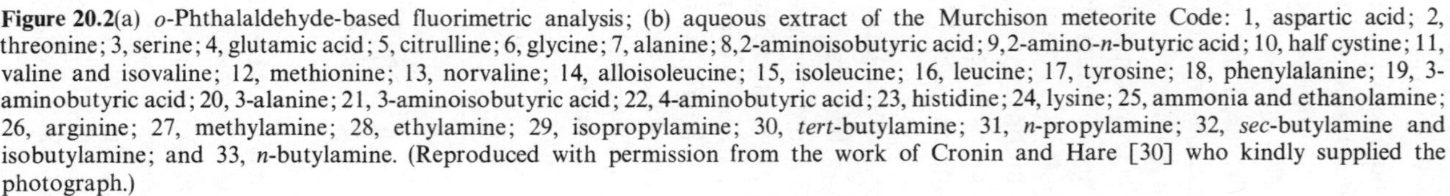

Figure 20.2(a) *o*-Phthalaldehyde-based fluorimetric analysis; (b) aqueous extract of the Murchison meteorite Code: 1, aspartic acid; 2, threonine; 3, serine; 4, glutamic acid; 5, citrulline; 6, glycine; 7, alanine; 8, 2-aminoisobutyric acid; 9, 2-amino-*n*-butyric acid; 10, half cystine; 11, valine and isovaline; 12, methionine; 13, norvaline; 14, alloisoleucine; 15, isoleucine; 16, leucine; 17, tyrosine; 18, phenylalanine; 19, 3-aminobutyric acid; 20, 3-alanine; 21, 3-aminoisobutyric acid; 22, 4-aminobutyric acid; 23, histidine; 24, lysine; 25, ammonia and ethanolamine; 26, arginine; 27, methylamine; 28, ethylamine; 29, isopropylamine; 30, *tert*-butylamine; 31, *n*-propylamine; 32, *sec*-butylamine and isobutylamine; and 33, *n*-butylamine. (Reproduced with permission from the work of Cronin and Hare [30] who kindly supplied the photograph.)

butyric acid, gave less than 10% of the fluorescent response of equimolar glycine. The final group produced a fluorescence yield of at least 65% of that obtained with glycine. β-Amino acids such as 3-aminobutyric acid generated compounds with a particularly high fluorescent response relative to glycine.

Lee *et al.* [31] also reported on a system for *o*-phthalaldehyde detection linked to automated amino acid analysis. As little as 0.2 μg of hydrolysed protein could be analysed with good precision. No attempt was made to evaluate the imino acid fraction of the protein. The fluorescence of the protein amino acids studied relative to that of glycine deviated from the values reported by Cronin and Hare (Table 20.3). This discrepancy may reflect the loss in the intrinsic fluorescence of the *o*-phthalaldehyde–amino acid complex caused by aging of the *o*-phthalaldehyde reagent [30].

Lindroth and Mopper [32] have reported on a reversed phase HPLC procedure associated with *o*-phthalaldehyde that separates and ably detects 24 amino acids and ammonia in 30 minutes. They report a 0.5% precision with a sample consisting of some 80 picomoles per amino acid and recoveries of known materials from human urine of greater than 98%. A detection limit of approximately 50 femtomoles (5×10^{-14} moles) was obtained. Indeed, a potential difficulty with such a procedure may reside in its marked sensitivity

Table 20.3 The relative molar fluorescence of amino acids[a]

	Relative fluorescence	
Amino acid	*Lee et al.*	*Cronin and Hare*
Cysteic acid	1.04	—
Carboxymethylcysteine	0.97	—
Aspartate	1.14	0.67
Threonine	1.10	0.75
Serine	1.10	0.98
Glutamate	1.06	0.72
Glycine	1.00	1.00
Alanine	0.96	0.96
Cystine	0.10	0.03
Valine	1.18	0.71
Methionine	1.02	0.68
Isoleucine	1.14	0.78
Leucine	1.15	0.78
Tyrosine	0.77	0.71
Phenylalanine	1.01	0.64
Histidine	1.03	1.12
Lysine	0.83	—
Arginine	0.92	0.88

[a] The peak areas determined after reaction with *o*-phthalaldehyde were calculated per nanomole of each amino acid and are given relative to those for glycine. (Reproduced with permission from the research of Lee *et al.* [31]).

which can make analysis of certain biological samples a virtual nightmare of complexity.

o-Phthalaldehyde detection has also been applied to the identification of amino acids separated by thin-layer chromatography [33]. The amino acid-laden plates are sprayed with 0.1% *o*-phthalaldehyde and 0.1% 2-mercaptoethanol in acetone and 5 minutes later with 1.0% triethylamine in acetone. In this way, 50 to 100 picomoles of many amino acids can be detected by viewing the developed plates with a 350 nm UV lamp. *o*-Phthalaldehyde and 2-mercaptoethanol exhibit limited reactivity toward such amino acids as lysine and cystine [28, 34]. However, Benson and Hare [35] found that augmenting the concentration of 2-mercaptoethanol ten-fold and adding Brij-35 permits detection of these amino acids.

A serious limitation in the use of both fluorescamine and *o*-phthalaldehyde is their inability to form fluorescent compounds with imino acids. With fluorescamine, proline can be detected if this imino acid is initially reacted with *N*-chlorosuccinimide and oxidatively decarboxylated to 4-aminobutyraldehyde [36]. Details of the reaction of *N*-chlorosuccinimide with proline [37] and hydroxyproline [38] and their subsequent analysis with fluorescamine have been published. The ability of fluorescamine to determine proline after treatment with *N*-chlorosuccinimide has made possible total fluorimetric automated amino acid analysis [37]. In this procedure, a solution of *N*-chlorosuccinimide is pumped into the column effluent of the amino acid analyser. Another pump introduces the boric acid buffer just before the final addition of fluorescamine in acetone. This analytical system thus requires four pumps and is somewhat complex. Nevertheless, the fluorescence of proline is linear to at least 10 nanomoles, detection of only 100 picomoles of proline is feasible, and the total analysis of protein amino acid is effectively achieved. The assay of hydroxyproline is linear to 5 nanomoles with a detection limit of around 250 picomoles [38].

The above modification is not applicable to *o*-phthalaldehyde detection of imino acids since this reagent is affected adversely by *N*-chlorosuccinimide [35, 39]. A recent description of a procedure in which the imino acids separated by automated amino acid analysis are reacted with hypochlorite has circumvented this deficiency [40]. In this method, the amino acid analyser column effluent is mixed with hypochlorite and heated to 60°C prior to its reaction with *o*-phthalaldehyde at ambient temperature (Fig. 20.3). Detection limits of 3 picomoles for proline and 5 picomoles for hydroxyproline were secured. A serious drawback to this procedure is the destruction of amino acids by hypochlorite. This situation necessitated a three-way valve that can be programmed to direct the hypochlorite solution either into the column effluent or back to the reagent reservoir. In this way, hypochlorite is introduced only with the elution of the imino acid. Use of hypochlorite also required a doubling of the re-equilibrium period to 60 min prior to the introduction of the subsequent sample for analysis.

Another difficulty arising from the use of *o*-phthalaldehyde is its low

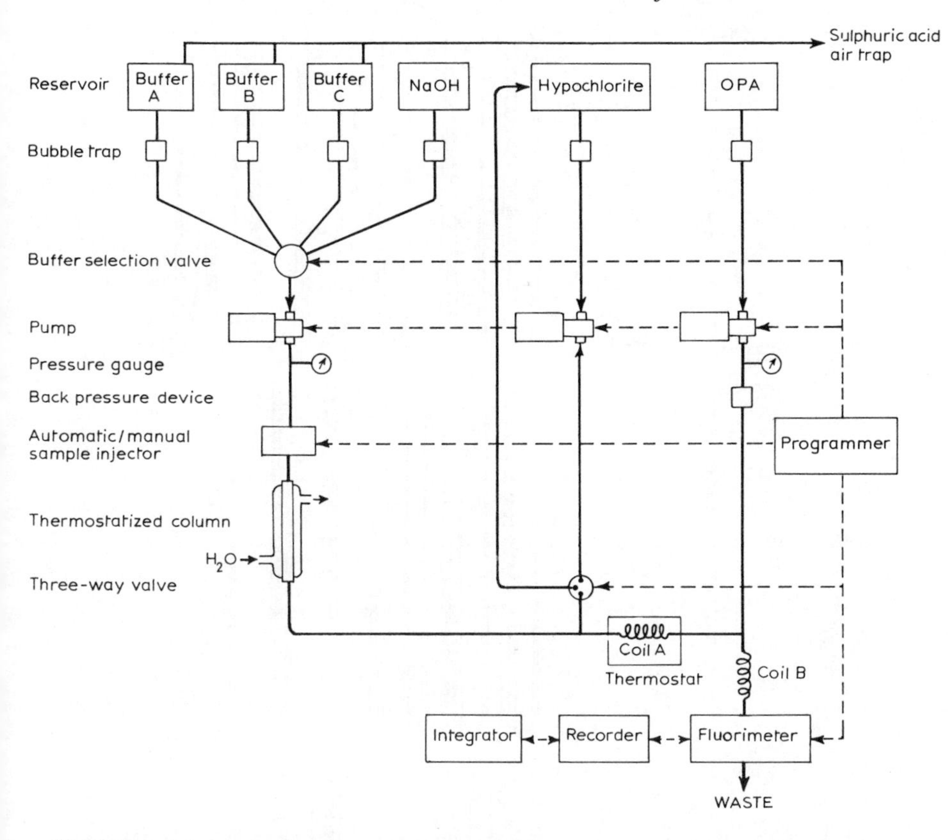

Figure 20.3 Schematic representation of an automated amino acid analyser linked to *o*-phthalaldehyde detection of the eluted compounds. The shaded area denotes modifications made specifically for imino acid detection. (Reproduced with permission from the work of Böhlen and Mellet [40].)

fluorescence yield with amino acids having a fully substituted carbon atom alpha to the carboxylic acid group such as with 2-aminoisobutyric acid. The fluorescence of such α-dialkylated amino acids can be augmented if the reaction time with *o*-phthalaldehyde and 2-mercaptoethanol is increased [41]. This interaction was optimized by passing the column effluent through a 2 m reaction coil maintained at 100°C. An observation that either methanethiol or ethanethiol reacted with primary amines and *o*-phthalaldehyde to generate a more stable fluorophore than with 2-mercaptoethanol [42] resulted in the application of these thiols. An elevated reaction temperature coupled to the use of these novel thiols substantially increased the fluorescence of fully substituted amino acids.

A direct comparison has been made of ninhydrin detection [35] with that of *o*-phthalaldehyde- and fluorescamine-mediated fluorimetric detection (Fig. 20.4). *o*-Phthalaldehyde results in nearly 1 order of magnitude greater fluorescence intensity. In addition, *o*-phthalaldehyde has the added advantage of being soluble

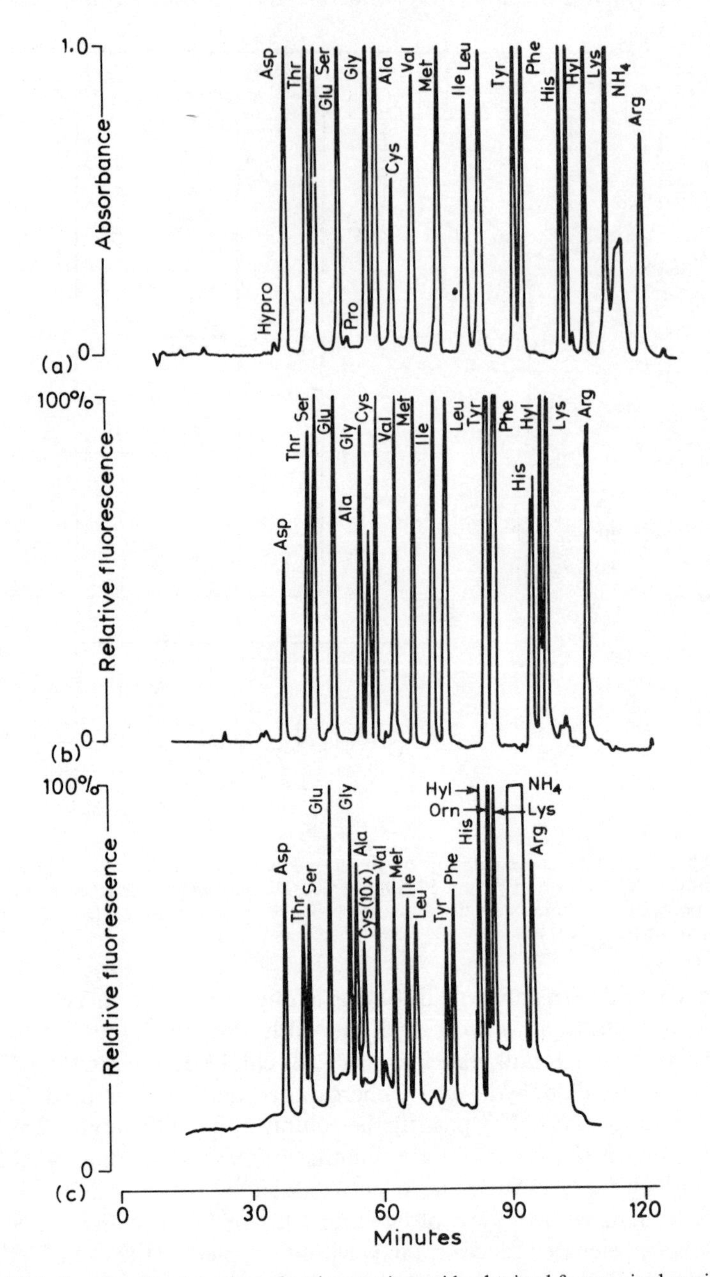

Figure 20.4 Comparative detection of various amino acids obtained from a single resin bed. One nanomole of each amino acid was detected with ninhydrin (a) or fluorescamine (b) or 100 picomoles of each amino acid were detected with *o*-phthalaldehyde (c). See original for additional experimental details. (Reproduced with permission from the research of Benson and Hare [5].)

in aqueous buffers which adds convenience to its use and aids in achieving more stable baseline values. Finally, *o*-phthalaldehyde is considerably more economical to use than fluorescamine. On the other hand, *o*-phthalaldehyde requires a longer reaction time to form fluorescent amino acid derivatives and the reagent must be made daily. Fluorescamine, however, is stable for at least three months in acetone or acetonitrile but for only two weeks in dioxane [43].

(c) Sulphonyl chlorides

In an excellent review of the chromatography of biogenic amines, Seiler [44] lists twelve compounds that have proved of value in the detection of primary and/or secondary amines. Many of these have found significant application in the detection of amino acids. Among the better known of this group are the fluorescent derivatives of 5-dimethylaminonaphthalene-1-sulphonyl chloride (commonly known as dansyl chloride). More recently, 5-di-*n*-butyl-aminonaphthalene-l-sulphonyl chloride, 6-methylanilinonaphthalene-2-sulphonyl chloride and 2-*p*-chlorosulphophenyl-3-phenylindenone are other sulphonyl chlorides that have been used for preparing amino acid fluorophores. These sulphonyl chlorides are very similar in their application and share a reactivity toward the amino groups to produce the corresponding sulphonamides, phenols to yield phenol esters, imidazoles to give *S*-sulphonyl derivatives, as well as certain alcohols [44]. Sulphonyl chlorides are noteworthy for yielding detectable fluorescence with nanogram amounts of amino acids [45–47].

Dansyl chloride

The predominant application of dansylated amino acid derivatives is in the identification of the *N*-terminal residue of proteins and polypeptides. Separation and quantitative determination of 10^{-9} to 10^{-10} moles of dansyl amino acid can be obtained by thin-layer chromatography [48, 49]. Once again, it is enhanced detectability that accounts for the preferential use of a fluorescent derivative since dansyl chloride derivatives are detected with greater sensitivity than are those of 1-fluoro-2,4-dinitrobenzene or phenylisothiocyanate.

In an extension of customary thin-layer chromatographic procedures, Airhart *et al.* [50] eluted radioactive, dansylated amino acids from polyamide sheets with redistilled chloroform (Fig. 20.5). Their methods permit determination of as little as 3×10^{-12} moles of amino acid. They claim the same accuracy but 10 000 times greater sensitivity than for automated amino acid analysis and ninhydrin detection [51, 52].

It stands to reason that HPLC can be a powerful tool for the separation of amino acids previously dansylated. Such an automated system, involving pre-column amino acid dansylation, has been developed with polyvinyl acetate gel columns [53]. Bayer *et al.* [54] present several cogent arguments for the desirability of HPLC with silica gel columns for amino acid separation as compared to ion-exchange columns, because much more rapid separations are

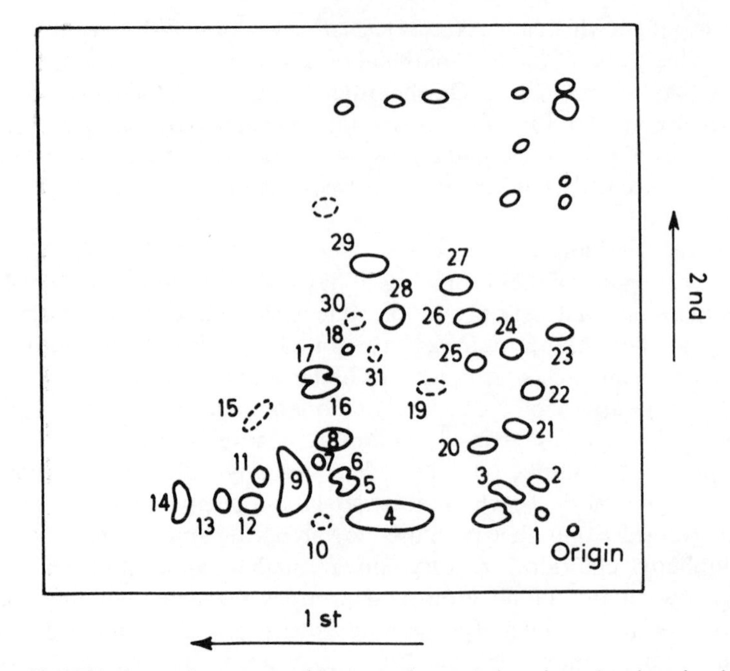

Figure 20.5 Thin-layer chromatographic separation and detection of a dansylated, acid-soluble extract of rat muscle. First dimension, formic acid: water (3:200, v/v); second dimension, benzene:acetic acid (90:10, v/v). See original for additional experimental details. Code: 1, cystine; 2, tryptophan; 3, taurine; 4, dansyl OH (blue); 5, aspartic acid; 6, glutamic acid; 7, threonine; 8, glycine; 9, dansyl-OH (blue); 10, phosphoserine; 11, serine; 12, glutamine; 13, asparagine; 14, arginine; 15, hydroxyproline; 16, dansyl-NH_2 (blue); 17, alanine; 18, 3-alanine; 19, methionine; 20, ornithine; 21, lysine; 22, carnosine; 23, tyrosine (orange); 24, histidine (yellow-orange); 25, phenylalanine; 26, leucine; 27, isoleucine (bluish); 28, valine; 29, proline (bluish); 30, cystathione; 31, 4-aminobutyric acid. All of the dansyl derivatives fluoresce yellow-green under ultraviolet radiation unless otherwise indicated. Creatine and phosphoethanolamine travelled with the second-dimension solvent front. (Reproduced with permission from the work of Airhart *et al.* [50].)

achieved with the former. Their detection system was predicated upon dansyl chloride rather than fluorescamine owing to the latter's lack of reactivity with imino acids. Once again, the dansylated amino acids were prepared prior to column chromatography. Their dual system can separate and detect 18 amino acids in 30 minutes using both gradient and isocratic elution. A reversed phase chromatographic system that separated protein amino acids in under 35 minutes was also described (Fig. 20.6). Chromatographic evidence was provided for the detection of 68 femtomoles of dansylated lysine.

5-Di-n-butylaminonaphthalene-1-sulphonyl chloride

A detailed consideration of the fluorescent labelling of amino acids with 5-di-*n*-butylaminonaphthalene-1-sulphonyl chloride (BANS-Cl) has been provided by Seiler *et al.* [55]. The BANS derivatives are less polar than the corresponding

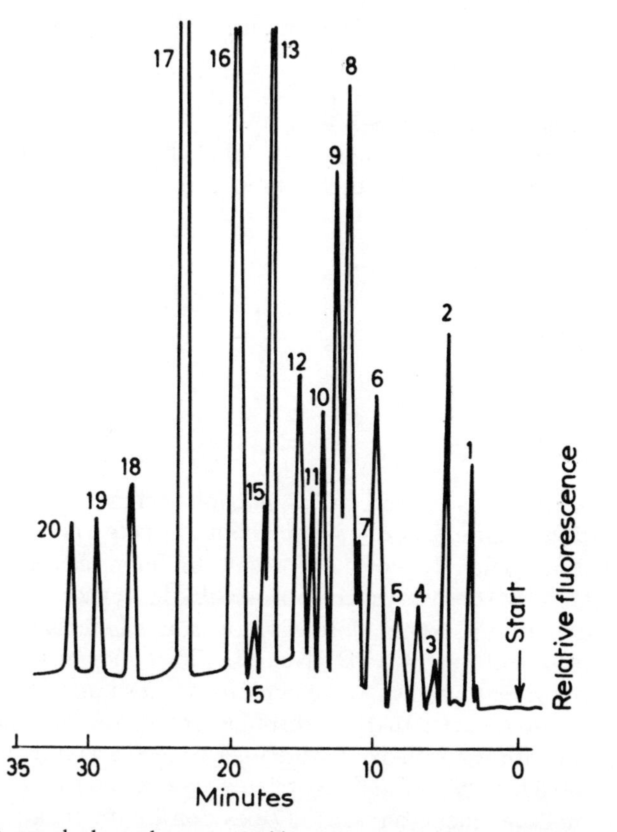

Figure 20.6 Reversed phase chromatographic separation of various dansyl amino acids and other reaction products. See original for experimental details. Code: 1, dimethylaminonaphthalene-5-sulphonic acid; 2, histidine; 3, serine; 4, glycine; 5, threonine; 6, alanine; 7, arginine; 8, proline; 9, valine; 10, methionine; 11, phenylalanine; 12, isoleucine; 13, leucine; 14, cystine; 15, aspartic acid; 16, tryptophan; 17, lysine (didansyl derivative); 18, tyrosine (didansyl derivative); 19 dansyl amide; 20, dansyl chloride. (Reproduced with permission from the research of Bayer *et al.* [54].)

dansyl derivatives and can be extracted more effectively by such solvents as ethyl acetate and chromatographically separated in more non-polar solvent systems. In ethyl acetate, the fluorescence intensity of BANS derivatives is about 15% greater than the corresponding dansyl derivatives. The basic reaction rates and the spectral excitation and emission characteristics for amino acid derivatives prepared from these two sulphonyl chlorides are much the same.

2-p-Chlorosulphophenyl-3-phenylindenone

Amino acids react with 2-*p*-chlorosulphophenyl-3-phenylindenone to form diphenylindenone sulphonyl derivatives (trivially called disyl derivatives). Treating disyl derivatives with sodium ethoxide *in situ* on silica gel chromatograms yields intense yellow-green fluorescent spots (UV light at 365 nm) that permit picomole-level detection. Little direct application of this compound has been made [56].

(d) 7-Chloro-4-nitrobenzo-2-oxa-1,3-diazole

7-Chloro-4-nitrobenzo-2-oxa-1,3-diazole (NBD chloride (*20.5*), not fluorescent) reacts with amino acids and other amines to produce highly fluorescent derivatives (e.g. *20.6*).

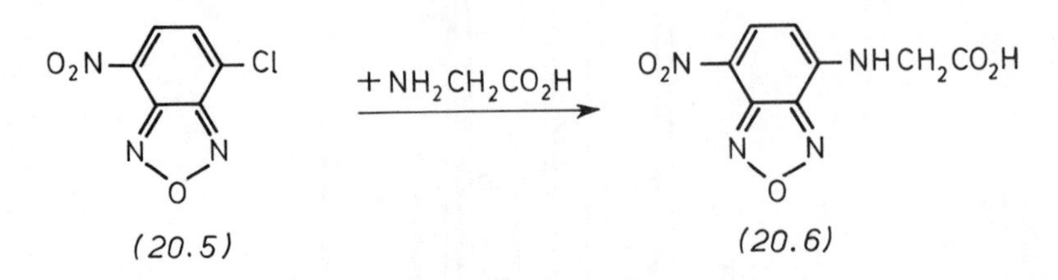

The fluorescence of NBD derivatives, maximized in solvents of low polarity, has an excitation maximum near 464 nm. This is distinctive because absorption of visible radiation generally does not result in fluorescence of amino acid derivatives. The NBD derivatives are more soluble in aqueous solution than their dansyl counterparts and NBD chloride is more stable to moisture than is dansyl chloride. Detection of 1 ng of NBD-glycine in direct fluorimetric assay and 1 μg on thin-layer chromatograms can be achieved (Ghosh and Whitehouse [57]). These workers also reported that (a) substituents on the amino group that restrict its conjugation with the benzoxadiazole nucleus (e.g. acetyl or phenyl) cause a drastic loss of fluorescence, and (b) related compounds that are not amine derivatives, but have other groups in the 7-position able to donate electrons to the nucleus, e.g. –OR or –SR, also fluoresce, but to a much smaller extent than the 7-amino derivatives. NBD derivatives of sulphydryl-containing amino acids are also easily formed [58]. The toxicity of NBD chloride is mentioned on p. 458.

(e) Pyridoxal derivatives

Under alkaline conditions, amino acids are known to react with pyridoxal (*20.7*) to form a Schiff base (*20.8*) that can be reduced chemically with tetrahydroborate to produce pyridoxyl-amino acid derivatives (*20.9*):

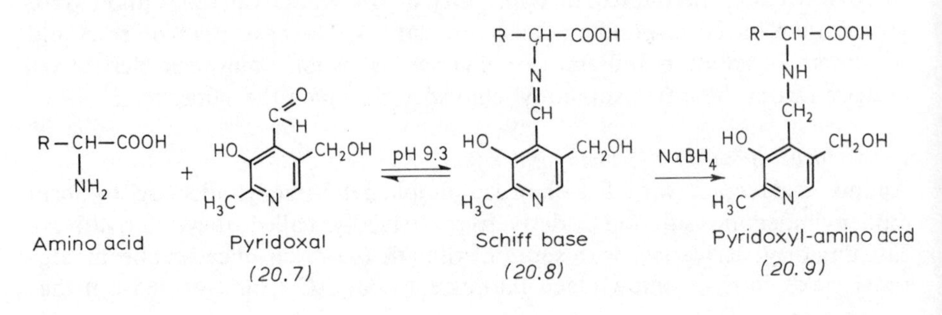

These derivatives can be detected at a concentration of 5×10^{-10} M and exhibit a λ_{ex} of 322 nm and a λ_{em} of 440 nm [59, 60].

In a modification of this procedure, pyridoxal phosphate and zinc acetate, dissolved in pyridinemethanol, interact with amino acids to form easily measured fluorophores. These fluorescent derivatives have a λ_{ex} of 390 ± 5 nm and a λ_{em} of 470 ± 5 nm. Neither tryptophan nor proline produce a fluorescent derivative by this procedure [61]. It was suggested that the zinc chelates of the Schiff bases (*20.10*) are the actual fluorescent species.

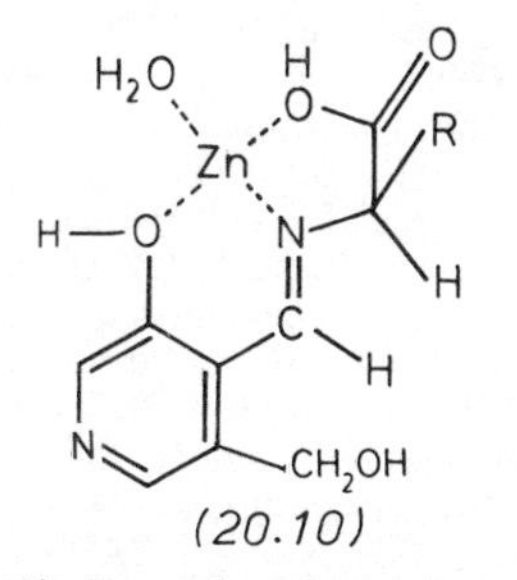

(20.10)

This reaction has been applied to the detection of amino acids eluted from automated amino acid analysers. The pyridoxal–zinc reagent is mixed with the column effluent and heated between 65° and 70°C for 10 minutes prior to its passage to the fluorimeter [62].

The above investigations provide ample evidence of the emerging and important role of fluorimetry in amino acid detection. They reveal a potential for fluorimetric amino acid analysis that fully rivals that of present day colorimetric procedures.

ACKNOWLEDGEMENTS

The author gratefully acknowledges the helpful suggestions of Dr James E. O'Reilly of the Chemistry Department of the University of Kentucky.

REFERENCES

1. McCaldin, D.J. (1960) *Chem. Rev.*, **60**, 39.
2. Lamothe, P.J. and McCormick, P.G. (1973) *Anal. Chem.*, **45**, 1906.
3. Greenstein, J.P. and Winitz, M. (1961) *Chemistry of the Amino Acids*, (Vol. 2) John Wiley and Sons, Inc., New York.
4. Rokushika, S., Murakami, F. and Hatano, H. (1977) *J. Chromatogr.*, **130**, 324.
5. Charlwood, B.V. and Bell, E.A. (1977) *J. Chromatogr.*, **135**, 377.
6. Moffat, E.D. and Lytle, R.I. (1959) *Anal. Chem.*, **31**, 926.
7. Dawson, R.M.C., Elliott, D.C., Elliott, W.H. and Jones, K.M. (1969) *Data for Biochemical Research*, Oxford University Press, Oxford.
8. Brenner, A., Niederwieser, A. and Pataki, G. (1962) in *Dunnschicht-Chromatographie* (ed. E. Stahl) Springer-Verlag, Heidelberg, pp. 403–452.
9. McCaman, M.W. and Robins, E. (1962) *J. Lab. Clin. Med.*, **59**, 885.
10. Samejima, K., Dairman, W. and Udenfriend, S. (1971) *Anal. Biochem.*, **42**, 222.
11. Samejima, K., Dairman, W., Stone, J. and Udenfriend, S. (1971) *Anal. Biochem.*, **42**, 237.

12. Weigele, M., Blount, J.F., Tengi, J.P., Czaikowski, R.C. and Leimgruber, W. (1972a) *J. Amer. Chem. Soc.*, **94**, 4052.
13. Weigele, M., DeBernardo, S.L., Tengi, J.P. and Leimgruber, W. (1972b) *J. Amer. Chem. Soc.*, **94**, 5927.
14. Stein, S., Böhlen, P., Dairman, W. and Udenfriend, S. (1973) *Arch. Biochem. Biophys.*, **155**, 203.
15. Stein, S., Böhlen, P. and Udenfriend, S. (1974a). *Arch. Biochem. Biophys.*, **163**, 400.
16. Udenfriend, S., Stein, S., Bohlen, P., Dairman, W., Leimgruber, W. and Weigele, M. (1972b). *Science*, **178**, 871.
17. Furlan, M. and Beck, E.A. (1974) *J. Chromatogr.*, **101**, 244.
18. Udenfriend, S., Stein, S., Böhlen, P. and Dairman, W. (1972a) in *Chemistry and Biology of Peptides* (ed. J. Meienhofer) Ann Arbor Science, Ann Arbor, MI.
19. Abe, F. and Semejima, K. (1975) *Anal. Biochem.*, **67**, 298.
20. Sterling, J. and Haney, W.G. (1974) *J. Pharm. Sci.*, **63**, 1448.
21. McHugh, W., Sandmann, R.A., Haney *et al.* (1976) *J. Chromatogr.*, **124**, 376.
22. Felix, A.M. and Jimenez, M.H. (1974) *J. Chromatogr.*, **89**, 361.
23. Touchstone, J.C., Sherman, J., Dobbins, M.F. and Hansen, G.R. (1976) *J. Chromatogr.*, **124**, 111.
24. Imai, K., Böhlen, P., Stein, S. and Udenfriend, S. (1974) *Arch. Biochem. Biophys.*, **161**, 161.
25. Nakamura, H. and Pisano, J.J. (1976) *J. Chromatogr.*, **121**, 79.
26. Stein, S., Chang, C.H., Böhlen, *et al.* (1974b) *Anal. Biochem.*, **60**, 272.
27. Vandekerckhove, J. and Von Montagu, M. (1974) *Eur. J. Biochem.*, **44**, 279.
28. Roth, M. (1971) *Anal. Chem.*, **43**, 880.
29. Cohn, V.H. and Lyle, J. (1966) *Anal. Biochem.*, **14**, 434.
30. Cronin, J.R. and Hare, P.E. (1977) *Anal. Biochem.*, **81**, 151.
31. Lee, H.-M., Forde, M.D., Lee, M.-C. and Bucher, D.J. (1979) *Anal. Biochem.*, **96**, 298.
32. Lindroth, P. and Mopper, K. (1979) *Anal. Chem.*, **51**, 1667.
33. Lindeberg, E.G.G. (1976) *J. Chromatogr.*, **117**, 439.
34. Roth, M. and Hampai, A. (1973) *J. Chromatogr.*, **83**, 353.
35. Benson, J.R. and Hare, P.E. (1975) *Proc. Natl. Acad. Sci. USA*, **72**, 619.
36. Weigele, M., DeBernardo, S. and Leimgruber, W. (1973) *Biochem. Biophys. Res. Comm.*, **50**, 352.
37. Felix, A.M. and Terkelsen, G. (1973a) *Arch. Biochem. Biophys.*, **157**, 177.
38. Felix, A.M. and Terkelsen, G. (1973b) *Anal. Biochem.*, **56**, 610.
39. Benson, J.R. (1973) in *Instrumentation in Amino Acid Sequence Analysis* (ed. R.N. Perham), Academic Press, New York, p. 1.
40. Böhlen, P. and Mellet, M. (1979) *Anal. Biochem.*, **94**, 313.
41. Cronin, J.R., Pizzarello, S. and Gandy, W.E. (1979) *Anal. Biochem.*, **93**, 174.
42. Simons, J.R., Pizzarello, S. and Gandy, W.E. (1979) *Anal. Biochem.*, **93**, 174.
43. Böhlen, P., Stein, S., and Udenfriend, S. (1974) *Arch. Biochem. Biophys.*, **163**, 390.
44. Seiler, N. (1977) *J. Chromatogr.*, **143**, 221.
45. Gray, W.R. and Smith, J.F. (1970) *Anal. Biochem*, **33**, 36.
46. Seiler, N. (1970) *Meth. Biochem. Anal.*, **18**, 359.
47. Deyl, Z. and Rosmus, J. (1972) *J. Chromatogr.*, **69**, 129.
48. Gros, C. and Labouesse, B. (1969) *Eur. J. Biochem.*, **7**, 463.
49. Schmer, G. (1967) *Z. Physiol. Chem.*, **348**, 199.
50. Airhart, J., Sibiga, S., Sanders, H. and Khairallah, E.A. (1973) *Anal. Biochem.*, **53**, 132.
51. Seiler, N. and Knödgen, B. (1977) *J. Chromatogr.*, **131**, 109.
52. Spivak, V.A., Fedoseev, V.A., Orlov, V.M. and Yarshavsky, J.A.M. (1971) *Anal. Biochem.*, **44**, 12.
53. Yambe, T., Takai, N. and Nakamura, H. (1925) *J. Chromatogr.*, **104**, 359.
54. Bayer, E, Grom, E., Kaltenegger, B. and Uhmann, R. (1976) *Anal. Chem.*, **48**, 1106.
55. Seiler, N., Schmidt-Glenewickel. T. and Schneider, H.H. (1973) *J. Chromatogr.*, **84**, 95.
56. Ivanov, CH. P. and Vladovska-Yuknovska, Y. (1972) *J. Chromatogr.*, **71**, 111.
57. Ghosh, P.B. and Whitehouse, M.W. (1968) *Biochem. J.*, **108**, 155.
58. Birkett, D.J., Price, N.C., Radda, G.K. and Salmon, A.G. (1970) *FEBS Letts.*, **6**, 346.
59. Lustenberger, N., Lange, H.W. and Hempel, K. (1972) *Angew. Int. Ed. Engl.*, **11**, 227.
60. Lange, H.W., Lustenberger, N. and Henifel, K. (1972) *Z. Anal. Chem.*, **261**, 337.
61. Maeda, M. and Tsuji, A. (1973) *Anal. Biochem.*, **52**, 555.
62. Maeda, M., Tsuji, A., Ganno, S., and Onishi, Y. (1973) *J. Chromatogr.*, **77**, 434.

CHAPTER TWENTY-ONE

Physical Properties of Amino Acid Solutions

T.H. LILLEY

21.1 INTRODUCTION

There is currently considerable and renewed interest in what may be termed the 'non-bonding' interactions which occur between atoms and molecules and much of this interest stems from an appreciation of the role such interactions play in biological systems [1–5]. A significant amount of the work in this area has been performed where the principal objective has been the investigation of relatively simple systems containing small molecules which incorporate some of the molecular features associated with biopolymers. It is not unfair to say that most of our, admittedly sketchy, knowledge of the nature and magnitudes of non-bonding interactions, has been derived from studies on such 'model' compounds. The amino acids have not surprisingly attracted considerable interest in this regard and the objective of this chapter is to present a summary of some aspects of the physicochemical behaviour of such molecules in aqueous systems. The approach adopted is largely based upon pictorial and diagrammatic representations rather than the algebraic approach frequently, and often necessarily, used by investigators with a physical chemistry background. One consequence of the decision to use this type of approach is that it is sometimes necessary to oversimplify situations, but this is not uncommon in chemistry. Many of the ideas used here are applicable to solutions containing solutes other than amino acids and occasionally we shall draw on information from other systems if the appropriate information is not available for amino acids.

Generally speaking, investigations into the properties of solutes in condensed media fall into two categories. There are those investigations which are directed towards explaining the properties of isolated molecules, or frequently the behaviour of a molecule surrounded by solvent, and there are those studies which are concerned with the interactions occurring between solute molecules, usually in a solvent. From a chemical viewpoint the latter types of investigation are the more important and the principal utility of the results obtained from studies in the first category is the insight they give towards interactive problems. In view of this broad subdivision of properties the two types of investigation will be surveyed separately.

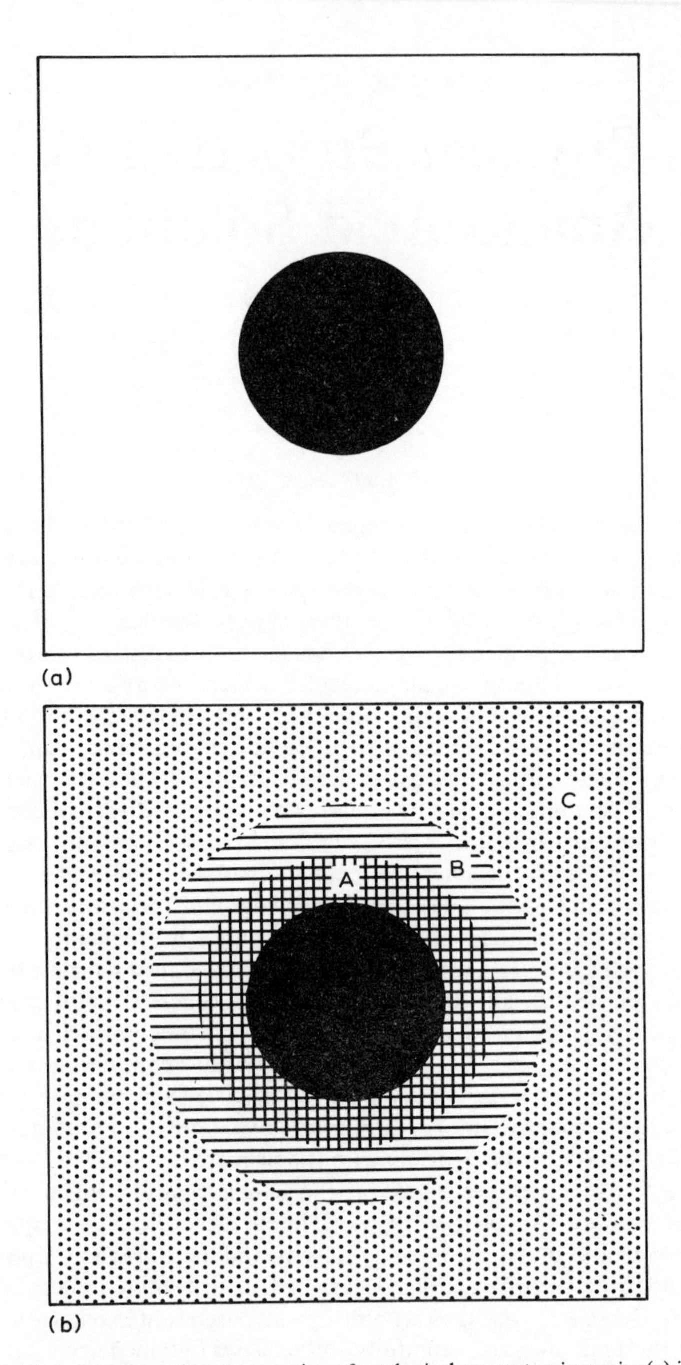

Figure 21.1 A diagrammatic representation of a spherical monoatomic species (a) in the gas phase and (b) in a solvent. The primary solvation region (region A – see text) is represented by cross-hatching, the bulk solvent (region C) by dots and the interfacial region (region B) by horizontal lines. Regions A and B constitute the co-sphere.

21.2 ISOLATED MOLECULES

For the purpose of illustration let us consider a very simple situation. Suppose we have an isolated monoatomic species in the gas phase and in some solvent. What happens at the molecular level is illustrated in Fig. 21.1. No intermolecular interactions of any type occur with the molecule in the gas phase (Fig. 21.1(a)) but this is manifestly not so when the molecule resides in a solvent (Fig. 21.1(b)). The solute interacts to a greater or lesser extent with the solvent in its immediate vicinity (region A) and there is a concomitant perturbation in the properties of the solute. Region A may be considered as the primary solvation region and there is strong evidence from neutron scattering studies [6, 7], that for some solutes at least, this region is well defined. One knows that at some distance from the solute, the solvent will have the properties of pure solvent (region C) and one can imagine that there will be a region of the solvent (B) where the solvent molecules are under the opposing influences of the primary solvation region and the bulk region. For spherical solutes regions A and B, where the solvent has different properties to that of bulk solvent, have together been called the co-sphere [9, 10].

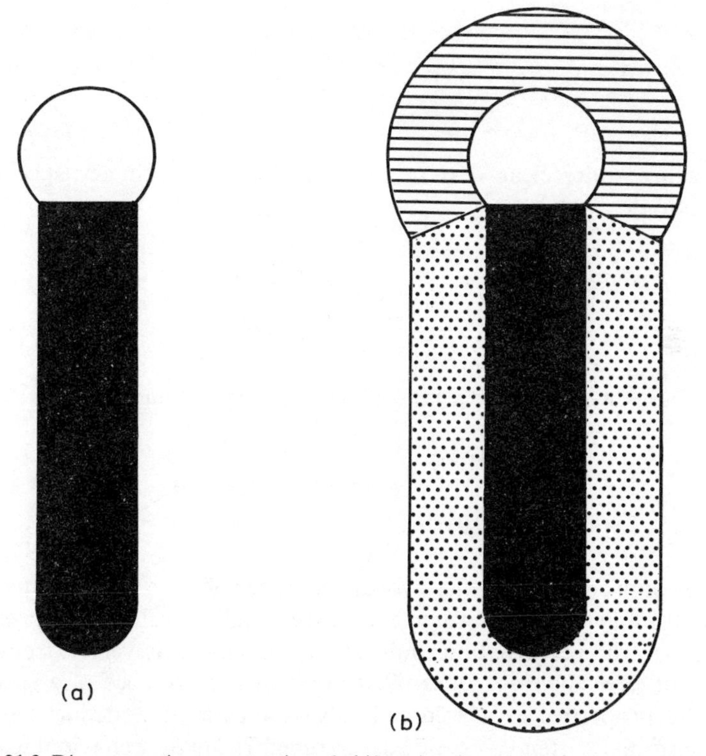

Figure 21.2 Diagrammatic representation of a bifunctional species such as a primary amine in (a) the gas phase and (b) in a solvent. The co-regions about the two functionalities are indicated by dots and horizontal lines.

If we now consider the more complex situation of a bifunctional solute dissolved in a solvent the ideas presented in Fig. 21.1 may be extended. Fig. 21.2 is a corresponding representation of a single molecule, such as an aliphatic primary amine, first in the gas phase and then in a solvent (we have neglected intramolecular effects such as, for example, possible conformational features arising from the backbone folding). For convenience we have chosen to represent the regions corresponding to A and B in Fig. 21.1 as a single region, which by analogy with the terminology for spherical solutes we call the co-region. It seems chemically sensible that the co-regions peripheral to the two functionalities should have different characters and extents and that there should be some mismatch in the vicinity of junctions of adjacent co-regions. It is clear that if the solute and the solvent are interacting, as indeed they must, then the chemistry of the solute in a solvent must be different, to some degree at least, to the chemistry of the solute in the truly isolated gas phase situation.

The perturbation in the chemical behaviour induced by solvent is illustrated rather nicely and dramatically by some recent investigations [11–13] on glycine in the gas phase and in water. It has been known for many years that glycine exists in water largely as the zwitterionic species.

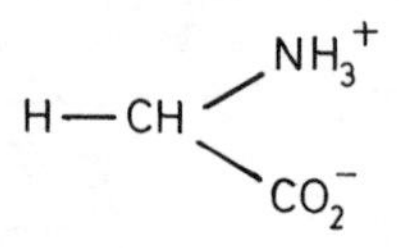

But it is now quite clear that in the gas phase the primary species is the nonzwitterionic form

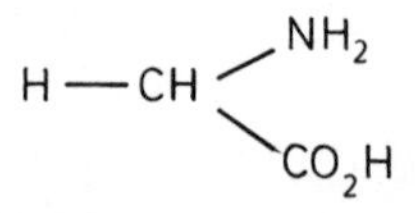

In both phases we can write a tautomerization equilibrium

The equilibrium constant for this process in the gas phase may be estimated to be roughly 10^{-21} whereas in water the corresponding value is approximately 4×10^{6} [13]. The perturbation induced by the solvent is very extreme in this instance but nevertheless it does illustrate how the presence of a solvent can modify the properties of a solute. It also indicates that deductions on the behaviour of 'real' systems (i.e. solutes in a solvent) from the behaviour (calculated or experimental) of solutes in the gas phase, need to be considered with some caution.

21.3 THE SOLVATION PROPERTIES OF AMINO ACIDS

There are several experimental procedures which can in principle be used to obtain information on the solvation of molecules. In many ways the most direct thermodynamically based procedure would be the free energy of transfer of the substance from the gas phase to a solvent. There are some problems regarding the choice of standard states [14–16] but in a relative sense such information would be most useful. It has not, however, been possible to perform such experiments on amino acids because of their extremely low volatility although an attempt has been made [17] via the properties of other compounds to correlate the distribution of molecules in the gas phase relative to water, with the free energies of transfer of amino acid side-chains from the interior to the exterior of globular proteins. It is possible to obtain the free energy, enthalpy and entropy of transfer of amino acids from the solid to water and some recent work [18] indicates how this may be done. However, it is at present not possible to draw any conclusions regarding the solvation of the amino acids from such investigations since each property contains specific and non-quantifiable contributions from the solid, crystalline state.

Fortunately there are other methods which can be used to obtain information on the solvation of molecules and two of these, partial molar volumes and partial molar heat capacities, have attracted considerable attention in recent years largely due to advances in experimental techniques [19, 20]. It is probably also true to state that these properties have been studied because, of the various thermodynamic quantities, both are amenable to imaginative discussion.

Since we are concerned with solvation properties, the quantities which we need to discuss are the partial molar properties at infinite dilution of the solute so that solute–solute interactions make no contribution. A pictorial representation of the partial molar volume at infinite dilution, $V^{\ominus}$, (often written as $\varphi_v{}^0$) may be obtained by considering the following thought experiment carried out under conditions of constant temperature and pressure. Suppose we take one molecule of a solute and immerse this in a large volume of solvent then the difference between the volume of the solution and the initial volume of the solvent may be termed the partial molecular volume of the solute and if this volume change is multiplied by Avogadro's constant, we obtain the partial molar volume of the solute. In practice, partial molar volumes are obtained indirectly from precise density measurements but the experimental methods need not concern us here. As was mentioned earlier there have been several investigations [21–27] in this area and a compilation of recent measurements is given in Table 21.1. In Fig. 21.3 we illustrate the variation of partial molar volume with backbone chain length for the α, ω-amino acids from glycine to 11-amino undecanoic acid and α-amino acids with unbranched side-chains from glycine to norleucine (see Table 21.1 for references). It is apparent from these data that for both sets of molecules an approximately linear correlation exists between the partial molar volume and the number of carbon atoms in the backbone. The slopes of the two lines are about

Table 21.1 Partial molar volumes at infinite dilution of amino acids and peptides in water at 25°C

Compound	$V^{\ominus}$ (*cm*^{3} *mol*$^{-1}$)
Glycine	43.25[a]
Alanine	60.46[b]
α-Aminobutyric acid	75.66[c]
Valine	90.78[d]
Norvaline	92.7[e]
Leucine	107.75[f]
Norleucine	107.72[g]
Phenylalanine	121.48[d]
Serine	60.62[h]
Tyrosine	123.6[i]
Asparagine	78.0[i]
Glutamine	93.9[i]
Aspartic acid	73.83[d]
Methionine	105.35[d]
Histidine	98.79[d]
Arginine	127.34[d]
Cysteine	73.44[d]
Glutamic acid	85.88[d]
Proline	82.83[d]
Tryptophan	143.91[d]
Diglycine	76.27[h]
Triglycine	111.81[h]
Tetraglycine	149.7[h]
Pentaglycine	187.1[h]
Dialanine	110.30[h]
Trialanine	163.80[h]
Tetraalanine	220.1[h]
Diserine	111.8[h]
Triserine	166.0[h]
α-Aminobutyric acid	75.66[c]
β-Aminobutyric acid	76.21[c]
γ-Aminobutyric acid	73.23[c]
5-Aminopentanoic acid	87.65[c]
6-Aminohexanoic acid	104.09[c]
7-Aminoheptanoic acid	120.0[c]
8-Aminooctanoic acid	136.03[c]
9-Aminononanoic acid	151.3[i]
10-Aminodecanoic acid	167.3[i]
11-Aminoundecanoic acid	183.0[f]

[a] Mean of values given in references 21, 22 and 24.
[b] Mean of values given in references 22 and 24.
[c] Reference 21.
[d] Reference 22.
[e] Cohn, E.J., Meekin, T.L., Edsall, J.T. and Blanchard, M.H. (1934) *J. Am. Chem. Soc.*, **56**, 784.
[f] Mean of values given in references 22 and 25.
[g] Reference 25.
[h] Reference 24.
[i] Reference 27.

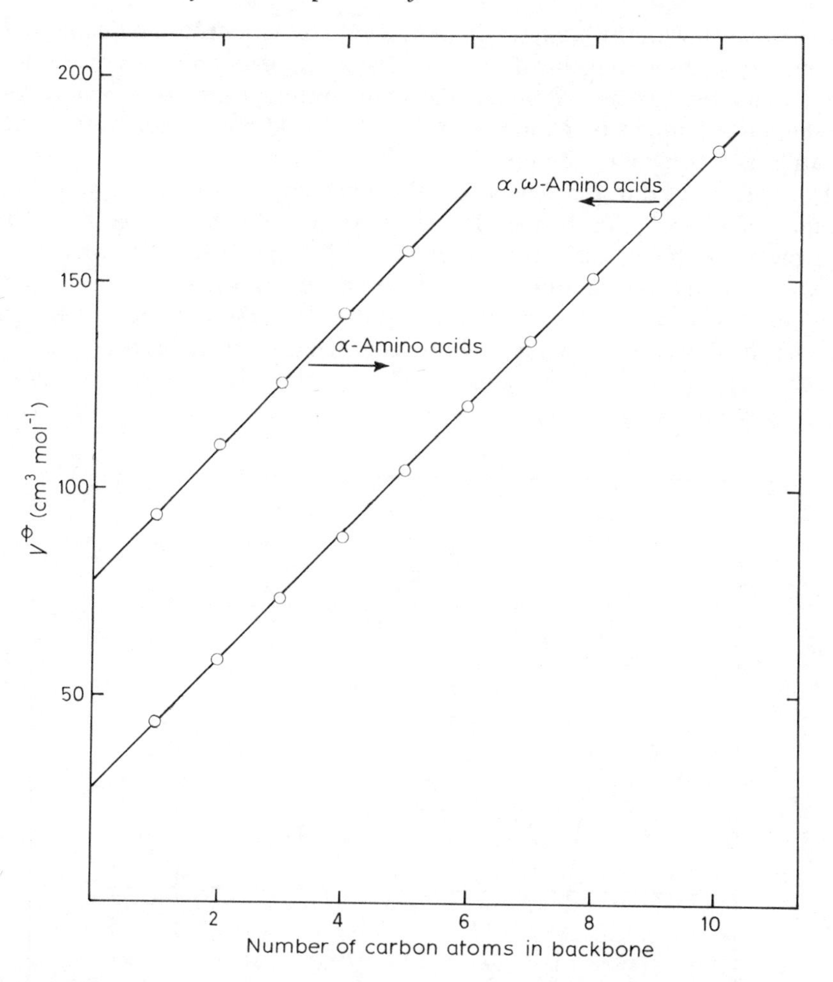

Figure 21.3 The variation of the partial molar volume at infinite dilution of α- and α,ω-amino acids in water at 25°C.

the same, corresponding to a volume contribution of approximately 16 cm^3 mol^{-1} of CH_2 groups and the intercepts indicate a volume contribution from the head group (NH_3^+, CO_2^-) of about 28 cm^3 mol^{-1}. It is instructive to compare these volumes with the 'intrinsic' volumes of the two groups, using Edwards' [28] compilation of atomic contributions. The methylene group volume calculated from this source is 10.2 cm^3 mol^{-1} and so there is an apparent increase in volume associated with the methylene group of some 6 cm^3 mol^{-1} (approximately 60%) which is indicative of packing problems [29–31] of the group in water and perhaps structure formation [22, 32] of the co-region of the group. In contrast the head group intrinsic volume is 24.8 cm^3 mol^{-1} and so given that packing problems will always be present, the relatively small volume increase of about

13% associated with the group when in water almost certainly is indicative of the volume contraction associated with the strong interaction between the ionic groups and the solvent. Although the above rationalizations probably have considerable validity it is apparent that if one looks at the data more carefully, the situation is rather more complex.

This is illustrated in Fig. 21.4 where the difference between the partial molar volumes of adjacent members in the above mentioned two series is shown. The data on the α-amino acids with unbranched side-chains are somewhat sparse because of solubility limitations of the higher members, whereas the data on the α, ω-amino acids are more extensive although for the higher members, at least, are of somewhat lower precision. It might be expected from what has been suggested earlier (p. 594) that once the region of influence of the head group over the co-region of methylene groups is exceeded then a constant increment in the partial

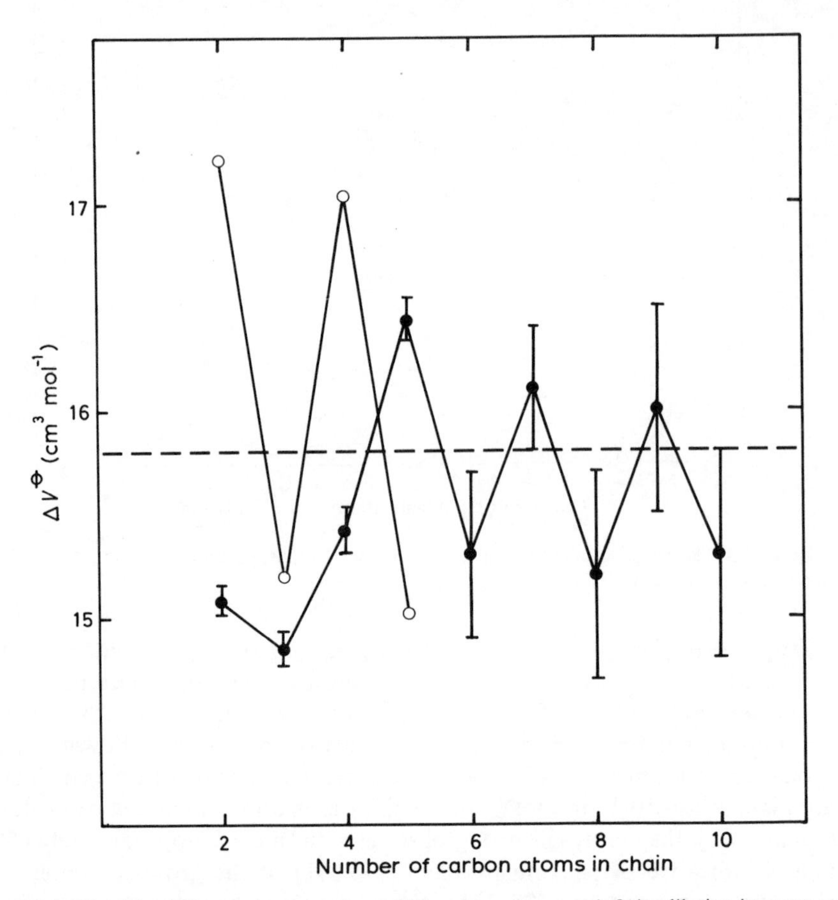

Figure 21.4 The difference between the partial molar volumes at infinite dilution in water at 25°C, of adjacent amino acids: α-amino acids (○); α,ω-amino acids (●) The dashed line corresponds to an increment of 15.8 $cm^3 mol^{-1}$ (see text).

molar volume of some 15.8 $cm^3\,mol^{-1}$ for each methylene group added would result [21]. It is possible for both series considered that this situation might be so and it has been suggested [21] that for the α-amino acids the co-region perturbation extends to C-4 whereas for the α, ω-amino acids the independence of methylene co-regions does not prevail until some six carbon atoms separate the amino and carboxylate groups. However it is not yet possible to be definite about this since there are indications of oscillatory behaviour in the volumes of the two series shown in Fig. 21.4 which is reminiscent of the variation of melting temperature with chain length in homologous series [33]. These oscillations, which are approximately covered by the estimated experimental error, might well indicate discontinuities in the packing of the solvent around the puckered hydrocarbon backbone.

The eventual additivity of volumes of group contributions distant from head groups is illustrated more clearly in a recent investigation [24] of the volumetric properties of some glycine, alanine and serine containing homopeptides. The results are summarized in Fig. 21.5 for the glycine and alanine series. The former series seems to indicate that, for peptides greater than triglycine, volumetric effects are additive and each glycyl residue contributes some 37.7 $cm^3\,mol^{-1}$ to the partial molar volume of the peptide. The more limited data set from the

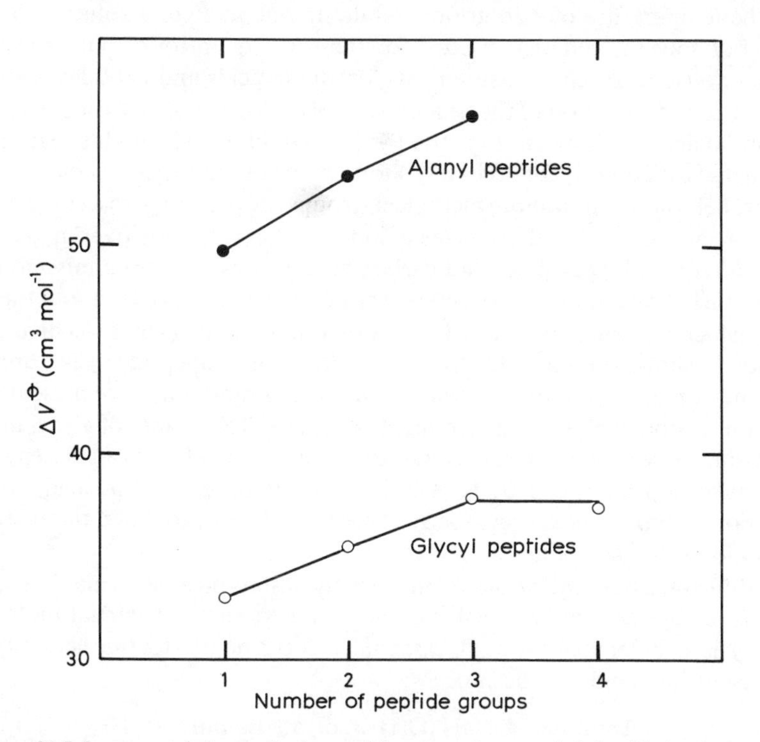

Figure 21.5 Incremental changes in the partial molar volumes at infinite dilution in water at 25°C of glycine homopeptides and alanine homopeptides.

alanine homopeptides indicate a corresponding volume change for an independent alanyl residue of about 56.5 $cm^3\ mol^{-1}$.

Notwithstanding the above details, there have been some semi-empirical attempts to describe the partial molar volumes of amino acids and peptides in terms of group contributions. One of the more successful of these from a predictive sense is that of Farrell and coworkers [26, 27, 34] in which the partial molar volumes of a species in a given homologous series is expressed by

$$V^{\circ} = aV_w + b \qquad (21.1)$$

where a and b are constants for the series and V_w is the intrinsic volume [28] of the species. Since there is a constant increment in the intrinsic volume as a series is ascended then the relationship in Equation (21.1) is essentially equivalent to that shown in Fig. 21.3.

There have also been recent attempts [21, 22] to construct group partial molar volumes. Although they give little molecular information or insight they nevertheless are useful if reasonably reliable estimates of the partial molar volumes of amino acids are required.

Consideration is now given to partial molar heat capacities at infinite dilution. These give a measure of the difference in the heat energy which must be added to raise the temperature of a solution containing one mole of a solute, in a large volume of solvent, and that needed for the same quantity of pure solvent. A summary of some recent measurements for amino acids and peptides is given in Table 21.2. The variation of the heat capacities with chain length for some α- and α,ω-amino acids is shown in Fig. 21.6 [35]. It would seem from this that the two series behave differently for relatively short chains but the α,ω-line and the α-line run parallel when four or more methylene groups are present with a slope of some 95 $J\ K^{-1}$ (mol of CH_2)$^{-1}$. This seems to indicate that co-region overlap occurs as far as C-3 for both the amino and carboxylate groups and is reminiscent of the volumetric behaviour of the two series. The increment observed is about the same as that which has been obtained [36] for other homologous series. A corresponding plot is shown in Fig. 21.7 for two series of homopeptides [24] and it is apparent from the glycine series that when three or more units are present linear behaviour is obtained with an increment of some 98 $J\ K^{-1}$ (mol of glycyl unit)$^{-1}$. The alanine series is less extensive but the general trend indicates an increment for the alanyl group of about 210 $J\ K^{-1}\ mol^{-1}$. As with the α,ω-amino acid series the indications are that the co-regions are only fully developed when there are four atoms along the backbone.

It is difficult, if not impossible, to quantify the results on a molecular basis, but a rational picture can be obtained if one remembers that the partial molar heat capacity is a difference term. It essentially corresponds to the heat capacity of the right hand side of the process:

$$\text{Amino acid} + nH_2O(1) \rightarrow \text{solvated amino acid}$$

There will be a contribution from skeletal motions such as vibrations, rotations

Table 21.2 Partial molar heat capacities at infinite dilution of amino acids and peptides in water at 25°C

Compound	$C_p^\ominus$ $(JK^{-1}\ mol^{-1})$
Glycine	39[a]
Alanine	141.4[a]
α-Aminobutyric acid	224[b]
Valine	307[b]
Leucine	382[b]
Phenylalanine	391[b]
Serine	117.4[a]
Proline	170[b]
Threonine	211[c]
Hydroxyproline	197[c]
Asparagine	158[c]
Glutamine	246[c]
Norvaline	335[c]
Norleucine	400[d]
β-Alanine	91[c]
β-Aminobutyric acid	182[c]
γ-Aminobutyric acid	154[c]
α-Aminoisobutyric acid	236[c]
β-Aminoisobutyric acid	174[c]
5-Aminopentanoic acid	219[c]
6-Aminohexanoic acid	278[d]
7-Aminoheptanoic acid	401[c]
8-Aminooctanoic acid	496[c]
Diglycine	105.0[a]
Triglycine	185.9[a]
Tetraglycine	283[a]
Pentaglycine	373[a]
Dialanine	330.8[a]
Trialanine	526[a]
Tetraalanine	733.4[a]
Diserine	259.9[a]
Triserine	398[a]
Glycyl-α-aminobutyric acid	331[e]
Glycylvaline	406[e]
Glycylleucine	497[e]
Alanylglycylglycine	329[c]

[a] Reference 24.
[b] Spink, C.H. and Wadsö, I. (1975) *J. Chem. Thermodynamics*, **7**, 561.
[c] Reference 35.
[d] Ahluwalia, J.C., Ostiguy, C., Perron, G. and Desnoyers, J.E. (1977) *Canad. J. Chem.*, **55**, 3364.
[e] Prasad, K.P. and Ahluwalia, J.C. (1980) *Biopolymers*, **19**, 265.

and translations of the amino acid and one can state that qualitatively vibrational contributions from this source would be very small since the energy needed to excite such motions would be too large. The rotational and translational contributions of the amino acid skeleton would be expected to be considerably smaller than those present in the gas phase (some 29 J K^{-1} mol^{-1}). Consequently most of the partial molar heat capacity will arise from solvation effects. If one

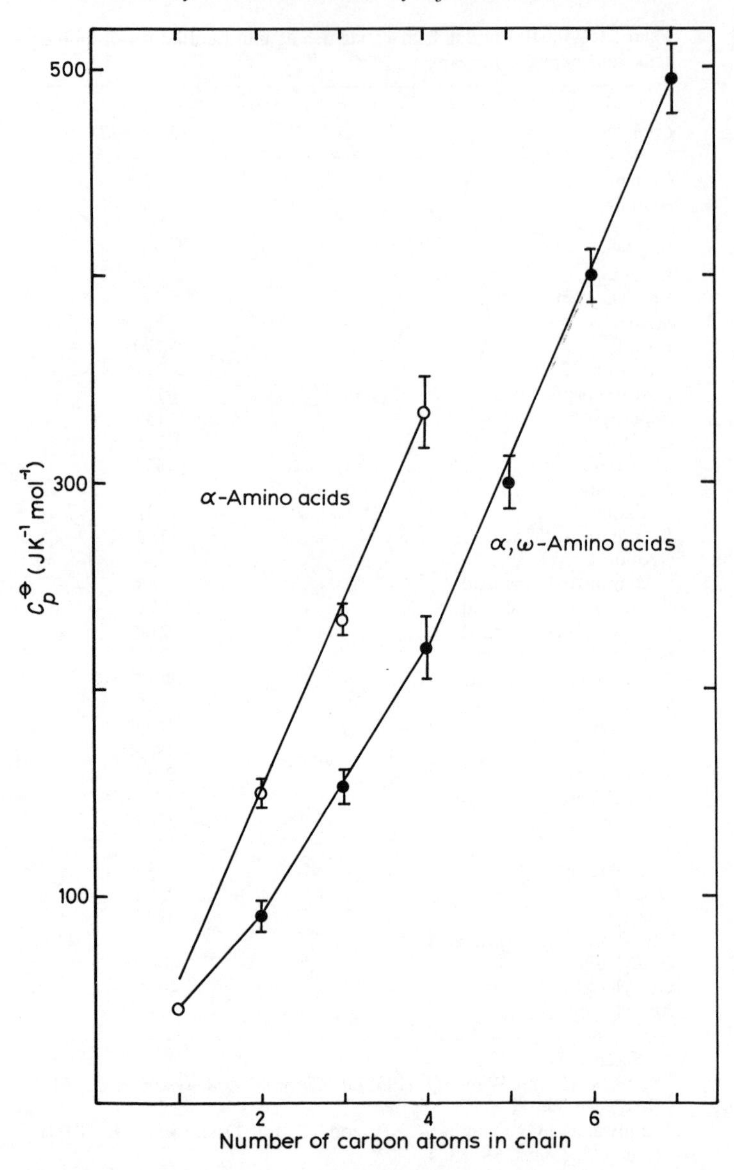

Figure 21.6 The variation with chain length of the partial molar heat capacity, at infinite dilution and at 30°C, of α- and α,ω-amino acids.

considers the contribution from an isolated head group, in a crude way we might expect that there would be a loss of motion from those water molecules solvating the ionic groups and this would in turn lead to a low (perhaps negative) partial molar heat capacity. It is apparent from Fig. 21.6 that extrapolation to the situation with no carbon atoms on the chain would give a negative value for the

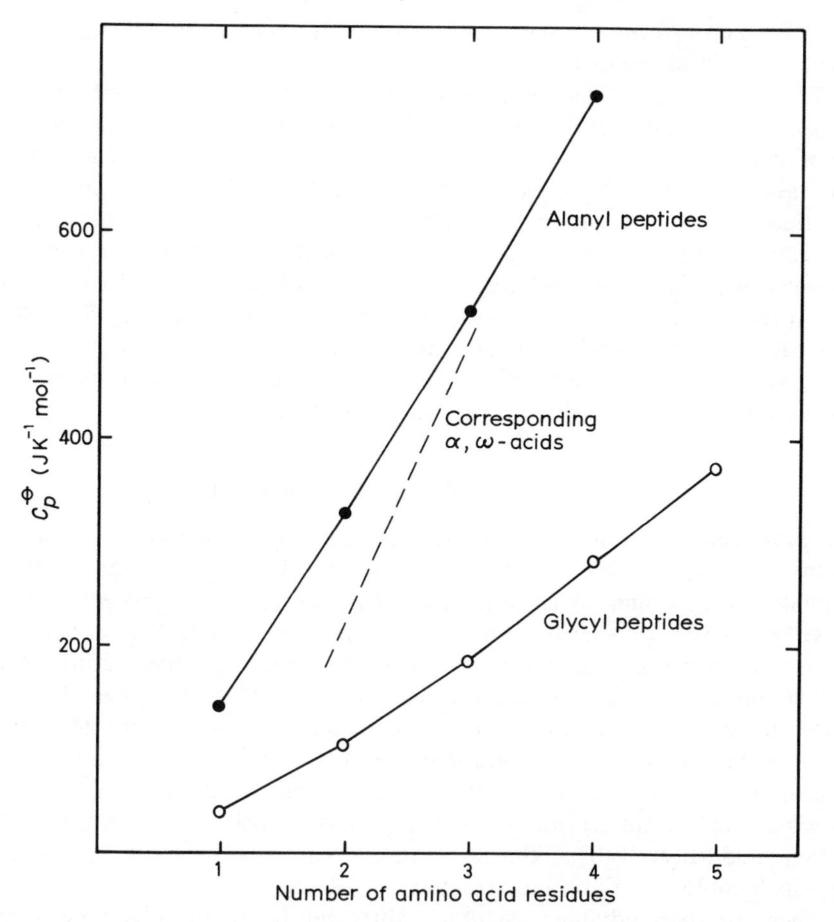

Figure 21.7 The variation with the number of amino acid residues of the partial molar heat capacity at infinite dilution and at 25°C of glycine and alanine homopeptides. The dashed line indicates the results for the corresponding α,ω-amino acids.

head group contribution. Suppose we now consider the other extreme of having an isolated methylene group. The experimental observation is that this would give a partial molar heat capacity of some $95\,\mathrm{J\,K^{-1}\,mol^{-1}}$. There is considerable evidence [37–39] which indicates that hydrophobic groups when present in water lead to the formation of clathrate-like structures. Such structures would lead to a diminution of the heat capacity due to loss of motion of the water, but presumably the structures formed would be held by relatively weak hydrogen bonds and these would have low energy vibrational manifolds associated with them. Since generally, excitable vibrations give a greater contribution to heat capacities than other sources, these would dominate in the experimental observations. The non-linearity observed, for example, with the α,ω-amino acids probably arises because of the incompatible nature of the co-regions about the ionic head groups and the methylene group, i.e. the head group solvation inhibits the formation of the methylene solvation region.

There is a marked difference between the dependence of the partial molar heat capacity on the number of atoms along the backbone for the glycine homopeptides and the α,ω-amino acids. This is illustrated in Fig. 21.7 where it seems that the values increase more rapidly with chain length for the amino acids than the peptides. It would seem that this arises because with the peptides clathrate formation is minimal and the formation of relatively strong hydrogen bonds between peptide groups and water molecules leads to decreased translational and rotational motion with a consequent reduction in heat capacity. The alanine peptides have considerably higher values for heat capacity than the corresponding glycine compounds and this would indicate that at least vestigial clathrate formation about the methyl groups occurs in the alanine series.

21.4 SOLUTE–SOLUTE INTERACTIONS

In many chemical systems the route into the investigation of the interaction between species has been via physical methods which can be analysed to give equilibrium constants. A large number of the experimental procedures which have been developed are based on spectroscopic probes of one sort or another but for many systems there are major difficulties in finding a suitable technique which can monitor molecular interactions. The problems which arise can stem from either the intrinsic insensitivity of the spectroscopic probe or from the fact that many of the interactions which are of interest are weak. Favourable conditions do exist in the amino acid area where association can be followed by changes in spectroscopic parameters and such a situation is illustrated by the NMR work of Kumar and Roeske [40] on the interactions occurring between amino acids, one of which contains an aromatic residue.

There are thermodynamic methods which can be used to give a measure of molecular interactions and one such procedure which has attracted considerable attention of late [41, 42] is that based upon the excess thermodynamic properties. Before outlining this it is important to recognize that the term 'solute–solute interaction' is a composite term and includes not only the direct interactions which occur between solute species but also embraces contributions arising from changes in the co-regions when solute molecules are in proximity (see Fig. 21.8).

The excess property of a solution (X^{ex}) is simply the difference between the value of the property of the solution (X) and the value the property would take if the solution were ideal (X^{id}).

$$X^{ex} = X - X^{id} \tag{21.2}$$

This is illustrated in molecular terms in Fig. 21.9 where the ideal solution is considered to be a dilute solution in which the co-regions of the molecules do not overlap. This is in contrast to the real solution where such overlap does occur to varying degrees. As might be expected in solutions which are not too con-

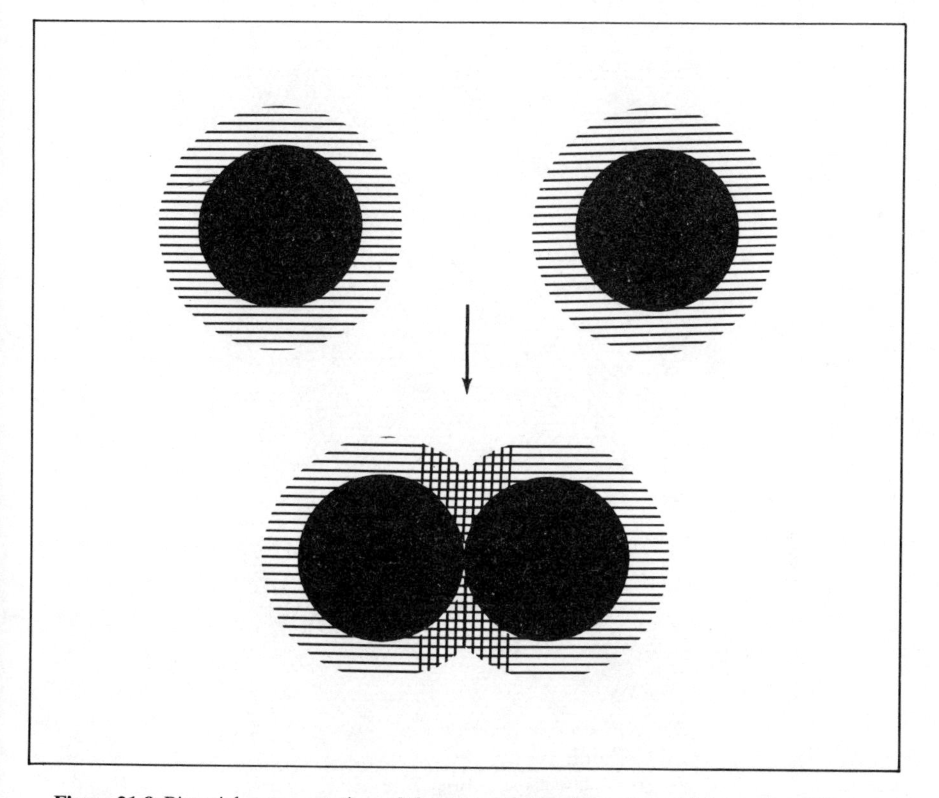

Figure 21.8 Pictorial representation of the interaction between two solute species. The co-sphere is indicated by horizontal lines and the cross-hatched region indicates the perturbation induced into co-spheres when the solutes are in proximity.

centrated, molecular events in which pairs of solute molecules are in proximity are more likely than situations with three or more solute species in close contact. It is clear from Fig. 21.9 that for any measurable property the difference observed between the two solutions will reflect molecular associations.

If one considers relatively dilute solutions of non-electrolytic species then one can invoke the above idea in a formal sense to describe the interactions which occur in real solutions by the use of a chemical association model [43, 44].

In this approach it is assumed that all deviations from ideality arise from association between solute molecules. Suppose we have a solution which consists of 1 kg of solvent and m_A^o moles of a solute A (i.e. an m_A^o molal solution), then we can postulate that the following associative equilibria occur

$$2A = A_2$$
$$3A = A_3 \text{ etc.}$$

For purposes of illustration and simplicity let us assume that only the first of these equilibria need be considered and that the equilibrium (association)

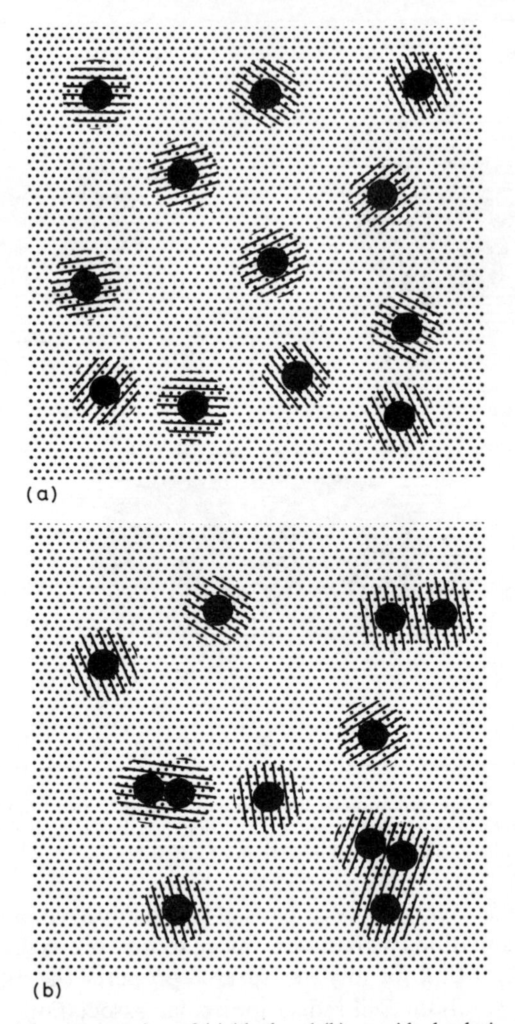

Figure 21.9 Pictorial representation of (a) ideal and (b) non-ideal solutions. The co-spheres are shown by cross-hatching and the unperturbed solvent by the dotted region. Notice the way the ideal solution corresponds to the (hypothetical) situation where the co-spheres are distinct and separate.

constant for this is given by

$$K_{A_2} = m_{A_2}/m_A^2 \tag{21.3}$$

where m_{A_2} and m_A are the molalities of dimer and monomer respectively. Under these circumstances the Gibbs free energy of the solution is simply the sum of the component terms, i.e.

$$G = m_A\mu_A + m_{A_2}\mu_{A_2} + n_w\mu_w \tag{21.4}$$

where μ_i is the chemical potential of i and n_w is the number of moles of solvent (w) in 1 kg. At equilibrium we have, from the condition given by the equilibrium constant

$$\mu_{A_2} = 2\mu_A \tag{21.5}$$

and so since $m_A^o = m_A + 2m_{A_2}$ (from mass balance considerations)

$$G = m_A\mu_A + n_w\mu_w \tag{21.6}$$

The chemical potential of the solvent depends upon the total molality of solute species and $n_w\mu_w$ may consequently be written as [45]

$$n_w\mu_w = n_w\mu_w^\ominus - \boldsymbol{R}T(m_A + m_{A_2}) \tag{21.7}$$

$\mu_w^\ominus$ being the molar free energy of pure water, whereas the chemical potential of the solute is expressed by

$$\mu_A = \mu_A^\ominus + \boldsymbol{R}T \ln m_A \tag{21.8}$$

Consequently the free energy of the solution is given by

$$G = m_A^o\mu_A^\ominus + m_A^o\boldsymbol{R}T \ln m_A + n_w\mu_w^\ominus - \boldsymbol{R}T(m_A + m_{A_2}) \tag{21.9}$$

If we consider the ideal solution of A in the solvent, this corresponds to the above situation with the association 'switched off' and the free energy under this circumstance would be

$$G^{id} = m_A\mu_A^o + m_A^o\,\boldsymbol{R}T \ln m_A^o + n_w\mu_w^\ominus - \boldsymbol{R}Tm_A^o \tag{21.10}$$

The free energy difference between the real and ideal solutions is then

$$G^{ex} = G - G^{id} = m_A^o\boldsymbol{R}T \ln(m_A/m_A^o) + \boldsymbol{R}Tm_{A_2} \tag{21.11}$$

This can be simplified if only weak association is considered i.e. the association constant is small, since then

$$K_{A_2} \simeq m_{A_2}/m_A^{o2} \tag{21.12}$$

and so

$$G^{ex} = m_A^o\boldsymbol{R}T \ln(1 - 2K_{A_2}m_A^o) + \boldsymbol{R}T\,K_{A_2}m_A^{o2} \tag{21.13}$$

In dilute solutions the logarithmic term may be expanded and only the leading term retained which leads to

$$G^{ex} = -\,\boldsymbol{R}T\,K_{A_2}m_A^{o2} \tag{21.14}$$

Consequently the excess free energy of the solution is proportional to the square of the molality of the solute.

The restriction of considering only pairwise association can be lifted and, by application of the above procedure, what is found is that further terms are obtained, in rising powers of molality, in the expression for the excess free energy and these terms represent interactions between triplets, quartets and higher groups of solute molecules.

It has become more customary [46] to represent the excess free energy by a polynomial equation of the form

$$G^{ex} = g_{AA}m_A^{o2} + g_{AAA}m_A^{o3} + \cdots \quad (21.15)$$

where the g terms may be called 'interaction coefficients' and so we can identify the first term in this expression with $-\boldsymbol{RT}K_{A_z}$. Various manipulations are possible [47] from the excess free energy polynomial expression. For example, the osmotic coefficient, (which is related to and measured via the solvent properties in solutions) is given by

$$\phi = 1 + (g_{AA}m_A^o + 2g_{AAA}m_A^{o2} + \cdots)/\boldsymbol{RT} \quad (21.16)$$

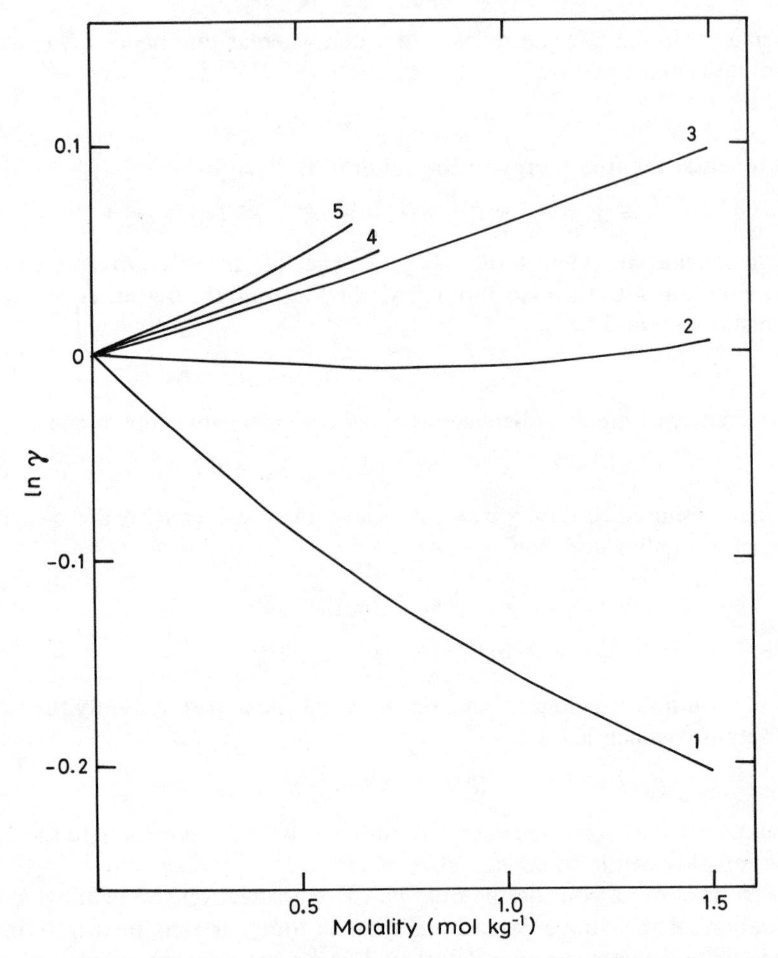

Figure 21.10 The variation of the logarithm of the molal activity coefficient of some amino acids in water at 25°C with molality: 1, glycine; 2, α-alanine; 3, α-aminobutyric acid; 4, norvaline; 5, valine.

and the logarithm of the activity coefficient (γ) of the solute is given by

$$\ln \gamma = (2g_{AA}m_A^o + 3g_{AAA}m_A^{o2} + \cdots)/RT \quad (21.17)$$

In experimental situations it is the coefficients in these equations which are obtained, either directly or indirectly, and it is from the concentration dependence of either the osmotic coefficient or the activity coefficient that the interaction coefficients are obtained.

Although this association model gives a simple molecular picture of the way excess properties are related to molecular interactions, it is only valid when there is positive association i.e. when the association constant is positive. This implies that G^{ex} should always be negative and, for example, activity coefficients should always be less than unity. There are many experimental situations where this is not so as is illustrated in Fig. 21.10. As the approach stands it cannot be applied to systems where, for example, the pairwise interaction coefficient is positive since this would then imply a negative association constant which in turn would indicate a negative concentration for the molality of the dimeric species! Consequently, although it is possible to rationalize the use of a polynomial expansion to express solution properties using a simple chemical model, such an approach can lead to conflict between the model and experimental results. However, as was mentioned, the polynomial approach does adequately represent experimental data and in Table 21.3 are presented activity coefficient data as a function of molality obtained from a polynomial least squares fitting method for all of the amino acids and peptides which have been investigated.

The fundamental problem of using chemical equilibrium models lies in the fact that they do not recognize that in molecular interactions, the nett effect can be repulsive. It has however been known for some time [47] that such a situation can prevail in the gas phase and procedures have been developed to cope with such systems. It is worthwhile at this point to give a simplified approach to a very simple interacting system to illustrate the approach which can be used to deal with repulsive interactions between solutes.

Suppose we have a system at constant temperature and volume containing a number N_A^o of spherical molecules in the gas phase. Let us also assume that the system is sufficiently dilute so that three-body collisions occur so infrequently as to be of negligible importance. We further take one molecule of A as the centre of our co-ordinate system and then look at the probability of finding other molecules at certain distances from the central molecule. If there are no attractions or repulsions between the molecules then the number of molecules in a spherical shell of thickness dr at a distance r from the central species (see Fig. 21.11) is simply equal to the number of molecules in the system (excluding the central one) multiplied by the ratio of the volume of the shell ($4\pi r^2 \mathrm{d}r$) to the volume of the system, i.e.

$$n_{AA}^{id}(r, r + \mathrm{d}r) = (N_A^o - 1)(4\pi r^2 \mathrm{d}r/V) \quad (21.18)$$

Table 21.3 Molal activity coefficients of amino acids and peptides in water at 25°C

Molality (mol kg^{-1})	Activity coefficient									
	Glycine[a]	*α-Alanine*[b]	*Valine*[a]	*α-Amino-butyric acid*[a]	*α-Amino-valeric acid*[c]	*β-Alanine*[d]	*β-Amino-butyric acid*[e]	*ε-Amino-caproic acid*[d]	*γ-Amino-butyric acid*[d]	*β-Amino-valeric acid*[e]
0.1	0.981	0.996	1.008	1.005	1.007	0.987	0.990	0.981	0.995	1.006
0.2	0.963	0.993	1.016	1.011	1.014	0.977	0.982	0.967	0.992	1.013
0.3	0.947	0.991	1.025	1.016	1.021	0.968	0.976	0.957	0.991	1.021
0.4	0.931	0.989	1.036	1.022	1.028	0.961	0.971	0.950	0.992	1.029
0.5	0.917	0.989	1.048	1.028	1.035	0.955	0.967	0.947	0.995	1.037
0.6	0.903	0.989	1.062	1.035	1.042	0.950	0.965	0.945	0.998	1.046
0.7	0.891	0.989		1.042	1.049	0.947	0.964	0.946	1.003	1.055
0.8	0.879	0.989		1.049		0.945	0.965	0.949	1.009	1.065
0.9	0.868	0.990		1.056		0.943	0.966	0.953	1.015	1.075
1.0	0.858	0.990		1.063		0.942	0.968	0.959	1.022	1.085
1.5	0.815	0.990		1.102		0.947	0.992	1.002	1.061	1.142
2.0	0.785	0.997		1.144		0.962	1.032	1.061	1.106	1.206
2.5	0.761					0.984	1.084	1.132	1.155	1.275
3.0	0.741					1.012	1.144	1.211	1.210	1.349
3.5						1.044	1.211	1.300	1.271	1.424
4.0						1.080	1.281	1.397	1.338	1.501
4.5						1.119	1.356	1.500	1.408	1.576
5.0						1.61	1.435	1.067	1.477	1.650
5.5						1.205	1.519		1.545	1.722
6.0						1.251	1.607		1.612	1.791
6.5						1.297	1.701		1.682	1.859
7.0						1.346	1.796		1.763	1.926
7.5						1.395	1.887		1.857	
8.0						1.444				

Table 21.3 (*Contd.*)

Molality (*mol kg*$^{-1}$)	*Activity coefficient*									
	γ-Amino-valeric acid[e]	*Serine*[f]	*Threonine*[f]	*Proline*[f]	*Hydroxy-proline*[f]	*Glycyl-glycine*[a]	*Alanyl-alanine*[g]	*Glycyl-alanine*[g]	*Alanyl-glycine*[g]	*Triglycine*[h]
0.1	0.994	0.983	0.991	1.006	0.998	0.951	0.980	0.964	0.959	0.913
0.2	0.990	0.963	0.983	1.013	0.995	0.911	0.970	0.934	0.927	0.848
0.3	0.987	0.941	0.976	1.019	0.994	0.877	0.967	0.911	0.903	0.797
0.4	0.986	0.919	0.970	1.027	0.992	0.848	0.968	0.892	0.884	
0.5	0.987	0.901	0.964	1.034	0.991	0.823	0.972	0.877	0.870	
0.6	0.988		0.959	1.042	0.990	0.801	0.978	0.865	0.860	
0.7	0.991		0.954	1.050	0.989	0.782	0.984	0.857	0.853	
0.8	0.995		0.950	1.059	0.989	0.764	0.991	0.850	0.848	
0.9	1.000		0.946	1.068	0.989	0.749	0.999	0.844	0.845	
1.0	1.006		0.943	1.077	0.989	0.735	1.011	0.840	0.842	
1.5	1.048		0.932	1.128	0.994	0.688				
2.0	1.106		0.924	1.187	1.006					
2.5	1.174			1.253	1.025					
3.0	1.251			1.325						
3.5	1.334			1.402						
4.0	1.425			1.484						
4.5	1.523			1.568						
5.0	1.630			1.651						
5.5	1.744			1.732						
6.0				1.808						
6.5										
7.0										
7.5										
8.0										

[a] Ellerton, H.D., Reinfelds, G., Mulcahy, D.E. and Dunlop, P.J. (1964) *J. Phys. Chem.*, **68**, 398.
[b] Robinson, R.A. (1952) *J. Biol. Chem.*, **199**, 71.
[c] Reanalysed data of Smith, P.K. and Smith, E.R.B. (1937) *J. Biol. Chem.*, **121**, 607.
[d] Reference 62.
[e] Reanalysed data of Smith, E.R.B. and Smith, P.K. (1940) *J. Biol. Chem.*, **132**, 47.
[f] Reanalysed data of Smith, P.K. and Smith, E.R.B. (1940) *J. Biol. Chem.*, **132**, 57.
[g] Reanalysed data of Smith, E.R.B. and Smith, P.K. (1940) *J. Biol. Chem.*, **135**, 273.
[h] Reference 61.

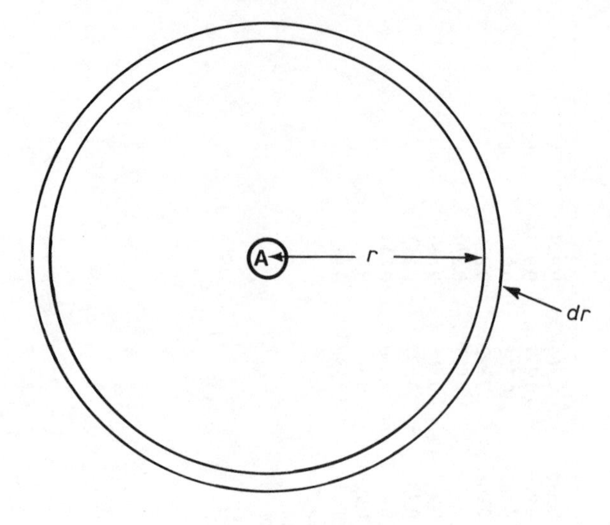

Figure 21.11

If one looks at how this number varies with distance it is found to take the form shown in Fig. 21.12(a). We now add the simplest possible complication namely that the molecules are hard spheres with a distance of closest approach between the molecules corresponding to their diameter (a). This means that two molecules can approach each other no closer than a and consequently the variation of the number of A molecules about the central species with distance is now as shown in Fig. 21.12(b). We now introduce some physical reality into the picture and assume that there exists both attractions and repulsions between the moleclues and these are expressed via the intermolecular potential function $u(r)$ which varies with distance in some way, perhaps in the form indicated by Fig. 21.13(c). (For the

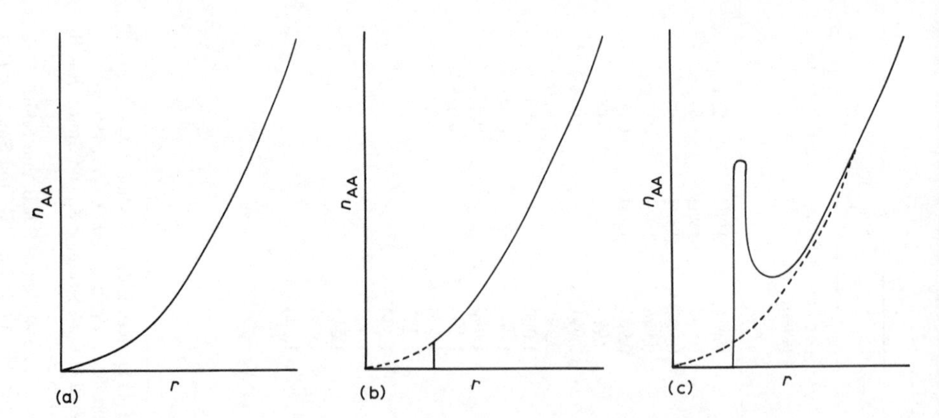

Figure 21.12 The variation in the number of molecules (n_{AA}) about a central A species (see Fig. 21.11) for various forms of the intermolecular potential: (a) the ideal situation; (b) the situation when the species are hard spheres; (c) a possible real intermolecular potential.

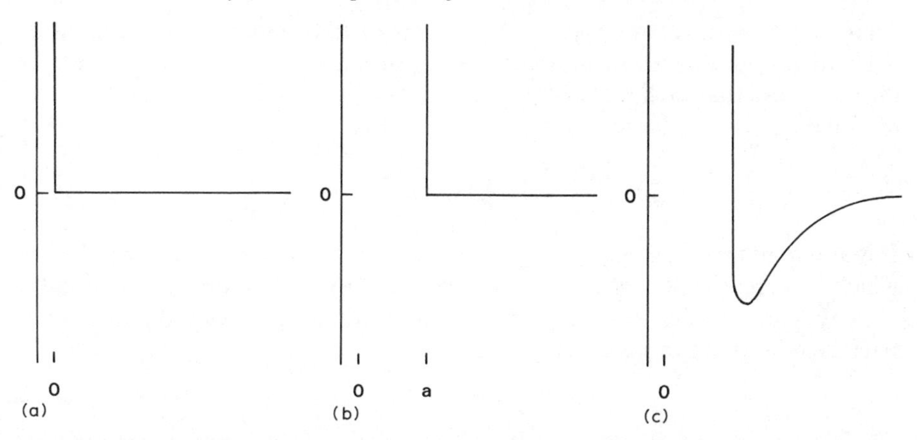

Figure 21.13 Intermolecular potentials corresponding to the distributions in Fig. 21.12. The ordinate is the intermolecular potential ($u(r)$) and the abscissa the intermolecular separation (r).

purpose of comparison we have included diagrammatic representations of the intermolecular potential functions which correspond to Fig. 21.12(a) and (b).) The qualitative consequences of this potential function are that at short distances there will be fewer molecules about the central molecule (because of closed-shell repulsions), at large distances the number of molecules in a shell will be the same as that indicated in Fig. 21.12(a) (since $u(r)$ approaches zero) and at some intermediate separations there will be more molecules in any shell because of the attractions between the molecules. The number–distance profile would then take the form shown in Fig. 21.12(c). The above can be quantified by recognizing that the scaling factor which makes the real number distribution different to the ideal number distribution is simply a Boltzmann term. In other words the expression corresponding to the equation for the real system is

$$n_{\mathrm{AA}}^{\mathrm{real}}(r,r+\mathrm{d}r) = n_{\mathrm{AA}}^{\mathrm{id}}(r,r+\mathrm{d}r) \times \exp\left\{\frac{-u(r)}{kT}\right\} \tag{21.19}$$

where k and T have their usual meanings. The Boltzmann factor essentially expresses the fact that if the molecules are attracting each other at the separation r then there is greater than random chance of finding them at this distance and conversely if the molecules are repelling each other there is a less than random chance of the two species being at this separation.

Suppose we now compare this with the ideal number and call the difference, the excess number (it can be positive or negative), we then get

$$n_{\mathrm{AA}}^{\mathrm{ex}}(r,r+\mathrm{d}r) = n_{\mathrm{AA}}^{\mathrm{id}}(r,r+\mathrm{d}r)\exp\left\{\left\{\frac{-u(r)}{kT}\right\}-1\right\} \tag{21.20}$$

This expression gives a measure of the *non-random* probability that two species will be in propinquity at a certain separation. The total excess number (which again may be positive or negative), N_{AA}^{ex}, is obtained by integrating this equation over all distances and all molecules, recognizing that a factor of $\frac{1}{2}$ must be introduced so that each molecule is counted only once and that the total number of molecules is very large ($N_A^0 \gg 1$). The result is

$$N_{AA}^{ex} = \frac{2\pi N_1^{02}}{V} \int_0^\infty \left\{ \exp\left\{\frac{-u(r)}{\boldsymbol{R}T}\right\} - 1 \right\} r^2 \mathrm{d}r. \qquad (21.21)$$

It is apparent from this that for a given system the total excess number of species which is present is proportional to the square of the total number of molecules. Now we can consider this excess number to be 'dimeric' species and from mass balance considerations we have

$$N_A^o = N_A + 2N_{AA} \qquad (21.22)$$

where N_A represents those molecules which are so far from other molecules that no interactions are occurring i.e. they are monomeric. This equation can be expressed in terms of the molarities of species by dividing by LV (L is Avogadro's constant) to give

$$c_A^o = c_A + 2c_{AA}$$

and Equation (21.21) can be similarly treated to give

$$c_{AA} = c_A^{o2} K_S \qquad (21.23)$$

where K_S (which Guggenheim [48] has called the *sociation* constant) is given by

$$K_S = 2\pi L \int_0^\infty \left\{ \exp\left[\frac{-u(r)}{kT}\right] - 1 \right\} r^2 \mathrm{d}r \qquad (21.24)$$

If we now consider how this is related to a measurable property namely the pressure exerted by the gas we have

$$\begin{aligned} P &= \boldsymbol{R}T(c_A + c_{AA}) \\ &= \boldsymbol{R}T(c_A^o - c_{AA}) \end{aligned} \qquad (21.25)$$

or using Equation (21.23)

$$P/\boldsymbol{R}T = c_A^o - K_S c_A^{o2} \qquad (21.26)$$

The first term of this expression represents the ideal gas law and the second term expresses the derivations from this law arising from molecular interactions. It should be noted that the sociation constant can take either positive or negative values, the sign depending upon the difference between the two terms in Equation (21.24). A positive value results if the area above the random distribution line in Fig. 21.12(c) is greater than the area below the line – this corresponds to a nett molecular attraction. A negative value of the sociation constant occurs if the repulsive contribution exceeds the attractive contribution.

The importance of the above discussion arises from the fact that for complex polyfunctional molecules in solution, a precisely analogous expression is obtained if the pressure of the gas is replaced by the osmotic pressure (π) of the solution [49]

i.e.
$$\pi/RT = c_A^o - K_S c_A^{o2} \tag{21.27}$$

and consequently this gives a means of obtaining a link between the interaction between solute molecules (which are contained in the sociation constant) and thermodynamic measurements. The osmotic pressure is not measured too frequently these days but it can be linked via conventional thermodynamic manipulations to, for example, osmotic coefficients and activity coefficients, both of which are readily accessible and measurable properties. It is worth mentioning that when solutions are considered, the intermolecular potential is replaced by the potential of mean force which is effectively a measure of how two solute species interact with each other *in the solvent.*

The molar osmotic coefficient [50–52] (ϕ') is defined by

$$\phi' = \pi/\pi^{id} \tag{21.28}$$

and so using Equation (21.27) we get

$$\phi' = 1 - K_S c_A^o \tag{21.29}$$

which should be compared with Equation (21.16), which expresses the molality dependence of the molal osmotic coefficient. It is this latter property which is usually measured but it can be shown that the sociation constant (which gives the direct link to intermolecular forces) can be linked to the molal pairwise

Table 21.4 Pairwise interaction coefficients, sociation constants and repulsive and attractive contributions to the latter for amino acids and peptides in water at 25°C

Compound	g_{AA} (*J kg mol*$^{-2}$)	K_S (*cm*^{3} *mol*$^{-1}$)	K_S^{rep} (*cm*^{3} *mol*$^{-1}$)	K_S^{att} (*cm*^{3} *mol*$^{-1}$)
Glycine	−238	52	−140	192
α-Alanine	−61	−35	−181	146
α-Aminobutyric acid	62	−100	−222	122
Norvaline	85	−126	−262	136
Valine	95	−128	−262	134
β-Alanine	−136	−2	−181	179
β-Aminobutyric acid	−85	−41	−222	181
β-Aminovaleric acid	77	−123	−262	139
γ-Aminobutyric acid	−137	−17	−222	205
γ-Aminovaleric acid	−88	−53	−262	209
ε-Aminocaproic acid	−249	−2	−303	301
Serine	−168	8	−200	208
Threonine	−115	−28	−240	212
Proline	−74	−52	−243	191
Hydroxyproline	−32	−70	−262	192
Glycylglycine	−665	193	−308	501
Alanylalanine	−331	24	−390	414
Glycylalanine	−498	108	−349	457
Alanylglycine	−566	135	−349	484
Glycylglycylglycine	−1260	398	−476	874

interaction coefficient (g_{AA} in Equation (21.16)) by the transformation

$$K_S = \{\tfrac{1}{2}\alpha \boldsymbol{RT} - V^{\ominus} - (g_2 V_w^{\ominus})/\boldsymbol{RT}\} \qquad (21.30)$$

in which α is the isothermal compressibility of the pure solvent and $V_w^{\ominus}$ is the volume occupied by 1 kg of pure solvent.

In Table 21.4 we list the pairwise interaction coefficients for those amino acids

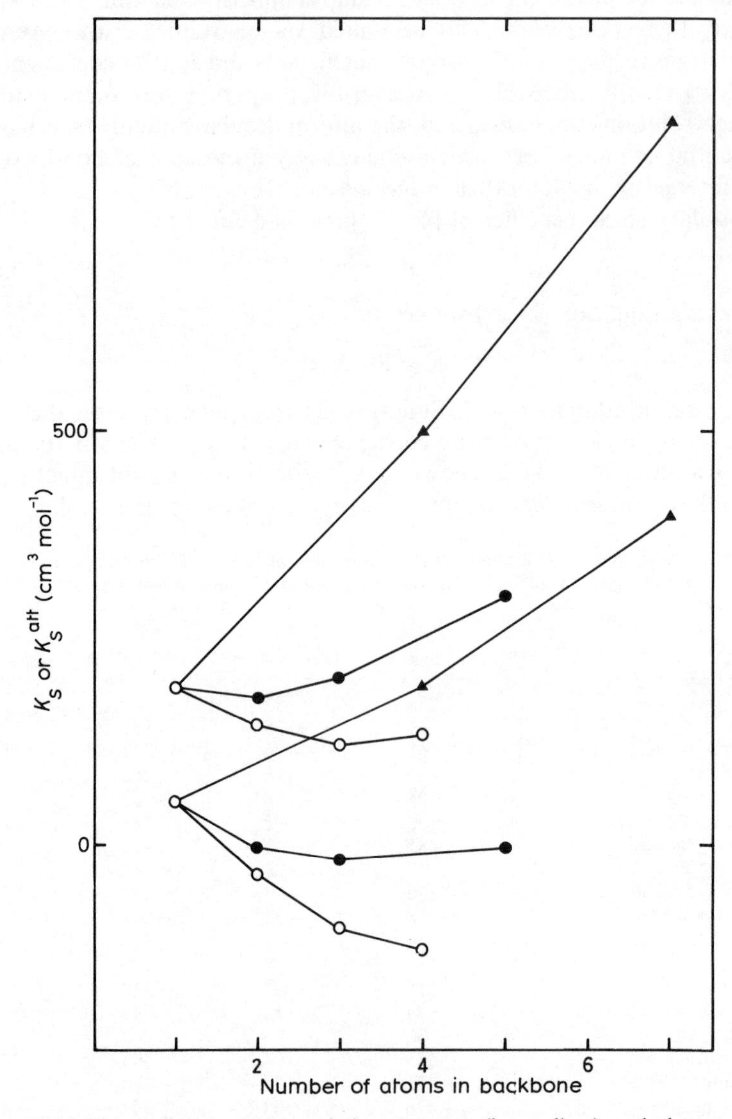

Figure 21.14 The variation of the sociation constant (lower lines) and the attractive contribution to the sociation constant (upper lines) for α-amino acids (○) α,ω-amino acids (●) and glycine homopeptides (▲) with chain length.

and peptides which have been investigated. In some instances there is considerable uncertainty in the values and some experimental reinvestigations are required. Also included in this table are the corresponding sociation constants, calculated from the interaction coefficients and the known or estimated (see earlier) partial molar volumes.

The dependence of sociation constant on chain length for some amino acids and peptides is shown by the lower three sets of lines in Fig. 21.14. These indicate that for both the α- and the α,ω-amino acids, members in the two homologous series, higher than glycine, have negative sociation constants and consequently show a nett *repulsion* between molecules. The glycine peptides have positive sociation constants and these become larger as the number of residues increases. The superficial results are somewhat misleading, since part of the sociation constant arises simply from excluded volume effects [51].

In view of the general shape of the potential of mean force as a function of intermolecular separation (see Fig. 21.13(c)), it is convenient to deconvolute the sociation constant into two contributions, one arising from attractive sources and the other from repulsive sources. A rigorous breakdown into these two factors is prohibitively difficult but an approximate procedure can be developed [53, 54] by assuming that the repulsive contribution is that which would result if the solute was a hard-sphere. The sociation constant then becomes

$$K_S^{HS} = K_S^{rep} = -4V_A^o \tag{21.31}$$

where V_A^o is the molar volume (note *not* the partial molar volume) of the solute. The attractive contribution to the sociation constant is then

$$K_S^{att} = K_S - K_S^{rep} \tag{21.32}$$

The attractive and repulsive contributions to the sociation constants are included in Table 21.4 and the upper set of lines in Fig. 21.14 shows the attractive terms for the acids and peptides. The following points can be concluded regarding the pairwise interactions of the series of molecules included in this figure:

(i) For the α-amino acids the attractive contribution changes little with increasing chain length. Glycine molecules interact the most favourably (presumably through the head groups) and as the series is ascended the initial decrease is followed by a slight increase. Actually the increase, although not marked probably indicates that the association between molecules tends to be 'side-on' rather than 'end-on' (see later with regard to heats of interaction).

(ii) The α,ω-amino acids show a marked variation with increasing chain length. The initial downturn is followed by a pronounced increase. One imagines that the α,ω-amino acids have a propensity to associate in a head-to-tail arrangement and the increase with chain length probably stems from two sources viz. an increasingly attractive hydrophobic association effect and an electrostatic effect which decreases with increasing chain length from the similarly charged ends of the associating molecules.

(iii) The attraction between peptides is comparatively strong (much stronger than the equivalent α,ω-amino acid) and although there will be an electrostatic component like that in (ii), there must also be a major contribution from intermolecular peptide–peptide bond interactions.

The fact that one needs volumetric data to transpose experimental free energy coefficients to intermolecular information is sometimes limiting since in many instances the required partial molar volume data are not available. It is even more restricting when derivatives of the free energy such as the pairwise enthalpy coefficient are investigated. The pairwise enthalpy coefficient is defined by [44]

$$h_{\mathrm{AA}} = (\partial(g_{\mathrm{AA}}/T)/\partial T^{-1})_P \tag{21.33}$$

and it qualitatively corresponds to the enthalpy change associated with bringing two A solute species from infinite separation in the solvent to proximity. These parameters can be obtained fairly readily from heat of dilution measurements but the transposition of these to molecular events requires knowledge of the temperature derivative of the partial molar volume [54] of the solutes considered. Not surprisingly it is only rarely that such measurements have been made prior to the experimental enthalpy measurements and generally obtaining such data is a tiresome and time-consuming task. However, it is possible to conclude from the free energy data on amino acids and peptides, that there is a reasonably good correlation between the experimental pairwise free energy coefficients with both the sociation constant and the attractive contribution (see Fig. 21.15). Given this it is not unreasonable to assume that a similar correlation will probably exist

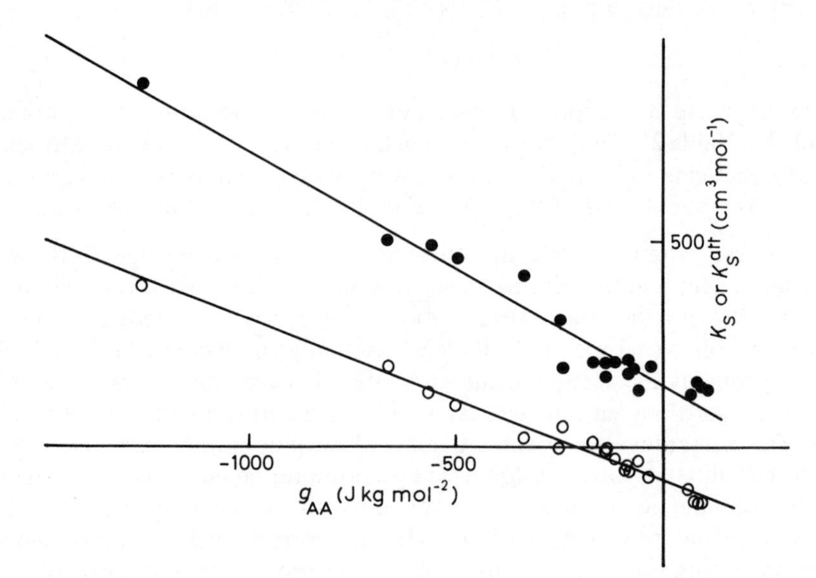

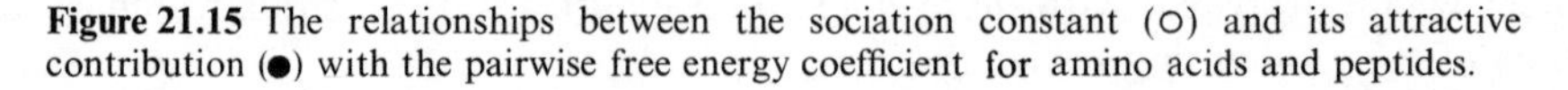
Figure 21.15 The relationships between the sociation constant (○) and its attractive contribution (●) with the pairwise free energy coefficient for amino acids and peptides.

Table 21.5 Pairwise enthalpy interaction coefficients for amino acids and peptides in water at 25°C

Compound	h_{AA} ($J\ kg\ mol^{-2}$)
Glycine	− 410,[a] − 439[b]
α-Alanine	213,[c] 217[b]
α-Aminobutyric acid	586[d]
Norvaline	1280[e]
Norleucine	1373[e]
Arginine	− 1764[b]
Cysteine	− 113[b]
Serine	− 720[b]
Valine	858[b]
β-Alanine	203[c]
γ-Aminobutyric acid	586[d]
ε-Aminocaproic acid	855[e]
Glycylglycine	− 796[f]
α-Aminoisobutyric acid	769[d]
Glycylglycylglycine	− 1899[f]

[a] Reanalysed data from Gucker, F.T., Pickard, H.B. and Ford, W.L. (1940) *J. Am. Chem. Soc.*, **62**, 2698; Wallace, W.E., Offutt, W.F. and Robinson, A.L. (1943) *J. Am. Chem. Soc.*, **65**, 347; Sturtevant, J.M. (1940) *J. Am. Chem. Soc.*, **62**, 1879.
[b] Humphrey, R.S., Hedwig, G.R., Watson, I.D. and Malcolm, G.N. (1980) *J. Chem. Thermodynamics*, **12**, 595.
[c] Reanalysed data from Benesi, H.A., Mason, L.S. and Robinson, A.L. (1946) *J. Am. Chem. Soc.*, **68**, 1755.
[d] Reanalysed data from Mason, L.S. and Robinson, A.L. (1947) *J. Am. Chem. Soc.*, **69**, 880.
[e] Reanalysed data from Mason, L.S., Offutt, W.F. and Robinson, A.L. (1949) *J. Am. Chem. Soc.*, **71**, 1463.
[f] Davis, K.G. (1984). Thesis, University of Sheffield.

between the corresponding heat terms and we will make this assumption in the brief comments which follow.

Table 21.5 lists pairwise enthalpy of interaction coefficients for some amino acids and peptides and Fig. 21.16 shows the variation with chain length for some free energy, enthalpy and entropy pairwise interaction coefficients at 25°C. (For convenience of scales we have plotted Ts_{AA} where

$$Ts_{AA} = h_{AA} - g_{AA}.)$$

The most notable feature for both the α- and the α,ω-series is that the free energy terms are relatively small but the component enthalpy and entropy terms are both large and in opposition. This seems to be a general feature of aqueous systems [55]. The overall increase in entropy with increasing size is notable and almost certainly reflects the loss of structured co-region solvent to the bulk solvent when the molecules interact. The trend with chain length for the α-amino acids indicates, as suggested earlier, that association for those molecules occurs in a side by side arrangement.

There is little information available on peptides but diglycine has a con-

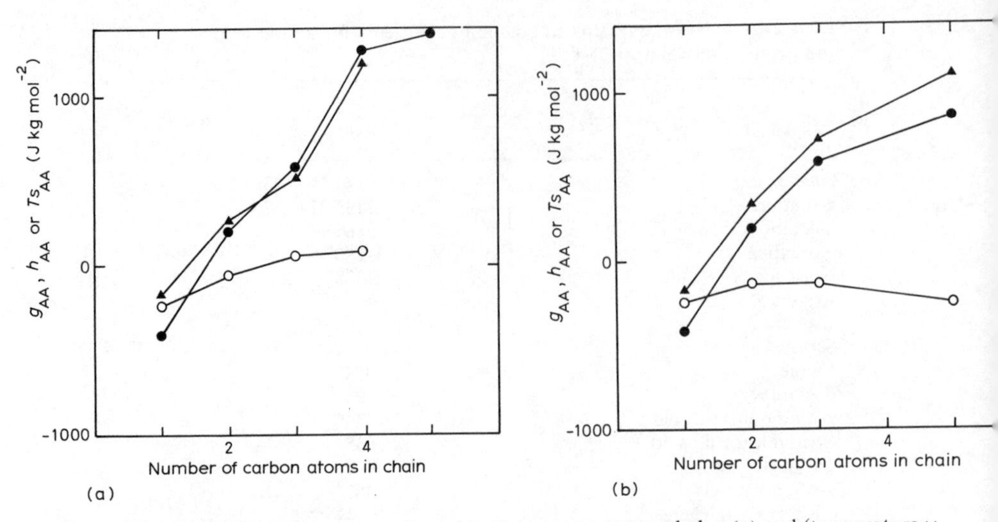

Figure 21.16 The dependence of pairwise free energy (○), enthalpy (●) and (temperature × entropy) (▲) coefficients for (a) α-amino acids and (b) α,ω-amino acids.

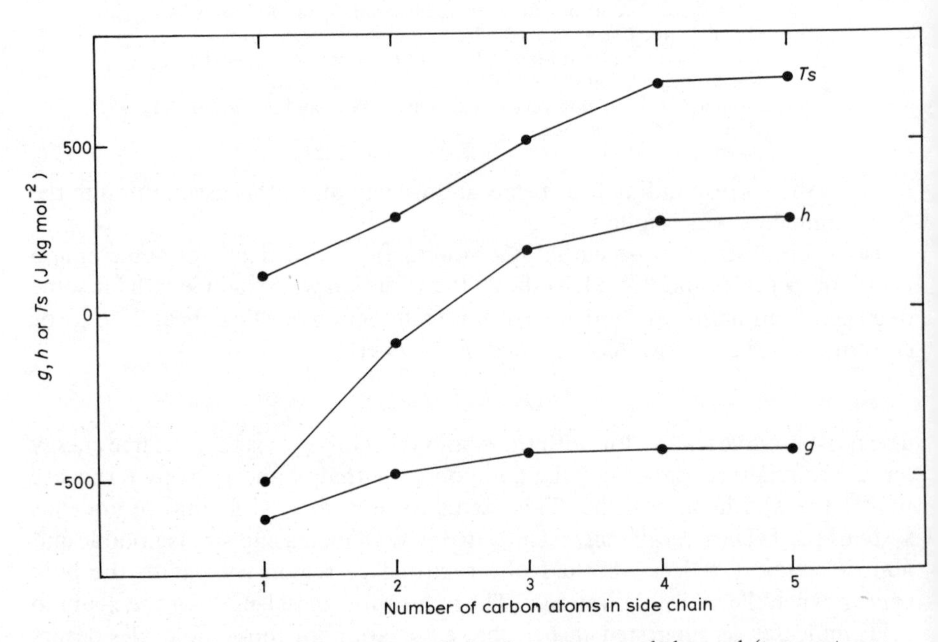

Figure 21.17 The dependence of the pairwise free energy, enthalpy and (temperature × entropy) coefficients for the interaction of the straight chain α-amino acids with sodium chloride in water at 25°C.

siderably more negative pairwise entropy term than has glycine which indicates that the solvation contribution for the former is dominated by the entropy loss resulting from the decreasing mobility of the associating species.

It is apparent from the above that the zwitterionic head group of the amino acids plays an important role in their interactions. This is reinforced by studies on the interactions occurring between amino acids and salts. A considerable amount of information is available [54, 56–63] on such interactions and Fig. 21.17 illustrates the variation of the interaction parameters for α-amino acids with sodium chloride in water [60]. These parameters are analogous to those for self-interactions but because it is not possible to separate ionic contributions, the free energy parameter, for example, is the sum of two terms, one from the cation and the other from the anion. It seems from these data that the ions essentially concentrate their attention on the polar head group region and that when the hydrocarbon chain is sufficiently long (about four carbon atoms) further increase in length has little or no effect on the interaction. The envisaged molecular situation is shown in Fig. 21.18.

When the charged nature of the amino acids is removed by appropriate substitution, the overall effects are considerably simpler in many ways. A recent investigation [64] of the energetics of the interactions occurring between some *N*-acetyl amino acid amides and some *N*-acetyl peptide amides shows that it is possible to rationalize a large body of experimental data using a simple model based on group-additivity [44, 65–67]. In essence the basic assumption is that every functional group on one molecule can interact with every functional group

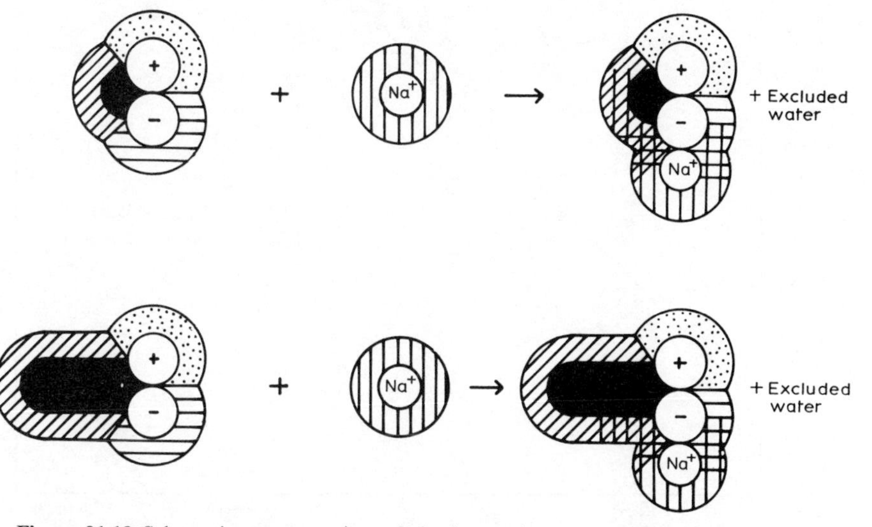

Figure. 21.18 Schematic representation of the interaction of a sodium ion with glycine (upper) and norleucine (lower). The various shadings correspond to the solvation regions around the groups and ion. The cation is shown adjacent to the carboxylate group during the interaction. This is of course an extreme situation and many other configurations must necessarily exist.

on another molecule. For example, if one considers molecules containing only peptide groups and methylene groups (methyl and methine groups are assumed to be equivalent to one and one half and half a methylene group respectively), then the pairwise enthalpy coefficient for the interaction between an A molecule and a B molecule is given by

$$h_{AB} = n^{A}_{CH_2} n^{B}_{CH_2} H_{CH_2-CH_2} + (n^{A}_{CH_2} n^{B}_{Pep} + n^{B}_{CH_2} n^{A}_{Pep}) H_{CH_2-Pep} + n^{A}_{Pep} n^{B}_{Pep} H_{Pep-Pep} \tag{21.34}$$

In this expression, for example $n^{A}_{CH_2}$ is the number of methylene groups on molecule A and H_{CH_2-Pep} represents the enthalpy change on bringing one mole of methylene groups towards one mole of peptide groups. The only unknowns in

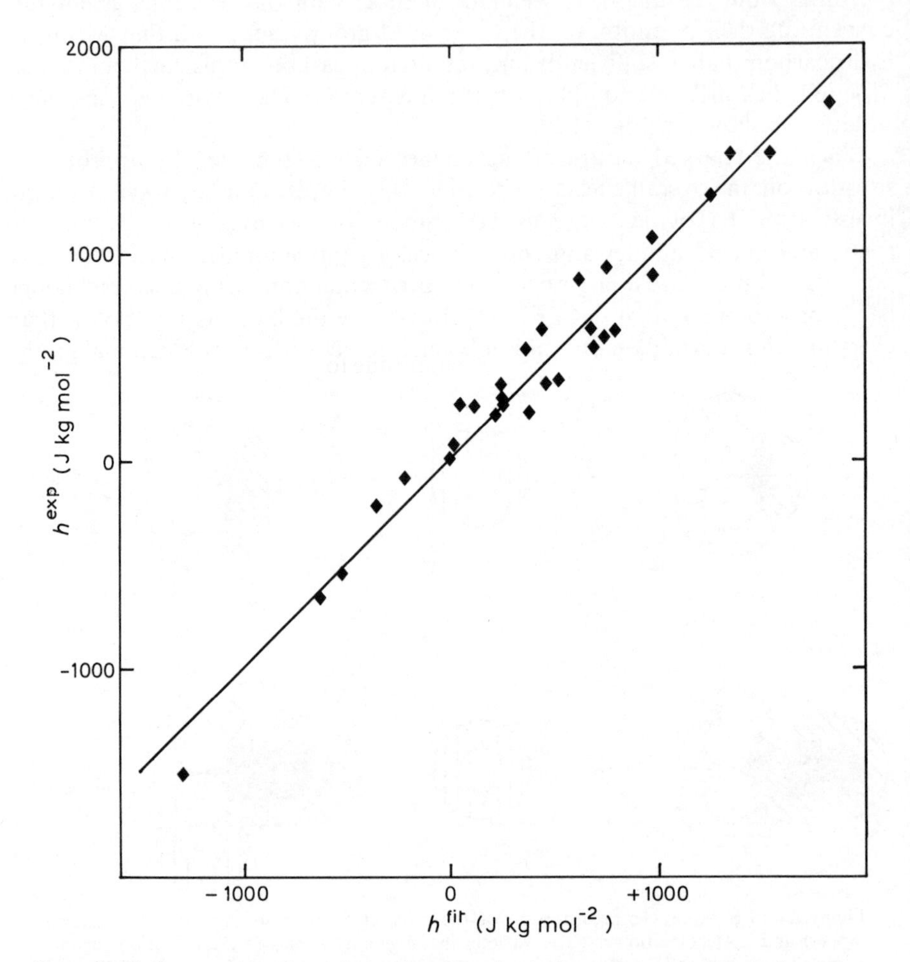

Figure 21.19 Comparison of experimental and fitted pairwise enthalpy coefficients for some amides and peptide derivatives. The fitted coefficients were obtained using the data in Table 21.6.

Table 21.6 Free energies, enthalpies and entropies of interaction of functional groups; units are J kg mol^{-2}

i/j		*CH$_2$*	*Peptide*
CH_2	G_{ij}	−29	
	H_{ij}	+14	
	TS_{ij}	+43	
Peptide	G_{ij}	+28	−48
	H_{ij}	+95	−311
	TS_{ij}	+67	−263

such a data set are the H terms and in Fig. 21.19 the experimental and fitted enthalpy coefficients are compared. It is clear from this that three terms ($H_{CH_2-CH_2}$, H_{CH_2-Pep} and $H_{Pep-Pep}$) are then required to obtain a reasonably good representation of quite a large body of data. An analogous approach to other properties can be made and a summary of the free energy, enthalpy and entropy terms for the various possible group interactions is given in Table 21.6. The CH_2–CH_2 pair interaction is qualitatively consistent with the established view [37–39] of the hydrophobic interaction in that it is dominated by the positive entropy term. This is in contrast to the peptide–peptide pair interaction for which the large and negative entropy term is outweighed by the more negative enthalpy term. As might be expected the interaction between methylene and peptide groups is repulsive but it is noteworthy that the numerical value of the three terms for such interactions are comparable in magnitude to the corresponding terms for methylene–methylene and peptide–peptide interactions. This has fairly major consequences on experimental observations and the superficial but attractive view that the interaction between peptide molecules would be dominated by intermolecular peptide–peptide bonds is negated by the observation that the enthalpy terms for the self interaction of *N*-acetyl glycylglycinamide and *N*-acetyl alanylalaninamide are respectively −646 and +939 J kg mol^{-2}. These results alone highlight the subtleties of the interactions of polyfunctional solutes in water and indicate the care which must be taken in interpreting them in molecular terms.

REFERENCES

1. Nemethy, G. and Scheraga, H.A. (1977) *Q. Rev. Biophys.*, **10**, 239.
2. Clementi, E. (1980) *Lecture Notes in Chemistry* (Vol. 19), Springer-Verlag, New York.
3. Franks, F. (ed.) (1973) *Water, A Comprehensive Treatise* (Vol. 2), Plenum Press, New York.
4. Franks, F. (ed.) (1975) *Water, A Comprehensive Treatise* (Vol. 4), Plenum Press, New York.
5. Nemethy, G., Peer, W.J. and Scheraga, H.A. (1981) *Ann. Rev. Biophys. Bioeng.*, **10**, 459.
6. Enderby, J.E. and Neilson, G.W. (1979) in *Water, A Comprehensive Treatise* (ed. F. Franks) (Vol. 6), Plenum Press, New York, p. 1.
7. Sandstrom, D.R. and Lytle, F.W. (1979) *Ann. Rev. Phys. Chem.*, **30**, 215.
8. Frank, H.S. and Wen, W-Y. (1957) *Disc. Faraday Soc.*, **24**, 133.
9. Gurney, R.W. (1953) *Ionic Processes in Solution*, McGraw Hill New York.
10. Friedman, H.L. and Krishnan, C.V. (1973) in *Water, A Comprehensive Treatise*, (ed. F. Franks) (Vol. 3) Plenum Press, New York.
11. Tse, Y.-C., Newton, M.D., Vishveshwara, S. and Pople, J.A. (1978) *J. Am. Chem. Soc.*, **100**, 4329.

12. Locke, M.J., Hunter, R.L. and McIver, R.T. (1979) *J. Amer. Chem. Soc.*, **101**, 272.
13. Haberfield, P. (1980) *J. Chem. Ed.*, **57**, 346.
14. Ben-Naim, A. (1978) *J. Phys. Chem.*, **82**, 792.
15. Ben-Naim, A. (1979) *J. Phys. Chem.*, **83**, 1803.
16. Tanford, C. (1979) *J. Phys. Chem.*, **83**, 1802.
17. Wolfenden, R., Andersson, L., Cullis, P.M. and Southgate, C.C.B. (1981) *Biochemistry*, **20**, 849.
18. Bull, H.B., Breese, K. and Swenson, C.A. (1978) *Biopolymers*, **17**, 1091. (This also gives references to earlier compilations of amino acid solubilities.)
19. Picker, P., Tremblay, E. and Jolicoeur, C. (1974) *J. Solution Chem.*, **3**, 377.
20. Picker, P., Leduc, P.-A., Philip, P.R. and Desnoyers, J.E. (1971) *J. Chem. Thermodynamics*, **3**, 631.
21. Lepori, L. and Mollica, V. (1980) *Z. Phys. Chem.* (Neue Folge), **123**, 51.
22. Millero, F.J., Lo Surdo, A. and Schin, C. (1978) *J. Phys. Chem.*, **82**, 784.
23. DiPaola, G. and Belleau, B. (1978) *Canad. J. Chem.*, **56**, 1827.
24. Jolicoeur, C. and Boileau, J. (1978) *Canad. J. Chem.*, **56**, 2707.
25. Ahluwalia, J.C., Ostigoy, C., Perron, G. and Desnoyers, J.E. (1977) *Canad. J. Chem.*, **55**, 3364.
26. Kirchnerova, J., Farrell, P.G. and Edward, J.T. (1976) *J. Phys. Chem.*, **80**, 1974.
27. Shahidi, F. and Farrell, P.G. (1978) *J.C.S. Faraday Trans I*, **74**, 858.
28. Edward, J.T. (1970) *J. Chem. Ed.*, **47**, 261.
29. Scott, G.D. (1960) *Nature*, **185**, 68.
30. Bernal, J.D. and Finney, J.L. (1967) *Disc. Faraday Soc.*, **43**, 62.
31. Duer, W.C., Greenstein, J.R., Oglesby, G.B. and Millers, F.J. (1977) *J. Chem. Ed.*, **54**, 139.
32. Gucker, F.T., Ford, W.L. and Moser, C.E. (1939) *J. Phys. Chem.*, **43**, 153.
33. Turner, W.R., (1971) *I. and E.C. Prod. Res. and Development*, **10**, 238.
34. Shahidi, F. and Farrell, P.G. (1978) *J.C.S. Faraday Trans I*, **74**, 1268.
35. Prasad, K.P. and Ahluwalia, J.C. (1976) *J. Solution Chem.*, **5**, 491.
36. Nichols, N., Sköld, R., Spink, C., *et al.* (1976) *J. Chem. Thermodynamics*, **8**, 1081.
37. Kauzmann, W. (1959) *Adv. Protein Chem.*, **14**, 1.
38. Nemethy, G. and Scheraga, H.A. (1962) *J. Phys. Chem.*, **66**, 1773.
39. Tanford, C. (1973) *The Hydrophobic Effect*, John Wiley, New York.
40. Kumar, N.G. and Roeske, R.W. (1978) *Org. Magnetic Resonance*, **11**, 291.
41. Lilley, T.H. and Scott, R.P. (1976) *J.C.S. Faraday Trans I*, **72**, 184.
42. Franks, F. (1979) in *Biochemical Thermodynamics*, (ed. M.N. Jones), Elsevier, Amsterdam.
43. Gill, S.J. and Noll, L. (1972) *J. Phys. Chem.*, **76**, 3065.
44. Okamoto, B.Y., Wood, R.H. and Thompson, P.T. (1978) *J.C.S. Faraday Trans I*, **74**, 1890.
45. Friedman, H.L. (1962) *Ionic Solution Theory*, Interscience, New York.
46. Blackburn, G.M., Lilley, T.H. and Walmsley, E. (1980) *J.C.S. Faraday Trans I*, **76**, 915.
47. Wood, R.H., Lilley, T.H. and Thompson, P.T. (1978) *J.C.S. Faraday Trans I*, **74**, 1990.
48. Guggenheim, E.A. (1960) *Trans Faraday Soc.*, **56**, 1159.
49. McMillan, W.G. and Mayer, J.E. (1945) *J. Chem. Phys.*, **13**, 276.
50. Friedman, H.L. (1972) *J. Solution Chem.*, **1**, 387.
51. Friedman, H.L. (1972) *J. Solution Chem.*, **1**, 413.
52. Friedman, H.L. (1972) *J. Solution Chem.*, **1**, 419.
53. Kozak, J.J., Knight, W.S. and Kauzmann, W. (1968) *J. Chem. Phys.*, **48**, 675.
54. Lilley, T.H., Moses, E. and Tasker, I.R. (1980) *J.C.S. Faraday Trans I*, **76**, 906.
55. Lumry, R. and Rajender, S. (1970) *Biopolymers*, **9**, 1125.
56. Lilley, T.H. and Scott, R.P. (1976) *J.C.S. Faraday Trans I*, **72**, 197.
57. Kelley, B.P. and Lilley, T.H. (1978) *J. Chem. Thermodynamics*, **10**, 703.
58. Kelley, B.P. and Lilley, T.H. (1978) *J.C.S. Faraday Trans I*, **74**, 2771.
59. Kelley, B.P. and Lilley, T.H. (1978) *J.C.S. Faraday Trans I*, **74**, 2779.
60. Lilley, T.H. and Tasker, I.R. (1982) *J.C.S. Faraday Trans I*, **78**, 1.
61. Schrier, E.E. and Robinson, R.A. (1974) *J. Solution Chem.*, **3**, 493.
62. Schrier, E.E. and Robinson, R.A. (1971) *J. Biol. Chem.*, **246**, 2870.
63. Briggs, C.C., Lilley, T.H., Rutherford, J. and Woodhead, S. (1974) *J. Solution Chem.*, **3**, 649.
64. Blackburn, G.M., Lilley, T.H. and Walmsley, E. (1982) *J.C.S. Faraday Trans I*, **78**, 1641.
65. Savage, J.J. and Wood, R.H. (1976) *J. Solution Chem.*, **5**, 733.
66. Lilley, T.H. and Wood, R.H. (1980) *J.C.S. Faraday Trans I*, **76**, 901.
67. Wood, R.H. and Hiltzik, L.H. (1980) *J. Solution Chem.*, **9**, 45.

CHAPTER TWENTY-TWO

X-Ray Crystal Structures of Amino Acids and Selected Derivatives

V. CODY

22.1 INTRODUCTION

The past two decades have seen a tremendous increase both in the number and accuracy of amino acid, peptide, and protein crystal structure determinations. The recent strides made in computer and diffractometer technologies and methods of crystal structure determination have been primarily responsible for this increase, which has in turn made possible the study of more complex biological systems. Since protein folding is determined by amino acid sequence and residue side-chain conformation is influenced by the protein main chain conformation, knowledge of amino acid geometry and conformational flexibility is of key importance in elucidating protein structure–activity relationships. Consequently, the goal of most amino acid structural studies has been to compile accurate information concerning amino acid residue geometry, conformational flexibility among families of amino acids in various environments, and their hydrogen bonding interactions. These data are also of considerable value in energy calculations of conformational flexibility. Ultimately, this approach should provide a means of predicting three-dimensional protein structures and understanding the relationships among amino acid sequence, structure and biological properties.

Since the focus of this book is on amino acids, this chapter will describe the three-dimensional crystal structure determinations of amino acids only, in particular those commonly found in proteins, as well as selected amino acid derivatives. Metal–amino acid complexes will be mentioned only briefly as a full description of these interactions is beyond the scope of this work. Dipeptides will not be described although they can be considered amino acid derivatives. Because the number of crystal structure reports of amino acids has steadily increased in recent years, now over 1000, this review should not be considered exhaustive. For the most recent compilations of amino acids structures the reader is directed to read the current volumes in the series *Molecular Structures and Dimensions* [1], which lists all crystal structure reports including those reported in abstract form and is available in computer-search form on the Cambridge Data Files [2]; volumes of *Molecular Structure by Diffraction Methods* [3] and volumes of *Amino Acids, Peptides and Proteins* [4], both specialist periodical reports published by The Chemical Society.

X-Ray crystal structure determinations provide the most precise details possible of molecular composition, configuration and conformation. Also, crystallographic techniques can be used to identify intermediates, metabolites, and their reaction products as well as to delineate active comformations. As illustrated with thyroxine [5–7], an iodinated amino acid derivative (Fig. 22.1), composition defines the number and kinds of atoms that make up the molecule and is represented by its chemical formula. Constitution refers to the connectivity of the molecule and is illustrated by a line drawing showing which atoms are bonded to one another. Molecular configuration defines the chirality of all asymmetric centres in the molecule. Conformation refers to the total geometric description and disposition of the atoms in three dimensions and is usually described in terms of torsion angles [8].

Multiple crystallographic observations of the same compound, crystallized under different conditions (acids, bases, aqueous salts, etc.), as metal complexes, or as derivatives of the parent molecule, provide a measure of the conformational flexibility inherent within the molecule and also indicate the presence of patterns

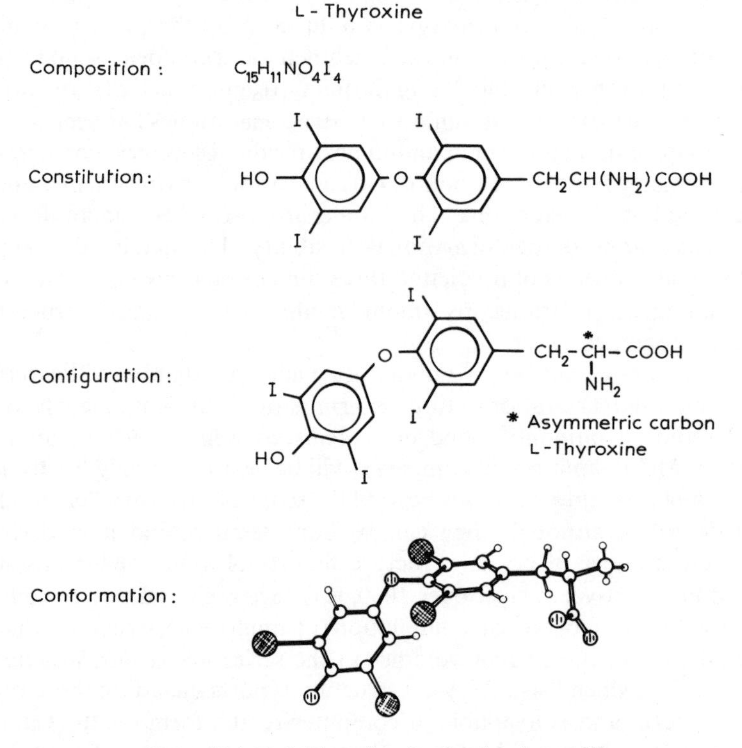

Figure 22.1 Definition of molecular descriptors [5].

in conformational isomerism. In addition to knowing the details of molecular conformation, it is important to understand how these molecules interact with their environment. Since many protein–substrate interactions take place via hydrogen bonds, an analysis of the hydrogen bonding and molecular packing of the amino acids in their crystalline structure offers an insight into the molecular details of hydrogen-bond strength and directionality at the protein–substrate site. However, to interpret the local hydrogen bonding environment, precise knowledge of the hydrogen atom positions is required. The availability of neutron diffraction determinations of amino acids and the increased accuracy of X-ray crystal structure results with reliability indices, R, around 3–5%, now provide these data.

While it is not within the scope of this chapter to describe the details of X-ray and neutron diffraction techniques [9–13], important differences in their properties should be emphasized. The difficulty in locating precise hydrogen atom positions from X-ray crystal structure determinations is a result of the small scattering power of hydrogen relative to the rest of the atoms in the structure. This comes about because the diffraction of X-rays is caused by an interaction with the electrons of the atoms in the crystal, thus the scattering power is proportional to the number of electrons present in the atoms. The accuracy of earlier X-ray crystal structure determinations, as indicated by reliability indices of $\sim$ 10–15%, was such that hydrogen atom positions were not discernible from difference electron density maps. Therefore, potential hydrogen bonding interactions were derived from the closest approach of acceptable donor and acceptor atoms with the hydrogen atom presumed to be located on the line joining the atom-centres of these atoms. Even as X-ray crystal structure determinations became more accurate ($R \simeq 3$–5%) and heavy atoms such as iodine or bromine were not present in the crystal structure, thereby permitting the location of reasonable hydrogen atom positions, those involved in the hydrogen bonding interactions were not always found.

Neutron diffraction, on the other hand, is capable of providing hydrogen atom positions as precisely as the other atoms since hydrogen or deuterium atoms have neutron scattering lengths comparable in magnitude to other elements, even heavy ones. In practice, however, it is more difficult to collect neutron diffraction data since the neutron experiment requires (i) access to a neutron facility with high neutron beam flux, (ii) longer exposure times and (iii) larger crystal sizes. Growing suitably large crystals of most biological molecules is still a problem; however the availability of neutron beams with higher flux permits the use of smaller crystals.

22.2 AMINO ACID DEFINITIONS

The general classes of amino acid residues that will be considered are illustrated in Fig. 22.2. Comprehensive reviews [14–21] of the early X-ray diffraction studies have defined the general geometrical (Fig. 22.3) and conformational (Table 22.1)

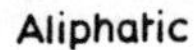
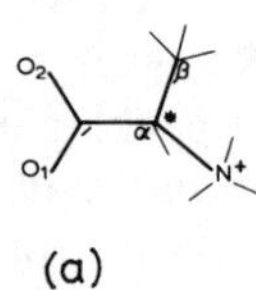
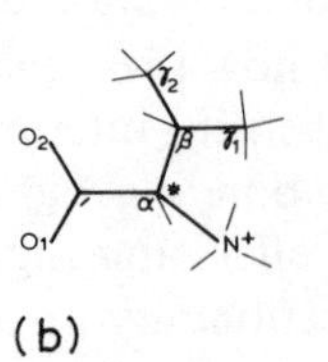
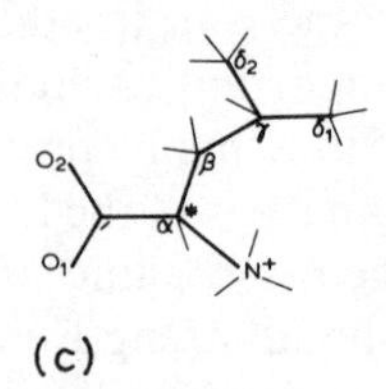
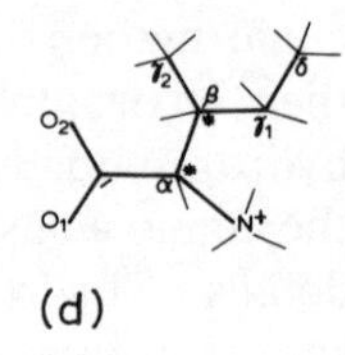
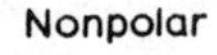
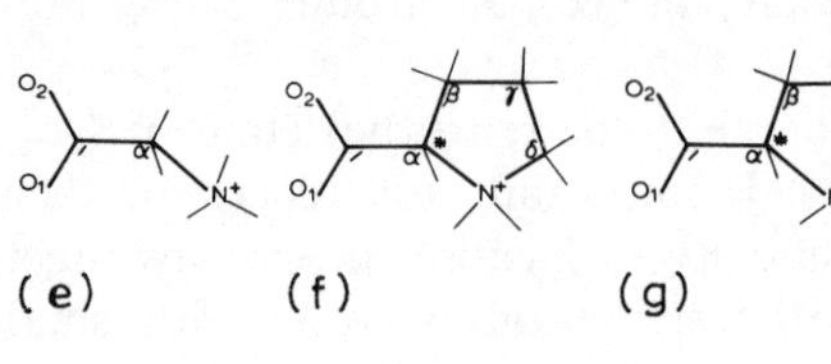
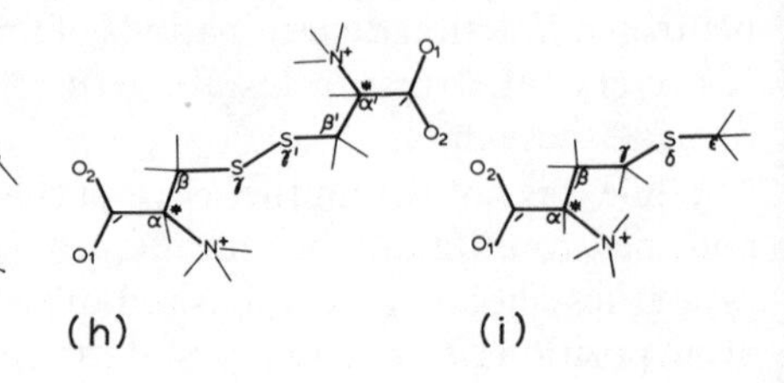
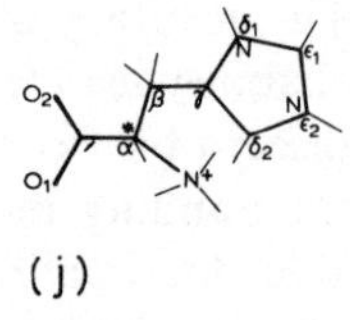
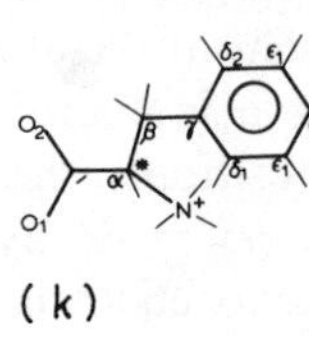
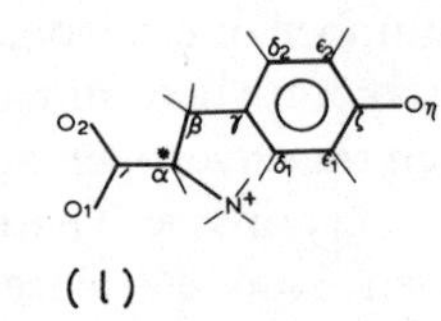
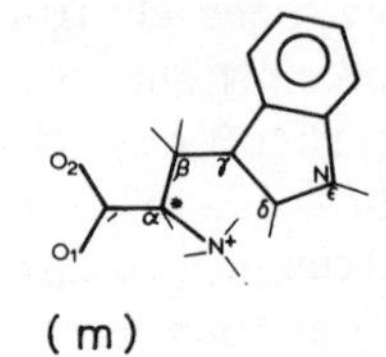
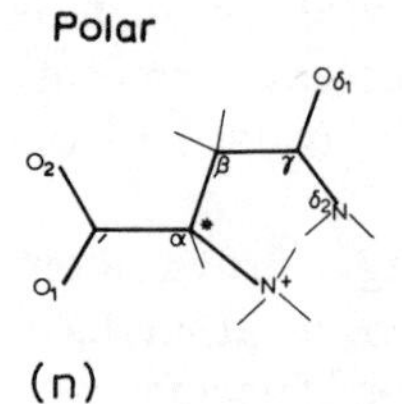
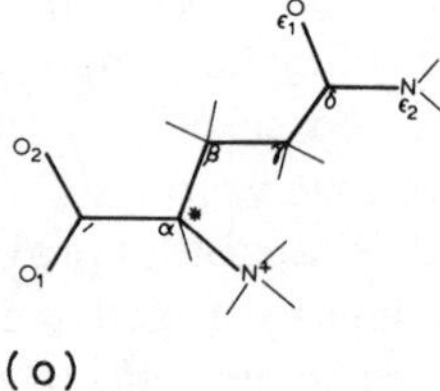
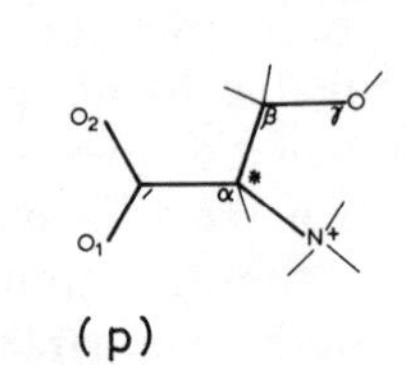
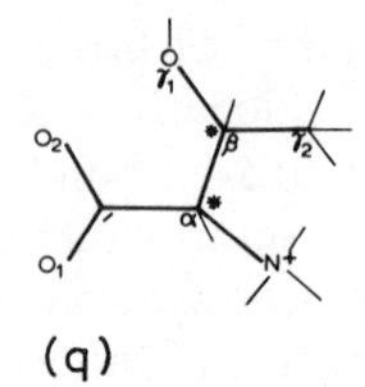
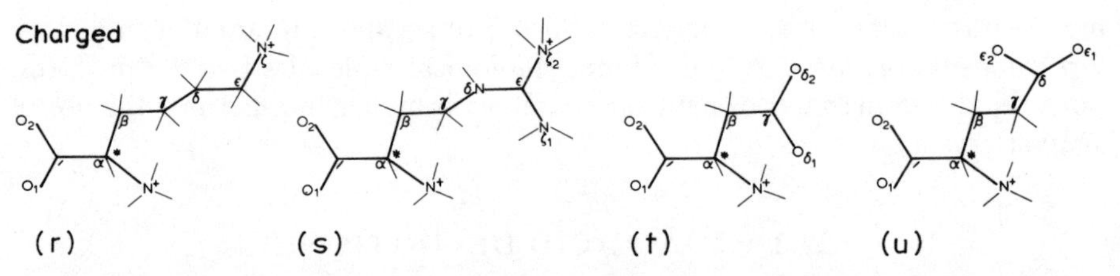

Figure 22.2 Classifications of naturally occurring amino acid residues. The asterisk indicates an asymmetric carbon; all figures denote L-configuration; all carboxylic groups are COO^- representations. (a) Ala; (b) Val; (c) Leu; (d) Ile; (e) Gly; (f) Pro; (g) Cys; (h) Cys–Cys; (i) Met; (j) His; (k) Phe; (l) Tyr; (m) Trp; (n) Asn; (o) Gln; (p) Ser; (q) Thr; (r) Lys; (s) Arg; (t) Asp; (u) Glu.

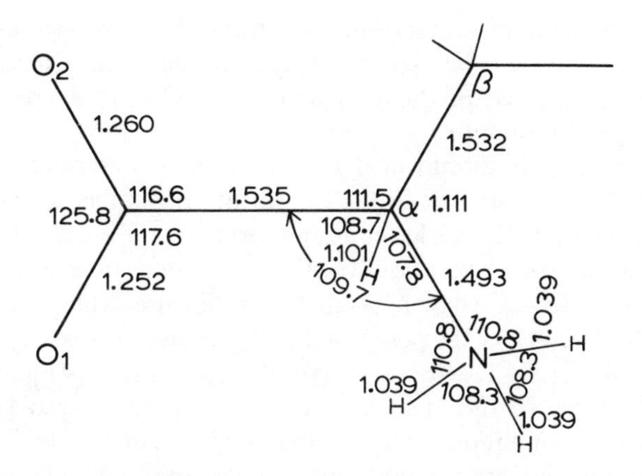

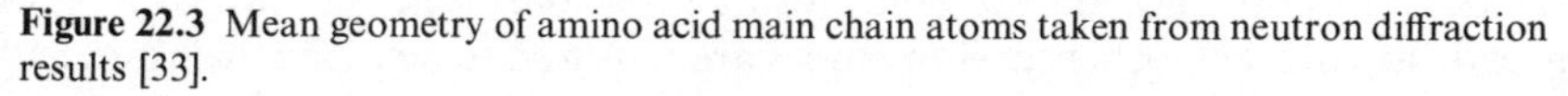
Figure 22.3 Mean geometry of amino acid main chain atoms taken from neutron diffraction results [33].

preferences of these structures. In addition, detailed studies of metal co-ordination with amino acids have been carried out [16–18] that describe the metal atom co-ordination number and the type of chelation geometry preferred by the various amino acids. These observations [16] are also illustrated with stereo drawings of the metal co-ordination spheres. Updates of the review by Freeman [16] have been made periodically and can be requested. Conformational energy calculations carried out on amino acids and peptides [22–25] indicate that only certain conformations are energetically favourable and are generally those observed in the X-ray crystal structures. With the availability of more accurate X-ray diffraction data [26–31], as well as neutron diffraction data [32–35], attention has focused on defining accurate hydrogen bonding geomet-

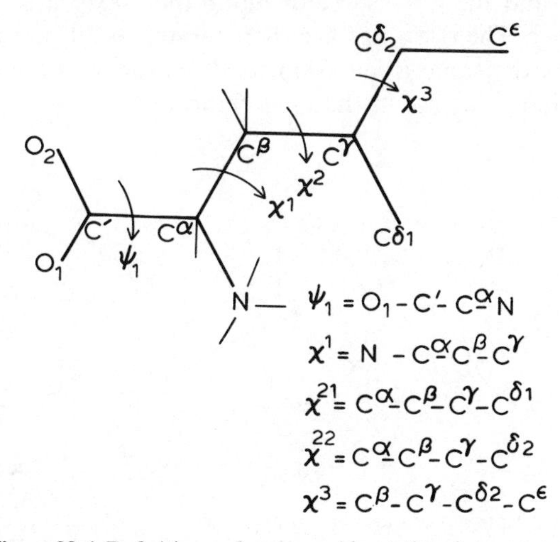

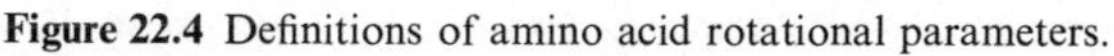
Figure 22.4 Definitions of amino acid rotational parameters.

ries based on these amino acid crystal structures. Also, the availability of three-dimensional protein data has permitted the comparison of residue side-chain conformations with those predicted from energy calculations or observed from amino acid crystal structures [36–40].

The conformation of amino acid residues can be described by the torsion parameters [41] illustrated in Fig. 22.4. All atom numbering begins with the alpha carbon, using the Greek letter designations for the atoms along the residue (Fig. 22.2). The conformation about the $C^\alpha - C^\beta$ bond is described by the torsion angle χ^1 ($N - C^\alpha - C^\beta - C^\gamma$) which has three equilibrium values [19, 21]: $\chi^1 \approx 60°$, 180° and −60°. Because of the planar nature of the carboxyl group, the difference between ψ^1 and ψ^2 ($N - C^\alpha - C' - O_2$) is 180°. Convention [41] defines O_2 to be *trans* to the amine. In the ideal geometry ψ should be zero [19]; however, inspection of all the amino acid structures reveals that this seldom is the case.

Depending on the amino acid residue or the type of carboxyl group substitution present, other torsion angles are required to describe their molecular conformations. For those amino acids with branched side-chains, the torsion parameters χ^{21} and χ^{22} define the rotations about the $C^\beta - C^\gamma$ bond. In the case of aromatic residues such as tyrosine or phenylalanine, $\chi^{21} - \chi^{22} = 180°$ since the C^β, C^γ, C^{δ_1}, C^{δ_2} group is planar. In the case of the tetrahedrally branched side-chains such as leucine, there are three staggered rotamers for χ^{11} and χ^{12} that are represented (Fig. 22.5) as *trans* ($\chi^{11} = 180°$, $\chi^{12} = -60°$), *gauche I* ($\chi^{11} = 60°$, $\chi^{12} = 180°$) or *gauche II* ($\chi^{11} = -60°$, $\chi^{12} = 60°$). Likewise, ψ^3 etc. describes the carboxyl substitutions and ϕ' those on the amine (Fig. 22.4).

The various amino acid residues that will be discussed here are classified in Fig. 22.2 which shows these amino acids in the same orientation. The asymmetric carbon in each molecule is also indicated as well as the standard amino acid numbering. The orientations shown are defined by placing the $C' - C^\alpha$ vector along the *x*-axis and the $C^\alpha - N$ vector along the *y*-axis, assumed to lie in the plane of the paper. The drawings are then rotated −30° around *x* and −10° around *y*. This is the same orientation used for the stereo pairs [42]. Unless specified, all amino acids are in the L-configuration.

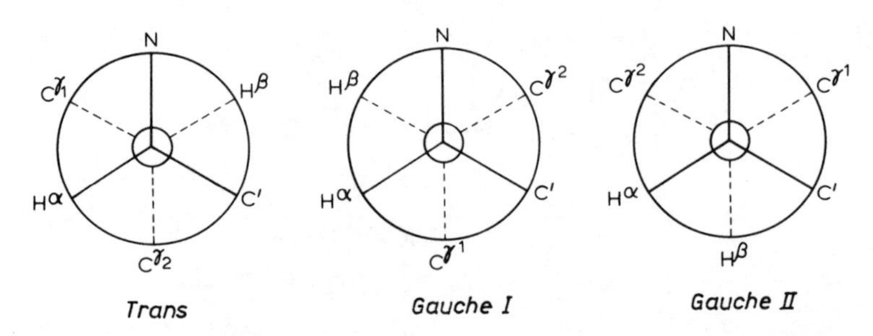

Figure 22.5 Newman projections looking from $C^\alpha - C^\beta$, for isoleucine illustrating the three staggered energy minimum conformations, *trans*, *gauche I* and *gauche II*, respectively.

22.3 AMINO ACID RESIDUE CONFORMATIONS

22.3.1 Aliphatic (Ala, Val, Leu, Ile)

The amino acids alanine, valine, leucine, and isoleucine (Fig. 22.2(a)–(d)) are aliphatic in character. Of the structural reports (> 100) containing these simple amino acids listed in the *Cambridge Data Files* [1, 2], many have been studied as halide salts of the *N*-acetyl derivatives of either the L-enantiomorph or the DL pair. When these determinations were carried out in the early 1970s, these structures delineated the preferred conformations likely to be present in large peptides and proteins. Another purpose of these studies was to understand the packing of these molecules [43] as it had been suggested that the structure of racemic and optically active structure pairs are often related in such a way that a plane of close-packed molecules of the same configuration (i.e., L- vs L of DL pair) generates the two structures through appropriate symmetry elements [44].

In this group of amino acids only alanine [45] and valine [46] have been studied using neutron diffraction. Thus for these structures precise values for the hydrogen atom positions and hydrogen bonding interactions are known.

Analysis of the alanine-containing structures (> 25) shows that their ψ values cluster near 5° and $-19°$ (Table 22.1). It has been suggested that the presence of a C^{β} atom influences the orientation of the carboxyl plane making ψ negative [19, 47]. The structure of L-alanine [45] is shown in Fig. 22.6.

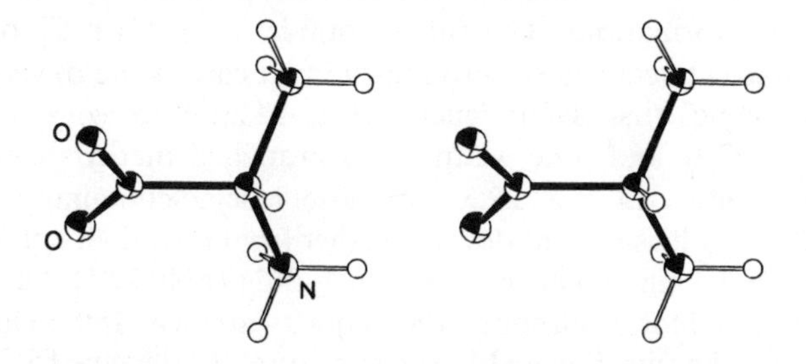

Figure 22.6 Stereo conformation of L-alanine [45] with $\Psi = -10°$.

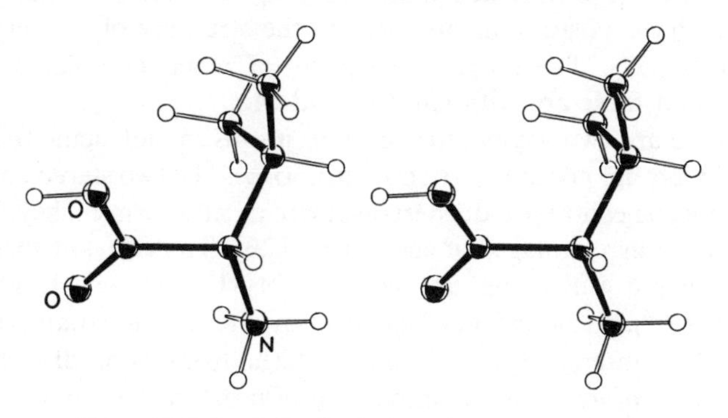

Figure 22.7 Stereo conformation of L-valine HCl [46].

The Cambridge Data Files [1,2] also list alanine–metal complexes (>20) in which Cu and Co are equally preferred as a chelator. As demonstrated in the structure of 1-mer-tris(L-alaninato)cobalt (III) monohydrate [48], the alanine chelates through the amine and *cis*-carboxyl oxygen with ψ values near 25°.

In addition to ψ, the conformation of valine (Fig. 22.2(b)) is further defined by the parameters χ^{11} and χ^{12} ($N-C^{\alpha}-C^{\beta}-C^{\gamma_2}$) (Fig. 22.4). Compilation of these parameters (Table 22.1) for the valine-containing structures (>15) shows ψ to cluster near $-11°$ and $-35°$ with a greater variation in these values than seen in the alanine structures. Of the three possible rotamers of χ^{11} and χ^{12}, the majority of valine structures are in the *gauche I* conformation [49] (Fig. 22.5), as depicted in L-valine HCl [46] (Fig. 22.7). The more unusual conformation is *gauche II* as observed in the structures of L-valine HCl H_2O [50] and an organophosphorous valine derivative [51].

A series of *N*-acylated derivatives of norvaline [52, 53] has been studied in which the acetyl moiety was replaced systematically by different substituents in order to determine the influence of C^{α}-substituent orientation and van der Waals radii on their rates of enzyme hydrolysis. The results of these diffraction studies are in agreement with the conformational parameters predicted from reaction kinetics data.

There are less than ten structure reports of metal chelates with valine. The chelating metals include Cu, Ni, Zn, Pd, Mn and Ca. The chelation of valine Pd [54] is through the N and O with ψ values near $-10°$ and χ^{11}, χ^{12} in the *gauche I* and *gauche II* conformations. The valine complex with Ca [55], on the other hand, shows that valine chelates through the two carboxylic oxygens.

The torsion angles that define leucine (Fig. 22.2(c)) are ψ, χ^{1} and χ^{21} and χ^{22} ($C^{\alpha}-C^{\beta}-C^{\gamma}C^{\delta_1}$), and, like valine, the branched methyls can adopt the three rotamers shown in Fig. 22.5. Of those leucine-containing compounds reported (>20), ψ values tend to deviate further from zero than either alanine or valine and have local minima near $-18°$ and $\pm 36°$ (Table 22.1). All three energy minima are observed for χ^{1} although the majority are near 180°. The preference for χ^{21}, χ^{22} is *gauche I* as depicted in the structure of L-leucine [56] (Fig. 22.8). However, when there is an *N*-acetyl substitution, χ^{1} tends to be near $-60°$ and χ^{21}, χ^{22} in the *trans* position as observed in the structure of *N*-acetyl-L-leucyl-isopropylamide [57]. There are few reports of metal complexes containing leucine (<5) and these are with Cu, Co, and Mo.

Because there are two asymmetric carbon atoms in isoleucine (Fig. 22.2(d)), four stereoisomers are possible (L,L; L,D; D,L; D,D). The two stereoisomers of the second asymmetric centre are diastereoisomers or allo forms. They, like the DL forms of C^{α}, are mirror images of each other [26]. The relevant torsion angles required to define isoleucine are ψ, χ^{11} ($N-C^{\alpha}-C^{\beta}-C^{\gamma_1}$), χ^{12}, and χ^{2} ($C^{\alpha}-C^{\beta}-C^{\gamma_1}-C^{\delta}$). As indicated, C^{γ} can occupy one of three rotameric positions (Fig. 22.5) and by interchanging C^{δ_1} and C^{δ_2}, the isomers of all-isoleucine will be described. A conformational comparison of most of the reported structures (>10) of isoleucine or allo-isoleucine has been carried out by Varughese [58]. His

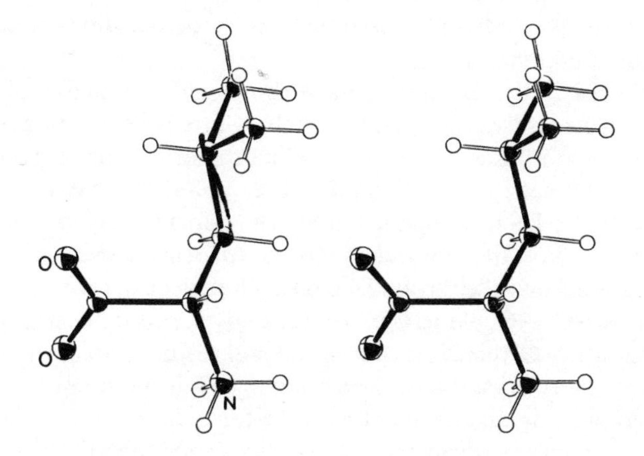

Figure 22.8 Stereo conformation of L-leucine [56].

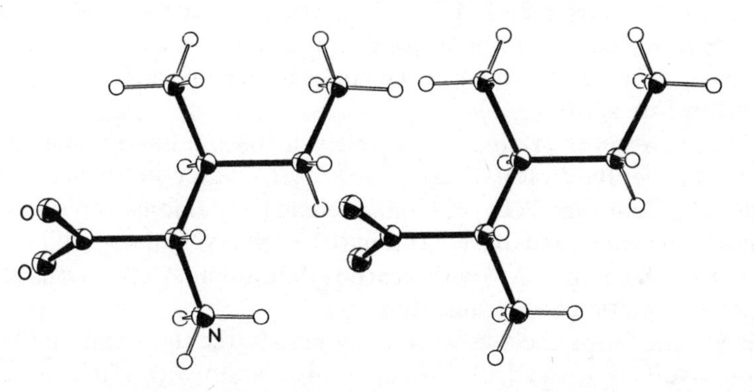

Figure 22.9 Stereo conformation of L-isoleucine [59].

analysis shows that ψ tends toward $-15°$, but when there are two molecules in the asymmetric unit [59, 60], one molecule has ψ near $-45°$. The author suggests that the side-chain C^{γ} atom causes ψ to deviate from zero in such a way as to maximize the O_2 (*trans* to N)...C^{γ} distance. All three rotamers of χ^{11}, χ^{12} (Fig. 22.5) have been observed with the majority in the *gauche I* conformation. The C^{δ} atom on the isoleucine side-chain [61] tends to be in an extended conformation ($\chi^2 \approx 180°$) *trans* to C^{α}. These features are illustrated in Fig. 22.9 for one of the two independent molecules in the structure of L-isoleucine [59]. There is only one report of a Cu complex involving isoleucine [62], and here metal chelation is through the amino nitrogen and carboxylic oxygen.

22.3.2 Non-polar (Gly, Pro, Cys, Met)

The amino acid residues glycine, proline, cysteine and methionine (Fig. 22.2(e)–(i)), while seemingly dissimilar, are all non-polar in character. Included in the structure reports (> 200) on this series are neutron diffraction studies of glycine, proline and cysteine.

Glycine (Fig. 22.2(e)) is by far the most thoroughly studied amino acid with over 100 crystal structures reported of which 11 involve neutron diffraction and 60 involve metal complexes. Glycine is unlike the other amino acid residues in that it has no asymmetric carbon and has three polymeric crystalline forms, α, β and γ, that have also been studied in detail. The α-form is the most common and has been shown to have the molecules oriented to form antiparallel double layers with alternating antiparallel layers [63, 64]. The tilt of the carboxyl plane with respect to the $N-C^{\alpha}-C'$ plane, ψ, is 19° for α-glycine as determined by neutron diffraction studies. β-Glycine is unstable and will readily transform to the α-form in air. The one X-ray crystal structure report [65], using film data, shows it to have the same packing arrangement as the α-form, and $\psi = 27°$. The γ-form of glycine is strongly piezoelectric and its packing is such that all the molecules are oriented in the same fashion forming helices around the crystallographic 3_2 screw axes [66, 67]. A neutron study [67] was carried out at two temperatures and shows significant changes in the lattice due to temperature changes. The torsion angle ψ is $-15°$ in these structures. The molecular conformation of γ-glycine is illustrated in Fig. 22.10.

The conformation of glycine as a simple salt [68, 69] has a planar backbone ($\psi = -1°$), as does the *N*-acetyl derivative [70]. However, in triglycine salts with sulphate [71], fluoroberyllate [72] and selenate [73] anions, two of the glycine molecules are planar and have protonated carboxyl groups while the third glycine is non-planar ($\psi = 20°$) with a carboxylate anion. All these triglycine salts undergo ferroelectric phase transitions.

While glycine forms chelates with many metals, the most commonly studied are those involving Cu and Co. Metal co-ordination with glycine can involve either the amine nitrogen and carboxyl oxygen [74, 75] or both carboxylate oxygens [76, 77].

Proline (Fig. 22.2(f)) and hydroxyproline (imino acid) are very different from the other amino acids in that they contain the pyrrolidine five-membered ring. Also, because of proline's unique conformation, its presence in a protein chain can disrupt the α-spiral and give rise to a collagen spiral. For this reason it is referred to as a helix breaker. Furthermore, there are steric interactions of the proline carboxyl group with neighbouring polypeptide residues. Because of its

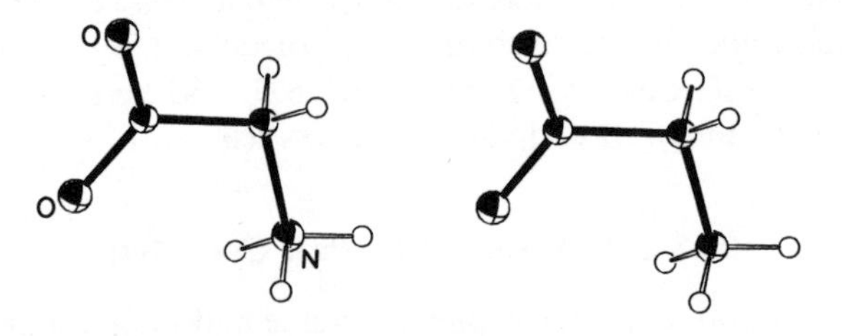

Figure 22.10 Stereo conformation of γ-glycine [67].

unique character, many proline derivatives (> 35), particularly the *t*-butoxycarbonyl and acetyl amides, have been studied as model systems to delineate the neighbouring group effects of proline in polypeptide chains [78–82].

In addition to the normal conformational descriptors the conformation of proline is described by the mode of puckering of the pyrrolidine ring. While there are many conformational descriptors of five-membered rings [83–86], most denote the five-membered ring according to the symmetry element present. The envelope form (C_S-mirror symmetry) has four atoms in a plane with the fifth above or below this plane, and the half-chair form (C_2-two-fold symmetry) has three atoms in a plane with the other two above and below it. Proline ring conformations are most often expressed [87] in terms of the approximate symmetry (C_S or C_2), whether C^{α}, C^{β}, or C^{γ} is outside the plane and the direction of shift of the atom from the plane in relation to C′ (*endo* or *exo*). In addition, the conformation is further defined by whether χ^1 is negative (A) or positive (B) [20]. Therefore, the torsion parameters χ^1 ($N-C^{\alpha}-C^{\beta}-C^{\gamma}$), χ^2 ($C^{\alpha}-C^{\beta}-C^{\gamma}-C^{\delta}$), χ^3 ($C^{\beta}-C^{\gamma}-C^{\delta}-N$), χ^4 ($C^{\gamma}-C^{\delta}-N-C^{\alpha}$) and χ^5 ($C^{\delta}-N-C^{\alpha}-C^{\beta}$) are required in order to ascertain the pucker of the pyrrolidine ring.

For most proline derivatives, ψ is near $\pm 10°$ with a few values near $\pm 25°$. There is an equal distribution of the A and B conformations as designated by the sign of χ^1. Detailed comparisons of several protected proline derivatives, as well as free amino acids, have been carried out by Benedetti and his coworkers [49, 79, 80] in order to delineate the preferred conformations of these molecules and their mode of packing. In general, the pyrrolidine ring adopts a C_S or envelope conformation with a β-pucker (C^{β}-*exo* or *endo*) or *N*-pucker. Those that have C_2 or half-chair conformations have C^{β}-*exo*-C^{γ}-*endo* as their preferred conformation. These analyses [49, 79, 80] suggest that the conformation of the flexible five-membered proline ring in these structures is determined, for the most part, by crystal-packing considerations and all possible puckerings involving C^{α}, C^{β}, C^{γ} and C^{δ} are observed. Puckering involving C^{δ} is usually not found because this places the C^{δ} hydrogens in unfavourable non-bonded interactions by eclipsing the carbonyl of an adjacent residue. In the neutron diffraction study of 4-hydroxyproline [88] (Fig. 22.11) the five-membered ring adopts the C_S envelope with a C^{γ}-pucker.

There are relatively few proline–metal chelates reported (< 20) and most

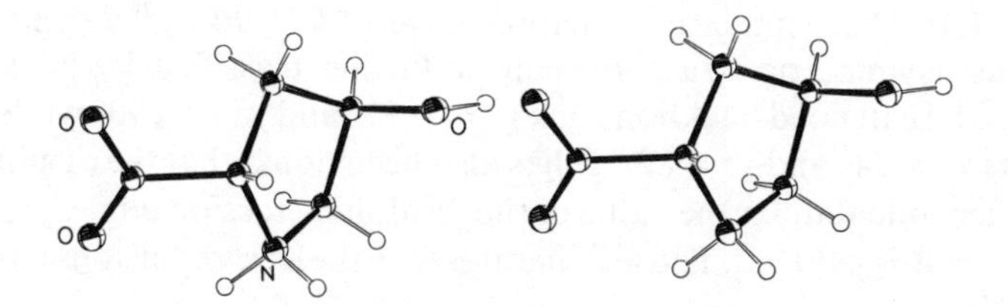

Figure 22.11 Stereo conformation of 4-hydroxyproline [88].

involve Cu in which the co-ordination is through the nitrogen and carboxylic oxygen [89, 90]. The proline ring is observed in both C_2 and C_S conformations, although C_S is more common. In other metal chelates, such as Mn [91], chelation is through the carboxylic oxygens.

The sulphur-containing amino acids cysteine (Fig. 22.2(g)) and cystine (Fig. 22.2(h)) have attracted much attention recently because the cysteine thiol group of specific protein residues is important in the active or catalytic sites of enzymes. Studies have also suggested that the thiol can act as a hydrogen bond donor or acceptor. In addition, cystine plays a special role as a cross-linking agent in protein structure since the two cysteine half-residues are joined by a covalent disulphide linkage.

There are only two torsion angles needed to describe cysteine: ψ and χ^2 ($N-C^{\alpha}-C^{\beta}-S^{\gamma}$). In the more than 20 cysteine derivatives reported, ψ tends to have values near zero or $-30°$ (Table 22.1). Most of these structures [92] have χ^1 values near $\pm 60°$, while there is only one example with $\chi^1 = 180°$. This unusual *trans* conformation occurs in one of the two independent molecules in the crystal structure of L-cysteine [93] (Fig. 22.12(a)). The structure of *N*-acetyl-L-cysteine has been reported from both X-ray [93, 94] and neutron diffraction data [95]. It is of interest to note that while these two determinations give the same conformation, different triclinic crystal lattices are reported and the crystal packing arrangements appear to be different. Structures of cysteic acid [96, 97] show the same conformational preferences as cysteine although the SO_3-moiety can participate in many more hydrogen-bonding interactions.

Because cystine (Fig. 22.2(h)) joins two cysteine residues by a disulphide bond, the added asymmetric centre results in only three stereoisomeric forms rather than four [26]. This arises because the two halves of cystine, though geometrically independent, are chemically identical and thus only L-cystine (L-configuration at both C^{α} atoms), D-cystine (D- at both C^{α} atoms) and *meso*-cystine (L- at one C^{α}, and D-at the other) are found. The molecular conformation is defined by identical labelling of each cysteine residue resulting in the parameters χ^{11}, χ^{12} ($N-C^{\alpha}-C^{\beta}-S^{\gamma}$), χ^{21}, χ^{22} ($C^{\alpha}-C^{\beta}-S^{\gamma}-S^{\gamma'}$) and χ^3 ($C^{\beta}-S^{\gamma}-S^{\gamma'}-C^{\beta'}$). Since the value of χ^3 is usually near 90°, two configurations are possible [26] corresponding to $\chi^3 \approx \pm 90°$. This chirality of disulphides gives rise to dissymmetric chromophores. Most cystine structures fall into one of two conformational types; those with approximate two-fold symmetry between the molecular halves [98], (Fig. 22.12(b)) and those that are asymmetric [99–103] (Fig. 22.12(c)). The symmetric structures have $\chi^3 \approx \pm 90°$, $\chi^2 \approx \pm 60°$ and $\chi^1 \approx \pm 60°$. The asymmetric structures can be further classified by backbone type [26, 102]. The extended backbone has $\chi^3 = -81°$ and $\chi^2 = -90°$ while the coiled S-type has $\chi^3 = 74°$ and $\chi^2 = 82°$. It has also been noted that the disulphide bond length is dependent upon the value of the disulphide torsion angle, χ^3, because of lone-pair repulsions [103]. Those structures with the least repulsions have $\chi^3 \approx 90°$ (Fig. 22.12(c)).

Metal complexes of sulphur-containing amino acids have an important role because of their potential therapeutic action against toxic heavy metals. Cysteine

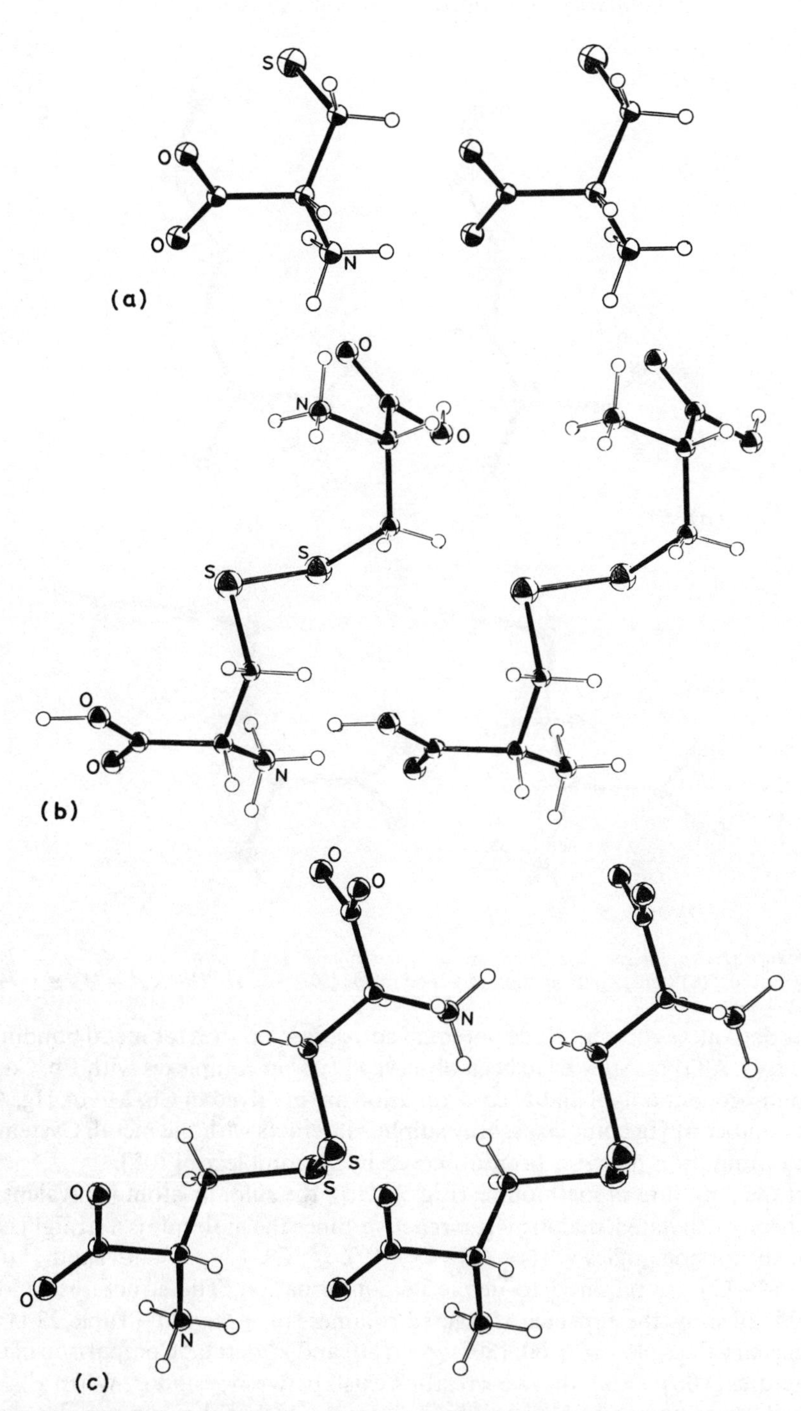

Figure 22.12(a) Stereo conformation of L-cysteine [93]; (b) stereo conformation of L-cystine dihydrochloride [103]; (c) stereo conformation of L-cystine dihydrobromide dihydrate [99].

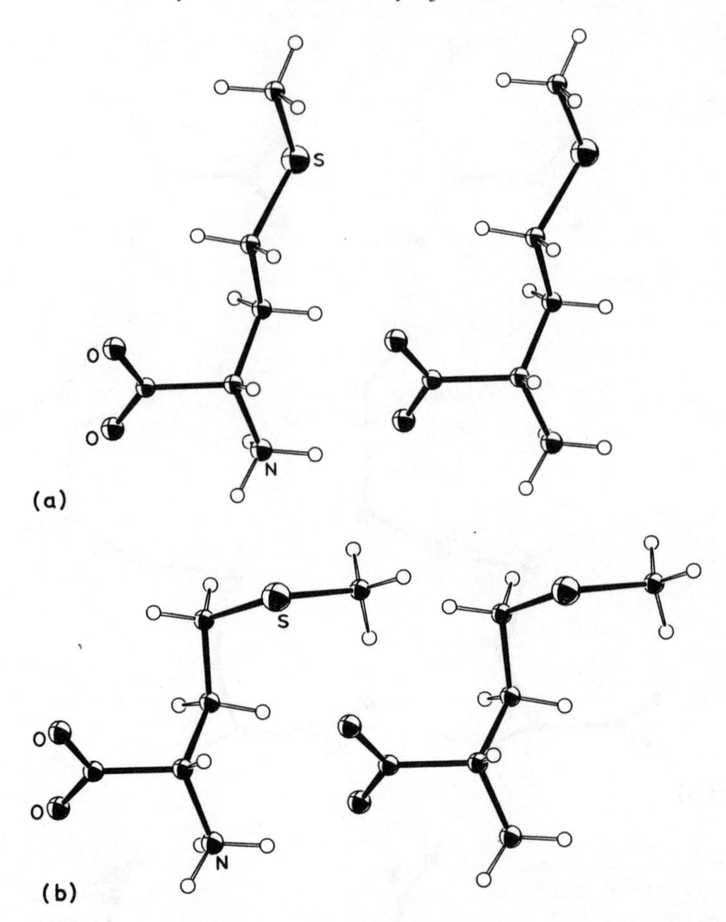

Figure 22.13(a) Stereo conformation of L-methionine [107] with $\chi^1 = 194°$, $\chi^2 = 174°$, $\chi^3 = 180°$; (b) stereo conformation of L-methionine [107] with $\chi^1 = 194°$, $\chi^2 = 74°$, $\chi^3 = 74°$.

or its derivatives present three potential co-ordination sites for metal bonding: S, N and O. All three sites have been observed [104] in complexes with Pb, Cd, Co, Cr and Mo, but only S and N co-ordination are involved in Cu, Mo or Hg. Also, in a number of Hg complexes, only sulphur interacts with the metal. Cysteine, S, N co-ordination has also been observed in Sn complexes [105].

In the structure of methionine (Fig. 22.2(i)), the sulphur atom is bivalent and has been methylated, making it less reactive. Since the molecule is a straight chain only the torsion angles χ^1 ($N-C^{\alpha}-C^{\beta}-C^{\gamma}$), χ^2 ($C^{\alpha}-C^{\beta}-C^{\gamma}-C^{\delta}$) and χ^3 ($C^{\beta}-C^{\gamma}-S^{\delta}-C^{\varepsilon}$) are required to define its conformation. The structures reported (>20) all show the expected staggered rotamers of $\pm 60, 180°$ (Table 22.1) with preferred values of $\chi^1 = \pm 60, 180°$; $\chi^2 = 180°$ and $\chi^3 = 180°$. Comparison of these structures [106] reveals that a correlation exists between χ^1 and χ^2. When $\chi^1 = 60°$, $\chi^2 = \pm 60°$ is unfavourable because it places the carboxyl group too close to the sulphur; likewise when $\chi^1 = 180°$ and $\chi^2 = -60°$. Two conformations [107] of L-methionine are illustrated in Fig. 22.13.

The metal complexes of methionine (< 20) that have been reported involve Cu, Zn, Pt, Pd and Hg. Those with Pt or Pd form strong metal–S bonds with S, N co-ordination [108], whereas Cu has no sulphur co-ordination [109].

22.3.3 Aromatic (His, Phe, Tyr, Trp)

The aromatic amino acid residues (Fig. 22.2(j)–(m)) histidine, phenylalanine, tyrosine and tryptophan are all characterized by having planar groups attached at the C^β position. Histidine plays a special role in the active sites of many enzymes because of the basic properties of the imidazole ring which may or may not be protonated depending on the pH of its environment. The geometry and conformation of the histidine moiety have been observed in both its protonated and unprotonated states as well as in complexes with several metals. Because the imidazole ring is planar, only the torsion angles ψ, χ^1 and χ^{21}, χ^{22} ($C^\alpha - C^\beta - C^\gamma - B^\delta$) are required to define the histidine conformation. Analysis of the structure reports (> 20), including neutron diffraction studies, shows ψ to be near $-25°$, χ^1 is $\pm 60°$ with no examples of $\chi^1 = 180°$, and χ^2 is near $\pm 90°$, as illustrated in the structure of L-histidine HCl [110] (Fig. 22.14). In the crystal of L-N-acetylhistidine H_2O [111], the two independent molecules in the structure have $\chi^1 = 60°$ and $-60°$, and are stabilized by the hydrogen bonding network. The neutron diffraction study of L-histidine [112] shows the molecule is a zwitterion ($NH_3^+COO^-$) with an intramolecular hydrogen bond from the alpha amino group to the ring imidazole nitrogen, stabilizing the $\chi^1 = 60°$ conformation. This same conformation is observed in histidine dihydrochloride [113] in which the imidazole ring is protonated as well. The structure of 5-nitro-L-histidine H_2O [114] also has a protonated imidazole ring, $\chi^1 = -60°$, and no intramolecular hydrogen bond.

Most of the metal chelates of histidine involve Cu and Co, with complexes of Ni, Cd, Cr, Mo and Hg also reported. Characteristic of many of these complexes

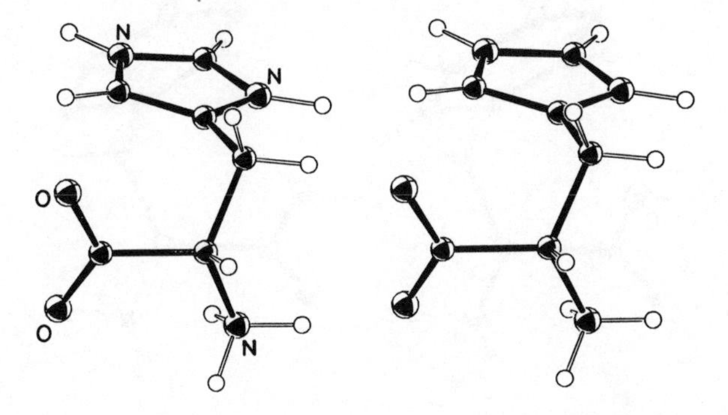

Figure 22.14 Stereo conformation of L-histidine hydrochloride monohydrate [110].

is tridentate co-ordination [115] with the carboxylic oxygen, alpha nitrogen and the imidazole nitrogen. Other complexes are bidentate with co-ordination through the amino acid and imidazole nitrogens [116].

Phenylalanine (Fig. 22.2(k)) can be described by the same torsion angles as shown for histidine. However, because it lacks any functional groups in its ring, it does not possess the same range of interactions. As pointed out in reviews [19, 21, 117], phenylalanine has been observed in all preferred χ^1 conformations ($\pm 60°$, $180°$) and χ^2 values near 90° as illustrated in the structure of phenylalanine HCl [118] (Fig. 22.15). There have been relatively few (<5) reports of phenylalanine–metal complexes. These involve Cu [119, 120], although there are unpublished reports of Pt and Co complexes [16].

The conformation of tyrosine (Fig. 22.2(l)) has been of great interest because it is involved in the catalytic site reaction mechanisms of many enzymes, and it is a metabolic precursor of both the neurotransmitters of the adrenergic system and of the thyroid hormones (Fig. 22.1). The molecular conformations of tyrosine have been reviewed [21] and like phenylalanine, all the energetically favourable conformers [121] of χ^1 and χ^2 have been observed. While the preferred conformation for ψ is relatively close to zero, i.e. the carbonyl oxygen is *cis* to the nitrogen, there are a few examples [122–124] in which the carbonyl oxygen is *trans* to the nitrogen. Another molecule that showed unusual structural characteristics is the diaquasodium salt of 3,5-dinitrotyrosine [125]. In the two crystallographically independent tyrosine molecules in the structure, not only do the two amino acid side-chain conformations differ ($\chi^1 = 57°/-71°$), but the phenolic bonds differ significantly as well; one is a hydroxyl with an intramolecular hydrogen bond and the other is a phenoxide ion. There are surprisingly few reports (<5) of tyrosine–metal complexes. Those reported involve co-ordination through the N and O of the amino acid terminus to Cu [126] Pd [127], Ni [128] and Hg [129].

The molecular conformation of the halogenated tyrosine derivatives, mono- and diiodotyrosine, dibromotyrosine and the halogenated thyronines (thyroid

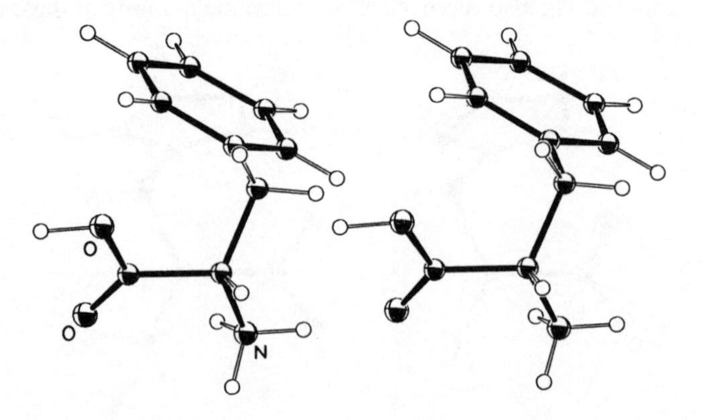

Figure 22.15 Stereo conformation of L-phenylalanine hydrochloride [118].

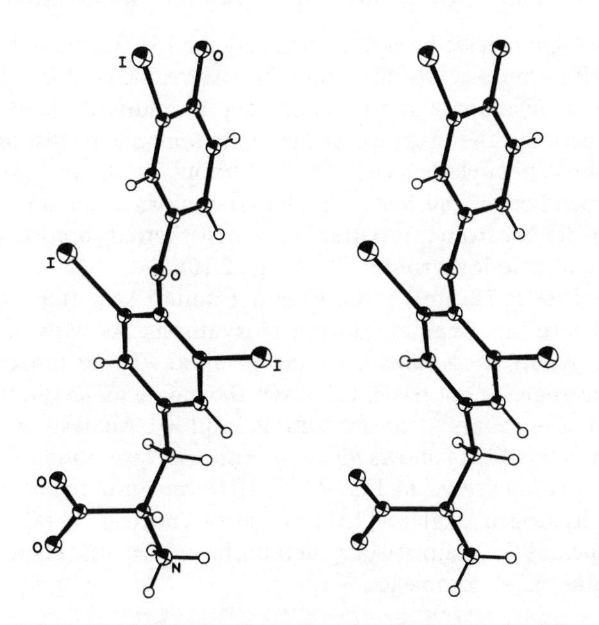

Figure 22.16 Stereo conformation of L-triiodothyronine [6].

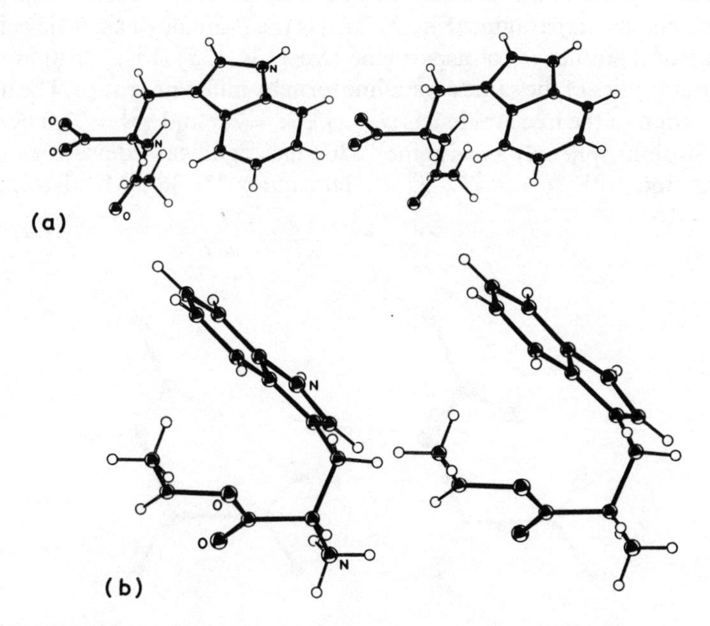

Figure 22.17(a) Stereo conformation of *N*-acetyl-L-tryptophan [132]; (b) stereo conformation of DL-tryptophan ethyl ester HCl [133].

hormone analogues), have been described in detail [5–7]. Those thyroid hormone analogues with amino acid side-chains show the same conformational patterns revealed for phenylalanine and tyrosine [21]. In addition, there is an apparent correlation between the presence of the alpha amino nitrogen and the conformation of the diphenylether [6]. Those without the amino group adopt the skewed (perpendicular and bisecting planes) conformation, and those with the amino group deviate from this pattern to a twist-skewed conformation as seen in the structure of triiodothyronine [6] (Fig. 22.16)

Tryptophan (Fig. 22.2(m)) is an essential amino acid that is known to be transformed into nicotinamide during biosynthesis. As with the other amino acids in this group, it contains a planar group at C^{β}, but unlike histidine, the indole ring nitrogen is not basic. However, the indole moiety in tryptophan can participate in $\pi - \pi$ charge-transfer complexes [130]. Analysis of the tryptophan torsion parameters [131] shows all the χ^1 rotamers are observed and χ^2 values are near 90°, as illustrated in Fig. 22.17. However only in the structure of D-tryptophan hydrogen oxalate [131] is the extended $\chi^1 = 180°$ conformation observed, whereas the majority of structures have $\chi^1 = 60°$. There are no reports of tryptophan–metal complexes.

22.3.4 Polar (Asn, Gln, Ser, Thr)

The amino acid residues asparagine, glutamine, serine and threonine (Fig. 22.2(n)–(q)) are all characterized by having polarizable functional groups in their side-chains. Asparagine (Fig. 22.2(n)) is the β-amide of aspartic acid. Of the few structural studies (4) of asparagine two [134, 135] (Fig. 22.18) are neutron diffraction studies of the same crystalline form by different groups. The molecular conformation of the free amide has $\psi = 11°$, $\chi^1 = 72°$ and $\chi^{21} = 3°$. The structure of N-(2,4-dinitrophenyl) asparagine [136] has significant deviations from this conformation with $\psi = -47°$, $\chi^1 = -60°$ and $\chi^{21} = 36°$. Metal complexes of

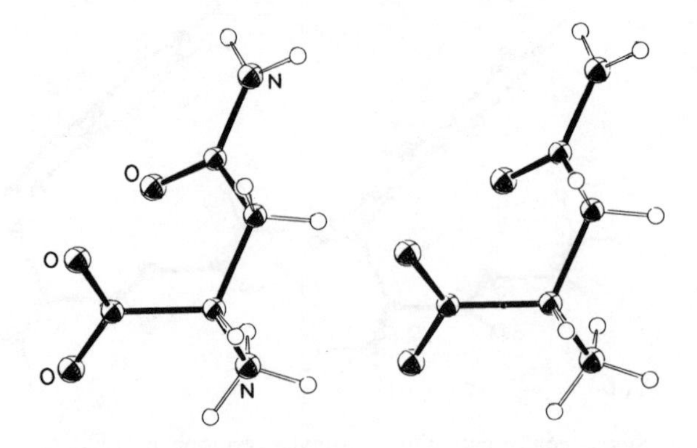

Figure 22.18 Stereo conformation of asparagine [134].

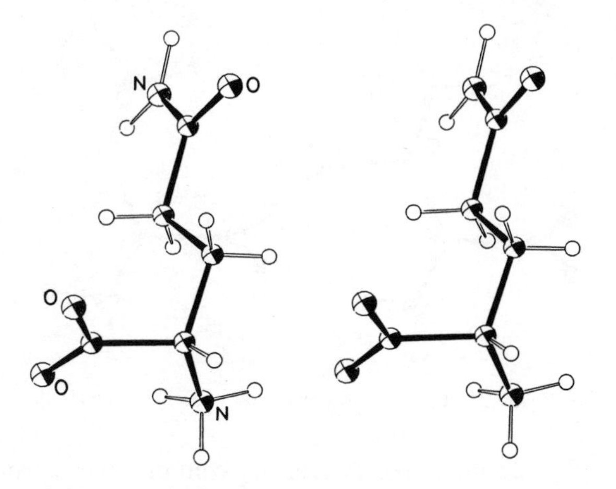

Figure 22.19 Stereo conformation of L-glutamine [139].

asparagine [137] with Cu, Zn, Cd and Co have been of interest for their use as potential cancer chemotherapy agents.

Glutamine (Fig. 22.2(o)) is the γ-amide of glutamic acid and plays an essential role in metabolic processes. The amide in this molecule is also more susceptible to hydrolysis in buffered solutions and will decompose under conditions that leave other amides unchanged. The torsion parameters that describe its conformation are $\psi \approx -18°$, $\chi^1 \simeq \pm 60°$, $\chi^2 \simeq 180°$ (with two exceptions [138]), and χ^{31}, χ^{32} near $\pm 15°$, indicating a planar amide group. The neutron diffraction structure [139] (Fig. 22.19) provides accurate parameters for the amide group. There are no reports of metal chelates of glutamine.

The conformation of serine (Fig. 22.2(p)), a β-hydroxylated alanine structure, is described by the torsion angles ψ and χ^1 ($N-C^\alpha-C^\beta-O^\gamma$). Of the reported structures (>15) all but one have the same conformation with $\psi \simeq 5°$, $\chi^1 \simeq 60°$. The structure of the 1:1 complex of L-serine-L-ascorbic acid [140] (Fig. 22.20) has $\psi = -28°$, $\chi^1 = -61°$. Structures of serine co-ordinated with Cu, Zn, Ni, Co and Pd have been studied. When co-ordination is only through the amino acid functional groups, $\chi^1 \approx 60°$, except in a Ni complex [141] when χ^1 is $-72°$.

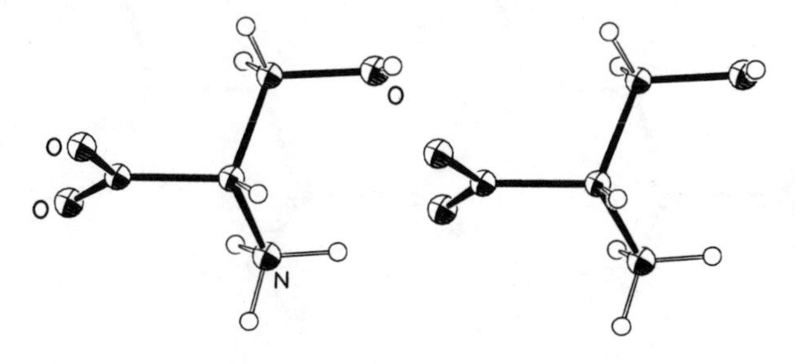

Figure 22.20 Stereo conformation of L-serine [140].

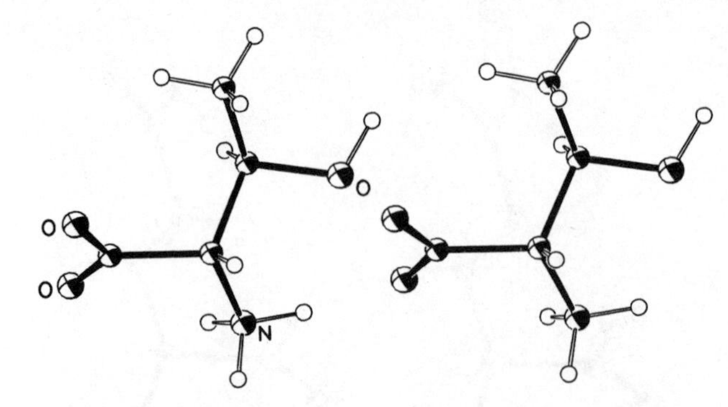

Figure 22.21 Stereo conformation of L-threonine [142].

Threonine (Fig. 22.2(q)), like isoleucine, contains two asymmetric carbon atoms and thus has four structural isomers [26]. The torsion parameters χ^{11} ($N-C^{\alpha}-C^{\beta}-O^{\gamma}$) and χ^{12} ($N-C^{\alpha}-C^{\beta}-C^{\gamma}$) describe the positions of the hydroxyl and methyl groups. For most structures (>10) $\psi \simeq -25°, \chi^{11} \simeq -60°$ with $\chi^{12} \simeq 180°$ or $\chi^{11} \simeq 60°, \chi^{12} \simeq -60°$, as illustrated by Fig. 22.21. To date, the only structures with $\chi^{11} \simeq 60°$ are part of modified nucleoside structures [143] in which the hydroxyl participates in an intramolecular hydrogen bond. There are only a few (<5) reports of threonine–metal complexes, and these involve Cu and Zn.

22.3.5 Charged (Lys, Arg, Asp, Glu)

The amino acid residues lysine and arginine (Fig. 22.2(r) and (s)) have a formal positive charge on their side-chains, and aspartic and glutamic acids (Fig. 22.2(t) and (u)) have a negative charge. Lysine is one of four basic amino acids and is

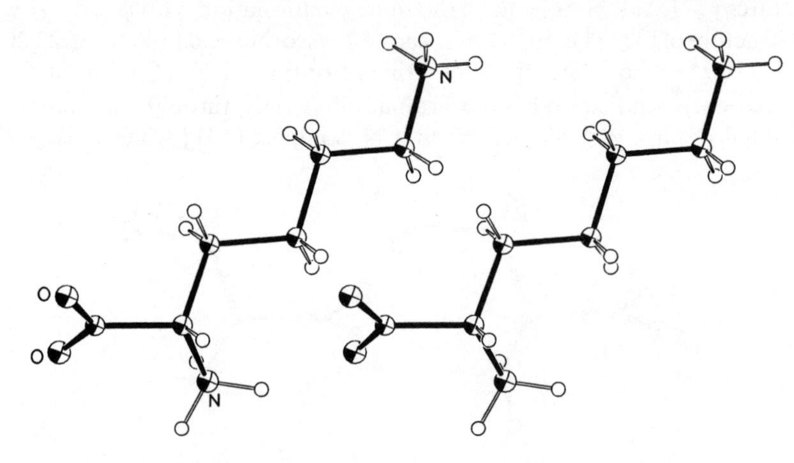

Figure 22.22 Stereo conformation of L-lysine [144].

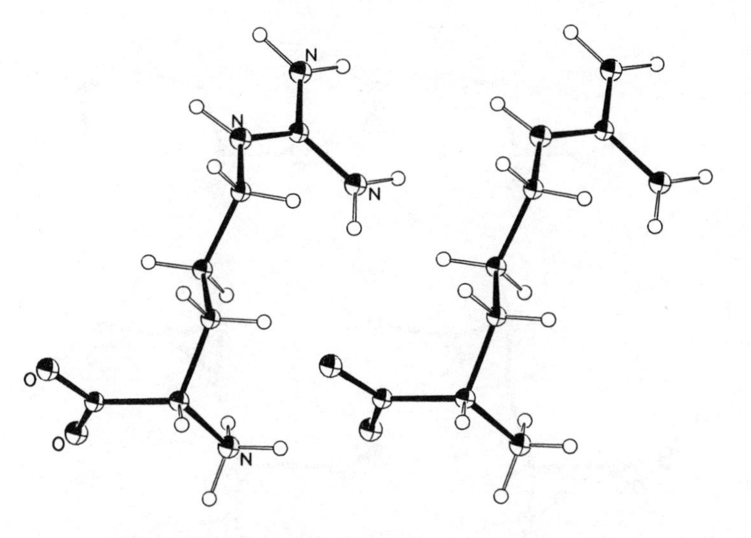

Figure 22.23 Stereo conformation of L-arginine [146].

characterized by having an amine group at C^{ε}. As a simple straight chain, the preferred conformation is fully extended [144], as shown in Fig. 22.22 with $\psi = -20°$, $\chi^1 = -60°$, $\chi^2 = \chi^3 = \chi^4 = 180°$. When co-ordinated to Pt, the only reported metal complex of lysine [145], the conformation twists ($\chi^1 = -179°$, $\chi^2 = 168°$, $\chi^3 = -171°$, $\chi^4 = -73°$) so as to co-ordinate the terminal amine.

Arginine (Fig. 22.2(s)) is a basic amino acid with a guanidyl side-chain functional group. It is usually observed in the extended conformation [146] with $\psi = -15°$, $\chi^1 = \pm 60°$, $\chi^2 = 180°$, $\chi^3 = 180°$ and $\chi^4 = \pm 10°$, demonstrated in Fig. 22.23. It can also adopt the conformation with $\chi^1 = 180°$, $\chi^2 = 180°$, and $\chi^3 = 180°$, and in a structure where arginine is complexed with diethylphosphoric acid [147], $\chi^3 = -60°$. There is one report of an arginine–Co complex [148] which shows the Co co-ordinated by the N and O of the amino acid and the nitrogen of the guanine.

While aspartic acid (Fig. 22.2(t)) has been the least studied (< 4) amino acid, its conformation is the most varied. The χ^1 value is $-60°$ for all the structures, but ψ

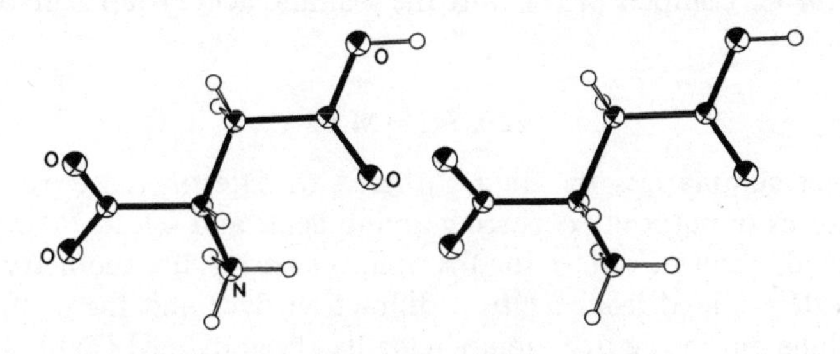

Figure 22.24 Stereo conformation of DL-aspartic acid [150].

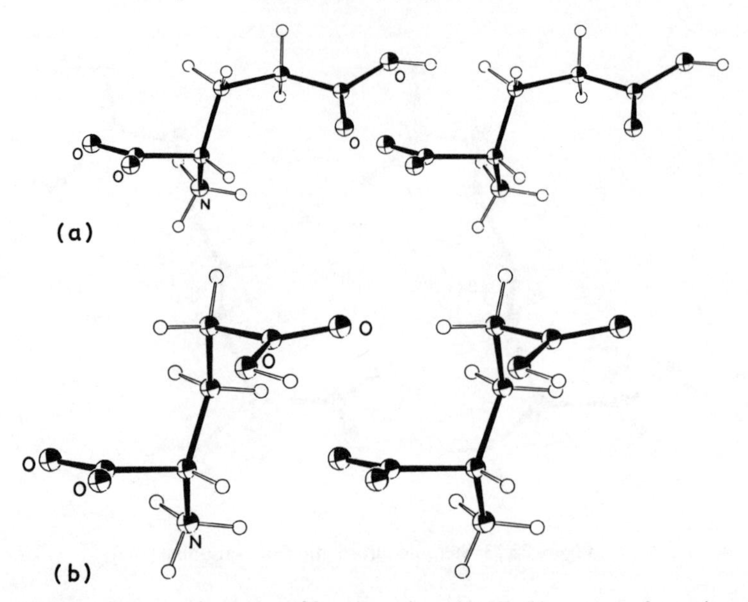

Figure 22.25 (a) Stereoconformation of β-L-glutamic acid [155]; (b) stereo conformation of α-L-glutamic acid [154].

and χ^{21}, the two carboxylic acid descriptors, differ considerably: 41°, 72°; −7°, 3°; −38°, 131° for the salt [149], DL- [150] and L-forms [151], respectively. The structure of DL-aspartic acid [150] is shown in Fig. 22.24. Metal complexes with Zn, Ni, Co [152] show tridentate co-ordination resulting in a folded aspartate conformation.

Glutamic acid (Fig. 22.2(u)), like glycine, crystallizes in more than one polymeric form. The more stable α-form will convert to the β-form on long-term standing in solution. The conformation observed for most glutamic acid structures [153, 154] (> 25) has $\psi \approx 0°$ or $-30°$, χ^1 and $\chi^2 = \pm 60°$, 180°, and $\chi^3 = \pm 20°$ (Fig. 22.25 (a)), except in a few structures where χ^3 is 75° (Fig. 22.25(b)). When complexed with Cu, Zn, Co, and Cd [156], all of the glutamic acid functional groups are involved in the co-ordination, but not as a tridentate ligand. The γ-carboxylic group co-ordinates to another metal in the lattice. In the Ca complex [157], only the α-amino acid group is involved in the coordination.

22.4 SUMMARY

This chapter summarizes the data gathered to date on the crystal structure determinations of naturally occurring amino acids and selected derivatives. As demonstrated, accurate values for the amino acid residue geometry (Fig. 22.3) have been determined from neutron diffraction data and the conformational flexibility inherent in the free amino acids has been defined (Table 22.1) as the result of multiple structure determinations of the same amino acid in various crystalline environments and with various functional groups derivatized. In addition, hydrogen bond geometries have been accurately described, and patterns in hydrogen bond directionality delineated. These data can now provide the basis for interpretation of residue side-chain conformation in protein structures.

Since these data were restricted exclusively to the simple amino acid structures, the concern is how well these results correlate with observed side-chain conformations in peptide and protein environments. With the availability of many accurate protein crystal structures [158–159], as well as numerous peptide structures [1, 2], efforts have been made to study these correlations [39, 40, 49]. The observed distributions of χ^1 and χ^2 in protein cyrstal structures have been compared with predictions made on the basis of energy calculations and with free amino acid structural data, and the results are in good agreement.

It has been shown [39] that steric hindrance limits the range of χ^1 and χ^2 for most residues, and the least-favourable conformations observed in the free amino acid structures are rarely found in proteins. Thus, $\chi^1 = 60°$ has been observed for some valine, tyrosine and phenylalanine structures [21, 49], but is seldom observed in protein structures because of the unfavourable side-chain interactions with functional groups of adjacent peptide residues. Also, the range of observed conformations is small and the distribution among permitted configurations varies only slightly with the main-chain conformation and with the residue position relative to the protein surface [39].

In general, the χ^1 and χ^2 distributions follow the same pattern observed in the free amino acids (Table 22.1). Of the three χ^1 minima ($\pm 60°, 180°$), $\chi^1 = -60°$ is the most favourable. Once the initial choice of χ^1 is made during protein folding, small variations ($\sim 15°$) around the preferred values occur to permit optimal packing of the side-chains in the protein interior [39]. The preferred χ^2 values are dependent upon the type of C^γ present: 180° for tetrahedral substituents and $\pm 90°$ for trigonal or planar groups.

Other studies of the main-chain conformational similarities among amino acid residues [40] show that amino acids with different chemical properties can have similar (ϕ, ψ) probability distributions indicating that more than the amino acid structural data are required to predict protein three-dimensional conformation. These data show the residues proline and glycine take unique main-chain conformations in proteins. These probability distributions also indicate that certain amino acid residue pairs exhibit main-chain conformational similarities which are reciprocal (i.e., Ala–Ser) whereas others are not (Leu–Ile). This means that when the probability distribution of leucine is compared with those of other residues, it is most similar to isoleucine, but when isoleucine is compared, it is found to be most similar to valine [40].

Two examples of how knowledge of amino acid conformational flexibility and hydrogen bonding directionality have been used in the interpretation of protein active site requirements involve thyroid hormone studies [160–162]. In order to investigate likely orientations of thyroid hormones at their protein binding sites, theoretical energy calculations [160] of inter- and intramolecular hydrogen bond strengths of model compounds were computed. The hydrogen bonding (O... O,N) preferences and geometry were in general agreement with the hydrogen bonding patterns observed from thyroid hormone analogue structural studies [161]. Additionally, the relationships among amino acid sequence, structure, and biological properties were exploited to permit an evaluation of the conformational factors that could explain the peptide sequence specificity observed in the diiodotyrosine coupling reaction to form thyroxine in thyroglobulin [162].

These studies indicate the progress that has been made in the understanding of the factors that control peptide and protein conformation and ultimately their structure–activity relationships in biological processes.

Table 22.1 Preferential conformations of amino acid residues*

Residue	ψ	χ^{11}	χ^{12}	χ^{21}	χ^{22}	χ^{3}	χ^{4}	χ^{5}
Alanine	5°; – 19°							
Valine	– 11; – 35	– 60°	180°					
Leucine	– 18; ± 36	180		– 60°	180°			
		– 60		– 60	180			
		60		60	180			
Isoleucine	– 15; – 45	60	180	180				
		180	– 60	180				
		– 60	180	180				
Glycine	0;25							
Proline	± 10	± 35		∓ 35		± 25°	∓ 20°	± 10°
Cysteine	– 30	– 60						
		60						
Methionine	± 20	180		180		– 60		
		– 60		180		180		
		60		180		180		
Histidine	– 25	– 60		60	– 120			
		60		60	– 120			
Phenylalanine	± 20	180		90				
		– 60		90				
		60		90				
Tyrosine	± 20	– 60		90				
		180		90				
		60		90				
Tryptophan	– 10	– 60		90				
		60		90				
Asparagine	11	60		0				

Table 22.1 (*Contd.*)

Residue	ψ	χ^{11}	χ^{12}	χ^{21}	χ^{22}	χ^{3}	χ^{4}	χ^{5}
Glutamine	−20	60		180		−15		
		−60		180		25		
Serine	5	60						
Threonine	−25	−60	180					
		60	−60					
Lysine	−20	−60		180		180	180	
Arginine	−15	60		180		180	±10	
		−60		180		180	±10	
Aspartic	−7; ±35	−60		±5				
Glutamic	0; ±35	180		180		±15		
		−60		180		±15		
		60		180		±15		

* These values approximate the more commonly observed conformations with the most frequent listed first.

ACKNOWLEDGEMENTS

The author is deeply grateful to Dr John C. Huffman, Indiana University, for his generosity in providing all the stereo drawings for this chapter and for his tireless efforts to make the software changes necessary to access the *Cambridge Data Files* used in this compilation. Molecules were located using the Indiana University Molecular Structure Center XTEL Library version of the Cambridge Files and the molecular graphics programs of J.C. Huffman. The services of the Indiana University Computing Facilities are also acknowledged. The critical review by Dr Jane Griffin and Dr R. Parthasarathy is appreciated as are the efforts of Mrs C. Devine and Miss Danielle Smith to compile the literature survey, Miss Melda Tugac and Miss Gloria Del Bel for graphic illustration and Mrs Brenda Giacchi for expert secretarial assistance. This work was supported in part by grant DHEW AM-15051.

REFERENCES

1. Kennard, O., Watson, D.G., Allen, F.H. and Weeds, S.M. (eds), *Molecular Structures and Dimensions* (Vol. 1–11) D. Reidel, Dordrecht, The Netherlands.
2. *Cambridge Crystallographic Data File*, Crystallographic Data Centre, University Chemical Laboratory, Cambridge, UK.
3. Sim, G.A. and Sutton, L.E., *Molecular Structure by Diffraction Methods*, A Specialist Periodical Report (Vol. 1–6) The Royal Society of Chemistry, London.

4. Sheppard, R.C., *Amino Acids, Peptides and Proteins*, A Specialist Periodical Report (Vol. 1–13) The Royal Society of Chemistry, London.
5. Cody, V. (1978) in *Recent Progress in Hormone Research*, (ed. R.O. Greep) (Vol. 34) Academic Press, New York, 1978, p. 437.
6. Cody, V. (1980) *Endocrine Rev.*, **1**, 140.
7. Cody, V. (1981) *Acta Crystallogr.*, **B37**, 1685.
8. Klyne W. and Prelog, V. (1960) *Experientia*, **16**, 521.
9. Cantor, C.R. and Schimmel, P.R. (1980) *Biophysical Chemistry*, Part II, Freeman and Co., San Francisco, pp. 687–846.
10. Glusker, J.P. and Trueblood, K.N. (1972) *Crystal Structure Analysis: A Primer*, Oxford University Press, London.
11. Stout, G.H. and Jensen, L.H. (1968) *X-Ray Structure Determination: A Practical Guide*, Macmillan, New York.
12. Bacon, G.E. (1977) *Neutron Scattering in Chemistry*, Butterworths, London.
13. Dachs, H. (ed.) (1978) *Neutron Diffraction*, Springer-Verlag, New York.
14. Gurskaya, G.V. (1968) *The Molecular Structure of Amino Acids*, Consultants Bureau, New York.
15. March, R.E. and Donohue, J. (1967) *Adv. Prot. Chem.*, **22**, 235.
16. Freeman, H.C. (1967) *Adv. Prot. Chem.*, **22**, 257.
17. Osterberg, R. (1974) *Coord. Chem. Rev.*, **12**, 309.
18. Saito, Y. (1974) *Coord. Chem. Rev.*, **13**, 305.
19. Lakshminarayanan, A.V., Sasisekharan, V. and Ramachandran, G.N. (1967) in *Conformation of Biopolymers*, (ed. G.N. Ramachandran) (Vol. 1) Academic Press, New York, p. 61.
20. Balasubramanian, R., Lakshminarayanan, A.V., Sebesan, M.N. *et al.* (1971) *Int. J. Protein Res.* **III**, 25.
21. Cody, V., Duax, W.L. and Hauptman, H.A. (1973) *Int. J. Peptide Protein Res.*, **5**, 297.
22. Ponnuswamy, P.K., Lakshminarayanan, A.V. and Sasisekharan, V. (1971) *Biochim. Biophys. Acta*, **229**, 596 and other papers in this series.
23. Perahia, D., Maigret, B. and Pullman, B. (1970) *Theor. Chim. Acta*, **19**, 121 and other papers in this series.
24. Ramachandran, G.N. and Ramakrishnan, C. (1968) in *Symposium on Fibrous Proteins* (ed. W.G. Grewther), Plenum Press, New York.
25. Zimmerman, S.S., Pottle, M.S., Memethy, G. and Scheraga, H.A. (1977) *Macromolecules*, **10**, 1.
26. Parthasarathy, R. (1977) in *Stereochemistry Fundamentals and Methods*, (ed. H.B. Kagan) (Vol. 1) Georg Thieme Publishers, Stuttgart, p. 182.
27. Chen, C.-S. and Parthasarathy, R. (1978) *Int. J. Peptide Protein Res.*, **11**, 9.
28. Mitra, J. and Ramakrishnan, C. (1977) *Int. J. Peptide Protein Res.*, **9**, 27.
29. Mitra, J. and Ramakrishnan, C. (1981) *Int. J. Peptide Protein Res.*, **17**, 401.
30. Zefirov, YU.V. (1977) *J. Gen. Chem. USSR*, **46**, 2519.
31. Vinogradov, S.N. (1979) *Int. J. Peptide Protein Res.*, **14**, 281.
32. Hamilton, W.C., Frey, M.N., Golic, L. *et al.* (1972) *Mater. Res. Bull.*, **7**, 1225.
33. Koetzle, T.F. and Lehmann, M.S. (1976) in *The Hydrogen Bond – Recent Developments in Theory and Experiments* (ed. P. Schuster), North-Holland Publishing Co., Amsterdam, p. 459.
34. Ramanadham, M. and Chidambaram, R. (1978) in *National Conference in Crystallography*, New Delhi, Oxford and IBH, Oxford.
35. Jeffrey, G.A. and Maluszynska, H. (1982) *Int. J. Biol. Macromol.*, **4**, 173.
36. Burgess, A.W., Ponnuswamy, P.K. and Scheraga, H.A. (1974) *Israel J. Chem.*, **12**, 239.
37. Nemethy, G. and Scheraga, H.A. (1977) *Quart. Rev. Biophys.*, **10**, 239.
38. Finkelstein, A.V. and Ptitsyn, O.B. (1977) *Biopolymers*, **16**, 469.
39. Janin, J., Wodak, S., Levitt, M. and Maigret, B. (1978) *J. Mol. Biol.*, **125**, 357.
40. Kolaskar, A.S. and Ramabrahmam, V. (1981) *Int. J. Biol. Macromol.*, **3**, 171.
41. IUPAC-IUB Commission on Biochemical Nomenclature (1970), *J. Mol. Biol.*, **52**, 1.
42. Johnson, C.K. (1965) *ORTEP, Oak Ridge National Laboratory Report*, ORNL-3794.
43. Pedone, C. and Benedetti, E. (1972) *Acta Crystallogr.*, **B28**, 1970.
44. DiBlasio, B., Napolitano, G. and Pedone, C. (1977) *Acta Crystallogr.*, **B33**, 542.
45. Lehmann, M.S., Koetzle, T.F. and Hamilton, W.C. (1972) *J. Am. Chem. Soc.*, **94**, 2657.
46. Koetzle, T.F., Golic, L., Lehmann, M.S. *et al.* (1974) *J. Chem. Phys.*, **60**, 4690.
47. Ponnuswamy, P.K. and Sasisekharan, V. (1970) *Int. J. Protein Res.*, **II**, 37.

48. Herak, R., Prelenik, B. and Krstanovic, I. (1978) *Acta Crystallogr.*, **B34**, 91.
49. Benedetti, E. (1977) in *Peptides* (eds M. Goodman and J. Meienhofer), Halsted Press, New York, p. 257.
50. Rao, S.T. (1969) *Z. Kristallogr.*, **128**, 339.
51. Andrianov, Y.G., Kalinin, A.E., Struchkov, Yu.T. *et al.* (1981) *J. Struct. Chim. USSR*, **22**, 49.
52. Lovas, Gy., Kalman, A. and Argay, Gy. (1974) *Acta Crystallogr.*, **B30**, 2882.
53. Czugler, M., Moravcsik, E., Tudos, H. and Lovas, Gy. (1976) *Cryst. Struct. Comm.*, **5**, 651.
54. Jarzab, T.C., Hare, C.R. and Langs, D.A. (1973) *Cryst. Struct. Comm.*, **3**, 395.
55. Glowiak, T. and Ciunik, Z. (1978) *Bull. Acad. Polon. Sci.*, **26**, 43.
56. Harding, M.M. and Howieson, R.M. (1976) *Acta Crystallogr.*, **B32**, 633.
57. Aubry, A., Marraud, M., Cung, M.-T. and Protas, J. (1975) *C. R. Acad. Sci. Paris*, **281**, 861.
58. Varughese, K.I. (1977) *Int. J. Peptide Protein Res.*, **9**, 81.
59. Torri, K. and Iitaka, Y. (1971) *Acta Crystallogr.*, **B27**, 2237.
60. Varughese, K.I. and Srinivasan, R. (1975) *J. Cryst. Mol. Struct.*, **5**, 317.
61. Ponnuswamy, P.K. and Sasisekharan, V. (1971) *Int. J. Protein Res.*, **III**, 9.
62. Weeks, C.M., Cooper, A. and Norton, D.A. (1969) *Acta Crystallogr.*, **B25**, 443.
63. Jonsson, P.G. and Kvick, A. (1972) *Acta Crystallogr.*, **B28**, 1827.
64. Power, L.F., Turner, K.E. and Moore, F.H. (1976) *Acta Crystallogr.*, **B32**, 11.
65. Iitaka, Y. (1960) *Acta Crystallogr.*, **13**, 35.
66. Iitaka, Y. (1961) *Acta Crystallogr.*, **14**, 1.
67. Kvick, A., Canning, W., Koetzle, T.F. and Williams, G.J.B. (1980) *Acta Crystallogr.*, **B36**, 115.
68. Al-Karaghouli, A.R., Cole, F.E., Lehmann, M.S. *et al.* (1975) *J. Chem. Phys.*, **63**, 1360.
69. DiBlasio, B., Pavone, V. and Pedone, C. (1977) *Cryst. Struct. Comm.*, **6**, 745.
70. Mackay, M.F. (1975) *Cryst. Struct. Comm.*, **4**, 225.
71. Fletcher, S.R., Keve, E.T. and Skapski, A.C. (1976) *Ferroelectrics*, **14**, 789.
72. Waskowska, A., Olejnik, S., Lukaszewicz, K. and Ciechanowicz-Rutkowska, M. (1979) *Ferroelectrics*, **22**, 855.
73. Olejnik, S. and Lukaszewicz, K. (1977) *Acta U. Wratislava, Mat. F. As*, **43**, 450.
74. Miyamae, H. and Saito, Y. (1978) *Acta Crystallogr.*, **B34**, 937.
75. Neitzel, C.J. and Desiderato, R. (1975) *Cryst. Struct. Comm.*, **4**, 333.
76. Glowiak, T. and Ciunik, Z. (1978) *Acta Crystallogr.*, **B34**, 1980.
77. Corradi, A.B. (1976) *Cryst. Struct. Comm.*, **5**, 923.
78. Kamwaya, M.E., Oster, O. and Bradaczek, H. (1981) *Acta Crystallogr.*, **B37**, 364; 1391.
79. Benedetti, E., Ciajolo, A., DiBlasio, B. *et al.* (1979) *Int. J. Peptide Protein Res.*, **14**, 130.
80. Benedetti, E., DiBlasio, B., Pavone, V. and Pedone, C. (1981) *Biopolymers*, **20**, 283.
81. Fujinaga, M. and James, M.N.G. (1980) *Acta Crystallogr.*, **B36**, 3196.
82. Druck, U., Littke, W. and Main, P. (1979) *Acta Crystallogr.*, **B35**, 253.
83. Kilpatrick, J.E., Pitzer, K.S. and Spitzer, R. (1947) *J. Am. Chem. Soc.*, **69**, 2483.
84. Pitzer, K.S. and Donath, W.E. (1960) *J. Am. Chem. Soc.*, **81**, 3213.
85. Altona, C., Geise, H.J. and Romers, C. (1968) *Tetrahedron*, **24**, 13.
86. Duax, W.L. and Norton, D.A. (1975) *Atlas of Steroid Structure*, (Vol. 1) Plenum Press, New York.
87. Ashida, T. and Kakudo, M. (1974) *Bull. Chem. Soc. Jpn.*, **47**, 1129.
88. Koetzle, T.F., Lehmann, M.S. and Hamilton, W.C. (1973) *Acta Crystallogr.*, **B29**, 231.
89. Shamala, N. and Venkatesan, K. (1973) *Cryst. Struct. Comm.*, **2**, 5.
90. Aleksandrov, G.G., Struchkov, Yu.T. and Kurganov, A.H. (1973) *Zh. Strukt. Khim.*, **14**, 492.
91. Glowiak, T. and Ciunik, Z. (1977) *Acta Crystallogr.*, **B33**, 3237.
92. Ramachandra Ayyar R. (1968) *Z. Kristallogr.*, **26**, 227.
93. Harding, M.M. and Long, H.A. (1968) *Acta Crystallogr.*, **B24**, 1096.
94. Lee, Y.J. and Suh, I.H. (1980) *J. Korean Chem. Soc.*, **24**, 193.
95. Takusagawa, F., Koetzle, T.F., Kou, W.W.H. and Parthasarathy, R. (1981) *Acta Crystallogr.*, **B37**, 1591.
96. Hendrickson, W.A. and Karle, J. (1971) *Acta Crystallogr.*, **B27**, 427.
97. Ramanadham, M., Sikka, S.K. and Chidambaram, R. (1973) *Acta Crystallogr.*, **B29**, 1167.
98. Rosenfield, R.E. Jr. and Parthasarathy, R. (1975) *Acta Crystallogr.*, **B31**, 462.
99. Rosenfield, R.E. Jr. and Parthasarathy, R. (1975) *Acta Crystallogr.*, **B31**, 816.
100. Vijayalakshmi, B.K. and Srinivasan, R. (1975) *Acta Crystallogr.*, **B31**, 993.

101. Kominani, S., Riesz, P., Akiyama, T. and Silverton, J.V. (1976) *J. Phys. Chem.*, **80**, 203.
102. Gupta, S.C., Sequeira, A. and Chidambaram, R. (1974) *Acta Crystallogr.*, **B30**, 562.
103. Jones, D.D., Bernal, I. Frey, M.N. and Koetzle, T.F. (1974) *Acta Crystallogr.*, **B30**, 1220.
104. Meester, P.de and Hodgson, D.J. (1977) *J. Am. Chem. Soc.*, **99**, 6884.
105. Domazetis, G., Mackay, M.F., Magee, R.J. and James, B.D. (1979) *Inorg. Chim Acta*, **34**, L247.
106. Chen, C. and Parthasarathy, R. (1977) *Acta Crystallogr.*, **B33**, 3332.
107. Torii, K. and Iitaka, Y. (1973) *Acta Crystallogr.*, **B29**, 2799.
108. Freeman, W.A. (1977) *Acta Crystallogr.*, **B33**, 191.
109. Bear, C.A. and Freeman, H.C. (1976) *Acta Crystallogr.*, **B32**, 2534.
110. Oda, K. and Koyama, H. (1972) *Acta Crystallogr.*, **B28**, 639.
111. Kistenmacher, T.J., Hunt, D.J. and Marsh, R.E. (1972) *Acta Crystallogr.*, **B28**, 3352.
112. Lehmann, M.S., Koetzle, T.F. and Hamilton, W.C. (1972) *Int. J. Peptide Protein Res.*, **4**, 229.
113. Kistenmacher, T.J. and Sorrell, T. (1974) *J. Cryst. Mol. Struct.*, **4**, 419.
114. Solans, X. and Font-Altaba, M. (1981) *Acta Crystallogr.*, **B37**, 2111.
115. de Meester, P. and Hodgson, D.J. (1977) *J. Am. Chem. Soc.*, **99**, 101.
116. Camerman, N., Fawcett, J.K., Kruck, T.P.A., *et al.* (1978) *J. Am. Chem. Soc.*, **100**, 2690.
117. Harada, Y. and Iitaka, Y. (1977) *Acta Crystallogr.*, **B33**, 247.
118. Al-Karaghouli, A.R. and Koetzle, T.F. (1975) *Acta Crystallogr.*, **B31**, 2461.
119. van der Helm, D., Lawson, M.B. and Enwall, E.L. (1971) *Acta Crystallogr.*, **B27**, 2411.
120. Casella, L., Gullotti, M., Pasini, A., *et al.* (1978) *Inorg. Chim. Acta*, **26**, L1.
121. Frey, M.N., Koetzle, T.F., Lehmann, M.S. and Hamilton, W.C. (1973) *J. Chem. Phys.*, **58**, 2547.
122. Michel, A.G. and Durant, F. (1976) *Acta Crystallogr.*, **B32**, 1574.
123. Koszelak, S.N. and van der Helm, D. (1981) *Acta Crystallogr.*, **B37**, 1122.
124. Cody, V. unpublished results.
125. Cody, V., Langs, D.A. and Hazel, J.P. (1979) *Acta Crystallogr.*, **B35**, 1829.
126. van der Helm, D. and Tatsch, C.E. (1972) *Acta Crystallogr.*, **B28**, 2307.
127. Jarzab, T.C., Hare, C.R. and Langs, D.A. (1973) *Cryst. Struct. Comm.*, **3**, 399.
128. Hamalainen, R., Ahlgren, M., Turpeinen, U. and Raikas, T. (1978) *Cryst. Struct. Comm.*, **7**, 379.
129. Alcock, N.W., Lampe, P.A. and Moore, P. (1978) *J. Chem. Soc., Dalton Trans.*, 1324.
130. Ash, R.P., Herriott, J.R. and Deranleau, D.A. (1977) *J. Am. Chem. Soc.*, **99**, 13.
131. Bakke, O. and Mostak, A. (1980) *Acta Chem. Scand.*, **B34**, 559.
132. Yamane, T., Andou, T. and Ashida, T. (1977) *Acta Crystallogr.*, **B33**, 1650.
133. Vijayalakshmi, B.K. and Srinivasan, R. (1975) *Acta Crystallogr.*, **B31**, 999.
134. Ramanadham, M., Sikka, S.K. and Chidambaram, R. (1972) *Acta Crystallogr.*, **B28**, 3000.
135. Verbist, J.J., Lehmann, M.S., Koetzle, T.F. and Hamilton, W.C.(1972) *Acta Crystallogr.*, **B28**, 30
136. Maugen, Y., Brunie, S., Tsoucaris, G. and Knossow, M. (1976) *Cryst. Struct. Comm.*, **5**, 723.
137. Stephens, F.S., Vagg, R.S. and Williams, R.A. (1977) *Acta Crystallogr.*, **B33**, 433.
138. Aubry, A., Protas, J. and Marraud, M. (1977) *Acta Crystallogr.*, **B33**, 2534.
139. Koetzle, T.F., Frey, M.N., Lehmann, M.S. and Hamilton W.C. (1973) *Acta Crystallogr.*, **B29**, 2571.
140. Sudhakar, V., Bhat, T.N. and Vijayan, M. (1980) *Acta Crystallogr.*, **B36**, 125.
141. van der Helm, D. and Hossain, M.B. (1969) *Acta Crystallogr.*, **B25**, 457.
142. Ramanadham, M., Sikka, S.K. and Chidambaram, R. (1973) *Pramana*, **1**, 247.
143. Parthasarathy, R., Ohrt, J.M. and Chedda, G.B. (1977) *Biochem.*, **16**, 4999.
144. Koetzle, T.F., Lehmann, M.S., Verbist, J.J. and Hamilton, W.C. (1972) *Acta Crystallogr.*, **B28**, 3207.
145. L'Haridon, P., Lang, J., Pastuszak, R. and Dobrowolski, J. (1978) *Acta Crystallogr.*, **B34**, 2436.
146. Sudhakar, V. and Vijayan, M. (1980) *Acta Crystallogr.*, **B36**, 120.
147. Furberg, S. Solbakk, J. (1973) *Acta Chem. Scand.*, **27**, 1226.
148. Watson, W.H., Johnson, D.R., Celap, M.B. and Kamberi, B. (1972) *Inoorg. Chim. Acta*, **6**, 591.
149. Dawson, B. (1977) *Acta Crystallogr.*, **B33** 882.
150. Rao, S.T. (1973) *Acta Crystallogr.*, **B29**, 1718.
151. Derissen, J.L., Endeman, H.J. and Peerdeman, A.F. (1968) *Acta Crystallogr.*, **B24**, 1349.
152. Oonishi, I., Sato, S. and Saito, Y. (1975) *Acta Crystallogr.*, **B31**, 1318.
153. Hirayama, N., Shirahata, K., Ohashi, Y. and Sasada, Y. (1980) *Bull. Chem. Soc. Jpn*, **53**, 30.
154. Lehmann, M.S. and Nunes, A.C. (1980) *Acta Crystallogr.*, **B36**, 1621.
155. Lehmann, M.S., Koetzle, T.F. and Hamilton, W.C. (1972) *J. Cryst. Mol. Struct.*, **2**, 225.

156. Flook, R.J., Freeman, H.C. and Scudder, M.L. (1977) *Acta Crystallogr.*, **B33**, 801.
157. Einspahr, H., Gartland, G.L. and Bugg, C.E. (1977) *Acta Crystallogr.*, **B33**, 3385.
158. Feldman, R.I. (1977) *Atlas of Molecular Structure on Microfiche*, Tracor-Jitco, Rockville, Maryland.
159. Protein Data Bank, Brookhaven National Laboratory.
160. Dietrich, S.W., Bolger, M.B., Kollman, P.A. and Jorgensen, E.C. (1977) *J. Med. Chem.*, **20**, 863.
161. Cody, V. (1979) in *Computer-Assisted Drug Design* (eds E.C. Olson and R.E. Christoffersen), ACS Symposium No. 112, American Chemical Society, Washington, DC, p. 281.
162. Cody, V. (1981) *Endocrine Society Abstracts*, No. 432, Cincinnati, Ohio.

Index